Millions of Years Ago	Period	Life Forms	Climates and Major Physical Events
Paleozoic Era			
248–286	Permian	Reptiles diversify. Origin of conifers, cycads, and ginkgos; possible origin of flowering plants; earlier forest types wane.	Extensive glaciation in Southern Hemisphere. Seas drain from land; worldwide aridity. Appalachians formed by end of Paleozoic.
286–360	Carboniferous Pennsylvanian Mississippian	Age of amphibians. First reptiles. Variety of insects. Sharks abundant. Great swamps, forests of ferns, gymnosperms, and horsetails.	Warm; conditions like those in temperate or subtropical zones—little seasonal variation, water plentiful. Lands low, covered by shallow seas or great coal swamps. Mountain building in eastern U.S., Texas, Colorado.
360–408	Devonian	Age of fishes. Amphibians appear. Mollusks abundant. Lunged fishes. Extinction of primitive vascular plants. Origin of modern groups of vascular plants.	Europe mountainous with arid basins. Mountains and volcanoes in eastern U.S. and Canada. Rest of North America low and flat. Sea covers most of land.
408–438	Silurian	Rise of fishes and reef-building corals. Shell-forming sea animals abundant. Invasion of land by arthropods. Earliest vascular plants. Modern groups of algae and fungi.	Mild. Continents generally flat. Again flooded. Mountain building in Europe.
438–505	Ordovician	First primitive fishes. Invertebrates dominant. First fungi. Possible invasion of land by plants.	Mild. Shallow seas, continents low; sea covers U.S. Limestone deposits.
505–590	Cambrian	Shelled marine invertebrates. Explosive diversification of eukaryotic organisms.	Mild. Extensive seas, spilling over continents.
Precambrian Era			
590–4,500		Origin of life. Prokaryotes. Eukaryotic cells and multicellularity by close of era. Earliest known fossils, including soft-bodied marine invertebrates.	Dry and cold to warm and moist. Planet cools. Formation of Earth's crust. Extensive mountain building. Shallow seas. Accumulation of free oxygen.

Invitation to Biology

Summer is brief within the Arctic Circle, where these snowy owl chicks have made their appearance. The sun never sets, moving in a circle, never directly overhead but never dipping below the horizon. Its glancing rays melt the snow and the upper inches of frozen soil, but just beneath the surface the soil remains permanently frozen. Water cannot penetrate the solid ice and so the tundra—as this land of permafrost is called—stays boggy wet all summer long, even though there is as little rainfall as in a desert. The roots of plants cannot penetrate the permafrost either, so all the plants, even the trees, are only a few inches tall. When you walk across the tundra, you are walking on a forest of birches and willows, some hundreds of years old. Flowering plants, racing to set seed before the darkness returns, sometimes bloom when they are only an inch in height. Snowy owls and other land birds nest on the treeless expanse, where the bright, short-stemmed flowers and stunted vegetation barely conceal these curious chicks.

Among the largest of the owls, snowies are almost two feet tall, with a five-foot wingspan. As you can see, the chicks are soot-colored, but most adults are a dazzling white. When courting, it is reported, the male dances stiffly on the ground while holding a dead lemming. Who could resist?

Every year, birds of many species migrate by the millions to the Arctic Circle to take advantage of the abundant fish, the swarms of insects that emerge each season, and most of all, the 24-hour daylight for feeding their young. When fall comes, most head south. The snowy owl is one that remains, seeking lemmings and Arctic hares rustling beneath the snow cover. During the long winter night, the air is cold and clear, and the starlight and moonlight reflect off the surface of the snow. By then, these young snowies will be full grown and airborne—huge white owls gliding as silent as the snow through the cold air.

Invitation to Biology

Fifth Edition

HELENA CURTIS

N. SUE BARNES

Illustrated by Shirley Baty

WORTH
FREEMAN

Invitation to Biology, Fifth Edition—Complete book (casebound)
Invitation to Biology, Fifth Edition—Part 1: Sections 1, 2, 3, 4, 8 (paperback)
Invitation to Biology, Fifth Edition—Part 2: Sections 5, 6, 7 (paperback)

Copyright 1994, 1985, 1981, 1977, 1972 by Worth Publishers, Inc.

Printed in the United States of America

Library of Congress Catalog Card Number 93-61158—Complete book (casebound)
Library of Congress Catalog Card Number 94-60032—Parts 1 and 2 (paperbacks)

ISBN: 0-87901-679-5—Complete book (casebound)
ISBN: 0-87901-734-1—Part 1: Sections 1, 2, 3, 4, 8 (paperback)
ISBN: 0-87901-735-X—Part 2: Sections 5, 6, 7 (paperback)

Printing: 3   4   5—98

Editor: Sally Anderson

Design: Malcolm Grear Designers

Art Director: George Touloumes

Production Supervisor: Sarah Segal

Layout: Matthew Dvorozniak

Picture Editor: Lana Berkovich

Illustration Consultant: John Norton

Typographer: CRWaldman Graphic Communications

Color Separations and Stripping: Creative Graphic Services Corporation

Printing and Binding: Von Hoffmann Press, Inc.

Covers and Frontispiece: Snowy owlets, © Art Wolfe

Worth Publishers
33 Irving Place
New York, New York 10003

Preface

Preface to the Instructor

As even a glance at the Table of Contents or a few minutes spent thumbing through the text itself will reveal, this Fifth Edition of *Invitation to Biology* is a major revision, the result of a thorough reexamination and rethinking of all aspects of the book.

The most obvious changes are in the design and the illustration program. Our goal was twofold: to make the book as inviting as possible to today's students, who live in a world dominated by visual cues, and, simultaneously, to use art and photographs not simply to enhance the text but also to clarify difficult concepts and actively teach. With the able assistance of our art consultant, John Norton, who provided us with a multitude of suggestions, and the dedicated labors of our gifted illustrator, Shirley Baty, who translated ideas into reality, more than 500 new pieces of art now grace the pages of *Invitation*. A similar effort, involving both our photo researcher, Lana Berkovich, and our editor, Sally Anderson, has resulted in the many marvelous new photographs that appear throughout the book. Of particular importance are the full-page photographs with which each chapter begins. Finding exactly the right photograph for each chapter—a photograph that would, in and of itself, appeal to students and, at the same time, lend itself to a brief essay that would intrigue students and set the stage for the chapter—was a daunting but rewarding task.

Adding more illustrations and increasing their size in order to maximize their appeal and pedagogical effectiveness does, of course, require more space. In order to accommodate the new illustration program without greatly increasing the length of the book—or having to compress explanatory material at the expense of clarity—we have, with this edition, moved from our traditional one-column format to a two-column format. The new design, created by Pat Appleton of Malcolm Grear Designers, has provided great flexibility in both the use of color and the layout of the individual pages. This flexibility and the enormous effort devoted by art director George Touloumes and his colleague, Matthew

Dvorozniak, to the detailed layout of the book, page by page, has ensured that the two-column format has not become so tight or "busy" that it is intimidating to students. Working with the new design has been exciting for us, and we believe that it will help draw students into the wonders of biology.

As in all previous editions, the text itself has been subjected to an intense reconsideration and reworking—section by section, chapter by chapter, and sentence by sentence—and then once again thoroughly reviewed. As always, we have had to face the questions of how to organize the book, what to put in, and, most difficult of all, what to leave out. In terms of the depth and breadth of the coverage, it has seemed to us wisest to maintain—and where possible, to strengthen—the firm emphasis on the basic principles of biology that has characterized previous editions of *Invitation*. Specific topics for detailed treatment have been selected on the basis of their centrality to modern biology, their utility in illuminating basic principles, their importance as part of the requisite store of knowledge of an educated adult, and their inherent appeal to students. Throughout, we have tried to lay the foundations and arouse student curiosity so that the ground is prepared for those areas—very diverse—in which individual instructors may wish to give more extensive coverage than is possible in any "short" text.

The essential foundation of biology is, of course, evolution, the major organizing theme of this and all other modern biology texts. The stage is set in the Introduction, which begins with the fascinating organisms of one particular "living laboratory of evolution"—the island of Madagascar—and then moves on to the development and key features of the Darwinian theory. The Introduction concludes with a discussion of the nature of science, where the story of the peppered moth provides both a fully worked-out example of scientific methodology and a concrete example of natural selection in action.

With this edition, we have moved the detailed treatment of evolution much earlier in the book, so that it is now Section 4, following immediately after genetics. This repositioning makes it possible for students to consider the complexities of population genetics while genetics itself is still fresh in their minds. It also makes possible a stronger and more explicit evolutionary theme in Section 5 (The Diversity of Life), Section 6 (Biology of Animals), and Section 7 (Biology of Plants), all of which, in previous editions, preceded the evolution section.

Moving evolution earlier in the book led to other changes, some of which we anticipated from the outset and some of which we discovered only as we worked our way through the revision. In previous editions, for example, human evolution was covered in the evolution section. With the new organization, however, the necessary foundation for a discussion of human evolution—namely the evolution of the vertebrates in general and the mammals in particular—is not covered until after the evolution section. Thus, we moved human evolution into the diversity section. What was previously one chapter on animal diversity is now two chapters: the first covers the invertebrates, and the second covers the vertebrates, culminating with the story of human evolution.

With that change made, it then seemed wise to move directly into animal physiology, for which our representative organism is *Homo sapiens*. Having Section 6, Biology of Animals, follow immediately after the two chapters on animal diversity should make the various examples of comparative animal physiology in that section more meaningful to students, once again because the relevant material will be fresh in their minds. We have also added two new chapters to Section 6: Chapter 33, The Immune Response, and Chapter 38, Development. Both immunology and developmental biology are currently undergoing an explosive growth of knowledge and are also, we believe, of great interest to students.

Section 7, Biology of Plants, which now follows the section on the biology of animals, begins with a chapter on reproduction in the flowering plants, which is followed, in turn, by a chapter on the plant body and its development. This provides a clear and immediate counterpoint to the last two chapters of Section 6, and it should enable students to see more easily the similarities and differences in the ways in which the organisms of the two largest kingdoms accomplish the fundamental tasks of reproduction and development. Beginning Section 7 with the flower and reproduction—rather than with a chapter on plant structure—also enables us to draw students into the fascinating world of plants with that which is familiar and beautiful.

From the plants, we then move into Section 8, Ecology. Here too, the new organization should enable students to have fresh in their minds information that will enhance their exploration of a new topic. This section has been extensively revised, as we attempt to track the continual shifts, rethinkings, and controversies that characterize this most vibrant science. As in the Fourth Edition, the section moves from population dynamics, through the interactions of populations in communities and ecosystems, to the overall organization and distribution of life on Earth. The book ends with a consideration of the tropical forests—the most complex and most seriously threatened of all ecological systems.

All of the feedback we have received—in letters and in surveys—has indicated that, with a few exceptions, the previous organization and sequence of the first three sections of the book—Section 1 (The Unity of Life), Section 2 (Energetics), and Section 3 (Genetics)—was sound and should not be altered. The exceptions were the placement of the chapter on the reproduction of cells (mitosis and cytokinesis), the sequence of chapters at the beginning of the genetics section, and the placement of human genetics in a separate chapter.

In response to your suggestions, we have made several changes. First, the chapter on cell reproduction has been moved out of the first section and is now Chapter 10, the first chapter of genetics. Second, we have resequenced the beginning of Section 3 so that it moves, in three consecutive chapters, from mitosis to meiosis to Mendel. Third, we have eliminated the separate chapter on human genetics and integrated that material throughout Section 3. Because of the inherent interest of human genetics, this change should make the other chapters of the section more appealing. Also, it avoids the problems that are involved in positioning a human genetics chapter that, if it is to be thorough, must range from Mendelian genetics to the cutting edge of molecular genetics. If such a chapter is positioned immediately after the chapters on classical genetics, either it must omit new developments in the diagnosis and treatment of human genetic disorders, or, alternatively, it must attempt to cover those developments before a proper foundation has been laid. If, on the other hand, such a chapter is positioned at the end of the genetics section, students are compelled to make a sudden mental switch from molecular genetics back to classical genetics—and then back again—all within the same chapter. The new organization should lead to both increased student understanding and increased student comfort.

One of the most striking aspects of the enormous burst of new discoveries in molecular and cell biology in recent years is the power of these discoveries to explain processes that previously could only be described.

The immune response, conduction of the nerve impulse, events at the synapse, the summing of information by individual neurons, and differentiation and morphogenesis in animal development are just a few of the many phenomena whose secrets are being revealed by studies at the molecular and cellular level. These revelations depend, in large part, on what is now a flood tide of reports identifying specific membrane proteins, their amino acid sequences, their three-dimensional structures, and, in many cases, the sequences and locations within the genome of the genes coding for these proteins. Because these discoveries, although fascinating in and of themselves, are of such value in explaining organismal phenomena, we have generally chosen to defer their discussion to later sections of the text, where their significance will be most readily grasped. Molecular and cell biology have, in many ways, come of age, and it seems to us that the essential task in the early sections of the book is the clear communication of the underlying principles on which so much is now being built—rather than a catalog of the latest new discoveries, which will soon be superseded by other new discoveries. We have endeavored to explain the fundamental principles as clearly as possible, and the many new illustrations should make these early sections easier to teach and more understandable for students.

Increased student understanding of the phenomena of the living world—coupled with the sheer pleasure of discovering the wonder of it all—is, of course, the whole point of creating a textbook and of teaching biology. As an aid toward the accomplishment of that goal, a number of supplements have been revised or newly created to accompany this edition of *Invitation to Biology*. The excellent *Study Guide* by David J. Fox of the University of Tennessee, Knoxville, makes students active participants in the learning process. A new edition of the well-received *Biology in the Laboratory*, by Doris R. Helms of Clemson University, engages students in critical thinking and problem solving; it is accompanied by a revised *Preparator's Guide* and *Answer Guide* that will make the laboratory course easier to plan and to teach. A set of 200 full-color acetate transparencies and a videodisc provide excellent visual enhancement for lectures. A *Test Bank*, of more than 2,500 multiple-choice, true-false, and brief essay questions, prepared by David Fox, is available on both paper and computer disks.

Of particular interest is *Instructor's Resources* by Robert C. Evans of Rutgers University, Camden. This new supplement includes suggestions for using all of the materials in the package, lecture guides, learning objectives, answers to the questions in the textbook, and a list of appropriate resources for each chapter. Instruc-

tions for outlining and concept mapping, as well as listings of available films and videos, computer programs, and videodiscs, with sources, are also included. The *Lecture Guides* from this supplement plus *Focus Points*, prepared by Bruce Chorba of Mercer County Community College, are also available together on computer disk for preparing your own handouts. Another supplement, *Student Activities*, provides three-hole-punched master copies of critical thinking, extension, and application activities for each chapter.

As with previous editions, we have been greatly assisted by the advice of our consultants and reviewers. In addition to his work on *Instructor's Resources*, Bob Evans reviewed the entire manuscript. His sensitivity to students, his good judgment, and his steady, calm counsel have been a great blessing. The entire manuscript was also expertly reviewed by both Janet Wagner, whose suggestions were invaluable, particularly in the reworking of the genetics section, and Linda Hansford, who has, in addition, prepared the Index. Manuel C. Molles, Jr., of the University of New Mexico, and Andrew Blaustein, of Oregon State University, both made major contributions to our revision of the evolution and ecology sections. Rick Weiss, formerly of *Science News*, and Judith Wilson, of Worth Publishers, contributed to the new essays in Section 6.

In addition, we are grateful for the constructive criticism and many helpful suggestions we received from the following reviewers of this edition:

William L. Bischoff, University of Toledo

Richard Blazier, Parkland College

Maren Brown, Onondaga Community College

Rita Calvo, Cornell University

Mary Colavito, Santa Monica College

Jean DeSaix, University of North Carolina, Chapel Hill

Mark Dubin, University of Colorado

Ray F. Evert, University of Wisconsin

William Greenhood, Santa Fe Community College

Mart Gross, University of Toronto

Thomas Hanson, Temple University

Jean B. Harrison, Mount St. Mary's College

George Hennings, Kean College of New Jersey

Merrill Hille, University of Washington

Ronald Hoham, Colgate University

Norma Johnson, University of North Carolina, Chapel Hill

Stephen W. Kress, National Audubon Society

Melanie Loo, California State University, Sacramento

Joyce Maxwell, California State University, Northridge

Dorothy McMeekin, Michigan State University

Richard Milner, American Museum of Natural History

Douglas W. Morrison, Rutgers University, Newark

David Netzly, Hope College

Bette Nicotri, University of Washington

Frank E. Price, Hamilton College

Julia Riggs, Victoria College

Edward S. Ross, California Academy of Sciences

Anthony P. Russell, University of Calgary

Robert W. Schuhmacher, Kean College of New Jersey

Ian Tattersall, American Museum of Natural History

Robin Tyser, University of Wisconsin, La Crosse

Helen Young, Barnard College

The preparation of this new edition has been a complex and all-consuming task, involving the collaborative efforts of a large number of talented people, many of whom have been noted already. Particular thanks are also due to Bob Worth, whose vision and support through the years have made it all possible; to Sarah Segal, who has managed the production process and somehow kept us all on course—often a thankless and frustrating task; to Demetrios Zangos, who labeled all of the new art and prepared and incorporated structural formulas into the illustrations in which they appear; to Timothy Prairie, who coordinated and edited the supplements; to Brian Donnelley and Treë, both of whom assisted with the photo research; and to Lindsey Bowman, Mary Mazza, and Alexandra Swieconek, who provided expert editorial assistance at different stages of the project. Our greatest debt, however, is to our extraordinary editor, Sally Anderson. Her dedication, her thorough knowledge of biology, her editorial expertise, and her long experience in working with us both have once again seen us through to the successful completion of an enormous project.

Finally, we want to thank all of the instructors and students who have written to us, some with criticisms, some with suggestions, some with questions, and some simply because they enjoyed the book. Those letters remind us of how privileged we are to be writing for students. We continue to appreciate their curiosity, their energy, their imaginativeness, their sense of humor, and their dislike of the pompous and pedantic. We hope we serve them well.

Helena Curtis
Sue Barnes

Long Island, New York
January, 1994

Leaf-tailed gecko, Madagascar

Contents in Brief

Contents

SECTION 1

The Unity of Life   18

1

2

SECTION 2

SECTION 3

SECTION 4

Evolution   318

19

The Genetic Basis of Evolution   318

SECTION 5

SECTION 6

Biology of Animals 506

SECTION 7

SECTION 8

Distant islands are the homes of many strange inhabitants—plants and animals found nowhere else on our planet. Lemurs, such as the sifakas shown here, are found only on the island of Madagascar, which lies off the southeast coast of Africa. Some 45 different species of lemurs have been discovered, of which 17 are now extinct. No monkeys, or apes, or other primates (except for humans, who were late arrivals) have been found on Madagascar—only the lemurs. Why are lemurs here and nowhere else?

This same sort of question was repeated over and over in different forms as the naturalists of the eighteenth and nineteenth centuries explored widely separated regions of the world—South America, the Galapagos Islands, Australia, the Malay Archipelago, and Madagascar. These travelers—Charles Darwin among them—were astonished by the vast numbers and extraordinary diversity of the organisms they encountered. The doctrine of special creation held that each kind of plant and animal was created in its present form and placed in the location in which it was subsequently found. But, the naturalist explorers wondered, was it possible, perhaps, that groups of plants and animals, separated from one another—as on islands—change and diversify over time into new kinds of organisms?

The lemur shown here, known to scientists as *Propithecus verreauxi,* is about the size of a small house cat. It is among the largest of the surviving lemurs. Sifakas are the acrobats of this primate world, using their long tails for balance as they fling themselves through the air. They can cover a distance as great as 25 feet in a single leap. These two sifakas are resting in a tamarind tree, found in the dry forest region of Madagascar. When faced with an enemy, they hiss the alarm call that gives them their name: Shifakh!

Introduction

Madagascar, the world's fourth largest island, lies in the Indian Ocean 250 miles east of the African mainland. Slightly smaller than the state of Texas, the island is located almost entirely within the tropics. Two principal types of vegetation are found on Madagascar: dry forest to the west and moist forest to the east, separated by a treeless mountain chain that runs the length of the island from north to south.

Some 200 million years ago, when today's continents were joined together in a single supercontinent, the fragment that would become Madagascar was located between the regions that would ultimately form Africa and India. When Africa and India began to separate, about 160 million years ago, Madagascar was split off. As it gradually moved away from the larger land masses, its plants and animals became increasingly isolated from those on the mainland of both Africa and India. Although occasional plant and animal colonists may have arrived on floating clumps of vegetation carried by the ocean currents, by about 50 million years ago the isolation was complete. The plants and animals of the island remained undisturbed until the arrival of the first human colonists some 2,000 years ago.

In 1771, the French naturalist Philibert de Commerson wrote of his first visit to Madagascar: "Nature seems to have retreated there into a private sanctuary, where she could work on models different from any she has used elsewhere. There, you meet bizarre and marvellous forms at every step." On Madagascar today there are literally thousands of **species,** or kinds, of plants and animals found nowhere else on Earth. For example, of the 8,000 species of flowering plants found on Madagascar, 80 percent grow only on this island. Of the world's 85 species of chameleons, two-thirds are here alone. Similarly, all but two of the 150 species of frogs on Madagascar are unique to the island. Here are found 3,000 different species of butterflies and moths, 97 percent of which live only on Madagascar.

1

I–1 *The island of Madagascar broke off from Africa some 160 million years ago and then, over a period of approximately 70 million years, moved slowly south to its present position. Madagascar is about 1,000 miles long and 350 miles across at its widest point. Conservationist Gerald Durrell recently described its shape as resembling "a badly made omelet."*

Among the most unusual inhabitants of Madagascar are the small, furry animals known as lemurs. Lemurs are classified by biologists with the primates, a group of mammals that includes monkeys, apes, and ourselves. At one time there were at least 45 species of lemurs, all of which inhabited Madagascar and Madagascar alone. In the last 2,000 years, however, almost half of the lemur species have vanished. Some of the larger species, with adults about the size of a Saint Bernard, were hunted to extinction before the first European explorers arrived on Madagascar. Other species have become extinct over the course of the past two centuries, as the rapid growth of the human population has led to the destruction of the island's forests. Today less than 15 percent of the lemurs' original habitat remains, and only 28 species survive (Figure I–2).

Madagascar has long been of interest to biologists not only because of its array of unique plants and animals but also because of the plants and animals that are *not* found on the island. Totally absent are the monkeys, apes, antelope, gazelles, zebras, giraffes, wildebeest, elephants, and lions of the neighboring African mainland. As European naturalists explored widely separated regions of the world in the eighteenth and nineteenth centuries, they found that other islands and island chains resembled Madagascar in two important respects: (1) distant islands were typically home to an enormous variety of organisms unique to that island or island group, and (2) such islands often lacked many of the plants and animals characteristic of the nearest mainland.

These observations raised an important question for the explorers: How does one make sense of the bewildering diversity of living things and the patterns in which they are scattered in such profusion over the surface of the Earth? One answer was that each species—each different kind of living organism—was specially created and placed by the creator in its present location. This was the traditional explanation. Another idea was in the air, however, and that idea is the subject of this Introduction.

The Road to Evolutionary Theory

In 1831, the young Charles Darwin set sail from England on what was to prove the most consequential voyage in the history of biology. Not yet 23, Darwin had already abandoned a proposed career in medicine—he described himself as once fleeing a surgical theater in which an operation was being performed on an unanesthetized child—and was a reluctant candidate for the clergy, a profession deemed suitable for the younger son of an English gentleman. An indifferent student, Darwin was an ardent hunter and horseman, a collector of beetles, mollusks, and shells, and an amateur botanist and geologist. When the captain of the surveying ship H.M.S. *Beagle,* himself only a little older than Darwin, offered passage for a young gentleman naturalist who would volunteer to go without pay, Darwin eagerly seized the opportunity to pursue his interest in natural history. The voyage, which lasted five years, shaped the course of Darwin's future work. He returned to an inherited fortune, an estate in the English countryside, and a lifetime of independent work and study that radically changed our view of life and of our place in the living world.

That Darwin was the founder of the modern theory of evolution is well known. Although he was not the first to propose that organisms **evolve**—or change—through time, he was the first to gather a large body of supporting evidence and the first to propose a valid mechanism by which evolution might occur. In order to understand the meaning and significance of Darwin's theory, it is useful to look at the intellectual climate in which it was formulated.

I–2 *Six of the 28 surviving lemur species on Madagascar.* (a) *Indri, the largest of the living lemurs. In Malagasy, the principal language of Madagascar, the name of this lemur is "babakoto," the word for "grandfather." However, when one of the local inhabitants used the expression "indri," which means "look at that!" to point out the animal to an early European explorer, the explorer misinterpreted it as the animal's name. With the bodies of monkeys, the black-and-white fur of pandas, the heads of dogs, and no tails, the indri are indeed a sight to behold. They are also noted for their eerie and deafening wail, in which a single song may last as long* as 4 minutes. (b) *An aye-aye, the rarest of the lemurs. This small, solitary, and nocturnal animal may have been the original source for the name lemur, which means "ghost." Aye-ayes have bat-like ears, large eyes, and a long third finger that they drum against trees to determine whether they are hollow. If the sound is right, the aye-aye (like the familiar woodpecker) proceeds to extract insects from the decaying wood. Aye-ayes also have a taste for sweet foods, such as sugar cane and coconuts. Malagasy legends say that if an aye-aye should point its long third finger at you, you are destined to die, horribly and* swiftly. (c) *The tiny mouse lemur, which is thought to resemble the ancestral primate from which the diverse species of lemurs arose.* (d) *The golden bamboo lemur, first discovered in 1986, is one of the world's most seriously endangered species. The number of surviving individuals is unknown but is thought to be only a few dozen.* (e) *A female black lemur. Unlike the males, for which the species is named, the females are a rusty brown.* (f) *A ring-tailed lemur. The second toe of each foot has a special grooming claw with which the lemur combs its long tail.*

I–3   *Charles Darwin in 1840, four years after he returned from his five-year voyage on H.M.S.* Beagle. *In his autobiography, written in 1876 for his children, Darwin made the following comments about his selection for the voyage: "Afterwards, on becoming very intimate with Fitz Roy [the captain of the Beagle], I heard that I had run a very narrow risk of being rejected on account of the shape of my nose! He . . . was convinced that he could judge of a man's character by the outline of his features; and he doubted whether anyone with my nose could possess sufficient energy and determination for the voyage. But I think he was afterwards well satisfied that my nose had spoken falsely."*

Aristotle (384–322 B.C.), the first great biologist, believed that all living things could be arranged in a hierarchy. This hierarchy became known as the *Scala Naturae,* or ladder of nature, in which the simplest creatures had a humble position on the bottommost rung, man occupied the top rung, and all other organisms had their proper places in between. Until the late nineteenth century, many biologists believed in such a natural hierarchy. But whereas to Aristotle living organisms had always existed, the later biologists (at least those of the Western world) believed, in harmony with the teachings of the Old Testament, that all living things were the products of a divine creation. They believed, moreover, that most were created for the service or pleasure of mankind.

That each type of living thing came into existence in its present form—specially and specifically created—was a compelling idea. How else could one explain the astonishing extent to which every living thing was adapted to its environment and to its role in nature? It was not only the authority of the Bible but also, so it seemed, the evidence before one's own eyes that gave such strength to the concept of special creation.

Among those who believed in divine creation was Carolus Linnaeus (1707–1778), the great Swedish naturalist who devised our present system for naming species of organisms. In 1753, Linnaeus published two encyclopedic volumes describing every species of plant known at the time. Even as Linnaeus was at work on this massive project, explorers were returning to Europe from Africa and the New World with thousands of previously undescribed plants and animals. Linnaeus revised edition after edition to accommodate these findings, but he did not change his opinion that all species now in existence were created by the sixth day of God's labor and have remained the same ever since. During Linnaeus's time, however, it became clear that the pattern of creation was far more complex than had been previously imagined.

Evolution before Darwin

The idea that organisms might evolve through time, with one type of organism giving rise to another type of organism, is an ancient one, predating Aristotle. A school of Greek philosophy, founded by Anaximander (611–547 B.C.) and culminating in the writings of the Roman Lucretius (99–55 B.C.), developed not only an atomic theory but also an evolutionary theory, both of which are strikingly similar to modern conceptions. However, as the science of biology began to take form in the eighteenth and nineteenth centuries, the work of this school was largely unknown, as were the ideas of other, non-European cultures.

In the eighteenth century, the French scientist Georges-Louis Leclerc de Buffon (1707–1788) was among the first to propose that species might undergo changes in the course of time. He suggested that, in addition to the numerous creatures that were produced by divine creation at the beginning of the world, "there are lesser families conceived by Nature and produced by Time." Buffon believed that these organisms were produced by a process of degeneration. For example, although an "ideal" type of feline had been created, its descendants had degenerated into a variety of forms, including lions, jaguars, cheetahs, and house cats. Buffon's hypothesis, although vague as to the way in which such changes might occur, did attempt to explain the bewildering variety of creatures in the modern world.

Another doubter of fixed and unchanging species was Erasmus Darwin (1731–1802), Charles Darwin's grandfather. Erasmus Darwin was a physician, a gentleman naturalist, and a prolific writer, often in verse, on both botany and zoology. He suggested, largely in asides and footnotes, that species have historical connections with one another, that animals may change in response to their environment, and that their offspring may inherit these changes. He maintained, for instance, that a polar bear is an "ordinary" bear that, by living in the Arctic, became modified and passed the modifications along to its cubs. These ideas were never convincingly formulated but are interesting because of their possible effects on Charles Darwin.

The Age of the Earth

It was geologists, more than biologists, who paved the way for modern evolutionary theory. One of the most influential of these was James Hutton (1726–1797). Hutton proposed that the Earth had been molded not by sudden, violent events but by slow and gradual processes—wind, weather, and the flow of water—the same processes that can be seen at work in the world today. This theory of Hutton's, which later became known as **uniformitarianism,** was important for three reasons. First, it implied that the Earth has a long history, which was a new idea to eighteenth-century

Europeans. Christian theologians, by counting the successive generations since Adam (as recorded in the Bible), had calculated the maximum age of the Earth at about 6,000 years. Yet 6,000 years is far too short for major evolutionary changes to take place, by any theory. Second, the theory of uniformitarianism stated that change is itself the *normal* course of events, as opposed to a static system interrupted by an occasional unusual event, such as an earthquake. Third, although this was never explicit, uniformitarianism suggested that there might be alternatives to the literal interpretation of the Bible.

The Fossil Record

During the latter part of the eighteenth century, there was a revival of interest in **fossils,** which are the preserved remains of organisms long since dead (Figure I–4). In previous centuries, fossils had been collected as curiosities, but they had generally been regarded either as accidents of nature—stones that somehow looked like shells—or as evidence of great catastrophes, such as the Flood described in the Old Testament.

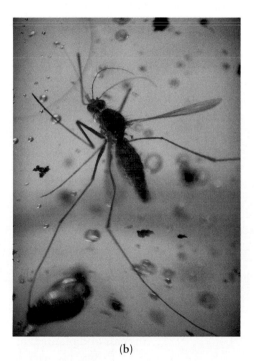

(a)   (b)   (c)

I–4   *A fossil is a remnant or trace of an organism that died many, many years earlier. (a) Minerals may seep into and fill the hollows left by the decay of soft tissues, as in this cycad leaf dating from the Jurassic period (144 to 213 million years ago). A great diversity of cycads were contemporary with the dinosaurs; about 100 species survive today. (b) A primitive mosquito caught in amber formed from the resin of a tree that lived some 25 to 40 million years ago. The blood from this mosquito's last meal is still visible in her digestive tract. (c) Among the most common early fossils are the outer skeletons of trilobites, marine animals that first appear in the fossil record during the Cambrian period (505 to 590 million years ago). This fossil, with its well-preserved antennae and legs, is filled with fool's gold (pyrite). Trilobites flourished for more than 300 million years before they became extinct.*

I-5   *An outcrop at Olduvai Gorge, in Tanzania, that clearly shows different layers, or strata, in the Earth's surface. The work of geologists who discovered that different strata contain different types of fossils was important in the development of evolutionary theory. The strata at Olduvai have yielded a number of fossils that have contributed to our growing understanding of human evolution.*

The English surveyor William Smith (1769–1839) was among the first to study the distribution of fossils scientifically. Whenever his work took him down into a mine or along canals or cross-country, he carefully noted the order of the different layers of rock, known as **strata,** and collected fossils from each layer. He eventually established that each stratum, no matter where he came across it in England, contained characteristic kinds of fossils. Moreover, these "index fossils" were actually the best way to identify a particular stratum in a number of different geographic locations. Smith, who was nicknamed "Strata" by his contemporaries, did not interpret his findings. However, the implication that the present surface of the Earth had been formed layer by layer over the course of time was an unavoidable one.

Like Hutton's world, the world seen and described by William Smith was a very ancient one. A revolution in geology was beginning; earth science was becoming a study of time and change rather than a mere cataloging of types of rocks. As a consequence, the history of Earth itself became inseparable from the history of living organisms, as revealed in the fossil record (Figure I–5).

Catastrophism

Although the way was being prepared by the revolution in geology, the time was not yet ripe for a parallel revolution in biology. The dominating force in European natural science in the early nineteenth century was Georges Cuvier (1769–1832). Cuvier was the founder of **vertebrate paleontology,** the scientific study of the fossil record of **vertebrates** (animals with backbones). An expert in anatomy and zoology, he applied his knowledge of the way in which animals are constructed to the study of fossil animals. From a few fragments of bone in a rock slab, he could identify an entire animal. To his students, Cuvier's brilliant deductions seemed almost magical.

Today we think of paleontology and evolution as so closely connected that it is surprising to learn that Cuvier was a staunch and powerful opponent of evolutionary theories. He did, however, recognize that many species that had once existed no longer did—a startling conclusion at the time. (In fact, according to modern estimates, considerably less than one percent of all species that have ever lived are represented on Earth today.) Cuvier explained the extinction of species by postulating a series of catastrophes. After each catastrophe, the most recent of which was the Flood, new species filled the vacancies.

Cuvier hedged somewhat on the source of the new animals and plants that appeared after the extinction of older forms. He was inclined to believe they moved in from parts unknown. Another major opponent of evolution, Louis Agassiz (1807–1873), America's leading nineteenth-century biologist, was more straightforward. According to Agassiz, the fossil record revealed 50 to 80 total extinctions of life, followed by an equal number of new, separate creations.

The Concepts of Lamarck

The first European scientist to work out a systematic concept of evolution was Jean Baptiste Lamarck (1744–1829). "This justly celebrated naturalist," as Darwin referred to him, boldly proposed in 1801 that all species, including *Homo sapiens,* are descended from other species. Lamarck, unlike most of the other zoologists of his time, studied one-celled organisms and **invertebrates** (animals without backbones). Undoubtedly it was his long study of these forms of life that led him to think of living things in terms of constantly increasing complexity, each species derived from an earlier, less complex one.

Like Cuvier and others, Lamarck noted that older rocks generally contained fossils of simpler forms of life. Unlike Cuvier, however, Lamarck interpreted this as meaning that the more complex forms had arisen from the simpler forms by a kind of progression. According to his hypothesis, this progression, or evolution, to use the modern term, is dependent on two main forces. The first is the inheritance of acquired characteristics, an idea that did not originate with Lamarck but was, in fact, widely accepted at the time by both scientists and the general public. It was believed that organs in animals became stronger or weaker, larger or smaller, through use or disuse, and that these changes were transmitted from the parents to the offspring. Lamarck's most famous example was the evolution of the giraffe. According to Lamarck, the modern giraffe evolved from ancestors that stretched their necks to reach leaves on high branches. These ancestors transmitted the longer necks—acquired by stretching—to their offspring, which stretched their necks even longer, and so on.

The second important force in Lamarck's concept of evolution was a universal creative principle, an unconscious striving upward on the *Scala Naturae* that moved every living creature toward greater complexity. Amoebas were on their way to *Homo sapiens.* Some might get waylaid—the orangutan, for instance, had been diverted from its course by being caught in an unfavorable environment—but the will was always present. Life in its simplest forms emerged spontaneously to fill the void left at the bottom of the ladder. In Lamarck's formulation, Aristotle's ladder of nature had been transformed into a steadily ascending escalator powered by a universal will.

Lamarck's contemporaries generally did not object to the idea of the inheritance of acquired characteristics, which we, with our present knowledge of genetics, know to be false. Nor did they criticize his belief in an upward-striving metaphysical force, which was a common element in many of the concepts of the time.

But these vague, untestable postulates provided a very shaky foundation for the radical proposal that more complex forms of life evolved from simpler forms. Moreover, Lamarck personally was no match for the brilliant and prestigious Cuvier, who relentlessly attacked his ideas. As a result, Lamarck's career was ruined, and scientists as well as the public became even less prepared to accept any evolutionary doctrine.

Development of Darwin's Theory

Earth Has a History

The person who most influenced Darwin, it is generally agreed, was Charles Lyell (1797–1875), a geologist who was Darwin's senior by 12 years. One of the books that Darwin took with him on his voyage was the first volume of Lyell's newly published *Principles of Geology,* and the second volume was sent to him while he was on the *Beagle.* On the basis of his own observations and those of his predecessors, Lyell opposed the theory of catastrophism. Instead, he produced new evidence in support of Hutton's earlier theory of uniformitarianism.

According to Lyell, the slow, steady, and cumulative effect of natural forces had produced continuous change in the course of the Earth's history. Since this process is demonstrably slow, its results being barely visible in a single lifetime, it must have been going on for a very long time. What Darwin's theory needed was time, and it was time that Lyell gave him. In the words of biologist Ernst Mayr of Harvard University, the discovery that the Earth was ancient "was the snowball that started the whole avalanche."

The Voyage of the *Beagle*

This, then, was the intellectual climate in which Charles Darwin set sail from England. As the *Beagle* moved down the Atlantic coast of South America, through the Strait of Magellan, and up the Pacific coast (Figure I–6), Darwin often left the ship and traveled the interior. He explored the rich fossil beds of South America (with the theories of Lyell fresh in his mind) and collected specimens of the many new kinds of plant and animal life he encountered. He was impressed most strongly during his long, slow trip down one coast and up the other by the constantly changing varieties of organisms he saw. The birds and other animals on the west coast, for example, were very different from those on the east coast, and even as he moved slowly up the western coast, one species would give way to another.

I–6   The Beagle's *voyage around South America. The ship left England in December of 1831 and arrived at Bahia, Brazil, in late February of 1832. About 3½ years were spent along the coast of South America, surveying and making inland explorations. The stop at the Galapagos Islands was for slightly more than a month, and, during that brief time, Darwin made the wealth of observations that were to change the course of the science of biology. The remainder of the voyage, across the Pacific to New Zealand and Australia, across the Indian Ocean to the Cape of Good Hope, back to Bahia once more, and at last home to England, occupied another year.*

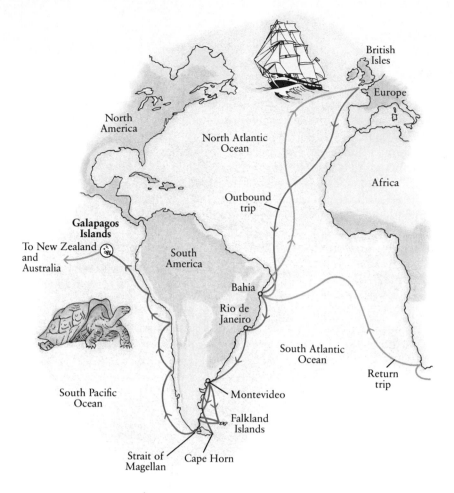

I–7   *A distinguishing feature of the Galapagos tortoise is the shape of its shell, or carapace, which varies according to its island of origin. (a) Tortoises found on islands with comparatively lush vegetation are characterized by a domed shell. The domed shell protects the tortoise's soft parts as it makes its way through the thick undergrowth. This tortoise is a native of Isabela Island. (b) Tortoises found on arid islands, on which the typical vegetation is thornbush and cactus, are characterized by a saddleback shell. The high arch at the front of the shell enables the tortoise to reach upward in search of food. This tortoise was photographed on the island of Española.*

Most interesting to Darwin were the animals and plants that inhabited a small, barren group of islands, the Galapagos, which lie some 580 miles off the coast of Ecuador. The Galapagos were named after the islands' most striking inhabitants, the tortoises *(galápagos* in Spanish), some of which weigh 200 pounds or more. Each island has its own type of tortoise; sailors who took the tortoises on board and kept them as convenient sources of fresh meat on their sea voyages could generally tell which island any particular tortoise had come from (Figure I–7).

(a)

(b)

I–8   *The woodpecker finch is a rare phenomenon in the bird world because it is a tool user. Like the true woodpecker, it feeds on grubs, which it digs out of trees using its beak as a chisel. Lacking the woodpecker's long, barbed tongue, however, it resorts to an artificial probe, a twig or cactus spine, to remove the grubs from their holes. The woodpecker finch shown here has selected a cactus spine, which it has (a) inserted into a grub hole on a dead tree and (b) used to remove the grub from the hole. If the particular tool is an efficient one, as this cactus spine appears to be, the bird will carry it from tree to tree in its search for grubs.*

(a)   (b)

In addition to the tortoises, there was a group of finchlike birds, 13 species in all, that differed from one another in the sizes and shapes of their bodies and beaks, and particularly in the type of food they ate. In fact, although clearly finches, they had many characteristics seen only in completely different types of birds on the mainland. One finch, for example, feeds—as does the familiar woodpecker and also the aye-aye of Madagascar (Figure I–2b)—by routing insects out of the bark of trees. The finch is not fully equipped for this, however, having neither the long tongue of the woodpecker nor the long third finger of the aye-aye. Instead, the woodpecker finch (Figure I–8) uses a small stick or cactus spine to pry the insects loose.

From his knowledge of geology, Darwin knew that the Galapagos, clearly of volcanic origin, were much younger than the mainland. Yet the plants and animals of the islands were different from those of the mainland. Moreover, the inhabitants of the different islands in the archipelago differed from one another. Were the living things on each island the product of a separate special creation? "One might really fancy," Darwin mused at a later date, "that from an original paucity of birds in this archipelago one species had been taken and modified for different ends." This problem continued, in his own word, to "haunt" him.

The Darwinian Theory

Not long after his return to England, Darwin read an essay by the Reverend Thomas Malthus that had first appeared in 1798. In this essay, Malthus warned, as economists have warned ever since, that the human population was increasing so rapidly that it would soon be impossible to feed all of Earth's inhabitants.

Darwin saw that Malthus's conclusion—that food supply and other factors hold populations in check—is true for all species, not just the human one. For example, Darwin calculated that a single breeding pair of elephants, which are among the slowest reproducers of all animals, would, if all their offspring lived and, in turn, reproduced the normal number of offspring over a normal life span, produce a standing population of 19 million elephants in 750 years. Yet the average number of elephants generally remains the same over the years. So, although a single breeding pair could have, in theory, produced 19 million descendants, it did, in fact, produce an average of only two. But why these particular two? The process by which the survivors are "chosen" Darwin called **natural selection.**

Natural selection, according to Darwin, was a process analogous to the type of selection exercised by breeders of cattle, horses, dogs, and pigeons (Figure I–9). In **artificial selection,** we humans choose individual specimens of plants or animals for breeding on the basis of characteristics that seem to us desirable. In natural selection, the action of the environment takes the place of human choice. As individuals with certain hereditary characteristics survive and reproduce, and as individuals with other hereditary characteristics are eliminated, the population will slowly change. If some horses were swifter than others, for example, these individuals would be more likely to escape predators and survive, and their offspring, in turn, might be swifter, and so on.

I–9   *As a pigeon fancier, Darwin knew of the wide variety of exotic breeds that had been produced through artificial selection. All of the birds shown in this illustration from a book published in Germany in the early 1900s were descended from the common rock dove, the familiar city pigeon. As Darwin wrote, "If these varieties can be produced by man's hand, what might Nature not achieve?"*

According to Darwin, inherited variations among individuals, which occur in every natural population, are a matter of chance. They are not produced by the environment, by a "creative force," or by the unconscious striving of the organism. In themselves, they have no goal or direction. They may, however, turn out to be more or less useful to an organism, enhancing or diminishing its survival and reproduction. It is the operation of natural selection—the interaction of individual organisms with their environment—over a series of generations that gives direction to evolution. A variation arises by chance, but if it gives an organism even a slight advantage, it makes it more likely that the organism will leave surviving offspring.

To return to the giraffe, an animal with a slightly longer neck may have an advantage in feeding and thus be healthier, survive longer, and leave more offspring than one with a shorter neck. If neck length is an inherited trait, some of these offspring will also have long necks, and if the long-necked animals in this generation have an advantage, the next generation will include more long-necked individuals. In time, the population of short-necked giraffes will have become a population of longer-necked giraffes (although there will still be variations in neck length).

The essential difference between Darwin's formulation and that of any of his predecessors is the central role he gave to variation. Others had thought of variations as mere disturbances in the overall design, whereas Darwin saw that variations among individuals are the raw material of the evolutionary process. Species arise, he proposed, when differences among individuals within a group are gradually converted, over many generations, into differences between groups. This can occur as subsets of the original group become separated in space and time and are subjected to different environmental forces.

Darwin's *The Origin of Species,* which resulted from more than 20 years of thought after his return to England, is, in his own words, "one long argument." Fact after fact, observation after observation, culled from the most remote Pacific island to a neighbor's pasture, is recorded, analyzed, and commented upon. Every objection is weighed, anticipated, and countered. *The Origin of Species* was published on November 24, 1859, and the Western world has not been the same since.

Acceptance of Darwin's argument revolutionized the science of biology. As Ernst Mayr wrote in 1963:

> The theory of evolution is quite rightly called the greatest unifying theory in biology. The diversity of organisms, similarities and differences between kinds of organisms, patterns of distribution and behavior, adaptation and interaction, all this was merely a bewildering chaos of facts until given meaning by the evolutionary theory. There is no area of biology in which that theory has not served as an ordering principle.*

* From Ernst Mayr, *Animal Species and Evolution,* Harvard University Press, Cambridge, Mass., 1963.

Darwin returned to England with the *Beagle* in 1836. Two years later, he read the essay by Malthus, and in 1842 he wrote a preliminary sketch of his theory, which he revised in 1844. On completing the revision, he wrote a formal letter to his wife requesting her, in the event of his death, to publish the manuscript. Then, with the manuscript and letter in safekeeping, he turned to other work. For more than 20 years following his return from the Galapagos, Darwin mentioned his ideas on evolution only in his private notebooks and in letters to his scientific colleagues.

In 1856, urged on by his friends Charles Lyell and botanist Joseph Hooker, Darwin set slowly to work preparing a manuscript for publication. In 1858, some 10 chapters later, Darwin received a letter from the Malay archipelago from another English naturalist, Alfred Russel Wallace, who had corresponded with Darwin on several previous occasions. Wallace presented a theory of evolution that exactly paralleled Darwin's own. Like Darwin, Wallace had traveled extensively and also had read Malthus's essay. Wallace, tossing in bed one night with a fever, had a sudden flash of insight. "Then I saw at once," Wallace recollected, "that the ever-present variability of all living things would furnish the material from which, by the mere weeding out of those less adapted to the actual conditions, the fittest alone would continue the race." Within two days, Wallace's 20-page manuscript was completed and in the mail to Darwin.

When Darwin received Wallace's letter, he turned to his friends for advice, and Lyell and Hooker, taking matters into their own hands, presented the theory of Darwin and Wallace at a scientific meeting just one month later. At that meeting, Lyell and Hooker read to the au-

Alfred Russel Wallace (1823–1913). As a young man, Wallace explored the Malay archipelago for eight years, covering about 14,000 miles by foot and native canoe. During his stay there, he collected 125,000 specimens of plants and animals, many of them previously unknown. His book about his Malay travels bears this inscription: "To Charles Darwin, Author of 'The Origin of Species,' I dedicate this book, not only as a token of personal esteem and friendship but also to express my deep admiration for his genius and his works."

dience four papers from Darwin's notes of 1844, excerpts from two letters written by Darwin, and Wallace's manuscript. Their presentation received little attention, but for Darwin the floodgates were opened. He finished his long treatise in little more than a year, and the book was finally published. The first printing was a mere 1,250 copies, but they were sold out the same day.

Why Darwin's long delay? His own writings, voluminous though they are, shed little light on this question. But perhaps his background does. He came from a conventionally devout family, and he himself had been a divinity student. Perhaps most important, his wife, to whom he was deeply devoted, was extremely religious. It is difficult to avoid the speculation that Darwin, like so many others, found the implications of his theory difficult to confront.

As we shall see throughout the pages that follow, evolution is the thread that links together all the diverse phenomena of the living world.

The theory of evolution has also deeply influenced our way of thinking about ourselves. With the possible exception of the new astronomy of Copernicus and Galileo in the sixteenth and seventeenth centuries, no revolution in scientific thought has had as much effect on human culture as this one. One reason is, of course, that evolution is in contradiction to the literal interpretation of the Bible. Another difficulty is that it seems to diminish human significance. The new astronomy had made it clear that Earth is not the center of the universe or even of our own solar system. Then the new biology asked us to accept the proposition that, as far as science can show, we are not fundamentally different from other organisms in either our origins or our place in the natural world.

Evolutionary Theory Today

Later on, in Chapters 19–27, we shall consider in more detail the process of evolution and the diversity of organisms it has produced. As you will see then, evolutionary theory is itself still evolving. Lively debates about various aspects of evolution occur almost constantly among its students (biologists, like all scientists, are students). Should we, for instance, measure evolutionary relationships—as between humans and chimps—by physical appearance, by the evidence of the fossil record, or by minute variations in the genetic information encoded in the chromosomes? Are animals capable of unselfish behavior, or is seemingly unselfish behavior always—in the evolutionary sense—really self-serving? (One biologist, calculating rapidly, announced that he would lay down his life "for two brothers or eight first cousins.") There will always be controversies of this sort, because that is the way science works.

That evolution has occurred is not, however, a matter of controversy among biologists. There is virtually complete agreement that Earth has had a long history—some 4.5 billion years is the current estimate—and that, during the course of that history, the complex living things present today, including ourselves, arose from simpler forms. There is also agreement that the particular features that suit an organism to its environment—the swiftness of the cheetah, the fragrance of the flower, the camouflage of the katydid, our own opposable thumb—have arisen by the process of natural selection as proposed almost 140 years ago by Charles Darwin.

There are, however, controversies between scientists and religious fundamentalists of various faiths. The fundamentalists seek to explain the world in supernatural or spiritual terms, in accordance with their particular religious doctrines. Scientists do not deny the possibility of supernatural phenomena or the importance of religious belief. They maintain, however, that phenomena such as special creation, whether or not they exist, lie outside the realm of science. Such questions cannot be answered, or even argued, by scientific means because the answers depend ultimately on one's inner faith, rather than on observations or experiments involving the external, material world.

The Nature of Science

Science, including biology, is a way of seeking principles of order. Art is another way, as are religion and philosophy. Science differs from these others in that it limits its search to the natural world, the physical universe. It also differs in the way it conducts the search for understanding.

I–10   *The Christmas star orchid of Madagascar is characterized by a tubular nectary, generally about 11½ inches long. A little more than an inch of nectar is found at the very bottom of the tube. Darwin's study of this orchid led him to predict the existence of a moth with a tongue long enough to reach the nectar. Although he was ridiculed at the time by biologists who specialized in the study of insects, such a moth, which feeds on the nectar and pollinates the flowers, was later observed and identified. The young man in the photograph is Bedo Jaosolo, a talented and largely self-taught naturalist and guide who worked closely with scientists studying the rain forest of Madagascar. In 1989, at the age of 20, he was murdered, possibly because of his conservation activities.*

Observation, Hypothesis, Prediction, and Testing

The scientific process begins with an observation that leads to a question. (One of the most important in the history of biology was the question with which we began this Introduction: How does one make sense of the bewildering diversity of living things and the patterns in which they are scattered over the surface of the Earth?) A scientist seeks to answer the question by accumulating further information. Such information can be the product of concentrated thought. Or the additional information may be in the form of further observations, either original observations or reports from other observers (usually a combination of the two). Darwin, for example, spent decades gathering information—and thinking about that information—before proposing his concept of evolution by means of natural selection.

Early on, a scientist usually formulates a **hypothesis**—a tentative explanation. On the basis of the hypothesis, the scientist makes a **prediction** as to what will or will not be observed if the hypothesis is correct. The hypothesis is then subjected to a series of tests. For instance, birds migrate south in the fall, often flying at night, sometimes over open water, to a place they've never been before. How do they find their way? A tentative explanation: They navigate by the stars. A prediction: If birds navigate by the stars, they will not fly on cloudy nights. Testing by observation: Do migrating birds fly on cloudy nights? You might want to look for yourself.

Another example of prediction is provided by Darwin's study of the Christmas star orchid of Madagascar (Figure I–10). The long tubular structure trailing from each flower contains, at the bottom of the tube, a small amount of sugary nectar. On the basis of this structure, Darwin predicted that a moth would be found on Madagascar with a tongue between 10 and 11 inches long—because that length would be required to reach the nectar. Such a moth was discovered, some 40 years later, confirming Darwin's prediction.

Often, the tests of a hypothesis take the form of experiments. These are carefully structured observations that involve some form of manipulation or intervention by the investigator. As an example of the sequence from puzzling observation to hypothesis to prediction to testing, let us look at a study in England that also provides an example of the process of natural selection.

The Case of the Peppered Moth

The peppered moth, known scientifically as *Biston betularia,* was well known to British naturalists of the nineteenth century, who remarked that the moths were usually found on lichen-covered trees and rocks. Against this background, the light coloring of the moths made them practically invisible (Figure I–11). Until 1845, all reported specimens of *Biston betularia* were light-colored, but in that year one black moth of this species was captured in the growing industrial center of Manchester.

With the increasing industrialization of England, smoke particles began to pollute the foliage in the vicinity of industrial towns, killing the lichens and leaving the tree trunks bare. In heavily polluted districts, the trunks and even the rocks and ground became black. During this period, more and more black *Biston betularia* were found. Replacement of light-colored moths by dark ones proceeded briskly. By the 1950s, only a few of the light-colored population could be found, and these were far from industrial centers. Because of the prevailing westerly wind in England, pollutants were carried to the east of industrial towns, and the moths

tended to be of the black variety right up to the east coast of England. The few light-colored populations were concentrated in the west, where lichens still grew.

Where did the black *Biston betularia* come from? Eventually, it was demonstrated that the black color was a natural variation in the population. The black moths had always been there, in very small numbers. But why had their numbers increased so dramatically? In the late 1950s, H. B. D. Kettlewell, a physician who was also an amateur collector of moths and butterflies, hypothesized that the color of the moths protected them from insect-eating birds (Figure I–12).

In the face of strong opposition from biologists who specialized in the study of insects—all of whom claimed they had never seen a bird eat a *Biston betularia* of any color—Kettlewell set out to test his hypothesis. He predicted that if he released both light and dark moths in a polluted area and then recaptured them later, he would recapture more of the dark moths than of the light moths. Why? Conversely, he predicted that if he released both types of moths in an unpolluted area, he would later recapture more of the light moths than of the dark ones.

As an experimental test, Kettlewell marked a sample of moths of each color by carefully putting a spot on the underside of the wings, where it could not be seen by birds. Then he released known numbers of marked individuals into a bird reserve near Birmingham, an industrial area where 90 percent of the local *Biston betularia* population consisted of black moths. Another

I–11 Biston betularia, *the peppered moth, resting on a lichen-covered tree trunk in an unpolluted English countryside. Once the moth has positioned itself on a patch of lichen, it remains absolutely motionless, further enhancing its camouflage.*

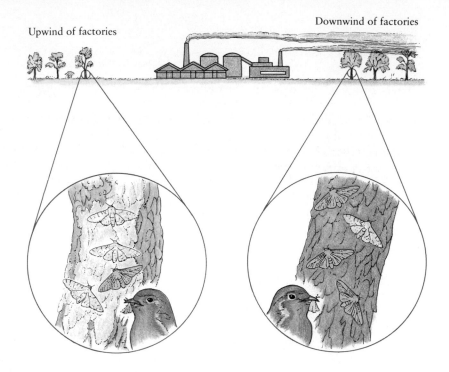

Upwind of factories

Downwind of factories

I-12   *Kettlewell hypothesized that the distribution and abundance of the light and dark varieties of the peppered moth were determined by the visibility of the moths to insect-eating birds. In unpolluted areas upwind of industrial zones, where the trees were covered with lichens, the light-colored moths would be less visible than the black moths and thus would be less likely to be eaten. In polluted areas downwind, however, where the trees were covered with soot, the black moths would be less visible and thus less likely to be eaten.*

sample was released into an unpolluted Dorset countryside, where no black moths were ordinarily found. He returned at night with light traps to recapture the marked moths. From the area around Birmingham, he recovered 40 percent of the black moths, but only 19 percent of the light ones. From the area in Dorset, 6 percent of the black moths and 12.5 percent of the light moths were recaptured.

To clinch the argument, Kettlewell placed moths on tree trunks in both locations, focused hidden movie cameras on them, and was able to record birds actually selecting and eating the moths. As his films revealed, they do this so rapidly that it is not surprising it was not previously observed. The photographic record showed that when equal numbers of black and light moths were available near Birmingham, the birds seized 43 light-colored moths and only 15 black ones; in Dorset, they took 164 black moths but only 26 light-colored forms.

Note that Kettlewell did not just think about the puzzling observation (the increase in the number of black moths) and a possible explanation (in heavily polluted districts, black moths were less likely to be seen and eaten by birds). He also devised tests of his hypothesis. And, when his first test, which was an indirect one, supported the hypothesis, he devised a second, more direct test. Repeated testing, typically using several different methods, is an essential characteristic of the scientific process.

When a scientist has collected enough information to support a particular hypothesis, he or she then reports the results to other scientists. Such a report may take place at a scientific meeting, or in a scientific publication, such as a journal or book (for example, Darwin's

The Origin of Species). If the information is sufficiently interesting or the hypothesis is an important one, other scientists will attempt to repeat the observations or experiments in order to confirm, refute, or extend them.

In the case of the moths, investigations by other scientists in England and the United States have demonstrated that the replacement of light-colored moths by dark-colored ones is a common occurrence in heavily polluted areas. It has been observed among some 70 other moth species in England and some 100 species of moths in the area of Pittsburgh, Pennsylvania. It has also been observed in many species of butterflies.

From Hypothesis to Theory

Once a hypothesis of broad, fundamental importance has survived a number of tests, it is generally referred to as a **theory.** Thus a theory in science has a different meaning from the word "theory" in everyday usage, in which "just a theory" carries with it the implication of a flight of fancy, a hunch, or some abstract, impractical notion. In science, a theory is a carefully formulated, well-tested proposition that describes the relationship among a number of different observations. The theory of evolution, for example, is built on an enormous number and variety of observations.

As new observations are made, however, theories are challenged. For example, various theories describing the structure of the atom have superseded one another as powerful new instruments have made new observations possible. However, the existence of atoms has not been questioned. Similarly, over the past 135 years, the theory of evolution has been tested, retested, amended, and fine-tuned. Some of the new discoveries in genetics—how traits are passed from parents to offspring—seemed at first to challenge the theory of evolution, but, as we shall see in later chapters, they have now become an important part of it. Such challenges will always continue, because, as long as there are curious human beings, there will always be new observations leading to new hypotheses, new predictions, and new tests. That is the way science works.

The Limits of Science

The raw materials of science are our observations of the phenomena of the natural universe. Science—unlike art, religion, or philosophy—is limited to what is observable and measurable and, in this sense, is rightly categorized as materialistic. Hunches are abandoned, hypotheses superseded, theories revised—and occasionally, shattered—but observations endure, and, moreover, they are used over and over again, sometimes in wholly new ways. It is for this reason that scientists stress and seek objectivity. In the arts, by contrast, the emphasis is on subjectivity—experience as filtered through the individual consciousness.

Because of this emphasis on objectivity, value judgments cannot be made in science in the way that such judgments are made in philosophy, religion, and the arts, and indeed in our daily lives. Whether or not something is good or beautiful or right in a moral sense, for example, cannot be determined by scientific methods.

Such judgments, even though they may be supported by a broad consensus, are not subject to scientific testing.

At one time, the sciences, like the arts, were pursued solely for their own sake, for pleasure and excitement and satisfaction of the insatiable curiosity with which we are both cursed and blessed. In the twentieth century, however, the sciences have spawned a host of giant technological achievements—polio vaccine, pesticides, electronic communications, indestructible plastics, nuclear power, perhaps even ways to manipulate our genetic heritage. Yet science cannot give us insight into how to use these achievements wisely or guidance in how to solve the problems they have generated.

The reason that science cannot and does not solve many of the problems for which we seek solutions is inherent in its nature. Most of the problems we now confront can be solved only by value judgments. For example, science has given us nuclear power and can give us predictions as to the extent of the biological damage that would result from accidents that allowed radioactivity to escape into the environment. Yet science cannot help us, as citizens, in weighing our fear of nuclear accidents against our energy needs and our desire for cleaner air. Scientists can develop a synthetic hormone that will produce early abortions, but they are not the ones to decide whether such a hormone will be made available to pregnant women seeking abortions. Science can give us information on which to base our judgments, but it cannot make these judgments for us.

I–13   *In science, great stress is placed on objectivity. As a consequence, scientists, speaking as scientists, refrain from making value judgments. Thus, for example, they would not identify any organism as the ugliest or the most beautiful. As a private individual, however, a scientist may have his or her own opinion about (a) the fishing bat and (b) the sacred lotus.*

(a)   (b)

It is one of the ironies of this so-called "age of science and materialism" that probably never before have ordinary individual men and women, including scientists, been confronted with so many moral and ethical dilemmas. In this text, we shall discuss some of the dilemmas that have grown out of the achievements of modern science and technology. Our greater concern, however, is to provide you with the biological knowledge necessary to understand the relevant information as you make your own value judgments regarding the problems that confront us now and that will do so in the future.

You may have decided to study biology because of an interest in environmental problems or a desire to know more about the workings of your own body or a concern with the implications of genetic engineering—in short, because it is "relevant." The study of biology is, indeed, pertinent to many aspects of our day-to-day existence, but do not make this your main focus. Above all other considerations, study biology because it is "irrelevant"—that is, study it for its own sake, because, like art and music and literature, it is an adventure for the mind and nourishment for the spirit.

Summary

Until the eighteenth century, it was generally accepted that species (kinds of organisms) were the products of a special (divine) creation and had remained unchanged ever since. This concept came into question as a result of several developments: (1) the discovery of an enormous number of new species by the European naturalists who fanned out around the globe; (2) studies in geology that indicated a process of gradual, constant change in the surface of the Earth; (3) the accompanying recognition that the Earth has had a long history (without which evolution would have been impossible); and (4) the discovery of the real nature of fossils.

Darwin was not the first to propose a theory of evolution. His most notable predecessor was Lamarck, whose theory of evolution—now known to be in error—was based on the inheritance of characteristics acquired by an organism during its lifetime. Darwin's theory differed from others in that it envisioned evolution as a two-part process, depending upon (1) the existence in nature of inheritable variations among organisms and (2) the process of natural selection by which some organisms, by virtue of their inheritable variations, leave more offspring than others. Darwin's theory is regarded as the greatest unifying principle in biology.

Science is a way of seeking knowledge of the natural world. The essential components of the scientific process are an observation that leads to a question, the formulation of a hypothesis (a tentative answer to the question), the prediction of what would be observed under a particular set of circumstances if the hypothesis were correct, and the testing of the prediction through further observations. Hypotheses that describe the relationship among a number of different observations and that have been repeatedly tested in a variety of different ways are known as theories. Theories provide a framework for further observations, questions, hypotheses, predictions, and tests. They are, as a result, always subject to further revision.

Questions

1. What is the essential difference between Darwin's theory of evolution and that of Lamarck?

2. The chief predator of an English species of snail is the song thrush. Snails that inhabit woodland floors usually have dark shells, whereas those that live on grass have yellow shells, which are less clearly visible against the lighter background. Explain, in terms of Darwinian principles.

3. The phrase "chance and necessity" has been used to describe the Darwinian theory of evolution. Relate this to the fact that snails living on grass do not have green shells, but there are, for example, green frogs and green insects.

4. What tool-using animals other than the woodpecker finch can you think of?

5. When scientists report new findings, they are expected to reveal their methods and raw data as well as their conclusions. Why is such reporting considered essential?

Suggestions for Further Reading

Berra, Tim M.: *Evolution and the Myth of Creationism: A Basic Guide to the Facts in the Evolution Debate*, Stanford University Press, Stanford, Calif., 1990.*

> *An excellent presentation of the modern theory of evolution, the evidence in its support, and examples of its explanatory power. Written for the general reader, this well-illustrated book also provides an overview of the major events in evolutionary history. It concludes with a discussion of recent political controversies over evolution.*

Darwin, Charles: *The Origin of Species by Means of Natural Selection, or The Preservation of Favored Races in the Struggle for Life*, New American Library, Inc., New York, 1986.*

> *Darwin's "long argument." Every student of biology should, at the very least, browse through this book to catch its special flavor and to begin to understand its extraordinary force.*

Darwin, Charles: *The Voyage of the Beagle*, New American Library, Inc., New York, 1988.*

> *Darwin's own chronicle of the expedition on which he made the discoveries and observations that eventually led him to his theory of evolution. The sensitive, eager young Darwin that emerges from these pages is very unlike the image many of us have formed of him from his later portraits.*

Kettlewell, H. B. D.: "Selection Experiments on Industrial Melanism in the Lepidoptera," *Heredity*, vol. 10, pages 287–301, 1955.

> *The journal article in which Kettlewell reported his now classic studies of natural selection in the peppered moth.*

Lanting, Frans: *A World out of Time: Madagascar*, Aperture Foundation, Inc., New York, 1990.

> *This beautiful book is filled with more than 100 of Lanting's photographs of the peoples, plants, and animals of Madagascar. Essays by Alison Jolly, one of the world's leading authorities on lemurs, John Mack, a cultural anthropologist, and conservationist Gerald Durrell make clear not only the uniqueness of this island paradise but also its imminent peril. Highly recommended.*

* Available in paperback.

Lewin, Roger: *Thread of Life: The Smithsonian Looks at Evolution*, Smithsonian Books, Washington, D.C., 1991.*

> *An absorbing account of the historical development of evolutionary thought, with current applications from biochemistry, paleontology, and geology. Well written and beautifully illustrated with a rich assortment of color photographs.*

Mayr, Ernst: *The Growth of Biological Thought: Diversity, Evolution, and Inheritance*, Harvard University Press, Cambridge, Mass., 1982.*

> *The first of two volumes on the history of biology and its major ideas, written by one of the leading figures in the study of evolution. The introductory chapters provide an outstanding analysis of the philosophy and methodology of the biological sciences. This book, like Darwin's* The Origin of Species, *should at least be sampled by every serious student of biology.*

Milner, Richard: *The Encyclopedia of Evolution: Humanity's Search for Its Origins*, W. W. Norton & Company, New York, 1993.*

> *With more than 600 entries, this engagingly written reference covers a wide range of topics on the theory of evolution and its influence on Western culture—from Darwin and Wallace to sperm banks, Chief Red Cloud, and* Fantasia. *Well illustrated, authoritative, and rich with fascinating detail.*

Moorehead, Alan: *Darwin and the Beagle*, Penguin Books, New York, 1978.*

> *A delightful narrative of Darwin's journey, beautifully illustrated with contemporary or near contemporary drawings, paintings, and lithographs.*

Simpson, George Gaylord: *Fossils and the History of Life*, W. H. Freeman and Company, New York, 1983.

> *A beautifully written and illustrated summary of the fossil evidence on which much of our knowledge of evolution rests, by one of the major figures in twentieth-century evolutionary biology.*

Tattersall, Ian: "Madagascar's Lemurs," *Scientific American*, January 1993, pages 110–117.

Wallace, Alfred Russel: *The Malay Archipelago*, Oxford University Press, New York, 1990.*

> *A reprint of Wallace's classic work, accompanied by an excellent introduction that provides much additional information about Wallace's life and scientific work.*

The Unity of Life

1

Atoms and Molecules

The universe in which we live is composed of stars that, like our Sun, are born and die. In the birth and death of stars, all of the atoms that make up our solar system, our planet, and ourselves had their beginnings.

The birth of a star begins in clouds of gases, mainly hydrogen and helium, that surge and eddy through the spaces between previously formed stars. Moving at random and accompanied by very fine dust, the gases form accidental pockets, their atoms held together by gravity. As a cloud contracts, its interior becomes more dense and its temperature rises. From this mass, a star will form.

As the embryonic star continues to contract, the temperature rises millions of degrees, and nuclear fusion reactions take place. In these reactions, hydrogen is transformed to helium and energy is released. At this moment, the star is born. "Fireflies tangled in a silver braid," wrote the poet Tennyson about the Pleiades, the group of young stars shown here. The haloes and blue patches of haze surrounding the stars are remnants of the dust and gases from which they formed. The Pleiades can be seen in the "shoulder" of the constellation Taurus (the bull).

During a star's lifetime, hydrogen is transformed to helium, and helium to carbon and oxygen, which themselves fuse to form the heavier elements. These reactions release enormous quantities of energy, visible to us as the light of distant stars and as the light and heat of the nearest star—our Sun—which formed some 5 billion years ago.

Eventually a star uses up its fuel, the hydrogen atoms at its core. At its death, there is a sudden great explosion in the sky. During the few minutes of the explosion, the old, dying star releases energy comparable to the total amount emitted by our Sun in a billion years. This enormous blast blows the outer layers of the star into space, creating new clouds of gases and dust. Over time, these will cool and merge with the interstellar clouds of dust and gas from which new stars will be born.

Our universe began, according to current theory, with the "big bang," a tremendous explosion that filled all space. Prior to this, all of the energy and matter of the present universe is thought to have existed in the form of pure energy, compressed into an infinitesimally small point. This energy was released by the "big bang," and every particle of matter formed from the energy was hurled away from every other particle. The temperature at the time of the explosion—some 10 to 20 billion years ago—was about 100,000,000,000 degrees Celsius (10^{11} °C). At this temperature, not even atoms could hold together; all matter was in the form of subatomic particles. Moving at enormous velocities, these particles had fleeting lives. Colliding with great force, they annihilated one another, creating new particles and releasing more energy.

As the universe expanded and gradually cooled, more matter began to form from energy. When the universe reached a temperature of 2500°C (about the temperature of a white-hot wire in an incandescent light bulb), two types of stable particles, previously present only in relatively small amounts, began to assemble. These particles—protons and neutrons—are very heavy as subatomic particles go. Held together by forces that are still not completely understood, they formed the central cores, or nuclei, of atoms. These nuclei, with their positively charged protons, attracted small, light, negatively charged particles—electrons—which moved rapidly around them. Thus, atoms came into being.

It is from these atoms—blown apart, formed, and re-formed over the course of several billion years—that all the stars and planets of our universe are formed, including our particular star and planet. And it is from the atoms present on this planet that living systems assembled themselves and evolved. Each atom in our own bodies had its origin in that enormous explosion so long ago. You and I are flesh and blood, but we are also stardust.

This text begins where life began, with the atom. At first, it might appear that lifeless atoms have little to do with biology. Bear with us, however. A closer look reveals that the activities we associate with being alive depend on combinations and exchanges between atoms. Moreover, the force that binds the electron to the atomic nucleus stores the energy that powers living systems.

Atoms

All matter, including the most complex living organisms, is made up of combinations of **elements.** Some 92 elements occur naturally on Earth. Many of them are familiar to you; for example, carbon, found in its purest forms in diamond and graphite; oxygen, present in the air we breathe; calcium, incorporated by a great variety of organisms into sea shells, egg shells, bones, and teeth; and iron, the element responsible for the red color of our blood. Elements are, by definition, substances that cannot be broken down into other substances by ordinary chemical reactions. The smallest particle of an element that possesses the properties of that element is an **atom.**

Every atom has a core, or **nucleus,** that contains one or more positively charged particles, called **protons.** The number of protons distinguishes the atoms of different elements from one another. For example, an atom of hydrogen, the simplest of the elements, has 1 proton in its nucleus; an atom of carbon has 6 protons in its nucleus. The number of protons in the nucleus of a particular atom is called its **atomic number.** The atomic number of hydrogen is therefore 1, and the atomic number of carbon is 6.

Outside the nucleus of an atom are negatively charged particles, the **electrons,** which are attracted by the positive charge of the protons. The number of electrons in an atom is determined by and is equal to the number of protons in its nucleus. The electrons, in turn, determine whether an atom will react with other atoms and how it will do so. As we shall see, chemical reactions involve changes in the numbers and energy of electrons.

Atoms also contain **neutrons,** which are uncharged particles of about the same weight as protons. These, too, are found in the nucleus of the atom, where they seem to play a role in stabilizing the nucleus. The **atomic weight** of an element is essentially equal to the number of protons plus neutrons in the nuclei of its atoms. As you can calculate from the data in Table 1–1, the atomic weight of carbon is 12, whereas that of hydrogen, which contains no neutrons, is 1. Electrons are so

light by comparison that they are usually disregarded in calculations of atomic weight. When you weigh yourself, only about 30 grams—approximately 1 ounce—of your total weight is made up of electrons.

Isotopes

All atoms of a particular element have the same number of protons in their nuclei. Sometimes, however, different atoms of the same element contain different numbers of neutrons. These atoms, which differ from one another in their atomic weights but not in their atomic numbers, are known as **isotopes** of the element.

Hydrogen, for example, exists in three different isotopic forms (Table 1–2). The common form of hydro-

Table 1–1	Atomic Structure of Some Familiar Elements			
		Nucleus		
Element	Symbol	Number of Protons	Number of Neutrons*	Number of Electrons
Hydrogen	H	1	0	1
Helium	He	2	2	2
Carbon	C	6	6	6
Nitrogen	N	7	7	7
Oxygen	O	8	8	8
Sodium	Na	11	12	11
Magnesium	Mg	12	12	12
Phosphorus	P	15	16	15
Sulfur	S	16	16	16
Chlorine	Cl	17	18	17
Potassium	K	19	20	19
Calcium	Ca	20	20	20

* In most common isotope.

Table 1–2	Isotopes of Hydrogen					
Isotope						
Name	Symbol	Atomic Number	Atomic Weight	Number of Protons	Number of Neutrons	Number of Electrons
Hydrogen	^1H	1	1	1	0	1
Deuterium	^2H	1	2	1	1	1
Tritium	^3H	1	3	1	2	1

gen, with its one proton, has an atomic weight of 1 and is symbolized as ^{1}H, or simply H. A second isotope of hydrogen contains one proton and one neutron and so has an atomic weight of 2; this isotope is symbolized as ^{2}H. A third, extremely rare isotope, ^{3}H, has one proton and two neutrons and so has an atomic weight of 3. The chemical behavior of the two heavier isotopes is essentially the same as that of ordinary hydrogen. All three isotopes have only one electron each, and it is the electrons that determine chemical properties.

Most elements have several isotopic forms. The differences in weight, although very small, are sufficiently great that they can be detected with modern laboratory apparatus. Moreover, many, but not all, of the less common isotopes are radioactive. This means that the nucleus of the atom is unstable and emits energy as it changes to a more stable form. The energy given off by the nucleus of a radioactive isotope can be detected with a Geiger counter or on photographic film.

Isotopes have a number of important uses in biological research and in medicine. They can be used, for example, to determine the age of fossils. Each type of radioactive isotope emits energy and changes into another kind of isotope at a characteristic and fixed rate. As a result, the relative proportions of different isotopes in a fossil or in a nearby rock sample give a good indication of how long ago it was formed.

Another use of radioactive isotopes is as "tracers." For example, an isotope of the element thallium, which is unreactive in the human body, can be used to identify blocked blood vessels in persons with symptoms of heart disease. Isotopes have other diagnostic uses in medicine, as well as an important role in the treatment of many forms of cancer.

Since isotopes of the same element all have the same chemical properties, a radioactive isotope will behave in an organism just as its more common nonradioactive isotope does. As a result, biologists have been able to use isotopes of a number of elements—especially carbon, nitrogen, and oxygen—to trace the course of many essential chemical reactions in living organisms.

Electrons and Energy

Atoms—the indivisible units of the elements—are the fundamental building blocks of all matter, both living and nonliving. Yet they are mostly empty space. As the electrons move around the nucleus at almost the speed of light, the distance from electron to nucleus is, on average, about 1,000 times the diameter of the nucleus. The electrons are so exceedingly small that the space is almost entirely empty.

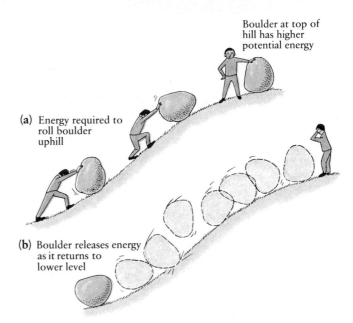

Boulder at top of hill has higher potential energy

(a) Energy required to roll boulder uphill

(b) Boulder releases energy as it returns to lower level

1-1   (a) *The energy used to push a boulder to the top of a hill (less the heat energy resulting from friction between the boulder and the hill) becomes potential energy, stored in the boulder as it rests at the top of the hill.* (b) *When the boulder rolls back down the hill, this potential energy is converted to energy of motion and released.*

The distance between an electron and the nucleus is determined by the amount of potential energy the electron possesses. **Potential energy**, often called "energy of position," is the stored energy that any object—not just an electron—possesses because of its position. The greater the amount of energy possessed by the electron, the farther it will be from the nucleus. Thus, an electron with a relatively small amount of energy is found close to the nucleus and is said to be at a low **energy level**. An electron with more energy is farther from the nucleus, at a higher energy level.

An analogy may be useful. A boulder resting on flat ground neither gains nor loses potential energy. If, however, you change its position by pushing it up a hill, you increase its potential energy. As long as it sits on the peak of the hill, the boulder once more neither gains nor loses potential energy. If it rolls down the hill, however, potential energy is converted to energy of motion and released (Figure 1-1). The electron is like the boulder in that an input of energy can move it to a higher energy level—farther away from the nucleus. As long as it remains at the higher energy level, it possesses the added energy. And, just as the boulder is likely to roll downhill, the electron also tends to go to its lowest possible energy level.

1–2   (a) *When an atom, such as the hydrogen atom in this diagram, receives an input of energy, an electron (shown here in red) may be boosted to a higher energy level. The electron thus gains potential energy.* (b) *When the electron returns to its previous level, this energy is released in the form of heat or light.*

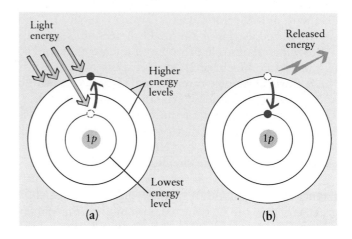

It takes energy to move a negatively charged electron farther away from a positively charged nucleus, just as it takes energy to push a boulder up a hill. However, unlike the boulder on the hill, the electron cannot be pushed part of the way up. With a sufficient input of energy, an electron can move from a lower energy level to any one of several higher energy levels, but it cannot move to an energy state somewhere in between. For an electron to move from one energy level to a higher one, it must absorb a discrete amount of energy, equal to the difference between the two particular energy levels. When the electron returns to its original energy level, that same amount of energy is released (Figure 1–2).

In the green cells of plants and algae, the radiant energy of sunlight raises electrons to higher energy levels. In a series of reactions, which will be described in Chapter 9, these electrons are passed "downhill" from one energy level to another until they return to their original energy level. During these transitions, the radiant energy of sunlight is transformed into the chemical energy on which life on Earth depends.

The Arrangement of Electrons

The way an atom reacts chemically is determined by the number and arrangement of its electrons. An atom is most stable when all of its electrons are at their lowest possible energy levels. Therefore the electrons of an atom fill the energy levels in order—the first is filled before the second, the second before the third, and so on.

The first energy level, closest to the nucleus, can hold a maximum of two electrons. Thus the single electron of hydrogen (atomic number 1) moves around the nucleus within the first energy level. Similarly, the two electrons of helium (atomic number 2) move within the first energy level.

1–3   *The leaves of these corn plants contain chlorophyll, which gives them their green color. When light strikes a molecule of chlorophyll, electrons in the molecule are raised to higher energy levels. As each electron returns to its previous energy level, a portion of the energy released is captured in the bonds of carbon-containing molecules.*

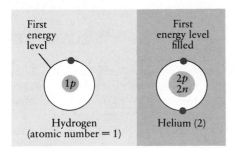

First energy level

Hydrogen (atomic number = 1)

1p

First energy level filled

2p
2n

Helium (2)

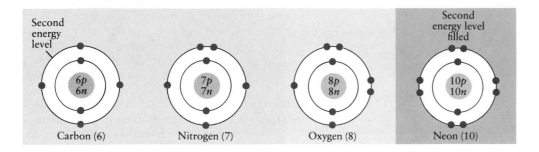

Second energy level

6p
6n

Carbon (6)

7p
7n

Nitrogen (7)

8p
8n

Oxygen (8)

Second energy level filled

10p
10n

Neon (10)

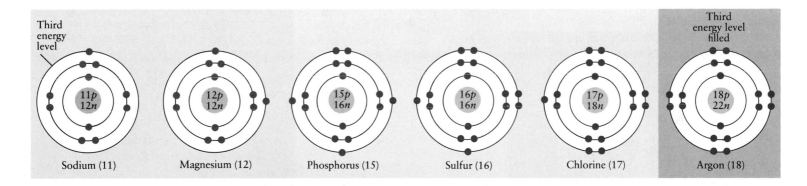

Third energy level

11p
12n

Sodium (11)

12p
12n

Magnesium (12)

15p
16n

Phosphorus (15)

16p
16n

Sulfur (16)

17p
18n

Chlorine (17)

Third energy level filled

18p
22n

Argon (18)

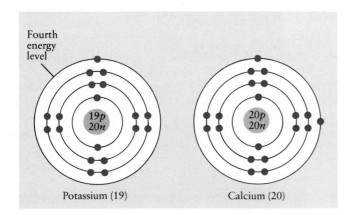

Fourth energy level

19p
20n

Potassium (19)

20p
20n

Calcium (20)

1–4 *Schematic diagrams of the electron arrangements in some familiar elements. In each of these diagrams, the electrons are at their lowest possible energy levels. Note how the energy levels are filled in sequence as the atomic number increases.*

The first energy level can hold a maximum of 2 electrons. The second energy level can hold a maximum of 8 electrons, as can the third energy level of the elements through calcium (atomic number 20). As you can see, potassium (atomic number 19) has one electron in the fourth energy level, and calcium (atomic number 20) has two electrons in the fourth energy level. In atoms of higher atomic number, additional electrons (up to a maximum of 10) fill the third energy level before any more electrons are added to the fourth energy level.

Atoms of higher atomic number than helium have more than two electrons. Since the first energy level is filled by two electrons, the additional electrons must occupy higher energy levels, farther from the nucleus. The second energy level can hold a maximum of eight electrons—and so can the third energy level of elements through atomic number 20 (calcium).

Figure 1–4 illustrates the electron arrangements in the atoms of 14 representative elements, 11 of which play major roles in living systems. Each diagram shows the energy levels occupied by the electrons when the atom is in its ground state—that is, when all of its electrons are at their lowest possible energy levels. You will find it helpful to refer to these diagrams as we consider the chemical behavior of specific elements.

Models of Atomic Structure

The concept of the atom as the indivisible unit of the elements is almost 200 years old. However, our ideas about atomic structure have undergone many changes over the years. These ideas are usually presented in the form of models.

The earliest model of the atom, emphasizing its indivisibility, portrayed the atom as a sphere like a billiard ball. When electrons were discovered and it was realized that they could be removed from the atom, the billiard-ball model gave way to the plum-pudding model. This model represented the atom as a solid, positively charged mass with negatively charged particles, the electrons, embedded in it.

Subsequently, however, physicists found that the electrons are located outside the nucleus, moving around it at very high speed. This discovery gave rise to the planetary model of the atom **(a)**, which depicted the electrons as moving in orbits around the nucleus. Although this model was of limited scientific use,

it gained wide currency as a symbol for atomic energy.

A more useful model is derived from the work of the Danish physicist Niels Bohr (1885–1962), who made the key discovery that different electrons have different amounts of energy and are at different distances from the nucleus. In the Bohr model **(b),** the energy lev-

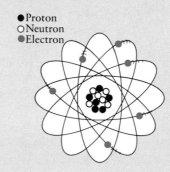

(a) A planetary model of the carbon atom.

els are depicted as concentric circles surrounding the nucleus. The Bohr model is not, in any sense, a true "picture" of the atom, and it has been superseded by another model. However, schematic diagrams derived from the Bohr model can help us visualize the energy levels of an atom, keep track of the number of electrons in each energy level, and see how atoms are likely to interact with one another.

The most recent model of atomic structure provides a more accurate picture of the atom. An electron is so small and moves so rapidly that it is theoretically impossible to determine, at any given moment, both its precise location and the exact amount of energy it possesses. As a result of this difficulty, the current model describes the pattern of an electron's motion rather than its position. The volume of space in which an electron will be found 90 percent of the time is defined as its **orbital.**

In any atom, the electrons at the lowest energy level—the first energy level—occupy a single spherical orbital, which can contain a

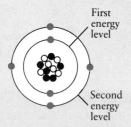

(b) A Bohr model of the carbon atom.

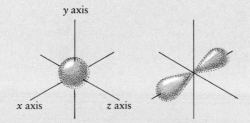

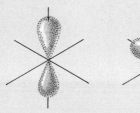

(c) The four orbitals of the second energy level. Each orbital can hold a maximum of two electrons. One orbital is spherical and encloses the spherical orbital of the first energy level, which, in turn, encloses the nucleus, located at the intersection of the axes. The other three orbitals

are dumbbell-shaped, with their axes perpendicular to one another. For clarity, the orbitals are shown individually in this diagram. In reality, the orbitals influence one another and determine the overall shape of the atom.

maximum of two electrons. At the second energy level, which can contain a maximum of eight electrons, there are four orbitals **(c)**. At the third energy level of elements through atomic number 20 (calcium), there are also four orbitals. In elements of higher atomic number, the pattern becomes more complex.

The electrons in each orbital traverse a particular region of three-dimensional space. Taken together, these patterns of electron movement give the atom as a whole a particular three-dimensional shape **(d)**. The atom, however, has no rigid boundaries but is instead defined by regions of charge. Even these are not always the same. About 10 percent of the time, the electrons of any given orbital are somewhere else in space, farther from the nucleus.

Such a "soft" structure would appear to provide an unlikely foundation for the material world that surrounds us and for the organisms with which it is filled. And, yet, according to modern physics, such is the case.

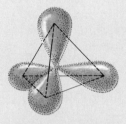

(d) An orbital diagram of the carbon atom. In this atom, each orbital of the second energy level contains only one electron. Each electron moves in a pattern that provides maximum separation from the other electrons. The resulting shape of the atom resembles four teardrops, extending out from the nucleus.

The Basis of Chemical Reactivity

As we have noted previously, an atom is most stable when all of its electrons are at their lowest possible energy levels. Moreover, an atom in which the outermost energy level is completely filled with electrons is more stable than one in which the outer energy level is only partially filled. For example, helium (atomic number 2) has two electrons at the first energy level, which means that its outer energy level (in this case, also its lowest energy level) is completely filled. Helium is therefore extremely stable and does not ordinarily participate in chemical reactions. Similarly, neon (atomic number 10) has two electrons at the first energy level and eight at the second energy level. Both energy levels are completely filled, and neon is unreactive.

In the atoms of most elements, however, the outer energy level is only partially filled. Atoms in which the outer energy level is partially filled tend to interact with other atoms in such a way that after the reaction both atoms have completely filled outer energy levels. Some atoms lose electrons. Others gain electrons. And, in many of the most important chemical reactions that occur in living systems, atoms share their electrons with each other.

Bonds and Molecules

When atoms interact with one another, resulting in filled outer energy levels, new, larger particles are formed. Discrete particles, consisting of two or more atoms, are known as **molecules.** The forces that hold atoms together within molecules are known as **chemical bonds.** Two principal types are ionic and covalent.

Ionic Bonds

For many atoms, the simplest way to attain a completely filled outer energy level is either to gain or to lose one or two electrons. For example, a chlorine atom (atomic number 17) needs one electron to complete its outer energy level. By contrast, a sodium atom (atomic number 11) has a single electron in its outer energy level. This electron is strongly attracted by the chlorine atom and jumps from the sodium to the chlorine (Figure 1–5). As a result of this transfer, both atoms have outer energy levels that are completely filled, and all the electrons are at their lowest possible energy levels.

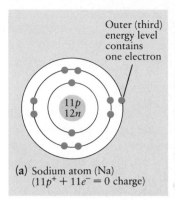

Outer (third) energy level contains one electron

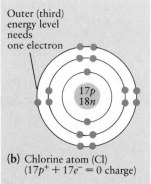

Outer (third) energy level needs one electron

(a) Sodium atom (Na)
$(11p^+ + 11e^- = 0$ charge$)$

(b) Chlorine atom (Cl)
$(17p^+ + 17e^- = 0$ charge$)$

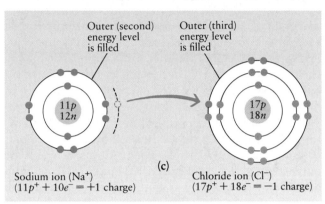

Outer (second) energy level is filled

Outer (third) energy level is filled

(c)

Sodium ion (Na⁺)
$(11p^+ + 10e^- = +1$ charge$)$

Chloride ion (Cl⁻)
$(17p^+ + 18e^- = -1$ charge$)$

1–5 *The interaction of sodium and chlorine.* **(a)** *The sodium atom has only one electron in its outer energy level.* **(b)** *The chlorine atom, by contrast, needs to gain one electron in order to complete its outer energy level.* **(c)** *When these two atoms come close to one another, the single electron in the outer energy level of the sodium atom jumps to the outer energy level of the chlorine atom, thereby completing that energy level. When sodium loses the electron, its second energy level—with a full complement of 8 electrons—becomes its outer energy level. Thus both atoms now have filled outer energy levels and are more stable than they were before the electron transfer. However, the atoms have become electrically charged. Sodium now has a charge of +1, and chlorine a charge of −1.*

In this process, however, the original atoms have become electrically charged. Such charged atoms are known as **ions**. The chlorine atom, having accepted an electron from sodium, now has one more electron than proton and is a negatively charged chloride ion, symbolized as Cl^-. Conversely, the sodium atom now has one less electron than proton and is a positively charged sodium ion, symbolized as Na^+.

Because of their charges, positive and negative ions attract one another. Thus the sodium ion (Na^+) with its single positive charge is attracted to the chloride ion (Cl^-) with its single negative charge. The resulting substance, sodium chloride (NaCl), is ordinary table salt (Figure 1–6). Similarly, when a calcium atom (atomic number 20) loses two electrons, the resulting calcium ion (Ca^{2+}) can attract and hold two Cl^- ions. Calcium chloride is identified in chemical shorthand as $CaCl_2$, with the subscript 2 indicating that two chloride ions are present for each ion of calcium.

Bonds that involve the mutual attraction of ions of opposite charge are known as **ionic bonds**. Such bonds can be quite strong, but, as we shall see in the next chapter, many ionic substances break apart easily in water, producing free ions. Small ions such as Na^+ and Cl^- make up less than 1 percent of the weight of most living matter, but they play crucial roles. Potassium ion (K^+) is the principal positively charged ion in most organisms, and many essential biological processes occur only in its presence. Calcium ion (Ca^{2+}), K^+, and Na^+ are all involved in the production and propagation of the nerve impulse. In addition, Ca^{2+} is required for the contraction of muscles and for the maintenance of a normal heartbeat. Magnesium ion (Mg^{2+}) forms a part of the chlorophyll molecule, the molecule in green plants and algae that traps radiant energy from the sun.

1–6 **(a)** *Oppositely charged ions, such as sodium and chloride ions, depicted here as spheres, attract one another. Table salt is crystalline NaCl, a latticework of alternating Na^+ and Cl^- ions held together by their opposite charges. Such bonds between oppositely charged ions are known as ionic bonds.*

(b) *The regularity of the latticework is reflected in the structure of salt crystals, magnified here about 30 times.*

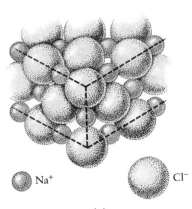

Na⁺   Cl⁻

(a)

(b)

1–7   *The acrobatic leap of the European tree frog, like the movement of all complex animals, is the result of coordinated muscle contractions triggered by nerve impulses. Sodium, potassium, and calcium ions are involved in producing and propagating nerve impulses, and calcium ions are required for the contraction of muscle fibers.*

Covalent Bonds

Another way for atoms to complete their outer energy levels is by sharing electrons with each other. Chemical bonds that result from the sharing of one or more pairs of electrons are known as **covalent bonds.** In a covalent bond, each electron spends part of its time around one nucleus and part of its time around the other. Thus the sharing of electrons completes the outer energy level of each atom and neutralizes the positive charge of its nucleus.

Atoms that need to gain electrons to achieve a filled, and therefore stable, outer energy level have a strong tendency to form covalent bonds. Thus, to take the simplest example, a hydrogen atom forms a covalent bond with another hydrogen atom (Figure 1–8), producing a molecule of hydrogen gas (H_2). A hydrogen atom can also form a covalent bond with any other atom that needs to gain an electron to complete its outer energy level.

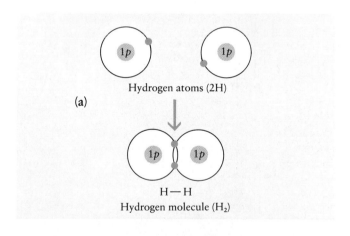

1–8   (a) *When a molecule of hydrogen is formed, each atom shares its single electron with the other atom. As a result, both atoms effectively have a filled first energy level, containing two electrons—a highly stable arrangement. This type of bond, in which electrons are shared, is known as a covalent bond. In structural formulas, it is represented as a single line: —.*

(b) *A representation of the volume of space traversed by the two electrons of the covalent bond as they move around the hydrogen nuclei. Because the electrons move so rapidly—at nearly the speed of light—the charges of both nuclei are effectively neutralized at all times.*

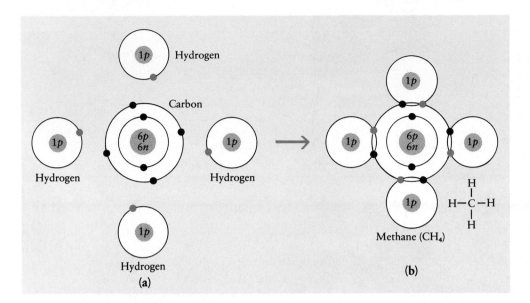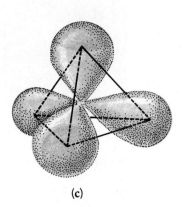

1–9   *A carbon atom, with four electrons in its outer energy level, can form covalent bonds with as many as four other atoms.*
(a) When a carbon atom reacts with four hydrogen atoms, each of the electrons in its outer energy level forms a covalent bond

with the single electron of one hydrogen atom, producing (b) a molecule of methane (CH_4).
(c) A three-dimensional representation of the methane molecule. The electrons forming the covalent bonds move rapidly in complex

patterns that encompass the hydrogen nuclei and also bring the electrons close to the carbon nucleus. The positive charge of each nucleus is neutralized, and the outer energy level of each atom is filled.

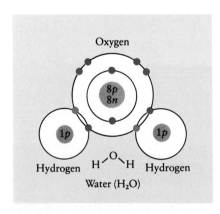

1–10   *A schematic diagram of the water molecule (H_2O). Each of the two single covalent bonds in the molecule consists of one electron contributed by oxygen and one electron contributed by hydrogen.*

Of extraordinary importance in living systems is the capacity of carbon atoms to form covalent bonds. A carbon atom has four electrons in its outer energy level (see Figure 1–4). It can share each of those electrons with another atom, forming covalent bonds with as many as four other atoms (Figure 1–9). The covalent bonds formed by a carbon atom may be with different atoms (most frequently hydrogen, oxygen, and nitrogen) or with other carbon atoms. As we shall see in Chapter 3, this tendency of carbon atoms to form covalent bonds with other carbon atoms gives rise to the large molecules that form the structures of living organisms and that participate in essential life processes.

Single and Double Covalent Bonds

There are various ways in which atoms can participate in covalent bonds and fill their outer energy levels. Oxygen, for example, has six electrons in its outer energy level (see Figure 1–4). Four of these electrons are grouped into two pairs and are generally unavailable for covalent bonding. The other two electrons are unpaired, and each can be shared with another atom in a covalent bond. In the water molecule (H_2O), one of these electrons participates in a covalent bond with one hydrogen atom, and the other in a covalent bond with a different hydrogen atom (Figure 1–10). Two **single bonds** are formed, and all three atoms have filled outer energy levels.

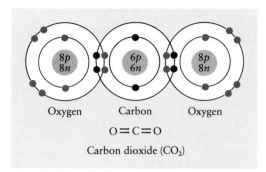

1-11 *A schematic diagram of the carbon dioxide molecule* (CO_2). *The carbon atom in the center of the molecule participates in two double covalent bonds, one with each oxygen atom. Each double bond consists of two pairs of electrons shared by the two atoms that participate in the bond. In structural formulas, a double bond is represented by two parallel lines:* =.

The bonding situation is different in another familiar substance, carbon dioxide (CO_2). In this molecule, the two available electrons of each oxygen atom participate with two electrons of a *single* carbon atom in the formation of *two* covalent bonds (Figure 1–11). Each oxygen atom is joined to the central carbon atom by two pairs of electrons (four electrons). Such bonds are called **double bonds.** Carbon atoms can form double bonds with each other as well as with other atoms, and so the variety of kinds of molecules that carbon can form is very large.

Single bonds are flexible, leaving atoms free to rotate in relation to one another. Double bonds, however, hold the atoms relatively rigid in relation to one another. The presence of double bonds in a molecule can make a significant difference in its properties. For example, both fats and oils are composed of carbon and hydrogen atoms covalently bonded together, but in fats the bonds are all single and in oils some of the bonds are double. The rigidity caused by these double bonds prevents the oil molecules from packing together, and, as a consequence, oils are liquids at room temperature. In fats, by contrast, the molecules can bend and twist, fitting closely together in a structure that is solid at room temperature.

Polar Covalent Bonds

The atomic nuclei of different elements have different degrees of attraction for electrons. As a consequence, in covalent bonds formed between atoms of different elements, the electrons are not shared equally between the atoms involved. Instead, the shared electrons tend to spend more time around the nucleus with the greater attraction. The atom around which the electrons spend more time has a slightly negative charge. The other atom has a slightly positive charge, since the electrons spend less time around it and its nuclear charge is not entirely neutralized.

Covalent bonds in which electrons are shared unequally are known as **polar covalent bonds.** Such bonds often involve oxygen atoms, to which electrons are strongly attracted. In molecules that are perfectly symmetrical, such as carbon dioxide, the unequal charges cancel out and the molecule as a whole is **nonpolar.** However, in asymmetrical molecules, such as water, the molecule as a whole is **polar,** with regions of partial positive charge and regions of partial negative charge. Many of the special properties of water, upon which life depends, derive largely from its polar nature, as we shall see in the next chapter.

Chemical Reactions

Chemical reactions—exchanges of electrons among atoms—can be compactly described by **chemical equations.** For example, the equation for the formation of sodium chloride from sodium and chloride ions is

$$\underset{\text{Reactants}}{Na^+ + Cl^-} \longrightarrow \underset{\text{Product}}{NaCl}$$

The arrow in the equation means "forms" or "yields," and it shows the direction of chemical change. Like algebraic equations, chemical equations "balance." The number and kinds of atoms in the products of the reaction must equal the number and kinds of atoms in the original reactants.

To take a slightly more complex example, hydrogen gas can combine with oxygen gas to produce water. As you know, hydrogen gas is H_2; similarly, oxygen gas is O_2. Each molecule of water contains two atoms of hydrogen and one of oxygen. Therefore the reactants must be in the proportion of two hydrogens to one oxygen:

$$\underset{\text{Reactants}}{2H_2 + O_2} \longrightarrow \underset{\text{Product}}{2H_2O}$$

Two molecules of H_2 plus one molecule of O_2 yield two molecules of water. The equation for a chemical reaction thus tells us the kinds of atoms that are present, their proportions, and the direction of the reaction.

A substance that contains atoms of two or more different elements, held together in a definite and constant proportion by chemical bonds, is known as a chemical **compound.** Examples of chemical compounds include water (H_2O), sodium chloride (NaCl), carbon dioxide (CO_2), methane (CH_4), and glucose ($C_6H_{12}O_6$).

The Signs of Life

What do we mean when we speak of "life" or "the evolution of life"? Actually, there is no simple definition. Life does not exist in the abstract. There is no "life," only living things. Moreover, there is no single, simple way to draw a sharp line between the living and the nonliving. Certain characteristics, however, taken together, distinguish living things from inanimate (that is, nonliving) objects.

(a) Living things are highly organized, as in this cross section of a stem of a sycamore sapling. This stem reflects the complicated organization of many different kinds of atoms into molecules and of molecules into complex structures. Such complexity of form, which is never found in inanimate objects of natural origin, makes possible the specialization of different parts of a living organism for different functions.

(b) Living organisms are homeostatic, which means simply "staying the same." Although they constantly exchange materials with the outside world, they maintain a relatively stable internal environment quite unlike that of their surroundings. Even this tiny, apparently fragile animal, a rotifer, has an internal chemical composition that differs from its changing environment.

(c) Living things reproduce themselves. They make more of themselves, generation after generation, with astonishing fidelity (and yet, as we shall see, with just enough variation to provide the raw material for evolution). Flowers, the familiar symbols of spring and romance, are the reproductive structures of the largest and most diverse group of plants.

(d) Living organisms grow and develop. Growth and development are the processes by which a single living cell, the fertilized egg, becomes, for example, a tree, or an elephant, or a young koala, shown here with its mother.

(e) Living things take energy from the environment and change it from one form to another. The processes of energy conversion are highly specialized and remarkably efficient. This European kingfisher has converted chemical energy stored in its body to the energy of motion used in catching a fish. After the kingfisher has eaten and digested the fish, the chemical energy stored in the body of the fish will be available for the kingfisher's use.

(f) Living things are adapted to their environment. The Arctic fox is protected from the harsh conditions of winter by a luxuriant coat of soft fur and a thick layer of body fat just below the skin. This insulation is so effective that the fox can comfortably tolerate actual or wind chill temperatures as low as $-60°$ Fahrenheit. When it is inactive or sleeping, the fox curls into a tight ball, providing added protection for tender areas, such as the paws, the belly, and the muzzle.

(g) Living organisms respond to stimuli. When this sphinx moth caterpillar sensed a threat—an approaching photographer—it inflated the front portions of its body, released its front legs from the branch on which it was resting, hung upside down, and waved its body at the camera. Although the caterpillar is only 6 centimeters long, its "fake snake" disguise is sufficiently convincing to protect it from attacks by would-be predators, such as hungry birds.

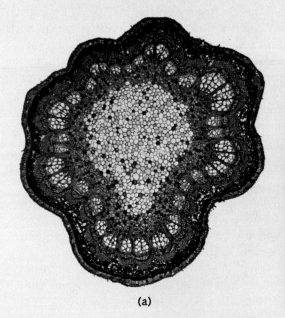

(a)

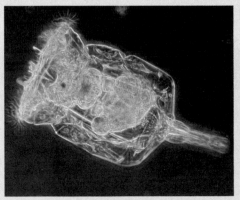

(b)

(c)

(d)

(e)

(f)

(g)

The Biologically Important Elements

Of the 92 naturally occurring elements, only six make up some 99 percent of all living tissue (Table 1–3). These six elements are carbon, hydrogen, nitrogen, oxygen, phosphorus, and sulfur, conveniently remembered as CHNOPS. These are not the most abundant of the elements of the Earth's surface. Why, as life assembled and evolved from stardust, were these of such importance?

One clue is that the atoms of all of these elements need to gain electrons to complete their outer energy levels (see Figure 1–4). Thus they generally form covalent bonds. Because these atoms are small, the shared electrons in the bonds are held closely to the nuclei, producing very stable molecules. Moreover, with the exception of hydrogen, atoms of these elements can all form bonds with two or more atoms. This makes possible the formation of the complex molecules essential for the structures and functions of living systems.

Levels of Biological Organization

In beginning our study of biology with an examination of how atoms and molecules are put together and how they interact, we have made two fundamental assumptions: (1) Living things are made up of the same chemical components—atoms and molecules—as nonliving things, and (2) living things obey the same chemical and physical laws as nonliving things. This does not mean, however, that organisms are "nothing but" the atoms and molecules of which they are composed. Indeed, there are recognizable differences between living and nonliving systems (see essay, pages 30–31). The key to these differences is organization.

A proton, as we have seen, is a positively charged particle, and an electron a negatively charged particle. Put them together and you have a hydrogen atom, an entity with quite different properties from those of an electron or a proton. Combine hydrogen atoms with each other in hydrogen molecules, and they form a colorless, highly flammable gas. Combine hydrogen atoms with the atoms of oxygen (whose molecules form a different colorless gas), and liquid water is produced—not a gas at all and with very different and remarkable properties, not a bit like those of isolated electrons and protons or of elemental hydrogen and oxygen.

Take just a few more kinds of atoms—carbon, nitrogen, a little sulfur—and put them together in a certain way, and you have the contractile machinery of a muscle cell, the basis for all the voluntary movements of our bodies. Add a little phosphorus, combine the atoms in a different way, and you can spell out the genetic code. Put together enough of these atoms in the right way, enclose them in a membrane (made up of the same kinds of atoms), and they form a living cell.

Each of these states represents a new level of organization. The characteristics of each level are not simply a combination of the characteristics of its components but are, in fact, totally different. At each new level of organization, new and different properties emerge.

The **cell** is the level of organization at which life can unarguably be said to appear as a new, emergent property. Other properties emerge when individual, specialized cells are organized at still higher levels, in the tissues and organs of a multicellular organism. Organized in one way, the cells form a liver; in another way, an intestinal tract; in yet another, a human brain, which represents an extraordinary degree of organizational complexity. Yet it is, in turn, only part of a larger entity whose characteristics are different from those of the brain, although they depend on them.

Nor is the individual organism the ultimate level of biological order. Living organisms interact with each other, individually and in groups. Finally, groups of living organisms are themselves part of an even vaster system of organization. This ultimate level of organization, the **biosphere,** involves not only the great diversity of plants and animals and microorganisms and their interactions with each other but also the physical characteristics of the environment and the planet Earth itself.

Table 1–3 Atomic Composition of Three Representative Organisms			
Element	Human	Alfalfa	Bacterium
Carbon	19.37%	11.34%	12.14%
Hydrogen	9.31	8.72	9.94
Nitrogen	5.14	0.83	3.04
Oxygen	62.81	77.90	73.68
Phosphorus	0.63	0.71	0.60
Sulfur	0.64	0.10	0.32
CHNOPS total:	97.90%	99.60%	99.72%

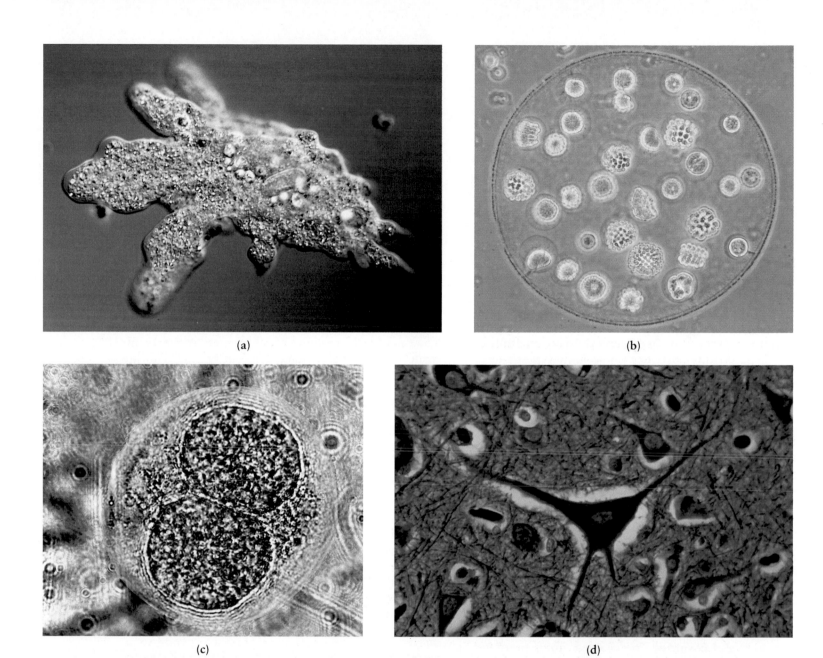

(a)

(b)

(c)

(d)

1–12 *A gallery of cells.* (**a**) Amoeba pro-teus, *a single-celled organism named for Pro-teus, a Greek god capable of changing his shape at will. Extensions of the cell, known as pseudopods, enable amoebas to move and to capture prey.*

(**b**) *This simple organism, called Eudorina, is made up of 32 cells, held together by a jelly-like substance, whose outlines you can see. When the organism reproduces, some—but* not all—*of the cells divide repeatedly. If you look closely you can see these cells, which appear to be miniature versions of the parent colony. When the cell divisions are com-pleted in the reproductive cells, the parent colony breaks apart and the daughter colo-nies begin their independent existence.*

(**c**) *The embryo of a mouse at the two-celled stage. Within each cell is a nucleus* that carries all the genetic information needed for every cell in the mature mouse.

(**d**) *These cells are from the cerebral cortex of a human brain—the most highly organ-ized structure on Earth. The actions of the cells of the cerebral cortex and the intercon-nections among them are responsible for consciousness, intelligence, dreams, and memory.*

Summary

Matter is composed of atoms, the smallest units of chemical elements. The nucleus of an atom contains positively charged protons and (except for hydrogen, ^1H) neutrons, which have no charge. The atomic number of an atom is equal to the number of protons in its nucleus. The atomic weight of an atom is the sum of the number of protons and neutrons in its nucleus. How an atom will react with other atoms is determined by its electrons—small, negatively charged particles found outside the nucleus. The number of electrons in an atom equals the number of protons.

The nuclei of different isotopes of the same element contain the same number of protons but different numbers of neutrons. Thus the isotopes of an element have the same atomic number but different atomic weights.

The electrons of an atom have differing amounts of energy. Electrons closer to the nucleus have less energy than those farther from the nucleus and thus are at a lower energy level. An electron tends to occupy the lowest available energy level, but with an input of energy, it can be boosted to a higher energy level. When the electron returns to a lower level, energy is released.

The chemical behavior of an atom is determined by the number and arrangement of its electrons. An atom is most stable when all of its electrons are at the lowest available energy levels and those energy levels are completely filled with electrons. The first energy level can hold two electrons, the second energy level can hold eight electrons, and so can the third energy level of the atoms of greatest interest in biology. Chemical reactions between atoms result from the tendency of atoms to reach the most stable electron arrangement possible.

Molecules consist of two or more atoms held together by chemical bonds. Two principal types of bonds are ionic and covalent. The charged atoms produced when electrons jump from one atom to another are known as ions. Ionic bonds are formed by the mutual attraction of ions of opposite charge. In covalent bonds, pairs of electrons are shared between atoms. In single covalent bonds, one pair of electrons is shared; in double covalent bonds, two pairs of electrons are shared. In some covalent bonds, known as polar covalent bonds, pairs of electrons are shared unequally, giving the molecule regions of positive and negative charge.

Chemical reactions—exchanges of electrons among atoms—can be represented by chemical equations. Substances that consist of the atoms of two or more different elements, in definite and constant proportions, are known as chemical compounds.

Six elements (CHNOPS) make up 99 percent of all living matter. The atoms of all of these elements are small and form tight, stable covalent bonds. With the exception of hydrogen, they can all form covalent bonds with two or more atoms, giving rise to the complex molecules that characterize living systems.

Living things are made up of the same chemical components—atoms and molecules—as nonliving things, and they obey the same chemical and physical laws. However, new properties emerge with each increasing level of organization. The properties of a complex molecule depend upon the organization of the atoms within the molecule. Similarly, the properties of a living cell depend upon the organization of molecules within the cell, and the properties of a multicellular organism depend upon the organization of the cells within its body. The ultimate level of biological organization, the biosphere, results from the interactions of the plants, animals, and microorganisms of the Earth with each other and with physical factors in the environment.

Questions

1. Describe the three types of particles of which atoms are composed. What is the atomic number of an atom? The atomic weight?

2. For each of the following isotopes, determine the number of protons and neutrons in the nucleus: (a) ^{11}C, ^{12}C, ^{14}C; (b) ^{31}P, ^{32}P, ^{33}P; (c) ^{32}S, ^{35}S, ^{38}S.

3. Consider the isotopes of phosphorus listed in Question 2. Would you expect all three of these isotopes to exhibit the same chemical properties in a living organism? Why or why not?

4. The street lights in many cities contain bulbs filled with sodium vapor. When electrical energy is passed through the bulb, a brilliant yellow light is given off. What is happening to the sodium atoms to cause this?

5. Determine the number of protons, the number of neutrons, the number of energy levels, and the number of electrons in the outermost energy level in each of the following atoms: oxygen, nitrogen, carbon, sulfur, phosphorus, chlorine, potassium, and calcium.

6. How many electrons does each of the atoms in Question 5 need to share, gain, or lose to acquire a completed outer energy level?

7. Magnesium has an atomic number of 12. How many electrons are in its first energy level? Its second energy level? Its third energy level? How would you expect magnesium and chlorine to interact? Write the formula for magnesium chloride.

8. Explain the differences between ionic, covalent, and polar covalent bonds. What tendency of atoms causes them to interact with each other, forming bonds?

9. Molecules that contain polar covalent bonds typically have regions of positive and negative charge and thus are polar. However, some molecules containing polar covalent bonds are nonpolar. Explain how this is possible.

10. Knowing that chemical reactions have to be balanced, fill the appropriate numbers into the underlined spaces (*hint:* from 1 to 3 in all cases):

(a) ____ H_2CO_3 $\longrightarrow$ ____ H_2O + ____ CO_2
Carbonic
acid

(b) ____ H_2 + ____ N_2 $\longrightarrow$ ____ NH_3
Ammonia

(c) ____ NaOH + ____ H_2CO_3 $\longrightarrow$
Sodium
hydroxide

____ Na_2CO_3 + ____ H_2O
Sodium
carbonate

11. What six elements make up the bulk of living tissue? What characteristics do the atoms of these six elements share?

Suggestions for Further Reading

Atkins, P. W.: *Molecules*, W. H. Freeman and Company, New York, 1987.*

> *In this beautifully illustrated book, Professor Atkins explores the structure and properties of 160 of the most familiar molecules of our everyday world.*

Powell, Corey S.: "The Golden Age of Cosmology," *Scientific American*, July 1992, pages 17–22.

Silk, Joseph: *The Big Bang: The Creation and Evolution of the Universe*, 2d ed., W. H. Freeman and Company, New York, 1988.

> *A discussion of modern evidence concerning the formation of the solar system and the planet Earth. An excellent, well-written introduction to cosmology.*

* Available in paperback.

Weinberg, Steven: *The Discovery of Subatomic Particles*, W. H. Freeman and Company, New York, 1990.*

> *In this handsome book, an introduction to the structure of the atom is combined with a lively history of twentieth-century physics. This revolution in physics profoundly influenced modern biology.*

Weinberg, Steven: *The First Three Minutes: A Modern View of the Origin of the Universe*, revised ed., Basic Books, Inc., New York, 1988.*

> *A wonderful story, written for the intelligent nonscientist (characterized by the author as a smart old attorney who expects to hear some convincing arguments before he makes up his mind).*

Weisskopf, Victor F.: "The Origin of the Universe," *American Scientist*, vol. 71, pages 473–480, 1983.

2

Water

Life on Earth depends on the presence of water. The first living systems came into being, according to present hypotheses, in the warm primitive seas, and for many organisms, ourselves included, each new individual begins life bathed and cradled in water.

These are developing embryos of the rain frog, which is native to the rain forests of Costa Rica. The eggs are laid on any convenient surface, including the mosses and other plants that grow on the trunks of the large trees of the forest. Within its gelatinous egg case, which contains an ample supply of water and nutrients, each embryo develops directly into a miniature adult without passing through the free-living tadpole stage commonly seen in frogs. Rain frogs can therefore lay their eggs on terrestrial surfaces, well away from puddles, ponds, and other bodies of water. These two froglets are almost ready to emerge from their egg cases.

Water is composed of atoms of hydrogen and oxygen, and yet its properties are very different from those of either hydrogen or oxygen. It is these special properties of water that are vital to life here on Earth.

In this chapter and the next, we are going to examine the molecules of which living organisms are composed. By far the most abundant of these molecules is water, which makes up 50 to 95 percent of the weight of any functioning living system.

Life on this planet began in water, and today, wherever liquid water is found, life is also present. There are one-celled organisms that eke out their entire existence in no more water than can cling to a grain of sand. Some kinds of algae are found only on the melting undersurfaces of polar ice floes. Certain bacteria thrive in the near-boiling water of hot springs. In the desert, plants race through an entire life cycle—seed to flower to seed—following a single rainfall. In the tropical rain forest, the water cupped in the leaves of a plant forms a miniature world in which a multitude of small organisms are born, mature, reproduce, and die.

Water is the most common liquid on Earth. Three-fourths of the surface of the Earth is covered by water. In fact, if the Earth's land surface were absolutely smooth, all of it would be under 2.5 kilometers* of water. But do not mistake "common" for "ordinary." Water is not an ordinary liquid at all. Compared with other liquids it is, in fact, quite extraordinary. If it were not, it is unlikely that life on Earth could ever have evolved.

The Structure of Water

In order to understand why water is so extraordinary and how, as a consequence, it can play its unique and crucial role in relation to living systems, we must look at its molecular structure. As you know, each water molecule is made up of two atoms of hydrogen and one

---

* A metric table with English equivalents can be found inside the back cover.

atom of oxygen (Figure 2–1). Each of the hydrogen atoms is held to the oxygen atom by a covalent bond; that is, the single electron of each hydrogen atom is shared with the oxygen atom, which also contributes an electron to each bond.

The water molecule as a whole is neutral in charge, having an equal number of electrons and protons. However, the molecule is polar (page 29). Because of the very strong attraction of the oxygen nucleus for electrons, the shared electrons of the covalent bonds spend more time around the oxygen nucleus than they do around the hydrogen nuclei. As a consequence, the region near each hydrogen nucleus has a weak positive charge. Moreover, the oxygen atom has four other electrons in its outer energy level. These electrons, which are not involved in covalent bonding to hydrogen, are paired in two orbitals (see essay, page 24). Each of these orbitals has a weak negative charge. Thus, the water molecule, in terms of its polarity, is four-cornered, with two positively charged "corners" and two negatively charged ones (Figure 2–2a).

When one of these charged regions comes close to an oppositely charged region of another water molecule, the force of attraction forms a bond between them, which is known as a **hydrogen bond**. Hydrogen bonds are found not only in water but also in many large molecules, where they help maintain structural stability. A

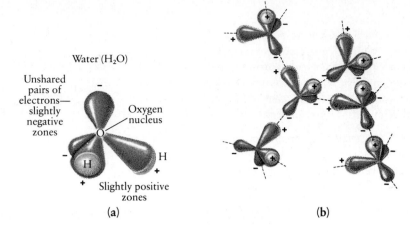

Water (H₂O)

2–2   *The polarity of the water molecule and its consequences.* (a) *As shown in this orbital model of water, four orbitals branch off from the oxygen nucleus. Two of the orbitals are formed by the shared pairs of electrons bonding the hydrogen atoms to the oxygen atom. These two orbitals have a slightly positive charge because their electrons spend more time around the oxygen nucleus than around the hydrogen nuclei. The other two orbitals, containing electrons that are not involved in covalent bonds, have a slightly negative charge.*

(b) *As a result of these positive and negative zones, each water molecule can form hydrogen bonds (dashed lines) with as many as four other water molecules. Under ordinary conditions of pressure and temperature, the hydrogen bonds are continually breaking and re-forming in a shifting pattern. Thus water is a liquid.*

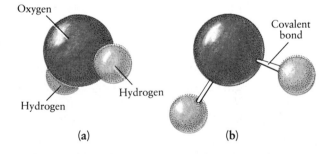

2–1   *The structure of the water molecule (H₂O) can be depicted in several different ways, illustrating different aspects of the structure. The schematic diagram in Figure 1–10 (page 28), for example, shows the electron structure of the atoms and the electron sharing within the molecule. By contrast, the space-filling model (a), in which the atoms are represented by color-coded spheres, provides a compact three-dimensional representation of the molecule. Because of its simplicity, this model is often used as a convenient symbol of the water molecule.*

(b) *The ball-and-stick model emphasizes the covalent bonds that join the atoms. It also gives some indication of the geometry of the molecule. A more accurate description of the molecule's shape is provided by the orbital model in Figure 2–2a.*

hydrogen bond can form between any hydrogen atom that is covalently bonded to an atom that has a strong attraction for electrons—usually oxygen or nitrogen—and the oxygen or nitrogen atom of another molecule. In water, a hydrogen bond forms between a negative "corner" of one water molecule and a positive "corner" of another. Every water molecule can establish hydrogen bonds with as many as four other water molecules (Figure 2–2b).

Any single hydrogen bond is significantly weaker than either a covalent or an ionic bond. Moreover, it has an exceedingly short lifetime. On average, each hydrogen bond in liquid water lasts approximately 1/100,000,000,000th of a second. But, as one is broken, another is made. All together, the hydrogen bonds have considerable strength, causing the water molecules to cling together as a liquid under ordinary conditions of temperature and pressure.

Now let us look at some of the consequences of these attractions among water molecules, especially as they affect living organisms.

(a)   **(b)**

2–3   *Some consequences of the surface tension of water.*
(a) A water strider, at rest on the calm surface of a pond. The weight of the insect's body is easily supported by the continuous sheet formed by the surface of the water. Note the elasticity of the surface, revealed by the slight depression below each of the insect's legs. The green objects are duckweed, a tiny flowering plant commonly found growing on the surface of ponds and lakes.

(b) When a drop of water strikes a water surface, or a rain drop falls on a pond, the water splashed into the air at impact forms a cylinder of water topped by tiny spheres, each held together by the cohesion of the water molecules. The point of impact, captured here by high-speed photography, resembles a jewel-studded crown.

Consequences of the Hydrogen Bond

Surface Tension

Look at water dripping from a faucet. Each drop clings to the rim and dangles for a moment by a thread of water. Then, just as the tug of gravity breaks it loose, its outer surface is drawn taut, to form a sphere as the drop falls free. Gently place a needle or a razor blade flat on the surface of the water in a glass. Although the metal is denser than water, it floats. Look at a pond in spring or summer. You will see water striders and other insects walking on its surface almost as if it were solid (Figure 2–3a). Rain drops create small crowns as they splash on the pond's surface (Figure 2–3b). These phenomena are all the result of **surface tension**—a taut yet elastic "skin" at the water's surface. Surface tension is produced by the clinging together of the water molecules, caused by their hydrogen bonding. Such a holding together of molecules of the same substance is known as **cohesion.**

Water, because of its negative and positive charges, also clings tightly to other charged molecules and to charged surfaces. Such a holding together of molecules of different substances is known as **adhesion.** The "wetting" capacity of water—that is, its ability to adhere to a surface—is another result of its polar structure.

Capillary Action and Imbibition

If you hold two dry glass slides together and dip one corner in water, the water will spread upward between the two slides. This movement of liquid through a narrow passage is known as **capillary action.** It results from the combined effects of the cohesion of water molecules to each other and their adhesion to the surface of the glass (Figure 2–4). Capillary action causes water to rise in very fine glass tubes, to creep up a piece of blotting paper, or to move slowly through the tiny spaces between soil particles and so become available to the roots of plants.

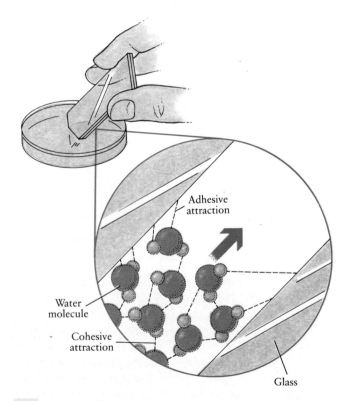

Adhesive attraction

Water molecule

Cohesive attraction

Glass

2–4   *The capillary movement of water between two glass slides. Because of the polarity of the water molecules, the molecules at the leading edge of the water surface are attracted to the charged surface of the glass. As the water molecules are pulled forward by this adhesive attraction, their cohesive attraction for other water molecules—a direct result of hydrogen bonding—pulls those molecules along behind. The combined effect of these two forces will continue to move the water forward until the forces of attraction are balanced by the pull of gravity on the increasing weight of the column of water.*

2–5   *The germination of seeds begins with changes in the seed coat that permit a massive uptake of water. The embryo and surrounding structures then swell, bursting the seed coat. In this acorn, photographed on a forest floor, the embryonic root emerged after the tough outer layers of the fruit split open.*

Imbibition ("drinking up") is the capillary movement of water molecules into substances such as wood or gelatin, which swell as a result. The pressures developed by imbibition can be astonishingly great. It is said that stone for the ancient Egyptian pyramids was quarried by driving wooden pegs into holes drilled in the rock face and then soaking the pegs with water. The swelling of the wood created a force great enough to break the stone slab free. Seeds imbibe water as they begin to germinate, swelling and bursting their seed coats (Figure 2–5).

Resistance to Temperature Changes

If you go swimming in the ocean or a lake on one of the first hot days of summer, you will quickly be aware of a striking difference between the air temperature and the water temperature. This difference occurs because a greater input of energy is required to raise the temperature of water than to raise the temperature of air. The amount of heat a given substance requires for a given increase in temperature is known as its **specific heat** (also called heat capacity). One calorie* is defined as the amount of heat that will raise the temperature of 1 gram (1 milliliter or 1 cubic centimeter) of water 1°C. The specific heat of water is about twice the specific heat of oil or alcohol; that is, only 0.5 calorie is needed to

---

* Dietary Calories are actually kilocalories (kcal); 1 kilocalorie equals 1,000 calories.

raise the temperature of 1 gram of oil or alcohol 1°C. The specific heat of water is four times the specific heat of air and 10 times that of iron. Only liquid ammonia has a higher specific heat.

Heat is a form of energy—the **kinetic energy,** or energy of motion, of molecules. Molecules are always moving. They vibrate, rotate, and shift position in relation to other molecules. Heat, which is measured in calories, reflects the *total* kinetic energy in a collection of molecules. It includes not only the magnitude of the molecular movements but also the mass and number of moving molecules present. By contrast, temperature, which is measured in degrees, reflects the *average* kinetic energy of the molecules in question. Thus, heat and temperature are not identical. For example, a lake may have a lower temperature than does a bird flying over it, but the lake contains more heat because it has many more molecules in motion.

The high specific heat of water is a consequence of hydrogen bonding. The hydrogen bonds in water tend to restrict the movement of the molecules. In order for the kinetic energy of water molecules to increase sufficiently for the temperature to rise 1°C, it is necessary first to rupture a number of the hydrogen bonds holding the molecules together. When you heat a pot of water, much of the heat energy added to the water is used in breaking the hydrogen bonds between the water molecules. Only a relatively small amount of heat energy is therefore available to increase molecular movement.

What does the high specific heat of water mean in biological terms? It means that for a given rate of heat input, the temperature of water will rise more slowly than the temperature of almost any other material. Conversely, the temperature will drop more slowly as heat is removed. Because so much heat input or heat loss is required to raise or lower the temperature of water, organisms that live in the oceans or large bodies of fresh water live in an environment where the temperature is relatively constant. Also, the high water content of terrestrial plants and animals helps them to maintain a relatively constant internal temperature. This constancy of temperature is critical because biologically important chemical reactions take place only within a narrow temperature range.

Vaporization

Vaporization—or evaporation, as it is more commonly called—is the change from a liquid to a gas. Water has a high **heat of vaporization.** At water's boiling point (100°C at sea level), it takes 540 calories to change 1 gram of liquid water into vapor, almost 60 times as much as for ether and twice as much as for ammonia.

(a)

(b)

(c)

2–6   *Animals have various devices for utilizing the heat-absorbing properties of evaporating water. (a) Dogs unload heat by panting, which involves short, shallow breaths and the production of a copious saliva that evaporates from the tongue. (b) Among the animals that can sweat over their entire body surface are horses and humans. (c) Elephants, which cannot perspire at all, hose themselves down to keep cool.*

Hydrogen bonding is also responsible for water's high heat of vaporization. Vaporization comes about because some of the most rapidly moving molecules of a liquid break loose from the surface and enter the air. The hotter the liquid, the more rapid the movement of its molecules and, hence, the more rapid the rate of evaporation. But, whatever the temperature, so long as a liquid is exposed to air that is less than 100 percent saturated with the vapor of that liquid, evaporation will take place, down to the last drop.

In order for a water molecule to break loose from its fellow molecules—that is, to vaporize—the hydrogen bonds have to be broken. This requires heat energy. As a consequence, when water evaporates, as from the surface of your skin or a leaf, the escaping molecules carry a great deal of heat away with them. Thus evaporation has a cooling effect. Evaporation from the surface of a land-dwelling plant or animal is one of the principal ways in which these organisms "unload" excess heat and so stabilize their temperatures.

Freezing

Water exhibits another peculiarity when it undergoes the transition from a liquid to a solid (ice). In most liquids, the **density**—that is, the weight of the material in a given volume—increases as the temperature drops. This greater density occurs because the individual molecules are moving more slowly and so the spaces between them decrease, leading to more molecules in the same volume. The density of water also increases as the temperature drops, until it nears 4°C. Then the water molecules come so close together and are moving so slowly that *every one* of them can form hydrogen bonds simultaneously with four other molecules—something they could not do at higher temperatures. However, the geometry of the water molecule is such that, as the temperature drops below 4°C, the molecules must move slightly apart from each other to maintain the maximum number of hydrogen bonds in a stable structure. At 0°C, the freezing point of water, this creates an open latticework (Figure 2–7). Thus water as a solid takes up more volume than water as a liquid. Ice is less dense than liquid water and therefore floats in it.

This increase in volume has occasional disastrous effects on water pipes but, on the whole, turns out to be enormously beneficial for life forms. If water continued to contract as it froze, ice would be heavier than liquid water. As a result, lakes and ponds and other bodies of water would freeze from the bottom up. Once ice began to accumulate on the bottom, it would tend not to melt, season after season. Spring and summer might stop the freezing process, but laboratory experiments have shown that if ice is held to the bottom of even a relatively shallow tank, water can be boiled on the top without melting the ice. Thus if water did not expand when it froze, it would continue to freeze from the bottom up, year after year, and never melt again. Eventually, the body of water would freeze solid and any life in it would be destroyed. By contrast, the layer of floating ice that actually forms tends to protect the organisms in the water. The ice layer effectively insulates the liquid water beneath it, keeping its temperature at or above the freezing point of water.

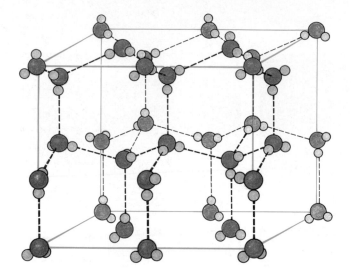

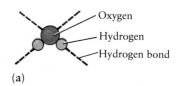

Oxygen
Hydrogen
Hydrogen bond

(a)

(b)

2–7   (a) *In the crystalline structure of ice, each water molecule is hydrogen-bonded to four other water molecules in a three-dimensional open latticework. As the water molecules link up in a hexagonal arrangement, the bond angles in some of the molecules are distorted. The hexagonal arrangement, shown here in a small section of the latticework, is repeated throughout the crystal and is responsible for the beautiful patterns seen in snowflakes and frost. The water molecules are actually farther apart in ice than they are in liquid water.*

(b) An ice "lens," photographed on the Alaskan tundra. When snow melts during the Arctic spring and summer, the meltwater seeps below the soil surface. It is unable to penetrate deeply, however, because of a thick, permanently frozen layer of subsoil known as permafrost. In the autumn, when the surface temperatures once more drop below 0°C, the trapped water expands as it freezes. The force generated by this expansion heaves the soil upward, revealing the ice lens.

When ice melts, it draws heat from its surroundings. For example, ice cubes in a glass of water gradually melt, cooling the water in the process. The heat energy absorbed by the ice breaks the hydrogen bonds of the latticework. Conversely, as water freezes, it releases heat into its surroundings. In this way, ice and snow also serve as temperature stabilizers, particularly during the transition periods of fall and spring. Moderation of sudden changes in temperature gives organisms time to make seasonal adjustments essential to survival.

Water as a Solvent

Many substances within living systems are found in aqueous solution. A **solution** is a uniform mixture of the molecules of two or more substances. The substance present in the greatest amount—usually a liquid—is called the **solvent,** and the substances present in lesser amounts are called **solutes.**

The polarity of water molecules is responsible for water's capacity as a solvent. The polar water molecules tend to separate ionic substances, such as sodium chloride (NaCl), into their constituent ions. As shown in Figure 2–8, the water molecules cluster around and segregate the charged ions.

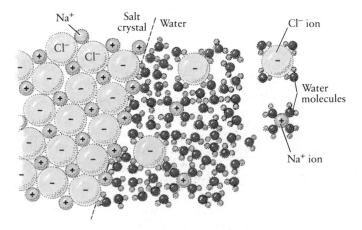

Na⁺   Salt crystal   Water   Cl⁻ ion

Cl⁻

Cl⁻

Water molecules

Na⁺ ion

2–8   *Because of the polarity of water molecules, water can serve as a solvent for ionic substances and polar molecules. This diagram shows table salt, sodium chloride (NaCl), dissolving in water. The polar water molecules cluster around the individual sodium ions (positively charged) and chloride ions (negatively charged), separating them from one another. Note the difference between the way the water molecules are arranged around the sodium ions and the way they are arranged around the chloride ions.*

The Four Seasons of a Lake

As we have seen, water increases in density as its temperature drops, until it reaches 4°C, the temperature of maximum density. Water either colder or warmer than 4°C is less dense and floats above water at 4°C. As a result, the water of temperate-zone lakes is stratified in the summer and winter but undergoes considerable mixing in the fall and spring. The stratifications of summer and winter enable lake-dwelling organisms to avoid life-threatening temperature extremes, while the mixing that occurs in fall and spring provides nutrients and oxygen to organisms at all levels of the lake.

In the summer (a), the top layer of water, called the epilimnion, is heated by the sun and the surrounding air, becoming warmer than the lower layers. Since it becomes less dense as it becomes warmer, this water remains at the surface. Only the water in the epilimnion circulates. In the middle layer, there is an abrupt drop in temperature, known as the thermocline. Since the water in this layer is progressively colder and therefore progressively more dense, it does not mix with the lighter water above. The water of the middle layer effectively cuts off the circulation of oxygen-laden water from the surface into the third layer, the hypolimnion. As the organisms of the hypolimnion gradually use up the available oxygen, the summer stagnation results.

In the fall (b), the temperature of the epilimnion drops until it is the same as that of the hypolimnion. As the surface water becomes more dense, it sinks, and the warmer water in the middle layer rises to the surface, producing the fall overturn. Aided by the fall winds (c), water begins to circulate throughout the lake. Oxygen is returned to the depths, and nutrients released by the activities of bottom-dwelling bacteria are carried to the upper layers of the lake.

As winter deepens (d), the surface water cools below 4°C, becoming lighter as it expands. This water remains on the surface and, in many areas, freezes. The result is the winter stratification.

In the spring (e), as the ice melts and the water on the surface warms to 4°C, it sinks to the bottom, producing the spring overturn. Aided by the spring winds (f), another thorough mixing of the water in the lake occurs.

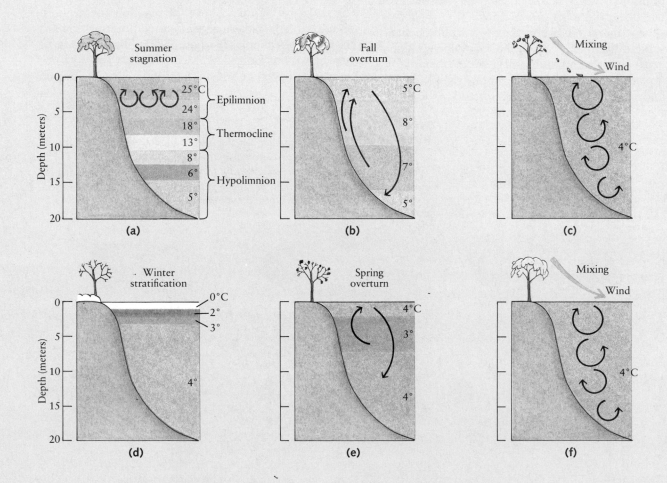

Many of the covalently bonded molecules important in living systems—such as sugars—are polar; that is, they have regions of weak positive and negative charge. (Such charged regions arise, as you will recall, because the nuclei of atoms of different elements exert differing degrees of attraction for electrons.) Because of their polarity, these molecules attract water molecules and dissolve in water. Polar molecules that readily dissolve in water are often called **hydrophilic** ("water-loving"). Such molecules slip into aqueous solution easily because their charged regions attract water molecules as much as or more than they attract each other. The polar water molecules thus compete with the attraction between the solute molecules themselves.

Molecules, such as fats, that lack polar regions tend to be very insoluble in water. The hydrogen bonding between the water molecules acts as a force to exclude the nonpolar molecules. As a result of this exclusion, nonpolar molecules tend to cluster together in water, just as droplets of fats tend to coalesce, for example, on the surface of chicken soup. Such molecules are said to be **hydrophobic** ("water-fearing"), and the clusterings are known as **hydrophobic interactions.**

We will encounter the properties of hydrophilic and hydrophobic molecules again in later chapters. These weak forces—hydrogen bonds and hydrophobic interactions—play crucial roles in determining the shape of large, biologically important molecules and, as a consequence, in dictating their properties.

The Ionization of Water

In liquid water, there is a slight tendency for the nucleus of one of the hydrogen atoms in a water molecule to leave the oxygen atom to which it is covalently bonded and jump to the oxygen atom to which it is hydrogen-bonded (Figure 2–9). In this reaction, two ions are produced: the hydronium ion (H_3O^+) and the hydroxide ion (OH^-). In any given volume of pure water, a small but constant number of water molecules will be ionized in this way. The number is constant because the tendency of water to ionize is offset by the tendency of the ions to reunite. Thus, even as some molecules are ionizing, an equal number of other molecules are being formed as ions reunite. This state, in which two opposite and equal processes are occurring simultaneously, is known as dynamic **equilibrium.**

The ionization of water is expressed by the following chemical equation:

$$2H_2O \rightleftharpoons H_3O^+ + OH^-$$

The two arrows indicate that the reaction goes in both directions. The fact that the arrow pointing toward $2H_2O$ is longer indicates that, at equilibrium, most of the water is not ionized. As a consequence, in any sample of pure water, only a small fraction exists in ionized form.

2–9   *When water ionizes, a hydrogen nucleus (that is, a proton) shifts from the oxygen atom to which it is covalently bonded to the oxygen atom to which it is hydrogen-bonded. The resulting ions are the negatively charged hydroxide ion and the positively charged hydronium ion.*

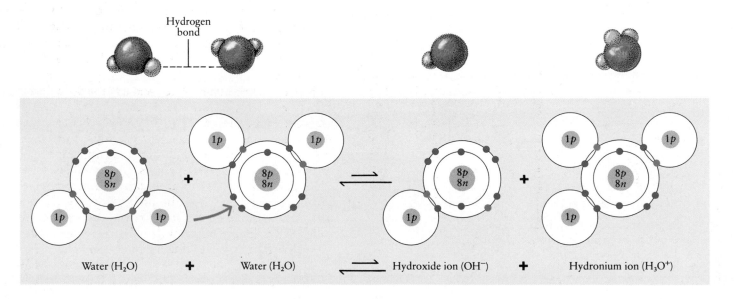

Water (H₂O)   **+**   Water (H₂O)   ⇌   Hydroxide ion (OH⁻)   **+**   Hydronium ion (H₃O⁺)

Although the positively charged ion formed when water ionizes is the hydronium ion (H_3O^+), it is, by convention, usually represented as a hydrogen ion (H^+). This convention, which we shall use in the remainder of this chapter and in subsequent chapters, simplifies the bookkeeping involved in balancing chemical equations.

Acids and Bases

In pure water, the number of H^+ ions exactly equals the number of OH^- ions. This is necessarily the case since neither ion can be formed without the other when only H_2O molecules are present. However, when an ionic substance or a substance with polar molecules is dissolved in water, it may change the relative numbers of H^+ and OH^- ions. For example, when hydrogen chloride (HCl) dissolves in water, it is almost completely ionized into H^+ and Cl^- ions. As a result, an HCl solution (hydrochloric acid) contains more H^+ ions than OH^- ions. Conversely, when sodium hydroxide (NaOH) dissolves in water, it forms Na^+ and OH^- ions. Thus, in a solution of sodium hydroxide in water, there are more OH^- ions than H^+ ions.

A solution is acidic when the number of H^+ ions exceeds the number of OH^- ions. Conversely, a solution is basic (alkaline) when the number of OH^- ions exceeds the number of H^+ ions. Thus, an **acid** is a substance that causes an increase in the relative number of H^+ ions in a solution, and a **base** is a substance that causes an increase in the relative number of OH^- ions.

Strong acids and bases are substances, like HCl and NaOH, that ionize almost completely in water, resulting in relatively large increases in the concentrations of H^+ and OH^- ions, respectively. Weak acids and bases, by contrast, are substances that ionize only slightly, resulting in relatively small increases in the concentration of H^+ or OH^- ions.

Chemists express degrees of acidity by means of the **pH scale** (Figure 2–10). (The symbol "pH" is derived from the German *potenz Hydrogen*, "power of hydrogen.") At pH 7, the concentrations of H^+ and OH^- are exactly the same, as they are in pure water. This is a neutral state. Any pH below 7 is acidic, and any pH above 7 is basic. A difference of one pH unit represents a tenfold difference in the concentration of H^+ ion. Most of the chemical reactions of living systems take place within a narrow range of pH that hovers around neutrality.

2–10   *The pH scale provides a measure of the degree of acidity of solutions. On this scale, 7 represents the neutral state, in which the concentrations of H^+ and OH^- ions are equal. Lower numbers represent increasingly acidic solutions, and higher numbers represent increasingly basic solutions. Notice that as the H^+ concentration increases, the OH^- concentration decreases, and vice versa.*

A difference of one pH unit reflects a tenfold difference in the concentration of H^+ ions. Cola, for instance, is 10 times as acidic as tomato juice, and gastric juices are about 100 times more acidic than cola drinks.

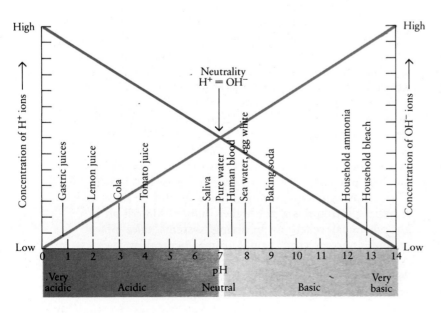

The Water Cycle

Most of the water on Earth—almost 98 percent—is in liquid form, in the oceans, lakes, and streams. Of the remaining 2 percent, some is frozen in polar ice and glaciers, some is in the soil, some is in the atmosphere in the form of vapor, and some is in the bodies of living organisms.

Water is made available to land organisms by processes powered by the sun. Solar energy evaporates water from the oceans, leaving the salt behind. Water is also evaporated, but in much smaller amounts, from lakes and ponds, from rivers and streams, from moist soil surfaces, from the leaves of plants, and from the bodies of other organisms. These molecules—now water vapor—are carried up into the atmosphere by air currents. At the cooler temperatures of the atmosphere, the molecules condense into clouds of liquid water or ice.

Eventually this water returns to the Earth's surface in the form of rain or snow. Most of the water falls on the oceans, since they cover most of the Earth's surface.

The water that falls on land is pulled toward the oceans by the force of gravity. Some of this runoff, reaching low ground, forms ponds or lakes and streams or rivers, which pour water back into the oceans.

Some of the water that falls on the land soaks in and then percolates down through the soil until it reaches a zone of saturation. In the zone of saturation, all pores and cracks in the rock are filled with water. The upper surface of the zone of saturation is known as the water table; the water within this zone is known as groundwater. Below the zone of saturation is solid rock, through which the water cannot penetrate. The deep groundwater, moving extremely slowly, eventually also reaches the ocean, thereby completing the water cycle.

As we have seen in this chapter, water, essential for life, is a most extraordinary substance. The Earth's supply of water is the permanent possession of our planet, held to its surface by the force of gravity. Through the movements of the water cycle, it is perpetually available to living organisms.

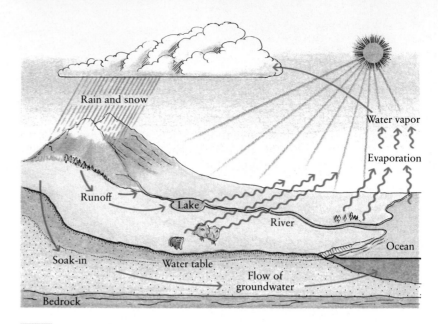

2–11   *The water cycle, powered by solar energy, provides a continuous circulation of the Earth's precious supply of water.*

Summary

Water, the most common liquid on the Earth's surface and the major component, by weight, of all living things, has a number of remarkable properties. These properties are a consequence of its molecular structure and are responsible for water's "fitness" for its roles in living systems.

Water is made up of two hydrogen atoms and one oxygen atom held together by covalent bonds. The water molecule is polar, with two weakly negative zones and two weakly positive zones. As a consequence, weak bonds form between water molecules. Such bonds, which link a somewhat positively charged hydrogen atom that is part of one molecule to a somewhat negatively charged atom that is part of another molecule, are known as hydrogen bonds. Each water molecule can form hydrogen bonds with as many as four other water molecules. Although individual bonds are weak and constantly shifting, the total strength of the bonds holding the molecules together is very great.

Because of the hydrogen bonds holding the water molecules together (cohesion), water has a high surface tension and a high specific heat (the amount of heat that a given amount of the substance requires for a given increase in temperature). It also has a high heat of vaporization (the heat required to change a liquid to a gas). Just before water freezes, it expands; thus, ice has a lower density than liquid water. As a result, ice floats in water.

The polarity of the water molecule is responsible for water's adhesion to other polar substances and hence its tendency for capillary movement. Similarly, water's polarity makes it a good solvent for ions and polar molecules. Molecules that dissolve readily in water are known as hydrophilic. Water molecules, as a consequence of their polarity, actively exclude nonpolar molecules from solution. Molecules that are excluded from aqueous solution are known as hydrophobic.

Water has a slight tendency to ionize, that is, to separate into H^+ ions and OH^- ions. In pure water, the number of H^+ ions and OH^- ions is equal. A solution that contains more H^+ ions than OH^- ions is acidic; one that contains more OH^- ions than H^+ ions is basic. The pH scale reflects the proportion of H^+ ions to OH^- ions. An acidic solution has a pH lower than 7; a basic solution has a pH higher than 7. Almost all of the chemical reactions of living systems take place within a narrow range of pH around neutrality.

Through the water cycle, the water above, on, and below the Earth's surface is recirculated. As a result, it is continuously available to living organisms.

Questions

1. (a) Sketch the water molecule and label the regions of weak positive and negative charge. (b) What are the major consequences of the polarity of the water molecule? (c) How are these effects important to living systems?

2. Distinguish among the following terms: cohesion/adhesion; capillary action/imbibition; heat/temperature; solution/solvent/solute; hydrophilic/hydrophobic; hydrogen bond/hydrophobic interaction; acidic/neutral/basic.

3. The trick with the razor blade (page 39) works better if the blade is a little greasy. Why?

4. Surfaces such as glass or raincoat cloth can be made "nonwettable" by application of silicone oils or other substances that cause water to bead up instead of spreading flat. What do you suppose is happening, in molecular terms, when a surface becomes nonwettable?

5. Generally, coastal areas have more moderate temperatures (not as cold in winter, nor as hot in summer) than inland areas at the same latitude. What explanation can you give for this phenomenon?

6. What is vaporization? Describe the changes that take place in water as it vaporizes. What is heat of vaporization? Why does water have an unusually high heat of vaporization?

7. Many of the molecules important in living systems contain the carboxyl group (—COOH). When these molecules are dissolved in water, the carboxyl group ionizes, as shown by the following equation:

$$-COOH \rightleftharpoons H^+ + -COO^-$$

At equilibrium, is the carboxyl group mostly ionized or only partially ionized? (*Hint:* look at the relative lengths of the two arrows.) Is the carboxyl group an acid or a base? Is it weak or strong?

8. Another group found in many molecules in living systems is the amino group (—NH$_2$). The amino group reacts with water as follows:

$$-NH_2 + H_2O \rightleftharpoons -NH_3^+ + OH^-$$

Is the amino group an acid or a base? Is it weak or strong?

9. Hydrochloric acid (HCl) is a major component of gastric juice, and the digestive processes in the human stomach take place at a pH of about 2. When the food being digested reaches the small intestine, a weak base, sodium bicarbonate (NaHCO$_3$), is released from the pancreas into the small intestine. What effect would you expect this to have on the pH of the partially digested food mass?

Suggestions for Further Reading

Gould, James L., and Carol Grant Gould (eds.): *Life at the Edge,* W. H. Freeman and Company, New York, 1989.*

> *A collection of articles from* Scientific American *on organisms that thrive in extraordinarily difficult environments—such as the frigid waters of the Antarctic or boiling volcanic vents in the ocean depths.*

La Rivière, J. W. Maurits: "Threats to the World's Water," *Scientific American,* September 1989, pages 80–94.

Milne, Lorus J., and Margery Milne: "Insects of the Water Surface," *Scientific American,* April 1978, pages 134–142.

Peterson, Ivars: "A Biological Antifreeze," *Science News,* November 22, 1986, pages 330–332.

Stillinger, Frank H.: "Water Revisited," *Science,* vol. 209, pages 451–457, 1980.

Storey, Kenneth B., and Janet M. Storey: "Frozen and Alive," *Scientific American,* December 1990, pages 92–97.

*Available in paperback.

Atoms, forged in the stars, are incorporated into living organisms through the process of photosynthesis. In this process, green plants extract carbon atoms from the carbon dioxide gas in the atmosphere and harness the energy of the sun to build the complex molecules of which living organisms are formed. These molecules, known as organic molecules, are all constructed on backbones of carbon atoms.

Among the principal types of organic molecules are sugars, such as those contained in the syrupy nectar that has lured this honey bee to this flower. Sugar molecules are a source of energy. Plants use this energy to power the synthesis of other kinds of molecules, such as the proteins and oils present in the pollen grains clinging to the foraging bee. Pollen grains also contain molecules of nucleic acid, which carry the hereditary blueprint of the plant. Rewarded by the energy-rich nectar offered by the flower, the bee transports the pollen grains—and the genetic instructions they contain— from flower to flower.

Soon after the emergence of flowering plants on Earth, an event that occurred more than 120 million years ago, flowers and insects began to adapt to one another. Competition among plants for more efficient means of pollination led to the evolution of nectar and of the brightly colored flowers that advertise its presence. Competition for nectar led to evolutionary changes among insects. The bee's long mouthparts, for example, enable her to probe deeply within the flower for its sugary nectar. Clusters of specialized hairs on the rear legs form baskets for carrying the nutritious pollen grains back to the hive to share with her queen and her sister bees.

3

Organic Molecules

In this chapter, we present some of the types of **organic molecules**—molecules containing carbon—that are found in living things. The molecular drama is a grand spectacular with, literally, a cast of thousands. A single bacterial cell contains some 5,000 different kinds of organic molecules, and an animal or plant cell has about twice that many. These thousands of molecules, however, are composed of relatively few elements (CHNOPS). Similarly, relatively few types of molecules play the major roles in living systems. Consider this chapter, if you will, an introduction to the principal characters. The plot begins to unfold in Chapter 4.

The Central Role of Carbon

As we noted in the previous chapter, water makes up from 50 to 95 percent of a living organism, and small ions, such as sodium (Na^+), potassium (K^+), and calcium (Ca^{2+}), account for no more than 1 percent. Almost all the rest of an organism, chemically speaking, is composed of organic molecules.

Four different types of organic molecules are found in large quantities in organisms. These four are **carbohydrates** (composed of sugars), **lipids** (nonpolar molecules, most of which contain fatty acids), **proteins** (composed of amino acids), and **nucleic acids** (composed of complex molecules known as nucleotides). All of these molecules—carbohydrates, lipids, proteins, and nucleic acids—contain carbon, hydrogen, and oxygen. In addition, proteins contain nitrogen and sulfur. Nucleic acids, as well as some lipids, contain nitrogen and phosphorus.

The Carbon Backbone

As we saw in Chapter 1, a carbon atom can form four covalent bonds with as many as four other atoms. Methane (CH_4), which is natural gas, is an example (see Figure 1–9, page 28). Even more important, in terms

49

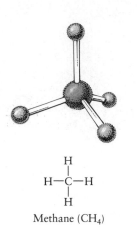
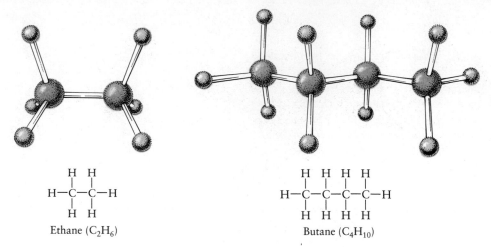

H
|
H—C—H
|
H

Methane (CH₄)

H H
| |
H—C—C—H
| |
H H

Ethane (C₂H₆)

H H H H
| | | |
H—C—C—C—C—H
| | | |
H H H H

Butane (C₄H₁₀)

3–1   *Ball-and-stick models and structural formulas of methane, ethane, and butane. The gray spheres represent carbon atoms,* *and the smaller blue spheres represent hydrogen atoms. The sticks in the models—and the lines in the structural formulas—repre-* *sent covalent bonds, each of which consists of a shared pair of electrons. Note that every carbon atom forms four covalent bonds.*

of carbon's biological role, carbon atoms can form bonds with each other. Ethane, for example, contains two carbons; propane, three; butane, four; and so on, forming long chains (Figure 3–1). In general, an organic molecule derives its overall shape from the arrangement of the carbon atoms that form the backbone, or skeleton, of the molecule. The shape of the molecule, in turn, determines many of its properties and its function within the living organism.

In the molecules shown in Figure 3–1, every carbon bond that is not occupied by another carbon atom is taken up by a hydrogen atom. Such compounds, consisting of only carbon and hydrogen, are known as **hydrocarbons.** Structurally, they are the simplest kind of organic compounds. Although hydrocarbons are relatively unimportant in living organisms, they are of great economic importance. The fuels on which we depend—natural gas, heating oil, gasoline, and diesel fuel—are all hydrocarbons. They are derived from the remains of organisms that died millions of years ago. Conditions were such that the bodies did not decay but were instead converted into coal and petroleum.

Carbohydrates:
Sugars and Polymers of Sugars

Carbohydrates are the primary energy-storage molecules in most living things. In addition, they form a variety of structural components of living cells. The walls of young plant cells, for example, are about 40 percent cellulose—a carbohydrate that is the most common organic compound in the biosphere.

Carbohydrates are formed from small molecules known as **sugars.** There are three principal kinds of carbohydrates, classified according to the number of sugar subunits they contain. **Monosaccharides** ("single sugars"), such as ribose, glucose, and fructose, consist of only one sugar molecule. **Disaccharides** ("two sugars") contain two sugar subunits linked covalently. Familiar examples are sucrose (table sugar), maltose (malt sugar), and lactose (milk sugar). **Polysaccharides,** such as cellulose and starch, contain many sugar subunits linked together. Large molecules, such as polysaccharides, that are made up of similar or identical subunits are known as **polymers** ("many parts"). The subunits are called **monomers** ("single parts").

Monosaccharides:
Ready Energy for Living Systems

Monosaccharides are organic compounds composed of carbon, hydrogen, and oxygen. The proportion is one carbon atom to two hydrogen atoms to one oxygen atom, as indicated by the shorthand formula for monosaccharides: $(CH_2O)_n$. These proportions gave rise to the term "carbohydrate" (meaning "carbon with water added") for sugars and the larger molecules formed from sugar subunits. In the formula, n may be as small as 3, as in $C_3H_6O_3$, or as large as 8, as in $C_8H_{16}O_8$.

Monosaccharides are the building blocks from which living cells construct polysaccharides and other essential molecules. Moreover, they are the principal energy source for most organisms. To understand how monosaccharides provide energy for living systems, we must look again at the covalent bond and consider what happens in a chemical reaction.

Representations of Molecules

As we saw in Chapters 1 and 2, chemists have developed various models to represent the structures of atoms and molecules. Each of these models is a way of organizing a set of scientific data and of focusing attention on particular characteristics of atoms and molecules.

Because the properties of a molecule depend on its three-dimensional shape, physical models are often the most useful. For example, ball-and-stick models of the kind shown in Figure 3–1 emphasize the geometry of a molecule and, in particular, the bonds between atoms. These models, however, fail to suggest the overall shape of the molecule created by the movement of electrons around the atomic nuclei.

A closer approximation of molecular shape is provided by space-filling models. Each atom is represented by the outermost edge of the region occupied by its electrons. Space-filling models are misleading, however, in that molecules do not fill space in the same way that we think of a table or a rock as filling space. The atoms that make up molecules consist mostly of empty space. If the outer perimeter of the region traversed by the electrons in an oxygen atom were the size of the perimeter of the Astrodome in Houston, the nucleus would be a ping-pong ball in the center of the stadium.

The space in molecules is "filled" with regions of charge, associated with the movements of the electrons around the nuclei. One molecule "sees" another molecule in terms of these regions of charge. As a consequence, for instance, a protein that transports glucose molecules into the living cell will not transport fructose molecules because of the differences in the shape of the regions of charge. All the intricate biochemistry that goes on in the cell is based on this ability of molecules to "recognize" one another.

Ball-and-stick and space-filling models are often used in the laboratory, but they are less useful on paper because it is necessary to see them from all angles to see all of the atoms and their bonds. The most accurate two-dimensional representations of molecular structure are orbital models, such as those shown in Figure 2–2 (page 38). For molecules containing more than a few atoms, however, orbital models become extremely complicated. Thus, when representing complex molecules, such as those found in living systems, chemists usually use molecular formulas or structural formulas. A **molecular formula** indicates the number of atoms of each kind within the molecule, while a **structural formula** shows how the atoms are bonded to one another.

Glucose, for example, has 6 carbon atoms, 12 hydrogen atoms, and 6 oxygen atoms. Its molecular formula is $C_6H_{12}O_6$. However, fructose also contains 6 carbons, 12 hydrogens, and 6 oxygens and has a similar structure—a chain of carbon atoms to which hydrogen and oxygen atoms are attached. The differences between glucose and fructose are determined by which carbon atoms the other atoms are attached to. The molecules can therefore be distinguished by their structural formulas:

In these formulas, the symbol — represents a single covalent bond, and the symbol = represents a double covalent bond.

When glucose and fructose are in solution, however, they tend to form rings and so are more accurately represented by these structural formulas:

The lower edges of the rings are drawn thicker to hint at a three-dimensional structure. The ring is perpendicular to the page, with the thick edges projecting toward you; the thin edges project behind the page. By convention, the carbon atoms at the intersections of the links in an organic ring structure are "understood" to be present and are not labeled. Although it is not necessary to number the carbon atoms, doing so often makes it easier to interpret the structural formulas.

Structural formulas give us less information than physical models, but you will find them a convenient tool as we examine the molecules involved in the structures and processes of living systems.

Space-filling models of the sugars glucose and fructose. The gray spheres, almost completely hidden at the center of each molecule, represent the carbon atoms. The red spheres at the surface of each molecule represent oxygen atoms, while the blue spheres represent hydrogen atoms.

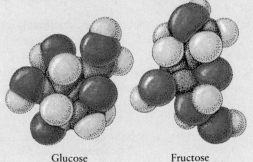

Glucose   Fructose

Covalent bonds—the bonds commonly found in organic molecules—are strong, stable bonds consisting of shared electrons moving around two or more atomic nuclei. As you will recall from the last chapter, molecules are always in motion—vibrating, rotating, and shifting position in relation to other molecules. The atoms within molecules are also in motion—vibrating and, often, rotating about the axes of their bonds. If this motion becomes great enough (that is, if the atoms acquire enough kinetic energy), the bond will "break" and the atoms will become separated from each other.

Different bonds have different characteristic strengths. Bond strengths are conventionally measured in terms of the amount of energy that must be supplied to break the bond under standard conditions of temperature and pressure. The units in which this energy is expressed are kilocalories per mole.* Table 3–1 compares the strengths of the covalent bonds that occur most frequently in organic molecules. The stronger the bond, the greater the amount of energy required to break it. As you can see, double bonds are considerably stronger than single bonds. Moreover, the single bond between carbon and hydrogen is stronger than the single bond between carbon and oxygen or between two carbon atoms.

When a covalent bond breaks, atoms (or, in some cases, groups of atoms) are released, and each atom usually takes its own complement of electrons with it. This results in atoms whose outer energy levels are only partially filled with electrons. For example, when the atoms of a methane molecule are vibrating and rotating so rapidly that the four carbon-hydrogen bonds break, one carbon atom and four hydrogen atoms are produced—and each of these atoms needs to gain one or more electrons to complete its outer energy level. Thus, the atoms tend to form new covalent bonds quite rapidly, restoring the stable condition of filled outer energy levels. Whether the new bonds that form are identical to those that were broken or are different depends on a number of factors—the temperature, the pressure, and, most important, what other atoms are available in the immediate vicinity.

Chemical reactions that produce new combinations of atoms—that is, new molecules—always involve the

* The mole is the principal unit of measure for quantities of substances involved in chemical reactions. The number of particles (whether ions, atoms, or molecules) in 1 mole of any substance is always exactly the same: 6.023×10^{23}. For example, 1 mole of hydrogen ions contains 6.023×10^{23} ions, 1 mole of carbon atoms contains 6.023×10^{23} atoms, and 1 mole of the monosaccharide glucose contains 6.023×10^{23} molecules.

Table 3–1   A Comparison of Bond Strengths

Bond	$\diagup$C=O	$\diagup$C=C$\diagdown$	—C—H	—C—O—	—C—C—
Energy needed to break bond (kcal/mole)	171	147	99	84	83

rupture of existing chemical bonds and the formation of new bonds. If the new bonds are weaker than the bonds that were broken, energy will be released. However, if the new bonds are stronger than the bonds that were broken, energy will be taken up from the surroundings. For example, in the process of **photosynthesis**, which we shall discuss in Chapter 9, radiant energy from sunlight is taken up by the green cells of plants and algae. These cells use the energy to break and rearrange covalent bonds in carbon dioxide and water, making possible the formation of new chemical combinations—sugar molecules and oxygen molecules. This process, which stores the energy captured from sunlight in the covalent bonds of the sugar molecules, involves many chemical reactions. It can, however, be summarized as follows:

Carbon dioxide + Water + Energy ⟶
  Sugar + Oxygen

In another series of reactions, living cells can break the covalent bonds of the sugar molecules. In the presence of oxygen, new combinations of atoms—carbon dioxide and water—are formed, and the energy stored in the bonds of the sugar molecules is released:

Sugar + Oxygen ⟶
  Carbon dioxide + Water + Energy

Other organic substances, such as wood, gasoline, and alcohol, also release their stored energy when they are oxidized (broken down in the presence of oxygen), as occurs when they are burned for fuel. The amount of stored energy an organic compound contains is measured in terms of the amount of heat energy, expressed in kilocalories per mole, released when the compound is oxidized.

The monosaccharide glucose is the principal energy source for most living cells. When 1 mole of glucose is broken down to carbon dioxide and water, 686 kilocalories of energy are released. The yield is the same

3–2 *Over the millennia, many organisms have evolved sensitive mechanisms for detecting sugars. Houseflies, for instance, have sugar detectors in their feet. When a housefly lights on droplets of a sugar solution, its tubular mouthparts are automatically extended and the fly begins to feed. We also have sensory receptors tuned for the detection of sugar, although our receptors, in keeping with our eating habits, are in our tongues. A housefly reacts positively to about the same types of sugars that we do, although its detection mechanism is about 10 million times more sensitive.*

In the course of consuming sugary plant products, houseflies, humans, and other animals obtain not only a rich energy supply but also other essential nutrients, such as plant proteins, lipids, vitamins, and minerals.

whether the reaction takes place in a living cell or in the laboratory. Inside the cell, however, almost 40 percent of this energy is repackaged in the bonds of other organic molecules, as we shall see in Chapter 8. In the laboratory, all of the released energy is given off as heat.

Disaccharides: Transport Forms

Although glucose is the common transport sugar for humans and other vertebrate animals, sugars are often transported in other organisms as disaccharides. Sucrose, often called cane sugar, is the form in which sugar is transported in plants from the photosynthetic cells (mostly in the leaves), where it is produced, to other parts of the plant body. Sucrose is composed of the monosaccharides glucose and fructose. Sugar is transported through the bodies of many insects in the form

of another disaccharide, known as trehalose, which consists of two glucose units linked together. Another common disaccharide is lactose, a sugar that occurs only in milk. Lactose is made up of glucose combined with another monosaccharide, galactose.

In the synthesis of a disaccharide molecule from two monosaccharide molecules, a molecule of water is removed in the process of forming the new bond between the two monosaccharides (Figure 3–3). This type of chemical reaction, which occurs in the formation of most organic polymers from their subunits, is known as **dehydration synthesis.**

When a disaccharide is split into its monosaccharide subunits, which happens when it is used as an energy source, the molecule of water is added again. This splitting is known as **hydrolysis,** from *hydro,* meaning "water," and *lysis,* "breaking apart." Hydrolysis is an energy-releasing reaction. The hydrolysis of sucrose, for example, yields 5.5 kilocalories per mole. Conversely, the formation of sucrose from glucose and fructose requires an energy input of 5.5 kilocalories per mole of sucrose.

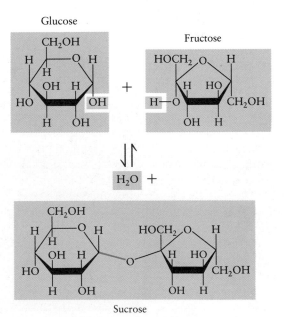

3–3 *Sucrose is a disaccharide made up of two monosaccharide subunits, one glucose and one fructose. The formation of sucrose involves the removal of a molecule of water (dehydration synthesis). The new chemical bond forged in the course of this reaction is shown in blue. The reverse reaction—splitting sucrose back into its constituent monosaccharides—requires the addition of a water molecule (hydrolysis).*

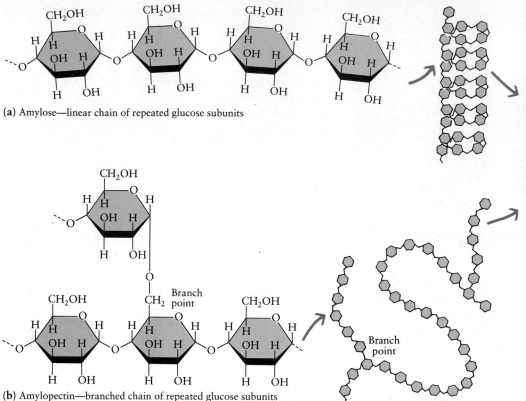

(a) Amylose—linear chain of repeated glucose subunits

(b) Amylopectin—branched chain of repeated glucose subunits

Branch point

Branch point

(c)

3–4 *In plants, sugars are stored in the form of starch. Starch is composed of two different types of polysaccharides, amylose and amylopectin. (a) A single molecule of amylose may contain 1,000 or more glucose subunits in a long unbranched chain, which winds to form a uniform coil. (b) A molecule of amylopectin may contain from 1,000 to 6,000 or more glucose subunits. Short chains of glucose subunits branch off from the main chain at frequent intervals. These side branches prevent the formation of a tight coil and produce a looser, more circular structure.*

(c) Starch molecules, perhaps because of their coiled nature, tend to cluster into granules. In this electron micrograph of a single storage cell of a potato, the spherical and egg-shaped objects are starch granules. They are magnified about 1,000 times.

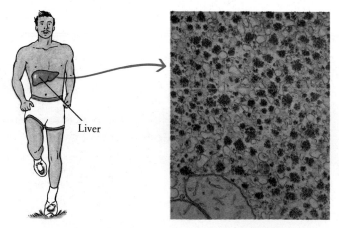

Liver

3–5 *In animal cells, sugars are commonly stored as glycogen, primarily in the liver and in skeletal muscle. Glycogen resembles the amylopectin found in plant cells, but glycogen is more highly branched and the individual branches are shorter. When glucose is needed, hydrolysis can occur at the ends of many branches simultaneously, providing abundant supplies of glucose. The dark granules in this liver cell, magnified about 55,000 times, are glycogen.*

Polysaccharides

Polysaccharides are made up of monosaccharides linked together in long chains. They constitute storage forms for sugars. The principal storage polysaccharide in plants is **starch** (Figure 3–4), and in animals and fungi it is **glycogen** (Figure 3–5). Both starch and glycogen are built up of many glucose units. The differences between them are in the length of the polysaccharide chains and in the ways in which the glucose molecules are linked. Polysaccharides must be hydrolyzed to monosaccharides or disaccharides before the sugar subunits can be used as energy sources or transported through living systems.

Polysaccharides also play structural roles. In plants, the principal structural molecule is the polysaccharide **cellulose.** Although cellulose, like starch, is a polymer of glucose, the bonds linking the glucose units in cellulose are slightly different from those in starch or glycogen. This small difference has a profound effect on the three-dimensional structure of the molecules and thus on their properties. Rather than forming granules,

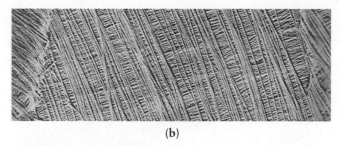

(a) Cellulose—linear chain of repeated glucose subunits

(b)

3–6 (a) *In the cellulose molecule, the bonds linking the glucose subunits are such that —OH groups (highlighted in blue) project from both sides of the chain. These —OH groups form hydrogen bonds with the —OH groups on adjacent cellulose molecules, producing bundles of cross-linked parallel chains. In the starch molecule (Figure 3–4), by contrast, most of the —OH groups capable of forming hydrogen bonds face toward the exterior of the coil, making it more readily soluble in the surrounding water.*

(b) *The wall of a young plant cell is about 40 percent cellulose. Hundreds of cellulose molecules are bundled together to form strands known as microfibrils, magnified here about 30,000 times. The microfibrils, as strong as an equivalent amount of steel, are embedded in other polysaccharides.*

3–7 *A cicada molting. The relatively hard outer coverings, or exoskeletons, of insects contain chitin, a modified polysaccharide that contains nitrogen. Because exoskeletons do not grow as the insect grows, they must be molted periodically. The discarded exoskeleton is at the bottom, below the insect, which is drying out and waiting for its new exoskeleton to harden. This cicada, a native of Australia, is commonly known as the "greengrocer."*

as do starch and glycogen molecules, cellulose molecules form long, rigid bundles (Figure 3–6).

Because of its structure, cellulose plays a biological role quite different from that of starch or glycogen. Cellulose is a major constituent of plant cell walls, but its glucose units are not readily available as an energy source, either for the plant or for other organisms. In fact, cellulose can be hydrolyzed by only a few microorganisms. Cows and other ruminants, termites, and cockroaches can use cellulose for energy only because of the microorganisms that inhabit their digestive tracts (see page 108).

Chitin, which is a major component of the exoskeletons (the hard, exterior coverings) of insects and other arthropods, is a tough, resistant, modified polysaccharide (Figure 3–7). Chitin is also found in the cell walls of fungi.

Lipids

Lipids are a group of organic substances that are insoluble in polar solvents, such as water, but that dissolve readily in nonpolar organic solvents, such as benzene. Typically, lipids serve as energy-storage molecules—usually in the form of fats or oils—and for structural purposes. Some lipids, however, play major roles as chemical "messengers," both within and between cells.

Fats and Oils: Energy in Storage

Unlike many plants, such as the potato, animals have only a limited capacity to store carbohydrates. In vertebrates, sugars consumed in excess of what can be stored as glycogen are converted into fats. Some plants also store food energy as oils, especially in seeds and fruits. Fats and oils contain a higher proportion of energy-rich carbon-hydrogen bonds than carbohydrates do and, as a consequence, contain more chemical energy. On average, the complete oxidation of fats yields about 9.3 kilocalories per gram* as compared to 3.8 kcal per gram of carbohydrate, or 3.1 kcal per gram of protein. Also, because fats are nonpolar, they do not attract water molecules and hence are not "weighted down" by them, as glycogen is. Taking into account the water factor, fats store six times as much energy, gram for gram, as glycogen, which is undoubtedly why in the course of evolution they came to play a major role in energy storage.

---

* 1,000 grams = 1 kilogram = 2.2 pounds, so oxidation of a pound of fat would yield about 4,200 kilocalories, more than the 24-hour requirement for a moderately active adult.

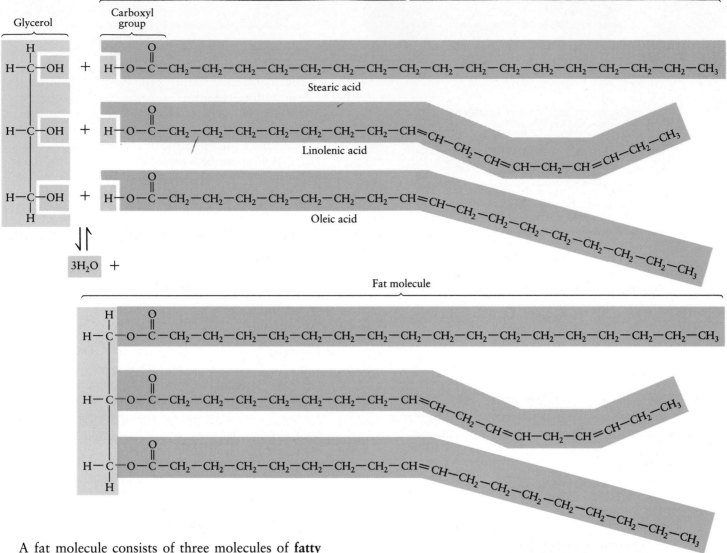

3-8 *A fat molecule consists of three fatty acids joined to a glycerol molecule (hence the term "triglyceride"). The long hydrocarbon chains of the fatty acids terminate in carboxyl (—COOH) groups, which become covalently bonded to the glycerol molecule. Each bond (dark blue) is formed when a molecule of water is removed (dehydration synthesis).*

Three different fatty acids are shown here. Stearic acid is saturated, whereas oleic acid and linolenic acid are unsaturated, as you can see by the double bonds in their structures. Oleic acid, with one double bond, is monounsaturated; linolenic acid, with three double bonds, is polyunsaturated.

The straight chains of saturated fatty acids allow saturated fat molecules to pack together, producing a solid, such as butter or lard. The kinks at the double bonds in unsaturated fatty acids tend to separate the fat molecules, producing a liquid, such as corn oil or peanut oil.

A fat molecule consists of three molecules of **fatty acid** covalently bonded to one **glycerol** molecule. As with the disaccharides and polysaccharides, each bond is formed by the removal of a molecule of water (dehydration synthesis), as shown in Figure 3–8. Fat molecules, which are also known as triglycerides, are said to be neutral because they contain no polar groups. As you would expect, they are extremely hydrophobic.

You have undoubtedly heard a lot about "saturated" and "unsaturated" fats. A fatty acid in which there are no double bonds between carbon atoms is said to be **saturated.** Each carbon atom in the chain has formed covalent bonds to four other atoms, and its bonding possibilities are therefore complete. By contrast, a fatty acid that contains carbon atoms joined by double bonds is said to be **unsaturated.** The double-bonded carbon atoms have the potential to form additional bonds with other atoms.

The physical nature of a fat is determined by the length of the carbon chains in the fatty acids and by whether the acids are saturated or unsaturated. Unsaturated fats tend to be liquid at room temperature. They are more common in plants than in animals; examples

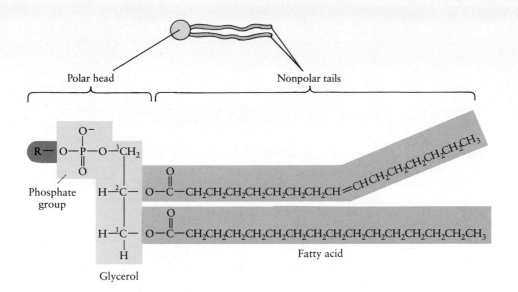

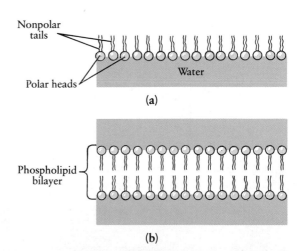

3-9   *A phospholipid molecule consists of two fatty acids linked to a glycerol molecule, as in a fat, with a phosphate group (shown in yellow) linked to the glycerol's third carbon. It also usually contains an additional chemical group (shown in red) attached to the phosphate group. This group is symbolized by the letter R; its chemical composition varies in different molecules.*

The fatty acid "tails" of a phospholipid molecule are nonpolar and therefore insoluble in water (hydrophobic). The polar "head" containing the phosphate and R groups is water-soluble (hydrophilic).

are olive oil, peanut oil, and corn oil. Animal fats and their derivatives, such as butter and lard, contain saturated fatty acids and are usually solid at room temperature.

Phospholipids and Glycolipids

Lipids, especially phospholipids and glycolipids, also play extremely important structural roles. Like fats, both phospholipids and glycolipids are composed of fatty acid chains attached to a glycerol backbone. In the **phospholipids,** however, the third carbon of the glycerol molecule is occupied not by a fatty acid but by a negatively charged phosphate group, PO_4^{3-} (Figure 3–9), to which another polar group is usually attached. The phosphate end of the molecule is hydrophilic, and the fatty acid portions are hydrophobic.

When phospholipids are added to water, they tend to form a film along its surface, with their hydrophilic "heads" under the water and their hydrophobic fatty acid "tails" protruding above the surface (Figure 3–10a). In the watery interior of the cell, phospholipids tend to align themselves in double rows, with their fatty acid tails oriented toward one another and their phosphate heads directed outward (Figure 3–10b). As we shall see in Chapter 5, this arrangement of phospholipid molecules, with their hydrophilic heads exposed and their hydrophobic tails clustered together, forms the structural basis of cell membranes.

In the **glycolipids** ("sugar lipids"), the third carbon of the glycerol molecule is occupied not by a phosphate group but by a short carbohydrate chain. Depending on the particular glycolipid, this chain may contain anywhere from 1 to 15 sugar subunits. Like the phosphate head of a phospholipid, the carbohydrate head of a glycolipid is hydrophilic, and the fatty acid tails are, of course, hydrophobic. In aqueous solution, glycolipids behave in the same fashion as phospholipids, and they are also important components of cell membranes.

3-10   (a) *Because phospholipids have polar heads and nonpolar tails, they tend to form a thin film on a water surface. The hydrophilic (water-loving) heads are in the water, and the hydrophobic (water-fearing) tails extend above the water. (b) Surrounded by water, phospholipid molecules spontaneously arrange themselves in two layers with their hydrophilic heads extending outward and their hydrophobic tails inward. This arrangement—a phospholipid bilayer—forms the structural basis of cell membranes.*

Regulation of Blood Cholesterol

Although cholesterol plays a number of essential roles in the animal body, it is also a principal villain in heart disease. Deposits containing cholesterol can narrow the arteries carrying blood to the heart muscle, and people with unusually large amounts of cholesterol in their blood have a high risk of heart attacks. How does the body regulate cholesterol levels? What goes wrong to cause elevated cholesterol levels? How does cholesterol cause heart attacks? Given the fact that heart disease is the major cause of death in this country, these questions are not only of biological interest but are also of personal importance to just about every one of us.

The key organ in cholesterol regulation is the liver, which not only synthesizes needed cholesterol from saturated fatty acids but also degrades excess cholesterol circulating in the blood—as a result, for example, of a diet rich in meat, cheese, and egg yolks. Cholesterol is transported to and from body cells, including those of the liver, by way of the bloodstream. Like other lipids, it is insoluble in plasma, the fluid portion of the blood. It is carried in particles consisting of a cholesterol interior and a lipid "wrapper" that has water-soluble proteins embedded in its outer surface. These large complexes exist in two principal forms: low-density lipoproteins (LDLs) and high-density lipoproteins (HDLs).

LDLs function as the delivery trucks of the system, carrying dietary cholesterol and newly synthesized cholesterol to various destinations in the body, including both the liver and the hormone-synthesizing organs. HDLs, however, function more like garbage trucks, carrying excess cholesterol on a one-way trip to the liver for degradation and excretion.

Normally the system is in balance, and the liver synthesizes or degrades cholesterol depending on the body's current needs and the amount of cholesterol in the blood. It can, however, be thrown out of balance by a number of factors. If, for example, the dietary intake of cholesterol is high, the liver becomes swamped and cannot degrade all of the excess. If the dietary intake of saturated fats is high, even in the absence of a high intake of cholesterol itself, the liver increases its synthesis of cholesterol.

Current evidence indicates that the liver monitors the level of cholesterol in the blood through its uptake of LDLs. If the quantities of LDLs taken up by the liver cells are low—a sign that blood cholesterol levels are low—the cells increase their synthesis and export of cholesterol. If the quantities of LDLs taken up by the cells are high—a sign that blood cholesterol levels are high—the cells stop synthesizing cholesterol.

When the quantities of circulating LDLs are greater than can be taken up by the liver and hormone-synthesizing organs, they are taken up by the cells lining the arteries supplying the heart. This can ultimately lead to total blockage of an artery and thus to a heart attack.

Heart disease often runs in families, suggesting that hereditary factors are involved in some cases. Normal liver cells have specialized receptors on their surface that participate in the transport of LDLs into the cells. In one type of hereditary heart disease, these receptors are absent. No LDLs can be taken up by the liver cells, and they continuously synthesize and export cholesterol. Persons with this disease have six to eight times the normal amount of cholesterol in their blood, usually have their first heart attack in childhood, and die of heart disease in their early twenties. Other families seem to be protected against heart disease, apparently because the individuals' bodies synthesize large quantities of HDLs, ensuring that all excess cholesterol makes a speedy one-way trip to the liver.

For most of us, however, the degree of risk depends on our behavior: whether we exercise regularly, which seems to increase HDL levels and thus protect against cholesterol buildup; whether we smoke cigarettes, which seems to decrease HDL levels; and the quantities of cholesterol and saturated fats that we consume.

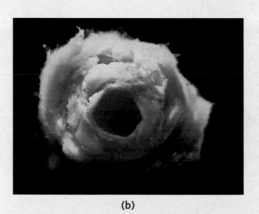

(a)

(b)

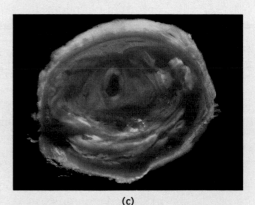

(c)

In one type of heart disease, atherosclerosis, cholesterol and other fatty substances accumulate in the walls of the arteries supplying the heart muscle. This accumulation triggers abnormal growth and the production of fibrous tissues by the cells of the walls. (a) A normal cor- *onary artery. Note the wide channel through which the blood can flow. (b) A coronary artery in which moderate atherosclerosis has developed. Fatty deposits have formed, and the space left for blood flow is significantly decreased. (c) A coronary artery in which the de-* *posits have become so great that only a very narrow channel remains open. Such a narrow channel can be completely blocked by a blood clot. The result is a heart attack and the death of the heart muscle supplied by the artery.*

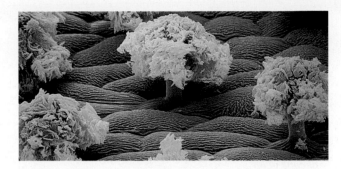

3–11 *All groups of land plants synthesize waxes, which protect exposed plant surfaces from water loss. This scanning electron micrograph reveals the wax-producing glands on the outer surface of a flower bud of a primrose. The stalked structures are the glands, and the fluffy material at the top of the stalks is the wax they have produced. The wax prevents water loss from the flower bud and insulates its delicate internal tissues from the cold. The wax deposits, magnified here 300 times, are sometimes so voluminous that the primrose plant appears to be dusted with flour.*

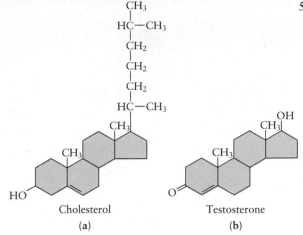

Cholesterol
(a)

Testosterone
(b)

3–12 *Two examples of steroids.* (a) *The cholesterol molecule consists of four carbon rings and a hydrocarbon chain.* (b) *Testosterone, a male sex hormone synthesized from cholesterol by cells in the testes, also has the characteristic four-ring structure but lacks the hydrocarbon tail.*

Waxes

Waxes are another type of structural lipid. They form protective, waterproof coatings on skin, fur, and feathers, on the leaves and fruits of land plants (Figure 3–11), and on the exoskeletons of many insects.

Cholesterol and Other Steroids

Cholesterol belongs to an important group of compounds known as the **steroids** (Figure 3–12). Although steroids do not resemble the other lipids structurally, they are grouped with them because they are insoluble in water. All the steroids have four linked carbon rings, and several of them have a hydrocarbon tail.

Cholesterol is found in animal cell membranes. About 25 percent (by dry weight) of the cell membrane of a red blood cell is cholesterol. It is synthesized in the liver from saturated fatty acids. Cholesterol is also obtained in the diet, principally in meat, cheese, and egg yolks. High concentrations of cholesterol in the blood are associated with atherosclerosis, in which cholesterol is found in fatty deposits on the interior lining of diseased blood vessels (see essay).

A number of hormones are also steroids. These include cortisol and the sex hormones estrogen and testosterone. In the body steroid hormones are synthesized from cholesterol.

Proteins

Proteins are among the most abundant organic molecules. In most living organisms they make up 50 percent or more of the dry weight. Only plants, with their high cellulose content, are less than half protein. Proteins perform an incredible diversity of functions in living systems. In their structure, however, they all follow the same simple blueprint: they are all polymers of nitrogen-containing molecules known as **amino acids,** arranged in a linear sequence. Some 20 different kinds of amino acids are used by living systems to build proteins.

Protein molecules are large, often containing several hundred amino acid monomers. Thus the number of different amino acid sequences, and therefore the possible variety of protein molecules, is enormous—about as enormous as the number of different sentences that can be written with our own 26-letter alphabet. Organisms, however, synthesize only a very small fraction of the proteins that are theoretically possible. The single-celled bacterium *Escherichia coli,** for example, contains 600 to 800 different kinds of proteins at any one time, and the cell of a plant or animal has several times that number. A typical cell in your body contains about 100 million protein molecules of about 10,000 different kinds. Each has a special function, and each, by its unique chemical nature, is specifically suited for that function.

---

* Biologists use a binomial ("two-name") system for designating organisms. Every different kind of organism has a unique two-part name. The first part of the name refers to the genus to which the organism belongs. The second part, in combination with the first part, refers to the particular species, a subdivision of the genus category. In this name, for example, *Escherichia* denotes the genus, while *coli* designates a particular kind, or species, of *Escherichia,* distinguished from all others by certain characteristics. By convention, with the second mention of a scientific binomial, it is permissible to abbreviate the first (genus) name. This is fortunate, particularly when dealing with such names as *Escherichia.*

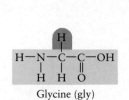

Glycine (gly)

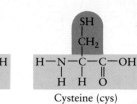

Valine (val)

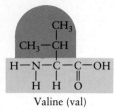

Cysteine (cys)

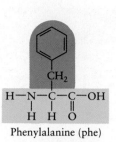

Serine (ser)

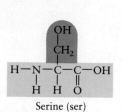

Phenylalanine (phe)

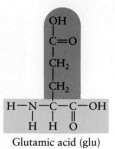

Glutamic acid (glu)

3–13 *Eight of the 20 different kinds of amino acids used in making proteins. As you can see, the fundamental structure is the same in all of the molecules, but the R groups differ. The other 12 amino acids—*

not shown here—are alanine (ala), glutamine (gln), histidine (his), leucine (leu), asparagine (asn), methionine (met), isoleucine (ile), aspartic acid (asp), lysine (lys), threonine (thr), tyrosine (tyr), proline (pro).

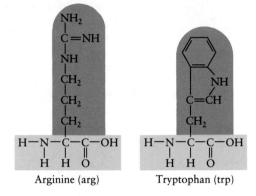

Arginine (arg)

Tryptophan (trp)

Amino Acids: The Building Blocks of Proteins

Amino acids are made up of carbon, hydrogen, and oxygen, as are sugars, and, in addition, they contain nitrogen. Every amino acid contains an amino group ($—NH_2$), a hydrogen atom, and a carboxyl group ($—COOH$) bonded to a central carbon atom:

$$H—N—C—C—OH$$

This is the fundamental structure of the molecule, and it is the same in all amino acids.

Different amino acids differ in the side group that occupies the fourth covalent bond of the central carbon atom. The side group is symbolized by the letter "R" (for "rest of molecule"). It has a different chemical structure in each different kind of amino acid (Figure 3–13). These differences are very important because they determine the different biological properties of the individual amino acids and, as a consequence, of the various kinds of proteins.

In yet another example of a dehydration synthesis, the amino "head" of one amino acid can be linked to the carboxyl "tail" of another by the removal of a molecule of water (Figure 3–14). The resulting covalent linkage is known as a **peptide bond,** and the molecule that is formed by the linking of many amino acids is called a **polypeptide** (Figure 3–15).

In order to assemble amino acids into proteins, a cell must have not only a large enough quantity of amino acids but also enough molecules of each kind. This fact is of great importance in human nutrition (see essay).

The Levels of Protein Organization

In a living cell, a protein is assembled in a long polypeptide chain, one amino acid at a time. In this process, the head of one amino acid is linked to the tail of an-

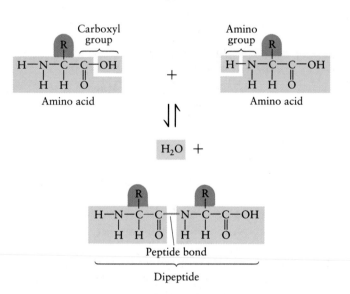

3–14 *A peptide bond is a covalent bond between two amino acids, formed by dehydration synthesis. The amino group of one amino acid is joined to the carboxyl group of the adjacent amino acid, and a molecule of water is split out.*

other, like a line of boxcars. The linear sequence of amino acids, which is dictated by the information stored in the cell for that particular protein, is known as the **primary structure** of the protein (Figure 3–16). Each different protein has a different primary structure. It determines the structural features of the molecule as a whole and thus its biological function. Even one small variation in the sequence may alter or destroy the way in which the protein functions.

Like fats, amino acids are formed within living cells using sugars as starting materials. But while fats contain only carbon, hydrogen, and oxygen atoms, all available in the sugar and water of the cell, amino acids also contain nitrogen. Most of the Earth's supply of nitrogen exists in the form of gas in the atmosphere. Only a few organisms, all microscopic, are able to incorporate nitrogen from the air into compounds—ammonia, nitrites, and nitrates—that can be used by living systems. Hence the proportion of the Earth's nitrogen supply available to the living world is very small.

Plants incorporate the nitrogen from ammonia, nitrites, and nitrates into carbon-hydrogen compounds to form amino acids. Animals are able to synthesize some of their amino acids, using ammonia as a nitrogen source. The amino acids they cannot synthe-size, the so-called **essential amino acids,** must be obtained in the diet, either from plants or from the meat of animals that have eaten plants. For adult human beings, the essential amino acids are lysine, tryptophan, threonine, methionine, phenylalanine, leucine, valine, and isoleucine.

People who eat meat usually get enough protein and the correct balance of amino acids. People who are vegetarians, whether for philosophical, esthetic, or economic reasons, have to be careful that they get enough protein and, in particular, all of the essential amino acids.

For many years, agricultural scientists concerned with the world's hungry people concentrated on developing plants with a high caloric yield. More recently, increasing recognition of the role of plants as a major source of amino acids for human populations has led to emphasis on the development of high-protein strains of food plants. Of particular importance has been the development of plants, such as "high-lysine" corn, with increased levels of one or more of the essential amino acids.

Another approach to obtaining the right balance of amino acids from plant sources is to combine certain foods. Beans, for instance, are likely to be deficient in tryptophan and in the sulfur-containing amino acids cysteine and methionine, but they are a good-to-excellent source of isoleucine and lysine. Rice is deficient in isoleucine and lysine but provides an adequate amount of the other essential amino acids. Thus rice and beans in combination make just about as perfect a protein menu as eggs or steak, as some nonscientists seem to have known for quite a long time.

3–15   *Polypeptides are polymers of amino acids linked together, one after another, by peptide bonds. The polypeptide chain shown here contains six different amino acids, but some chains may contain as many as 1,000 linked amino acid monomers.*

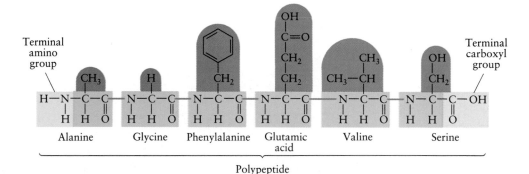

3–16   *The primary structure of a relatively small protein, human adrenocorticotropic hormone (ACTH). This was one of the first proteins for which the primary structure was determined. As you can see, it consists of a single polypeptide chain containing 39 amino acids. This hormone, secreted by the pituitary gland, stimulates the production of cortisol and related steroid hormones by the adrenal glands.*

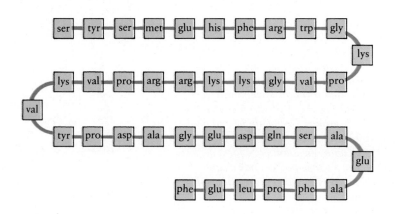

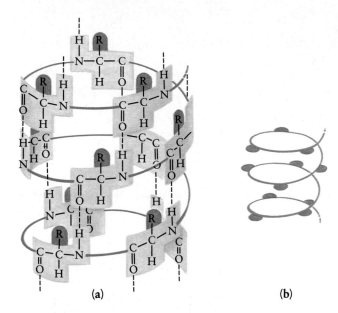

(a)   **(b)**

3–17   *In many different types of proteins, the polypeptide chain forms a secondary structure known as a helix. (a) The helix is held in shape by hydrogen bonds, indicated by the dashed lines. The hydrogen bonds form between the double-bonded oxygen atom in one amino acid and the hydrogen atom of the amino group in another amino acid that occurs four amino acids farther along the chain. The R groups, which appear flattened in this diagram, actually extend out from the helix, as shown in (b). In some proteins, virtually all of the molecule is in the form of a helix. In other proteins, only certain regions of the molecule have this secondary structure.*

As a polypeptide chain is assembled, interactions among the various amino acids along the chain cause it to fold in a simple pattern, known as its **secondary structure.** One common secondary structure is a coil or, as it is more commonly called by biochemists, a **helix.** The shape of the helix is maintained by hydrogen bonds (Figure 3–17). Because the hydrogen bonds break and re-form easily, helical proteins—or helical portions of proteins—are stretchable (elastic). Examples of such elastic proteins are myosin, which is one of the major components of muscle, and keratin, the protein that forms hair. The changes that occur when hair is washed and dried involve breaking hydrogen bonds and making new ones.

Other proteins, such as the silk fibers of cocoons and spider webs, are made of extended polypeptide chains lined up in parallel and linked to one another by hydrogen bonds. Proteins with this type of secondary structure, known as a **pleated sheet** (Figure 3–18), are smooth and supple but nonelastic.

Collagen, which is a principal component of cartilage, bones, and tendons, has yet another type of secondary structure. Collagen molecules are made by specialized cells called fibroblasts. As the long polypeptide chains are synthesized, three of them wrap around each other to form a cablelike molecule. These molecules are then packed into long, thin fibrils.

Proteins that exist for most of their length in a helical, pleated-sheet, or cablelike secondary structure are known as **fibrous proteins.** They play a variety of important structural roles in organisms.

3–18   *Another common secondary structure of proteins is the pleated sheet. (a) The pleats result from the alignment of the zigzag pattern of the atoms that form the backbone of polypeptide chains. The sheet is held together by hydrogen bonding between adjacent chains. The R groups extend above and below the pleats, as shown in (b). In some proteins, two or more polypeptide chains are aligned with one another to form a pleated sheet. In other proteins, a single polypeptide chain loops back and forth in such a way that adjacent portions of the chain form a pleated sheet.*

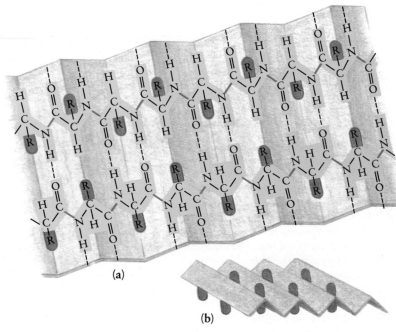

(a)

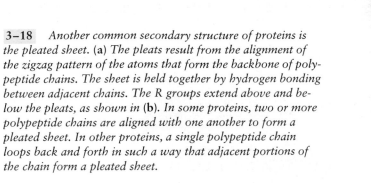

(b)

(a)

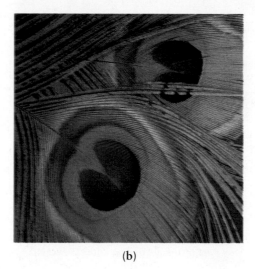

(b)

3-19 *The fibrous protein keratin is found in all vertebrates. It is the chief component of scales, wool, nails, and feathers. (a) The horn of a rhinoceros consists of tightly packed strands of keratin. Solid rhino horn is used for dagger handles, and powdered horn is sold as an aphrodisiac. A single horn can net a poacher far more than the average annual wage in many parts of Africa. (b) Feathers, such as these spectacularly colored peacock feathers, are made up of a shaft to which thousands of barbs—each with many tiny barbules—are attached.*

In other proteins, known as **globular proteins,** the secondary structure folds back on itself to make a complex **tertiary structure.** The tertiary structure forms as a result of complex interactions among the R groups in the individual amino acids. These interactions include attractions and repulsions among amino acids with polar R groups and repulsions between nonpolar R groups and the surrounding water molecules. In addition, the sulfur-containing R groups of molecules of the amino acid cysteine can form covalent bonds with each other. These bonds, known as **disulfide bridges,** lock portions of the molecule into a particular position.

A great diversity of globular proteins are synthesized by living organisms. They include enzymes, which regulate the chemical reactions occurring within organisms; antibodies, which are important components of the immune system; and receptors on the surface of the cell membrane that interact with a variety of other molecules. As we shall see in subsequent chapters, the three-dimensional structures of all of these proteins are of critical importance in determining their biological functions.

Many proteins are composed of more than one polypeptide chain. The polypeptide chains are held together by hydrogen bonds, disulfide bridges, attractions between positive and negative charges, and hydrophobic forces. This level of organization of proteins, which involves the interaction of two or more polypeptides, is called a **quaternary structure.** The hormone insulin, for example, has an intricate quaternary structure formed by the two polypeptide chains of which the molecule is composed (Figure 3–20).

The secondary, tertiary, and quaternary structures of a protein all depend on the primary structure—the linear sequence of amino acids.

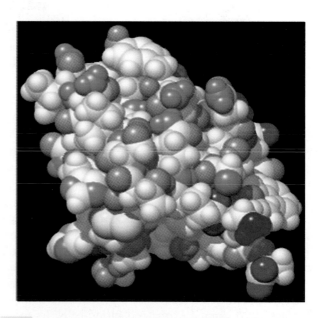

3-20 *A computer-generated model of the insulin molecule, which consists of two short polypeptide chains, folded together in an intricate three-dimensional structure. The individual atoms are identified by color. Carbon atoms are white, sulfur atoms are yellow, slightly hydrophilic nitrogen atoms are light blue, strongly hydrophilic nitrogen atoms are dark blue, slightly hydrophilic oxygen atoms are pink, and strongly hydrophilic oxygen atoms are red. Hydrogen atoms are the same color as the atoms to which they are bonded.*

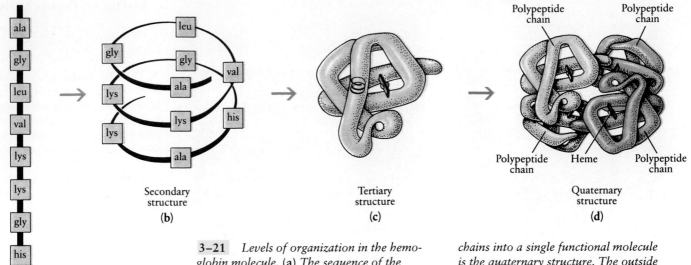

ala
gly
leu
val
lys
lys
gly
his
ala
lys

Primary
structure
(a)

Secondary
structure
(b)

Tertiary
structure
(c)

Quaternary
structure
(d)

Polypeptide chain   Polypeptide chain

Polypeptide chain   Heme   Polypeptide chain

3–21   *Levels of organization in the hemo-globin molecule. (a) The sequence of the amino acids in each chain is its primary structure. (b) The helical form assumed by any part of the chain as a consequence of hydrogen bonding between nearby*

$C=O$ *and* $—NH$ *groups is its secondary structure. (c) The folding of the chains in three-dimensional shapes is the tertiary struc-ture, and (d) the combination of the four*

chains into a single functional molecule is the quaternary structure. The outside of the molecule and the hole through the middle are lined by charged amino acids, and the uncharged amino acids are packed inside. Each of the four chains surrounds a heme group (red), which can hold a single oxygen molecule. A hemoglobin molecule is therefore capable of transporting four oxy-gen molecules.

Hemoglobin: An Example of Specificity

Fibrous proteins usually consist of a relatively small va-riety of amino acids in a repetitive sequence. Many globular proteins, by contrast, have extremely complex, irregular amino acid sequences, as complex and irreg-ular as the sequence of letters in a sentence on this page. Just as these sentences make sense (if they do) because the letters are the right ones and in the right order, the proteins make sense, biologically speaking, because their amino acids are the right ones in the right order.

Hemoglobin, for example, is a protein that is man-ufactured and carried in the red blood cells. Its mole-cules have the special property of being able to combine loosely with oxygen, collecting it in the lungs and re-

leasing it elsewhere in the body. The hemoglobin mol-ecule has a quaternary structure that consists of four polypeptide chains, each of which is combined with an iron-containing group known as heme. Hemoglobin has two identical chains of one type, designated alpha, and two identical chains of another type, designated beta. Each type of chain has a unique primary structure containing about 150 amino acids, for a total of about 600 amino acids in all (Figure 3–21).

Sickle-cell anemia is a disease in which the hemoglo-bin molecules are defective. When oxygen is released from them, these molecules change shape and combine with one another to form stiffened rodlike structures.

(a)

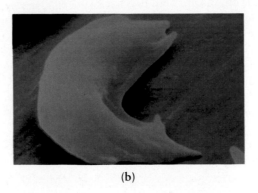

(b)

3–22   *Scanning electron micrographs of (a) a human red blood cell containing nor-mal hemoglobin, and (b) a red blood cell containing the abnormal hemoglobin associ-ated with sickle-cell anemia. When the oxy-gen concentration in the blood is low, the abnormal hemoglobin molecules stick to-gether, distorting the shape of the cells. As a result, the cells cannot pass readily through small blood vessels. These cells are magnified about 35,000 times.*

Normal hemoglobin

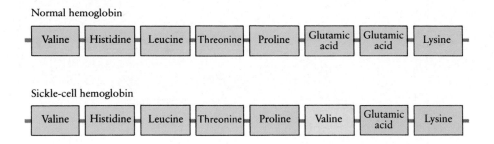

Sickle-cell hemoglobin

3–23 *An example of the remarkable precision of the "language" of proteins. Portions of the beta chains of the normal hemoglobin molecule and the sickle-cell hemoglobin molecule are shown. The entire structural difference between the normal molecule and the sickle-cell molecule (literally, a life-and-death difference) consists of one change in the sequence of each beta chain: one glutamic acid is replaced by one valine.*

Red cells containing large proportions of such hemoglobin molecules become stiff and deformed, taking on the characteristic sickle-cell shape (Figure 3–22). The deformed cells are more fragile than normal red blood cells and rupture easily. They also tend to clog small blood vessels, causing blood clots and depriving vital organs of their full supply of blood. The result is pain, intermittent illness, and, in many cases, a shortened life span.

Analysis of the hemoglobin molecules has revealed that the only difference between normal and sickle-cell hemoglobin is that at a precise location in each beta chain, one glutamic acid is replaced by one valine (Figure 3–23). When the beta chains assume their tertiary structure and become associated with the alpha chains in the quaternary structure of the completed protein, the location at which the substitution occurs is on the outside of the molecule. As you can see in Figure 3–13 (page 60), glutamic acid—present in the normal beta chain—is polar and thus hydrophilic. Valine, by contrast, is nonpolar and thus hydrophobic. The substituted valines in the two beta chains create two "sticky" regions on the surface of the hemoglobin molecule. It is the mutual attraction of these regions in neighboring molecules that causes sickle-cell hemoglobin to form rodlike structures.

In hemoglobin, a difference of only two amino acids in a total of almost 600 can be the difference between life and death. This gives you an idea of the precision and the importance of the arrangement of amino acids in a particular sequence in a protein. In living systems, which must perform many different activities simultaneously, the specificity of function that results from the structural precision of different protein molecules is of crucial importance.

Nucleic Acids

The information dictating the structures of the enormous variety of protein molecules found in living organisms is encoded in molecules known as nucleic acids. Just as proteins consist of long chains of amino acids, nucleic acids consist of long chains of molecules known as **nucleotides.** A nucleotide, however, is a more complex molecule than an amino acid.

As shown in Figure 3–24, a nucleotide consists of three subunits: a phosphate group, a five-carbon sugar, and a **nitrogenous base**—a molecule that has the properties of a base and contains nitrogen. The phosphate group (PO_4^{3-}) is an ion of phosphoric acid (H_3PO_4); it

3–24 *A nucleotide is made up of three different subunits: a phosphate group, a five-carbon sugar, and a nitrogenous base. The nitrogenous base in this nucleotide is adenine, and the sugar is ribose. The nucleotide is adenosine monophosphate, abbreviated AMP.*

Nitrogenous base

Phosphate group

Sugar

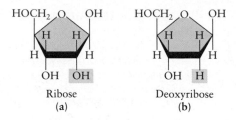

3–25   *The sugar subunit of a nucleotide may be either* (**a**) *ribose or* (**b**) *deoxyribose. The structural difference between the two sugars is highlighted in blue. As you can see, it is very slight. RNA is formed from nucleotides that contain ribose, and DNA from nucleotides that contain deoxyribose.*

is the source of "acid" in the term "nucleic acid." The sugar subunit of a nucleotide may be either ribose or deoxyribose, which contains one less oxygen atom than ribose (Figure 3–25). Five different nitrogenous bases occur in the nucleotides that are the building blocks of nucleic acids. In the nucleotide in Figure 3–24, the nitrogenous base is adenine. In adenine and other nitrogenous bases, each of the nitrogen atoms in the molecule has an unshared pair of electrons in the outer energy level. These electrons exert a weak attraction for hydrogen ions (H^+). Thus the molecule is a base, capable of combining with H^+ ions and thereby increasing the relative number of OH^- ions in a solution (see page 45).

Two types of nucleic acid are found in living organisms. In **ribonucleic acid (RNA)**, the sugar subunit in the nucleotides is ribose. In **deoxyribonucleic acid (DNA)**, it is deoxyribose. Like polysaccharides, fats, and proteins, RNA and DNA are formed from their subunits in dehydration synthesis reactions. The result is a linear molecule, consisting of one nucleotide after another (Figure 3–26).

3–26   *Nucleic acid molecules are long chains of nucleotides in which the sugar subunit of one nucleotide is linked to the phosphate group of the next nucleotide. The covalent bond linking one nucleotide to the next—shown here in blue—is forged by a dehydration synthesis reaction. This reaction is reversible; with the addition of a molecule of water in hydrolysis, the bond can be broken.*

RNA molecules consist of a single chain of nucleotides, as shown here. DNA molecules, by contrast, consist of two chains of nucleotides, coiled around each other in a double helix.

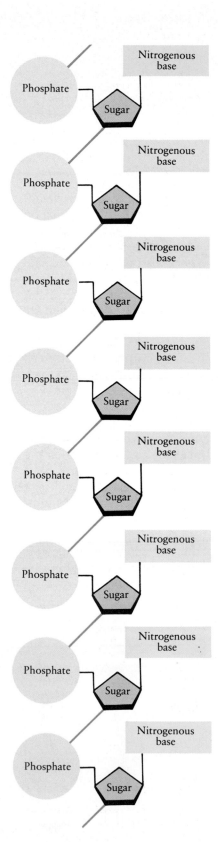

Although their chemical components are very similar, DNA and RNA generally play different biological roles. DNA is the carrier of the genetic message. It contains the information, organized in units known as **genes,** that we and other organisms inherit from our parents. RNA is a copy, or transcript, of the genetic message that serves as a blueprint for protein synthesis.

The discovery of the structure and function of DNA and RNA is undoubtedly the greatest triumph thus far of the molecular approach to the study of biology. In Section 3, we shall trace the events leading to the key discoveries and shall consider in some detail the marvelous processes—the details of which are still being worked out—by which the nucleic acids perform their functions.

ATP: The Cell's Energy Currency

In addition to their role as the building blocks of nucleic acids, nucleotides have an independent and crucial function in living systems. When modified by the addition of two more phosphate groups, they are the carriers of the energy needed to power the numerous chemical reactions occurring within the living cell. The energy in storage carbohydrates, such as starch and glycogen, and in lipids is like money in savings bonds or certificates of deposit—not readily accessible. The energy in glucose is like money in a checking account—accessible but not very handy for everyday transactions. The energy in modified nucleotides, however, is like money in your pocket—available in convenient amounts and universally accepted.

The principal energy carrier for most processes in living organisms is a molecule known as **adenosine triphosphate,** or **ATP.** This molecule is shown schematically in Figure 3–27. Note the three phosphate groups.

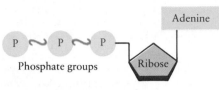

Phosphate groups · Ribose · Adenine

Adenosine triphosphate (ATP)

3–27 *A schematic diagram of the ATP (adenosine triphosphate) molecule. The only difference between this molecule and the AMP (adenosine monophosphate) molecule shown in Figure 3–24 is the addition of two phosphate groups. Although this structural difference may seem minor, it is the key to the function of ATP in living systems.*

The bonds linking these groups, designated by a squiggle, are relatively weak, and they can be broken quite readily by hydrolysis. The products of the most common reaction, illustrated in Figure 3–28, are **ADP** (adenosine diphosphate), a phosphate group, and energy. As this energy is released, it can be used to power other chemical reactions.

ADP is "recharged" to ATP when glucose is oxidized to carbon dioxide and water, just as the money in your pocket is "recharged" when you cash a check or visit an automatic teller machine. In Chapter 8, we shall consider this process in more detail. For now, however, the important thing to remember is that ATP is the molecule that is directly involved in providing energy for the living cell, to which we shall turn our attention in the next three chapters.

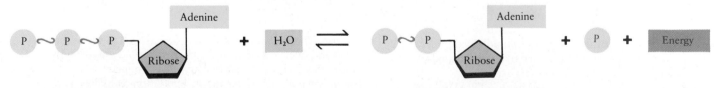

Adenosine triphosphate (ATP)   Adenosine diphosphate (ADP)

3–28 *The hydrolysis of ATP. With the addition of a molecule of water to ATP, one phosphate group is removed from the molecule. The products of this reaction are ADP, a free* *phosphate group, and energy. About 7 kilocalories of energy are released for every mole of ATP hydrolyzed. With an energy input of 7 kilocalories per mole, the reaction can be reversed.*

Summary

The chemistry of living organisms is, in essence, the chemistry of organic compounds—that is, compounds containing carbon. Carbon is uniquely suited to this central role by its ability to form multiple covalent bonds. Because of this ability, carbon can combine with carbon and other atoms to form a great variety of strong and stable chain and ring compounds. Organic molecules derive their three-dimensional shapes primarily from their carbon skeletons. Among the major types of organic molecules important in living systems are carbohydrates, lipids, proteins, and nucleic acids.

Carbohydrates serve as a primary source of chemical energy for living systems. The simplest carbohydrates are the monosaccharides ("single sugars"), such as glucose and fructose. Monosaccharides can be combined to form disaccharides ("two sugars"), such as sucrose, and polysaccharides (chains of many monosaccharides). The polysaccharides starch and glycogen are storage forms for sugar, whereas cellulose, another polysaccharide, is an important structural material in plants. Disaccharides and polysaccharides are formed by dehydration synthesis reactions in which monosaccharide units are covalently bonded with the removal of a molecule of water. They can be broken apart again by hydrolysis, with the addition of a water molecule.

Lipids are hydrophobic organic molecules that, like carbohydrates, play important roles in energy storage and as structural components. Compounds in this group include fats and oils, phospholipids, glycolipids, waxes, and cholesterol and other steroids. Phospholipids and glycolipids are major components of cell membranes.

Proteins are very large molecules composed of long chains of amino acids; these chains are known as polypeptides. The 20 different amino acids used in making proteins vary according to the properties of their side (R) groups. From these relatively few amino acids, an extremely large variety of different kinds of protein molecules can be synthesized, each of which has a highly specific function in living systems. Some proteins are fibrous and have important structural roles. Other proteins are globular and have regulatory and protective functions. The principal levels of protein organization are (1) primary structure, the linear amino acid sequence; (2) secondary structure, often a coiling of the polypeptide chain; (3) tertiary structure, the complex folding of the chain into various shapes; and (4) quaternary structure, the folding of two or more polypeptide chains around each other. The precise structure of a protein molecule is of critical importance in its capacity to carry out a specific biological function.

The building blocks of the nucleic acids are complex molecules known as nucleotides. A nucleotide consists of a phosphate group, a five-carbon sugar, and a nitrogenous base. The nucleic acid known as deoxyribonucleic acid (DNA) is the carrier of the genetic information. Another molecule, ribonucleic acid (RNA), is a copy of that information that functions as a blueprint for protein synthesis. Nucleotides also play key roles in the energy exchanges accompanying chemical reactions within living systems. The principal energy carrier for most processes is adenosine triphosphate (ATP).

Questions

1. Distinguish among the following: hydrocarbon/carbohydrate; glucose/fructose/sucrose; monomer/polymer; glycogen/starch/cellulose; saturated/unsaturated; mono-unsaturated/polyunsaturated; phospholipid/glycolipid; polysaccharide/polypeptide; primary structure/secondary structure/tertiary structure/quaternary structure; nitrogenous base/nucleotide/nucleic acid; ATP/ADP.

2. Many of the synthetic reactions in living systems take place by dehydration synthesis. What is a dehydration synthesis reaction? What types of molecules undergo dehydration synthesis reactions to form disaccharides and polysaccharides? To form fats? To form proteins? To form nucleic acids?

3. Disaccharides and polysaccharides, as well as lipids, proteins, and nucleic acids, can be broken down by hydrolysis. What is hydrolysis? What two types of products are released when a polysaccharide such as starch is hydrolyzed? How are these products important for the living cell?

4. What do we mean when we say that some polysaccharides are "energy-storage" molecules and that others are "structural" molecules? Give an example of each. In what sense should any polysaccharide be regarded as an "energy-storage" molecule?

5. Plants usually store energy reserves as polysaccharides, whereas, in most animals, lipids are the principal form of energy storage. Why is it advantageous for animals to have their energy reserves stored as lipids rather than as polysaccharides? (Think about the differences in "life style" between plants and animals.) What kinds of storage materials would you expect to find in seeds?

6. Sketch the arrangement of phospholipids when they are surrounded by water.

7. In pioneer days, soap was made by boiling animal fat with lye (potassium hydroxide). The bonds linking the fatty acids to the glycerol molecule were hydrolyzed, and the po-

tassium hydroxide reacted with the fatty acid to produce potassium stearate. A typical soap available today is sodium stearate. In water, it ionizes to produce sodium ions (Na^+) and stearate ions:

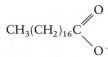

Explain how soap functions to trap and remove particles of grease and dirt.

8. Consider the R groups of the eight amino acids illustrated in Figure 3–13. Which of these R groups would you expect to be nonpolar? Polar? Are there any of the R groups that you would expect to function as weak acids in water solution? As weak bases?

Suggestions for Further Reading

Brown, Michael S., and Joseph L. Goldstein: "How LDL Receptors Influence Cholesterol and Atherosclerosis," *Scientific American*, November 1984, pages 58–66.

Karplus, Martin, and J. Andrew McCammon: "The Dynamics of Proteins," *Scientific American*, April 1986, pages 42–51.

Lehninger, Albert L., David L. Nelson, and Michael M. Cox: *Principles of Biochemistry*, 2d ed., Worth Publishers, Inc., New York, 1993.

This is an updated version of a classic introductory text, outstanding for both its clarity and its consistent focus on the living cell.

Mathews, Christopher K., and K. E. van Holde: *Biochemistry*, The Benjamin/Cummings Publishing Company, Redwood City, Calif., 1990.

Chapters 4–9 of this beautifully illustrated text describe the structure and properties of nucleic acids, proteins, carbohydrates, and lipids.

Olson, Arthur J., and David S. Goodsell: "Visualizing Biological Molecules," *Scientific American*, November 1992, pages 76–81.

Richards, Frederic M.: "The Protein Folding Problem," *Scientific American*, January 1991, pages 54–63.

Scientific American: *The Molecules of Life*, W. H. Freeman and Company, New York, 1986.*

This reprint of the October 1985 issue of Scientific American *includes 11 articles on molecules that play key roles in the living cell. The articles on proteins are of particular interest at this point in your study of biology. You will find the other articles in this collection useful at later points in the course.*

Sharon, Nathan: "Carbohydrates," *Scientific American*, November 1980, pages 90–116.

Stryer, Lubert: *Biochemistry*, 3d ed., W. H. Freeman and Company, New York, 1988.

An introductory text, with many examples of medical applications of biochemistry. Handsomely illustrated.

* Available in paperback.

Life first appeared on Earth more than 3.5 billion years ago, when the planet was still very young. The first step in the origin of life was the formation of complex organic molecules from the atoms and simple molecules present in the Earth's crust, waters, and atmosphere. This process required energy. One likely source was the internal heat of the Earth itself. Because of the high temperatures (2200°C to 2800°C) and pressures, rock in the interior of the Earth is molten. It is this molten material that erupts from volcanoes, such as Sakurajima in Japan, shown here in a photograph taken on May 18, 1991.

Volcanic eruptions, releasing enormous amounts of energy, are thought to have been far more frequent on the early Earth than they are today. The energy released takes several forms: dry heat, from the molten rock; damp heat, in the form of steam; and shock waves that travel through the air, water, and rock of the Earth's surface. Volcanic eruptions are also frequently accompanied by lightning. The lightning is caused by the imbalance of electric charge produced by the clouds of volcanic ash, steam, and other gases spewing up into the atmosphere. All of these forms of energy associated with volcanoes are thought to have played a role in the events leading to the origin of life.

The Earth's crust and the upper portion of its mantle are made up of separate plates on which the continents and oceans rest. Volcanoes occur most frequently in areas where different plates converge and one plate slips below another. Sakurajima and the other volcanoes of the Pacific "Ring of Fire" are located along the line where the Asian plate and the Pacific plate converge.

4

Cells: An Introduction

In the last three chapters, we have progressed from subatomic particles through atoms and molecules to large, complex molecules, such as proteins and nucleic acids. At each level of organization, new properties appear. For instance, water is not the sum of the properties of elemental hydrogen and oxygen. It is something more and also something different. In proteins, amino acids become organized into polypeptides, and polypeptide chains are arranged in new levels of organization, the secondary, tertiary, and, in some cases, quaternary structure of the complete protein molecule. Only at the final level of organization do the complex properties of the protein emerge, and only then can the molecule assume its function.

The characteristics of living systems, like those of atoms or of molecules, do not emerge gradually as the degree of organization increases. They appear quite suddenly and specifically, in the form of the living cell—something that is more than and different from the atoms and molecules of which it is composed. No one knows exactly when or how this new level of organization—the living cell—first came into being. However, increasing knowledge of the history of our planet and the results of numerous laboratory experiments provide strong evidence to support the hypothesis that living cells spontaneously self-assembled from molecules present in the primitive seas.

The Formation of the Earth

About 5 billion years ago, cosmologists calculate, the star that is our sun came into being. According to current theory, it formed, like other stars, from an accumulation of dust and hydrogen and helium gases whirling in space among the older stars.

The immense cloud that was to become the sun condensed gradually as the hydrogen and helium atoms were pulled toward one another by the force of gravity, falling into the center of the cloud and gathering speed

4–1   *Of the nine planets in our solar system, only one, as far as we know, has life on it. This planet, Earth, is visibly different from the others. From a distance, it appears blue and green and it shines a little. The blue is water, the green is chlorophyll, and the shine is sunlight reflected off the layer of gases surrounding the planet's surface. Life, at least as we know it, depends on these visible features of Earth.*

as they fell. As the cluster grew denser, the atoms moved more rapidly. More atoms collided with each other, and the gas in the cloud became hotter and hotter. As the temperature rose, the collisions became increasingly violent until the hydrogen atoms collided with such force that their nuclei fused, forming additional helium atoms and releasing nuclear energy. This thermonuclear reaction is still going on at the heart of the sun and is the source of the energy radiated from its turbulent surface.

The planets, according to current theory, formed from the remaining gas and dust moving around the newly formed star. At first, particles would have collected at random, but as each mass grew larger, other particles began to be attracted by the gravity of the largest masses, which would become the planets. The whirling dust and forming planetary spheres continued to revolve around the sun until finally each sphere had swept its own path clean, picking up loose matter like a giant snowball. The orbit nearest the sun was swept by Mercury, the next by Venus, the third by Earth, the fourth by Mars, and so on out to Neptune and Pluto,

the most distant of the planets. The planets, including Earth, are calculated to have come into being about 4.6 billion years ago.

During the time Earth and the other planets were being formed, the release of energy from radioactive materials kept their interiors very hot. When Earth was still so hot that it was mostly liquid, the heavier materials sank to the interior, forming a dense inner core whose diameter is about half that of the planet. As Earth's surface cooled, an outer crust, a skin as thin by comparison as the skin of an apple, solidified. The oldest known rocks in this layer have been dated by isotopic methods as about 4.1 billion years old. A mere 600 million years after these rocks formed—and perhaps even earlier—life on Earth began.

The Beginning of Life

Until very recently, biologists regarded the earliest events in the history of life as chapters that would probably remain forever closed to scientific investigation. Two developments, however, have greatly improved our long-distance vision. The first was the formulation of a testable hypothesis about the events preceding life's origins. This hypothesis generated questions for which answers could be sought experimentally. The results of the initial experimental tests led to the formulation of further hypotheses and to additional experiments, a process that continues today as scientists in many laboratories explore the question of life's origins. The second development was the discovery of fossilized cells more than 3 billion years old.

The testable hypothesis was offered by the Russian biochemist A. I. Oparin. According to Oparin, the appearance of life was preceded by a long period in which complex molecules formed and began to accumulate, a process that is sometimes called **chemical evolution.** The identity of the substances, particularly gases, present in the atmosphere and seas during this period is a matter of controversy. There is general agreement, however, on two critical issues. First, little or no free oxygen was present. The Earth's supply of oxygen was locked in simple compounds such as water, carbon monoxide, carbon dioxide, and oxides of silicon and various metals. Second, the four elements—hydrogen, oxygen, carbon, and nitrogen—that make up more than 95 percent of living tissues were available in some form in the atmosphere and waters of the primitive Earth.

In addition to these raw materials, energy abounded on the young Earth. Radioactive elements within the

Earth released their energy into the atmosphere, as did the frequent volcanic eruptions. There was heat energy, both boiling (moist) heat and baking (dry) heat. Water vapor spewed out of the primitive seas, cooled in the upper atmosphere, collected into clouds, fell back on the crust of the Earth, and steamed up again. Violent rainstorms were accompanied by lightning, which provided electrical energy. And, most important, the sun continuously bombarded the Earth's surface with high-energy particles and ultraviolet radiation.

Oparin hypothesized that under such conditions organic molecules were formed from the atmospheric gases and collected in a thin soup in the Earth's seas and lakes. In the absence of free oxygen, these organic molecules could persist. (Oxygen, as we noted in Chapter 1, has a powerful attraction for electrons; thus it tends to react with and break down organic molecules into simple substances such as carbon dioxide and water.) Some of the newly formed organic molecules might have become locally more concentrated by the drying up of a lake or by the adhesion of the molecules to solid surfaces, such as clay particles.

Oparin published his hypothesis in 1922, but the scientific community ignored his ideas. In the 1950s, the first test of the hypothesis (Figure 4–2) was performed by Stanley Miller, then a graduate student at the University of Chicago. Experiments of this sort, now repeated many times, have shown that almost any source of energy—lightning, ultraviolet radiation, or hot volcanic ash—would have converted molecules believed to have been present on the Earth's surface into a variety of complex organic compounds. With various modifications in the experimental conditions and in the mixture of gases placed in the reaction vessel, almost all of the common amino acids have been produced, as well as the nucleotides that are the essential components of DNA and RNA.

These experiments have not proved that such organic compounds were formed spontaneously on the primitive Earth, only that they could have formed. The accumulated evidence is nevertheless very great, and most biochemists now believe that, given the conditions existing on the young Earth, chemical reactions producing amino acids, nucleotides, and other organic molecules were inevitable.

As the concentrations of such molecules increased, bringing them into closer proximity to each other, they would have been subject to the same chemical forces that act on organic molecules today. As we saw in the last chapter, small organic molecules react with each other, typically in dehydration synthesis reactions, to form larger molecules. Moreover, such forces as hydro-

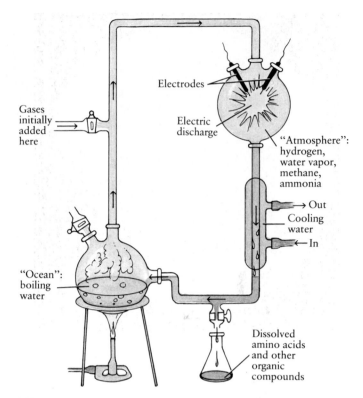

4–2 *In Miller's experiment, conditions believed to have existed on the primitive Earth were simulated in the laboratory. Hydrogen gas (H$_2$), water vapor, methane (CH$_4$), and ammonia (NH$_3$) were continuously circulated between a lower "ocean" and an upper "atmosphere." The "ocean" was heated, vaporizing the water and forcing the gases into the "atmosphere," through which electric discharges were transmitted. As the gases moved down the portion of the tubing surrounded by cooling water, the water vapor condensed to liquid water and carried with it any organic molecules that had formed. These molecules accumulated in the portion of the tubing leading back to the "ocean." At the end of 24 hours, about half of the carbon originally present in the methane gas was converted to amino acids and other organic molecules. This was the first test of Oparin's hypothesis.*

gen bonds and hydrophobic interactions cause these molecules to assemble themselves into more complex aggregates. In modern chemical systems—either in the laboratory or in the living organism—the more stable molecules and aggregates tend to survive, and the least stable are transitory. Similarly, the compounds and aggregates that had the greatest chemical stability under the prevailing conditions on the primitive Earth would have tended to survive. Hence a form of natural selection played a role in chemical evolution as well as in the biological evolution that was to follow.

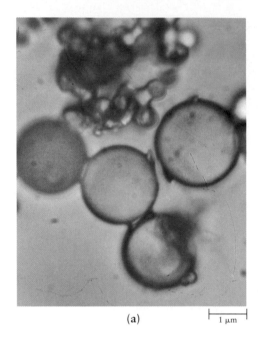

(a)   1 μm

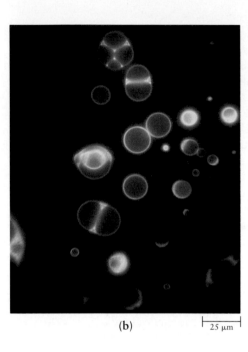

(b)   25 μm

4–3   (a) *When dry mixtures of amino acids are heated at moderate temperatures, polymers known as thermal proteinoids are formed. Each of these polymers may contain as many as 200 amino acid subunits. When the polymers are placed in water solution and maintained under suitable conditions, they spontaneously form proteinoid microspheres, shown here. The microspheres are separated from the surrounding solution by a membrane that appears to be two-layered.*

*(b) These membrane-bound structures, separated from the surrounding solution by a lipid bilayer, formed spontaneously from organic molecules extracted from the Murchison meteorite, which struck Australia in 1969. Samples were taken from the meteorite on the day it landed and then stored under conditions that would prevent their contamination by terrestrial materials. Since that time, samples of the meteorite have been made available to scientists around the world for a variety of investigations.**

The First Cells

Three characteristics distinguish the living cell from a mere aggregate of complex molecules: (1) the capacity to replicate itself, generation after generation; (2) the presence of enzymes, the complex proteins that are essential for the chemical reactions on which life depends; and (3) a membrane that separates the cell from the surrounding environment and enables it to maintain a distinct chemical identity. Which of these characteristics appeared first—and made possible the development of the others—remains an open question. However, as we shall see in Chapter 16, recently discovered functions of RNA suggest that a critical step in the evolution of the first cells may have been the self-assembly of RNA molecules from nucleotides produced by chemical evolution.

In studies simulating conditions during the Earth's first billion years, Sidney W. Fox and his co-workers at the University of Miami have produced membrane-bound structures, known as proteinoid microspheres, that can carry out a few chemical reactions resembling those of living cells. The microspheres (Figure 4–3a) grow slowly by the addition of proteinoid material from the solution and eventually bud off smaller microspheres. These structures are not living cells. Their formation, however, suggests the kinds of processes that could have given rise to self-sustaining protein entities, separated from their environment and capable of carrying out the chemical reactions necessary to maintain their physical and chemical integrity.

Although cell membranes contain large amounts of protein, their basic structure is formed by lipids. It has been hypothesized that comets and meteorites, which struck the primitive Earth with great frequency and sometimes contain carbonaceous material, might have been a source of organic compounds, including lipid-like substances. David W. Deamer of the University of California, Davis, and his co-workers extracted and analyzed organic molecules from carbon-containing meteorites. Some of the molecules spontaneously self-assemble into membrane-bound spheres (Figure 4–3b).

It is not known when the first living cells appeared on Earth, but we can establish some sort of time scale. The earliest fossils found so far (Figure 4–4), which resemble present-day bacteria, have been dated at 3.5 billion years—about 1.1 billion years after the formation of the Earth itself. These fossilized cells are sufficiently complex that it is clear that some little aggregation of chemicals had moved through the twilight zone separating the living from the nonliving millions of years before.

* The short straight lines at the bottom of these micrographs and those that follow are scale markers. A scale marker provides a reference for size. The same system is used to indicate distances on road maps. A micrometer, abbreviated μm, is 1/1,000,000 meter.

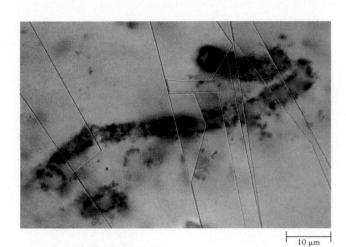

|— 10 μm —|

4–4   *This microfossil of a filament of bacteria-like cells was found in Western Australia in a deposit of a flintlike rock known as black chert. Dated at 3.5 billion years of age, it is one of the oldest fossils known.*

Why on Earth?

Of all the planets in our solar system, only Earth has been found to support life, a fact that holds clues to the requirements for life. A major factor is that Earth is neither too close to nor too distant from the sun. The chemical reactions on which life depends require liquid water, and they virtually cease at very low temperatures. Jupiter, where temperatures are about −150°C, and Saturn, where they are even lower—about −180°C—are far too cold to support life. Conversely, at high temperatures, water, which boils at 100°C, exists only in the form of vapor, and the complex chemical compounds essential for life become unstable. The upper limit for the stability of amino acids, for example, is between 200°C and 300°C. Life is clearly impossible on Mercury and Venus, where temperatures reach 430°C and 480°C, respectively.

Earth's size and mass are also important factors. Planets much smaller than Earth do not have enough gravitational pull to hold a protective atmosphere, and any planet much larger than Earth is likely to have so dense an atmosphere that light from the sun cannot reach its surface. Earth's atmosphere blocks out the most energetic radiations from the sun, which are capable of breaking the covalent bonds between carbon atoms. It does, however, permit the passage of visible light, which made possible the process of photosynthesis, one of the most significant steps in the evolution of complex living systems.

Heterotrophs and Autotrophs

The energy that produced the first organic molecules came from a variety of sources on the young Earth and in its atmosphere—heat, ultraviolet radiation, and electrical disturbances. When the first primitive cells or cell-like structures evolved, they required a continuing supply of energy to maintain themselves, to grow, and to reproduce. The manner in which these cells obtained energy is the subject of continuing discussion.

Modern organisms—and the cells of which they are composed—can meet their energy needs in one of two ways. **Heterotrophs** are organisms that are dependent upon outside sources of organic molecules for both their energy and their small building-block molecules. (*Hetero* comes from the Greek word meaning "other," and *troph* comes from *trophos,* "one that feeds.") All animals and fungi, as well as many single-celled organisms, are heterotrophs.

Autotrophs, by contrast, are "self-feeders." They do not require organic molecules from outside sources for energy or to use as small building-block molecules. They are, instead, able to synthesize their own energy-rich organic molecules from simple inorganic substances. Most autotrophs, including plants and several different types of single-celled organisms, are **photosynthetic,** meaning that the energy source for their synthetic reactions is the sun. Certain groups of bacteria, however, are **chemosynthetic.** These organisms capture the energy released by specific inorganic reactions to power their life processes, including the synthesis of needed organic molecules.

It has long been postulated that the first living cells were heterotrophs, utilizing organic molecules available in the primordial soup—the same molecules from which the cells were themselves assembled. As the primitive cells increased in number, according to this hypothesis, they began to use up the complex molecules on which their existence depended and which had taken millions of years to accumulate. As the supply of these molecules decreased, competition began. Under the pressures of competition, cells that could make efficient use of the limited energy sources now available were more likely to survive and reproduce than cells that could not. In the course of time, other cells evolved that were able to synthesize organic molecules out of simple inorganic materials.

Recent discoveries, however, have raised the possibility that the first cells may have been either chemosynthetic or photosynthetic autotrophs rather than heterotrophs. First, several different groups of chemosynthetic bacteria have been found that would have been well-suited to the conditions prevailing on the young Earth. Some of these bacteria (Figure 4–5) live in swamps, while others have been found in deep ocean trenches in areas where gases escape from fissures in the Earth's crust. There is evidence that these modern bacteria are the surviving representatives of very ancient groups of unicellular organisms. Second, some of the experiments simulating conditions on the young Earth have produced molecules that are the chemical precursors of chlorophyll. When these molecules are mixed with simple organic molecules in an oxygen-free environment and illuminated, primitive photosynthetic reactions occur. These reactions resemble the ones that occur in some types of photosynthetic bacteria.

Both heterotrophs and autotrophs seem to be represented among the earliest microfossils, and biologists are presently unable to resolve the question of which came first. However, it is certain that without the evolution of autotrophs, life on Earth would soon have come to an end. In the more than 3.5 billion years since life first appeared on Earth, the most successful autotrophs (that is, those that have left the most offspring and diverged into the greatest variety of forms) have been those that evolved a system for making direct use of the sun's energy in the process of photosynthesis. With the advent of photosynthesis, the flow of energy in the biosphere came to assume the predominant pattern seen today: radiant energy from the sun is channeled through photosynthetic autotrophs to all other forms of life.

1 μm

4–5   *Methanogens, such as the cells shown here, are chemosynthetic bacteria that produce methane (CH_4) and water from carbon dioxide and hydrogen gas. The chemosynthetic reaction releases about 33 kilocalories of energy for each mole of methane that is formed. This energy is used by the cells to power their life processes. Methanogens can live only in the absence of oxygen—a condition prevailing on the young Earth but occurring today only in isolated environments, such as the muck and mud at the bottom of swamps.*

The Cell Theory

One of the fundamental principles of biology is that all living organisms are composed of one or more similar cells. This concept is of tremendous and central importance to biology because it emphasizes the underlying sameness of all living systems. It therefore brings unity to widely varied studies involving many different kinds of organisms.

The word "cell" was first used in a biological sense some 300 years ago. In the seventeenth century, the English scientist Robert Hooke, using a microscope of his own construction, noticed that cork and other plant tissues are made up of small cavities separated by walls (Figure 4–6). He called these cavities "cells," meaning "little rooms." However, "cell" did not take on its present meaning—the basic unit of living matter—for more than 150 years.

In 1838, Matthias Schleiden, a German botanist, reported his observation that all plant tissues consist of organized masses of cells. In the following year, zoologist Theodor Schwann extended Schleiden's observation to animal tissues and proposed a cellular basis for all life. In 1858, the idea that all living organisms are composed of one or more cells took on an even broader significance when the pathologist Rudolf Virchow generalized that cells can arise only from preexisting cells: "Where a cell exists, there must have been a preexisting cell, just as the animal arises only from an animal and the plant only from a plant."

From the perspective provided by Darwin's theory of evolution, published in the following year, Virchow's concept takes on an even larger significance. There is an unbroken continuity between modern cells—and the organisms they compose—and the first primitive cells that appeared on Earth.

In its modern form, the **cell theory** states simply that (1) all living organisms are composed of one or more cells; (2) the chemical reactions of a living organism, including its energy-releasing processes and its biosynthetic reactions, take place within cells; (3) cells arise from other cells; and (4) cells contain the hereditary information of the organisms of which they are a part, and this information is passed from parent cell to daughter cell.

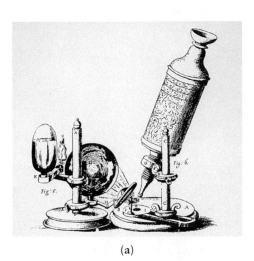

(a)

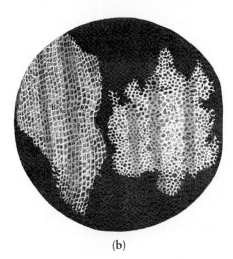

(b)

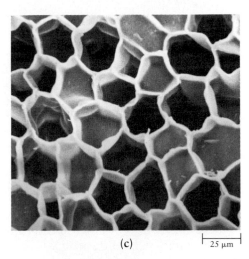

(c)

25 μm

4–6   (a) *Robert Hooke's drawing of his microscope, reproduced from a book he published in 1665. Light from an oil lamp was directed to the specimen through a water-filled glass globe that acted as a condenser. The specimen was mounted on a pin, just below the tip of the microscope. The microscope was focused by moving it up and* *down, using a screw held to the stand by a clamp. Among the many other illustrations in the book was a drawing of two slices of a piece of cork* (b). *The structure of cork is revealed in more detail in a modern electron micrograph* (c).

Hooke was the first to use the word "cells" to describe the tiny compartments *that together make up an organism. The cells in these pieces of cork have died—all that remain are the outer walls. The living cell, however, is filled with a variety of substances, organized into distinct structures and carrying out a multitude of essential processes.*

Prokaryotes and Eukaryotes

All cells—ancient and modern—share two essential features. One is an outer membrane, the **cell membrane** (known more precisely as the **plasma membrane**), that separates the cell from its external environment. The other is the genetic material—the hereditary information—that directs a cell's activities and enables it to reproduce, passing on the cell's characteristics to its offspring.

The organization of the genetic material is one of the characteristics that distinguish two fundamentally distinct kinds of cells, **prokaryotes** and **eukaryotes.** In prokaryotic cells, the genetic material is in the form of a large, circular molecule of DNA, with which a variety of proteins are loosely associated. This molecule is known as the **chromosome.** In eukaryotic cells, by contrast, the DNA is linear, forming a number of distinct chromosomes. Moreover, it is tightly bound to special proteins known as **histones,** which are an integral part of the chromosome structure.

Within the eukaryotic cell, the chromosomes are surrounded by a double membrane, the **nuclear envelope,** that separates them from the other cell contents in a distinct **nucleus** (hence the name, *eu,* meaning "true," and *karyon,* meaning "nucleus" or "kernel"). In prokaryotes ("before a nucleus"), the chromosome is not contained within a membrane-bound nucleus, although it is localized in a distinct region that is known as the **nucleoid.**

The remaining components of a cell (that is, everything within the plasma membrane except the nucleus or nucleoid and its contents) constitute the **cytoplasm.** The cytoplasm contains a large variety of molecules and molecular complexes. For example, both prokaryotes and eukaryotes contain complexes of protein and RNA, known as **ribosomes,** that play a crucial role in the assembly of protein molecules from their amino acid subunits. Eukaryotic cells also contain a variety of membrane-bound structures and compartments called **organelles.** These specialized structures carry out particular functions within the cell.

The plasma membrane of prokaryotes is surrounded by an outer **cell wall** that is manufactured by the cell itself. Some eukaryotic cells, including plant cells and fungi, have cell walls, although their composition and structure is different from that of prokaryotic cell walls. Other eukaryotic cells, including those of our own bodies and of other animals, do not have cell walls. Another feature distinguishing eukaryotes and prokaryotes is size: eukaryotic cells are usually larger than prokaryotic cells.

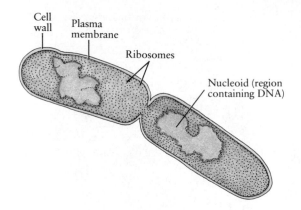

Cell wall
Plasma membrane
Ribosomes
Nucleoid (region containing DNA)

Modern prokaryotes include the bacteria (Figure 4–7) and the cyanobacteria (Figure 4–8), a group of photosynthetic prokaryotes that were formerly known as the blue-green algae.* According to the fossil record, the earliest living organisms were comparatively simple cells, resembling present-day prokaryotes. Prokaryotes were the only forms of life on this planet for almost 2 billion years until eukaryotes evolved.

Many biologists believe that the transition from the prokaryotic cell to the eukaryotic cell, a topic we shall explore in Chapter 24, was the most significant event in the history of life, second only in biological importance to the first appearance of living systems. Figure 4–9 gives an example of a modern single-celled photosynthetic eukaryote, the alga *Chlamydomonas*. It is a common inhabitant of freshwater ponds. These organisms are small and bright green (because of their chlorophyll), and they move very quickly with a characteristic darting motion. Being photosynthetic, they are usually found near the water's surface, where the light intensity is greatest.

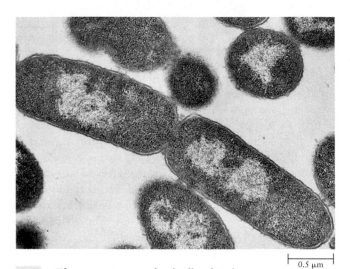

|—— 0.5 μm ——|

4–7 *Electron micrograph of cells of* Escherichia coli, *the heterotrophic prokaryote that is the most thoroughly studied of all living organisms. The genetic material (DNA) is in the lighter-appearing area in the center of each cell. This region, which is not enclosed by a membrane, is known as the nucleoid. The small, dense bodies in the cytoplasm are ribosomes. The two cells in the center have just finished dividing and have not yet separated completely.*

The Origins of Multicellularity

The first multicellular organisms, as far as can be told by the fossil record, made their appearance a mere 750 million years ago (Figure 4–10). The cells of modern multicellular organisms closely resemble those of single-celled eukaryotes. They are bound by a plasma membrane identical in appearance to the plasma membrane

---

* The Latin term *alga*, plural *algae*, means "seaweed." Algae is a general term applied to eukaryotic single-celled photosynthetic organisms and to many simple multicellular forms. Until recently, it was also applied to some photosynthetic prokaryotes.

4–8 *Electron micrograph of a single cell from a filament of cells of a photosynthetic prokaryote, the cyanobacterium* Anabaena azollae. *In addition to the genetic material, this cell contains a series of membranes in which chlorophyll and other photosynthetic pigments are embedded. Anabaena synthesizes its own energy-rich organic compounds in chemical reactions powered by the radiant energy of the sun.*

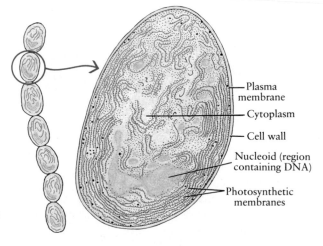

Plasma membrane
Cytoplasm
Cell wall
Nucleoid (region containing DNA)
Photosynthetic membranes

|—— 2 μm ——|

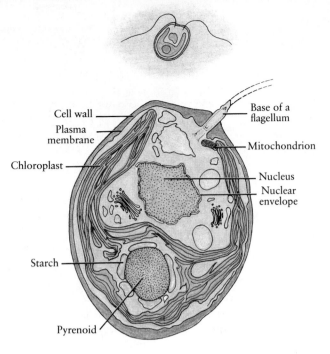

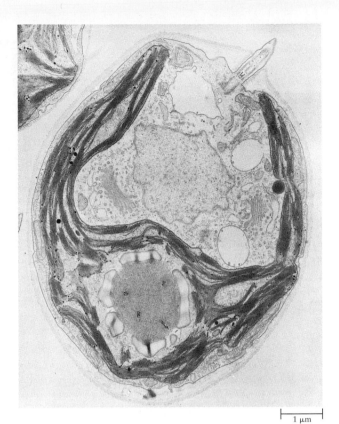

├─────┤ 1 μm

4–9 *Electron micrograph of* Chlamydomonas, *a photosynthetic eukaryotic cell, which contains a membrane-bound ("true") nucleus and numerous organelles. The most prominent organelle is the single, irregularly shaped chloroplast that fills most of the cell. It is surrounded by a double membrane and is the site of photosynthesis. Other membrane-bound organelles, the mitochondria, provide energy for cellular functions, including the flicking movements of the two flagella. These movements propel the cell through the water. The organism's food reserves are in the form of starch granules, stored around and within a structure known as the pyrenoid. The cytoplasm is enclosed by the plasma membrane, outside of which is a cell wall composed of polysaccharides.*

4–10 *The clockface of biological time, which shows when important events in the Earth's past would have occurred if the Earth's 4.6-billion-year history were condensed into one day. Life first appears relatively early, before 6:00 A.M. on a 24-hour time scale. The first multicellular organisms do not appear until the early evening of that 24-hour day, and* Homo, *the genus to which humans belong, is a late arrival—at about 30 seconds to midnight.*

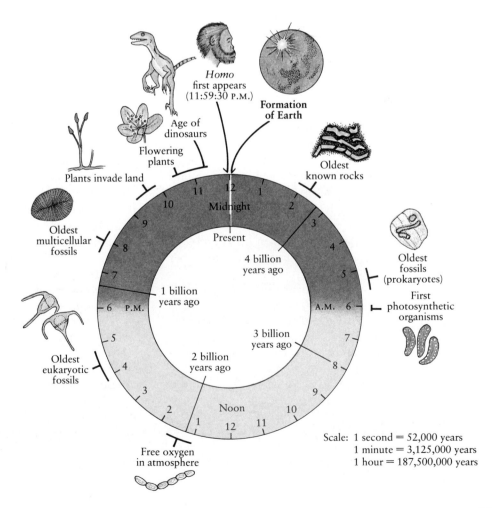

Scale: 1 second = 52,000 years
1 minute = 3,125,000 years
1 hour = 187,500,000 years

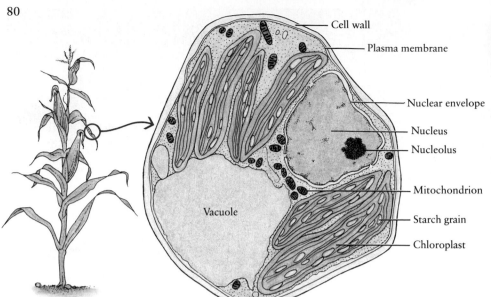

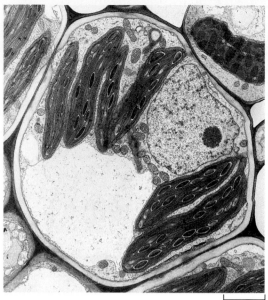

Cell wall

Plasma membrane

Nuclear envelope

Nucleus

Nucleolus

Mitochondrion

Vacuole

Starch grain

Chloroplast

0.5 µm

4–11 *Electron micrograph of cells from the leaf of a corn plant. The nucleus can be seen on the right side of the central cell. The granular material within the nucleus is chromatin. It contains DNA associated with his-* *tone proteins. The nucleolus is the region within the nucleus where the RNA components of ribosomes are synthesized. Note the many mitochondria and chloroplasts, all enclosed by membranes. The vacuole, a fluid-* *filled region enclosed by a membrane, and the cell wall are characteristic of plant cells but are generally not found in animal cells. As you can see, this cell closely resembles* Chlamydomonas, *shown in Figure 4–9.*

of a single-celled eukaryote. Their organelles are constructed according to the same design. The cells of multicellular organisms differ from single-celled eukaryotes in that each type of cell is specialized to carry out a relatively limited function in the life of the organism. However, each remains a remarkably self-sustaining unit.

Notice how similar a cell from the leaf of a corn plant (Figure 4–11) is to *Chlamydomonas*. This plant cell is also photosynthetic, supplying its own energy needs from sunlight. However, unlike the alga, it is part of a multicellular organism and depends on other cells for water, minerals, protection from drying out, and other necessities.

The human body, made up of trillions of individual cells, is composed of at least 200 different types of cells, each specialized for its particular function but all working as a cooperative whole. Figure 4–12 shows cells from the lining of an animal trachea (windpipe). They are part of an elaborate organ system involved in delivering oxygen to other cells in the body.

The Forms of Life

Although all organisms are made up of cells, there is incredible diversity in the living world. It is known that

we share this planet with at least 5 million different species of living organisms, and the actual number may be much higher. These different organisms exhibit great variety in the organization of their bodies, in their patterns of reproduction, growth, and development, and in their behavior.

Despite the seemingly overwhelming diversity of organisms, it is possible to group them in ways that reveal not only patterns of similarity and difference but also the evolutionary relationships among different groups. In Section 5, we shall consider these patterns and relationships in some detail. Before reaching that point, however, we shall encounter a marvelous variety of organisms. Thus you will find it helpful to be aware of the five major categories, or **kingdoms,** into which we group organisms in this text.

The first kingdom, **Monera,** includes all of the prokaryotes—the bacteria, the cyanobacteria, and their relatives. The second kingdom, **Protista,** includes the single-celled eukaryotes, as well as a number of simple multicellular eukaryotic organisms. The protists are an extremely varied collection of organisms, including both heterotrophs and autotrophs. Examples of protists are amoebas, paramecia, and the many forms of algae.

The three kingdoms of multicellular organisms—**Fungi, Plantae,** and **Animalia**—are thought to have

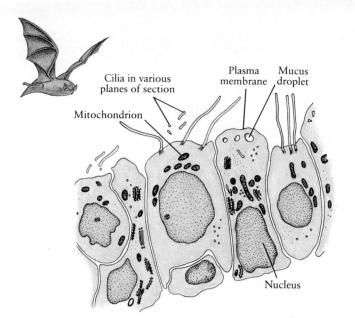

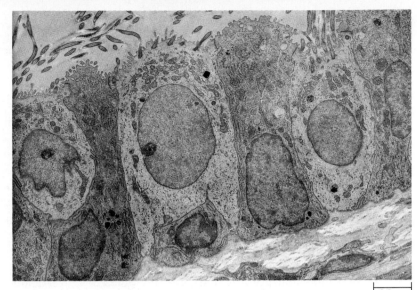

2.5 μm

4–12 *Cells from the surface of the trachea (windpipe) of a bat. The free surface of the larger cell is covered with cilia, which are essentially the same in structure as the flagella of Chlamydomonas. Next to the ciliated cells are cells that secrete mucus onto the cell surface. Mucus currents, swept by cilia, remove foreign particles from the trachea.*

evolved from different types of single-celled eukaryotes (Figure 4–13). The fungi include such organisms as molds, yeasts, and mushrooms. They are heterotrophs, absorbing nutrient molecules from the surrounding environment.

Plants are most concisely defined as multicellular photosynthetic autotrophs. They transform the energy of sunlight into the complex molecules of which their bodies are composed. These molecules, which include carbohydrates, proteins, and lipids, are the energy sources for animal life.

Animals are multicellular heterotrophs that must ingest other organisms—mostly plants or other animals—for their sustenance. From our anthropocentric point of view, animal usually means mammal, but actually most animals are invertebrates, animals without backbones. More than 1.5 million different kinds of animals have been recorded, of which 95 percent are invertebrates and more than a million of these are insects.

In the past century, our knowledge of the forms of life, past and present, of the processes occurring within their bodies, and of their interrelationships with one another has rapidly outstripped that gained in all previous centuries of human inquiry. Among the factors contributing to this explosion of knowledge have been significant improvements in microscopy.

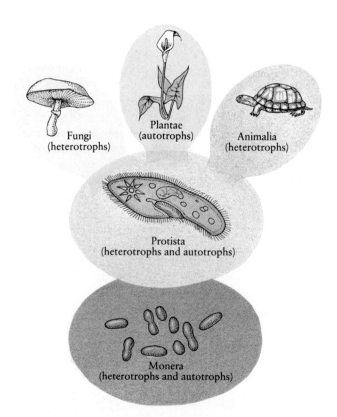

4–13 *The five kingdoms of living organisms. The organisms of kingdom Monera are all prokaryotes, whereas the organisms of the other four kingdoms are all eukaryotes. The first eukaryotic cells were derived from prokaryotes, and the three kingdoms of multicellular eukaryotes are thought to have evolved from different single-celled eukaryotes.*

81

Viewing the Cellular World

In the three centuries since Robert Hooke first observed the structure of cork through his simple microscope (Figure 4–6, page 77), our capacity to view the cell and its contents has increased dramatically. Unaided, the human eye is able to distinguish objects that are separated by a distance of at least 1/10 millimeter, or 100 micrometers (Table 4–1). This capacity is known as **resolving power.** It is expressed in terms of the minimum distance that must be between two objects for them to be perceived as separate objects. For example, if you look at two lines that are less than 100 micrometers apart, you will see a single, somewhat thickened line. Similarly, two dots less than 100 micrometers apart look like a single blurry dot. Conversely, if you look at two lines (or two dots) that are 120 micrometers apart, you can distinguish them from one another.

Most eukaryotic cells are between 10 and 30 micrometers in diameter—some 3 to 10 times below the resolving power of the human eye—and prokaryotic cells are smaller still. In order to distinguish individual cells, to say nothing of examining the structures of which they are composed, we must use instruments that provide greater resolution than does the naked eye. Most of our current knowledge of cell structure has been gained with the assistance of three different types of instruments: the **light microscope,** the **transmission electron microscope,** and the **scanning electron microscope** (Figure 4–14).

The best light microscopes have a resolving power of about 0.2 micrometer, or 200 nanometers, and so improve on the naked eye about 500 times. It is impossible to build a light microscope that will do better than this. The limiting factor is the wavelength of light; the shorter the wavelength, the greater the resolution. The shortest wavelength of visible light is about 0.4 micrometer, and this sets the limit of resolution with the light microscope.

Notice that resolving power and magnification are two different things. If you take a picture through the best light microscope of two lines that are less than 0.2 micrometer, or 200 nanometers, apart, you can enlarge that photograph indefinitely, but the two lines will continue to blur together. Similarly, by using more powerful microscope lenses, you can increase magnification, but this will not improve resolution; you will just see a larger blur.

With the transmission electron microscope, resolving power has been increased about 1,000 times over that provided by the light microscope. This is achieved by using "illumination" of a much shorter wavelength, consisting of electron beams instead of light rays. Areas in the specimen that permit more electrons to pass through—"electron-transparent" regions—show up bright, and areas that do not allow electrons to pass through—"electron-opaque" regions—are seen as dark. Transmission electron microscopy at present affords a resolving power of about 0.2 nanometer, roughly 500,000 times greater than that of the human eye. This is about twice the diameter of a hydrogen atom.

Although the resolving power of the scanning electron microscope is only about 10 nanometers, this instrument has become a valuable tool for biologists. In scanning electron microscopy, the electrons do not pass through the specimen but are instead deflected from its surface, providing an image of the surface contours. The electron beam is focused into a fine probe, which is rapidly passed back and forth over the specimen. Complete scanning from top to bottom usually takes a few seconds. Variations in the surface of the specimen affect the pattern in which the electrons are deflected from it. Holes and fissures appear dark, and knobs and ridges are light. The pattern produced by the electrons is amplified and transmitted to a television monitor, providing a visual image of the specimen. Scanning electron microscopy provides vivid three-dimensional representations of cells and cellular structures that more than compensate for its limited resolution.

Studies with these three types of microscopes, complemented by a variety of biochemical techniques, have revealed a wealth of knowledge about the structure of cells and the dynamic processes that characterize them. We shall begin to examine this knowledge in the next chapter.

Table 4–1   Measurements Used in Microscopy

1 centimeter (cm) = 1/100 meter = 0.4 inch*

1 millimeter (mm) = 1/1,000 meter = 1/10 cm

1 micrometer (μm)[†] = 1/1,000,000 meter = 1/10,000 cm

1 nanometer (nm) = 1/1,000,000,000 meter = 1/10,000,000 cm

or

1 meter = 10^2 cm = 10^3 mm = 10^6 μm = 10^9 nm

* A metric-to-English conversion table can be found inside the back cover.
† Micrometers were formerly known as microns (μ), and nanometers as millimicrons (mμ).

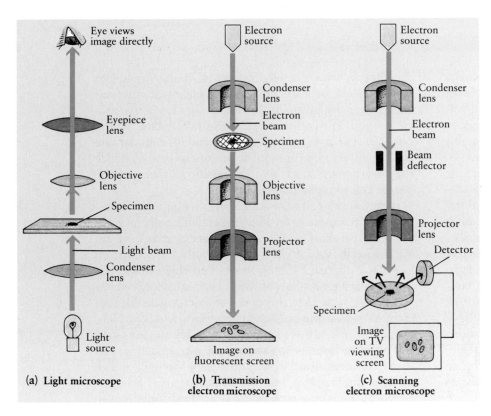

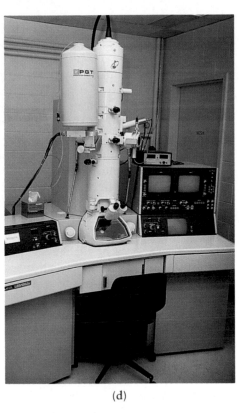

(d)

4–14   *A comparison of (a) the light microscope, (b) the transmission electron microscope, and (c) the scanning electron microscope. The focusing lenses in the light microscope are glass or quartz; those in the* *electron microscopes are magnetic coils. In both the light microscope and the transmission electron microscope, the illuminating beam passes through the specimen. In the scanning electron microscope, it is deflected* *from the surface of the specimen. (d) A modern electron microscope. Depending on how the controls are adjusted, it can function as either a transmission electron microscope or a scanning electron microscope.*

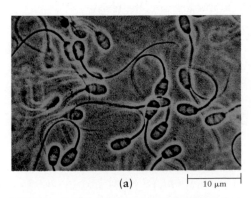

(a)   ⊢ 10 μm ⊣

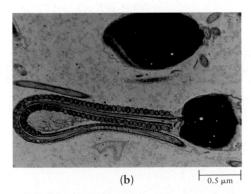

(b)   ⊢ 0.5 μm ⊣

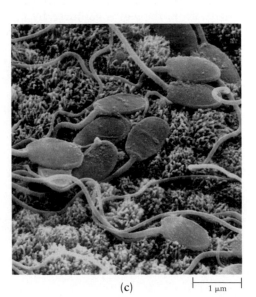

(c)   ⊢ 1 μm ⊣

4–15   *Rabbit sperm cells, as seen in (a) a light micrograph, (b) a transmission electron micrograph, and (c) a scanning electron micrograph. Note the dramatic increase in the resolution of structural detail in the electron micrographs.*

Electron microscopes produce images only in black and white. However, color is often added to electron micrographs to enhance their visual appeal, as in these micrographs, or to highlight specific structures.

Summary

The properties associated with living systems emerge at the cellular level of organization. Microfossils of bacteria-like cells have been discovered that are 3.5 billion years old. The complexity of these cells suggests that the first primitive cells arose early in the Earth's existence—sometime during the first billion years.

The primitive atmosphere held the raw materials of living matter—hydrogen, oxygen, carbon, and nitrogen—contained in molecules of water vapor and other simple compounds. The energy required to break apart the molecules and re-form them into more complex molecules was present in heat, lightning, radioactive elements, and high-energy radiation from the sun. Laboratory experiments have shown that under these conditions, the types of organic molecules characteristic of living systems can be formed. Other experiments have suggested the kinds of processes by which aggregations of organic molecules could have formed cell-like structures, separated from their environment by a membrane and capable of maintaining their structural and chemical integrity.

The earliest cells may have been heterotrophs (organisms that depend on outside sources for their energy-rich organic molecules) or autotrophs (organisms that can make their own organic molecules from inorganic substances). The first autotrophs may have been chemosynthetic (using the energy released by specific inorganic reactions to synthesize their own organic molecules) or photosynthetic (using the sun's energy to power their synthetic reactions). With the advent of photosynthesis, the flow of energy through the biosphere assumed its dominant modern form—radiant energy from the sun is captured by photosynthetic autotrophs and channeled through them to heterotrophic organisms.

One of the fundamental principles of biology is the cell theory, which states that (1) all living organisms consist of one or more cells; (2) the chemical reactions of a living organism, including energy-releasing processes and biosynthetic reactions, take place within cells; (3) cells arise from other cells; and (4) cells contain the hereditary information of the organisms of which they are a part, and this information is passed from parent cell to daughter cell.

There are two fundamentally distinct types of cells—prokaryotes and eukaryotes. Prokaryotic cells lack the membrane-bound nuclei and organelles found in eukaryotic cells. Prokaryotes were the only form of life on Earth for almost 2 billion years, and then, about 1.5 billion years ago, eukaryotic cells evolved. Multicellular organisms, which are composed of eukaryotic cells specialized to perform particular functions, evolved comparatively recently—only about 750 million years ago.

Five kingdoms of organisms are recognized in this text. The organisms of the kingdom Monera, which includes the bacteria and the cyanobacteria, are prokaryotes. The organisms of the other four kingdoms—Protista, Fungi, Plantae, and Animalia—are eukaryotes. The protists include single-celled organisms, both heterotrophic and autotrophic, as well as a number of simple multicellular forms. Fungi are multicellular heterotrophs that obtain organic molecules by absorption. Plants are multicellular photosynthetic autotrophs, and animals are multicellular heterotrophs that ingest other organisms to obtain needed organic molecules.

Because of the small size of cells and the limited resolving power of the human eye, microscopes are required to visualize cells and subcellular structures. The three principal types are the light microscope, the transmission electron microscope, and the scanning electron microscope.

Questions

1. Distinguish among the following: heterotroph/autotroph; chemosynthetic autotroph/photosynthetic autotroph; prokaryote/eukaryote; light microscope/transmission electron microscope/scanning electron microscope.

2. Why would energy sources have been necessary for the synthesis of simple organic molecules on the primitive Earth?

3. Although there is some uncertainty as to the exact mixture of gases that constituted the early atmosphere, there is general agreement that free oxygen was not present. What properties of oxygen would have made chemical evolution unlikely in an atmosphere containing O_2?

4. A key event in the origin of life was the formation of a membrane that separated the contents of primitive cells from their surroundings. Why was this so critical?

5. Some scientists think that other planets in our galaxy may well contain some form of life. If you were seeking such a planet, what characteristics would you look for?

Suggestions for Further Reading

Cloud, Preston: *Oasis in Space: Earth History from the Beginning*, W. W. Norton & Company, New York, 1989.*

> An accurate and up-to-date history of Earth's 4.6 billion years, written by a master storyteller. A major theme is the ongoing interplay between physical, chemical, and biological processes. Particular emphasis is placed on the early periods of Earth's history, during which life originated and many of the great diversifications of organisms occurred.

Groves, David I., John S. R. Dunlop, and Roger Buick: "An Early Habitat of Life," *Scientific American*, October 1981, pages 64–73.

Horgan, John: "In the Beginning . . .," *Scientific American*, February 1991, pages 116–125.

Kasting, James F., Owen B. Toon, and James B. Pollack: "How Climate Evolved on the Terrestrial Planets," *Scientific American*, February 1988, pages 90–97.

Oparin, A. I.: *The Origin of Life*, Dover Publications, Inc., New York, 1953.*

> In 1938, Oparin, a Russian biochemist, was the first to argue that life arose spontaneously in the oceans of the primitive Earth. Although his concepts have since been somewhat modified in detail, they form the basis for the present scientific theories on the origin of living things.

Scientific American: *Molecules to Living Cells*, W. H. Freeman and Company, New York, 1980.*

> A collection of outstanding articles from Scientific American. The first two chapters cover the origin of life and the evolution of the earliest cells. Highly recommended.

Taylor, D. Lansing, Michel Nederlof, Frederick Lanni, and Alan S. Waggoner: "The New Vision of Light Microscopy," *American Scientist*, vol. 80, pages 322–335, 1992.

Vidal, Gonzalo: "The Oldest Eukaryotic Cells," *Scientific American*, February 1984, pages 48–57.

York, Derek: "The Earliest History of the Earth," *Scientific American*, January 1993, pages 90–96.

* Available in paperback.

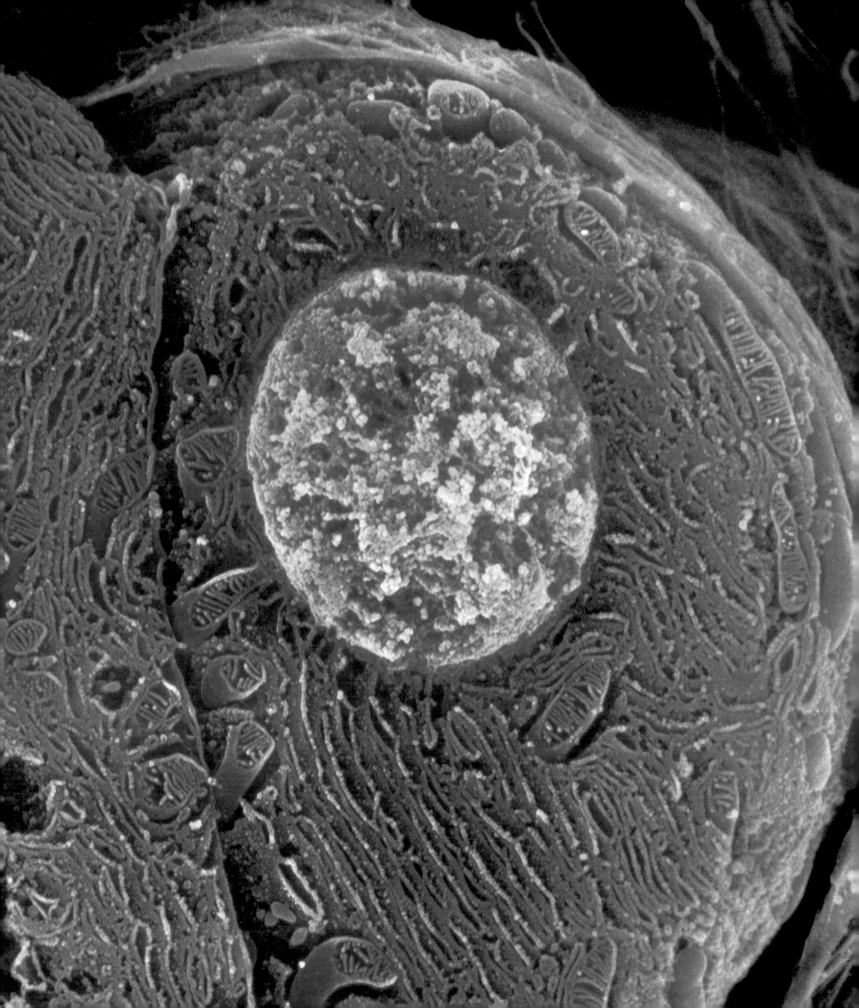

In multicellular organisms, cells are specialized to carry out the different functions necessary for the survival of the organism. This cell is from the human liver, a three-pound chemical "factory" that synthesizes, processes, and degrades many different types of molecules.

Color has been added to this scanning electron micrograph to highlight the various structures of the cell. The plasma membrane—that is, the outer membrane of the cell—has been colored green. It has been partially peeled away, revealing the contents of the cell.

The large body near the center of the cell, tinted yellow, is the nucleus. It contains the cell's chromosomes and is the control center for the cell's many activities.

Like other cells that produce proteins for export, this cell has a complex system of internal membranes known as the endoplasmic reticulum. These membranes, to which a greenish-tan color has been added, are involved in the synthesis of lipids and proteins.

Embedded in the endoplasmic reticulum and supplying the ATP molecules that power its activities are organelles known as mitochondria, highlighted in a reddish-brown. Mitochondria are also divided into specialized compartments by a complex membrane system.

The history of the evolution of cells can be traced in the increasing role of membranes in the life of the cell. This evolution began with a single, crucial outer membrane separating the earliest prokaryote from the warm seas in which it originated. From that simple beginning have come the complex membrane systems of modern eukaryotic cells.

5

How Cells Are Organized

There are many, many different kinds of cells. In a drop of pond water, you are likely to find a variety of protists, and in even a small pond, there are probably several hundred different kinds of these one-celled eukaryotes, plus a variety of prokaryotes. Plants are composed of cells that appear quite different from those of our bodies, and insects have many kinds of cells not found in either plants or vertebrates. Thus, the first remarkable fact about cells is their diversity.

The second, even more remarkable fact is their similarity. All cells have DNA as the genetic material, they perform the same types of chemical reactions, and they are all surrounded by an external membrane that conforms to the same general design in both prokaryotic and eukaryotic cells. Every cell must carry out essentially the same processes—acquire and assimilate nutrients, eliminate wastes, synthesize new cellular materials, and in many cases, be able to move and to reproduce. Just as the various organs of your body have a structure that enables them to carry out their functions, so your cells have a complex internal architecture that enables them to carry out their functions.

Although we can look at only one structure or process at a time, a cell is not a random assortment of parts. It is a dynamic, integrated entity. Most activities of a cell go on simultaneously and influence one another. For example, *Chlamydomonas* (Figure 4–9, page 79) is swimming, photosynthesizing, absorbing nutrients from the water, building its cell wall, making proteins, converting starch to sugar (or vice versa), and oxidizing food molecules for energy, all at the same time. It is also likely to be orienting itself in the sunlight, it is probably preparing to divide, it is possibly "looking" for a mate, and it is undoubtedly carrying out at least a dozen or more other important activities, many of which may still be unknown.

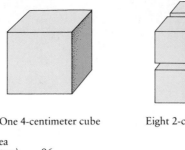 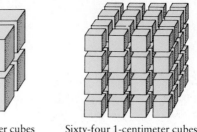

	One 4-centimeter cube	Eight 2-centimeter cubes	Sixty-four 1-centimeter cubes
Total surface area (square centimeters)	96	192	384
Total volume (cubic centimeters)	64	64	64
Surface-to-volume ratio	1.5 to 1	3 to 1	6 to 1

5–1 *The single 4-centimeter cube, the eight 2-centimeter cubes, and the sixty-four 1-centimeter cubes all have the same total volume. As the volume is divided up into smaller units, however, the total amount of surface area increases, as does the ratio of surface area to volume. For example, the sixty-four 1-centimeter cubes shown here have four times the total surface area of the single 4-centimeter cube. Moreover—and this is the important point—the ratio of surface area to volume in each 1-centimeter cube is four times that in the 4-centimeter cube. Similarly, smaller cells have a higher ratio of surface area to volume than larger cells. This means not only more membrane surface through which materials can move into or out of the cell but also less living matter to be serviced and shorter distances through which materials must move within the cell.*

Cell Size and Shape

Prokaryotic cells are, on average, about 2 micrometers long. By contrast, eukaryotic cells, such as the cells that make up a plant or animal body, are typically between 10 and 30 micrometers in diameter. A principal restriction on cell size is imposed by the relationship between volume and surface area. Figure 5–1 shows that as volume decreases, the ratio of surface area to volume increases rapidly.

It is through the membrane-bound surface of a cell that materials—such as oxygen, carbon dioxide, ions, food molecules, and waste products—enter and leave the cell. These substances are the raw materials and products of the cell's **metabolism,** which is the total of all of the chemical activities in which it is engaged. The more active the cell's metabolism, the more rapidly materials must be exchanged with the environment if the cell is to continue to function. In smaller cells, the ratio of surface area to volume is higher than in larger cells. Thus proportionately greater quantities of materials can move into, out of, and through smaller cells in a given period of time. A larger cell, by contrast, requires the exchange of greater quantities of materials in order to meet the needs of the larger volume of living matter—and yet, the larger the cell, the smaller the ratio of surface area to volume. It is not surprising, therefore, that the most metabolically active cells are usually small.

Another limitation on cell size appears to involve the capacity of the nucleus, the cell's control center, to provide enough copies of the information needed to regulate the processes occurring in a large, metabolically active cell. The exceptions seem to "prove" the rule. In certain large, complex one-celled protists, each cell has two or more nuclei, the additional ones apparently copies of the original.

Like drops of water and soap bubbles, cells have a tendency to be spherical. Often, however, cells have other shapes. For example, cell walls, found in plants, fungi, and many one-celled organisms, are generally in-

flexible. The shape of the enclosed cell is determined by the shape of the cell wall, which, depending on the organism and the cell type, may take any number of different forms. The shape of some cells is determined by attachments to and pressure from neighboring cells. This is the case, for example, with the columnar cells that line the surface of the trachea (Figure 4–12, page 81). And, as we shall see later in this chapter, the shape of still other cells is determined by arrays of structural filaments within the cell.

Cell Boundaries

The Plasma Membrane

When you look at a solitary cell—an amoeba, for instance—through a microscope, you can barely distinguish it from its watery surroundings. Although almost invisible and seemingly fragile, it is very different from the vast sea around it: it is alive. One of the crucial factors making possible this extraordinary difference is an external cell membrane—the **plasma membrane**—that separates the living material of the cell from the surrounding environment.

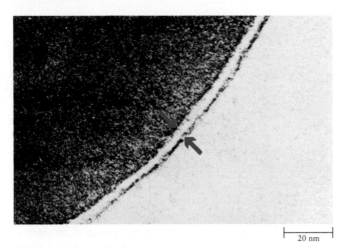

|— 20 nm —|

5–2 *Electron micrograph of a portion of a human red blood cell, showing a cross section of the plasma membrane (indicated by the arrows). In essence, the membrane is a "molecular sandwich." It consists of two layers of phospholipid molecules arranged with their electron-transparent (light) hydrophobic tails pointing inward, forming an inner "filling." The dark line on either side of the "filling" is formed by the electron-opaque hydrophilic heads of the phospholipid molecules. The darker material at the left of the micrograph is hemoglobin, which fills the red blood cell.*

The plasma membrane is only about 7 to 9 nanometers thick and therefore cannot be resolved by the light microscope. With the electron microscope, however, it can be visualized as a continuous, thin double line (Figure 5–2).

The fundamental structure of the plasma membrane is formed by a **phospholipid bilayer,** that is, a double layer of phospholipid molecules arranged with their hydrophobic fatty acid tails pointing inward (Figure 5–3). In animal cells, large numbers of cholesterol molecules are embedded in the hydrophobic interior of the bilayer.

5–3 *Model of the plasma membrane of an animal cell. The fundamental structure is formed from a network of phospholipid molecules, in which cholesterol molecules and large protein molecules are embedded. The phospholipid molecules are arranged in a bilayer with their hydrophobic tails pointing inward and their hydrophilic phosphate heads pointing outward. Cholesterol molecules nestle among the hydrophobic tails.*

The proteins embedded in the bilayer are known as integral membrane proteins. On the cytoplasmic face of the membrane, peripheral membrane proteins are bound to some of the integral proteins. The portion of a protein molecule's surface that is within the lipid bilayer is hydrophobic. The portion of the surface that is exposed outside the bilayer is hydrophilic. It is believed that pores with hydrophilic surfaces pass through some of the protein molecules.

Interspersed among the phospholipid molecules of the outer layer of the bilayer are glycolipid molecules. Their carbohydrate chains and the carbohydrate chains attached to the proteins protruding on the outside of the membrane are thought to be involved in the adhesion of cells to each other and with the "recognition" of molecules at the membrane surface.

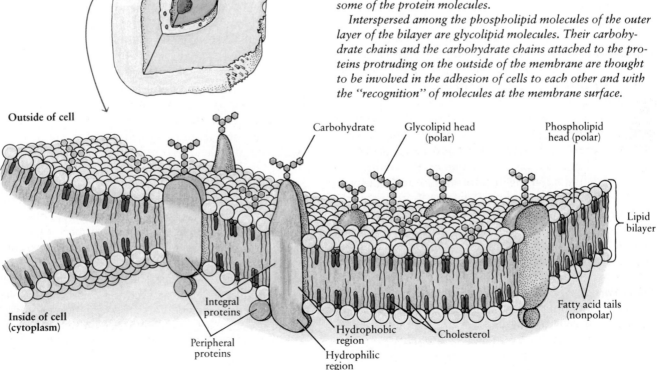

Outside of cell

Carbohydrate   Glycolipid head (polar)   Phospholipid head (polar)

Lipid bilayer

Inside of cell (cytoplasm)

Integral proteins   Hydrophobic region   Fatty acid tails (nonpolar)

Peripheral proteins   Cholesterol

Hydrophilic region

In prokaryotes and some protists and in the cells of plants and fungi, cholesterol is generally absent.

The plasma membrane also contains numerous protein molecules, suspended in the lipid bilayer. These proteins, known as **integral membrane proteins,** generally span the bilayer and protrude on either side. The portions embedded in the bilayer have hydrophobic surfaces, whereas the surfaces of those portions that extend beyond the bilayer are hydrophilic. On the cytoplasmic side of the membrane, additional protein molecules, known as **peripheral membrane proteins,** are bound to some of the integral proteins protruding from the bilayer.

Although many of the integral proteins appear to be anchored in place, either by peripheral proteins or by cytoplasmic protein filaments that are concentrated near the plasma membrane, the structure of the bilayer is generally quite fluid. The lipid molecules and at least some of the protein molecules can move laterally within the bilayer, forming different patterns (mosaics) that vary from time to time and place to place. Consequently, this widely accepted model of membrane structure is known as the **fluid-mosaic model.**

Membrane Specificity

The two surfaces of the plasma membrane—one facing the interior of the cell and the other facing the exterior of the cell—differ considerably in chemical composition. The two layers of the bilayer generally have different concentrations of specific types of lipid molecules. In many types of cells, the outer layer is particularly rich in glycolipid molecules. The carbohydrate chains of these glycolipid molecules are, like the phosphate heads of the phospholipid molecules, exposed on the surface of the membrane. The hydrophobic fatty acid tails of both types of molecules are within the membrane.

The protein composition of the outside and inside layers of the membrane also differ. The integral membrane proteins have a definite orientation within the bilayer, and the portions extending on either side are completely different in both amino acid composition and tertiary structure. On the outside of the membrane, short carbohydrate chains are attached to the protruding integral proteins, forming molecules known as **glycoproteins.** The carbohydrate chains of the glycoproteins, along with the carbohydrate chains of the glycolipids, form a carbohydrate coat on the outer surface of the membranes of many types of cells. The carbohydrates are thought to play a role in the adhesion of cells to one another and in the "recognition" of molecules (such as hormones, antibodies, and viruses) that interact with the cell.

In eukaryotes, all the membranes of a cell, including those surrounding the various organelles, have the same general structure as the plasma membrane. There are, however, differences in the types of lipids and, particularly, in the number and types of proteins and carbohydrates, which vary from membrane to membrane and also from place to place on the same membrane. These differences give the membranes of various cells and organelles unique properties that can be correlated with differences in function.

Membrane proteins, which are extremely diverse structurally, perform a variety of essential functions. Some are enzymes, regulating particular chemical reactions. Others are receptors, influencing the cell's uptake of hormones and other regulatory molecules. Still others are transport proteins, playing critical roles in the movement of ions and molecules across the membrane. As we shall see repeatedly in the course of this text, discoveries concerning the structure and function of specific membrane proteins are shedding new light on a wide variety of processes, ranging from photosynthesis to the response of the human body to infection to the transmission of the nerve impulse.

The Cell Wall

A principal distinction between plant and animal cells is that plant cells are surrounded by a cell wall. The wall is outside the plasma membrane and is constructed by the cell.

As a plant cell divides, a thin layer of gluey material forms between the plasma membranes of the two new cells. This layer becomes the **middle lamella.** Composed of pectins (the compounds that make jellies gel) and other polysaccharides, it holds adjacent cells together. Next, on either side of the middle lamella, each plant cell constructs its **primary cell wall** (Figure 5–4). This wall contains cellulose molecules bundled together in microfibrils and laid down in a matrix of gluey polymers. As you can see in Figure 3–6b on page 55, successive layers of microfibrils are oriented at right angles to one another in the completed cell wall. (Those of you familiar with building materials will note that the cell wall thus combines the structural features of both fiberglass and plywood.)

In plants, growth takes place largely by cell elongation. Studies have shown that the cell adds new materials to its walls throughout this elongation process. The cell, however, does not simply expand in all directions.

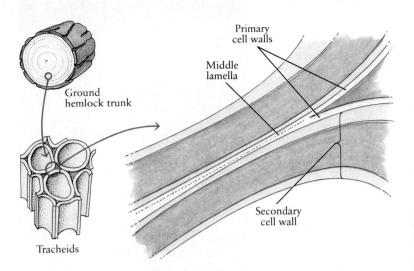

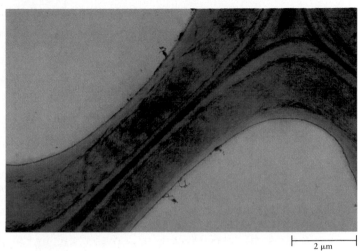

2 μm

5–4   *Two adjacent cell walls of tracheids, the cells through which water is conducted in plants. In the electron micrograph, you can see the middle lamella, the primary cell walls, and the layered secondary cell walls, deposited inside the primary walls. The cells, which are from the wood of a ground hemlock, have died. The transparent areas in the top left and lower right portions of the micrograph represent empty space, once filled by the living matter of the cells.*

Its final shape is determined by the structure of its cell wall (Figure 5–5).

As the cell matures, a **secondary wall** may be constructed. This wall is not capable of expansion, as is the primary wall. It often contains molecules that have stiffening properties. In such cells, the living material often dies, leaving only the outer wall, a monument to the cell's architectural abilities.

Cellulose-containing cell walls are also found in many algae. Fungi and prokaryotes also have cell walls, but they usually do not contain cellulose.

The Nucleus

In eukaryotic cells, the **nucleus** is a large, often spherical body, usually the most prominent structure within the cell. It is surrounded by the **nuclear envelope,** which is made up of two concentric membranes, each of which is a phospholipid bilayer. These two membranes are fused together at frequent intervals, creating small **nuclear pores** through which materials pass between the nucleus and the cytoplasm (Figure 5–6).

The chromosomes are found within the nucleus. When the cell is not dividing, the chromosomes are visible only as a tangle of fine threads, called **chromatin.** The most conspicuous body within the nucleus is the **nucleolus.** As we shall see in Chapter 16, the nucleolus is the site at which ribosomal subunits are constructed. Viewed with the electron microscope, the nucleolus appears to be a collection of fine granules and tiny fibers (Figure 5–7). These are thought to be parts of ribosomal subunits and threads of chromatin.

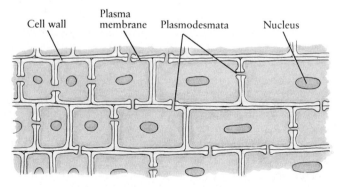

5–5   *Growth of plant cells is limited by the rate at which the cell walls expand. The walls control both the rate of growth and its direction. They do not expand in all directions but elongate in a single dimension. Cells at the left are newly formed. Cells farther to the right are older and have started to elongate. Plasmodesmata are channels connecting the cytoplasm of adjacent cells.*

5–6   *The nuclear envelope is formed from two concentric lipid bilayers. A surface view of the nuclear envelope of a guinea pig sperm cell is shown in (**a**). Nuclear pores, which are very dense in these cells, are clearly visible on its surface. Biochemical studies and electron micrographs of sections through the plane of the envelope (**b**) have revealed that the structure of each pore consists of eight protein-containing granules. The opening is a narrow channel in the center of the octagonal array. Passage of molecules into and out of the nucleus is thought to be regulated by the nuclear pores.*

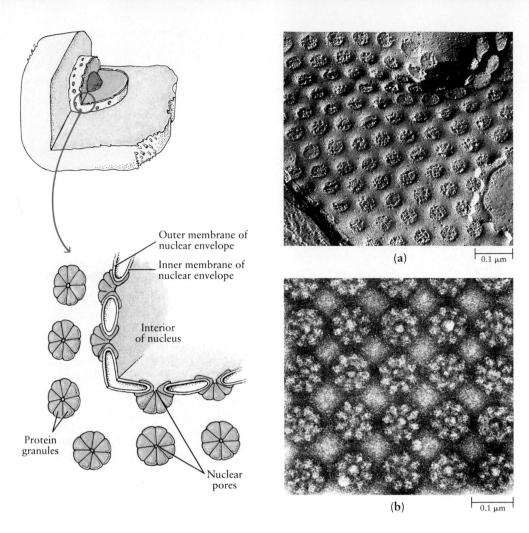

Outer membrane of nuclear envelope

Inner membrane of nuclear envelope

Interior of nucleus

Protein granules

Nuclear pores

(a)   0.1 μm

(b)   0.1 μm

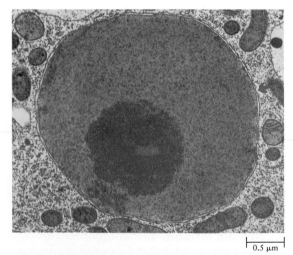

0.5 μm

5–7   *The nucleus of a root tip cell from* Arabidopsis thaliana, *as seen in an electron micrograph to which color has been added. The dark area in the lower left portion of the nucleus is the nucleolus. The major components of ribosomes are produced in the nucleolus. The dark granules around its periphery are partially formed ribosomes. Notice also the nuclear envelope with its nuclear pores, two of which are indicated by arrows. The structures highlighted in green are mitochondria.* Arabidopsis *is a small weed that has become an important experimental organism in the study of plant molecular genetics.*

The Functions of the Nucleus

Our current understanding of the role of the nucleus in the life of the cell began with some early microscopic observations. One of the most important of these observations was made more than a hundred years ago. A German embryologist, Oscar Hertwig, was observing the eggs and sperm of sea urchins (Figure 5–8). Sea urchins produce eggs and sperm in great numbers. The eggs are relatively large and so are easy to observe. They are fertilized in the open water, rather than internally, as is the case with land-dwelling vertebrates such as ourselves.

Watching the eggs being fertilized under his microscope, Hertwig observed that only a single sperm cell was required. Further, when the sperm cell penetrated the egg, its nucleus was released and fused with the nucleus of the egg. This observation, confirmed by other scientists and in other kinds of organisms, was important in establishing the fact that the nucleus is the carrier of the hereditary information: the only link between father and offspring is the nucleus of the sperm.

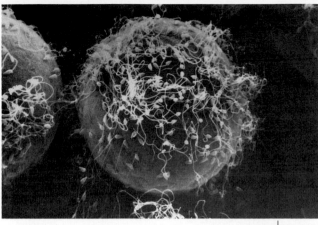

10 μm

5–8   *An egg from a sea urchin, surrounded by sperm cells. Despite the great differences in size of egg and sperm, both contribute equally to the hereditary characteristics of the individual. Since the nucleus is approximately the same size in both cells, the early microscopists postulated that this part of the cell must be the carrier of the hereditary information.*

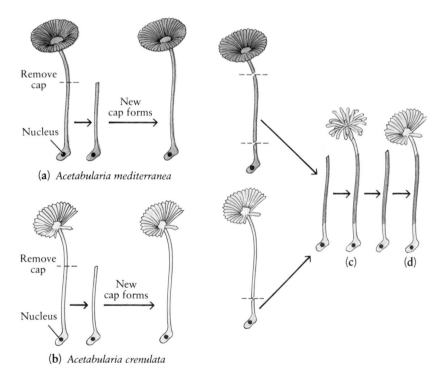

Remove cap

New cap forms

Nucleus

(a) *Acetabularia mediterranea*

Remove cap

New cap forms

Nucleus

(b) *Acetabularia crenulata*

(c)   (d)

5–9   (a) *One species of* Acetabularia *has an umbrella-shaped cap, and* (b) *another has a ragged, petal-like cap. If the cap is removed, a new cap forms, similar in appearance to the amputated one. However, if the "foot" (containing the nucleus) is removed at the same time as the cap and a new nucleus from the other species is transplanted, the cap* (c) *that forms will have a structure with characteristics of both species. If this cap is removed, the next cap* (d) *that grows will be characteristic of the cell that donated the nucleus, not of the cell that donated the cytoplasm.*

Since Hertwig's time, a number of experiments have explored the role of the nucleus in the cell. In one simple experiment, the nucleus was removed from an amoeba by microsurgery. The amoeba stopped dividing and, in a few days, it died. If, however, a nucleus from another amoeba was implanted within 24 hours after the original one was removed, the cell survived and divided normally.

In the early 1930s, Joachim Hämmerling studied the comparative roles of the nucleus and the cytoplasm by taking advantage of some unusual properties of the marine alga *Acetabularia*. The body of *Acetabularia* consists of a single huge cell 2 to 5 centimeters in height. Individuals have a cap, a stalk, and a "foot," all of which are differentiated portions of the single cell. If the cap is removed, the cell will rapidly regenerate a new one. Different species of *Acetabularia* have different kinds of caps. *Acetabularia mediterranea*, for example, has a compact umbrella-shaped cap, and *Acetabularia crenulata* has a cap of petal-like structures.

Hämmerling took the "foot," which contains the nucleus, from a cell of *A. crenulata* and grafted it onto a cell of *A. mediterranea*, from which he had first removed the "foot" and the cap. The cap that then formed had a shape intermediate between those of the two species. When this cap was removed, the next cap that formed was completely characteristic of *A. crenulata* (Figure 5–9).

Hämmerling interpreted these results as meaning that certain cap-determining substances are produced under the direction of the nucleus. These substances accumulate in the cytoplasm, which is why the first cap that formed after nuclear transplantation was of an intermediate type. By the time the second cap formed, however, the cap-determining substances present in the cytoplasm before the transplant had been exhausted, and the form of the cap was completely under the control of the new nucleus.

We can see from these experiments that the nucleus performs two crucial functions for the cell. First, it carries the hereditary information that determines whether a particular cell will develop into (or be a part of) a sea urchin, an oak, or a human—and not just any sea urchin, oak, or human, but one that resembles the parent or parents of that particular unique organism. Each time a cell divides, this information is passed on to the two new cells. Second, as Hämmerling's work with *Acetabularia* indicated, the nucleus exerts a continuing influence over the ongoing activities of the cell, ensuring that the complex molecules that the cell requires are synthesized in the number and of the kind needed. The way in which the nucleus performs these functions will be described in Section 3.

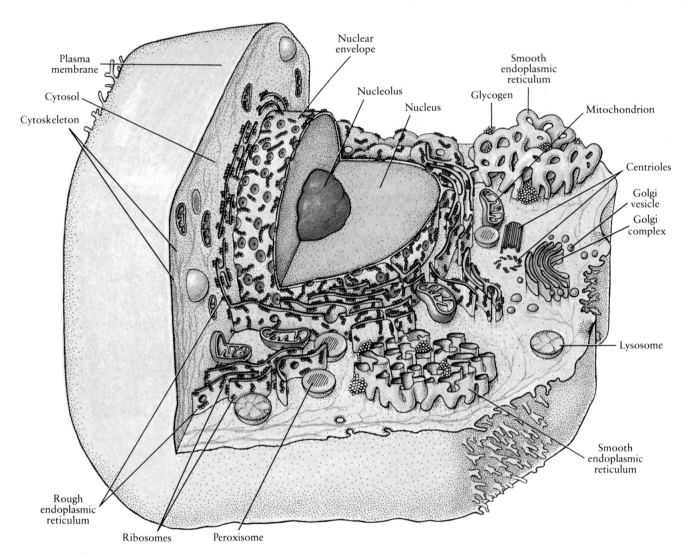

Plasma
membrane

Cytosol

Cytoskeleton

Nuclear
envelope

Nucleolus

Nucleus

Smooth
endoplasmic
reticulum

Glycogen

Mitochondrion

Centrioles

Golgi
vesicle

Golgi
complex

Lysosome

Smooth
endoplasmic
reticulum

Rough
endoplasmic
reticulum

Ribosomes

Peroxisome

5–10   *A representative animal cell, as interpreted from electron micrographs. Like all cells, this one is bounded by a* **plasma membrane.** *All materials that enter or leave the cell, including food, wastes, and chemical messages, must pass through this membrane.*

Surrounded by the plasma membrane is the cytoplasm, which consists of a dense solution, the **cytosol,** *and numerous membrane-bound organelles. The cytosol is traversed and subdivided by the membranes of an elaborate organelle system, the* **endoplasmic reticulum,** *a portion of which is shown here. In some areas, the endoplasmic reticulum is covered with* **ribosomes,** *the structures on which amino acids are assembled into proteins. Ribosomes are also found elsewhere in the cytosol.*

Golgi complexes are packaging centers for molecules synthesized within the cell. **Lysosomes** *and* **peroxisomes** *are vesicles in which a number of different types of molecules are broken down to simpler constituents that can either be used by the cell*

or, in the case of waste products, be safely removed from it. The **mitochondria** *are the sites of the chemical reactions that produce large quantities of ATP to power cellular activities.*

The largest body in the cell is the **nucleus.** *It is surrounded by a double membrane, the* **nuclear envelope,** *the outer membrane of which is continuous with the endoplasmic reticulum. The nuclear envelope is studded with nuclear pores, which regulate the movement of substances into and out of the nucleus. Contained within the nucleus is the* **nucleolus,** *the site where the ribosomal subunits are formed, and the chromatin, which consists of long strands of DNA and protein.*

An elaborate, highly structured network of protein filaments, the **cytoskeleton,** *pervades the cytosol. The cytoskeleton, of which only a hint is shown here, maintains the cell's shape, anchors its organelles, and directs the intracellular molecular traffic.*

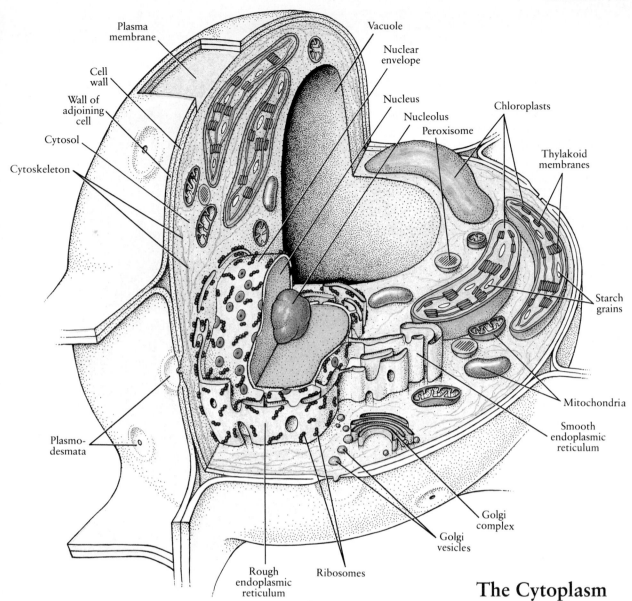

Plasma
membrane

Cell
wall

Wall of
adjoining
cell

Cytosol

Cytoskeleton

Plasmo-
desmata

Rough
endoplasmic
reticulum

Ribosomes

Vacuole

Nuclear
envelope

Nucleus

Nucleolus

Peroxisome

Chloroplasts

Thylakoid
membranes

Starch
grains

Mitochondria

Smooth
endoplasmic
reticulum

Golgi
complex

Golgi
vesicles

5–11   *A relatively young plant cell, as interpreted from electron micrographs. Like the animal cell, it is bounded by a*
plasma membrane. *Surrounding this membrane is a cellulose-
containing* **cell wall. Plasmodesmata,** *which are membrane-
lined channels through the cell walls, provide a cytoplasmic
connection between adjacent cells.*

The most prominent structure in many plant cells is a large
vacuole, *filled with a solution of salts and other substances. As
we shall see in the next chapter, the vacuole plays a key role in
keeping the cell wall stiff and the plant body crisp.*

Chloroplasts, *the large organelles in which photosynthesis
takes place, are generally concentrated near the surface of the
cell. Molecules of chlorophyll and the other substances involved in the capture of light energy from the sun are located
in the* **thylakoid membranes** *within the chloroplasts.*

*Like the animal cell, the living plant cell contains a promi-
nent* **nucleus,** *extensive* **endoplasmic reticulum,** *and many ribosomes* *and* **mitochondria.** *Especially numerous in the growing plant cell are* **Golgi complexes,** *which play an important
role in the assembly of materials for the expanding cell wall.*

The Cytoplasm

Not long ago, the cell was visualized as a bag of fluid
containing enzymes and other dissolved molecules
along with the nucleus and a few other structures. With
the development of electron microscopy, an increasing
number of complex structures have been identified
within the cytoplasm, which is now known to be highly
organized. The interior of a typical animal cell is shown
in Figure 5–10. Figure 5–11 shows a corresponding
view of a typical plant cell.

The fluid portion of the cytoplasm, known as the
cytosol, is a concentrated solution of ions, small molecules (such as amino acids, sugars, and ATP), and proteins. Most of these solutes are found only in low concentrations—or not at all—outside the cell. The cytosol
of both prokaryotic and eukaryotic cells also contains
numerous **ribosomes,** the RNA-protein complexes that
play a major role in the synthesis of protein molecules
from their amino acid subunits.

In most eukaryotic cells, about half of the cytoplasm consists of the membrane-bound compartments known as **organelles.** This name was coined by the early microscopists who viewed these structures, which they could barely distinguish, as comparable to the organs of multicellular animals. On closer inspection, the organelles turned out not to be miniature hearts or kidneys. However, like those organs, each type of organelle carries out its own specialized functions. The amount of membrane surface created within the cell by the organelles is enormous—vastly more than the total surface area of the plasma membrane.

The organelles are anchored within the cytosol by a network of protein filaments that form an internal **cytoskeleton.** The fibers of the cytoskeleton maintain the shape of the cell, enable it to move, and direct its molecular traffic.

In the pages that follow, we shall look at the individual structures of animal and plant cells and discuss their roles in the life of the cell.

Vacuoles and Vesicles: Support and Transport

Many cells, especially plant cells, contain structures known as **vacuoles.** A vacuole is a space in the cytoplasm surrounded by a single membrane and filled with water and solutes. Immature plant cells typically have many vacuoles, but as a plant cell matures, the numerous smaller vacuoles merge, forming one large, central, fluid-filled vacuole. When fully formed, this vacuole is a major supporting element of the cell (Figure 5–12).

Vesicles, which are found in all metabolically active eukaryotic cells, have the same general structure as vacuoles. They are, however, much smaller. One of the principal functions of vesicles is transport. As we shall see, they participate in the transport of materials both within the cell and into and out of the cell.

Ribosomes and Endoplasmic Reticulum: Protein and Lipid Biosynthesis

As we have noted previously, ribosomes are the sites at which amino acids are assembled into proteins. The more protein a cell is making, the more ribosomes it has. In eukaryotic cells, ribosomes are found within the cytosol and also attached to a complex system of internal membranes, the **endoplasmic reticulum.**

The endoplasmic reticulum forms a network of interconnecting flattened sacs, tubes, and channels. It occurs in two general forms, **rough** (with ribosomes attached) and **smooth** (without ribosomes). Rough endoplasmic reticulum (Figure 5–13) is continuous with the outer membrane of the nuclear envelope, which also has ribosomes attached. Similarly, smooth endoplasmic reticulum is continuous with the rough endoplasmic reticulum.

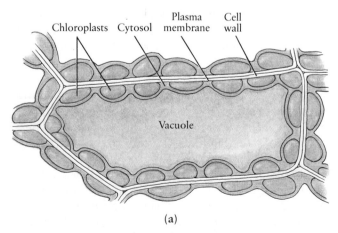

Chloroplasts   Cytosol   Plasma membrane   Cell wall

Vacuole

(a)

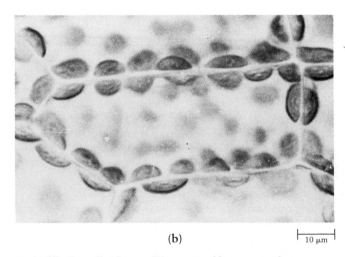

(b)   10 μm

5–12   (a) *Graphic interpretation and* (b) *electron micrograph of a vacuole. In this cell from the photosynthetic structure of a moss, the vacuole has expanded until it almost entirely fills the cell. The small amount of living cytoplasm, containing the chloroplasts, has been forced to the edges of the cell, up against the plasma membrane.*

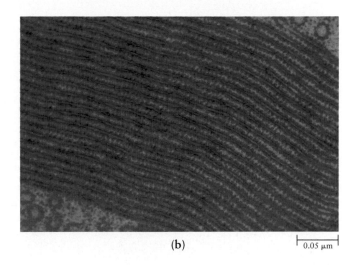

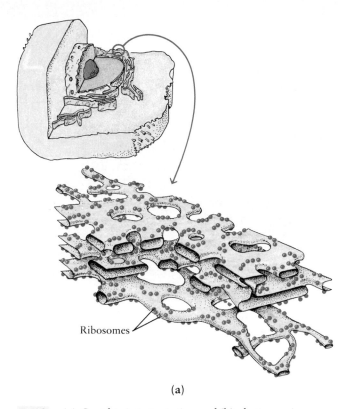

(b)

0.05 µm

Ribosomes

(a)

5–13   (a) *Graphic interpretation and* (b) *electron micrograph of rough endoplasmic reticulum. The endoplasmic reticulum is a system of membranes that separates the cell into channels and compartments and provides surfaces on which chemical activities take place. The dense objects on the membrane surfaces are ribosomes. The narrow compartments formed by the membranes of the rough endoplasmic reticulum contain newly synthesized proteins. This endoplasmic reticulum is from a cell of the pancreas, an organ extremely active in the synthesis of digestive enzymes.*

The amount of endoplasmic reticulum in a cell—and the relative proportions of the two forms—are not fixed but vary according to the cell's activities. In most cells, rough endoplasmic reticulum is the predominant form. Only cells specialized for the synthesis or metabolism of lipids—such as the cells that make steroid hormones—have large amounts of smooth endoplasmic reticulum. "Routine" lipid synthesis, such as the synthesis of membrane lipids, generally occurs in the rough endoplasmic reticulum, which also plays a role in the synthesis of many proteins.

The synthesis of proteins always begins on ribosomes that are located within the cytosol. The subsequent location of the ribosomes and the growing polypeptide chains to which they are attached is determined by the way in which the newly synthesized proteins will be used. Some proteins—hemoglobin and certain enzymes, for example—perform their functions within the cytosol. In the case of these molecules, the entire process of protein synthesis, from beginning to end, occurs in the cytosol.

Other proteins, such as digestive enzymes, hormones, or mucus, are released outside the cell, sometimes carrying out their functions at a great distance (on a cellular scale) from their source. Still other proteins are essential components of organelles, organelle membranes, or the plasma membrane. The synthesis of all these proteins begins in the cytosol with the synthesis of a "leader" of hydrophobic amino acids. This portion of the molecule, known as the **signal sequence,** directs the newly forming proteins and their accompanying ribosomes to specific regions of the endoplasmic reticulum. The ribosomes then attach to the endoplasmic reticulum, and the amino acids of the signal sequence assist in the transport of the proteins through the lipid bilayer to the interior cavity of the endoplasmic reticulum.

As synthesis proceeds, the growing proteins continue to move into the endoplasmic reticulum, a process powered by energy from ATP. When the proteins are complete, the ribosomes are detached and return to the cytosol. Subsequently, the newly synthesized proteins move from the rough endoplasmic reticulum through a special transitional endoplasmic reticulum. Here they are packaged, often with newly synthesized membrane lipids, in transport vesicles destined for a Golgi complex.

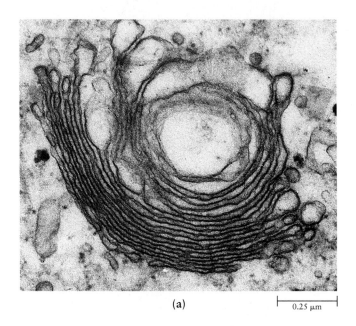

(a)

0.25 μm

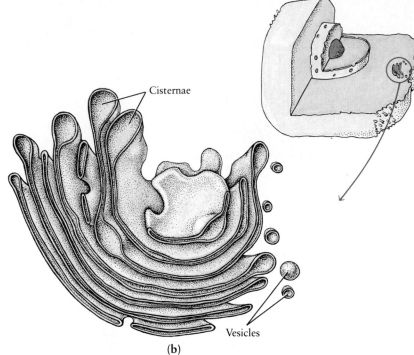

Cisternae

Vesicles

(b)

5–14 (a) *Electron micrograph and* (b) *graphic interpretation of a Golgi complex. A Golgi complex consists of four or more membrane-bound sacs, known as cisternae, arranged in a loose stack. Materials are packaged in membrane-enclosed vesicles at the Golgi complexes and are distributed within the cell or shipped to the cell surface. Note the vesicles pinching off from the edges of the flattened cisternae.*

Golgi Complexes: Processing, Packaging, and Distribution

Each **Golgi complex** consists of flattened, membrane-bound sacs stacked loosely on one another and surrounded by tubules and vesicles (Figure 5–14). The function of the Golgi complex is to accept vesicles from the endoplasmic reticulum, to modify the membranes and contents of the vesicles, and to incorporate these finished products into transport vesicles that deliver them to other parts of the cell and, especially, to the cell surface. Thus Golgi complexes serve as processing, packaging, and distribution centers. They are found in almost all eukaryotic cells. Animal cells usually contain 10 to 20 Golgi complexes, and plant cells may have several hundred.

One of the most critical products processed, packaged, and distributed by the Golgi complexes is new material for the plasma membrane and for the membranes of the cell's organelles. Membrane lipids and proteins, synthesized in the rough endoplasmic reticulum, are delivered to the Golgi complex in vesicles that fuse with it. Within the sacs of the Golgi complex, the final assembly of carbohydrates with proteins (forming glycoproteins) and with lipids (forming glycolipids) occurs. After the chemical processing is completed, the new membrane material is packaged in vesicles that are targeted to the correct location, whether it be the plasma membrane or the membrane of a particular organelle. In plant cells, Golgi complexes also bring together some of the components of the cell walls and export them to the cell surface where they are assembled.

In addition to their function in the assembly of cellular membranes, Golgi complexes have a similar role in the processing and packaging of materials that are released outside the cell. Figure 5–15 summarizes the way in which the ribosomes, the endoplasmic reticulum, and the Golgi complex and its vesicles interact to produce and deliver new material for cellular membranes and molecules for export. One of the intriguing questions about this process is how the various transport vesicles "know" where to go. The speed and efficiency with which each vesicle arrives at the proper location indicates the presence of a sophisticated mechanism for labeling and sorting. The nature of this mechanism is currently under investigation.

Lysosomes and Peroxisomes: Degradation and Recycling

One type of relatively large vesicle commonly formed from the Golgi complex is the **lysosome.** Lysosomes are essentially membranous bags that enclose enzymes involved in hydrolysis reactions. A single lysosome may

5–15 *The interaction of ribosomes, the endoplasmic reticulum, and the Golgi complex and its vesicles.* **(a)** *As proteins are synthesized on the ribosomes, they are fed into the rough endoplasmic reticulum. New membrane lipids are also synthesized in the endoplasmic reticulum.* **(b)** *The proteins then move through a transitional region of the endoplasmic reticulum and are released in vesicles that incorporate the newly synthesized membrane lipids. These vesicles move to the Golgi complex, where they fuse with its sacs.*

(c) *In the Golgi complex, carbohydrates, synthesized from sugar subunits available in the cytosol, are added to some of the proteins and lipids, producing glycoproteins and glycolipids. These large molecules are common components of cellular membranes. Molecules destined for export from the cell also undergo chemical processing in the Golgi complex. In some types of cells, this processing includes the addition of lipids to proteins, producing lipoproteins.*

Vesicles containing the finished molecules are released from the Golgi complex and move to other locations within the cell or to its exterior surface. If the vesicles contain new material for the plasma membrane or if their contents are to be secreted from the cell, the vesicle membrane fuses with the plasma membrane.

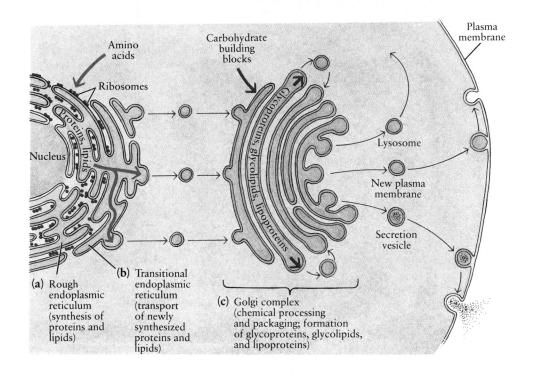

contain as many as 50 different kinds of hydrolytic enzymes. Within the lysosomes, these enzymes break down proteins, polysaccharides, and lipids from various types of cellular debris. The small molecules that result are then released back into the cytosol for reuse.

Another function of lysosomes is seen among white blood cells, which engulf bacteria in the human body. As the bacteria are taken up by the cell, they are wrapped in membrane-enclosed vacuoles. When this occurs, the lysosomes within the cell fuse with the vacuoles containing the bacteria and release their hydrolytic enzymes. The bacteria are quickly digested.

The membrane of a lysosome separates the hydrolytic enzymes it contains from the rest of the cell. If the lysosomes break open, the cell itself will be destroyed, since the enzymes they carry are capable of hydrolyzing the molecules of the living cell. Why the enzymes do not destroy the membranes of the lysosomes that carry them is an unanswered question. One clue is that the surface of the lysosome membrane facing the interior of the vesicle has a coating of carbohydrates that may perform a protective function.

Another type of enzyme-containing vesicle is the **peroxisome.** Peroxisomes are vesicles in which certain nitrogenous bases and other compounds are broken down by the cell. Some of the reactions that occur in the peroxisomes produce hydrogen peroxide, a compound that is extremely toxic to living cells. The peroxisomes, however, contain another enzyme that immediately breaks hydrogen peroxide into water and oxygen, preventing any damage to the cell.

Chloroplasts and Mitochondria: Cellular Power Plants

The activities of a cell require energy. As we have seen, some cells (autotrophs) manufacture their own energy-rich organic compounds from inorganic molecules. Other cells (heterotrophs) must obtain organic molecules from outside sources.

Photosynthetic autotrophs capture radiant energy from the sun and transform it to chemical energy stored in organic molecules. This process, **photosynthesis,** requires special pigments, of which chlorophyll is the most common. Photosynthesis takes place, however, only when the chlorophyll molecules are embedded in a membrane. In all photosynthetic eukaryotes, the chlorophyll-bearing membranes are organized within a membrane-bound organelle, the **chloroplast** (Figure 5–16).

5–16   (a) *Graphic interpretation and* (b) *electron micrograph of a chloroplast. A chloroplast is surrounded by two membranes, each of which is a lipid bilayer. In addition, it contains an elaborate internal membrane system in which chlorophyll and* *other photosynthetic pigments are embedded. The internal membranes form a series of flattened, interconnected sacs, which are called thylakoids. This chloroplast is from the leaf of a corn plant.*

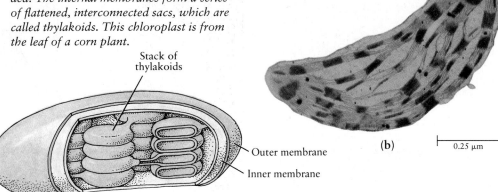

Stack of thylakoids

Outer membrane

Inner membrane

(b)

0.25 μm

(a)

Inner membrane

Outer membrane

Cristae

(a)

(b)

0.5 μm

5–17   (a) *Graphic interpretation and* (b) *electron micrograph of a mitochondrion. Like the chloroplast, the mitochondrion is surrounded by two distinct membranes. The inner membrane folds inward to make a series of shelves, or cristae. The membrane forming these shelves plays a crucial role in the energy-releasing chemical reactions that occur in the mitochondria.*

Virtually all eukaryotic cells (including photosynthetic ones) have **mitochondria** (singular, mitochondrion), which are also membrane-bound organelles. In the process of **cellular respiration,** which occurs in the mitochondria, energy-rich molecules are broken down. This process uses oxygen and releases energy. The energy is harnessed by the mitochondria to "recharge" ADP molecules to ATP, which is then available to power other cellular activities. As shown in Figure 5–17, mitochondria are always surrounded by two membranes, the inner one of which folds inward. These folds, known as **cristae,** are working surfaces for the reactions that occur in the mitochondria.

Chloroplasts and mitochondria are the essential power plants of eukaryotic cells. Without the energy they make available, most other cellular functions could not be carried out. We shall examine the structure and function of these organelles in more detail when we consider the processes of photosynthesis and respiration in Section 2.

The Cytoskeleton: Structural Support and Motility

The three-dimensional organization of the eukaryotic cell depends, in large part, on a network of filamentous proteins in the cytoplasm. These proteins form an internal cytoskeleton that maintains the shape of the cell, anchors its organelles, directs its traffic, and enables it to move. Three different types of filaments—**actin filaments, intermediate filaments,** and **microtubules**—have been identified as major participants in the cytoskeleton (Figure 5–18).

5–18   *These micrographs of whole cells from the rat kangaroo dramatically reveal the distribution of elements of the cytoskeleton. (a) Actin filaments are bundled together in stress fibers that run the length of the cell and are also concentrated just beneath the plasma membrane. (b) Intermediate filaments surround the nucleus and extend throughout the cytoplasm. (c) Microtubules radiate from an "organizing center" near the nucleus and end near the plasma membrane.*

These cells are from epithelial tissue, which provides a covering for body surfaces and lines the internal organs. Each cell was treated with fluorescent antibodies that would bind to one specific type of cytoskeletal protein. The pattern of fluorescence indicates the location of the protein.

(a)   20 μm

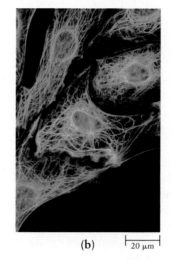

(b)   20 μm

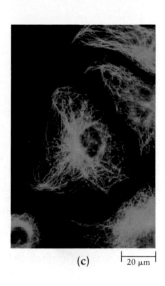

(c)   20 μm

Actin filaments are very fine protein threads, formed from the globular protein actin (Figure 5–19a). Actin is the most abundant protein in many animal cells, and its globular molecules can be rapidly assembled into filaments by the cell—and also rapidly disassembled—depending on the requirements of the cell at any given time. Actin filaments are often concentrated in bundles, known as stress fibers, that are stretched across the cytoplasm like cables or meshed together in the form of belts or webs. In many cells, the plasma membrane is supported by an internal skeleton of actin filaments. The filaments are anchored to particular integral proteins of the plasma membrane, and their assembly begins at these attachment points. Actin filaments also play an important role in cell division.

Intermediate filaments, which are intermediate in size between actin filaments and microtubules, are composed of fibrous proteins. In many cells, the intermediate filaments radiate out from the nuclear envelope.

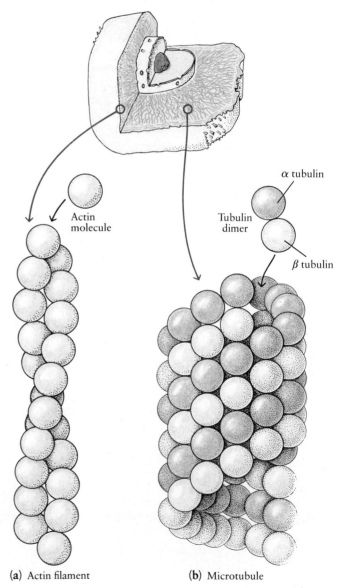

α tubulin

Tubulin dimer

β tubulin

Actin molecule

5–19   *Two components of the cytoskeleton, actin filaments and microtubules, are formed from globular protein subunits. (a) Actin filaments consist of two linear chains of identical actin molecules, coiled around one another to form a helix. (b) Microtubules are hollow tubes composed of two different types of molecules, alpha (α) tubulin and beta (β) tubulin. These molecules come together to form soluble dimers ("two parts"), which then self-assemble into insoluble hollow tubules. Both actin filaments and microtubules can be readily assembled and disassembled by the cell.*

(a) Actin filament   **(b)** Microtubule

Their function in the life of the cell is still poorly understood, but they are found in the greatest density in cells, such as those of the skin, that are subject to mechanical stress.

Microtubules, the largest structures of the cytoskeleton, are long, hollow tubes assembled by the cell from globular protein subunits (Figure 5–19b). They usually exist as individual tubules, radiating outward from an "organizing center" near the nucleus and ending near the cell surface. Microtubules determine the position of organelles within the cell, and they provide routes for the movement of both organelles and vesicles. Like actin filaments, microtubules play an important role in cell division. In addition, they seem to provide a temporary scaffolding for the construction of other cellular structures, such as the plant cell wall.

Numerous wisplike fibers interconnect all of the other structures within the cytoplasm. These fibers, formed from accessory proteins of the cytoskeleton, link the various filaments together in specific ways. Although the resulting network gives the cell a highly ordered three-dimensional structure, it is neither rigid nor permanent. The cytoskeleton is a dynamic framework, changing and shifting according to the activities of the cell and making possible its motility.

All cells exhibit some form of movement. Even plant cells, encased in a rigid cell wall, exhibit active movement of the cytoplasm within the cell as well as chromosomal movements and changes in shape during cell division. Embryonic cells migrate in the course of animal development. Amoebas pursue and engulf their prey, and little *Chlamydomonas* cells dart toward a light source.

Two different mechanisms of cellular movement have been identified. The first consists of assemblies of contractile proteins, in which actin filaments are a principal component. In addition to their structural role in the cytoskeleton, actin filaments participate in the internal movement of cellular contents and in the move-

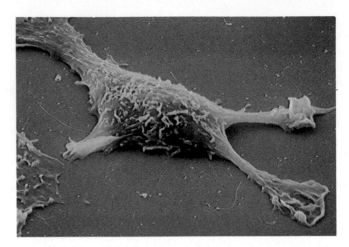

5–20   *A macrophage, a type of white blood cell, on the move across the surface of a culture dish. Its motility is made possible by the interaction of actin filaments with another protein known as myosin. As we shall see in Chapter 33, macrophages perform a number of vital functions in the body's response to invading microorganisms. They are also one of the principal cell types attacked by HIV, the virus that causes AIDS.*

ment of the cell itself (Figure 5–20). The second mechanism involves permanent locomotor structures, cilia and flagella, formed from assemblies of microtubules.

Cilia and Flagella

Cilia (singular, cilium) and **flagella** (singular, flagellum) are long, thin structures extending from the surface of many types of eukaryotic cells. They are essentially the same except for length. (Their names, which are derived from the Latin terms for "eyelash" and "whip," were given before their similarity was realized.) When they are shorter and occur in larger numbers, they are more likely to be called cilia. When they are longer and fewer, they are usually called flagella (Figure 5–21).

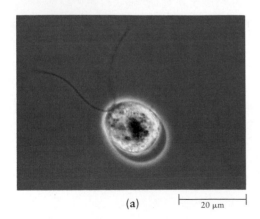

(a)   20 μm

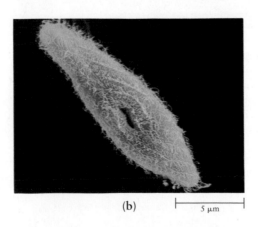

(b)   5 μm

5–21   (a) *The green alga* Chlamydomonas *propels itself through the water by means of its two flagella.* (b) *The heterotrophic protist* Paramecium, *by contrast, propels itself by means of numerous cilia.*

In one-celled protists and some small animals (such as a few types of flatworms), cilia and flagella are associated with movement of the organism. Similarly, the motile power of the mammalian sperm cell comes from its single powerful flagellum, or "tail."

Many of the cells that line the surfaces within our bodies and the bodies of other animals are also ciliated (see Figure 4–12, page 81). These cilia do not move the cells but, rather, serve to sweep substances across the cell surface. For example, cilia on the surface of cells of the respiratory tract beat upward, propelling a current of mucus that sweeps bits of soot, dust, pollen, tobacco tar—whatever foreign substances we have inhaled either accidentally or on purpose—to our throats, where they can be removed by swallowing.

Only a few large groups of eukaryotic organisms—most notably the flowering plants—have no cilia or flagella on any cells. Some bacteria move by means of flagella, but these prokaryotic flagella are so different in construction from those of eukaryotes that it would be useful if they had a different name.

Almost all eukaryotic cilia and flagella have the same internal structure. Nine fused pairs of microtubules form a cylinder that surrounds two additional, solitary microtubules in the center (Figure 5–22). The movement of cilia and flagella comes from within the structures themselves. If cilia are removed from cells and placed in a medium containing ATP, they beat or swim through the medium. The movement, according to the generally accepted hypothesis, is caused by each outer pair of microtubules moving tractor-fashion with respect to its nearest neighbor. The two "arms" that you can see on one microtubule of each outer pair are enzymes involved in ATP hydrolysis. Other proteins form spokes that extend from the nine pairs of outer microtubules toward the central pair, and still other proteins form more widely spaced links, rather like the hoops of a barrel, connecting the nine outer pairs to each other.

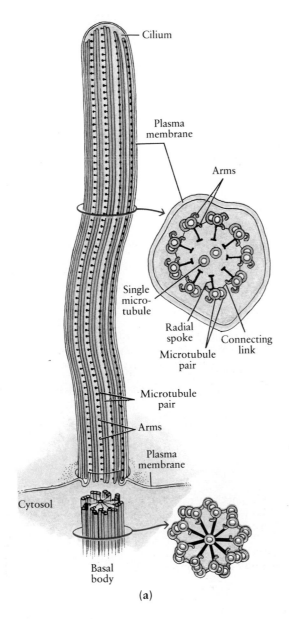

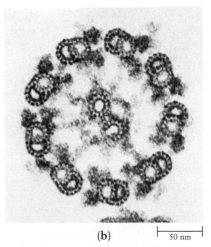

(b)   50 nm

5–22   (a) *Diagram of a cilium with its underlying basal body, and* (b) *an electron micrograph of the flagellum of* Chlamydomonas, *as seen in cross section. Virtually all eukaryotic cilia and flagella, whether they are found on protists or on the surfaces of cells within our own bodies, have this same internal structure, which consists of an outer cylinder of nine pairs of microtubules surrounding two additional microtubules in the center. The "arms," the radial spokes, and the connecting links are formed from different types of protein. The basal bodies from which cilia and flagella arise have nine outer triplets, with no microtubules in the center. The "hub" of the wheel in the basal body is not a microtubule, although it has about the same diameter.*

The spokes are thought to play a role in coordinating the tractorlike movements of the microtubules, whereas the links limit the amount of sliding possible and thus convert the movement into a bending motion.

Cilia and flagella arise from **basal bodies,** which are also made up of microtubules. Their number and arrangement are somewhat different, however, as you can see in Figure 5–22a. Many types of eukaryotic cells also contain structures identical to basal bodies that are known as **centrioles.** The distribution of centrioles within the cell is different from that of basal bodies, and, until recently, it appeared that their function was also different. Thus they were given a different name,

long before electron microscopy revealed their identical structure. Centrioles, like basal bodies, are found only in those groups of organisms that also have cilia and flagella (see Table 5–1). As we shall see in Chapter 10, centrioles also appear to play a role as organizers of microtubules.

The discovery of the complex internal structure of cilia and flagella, repeated over and over again throughout the living world, was one of the spectacular revelations of electron microscopy. For biologists, it is another glimpse down the long corridor of evolution, providing overwhelming evidence, once again, of the basic unity of Earth's living things.

Table 5–1   A Comparison of Cell Characteristics in the Five Kingdoms

	Monera	Protista	Fungi	Plantae	Animalia
Cell type	Prokaryotic	Eukaryotic	Eukaryotic	Eukaryotic	Eukaryotic
Plasma membrane	Present	Present	Present	Present	Present
Cell wall	Present (noncellulose polysaccharide plus protein)	Present in some forms, various types	Chitin and other noncellulose polysaccharides	Cellulose and other polysaccharides	Absent
Nuclear envelope	Absent	Present	Present	Present	Present
Chromosomes	Single, circular DNA molecule	Multiple, consisting of DNA and histone proteins	Multiple, consisting of DNA and histone proteins	Multiple, consisting of DNA and histone proteins	Multiple, consisting of DNA and histone proteins
Ribosomes	Present (smaller)	Present	Present	Present	Present
Endoplasmic reticulum	Absent	Present	Present	Present	Present
Golgi complexes	Absent	Present	Present	Present	Present
Lysosomes	Absent	Often present	Often present	Similar structures (lysosomal compartments) present	Often present
Peroxisomes	Absent	Often present	Absent	Often present	Often present
Vacuoles	Absent	Present	Present	Usually large single vacuole in mature cell	Small or absent
Mitochondria	Absent	Present	Present	Present	Present
Chloroplasts	Absent	Present (some forms)	Absent	Present	Absent
9 + 2 cilia or flagella	Absent	Often present	Absent	Absent (in flowering plants)	Often present
Centrioles	Absent	Often present	Absent	Absent (in flowering plants)	Present

Summary

Cells are the basic units of biological structure and function. The size of cells is limited by proportions of surface to volume; the greater a cell's surface area in proportion to its volume, the greater the quantity of materials that can move into and out of the cell in a given period of time. Cell size is also limited by the capacity of the nucleus to regulate cellular activities.

Cells are separated from their environment by the plasma membrane, which protects the cell's structural and functional integrity. According to the fluid-mosaic model, the plasma membrane and other cellular membranes are formed from phospholipid bilayers in which protein molecules are embedded. Different membrane proteins perform different functions; some are enzymes, others are receptors, and still others are transport proteins.

The two faces of the plasma membrane differ in chemical composition. The cytoplasmic face is characterized by additional protein molecules attached to the proteins embedded in the bilayer. The exterior face of the membrane is characterized by short carbohydrate chains. Some of these chains are the hydrophilic heads of glycolipid molecules that are interspersed among the phospholipid molecules; others are attached to the protruding portions of membrane proteins.

The cells of plants, most algae, fungi, and prokaryotes are further separated from the environment by a cell wall constructed by the cell itself.

The nucleus of eukaryotic cells is surrounded by a double membrane, the nuclear envelope, which contains pores through which molecules pass to and from the cytoplasm. The nucleus contains the chromosomes, which, when the cell is not dividing, exist in an extended form called chromatin. The nucleolus, visible within the nucleus, is involved in the formation of ribosomes. Interacting with the cytoplasm, the nucleus helps to regulate the cell's ongoing activities.

The cytoplasm consists of all the cell contents that are outside the nuclear envelope but within the plasma membrane. The cytosol, the fluid portion of the cytoplasm, is a dense solution of ions, small molecules, and proteins. Dispersed throughout the cytosol in both prokaryotic and eukaryotic cells are numerous ribosomes, complexes of protein and DNA that participate in protein synthesis.

In eukaryotic cells, the cytoplasm contains numerous membrane-bound compartments known as organelles, which are not found in prokaryotic cells (see Table 5–1). The simplest are vacuoles and vesicles, which are bounded by a single membrane.

The cytoplasm of eukaryotic cells is subdivided by a network of membranes known as the endoplasmic reticulum, which serves as a work surface for many of the cell's biochemical activities. Cells that are producing new membrane material or proteins for export have extensive systems of endoplasmic reticulum with ribosomes attached, known as rough endoplasmic reticulum.

Golgi complexes are processing and packaging centers for materials being moved through and out of the cell. Lysosomes, which contain hydrolytic enzymes, are involved in intracellular digestive activities. The enzymes for cellular reactions that produce hydrogen peroxide as a by-product are sequestered in the peroxisomes.

Chloroplasts and mitochondria are the organelles involved in energy capture and energy release, respectively. Photosynthesis occurs within the chloroplasts, and cellular respiration within the mitochondria.

The eukaryotic cytoplasm has a supporting cytoskeleton that includes three principal types of structures: actin filaments, intermediate filaments, and microtubules. The cytoskeleton maintains the shape of the cell, anchors its organelles, directs its traffic, and enables it to move. Assemblies of contractile proteins, including actin filaments, are associated with internal cellular movement and, in some types of cells, with the movement of the cell itself.

Cilia and flagella, which are formed from assemblies of microtubules, are associated with the external movement of cells or the movement of materials along cell surfaces. These whiplike appendages are found on the surface of many types of eukaryotic cells. They have a highly characteristic 9 + 2 structure, with nine pairs of microtubules forming a cylinder around two central microtubules. Cilia and flagella arise from basal bodies, which are cylindrical structures containing nine microtubule triplets with no inner pair. Centrioles have the same internal structure as basal bodies and are found only in those groups of organisms that also have cilia or flagella.

Questions

1. Distinguish among the following terms: plasma membrane/cell wall; nucleus/nucleolus; cytoplasm/cytosol/cytoskeleton; chloroplasts/mitochondria; cilia/flagella; basal body/centriole.

2. Describe the structure of the plasma membrane. How do the two faces of the membrane differ? What is the functional significance of these differences?

3. (a) Sketch an animal cell. Include the principal organelles and label them. (b) Prepare a similar, labeled sketch of a plant cell. (c) What are the major differences between the animal cell and the plant cell?

4. Why is the secondary wall of a plant cell *inside* the primary cell wall? Where is the plasma membrane in relation to the two cell walls?

5. Explain the functions of each of the following structures: ribosomes, endoplasmic reticulum, vesicles, Golgi complexes. How do they interact in the synthesis and delivery of new membrane material and in the export of proteins from the cell?

6. What are the functions of the cytoskeleton? Describe the similarities and differences between actin filaments, intermediate filaments, and microtubules.

7. Use a ruler and the scale marker at the bottom of each micrograph on pages 89 and 103 to determine: (a) the thickness (roughly) of the plasma membrane, (b) the diameter of a cilium, and (c) the diameter of a microtubule within a cilium. (This is how the sizes of cellular components are determined by microscopists.) Would a cilium be resolvable in a light microscope (that is, is its diameter more than $0.2 \ \mu m$)?

8. (a) Sketch a cross section of a cilium. (b) Sketch a cross section of the basal body of a cilium. (c) What are the differences between the two structures?

9. On the basis of what you know of the functions of each of the structures in Table 5–1, what components would you expect to find most prominently in each of the following cell types: muscle cells, sperm cells, green leaf cells, red blood cells, white blood cells?

10. Two brothers were under medical treatment for infertility. Microscopic examination of their semen showed that the sperm were immotile and that the little "arms" were missing from the microtubular arrays. The brothers also had chronic bronchitis and other respiratory difficulties. What is a plausible explanation?

Suggestions for Further Reading

Albersheim, Peter: "The Walls of Growing Plant Cells," *Scientific American*, April 1975, pages 81–95.

Alberts, Bruce, Dennis Bray, Julian Lewis, Martin Raff, Keith Roberts, and James D. Watson: *Molecular Biology of the Cell*, 3d ed., Garland Publishing, Inc., New York, 1994.

> *Progressing from the molecules of which cells are composed, through an examination of cellular structure and function, to the interactions of cells within tissues, this outstanding text describes not only our current knowledge and how it was attained but also the many areas still to be explored. It is clearly written and filled with wonderful micrographs and explanatory diagrams. Highly recommended.*

Allen, Robert Day: "The Microtubule as an Intracellular Engine," *Scientific American*, February 1987, pages 42–49.

Becker, Wayne M., and David W. Deamer: *The World of the Cell*, 2d ed., The Benjamin/Cummings Publishing Company, Redwood City, Calif., 1991.

> *An introductory textbook, designed for a first course in cell biology. Chapters 4–9 provide a concise but clear discussion of cell structure and function.*

Bretscher, Mark S.: "How Animal Cells Move," *Scientific American*, December 1987, pages 72–90.

Darnell, James, Harvey F. Lodish, and David Baltimore: *Molecular Cell Biology*, 2d ed., W. H. Freeman and Company, New York, 1990.

> *A comprehensive treatment of modern cell biology, richly illustrated with diagrams and micrographs. This text places particular emphasis on membrane structure and function, cytoplasmic organelles, and the cytoskeleton, as well as on the techniques used in the contemporary study of cell biology.*

de Duve, Christian: *A Guided Tour of the Living Cell*, W. H. Freeman and Company, New York, 1984.*

> *In this beautifully illustrated two-volume set, de Duve, one of the pioneers of modern cell biology, takes the reader—imagined to be a "cytonaut," a bacterium-sized tourist—on a journey through the eukaryotic cell. The first portion of the journey explores the cellular*

---

* Available in paperback.

membranes; the second, the cytoplasm and its organelles; and the third, the nucleus. At the conclusion of the journey, de Duve considers such key questions of modern biology as the origin of life and the mechanisms of evolution.

Dustin, Pierre: "Microtubules," *Scientific American*, August 1980, pages 67–76.

Ezzell, Carol: "Sticky Situations: Picking Apart the Molecules that Glue Cells Together," *Science News*, June 13, 1992, pages 392–395.

Lazarides, Elias, and Jean Paul Revel: "The Molecular Basis of Cell Movement," *Scientific American*, May 1979, pages 100–113.

Ledbetter, M. C., and Keith R. Porter: *Introduction to the Fine Structures of Plant Cells*, Springer-Verlag, New York, 1970.

An excellent atlas of electron micrographs of plant cells, with detailed explanations.

Porter, Keith R., and Mary A. Bonneville: *An Introduction to the Fine Structures of Cells and Tissues*, 4th ed., Lea & Febiger, Philadelphia, 1973.

An atlas of electron micrographs of animal cells, with detailed commentaries. These are magnificent micrographs, and the commentaries describe not only what the pictures show but also the experimental foundations of our knowledge of cell structure.

Prescott, David M.: *Cells: Principles of Molecular Structure and Function*, Jones and Bartlett Publishers, Boston, 1988.

An up-to-date, yet concise, textbook of cell biology, written for a first course at the undergraduate level. It is a wonderful introduction for any reader wishing to gain an overview of the exciting developments in contemporary cell biology.

Rothman, James E.: "The Compartmental Organization of the Golgi Apparatus," *Scientific American*, September 1985, pages 74–89.

Satir, Peter: "How Cilia Move," *Scientific American*, October 1974, pages 44–52.

Scientific American: *The Molecules of Life*, W. H. Freeman and Company, New York, 1986.*

This reprint of the October 1985 issue of Scientific American *includes 11 articles on molecules that play key roles in the living cell. You will find the articles on the molecules of the plasma membrane and the cytoskeleton of particular interest at this point in your study of biology.*

Scientific American: *Molecules to Living Cells*, W. H. Freeman and Company, New York, 1980.*

A collection of outstanding articles from Scientific American. *Two articles are devoted to cell membranes and their assembly.*

Sharon, Nathan, and Halina Liss: "Carbohydrates in Cell Recognition," *Scientific American*, January 1993, pages 82–89.

Sloboda, Roger D.: "The Role of Microtubules in Cell Structure and Cell Division," *American Scientist*, vol. 68, pages 290–298, 1980.

Stossel, Thomas P.: "How Cells Crawl," *American Scientist*, vol. 78, pages 408–423, 1990.

Unwin, Nigel, and Richard Henderson: "The Structure of Proteins in Biological Membranes," *Scientific American*, February 1984, pages 78–94.

* Available in paperback.

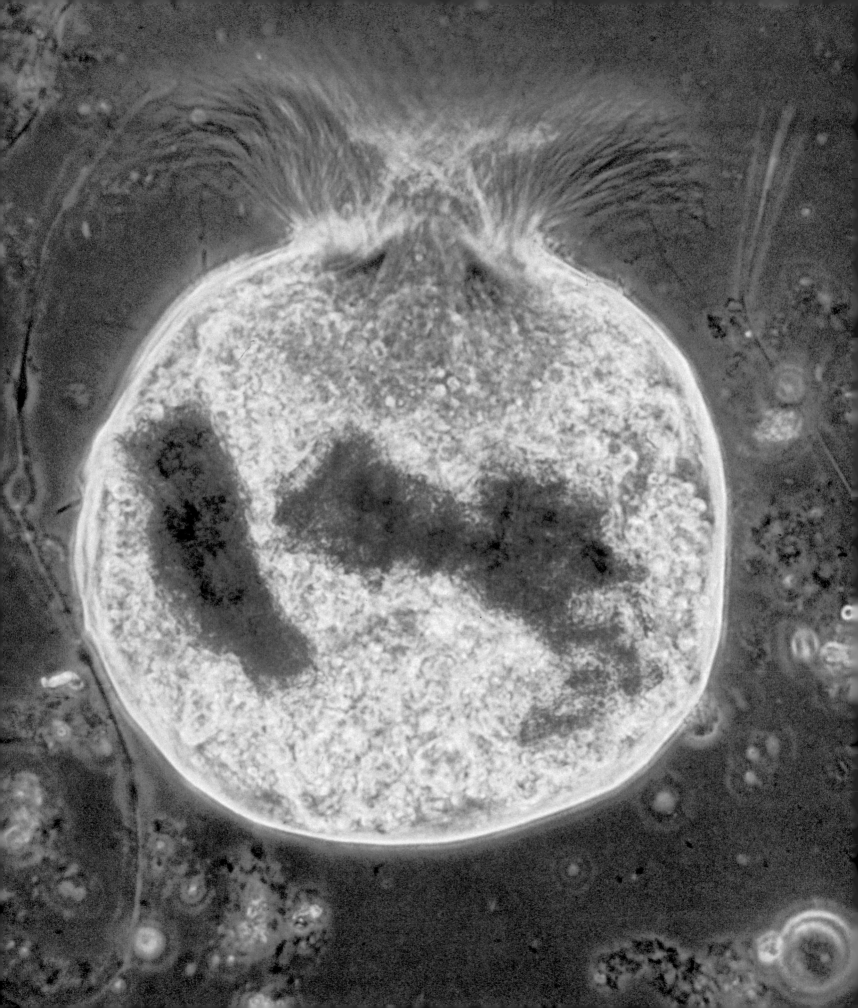

The light brown objects inside this remarkable cell are just what they look like: chips of wood. They were taken into the cell by a process known as phagocytosis—literally "cell-eating."

This organism, a protist bearing the melodic name *Barbulanympha ufalula,* lives in the intestine of the wood roach *Cryptocercus punctulatus.* The roach lives on and among dead logs, feeding on wood chips. Its digestive system breaks up the wood chips into protist-sized pieces, which are then engulfed and digested by *Barbulanympha.* Neither organism can live without the other.

Barbulanympha is about 0.3 millimeter in diameter, barely visible to the unaided eye. The tufts at the top of the cell are flagella—some 13,000 in all—that enable it to move about within the roach's intestine. The blue spherical region, visible below the flagella, is the nucleus. At the opposite end of the cell is a "back door"—a sensitive region of the plasma membrane through which wood chips are brought into the cell.

When the sensitive portion of the plasma membrane makes contact with a wood chip in the roach's intestine, the membrane extends around the wood chip, enclosing it in a vacuole. The vacuole then detaches from the plasma membrane and moves into the interior of the cell. Enzymes synthesized by *Barbulanympha* digest the cellulose in the wood into glucose and then into a smaller breakdown product, acetate. Both digestive enzymes and acetate are exported out of the cell, back into the intestine of the wood roach. In the mitochondria of the roach intestinal cells, the acetate is broken down to carbon dioxide and water. The energy released in this final stage of the process is harnessed to produce ATP.

Although not all cells are as photogenic as *Barbulanympha,* they are all equally dependent on their capacity to transport substances—ions, molecules, food particles, and even other cells—across the plasma membrane.

How Things Get into and out of Cells

Living matter is surrounded on all sides by nonliving matter with which it constantly exchanges materials. Yet living systems differ from their nonliving surroundings in the kinds and amounts of chemical substances they contain. Without this difference, living systems would be unable to maintain the organization on which their existence depends.

In all living systems, from prokaryotes to the most complex multicellular eukaryotes, exchanges of substances between the living organism and the nonliving world occur at the level of the individual cell. These exchanges are regulated by the plasma membrane. In multicellular organisms, the plasma membrane has the additional task of regulating exchanges of substances among the various specialized cells that constitute the organism. Control of these exchanges is essential to (1) protect each cell's integrity, (2) maintain the conditions at which its metabolic activities can take place, and (3) coordinate the activities of the different cells.

In addition to the plasma membrane, which controls the passage of materials between the cell and its environment, internal membranes, such as those surrounding mitochondria, chloroplasts, and the nucleus, control the passage of materials among intracellular compartments. This makes it possible for the cell to maintain the specialized chemical environments necessary for the processes occurring in the different organelles.

Maintenance of the internal environment of the cell and its various parts requires that the plasma membrane and the organelle membranes perform a complex double function: they must keep certain substances out while letting others in, and, conversely, they must keep certain substances in while letting others out. The capacity of a membrane to accomplish this function depends not only on the physical and chemical properties that result from its lipid and protein structure, but also on the physical and chemical properties of the ions and molecules that interact with the membrane.

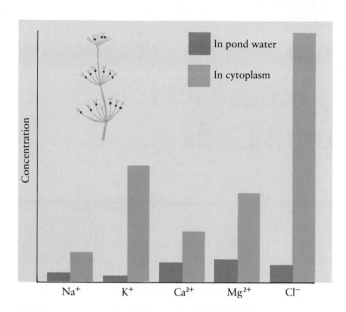

6–1   *Among the most important ions in many living cells are sodium (Na⁺), potassium (K⁺), calcium (Ca²⁺), magnesium (Mg²⁺), and chloride (Cl⁻). This graph shows the relative concentrations of these ions in the cytoplasm of the green alga* Nitella *and in the surrounding pond water. Such differences in ion concentration between the interior of living cells and their nonliving environment indicate that cells regulate their exchanges of materials with the surroundings. This regulation is accomplished by the plasma membrane.*

Of the many kinds of molecules surrounding and contained within the cell, by far the most common is water. Further, the many other molecules and ions important in the life of the cell (Figure 6–1) are dissolved in water. Therefore, let us begin our consideration of transport across cellular membranes by looking again at water, focusing our attention this time on how it moves.

The Movement of Water and Solutes

Except when it is locked in ice, water is constantly moving. It moves across the continents in rivers and streams, from the soil through the bodies of plants into the atmosphere, through the human body by way of the bloodstream, and into and out of living cells. In both the living and nonliving worlds, water molecules move from one place to another because of differences in potential energy. As you will recall (page 21), potential

energy is the stored energy an object—or a collection of objects, such as a collection of water molecules—possesses because of its position. The potential energy of water is usually referred to as the **water potential.**

Water moves from a region where water potential is higher to a region where water potential is lower, regardless of the reason for the difference in water potential (Figure 6–2). A simple example is water running downhill in response to gravity. Water at the top of a hill has more potential energy (that is, a higher water potential) than water at the bottom of a hill. As the water runs downhill, its potential energy is converted to kinetic energy. This, in turn, can be converted to mechanical energy doing useful work if, for example, a water wheel is placed in the path of the moving water.

Pressure is another source of water potential. If we fill an eyedropper with water and then squeeze the bulb, water will squirt out. Like water at the top of a hill, this water has been given a high water potential and will move to a lower one. Can we make the water that is running downhill move uphill by means of pressure? Obviously we can move water uphill—but only so long as the water potential produced by the pressure exceeds the water potential produced by gravity.

In solutions, water potential is affected by the concentration of dissolved particles (solutes). As the concentration of solute particles (that is, the number of solute particles per unit volume of solution) increases, the water potential decreases. Conversely, as the concentration of solute particles decreases, the water potential increases. In the absence of other factors (such as pressure) affecting the water potential, water molecules in solutions move from regions of lower solute concentration (higher water potential) to regions of higher solute concentration (lower water potential). This fact is of great importance for living systems.

The concept of water potential is a useful one because it enables us to predict the way that water will move under various combinations of circumstances. Measurements of water potential are usually made in terms of the pressure required to stop the movement of water—that is, the hydrostatic (water-stopping) pressure—under the particular circumstances. The unit usually used to measure this pressure is the atmosphere. One atmosphere is the average pressure of the air at sea level, about 1 kilogram per square centimeter (14.7 pounds per square inch).

Two mechanisms are involved in the movement of water and solutes: bulk flow and diffusion. In living systems, bulk flow is the process that moves water and solutes from one part of a multicellular organism to

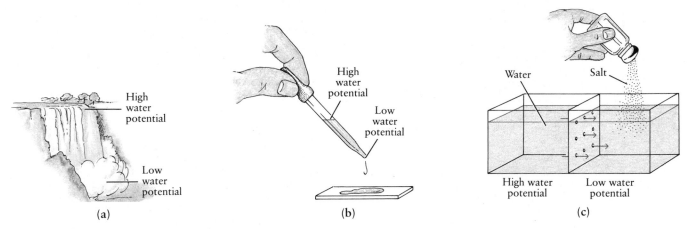

6–2 *The three factors that most commonly determine water potential are* (**a**) *gravity,* (**b**) *pressure, and* (**c**) *the concentration of dissolved solutes. Water moves from the region of higher water potential to the region of lower water potential, regardless of the reason for the difference in water potential.*

Often, a combination of factors are involved. In (**b**), *for example, the pressure applied when the bulb is squeezed is not the only factor contributing to the difference in water potential. Given the angle at which the eyedropper is held, gravity is also a factor.*

another part. Diffusion, by contrast, plays a major role in the movement of many molecules and ions into, out of, and through cells. A particular instance of diffusion—that of water moving through a membrane that separates solutions of different concentration—is known as osmosis.

Bulk Flow

Bulk flow is the overall movement of a fluid. The molecules move all together and in the same direction. For example, water runs downhill by bulk flow in response to the differences in water potential at the top and the bottom of a hill. Blood moves through your blood vessels by bulk flow as a result of the water potential (blood pressure) created by the pumping of your heart. Sap—an aqueous solution of sucrose and other solutes—moves by bulk flow from the leaves of a plant to other parts of the plant body.

Diffusion

Diffusion is a familiar phenomenon. If you sprinkle a few drops of perfume in one corner of a room, the scent will eventually permeate the entire room even if the air is still. If you put a few drops of dye in one end of a glass tank full of water, the dye molecules will slowly become evenly distributed throughout the tank. The process may take a day or more, depending on the size of the tank, the temperature, and the relative size of the dye molecules.

Why do the dye molecules move apart? If you could observe the individual dye molecules in the tank (Figure 6–3), you would see that each one of them moves individually and at random. Looking at any single molecule—at either its rate of motion or its direction of motion—gives you no clue at all about where the molecule is located with respect to the others. So how do the dye molecules get from one end of the tank to the other?

Imagine a thin section through the tank, running from top to bottom. Dye molecules will move in and out of this section, some moving in one direction, some moving in the other. But you will see more dye molecules moving from the side of greater dye concentration. Why? Simply because there are more dye molecules at that end of the tank. If there are more dye molecules on the left, more dye molecules, moving at random, will move to the right, even though there is an equal probability that any one molecule of dye will move from right to left. Consequently, the *net* movement of dye molecules will be from left to right. Similarly, if you could see the movement of the individual water molecules in the tank, you would see that their *net* movement is from right to left.

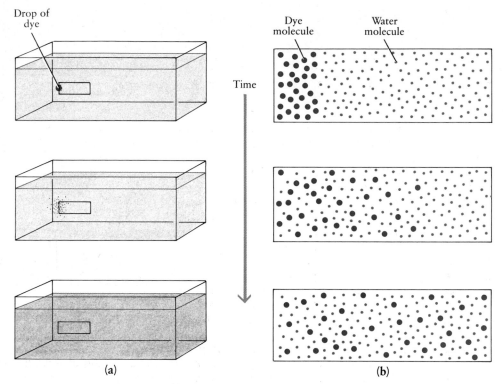

Drop of dye

Time

Dye molecule   Water molecule

(a)   (b)

6–3   *The diffusion process, as revealed by the addition of a drop of dye to a tank of water (top left). The outlined section within each panel in* (a) *is shown in close-up in* (b). *Diffusion is the result of the random movement of individual molecules (or ions), which produces a net movement of particles from a region where they are more concentrated to a region where they are less concentrated. This movement is* down *the concentration gradient. Notice that as the dye molecules (red) diffuse to the right, the water molecules (blue) diffuse to the left. The ultimate result is an even distribution of both types of molecules. Can you see why the* net *movement of molecules will slow down as equilibrium is reached?*

Substances that are moving from a region of higher concentration of their own molecules to a region of lower concentration are said to be moving *down a gradient,* which is analogous to flowing downhill. (A substance moving in the opposite direction, toward a region where it is more highly concentrated, moves *against a gradient,* which is analogous to being pushed uphill.) Diffusion occurs only *down* a gradient. The steeper the downhill gradient—that is, the larger the difference in concentration—the more rapid the diffusion.

In our imaginary tank, there are two gradients, one of dye molecules and one of water molecules. The dye molecules are moving in one direction, down their gradient, and the water molecules are moving in the opposite direction, down their gradient. In each case, the molecules are moving from a region of higher potential energy to a region of lower potential energy.

What happens when all the molecules are distributed evenly throughout the tank? The even distribution does not affect the behavior of the molecules as individuals; they still move at random. And, since the movements are random, just as many molecules go to the left as to the right. But because there are now as many molecules of dye and as many molecules of water on one side of the tank as on the other, there is no *net* movement of either. There is, however, just as much random motion as before, provided the temperature has not changed. When the molecules have reached a state of equal distribution, that is, when there are no more gradients, they are said to be in **dynamic equilibrium.**

The essential characteristics of diffusion are (1) that each molecule or ion moves independently of the others and (2) that these movements are random. The net result of diffusion is that the diffusing substances become evenly distributed.

Sensory Responses in Bacteria: A Model Experiment

Concentration gradients are, as we have seen, of vital importance in the diffusion of substances into, out of, and through cells. In the case of many single-celled organisms, they are also important in the movement of the cell itself through the surrounding medium.

Many types of bacteria are able to swim toward a food source or away from a noxious chemical. They accomplish this by moving along a concentration gradient, from a lower concentration of a particular type of molecule to a higher one, or vice versa. Such directed movements in bacteria are extremely sensitive and highly specific. The bacteria can sense only certain molecules and can sense them at very low concentrations. The sensory abilities, it has been shown, depend on receptors in the plasma membrane that detect the molecules in question.

If you watch flagellated bacteria swimming freely, you will see two types of movement. When the flagella are rotating, they drive the cell through the water in much the same way a propeller drives a boat. When the flagella stop, the cell tumbles wildly for perhaps a tenth of a second. Then the propellerlike motion begins again, and the cell moves off in a new direction. When the concentration of chemicals in the water is uniform, the cell tumbles often, changing direction every time. By contrast, when the cell is moving along a gradient, there are fewer tumbles, so the cell continues longer in the same direction.

How do bacterial cells "decide" to move in a particular direction? In other words, how do they detect the concentration gradient? For many years, the most widely held hypothesis was that a bacterial cell could detect the difference in concentration between its front end and its rear end. However, when Daniel E. Koshland, Jr., of the University of California, calculated the concentrations of molecules to which a cell could respond, he began to question this concept. A bacterial cell is so small that, even in a gradient steep enough to produce a strong response, the difference in concentration between the front end of the cell

and the rear end would be only on the order of one molecule in 10,000. Compounding the problem is the probability that the gradient would not be exactly uniform. In short, the analytical task confronting the cell on its journey would seem virtually impossible.

Koshland then formulated an alternative hypothesis and, more important, figured out a way to test between the two. Koshland's hypothesis was that the bacterial cells were making a comparison not in space—between the concentrations at their front end and at their rear end—but in time. In other words, they were comparing the concentration of the solution in which they were currently swimming with the concentration of the solution they had left behind just a few microseconds ago.

In order to choose between the alternatives, Koshland formulated an ingeniously simple experiment. Using a strain of the common bacterium *Salmonella,* he set up an apparatus with which he could transfer a sample of cells almost instantaneously from one liquid medium to another and compare the motility of the cells in the two environments.

First, he put *Salmonella* in a medium that contained no chemical attractants. The cells exhibited their normal tumble-and-run pattern of behavior. He transferred them to a new medium, also containing no chemical attractants. They did not change their pattern of movement. (This part of the experiment—the control—showed that moving the cells, by itself, did not affect their motility.)

Next, in the crucial part of the experiment, he placed *Salmonella* cells in a medium containing a uniform concentration of an attractant, the amino acid serine. The cells behaved just as they had when no attractant was present. Then he transferred them to a medium with a slightly higher concentration of serine. There was an immediate change. For a few seconds, the cells ran more than they tumbled. Then he transferred them to a medium with a lower concentration of serine. For a few seconds, they tumbled more and ran less. In other words, although the bacteria were actually

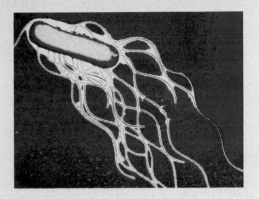

A *color-enhanced electron micrograph of a* Salmonella *cell.*

moving from one uniform concentration to another, they behaved as if they were moving up or down a gradient. Koshland had tricked the bacteria into revealing their secret and so was able to choose between the alternative hypotheses. The bacteria were analyzing differences in time, not space.

This experiment is a minilesson in how scientists go about their business. They formulate a testable hypothesis and then they challenge it. The test of the hypothesis can take the form of a clever, well-designed experiment, as in this example, of accumulated observations, or of the analysis of reports made by other observers. However, two components are always necessary: the testable hypothesis and the data with which to test it.

Of course, many questions remain. Exactly how do the receptors on the plasma membrane recognize particular substances? How does the cell "remember" the concentration from one moment to the next? How does the sensory response (the detection of the chemical) trigger the motor response (the movement of the flagella)? Here again is a characteristic of the scientific process: the answer to one question nearly always raises still more questions.

Cells and Diffusion

Water, oxygen, carbon dioxide, and a few other simple molecules diffuse freely across the plasma membrane. Carbon dioxide and oxygen, which are both nonpolar, are soluble in lipids and move easily through the lipid bilayer. Despite their polarity, water molecules also move through the membrane without hindrance, apparently through momentary openings created by spontaneous movements of the membrane lipids. Other polar molecules, provided they are small enough, also diffuse through these openings. The permeability of the membrane to these solutes varies inversely with the size of the molecules, indicating that the openings are small and that the membrane acts like a sieve in this respect.

Diffusion is also a principal way in which substances move within cells. One of the major factors limiting cell size is this dependence upon diffusion, which is essentially a slow process, except over very short distances. It becomes increasingly slower and less efficient as the distance traveled by the diffusing molecules increases.

Efficient diffusion requires not only a relatively short distance but also a steep concentration gradient. Cells maintain such gradients by their metabolic activities, thereby hastening diffusion. For example, carbon dioxide is constantly produced as the cell oxidizes fuel molecules for energy. As a result, there is a higher concentration of carbon dioxide inside the cell than out. Thus a gradient is maintained between the inside of the cell and the outside, and carbon dioxide diffuses out of the cell down this gradient. Conversely, oxygen is used up by the cell in the course of its activities, so oxygen present in air or water or blood tends to move into the cells by diffusion, again down a gradient. Similarly, within a cell, molecules or ions are often produced at one place and used at another. Thus a concentration gradient is established between the two regions, and the substance diffuses down the gradient from the site of production to the site of use.

Osmosis: A Special Case of Diffusion

A membrane that permits the passage of some substances, while blocking the passage of others, is said to be **selectively permeable.** The movement of water molecules through such a membrane is a special case of diffusion, known as **osmosis.** Osmosis results in a net transfer of water from a solution that has higher water potential to a solution that has lower water potential. In the absence of other factors that influence water potential (such as pressure), the movement of water in osmosis will be from a region of lower solute concentration (and therefore of higher water potential) to a region of higher solute concentration (lower water potential).

The diffusion of water is not affected by *what* is dissolved in the water, only by *how much* is dissolved—that is, the concentration of particles of solute (molecules or ions) in the water. A small solute particle, such as a sodium ion, counts just as much as a large solute particle, such as a sugar molecule.

Two or more solutions that have equal numbers of dissolved particles per unit volume—and therefore the same water potential—are said to be **isotonic** (from the Greek *isos,* meaning "equal," and *tonos,* "tension"). There is no net movement of water across a membrane separating two solutions that are isotonic to one another, unless, of course, pressure is exerted on one side.

In comparing solutions of different concentration, the solution that has less solute (and therefore a higher water potential) is known as **hypotonic,** and the one that has more solute (a lower water potential) is known as **hypertonic.** (Note that *hyper* means "more"—in this case, more particles of solute; and *hypo* means "less"—in this case, fewer particles of solute.) In osmosis, water molecules diffuse from a hypotonic solution (or from pure water), through a selectively permeable membrane, into a hypertonic solution (Figure 6–4).

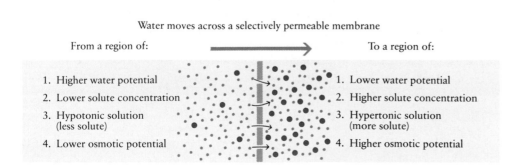

Water moves across a selectively permeable membrane

From a region of: → To a region of:

1. Higher water potential
2. Lower solute concentration
3. Hypotonic solution (less solute)
4. Lower osmotic potential

1. Lower water potential
2. Higher solute concentration
3. Hypertonic solution (more solute)
4. Higher osmotic potential

6–4   *The direction of water movement in osmosis.*

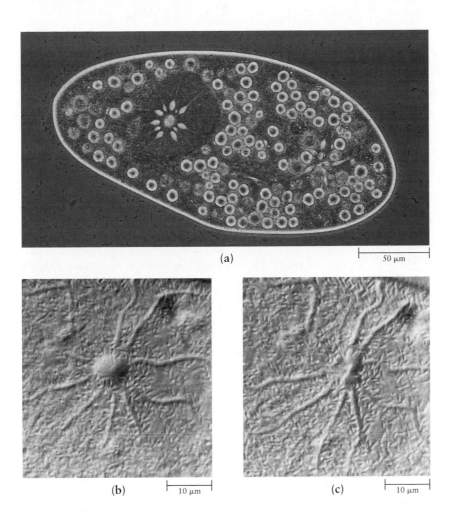

(a)

50 μm

(b)   10 μm

(c)   10 μm

6–5 *A* Paramecium *is hypertonic in relation to its environment, and hence water tends to move into the cell by osmosis. Excess water is expelled through its contractile vacuoles.* **(a)** *Photomicrograph of a living* Paramecium *in which one of its two rosette-like contractile vacuoles is clearly visible at the left. If you look very closely, you can find the other contractile vacuole at the opposite end of the cell. The numerous small, spherical objects in the cell are food vacuoles, filled with bacteria that are being digested by the* Paramecium.

A contractile vacuole, **(b)** *filled and* **(c)** *empty, as revealed by the scanning electron microscope. A group of long, thin collecting tubules converge on a central vacuole. Water moves from the cytosol into these tubules and through them into the central vacuole. When the central vacuole is filled, it contracts, propelling the water out of the cell through a small pore in the plasma membrane. Actin filaments (page 101) are involved in the contraction of the vacuole.*

Osmosis and Living Organisms

The osmotic movement of water across the selectively permeable plasma membrane causes some crucial problems for living systems. These problems vary according to whether the cell or organism is hypotonic, isotonic, or hypertonic in relation to its environment.

The body fluids of many marine fishes, for example, are hypotonic in relation to the surrounding salt water. Thus water tends to move out of their bodies by osmosis. As we shall see in Chapter 32, a combination of mechanisms enable these fishes to drink sea water and then excrete the excess solutes, maintaining their internal solute concentrations at the proper levels.

One-celled organisms that live in salt water are usually isotonic with the medium they inhabit, which is another way of solving the problem. The cells of most marine invertebrates are also isotonic with sea water. Similarly, the cells of vertebrate animals are isotonic with the blood and lymph that constitute the watery medium in which they live.

Many types of cells, however, live in a hypotonic environment. In all single-celled organisms that live in fresh water, such as *Paramecium*, the interior of the cell is hypertonic to the surrounding water. Consequently, water tends to move into the cell by osmosis. If too much water were to move into the cell, it could dilute the cell contents to the point of interfering with function and could even eventually rupture the plasma membrane. This is prevented by a specialized organelle known as a contractile vacuole, which collects water from various parts of the cell and pumps it out with rhythmic contractions (Figure 6–5). As you might expect, this bulk transport process requires an expenditure of energy by the cell.

Osmotic Potential

The water potential of two solutions separated by a selectively permeable membrane will become equal if enough water moves from the hypotonic solution into the hypertonic solution to equalize the solute concentrations—that is, to make the solutions isotonic. If, however, physical barriers prevent the expansion of the hypertonic solution as water moves into it by osmosis, there will be increasing resistance as water molecules continue to move across the membrane. This resistance is caused by a buildup of pressure that gradually increases the water potential of the hypertonic solution, decreasing the gradient of water potential between the two solutions. As the pressure increases, the *net* flow of water molecules will slow and then cease as the gradient of water potential disappears. Individual water molecules continue to move back and forth across the membrane, but these movements are in equilibrium and there is no *net* movement of water.

The pressure that is required to stop the osmotic movement of water into a solution (Figure 6–6) is a measure of the **osmotic potential** of the solution—that is, of the tendency of water to move across a membrane into the solution. The greater the tendency of water to move into the solution, the higher the pressure required to stop its movement—and thus the higher the osmotic potential.

Turgor   Plant cells are usually hypertonic to their surrounding environment, and so water tends to diffuse into them. This movement of water into the cell creates pressure within the cell against the cell wall. In young cells, this pressure causes the cell wall to expand and the cell to enlarge. The elongation that occurs as a plant cell matures (Figure 5–5, page 91) is a direct result of the osmotic movement of water into the cell.

As the plant cell matures, the cell wall stops growing. Moreover, mature plant cells typically have large central vacuoles that contain solutions of salts and other materials. (In citrus fruits, for example, they contain the acids that give the fruits their characteristic sour taste.) Because of these concentrated solutions, plant cells have a high osmotic potential—that is, water has a strong tendency to move into the cells. In the mature cell, however, the cell wall does not expand further. Its resistance to expansion results in an inward directed pressure, analogous to the pressure exerted by the depressed piston in Figure 6–6. This pressure prevents the net movement of additional water into the cell. Consequently, equilibrium of salt concentration is not reached and water continues to "try" to move into the cell, maintaining a constant pressure on the cell wall from the inside (Figure 6–7). This internal pressure on the cell wall is known as **turgor,** and it keeps the cell walls stiff and the plant body crisp. When turgor is reduced, as a consequence of water loss, the plant wilts.

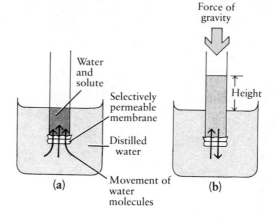
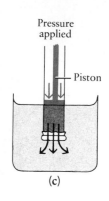

6–6   *Osmosis and the measurement of the osmotic potential of a solution.* (**a**) *The tube contains a solution and the beaker contains distilled water. A selectively permeable membrane at the base of the tube permits the passage of water molecules but not of solute particles.* (**b**) *The diffusion of water into the solution causes its volume to increase, and thus the column of liquid to rise in the tube. However, the downward pressure created by the force of gravity acting on the column of solution is proportional to the height of the column and the density of the solution. Thus, as the column of solution rises in the tube, the downward pressure gradually increases until it becomes so great that it counterbalances the tendency of water to move into the solution. In other words, the water potential on the two sides of the membrane becomes equal. At this point, there is no further net movement of water.* (**c**) *The pressure that must be applied to force the column of solution back to the level of the water in the beaker provides a quantitative measure of the osmotic potential of the solution—that is, of the tendency of water to diffuse across the membrane into the solution.*

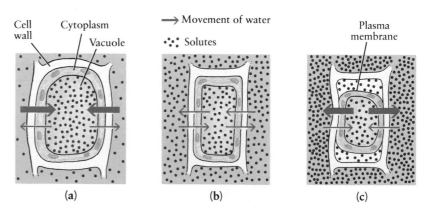

6–7   (**a**) *A turgid plant cell. The central vacuole is hypertonic in relation to the fluid surrounding it and so gains water. The expansion of the cell is held in check by the cell wall.* (**b**) *A plant cell begins to wilt if it is placed in an isotonic solution, so that water pressure no longer builds up within the vacuole.* (**c**) *A plant cell in a hypertonic solution loses water to the surrounding fluid and so collapses, with its plasma membrane pulling away from the cell wall. Such a cell is said to be plasmolyzed.*

Carrier-Assisted Transport

Cell membranes, as we have seen, are permeable to such substances as water, oxygen, and carbon dioxide, which move through them easily by diffusion. Other molecules used or produced by the cell are unable to diffuse through cell membranes, because of either their size or their polarity. Most ions and polar molecules, for example, cannot pass through the hydrophobic interior of the lipid bilayer. The transport of such substances depends upon integral membrane proteins that act as carriers, ferrying molecules and ions back and forth.

The transport proteins of the plasma membrane and of organelle membranes are highly selective. A particular protein may accept one molecule while it excludes a nearly identical one. It is the configuration of the protein molecule—that is, its tertiary or, in some cases, quaternary structure—that determines what molecules it can transport.

Some transport proteins can move substances across a membrane only if there is a favorable concentration gradient—that is, they can only move substances *down* a concentration gradient. Such carrier-assisted transport is known as **facilitated diffusion.** Facilitated diffusion, like simple diffusion, is driven by the potential energy of a concentration gradient. It is a passive process, requiring no energy outlay by the cell.

Other transport proteins can move molecules *against* a concentration gradient, a process known as **active transport.** Unlike facilitated diffusion, active transport requires the expenditure of energy by the cell (Figure 6–8).

Depending on the situation, the transport of a par-ticular molecule into a cell may involve either facilitated diffusion or active transport. For example, most cells oxidize glucose to meet their energy needs. A steady supply of glucose is carried into these cells, down the concentration gradient, by facilitated diffusion. Liver cells, however, store glucose, readily converting it to and from glycogen. Although these cells have a high internal concentration of glucose, additional glucose is moved into them, against the concentration gradient, by active transport.

An Example of Active Transport: The Sodium-Potassium Pump

Most animal cells maintain steep concentration gradients of sodium ions (Na^+) and potassium ions (K^+) across the plasma membrane. The concentration of Na^+ outside the cell is as much as 14 times higher than the concentration inside the cell. By contrast, the concentration of K^+ inside the cell is, depending on the cell type, anywhere from 10 to 30 times higher than the concentration on the outside. These concentration gradients, which are important in maintaining osmotic balance and in controlling cell volume, are produced by an active-transport system known as the **sodium-potassium pump.**

The sodium-potassium pump is powered by energy supplied by ATP, which, as you will recall from Chapter 3, is the cell's ready energy currency. A measure of the importance of the sodium-potassium pump to the organism is that more than a third of the ATP used by a resting animal is consumed by this one ion-pumping mechanism.

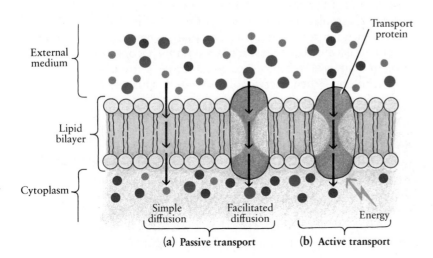

6–8 *Modes of transport through the plasma membrane.* (a) *In simple diffusion (tan spheres) and facilitated diffusion (blue-green spheres), molecules or ions move* down *a concentration gradient. The potential energy of the concentration gradient drives these processes, which are, from the standpoint of the cell, passive.* (b) *In active transport (red spheres), by contrast, molecules or ions are moved* against *a concentration gradient. Energy, most often supplied by ATP, is required to power active transport. Both facilitated diffusion and active transport require the presence of integral membrane proteins, specific for the substance being transported.*

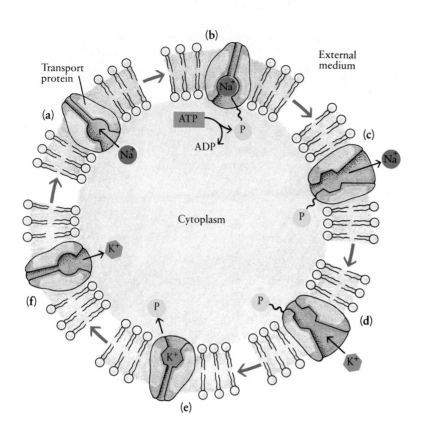

6–9   *A model of the sodium-potassium pump.* (a) *An Na⁺ ion in the cytosol fits precisely into the transport protein (top left).* (b) *The hydrolysis of an ATP molecule then attaches a phosphate group (P) to the protein, releasing ADP. This process results in* (c) *a change in the shape of the protein that causes the Na⁺ to be released outside the cell.* (d) *A K⁺ ion outside the cell is then bound to the transport protein, which in this form provides a better fit for K⁺ than for Na⁺.* (e) *The phosphate group is released from the protein, inducing conversion back to the other shape* (f) *and the release of the K⁺ ion into the cytosol. The protein is now once more ready to transport Na⁺ out of the cell.*

For clarity, only single ions are shown in this diagram. Quantitative studies have shown, however, that each complete pumping sequence, using only a single molecule of ATP, transports three Na⁺ ions out of the cell and two K⁺ ions into the cell. The pump works so rapidly that it can use as many as 100 molecules of ATP per second.

The pumping of Na⁺ and K⁺ ions is accomplished by a transport protein thought to exist in two alternative configurations. One configuration has cavities opening to the inside of the cell, into which Na⁺ ions can fit. The other configuration has cavities opening to the outside, into which K⁺ ions fit. As shown in Figure 6–9, Na⁺ within the cell binds to the transport protein. At the same time, the hydrolysis of ATP to ADP and phosphate (see Figure 3–28, page 67) results in the attachment of a phosphate group to the protein. This triggers its shift to the alternative configuration and the release of the Na⁺ to the outside of the cell. The transport protein is now ready to pick up K⁺, which results in the release of the phosphate group from the protein. This, in turn, causes it to return to the first configuration and to release the K⁺ to the inside of the cell. As this process is repeated, over and over, it maintains the observed gradients of Na⁺ and K⁺ ions across the membrane.

Types of Transport Proteins

Many ingenious models have been proposed to show how transport proteins, such as the sodium-potassium pump, might accept and eject their passengers. A cur-rent model hypothesizes that transport proteins have hydrophilic cores, or channels, through which the transported molecules are squeezed, propelled by changes in the configuration of the protein. These channels are thought to be regions of the tertiary (or, in some cases, quaternary) structure of the proteins formed by amino acids with hydrophilic R groups.

The changes in configuration that accomplish the transport may be triggered directly, by the binding of the molecule to be transported, or indirectly, by the interaction of a receptor on the membrane surface with some other molecule or ion that is not actually transported through the membrane. The plasma membrane of most nerve cells, for example, contains a complex protein that is a receptor for a molecule known as acetylcholine. When acetylcholine binds to this receptor, a channel is opened in another membrane protein, closely associated with the acetylcholine receptor. This channel allows sodium ions (Na⁺) to flow into the cell, down the concentration gradient created and maintained by the action of the sodium-potassium pump. As we shall see in Chapter 35, this flow of Na⁺ ions into the cell is a key event in the propagation of the nerve impulse.

Current evidence indicates that there are at least three general types of transport proteins (Figure 6–10).

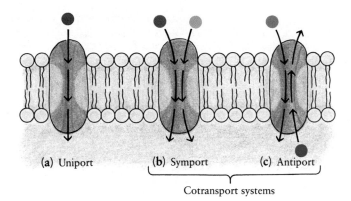

(a) Uniport   **(b)** Symport   **(c)** Antiport

Cotransport systems

6–10   *Three types of transport molecules. (a) In the simplest, known as a uniport, one particular solute is moved directly across the membrane in one direction. (b) In the type of cotransport system known as a symport, two different solutes are moved across the membrane, simultaneously and in the same direction. Often, a concentration gradient involving one of the transported solutes powers the transport of the other solute. (c) In another type of cotransport system, known as an antiport, two different solutes are moved across the membrane, either simultaneously or sequentially, but in opposite directions. The sodium-potassium pump is an example of a cotransport system involving an antiport.*

The simplest transfers one particular kind of molecule or ion directly across the membrane. More complex proteins function as **cotransport systems,** in which the transport of a particular molecule or ion depends upon the simultaneous or sequential transport of a different molecule or ion. In some cotransport systems, both solutes are transported in the same direction. In others, such as the sodium-potassium pump, the two different solutes are transported in opposite directions.

Vesicle-Mediated Transport

The transport proteins that ferry ions and small polar molecules across the plasma membrane cannot accommodate large molecules, such as proteins and polysaccharides, or large particles, such as microorganisms or bits of cellular debris. These large molecules and particles are transported by means of vesicles or vacuoles that bud off from or that fuse with the plasma membrane. For example, many substances are exported from cells in vesicles formed by the Golgi complexes. As we saw in Figure 5–15 (page 99), vesicles move from the Golgi complexes to the surface of the cell. When a vesicle reaches the cell surface, its membrane fuses with the plasma membrane, thus expelling its contents to the outside (Figure 6–11). This process is known as **exocytosis.**

Transport by means of vesicles or vacuoles can also work in the opposite direction. In **endocytosis,** material to be taken into the cell induces the plasma membrane to bulge inward, producing a vesicle enclosing the substance. This vesicle is released into the cytoplasm. Three different forms of endocytosis are known: phagocytosis ("cell-eating"), pinocytosis ("cell-drinking"), and receptor-mediated endocytosis.

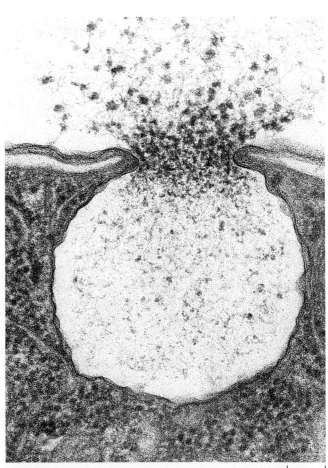

0.1 μm

6–11   *Exocytosis. A secretion vesicle, formed by a Golgi complex of the protist* Tetrahymena furgasoni, *discharges mucus at the cell surface. Notice how the membrane enclosing the vesicle has fused with the plasma membrane.*

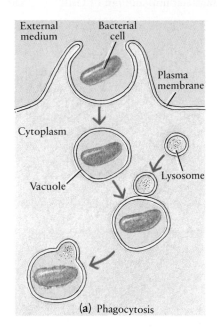

(a) Phagocytosis

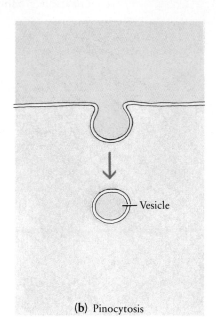

(b) Pinocytosis

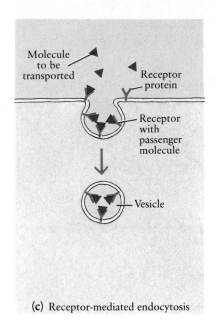

(c) Receptor-mediated endocytosis

6–12   *Three types of endocytosis. (a) In phagocytosis, contact between the plasma membrane and particulate matter, such as a bacterial cell, causes the plasma membrane to extend around the particle, engulfing it in a vacuole. One or more lysosomes then fuse with the vacuole, spilling hydrolytic enzymes into it.*

(b) In pinocytosis, the plasma membrane pouches inward, forming a vesicle around liquid from the external medium that is to be taken into the cell.

(c) In receptor-mediated endocytosis, the molecules to be transported into the cell must first bind to specific receptor proteins. The receptors are either localized in indented areas of the plasma membrane, known as pits, or migrate to such areas after binding the molecules to be transported. When filled with receptors carrying their particular molecules, the pit buds off as a vesicle.

When the substance to be taken into the cell in endocytosis is a solid particle, such as a wood chip (page 108) or a bacterial cell, the process is usually called **phagocytosis** (Figure 6–12a). Many heterotrophic protists, such as amoebas, feed in this way. Similarly, macrophages (Figure 5–20, page 102) and other types of white blood cells in our own bloodstreams engulf bacteria and other invaders in phagocytic vacuoles.

The taking in of liquids, as distinct from particulate matter, is given the special name of **pinocytosis** (Figure 6–12b). It is the same in principle as phagocytosis. Unlike phagocytosis, however, which is carried out only by certain specialized cells and involves specific recognition processes, pinocytosis occurs in all eukaryotic cells, as the cells continually and indiscriminately sip small amounts of fluid from the surrounding medium.

In **receptor-mediated endocytosis**, currently the subject of a great deal of research, particular membrane proteins serve as receptors for specific molecules that are to be transported into the cell (Figure 6–12c). Cholesterol, for example, is carried into animal cells by receptor-mediated endocytosis. As we noted earlier (page 58), cholesterol circulates in the bloodstream in the form of LDL particles, which interact with specific receptors on the cell surface. Binding of LDL particles to the receptor molecules triggers the formation of a ves-

icle that transports the cholesterol molecules into the cell.

Receptors for some substances, such as the hormone insulin, are apparently free to move laterally in the plasma membrane and, when unoccupied, are scattered at random locations on its surface. As the molecules to be transported into the cell bind to the receptors, the receptors move close together. A vesicle forms, and the hormone-laden receptors are carried into the cell. Receptors for other substances, such as LDL particles, appear to be localized in groups in specific areas of the plasma membrane even before binding of the substance to be transported.

In the areas where specific receptors are localized—or to which they migrate, as in the case of insulin receptors—the inner, or cytoplasmic, face of the plasma membrane is densely coated with a peripheral membrane protein known as clathrin. These areas, which are slightly indented, are known as **coated pits.** The vesicles that form from them, containing receptor molecules and their passengers, thus acquire an external, cagelike coating of clathrin. The formation of such a coated vesicle is illustrated in Figure 6–13.

As you can see by studying Figures 6–12 and 6–13, the surface of the membrane facing the interior of a vesicle or vacuole is equivalent to the surface of the

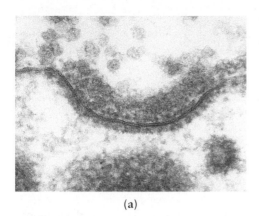

(a)

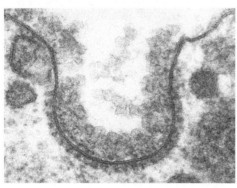

(b)

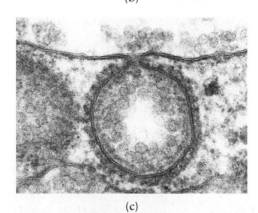

(c)

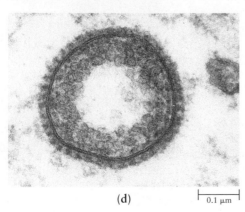

(d)   0.1 μm

6–13   *The formation of a coated vesicle in the developing egg cell of a hen. (a) A coated pit in the plasma membrane is covered on the cytoplasmic face with a latticework of clathrin molecules. The large particles clustered in the shallow pit on the external face are lipoprotein molecules, gathered from the surrounding medium and bound to specific membrane receptors that are associated with the underlying clathrin layer. (b) The pit deepens, and then (c) the plasma membrane closes around the pit to form the vesicle. (d) The completed vesicle with its outer coating of clathrin buds off and moves into the cell. The lipoproteins carried by this coated vesicle will be incorporated into the egg yolk.*

plasma membrane facing the exterior of the cell. Similarly, the surface of the vesicle or vacuole membrane facing the cytoplasm is equivalent to the cytoplasmic surface of the plasma membrane. As we noted in the last chapter, new material needed for expansion of the plasma membrane is transported, ready-made, from the Golgi complexes to the membrane by a process similar to exocytosis. Current evidence indicates that the portions of the plasma membrane used in forming endocytic vesicles or vacuoles are also returned to the membrane in exocytosis, thus recycling the membrane lipids and proteins, including the specific receptor molecules.

Cell-to-Cell Communication

Thus far in our consideration of the transport of substances into and out of cells, we have assumed that individual cells exist in isolation, surrounded by a watery environment. In multicellular organisms, however, this is generally not the case. Cells are organized into **tissues,** groups of specialized cells with common functions. Tissues are further organized in concert to form **organs,** each of which has a structure that suits it for a specific function.

As you might imagine, in multicellular organisms it is essential that individual cells communicate with one another so that they can collaborate to create a harmonious tissue or organ. This communication is accomplished by means of chemical signals—that is, by substances that are transported out of one cell and travel to another cell. When they reach the plasma membrane of the target cell, they may be transported into the cell, by any one of the processes we have considered. Alternatively, they may bind to specific membrane receptors at the surface of the target cell, thereby triggering chemical changes within that cell.

Communication in the Cellular Slime Mold

A cellular communication system of particular interest to biologists, because of the comparative ease with which it can be studied, is seen among a group of organisms known as the cellular slime molds. The slime mold *Dictyostelium discoideum* is an example. At one stage in its life cycle, it exists as a swarm of small individual amoeba-like cells, which divide and grow and feed until their food supply (mostly bacteria) gives out.

At this point, the cells alter both their shape and behavior: they become sausage-shaped and begin to migrate toward the center of the group **(a)**. Eventually, they pile up in a heap.

The heap gradually takes on the form of a multicellular mass somewhat resembling a garden slug **(b)**, slowly migrating and depositing a thick slime sheath that collapses behind it.

The sluglike mass soon stops its migration, gathers itself into a mound **(c)**, and sends up a long stalk at the tip of which a small, shimmering fruiting body forms **(d)**. The fruiting body matures **(e)** and eventually bursts open, releasing a new swarm of tiny cells, and the cycle begins again.

The chemical that spreads from cell to cell to initiate this remarkable sequence of events was first called acrasin, after Acrasia, the cruel witch in Spenser's *Faerie Queene* who attracted men and turned them into beasts. Acrasin was later identified as the chemical compound cyclic AMP (adenosine monophosphate). In recent years, it has become clear that many of the communications among cells in the human body also involve cyclic AMP. As we shall see in Chapter 34, the interaction of a number of vertebrate hormones with their specific receptors in the plasma membrane triggers a sequence of events within the cell in which cyclic AMP plays a central role.

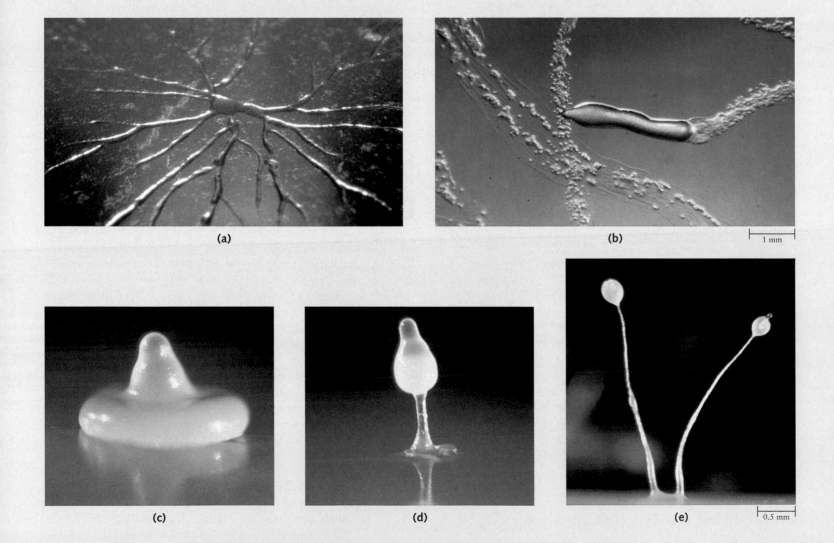

(a)

(b) 1 mm

(c)

(d)

(e) 0.5 mm

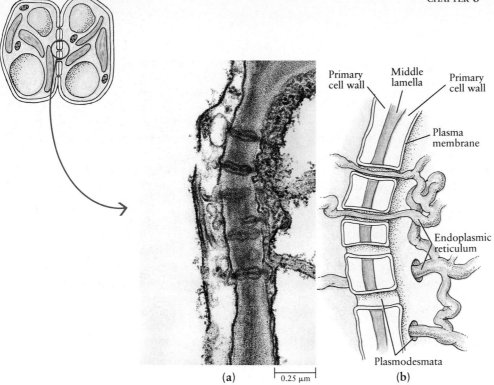

6–14 (a) *Electron micrograph and* (b) *graphic interpretation of plasmodesmata connecting two leaf cells from a corn plant. The adjacent primary walls of the two cells form the wide gray area running vertically through the micrograph. The walls are traversed by the plasmodesmata, each of which is lined by plasma membrane and filled with cytosol. The dark line extending through the center of each plasmodesma is an extension of the endoplasmic reticulum.*

Primary cell wall
Middle lamella
Primary cell wall
Plasma membrane
Endoplasmic reticulum
Plasmodesmata

(a)   0.25 μm   (b)

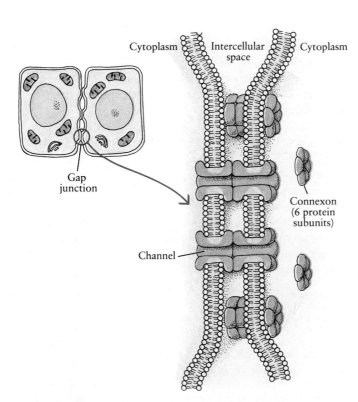

Cytoplasm   Intercellular space   Cytoplasm
Gap junction
Connexon (6 protein subunits)
Channel

6–15 *A model of a gap junction. Embedded in the plasma membranes of two adjacent cells are structures that have been appropriately named "connexons." Each connexon consists of six identical membrane protein subunits, arranged in a hexagonal pattern with a space through the center. Connexons in adjacent plasma membranes abut each other in perfect alignment, providing a channel connecting the cytosol of the two cells.*

Often, however, the cells within a tissue or organ are tightly packed, allowing for direct and intimate contacts of various types between the cells. Among plant cells, which are separated from one another by cell walls, channels called **plasmodesmata** (singular, plasmodesma) traverse the walls, directly connecting the cytoplasm of adjacent cells (Figure 6–14).

In animal tissues, structures known as **gap junctions** permit the passage of materials between cells. These junctions appear as fixed clusters of very small channels surrounded by an ordered array of proteins (Figure 6–15). Ions and small molecules, such as amino acids and ATP, readily pass through these channels from one cell to another.

The transport of materials into and out of cells through the channels of plasmodesmata or gap junctions, through integral membrane proteins, and by means of endocytosis and exocytosis appear superficially to be three quite different processes. They are fundamentally similar, however, in that they all depend on the precise, three-dimensional structure of a great variety of specific protein molecules. These protein molecules not only form channels through which transport can occur but also endow the plasma membrane with the capacity to "recognize" particular molecules. This capacity is the result of billions of years of an evolutionary process that began, as far as we are able to discern, with the formation of that first fragile film around a few organic molecules.

Summary

The plasma membrane regulates the passage of materials into and out of the cell, a function that makes it possible for the cell to maintain its structural and functional integrity. This regulation depends on interactions between the membrane and the materials that pass through it.

One of the principal substances passing into and out of cells is water. Water potential determines the direction in which water moves; that is, water moves from where the water potential is higher to where it is lower.

Water movement takes place by bulk flow and diffusion. Bulk flow is the overall movement of water molecules and dissolved solutes as a group, as when water flows in response to gravity or pressure. The circulation of blood through the human body is an example of bulk flow.

Diffusion involves the random movement of individual molecules or ions and results in net movement down a concentration gradient. It is most efficient when the surface area through which diffusion is occurring is large in relation to volume, when the distance involved is short, and when the concentration gradient is steep. By their metabolic activities, cells maintain steep concentration gradients of many substances across the plasma membrane and between different regions of the cytoplasm.

Osmosis is the diffusion of water through a membrane that permits the passage of water but inhibits the movement of most solutes. Such a membrane is said to be selectively permeable. In the absence of other forces, the net movement of water in osmosis is from a region of lower solute concentration (a hypotonic medium), and therefore of higher water potential, to one of higher solute concentration (a hypertonic medium), and so of lower water potential. Turgor in plant cells is a consequence of osmosis.

Molecules cross the plasma membrane by simple diffusion or are transported by integral membrane proteins. If carrier-assisted transport is driven by the concentration gradient, the process is known as facilitated diffusion. If the transport requires the expenditure of energy by the cell, it is known as active transport. Active transport can move substances against their concentration gradients. One of the most important active-transport systems is the sodium-potassium pump, which maintains sodium ions at relatively low concentration and potassium ions at relatively high concentration in the cytoplasm.

Controlled movement into and out of a cell may also occur by endocytosis or exocytosis, in which the substances are transported in vacuoles or vesicles composed of portions of the plasma membrane. Three forms of endocytosis are phagocytosis, in which solid particles are taken into the cell; pinocytosis, in which liquids are taken in; and receptor-mediated endocytosis, in which molecules or ions to be transported into the cell are bound to specific receptors in the plasma membrane.

In multicellular organisms, communication among cells is essential for coordination of the different activities of the cells in the various tissues and organs. Much of this communication is accomplished by chemical agents that either pass through the plasma membrane or interact with receptors on its surface. Communication may also occur directly, through the channels of plasmodesmata (in plant tissues) or gap junctions (in animal tissues).

Questions

1. Distinguish among the following: bulk flow/diffusion/osmosis; hypotonic/hypertonic/isotonic; facilitated diffusion/active transport; endocytosis/exocytosis; phagocytosis/pinocytosis/receptor-mediated endocytosis; plasmodesmata/gap junctions.

2. What is a concentration gradient? How does a concentration gradient affect diffusion? How does a concentration gradient affect osmosis?

3. When diffusion of dye molecules in a tank of water is complete, random movement of molecules continues at the same rate (as long as the temperature remains the same). However, net movement stops. How do you reconcile these two facts?

4. Why is diffusion more rapid in gases than in liquids? Why is it more rapid at higher temperatures than at lower temperatures?

5. Three funnels have been placed in a beaker containing a solution (see the figure below). What is the concentration of the solution in the beaker? Explain your answer.

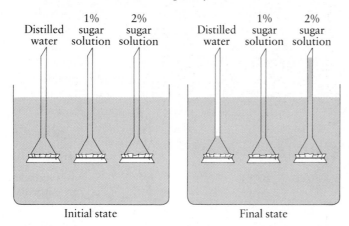

Initial state   Final state

6. Imagine a pouch with a selectively permeable membrane containing a saltwater solution. It is immersed in a dish of fresh water. Which way will the water move? If you add salt to the water in the dish, how will this affect water movement? What living systems exist under analogous conditions? How do you think they maintain water balance?

7. When you forget to water your house plants, they wilt and the leaves (and sometimes the stems) become very limp. What has happened to the plants to cause this change in appearance and texture? Within a few hours after you water your plants, they resume their normal, healthy appearance. What has occurred within the plants to cause this restoration? Sometimes, if you wait too long to water your plants, they never revive. What do you suppose has happened?

8. What limits the passage of water and other polar molecules and ions through the plasma membrane? How do such molecules get into and out of the cell? Describe four possible routes.

9. In what three ways does active transport differ from simple diffusion? How does it differ from facilitated diffusion?

10. Justify the conclusion that differences in ion concentration between cells and their surroundings (see Figure 6–1) indicate that cells regulate the passage of materials across the plasma membrane.

Suggestions for Further Reading

Brown, Michael S., and Joseph L. Goldstein: "How LDL Receptors Influence Cholesterol and Atherosclerosis," *Scientific American,* November 1984, pages 58–66.

Dautry-Varsat, Alice, and Harvey F. Lodish: "How Receptors Bring Proteins and Particles into Cells," *Scientific American,* May 1984, pages 52–58.

Kessin, Richard H., and Michiel M. Van Lookeren Campagne: "The Development of a Social Amoeba," *American Scientist,* vol. 80, pages 556–565, 1992.

Lienhard, Gustav E., Jan W. Slot, David E. James, and Mike M. Mueckler: "How Cells Absorb Glucose," *Scientific American,* January 1992, pages 86–91.

Linder, Maurine E., and Alfred G. Gilman: "G Proteins," *Scientific American,* July 1992, pages 56–65.

Satir, Birgit: "The Final Steps in Secretion," *Scientific American,* October 1975, pages 28–37.

Staehelin, L. Andrew, and Barbara E. Hull: "Junctions between Living Cells," *Scientific American,* May 1978, pages 141–152.

Energetics

The Flow of Energy

Sea otters are solar-powered. Long strands of kelp, the common marine alga in which this otter is floating, capture the sun's rays and convert them to the energy-rich chemicals of the algal cells. Certain animals, such as the sea urchin seen here cradled in the otter's paws, feed on those algal cells. When the otter eats the urchin—an event about to take place—the sun's energy is converted into the molecules, organelles, cells, organs, and activities that, all together, constitute a sea otter.

It takes a lot of sunshine to make an otter. Only 1 to 3 percent of the sunlight that reaches the alga is captured by its cells. In turn, only about 10 percent of the chemical energy stored in the alga is transferred to and stored in the tissues of the sea urchin. And, only 10 percent of that 10 percent is transferred to and stored in the tissues of the otter. As a consequence, you can always expect to find more algae off the coast of California than sea urchins and more sea urchins than sea otters. Similarly, in the forests of eastern North America, you can expect to find more maple leaves than caterpillars of the maple leaf-cutter moth and more caterpillars than birds that feed upon them. And, in East Africa, more grass than zebras and more zebras than lions.

Each of these—alga, sea urchin, sea otter, leaf, caterpillar, bird, grass, zebra, and lion—owes its existence to continued thermonuclear events taking place at the heart of a middle-aged, mid-sized star—our sun—and to the evolution on Earth of mechanisms and devices to capture and use its energy.

Life on Earth depends on a steady flow of energy from the sun. The amount of energy delivered to the Earth by the sun is about 13×10^{23} (the number 13 followed by 23 zeros) calories per year. It is a difficult quantity to imagine. For example, the amount of solar energy striking the Earth every day is about 1.5 billion times greater than the amount of electricity generated in the United States each year.

About 30 percent of this solar energy is reflected back into space as light, and about 20 percent is absorbed by the atmosphere. Most of the remaining 50 percent, which reaches the surface of the Earth, is absorbed and converted to heat. Some of this absorbed heat energy serves to evaporate the waters of the oceans, producing the clouds that, in turn, produce rain and snow. Solar energy, in combination with other factors, is also responsible for the movements of air and water that help set patterns of climate over the surface of the Earth.

A small fraction—less than 1 percent—of the solar energy reaching the Earth's surface becomes, through a series of operations performed by the cells of plants and other photosynthetic organisms, the energy that drives all the processes of life. Living systems change energy from one form to another, transforming the radiant energy captured from the sun into the chemical and mechanical energy used by everything that is alive.

This flow of energy is the essence of life. In fact, one way of viewing evolution is as a competition among organisms for the most efficient use of energy resources. A cell can be best understood as a complex system for transforming energy. At the other end of the biological scale, the structure of an ecosystem (that is, all the living organisms in a particular locale and the nonliving factors with which they interact) or of the biosphere itself is determined by the energy exchanges occurring among the groups of organisms within it.

In this chapter, we shall look first at the general principles governing all energy transformations. Then we shall turn our attention to the characteristic ways in which cells regulate the energy transformations that take place within living systems. In the chapters that follow, the principal and complementary processes of energy flow through the biosphere will be examined—glycolysis and respiration in Chapter 8 and photosynthesis in Chapter 9.

The Laws of Thermodynamics

"Energy" is such a familiar term today that it is surprising to learn that the word first came into common use less than 200 years ago, at the time of the development of the steam engine. It was only then that scientists and engineers began to understand that heat, motion, light, electricity, and the forces holding atoms together in molecules are all different forms of energy—which can be defined most simply as the capacity to cause change, or, as it is often expressed, to do work. This new understanding led to the study of **thermodynamics**—the science of energy transformations—and to the formulation of its laws.

The First Law

The **first law of thermodynamics** states, quite simply: *Energy can be changed from one form to another, but it cannot be created or destroyed.*

Electricity is a form of energy, as is light. Electrical energy can be changed to light energy (for example, by letting an electric current flow through the tungsten wire in a light bulb). Conversely, light energy can be changed to electrical energy, a transformation that is the essential first step of photosynthesis, as we shall see in Chapter 9.

Energy can be stored in various forms and then changed into other forms. In automobile engines, for example, the energy stored in the chemical bonds of gasoline is converted to heat (kinetic energy), which is then partially converted to mechanical movements of the engine parts. Some of the energy is converted back to heat by the friction of the moving engine parts, and some of it leaves the engine in the exhaust products. Similarly, when organisms break down carbohydrates, they convert the energy stored in chemical bonds to other forms. On a summer evening, for example, a firefly converts chemical energy to mechanical energy, to heat, to flashes of light, and to electrical impulses that travel along the nerves of its body. Birds and mammals convert chemical energy into the heat necessary to maintain their body temperature, as well as into mechanical energy, electrical energy, and other forms of chemical energy. According to the first law of thermo-

dynamics, in these energy conversions, and in all others, energy is neither created nor destroyed.

In all energy conversions, however, some useful energy is converted to heat and dissipates. In an automobile engine, for example, the heat produced by friction and lost in the exhaust, unlike the heat confined in the engine itself, cannot produce work—that is, it cannot drive the pistons and turn the gears—because it is dissipated into the surroundings. But it is nevertheless part of the total equation. In a gasoline engine, about 75 percent of the energy originally present in the fuel is transferred to the surroundings in the form of heat—that is, it is converted to increased motion of atoms and molecules in the air. Similarly, the heat produced by the metabolic processes of animals is dissipated into the surrounding air or water.

Thus, another way of stating the first law of thermodynamics is that *in all energy exchanges and conversions, the total energy of the system and its surroundings after the conversion is equal to the total energy before the conversion.* A "system" can be any clearly defined entity—for example, an exploding stick of dynamite, an idling automobile engine, a mitochondrion, a living cell, a sea otter, a forest, or the Earth itself. The "surroundings" consist of everything outside the system.

7–1 *Electrical energy can be converted to light energy, as in these Hong Kong signs, for example. The energy emitted as an electron falls from one energy level to another is a discrete amount, characteristic for each atom. When electricity is passed through a tube of gas, electrons in the atoms of the gas are boosted to higher energy levels. As they fall back, light energy is emitted, producing, for example, the red glow characteristic of neon and the yellow glow characteristic of sodium vapor.*

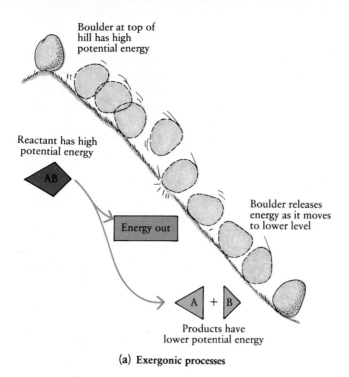

(a) Exergonic processes

(b) Endergonic processes

7–2   (a) *In exergonic processes, whether physical or chemical, the potential energy of the final state is lower than the potential energy of the initial state, and energy is released. Exergonic processes occur spontaneously. Although "spontaneously" has an explosive sound, the word says nothing about the speed of the process—just that it can take place* *without an input of energy.* (b) *In endergonic processes, by contrast, the potential energy of the final state is higher than the potential energy of the initial state. Such processes cannot take place spontaneously. For them to occur, an input of energy is required.*

Stated in this form, the first law of thermodynamics can be used as a simple bookkeeping rule: when we add up the energy income and expenditures for any physical process or chemical reaction, the books must always balance.

The Second Law

The energy that is released as heat in an energy conversion has not been destroyed—it is still present in the random motion of atoms and molecules—but it has been "lost" for all practical purposes. It is no longer available to do useful work. This brings us to the **second law of thermodynamics,** which is the more interesting one, biologically speaking. It predicts the direction of all events involving energy exchanges. Thus it has been called "time's arrow."

The second law states that *in all energy exchanges and conversions—if no energy leaves or enters the system under study—the potential energy of the final state will always be less than the potential energy of the initial state.* The second law is entirely in keeping with everyday experience. A boulder will roll downhill but never uphill. Heat will flow from a hot object to a cold one and never the other way. A ball that is dropped will

bounce—but not back to the height from which it was dropped.

A process in which the potential energy of the final state is less than that of the initial state is one that releases energy. Such a process is said to be **exergonic** ("energy-out"). Only exergonic processes can take place spontaneously—that is, without an input of energy from outside the system (Figure 7–2a). By contrast, a process in which the potential energy of the final state is greater than that of the initial state is one that requires energy. Such processes are said to be **endergonic** ("energy-in"). In order for them to proceed, there must be an input of energy to the system (Figure 7–2b).

Molecules, as you know, contain potential energy, stored in the chemical bonds holding the atoms together. When these bonds are broken in a chemical reaction, the energy they contain can be used to form new chemical bonds or can be released in the form of heat. If the potential energy of the products of the reaction is lower than the potential energy of the reactants, the reaction is exergonic. Energy will be released. If, however, the potential energy of the products is higher than the potential energy of the reactants, the reaction is endergonic. It will occur only with an input of energy.

One factor determining the difference in potential energy between the reactants and the products of a chemical reaction is their heat content. The lower the heat content, the lower the potential energy. However, another factor—the degree of disorder or randomness of the reactants and the products—is also involved. The greater the disorder or randomness, the lower the potential energy.

Previously, we stated the second law of thermodynamics in terms of the energy change between the initial and final states of a process. The second law can also be stated in another, simpler way: *All natural processes tend to proceed in such a direction that the disorder, or randomness, of the universe increases* (Figure 7–3). This disorder, or randomness, is known as **entropy**.

Living Systems and the Second Law

The universe, according to the present model, is a closed system—that is, neither matter nor energy enters or leaves the system. The matter and energy present in the universe at the time of the "big bang" are all the matter and energy it will ever have. Moreover, after each and every energy exchange and transformation, the universe as a whole has less potential energy and more entropy than it did before. In this view, the universe is running down. The stars will flicker out, one by one. Life—any form of life on any planet—will come to an end. Finally, even the motion of individual molecules will cease. However, even the most pessimistic among us do not believe this will occur for another 20 billion years or so.

In the meantime, life can exist *because* the universe is running down. Although the universe as a whole is a closed system, the Earth is not. It is an open system (Figure 7–4), receiving an energy input of about 13×10^{23} calories per year from the sun. Photosynthetic organisms are specialists at capturing the light energy released by the sun as it slowly burns itself out. They use this energy to organize small, simple molecules (water and carbon dioxide) into larger, more complex molecules (sugars). In the process, the captured light energy is stored as chemical energy in the bonds of sugars and other molecules.

Living cells—including photosynthetic cells—can convert this stored energy into motion, electricity, light, and, by shifting the energy from one type of chemical bond to another, into more convenient forms of chemical energy. At each transformation, energy is lost to the surroundings as heat. But before the energy captured from the sun is completely dissipated, organisms use it to create and maintain the complex organization of structures and activities that we know as life.

Initial State   Final State
Copper blocks

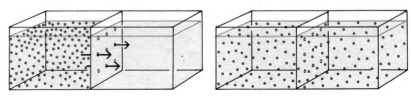

Heat flows from hot body to cold body

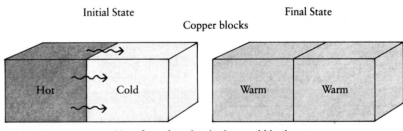

Solute particles move from region of high concentration to region of low concentration

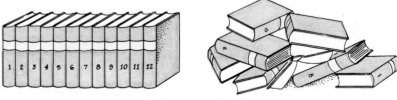

Order becomes disorder

7–3   *Some illustrations of the second law of thermodynamics. In each case, a concentration of energy—in the hot copper block, in the solute particles on one side of a tank, and in the neatly organized books—is dissipated. In nature, processes tend toward randomness, or disorder. Only an input of energy can reverse this tendency and reconstruct the initial state from the final state. Ultimately, however, disorder will prevail, since the total amount of energy in the universe is finite.*

7-4 (a) *A fish bowl is, like the Earth, an open system—matter and energy enter and leave the system. Sunlight passes through the glass, oxygen diffuses into the water at its surface, and food is added by a human caretaker. Heat leaves the system through the glass and through the opening at the top, carbon dioxide diffuses out of the water at its surface, and the waste products of the animals are removed when the bowl is cleaned. Although energy is lost from the system with each and every energy exchange, a steady supply of energy from outside the system maintains its order.*

(b) *If, however, the fish bowl is placed in an opaque container that is sealed and insulated, it becomes a closed system, consisting of the fish bowl, its contents, and the air within the container. Neither matter nor energy can enter or leave the system. For a period of time after the container is sealed, energy continues to be converted from one form to another by the organisms in the fish bowl. With each and every conversion, however, some of the energy is given off as heat and dissipated into the water, the glass, and the air within the container. In time, the system will run down—the organisms will die and their bodies will disintegrate. The order originally present in the system will have become the disorder of individual atoms and molecules moving at random.*

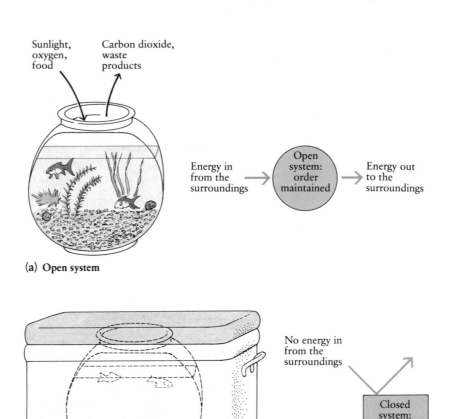

(a) **Open system**

(b) **Closed system**

(a)   (b)   (c)

7-5 *Although energy is dissipated in every conversion from one form to another, considerable work can be accomplished in the process. For example, the transformation of stored chemical energy to mechanical energy can be used by an organism to move into a position to obtain a meal—and thus a* new supply of chemical energy. (a) *A rainbow nudibranch (right), a type of mollusk, encounters a sea anemone (left) that has fully extended its tentacles from the tube that forms the base of its body.* (b) *The nudibranch slowly rears up and prepares to strike. When the nudibranch grasps the anemone's* tentacles with its large jaws, the anemone rapidly retreats into its tube, pulling the nudibranch down with it (c). *In this position, the nudibranch will consume the anemone at its leisure.*

Oxidation-Reduction

You will recall from Chapter 1 that electrons possess differing amounts of potential energy depending on their distance from the atomic nucleus and the attraction of the nucleus for electrons. An input of energy will boost an electron to a higher energy level, but without added energy an electron will remain at the lowest energy level available to it.

Chemical reactions are essentially energy transformations in which the potential energy stored in the chemical bonds of one substance is transferred to other, newly formed chemical bonds of a different substance. In such transfers, electrons shift from one energy level to another. In many reactions, electrons also pass from one atom or molecule to another. These reactions, which are of great importance in living systems, are known as oxidation-reduction (or redox) reactions.

The *loss* of an electron is known as **oxidation,** and the atom or molecule that loses the electron is said to be oxidized. The reason electron loss is called oxidation is that oxygen, which attracts electrons very strongly, is most often the electron acceptor.

Reduction is, conversely, the *gain* of an electron. Oxidation and reduction always take place simultaneously because an electron that is lost by the oxidized atom is accepted by another atom, which is itself reduced in the process.

Redox reactions may involve only a solitary electron, as when sodium loses an electron and becomes oxidized to Na^+, and chlorine gains an electron and is reduced to Cl^- (Figure 7–6a). Often, however, the electron travels with a proton, that is, as a hydrogen atom. In such cases, oxidation involves the removal of hydrogen atoms, and reduction the gain of hydrogen atoms (Figure 7–6b). For example, when glucose is oxidized, hydrogen atoms are lost by the glucose molecule and gained by oxygen:

$$C_6H_{12}O_6 + 6O_2 \longrightarrow 6CO_2 + 6H_2O + \text{Energy}$$

Glucose   Oxygen   Carbon   Water
dioxide

The electrons are moving to a lower energy level, and energy is released.

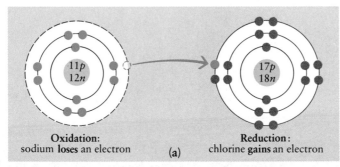

Oxidation:
sodium **loses** an electron   **(a)**   **Reduction:**
chlorine **gains** an electron

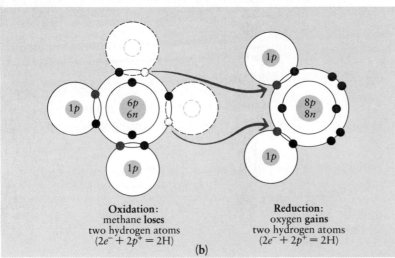

Oxidation:
methane **loses**
two hydrogen atoms
$(2e^- + 2p^+ = 2H)$ **(b)**   **Reduction:**
oxygen **gains**
two hydrogen atoms
$(2e^- + 2p^+ = 2H)$

7–6   (a) *In some oxidation-reduction reactions, such as the oxidation of sodium and the reduction of chlorine, a solitary electron is transferred from one atom to another. Such simple reactions typically involve elements or inorganic compounds.* (b) *In other oxidation-reduction reactions, such as the partial oxidation of methane (CH₄), the electrons are accompanied by protons. In these reactions, which often involve organic molecules, oxidation is the loss of hydrogen atoms, and reduction is the gain of hydrogen atoms. When an oxygen atom gains two hydrogen atoms, as shown here, the product is, of course, a water molecule.*

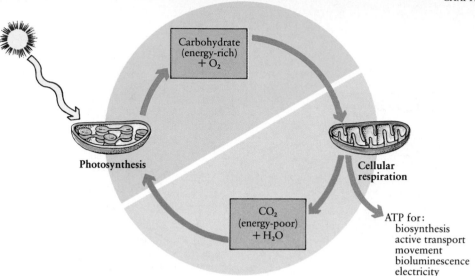

7-7   *The flow of energy through the biosphere. Chloroplasts, present in all photosynthetic eukaryotic cells, capture the radiant energy of sunlight and use it to convert water and carbon dioxide into carbohydrates, such as glucose, starch, and other foodstuff molecules. Oxygen is released as a product of the photosynthetic reactions.*

Mitochondria, present in all eukaryotic cells, carry out the final steps in the breakdown of these carbohydrates and capture their stored energy in ATP molecules. This process, cellular respiration, consumes oxygen and produces carbon dioxide and water, completing the cycling of the molecules.

With each transformation, some energy is dissipated to the environment in the form of heat. Thus the flow of energy through the biosphere is one-way. It can continue only so long as there is an input of energy from the sun.

Conversely, in the process of photosynthesis, hydrogen atoms are transferred from water to carbon dioxide, thereby reducing the carbon dioxide to form glucose:

$$6CO_2 + 6H_2O + Energy \longrightarrow C_6H_{12}O_6 + 6O_2$$

Carbon   Water   Glucose   Oxygen
dioxide

In this case, the electrons are moving to a higher energy level, and an energy input is required to make the reaction occur.

In living systems, the energy-capturing reactions (photosynthesis) and energy-releasing reactions (glycolysis and respiration) are oxidation-reduction reactions. The reduction of carbon dioxide to form a mole of glucose stores 686 kilocalories of energy in the chemical bonds of glucose. Conversely, the complete oxidation of a mole of glucose releases 686 kilocalories of energy.

If this energy were to be released all at once, most of it would be dissipated as heat. Not only would it be of no use to the cell, but the resulting high temperature would be lethal. However, mechanisms have evolved in living systems that regulate these chemical reactions—and a multitude of others—in such a way that energy is stored in particular chemical bonds from which it can be released in small amounts as the cell needs it. These mechanisms, which require only a few kinds of molecules, enable cells to use energy efficiently, without disrupting the delicate balances that characterize a living system. To understand how they work, we must look more closely at the proteins known as enzymes and at the molecule known as ATP.

Enzymes

Most chemical reactions require an initial input of energy to get started. This is true even for exergonic reactions such as the oxidation of glucose or the burning of natural gas. The added energy increases the kinetic energy of the molecules, enabling them to collide with enough force (1) to overcome the repulsion between the electrons surrounding one molecule and those surrounding the other molecule and (2) to break existing chemical bonds within the molecules, making possible the formation of new bonds. The energy that must be possessed by the molecules in order to react is known as the **energy of activation.**

In the laboratory, the energy of activation is usually supplied as heat. But, in a cell, many different reactions are going on at the same time, and heat would affect all of these reactions indiscriminately. Moreover, heat would break the hydrogen bonds that maintain the structure of many of the molecules within the cell and would have other generally destructive effects. Cells get around this problem by the use of **enzymes,** molecules that are specialized to serve as catalysts.

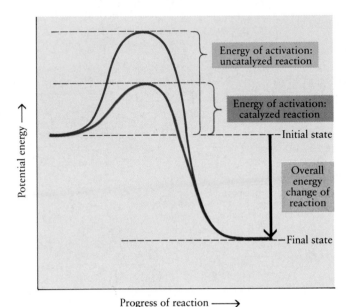

7–8   *In order to react, molecules must possess enough energy—the energy of activation—to collide with sufficient force to overcome their mutual repulsion and to break existing chemical bonds. An uncatalyzed reaction requires more activation energy than a catalyzed one, such as an enzymatic reaction. The lower activation energy in the presence of the catalyst is often within the range of energy possessed by the molecules, and so the reaction can occur at a rapid rate with little or no added energy. Note, however, that the overall energy change from the initial state to the final state is the same with and without the catalyst.*

A **catalyst** is a substance that lowers the activation energy required for a reaction by forming a temporary association with the molecules that are reacting (Figure 7–8). This temporary association brings the reacting molecules close to one another and may also weaken existing chemical bonds, making it easier for new ones to form. As a result, little, if any, added energy is needed to start the reaction, and it goes more rapidly than it would in the absence of the catalyst. The catalyst itself is not permanently altered in the process, and so it can be used over and over again. (In Chinese, the word for "catalyst" is the same as the word for "marriage broker," and the functions are indeed similar.)

Because of enzymes, cells are able to carry out chemical reactions at great speed and at comparatively low temperatures. A single enzyme molecule may catalyze the reaction of tens of thousands of identical molecules in a second. Thus enzymes are typically effective in very small amounts.

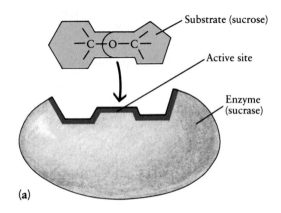

(a)

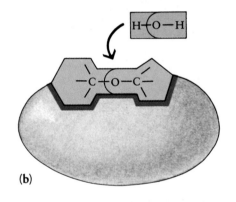

(b)

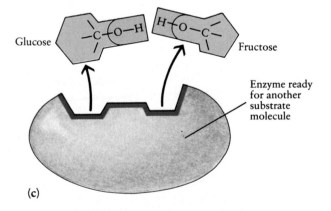

(c)

7–9   *A model of enzyme action. (a, b) Sucrose, a disaccharide, is hydrolyzed to yield a molecule of glucose and a molecule of fructose (c). The enzyme involved in this reaction, sucrase, is specific for this process. As you can see, the active site of the enzyme fits the opposing surface of the sucrose molecule. The fit is so exact that a molecule composed, for example, of two subunits of glucose would not be affected by this enzyme.*

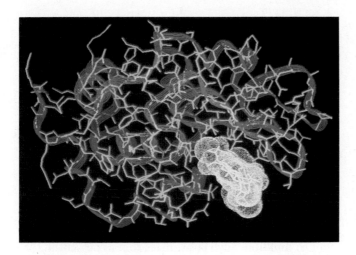

7–10 *A computer-generated model of the enzyme lysozyme. The blue lines represent the covalent bonds linking the atoms in the molecule to one another. The purple ribbon represents the backbone of the polypeptide chain, revealing the folding and bending that produce the tertiary structure. The substrate, a complex sugar, is shown in yellow, nestled into the active site.*

Lysozyme acts on the polysaccharides found in the cell walls of many types of bacteria. It was first detected by Sir Alexander Fleming, better known for his discovery of penicillin. Fleming was suffering from a bad cold, and a droplet fell from his nose onto a culture dish containing bacteria. The bacteria in the vicinity of the droplet were destroyed. Lysozyme was subsequently discovered in tears, saliva, milk, egg white, and other body fluids of many different animals.

Almost 2,000 different enzymes are now known, each of them capable of catalyzing a specific chemical reaction. No cell, however, contains all the known enzymes—different types of cells manufacture different types of enzymes. The particular enzymes that a cell manufactures are a major factor in determining the biological activities and functions of that cell. A cell can carry out a given chemical reaction at a reasonable rate only if it has a specific enzyme that can catalyze that reaction.

The molecule (or molecules) on which an enzyme acts is known as its **substrate.** For example, in the reaction diagrammed in Figure 7–9, sucrose is the substrate, and sucrase is the enzyme.

Enzyme Structure and Function

A few enzymes are RNA molecules. All other enzymes, however, are large, complex globular proteins consisting of one or more polypeptide chains. The polypeptide chains of an enzyme are folded in such a way that they form a groove or pocket on the surface (Figure 7–10). The substrate fits into this groove, which is the site of the reactions catalyzed by the enzyme. This portion of the molecule is known as the **active site.**

Only a few amino acids of the enzyme are involved in any particular active site. Some of these may be adjacent to one another in the primary structure, but often the amino acids of the active site are brought close to one another by the intricate folding of the polypeptide chain that produces the tertiary structure. In an enzyme with a quaternary structure, the amino acids of the active site may even be on different polypeptide chains (Figure 7–11).

7–11 *A model of the digestive enzyme chymotrypsin. This enzyme is composed of three polypeptide chains. The three-dimensional shape of the molecule is a result of a combination of disulfide bridges (yellow) and interactions among the chains and between the chains and the surrounding water molecules. These interactions are based on the polarity of the R groups of the various amino acids. As a result of the bending and twisting of the polypeptide chains, particular amino acids come together in a highly specific configuration to form the active site of the enzyme. Three amino acids known to be part of the active site are shown in blue.*

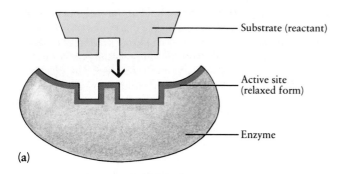

Substrate (reactant)

Active site
(relaxed form)

Enzyme

(a)

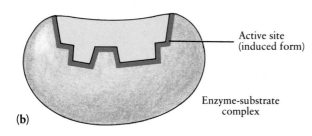

Active site
(induced form)

Enzyme-substrate
complex

(b)

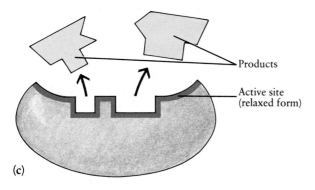

Products

Active site
(relaxed form)

(c)

7–12   *The induced-fit hypothesis.* (a) *The active site is believed to be flexible and* (b) *to adjust its shape to that of the
substrate. This induces a close fit between the active site and
the substrate and may also put some strain on the substrate
molecule, facilitating the reaction* (c).

Recent studies of enzyme structure have suggested
that the active site is flexible. The binding between enzyme and substrate appears to alter the shape of the
enzyme, thus inducing a close fit between the active site
and the substrate (Figure 7–12). It is believed that this
induced fit may put some strain on the reacting molecules and so further facilitate the reaction.

Cofactors in Enzyme Action

The catalytic activity of some enzymes appears to depend only on the interaction between the amino acids
of the active site and the substrate. Many enzymes,
however, require additional nonprotein substances in
order to function. Such substances that are essential for
enzyme activity are known as **cofactors.**

Certain ions are cofactors for particular enzymes.
For example, the magnesium ion (Mg^{2+}) is required in
all enzymatic reactions involving the transfer of a phosphate group from one molecule to another. The phosphate group is usually negatively charged in solution,
and its two negative charges are attracted by the two
positive charges of the magnesium ion. This attraction
holds it in position at the active site of the enzyme.

Nonprotein organic molecules may also function as
enzyme cofactors. Such molecules are called **coenzymes.**
They are bound, either temporarily or permanently, to
the enzyme, usually fairly close to the active site. Many
vitamins—the compounds that humans and other animals cannot synthesize and so must obtain in their
diets—are coenzymes or parts of coenzymes.

Some coenzymes function as electron acceptors in
oxidation-reduction reactions, receiving electrons—
often a pair of electrons accompanied by a hydrogen
ion (that is, a proton)—and then passing them on to
another molecule. There are several different kinds of
electron-accepting coenzymes in any given cell, each
able to hold electrons at a slightly different energy level.
A coenzyme can accept electrons in one particular reaction, and then release them in another reaction. This
enables the cell to capture energy efficiently as electrons
move from high energy levels to lower ones.

Biochemical Pathways

Enzymes characteristically work in series, like workers
on an assembly line. Each enzyme catalyzes one small
step in an ordered series of reactions that together form
a **biochemical pathway** (Figure 7–13). Different biochemical pathways serve different functions in the life
of the cell. For example, one pathway may be involved
in the breakdown of the polysaccharides in bacterial cell
walls, another in the breakdown of glucose, and another in the synthesis of a particular amino acid.

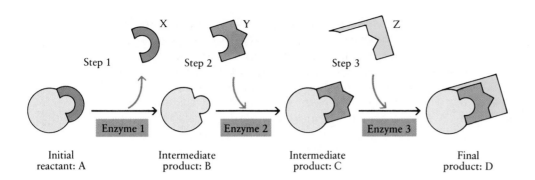

Step 1   Step 2   Step 3

Enzyme 1   Enzyme 2   Enzyme 3

Initial reactant: A   Intermediate product: B   Intermediate product: C   Final product: D

7–13 *A schematic representation of a biochemical pathway. In order to produce the final product (D) from the initial reactant (A), a series of reactions is required. Each reaction is catalyzed by a different enzyme, and each results in a small but significant modification in the substrate molecule. If any step of the pathway is inhibited—either because of a nonfunctioning enzyme or because a substrate is unavailable—the pathway will shut down, and the subsequent reactions of the series will not occur.*

Cells derive several advantages from this sort of arrangement. First, the groups of enzymes making up a common pathway can be segregated within the cell. Some are found in solution, as in the lysosomes, whereas others are embedded in the membranes of particular organelles. The enzymes located in membranes appear to be lined up in sequence, so the product of one reaction moves directly to the adjacent enzyme for the next reaction of the series. A second advantage is that there is little accumulation of intermediate products, since each product tends to be used up in the next reaction along the pathway. A third advantage is a thermodynamic one. If any of the reactions along the pathway are highly exergonic (that is, energy-releasing), they will rapidly use up the products of the preceding reactions, pulling those reactions forward. Similarly, the accumulation of products from the exergonic reactions will push the subsequent reactions forward by increasing the concentrations of the reactants.

The Cell's Energy Currency: ATP Revisited

Glucose and other carbohydrates are storage forms of energy. They are also the forms in which energy is transferred from cell to cell and organism to organism. As we noted in Chapter 3, they are like money in the bank. **ATP** (adenosine triphosphate), however, is like the change in your pocket—it is the cell's immediately spendable energy currency. Almost every energy-requiring transaction in the cell involves ATP. For this reason, ATP is often referred to as the "universal currency" of the cell.

As you know, ATP is made up of the nitrogenous base adenine, the five-carbon sugar ribose, and three phosphate groups (Figure 7–14). The covalent bonds linking these three phosphates to each other, symbolized by a squiggle, ~, are the key to ATP's function.

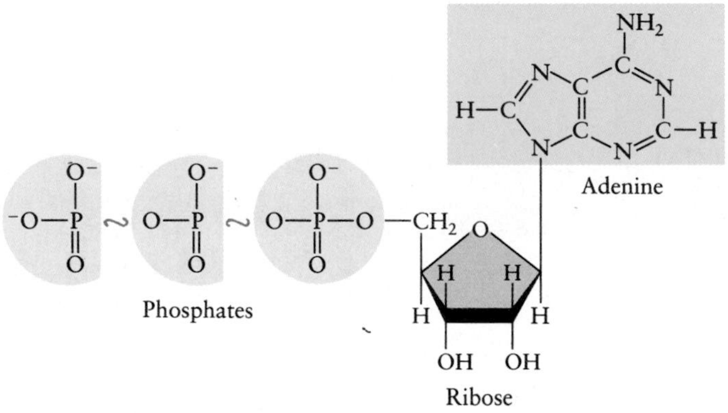

7–14 *The structure of adenosine triphosphate (ATP), the cell's chief energy currency. ATP function depends on the bonds, shown here in blue, between the three phosphate groups in the molecule.*

NH_2

Adenine

Phosphates

CH_2

H   H   H   H

OH   OH

Ribose

These particular covalent bonds, often called "high-energy" bonds, are easily broken, releasing an amount of energy adequate to drive many of the essential reactions of the cell.

When one phosphate group is removed by hydrolysis, the ATP molecule becomes ADP, adenosine diphosphate (see Figure 3–28, page 67). About 7 kilocalories of energy per mole are released. If another phosphate group is removed by hydrolysis, the ADP molecule becomes AMP, adenosine monophosphate, and another 7 kilocalories of energy per mole are released. In most cellular reactions, however, only the first phosphate group is removed. The resulting ADP molecule can be "recharged" to ATP with the addition of a phosphate group and an energy input of 7 kilocalories per mole.

ATP in Action

Coupled Reactions

To understand the role of ATP in living systems, we must return briefly to the concept of the chemical bond. Because a chemical bond is a stable configuration of electrons, reacting molecules must possess a certain amount of energy before existing chemical bonds can be broken and new bonds formed. This is the energy of activation. Because of enzymes, which reduce the required energy of activation, the reactions essential to life are able to proceed at an adequate rate. However,

in biosynthetic reactions, in which larger, more complex molecules are synthesized from smaller, simpler molecules, additional energy is required.

In a biosynthetic reaction—for example, the formation of a disaccharide from two monosaccharide molecules—the electrons forming the chemical bonds of the product are at a higher energy level than the electrons in the bonds of the reactants. In other words, the potential energy of the product is greater than the potential energy of the reactants. Thus the reaction is endergonic—it can occur only with an input of energy. Cells get around this difficulty by means of **coupled reactions,** in which endergonic reactions are linked to exergonic reactions that provide a surplus of energy, enabling the overall process to proceed (Figure 7–15). The molecule that most frequently supplies the energy in coupled reactions is ATP.

Hydrolysis and Phosphorylation

In living cells, ATP is sometimes directly hydrolyzed to ADP plus phosphate, releasing energy for a variety of activities. ATP hydrolysis provides, for example, a means for producing heat, as in those animals, such as birds and mammals, that generally maintain a high and constant body temperature.

Enzymes catalyzing the hydrolysis of ATP are known as **ATPases.** A variety of different ATPases have been identified. The protein "arms" on the microtubules in cilia and flagella (page 103), for example, are ATPase

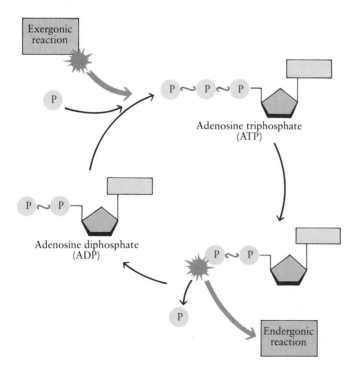

7–15 *In living systems, endergonic reactions, such as biosynthetic reactions, are powered by the energy released in exergonic reactions to which they are coupled. In most coupled reactions, ATP is the intermediate that carries energy from one reaction to the other.*

Exergonic reaction

Adenosine triphosphate (ATP)

Adenosine diphosphate (ADP)

Endergonic reaction

(a)

(b)

7–16 *Living organisms use the energy stored in the phosphate bonds of ATP for a variety of purposes. (a) In the skunk cabbage, a common plant in bogs and marshy areas of the northeastern United States, ATP is hydrolyzed to ADP, releasing energy in the form of heat. The plant produces enough heat to melt surrounding snow or ice, while maintaining a nearly constant internal temperature of about 22°C (72°F). (b) Flashlight fish contain specialized organs, located just below the eyes, that are filled with bioluminescent bacteria—as many as 10 billion cells in each milliliter of fluid. These bacteria transform ATP energy into light energy, thereby glowing in the dark. The French divers who discovered the first flashlight fish, with its blinking headlights, called it "le petit Peugeot." Flashlight fish live in reefs in the Indian Ocean and the Red Sea. These fish were photographed off the Comoro Islands, which are located at the northern end of the channel separating Madagascar from the African mainland.*

molecules, catalyzing the energy release that causes the microtubules to move past one another. Many of the proteins that move molecules and ions through cell membranes against a concentration gradient (page 117) are not only transport proteins but also ATPases, releasing energy to power the transport process.

Usually, however, the terminal phosphate group of ATP is not simply removed but is transferred to another molecule. This addition of a phosphate group is known as **phosphorylation.** Phosphorylation reactions transfer some of the energy of the phosphate group in the ATP molecule to the phosphorylated compound, which, thus energized, participates in a subsequent reaction.

Consider, for example, the formation of sucrose, a disaccharide synthesized in plants from the monosaccharides glucose and fructose. This process requires an input of energy to raise electrons to the higher energy level of the new chemical bond. The energy is supplied by the phosphorylation of the glucose and fructose molecules:

$$ATP + Glucose \longrightarrow Glucose\ phosphate + ADP$$

$$ATP + Fructose \longrightarrow Fructose\ phosphate + ADP$$

In each of these reactions, some of the energy made available by the conversion of ATP to ADP is conserved by the transfer of the terminal phosphate group of an ATP molecule to a sugar molecule. The glucose and fructose molecules thus become energized.

Next, glucose phosphate reacts with fructose phosphate to form sucrose:

$$Glucose\ phosphate + Fructose\ phosphate \longrightarrow$$
$$Sucrose + 2\ Phosphates$$

In this second step, the phosphate groups are released from the glucose and fructose. Much of the energy made available by their release (energy originally derived from ATP) is used to form the bond between glucose and fructose, producing sucrose.

Where does the ATP come from? As we shall see in the next chapter, energy released in the cell's oxidation reactions, such as the breakdown of glucose, is used to "recharge" the ADP molecule to ATP by the addition of a phosphate group. Thus the ATP/ADP system serves as a universal energy-exchange system, shuttling between energy-releasing processes and energy-requiring ones.

Summary

Living systems convert energy from one form to another as they carry out essential functions of maintenance, growth, and reproduction. In these energy conversions, as in all others, some useful energy is lost to the surroundings at each step.

The laws of thermodynamics govern transformations of energy. The first law states that energy can be converted from one form to another but cannot be created or destroyed. The second law of thermodynamics states that in the course of energy conversions, the potential energy of the final state will always be less than the potential energy of the initial state. Stated differently, all natural processes tend to proceed in such a direction that the disorder, or randomness, of the universe increases. To maintain the organization on which life depends, living systems must have a constant supply of energy to overcome the tendency toward increasing disorder. The sun is the original source of this energy.

The energy transformations in living cells involve the movement of electrons from one energy level to another and, often, from one atom or molecule to another. Reactions in which electrons move from one atom to another are known as oxidation-reduction reactions. An atom or molecule that loses electrons is oxidized, and one that gains electrons is reduced.

Enzymes are the catalysts of biological reactions, lowering the energy of activation and thus enormously increasing the rate at which reactions take place. With a few exceptions, enzymes are large, globular protein molecules folded in such a way that particular groups of amino acids form an active site. The reacting molecule or molecules, known as the substrate, fit precisely into this active site. Although the shape of an enzyme may change temporarily in the course of a reaction, it is not permanently altered.

Many enzymes require cofactors, which may be simple ions or nonprotein organic molecules known as coenzymes. Coenzymes often serve as electron carriers, with different coenzymes holding electrons at slightly different energy levels.

Enzymes generally work in series. Each enzyme in the series catalyzes one specific reaction in the sequence of reactions that constitute a biochemical pathway. The stepwise reactions of biochemical pathways enable cells to carry out their chemical activities with remarkable efficiency in terms of both energy and materials.

ATP supplies the energy for most of the activities of the cell. The ATP molecule consists of the nitrogenous base adenine, the five-carbon sugar ribose, and three phosphate groups. The three phosphate groups are linked by two covalent bonds that are easily broken, each yielding about 7 kilocalories of energy per mole. Cells are able to carry out endergonic (energy-requiring) reactions by coupling them with exergonic (energy-yielding) reactions that provide a surplus of energy. Such coupled reactions usually involve ATP.

Questions

1. Distinguish among the following: the first law of thermodynamics/the second law of thermodynamics; exergonic reaction/endergonic reaction; oxidation/reduction; active site/substrate; ATP/ADP/AMP.

2. At present, at least four types of energy conversions are going on in your body. Name them.

3. The laws of thermodynamics apply only to closed systems, that is, to systems in which no energy is entering or leaving. Is a terrarium ordinarily a closed system? Could you convert it to one? A spaceship may or may not be a closed system, depending on certain features of its design. What would these features be? Is the Earth a closed system?

4. Explain why it is that living systems, despite appearances, are not in violation of the second law of thermodynamics.

5. What is there about the orderliness of a living organism that most significantly distinguishes it from the orderliness of a machine, such as a computer or the telephone system?

6. What is the basis for the specificity of enzyme action? What is the advantage to the cell of such specificity? What might be its disadvantages to the cell?

7. When a plant does not have an adequate supply of an essential mineral, such as magnesium, it is likely to become sickly and may die. When an animal is deprived of a particular vitamin in its diet, it too is likely to become ill and may die. What is a reasonable explanation of such phenomena?

8. Some human societies use a barter system for the exchange of goods and services. However, all complex societies have some form of monetary exchange. What are the advantages of a monetary exchange? Relate your answer to the ATP/ADP system.

9. In the photo at the right, why are there more plants than wildebeest and more wildebeest than cheetahs? (Explain in terms of thermodynamics.)

Suggestions for Further Reading

Alberts, Bruce, Dennis Bray, Julian Lewis, Martin Raff, Keith Roberts, and James D. Watson: *Molecular Biology of the Cell*, 3d ed., Garland Publishing, Inc., New York, 1994.

This outstanding cell biology text includes a clear, up-to-date discussion of our current understanding of the processes that occur in mitochondria and chloroplasts. The authors' discussion of the experimental procedures used to study these processes is especially helpful.

Darnell, James, Harvey F. Lodish, and David Baltimore: *Molecular Cell Biology*, 2d ed., W. H. Freeman and Company, New York, 1990.

A comprehensive treatment of modern cell biology, richly illustrated with diagrams and micrographs. Chapters 15 and 16 of this outstanding text are devoted to a thorough explication of our current understanding of the processes that occur in mitochondria and chloroplasts.

Dressler, David, and Huntington Potter: *Discovering Enzymes*, W. H. Freeman and Company, New York, 1990.

This beautifully illustrated volume from the Scientific American Library *provides both a history of the science of biochemistry and a clear presentation of our current knowledge of enzyme structure and function in living organisms.*

Karplus, Martin, and J. Andrew McCammon: "The Dynamics of Proteins," *Scientific American*, April 1986, pages 42–51.

Knowles, Jeremy R.: "Tinkering with Enzymes: What Are We Learning?" *Science*, vol. 236, pages 1252–1258, 1987.

Kornberg, Arthur: *For the Love of Enzymes: The Odyssey of a Biochemist*, Harvard University Press, Cambridge, Mass., 1991.*

Kornberg, who began his scientific career with studies of jaundice in World War II army recruits, went on to make major contributions to our understanding of enzymes. He is a wonderful writer, and this book is enriched by the clarity of his explanations, by his wise reflections on the scientific enterprise, and by his sense of humor.

Lehninger, Albert L., David L. Nelson, and Michael M. Cox: *Principles of Biochemistry*, 2d ed., Worth Publishers, Inc., New York, 1993.

Lehninger was one of the foremost experts on cellular energetics. This updated version of his classic introductory text is enriched by his vast experience and thorough understanding of the processes by which cells provide themselves with energy.

Mathews, Christopher K., and K. E. van Holde: *Biochemistry*, The Benjamin/Cummings Publishing Company, Redwood City, Calif., 1990.

A clearly written and beautifully illustrated textbook. Chapter 3 covers the principles of cellular energetics, and Chapters 10 and 11 are devoted to enzymes and the regulation of their activity.

Neurath, Hans: "Evolution of Proteolytic Enzymes," *Science*, vol. 224, pages 350–357, 1984.

Prescott, David M.: *Cells: Principles of Molecular Structure and Function*, Jones and Bartlett Publishers, Boston, 1988.

An up-to-date, yet concise, textbook of cell biology, written for a first course at the undergraduate level. Chapter 4 is devoted to energy flow and metabolism.

Scientific American: *Molecules to Living Cells*, W. H. Freeman and Company, New York, 1980.*

A collection of articles from Scientific American. *Chapters 5 and 6 are concerned with the structure and function of enzymes.*

Stryer, Lubert: *Biochemistry*, 3d ed., W. H. Freeman and Company, New York, 1988.

A good introduction to cellular energetics, handsomely illustrated.

Weiss, Rick: "Blazing Blossoms: Investigating the Metabolic Machinations of Heat-generating Plants," *Science News*, June 24, 1989, pages 392–394.

* Available in paperback.

ATP is the universal energy currency in living systems. In eukaryotes, such as this winning example, most of the body's ATP is generated in specialized organelles, the mitochondria. Muscle cells, like other cells with high energy requirements, abound in mitochondria. Powered by the ATP produced in their mitochondria, muscle cells are capable of contraction. It is their contractions that make possible the movements of walking, running, swimming, flying—and diving.

With the remarkable airborne performance shown here, Fu Mingxia of China won the 1992 Olympic gold medal in the women's 10-meter platform diving event. At the time, she was only thirteen years old.

In addition to providing the energy that made Fu Mingxia's dive possible, ATP was also simultaneously powering the many chemical activities taking place within the other cells of her body and providing the heat that kept her internal temperature well above that of the surrounding air and water.

The chemical energy stored in Fu Mingxia's ATP molecules came from the oxidation of glucose and other organic molecules. Ultimately, however, Fu Mingxia, like the rest of us, operates on solar power. And, evolution, somewhat like the Olympics, can be viewed as a contest in which the victors are those individuals who capture and use energy most efficiently.

How Cells Make ATP: Glycolysis and Respiration

In living organisms, the principal energy carrier is ATP. It participates in a great variety of cellular events, from the biosynthesis of organic molecules, to the flick of a cilium, the flashing of a firefly on a summer evening, the twitch of a muscle, or the active transport of a molecule across the plasma membrane. In this chapter, we shall show how a cell breaks down carbohydrates and captures a portion of the released energy in the terminal phosphate bonds of ATP. The process begins with the glucose molecule, the form in which carbohydrates usually enter the cell.

An Overview of Glucose Oxidation

The oxidation of glucose (and other carbohydrates) is complicated in detail, but simple in its overall design. **Oxidation,** as we saw in the last chapter, is the loss of electrons. **Reduction** is the gain of electrons. In the oxidation of glucose, the molecule is split apart, and the hydrogen atoms (that is, electrons and their accompanying protons) are removed from the carbon atoms and combined with oxygen, which is thereby reduced. The electrons go from higher energy levels to lower energy levels, and energy is released. The summary equation for this process is

Glucose + Oxygen $\longrightarrow$
$$\text{Carbon dioxide} + \text{Water} + \text{Energy}$$

Or,

$$C_6H_{12}O_6 + 6O_2 \longrightarrow$$
$$6CO_2 + 6H_2O + 686 \text{ kilocalories/mole}$$

Each time we inhale, we bring into our bodies the oxygen required for this process. Each time we exhale, we expel the carbon dioxide that is formed as a by-product.

About 40 percent of the energy released by the oxidation of glucose is used to convert ADP to ATP. As you will recall, about 75 percent of the energy in gasoline is "lost" as heat by an automobile engine, and only 25 percent is converted to useful forms of energy. The living cell is significantly more efficient. Its greater efficiency is largely due to the fact that the energy release occurs over the course of a series of reactions, each of which involves only a small change in energy.

The oxidation of glucose takes place in two major stages (Figure 8–1). The first is known as **glycolysis**— the lysis, or splitting, of glucose. The second is **respiration,** which, in turn, consists of two stages: the **Krebs cycle** and **terminal electron transport.** Glycolysis occurs in the cytosol, and the two stages of respiration take place within the mitochondria.

In glycolysis and the Krebs cycle, hydrogen atoms are removed from the carbon backbone of the glucose molecule and are passed to coenzymes that function as electron carriers. The first of these is nicotinamide adenine dinucleotide, abbreviated NAD^+. (As you can tell from its name, the molecule consists of two nucleotides, each containing a nitrogenous base, a sugar, and a phosphate group.) NAD^+ can accept a proton and two electrons, becoming reduced to **NADH.** The second molecule is flavin adenine dinucleotide, abbreviated **FAD.** FAD can accept two hydrogen atoms (that is, two protons and two electrons), being reduced to $FADH_2$. In glycolysis and the Krebs cycle, NAD^+ and FAD accept electrons and protons and are thus reduced.

In the final stage of respiration, NADH and $FADH_2$ give up their electrons to an **electron transport chain.**

8–1   *An overview of the oxidation of glucose. In glycolysis, glucose is split into pyruvic acid. A small amount of ATP is synthesized from ADP and phosphate, and a few electrons (e^-) and their accompanying protons (H^+) are transferred to coenzymes that function as electron carriers.*

In the presence of oxygen, the pyruvic acid is fed into the Krebs cycle. In the course of this cycle, additional ATP is synthesized and more electrons and protons are transferred to electron carriers. The electron carriers then transfer the electrons to an electron transport chain in which the electrons drop, step by step, to lower energy levels. As they do so, considerably more ATP is synthesized. At the end of the electron transport chain, the electrons reunite with protons and combine with oxygen to form water.

In the absence of oxygen, the pyruvic acid is converted to either lactic acid or ethanol. This process, known as fermentation, produces no ATP, but it does regenerate the electron carriers that are necessary for glycolysis to continue.

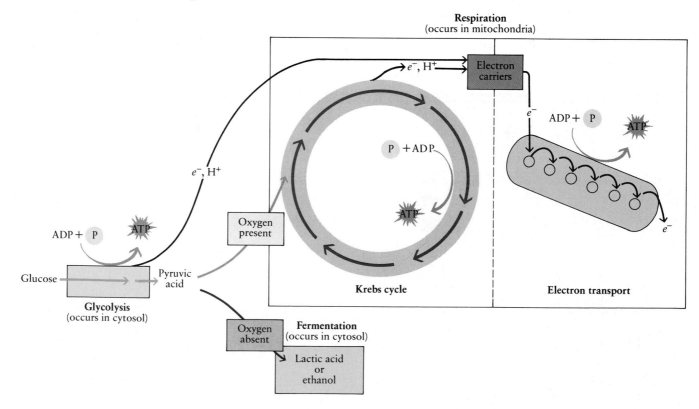

8–2 *Ancient Egyptian wall paintings, such as the one shown here, are the earliest historical record of wine-making. They have been dated to about 5,000 years ago. However, recently discovered pottery fragments, stained with wine, suggest that the Sumerians had mastered the art of wine-making at least 500 years before the Egyptians.*

The electrons are then passed "downhill" along a series of electron carriers. As the electrons drop, step by step, to lower energy levels, the energy released is used to power the formation of ATP from ADP and phosphate. When the electrons reach their lowest energy level, they combine with protons (H^+) and oxygen to form water.

In the presence of oxygen (an **aerobic** environment), the complete oxidation of a molecule of glucose produces about 38 molecules of ATP. These convenient packages of energy are then available to power the many activities of the cell.

In the absence of oxygen (an **anaerobic** environment), respiration cannot occur. Such a situation exists, for example, in muscle cells depleted of oxygen by strenuous exercise and in kegs containing yeast cells and fruit juices. Glycolysis, however, can occur, in conjunction with a process known as **fermentation**. The energy yield is only two molecules of ATP for each molecule of glucose, but this is enough to supply the immediate needs of muscle cells until the oxygen supply can be replenished. And, it is enough to keep yeast cells alive and well while their enzymes carry out the reactions that produce wine, appreciated by *Homo sapiens* (and perhaps other animals as well) since the dawn of history (Figure 8–2).

Now that you have the general outline in mind, let us look at the details of glycolysis, respiration, and fermentation.

Glycolysis

In glycolysis, the six-carbon glucose molecule is split into two molecules of a three-carbon compound known as **pyruvic acid** (Figure 8–3). Four hydrogen atoms (that is, four electrons and four protons) are removed from the glucose molecule in this process. The electrons

and two of the protons are accepted by NAD^+ molecules, while the other two protons remain in solution as hydrogen ions (H^+). Approximately 143 kilocalories are released per mole of glucose during glycolysis. The other 543 kilocalories remain stored in the bonds of the pyruvic acid molecules.

Glycolysis exemplifies the way the biochemical processes of a living cell proceed in small sequential steps. It takes place in a series of nine reactions, each catalyzed by a specific enzyme. Do *not* try to memorize the steps of glycolysis, but, as we describe each step (beginning at the top of the next page), look closely at Figure 8–4 to see how a series of small changes can add up to a major transformation. Notice especially how the carbon backbone of the glucose molecule is dismembered and its atoms rearranged step by step. Note also the formation of ATP from ADP and of NADH and H^+ from NAD^+. *ATP and NADH represent the cell's net energy harvest from glycolysis.*

8–3 *In glycolysis, the six-carbon glucose molecule is split into two molecules of a three-carbon compound known as pyruvic acid. In the course of glycolysis, four hydrogen atoms are removed from the original glucose molecule.*

8–4   *The steps of glycolysis.*

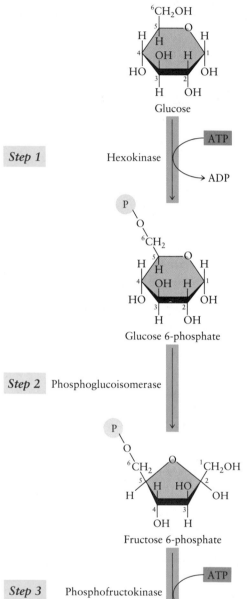

Step 1

Hexokinase

ATP

ADP

Glucose

Glucose 6-phosphate

Step 2   Phosphoglucoisomerase

Fructose 6-phosphate

Step 3   Phosphofructokinase

ATP

ADP

Fructose 1,6-bisphosphate

Step 1. The first steps in glycolysis require an input of energy. This energy is supplied by coupling these steps to the ATP/ADP system. The terminal phosphate group is transferred from an ATP molecule to the carbon in the sixth position of the glucose molecule, to make glucose 6-phosphate. The reaction of ATP with glucose to produce glucose 6-phosphate and ADP is an exergonic reaction. Some of the energy it releases is conserved in the chemical bond linking the phosphate to the glucose molecule, which thus becomes energized. This reaction is catalyzed by a specific enzyme (hexokinase), and each of the reactions that follows is similarly catalyzed by a specific enzyme.

Step 2. The molecule is reorganized. The six-sided ring characteristic of glucose becomes the five-sided fructose ring. (As you know, glucose and fructose both have the same number of atoms—$C_6H_{12}O_6$—and differ only in the arrangements of these atoms.) This reaction is pushed forward by the accumulation of glucose 6-phosphate and the removal of fructose 6-phosphate as the latter enters Step 3.

Step 3. In this step, which is similar to Step 1, fructose 6-phosphate gains a second phosphate by the investment of another ATP. The added phosphate is bonded to the first carbon, producing fructose 1,6-bisphosphate, that is, fructose with phosphates in the 1 and 6 positions. Note that in the course of the reactions thus far, two molecules of ATP have been converted to ADP and no energy has been recovered.

Step 4. The six-carbon sugar molecule is split in half, producing two three-carbon molecules, dihydroxyacetone phosphate and glyceraldehyde phosphate. These two molecules can be converted from one to the other by the enzyme isomerase. However, because the glyc-

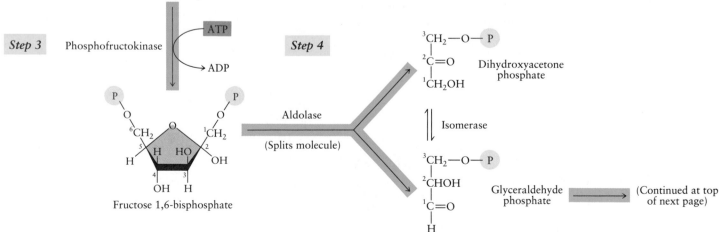

Step 4

Aldolase

(Splits molecule)

Dihydroxyacetone phosphate

Isomerase

Glyceraldehyde phosphate   (Continued at top of next page)

eraldehyde phosphate is used up in subsequent reactions, all of the dihydroxyacetone phosphate is eventually converted to glyceraldehyde phosphate. Thus, *the products of all subsequent steps must be counted twice to account for the fate of one glucose molecule.* With the completion of Step 4, the preparatory reactions are complete.

Step 5. Glyceraldehyde phosphate molecules are oxidized—that is, hydrogen atoms with their electrons are removed—and NAD^+ is reduced to NADH and H^+ (a total of two molecules of NADH and two H^+ ions per molecule of glucose). This is the first reaction from which the cell harvests energy. Some of the energy from this oxidation reaction is also conserved in the attachment of a phosphate group to the 1 position of the molecule, forming 1,3-bisphosphoglycerate. The properties of this new bond are similar to those of the phosphate bonds of ATP, as indicated by the squiggle.

Step 6. The phosphate in the 1 position is released from the bisphosphoglycerate molecule and used to recharge a molecule of ADP (a total of two molecules of ATP per molecule of glucose). This is an exergonic reaction, and it pulls all the preceding reactions forward.

Step 7. The remaining phosphate group is enzymatically transferred from the 3 position to the 2 position.

Step 8. In this step, a molecule of water is removed from the three-carbon compound. The resulting internal rearrangement of the molecule concentrates energy in the vicinity of the phosphate group.

Step 9. The phosphate is transferred to a molecule of ADP, forming another molecule of ATP (again, a total of two molecules of ATP per molecule of glucose). This is also an exergonic reaction, and it pulls forward the preceding two reactions (Steps 7 and 8).

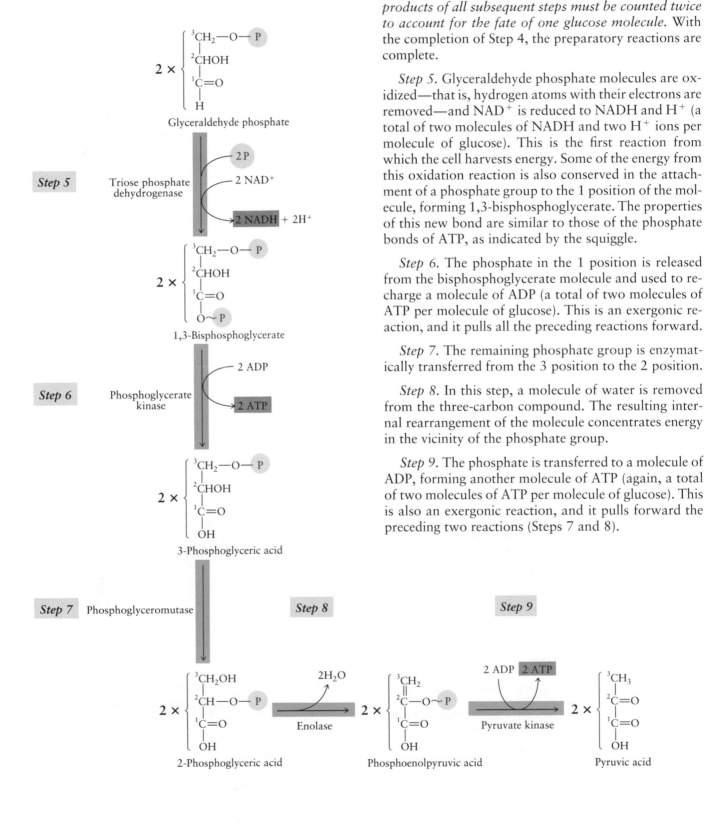

Summary of Glycolysis

The series of reactions we have just described is carried out by virtually all living cells—from the simplest prokaryote to the eukaryotic cells of our own bodies. The complete sequence, which is summarized in Figure 8–5, begins with one molecule of glucose. Energy is invested at Steps 1 and 3 by the transfer of a phosphate group from an ATP molecule—one at each step—to the sugar molecule. The six-carbon molecule splits at Step 4, and from this point on, the sequence yields energy. At Step 5, NAD$^+$ takes energy from the system and is reduced to NADH and H$^+$. At Steps 6 and 9, ADP takes energy from the system, becoming phosphorylated to ATP.

To sum up: The energy from the phosphate bonds of two ATP molecules is needed to initiate the glycolytic sequence. Subsequently, two NADH molecules are produced from two NAD$^+$ and four ATP molecules from four ADP plus four phosphates ("P$_i$" in the following equation—the "i" stands for "inorganic").

$$\text{Glucose} + 2\text{ATP} + 4\text{ADP} + 2\text{P}_i + 2\text{NAD}^+ \longrightarrow$$
$$2 \text{ Pyruvic acid} + 2\text{ADP} + 4\text{ATP} + 2\text{NADH} + 2\text{H}^+$$

Thus the net harvest—the energy recovered—is two molecules of ATP and two molecules of NADH per molecule of glucose. Simultaneously, the glucose molecule has been converted to two molecules of pyruvic acid. These molecules still contain a large proportion of the energy that was stored in the original glucose molecule.

Respiration

In the presence of oxygen, the pyruvic acid produced in glycolysis is broken down completely, to carbon dioxide and water. This process, respiration, takes place in two stages: the Krebs cycle and terminal electron transport. In eukaryotic cells, these reactions take place within the mitochondria.

8–5   *A summary of the two stages of glycolysis. The first stage requires an energy investment of 2 ATP. This stage ends with the splitting of the six-carbon sugar molecule into two three-carbon sugar molecules. The second stage produces an energy yield of 4 ATP and 2 NADH—a substantial return on the original investment.*

Compounds other than glucose, including glycogen, starch, various disaccharides, and a number of monosaccharides, can undergo glycolysis once they have been converted to glucose 6-phosphate or fructose 6-phosphate.

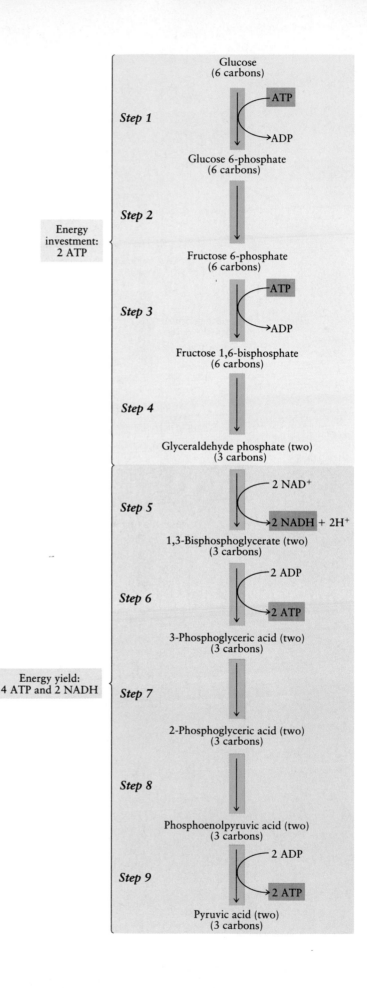

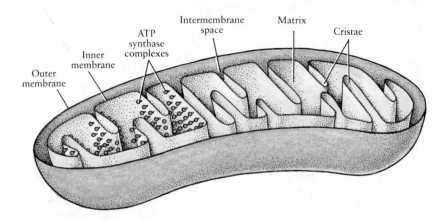

8–6 *A mitochondrion is surrounded by two membranes. The inner membrane folds inward to make a series of shelves, the cristae. Many of the enzymes and electron carriers involved in respiration are built into the membrane forming the cristae. Among these enzymes are the ATP synthase complexes, which, as we shall see, play a critical role in the formation of ATP in the final stage of respiration. The inner membrane encloses a dense solution, the matrix, which contains enzymes involved in the earlier stages of respiration, plus coenzymes, phosphates, and other solutes. The space between the inner membrane and the outer membrane (the intermembrane space) contains a solution of different composition.*

The Structure of the Mitochondrion

A mitochondrion, as we saw in Chapter 5, is surrounded by two membranes, each of which is a phospholipid bilayer. The outer membrane is smooth, but the inner membrane folds inward, forming a convoluted series of shelves known as **cristae** (Figure 8–6). Within the inner compartment of the mitochondrion, bathing the cristae, is a dense solution known as the **matrix.** The matrix contains enzymes, coenzymes, water, phosphates, and other molecules involved in respiration. Some of the enzymes of the Krebs cycle are in solution in the matrix. Other Krebs cycle enzymes and the enzymes and other components of the electron transport chain are built into the cristae.

The outer membrane of the mitochondrion is permeable to most small molecules, and the solution in the space between the inner and outer membranes is similar in composition to the cytosol. The inner membrane, however, permits the passage of only certain molecules, such as pyruvic acid, ADP, and ATP. It restrains the passage of other molecules and ions, including H^+ ions (protons). As we shall see, this selective permeability of the inner membrane is critical to the ability of the mitochondria to harness the power of respiration to the production of ATP.

A Preliminary Step: The Oxidation of Pyruvic Acid

Pyruvic acid passes from the cytosol, where it is produced by glycolysis, to the matrix of the mitochondrion, crossing the outer and inner membranes. Before entering the Krebs cycle, the three-carbon pyruvic acid molecule is oxidized (Figure 8–7). The first carbon and its attached oxygen atoms are removed in the form of carbon dioxide, and a two-carbon acetyl group (CH_3CO) remains. In the course of this reaction, a molecule of NADH is formed from NAD^+. Because the oxidation

of one glucose molecule produces two pyruvic acid molecules, this step yields two NADH molecules for each glucose molecule. The original glucose molecule has now been oxidized to two CO_2 molecules and two acetyl groups.

Each acetyl group is momentarily accepted by a compound known as **coenzyme A.** Like many other coenzymes, it is a large molecule, a portion of which is a nucleotide. The combination of an acetyl group and coenzyme A is abbreviated as **acetyl CoA.** Its formation is the link between glycolysis and the Krebs cycle. Acetyl groups from sources other than the oxidation of pyruvic acid—for example, from the breakdown of fats and amino acids—can also combine with coenzyme A and enter the respiratory sequence at this point.

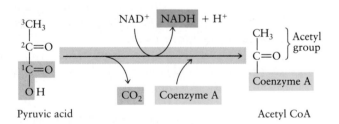

8–7 *The three-carbon pyruvic acid molecule is oxidized to the two-carbon acetyl group, which is combined with coenzyme A to form acetyl CoA. Following this reaction, acetyl CoA enters the Krebs cycle. The oxidation of the pyruvic acid molecule is coupled to the reduction of NAD^+. One molecule of NADH is produced for each molecule of pyruvic acid oxidized. Thus two molecules of NADH are produced for each glucose molecule that enters glycolysis.*

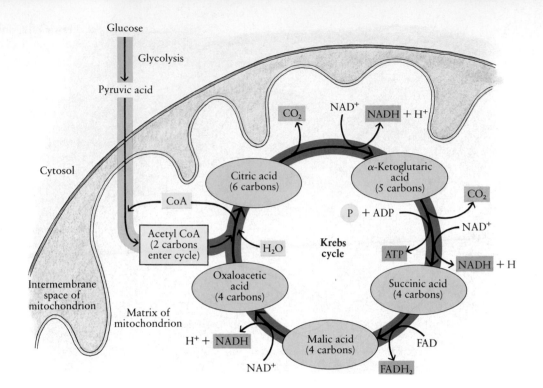

8–8 *The principal steps of the Krebs cycle. In the course of this reaction sequence, the carbons donated by the acetyl group of acetyl CoA are oxidized to carbon dioxide, and the hydrogen atoms are passed to the electron carriers NAD$^+$ and FAD. As in glycolysis, a specific enzyme catalyzes each step. Coenzyme A shuttles back and forth between the oxidation of pyruvic acid and the Krebs cycle, linking these two stages of respiration.*

One molecule of ATP, three molecules of NADH, and one molecule of FADH$_2$ represent the energy yield of one turn of the cycle. Two turns are required to complete the oxidation of one molecule of glucose.

This cycle is named after its discoverer, Sir Hans Krebs. It is also known as the citric acid cycle.

The Krebs Cycle

The Krebs cycle, like other biochemical pathways, consists of a series of reactions in which each step involves only a small change but the process as a whole produces major change. The principal steps of the Krebs cycle are shown in Figure 8–8.

The process begins when the two-carbon acetyl group is combined with a four-carbon compound (oxaloacetic acid) to form a six-carbon compound (citric acid). In the course of the Krebs cycle, two of the six carbons are removed and oxidized to carbon dioxide. And, a molecule of oxaloacetic acid is regenerated—making this series literally a cycle. Each turn around the cycle uses up one acetyl group and regenerates a molecule of oxaloacetic acid, which is then ready to begin the sequence again.

In the course of these reactions, some of the energy released by the oxidation of the carbon atoms is used to convert ADP to ATP (one molecule per cycle), some is used to produce NADH and H$^+$ from NAD$^+$ (three molecules per cycle), and some is used to produce FADH$_2$ from FAD (one molecule per cycle).

Oxaloacetic acid + Acetyl CoA + H$_2$O +
$$\text{ADP} + \text{P}_i + 3\text{NAD}^+ + \text{FAD} \longrightarrow$$
Oxaloacetic acid + 2CO$_2$ + CoA +
$$\text{ATP} + 3\text{NADH} + 3\text{H}^+ + \text{FADH}_2$$

Two turns of the cycle are required to complete the oxidation of one molecule of glucose. Thus the total energy yield of the Krebs cycle for one glucose molecule is two molecules of ATP, six molecules of NADH, and two molecules of FADH$_2$.

No oxygen is required for the Krebs cycle. The electrons and protons removed in the oxidation of carbon are accepted by NAD$^+$ and FAD. However, oxygen *is* required for the next—and last—stage of respiration, which begins when NADH and FADH$_2$ give up their electrons and protons, regenerating NAD$^+$ and FAD. If oxygen is not available, the supply of NAD$^+$ and FAD is rapidly depleted and the Krebs cycle shuts down.

Terminal Electron Transport

The glucose molecule is now completely oxidized. Some of its energy has been used to produce ATP from ADP and phosphate. Most of its energy, however, remains in the electrons removed from the carbon atoms and passed to the electron carriers NAD$^+$ and FAD. These electrons—from glycolysis, the oxidation of pyruvic acid, and the Krebs cycle—are still at a high energy level.

In terminal electron transport, which is the final stage of respiration, these high-energy-level electrons are passed step-by-step to the low energy level of oxygen. This stepwise passage is made possible by the electron transport chain (Figure 8–9), a series of electron carriers, each of which holds the electrons at a slightly lower energy level. Among the principal components of the electron transport chain are molecules known as **cytochromes.** Although the structures of the individual cytochromes are similar, they differ enough to enable them to hold electrons at different energy levels.

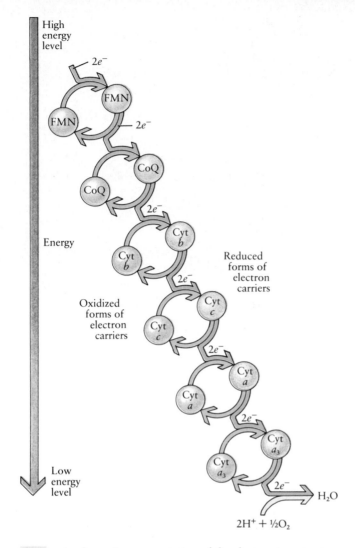

High
energy
level

$2e^-$

FMN

FMN

$2e^-$

CoQ

CoQ

Energy

$2e^-$

Cyt
b

Reduced
forms of
electron
carriers

Cyt
b

$2e^-$

Oxidized
forms of
electron
carriers

Cyt
c

Cyt
c

$2e^-$

Cyt
a

Cyt
a

$2e^-$

Cyt
a_3

Cyt
a_3

Low
energy
level

$2e^-$ H_2O

$2H^+ + \frac{1}{2}O_2$

8–9   *A schematic representation of the electron transport chain. The molecules shown here—flavin mononucleotide (FMN), coenzyme Q (CoQ), and cytochromes b, c, a, and a₃—are the principal electron carriers of the chain. At least nine other molecules function as intermediates between these electron carriers.*

Electrons carried by NADH enter the chain when they are transferred to FMN, which is thus reduced (blue). Almost instantaneously, FMN passes the electrons on to CoQ. FMN returns to its oxidized form (gray), ready to receive another pair of electrons, and CoQ is reduced. CoQ then passes the electrons on to the next carrier and returns to its oxidized form, and so on down the line. As the electrons move down the chain, they drop to successively lower energy levels. The electrons are ultimately accepted by oxygen, which combines with protons (hydrogen ions) to form water.

Electrons carried by FADH₂ are at a slightly lower energy level than those carried by NADH. They enter the electron transport chain farther down the line at CoQ.

As the electrons move through the electron transport chain, dropping to lower energy levels, energy is released. This energy is harnessed by the mitochondrion to power the synthesis of ATP from ADP, in a process known as **oxidative phosphorylation.** Quantitative measurements show that for every two electrons that pass from NADH to oxygen, three molecules of ATP are formed from ADP and phosphate. For every two electrons that pass from FADH₂, which holds them at a slightly lower energy level than NADH, two molecules of ATP are formed.

With the synthesis of ATP in oxidative phosphorylation, the process that began with the glucose molecule is complete. However, before we sum up the total energy harvest, let us look more closely at how ATP is synthesized in conjunction with terminal electron transport.

The Mechanism of Oxidative Phosphorylation: Chemiosmotic Coupling

The mechanism of oxidative phosphorylation was, until the early 1960s, one of the most baffling puzzles in all of cell biology. As a result of the insight and experimental creativity of the British biochemist Peter Mitchell (1920–1992)—and the subsequent work of many other investigators—much of the puzzle has now been solved. Oxidative phosphorylation depends on a gradient of protons (H^+ ions) across the mitochondrial membrane and the subsequent use of the potential energy stored in that gradient to form ATP from ADP and phosphate.

As shown in Figure 8–10, the components of the electron transport chain are arranged sequentially in the inner membrane of the mitochondrion. Most of the electron carriers are tightly associated with proteins embedded in the membrane, forming three distinct complexes. According to current evidence, these complexes are locked into place in the membrane. Within each complex, the electron carriers are held in the proper positions in relation to one another.

The protein complexes also have another crucial function: they are proton pumps. As the electrons drop to lower energy levels during their transit through the electron transport chain, the released energy is used by the protein complexes to pump protons from the mitochondrial matrix into the intermembrane space. It is thought that for each pair of electrons moving down the electron transport chain from NADH to oxygen, 10 protons are pumped out of the matrix.

The inner membrane of the mitochondrion is, as we noted earlier, impermeable to protons. Thus, the protons that are pumped into the intermembrane space cannot easily move back across the membrane into the

8–10   *The arrangement of the components of the electron transport chain in the inner membrane of the mitochondrion. Three complex protein structures (here labeled I through III) are embedded in the membrane. They contain electron carriers and the enzymes required to catalyze the transfer of electrons from one carrier to the next. Complex I contains the electron carrier FMN and receives electrons from NADH. CoQ, which is located in the lipid interior of the membrane, ferries electrons from complex I to complex II, which contains cytochrome b. From complex II, the electrons move to cytochrome c, which is a peripheral membrane protein that shuttles back and forth between complexes II and III. The electrons then move through cytochromes a and a_3, located in complex III, back into the matrix where they combine with H^+ ions and oxygen, forming water.*

As the electrons make their way down the electron transport chain, protons (H^+ ions) are pumped through the three protein complexes from the matrix to the intermembrane space. This transfer of protons from one side of the membrane to the other establishes the proton gradient that powers the synthesis of ATP.

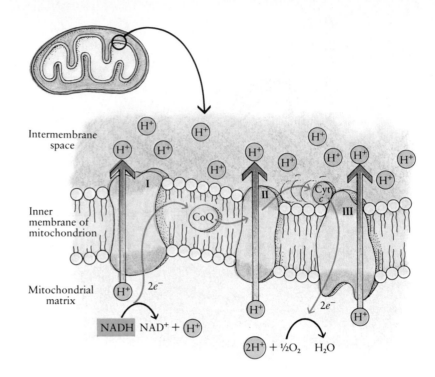

matter. The result is a concentration gradient of protons across the inner membrane of the mitochondrion, with a much higher concentration of protons in the intermembrane space than in the matrix.

Like a boulder at the top of a hill or water at the top of a falls, the difference in the concentration of protons between the intermembrane space and the matrix represents potential energy. This potential energy results not only from the actual concentration difference (more hydrogen ions outside the matrix than inside) but also from the difference in electric charge (more + charges outside than inside). The potential energy is thus in the form of an **electrochemical gradient.** It is available to power any process that provides a channel allowing the protons to flow down the gradient back into the matrix.

Such a channel is provided by a large enzyme complex known as **ATP synthase** (Figure 8–11). This enzyme complex, which is embedded in the inner membrane of the mitochondrion, has binding sites for ATP and ADP. It also has an inner channel, or pore, through which protons can pass. When protons flow through this channel, moving down the electrochemical gradient from the intermembrane space back into the matrix, the energy released powers the synthesis of ATP from ADP and phosphate.

This mechanism of ATP synthesis, summarized in Figure 8–12, is known as **chemiosmotic coupling.** The term "chemiosmotic," which was coined by Peter Mitchell, reflects the fact that the production of ATP in

oxidative phosphorylation includes both chemical processes and transport processes across a selectively permeable membrane. As we have seen, two distinct events take place in chemiosmotic coupling: (1) a proton gradient is established across the inner membrane of the mitochondrion, and (2) potential energy stored in the gradient is used to generate ATP from ADP and phosphate.

Chemiosmotic power also has other uses in living systems. For example, it provides the power that drives the rotation of bacterial flagella. In photosynthetic cells, as we shall see in the next chapter, it is involved in the formation of ATP using energy supplied to electrons by the sun. And, it can be used to power other transport processes. In the mitochondrion, for example, the energy stored in the proton gradient is also used to carry other substances through the inner membrane. Both phosphate and pyruvic acid are carried into the matrix by membrane proteins that simultaneously transport protons down the gradient.

We noted earlier (page 145) that "about" 38 molecules of ATP are formed for each molecule of glucose oxidized to carbon dioxide and water. The exact amount of ATP formed depends on how the cell apportions the energy made available by the proton gradient. When more of this energy is used in other transport processes, less of it is available for ATP synthesis. The needs of the cell vary according to the circumstances, and so does the amount of ATP synthesized.

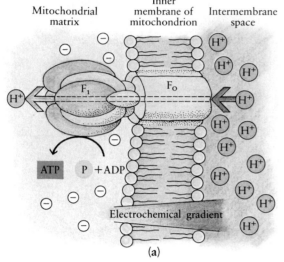

Mitochondrial matrix

Inner membrane of mitochondrion

Intermembrane space

F_1

F_o

H^+

ATP   P +ADP

Electrochemical gradient

(a)

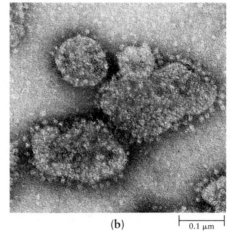

(b)   0.1 μm

8–11 (a) *The ATP synthase complex. This enzyme complex consists of two major portions, F_o, which is contained within the inner membrane of the mitochondrion, and F_1, which extends into the matrix. Binding sites for both ATP and ADP are located on the F_1 portion, which consists of nine separate protein subunits. A channel, or pore, connecting the intermembrane space to the mitochondrial matrix, passes through the entire complex. When protons flow through this channel, moving down the electrochemical gradient, ATP is synthesized from ADP and phosphate.*

(b) The knobs protruding from the vesicles in this electron micrograph are the F_1 portions of ATP synthase complexes. The F_o portions to which they are attached are embedded in the membrane and are not visible. The vesicles were prepared by disrupting the inner mitochondrial membrane with ultrasonic waves. When the membrane is disrupted in this way, the fragments immediately reseal, forming closed vesicles. The vesicles are, however, inside-out. The outer surface here is the surface that faces the matrix in the intact mitochondrion.

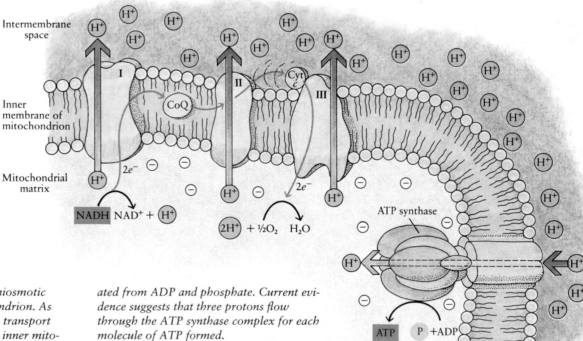

Intermembrane space

Inner membrane of mitochondrion

Mitochondrial matrix

$2e^-$

I

CoQ

II

Cyt c

III

NADH NAD$^+$ + H^+

$2H^+$ + ½O$_2$ H$_2$O

$2e^-$

ATP synthase

H^+

ATP   P +ADP

8–12 *A summary of the chemiosmotic synthesis of ATP in the mitochondrion. As electrons pass down the electron transport chain, which forms a part of the inner mitochondrial membrane, protons are pumped out of the mitochondrial matrix into the intermembrane space. This creates an electrochemical gradient. The subsequent movement of protons down the gradient as they pass through the ATP synthase complex provides the energy by which ATP is regener-* *ated from ADP and phosphate. Current evidence suggests that three protons flow through the ATP synthase complex for each molecule of ATP formed.*

It is estimated that the inner membrane of each mitochondrion in a liver cell has more than 10,000 copies of the electron transport chain and of the ATP synthase complex. Each mitochondrion in a heart muscle cell, which uses enormous quantities of ATP, is estimated to have more than 30,000 copies.

8–13   *A summary of the maximum energy yield from the complete oxidation of one molecule of glucose.*

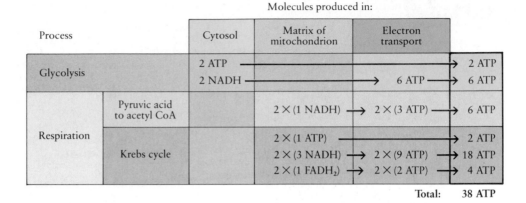

Process		Molecules produced in:			
		Cytosol	Matrix of mitochondrion	Electron transport	
Glycolysis		2 ATP			2 ATP
		2 NADH		6 ATP	6 ATP
Respiration	Pyruvic acid to acetyl CoA		2 × (1 NADH) →	2 × (3 ATP) →	6 ATP
	Krebs cycle		2 × (1 ATP)		2 ATP
			2 × (3 NADH) →	2 × (9 ATP) →	18 ATP
			2 × (1 FADH₂) →	2 × (2 ATP) →	4 ATP

Total: 38 ATP

Overall Energy Harvest

We are now in a position to see how much of the energy originally present in the glucose molecule has been recovered in the form of ATP. Because the proton gradient across the inner membrane of the mitochondrion can be used for purposes other than ATP synthesis, the amount of ATP formed varies. The numbers shown in Figure 8–13 represent the maximum energy harvest.

Glycolysis yields two molecules of ATP directly and two molecules of NADH. When oxygen is present, the electrons held by these NADH molecules are "shuttled across" the inner membrane of the mitochondrion and then enter the electron transport chain. In most cells, the energy cost of the shuttle is quite low, and each NADH formed in glycolysis ultimately results in the synthesis of three molecules of ATP. In these cells, the total gain from glycolysis is 8 ATP. In other cells, including those of brain, skeletal muscle, and insect flight muscle, the energy cost of the shuttle is higher. In such cells, the total gain from glycolysis is only 6 ATP.

The conversion of pyruvic acid to acetyl CoA, which occurs in the matrix of the mitochondrion, yields two molecules of NADH for each molecule of glucose. When the electrons held by these two NADH molecules pass down the electron transport chain, 6 ATP are produced.

The Krebs cycle, which also occurs in the matrix of the mitochondrion, yields two molecules of ATP, six of NADH, and two of FADH₂. The passage down the electron transport chain of the electrons held by these NADH and FADH₂ molecules yields 22 ATP. Thus, for each molecule of glucose, the total yield of the Krebs cycle is 24 ATP.

As Figure 8–13 shows, the complete yield from a single molecule of glucose is a maximum of 38 molecules of ATP. Note that all but 2 of the 38 molecules of ATP have come from reactions taking place in the mitochondrion. And, all but 4 result from the passage down the electron transport chain of electrons carried by NADH or FADH₂.

In the course of glycolysis and respiration, 686 kilocalories are released per mole of glucose. About 266 kilocalories (7 kilocalories per mole of ATP × 38 moles of ATP) have been captured in the phosphate bonds of ATP, an efficiency of almost 40 percent.

The ATP molecules, once formed, are exported across the inner membrane of the mitochondrion by a shuttle system that simultaneously brings in one molecule of ADP for each ATP exported.

Fermentation

The pathway of respiration, which we have just examined, is the principal pathway of energy metabolism for most cells in the presence of oxygen. In the absence of oxygen, however, fermentation occurs. The pyruvic acid formed in glycolysis is converted to ethanol (ethyl alcohol) or to one of several organic acids, of which lactic acid is the most common. The reaction product depends on the type of cell.

Yeast cells, which are often present as a "bloom" on the skin of grapes, can grow either with or without oxygen. When the sugar-filled juices of grapes are extracted and stored under anaerobic conditions, the yeast cells turn the fruit juice to wine by converting glucose into ethanol (Figure 8–14). When the sugar is exhausted, the yeast cells cease to function. At this point, the alcohol concentration is between 12 and 17 percent, depending on the variety of the grapes and the season at which they were harvested.

Lactic acid is formed from pyruvic acid by a variety of microorganisms and also by some animal cells when oxygen is scarce or absent (Figure 8–15). As we noted on page 145, it is produced in muscle cells during strenuous exercise, as when an athlete runs a sprint. We breathe hard when we run fast, thereby increasing the supply of oxygen, but even this increase may not be

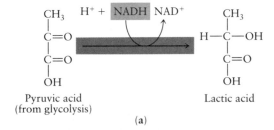

Pyruvic acid
(from glycolysis)   Acetaldehyde   Ethanol

(a)

(b)

8–14   (a) *The steps by which pyruvic acid, formed in glyco-lysis, is converted to ethanol (ethyl alcohol) in the absence of oxygen. In the first step, carbon dioxide is released. In the sec-ond, NADH is oxidized, and acetaldehyde is reduced. Most of the energy in the original glucose molecule remains in the alco-hol molecule, which is the end product of the sequence. How-ever, by regenerating* NAD^+, *these steps allow glycolysis to continue, with its small but sometimes vitally necessary yield of ATP.*

(b) *Yeast cells on the skins of these grapes give them their dustlike "bloom." When the grapes are crushed, the yeast cells mix with the juice. Storing the mixture under anaerobic condi-tions causes the yeast to break down the glucose in the grape juice to alcohol.*

Pyruvic acid
(from glycolysis)   Lactic acid

(a)

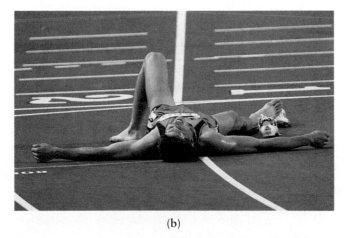

(b)

8–15   (a) *The enzymatic reaction that produces lactic acid from pyruvic acid anaerobically in muscle cells. In the course of this reaction, NADH is oxidized and pyruvic acid is re-duced. The* NAD^+ *molecules produced in this reaction and the one shown in Figure 8–14 are recycled in glycolysis. Without this recycling, glycolysis cannot proceed.*

(b) *Later, when ample oxygen is available and ATP demand is reduced, the reaction shown in* (a) *is reversed. The accumu-lated lactic acid is converted back to pyruvic acid, which is then used in the resynthesis of glucose and glycogen. This re-stores the chemical energy reserves of the muscle cells, which are then ready, once more, to meet whatever demands are placed upon them.*

enough to meet the immediate needs of the muscle cells. These cells, however, can continue to work by accu-mulating what is known as an **oxygen debt.** Glycolysis continues, using glucose released from glycogen stored in the muscle, but the resulting pyruvic acid is not ox-idized to acetyl CoA. Instead, it is converted to lactic acid, which, as it accumulates, produces the sensations of muscle fatigue.

Later, when oxygen is more abundant (as a result of the deep breathing that follows strenuous exercise) and ATP demand is reduced, the lactic acid is converted back to pyruvic acid. The pyruvic acid is then used in the resynthesis of glucose or glycogen.

Why is pyruvic acid converted to lactic acid, only to be converted back again? The function of the initial conversion is simple: it uses NADH and regenerates the NAD^+ without which glycolysis cannot go forward (see Step 5, page 147). Even though the overall process seems to be wasteful in terms of energy consumption, the regeneration of NAD^+ may spell the difference be-tween life and death when an animal "out of breath" needs one last burst of ATP to escape from a predator or to capture a prey.

Ethanol, NADH, and the Liver

The human body can dispose fairly readily of most toxic products of its own manufacture, such as carbon dioxide and the end products of nitrogen metabolism. In contrast, most ingested toxic substances, such as ethanol (beverage alcohol), must first be broken down by the liver, which possesses special enzymes not present in other tissues.

It has been known for many years that heavy drinkers are at great risk for severe, and often fatal, liver disease. Biochemical studies have demonstrated that the origin of the problem lies in the simple chemical steps involved in the breakdown of ethanol. Enzymes in the liver first oxidize ethanol (CH_3CH_2OH) to acetaldehyde (CH_3CHO), removing two hydrogen atoms and reducing a molecule of NAD^+. This is the reverse of the second reaction shown in Figure 8–14a. The acetaldehyde is then oxidized to acetic acid, which is, in turn, oxidized to carbon dioxide and water and eliminated from the body.

The intoxicating effects of alcohol are due mostly to the acetaldehyde, which stimulates the release of adrenalinelike agents. The chief culprits in the development of liver disease, however, are the hydrogen atoms (electrons and protons) removed from ethanol. These "extra" hydrogens—carried by NADH—follow two principal pathways within the cell.

Most of the hydrogen atoms are fed directly into the electron transport chain, producing water and ATP. Because of the high levels of NADH present in the cell from the oxidation of ethanol, the production of NADH by glycolysis and the Krebs cycle is reduced. As a result, sugars, amino acids, and fatty acids are not broken down but are instead converted to fats. The fats accumulate in the liver. The mitochondria also swell, presumably as a result of the distortion of their normal function—the electron transport chain is doing very heavy duty, while the Krebs cycle is effectively shut down.

Other hydrogen atoms are used in the synthesis of fatty acids from the sugars and amino acids that are not being processed in glycolysis and the Krebs cycle. More fats accumulate. It does not take long. In human volunteers fed a good high-protein, low-fat diet, six drinks (about 10 ounces) a day of 86 proof alcohol produced an eightfold increase in fat deposits in the liver in only 18 days. Fortunately, these early effects are completely reversible.

The liver cells work hard to get rid of the excess fats, which are not soluble in blood plasma. Before being released into the bloodstream, the fats are coated with a thin layer of protein in a process carried out on the membranes of the endoplasmic reticulum. The liver cells of heavy drinkers show enormous proliferation of the endoplasmic reticulum.

After a few years—depending on how much alcohol is consumed—liver cells, engorged with fat, begin to die, triggering the inflammatory process known as alcoholic hepatitis. Liver function becomes impaired. Cirrhosis is the next step. It is the formation of scar tissue, which interferes with the function of the individual cells and also with the supply of blood to the liver. This leads to the death of more cells. The liver can no longer carry out its normal activities—such as breaking down nitrogenous wastes—which is why cirrhosis is a cause of death. In fact, cirrhosis of the liver is the ninth leading cause of death in the United States for the population as a whole, and the seventh leading cause of death among those who are 25 to 44 years old.

Not so long ago, it was commonly believed

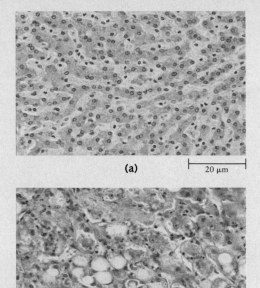

(a) 20 μm

(b) 20 μm

(a) *Normal cells from a human liver.*
(b) *Abnormal cells from the liver of a person with cirrhosis of the liver. Note the many globular fat droplets that have accumulated in the cells.*

that a good diet was all that was required to protect even a heavy drinker from the deleterious effects of alcohol. In fact, if one were just to add a few vitamins to the alcohol itself, some sophisticates maintained, most of the long-term physical damage of alcohol would disappear. This evidence refutes such comforting notions, and it comes at a time when alcohol is enjoying a resurgence of popularity among persons of high school and college age.

The fact that glycolysis does not require oxygen suggests that the glycolytic sequence evolved early, before free oxygen was present in the atmosphere. Primitive one-celled organisms presumably used glycolysis (or something very much like it) to extract energy from organic compounds they absorbed from their watery surroundings. Although anaerobic glycolysis generates only two molecules of ATP for each glucose molecule processed (a small fraction of the ATP that is generated when oxygen is present), it was and is adequate for the needs of many organisms.

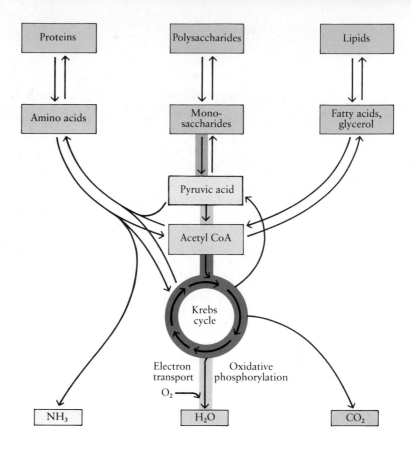

8-16 *An outline of the major pathways of catabolism and anabolism in the living cell. Catabolic pathways (arrows pointing down) are exergonic. A significant portion of the energy released in these pathways is, as we have seen, captured in the synthesis of ATP. Anabolic pathways (arrows pointing up) are endergonic. The energy that powers the reactions in these pathways is supplied primarily by ATP and NADH.*

The Strategy of Energy Metabolism

Most organisms do not feed directly on glucose. How do they extract energy from, for example, fats or proteins? The answer lies in the fact that the Krebs cycle is a Grand Central Station for energy metabolism. Other foodstuffs are broken down and converted into molecules that can feed into this central pathway.

Polysaccharides, such as starch, are broken down to their constituent monosaccharides and phosphorylated to glucose 6-phosphate or fructose 6-phosphate. In these forms, they enter the glycolytic pathway at Steps 2 and 3, respectively (see page 146).

Fats are first split into their glycerol and fatty acid components. Glycerol is subsequently broken down in a series of reactions that ultimately result in the transfer of electrons into the electron transport chain at CoQ. The fatty acids are chopped up into two-carbon fragments and slipped into the Krebs cycle as acetyl CoA.

Proteins are similarly broken down into their constituent amino acids, and the amino groups are removed. The residual carbon skeleton is converted either to an acetyl group, which enters the Krebs cycle as acetyl CoA, or to one of the larger compounds that are intermediates in glycolysis or in the Krebs cycle. If the amino groups are not reused, they are excreted as nitrogenous wastes.

These various pathways by which foods are broken down to yield energy are known collectively as **catabolism.** They are also central to the biosynthetic processes of life. These processes, known collectively as **anabolism,** are the pathways by which cells synthesize the diversity of molecules that constitute a living organism. Since many of these molecules, such as proteins and lipids, can be broken down and fed into the central pathway of glucose metabolism, you might guess that the reverse process can occur—namely, that the various intermediates of glycolysis and the Krebs cycle can serve as precursors for biosynthesis. This is indeed the case, as outlined in Figure 8–16.

For the reactions of the catabolic and anabolic pathways to occur, there must be a steady supply of organic molecules that can be broken down to yield energy and building-block molecules. Without a supply of such molecules, the metabolic pathways cease to function and the life of the organism ends. Heterotrophic cells (including the heterotrophic cells of plants, such as root cells) are dependent on external sources—specifically, autotrophic cells—for the organic molecules that are essential for life. Autotrophic cells, however, are able to synthesize monosaccharides from simple inorganic molecules and an external energy source. These monosaccharides are then used not only to supply energy but also as building-block molecules.

By far the most important autotrophic cells are the photosynthetic cells of algae and plants. In the next chapter, we shall examine how these cells capture the energy of sunlight and use it to synthesize the monosaccharide molecules on which life on this planet depends.

Summary

The oxidation of glucose is a chief source of energy in most cells. As the glucose is broken down in a series of enzymatic reactions, a significant portion of the energy in the molecule is repackaged in the phosphate bonds of ATP molecules.

The first phase in the breakdown of glucose is glycolysis, in which the six-carbon glucose molecule is split into two three-carbon molecules of pyruvic acid. A net yield of two molecules of ATP and two of NADH results from the process. Glycolysis takes place in the cytosol of the cell.

The second phase in the breakdown of glucose and other fuel molecules is respiration. It requires oxygen and, in eukaryotic cells, takes place in the mitochondria. It occurs in two stages: the Krebs cycle and terminal electron transport.

In the course of respiration, the three-carbon pyruvic acid molecules from glycolysis are broken down to two-carbon acetyl groups, which then enter the Krebs cycle. In a series of reactions in the Krebs cycle, the two-carbon acetyl group is oxidized completely to carbon dioxide. In the course of the oxidation of each acetyl group, four electron acceptors (three NAD^+ and one FAD) are reduced, and another molecule of ATP is formed.

The final stage of respiration is terminal electron transport, which involves a chain of electron carriers and enzymes embedded in the inner membrane of the mitochondrion. Along this series of electron carriers, the high-energy electrons carried by NADH from glycolysis and the oxidation of pyruvic acid and by NADH and $FADH_2$ from the Krebs cycle pass downhill to oxygen. During their passage down the electron transport chain, the energy released powers the pumping of protons (H^+ ions) out of the mitochondrial matrix. This creates a gradient of potential energy across the inner membrane of the mitochondrion. When protons pass through the ATP synthase complex as they flow down the gradient back into the matrix, the energy released powers the formation of ATP molecules from ADP and phosphate. This mechanism, by which oxidative phosphorylation is accomplished, is known as chemiosmotic coupling.

In the course of the breakdown of the glucose molecule, a maximum of 38 molecules of ATP can be formed, most of them from processes occurring in the mitochondrion.

In the absence of oxygen, the pyruvic acid produced by glycolysis is converted to either ethanol or lactic acid by the process of fermentation. NAD^+ is regenerated, allowing glycolysis to continue, producing a small but vital supply of ATP for the organism.

Organisms can also obtain energy from food molecules other than glucose. Polysaccharides, fats, and proteins are degraded to molecules that can enter the pathways of glucose metabolism at various steps. The biosynthesis of these substances also originates with molecules derived from intermediates in the respiratory sequence and is driven by the energy derived from those processes.

Questions

1. Distinguish among the following: oxidation of glucose/glycolysis/respiration/fermentation; NAD^+/NADH; FAD/$FADH_2$; Krebs cycle/electron transport; aerobic pathways/anaerobic pathways.

2. Sketch the structure of a mitochondrion. Describe where the various stages in the breakdown of glucose take place in relation to mitochondrial structure. What molecules and ions cross the mitochondrial membranes during these processes?

3. (a) As we have seen, a cell can obtain a maximum of 38 molecules of ATP from each molecule of glucose that is completely oxidized. Account for the production of each molecule of ATP. (b) In the course of glycolysis, the Krebs cycle, and electron transport, 40 molecules of ATP are actu- ally formed. Why is the net yield for the cell only 38 molecules? (c) What other factors can reduce the yield of ATP?

4. Cyanide can combine with—and thereby deactivate—cytochrome a and cytochrome a_3. In our bodies, however, cyanide tends to react first with hemoglobin and to make it impossible for oxygen to bind to the hemoglobin. Either way, cyanide poisoning has the same effect: it inhibits the synthesis of ATP. Explain how this is so.

5. Certain chemicals function as "uncoupling" agents when they are added to respiring mitochondria. The passage of electrons down the electron transport chain to oxygen continues, but no ATP is formed. One of these agents, the antibiotic valinomycin, is known to transport K^+ ions through the inner membrane into the matrix. Another, 2,4-dinitro-

phenol, transports H^+ ions through the membrane. How do these substances prevent the formation of ATP?

6. In the cells of a specialized tissue known as brown fat, the inner membrane of the mitochondrion is permeable to H^+ ions. These cells contain large stores of fat molecules, which are gradually broken down to acetyl groups. These acetyl groups are fed into the Krebs cycle. The electrons captured by NADH and $FADH_2$ are, in turn, fed into the electron transport chain and ultimately accepted by oxygen. No ATP is synthesized, however. Why not? Brown fat tissue is found in some hibernating animals and in mammalian infants that are born hairless, including human infants. What do you suppose the function of brown fat tissue is?

7. If aerobic (oxygen-utilizing) organisms are so much more efficient than anaerobes in extracting energy from organic molecules, why are there any anaerobes left on this planet? Why didn't they all become extinct long ago?

8. Describe the process of fermentation. What conditions are essential if it is to occur? With some strains of yeast, fermentation stops before the sugar is exhausted, usually at an alcohol concentration in excess of 12 percent. What is a plausible explanation?

9. Describe how the cell is adapted to the efficient use of a variety of foodstuffs, and to the efficient production of the variety of molecules that it needs to manufacture for its own use.

10. In terms of the cell's economy, what do catabolic processes provide for the cell? What do anabolic processes provide? How are they dependent on each other?

Suggestions for Further Reading

Alberts, Bruce, Dennis Bray, Julian Lewis, Martin Raff, Keith Roberts, and James D. Watson: *Molecular Biology of the Cell,* 3d ed., Garland Publishing, Inc., New York, 1994.

This outstanding cell biology text includes a clear, up-to-date discussion of our current understanding of the processes that occur in the mitochondria.

Darnell, James, Harvey F. Lodish, and David Baltimore: *Molecular Cell Biology,* 2d ed., W. H. Freeman and Company, New York, 1990.

A comprehensive treatment of modern cell biology, richly illustrated with diagrams and micrographs. Chapter 15 of this outstanding text is devoted to a thorough explication of our current understanding of glycolysis and respiration.

Dickerson, Richard E.: "Cytochrome *c* and the Evolution of Energy Metabolism," *Scientific American,* March 1980, pages 136–153.

Kolata, Gina: "How Do Proteins Find Mitochondria?" *Science,* vol. 228, pages 1517–1518, 1985.

Lane, M. Daniel, Peter L. Pedersen, and Albert S. Mildvan: "The Mitochondrion Updated," *Science,* vol. 234, pages 526–527, 1986.

Lehninger, Albert L., David L. Nelson, and Michael M. Cox: *Principles of Biochemistry,* 2d ed., Worth Publishers, Inc., New York, 1993.

Lehninger, who made the key discovery that oxidative phosphorylation occurs in the mitochondria, was one of the foremost experts on cellular energetics. This updated version of his classic introductory text is enriched by his vast experience and thorough understanding of

the processes by which cells provide themselves with energy.

Newsholme, Eric, and Tony Leech: *The Runner: Energy and Endurance,* Fitness Books, Roosevelt, N.J., 1984. *

A lively introduction to the energetics of the working human body. Equally useful as a primer for runners who want to know more about their own physiology and as an introduction to the biochemistry of carbohydrate and fat metabolism.

Prescott, David M.: *Cells: Principles of Molecular Structure and Function,* Jones and Bartlett Publishers, Boston, 1988.

An up-to-date, yet concise, textbook of cell biology, written for a first course at the undergraduate level. Glycolysis and respiration are covered in Chapter 4.

Scientific American: *Molecules to Living Cells,* W. H. Freeman and Company, New York, 1980. *

A collection of articles from Scientific American. *Chapter 12 is an excellent presentation of the chemiosmotic synthesis of ATP in both mitochondria and chloroplasts.*

Stryer, Lubert: *Biochemistry,* 3d ed., W. H. Freeman and Company, New York, 1988.

Glycolysis, the Krebs cycle, and oxidative phosphorylation are covered in Chapters 15, 16, and 17 of this outstanding text.

Webb, A. Dinsmoor: "The Science of Making Wine," *American Scientist,* vol. 72, pages 360–367, 1984.

* Available in paperback.

We are here on Earth—you and I—because of events taking place at this very moment in the cells of plants and algae. These events—known as photosynthesis—capture the radiant energy of the sun and use it to form carbohydrates and other organic molecules. These molecules, which are the building blocks of the tissues of plants and algae, are incorporated into the bodies of animals and other organisms that eat plants and algae. The molecules produced in photosynthesis also serve as the power source—the fuel—for the nonphotosynthetic inhabitants of our planet.

The three great staple foods for the human population are wheat, which had its origins in the temperate regions of the Old World; corn, which had its beginnings in the wild plants of the New World; and rice, found in the tropics and subtropics. Much of the world's rice crop—almost 500 million metric tons each year—is harvested by hand, as in these rice paddies on the island of Sumatra, in Indonesia.

Learning to plant, cultivate, and harvest crops such as rice was the most important event in the cultural evolution of *Homo sapiens*. It led to the end of the hunter-gatherer way of life for most of the human population. For better or worse, people began to have permanent dwelling places and to accumulate food, land, and other wealth. This transition led to a human situation that endures to this day—a large and still growing population, divided between rich and poor.

Photosynthesis, Light, and Life

The first photosynthetic organisms probably appeared 3 to 3.5 billion years ago. Before the evolution of photosynthesis, the physical characteristics of Earth and its atmosphere were the most powerful forces shaping the course of natural selection. With the evolution of photosynthesis, however, organisms began to change the face of our planet and, as a consequence, to exert strong influences on each other. Organisms have continued to change the environment, at an ever-increasing rate, up to the present day.

As we saw earlier, the atmosphere in which the first cells evolved lacked free oxygen. These earliest organisms were, of course, adapted to living in that environment. In fact, free oxygen, with its powerful electron-attracting capacities, would have been poisonous to them (as it is to many modern anaerobes). Their energy came from anaerobic processes, most likely glycolysis and fermentation, which would have released carbon dioxide into the atmosphere. The organic molecules they used as fuel may have been formed by nonbiological processes (page 73), or they may have been produced by chemosynthetic autotrophs (page 75) or by primitive photosynthetic cells that, like some modern photosynthetic bacteria, did not release oxygen to the environment.

Then, it is hypothesized, there evolved photosynthetic organisms that used carbon dioxide as their carbon source and released oxygen, as do most modern photosynthetic forms. As these photosynthetic organisms multiplied, they provided a new supply of organic molecules, and free oxygen began to accumulate. In response to these changing conditions, cell species arose for which oxygen was not a poison but rather a requirement for existence.

As we saw in the last chapter, oxygen-utilizing organisms have an advantage over those that do not use oxygen. A higher yield of energy can be extracted per molecule from the aerobic breakdown of carbon-containing compounds than from anaerobic processes, in which fuel molecules are not completely oxidized.

161

9–1   *The bubbles on the leaves of this sprig of* Elodea, *a common pondweed and aquarium plant, are oxygen. As we shall see, the oxygen is released when water molecules are split apart in the first step of photosynthesis.*

Energy released in cells by reactions using oxygen made possible the development of increasingly active, increasingly complex organisms. Without oxygen, the complex forms of life that now exist on Earth could not have evolved.

Life on Earth continues to be dependent on photosynthesis both for its oxygen and for its carbon-containing fuel molecules. Photosynthetic organisms capture light energy and use it to form carbohydrates and free oxygen from carbon dioxide and water, in a complex series of reactions. The overall equation for photosynthesis can be summarized as:

Carbon dioxide + Water + Light energy $\longrightarrow$
$$\text{Glucose + Oxygen}$$

Or,

$$6CO_2 + 6H_2O + 686 \text{ kilocalories/mole} \longrightarrow$$
$$C_6H_{12}O_6 + 6O_2$$

To understand how organisms are able to capture light energy and convert it into stored chemical energy, we must look first at the characteristics of light itself.

The Nature of Light

Over 300 years ago, the English physicist Sir Isaac Newton (1642–1727) separated visible light into a spectrum of colors by passing it through a prism. Then by passing the light through a second prism, he recombined the colors, producing white light once again. By this experiment, Newton showed that white light is actually made up of a number of different colors, ranging from violet at one end of the spectrum to red at the other (Figure 9–2). Their separation is possible because light of different colors is bent at different angles in passing through the prism.

In the nineteenth century, through the genius of James Clerk Maxwell (1831–1879), it became known that what we experience as light is in truth a very small part of a vast continuous spectrum of radiation, the **electromagnetic spectrum.** All the radiations in this spectrum act as if they travel in waves. The **wavelengths**—that is, the distances from the crest of one wave to the crest of the next—range from those of gamma rays, which are measured in fractions of a nanometer (1 nanometer = 10^{-9} meter), to those of low-frequency radio waves, which are measured in kilometers (1 kilometer = 10^3 meters). Radiation of each particular wavelength has a characteristic amount of energy associated with it. The longer the wavelength, the lower the energy; conversely, the shorter the wavelength, the higher the energy. Within the spectrum of visible light, violet light has the shortest wavelengths, and red light the longest. The shortest rays of violet light have almost twice the energy of the longest rays of red light.

The Fitness of Light

From the physicist's point of view, the difference between radiations we can see and radiations we cannot see—so dramatic to the human eye—is only a few nanometers of wavelength, or, expressed differently, a small amount of energy. Why does this particular group of radiations, rather than some other, make the leaves grow and the flowers burst forth, cause the mating of fireflies and palolo worms, and, when reflecting off the surface of the moon, excite the imagination of poets and lovers? Why is this tiny portion of the electromagnetic spectrum responsible for vision, for the rhythmic day-night regulation of many biological activities, for the bending of plants toward the light, and—most important of all—for photosynthesis, on which life depends? Is it an amazing coincidence that all these biological activities are dependent on these same wavelengths?

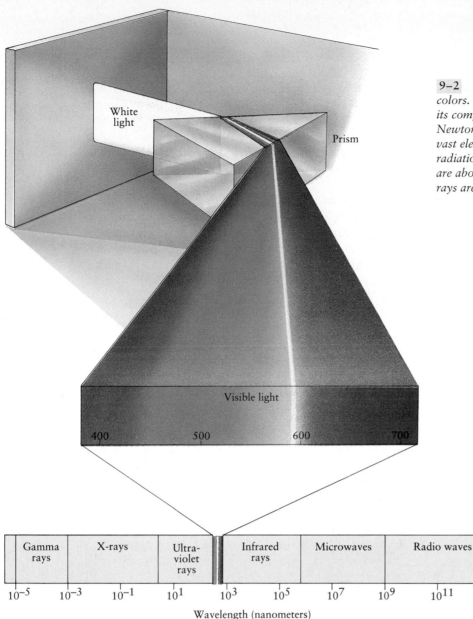

9–2   *White light is actually a mixture of light of different colors. When it is passed through a prism, it is separated into its components—"the celebrated phaenomena of colors," as Newton referred to it. Visible light is only a small portion of a vast electromagnetic spectrum. For the human eye, the visible radiations range from violet light, in which the shortest rays are about 380 nanometers, to red light, in which the longest rays are about 750 nanometers.*

George Wald of Harvard University, an expert on the subject of light and life, says no. He thinks that if life exists elsewhere in the universe, it is probably dependent on the same fragment of the electromagnetic spectrum. Wald bases this conjecture on two points. First, living things are composed of large, complicated molecules held in special configurations and relationships to one another by hydrogen bonds and other weak bonds. Radiation of even slightly higher energies than the energy of violet light breaks these bonds and so disrupts the structure and function of the molecules. DNA molecules, for example, are particularly vulnerable to such disruption. Radiations with wavelengths less than 200 nanometers—that is, with still higher energies— drive electrons out of atoms. On the other hand, light of wavelengths longer than those of the visible band—

that is, with less energy than red light—is absorbed by water, which makes up the great bulk of all living things on Earth. When this light is absorbed by molecules, its energy causes them to increase their motion (increasing heat), but it does not trigger changes in their electron configurations. Only those radiations within the range of visible light have the property of exciting molecules—that is, of moving electrons into higher energy levels—and so of producing chemical and, ultimately, biological change.

The second reason that the visible band of the electromagnetic spectrum has been "chosen" by living things is simply that it is what is available. The bulk of the radiation reaching the surface of the Earth from the sun is within this range. Most of the higher-energy wavelengths are screened out by oxygen and ozone in the atmosphere. Much infrared radiation is screened out by water vapor and carbon dioxide before it reaches the Earth's surface.

This is an example of what has been termed "the fitness of the environment." The suitability of the environment for life and that of life for the physical world are exquisitely interrelated. If they were not, life could not exist.

Chlorophyll and Other Pigments

In order for light energy to be used by living systems, it must first be absorbed. A substance that absorbs light is known as a **pigment.** Some pigments absorb all wavelengths of light and so appear black. Others absorb only certain wavelengths, transmitting or reflecting the wavelengths they do not absorb (Figure 9–3). For example, chlorophyll, the pigment that makes leaves green, absorbs light in the violet and blue wavelengths and also in the red. Because it reflects and transmits green light, it appears green.

Different groups of photosynthetic organisms use various pigments in photosynthesis. There are several different kinds of chlorophyll that vary slightly in their molecular structure. In photosynthetic eukaryotes (plants and algae), **chlorophyll** *a* is the pigment directly involved in the transformation of light energy to chemical energy. Most photosynthetic cells also contain a second type of chlorophyll—in plants and green algae, it is **chlorophyll** *b*—and representatives of another group of pigments called the **carotenoids.** The carotenoids are red, orange, or yellow pigments. In the green leaf, their color is masked by the chlorophylls, which are more abundant. In some tissues, however, such as those of a ripe tomato, the carotenoid colors predominate, as they do also when leaf cells stop synthesizing chlorophyll in the fall.

Chlorophyll *b* and the carotenoids are able to absorb light at wavelengths different from those absorbed by chlorophyll *a* (Figure 9–4). They apparently can pass the energy on to chlorophyll *a*, thus extending the range of light available for photosynthesis.

9–3 *When light strikes a pigmented object, some wavelengths are absorbed, whereas others are transmitted or reflected. The colors we perceive are the wavelengths of light that are transmitted and reflected. For example, a ripe tomato appears red because it reflects light in the red portion of the spectrum; light in all the other portions of the visible spectrum is absorbed. Similarly, the leaves of the tomato plant appear green because they reflect light in the green portion of the spectrum.*

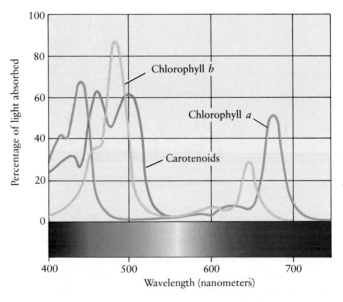

9–4 *Different pigments absorb light energy of different wavelengths. The percentage of the light of each wavelength that is absorbed by a pigment can be measured and the results presented in the form of a graph. Combining the graphs for different pigments makes it easy to compare the absorption patterns. The three curves shown here reveal the absorption patterns of the principal pigments found in the chloroplast: chlorophyll a, chlorophyll b, and the carotenoids. The two forms of chlorophyll absorb violet, blue, orange, and red wavelengths of light. The carotenoids, by contrast, absorb violet, blue, and green wavelengths.*

When a pigment absorbs light, electrons within the pigment molecules are boosted to higher energy levels. In most cases, the electrons fall back to their original energy levels almost immediately. The energy released as they fall back may (1) be absorbed by a neighboring molecule, boosting its electrons to higher energy levels, or (2) be dissipated as heat, or (3) be re-emitted as light of a longer wavelength, a phenomenon known as fluorescence.

In other cases, however, the absorbed energy triggers a chemical reaction. The energy absorbed by the pigment drives electrons from the molecule, which is thus oxidized. These high-energy electrons are immediately accepted by another molecule, which is thus reduced. Whether a chemical reaction occurs depends not only on the structure of the particular pigment but also on its relationships with neighboring molecules. Chlorophyll can convert light energy to chemical energy—a process that begins with a simple oxidation-reduction reaction—only when it is associated with certain proteins and is embedded in a specialized membrane.

Chloroplasts

The specialized membranes in which chlorophyll and other pigments are embedded are known as **thylakoids**. They usually take the form of flattened sacs, or vesicles. In eukaryotes, the thylakoids form a part of the internal membrane structure of specialized organelles, the chloroplasts (Figure 9–5). The alga *Chlamydomonas*, for

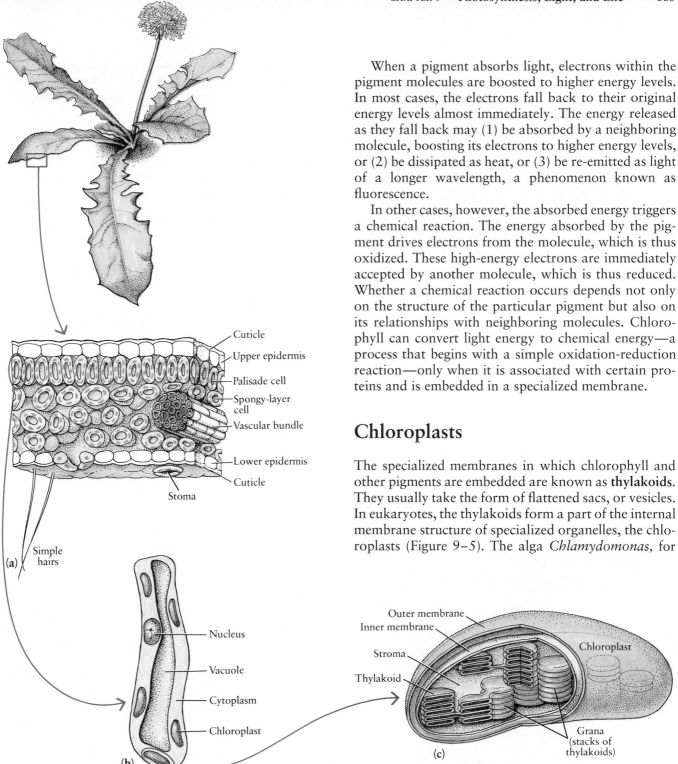

9–5 *Journey into the chloroplast of a dandelion leaf.* (a) *The inner tissues of the leaf are completely enclosed by transparent epidermal cells that are coated with a waxy layer, the cuticle. Oxygen, carbon dioxide, and other gases enter the leaf largely through special openings, the stomata (singular, stoma). These gases and water vapor fill the* spaces between cells in the spongy layer, leaving and entering cells by diffusion. Water, taken up by the roots, enters the leaf by way of the vascular bundles, and sugars, the products of photosynthesis, move out of the leaf by this same route, traveling to nonphotosynthetic parts of the plant. Much of the photosynthesis takes place in the palisade *cells* (b), *elongated cells directly beneath the upper epidermis. Palisade cells have a large central vacuole and numerous chloroplasts* (c) *that move within the cytosol, orienting themselves with respect to the light. Light is captured in the membranes of the disk-shaped thylakoids within the chloroplast.*

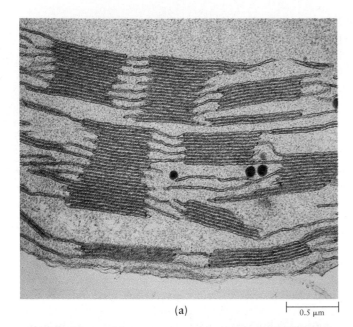

(a)   0.5 μm

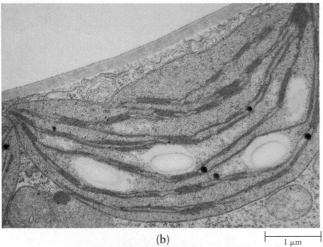

(b)   1 μm

9-6   *The first stage of photosynthesis takes place in the thylakoid, a flattened sac, whose membranes contain chlorophyll and other pigments. In plants and algae, thylakoids are part of an elaborate membrane system enclosed in a special organelle, the chloroplast. (a) Stacks of thylakoids (grana) in a chloroplast from a leaf cell of a corn plant. The inner compartments of the thylakoids are interconnected, forming the thylakoid space, which contains a solution whose composition differs from that of the stroma and of the cytosol. The black particles are droplets of lipid, available for the synthesis of new membrane material. (b) A chloroplast, showing the elaborate system of internal membranes comprising interconnected stacks of thylakoids. This chloroplast is from the leaf of a soybean plant.*

instance, has a single very large chloroplast. The cell of a leaf characteristically has 40 to 50 chloroplasts, and there are often 500,000 chloroplasts per square millimeter of leaf surface.

Chloroplasts, like mitochondria, are surrounded by two membranes. The thylakoids, in the interior of the chloroplast, constitute a third membrane system. Surrounding the thylakoids, and filling the interior of the chloroplast, is a dense solution, the **stroma,** which is different in composition from the cytosol. The thylakoids enclose an additional compartment, known as the **thylakoid space,** which contains a solution of still different composition.

With the light microscope, it is possible to see little spots of green within the chloroplasts of leaves. The early microscopists called these green specks **grana** ("grains"), and this term is still in use. With the electron microscope, it can be seen that the grana are stacks of thylakoids (Figure 9–6). All the thylakoids in a chloroplast are oriented parallel to each other. Thus, as the chloroplast swings toward the light, all of its millions of pigment molecules can be simultaneously aimed for optimum reception, as if they were miniature electromagnetic antennae (which, indeed, they are).

The Stages of Photosynthesis

The reactions of photosynthesis take place in two stages (Figure 9–7). In the first stage—the **light-dependent reactions**—light strikes chlorophyll *a* molecules that are packed in a special way in the thylakoid membranes. Electrons from the chlorophyll *a* molecules are boosted to higher energy levels, and the chlorophyll *a* molecules are oxidized. In a stepwise series of reactions, the energy carried by these electrons is used to form ATP from ADP and to reduce a molecule known as **NADP⁺**. NADP⁺ closely resembles NAD⁺, and it too is reduced by the addition of two electrons and a proton, forming **NADPH.** Water molecules are also broken apart in this stage of photosynthesis, supplying electrons that replace those boosted from the chlorophyll *a* molecules.

In the second stage of photosynthesis—the **light-independent reactions**—the ATP and NADPH formed in the first stage are used to reduce carbon dioxide to a simple sugar. Thus the chemical energy temporarily stored in ATP and NADPH molecules is transferred to molecules suitable for transport and storage in the algal cell or plant body. At the same time, a carbon skeleton is formed, from which other organic molecules can be built. This incorporation of carbon dioxide into organic compounds, known as **carbon fixation,** occurs in the stroma of the chloroplast.

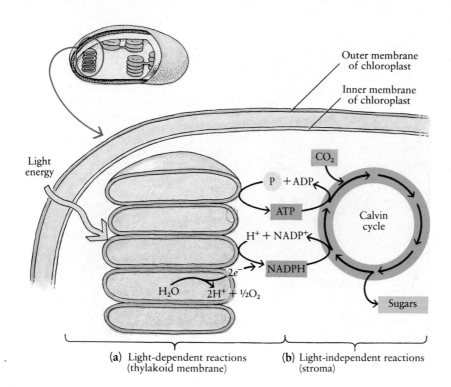

(a) Light-dependent reactions
(thylakoid membrane)

(b) Light-independent reactions
(stroma)

9–7 *An overview of photosynthesis. Photosynthesis takes place in two stages: the light-dependent reactions and the light-independent reactions.* **(a)** *In the light-dependent reactions, light energy absorbed by chlorophyll a molecules in the thylakoid membrane is used to power the synthesis of ATP. Simultaneously, in the interior of the thylakoid, water is split into oxygen gas and hydrogen atoms (electrons and protons). The electrons are ultimately accepted by $NADP^+ + H^+$, producing NADPH.* **(b)** *In the light-independent reactions, which occur in the stroma of the chloroplast, sugars are synthesized from carbon dioxide and the hydrogen carried by NADPH. This process is powered by the ATP and NADPH produced in the light-dependent reactions. As we shall see, it involves a series of reactions, known as the Calvin cycle, that are repeated over and over.*

The Light-Dependent Reactions

In the thylakoids, chlorophyll and other molecules are, according to the present model, packed into units called **photosystems.** Each unit contains from 250 to 400 molecules of pigment, which serve as light-trapping antennae. Once light energy is absorbed by one of the antenna pigments, it is bounced around (like a hot potato) among the other pigment molecules of the photosystem until it reaches a particular form of chlorophyll *a,* which is the reaction center of the photosystem.

Present evidence indicates that there are two different photosystems. In **Photosystem I,** the reactive chlorophyll *a* molecule is known as P_{700} (P is for pigment) because one of the peaks of its absorption curve is at 700 nanometers, a slightly longer wavelength than the usual chlorophyll *a* peak. P_{700} is not a different kind of chlorophyll but rather two chlorophyll *a* molecules that are bound together. Its unusual properties result from its association with specific proteins in the thylakoid membrane and its position in relation to other molecules. **Photosystem II** also contains a reactive chlorophyll *a* molecule, P_{680}.

Figure 9–8 shows how the two photosystems are thought to work together in photosynthesis. Light energy enters Photosystem II, where it is trapped by the reactive chlorophyll *a* molecule P_{680}, causing an electron from the P_{680} molecule to be boosted to a higher energy level. The electron is then transferred to a primary electron-acceptor molecule that holds electrons at a higher energy level than does chlorophyll *a.* Next, the electron passes downhill along an electron transport chain to Photosystem I. As electrons travel down this chain, the energy they release is used to pump protons from the stroma into the thylakoid space. A proton gradient is thus established across the thylakoid membrane. ATP synthase complexes, embedded in the thylakoid membrane, provide a channel through which protons can flow down the gradient, back into the stroma. As they do so, the potential energy of the gradient drives the synthesis of ATP from ADP. For each molecule of ATP synthesized, two electrons must travel down the electron transport chain from Photosystem II to Photosystem I.

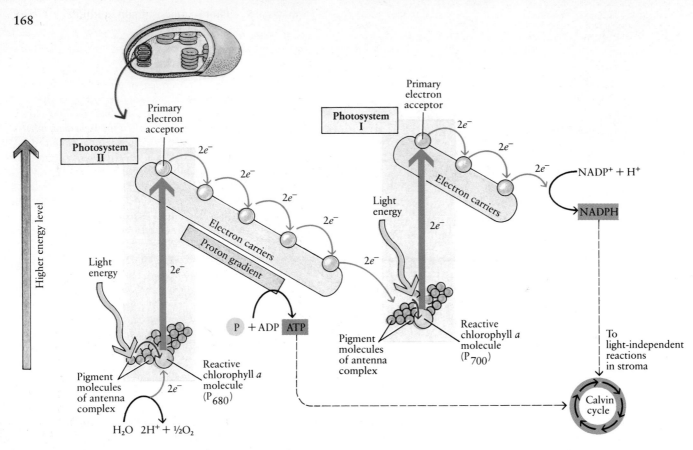

9–8 *A schematic diagram of the light-dependent reactions of photosynthesis. In Photosystem II, light energy absorbed by pigment molecules of the antenna complex is transferred to the reactive chlorophyll a molecule (P_{680}). This energy boosts electrons from the P_{680} molecule to a primary electron acceptor at a higher energy level. The electrons are then passed from the primary electron acceptor down an electron transport chain to a lower energy level, the reactive chlorophyll a molecule (P_{700}) of Photosystem I. As the electrons travel down this transport chain, the energy they release is used to power the synthesis of ATP. To generate one molecule of ATP, two electrons must be boosted from Photosystem II and pass down the electron transport chain to Photosystem I. The electrons removed from Photosystem II are replaced by electrons released when water molecules are split into protons (H^+ ions) and oxygen.*

In Photosystem I, light energy boosts electrons from the P_{700} molecule to another primary electron acceptor. From this electron acceptor, they are passed via other carriers to $NADP^+$. Two electrons and one proton (H^+ ion) combine with one molecule of $NADP^+$ to form one molecule of NADPH. The additional H^+ ion released from each water molecule that was split by the reaction in Photosystem II remains in solution in the thylakoid space. The electrons removed from Photosystem I are replaced by those from Photosystem II.

ATP and NADPH represent the net gain from the light-dependent reactions. To generate one molecule of NADPH, two electrons must be boosted from Photosystem II and two from Photosystem I.

This chemiosmotic process, illustrated in Figure 9–9, is known as **photophosphorylation.** It is similar to the oxidative phosphorylation that occurs in the mitochondria (page 151).

During photophosphorylation, three other processes are taking place simultaneously:

1. The P_{680} chlorophyll molecule, having lost two electrons, is avidly seeking replacements. It finds them in a water molecule, which is stripped of two electrons and then broken into protons and oxygen.

2. Additional light energy is trapped in the reactive chlorophyll molecule (P_{700}) of Photosystem I. The molecule is oxidized, and electrons are boosted to a primary electron acceptor, from which they go downhill to $NADP^+$. Two electrons and one proton combine with $NADP^+$ to form NADPH.

3. The electrons removed from the P_{700} molecule of Photosystem I are replaced by the electrons that moved downhill from the primary electron acceptor of Photosystem II.

Thus in the light there is a continuous flow of electrons from water to Photosystem II to Photosystem I to $NADP^+$. In the words of the late Nobel laureate Albert Szent-Györgyi: "What drives life is . . . a little electric current, kept up by the sunshine."

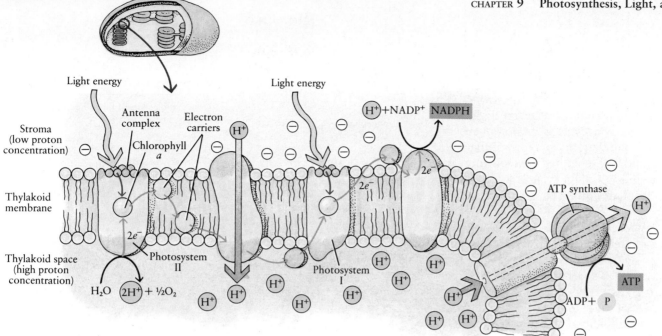

9–9 *Embedded in the thylakoid membrane, or closely associated with it, are the molecules and molecular complexes that participate in the light-dependent reactions of photosynthesis. These include the pigments and proteins of Photosystems I and II, the carriers of their respective electron transport chains, and all necessary enzymes, including ATP synthase.*

The arrangement of these molecules in the thylakoid membrane makes possible the

chemiosmotic synthesis of ATP in photophosphorylation. In this process, electrons from the reactive chlorophyll a molecule of Photosystem II are boosted to a high energy level by sunlight. As they flow down a series of electron carriers that lead to the reactive chlorophyll a molecule of Photosystem I, the energy they release is used to pump protons (H^+ ions) through one of the carrier complexes. The protons are pumped from the stroma into the thylakoid space. This creates

an electrochemical gradient of potential energy. As the protons flow down this gradient through the ATP synthase complexes, moving from the thylakoid space back into the stroma, ADP is phosphorylated to ATP.

The chemical structures of the electron carriers and enzymes of the thylakoid membrane (including ATP synthase) are only slightly different from those of the mitochondrial membrane.

The Light-Independent Reactions

In the first stage of photosynthesis, light energy is converted to electrical energy—the flow of electrons—and the electrical energy is converted to chemical energy stored in the bonds of NADPH and ATP. In the second stage of photosynthesis, this energy is used to reduce carbon and synthesize simple sugars.

Carbon is available to photosynthetic cells in the form of carbon dioxide. Algal cells obtain dissolved carbon dioxide directly from the surrounding water. In plants, however, carbon dioxide reaches the photosynthetic cells through specialized openings, called **stomata** (Figure 9–10), that are found in leaves and green stems.

The reactions in the second stage of photosynthesis require molecules—ATP and NADPH—that are synthesized in the chloroplast only in the presence of light. However, once these molecules are available, the subsequent reactions can take place regardless of whether light is or is not present. Thus, these reactions are said to be "light-independent."

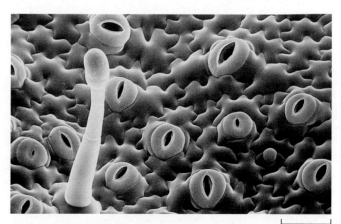

50 µm

9–10 *Scanning electron micrograph of stomata on the lower surface of a tobacco leaf. The plant can open or close its stomata, as needed; in this micrograph, most of them are open. It is through the stomata that the carbon dioxide required for photosynthesis diffuses into the interior of the leaf and the oxygen produced as a by-product of photosynthesis diffuses out. The structure extending from the surface at the lower left is a leaf hair.*

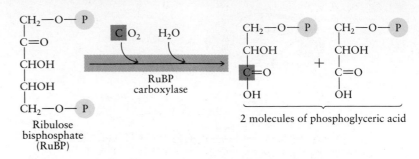

Ribulose
bisphosphate
(RuBP)

2 molecules of phosphoglyceric acid

The Calvin Cycle: The Three-Carbon Pathway

The reduction of carbon takes place in the stroma in a cyclical series of reactions named after its discoverer, Melvin Calvin. The **Calvin cycle** is analogous to the Krebs cycle (page 150) in that, in each turn of the cycle, the starting compound is regenerated. The starting (and ending) compound in the Calvin cycle is a five-carbon sugar with two phosphates attached, **ribulose bisphosphate (RuBP)**.

The cycle begins when carbon dioxide is bound to RuBP, which immediately splits to form two molecules of phosphoglyceric acid, or PGA (Figure 9–11). This reaction is catalyzed by a specific enzyme, **RuBP carboxylase** (known also as "rubisco"), which constitutes more than 15 percent of the chloroplast protein. In fact, RuBP carboxylase is thought to be the most abundant protein on Earth. Each of the PGA molecules produced in this initial reaction contains three carbon atoms; thus the Calvin cycle is also known as the three-carbon pathway.

The complete cycle is diagrammed in Figure 9–12. As in the Krebs cycle, each step is catalyzed by a specific enzyme. At each full turn of the cycle, one molecule of carbon dioxide enters the cycle and is reduced, and one molecule of RuBP is regenerated. Three turns of the cycle introduce three molecules of carbon dioxide, enough to produce one molecule of a three-carbon sugar. Six revolutions of the cycle, with the introduction of six molecules of carbon dioxide, are necessary to produce the equivalent of a six-carbon sugar, such as glucose. The overall equation is

$$6RuBP + 6CO_2 + 12NADPH + 12H^+ + 18ATP \longrightarrow$$
$$6RuBP + Glucose + 12NADP^+ +$$
$$18ADP + 18P_i + 6H_2O$$

The immediate product of the Calvin cycle is glyceraldehyde phosphate. This same three-carbon sugar-phosphate molecule is formed when the fructose bisphosphate molecule is split at the fourth step in glycolysis (page 146).

The Problem of Photorespiration

In the presence of ample carbon dioxide, the enzyme RuBP carboxylase feeds carbon dioxide into the Calvin cycle with great efficiency. However, when the carbon dioxide concentration in the leaf is low in relation to the oxygen concentration, this same enzyme catalyzes a reaction of RuBP with oxygen, rather than with carbon dioxide. This reaction is the first step in a process known as **photorespiration**, in which carbohydrates are oxidized to carbon dioxide and water in the presence of light. Unlike mitochondrial respiration, however, photorespiration is a wasteful process, yielding neither

9–11   *In order to trace the sequence of reactions in which carbon is reduced, Calvin and his collaborators briefly exposed photosynthesizing algae to carbon dioxide in which the carbon atom was radioactive (shown in red). The first product in which they could detect the radioactive carbon atom was phosphoglyceric acid (PGA). This indicated that when carbon dioxide binds to RuBP, the molecule immediately splits to form two molecules of PGA. This is the first step of the Calvin cycle.*

ATP nor NADH. In some plants, as much as 50 percent of the carbon fixed in photosynthesis may be reoxidized to carbon dioxide during photorespiration.

Conditions that can set the stage for photorespiration are quite common. Carbon dioxide is not continuously available to the photosynthesizing cells of a plant. As we have seen, it enters the leaf by way of the stomata, specialized pores that open and close, depending on, among other factors, water stress. When a plant is subjected to hot, dry conditions, it must close its stomata to conserve water. This cuts off the supply of carbon dioxide and also allows the oxygen produced by photosynthesis to accumulate. The resulting low carbon dioxide and high oxygen concentrations lead to photorespiration.

Also, when plants are growing in close proximity to one another, the air surrounding the leaves may be quite still, with little gas exchange between the immediate environment and the atmosphere as a whole. Under such conditions, the concentration of carbon dioxide in the air closest to the leaves may be rapidly reduced to low levels by the photosynthetic activities of the plant. Even if the stomata are open, the concentration gradient between the outside of the leaf and the inside is so slight that little carbon dioxide diffuses into the leaf. Meanwhile, oxygen accumulates, and photorespiration is likely to occur, greatly reducing the photosynthetic efficiency of the plants.

A Solution: The Four-Carbon Pathway

The problem of photorespiration is solved in some plants by an alternative pathway for the capture of carbon dioxide. In these plants, the first step in the fixation of carbon is the binding of carbon dioxide to a compound known as phosphoenolpyruvic acid (PEP), forming the four-carbon compound oxaloacetic acid (Figure

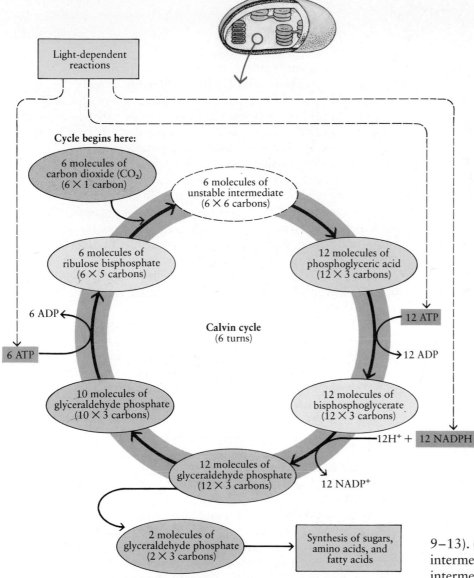

Cycle begins here:

6 molecules of
carbon dioxide (CO₂)
(6 × 1 carbon)

6 molecules of
unstable intermediate
(6 × 6 carbons)

6 molecules of
ribulose bisphosphate
(6 × 5 carbons)

12 molecules of
phosphoglyceric acid
(12 × 3 carbons)

12 ATP

12 ADP

6 ADP

6 ATP

Calvin cycle
(6 turns)

10 molecules of
glyceraldehyde phosphate
(10 × 3 carbons)

12 molecules of
bisphosphoglycerate
(12 × 3 carbons)

12H⁺ + 12 NADPH

12 molecules of
glyceraldehyde phosphate
(12 × 3 carbons)

12 NADP⁺

2 molecules of
glyceraldehyde phosphate
(2 × 3 carbons)

Synthesis of sugars,
amino acids, and
fatty acids

Light-dependent
reactions

9–12 *Summary of the Calvin cycle. At each full "turn" of the cycle, one molecule of carbon dioxide enters the cycle. Six turns are summarized here—the number required to make two molecules of glyceraldehyde phosphate, the equivalent of one molecule of a six-carbon sugar. The energy that drives the Calvin cycle is in the form of ATP and NADPH, produced by the light-dependent reactions in the first stage of photosynthesis.*

The cycle begins at the upper left. Six molecules of ribulose bisphosphate (RuBP), a five-carbon compound, are combined with six molecules of carbon dioxide. This produces six molecules of an unstable intermediate that splits apart immediately, yielding twelve molecules of phosphoglyceric acid (PGA), a three-carbon compound. In the next two steps, these molecules are reduced to twelve molecules of glyceraldehyde phosphate. Ten of these three-carbon molecules are combined and rearranged to form six five-carbon molecules of RuBP. The two "extra" molecules of glyceraldehyde phosphate represent the net gain from the Calvin cycle. Using these molecules as the starting point, the cell can synthesize a variety of sugars, amino acids, and fatty acids.

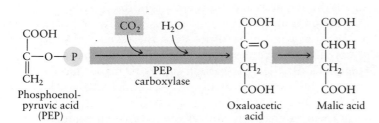

9–13 *In C₄ plants, carbon dioxide is first combined with phosphoenolpyruvic acid (PEP) to yield the four-carbon oxaloacetic acid. This reaction is catalyzed by the enzyme PEP carboxylase. The oxaloacetic acid is then converted either to malic acid, as shown here, or to aspartic acid. Carbon dioxide is subsequently released from the malic acid (or aspartic acid) for use within the Calvin cycle. Sugarcane, corn, and sorghum are among the best-known C₄ plants.*

9–13). (Phosphoenolpyruvic acid, you may recall, is an intermediate in glycolysis, and oxaloacetic acid is an intermediate in the Krebs cycle.) Plants that utilize this pathway are commonly called **C₄ plants,** as distinct from the **C₃ plants,** in which carbon is bound first into the three-carbon compound phosphoglyceric acid (PGA).

PEP carboxylase, the enzyme that catalyzes the formation of oxaloacetic acid in C₄ plants, has a much higher affinity for carbon dioxide than does RuBP carboxylase, the enzyme that catalyzes the formation of PGA. Compared with RuBP carboxylase, PEP carboxylase fixes carbon dioxide faster and at lower levels, thereby keeping the carbon dioxide concentration lower within the cells near the surface of the leaf. This maximizes the gradient of carbon dioxide between the outside air and the leaf interior. Thus, when the stomata are open, carbon dioxide readily diffuses down the concentration gradient into the leaf. If the stomata must be closed much of the time—as they must be to conserve water in a hot, dry climate—the plant with C₄ metabolism will take up more carbon dioxide with each gasp (so to speak) than the plant that has only C₃ metabolism.

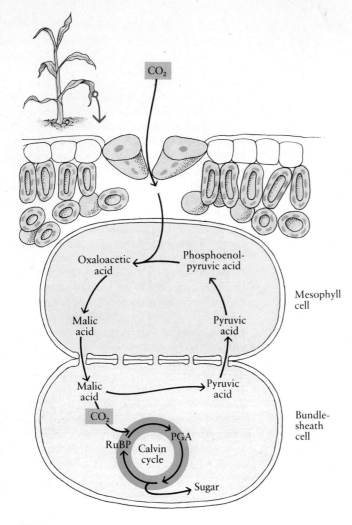

9–14 *A pathway for carbon fixation in C_4 plants. In cells in the leaf interior, known as mesophyll ("middle leaf") cells, CO_2 is initially fixed in oxaloacetic acid, which is then converted to malic acid. By way of the plasmodesmata, the malic acid passes to bundle-sheath cells, which form sheaths around the vascular bundles. Here the malic acid is broken down to pyruvic acid and CO_2, which enters the Calvin cycle. The pyruvic acid moves through the plasmodesmata into the mesophyll cells, where it is converted back to phosphoenolpyruvic acid (PEP), ready to receive another molecule of CO_2.*

The carbon dioxide bound into oxaloacetic acid by C_4 plants is ultimately released and then bound to RuBP, entering the Calvin cycle. This does not occur, however, until it has passed through a series of reactions (Figure 9–14) that transport it to other cells, known as **bundle-sheath cells**, that surround the vascular bundles of the leaf. In the bundle-sheath cells, the carbon dioxide is released at a high concentration, enabling RuBP carboxylase and the other enzymes of the Calvin cycle to function with maximum efficiency.

C_4 plants evolved primarily in the tropics and are especially well adapted to high light intensities, high temperatures, and dryness. The optimal temperature range for C_4 photosynthesis is much higher than that for C_3 photosynthesis, and C_4 plants flourish at temperatures that would eventually be lethal to many C_3 species. Because of their more efficient use of carbon dioxide, C_4 plants can attain the same photosynthetic rate as C_3 plants but with smaller stomatal openings and, hence, with considerably less water loss.

A familiar example of the competitive capacity of C_4 plants is seen in lawns in the summertime. In most parts of the United States, lawns consist mainly of C_3 grasses such as Kentucky bluegrass. As the summer days become hotter and drier, these dark green, fine-leaved grasses are often overwhelmed by rapidly growing crabgrass, which disfigures the lawn as its yellowish-green, broader-leaved plants slowly take over. Crabgrass, you will not be surprised to hear, is a C_4 plant.

The list of plants known to utilize the four-carbon pathway has grown to over 1,000 species. Current evidence indicates that this pathway has arisen independently many times in the course of evolution. It is another example of the exquisite adaptation of living systems to their environment.

The Products of Photosynthesis

Glyceraldehyde phosphate, the three-carbon sugar produced by the Calvin cycle, may seem an insignificant reward, both for all the enzymatic activity on the part of the cell and for the intellectual effort required to understand photosynthesis. However, this molecule and those derived from it provide (1) the energy source for virtually all living systems, and (2) the basic carbon skeleton from which the great diversity of organic molecules can be synthesized. Carbon has been fixed—that is, it has been brought from the inorganic world into the organic one.

Molecules of glyceraldehyde phosphate may flow into a variety of different metabolic pathways, depending on the activities and requirements of the cell. Often they are built up to glucose or fructose, following a sequence that is in many of its steps the reverse of the glycolytic sequence described in the previous chapter. Plant cells use these six-carbon sugars to make starch and cellulose for their own purposes and sucrose for export to other parts of the plant body. Animal cells store them as glycogen. All cells use sugars, including glyceraldehyde phosphate and glucose, as the starting point for the manufacture of other carbohydrates, fats and other lipids, and, with the addition of nitrogen, amino acids and nitrogenous bases. Finally, as we saw in the preceding chapter, the carbon fixed in photosynthesis is the source of ATP energy for heterotrophic organisms and for the heterotrophic cells of plants.

The Carbon Cycle

In photosynthesis, living systems incorporate carbon dioxide from the atmosphere into organic compounds. In respiration, organic compounds are broken down again into carbon dioxide and water. These processes, viewed on a worldwide scale, result in the carbon cycle.

The principal photosynthesizers in the carbon cycle are plants and microscopic marine algae. These organisms synthesize carbohydrates from carbon dioxide and water and release oxygen into the atmosphere. About 100 billion metric tons of carbon per year are bound into carbon compounds by photosynthetic organisms.

Some of the carbohydrates are used by the photosynthesizers themselves in glycolysis and respiration. Plants release carbon dioxide from their roots and leaves, and marine algae release it into the water. The carbon dioxide dissolved in the waters of the Earth is in equilibrium with the carbon dioxide in the atmosphere. Some 500 billion metric tons of carbon are "stored" as dissolved carbon dioxide in surface waters, and some 740 billion metric tons in the atmosphere.

Some of the carbohydrates produced in photosynthesis are oxidized in the cells of animals that feed on the living plants, on algae, and on one another, releasing carbon dioxide. An enormous amount of carbon, however, is contained in the dead bodies of plants and other organisms plus discarded leaves and

shells, feces, and other waste materials that settle into the soil or sink to the ocean floors where they are consumed by small invertebrates, bacteria, and fungi. The metabolic processes of these organisms also release carbon dioxide into the reservoir of the atmosphere and the oceans. Another, even larger store of carbon lies below the surface of the Earth in the form of coal and oil, deposited there some 300 million years ago.

The natural processes of photosynthesis and respiration generally balance one another out. Over the long span of geologic time, the carbon dioxide concentration of the atmosphere has varied, but for the last 10,000 years it has remained relatively constant. By volume, it is a very small proportion of the atmosphere, only about 0.03 percent. It is important, however, because carbon dioxide, unlike most other components of the atmosphere, acts as a blanket, trapping the heat produced when sunlight is absorbed at the Earth's surface. Since 1850, carbon dioxide concentrations in the atmosphere have been increasing, owing in large part to our use of fossil fuels, to our plowing of the soil, and to our destruction of forests, particularly in the tropics.

At the beginning of the 1980s, a major study by the Environmental Protection Agency predicted that this increase in the carbon dioxide "blanket" would significantly increase the average temperatures here on Earth, be-

ginning by the turn of the century. A number of subsequent studies, coupled with worldwide temperatures in the 1980s that were higher than in any decade since records have been kept, have now convinced not only members of the scientific community but also political leaders throughout the world that the increase is real and has already begun.

The consequences of this increase cannot be known with certainty. In some parts of the world, there may be lengthened growing seasons, increased precipitation, and, in conjunction with the increased levels of carbon dioxide available to plants, greater agricultural productivity. In other parts of the world, however, it is thought that precipitation will be reduced, lowering crop yields and, in already arid areas, accelerating the spread of the great deserts of the world. Rises in sea level, resulting from the melting of polar ice, pose a potential threat not only to human inhabitants of coastal regions but also to the multitude of marine organisms that live or reproduce in shallow waters at the continents' edges.

Although it now seems apparent that a certain amount of global warming is the inevitable consequence of past and present human activities, national and international efforts are currently underway to develop strategies for agriculture, energy use, and manufacturing that will slow—and perhaps eventually reverse—the process. It remains to be seen if we have awakened in time.

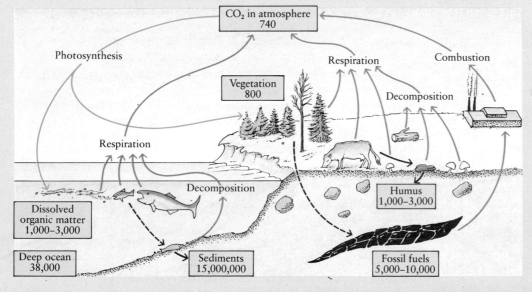

The carbon cycle. The arrows indicate the movement of carbon atoms. The numbers are all estimates of the amount of carbon, expressed in billions of metric tons, stored in various reservoirs as of 1988. The amount of carbon released by respiration and combustion has begun to exceed the amount fixed by photosynthesis. The amount of carbon stored in the atmosphere is currently increasing by 3 billion metric tons each year.

Summary

In photosynthesis, light energy is converted to chemical energy, and carbon is fixed into organic compounds.

Light energy is captured by the living world by means of pigments. The pigments involved in photosynthesis in eukaryotes include the chlorophylls and the carotenoids. Light absorbed by antenna pigments boosts their electrons to higher energy levels. Because of the way the pigments are packed into membranes, they are able to transfer this energy to reactive chlorophyll *a* molecules.

In eukaryotes, photosynthesis takes place within organelles known as chloroplasts, which are surrounded by two membranes. Contained within the membranes of the chloroplast are a solution of organic compounds and ions known as the stroma and a complex internal membrane system that forms sacs called thylakoids. The pigments and other molecules responsible for capturing light are located in and on the thylakoid membranes.

Photosynthesis takes place in two stages: (1) the light-dependent reactions, in which light energy trapped by chlorophyll is converted to the chemical energy of ATP and NADPH; and (2) the light-independent reactions, in which carbon dioxide is reduced and carbohydrates formed.

In the currently accepted model of the light-dependent reactions, light energy strikes antenna pigments of Photosystem II. Electrons are boosted uphill from the reactive chlorophyll *a* molecule P_{680} to a primary electron acceptor. As the electrons are removed, they are replaced by electrons from water molecules, with the simultaneous production of free oxygen and protons (H^+ ions). The electrons then pass downhill to Photosystem I along an electron transport chain. This passage generates a proton gradient that drives the synthesis of ATP from ADP (photophosphorylation). Light energy absorbed in antenna pigments of Photosystem I and passed to chlorophyll P_{700} results in the boosting of electrons to another primary electron acceptor. The electrons removed from P_{700} are replaced by the electrons from Photosystem II. The electrons are ultimately accepted by the electron carrier $NADP^+$. The energy yield from this series of reactions is contained in molecules of NADPH and ATP.

In the light-independent reactions, which take place in the stroma, NADPH and ATP produced in the light-dependent reactions are used to reduce carbon dioxide to organic carbon. This is accomplished by means of the Calvin cycle. In the Calvin cycle, a molecule of carbon dioxide is combined with the starting material, a five-carbon sugar called ribulose bisphosphate (RuBP). At each turn of the cycle, one carbon atom enters the cycle. Three turns of the cycle produce a three-carbon molecule, glyceraldehyde phosphate. Two molecules of glyceraldehyde phosphate (six turns of the cycle) can combine to form a glucose molecule. With each turn of the cycle, RuBP is regenerated. The glyceraldehyde phosphate can also be used as a starting material for other organic compounds needed by the cell.

When the carbon dioxide concentration in photosynthetic cells is low in relation to the oxygen concentration, photorespiration occurs. In this process, which requires sunlight, carbohydrates are oxidized to carbon dioxide and water, reducing the photosynthetic efficiency of the plant. In C_4 plants, photorespiration is minimized. Carbon dioxide is initially accepted by a compound known as PEP to yield the four-carbon oxaloacetic acid. Following a series of reactions that transport it to cells that surround the vascular bundles of the leaf, the carbon dioxide is released, binding with RuBP and entering the Calvin cycle. The C_4 pathway enables plants to minimize water loss and maximize photosynthetic efficiency. Under conditions of intense sunlight, high temperatures, or drought, C_4 plants are more efficient than C_3 plants.

Questions

1. Distinguish between the following: grana/thylakoid; stroma/thylakoid space; light-dependent reactions/light-independent reactions; Photosystem I/Photosystem II; C_3 photosynthesis/C_4 photosynthesis.

2. Why is it plausible to argue, as the Nobel laureate George Wald does, that wherever in the universe we find living organisms, we will find them (or at least some of them) to be colored?

3. Predict what colors of light might be most effective at stimulating plant growth.

4. Sketch a chloroplast and label its structures. Compare your sketch with Figure 9–5.

5. Describe in general terms the events of photosynthesis.

6. What is the source of the oxygen released in photosynthesis?

7. The experiment shown in Figure 4–2 on page 73 was attempted in Stockton, California, in 1973, but instead of making amino acids, the electric sparks caused the apparatus to explode. (No one was hurt, fortunately.) What is present in today's atmosphere that was not present in the primitive atmosphere and that would account for the explosion?

8. Some plants have acquired C_4 photosynthesis in the course of evolution. How is this adaptation advantageous to these plants? Many plants, however, have not evolved C_4 photosynthesis. Why is it advantageous to such plants *not* to have C_4 photosynthesis?

9. Trace a carbon atom through a series of biological events, such as those illustrated on page 126.

Suggestions for Further Reading

Alberts, Bruce, Dennis Bray, Julian Lewis, Martin Raff, Keith Roberts, and James D. Watson: *Molecular Biology of the Cell,* 3d ed., Garland Publishing, Inc., New York, 1994.

This outstanding cell biology text includes a clear, up-to-date discussion of our current understanding of the processes that occur in the chloroplast.

Bazzaz, Fakhri A., and Eric D. Fajer: "Plant Life in a CO_2-Rich World," *Scientific American,* January 1992, pages 68–74.

Berner, Robert A., and Antonio C. Lasaga: "Modeling the Geochemical Carbon Cycle," *Scientific American,* March 1989, pages 74–81.

Conant, James Bryant (ed.): *Harvard Case Histories in Experimental Science,* vol. 2, Harvard University Press, Cambridge, Mass., 1964.

Case #5, Plants and the Atmosphere, *edited by Leonard K. Nash, describes the early work on photosynthesis, often presented in the words of the investigators themselves. The narrative illuminates the historical context in which the discoveries were made.*

Detwiler, R. P., and C. A. S. Hall: "Tropical Forests and the Global Carbon Cycle," *Science,* vol. 239, pages 42–47, 1988.

Govindjee, and William J. Coleman: "How Plants Make Oxygen," *Scientific American,* February 1990, pages 50–58.

Houghton, Richard A., and George M. Woodwell: "Global Climatic Change," *Scientific American,* April 1989, pages 36–44.

Jones, Philip D., and Tom M. L. Wigley: "Global Warming Trends," *Scientific American,* April 1990, pages 84–91.

Kasting, James F., Owen B. Toon, and James B. Pollack: "How Climate Evolved on the Terrestrial Planets," *Scientific American,* February 1988, pages 90–97.

Ledbetter, M. C., and Keith R. Porter: *Introduction to the Fine Structures of Plant Cells,* Springer-Verlag, New York, 1970.

An excellent atlas of electron micrographs of plant cells, with detailed explanations.

Nassau, Kurt: "The Causes of Color," *Scientific American,* October 1980, pages 124–154.

Post, Wilfred M., et al.: "The Global Carbon Cycle," *American Scientist,* vol. 78, pages 310–326, 1990.

Scientific American: *Managing Planet Earth,* W. H. Freeman and Company, New York, 1990.*

This reprint of the September 1989 issue of Scientific American *contains 11 articles on the environmental crises currently threatening our planet, the effect of continuing human population growth, and strategies for the future to sustain economic growth without further damage to the environment. Chapters 2 and 3 examine the changes now occurring in the atmosphere and in worldwide climate as a consequence of the imbalance in the carbon cycle caused by human activities.*

Scientific American: *Molecules to Living Cells,* W. H. Freeman and Company, New York, 1980.*

A collection of articles from Scientific American. *Chapter 12 is an excellent presentation of the chemiosmotic synthesis of ATP in both mitochondria and chloroplasts, and Chapter 13 examines the photosynthetic membrane.*

Waggoner, Paul E.: "Agriculture and Carbon Dioxide," *American Scientist,* vol. 72, pages 179–184, 1984.

Youvan, Douglas C., and Barry L. Marrs: "Molecular Mechanisms of Photosynthesis," *Scientific American,* June 1987, pages 42–48.

*Available in paperback.

Genetics

The Reproduction of Cells

"Of everything that creepeth on the Earth, there went in two and two . . ."

Knowledge that like begets like is at least as old as recorded history. Biological inheritance has been perhaps the single most important factor in shaping the social evolution of humankind. Since the beginning of civilization, the ties of maternity and paternity have governed the transfer of power and wealth (and also of powerlessness and poverty). Today, throughout the world, the biological relationships of race and tribe and royalty still play a major role in determining the fates of nations and of individuals.

What is the source of this great power? What exactly do we mean by inheritance? What is passed from generation to generation? Why is it necessary to have two lions—and lions of opposite sexes, at that— in order to ensure a continuity of lions?

The way in which the continuity of life is preserved from generation to generation, and the way that male and female each contribute to the characteristics of their offspring were fundamental questions of biological science from its earliest beginnings. These ancient questions have now been largely answered, as the result of an extraordinary period of research that began in the middle of the nineteenth century and that has continued, at an ever-accelerating pace, to the present.

Our story begins with a seemingly simple event—the reproduction of the individual cell.

Cells reproduce in a process known as **cell division,** in which the cellular contents are divided between two new **daughter cells.** In one-celled organisms, such as bacteria and many protists, cell division increases the number of individuals in the population. In many-celled organisms, such as plants and animals, cell division is the means by which the organism grows, starting from one single cell. It is also the means by which injured or worn-out tissues are repaired and replaced.

An individual cell grows by taking in materials from its environment and synthesizing these materials into new structural and functional molecules. When the cell reaches a certain critical size and metabolic state, it divides. The two daughter cells, each of which has received about half of the mass of the parent cell, then begin growing again. A bacterial cell may divide as often as every six minutes. In a one-celled eukaryote, such as *Paramecium,* cell division may occur every few hours.

The new cells are structurally and functionally similar both to the parent cell and to one another. They are similar, in part, because each new cell usually receives about half of the parent cell's cytosol and organelles. More important, in terms of structure and function, each new cell inherits an exact replica of the hereditary information of the parent cell.

Cell Division in Prokaryotes

The distribution of exact replicas of the hereditary information is relatively simple in prokaryotic cells. In such cells, most of the hereditary material is in the form of a single, long, circular molecule of DNA, with which a variety of proteins are associated. This molecule, the cell's **chromosome,** is replicated before cell division.

According to present evidence, each of the two daughter chromosomes is attached to a different spot on the interior of the plasma membrane. As the membrane elongates, the chromosomes move apart

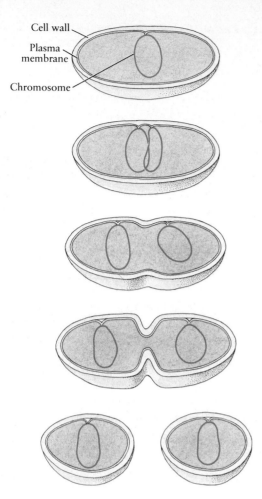

Cell wall
Plasma membrane
Chromosome

10–1   *Cell division in a bacterium. Attachment of the chromosome to an inward fold of the plasma membrane ensures that one chromosome replica is distributed to each daughter cell as the plasma membrane elongates.*

(Figure 10–1). When the cell has approximately doubled in size and the chromosomes are completely separated, the plasma membrane pinches inward, and a new cell wall forms that separates the two new cells and their chromosome replicas.

Cell Division in Eukaryotes

In eukaryotic cells, the problem of dividing the genetic material is much more complex. A typical eukaryotic cell contains about a thousand times more DNA than a prokaryotic cell, and this DNA is linear, forming a number of distinct chromosomes. For instance, human body cells have 46 chromosomes, each different from the others. When these cells divide, each daughter cell has to receive one copy—and only one copy—of each of the 46 chromosomes.

The solution to this problem is, as you will see, ingenious and elaborate. In a series of steps, known collectively as **mitosis,** a complete set of chromosomes is allocated to each of two daughter nuclei. Mitosis is usually followed by **cytokinesis,** a process that divides the cell into two new cells. Each new cell contains not only a nucleus with a full chromosome complement but also approximately half of the cytosol and organelles of the parent cell.

Although mitosis and cytokinesis are the culminating events of cell division, they represent only two stages of a larger process.

The Cell Cycle

Dividing eukaryotic cells pass through a regular, repeated sequence of cell growth and division, known as the **cell cycle** (Figure 10–2). The cycle consists of five major phases: G_1, S, G_2, mitosis, and cytokinesis. Completion of the cycle requires varying periods of time from a few hours to several days, depending on both the type of cell and external factors, such as temperature or available nutrients.

Before a cell can begin mitosis and actually divide, it must replicate its DNA, synthesize more of the proteins associated with the DNA in the chromosomes, produce a supply of organelles adequate for two daughter cells, and assemble the structures needed to carry out mitosis and cytokinesis. These preparatory processes occur during the G_1, S, and G_2 phases of the cell cycle, which are known collectively as **interphase.**

The key process of DNA replication occurs during the **S phase** (synthesis phase) of the cell cycle, a time in which many of the DNA-associated proteins are also synthesized. G (gap) phases precede and follow the S phase.

The **G_1 phase,** which precedes the S phase, is a period of intense biochemical activity. The cell doubles in size, and its enzymes, ribosomes, mitochondria, and other cytoplasmic molecules and structures also increase in number. Some of the structures can be synthesized entirely *de novo* ("from scratch") by the cell. These include microtubules, actin filaments, and ribosomes, all of which are composed, at least in part, of protein subunits. Membranous structures, such as Golgi complexes, lysosomes, vacuoles, and vesicles, are all apparently derived from the endoplasmic reticulum, which is renewed and enlarged by the synthesis of lipid and protein molecules.

In those cells that contain centrioles (that is, most eukaryotic cells except those of the flowering plants),

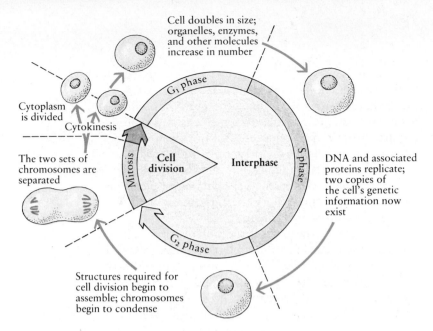

Cell doubles in size; organelles, enzymes, and other molecules increase in number

Cytoplasm is divided

Cytokinesis

The two sets of chromosomes are separated

G₁ phase

Mitosis

Cell division

Interphase

S phase

DNA and associated proteins replicate; two copies of the cell's genetic information now exist

G₂ phase

Structures required for cell division begin to assemble; chromosomes begin to condense

10–2 *The cell cycle. Cell division, which consists of mitosis (the division of the nucleus) and cytokinesis (the division of the cytoplasm), takes place after the completion of the three preparatory phases (G_1, S, and G_2) that constitute interphase. After the G_2 phase comes mitosis, which is usually followed immediately by cytokinesis. In cells of different species or of different tissues within the same organism, the different phases occupy different proportions of the total cycle.*

the two centrioles begin to separate from each other and to replicate. Mitochondria and chloroplasts, which are produced only from previously existing mitochondria and chloroplasts, also replicate. Each mitochondrion and chloroplast has its own chromosome, which is organized much like the single circular chromosome of the bacterial cell.

During the **G_2 phase**, which follows the S phase and precedes mitosis, the final preparations for cell division occur. The newly replicated chromosomes, which are dispersed in the nucleus in the form of threadlike strands of chromatin, slowly begin to coil and condense into a compact form. Replication of the centriole pair is completed, with the two mature centriole pairs lying just outside the nuclear envelope, somewhat separated from each other (Figure 10–3). Also during this period, the cell begins to assemble the structures required for the allocation of a complete set of chromosomes to each daughter cell during mitosis and for the separation of the two daughter cells during cytokinesis.

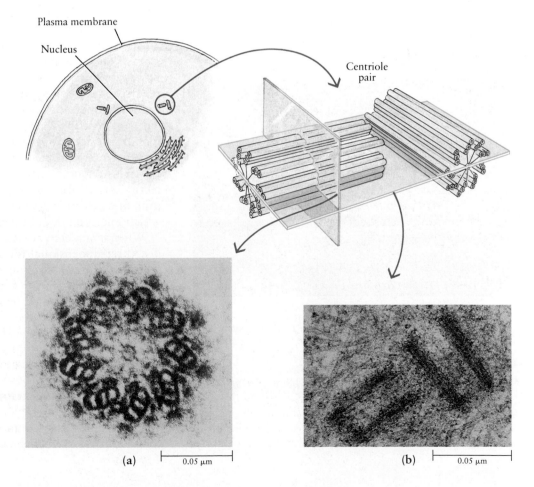

Plasma membrane

Nucleus

Centriole pair

10–3 *By the end of the G_2 phase of the cell cycle, two centriole pairs lie outside the nucleus. (a) As this cross section of a centriole from a human cell reveals, the structure of a centriole is identical to that of a basal body (page 103). (b) A centriole pair from a human cell, as seen in longitudinal section. Centrioles are formed either from preexisting centrioles, with the newly formed centriole appearing at right angles to the previously existing one, or from basal bodies.*

(a)   0.05 μm

(b)   0.05 μm

Regulation of the Cell Cycle

Some cell types pass through successive cell cycles throughout the life of the organism. This group includes the one-celled organisms and certain types of cells in both plants and animals. An example is provided by the **stem cells** in the human bone marrow that give rise to red blood cells. The average red blood cell lives only about 120 days, and there are about 25 trillion (2.5×10^{13}) of them in an adult. To maintain this number, about 2.5 million new red blood cells must be produced each second by the division of stem cells. The red blood cells themselves, like certain other highly specialized cells, such as nerve cells, lose the capacity to divide once they are mature.

Some cell types retain the capacity to divide after they mature but do so only under special circumstances. Cells in the human liver, for example, do not ordinarily divide. However, if a portion of the liver is removed surgically, the remaining cells (even if as few as a third of the original number) continue to divide until the liver reaches its former size. Then they stop. All told, about 2 trillion (2×10^{12}) cell divisions occur in an adult human every 24 hours, or about 25 million per second.

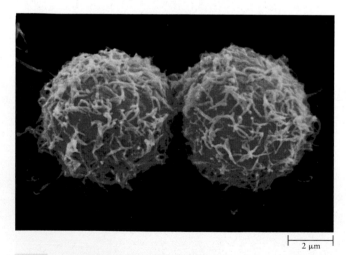

10–4   *These dividing malignant cells, of a type known as HeLa cells, are just at the point of separation. HeLa cells are derived from a tissue sample obtained in 1951 from an extremely aggressive cervical cancer that afflicted Henrietta Lacks, a Baltimore mother of four. Since that time, HeLa cells have been studied in laboratories around the world and have continued to divide, over and over and over again. By contrast, when normal human cells are cultured in the laboratory, they typically divide about 50 times and then die.*

In a multicellular organism, it is of critical importance that cells of the various different types divide at a sufficient rate to produce as many cells as are needed for growth and replacement—and only that many. If any particular cell type divides more rapidly than is necessary, the normal organization and functions of the organism may be disrupted as specialized tissues are invaded and overwhelmed by the rapidly dividing cells. Such is the course of events in cancer.

The nature of the control or controls that regulate the cell cycle is currently the subject of intense research, not only because of its biological interest but also because of its potential importance in the control of cancer. A number of different substances that influence the growth and division of cells have been discovered in the last few years. Work is now underway to determine their molecular structure, the structure of the membrane receptors to which they bind, and the events triggered within the cell in response to that binding.

Mitosis

The function of mitosis is to maneuver the replicated chromosomes so that each new cell gets a full complement—one of each. The capacity of the cell to accomplish this distribution depends on (1) the condensed state of the chromosomes during mitosis and (2) an assembly of microtubules known as the **spindle.**

By the beginning of mitosis, the chromosomes have become sufficiently condensed to be visible under the light microscope. Each chromosome consists of two replicas, called **chromatids** (Figure 10–5), joined together by a constricted area common to both chromatids. This region of attachment is known as the **centromere.** Within the constricted region are disk-shaped protein-containing structures, the **kinetochores,** to which microtubules of the spindle are attached.

When completely formed, the spindle (Figure 10–6) is a three-dimensional football-shaped structure, consisting of at least two groups of microtubules: (1) **polar fibers,** which reach from each pole of the spindle (analogous to the ends of the football) to a central region midway between the poles, and (2) **kinetochore fibers,** which are attached to the kinetochores of the replicated chromosomes. These two groups of spindle fibers are responsible for separating the sister chromatids during mitosis.

In those cells that contain centrioles, each pole of the spindle is marked by a pair of newly replicated centrioles. Such cells also contain a third group of shorter spindle fibers, extending outward from the centrioles.

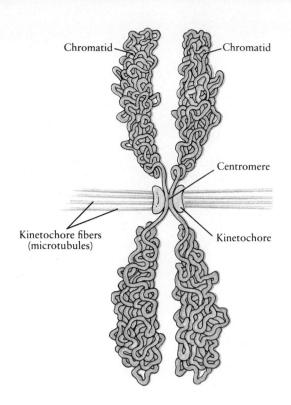

Chromatid — Chromatid

Centromere

Kinetochore fibers (microtubules)

Kinetochore

10–5   *A replicated, fully condensed chromosome. The DNA and proteins that form the chromosome were replicated during the S phase of the cell cycle. Each chromosome now consists of two identical parts, called chromatids. The centromere, the constricted area at the center, is the site of attachment of the two chromatids. The kinetochores are protein-containing structures, one on each chromatid, associated with the centromere. Attached to the kinetochores are microtubules that form part of the spindle.*

These additional fibers are known collectively as the **aster.** It has been hypothesized that the fibers of the aster may brace the poles of the spindle against the plasma membrane during the movements of mitosis. In cells that lack centrioles and asters, the rigid cell wall may perform a similar function.

Most of the tubulin molecules from which the microtubules of the spindle are formed (see Figure 5–19b, page 101) are apparently borrowed from the cytoskeleton. As a consequence, dividing cells take on a characteristic rounded appearance. Following cell division, the spindle is disassembled, the cytoskeletal network of microtubules is reassembled, and the cell assumes its nondividing shape.

The Phases of Mitosis

The process of mitosis is conventionally divided into four phases: **prophase, metaphase, anaphase,** and **telophase.** Of these, prophase is usually by far the longest. If a mitotic division takes 10 minutes (which is about the minimum time required), during about six of these minutes the cell will be in prophase.

10–6   *The mitotic spindle. (a) This micrograph of a dividing cell from the lung of an amphibian illustrates the spindle's three-dimensional quality. The red fibers are the spindle microtubules. The large blue bodies near the equator of the spindle are the chromosomes.*

The basic framework of the spindle in (b) an animal cell and (c) a plant cell. For clarity, only one replicated chromosome is shown in each cell. In the animal cell, a centriole pair is present at each pole. The polar fibers, which form the bulk of the spindle, are sharply focused on the centrioles, and additional fibers radiate outward from the centrioles, forming the aster. In plant cells, by contrast, centrioles are absent, and the spindle is less sharply focused at the poles. Although there are a few shorter fibers near the poles, no aster is formed.

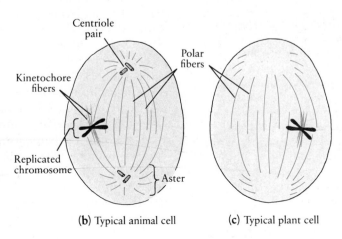

Centriole pair

Polar fibers

Kinetochore fibers

Replicated chromosome

Aster

(a)   10 μm

(b) Typical animal cell

(c) Typical plant cell

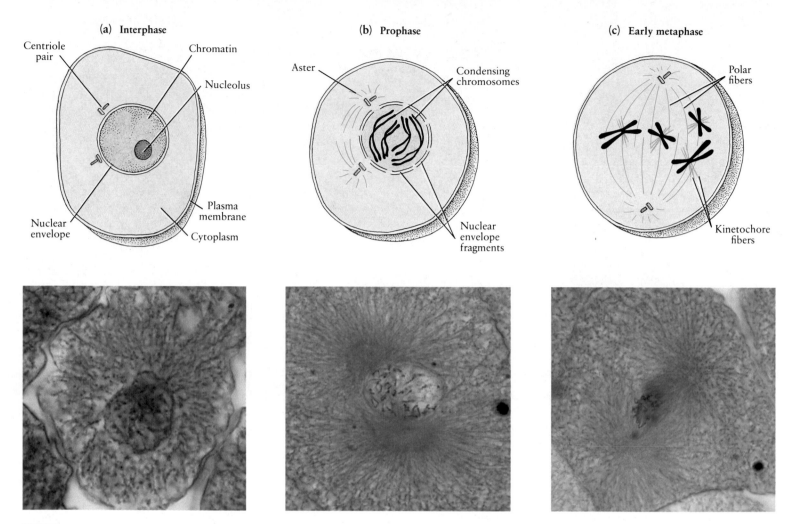

(a) Interphase

Centriole pair
Chromatin
Nucleolus
Plasma membrane
Nuclear envelope
Cytoplasm

(b) Prophase

Aster
Condensing chromosomes
Nuclear envelope fragments

(c) Early metaphase

Polar fibers
Kinetochore fibers

10–7   *The phases of mitosis as they occur in an animal cell. The accompanying micrographs show the phases of mitosis in embryonic cells of a whitefish. (a) Interphase. The chromatin, although replicated, has not yet begun to condense. Two centriole pairs lie just outside the nuclear envelope. (b) Prophase. The centrioles begin to move toward opposite poles of the cell, the condensing chromosomes become visible, the nuclear envelope breaks down, and the spindle apparatus begins to form. In cells with centrioles, the aster becomes visible. (c) Early metaphase. The chromatid pairs are tugged back and forth as they are moved into position, maneuvered by the polar and kinetochore fibers of the spindle. (d) Late metaphase. The chromatid pairs are lined up at the equator of the cell. (e) Anaphase. The two sets of newly separated chromosomes are pulled toward opposite poles of the cell. (f) Telophase. Nuclear envelopes form around the two sets of chromosomes, and the chromosomes uncoil, becoming diffuse once more. The spindle apparatus disappears, and the plasma membrane pinches in, a process that will ultimately separate the two daughter cells.*

During the interphase portions of the cell cycle, little can be seen in the nucleus (Figure 10–7a). The two centriole pairs can be seen at one side of the nucleus, outside the nuclear envelope.

By early prophase (Figure 10–7b), the chromatin has condensed sufficiently that the individual chromosomes, each consisting of two chromatids, become visible under the light microscope. (The threadlike appearance of the chromosomes when they first become visible is the source of the name "mitosis"; *mitos* is the Greek word for "thread.") The cell becomes more spherical and the cytoplasm more viscous at this stage, as the microtubules of the cytoskeleton are disassembled in preparation for the formation of the spindle.

During prophase, the centriole pairs move apart. Between the centriole pairs, forming as the centriole pairs separate (or, more likely, separating the centriole pairs as they form), are the microtubules that become the polar fibers of the spindle. In those cells that have centrioles, the microtubules that form the aster radiate outward from the centrioles. By this time the nucleolus has

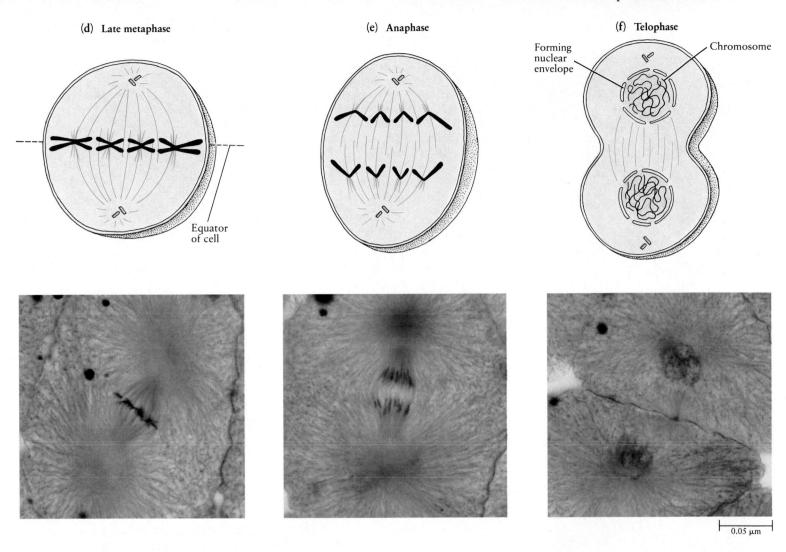

(d) Late metaphase

Equator of cell

(e) Anaphase

(f) Telophase

Forming nuclear envelope

Chromosome

0.05 μm

usually disappeared from view. As the chromosomes continue to condense, the nuclear envelope breaks down, dispersing into membranous fragments similar to fragments of endoplasmic reticulum.

By the end of prophase, the chromosomes are fully condensed. And, with the dispersion of the nuclear envelope, they are no longer separated from the cytoplasm. The centriole pairs have reached the poles of the cell, and the members of each pair are of equal size. The polar fibers of the spindle are fully formed, and the kinetochore fibers, attached to the kinetochores of the chromosomes, have also formed.

During early metaphase (Figure 10–7c), the chromatid pairs move back and forth within the spindle, apparently maneuvered by the spindle fibers. They appear to be tugged first toward one pole and then toward the other. Finally they become arranged precisely at the midplane (equator) of the cell (Figure 10–7d). This marks the end of metaphase.

At the beginning of anaphase, the most rapid stage of mitosis, the two chromatids of each pair, until this point attached at the centromere, separate. This happens simultaneously in all of the chromatid pairs. The chromatids of each pair then move apart, each chromatid now a separate chromosome, each apparently drawn toward the opposite pole by the kinetochore fibers. The centromeres move first, while the arms of the newly separated chromosomes seem to drag behind. As anaphase continues (Figure 10–7e), the two identical sets of chromosomes move rapidly toward the opposite poles of the spindle.

By the beginning of telophase, the chromosomes have reached the opposite poles and the spindle begins to disperse into tubulin molecules. During late telophase (Figure 10–7f), a nuclear envelope re-forms around each set of chromosomes, which once more become diffuse. In each nucleus, the nucleolus reappears. Often, a new centriole begins to form adjacent to each of the previous ones. As we saw earlier, replication of the centrioles continues during the subsequent cell cycle, so that each daughter cell has two centriole pairs by prophase of the next mitotic division.

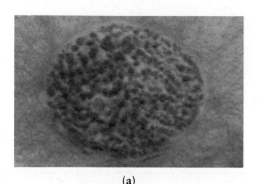

(a)

(b)

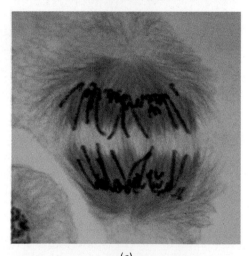

(c)

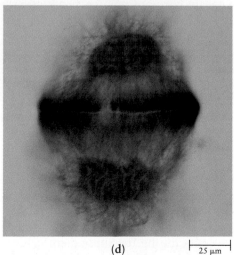

(d)   ⊢——25 μm——⊣

10–8   *Dividing cells in a seed of the African globe lily illustrate the phases of mitosis in a plant cell. (a) Early prophase. The microtubules of the cytoskeleton are in a meshwork surrounding the nucleus and have not yet become reorganized into a spindle. The chromosomes are condensing. (b) Metaphase. The condensed chromosomes are lined up at the equator of the cell. (c) Anaphase. The chromosomes are moving toward the poles. (d) Telophase. Note the unstained region at the equator of the cell. During cytokinesis, the cell plate will form here, completing the separation of the two cells.*

Cytokinesis

Cytokinesis, the division of the cytoplasm, usually but not always accompanies mitosis, the division of the nucleus. The visible process of cytokinesis generally begins during telophase of mitosis, and it usually divides the cell into two nearly equal parts.

Cytokinesis differs significantly in plant and animal cells. In animal cells, during early telophase, the plasma membrane begins to constrict along the circumference of the cell in the plane of the equator of the spindle. At first, a furrow appears on the surface, and this gradually deepens into a groove (Figure 10–9). Eventually, the connection between the daughter cells dwindles to a slender thread, which soon parts. The constriction of the plasma membrane is caused by a belt of actin filaments (page 101). These filaments are believed to act as a sort of "purse string," gathering in the plasma membrane of the parent cell at its midline, thus pinching apart the two daughter cells.

In plant cells, the cytoplasm is divided at the midline by a series of polysaccharide-containing vesicles produced from the Golgi complexes (Figure 10–10). These vesicles eventually fuse to form a flat, membrane-bound region, the **cell plate**. As more vesicles fuse, the edges of the growing plate fuse with the plasma membrane, completing the separation of the two daughter cells. During the fusion process, the polysaccharides within the vesicles are deposited between the two cells. This polysaccharide layer becomes impregnated with pectins and ultimately forms the middle lamella (page 91). Each new cell then constructs its own cell wall, laying down cellulose and other polysaccharides against the outer surface of its plasma membrane.

When cell division is complete, two daughter cells are produced, smaller than the parent cell but otherwise indistinguishable from it and from each other.

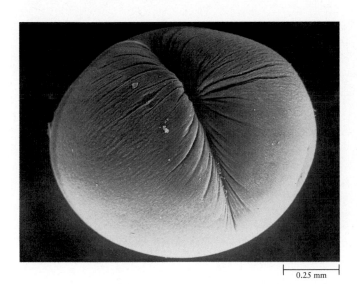

10–9   *Cytokinesis in an animal cell, the egg of a frog. In this scanning electron micrograph, note the puckering of the plasma membrane as the constriction furrow deepens into a groove. Eventually the two daughter cells will be connected by only a slender thread, similar to that connecting the two HeLa cells shown in Figure 10–4 (page 180). When that connection is severed, cytokinesis is complete.*

0.25 mm

Cell Division and the Reproduction of the Organism

In one-celled organisms, mitosis is the key event in reproduction. It is the means by which exact replicas of the chromosomes are transmitted from parent to offspring. Mitosis plays the same essential role in the reproduction of some large organisms. Sea anemones and sponges, for example, may break apart to form new sea anemones and sponges. Plants may produce roots or runners from which new individuals arise. Reproduction in which exact replicas of the chromosomes are transmitted from parent to offspring by mitosis is known as **asexual reproduction.** It is also sometimes called vegetative reproduction, since it is particularly common among plants.

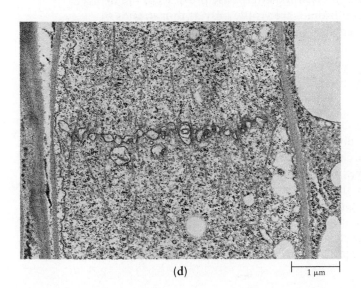

Plasma membrane

Cell wall

Vesicles

(a)

Cell plate

(b)

Middle lamella

New plasma membranes

Primary cell walls of daughter cells

(c)

(d)   1 µm

10–10   *In plants, the separation of the two daughter cells is effected by the formation of a structure known as the cell plate. (a) During telophase, vesicles containing polysaccharides become aligned across the equatorial plane of the cell. (b) The vesicles gradually fuse, forming a flat, membrane-bound region, the cell plate. This region is filled with the polysaccharides originally contained within the vesicles. The cell plate extends outward until it reaches the plasma membrane of the parent cell. As the last vesicles fuse with the plasma membrane, the two daughter cells become completely separated. (c) Each daughter cell then constructs a new primary cell wall on its side of the cell plate, which becomes the middle lamella.*

   (d) Cell plate formation in a seedling of the soft maple (Acer saccharinum).

10–11   *This green alga of the genus* Microasterias *can, like many one-celled organisms, reproduce either asexually or sexually. The mature cell consists of two halves separated by a narrow waist, which, in this micrograph, runs from top to bottom. When the cell divides, it splits at the waist. Each half then generates a perfect replica of itself, restoring the original shape. The new half-cell (at the right) is almost full size, and the cell will soon divide again.*

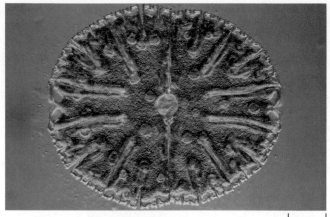

25 μm

All multicellular organisms produce new cells—and thus grow—by mitosis and cytokinesis, and some produce entirely new individuals in this fashion. Most multicellular organisms, however, arise from a unique single cell—the fertilized egg. This cell is formed by the fusion of two cells, the sperm and the egg. How are these cells formed? And what process ensures that the new organism receives a full complement of chromosomes, with exactly one half from the father and one half from the mother? In the next chapter, we shall turn our attention to these questions.

Summary

Cell division in prokaryotes is a relatively simple process, in which two daughter chromosomes are attached to different spots on the interior of the plasma membrane. As the membrane elongates, the chromosomes are separated. The plasma membrane then pinches inward, and a new cell wall forms, completing the division of the daughter cells.

Cell division is a more complex process in eukaryotes, which contain a vast amount of genetic material, organized into a number of different chromosomes. Dividing eukaryotic cells pass through a regular sequence of growth and division known as the cell cycle. The cycle consists of a G_1 phase, during which cytoplasmic molecules and structures increase in number; an S phase, during which the chromosomes are replicated; a G_2 phase, during which condensation of the chromosomes and assembly of the structures required for mitosis and cytokinesis begin; mitosis, during which the replicated chromosomes are apportioned between the two daughter nuclei; and cytokinesis, during which the cytoplasm is divided, separating the parent cell into two daughter cells. The first three phases of the cell cycle are known collectively as interphase. Regulation of the cell cycle is thought to involve a number of interacting chemical factors.

When the cell is in the interphase portions of the cycle, the chromosomes are visible only as thin strands of threadlike material (chromatin) within the nucleus. As prophase of mitosis begins, the condensing chromosomes, previously replicated during the S phase, become visible under the light microscope. They consist of pairs of identical replicas, called chromatids, held together at the centromere. Simultaneously, the spindle is forming. Prophase ends with the breakdown of the nuclear envelope and disappearance of the nucleolus. During metaphase, the chromatid pairs, maneuvered by the spindle fibers, move toward the center of the cell. At the end of metaphase, they are arranged on the equatorial plane. During anaphase, the sister chromatids separate, and each chromatid, now an independent chromosome, moves to an opposite pole. During telophase, a nuclear envelope forms around each group of chromosomes.

The spindle begins to break down, the chromosomes uncoil and once more become extended and diffuse, and nucleoli reappear.

Cytokinesis in animal cells results from constrictions in the plasma membrane between the two nuclei. In plant cells, the cytoplasm is divided by the fusion of vesicles to form the cell plate, within which new cell walls are subsequently laid down. In both cases, the result is the production of two new, separate cells. As a result of mitosis, each cell has received an exact copy of the chromosomes of the parent cell and, as a result of cytokinesis, approximately half of the cytosol and organelles.

In multicellular organisms, mitosis and cytokinesis produce new cells for growth and replacement. In unicellular organisms and some multicellular organisms, mitosis and cytokinesis produce new individuals, a process known as asexual reproduction.

Questions

1. Distinguish among the following: cell cycle/cell division; mitosis/cytokinesis; chromatid/chromosome; centriole/centromere/kinetochore.

2. Describe the activities occurring during each phase of the cell cycle and the role of each phase in the overall process of cell division.

3. What is a chromosome? How is it related to chromatin?

4. Why do we often refer to chromatids as sister chromatids? When are sister chromatids formed? How? When do they first become visible under the microscope?

5. The drawings at the right show stages in mitosis in a plant cell. Identify each stage and describe what is happening.

6. In what ways does cell division in plant cells differ from that in animal cells?

7. What is the function of cell division in the life of an organism? Suppose you, as an organism, were made up of a large single cell rather than trillions of small ones. How would you differ from your present self?

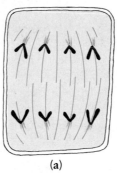

(a)

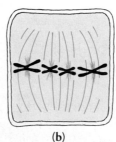

(b)

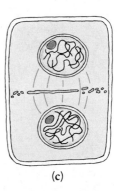

(c)

Suggestions for Further Reading

Aaronson, Stuart A.: "Growth Factors and Cancer," *Science,* vol. 254, pages 1146–1153, 1991.

Balter, Michael: "Cell Cycle Research: Down to the Nitty Gritty," *Science,* vol. 252, pages 1253–1254, 1991.

Glover, David M., Cayetano Gonzalez, and Jordan W. Raff: "The Centrosome," *Scientific American,* June 1993, pages 62–68.

Marx, Jean: "The Cell Cycle: Spinning Farther Afield," *Science,* vol. 252, pages 1490–1492, 1991.

McIntosh, J. Richard, and Michael P. Koonce: "Mitosis," *Science,* vol. 246, pages 622–628, 1989.

McIntosh, J. Richard, and Kent L. McDonald: "The Mitotic Spindle," *Scientific American,* October 1989, pages 48–56.

Murray, Andrew W., and Marc W. Kirschner: "What Controls the Cell Cycle," *Scientific American,* March 1991, pages 56–63.

Sloboda, Roger D.: "The Role of Microtubules in Cell Structure and Cell Division," *American Scientist,* vol. 68, pages 290–298, 1980.

This passionate embrace of the female golden toad by the male (from whose brilliant color the species derives its common name) often lasts for several hours. During that time, the male kicks his hind legs furiously to repel rivals. Eventually the female releases a procession of some 200 eggs. As they are laid, the male deposits his sperm on them. The sperm cells, which are too small to be seen in this photograph, then penetrate the outer layers of the eggs and fertilize them.

Each of the egg cells visible here contains just half the number of chromosomes present in the body cells of the female. Similarly, each of the sperm cells contains just half the number of chromosomes present in the body cells of the male. The process by which the chromosome number is halved is known as meiosis. At fertilization, when the sperm and egg unite, the number of chromosomes characteristic of the body cells is restored.

For reasons that remain unclear, these handsome amphibians, whose scientific name is *Bufo periglenes,* have disappeared from the few square miles of the Monteverde Cloud Forest Reserve in Costa Rica where they were previously found in great abundance. In the past, the entire population of golden toads emerged from their hiding places once a year when the first heavy spring rains filled the breeding pools. Within the space of a single week, all of the toads would congregate in the breeding pools, with thousands of eggs laid by the females and fertilized by the males. However, only a few of the fertilized eggs completed the journey to adulthood, largely because the little pools of water in which the eggs were laid dried up so quickly.

Each year from 1972 through 1987, some 1,500 golden toads were observed during the mating period. In 1988 and 1989, however, only a few were seen, and, since 1990, no golden toads have been observed at all. No one knows if the toads are alive and in hiding, perhaps waiting until an unusually long period of severe drought ends and pools of water suitable for breeding form once more—or if they have disappeared forever.

11

Meiosis and Sexual Reproduction

Most multicellular eukaryotic organisms—for example, golden toads, sea urchins, whitefish, pea plants, and human beings—reproduce sexually. Many unicellular eukaryotes—even those that typically reproduce asexually by mitosis and cytokinesis—are also capable of sexual reproduction.

Sexual reproduction generally requires two parents, and it always involves two events: **fertilization** and **meiosis.** Fertilization is the means by which the different genetic contributions of the two parents are brought together to form the genetic identity of the offspring. Meiosis is a special kind of nuclear division that is believed to have evolved from mitosis and that uses much of the same cellular apparatus. As we shall see, however, meiosis differs from mitosis in a number of important respects.

Haploid and Diploid

To understand meiosis, we must look once more at the chromosomes, focusing this time on their numbers. Every organism has a chromosome number characteristic of its particular species. In a mosquito, for example, each **somatic** (body) cell contains 6 chromosomes; in a cabbage, 18; in corn, 20; in a cat, 38; in a human being, 46; in a potato, 46; and in a goldfish, 94. However, in these organisms and most other familiar plants and animals, the sex cells—or **gametes**—have exactly half the number of chromosomes that is characteristic of the somatic cells of the organism.

The number of chromosomes in the gametes (from the Greek word *gamein,* "to marry") is referred to as the **haploid** ("single set") number. And, as you might expect, the number in the somatic cells is known as the **diploid** ("double set") number. Cells that have more than two sets of chromosomes are said to be **polyploid** ("many sets"). Polyploid cells rarely occur in animals, but the somatic cells of a wide variety of flowering plants are polyploid.

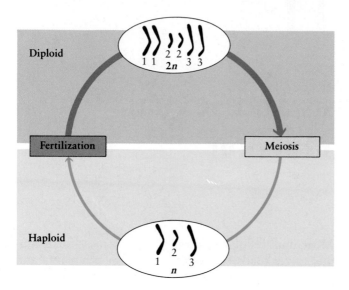

11–1 *Sexual reproduction is characterized by two events: the coming together of the gametes (fertilization) and the halving of the number of chromosomes (meiosis). Following meiosis, there is a single set of chromosomes, that is, the haploid number* (n). *Following fertilization, there is a double set of chromosomes, that is, the diploid number* (2n).

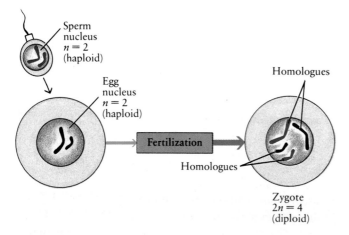

11–2 *During gamete formation, individual homologues (members of a homologous pair of chromosomes) are parceled out by meiosis so that a haploid* (n) *gamete, which is produced from a diploid* (2n) *cell, carries only one member of each homologous pair. At fertilization, the chromosomes in the sperm nucleus and the egg nucleus come together in the zygote, producing, once again, pairs of homologous chromosomes. Each pair consists of one homologue from the father (paternal chromosome) and one from the mother (maternal chromosome). Here, and in subsequent diagrams, red and black are used to indicate the paternal and maternal chromosomes of a homologous pair.*

In biological shorthand, the haploid number is designated as n and the diploid number as $2n$. In humans, for example, $n = 23$ and $2n = 46$. When a sperm fertilizes an egg, the two haploid nuclei fuse, $n + n = 2n$, and the diploid number is restored (Figure 11–1). A diploid cell produced by the fusion of two gametes is known as a **zygote** (from the Greek *zygotos,* "a pair").

In every diploid cell, each chromosome has a partner. These pairs of chromosomes are known as homologous pairs, or **homologues.** The two resemble each other in size and shape and also, as we shall see, in the kinds of hereditary information each contains. One homologue comes from the gamete of one parent, and its partner is from the gamete of the other parent. After fertilization, both homologues are present in the zygote (Figure 11–2).

In meiosis, the diploid set of chromosomes, which contains the two homologues of each pair, is reduced to a haploid set, which contains only one homologue of each pair. Meiosis thus counterbalances the effects of fertilization, ensuring that the number of chromosomes remains constant from generation to generation. Meiosis is also a source of new combinations within the chromosomes themselves.

Meiosis and the Life Cycle

Meiosis occurs at different times during the life cycle of different organisms (Figure 11–3). In many protists, such as the alga *Chlamydomonas,* it occurs shortly after fusion of the mating cells (Figure 11–4). The cells are ordinarily haploid, and meiosis restores the haploid number after fertilization.

In plants, a haploid phase typically alternates with a diploid phase. For example, in ferns (Figure 11–5), the form of the plant commonly seen is the **sporophyte,** the diploid organism. By meiosis, fern sporophytes produce spores, usually on the undersides of their fronds (leaves). These spores have only the haploid number of chromosomes. They germinate to form tiny plants, typically only a few cell layers thick, that are known as **gametophytes.** In these plants, all the cells are haploid. The small, haploid gametophytes produce gametes by mitosis. The gametes fuse and then develop into a new, diploid sporophyte. This process, in which a haploid phase is followed by a diploid phase and again by a haploid phase, is known as **alternation of generations.** As we shall see in Chapter 25, alternation of generations occurs in all sexually reproducing plants, although not always in the same form.

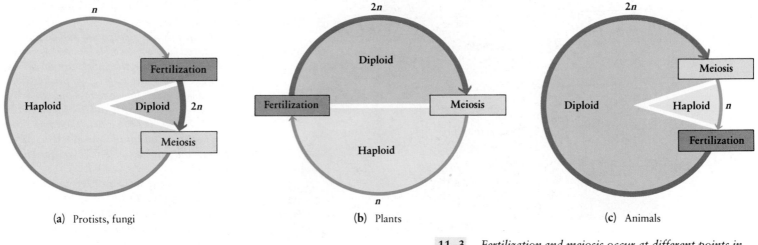

(a) Protists, fungi   (b) Plants   (c) Animals

11–3   *Fertilization and meiosis occur at different points in the life cycle of different organisms. (a) In many—but not all—protists and fungi, meiosis occurs almost immediately after fertilization. Most of the life cycle is spent in the haploid state (signified by the thin green arrow). (b) In plants, fertilization and meiosis are separated in time. The life cycle of the organism consists of both a diploid phase and a haploid phase. (c) In animals, completion of meiosis is followed almost immediately by fertilization. As a consequence, during most of the life cycle the organism is diploid (signified by the thick blue arrow).*

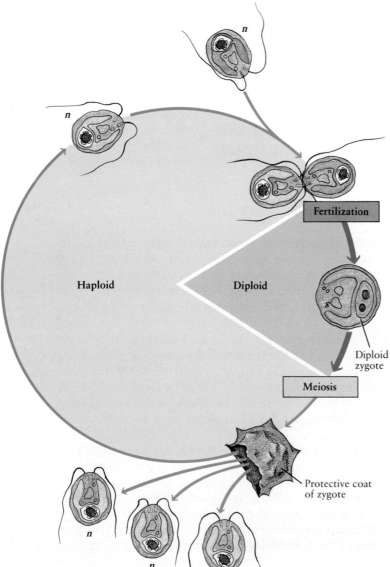

11–4   *The life cycle of* Chlamydomonas *is of the type shown in Figure 11–3a. The organism is haploid for most of its life cycle. Fertilization, the fusion of cells of different mating strains, produces the diploid zygote. The zygote manufactures a thick coat that allows it to remain dormant during harsh conditions. Following dormancy, the diploid zygote divides meiotically, forming four new haploid cells. Each haploid cell can reproduce asexually (by mitosis and cytokinesis) to form either more haploid cells or, in periods of environmental stress, haploid cells of a particular mating strain. These cells can fuse with cells of a different mating strain, and another sexual cycle is underway.*

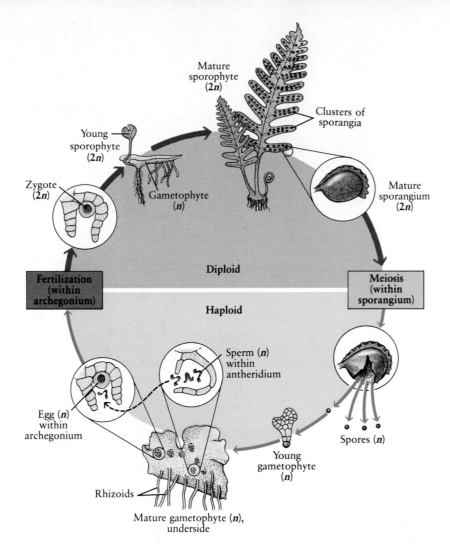

Mature sporophyte (2*n*)

Clusters of sporangia

Young sporophyte (2*n*)

Zygote (2*n*)

Gametophyte (*n*)

Mature sporangium (2*n*)

Diploid

Fertilization (within archegonium)

Meiosis (within sporangium)

Haploid

Sperm (*n*) within antheridium

Egg (*n*) within archegonium

Young gametophyte (*n*)

Spores (*n*)

Rhizoids

Mature gametophyte (*n*), underside

11–5 *The life cycle of a fern is of the type shown in Figure 11–3b. Following meiosis (far right), spores, which are haploid, are produced in the sporangia. As the sporangia dry and split open, the spores are shed. The spores develop into haploid gametophytes. In many species, the gametophytes are only one layer of cells thick and are somewhat heart-shaped, as shown here (bottom). From the lower surface of the gametophyte, filaments, the rhizoids, extend downward into the soil.*

On the lower surface of the gametophyte are borne the flask-shaped archegonia, which enclose the egg cells, and the antheridia, which enclose the sperm. When the sperm are mature and there is an adequate supply of water, the antheridia burst, and the sperm cells, which have numerous flagella, swim to the archegonia and fertilize the eggs.

From the zygote, the diploid (2n) sporophyte develops, growing out of the archegonium within the gametophyte. After the young sporophyte becomes rooted in the soil, the gametophyte disintegrates. The sporophyte matures, develops sporangia, in which meiosis occurs, and the cycle begins again.

Human beings have the typical animal life cycle, in which the diploid individual produces haploid gametes by meiosis shortly before fertilization. Fusion of male and female gametes at fertilization then restores the diploid number. Virtually all of the life cycle is spent in the diploid state (Figure 11–6).

Note that although meiosis in animals produces gametes, meiosis in plants produces **spores**. A spore is a haploid reproductive cell that, unlike a gamete, can develop into a whole organism (a haploid organism) without first fusing with another cell. With the formation of either gametes or spores, however, meiosis has the same result: at some point during the life cycle of a sexually reproducing organism, it reduces the diploid chromosome set to the haploid chromosome set.

The Preparations for Meiosis

As we saw in the last chapter, mitosis consists of a single nuclear division and results in the formation of two daughter nuclei. Each of these nuclei receives an exact copy of the parent cell's chromosomes. Meiosis, by con-

trast, consists of two successive nuclear divisions, producing a total of four daughter nuclei. Each of these daughter nuclei contains half the number of chromosomes present in the original nucleus. Moreover, each daughter nucleus receives just one member of each pair of homologous chromosomes.

The key events in meiosis—on which all else depends—occur in the interphase preceding meiosis and in prophase of the first meiotic division. During interphase, the chromosomes are replicated, so that by the beginning of meiosis each chromosome consists of two identical sister chromatids held together at the centromere.

Early in prophase, after the chromosomes are replicated, the homologous chromosomes come together in pairs (Figure 11–7a). Once contact is made at any point between two homologues, pairing extends, zipperlike, along the length of the chromatids. Since each chromosome consists of two identical chromatids, the pairing of the homologous chromosomes actually involves four chromatids. Each complex of paired homologous chromosomes is thus known as a **tetrad** (from the Greek *tetra*, meaning "four").

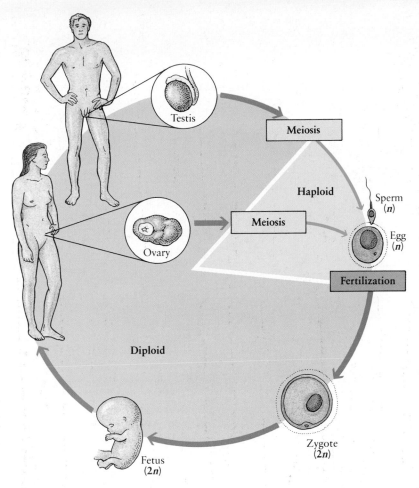

11–6 *The life cycle of* Homo sapiens. *Gametes—egg cells and sperm cells—are produced by meiosis. At fertilization, the haploid gametes fuse, restoring the diploid number in the fertilized egg. The zygote develops into a mature man or woman, who again produces haploid gametes. As is the case with most other animals, the cells are diploid during almost the entire life cycle, the only exception being the gametes. This is the type of life cycle diagrammed in Figure 11–3c.*

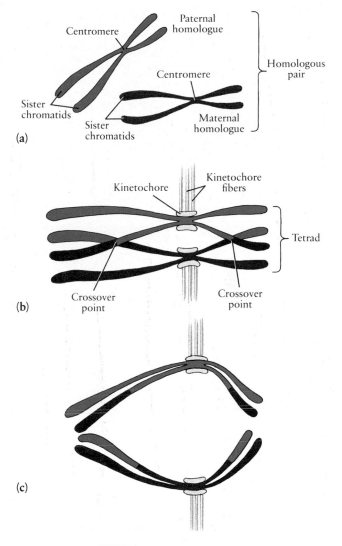

11–7 (a) *A homologous pair of chromosomes, prior to meiosis. One member of the pair is of paternal origin, and the other of maternal origin. Each of these chromosomes has replicated and consists of two sister chromatids.* (b) *In prophase of the first meiotic division, the two homologues of each pair come together and become closely associated with one another. Each homologous pair consists of four chromatids and is therefore also known as a tetrad. Within the tetrad, chromatids of the two homologues intersect at a number of points, making possible the exchange of chromatid segments. This phenomenon is known as crossing over, and the locations at which it occurs are called crossover points.* (c) *The result of crossing over is a recombination of the genetic material of the two homologues. The sister chromatids of each homologue are no longer identical.*

At this point, a crucial process occurs that can alter the genetic makeup of the chromosomes. This process, known as **crossing over,** involves the exchange of segments of one chromosome with corresponding segments from its homologous chromosome (Figure 11–7b). At the sites of crossing over, portions of the chromatids of one homologue are broken and exchanged with corresponding portions of the chromatids of the other homologue. The breaks are resealed, and the result is that the sister chromatids of a single homologue no longer contain identical genetic material (Figure 11–7c). The maternal homologue now contains portions of the paternal homologue, and vice versa. Thus crossing over is an important mechanism for recombining the genetic material from the two parents.

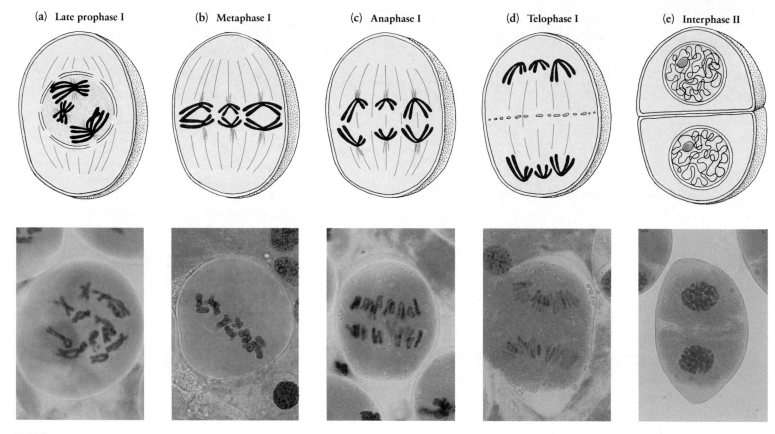

(a) Late prophase I   (b) Metaphase I   (c) Anaphase I   (d) Telophase I   (e) Interphase II

11–8   *The phases of meiosis as they occur in a plant cell in which the diploid number is 6 (n = 3). The accompanying micrographs show meiosis in the anther (the pollen-producing structure) of a lily.*

The Phases of Meiosis

The two successive nuclear divisions of meiosis are designated as meiosis I and meiosis II. In meiosis I, homologous chromosomes pair and then separate from one another. In meiosis II, the chromatids of each homologue separate. In this discussion, we shall describe meiosis in a plant cell in which the diploid number is 6 *(n = 3)*. Three of the six chromosomes were originally derived from one parent and three from the other parent. For each chromosome from one parent there is a homologous chromosome, or homologue, from the other parent.

Following replication of the chromosomes, the first of the two nuclear divisions of meiosis begins. It proceeds through the stages of prophase, metaphase, anaphase, and telophase. All of these are designated "I" to indicate that they are substages of meiosis I.

During **prophase I** (Figure 11–8a), the condensed,

replicated chromosomes first become visible under the light microscope. The spindle microtubules assemble and can be seen radiating out from the two poles of the cell. During this phase, the nucleolus disappears and the nuclear envelope breaks down. Also, and most important, the pairing of the homologous chromosomes and crossing over take place. If you remember one essential point—that the homologous chromosomes are arranged in pairs by the end of prophase I—you will be able to remember all of the subsequent events with little difficulty.

In **metaphase I** (Figure 11–8b), the homologous pairs line up along the equatorial plane of the cell. (By contrast, in metaphase of mitosis, replicated chromosomes line up at the equator with no pairing of the homologues.) Spindle fibers have become associated with the kinetochores (see page 180).

During **anaphase I** (Figure 11–8c), the homologues, each consisting of two sister chromatids, separate, as if pulled apart by the kinetochore fibers. However, the two sister chromatids of each homologue do not separate as they do in mitosis.

By the end of the first meiotic division, **telophase I**

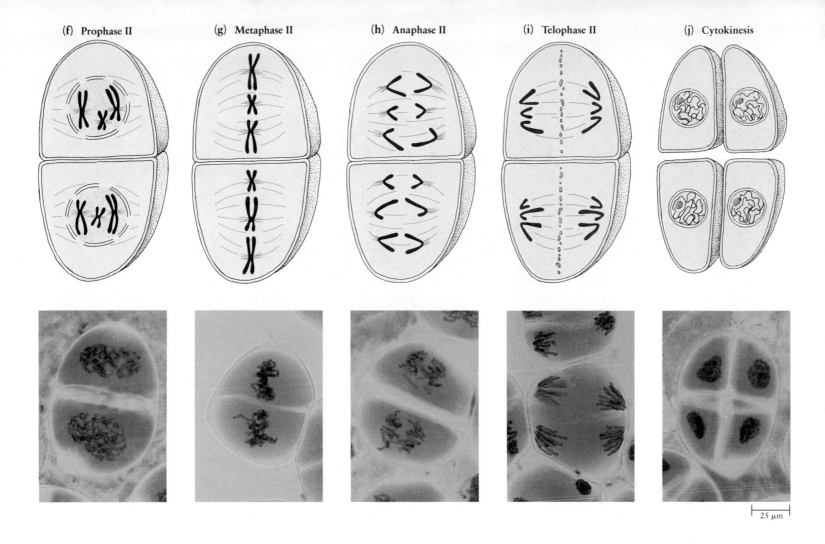

(f) Prophase II   **(g) Metaphase II**   **(h) Anaphase II**   **(i) Telophase II**   **(j) Cytokinesis**

25 µm

(Figure 11–8d), the homologues have moved to the poles. Each chromosome group now contains only half the number of chromosomes as the original nucleus.* Moreover, because of exchanges that took place during crossing over, these chromosomes may be different from any of those present in the original cell. Depending on the species, new nuclear envelopes may or may not form, and cytokinesis may or may not take place.

Meiosis, however, does not end here. Although each nucleus contains the haploid *number* of chromosomes, it nevertheless contains double the haploid *amount* of hereditary material. Why? Because each chromosome consists of two chromatids that have not yet separated.

Meiosis II resembles mitosis except that it is not preceded by replication of the chromosomal material. A short interphase (Figure 11–8e) may occur, during which the chromosomes partially uncoil, but meiosis in many species proceeds from telophase I directly to prophase II.

During **prophase II** (Figure 11–8f), the chromosomes, if dispersed, condense fully again. The nuclear envelopes, if present, disintegrate, and new spindle fibers begin to appear. Remember that in this example there are three chromosomes in each nucleus (the haploid number), and each is still in the form of two chromatids held together at the centromere.

During **metaphase II** (Figure 11–8g), the three chromosomes in each nucleus line up on the equatorial plane. Each consists of two sister chromatids. At **anaphase II** (Figure 11–8h), as in anaphase of mitosis, the sister chromatids of each chromosome separate from one another. Each individual chromatid, which can now be called a chromosome, moves toward one of the poles.

During **telophase II** (Figure 11–8i), the spindle microtubules disappear and a nuclear envelope forms around each set of chromosomes. There are now four nuclei in all, each containing the haploid number of chromosomes. Cytokinesis (Figure 11–8j) proceeds as it does following mitosis. Cell walls form, dividing the cytoplasm, and these haploid plant cells begin to differentiate into spores.

* In counting, it is often difficult to know whether to count a chromosome that has replicated but has not divided as one or two chromosomes. It is customary to count such a chromosome as one. The trick is to count centromeres.

195

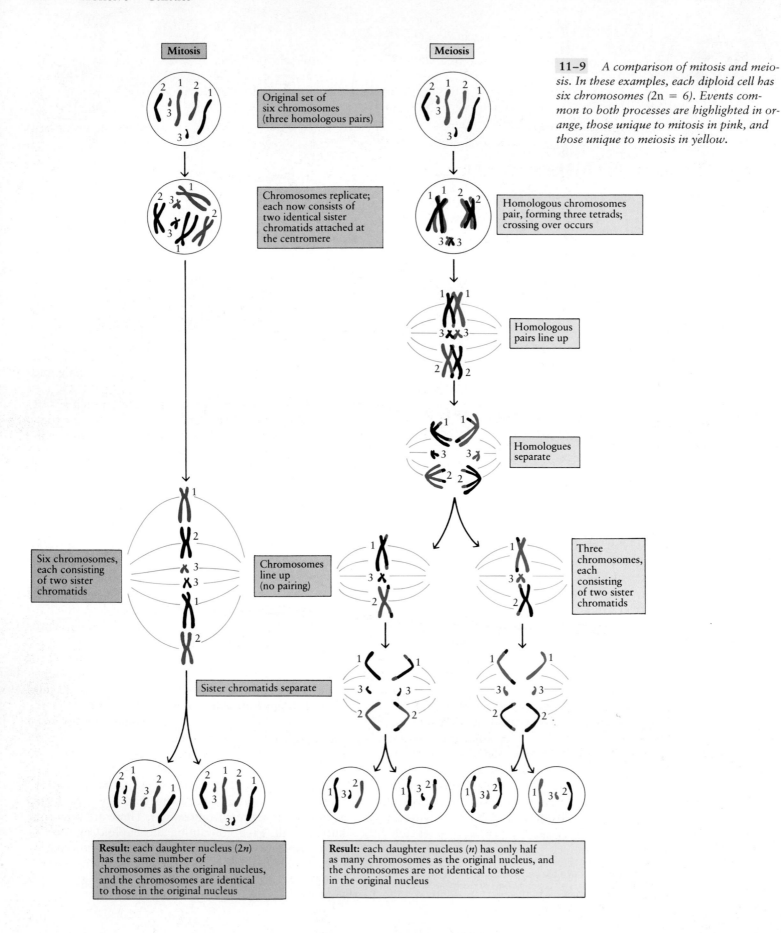

Mitosis

Meiosis

Original set of
six chromosomes
(three homologous pairs)

Chromosomes replicate;
each now consists of
two identical sister
chromatids attached at
the centromere

Homologous chromosomes
pair, forming three tetrads;
crossing over occurs

Homologous
pairs line up

Homologues
separate

Six chromosomes,
each consisting
of two sister
chromatids

Chromosomes
line up
(no pairing)

Three
chromosomes,
each
consisting
of two sister
chromatids

Sister chromatids separate

Result: each daughter nucleus (2*n*)
has the same number of
chromosomes as the original nucleus,
and the chromosomes are identical
to those in the original nucleus

Result: each daughter nucleus (*n*) has only half
as many chromosomes as the original nucleus, and
the chromosomes are not identical to those
in the original nucleus

11–9   *A comparison of mitosis and meiosis. In these examples, each diploid cell has six chromosomes (2n = 6). Events common to both processes are highlighted in orange, those unique to mitosis in pink, and those unique to meiosis in yellow.*

Thus, beginning with one cell containing six chromosomes (three homologous pairs), we end with four cells, each with three chromosomes (no homologous pairs). The chromosome number has been reduced from the diploid number to the haploid number.

Meiosis and Mitosis Compared

As we have seen, the events that take place during meiosis resemble those of mitosis. However, a direct comparison of the two processes (Figure 11–9) reveals a number of differences. The most crucial are the pairing of the homologous chromosomes in prophase I of meiosis and their alignment in metaphase I, followed by the separation of the homologues in anaphase I. These events, which do not occur in mitosis, are the key to the reduction in chromosome number.

Other important points to remember:

1. Meiosis can occur only in cells with the diploid (or polyploid) number of chromosomes—that is, in cells in which each chromosome has a partner. By contrast, mitosis, which does not require any pairing of the chromosomes, can occur in either haploid or diploid cells.

2. During meiosis, each diploid nucleus divides twice, producing a total of four nuclei. The chromosomes, however, replicate only once—prior to the first nuclear division.

3. Thus, each of the four nuclei produced contains half the number of chromosomes present in the original nucleus.

4. The haploid nuclei produced by meiosis contain new combinations of chromosomes. The homologous chromosomes, originally derived from the organism's parents, are assorted randomly among the four new haploid nuclei. Moreover, as a result of crossing over, these chromosomes are not identical to those with which meiosis began.

Meiosis in the Human Species

In all vertebrates, including humans, meiosis takes place in the reproductive organs, the **testes** of the male and the **ovaries** of the female.

In males, the diploid cells that undergo meiosis are known as **primary spermatocytes.** Each primary spermatocyte divides twice, producing four haploid **spermatids.** The spermatids then mature into motile sperm cells (Figure 11–10).

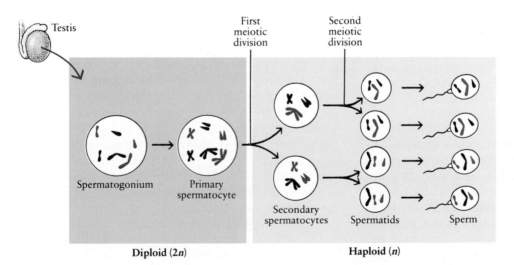

Diploid (2n)   **Haploid (n)**

11–10   *The series of changes resulting in the formation of sperm cells begins with the growth of spermatogonia into large cells known as primary spermatocytes. For clarity, only six (n = 3) of the 46 human chromosomes are shown. At the first meiotic division, each primary spermatocyte divides into two haploid secondary spermatocytes.*

The second meiotic division results in the formation of four haploid spermatids. The spermatids then mature into functional sperm cells. This process occurs continuously after puberty; the normal ejaculate of an adult human male contains between 300 and 400 million sperm cells.

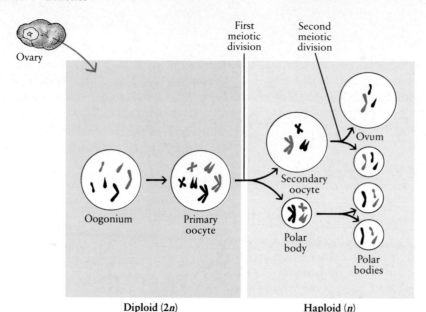

Ovary

First meiotic division

Second meiotic division

Oogonium

Primary oocyte

Secondary oocyte

Polar body

Ovum

Polar bodies

Diploid (2*n*)

Haploid (*n*)

11–11   *Formation of the ovum begins with the growth of an oogonium into a primary oocyte. In the first meiotic division, this cell divides into a secondary oocyte and a polar body. In human females, the first meiotic division begins during the third month of fetal development and ends at ovulation, which, in the case of some cells, may not take place until 50 years later. The second meiotic division, which produces the ovum and a second polar body, does not take place until after the fertilizing sperm cell has penetrated the secondary oocyte. The first polar body may also divide.*

In females, diploid cells known as **primary oocytes** undergo meiosis. Haploid nuclei are also produced by the meiotic divisions of these cells, but the cytoplasm is apportioned unequally during cytokinesis in both meiosis I and II. From each primary oocyte, only one **ovum** (egg cell) is produced, along with two or three **polar bodies** (Figure 11–11). The polar bodies contain the other nuclei formed by meiosis and usually disintegrate. As a result of this unequal division of the cytoplasm, the ovum is well supplied with ribosomes, mitochondria, enzymes, and stored nutrients, all important for the development of the embryo.

Mistakes in Meiosis

As you know, the diploid number of chromosomes in the human species is 46, and the haploid number is 23. In all of the homologous pairs except one, the chromosomes appear to be identical in both males and females; these chromosomes are known as **autosomes.** The structure of one pair, however, differs between males and females. The chromosomes of this pair are known as **sex chromosomes.** In females, the two sex chromosomes are identical, but in males they are dissimilar. One of the male sex chromosomes is the same as the female sex chromosomes, but the other is much smaller. The chromosome that is the same in the cells of both males and females is known as the **X chromosome,** and the unlike chromosome characteristic of the cells of males is known as the **Y chromosome** (Figure 11–12)

A number of genetic disorders are caused by abnormalities in the number or structure of either the autosomes or the sex chromosomes. These abnormalities result from "mistakes" during meiosis (or, less often, during mitotic divisions early in embryonic development). For example, from time to time, homologous chromosomes fail to separate during the first division of meiosis. Similarly, the chromatids occasionally fail to separate during the second division of meiosis. This phenomenon is known as **nondisjunction.**

When nondisjunction occurs in meiosis, the results are gametes with one or more chromosomes too many and other gametes with one or more chromosomes too few. A gamete with too few chromosomes (unless the missing chromosome is a sex chromosome) cannot produce a viable embryo. However, sometimes a gamete with too many chromosomes does produce a viable embryo. The result is an individual with one or more extra chromosomes in every cell of his or her body. In the vast majority of such cases, however, the fetus is spontaneously aborted early in pregnancy, an event (commonly known as a miscarriage) that occurs in 15 to 20 percent of recognized pregnancies.

Individuals with additional autosomes always have widespread abnormalities. With the exception of those with Down syndrome, those who are not stillborn typically survive only a few months. Among the few who survive, most are mentally retarded and those who survive to maturity are usually sterile. They frequently have abnormalities of the heart and other organs as well.

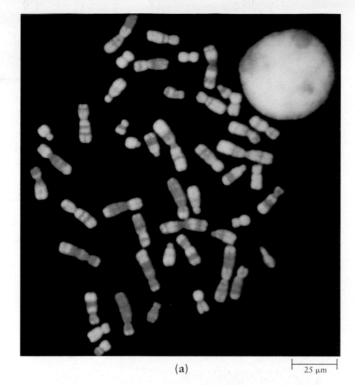

(a)

25 μm

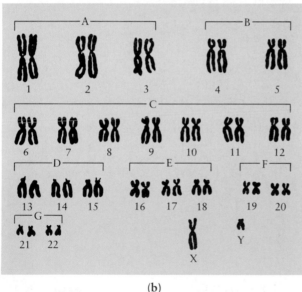

(b)

11–12 (a) *The chromosomes of a dividing somatic cell of a human female, photographed after replication and condensation. The chromosomes have been stained, revealing banding patterns that make it relatively easy to identify homologues. The large, round yellow-stained structure is the nucleus from which the chromosomes were released.*

(b) *Chromosomes from a single human somatic cell, arranged in the kind of graphic display known as a karyotype (see essay). In a karyotype, the autosomes are grouped by size (A, B, C, etc.), and then the probable homologues are paired. The normal diploid chromosome number of a human being is 46, 22 pairs of autosomes and two sex chromosomes. A normal woman has two X chromosomes and a normal man has an X and a Y, as shown here.*

Nondisjunction may also produce individuals with unusual numbers of sex chromosomes. A normal XY combination in the twenty-third pair produces maleness, but so do XXY, XXXY, and even XXXXY. These latter males, however, are usually sexually underdeveloped and sterile. XXX combinations sometimes produce normal females, but many of the XXX women and almost all women with only one X chromosome are sterile.

Nondisjunction is not the only type of "mistake" in meiosis that can cause abnormalities. For example, a **deletion** occurs when a segment of a chromosome breaks off and is not replaced by the corresponding segment from its homologue. In some cases, the segment is incorporated into the homologue, in which the segment then occurs twice. This abnormality is known as a **duplication.** Sometimes a deleted segment of one chromosome is transferred to and becomes part of another, nonhomologous chromosome, a phenomenon known as **translocation.** Depending on the size of the segment and the hereditary information it carries, the deletion of a portion of a chromosome can be as lethal as the loss of the entire chromosome. Similarly, duplication and translocation can, in some cases, have the same consequences as the presence of an extra chromosome.

Down Syndrome

One of the most familiar conditions resulting from an abnormality in the number of autosomal chromosomes is Down syndrome, named after the physician who first described it. Because it usually involves more than one defect, it is referred to as a **syndrome,** a group of disorders that occur together. Down syndrome includes, in most cases, a short, stocky body type with a thick neck; mental retardation, ranging from mild to severe in different individuals; a large tongue, resulting in speech defects; an increased susceptibility to infections; and, often, abnormalities of the heart and other organs. Individuals with Down syndrome who survive into their thirties or forties also have a high probability of developing a form of senility similar to Alzheimer's disease (to be discussed in Chapter 36).

Down syndrome arises when an individual has three, rather than two, copies of chromosome 21. In about 95 percent of the cases, the cause of the abnormality is nondisjunction during formation of a parental gamete, resulting in 47 chromosomes, with an extra copy of chromosome 21 in the cells of the affected individual (Figure 11–13).

(a)

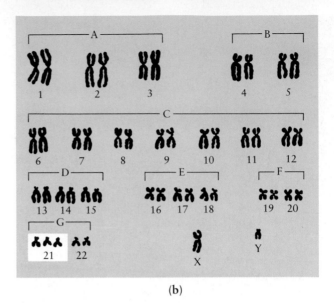

(b)

11–13   (a) *Although children with Down syndrome share certain physical characteristics, there is a wide range of mental capacity among these individuals.* (b) *The karyotype of a male with Down syndrome caused by nondisjunction. Note that there are three copies of chromosome 21.*

Down syndrome may also result from a translocation in the chromosomes of one of the parents. The person with Down syndrome caused by translocation usually has a third chromosome 21 (or, at least, most of it) attached to a larger chromosome, most often chromosome 14. Such an individual, although he or she has only 46 chromosomes, has the functional equivalent of a third chromosome 21.

When cases of Down syndrome due to translocation are studied, it is usually found that one parent, although normal, has only 45 separate chromosomes. One chromosome is usually composed of most of chromosomes 14 and 21 joined together. Thus, parents who have a child with Down syndrome are advised to have their karyotypes prepared. If either parent has the translocation, they are at a high risk of having another child with Down syndrome and, moreover, half of their children without Down syndrome are likely to be carriers of the translocation.

It has been known for many years that Down syndrome and a number of other disorders involving nondisjunction are more likely to occur among infants born to older women (Figure 11–14). The reasons for this are not known. Recent studies have also indicated that in about 5 percent of the cases of Down syndrome due to nondisjunction, the extra chromosome comes from the father rather than the mother.

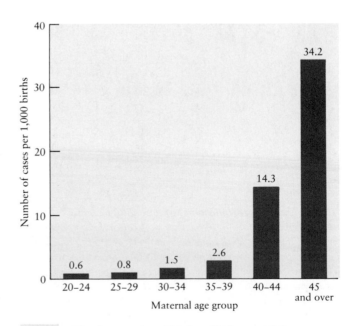

11–14   *The frequencies of births of infants with Down syndrome in relation to the ages of the mothers. The number of cases shown for each age group represents the occurrence of Down syndrome in every 1,000 live births by mothers in that group. As you can see, the risk of having a child with Down syndrome increases rapidly after the mother's age exceeds 40. An increased risk is also thought to occur after the father's age exceeds 55.*

Preparation of a Karyotype

A karyotype is a graphic display of the complete chromosome complement of a particular organism. Although the karyotypes of some organisms have been known since the 1920s, it was not until 1956 that the human karyotype was determined. This established, after a flurry of controversy, that the diploid number of human chromosomes is 46: 44 autosomes and two sex chromosomes. Interest in karyotyping received great impetus three years later with the discovery that the presence of an extra chromosome could be associated with a major medical problem, Down syndrome. Karyotypes are now prepared routinely for genetic counseling of couples at risk for having children with Down syndrome or other conditions associated with gross chromosome abnormalities.

The chromosomes shown in a karyotype are mitotic metaphase chromosomes, each consisting of two sister chromatids held together at the centromere. To prepare a karyotype, cells in the process of dividing are interrupted at metaphase by the addition of colchicine, a drug that interferes with the spindle microtubules and prevents the subsequent steps of mitosis from taking place. After treating and staining, the chromosomes are photographed, enlarged, cut out, and arranged according to size. Chromosomes of the same size are paired according to centromere position, which results in different "arm" lengths. From the karyotype, certain abnormalities, such as an extra chromosome or piece of a chromosome, can be detected.

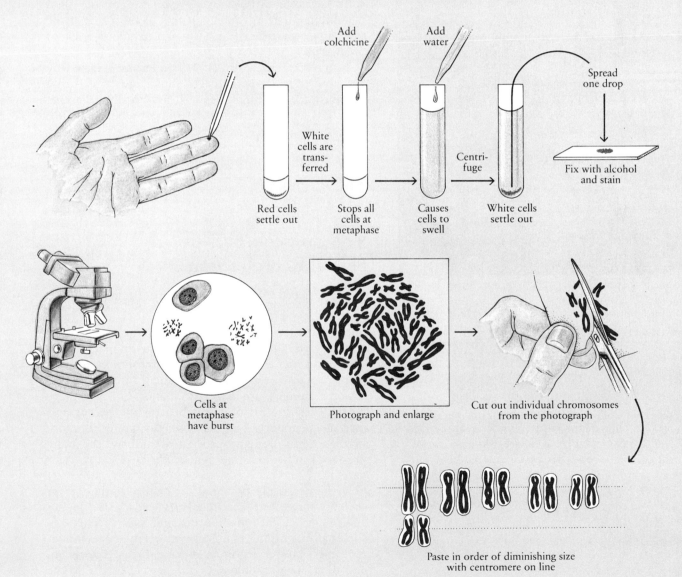

Add colchicine

Add water

Spread one drop

White cells are transferred

Centrifuge

Fix with alcohol and stain

Red cells settle out

Stops all cells at metaphase

Causes cells to swell

White cells settle out

Cells at metaphase have burst

Photograph and enlarge

Cut out individual chromosomes from the photograph

Paste in order of diminishing size with centromere on line

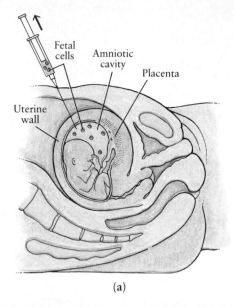

(a)

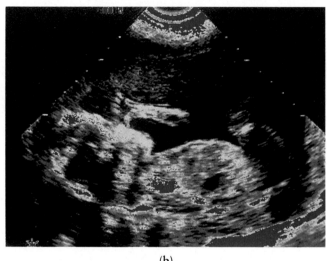

(b)

11–15 (a) *Amniocentesis. The position of the fetus is first determined by ultrasound, with the sonogram continuously displayed on a video monitor. Then a needle is inserted into the amniotic cavity, and fluid containing fetal cells is withdrawn into a syringe. The cells are grown in tissue culture and then are analyzed for chromosome abnormalities and other genetic disorders. The procedure is usually not performed until the fourteenth week of pregnancy, to ensure both that there are enough fetal cells in the amniotic cavity to make detection possible and that there is sufficient amniotic fluid that removal of the small amount necessary for the test will not endanger the fetus.*

(b) *A sonogram of the uterus of a pregnant woman carrying a four-month-old fetus. The solid black regions enclosed by the muscular uterine walls are the fluid surrounding the fetus. The fetus is lying on its back, with its head at the left, and appears to be sucking its thumb.*

Prenatal Detection

A procedure known as **amniocentesis** makes possible the prenatal detection of Down syndrome and a number of other genetic disorders in the fetus. A thin needle is inserted through the mother's abdominal wall and through the membranes that enclose the fetus, and a sample of the amniotic fluid surrounding the fetus is withdrawn (Figure 11–15). The amniotic fluid contains living cells sloughed off from the skin of the fetus. These cells can be induced to divide, thus providing mitotic cells from which a karyotype can be made. Although amniocentesis must be done with great care, it is simple, quick, and usually harmless.

More recently, a technique has been developed for collecting cells from the chorion, one of the fetal membranes. The advantage of this method is that the test can be performed as early as the eighth week of pregnancy.

In most cases, treatment of conditions detected by prenatal testing—particularly those involving gross chromosome abnormalities—is not possible. Parents are faced with the difficult decision of whether or not to abort the affected fetus. However, as we shall see in Chapter 18, current developments in gene therapy are providing the first rays of hope that successful treatment of other types of genetic disorders is not an impossible dream.

The Consequences of Sexual Reproduction

Many organisms can reproduce both asexually (by mitosis and cytokinesis) and sexually. For example, most one-celled eukaryotes have a life cycle in which either sexual or asexual reproduction may take place, often depending on environmental circumstances. Many plants can also reproduce both sexually and asexually. In some animals, asexual reproduction can take place by the breaking off of a fragment of the parent animal, as occurs in sponges and sea anemones. Because of the careful copying process of mitosis, asexually produced individuals are genetically identical to their parents.

By contrast, the potential for genetic variability in sexually produced individuals is enormous. Figure 11–16 shows the possible distributions of chromosomes at meiosis in organisms with relatively few chromosomes. The red chromosomes are of paternal origin, and the black chromosomes of maternal origin. In the course of meiosis, these chromosomes are distributed among the haploid cells. As you can see, chromosomes

(a)   (b)   (c)

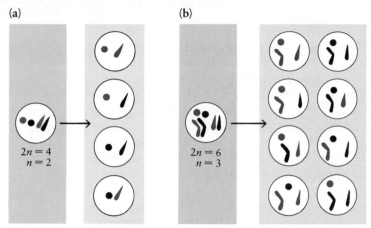

11–16   *The different chromosome combinations possible in the gametes of three organisms with relatively few chromosomes. Red represents chromosomes of paternal origin, and black represents chromosomes of maternal origin. (a) An organism in which the diploid number is 4, and the haploid number is 2. (b) An organism in which the diploid number is 6, and the haploid number is 3. (c) An organism in which the diploid number is 8, and the haploid number is 4.*

of maternal or paternal origin do not stay together but are assorted independently.

The number of possible chromosome combinations in the gametes is 2^n; 2 is the number of homologues in a pair, and n equals the haploid chromosome number. For example, if the haploid number of chromosomes is 2, the number of possible combinations of chromosomes is 2^2, or 4. If the haploid number is 3, the number of possible combinations is 2^3, or 8. If it is 4, then 16 different combinations (2^4) are possible.

A human male with his 46 chromosomes is capable of producing 2^{23} kinds of sperm cells—8,388,608 different combinations of chromosomes, a number larger than the population of New York City. Similarly, a human female is capable of producing 2^{23} kinds of egg cells—8,388,608 different combinations of chromosomes. And this does not take into account the additional variations that may be introduced by crossing over.

As you know from your own observations, the similarities and differences between parents and children and among brothers and sisters vary greatly from child to child and from trait to trait. Some children closely resemble one parent in certain traits but resemble the other parent in other traits. Other children resemble a grandparent more closely than they resemble either parent. And, in some cases, cousins resemble each other more strongly than they resemble their own brothers and sisters. This variability in inheritance, which results primarily from the enormous diversity of possible gametes and the random meeting of two particular gametes in fertilization, was a complete puzzle until the mid-nineteenth century. Then, as we shall see in the next chapter, an Augustinian monk began a series of experiments that revealed the fundamental principles governing inheritance.

Summary

Sexual reproduction involves a special kind of nuclear division called meiosis. Meiosis is the process by which the chromosomes are reassorted and cells are produced that have the haploid chromosome number *(n)*. The other principal component of sexual reproduction is fertilization, the coming together of haploid cells to form the zygote. Fertilization restores the diploid number *(2n)*. There are characteristic differences among major groups of organisms as to when in the life cycle these events take place.

During interphase preceding meiosis, the chromosomes replicate. During prophase of the first meiotic division, they arrange themselves in homologous pairs. One homologue of each pair is of maternal origin, and one is of paternal origin. Each homologue consists of two identical sister chromatids. While the chromosomes are tightly paired, crossing over occurs between homologues, resulting in exchanges of chromosomal material.

In the first stage of meiosis, meiosis I, the homologues are separated. Two nuclei are produced, each with a haploid number of chromosomes, which, in turn, consist of two chromatids each. The nuclei may enter interphase, but the chromosomal material is not replicated. In the second stage of meiosis, meiosis II, the sister chromatids of each chromosome separate as in mitosis. The two nuclei divide, cytokinesis occurs, and four haploid cells are formed.

Meiosis in the human male produces four haploid spermatids from each diploid primary spermatocyte. The spermatids mature into sperm cells. In meiosis in the human female, however, the cytoplasm is unequally divided. The result is that each diploid primary oocyte yields only one haploid ovum; the other haploid nuclei form polar bodies, which soon disintegrate.

Chromosome abnormalities resulting from "mistakes" in meiosis include extra chromosomes (usually a result of nondisjunction—the failure of two homologues or of two chromatids to separate), translocations, deletions, and duplications. Down syndrome is among the disorders associated with an extra chromosome; it may be caused either by nondisjunction or, less commonly, by translocation. Extra sex chromosomes can also result from nondisjunction and are usually associated with sterility. Many genetic disorders can now be detected in the fetus by the use of amniocentesis, the collection of fetal cells from the amniotic fluid.

Each of the haploid cells produced by meiosis contains a unique assortment of chromosomes due to crossing over and random assortment of homologues. Thus meiosis is a source of variation in the offspring.

Questions

1. Distinguish among the following: haploid/diploid/polyploid; gamete/zygote/spore; sporophyte/gametophyte; homologues/tetrad; meiosis I/meiosis II; sex chromosome/autosome; nondisjunction/translocation; deletion/duplication.

2. (a) Dogs have a diploid chromosome number of 78. How many chromosomes would you expect to find in a gamete? In a kidney cell? (b) Plums have a haploid chromosome number of 24. How many chromosomes would you expect to find in the cell of a leaf? In the nucleus of a pollen grain?

3. Draw a diagram of a cell with six chromosomes (n = 3) at meiotic prophase I. Label each pair of chromosomes differently (for example, label one pair A^1 and A^2, and another B^1 and B^2, etc.).

4. Diagram the eight possible gametes resulting from meiosis in a plant cell with six chromosomes (n = 3). Label each chromosome differently, as in Question 3. Assume that crossing over does not occur.

5. Identify the stages of meiosis in the lily cells shown in the micrographs at the right.

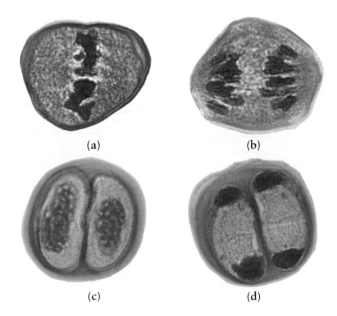

(a)   (b)

(c)   (d)

6. (a) Compare metaphase of mitosis and metaphase II of meiosis. (b) Compare anaphase of mitosis with anaphase I and anaphase II of meiosis. In your answers, consider both the positions and composition of the chromosomes, as well as the consequences.

7. Compare and contrast the processes and the genetic consequences of meiosis and mitosis.

8. In some species, sexual reproduction—that is, fertilization and meiosis—does not require two parents. Explain how this is possible.

9. Nondisjunction can occur at the first meiotic division or the second. How do the effects differ? Include diagrams with your answers.

10. Describe the two types of chromosome abnormalities that can cause Down syndrome. With which type is it possible to unequivocally identify prospective parents who are at a higher-than-average risk of having a child with Down syndrome? How?

11. In our bodies and in those of most other animals, both mitosis and meiosis occur. What are the end products of these two processes? Where in our bodies do these two processes occur?

Suggestions for Further Reading

Alberts, Bruce, Dennis Bray, Julian Lewis, Martin Raff, Keith Roberts, and James D. Watson: *Molecular Biology of the Cell,* 3d ed., Garland Publishing, Inc., New York, 1994.

This outstanding cell biology text provides an excellent discussion of our current understanding of meiosis and fertilization.

Anderson, Alun: "The Evolution of Sexes," *Science,* vol. 257, pages 324–326, 1992.

Mange, Arthur P., and Elaine Johansen Mange: *Genetics: Human Aspects,* 2d ed., Sinauer Associates, Inc., Sunderland, Mass., 1990.

A thorough account of human genetics. Chapters 7, 8, and 9 provide extensive coverage of chromosome abnormalities in both autosomes and sex chromosomes.

Patterson, David: "The Causes of Down's Syndrome," *Scientific American,* April 1987, pages 52–60.

Raghavan, V.: "Germination of Fern Spores," *American Scientist,* vol. 80, pages 176–185, 1992.

Weiss, Rick: "Uneven Inheritance: A Genetic Quirk Leaves Some People with a Chromosomal Odd Couple," *Science News,* July 7, 1990, pages 8–11.

The great contribution of Gregor Mendel, the "father of genetics," was to show that inheritance follows predictable patterns. His experimental subject was very ordinary—the common garden pea—and the traits he selected for study were unremarkable—flower color and seed shape, for example. But the principles he discovered opened the floodgates for the studies that have come to be known as classical genetics.

Mendel's discoveries have important modern applications in medical genetics—the study of inherited disorders. One human disorder that follows a Mendelian pattern of inheritance is Huntington's disease, which causes a progressive destruction of brain cells, followed by death within 10 to 20 years of the onset of symptoms. Most human hereditary disorders are evident at birth or in early infancy, but Huntington's usually first manifests itself among individuals in their 30s and 40s. One tragic consequence of this late onset is that the individual often already has children who, in turn, carry the same lethal, hidden time bomb. According to Mendelian principles, any child of a man or woman who develops Huntington's has a fifty-fifty chance of also developing the disorder.

Research into Huntington's disease has led geneticists to the shores of Lake Maracaibo in Venezuela. Among the inhabitants of a remote village, reachable only by small fishing boats, are more than 100 individuals with Huntington's and another 1,000 who, like the children shown here, are at high risk for the disorder. All are descendants of a single woman, Maria Concepción Soto, who lived in the early nineteenth century and is thought to have inherited the disorder from a German sailor who was her father.

In recent years, geneticists have compiled a pedigree (a detailed chart) tracing the family relationships and incidence of Huntington's among almost 10,000 individuals. This pedigree has made it possible to predict which children in the population are at a high risk of eventually developing the disorder.

From an Abbey Garden: The Beginning of Genetics

For more than 3 billion years, living things have been reproducing, generation after generation, with each parent passing on to its offspring all the biological instructions necessary for the offspring to develop into the same kind of organism as the parent. Although biological inheritance has been the object of wonder—and considerable thought—since early in human history, only rather recently have we begun to understand how it works. In fact, the scientific study of heredity—known as **genetics**—did not really begin until the second half of the nineteenth century.

The questions that are the focus of genetics are among the most fundamental in biology, including not only the transmission of hereditary information from generation to generation but also the translation of that information into the specific characteristics of a particular organism. Our present understanding of these processes is based on the work of some of the most brilliant scientists our civilization has known. In this chapter and the ones that follow, we are going to trace their discoveries.

The Concept of the Gene

In the mid-nineteenth century, well before the developments in microscopy that made possible the discovery of chromosomes and their movements in mitosis and meiosis, biologists were attempting to answer some especially puzzling questions about heredity. For example, why does a child resemble its mother in certain features and its father in other features? Why do some features seem to skip a generation, with the result that a child resembles a grandparent more closely than either parent? Similar questions were of considerable practical importance to breeders of plants and animals, who were attempting to develop varieties with particular desirable characteristics.

At about the same time that Darwin was writing *The Origin of Species* (page 10) and Hertwig was observing

207

the fusion of sea urchin egg and sperm (page 92), Gregor Mendel was beginning a series of experiments that provided the first useful answers to these basic questions about heredity. His work, pursued in a quiet monastery garden in what was then the city of Brünn in the Austro-Hungarian Empire (now Brno in the Czech Republic), and ignored until after his death, marks the beginning of modern genetics.

By Mendel's time, breeding experiments with domestic plants and animals had shown that both parents contribute to the characteristics of their offspring. Further, it was known that these contributions are carried in the gametes—that is, in the sperm and the eggs.

Mendel's great achievement was to demonstrate that inherited characteristics are determined by discrete factors, which are passed from one generation to the next and are parceled out separately (reassorted) in each generation. These discrete factors, which Mendel called *Elemente*, eventually came to be known as **genes**.

Mendel's Experimental Method

For his experiments in heredity, Mendel chose the common garden pea. It was a good choice. The plants were commercially available, easy to cultivate, and grew rapidly. Different varieties had clearly different characteristics that "bred true," appearing unchanged from one crop to the next. Further, the reproductive structures of the pea flower are entirely enclosed by petals, even when they are mature (Figure 12–1). Consequently, the flower normally self-pollinates; that is, sperm cells from the flower's own pollen fertilize its egg cells. Although the plants could be crossbred experimentally, accidental crossbreeding could not occur to confuse the experimental results. As Mendel said in his original paper, "The value and utility of any experiment are determined by the fitness of the material to the purpose for which it is used."

Mendel's choice of the pea plant for his experiments was not original. However, he was successful in formulating the fundamental principles of heredity—where others had failed—because of his approach to the problem. First, he tested a very specific hypothesis in a series of logical experiments. He planned his experiments carefully and imaginatively, choosing for study only clear-cut hereditary differences and avoiding characteristics that could be "more or less" apparent in the offspring. Second, he studied the offspring of not only the first generation but also of the second and subsequent generations. Third, and most important, he counted the different types of offspring resulting from each cross and then analyzed the results mathematically. Even though his mathematics was simple, the idea

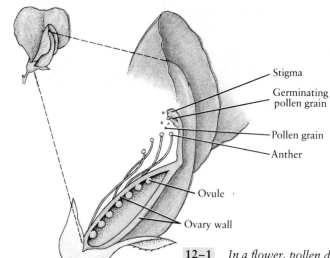

(a) Self-pollination

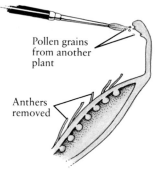

Pollen grains from another plant

Anthers removed

(b) Artificial cross-pollination

12–1   *In a flower, pollen develops in the anthers and the egg cells in the ovules. Pollination occurs when pollen grains land on the stigma, germinate, and grow down to the ovules, where they release the sperm nuclei. The nuclei of egg and sperm unite, and the fertilized eggs develop within the ovules, which are attached to the ovary wall. In the garden pea, the ovules with their enclosed embryos form the peas (the seeds), while the ovary wall becomes the pod.*

Pollination in most species of flowering plants involves the pollen from one plant (often carried by an insect) being caught on the stigma of another plant. This is called cross-pollination.

(a) In the pea flower, however, the stigma and anthers are completely enclosed by petals, and the flower, unlike most, does not open until after fertilization has taken place. Thus self-pollination is the normal course of events in the pea flower—that is, pollen is deposited on the stigma of the same flower in which it is produced. (b) In his crossbreeding experiments, Mendel pried open the flower bud before the pollen matured and removed the anthers with tweezers, preventing self-pollination. Then he artificially cross-pollinated the flower by dusting the stigma with pollen collected from another plant.

Labels for figure (a): Stigma, Germinating pollen grain, Pollen grain, Anther, Ovule, Ovary wall

that a biological problem could be studied quantitatively was startlingly new. Fourth, and last, he organized his data in such a way that his results could be evaluated simply and objectively. The experiments themselves were described so clearly that they could be repeated and checked by other scientists, as eventually they were.

Table 12–1	Results of Mendel's Experiments with Pea Plants			
	Original Crosses		**Second Filial Generation (F$_2$)**	
Trait	**Dominant**	**× Recessive**	**Dominant**	**Recessive**
Seed form	Round	× Wrinkled	5,474	1,850
Seed color	Yellow	× Green	6,022	2,001
Flower position	Axial	× Terminal	651	207
Flower color	Purple	× White	705	224
Pod form	Inflated	× Constricted	882	299
Pod color	Green	× Yellow	428	152
Stem length	Tall	× Dwarf	787	277

The Principle of Segregation

Mendel began with 32 different types of pea plants, which he studied for several years before he began his quantitative experiments. As a result of his preliminary observations, Mendel selected for study seven traits, each of which appeared in two conspicuously different forms in different varieties of plants. One variety of plant, for example, always produced yellow peas (seeds), while another always produced green ones. In one variety, the seeds, when dried, had a wrinkled appearance; in another variety, they were smooth. The complete list of traits is given in Table 12–1.

Mendel performed experimental crosses, removing the pollen-containing anthers from flowers and dusting their stigmas with pollen from a flower of another variety. He found that in every case in the first generation (now known in biological shorthand as the F_1, for "first filial generation"), all of the offspring showed only one of the two alternative characteristics. The other characteristic disappeared completely. For example, all of the seeds produced as a result of a cross between true-breeding yellow-seeded plants and true-breeding green-seeded plants were yellow. Similarly, all of the flowers produced by plants resulting from a cross between a true-breeding purple-flowered plant and a true-breeding white-flowered plant were purple. Characteristics that appeared in the F_1 generation, such as yellow seeds and purple flowers, Mendel called **dominant.**

The interesting question was: What had happened to the alternative characteristic—the greenness of the seed or the whiteness of the flower—that had been passed on so faithfully for generations by the parent stock? Mendel let the pea plant itself carry out the next stage of the experiment by permitting the F_1 plants to self-pollinate (Figure 12–2). The characteristics that had disappeared in the first generation reappeared in the second, or F_2, generation. These characteristics, which were present in the parental (**P**) generation and reappeared in the F_2 generation, must also have been present somehow in the F_1 generation, although not apparent there. Mendel called these characteristics **recessive.**

Mendel's actual counts are shown in Table 12–1. If you plug these numbers into your calculator, you will notice, as Mendel did, that the dominant and recessive characteristics appear in the second, or F_2, generation in ratios of approximately 3:1. How do the recessives disappear so completely and then reappear again, and always in such constant proportions? It was in answering these questions that Mendel made his greatest contribution. He saw that the appearance and disappearance of alternative characteristics, as well as their constant proportions in the F_2 generation, could be explained if hereditary characteristics are determined by

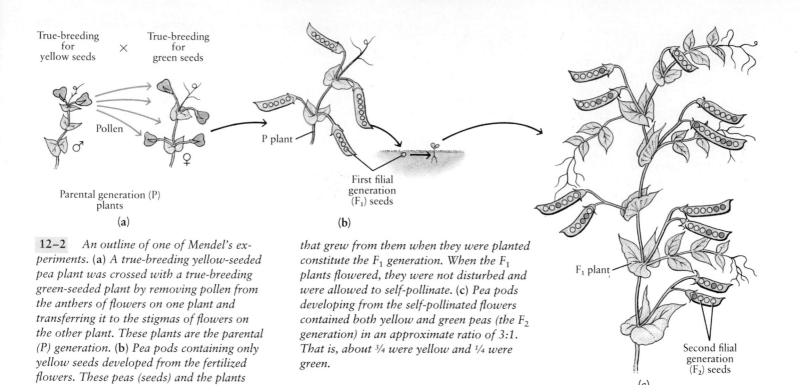

12–2 *An outline of one of Mendel's experiments.* (a) *A true-breeding yellow-seeded pea plant was crossed with a true-breeding green-seeded plant by removing pollen from the anthers of flowers on one plant and transferring it to the stigmas of flowers on the other plant. These plants are the parental (P) generation.* (b) *Pea pods containing only yellow seeds developed from the fertilized flowers. These peas (seeds) and the plants that grew from them when they were planted constitute the F_1 generation. When the F_1 plants flowered, they were not disturbed and were allowed to self-pollinate.* (c) *Pea pods developing from the self-pollinated flowers contained both yellow and green peas (the F_2 generation) in an approximate ratio of 3:1. That is, about 3/4 were yellow and 1/4 were green.*

discrete (separable) factors. These factors, Mendel realized, must have occurred in the F_1 plants in pairs, with one member of each pair inherited from the maternal parent and the other from the paternal parent. The paired factors separated again when the mature F_1 plants produced gametes, resulting in two kinds of gametes, with one member of the pair in each.

The hypothesis that every individual carries pairs of factors for each trait and that the members of a pair segregate (separate from each other) during the formation of gametes has come to be known as Mendel's first law, or the **principle of segregation.**

Consequences of Segregation

We now recognize that any given gene, for instance, the gene for seed color, can exist in different forms. These different forms of a gene are known as **alleles.** For example, yellow seededness and green seededness are determined by different alleles—that is, different forms—of the gene for seed color. The alleles are represented in biological shorthand by letters. By convention, capital letters are used for alleles for dominant characteristics and lowercase letters for alleles for recessive characteristics. Thus, for example, the allele for yellow seededness is represented by Y, and the allele for green seededness by y.

How a given trait is expressed in an organism is determined by the particular combination of two alleles for that trait present in the cells of the organism. If the two alleles are the same (for example, YY or yy), then the organism is said to be **homozygous** for that particular trait. (The term is from the Greek *homos*, meaning "same" or "similar," plus *zygotos*, "a pair.") If the two alleles are different from one another (for example, Yy), then the organism is **heterozygous** for that trait (from the Greek *heteros*, meaning "other" or "different").

When gametes are formed, alleles are passed on to them, but each gamete receives only one allele for any given gene (Figure 12–3). When two gametes combine to form a fertilized egg, the alleles occur in pairs again. If the two alleles in a given pair are the same (a homozygous state), the characteristic they determine will be expressed. If the alleles are different (a heterozygous state), one may be dominant over the other. A dominant allele is one that produces its particular characteristic in the heterozygous as well as in the homozygous state. A recessive allele, by contrast, is one that produces its particular characteristic only in the homozygous state.

The outward appearance and other observable characteristics of an organism constitute its **phenotype.** Even though a recessive allele may not be expressed in the phenotype, each allele of a pair still exists independently and as a discrete unit in the genetic makeup, or **genotype,** of the organism. The two alleles of a pair will separate from each other when gametes are again formed. Only if two recessive alleles come together in the fertilized egg—one from the female gamete and one from the male gamete—will the phenotype show the recessive characteristic.

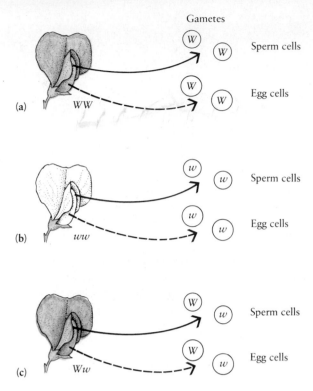

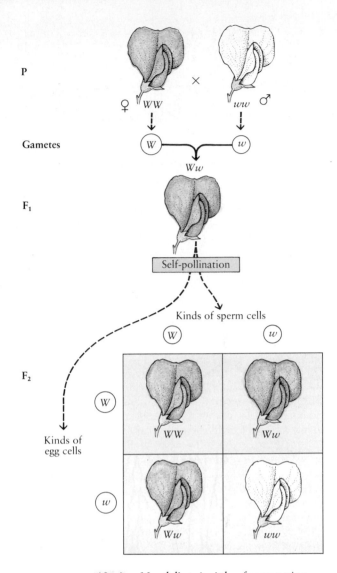

12–3 *An example of the segregation of alleles in gamete formation.* (a) *In a pea plant homozygous for purple flowers* (WW), *all of the gametes, both sperm and egg, will contain a purple-flower* (W) *allele.* (b) *Similarly, in a pea plant homozygous for white flowers* (ww), *all of the gametes will contain a white-flower* (w) *allele.* (c) *However, in a pea plant that is heterozygous for flower color, half of the sperm cells will contain a W allele and half will contain a w allele. Similarly, half of the egg cells will contain a W allele and half will contain a w allele.*

The letters W and w are used here because of a convention by which geneticists derive the allele symbol from the first letter of the name of the form that is less common in nature (in this case, white pea flowers).

12–4 *Mendel's principle of segregation, as exemplified in the F_1 and F_2 generations following a cross between two homozygous parent (P) pea plants, one with two dominant alleles for purple flowers* (WW) *and the other with two recessive alleles for white flowers* (ww). *The female symbol ♀ identifies the plant that contributes the egg cells (female gametes), and the male symbol ♂ indicates the plant that contributes the sperm cells (male gametes).*

The phenotype of the offspring in the F_1 generation is purple, but note that the genotype is Ww. The F_1 heterozygote produces four kinds of gametes, ♀ W, ♀ w, ♂ W, ♂ w, in equal proportions. When this plant self-pollinates, the W and w egg and sperm cells combine randomly to form, on the average, ¼ WW (purple), ²⁄₄ (or ½) Ww (purple), and ¼ ww (white) offspring. It is this underlying 1:2:1 genotypic ratio that accounts for the phenotypic ratio of 3 dominants (purple) to 1 recessive (white).

When pea plants homozygous for purple flowers are crossed with pea plants homozygous for white flowers, only pea plants with purple flowers are produced. However, each plant in this F_1 generation carries both an allele for purple and an allele for white. What happens in the F_2 generation when the F_1 generation self-pollinates? One of the simplest ways to predict the types of offspring that will be produced is to diagram the cross as shown in Figure 12–4. This sort of checkerboard diagram is known as a **Punnett square,** after the English geneticist who first used it for the analysis of genetically determined traits.

As you study Figure 12–4, notice that the result would be exactly the same if an F_1 individual were crossed with another F_1 individual. That is how these breeding experiments are performed with animals and with plants that are not self-pollinating.

211

From Pea Plants to Humans

The principles of genetics are, of course, the same for humans as they are for pea plants or members of any other diploid eukaryotic species. Historically, however, there have been important differences in the methodology of genetic studies. Breeding experiments, so readily performed with Mendel's pea plants, are not possible with humans. Thus, until very recently, most of our knowledge about human genetics has come from the analysis of inheritance patterns. Family pedigrees, which reveal these patterns, have usually been worked out only when a particular abnormality is associated with a medical problem.

Huntington's Disease

In terms of the numbers of individuals affected, serious medical problems caused by dominant alleles are rare, simply because severely afflicted individuals are typically unable to reproduce. Perhaps the most familiar disorder caused by a dominant allele is Huntington's disease (see page 207). Although the incidence of Huntington's is low (about 5 to 10 persons in every 100,000), the devastating physical and mental decline that it causes, coupled with its late onset, means that it affects the lives of many others.

Phenylketonuria (PKU)

Many inherited disorders are, like the white flowers of Mendel's pea plants, the result of the coming together of two recessive alleles. Individuals heterozygous for the gene are usually symptom-free. One of the best-studied examples of a disorder inherited as a Mendelian recessive is phenylketonuria, or PKU. Individuals with PKU lack an enzyme that plays a critical role in the breakdown of the amino acid phenylalanine. When this enzyme is missing or deficient, phenylalanine and its abnormal breakdown products accumulate in the bloodstream and urine. These substances are harmful to the cells of the developing nervous system. Without treatment, the result is profound mental retardation and a life span of little more than 30 years. About 1 in every 15,000 infants born in the United States is homozygous for this allele, which is most common among individuals of Northern European extraction.

It is not yet known how the high levels of phenylalanine and its abnormal breakdown products bring about the tragic neurological symptoms. However, the knowledge we do have is enough to effectively treat infants with PKU and prevent the symptoms from appearing. Most states now require routine tests of all newborn babies in order to detect PKU homozygotes. Those identified at birth are put on a special diet containing low amounts of phenylalanine—enough to supply dietary needs but not enough to permit toxic accumulations. After more than 30 years of experience, it is clear that low phenylalanine diets, maintained for at least the first six years of life (when brain development is still in progress), allow PKU homozygotes to develop normally.

More recently, a biochemical test has been developed that, in conjunction with amniocentesis, allows the detection of PKU in the developing fetus. In such cases, the mother-to-be is placed on a low phenylalanine diet for the remainder of her pregnancy, thus protecting her child from any early effects of PKU.

Tay-Sachs Disease

Another neurological disorder that occurs only in individuals homozygous for a recessive allele is Tay-Sachs disease. As in PKU, Tay-Sachs homozygotes appear normal at birth and through the early months. However, by about eight months, symptoms of severe listlessness become evident. Blindness usually occurs within the first year, and afflicted children rarely survive past their fifth year. Homozygous individuals lack an enzyme, normally found in the lysosomes of brain cells, that breaks down a specific lipid in the cells. In the child lacking this enzyme, the lysosomes of the brain cells fill with the lipid and swell, and the cells die. There is no therapy yet available for Tay-Sachs disease.

While Tay-Sachs disease is a rare disorder in the general population (1 in 300,000 births), until recently it has had a much higher incidence (1 in 3,600 births) among Jews of Eastern and Central European extraction, who make up more than 90 percent of the American Jewish population. It is estimated that among this population approximately 1 in 28 individuals is a heterozygous carrier of the Tay-Sachs allele. The development of a blood test measuring levels of the enzyme that is deficient in Tay-Sachs disease has made it possible for prospective parents to determine if they are carriers. Since its development, this test has been so extensively utilized by prospective parents in the American Jewish population that the incidence of Tay-Sachs births has now dropped dramatically. In fact, in the United States today, the majority of the Tay-Sachs babies born are to non-Jewish parents.

Sickle-Cell Anemia and Cystic Fibrosis

Other serious conditions caused by a recessive allele in the homozygous state are sickle-cell anemia (page 64) and cystic fibrosis. Sickle-cell anemia has a high incidence among African-Americans (1 in 400 births) and also occurs with some frequency among individuals of Mediterranean extraction. Cystic fibrosis is the most common genetic disorder among individuals of European descent (1 in 2,000 births). Affecting cells specialized for secretion, its most serious consequence is the production of copious amounts of thick mucus in the lungs. Although the resulting congestion can be relieved, the tissues of the lungs are gradually destroyed, leading to death. The average life span of cystic fibrosis victims is presently about 24 years.

As we shall see in subsequent chapters, the underlying basis of both sickle-cell anemia and cystic fibrosis is now understood, and effective treatments are, at long last, being developed.

A number of familiar human characteristics are governed by simple Mendelian inheritance of dominant alleles. Among these are dimples and cleft chin, as illustrated by the actor Kirk Douglas and his son Michael. Although this allele has no medical consequences for the individual who carries it, it has undoubtedly caused the hearts of many others to beat a bit faster.

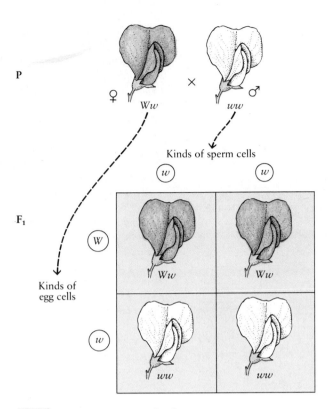

P

Kinds of sperm cells

F₁

Kinds of
egg cells

12–5 *A testcross. In order for a pea
flower to be white, the plant must be homozygous for the recessive allele (ww). But a
purple pea flower can be produced by a
plant with either a Ww or a WW genotype.
How can you determine the genotype of a
purple-flowered plant? Geneticists solve this
problem by crossing such plants with homozygous recessives. This sort of experiment is
known as a testcross. As shown here, a phenotypic ratio in the F₁ generation of one purple to one white indicates that the purple-
flowering parent used in the testcross must
have been heterozygous.*

*What would have been the result if the
plant being tested had been homozygous for
the purple-flower allele?*

A Testcross

In order to test the hypothesis that alleles occur in pairs
and that the two alleles of a pair segregate during gamete formation, it is necessary to perform an additional
experiment: cross purple-flowering F₁ plants (the result
of a cross between true-breeding purple- and white-
flowering plants) with white-flowering plants. To the
casual observer, it would appear as if this were simply
a repeat of Mendel's first experiment, crossing plants
having purple flowers with plants having white flowers.
But if Mendel's hypothesis is correct, the results will be
different from those of his first experiment. Can you
predict the results of such a cross? Stop a moment and
think about it.

The easiest way to analyze the possible results of
such a cross is with a Punnett square, as in Figure 12–5.
This type of experiment, which reveals the genotype of
the parent with the dominant phenotype, is known as
a **testcross**. A testcross is an experimental cross between
an individual with the dominant phenotype (genotype
unknown) for a given trait and another individual with
the recessive phenotype (and thus homozygous for the
recessive allele). Whether only one phenotype or two
different phenotypes are produced in the offspring indicates whether the parent with the dominant phenotype is homozygous or heterozygous for the trait being
studied. The testcross shown in Figure 12–5 reveals
that the genotype of the plant being tested is *Ww* rather
than *WW*.

The Principle of
Independent Assortment

In a second series of experiments, Mendel studied
crosses between pea plants that differed in two characteristics. For example, one parent plant produced
peas that were round and yellow, and the other had
peas that were wrinkled and green. Both the round and
yellow characteristics are dominant, and the wrinkled
and green are recessive (see Table 12–1). As you would
expect, all the F₁ seeds produced by a cross between the
true-breeding parental types were round and yellow.

When these F₁ seeds were planted and the resulting
flowers allowed to self-pollinate, 556 seeds were produced in the F₂ generation. Of these, 315 showed the
two dominant characteristics, round and yellow, but
only 32 combined the recessive characteristics, green
and wrinkled. All the rest of the seeds were unlike either
parent; 101 were wrinkled and yellow, and 108 were
round and green. Totally new combinations of characteristics had appeared.

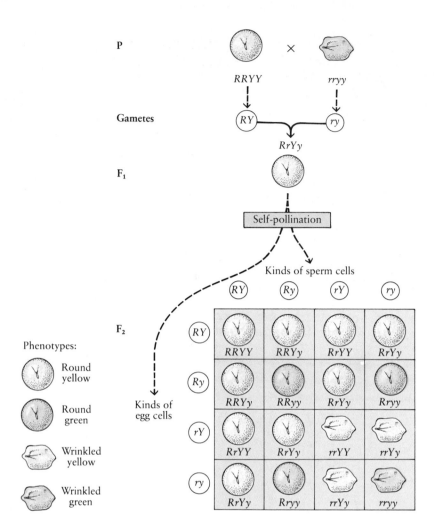

P

$RRYY$ $rryy$

Gametes RY ry

$RrYy$

F₁

Self-pollination

Kinds of sperm cells

RY Ry rY ry

F₂

Phenotypes:

Round yellow

Round green

Wrinkled yellow

Wrinkled green

Kinds of egg cells

RY
| $RRYY$ | $RRYy$ | $RrYY$ | $RrYy$ |

Ry
| $RRYy$ | $RRyy$ | $RrYy$ | $Rryy$ |

rY
| $RrYY$ | $RrYy$ | $rrYY$ | $rrYy$ |

ry
| $RrYy$ | $Rryy$ | $rrYy$ | $rryy$ |

12–6 *One of the experiments from which Mendel derived his principle of independent assortment. A plant homozygous for round (RR) and yellow (YY) peas is crossed with a plant having wrinkled (rr) and green (yy) peas. The F₁ peas are all round and yellow, but notice how the characteristics will, on the average, appear in the F₂ generation. Of the 16 possible combinations in the offspring, nine show the two dominant characteristics (round and yellow), three show one combination of dominant and recessive (round and green), three show the other combination (wrinkled and yellow), and one shows the two recessives (wrinkled and green). This 9:3:3:1 distribution of the phenotypes is always the expected result from a cross involving two independently assorting genes, each with one dominant and one recessive allele in each of the parents.*

This experiment did not, however, contradict Mendel's previous results. If the two traits, seed color and seed shape, are considered independently, round and wrinkled still appeared in a ratio of approximately 3:1 (423 round to 133 wrinkled), and so did yellow and green (416 yellow to 140 green). But the seed shape and the seed color characteristics, which had originally been combined in a certain way (round only with yellow and wrinkled only with green), behaved as if they were entirely independent of one another (yellow could now be found with wrinkled and green with round). From this, Mendel formulated his second law, the **principle of independent assortment.** This principle states that when gametes are formed, the alleles of one gene segregate independently of the alleles of another gene.

Figure 12–6 diagrams Mendel's interpretation of these results. It shows why, in a cross involving two independently assorting genes, each with one dominant and one recessive allele, the phenotypes in the offspring will, on the average, be in the ratio of 9:3:3:1. The fraction ⁹⁄₁₆ represents the proportion of the F₂ offspring expected to show the two dominant characteristics, ¹⁄₁₆ the proportion expected to show the two recessive characteristics, and ³⁄₁₆ and ³⁄₁₆ the proportions expected to show the two alternative combinations of dominant and recessive characteristics.

The 9:3:3:1 ratio in the F₂ offspring holds true when one of the original parents is homozygous for both recessive characteristics and the other parent is homozygous for both dominant characteristics, as in the experiment just described *(RRYY × rryy)*. It also holds true when each original parent is homozygous for one recessive and one dominant characteristic *(rrYY × RRyy).*

The Influence of Mendel

Mendel's experiments were first reported in 1865 before a small group of people at a meeting of the Brünn Natural History Society. None of them, apparently, understood the significance of Mendel's results. His paper was, however, published the following year in the *Proceedings* of the Society, a journal that was circulated to libraries all over Europe. In spite of this, his work was ignored for 35 years, during most of which he devoted himself to the duties of an abbot, and he received no scientific recognition until after his death.

It was not until 1900 that biologists were finally able to understand and accept Mendel's findings. Within a single year, his paper was independently rediscovered by three scientists, each working in a different European country. Each of the scientists had done similar breeding experiments and was searching the scientific literature for related work that would provide confirmation of his results. And each found, in Mendel's brilliant analysis, that much of his own work had been anticipated.

During the 35 years that Mendel's work remained in obscurity, great improvements were made in microscopy and, as a consequence, in **cytology**—the study of cell structure. It was during this period that chromosomes were discovered and their movements during mitosis and meiosis were first observed and recorded.

Cytology and Genetics Meet: Sutton's Hypothesis

In 1902, shortly after the rediscovery of Mendel's work, Walter Sutton, a graduate student at Columbia University, was studying the formation of sperm cells in male grasshoppers. Observing the process of meiosis, Sutton noticed that the chromosomes were paired early in the first meiotic division. He also noticed that the two chromosomes of any one pair had physical resemblances to one another. In diploid cells, he noted, chromosomes apparently come in pairs. The pairing was obvious only at meiosis, although the discerning eye might also find the matching, but unpaired, homologues during metaphase of mitosis (see, for example, Figure 11–12a on page 199).

Sutton was struck by the parallels between what he was seeing and the first principle of Mendel—the principle of segregation. Suddenly the facts fell into place. Suppose chromosomes carried genes, the factors described by Mendel, and suppose the alleles of a gene occurred on homologous chromosomes. Then the alleles would always remain independent and so would be separated at meiosis I as homologous chromosomes separated. New combinations of alleles would be formed as gametes fused at fertilization. Mendel's principle of the segregation of alleles could thus be explained by the segregation of the homologous chromosomes at meiosis.

12–7   *Gregor Mendel, holding a fuchsia, is third from the right in this photograph of members of the Augustinian monastery in Brünn in 1862. As a teenager, Mendel received a thorough training in agricultural science. He then spent two years at the University of Vienna, where he studied physics, chemistry, mathematics, and botany. Although Mendel published only two scientific papers during his lifetime, he performed breeding experiments with a variety of plants until his election as abbot of the monastery in 1871. Unfortunately, almost all of Mendel's papers relating to his scientific work were destroyed shortly before or after his death in 1884.*

Mendel and the Laws of Probability

In applying mathematics to the study of heredity, Mendel was asserting that the laws of probability apply to biology as they do to the physical sciences. Toss a coin. The probability that it will turn up heads is fifty-fifty, that is, one chance in two, or ½. The probability that it will turn up tails is also one chance in two, or ½. The probability that it will turn up one or the other is certain, or one chance in one.

Now toss two coins. The probability that one will turn up heads is again ½. The probability that the second will turn up heads is also ½. The probability that *both* coins will turn up heads is ½ × ½, or ¼. The probability that both coins will turn up tails is also ½ × ½. Similarly, the probability of the first turning up tails and the second turning up heads is ½ × ½, and the probability of the second turning up tails and the first heads is ½ × ½. We can diagram these four possible outcomes of a two-coin toss in a Punnett square (see figure), which indicates that the combination in each square has an equal probability of occurring.

The two-coin toss provides an example of what is known as the **product rule of probability.** This rule states that the probability of two independent events occurring together is simply the probability of one occurring alone multiplied by the probability of the other occurring alone. For instance, in Mendel's experiment diagrammed in Figure 12–4, the probability that a gamete produced by an F_1 plant of *Ww* genotype will carry the *W* allele

is ½, and the probability that it will carry the *w* allele is ½. Therefore the probability of any specific combination of the two alleles in the offspring—that is, *WW, Ww, wW,* or *ww*—is ½ × ½, or ¼. It was undoubtedly the observation that one-fourth of the offspring in the F_2 generation showed the recessive phenotype that indicated to Mendel that he was dealing with a simple case of the laws of probability.

Returning to our coin toss, if there were three coins involved, the probability of any given combination would be simply the product of all three individual probabilities: ½ × ½ × ½, or ⅛. Similarly, with four coins, the probability of any specific combination is ½ × ½ × ½ × ½, or ⅟₁₆. The Punnett square in Figure 12–6 expresses the probability of each of any one of four possible phenotype combinations.

When there is more than one possible arrangement of the events producing a specified outcome, the individual probabilities are added. For instance, what is the probability of throwing a head and a tail, in either order? There are two ways you could do this: by throwing a head first and then a tail (HT), or a tail first and then a head (TH). The probability of throwing a head first and then a tail (HT) is ½ × ½, or ¼. The probability of throwing a tail first and then a head (TH) is also ½ × ½, or ¼. Thus, the probability that a head and a tail will be thrown, in whichever

order, is the sum of their individual probabilities: (½ × ½) + (½ × ½) = ¼ + ¼ = ½. This is known as the **sum rule of probability.**

In the cross diagrammed in Figure 12–4, a heterozygote is produced by either *Ww* or *wW*. The probability of a heterozygote in the F_2 generation is the sum of the probability of each of the two possible combinations: ¼ + ¼ = ½.

The sum rule of probability, like the product rule, applies in more complex cases as well. For example, if you were asked the probability of throwing two heads and a tail, the answer would be ⅜. Three combinations are possible: HHT, HTH, and THH. For each of these combinations, the probability is ½ × ½ × ½ = ⅛, that is, the product of three independent throws. Thus, the probability of throwing two heads and a tail is the sum of the probability of each of the three possible combinations: ⅛ + ⅛ + ⅛ = ⅜.

Notice that in planning his experiments, Mendel made several assumptions: (1) of the male gametes produced, one-half contain one paternal allele and one-half contain the other paternal allele for each gene; (2) of the female gametes produced, one-half contain one maternal allele and one-half contain the other maternal allele for each gene; (3) the male and female gametes combine at random. Thus, the laws of probability could be employed—an elegant marriage of biology and mathematics.

If you toss two coins 4 times, it is unlikely that you will get the precise results diagrammed here. However, if you toss two coins 100 times, you will come close to the proportions predicted in the Punnett square, and if you toss two coins 1,000 times, you will be very close indeed. As Mendel knew, the ratio of dominants to recessives in the F_2 generation might well not have been so clearly visible if he had been dealing with a small sample. The larger the sample, however, the more closely it will conform to the results predicted by the laws of probability.

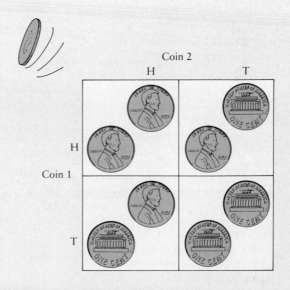

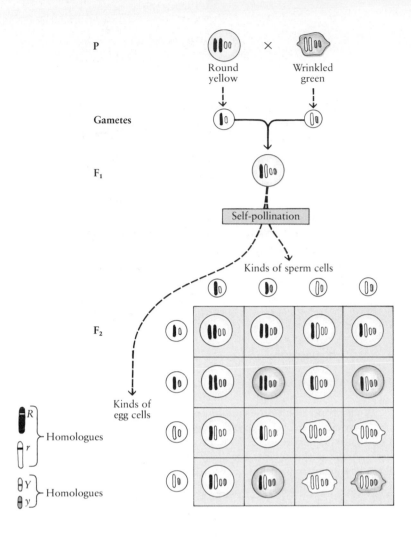

12–8   *The chromosome distributions in Mendel's cross of round yellow and wrinkled green peas, according to Sutton's hypothesis. Although the pea has 14 chromosomes (n = 7), only four are shown here, the homologous pair carrying the alleles for round or wrinkled and the homologous pair carrying the alleles for yellow or green. (This selection of specific pairs of homologous chromosomes is analogous to Mendel's selection of specific traits to study.) As you can see, one parent plant is homozygous for the dominants, and the other parent plant is homozygous for the recessives. Therefore, the gametes of one parent can contain only R and Y, and the gametes of the other parent can contain only r and y. The F₁ generation, therefore, must be Rr and Yy. When a cell of this generation undergoes meiosis, R is separated from r and Y from y when the members of each homologous pair separate at anaphase I. Because the alleles of the two genes are on different pairs of homologous chromosomes, they are assorted independently in meiosis. Four different types of haploid egg nuclei are possible, as the diagram reminds us, and also four different types of haploid sperm nuclei. These can combine in 4 × 4, or 16 different ways, as illustrated in the Punnett square.*

What about Mendel's second principle in relation to the movement of the chromosomes during meiosis? This principle, as we have seen, states that the alleles of different genes assort independently. We can see that this can be true if—and this is an important point—the genes are on different pairs of homologous chromosomes, as shown in Figure 12–8.

On the basis of these parallels, Sutton proposed that the factors described by Mendel are carried on the chromosomes. Although Sutton's proposal was widely accepted as a working hypothesis, proof of the physical location of the gene depended on further studies, which we shall consider in the next chapter.

Summary

How are hereditary characteristics passed from generation to generation? Gregor Mendel provided the first clear answers to this question. His breeding experiments in pea plants revealed that hereditary characteristics are determined by discrete factors (now called genes) that occur in pairs. According to Mendel's principle of segregation, the members of each pair segregate (that is, separate from one another) during gamete for-

mation. When two gametes come together in fertilization, the offspring receives one member of each pair from each parent.

The members of a given pair may be the same, in which case the individual is homozygous for the trait determined by that gene, or they may be different, in which case the individual is heterozygous for that trait. Different forms of the same gene are known as alleles.

The genetic makeup of an organism is known as its genotype. Its observable characteristics are known as its phenotype. An allele that is expressed in the phenotype of a heterozygous individual to the exclusion of the other allele is a dominant allele. An allele whose effects are concealed in the phenotype of a heterozygous individual is a recessive allele. In crosses involving two individuals heterozygous for the same gene, the expected ratio of dominant to recessive phenotypes in the offspring is 3:1.

Mendel's other great principle—independent assortment—applies to the behavior of two or more different genes. This principle states that, during gamete formation, the alleles of one gene segregate independently of the alleles of another gene. When organisms heterozygous for each of two independently assorting genes are crossed, the expected phenotypic ratio in the offspring is 9:3:3:1.

Sutton was among the first to notice the analogy between the behavior of the chromosomes at meiosis and the segregation and assortment of the factors described by Mendel. On the basis of this observation, Sutton proposed that genes are carried on chromosomes.

Questions

1. Distinguish between the following terms: gene/allele; dominant/recessive; homozygous/heterozygous; genotype/phenotype; the F_1/the F_2.

2. State Mendel's two principles in your own words.

3. In the experiments summarized in Table 12–1, which of the alternative characteristics appeared in the F_1 generation?

4. (a) What is the genotype of a pea plant that breeds true for tall? (Use the symbol T for tall, t for dwarf.) What possible gametes can be produced by such a plant? (b) What is the genotype of a pea plant that breeds true for dwarf? What possible gametes can be produced by such a plant? (c) What will be the genotype of the F_1 generation produced by a cross between a true-breeding tall pea plant and a true-breeding dwarf pea plant? (d) What will be the phenotype of this F_1 generation? (e) What will be the probable distribution of characteristics in the F_2 generation? Illustrate with a Punnett square.

5. PKU, phenylketonuria, is a disorder caused by the presence of two recessive alleles of a particular gene. Individuals who are homozygous for the dominant allele of that gene or who are heterozygous show no signs of PKU. If two parents, neither of whom has the disorder, have a child with PKU, what are their genotypes with respect to PKU? What is the probability that their next child will have the same disorder?

6. The inheritance of sickle-cell anemia is similar to that of PKU—that is, it occurs when two recessive alleles of a particular gene come together in the same individual. If two healthy parents have a child with sickle-cell anemia, what are their genotypes with respect to this allele? (Use H^a to represent the normal hemoglobin allele, and h^s to represent the sickle-cell allele.) Having had one such child, what is the probability that their next child will have the same disorder?

7. What proportion of the children of the parents in Question 6 would be expected to be carriers of the sickle-cell allele (that is, heterozygous)? What proportion of their children would not be expected to carry the allele? (Draw a Punnett square to diagram this problem.)

8. Why is a homozygous recessive always used in a test-cross?

9. The ability to taste a bitter chemical, phenylthiocarbamide (PTC), is due to a dominant allele. In terms of tasting ability, what are the possible phenotypes of a man both of whose parents are tasters? What are his possible genotypes? (Use N for the allele for tasting ability, and n for the allele for the absence of tasting ability.)

10. If the man in Question 9 marries a woman who is a nontaster, what proportion of their children could be tasters? Suppose one of the children is a nontaster. What would you know about the father's genotype? Explain your results by drawing Punnett squares.

11. A taster and a nontaster have four children, all of whom can taste PTC. What is the probable genotype of the parent who is a taster? Is there another possibility?

12. A pea plant that breeds true for round, green seeds *(RRyy)* is crossed with a plant that breeds true for wrinkled, yellow seeds *(rrYY)*. Each parent is homozygous for one dominant characteristic and for one recessive characteristic. (a) What is the genotype of the F_1 generation? (b) What is the phenotype? (c) The F_1 seeds are planted and their flowers are allowed to self-pollinate. Draw a Punnett square to determine the ratios of the phenotypes in the F_2 generation. How do the results compare with those of the experiment shown in Figure 12–6?

13. In Jimson weed, the allele for violet petals *(W)* is dominant over the allele for white petals *(w)*, and the allele for prickly capsules *(S)* is dominant over the allele for smooth capsules *(s)*. A plant with white petals and prickly capsules was crossed with one that had violet petals and smooth capsules. The F_1 generation was composed of 47 plants with white petals and prickly capsules, 45 plants with white petals and smooth capsules, 50 plants with violet petals and prickly capsules, and 46 plants with violet petals and smooth capsules. What were the genotypes of the parents?

14. Suppose you have just flipped a coin five times and it has turned up heads every time. What is the probability that the next time you flip it, it will turn up tails?

15. What is the probability of drawing two aces out of a deck of 52 playing cards, one the ace of hearts and the other the ace of spades?

16. (a) Suppose you would like to have a family consisting of two girls and a boy. What are your chances, assuming you have no children now? (b) If you already have one boy, what are your chances of completing your family as you would like? (c) If you have two girls, what is the probability that the next child will be a boy?

17. Mendel did not know of the existence of chromosomes. Had he known, what change might he have made in his second principle?

18. Segregation of alleles can occur at either of two stages of meiosis. Identify the two stages, and explain what happens in each of them.

19. You do not look exactly like your mother or your father. Why is this so? Explain how you might have inherited some of your maternal grandfather's characteristics. (Start with a gamete produced by your grandfather, and end with one of your somatic cells.)

Suggestions for Further Reading

Jacob, François: *The Logic of Life: A History of Heredity,* Pantheon Books, New York, 1982.*

> *Jacob's principal theme concerns the changes in the way people have looked at the nature of living beings. These changes, which are part of our total intellectual history, determine both the pace and direction of scientific investigation. The opening chapters are particularly brilliant.*

Klug, William S., and Michael R. Cummings: *Concepts of Genetics,* 3d ed., Macmillan Publishing Company, New York, 1991.

> *In this textbook, designed for a first course in genetics, the emphasis is on fundamental concepts, which are illustrated by carefully and clearly explained examples. It also contains many sample problems and their solutions.*

Mange, Arthur P., and Elaine Johansen Mange: *Genetics: Human Aspects,* 2d ed., Sinauer Associates, Inc., Sunderland, Mass., 1990.

> *A thorough account of human genetics.*

Oldroyd, David: "Gregor Mendel: Founding-father of Modern Genetics," *Endeavour,* vol. 8, pages 29–31, 1984.

Peters, James A. (ed.): *Classic Papers in Genetics,* Prentice-Hall, Inc., Englewood Cliffs, N.J., 1959.*

> *Includes papers by most of the scientists, including Mendel and Sutton, who were responsible for the major developments in genetics from the 1850s through the 1950s. The authors are surprisingly readable, and the papers give a feeling of immediacy that no modern account can achieve.*

Pines, Maya: "In the Shadow of Huntington's," *Science 84,* May 1984, pages 30–39.

Strickberger, Monroe W.: *Genetics,* 3d ed., Macmillan Publishing Company, New York, 1986.

> *A cohesive account of the science, this book provides a broad coverage of classical genetics.*

*Available in paperback.

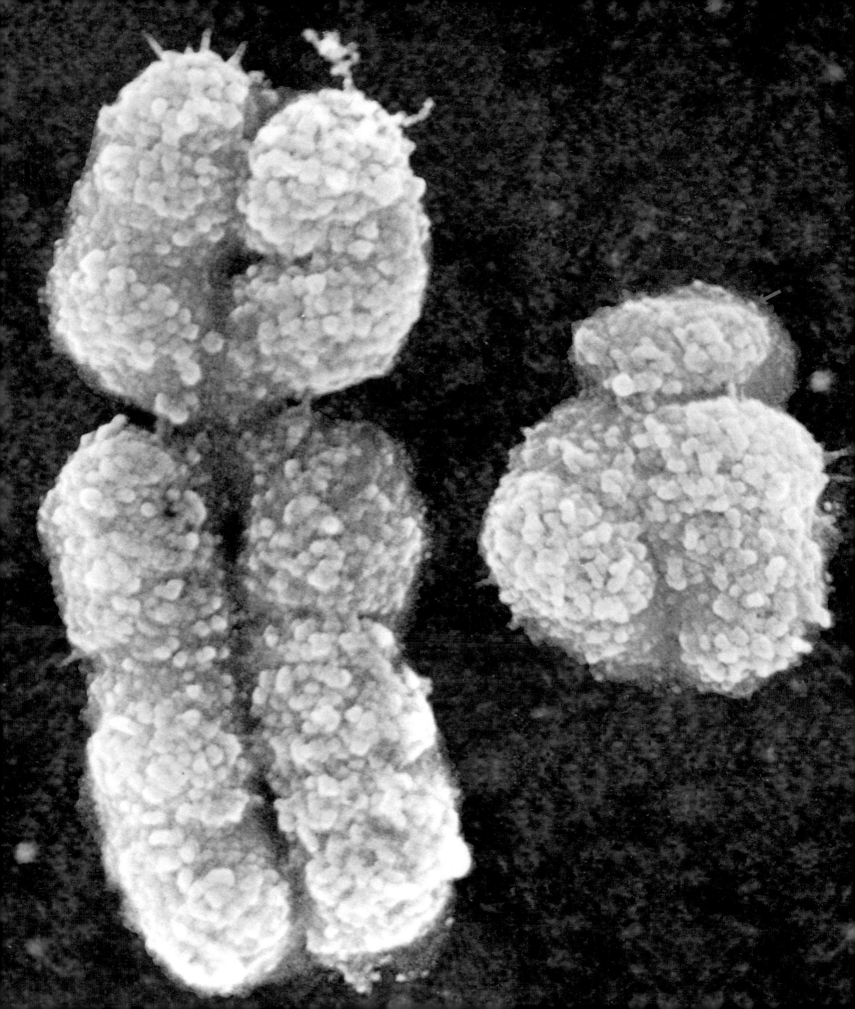

"Anatomy is destiny," according to one widely quoted source. Some of us may disagree with this statement and with its author. However, we are all likely to admit that whether we turned out male or female has played an important role in our life histories.

The determinants of this fundamental biological characteristic are two intensively studied microscopic structures: the human X and Y chromosomes. All human egg cells contain, as a result of meiosis, one X chromosome. Each human sperm cell contains either an X or a Y. Consequently, depending on which sperm wins the race to the egg, the zygote will be either XX (female) or XY (male)—and all else that follows from that.

As you can readily see in these scanning electron micrographs, the X chromosome (on the left) is much larger than the Y. In keeping with its size, the X carries many more genes than the Y. This difference is a significant factor in medical genetics. Suppose you have a defective allele on one X chromosome. If you are female, a normal allele on the other X chromosome can usually compensate for any deficiency. If you are male, however, you have only the one allele and so its effects are exposed. That is why some hereditary conditions, such as color blindness, hemophilia, and some forms of muscular dystrophy, are so much more common in males than in females.

The visible differences in the chromosome complements of males and females in many species provided some of the first evidence that the secrets of inheritance were to be found in the chromosomes. Differences in inheritance patterns in males and females noted by Mendel's successors lent strong support to the developing realization—now a commonplace fact—that particular chromosomes carry particular genes. Today, it is becoming possible to pinpoint the precise location of particular genes on particular chromosomes. An example is the human SRY gene (for "sex-determining region of the Y"), which is responsible for setting in motion the processes in embryonic development that result in maleness. Its location near the tip of the Y chromosome is marked by the arrow.

13

Mendel Rediscovered: Classical Genetics

In the decades following the rediscovery of Mendel's work in 1900, an enormous number of studies were carried out in genetics, as biologists sought to confirm and extend Mendel's observations. The investigations of this period included a wide variety of breeding experiments with many different types of organisms, as well as numerous cytological studies. The methods and the data that resulted from this span of about 50 years form the core of what is now known as **classical genetics.**

Classical genetics not only provided new understandings of the fundamental principles of heredity but also generated a wealth of data on the inheritance of specific traits in a diversity of plant and animal species. This data base has been—and continues to be—of great practical value to breeders of plants and animals and also to physicians concerned with human hereditary disorders. Moreover, both the data base and the methods of classical genetics are assuming a new importance as biologists try to relate the current avalanche of discoveries in molecular genetics to the phenotypic characteristics of living organisms.

Broadening the Concept of the Gene

The new studies that followed the rediscovery of Mendel's 1866 paper confirmed his work in principle. However, many of the studies showed that the patterns of inheritance are not always as simple and direct as Mendel's reported results had indicated. This is not surprising when you recall that Mendel had carefully selected certain traits for study, using only those that showed clear-cut differences.

The phenotypic effects of a particular gene, it was found, are influenced not only by the alleles of that gene present in the organism, but also by other genes and by the environment. And many, indeed most, traits are influenced by more than one gene, just as most genes can influence more than a single trait. Perhaps most surprising was the discovery that genes can change.

Mutations

In 1902, the Dutch botanist Hugo de Vries reported results of his studies on Mendelian inheritance in the evening primrose (Figure 13–1). Heredity in the primrose, he found, was generally orderly and predictable, as in the garden pea. Occasionally, however, a characteristic appeared that was not present in either parent or indeed anywhere in the lineage of that particular plant.

De Vries hypothesized that such characteristics came about as the result of abrupt changes in genes and that the characteristic produced by a changed gene was then passed along like any other hereditary characteristic. De Vries called these abrupt hereditary changes **mutations,** and organisms exhibiting such changes came to be known as **mutants.** Different alleles of a gene, de Vries proposed, arose as a result of mutations. For example, the allele for wrinkled peas is thought to have arisen as a mutation of the gene for round peas.

As it turned out, only about 2 of some 2,000 changes observed by de Vries in the evening primrose were actually mutations. The vast majority of the changes were simply the result of new allele combinations, produced in the course of meiosis, crossing over, and fertilization. However, de Vries's concept of mutation as the source of genetic variation proved of great importance, even though most of his examples were not valid.

Mutations and Evolutionary Theory

An important gap in Darwin's theory of evolution, first published in 1859, was the lack of explanation as to how variations can persist in populations. Mendel's work filled this gap. Segregation of alleles explained how variation is maintained from generation to generation. Independent assortment explained how individuals could have characteristics in combinations not present in either parent and so perhaps be better adapted, in evolutionary terms, than either parent.

Mendelian principles, however, presented new problems to the early evolutionists. If all hereditary variations were to be explained by the reshuffling process proposed by Mendel, there would be little or no opportunity for the kind of change in organisms envisioned by Darwin. As a result of mutations, however, there is a wide range of variability in natural populations. In a complex or shifting environment, a particular variation may give an individual or its offspring a slight edge. Although mutations seldom, if ever, determine the direction of evolutionary change, they are now recognized as the ultimate—and continual—source of the hereditary variations that make evolution possible.

13–1   *A flower of one species* (Oenothera biennis) *of evening primrose. Ironically, most of the genetic changes in the evening primrose observed by Hugo de Vries, one of the rediscoverers of Mendel's work, were not mutations. However, the concept that the basic genetic constitution of an organism is subject to change proved a valid and important one.*

Allele Interactions

Incomplete Dominance and Codominance

As the early studies in genetics proceeded, it soon became apparent that dominant and recessive characteristics are not always as clear-cut as in the seven traits studied by Mendel in the pea plant. Some characteristics appear to blend. For instance, a cross between a red-flowering snapdragon and a white-flowering snapdragon produces heterozygotes that are pink (Figure 13–2).

This phenomenon, in which the phenotype of the heterozygote is intermediate between those of the two homozygotes, is known as **incomplete dominance.** As we shall see in Chapter 15, it is a result of the combined effects of gene products—in this case, pigments synthesized by cells in the flower petals. When the heterozygous pink snapdragons are allowed to self-pollinate, the red and white characteristics are sorted out once again, showing that the alleles themselves, as Mendel had asserted, remain discrete and unaltered.

In other cases, alleles may act in a **codominant** manner, with heterozygotes expressing both homozygous phenotypes simultaneously. A familiar example is the human blood type AB, in which the red blood cells have the distinctive characteristics of both type A and type B.

P

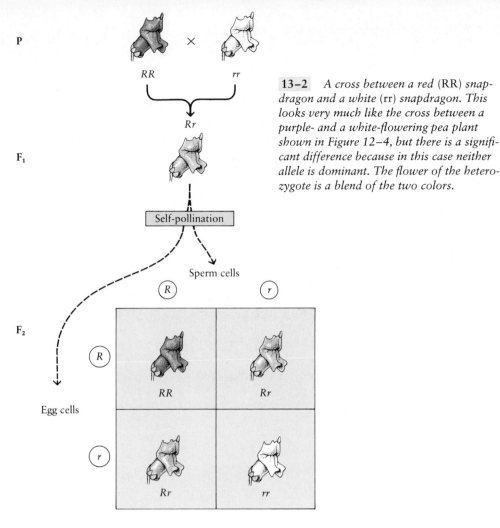

RR × rr

F₁

Rr

Self-pollination

Sperm cells

R   r

F₂

Egg cells

R

| RR | Rr |
| Rr | rr |

r

13–2 *A cross between a red (RR) snap-dragon and a white (rr) snapdragon. This looks very much like the cross between a purple- and a white-flowering pea plant shown in Figure 12–4, but there is a signifi-cant difference because in this case neither allele is dominant. The flower of the hetero-zygote is a blend of the two colors.*

(a)

(b)

(c)

Multiple Alleles

Although any individual diploid organism can have only two alleles of any given gene, it is possible that more than two forms of a gene—**multiple alleles**—may be present in a population of organisms (Figure 13–3). Multiple alleles result from different mutations of a single gene.

In humans, the four major blood groups (A, B, AB, and O) are determined by a gene with three alleles—*A, B,* and O. Phenotypically, the blood groups are char-acterized by particular polysaccharides on the surface of the red blood cells and by specific antibodies in the blood plasma.

13–3 *Coat color in rabbits is determined principally by a single gene, of which four different alleles are known. An individual rabbit, of course, carries only two alleles for this gene in its somatic (body) cells. Different combinations of alleles produce (a) the wild-type, or agouti, rabbit, (b) the chinchilla rab-bit, (c) the Himalayan rabbit, and (d) the albino rabbit.*

(d)

223

Table 13–1   The Genetic Basis of the ABO Blood Groups

Phenotype	Genotype (Alleles Present)	Polysaccharides on Surface of Red Blood Cells	Antibodies in Blood Plasma	Reaction with Antibodies	
				Antibody A	Antibody B
O	OO	—	Antibody A / Antibody B	No	No
A	AA, OA	A	Antibody B	Yes	No
B	BB, OB	B	Antibody A	No	Yes
AB	AB	A, B	—	Yes	Yes

The genetic basis of the blood groups is shown in Table 13–1. Alleles *A* and *B* are codominant, whereas allele *O* is recessive. Thus individuals with type A blood have either two *A* alleles or one *A* and one *O*, and their red blood cells bear the A polysaccharide. Their plasma does not contain antibodies against the A polysaccharide found on their own red blood cells, but it does have antibodies against the B polysaccharide. Individuals with type B blood have the B polysaccharide on their red blood cells and antibodies against the A polysaccharide in their plasma. Individuals with type AB blood have both polysaccharides but neither A nor B antibodies; conversely, type O individuals have neither polysaccharide but both A and B antibodies.

Gene Interactions

In addition to the interactions that occur between alleles of the same gene, interactions also occur among the alleles of different genes. Indeed, most of the characteristics (both structural and chemical) that constitute the phenotype of an organism are the result of the interaction of two or more distinct genes. These interactions may take a variety of forms.

Sometimes, when a trait is affected by two or more different genes, a completely novel phenotype may appear. For instance, comb shape in chickens is determined by two different genes, known as rose and pea, each with two alleles *(R, r* and *P, p)*. A genotype of *RR* or *Rr* results in rose comb, whereas *rr* produces single comb. Similarly, *PP* or *Pp* produces pea comb, and *pp* produces single comb. However, when *R* and *P* occur together in the same individual, a novel phenotype, walnut comb, results. Thus, four different types of combs are possible depending on the interaction of the alleles of these two genes (Figure 13–4).

In some cases, however, gene interaction does not produce a novel phenotype. Instead, one gene interferes with or masks the effect of another. This type of interaction is called **epistasis** ("standing upon"). A classic example of epistasis occurs in the sweet pea, a plant in which the flowers may be either purple or white. In order to produce purple flowers, a plant must have at least one dominant allele of each of two different genes. If a plant is homozygous recessive for either gene, the effects of the other gene are hidden and the flowers are white. Similarly, congenital deafness in humans can occur if either of two distinct genes is homozygous recessive.

(a) rrpp

(b) RRpp, Rrpp

(c) rrPP, rrPp

(d) RRPP, RRPp, RrPP, RrPp

13–4 *The four types of combs observed in chickens are single comb, rose comb, pea comb, and walnut comb. These comb types are determined by two different genes, of which the alleles are R, r and P, p. (a) Single comb occurs in chickens homozygous recessive for both genes. (b) Rose comb results from the presence of at least one dominant R allele coupled with two recessive p alleles. (c) Pea comb, by contrast, is produced by at least one dominant P allele coupled with two recessive r alleles. (d) A novel phenotype, walnut comb, is produced when at least one dominant allele of each gene is present in the same individual.*

Normal hearing requires at least one dominant allele of each of the two genes.

Polygenic Inheritance

Some traits, such as size or height, shape, weight, color, metabolic rate, and behavior, are not the result of interactions between one, two, or even several genes. Instead, they are the cumulative result of the combined effects of many genes. This phenomenon is known as **polygenic inheritance.**

A trait affected by a number of genes does not show a clear difference between groups of individuals—such as the differences tabulated by Mendel. Instead, it shows a gradation of small differences, which is known as **continuous variation.** If you make a chart of differences among individuals for any trait affected by a number of genes, for example, the height of adult men, you get a curve such as that shown in Figure 13–5.

One hundred years ago, the average height of men in the United States was less than it is now, but the shape of the curve was the same: the great majority fell within the middle range, and the extremes in height were represented by only a few individuals. Some of these height variations are produced by environmental factors, such as diet, but even if all the men in a population were maintained from birth on the same type of diet, there would still be a continuous variation in height in the population. This is due to genetically determined differences in hormone production, bone formation, and numerous other factors.

Table 13–2 illustrates a simple example of polygenic inheritance, color in wheat kernels, which is controlled by two genes, the four alleles of which exhibit cumulative quantitative effects. Human skin color is believed to be under a similar kind of genetic control (although involving more than two genes), as are many other traits.

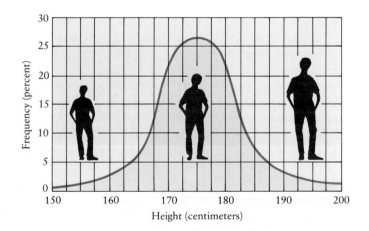

13–5 *Height distribution of males in the United States. Height is an example of polygenic inheritance; that is, it is affected by a number of genes. Such genetic traits are characterized by small gradations of difference. A graph of the distribution of such traits takes the form of a bell-shaped curve, as shown, with the mean, or average, usually falling in the center of the curve. The larger the number of genes involved, the smoother the curve.*

Multiple Effects of a Single Gene

Although most traits are influenced by a number of different genes, a single gene can often have multiple effects on the phenotype of an organism. This phenomenon is known as **pleiotropy** ("many turnings").

A striking example of pleiotropy is provided by a gene in rats that controls the production of a protein involved in the formation of cartilage. A mutation of this one gene causes a whole complex of birth defects, including thickened ribs, a narrowing of the passage through which air moves to and from the lungs, a loss of elasticity in the lungs, blocked nostrils, a blunt snout, and a thickening of the heart muscle. Needless to say, the mutation leads to a greatly increased mortality rate. Since cartilage is one of the most common structural substances of the body, the widespread effects of such a gene are not difficult to understand. Similarly, it is very likely that Mendel's allele for wrinkled peas affects other structural characteristics of the pea plant.

Genes and the Environment

The expression of a gene is always the result of its interaction with the environment. To take a common example, a seedling may have the genetic capacity to be green, to flower, and to fruit, but it will never turn green if it is kept in the dark, and it may not flower and fruit unless certain precise environmental requirements are met.

The water buttercup is a more striking example. It grows with half the plant body submerged in water. Although the leaves are genetically identical, the broad, floating leaves differ markedly in both form and physiology from the finely divided leaves that develop under water (Figure 13–6).

Temperature often affects gene expression. Primrose plants that are red-flowered at room temperature are white-flowered when raised at temperatures above 30°C (86°F). Similarly, as we saw in Figure 13–3c, Himalayan rabbits usually have black ears, forepaws, noses, and tails. However, when rabbits of the same genotype with respect to coat color are raised at high temperatures (above 35°C), their coats are entirely white.

The expression of a gene may be altered not only by factors in the external environment but also by factors in the internal environment of the organism, particularly during embryonic development. These factors include temperature, pH, ion concentrations, hormones, and a multitude of other influences, including the action of other genes.

When the expression of a gene is altered by environmental factors or by other genes, two outcomes are pos-

Table 13–2	The Genetic Control of Color in Wheat Kernels*		
Parents:	**RRSS** × **rrss**		
	(Dark Red)　(White)		
F₁:	**RrSs** (Medium Red)		

F₂:		Genotype		Phenotype	
1		RRSS		Dark red	
2	} 4	RRSs		Medium-dark red	
2		RrSS		Medium-dark red	
4		RrSs		Medium red	15 red
1	} 6	RRss		Medium red	to
1		rrSS		Medium red	1 white
2	} 4	Rrss		Light red	
2		rrSs		Light red	
1		rrss		White	

* Two genes are involved, each of which has two alleles, symbolized here as R and r for gene 1, and S and s for gene 2.

13–6　*The water buttercup,* Ranunculus peltatus, *grows with half the plant body submerged in water. Leaves growing above the water are broad, flat, and lobed. The genetically identical underwater leaves are thin and finely divided, appearing almost rootlike. These differences are thought to be related to differences in the turgor (page 116) of the immature leaf cells in the two environments. The degree of turgor affects the expansion of the cell walls and thus the ultimate size of the cells.*

sible. First, the degree to which a particular genotype is expressed in the phenotype may vary from individual to individual. In humans, this **variable expressivity** is seen in polydactyly, the presence of extra fingers and toes, which is caused by a dominant allele. Often, there is great variability in the expression of this allele among members of a family in which it is present. The result is that some individuals have both extra toes and fingers, while others have only a portion of an extra toe on one foot.

Second, the proportion of individuals that show the phenotype may be less than expected. The genotype shows **incomplete penetrance.** For example, individuals known to carry the allele for polydactyly may have absolutely normal hands and feet. Similarly, some individuals known to carry the dominant allele responsible for Huntington's disease (page 212) never develop the disorder.

These are all examples of a universal verity: The phenotype of any organism is the result of the interaction between its genes and their environment.

The Reality of the Gene

One of the most remarkable features of the earliest work in classical genetics was that the gene had no physical reality. It was a pure abstraction. The work of Sutton (page 215) and other cytologists was known, but, to most researchers, it seemed irrelevant to studies of inheritance. Several lines of study were to provide the evidence that conclusively linked genes and chromosomes.

Sex Determination

As Sutton had observed, the chromosomes of a diploid organism occur in pairs. In all of the pairs except one, the chromosomes in both males and females appear to be the same. As we noted in Chapter 11, these chromosomes are called autosomes. The structure of one pair, however, may differ between males and females. The chromosomes of this pair are known as the sex chromosomes.

In some animal groups, such as birds, the two sex chromosomes in males are structurally the same. These are, by convention, termed Z chromosomes, and the male is designated ZZ. The sex chromosomes of females in these groups consist of one Z chromosome (which is structurally identical to the male Z chromosome) and one W chromosome. Thus the females of these species are designated ZW.

By contrast, in mammals, including humans, and in many other groups of animals, it is the female in which

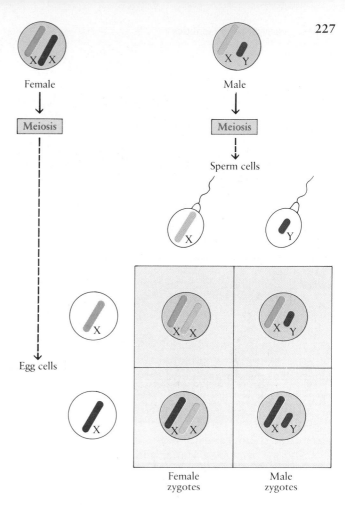

13–7 *Sex determination in organisms (such as humans and other mammals) in which the two sex chromosomes of the male are unlike. At meiosis, each egg cell receives an X chromosome from the mother. A sperm cell, however, may receive either an X chromosome or a Y chromosome.*

If a sperm cell carrying an X chromosome fertilizes an egg cell, the offspring will be female (XX). If, however, a sperm cell carrying a Y chromosome fertilizes an egg cell, the offspring will be male (XY). As the Punnett square predicts, male and female offspring are produced in approximately equal numbers.

the two sex chromosomes are the same. These are, by convention, termed X chromosomes, and the female is designated XX. The sex chromosomes of the male consist of one X chromosome (which is the same as the female X chromosome) and one Y chromosome. Thus the males of these species are designated XY.

In XY males, when the gametes are formed by meiosis, half of the sperm carry an X chromosome and half carry a Y chromosome. All the gametes formed by a female carry the X. The sex of the offspring is determined by whether the female gamete is fertilized by a male gamete carrying an X chromosome or a male gamete carrying a Y chromosome (Figure 13–7). Since equal numbers of X and Y sperm are produced, there is theoretically an even chance of producing male or female offspring.

The correlation of particular chromosomes with a particular trait—sex—gave strength to the hypothesis that genes are located on the chromosomes. Further evidence was to come from a variety of studies.

Sex Linkage

Introducing Drosophila

Much of the early work in classical genetics was carried out at a few research centers in Europe and the United States. Among the most important was the laboratory founded at Columbia University in 1909 by T. H. Morgan, a biologist from the United States who had visited de Vries's laboratory in Holland. By a remarkable combination of insight and good fortune, Morgan selected the fruit fly *Drosophila melanogaster* as his experimental organism.

Biologists have often used for their experiments "insignificant" plants and animals—such as pea plants, sea urchins, and fruit flies. Underlying this approach is the assumption that basic biological principles are universal, applying equally to all living things. As it turned out, the little fruit fly proved to be a "fit material" for a wide variety of genetic investigations. In the decades that followed, *Drosophila* was to become famous as the biologist's principal tool in studying animal genetics.

Drosophila means "lover of dew," although actually this useful animal is not attracted by dew but feeds on the fermenting yeast that it finds in rotting fruit. The fruit fly was an excellent choice for genetic studies since it is easy to breed and maintain. These tiny flies, each only 3 millimeters long, can produce a new generation every two weeks. Each female lays hundreds of eggs during her adult life, and very large numbers of flies can be kept in a half-pint bottle, as they were in Morgan's laboratory. Also, *Drosophila* has only four pairs of chromosomes (Figure 13–8), a feature that turned out to be particularly useful, although Morgan could not have foreseen that.

The White-Eyed Fruit Fly

The investigators in Morgan's laboratory were, at first, looking for genetic differences among individual flies that they could study in breeding experiments similar to those Mendel had carried out with pea plants. Shortly after Morgan established his *Drosophila* colony, such a difference appeared. One of the prominent and readily visible characteristics of fruit flies is their brilliant red eyes. One day, a white-eyed fly, a mu-

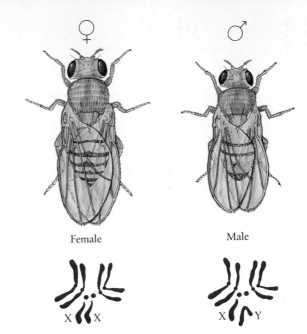

13–8   *The fruit fly* (Drosophila melanogaster) *and its chromosomes. Fruit flies have only four pairs of chromosomes (2n = 8), a fact that simplified Morgan's experiments. Six of the chromosomes (three pairs) are autosomes (including the two small spherical chromosomes in the center), and two are sex chromosomes.*

tant, appeared in the colony (Figure 13–9). This fly, a male, was mated with a red-eyed female, and all the F_1 offspring had red eyes. Apparently the white-eyed phenotype was recessive.

Morgan then crossbred the F_1 offspring, just as Mendel had done in his pea experiments. However, the offspring in the F_2 generation did not conform to the expected 3:1 ratio of dominant to recessive phenotypes (that is, of red-eyed to white-eyed individuals). And, as the detailed counts of the F_2 generation reveal, all of the white-eyed individuals were males:

Red-eyed females	2,459
White-eyed females	0
Red-eyed males	1,011
White-eyed males	782

Why were there no white-eyed females? To explore the situation further, Morgan crossed the original white-eyed male with one of the F_1 females. The following results were obtained from this testcross:

Red-eyed females	129
White-eyed females	88
Red-eyed males	132
White-eyed males	86

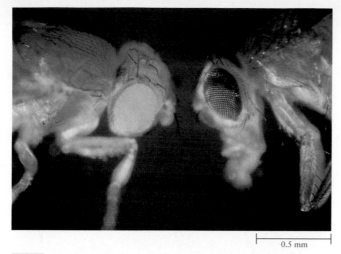

13–9 *A mutant white-eyed fruit fly (left) and a normal, or wild-type, red-eyed fruit fly (right). While looking for genetic differences among* Drosophila, *Calvin Bridges, a student in Morgan's laboratory, discovered a single white-eyed fruit fly in a population of thousands. This chance occurrence launched a series of studies that established the chromosomal basis of heredity.*

On the basis of these experiments, diagrammed in Figure 13–10, Morgan and his co-workers formulated the following hypothesis: The gene for eye color is carried only on the X chromosome. (In fact, as it was later shown, the Y chromosome of *Drosophila* carries very little genetic information of any kind.) The allele for white eyes must indeed be recessive, since all of the F_1 flies had red eyes. Thus a heterozygous female would have red eyes—which is why there were no white-eyed females in the F_2 generation. However, a male that received an X chromosome carrying the allele for white eyes would be white-eyed since no other allele would be present.

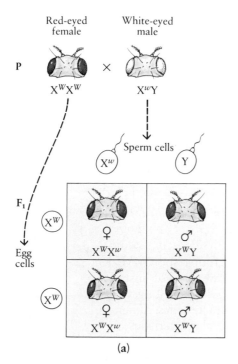

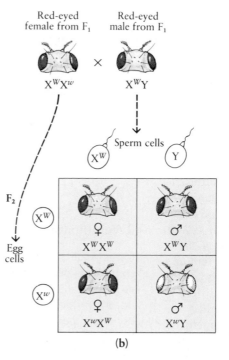

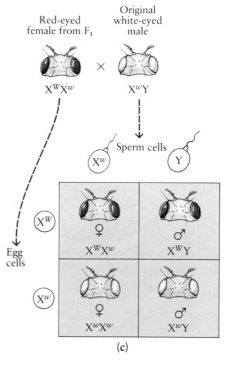

13–10 *Punnett square diagrams of the experiments Morgan performed after discovery of the white-eyed male* Drosophila. *The mutant allele for white eyes is represented here by w, and the wild-type allele for red eyes by W. Alleles located on the sex chromosomes are commonly designated by superscripts to the chromosome symbols.*
(a) Morgan first mated a homozygous red-eyed female to the white-eyed male. All of the F_1 offspring had red eyes. (b) Next, he mated an F_1 red-eyed female to an F_1 red-eyed male. Although both red-eyed and white-eyed males were produced in the F_2 generation, all F_2 females had red eyes, suggesting a relationship between the inheritance of eye color and the inheritance of the sex chromosomes.
(c) A testcross between a red-eyed F_1 female and the original white-eyed male produced both red-eyed and white-eyed flies of both sexes. This led to the conclusion that the gene for eye color must be carried on the X chromosome. The allele for red eyes (W) is dominant, and the allele for white eyes (w) is recessive.

Further experimental crosses, such as the one diagrammed in Figure 13–11, confirmed Morgan's hypothesis. They also revealed that white-eyed fruit flies are more likely to die before they reach adulthood than are red-eyed fruit flies. This suggested that the allele for white eyes has other effects as well, and it explained why the numbers of white-eyed fruit flies were lower than expected in both the F_2 generation and the test-cross.

These experiments introduced the concept of **sex-linked traits.** They also convinced Morgan, and many other geneticists as well, that Sutton's hypothesis was correct: Genes *are* on chromosomes. Conclusive demonstration of the physical location of the gene, however, depended on a subsequent series of experiments.

Linkage Groups

An important research tool became available when H. J. Muller, one of Morgan's collaborators, found that exposure to x-rays greatly increases the rate at which mutations occur in *Drosophila.* Other forms of radiation, such as ultraviolet light, and certain chemicals were also shown to act as **mutagens,** or agents that produce mutations. As increasing numbers of mutants were produced in Columbia University's *Drosophila* collection, it became possible to do breeding experiments in which the flies differed in more than one characteristic.

As you will recall, Mendel had demonstrated that certain pairs of alleles, such as those for round and wrinkled peas, assort independently of other pairs, such as those for yellow and green peas. And, as we noted at the end of the last chapter, the alleles of two different genes will always assort independently if the genes are on different pairs of homologous chromosomes (see Figure 12–8, page 217). If, however, the alleles of the two genes are on the same pair of homologous chromosomes, segregation of the alleles of one gene will *not* be independent of the segregation of the alleles of the other gene. In other words, if the alleles of two different genes are on the *same* chromosome, they should both be transmitted to the *same* gamete during meiosis. Genes that tend to stay together because they are on the same pair of homologous chromosomes are said to be in the same **linkage group.**

As the *Drosophila* breeding experiments continued, the mutations began to fall into four linkage groups, in accord with the four pairs of chromosomes visible in the cells. Indeed, in all organisms that have been studied in sufficient genetic detail, the number of linkage groups and the number of pairs of chromosomes have been the same, providing further support for Sutton's hypothesis that genes are on the chromosomes.

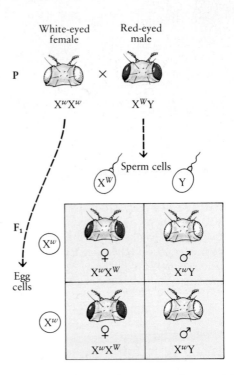

13–11   *Offspring of a cross between a white-eyed female fruit fly and a red-eyed male fruit fly, illustrating what happens when a recessive allele is carried on an X chromosome. The F_1 females, with one X chromosome from the mother and one from the father, are all heterozygous ($X^w X^W$) and so will be red-eyed. But the F_1 males, with their single X chromosome received from the mother carrying the recessive (w) allele, will all be white-eyed because the Y chromosome carries no gene for eye color. Thus the recessive allele on the X chromosome inherited from the mother will be expressed.*

Recombination

Large-scale studies of linkage groups soon revealed some unexpected difficulties, however. For instance, most fruit flies have light tan bodies and long wings, both of which are dominant characteristics. When individuals homozygous for these characteristics were bred with mutant fruit flies having black bodies and vestigial (short) wings, both of which are recessive characteristics, all the F_1 offspring had tan bodies and long wings, as would be expected. Then the F_1 generation was inbred. Two outcomes seemed possible:

1. The genes for body color and wing length would be assorted independently, giving rise to Mendel's 9:3:3:1 ratio in the phenotypes of the F_2 offspring. This would indicate that the genes for these two traits were on different pairs of homologous chromosomes.

Human Sex-Linked Traits

In humans, as in fruit flies, the Y chromosome carries much less genetic information than the X chromosome. Among the genes carried on the human X chromosome—but not on the Y—are the genes responsible for the capacity to distinguish red and green.

The ability to perceive color depends on three genes, producing three different visual pigments, each responsive to light in a different region of the spectrum. The genes for the pigments responsive to red and green light are both on the X chromosome.

In males, if the gene for green is defective, green cannot be distinguished from red, and, conversely, a defect in the gene for red results in red appearing as green (a). In heterozygous females, the defective alleles are recessive to the normal alleles on the other X chromosome, and so color vision is usually normal. Complete red-green color blindness in females occurs only when both X chromosomes carry the same defective allele.

If a woman carrying a defective allele on one X chromosome (b) transmits that X chromosome to a daughter, the daughter will also have normal color vision if she receives an X chromosome with the normal allele from her father (that is, if he is not color-blind). If, however, the X chromosome with the defective allele is transmitted from mother to son, he will be color-blind since, lacking a second X chromosome, he has only the defective allele.

Another classic example of sex-linked inheritance is the hemophilia that has afflicted some royal families of Europe since the nineteenth century. Hemophilia is a group of disorders in which the blood does not clot normally. Clotting occurs through a complex series of reactions in which each reaction depends on the presence of certain protein factors in the blood. Failure to produce one essential protein, known as Factor VIII, results in the most common form of hemophilia, which is associated with a recessive allele of a gene carried on the X chromosome.

For individuals with this form of hemophilia, even minor injuries carry the risk of bleeding to death. Until very recently, the only effective treatment was Factor VIII extracted from normal human blood. Unfortunately, this carried the risk of transmission of infectious agents, particularly the virus that causes AIDS. However, as we shall see in Chapter 18, a form of Factor VIII produced through genetic engineering is now available, eliminating the risk of contamination with infectious agents.

The original carrier of hemophilia among European royalty was probably Queen Victoria. One of her sons died of hemophilia at the age of 31. Because male-to-male inheritance of the disorder is impossible—as it is with all X-linked characteristics—Prince Albert, Victoria's husband, could not have been the source. Because none of Victoria's relatives other than her descendants were affected, it is thought that the mutation occurred on an X chromosome in one of her parents or in the cell line from which her own eggs were formed.

At least two of Victoria's daughters were carriers, since a number of *their* descendants were hemophiliacs. And so, through various intermarriages, the disease spread from throne to throne across Europe. Tsarevitch Alexis, the only son of Nicholas II and Alexandra, the last Tsar and Tsarina of Russia, inherited the allele for hemophilia from his mother, who was a granddaughter of Queen Victoria. The great concern of Alexis's parents for his health, which apparently distracted them from affairs of state, contributed to the turbulent events surrounding the Russian revolution.

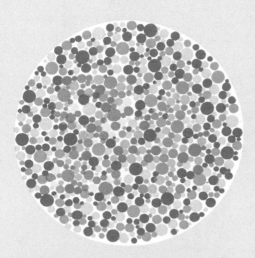

(a) *Individuals with normal color vision can easily read the two-digit number embedded in this dot pattern. Individuals with red-green color blindness—about 8 percent of human males and 0.04 percent of human females—cannot do so.*

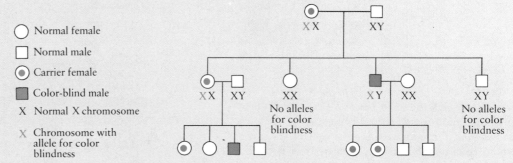

(b) *The pedigree of a family in which the mother has inherited one normal and one defective allele for red-green color vision. The normal allele is dominant, and she has normal color vision. However, half of her eggs (on the average) will carry the defective allele and half will carry the normal allele—and it is a matter of chance which kind is fertilized. Since her husband's Y chromosome, the one that determines a son rather than a daughter, carries no gene for color vision, the single allele the* mother contributes will determine whether or not a son is color-blind. Therefore, on average, half of her sons will be color-blind. Assuming that her children marry individuals with X chromosomes with the normal alleles, the expected distribution of the trait among her grandchildren will be as shown on the bottom line. Note that all the daughters of a color-blind man will be carriers of the defective allele, and all his sons will have normal color vision, unless their mother is a carrier or is herself color-blind.

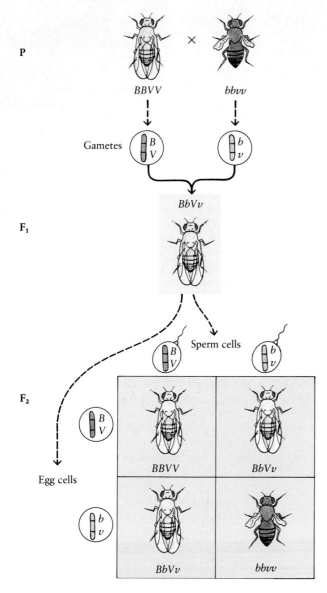

P

BBVV   ×   *bbvv*

Gametes

B
V   *b*
v

BbVv

F₁

Sperm cells

B
V   *b*
v

F₂

B
V

BBVV   *BbVv*

b
v

Egg cells

BbVv   *bbvv*

13–12   *The expected results in the F₂ generation of an initial cross between a fruit fly homozygous for tan body color and long wings and a fruit fly homozygous for black body color and short wings, if the two genes are located on the same pair of homologous chromosomes. The allele symbols are B for tan body, b for black body, V for long wings, and v for vestigial, or short, wings.*

When Morgan performed this experiment, the phenotypes in the F₂ generation were close to the 3:1 ratio shown in the Punnett square, but they did not conform exactly. Some of the F₂ flies were tan with vestigial wings, and some were black with long wings.

2. The genes for the two traits would be linked. This possibility is diagrammed in Figure 13–12. As you can see, 75 percent of the F₂ flies would be tan with long wings, and 25 percent, homozygous for both recessives, would be black with vestigial wings (a 3:1 ratio).

In the case of these particular traits, the results closely resembled the second possibility, but they did not conform exactly. In a few of the offspring, the genes for these traits appeared to assort independently. That is, some of the F₂ flies were tan with vestigial wings, and some were black with long wings. How could this be? Somehow alleles that were presumed to be on the same chromosome had become separated.

To find out what was happening, Morgan tried a testcross, breeding a member of the F₁ generation with a fly that was homozygous recessive for both genes. If the alleles for black and tan assorted independently of

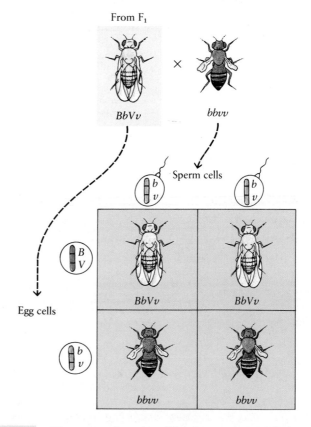

From F₁

BbVv   ×   *bbvv*

Sperm cells

b
v   *b*
v

B
V

BbVv   *BbVv*

Egg cells

b
v

bbvv   *bbvv*

13–13   *The expected results of a testcross of an F₁ fly from the cross in Figure 13–12 with a fly homozygous for black body and vestigial wings, if the two genes are located on the same pair of homologous chromosomes. Once again, the results were close to those shown in the Punnett square, but they did not conform exactly.*

the alleles for long and vestigial—that is, if the two genes were on different pairs of homologous chromosomes—25 percent of the offspring of this cross should be black with long wings, 25 percent tan with long wings, 25 percent black with vestigial wings, and 25 percent tan with vestigial wings (a ratio of 1:1:1:1).

On the other hand, if the genes for color and wing size were on the same pair of homologous chromosomes, the alleles of the two genes should move together. In this case, 50 percent of the testcross offspring should be tan with long wings and 50 percent should be black with vestigial wings, as shown in Figure 13–13. But actually, as it turned out, over and over, in counts of hundreds of fruit flies resulting from such crosses, 42 percent were tan with long wings, 42 percent were black with vestigial wings, 8 percent were tan with vestigial wings, and another 8 percent were black with long wings (Figure 13–14).

Morgan was convinced by this time that genes are located on chromosomes. It now seemed clear that the genes for these two traits were located on a single pair of homologous chromosomes, since the characteristics did not show up in the 1:1:1:1 ratio of independently assorted alleles. The only way in which the observed figures could be explained, Morgan reasoned, was if alleles could sometimes be exchanged between homologous chromosomes—that is, if they could be recombined.

As we noted in Chapter 11, it now has been established that exchange of portions of homologous chromosomes—crossing over—takes place in prophase I of meiosis (Figure 13–15). If crossing over takes place between the positions at which two different genes are located on a pair of homologues, then the alleles of the two genes can become separated as chromatids of the two homologues break and rejoin with each other (Figure 13–16).

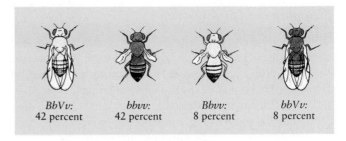

BbVv:
42 percent

bbvv:
42 percent

Bbvv:
8 percent

bbVv:
8 percent

13–14   *The actual results of the testcross of an F₁ fly from the cross illustrated in Figure 13–12 with a fly homozygous for black body and vestigial wings. These results suggested that alleles could sometimes be exchanged between homologous chromosomes. As we now know, this process—crossing over—occurs in prophase I of meiosis.*

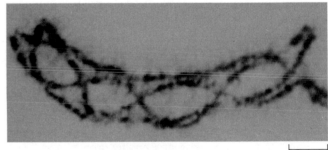

├─── 2 µm

13–15   *Homologous chromosomes of a grasshopper, as seen in prophase I. All four chromatids are visible. Crossing over—that is, the exchange of genetic material—has probably occurred at the points at which these chromatids intersect.*

13–16   (a, b) *Crossing over takes place when breaks occur in the chromatids of homologous chromosomes early in meiosis I, when the chromosomes are tightly paired. The broken end of each chromatid joins with the chromatid of a homologous chromosome. In this way, alleles are exchanged between chromosomes.* (c) *As a result of crossing over, two of the chromosomes have allele combinations different from those of the original chromosomes.*

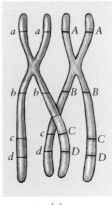

(a)

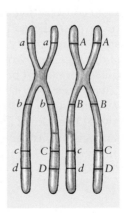

(b)

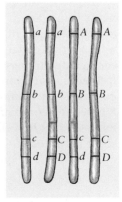

(c)

Mapping the Chromosome

With the discovery of crossing over, growing evidence clearly supported not only the premise that genes are carried on the chromosomes, but also that they must be at particular positions on the chromosomes. These positions are known as **loci** (singular, locus). It was also clear that the alleles of any given gene must occupy corresponding loci on homologous chromosomes. Otherwise, the exchange of sections of chromosomes would result in genetic chaos rather than in an exact exchange of alleles.

As other traits were studied, the data revealed that the percentage of recombinations between any two genes, such as those for body color and wing length, was different from the percentage of recombinations between two other genes, such as those for body color and leg length. In addition, as Morgan's experiments had shown, these percentages were fixed and predictable. It occurred to A. H. Sturtevant, then a student working in Morgan's laboratory, that the percentage of recombinations probably had something to do with the physical distances between the gene loci, or, in other words, with their spacing along the chromosome. This concept opened the way to the earliest "mapping" of chromosomes.

Sturtevant postulated that (1) genes are arranged in a linear series on chromosomes; (2) genes that are close together will be separated by crossing over less frequently than genes that are farther apart; and (3) it should therefore be possible to use the frequencies of various recombinations to plot the sequence of the genes along the chromosome and the relative distances between them. In Figure 13–16, for example, you can see that in a crossover, the chance that a strand would break and rejoin with its homologous strand somewhere between *A* and *C* should be more likely than this happening somewhere between *C* and *D*. This is the case simply because there is a greater distance between *A* and *C* in which crossing over can occur.

In 1913, Sturtevant began constructing chromosome maps using data from crossover studies in fruit flies. As a standard unit of measure, he arbitrarily defined one map unit as equal to the distance that would give (on the average) one recombinant organism per 100 fertilized eggs. For example, two genes with 10 percent recombination would be 10 units apart; two genes with 8 percent recombination would be 8 units apart. By this method, he and other geneticists constructed chromosome maps locating a variety of genes and their mutants in *Drosophila* (Figure 13–17).

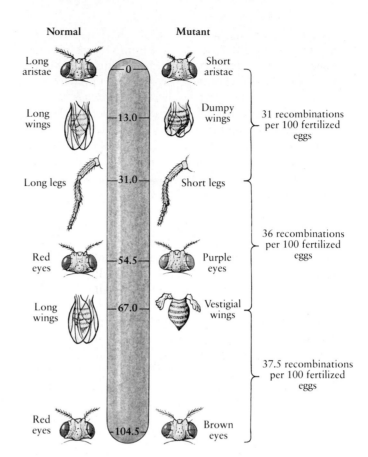

13–17   *A portion of the map of chromosome 2 of* Drosophila melanogaster, *showing the relative positions of some of the genes on that chromosome. The map distance between any two genes represents the number of recombinations of the alleles of those two genes per 100 fertilized eggs. As this map reveals, more than one gene may affect a single trait.*

Although Sturtevant's method of chromosome mapping has now been superseded by much more precise techniques, these studies were of critical importance in the history of genetics. They confirmed not only that genes are located on chromosomes, as Sutton had hypothesized, but also that they have fixed positions in a linear sequence. Moreover, the data from this work constitute a detailed atlas of the *Drosophila* chromosomes—an invaluable reference for current research workers.

Giant Chromosomes

Recombination does not usually affect the order of genes on a chromosome. However, as we noted in Chapter 11, when chromatid segments break during crossing over, they may sometimes be lost or they may join with a different chromosome. They may also rejoin the correct chromosome but in a different orientation. The outcome, in any of these cases, is a change in the sequence of genes on the affected chromosomes.

Evidence for such changes in chromosome structure, giving rise to changes in gene order and altered hereditary patterns, came from the study of giant chromosomes discovered in 1933 in the salivary glands of *Drosophila* larvae (immature forms). In *Drosophila*, as in many other insects, certain cells do not divide during the larval stages of the insect. In such cells, however, the chromosomes continue to replicate, over and over again, but since the daughter chromosomes of the nondividing cells do not separate from one another after replication, they simply become larger and larger. Ultimately, they may be composed of as many as a thousand copies. As you can see in Figure 13–18, giant chromosomes are characterized by very distinctive patterns of dark and light bands.

These banding patterns became another useful tool for geneticists, enabling them to detect structural changes in the chromosomes themselves. By observing changes in the banding patterns in the giant chromosomes, geneticists can see where changes in chromosome structure have occurred. Such changes include the deletions, duplications, and translocations discussed in Chapter 11 (page 199). They also include another type of chromosome abnormality, known as an **inversion**. This results when a double break in a chromosome occurs and a segment is turned 180° and then reincorporated into the chromosome. When one member of a homologous pair contains a large inverted segment, that chromosome must loop inside the other for close pairing to occur. In giant chromosomes, homologues are paired and such loops are greatly magnified, making it possible to readily identify regions with inverted segments.

The correlation of changes in the banding patterns of the giant chromosomes with observed genetic changes in the individual fruit flies provided conclusive confirmation of Sturtevant's hypothesis that genes occur in a linear sequence on the chromosomes.

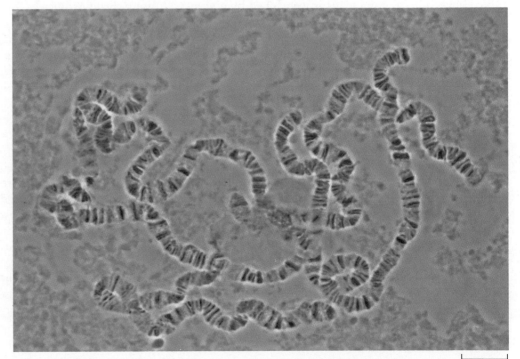

13–18   *Chromosomes from the salivary gland of a* Drosophila *larva. These chromosomes are hundreds of times larger than the chromosomes in ordinary body cells, and their details are therefore much easier to see. (This micrograph was taken with a light microscope, not an electron microscope.) Because of the distinctive banding patterns, it is possible in some cases to associate genes with specific regions in particular chromosomes. Each visible band represents several genes.*

25 μm

Summary

The rediscovery of Mendel's work in 1900 was the catalyst for many new discoveries in genetics, leading to the modification and extension of some of Mendel's conclusions and to the identification of chromosomes as the carriers of the genetic information.

Mutations are abrupt changes in the genotype. They are the ultimate source of the genetic variations studied by Mendel. Different mutations of a single gene increase the diversity of alleles of that gene in a population. As a consequence, mutation provides the variability among organisms that is the raw material for evolution.

Many traits are inherited according to the patterns revealed by Mendel. However, in others—perhaps the majority—the patterns are more complex. Although many alleles interact in a dominant-recessive manner, some show varying degrees of incomplete dominance and codominance. In a population of organisms, multiple alleles of a single gene may exist, but only two alleles can be present in any given diploid individual. The particular combination present determines both the genotype and the phenotype.

Different genes can also interact with one another. Novel phenotypes may result from these interactions, or genes may affect one another in an epistatic manner, such that one hides the effect of the other. The phenotypic expression of many traits is influenced by a number of genes. This phenomenon is known as polygenic inheritance. Such traits typically show continuous variation, as represented by a bell-shaped curve. Conversely, a single gene can affect two or more superficially unrelated traits. This property of a gene is known as pleiotropy.

Gene expression is also affected by factors in the external and internal environments. Variable expressivity or incomplete penetrance of particular alleles may result from environmental influences, interactions with other genes, or both.

Strong support for the hypothesis that genes are on the chromosomes came from work with the fruit fly *Drosophila*, which has been used in a wide variety of genetic studies. It has four pairs of chromosomes; three pairs (the autosomes) are structurally the same in both sexes, but the fourth pair, the sex chromosomes, is different. In fruit flies, as in many other species (including humans), the two sex chromosomes are XX in females and XY in males.

At the time of meiosis, the sex chromosomes, like the autosomes, segregate. Each egg cell receives an X chromosome, but half of the sperm cells receive an X chromosome and half receive a Y chromosome. Thus it is the sperm cell that determines the sex of the offspring in species with XY males.

In the early 1900s, breeding experiments with *Drosophila* showed that certain traits are sex-linked, that is, their genes are carried on the sex chromosomes. Because the X chromosome carries genes that are not present on the Y chromosome, a single recessive allele on the X chromosome of a male will result in a recessive phenotype—simply because no other allele is present. By contrast, a female heterozygous for a sex-linked trait will show the dominant phenotype. In humans, sex-linked characteristics include color blindness and hemophilia.

Some genes assort independently in breeding experiments, and others tend to remain together. Genes that do not assort independently are said to be in the same linkage group. A linkage group corresponds to a pair of homologous chromosomes.

Alleles are sometimes exchanged between homologous chromosomes as a result of crossing over in meiosis. Such recombinations can take place because: (1) the genes are arranged in a fixed linear array along the length of the chromosomes, and (2) the alleles of a given gene are at corresponding sites (loci) on homologous chromosomes. Chromosome maps, showing the relative positions of gene loci along the chromosomes, have been developed from recombination data provided by breeding experiments.

Genetic studies have shown that chromosome breaks other than those resulting in crossovers may sometimes occur. A portion of a chromosome may be lost, or deleted, it may be duplicated, it may be translocated to a nonhomologous chromosome, or it may be inverted. Studies of the giant chromosomes of *Drosophila* larvae provided visual confirmation of these changes, as well as the final, conclusive evidence that the chromosomes are the carriers of the genetic information.

Questions

1. Distinguish among the following: mutation/mutant/mutagen; incomplete dominance/codominance/epistasis; polygenic inheritance/pleiotropy; variable expressivity/incomplete penetrance.

2. The so-called "blue" (really gray) Andalusian variety of chicken is produced by a cross between the black and white varieties. Only a single pair of alleles is involved. What color chickens (and in what proportions) would you expect if you crossed two blues? If you crossed a blue and a black? Explain.

3. In snapdragons, the allele that produces tall stems is completely dominant to the allele for dwarf stems, while the allele that produces red flowers is only partially dominant to that for white flowers. (a) Describe the phenotype (height and flower color) of the F_1 plants resulting from a cross between a homozygous tall, red-flowered plant and a homozygous dwarf, white-flowered plant. (b) If one of these F_1 plants self-pollinates, what will be the appearance and proportions of phenotypes in the resulting F_2 generation? (c) Which two of these phenotypes will breed true?

4. If you are of blood type O and neither of your parents is, what are their possible genotypes? What is the probability that one of your siblings is also of type O? (The use of a Punnett square here may be helpful.)

5. Although blood types can never prove that someone is the father of a particular child, they can demonstrate that someone could *not* be the father. Fill in the table at the right to see how this is so. In the famous Charlie Chaplin paternity case in the 1940s, the baby's blood was B, the mother's A, and Chaplin's O. If you had been the judge, how would you have decided the case?*

6. In chickens, as we have seen, two pairs of alleles determine comb shape. *RR* or *Rr* results in rose comb, whereas *rr* produces single comb. *PP* or *Pp* produces pea comb, and *pp* produces single comb. When *R* and *P* occur together, they produce a new type of comb: walnut. (a) What would be the genotype of the F_1 generation resulting from *RRpp* × *rrPP*? The phenotype? (b) If F_1 hybrids were crossbred, what would be the probable distribution of genotypes? Of phenotypes? (Illustrate this cross with a Punnett square.)

* As a matter of fact, Chaplin was judged to be the father. In some states at that time, blood-group data were not admitted as evidence in cases of disputed parentage.

7. In Duroc-Jersey pigs, coat color is determined by two genes, *R* and *S*. The homozygous recessive condition, *rrss*, produces a white coat. The presence of at least one copy each of *R* and *S* produces red. The presence of one or the other allele (either *R* or *S)* produces a new phenotype, sandy. Give the phenotypes of the following genotypes:

RRSS	rrss
RrSs	rrSs
RRSs	rrSS
RrSS	RRss

8. Mating a red Duroc-Jersey boar to sow A (white) gave pigs in the ratio of 1 red: 2 sandy: 1 white. Mating this same boar to sow B (sandy) gave 3 red: 4 sandy: 1 white. When this boar was mated to sow C (sandy), the litter had equal numbers of red and sandy piglets. Using the information presented in Question 7, give the possible genotypes of the boar and the three sows.

9. In humans, either of two recessive alleles *(a or b)*, when homozygous, can cause congenital deafness. Hence, it is possible that all of the children of parents who are both congenitally deaf could have normal hearing. What would be the genotypes of the parents in such a situation?

Phenotypes of Parents		Phenotypes Possible in Children	Phenotypes Not Possible in Children
A	A		
A	B		
A	AB		
A	O		
B	B		
B	AB		
B	O		
AB	AB		
AB	O		
O	O		

10. A woman homozygous for the dominant alleles *A* and *B*, necessary for normal hearing, marries a man who is congenitally deaf. What are the possible genotypes of the man? What is the probability that this couple will have a deaf child?

11. You and a geneticist are looking at a mahogany-colored Ayrshire cow with a newly born red calf. You wonder if it is male or female, and the geneticist says it is obvious from the color which sex the calf is. She explains that in Ayrshires the genotype *AA* is mahogany and *aa* is red, but the genotype *Aa* is mahogany in males and red in females. What is she trying to tell you—that is, what sex is the calf? What are the possible phenotypes of the calf's father?

12. In one strain of mice, skin color is determined by five different pairs of alleles. The colors range from almost white to dark brown. Would it be possible for some pairs of mice to produce offspring darker or lighter than either parent? Explain.

13. The size of the eggs laid by one variety of hens is determined by three pairs of alleles. Hens with the genotype *AABBCC* lay eggs weighing 90 grams, and hens with the genotype *aabbcc* lay eggs weighing 30 grams. Each of the alleles *A*, *B*, or *C* adds 10 grams to the weight of the egg. When a hen from the 90-gram strain is mated with a rooster from the 30-gram strain, the hens of the F_1 generation lay eggs weighing 60 grams. If a hen and rooster from this F_1 generation are mated, what will be the weight of the eggs laid by hens of the F_2?

14. Height and weight in animals follow a distribution similar to that shown in Figure 13–5. By inbreeding large animals, breeders are usually able to produce some increase in size among their stock. But after a few generations, increase in size characteristically stops. Why?

15. Draw a diagram similar to Figure 13–7 indicating sex determination in a robin.

16. Why is male-to-male inheritance of red-green color blindness impossible? Under what conditions would such color blindness be found in a woman? If she married a man who was not color-blind, what proportion of her sons would be color-blind? Of her daughters?

17. What is the probability that a woman with normal color vision whose father is color-blind but whose husband has normal color vision will have a color-blind son? A color-blind daughter?

18. A woman whose maternal grandfather was hemophiliac has parents who are clinically normal. She too seems normal, as does her husband. What are the chances that her first son will *not* have hemophilia? *(Hint:* Determine the genotype of the woman's mother and then the possible genotypes of the woman herself.)

19. A man with a particular hereditary disorder marries a woman who does not have the disorder. They have six children, three girls and three boys. The girls all have the father's disorder but the boys do not. What type of inheritance pattern is suggested? (Although this situation was not discussed in the text, you should be able to deduce the answer.)

20. A particular diploid organism has 42 chromosomes per somatic cell. How many linkage groups does it have?

21. The first two pairs of genes studied simultaneously in crosses were *Rr* and *Yy* in pea plants. As it turned out, they were on different chromosomes. How do you think the development of the principles of genetics would have been affected had they been linked?

22. Does crossing over necessarily result in a recombination of alleles? Explain.

23. In a series of breeding experiments, a linkage group composed of genes *A*, *B*, *C*, *D*, and *E* was found to show approximately the recombination frequencies in the chart below. Using Sturtevant's standard unit of measure, "map" the chromosome.

		Gene				
		A	*B*	*C*	*D*	*E*
	A	—	8	12	4	1
	B	8	—	4	12	9
Gene	*C*	12	4	—	16	13
	D	4	12	16	—	3
	E	1	9	13	3	—

Recombinations per 100 Fertilized Eggs

Suggestions for Further Reading

Klug, William S., and Michael R. Cummings: *Concepts of Genetics*, 3d ed., Macmillan Publishing Company, New York, 1991.

> *In this textbook, designed for a first course in genetics, the emphasis is on fundamental concepts, which are illustrated by carefully and clearly explained examples. It also contains many sample problems and their solutions.*

Mange, Arthur P., and Elaine Johansen Mange: *Genetics: Human Aspects*, 2d ed., Sinauer Associates, Inc., Sunderland, Mass., 1990.

> *A thorough account of human genetics.*

Peters, James A. (ed.): *Classic Papers in Genetics*, Prentice-Hall, Inc., Englewood Cliffs, N.J., 1959.*

> *A collection of papers by the scientists—including Morgan and his collaborators at Columbia University—who were responsible for the major developments in genetics from the 1850s through the 1950s. The authors are surprisingly readable, and the papers give a feeling of immediacy that no modern account can achieve.*

Sapienza, Carmen: "Parental Imprinting of Genes," *Scientific American*, October 1990, pages 52–60.

Strickberger, Monroe W.: *Genetics*, 3d ed., Macmillan Publishing Company, New York, 1986.

> *A cohesive account of the science, this book provides a broad coverage of classical genetics.*

Weiss, Rick: "A Genetic Gender Gap," *Science News*, May 20, 1989, pages 312–315.

* Available in paperback.

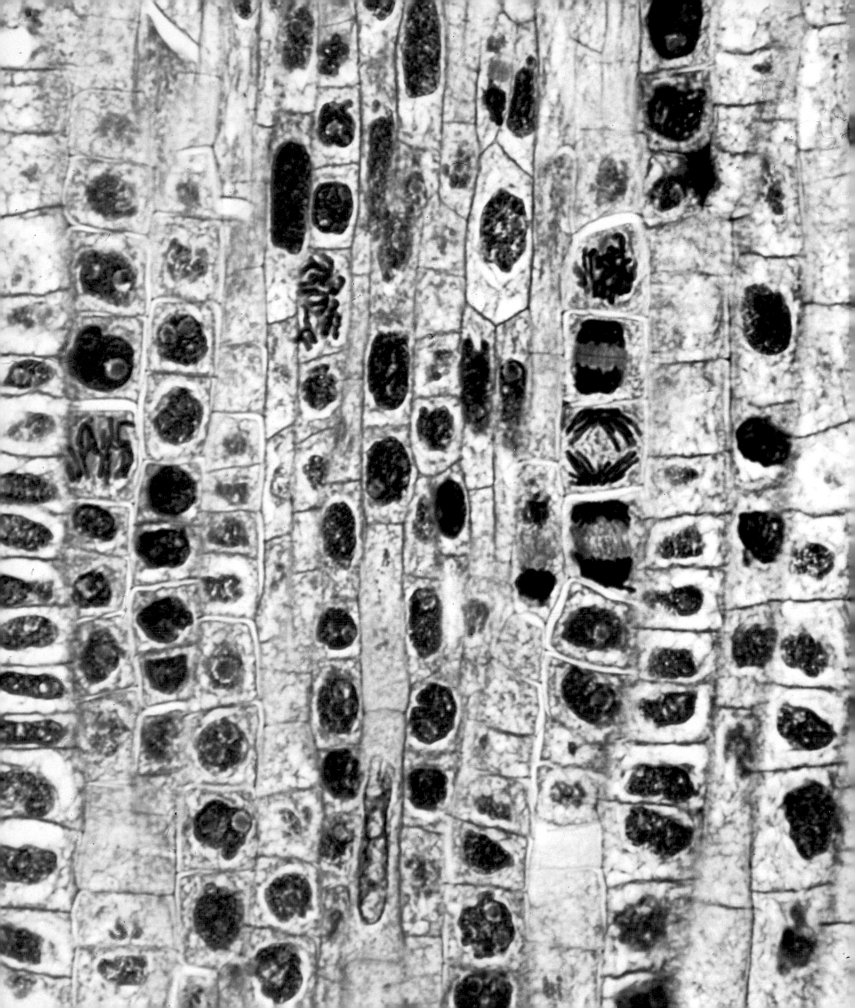

The Chemical Basis of Heredity: The Double Helix

This is a photomicrograph of dividing cells in the root tip of an onion. The cells, many of which are undergoing mitosis, have been stained to show the chromosomes.

As we have seen, chromosomes have two important functions in the life of a cell. First, they transmit the genetic information. When a cell undergoes mitosis and cytokinesis, each daughter cell receives a full and precise copy of all of the genes that were present in the parent cell. This everyday occurrence—taking place right now in almost every tissue of your body—is the basis for the survival of cells, of organisms, of generations—that is, of life on Earth.

The second important function of chromosomes is to instruct the cells in their day-to-day, moment-to-moment activities. In pea plants, for example, chromosomes direct cells in the flower petals to make pigment and cells in the roots to make the membrane proteins that allow them to absorb needed minerals from the soil.

The work of Mendel and his successors showed us that there were different genes for different functions and that genes were located on the chromosomes. But their work provided barely a clue as to how these microscopic scraps of matter could transmit information and direct cellular activities.

The answers to these crucial questions began to emerge when scientists turned their attention away from the microscopic examination of chromosomes and the counting of phenotypes toward a wholly new field of research—the chemistry of heredity.

By the early 1940s, the existence of genes and the fact that they are carried in the chromosomes were no longer in doubt. A turning point in the history of genetics came when scientists began to focus on the question of how it is possible for the chromosomes to carry what the scientists realized must be an enormous amount of extremely complex information.

The chromosomes, like all the other parts of a living cell, are composed of atoms arranged into molecules. Some scientists, a number of them eminent in the field of genetics, thought it would be impossible to understand the complexities of heredity in terms of the structure of "lifeless" chemicals. Others thought that if the chemical structure of the chromosomes were understood, we could then come to understand how chromosomes function as the bearers of the genetic information. This insight marked the beginning of the vast range of investigations that we know as **molecular genetics**.

The DNA Trail

Early chemical analyses revealed that the eukaryotic chromosome consists of both **deoxyribonucleic acid (DNA)** and protein, in about equal amounts. Thus, both were candidates for the role of the genetic material. Proteins seemed the more likely choice, because of their greater chemical complexity. As you will recall from Chapter 3, proteins are polymers of amino acids, of which there are some 20 different types in living cells. DNA, by contrast, is a polymer formed from only four different types of nucleotides.

Speculative thinkers in the field of biology were quick to point out that the amino acids, the number of which is so provocatively close to the number of letters in our own alphabet, could be arranged in a variety of different ways. The amino acids were seen as making up a sort of language—"the language of life"—that spelled out the directions for all the many activities of the cell.

Many prominent investigators, particularly those who had been studying proteins, believed that the genes themselves were proteins. They thought that the chromosomes contained master models of all the proteins that would be required by the cell and that enzymes and other proteins active in cellular life were copied from these master models. This was a logical hypothesis, but, as it turned out, it was wrong.

Sugar-Coated Microbes: The Transforming Factor

To trace the beginning of the other hypothesis—the one that ultimately proved correct—it is necessary to go back to 1928 and pick up an important thread in modern biological history. In that year, an experiment was performed that seemed at the time to have little relevance to the field of genetics. Frederick Griffith, a public health bacteriologist in England, was studying the possibility of developing vaccines against pneumococci, the bacteria that cause one kind of pneumonia. In those days, before the development of antibiotics, bacterial pneumonia was an extremely serious disease, often referred to as the grim "captain of the men of death."

Pneumococci, as Griffith knew, come in either **virulent** (disease-causing) forms with polysaccharide capsules or **nonvirulent** (harmless) forms without capsules (Figure 14–1). The production of the capsule and its composition are genetically determined—that is, they are inherited properties of the bacteria. Griffith thought that injections of heat-killed virulent bacteria or of liv-

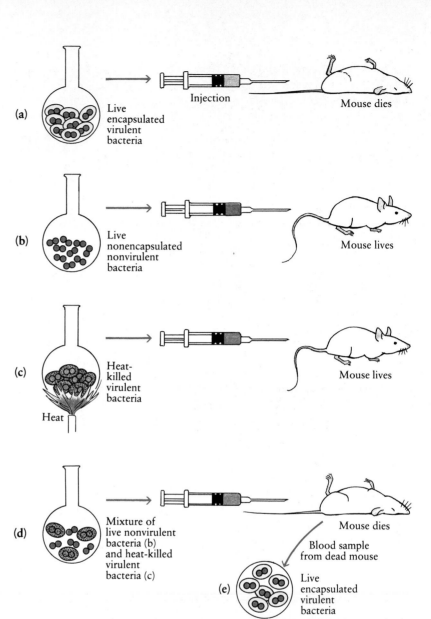

14–2   *Discovery of the transforming factor, a substance that can transmit genetic characteristics from one cell to another, resulted from studies of pneumococci. One strain of these bacteria has polysaccharide capsules; another does not. The capacity to make capsules and cause pneumonia is an inherited characteristic, passed from one bacterial generation to another as the cells divide.* (a) *Injection into mice of encapsulated pneumococci led to pneumonia, which killed the mice.* (b) *The nonencapsulated strain produced no infection.* (c) *If the encapsulated strain was heat-killed before injection, it too produced no infection.* (d) *If, however, heat-killed encapsulated bacteria were mixed with living nonencapsulated bacteria and the mixture was injected into mice, the mice developed pneumonia and died.* (e) *Blood samples from the dead mice revealed living encapsulated pneumococci. Something had been transferred from the dead bacteria to the living ones that endowed them with the capacity to make polysaccharide capsules and cause pneumonia. This "something" was later isolated and found to be DNA.*

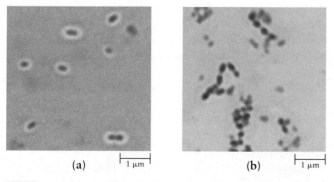

14–1   (a) *Encapsulated and* (b) *nonencapsulated forms of pneumococci. The capsules—the lighter colored zones surrounding the cells in* (a)—*are made up of polysaccharides deposited outside the cell walls. The encapsulated form, which is resistant to phagocytosis by white blood cells, can cause pneumonia. The nonencapsulated form is harmless.*

ing nonvirulent bacteria (without capsules) might immunize mice against living virulent pneumococci. As a result of his experiments, which are outlined in Figure 14–2, Griffith found that something can be passed from dead to living bacteria that changes their hereditary characteristics. This phenomenon became known as **transformation,** and the "something" that caused it was called the **transforming factor.**

In 1943, after almost a decade of patient chemical isolation and analysis, O. T. Avery and his co-workers at Rockefeller University demonstrated that the transforming factor is DNA. Subsequent experiments showed that a variety of genetic factors can be passed from bacterial cells of one strain to cells of another, similar strain by means of isolated DNA.

The Nature of DNA

DNA had first been isolated by a German physician named Friedrich Miescher in 1869—in the same remarkable decade in which Darwin published *The Origin of Species* and Mendel presented his results to the Brünn Natural History Society. The substance Miescher isolated was white, sugary, slightly acidic, and con-

tained phosphorus. Since it was found only in the nuclei of cells, it was called nucleic acid. This name was later amended to deoxyribonucleic acid, to distinguish it from a similar chemical also found in cells, ribonucleic acid (RNA).

By the time of Avery's discovery, it was known that DNA is made up of nucleotides. Each nucleotide consists of a nitrogenous base, a deoxyribose sugar, and a phosphate group (Figure 14–3a). The nitrogenous bases are of two kinds: **purines** (Figure 14–3b), which have two rings, and **pyrimidines** (Figure 14–3c), which have one ring. There are two kinds of purines in DNA, **adenine (A)** and **guanine (G),** and two kinds of pyrimidines, **cytosine (C)** and **thymine (T).** So DNA is made up of four types of nucleotides, differing only in their nitrogen-containing purine or pyrimidine.

Avery's results offered evidence for DNA as the genetic material, but his discovery was slow to gain full recognition. This was partly because bacteria, which are, of course, prokaryotes, were considered "lower" and "different." And, it was partly because the DNA molecule—made up of only four different nucleotides—seemed too simple for the enormously complex task of carrying the hereditary information.

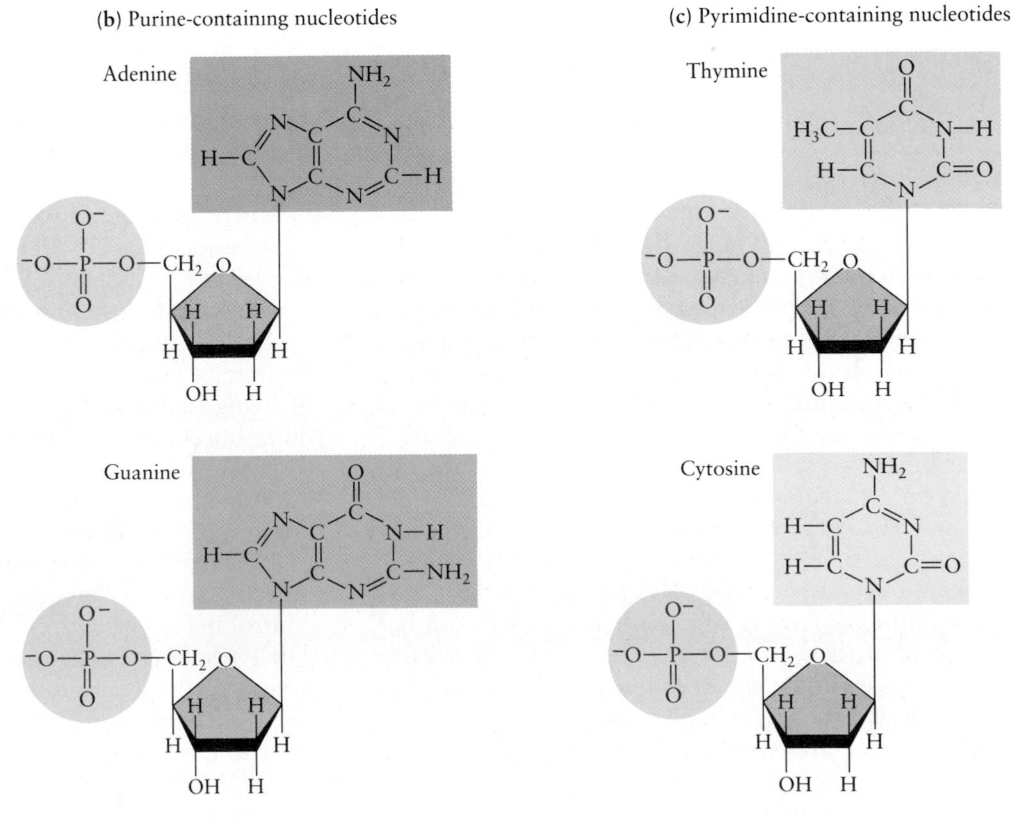

(a)

(b) Purine-containing nucleotides

(c) Pyrimidine-containing nucleotides

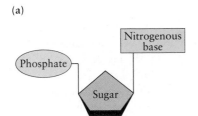

14–3   (a) *A nucleotide is made up of three different components: a nitrogenous base, a five-carbon sugar, and a phosphate group.* **(b)** *The two purine-containing nucleotides and* **(c)** *the two pyrimidine-containing nucleotides found in DNA. Each nucleotide consists of one of the four possible nitrogenous bases, a deoxyribose sugar, and a phosphate group.*

Adenine

Guanine

Thymine

Cytosine

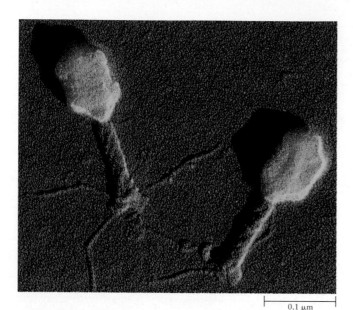

0.1 μm

14–4   *Bacteriophages of the type used in the studies that provided positive identification of DNA as the genetic material. Notice their highly distinctive shape. Each bacteriophage consists of a head, which appears hexagonal in electron micrographs, and a complex tail assembly. These bacteriophages attach to* E. coli *cells by means of the thin fibers extending from the tail assembly.*

The Bacteriophage Experiments

In 1940, a series of crucial experiments was begun with another "fit material," destined to become as important to genetic research as the garden pea and the fruit fly. The fit material was a group of viruses that attack bacteria and are therefore known as **bacteriophages** ("bacteria eaters"). The bacteriophages originally chosen for study were ones that attack *Escherichia coli,* the familiar bacterium found in the healthy human intestine.

These viruses, commonly called **phages,** were inexpensive, easy to maintain in the laboratory, and required little space or equipment. Moreover, they were phenomenal at reproducing themselves. About 25 minutes after a single virus infected a bacterial cell, the cell would burst open, releasing a hundred or more new viruses, all exact copies of the original virus. Another advantage of these phages was their highly distinctive shape (Figure 14–4), which enabled them to be readily identified with the electron microscope.

Chemical analysis of the bacteriophages revealed

that they consist quite simply of DNA and of protein, the two leading contenders for the role of the genetic material. The question of which type of molecule contains the viral genes—that is, the hereditary material that directs the synthesis of new viruses within the bacterial cells—was answered in 1952 by Alfred D. Hershey and Martha Chase. Their simple but ingenious experiments are summarized in Figure 14–5. Remember that protein contains sulfur (in the amino acids cysteine and methionine) but no phosphorus, and DNA contains phosphorus but no sulfur. You will then see how these experiments demonstrated that only the DNA of the bacteriophages was involved in the replication process and that the protein could not be the hereditary material.

Electron micrographs later confirmed that this type of bacteriophage attaches to the bacterial cell wall by its tail fibers and injects its DNA into the cell. The empty protein coat (the "ghost") is left on the outside (Figure 14–6). In short, the protein is just a container for the phage DNA. It is the DNA that enters the cell and carries the complete hereditary message of the virus, directing the formation of new viral DNA and new viral protein.

Further Evidence for DNA

The role of DNA in bacterial transformation and in viral replication formed very convincing evidence that DNA is the genetic material. Two other lines of work also helped to lend weight to the argument. First, Alfred Mirsky, in a long series of careful studies, showed that, in general, the somatic cells of any given species contain equal amounts of DNA. The gametes, however, contain just half as much DNA as the somatic cells.

A second important series of contributions was made by Erwin Chargaff. Chargaff analyzed the purine and pyrimidine content of the DNA of many different kinds of living things and found that the nitrogenous bases do *not* occur in equal proportions. The proportions of the four nitrogenous bases are the same in all cells of all individuals of a given species, but they vary from one species to another.

These variations suggested that the four bases might indeed provide a "language" in which the instructions controlling cell growth could be written. Some of Chargaff's results are reproduced in Table 14–1. Do you notice anything interesting about the relative proportions of the four bases? These ratios provided important clues to the molecular structure of DNA.

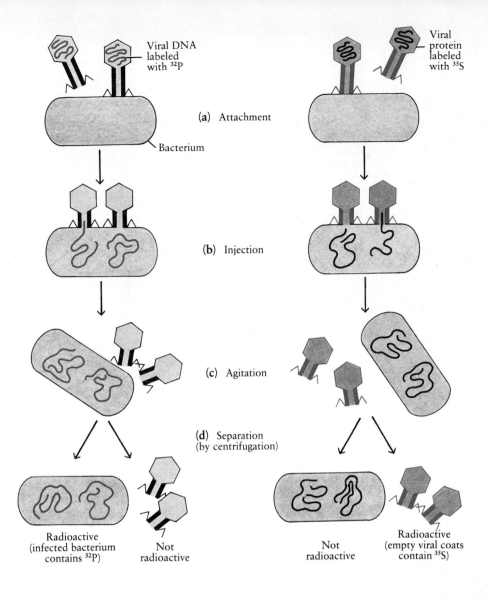

(a) Attachment

Bacterium

Viral DNA labeled with ^{32}P

Viral protein labeled with ^{35}S

(b) Injection

(c) Agitation

(d) Separation (by centrifugation)

Radioactive (infected bacterium contains ^{32}P)

Not radioactive

Not radioactive

Radioactive (empty viral coats contain ^{35}S)

14–5 *A summary of the Hershey-Chase experiments. Two separate samples of phages were prepared, one in which the DNA was labeled with a radioactive isotope of phosphorus, ^{32}P, and the other in which the protein was labeled with a radioactive isotope of sulfur, ^{35}S. (a) One culture of bacteria (left) was infected with the ^{32}P-labeled phage, and a similar culture (right) was infected with the ^{35}S-labeled phage. After the infection cycles had begun (b), the bacterial cells were agitated in a blender (c) and then spun down in a centrifuge (d) to separate the cells from any viral material remaining on the outside. Samples of extracellular material and of intracellular material from each culture were then tested for radioactivity. Hershey and Chase found that the ^{35}S had remained outside the bacterial cells with the empty viral coats and the ^{32}P had entered the cells, infected them, and caused the production of new viral particles. It was therefore concluded that the genetic material of the bacteriophage is DNA rather than protein.*

Table 14–1	Composition of DNA in Several Species			
	Purines		**Pyrimidines**	
Source	**Adenine**	**Guanine**	**Cytosine**	**Thymine**
Human being	30.4%	19.6%	19.9%	30.1%
Ox	29.0	21.2	21.2	28.7
Salmon sperm	29.7	20.8	20.4	29.1
Wheat germ	28.1	21.8	22.7	27.4
E. coli	24.7	26.0	25.7	23.6
Sea urchin	32.8	17.7	17.3	32.1

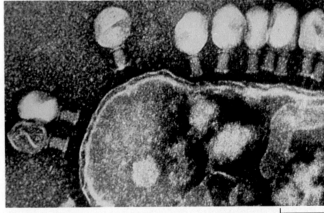

0.1 μm

14–6 *Electron micrograph of bacteriophages attacking a cell of E. coli. The viruses attach to the bacterial cell by their tail fibers. Viral DNA, contained within the head of the virus, is injected through the tail and into the cell. In this micrograph, the heads of some of the viruses are empty, indicating that the injection process has already occurred.*

The Watson-Crick Model

In the early 1950s, a young American scientist, James Watson, went to Cambridge, England, on a research fellowship to study problems of molecular structure. There, at the Cavendish Laboratory, he met physicist Francis Crick. Both were interested in DNA, and they soon began to work together to solve the problem of its molecular structure. They did not do experiments in the usual sense but rather undertook to examine all the data about DNA and to unify them into a meaningful whole.

The Known Data

By the time Watson and Crick began their studies, quite a lot of information on the subject had already accumulated. It was known that the DNA molecule is very large, very long and thin, and contains nucleotides, each consisting of either one purine or one pyrimidine plus a deoxyribose sugar and a phosphate group.

In 1950, Linus Pauling had shown that proteins sometimes take the form of a helix (see page 62) and that the helical structure is maintained by hydrogen bonding between successive turns in the helix. Pauling had suggested that the structure of DNA might be similar. Subsequent x-ray studies by Rosalind Franklin and Maurice Wilkins at Kings College, London, provided strong evidence that the DNA molecule was indeed a giant helix.

Finally, there were Chargaff's data, which indicated that the ratio of DNA nucleotides containing thymine to those containing adenine is approximately 1:1 and that the ratio of nucleotides containing guanine to those containing cytosine is also approximately 1:1.

Building the Model

From these data, Watson and Crick attempted to construct a model of DNA that would fit the known facts and explain the biological role of DNA. In order to carry the vast amount of genetic information, the molecules should be heterogeneous and varied. Also, there must be some way for them to replicate readily and with great precision so that faithful copies could be passed from cell to cell and from parent to offspring, generation after generation.

By piecing together the data, Watson and Crick were able to deduce that DNA is an exceedingly long, entwined double helix. If you were to take a ladder and twist it into a helix, keeping the rungs perpendicular, you would have a crude model of the DNA molecule (Figure 14–7). The two rails, or sides, of the ladder are made up of alternating sugar and phosphate molecules.

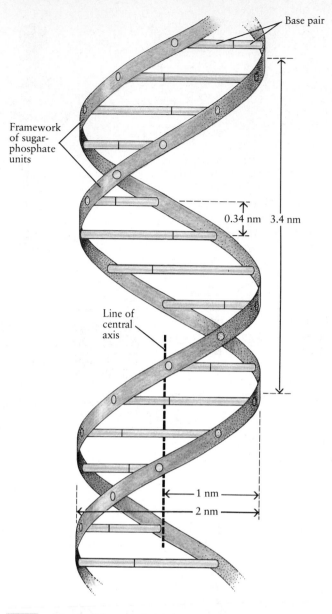

14–7   *The double-stranded helical structure of DNA, as first presented in 1953 by Watson and Crick. The framework of the helix is composed of the sugar-phosphate units of the nucleotides. The rungs are formed by the four nitrogenous bases adenine and guanine (the purines) and thymine and cytosine (the pyrimidines). Each rung consists of a pair of bases. Knowledge of the distances shown here was crucial in establishing the structure of the DNA molecule. These distances were determined from x-ray diffraction photographs of DNA taken by Rosalind Franklin.*

The perpendicular rungs of the ladder are formed by the nitrogenous bases—adenine (A), thymine (T), guanine (G), and cytosine (C)—a pair of bases to each rung. Each base is covalently bonded to a sugar subunit in the "rail" adjacent to it. The paired bases meet in the interior of the helix and are joined by hydrogen bonds, the relatively weak bonds that Pauling had demonstrated in his studies of protein structure.

Who Might Have Discovered It?

Then there is the question, what would have happened if Watson and I had not put forward the DNA structure? This is "iffy" history which I am told is not in good repute with historians, though if a historian cannot give plausible answers to such questions I do not see what historical analysis is about. If Watson had been killed by a tennis ball I am reasonably sure I would not have solved the structure alone, but who would? Olby has recently addressed himself to this question. Watson and I always thought that Linus Pauling would be bound to have another shot at the structure once he had seen the King's College x-ray data, but he has recently stated that even though he immediately liked our structure it took him a little time to decide finally that his own was wrong. Without our model he might never have done so. Rosalind Franklin was only two steps away from the solution. She needed to realise that the two chains must run in opposite directions and that the bases, in their correct tautomeric forms, were paired together. She was, however, on the point of leaving King's College and DNA, to work instead on TMV [tobacco mosaic virus] with Bernal. Maurice Wilkins had announced to us, just before he knew of our structure, that he was going to work full time on the problem. Our persistent propaganda for model building had also had its effect (we had previously lent them our jigs to build models but they had not used them) and he proposed to give it a try. I doubt myself whether the discovery of the structure could have been delayed for more than two or three years.

There is a more general argument, however, recently proposed by Gunther Stent and supported by such a sophisticated thinker as Medawar. This is that if Watson and I had not discovered the structure, instead of being revealed with a flourish it would have trickled out and that its impact would have been far less. For this sort of reason Stent had argued that a scientific discovery is more akin to a work of art than is generally admitted. Style, he argues, is as important as content.

I am not completely convinced by this argument, at least in this case. Rather than believe that Watson and Crick made the DNA structure, I would rather stress that the struc-

(a)

(a) *James Watson (left) and Francis Crick in 1953, with one of their models of DNA. At the time they announced their discovery of the DNA structure, Watson was 23 and Crick was 34.*

   (b) *A computer-generated model of a portion of a DNA molecule. The sugar-phosphate backbone is indicated by the blue ribbons and green dots. The purines are shown in yellow, and the pyrimidines in red. The hydrogen bonds linking*

ture made Watson and Crick. After all, I was almost totally unknown at the time and Watson was regarded, in most circles, as too bright to be really sound. But what I think is overlooked in such arguments is the intrinsic beauty of the DNA double helix. It is the molecule which has style, quite as much as scientists. The genetic code was not revealed all in one go but it did not lack for impact once it had been pieced together. I doubt if it made all that difference that it was Columbus who discovered America. What mattered much

(b)

the base pairs are represented by blue dashed lines.

   In 1993, 40 years after their discovery, Watson remarked, "The molecule is so beautiful. Its glory was reflected on Francis and me. I guess the rest of my life has been spent trying to prove that I was almost equal to being associated with DNA, which has been a hard task." Crick replied, "We were upstaged by a molecule."

more was that people and money were available to exploit the discovery when it was made. It is this aspect of the history of the DNA structure which I think demands attention, rather than the personal elements in the act of discovery, however interesting they may be as an object lesson (good or bad) to other workers.

Francis Crick: "The Double Helix: A Personal View," *Nature*, vol. 248, pages 766–769, 1974.

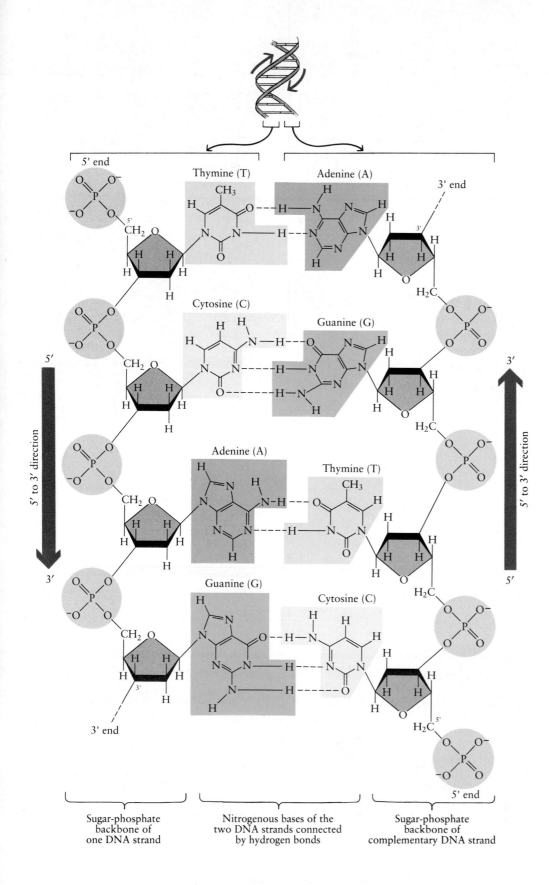

Sugar-phosphate
backbone of
one DNA strand

Nitrogenous bases of the
two DNA strands connected
by hydrogen bonds

Sugar-phosphate
backbone of
complementary DNA strand

14–8   *The double-stranded structure of a small portion of a DNA molecule. Each nucleotide consists of a phosphate group, a deoxyribose sugar, and a purine or pyrimidine base.*

Note the repetitive sugar-phosphate-sugar-phosphate sequence that forms the backbone of each strand of the molecule. Each phosphate group is attached to the 5' carbon of one sugar subunit and to the 3' carbon of the sugar subunit in the adjacent nucleotide. Each strand of the DNA molecule thus has a 5' end and a 3' end, determined by these 5' and 3' carbons. The strands are antiparallel—that is, the direction from the 5' to the 3' end of one strand is opposite to that of the other.

The strands are held together by hydrogen bonds (represented here by dashed lines) between the bases. Notice that adenine and thymine form two hydrogen bonds, whereas guanine and cytosine form three. Because of these bonding requirements, adenine can pair only with thymine, and guanine can pair only with cytosine. Thus the order of bases along one strand determines the order of bases along the other strand.

The sequence of bases varies from one DNA molecule to another. It is customarily written as the sequence in the 5' to 3' direction of one of the strands. Here, using the strand on the left, the sequence is TCAG.

According to x-ray measurements, the distance between the two sides, or rails, is 2 nanometers. Two purines in combination would take up more than 2 nanometers, and two pyrimidines would not reach all the way across. But if a purine paired in each case with a pyrimidine, there would be a perfect fit. The paired bases—the "rungs" of the ladder—would therefore always be purine-pyrimidine combinations.

As Watson and Crick worked their way through the data, they assembled actual tin-and-wire models of the molecules (see essay, page 247), testing where each piece would fit into the three-dimensional puzzle. As they worked with the models, they realized that the nucleotides along any one strand of the double helix could be assembled in any order. Since a DNA molecule may be thousands of nucleotides long, there is a possibility for great variety, one of the primary requirements for the genetic material.

The most exciting discovery, however, came when Watson and Crick set out to construct a matching strand. They encountered an interesting and important restriction. Not only could purines not pair with purines and pyrimidines not pair with pyrimidines, but also, because of the structures of the bases, adenine could pair only with thymine, forming two hydrogen bonds (A=T), and guanine could pair only with cytosine, forming three hydrogen bonds (G≡C). The paired bases were **complementary.** Look at Table 14–1 again and see how well these chemical requirements explain Chargaff's data.

The double-stranded structure of a small portion of a DNA molecule is shown in Figure 14–8. As you can see, the strands have direction. In each strand, each phosphate group is attached to one sugar at the 5′ position (the fifth carbon in the sugar ring) and to the other sugar at the 3′ position (the third carbon in the sugar ring). Thus, each strand has a 5′ end and a 3′ end. And, within the double helix, the two strands run in opposite directions—that is, the direction from the 5′ end to the 3′ end of each strand is opposite. The strands are therefore said to be **antiparallel.**

Although the nucleotides along one chain of the double helix can occur in any order, their sequence then determines the order of nucleotides in the other chain. This is necessarily the case, since the bases are complementary (G pairs only with C, and A pairs only with T).

DNA Replication

An essential property of the genetic material is the ability to provide for exact copies of itself. Implicit in the

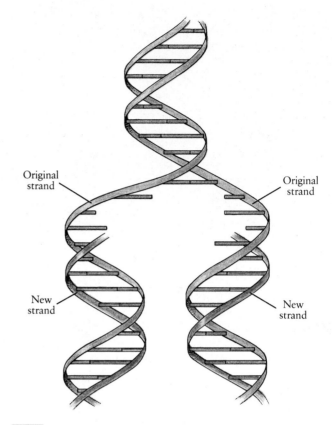

Original strand

Original strand

New strand

New strand

14–9   *Replication of the DNA molecule. The strands separate down the middle as the hydrogen bonds are broken and the paired bases separate. Each of the original strands then serves as a template along which a new, complementary strand forms from nucleotides available in the cell.*

double and complementary structure of the DNA helix is a mechanism by which it can reproduce itself.

At the time of chromosome replication, the molecule "unzips" down the middle, with the paired bases separating as the hydrogen bonds are broken. As the two strands separate, they act as **templates,** or guides. Each strand directs the synthesis of a new complementary strand along its length (Figure 14–9), using the raw materials in the cell. If a T is present on the original strand (the template), only an A can fit in the adjacent location of the new strand; a G will pair only with a C, and so on. In this way, each strand forms a copy of its original partner strand, and two exact replicas of the molecule are produced. The age-old question of how hereditary information is duplicated and passed on, generation after generation, had, in principle, been answered.

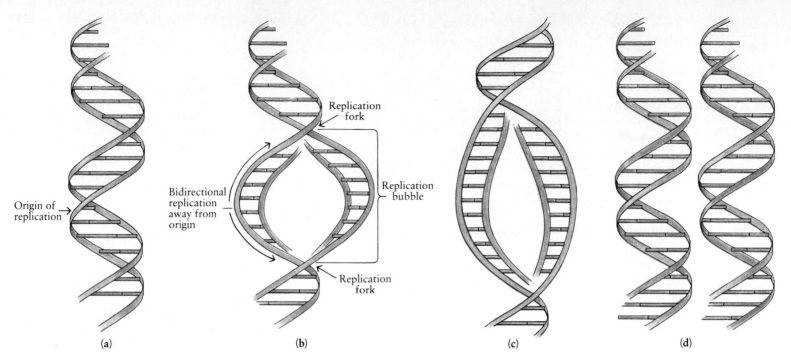

14-10   *An overview of DNA replication. The two strands of the DNA molecule (a) separate at the origin of replication as a result of the action of special initiator proteins and enzymes. (b, c) The two replication*

forks move away from the origin of replication in opposite directions, forming a replication bubble that expands in both directions (bidirectionally). (d) When synthesis of the new DNA strands is complete, the two

double-stranded chains separate into two new double helixes. Each consists of one old strand and one new strand.

The Mechanics of DNA Replication

Replication of DNA is a process that occurs only once in each cell generation, during the S phase of the cell cycle (page 178). It is the essential event in the replication of the chromosomes. In most eukaryotic cells, DNA replication leads ultimately to mitosis, but in the cells that give rise to gametes it leads instead to meiosis. It is a remarkably rapid process. For example, in humans and other mammals, the rate of synthesis is about 50 nucleotides per second. In prokaryotes, it is even faster—about 500 nucleotides per second.

The principle of DNA replication, in which each strand of the double helix serves as a template for the formation of a new strand, is relatively simple and easy to understand. However, the actual process by which the cell accomplishes replication is considerably more complex. Like other biochemical reactions of the cell, DNA replication requires a number of different enzymes, each catalyzing a particular step of the process. The identification of the principal enzymes, their precise functions, and the sequence of events in replication has required a number of years and the efforts of many scientists working in different laboratories. Although our understanding is still incomplete, the general outlines of the process are now clear.

Initiation of DNA replication always begins at a specific nucleotide sequence known as the **origin of repli-**

cation. It requires special initiator proteins and enzymes that break the hydrogen bonds linking the complementary bases at the replication origin, opening up the helix so replication can occur. As the two strands of the DNA double helix are separated, other proteins attach to the individual strands, holding them apart. This makes possible the next stage, the actual synthesis of the new strands, catalyzed by a group of enzymes known as **DNA polymerases.**

If replicating DNA is viewed under the electron microscope, the region of synthesis appears as an "eye," or **replication bubble.** At either end of the bubble, where the existing strands are being separated and the new, complementary strands are being synthesized, the molecule appears to form a Y-shaped structure. This is known as a **replication fork.** The two replication forks move in opposite directions away from the origin (Figure 14–10), and thus replication is said to be **bidirectional.**

In prokaryotes, there is a single replication origin in the chromosome. In eukaryotes, by contrast, there are many replication origins in each chromosome. Replication proceeds along the linear chromosome as each bubble expands bidirectionally (Figure 14–11) until it meets an adjacent bubble. As the bubbles merge, the entire length of the chromosome is replicated.

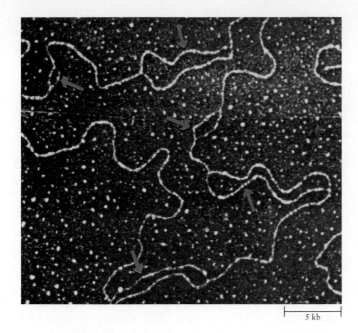

14–11 *Replication of eukaryotic chromosomes is initiated at multiple origins. The individual replication bubbles spread until ultimately they meet and join. In this electron micrograph of a replicating chromosome in an embryonic cell of* Drosophila, *the replication bubbles are indicated by the arrows.*

5 kb

Proofreading

One of the essential features of DNA replication is that DNA polymerase has a proofreading function. In the course of synthesis, errors are sometimes made, and the wrong nucleotide is added to the newly forming strand. In other words, the nucleotide added to the strand is not complementary to the nucleotide on the template strand. DNA polymerase, however, is able to add nucleotides to a strand only if the preceding nucleotides are correctly paired with their complementary nucleotides on the template strand. If a mistake has occurred, the enzyme backtracks, removing nucleotides until it encounters a correctly paired nucleotide. At that point, the enzyme begins moving forward once more, adding nucleotides to the growing strand as it goes. This ability to remove incorrectly paired nucleotides provides an important proofreading that ensures the accuracy of DNA replication.

In addition to the proofreading that occurs during DNA replication, other enzymes constantly monitor the DNA double helixes of the cell. Whenever an incorrectly paired nucleotide is encountered, DNA repair enzymes move in to snip it out and replace it with the correct nucleotide. This process is essential in maintaining the integrity of the genetic machinery.

The Energetics of DNA Replication

The nucleotides required for DNA replication are synthesized by the same biochemical pathways as the nucleotides required for other functions of the cell. However, they are supplied not in the form of the monophosphates shown in Figure 14–3, but rather as triphosphates. For example, adenine is supplied as deoxyadenosine triphosphate (dATP). The only differ-

ence between ATP and dATP is in the sugar subunit—in ATP, it is ribose, and in dATP, it is deoxyribose. Similarly, guanine is supplied as dGTP, cytosine as dCTP, and thymine as dTTP.

Just as ATP provides energy to power a variety of biosynthetic reactions, dATP, dGTP, dCTP, and dTTP provide energy to power the synthesis of DNA. As each nucleotide is attached to the growing DNA strand, the "extra" P~P group is removed and immediately split into two phosphate ions. The energy released in this process powers the reactions mediated by the DNA polymerases.

DNA as a Carrier of Information

You will recall that a necessary property of the genetic material is the capacity to carry information. The Watson-Crick model showed how the DNA molecule is able to do this. The information is carried in the sequence of the bases, and *any* sequence of bases is possible. Since the number of paired bases ranges from about 5,000 for the simplest known virus up to an estimated 6 billion in the 46 human chromosomes, the number of possible variations in sequence is astronomical.

The DNA from a single human cell—which if extended in a single thread would be almost 2 meters long—can contain information equivalent to some 600,000 printed pages of 500 words each, or a library of about a thousand books. Obviously, the DNA structure can well account for the endless diversity among living things.

Summary

Classical genetics had been concerned with the mechanics of inheritance—how the units of heredity are passed from one generation to the next and how changes in the hereditary material are expressed in individual organisms. New questions soon arose, however, and geneticists began to explore the nature of the gene—its structure, composition, and properties.

During the 1940s, many investigators believed that genes were proteins, but others were convinced that the hereditary material was deoxyribonucleic acid (DNA). Important, although not widely accepted, evidence for the genetic role of DNA was presented by Avery in his experiments to identify the transforming factor of pneumococci. Confirmation of Avery's hypothesis came from studies with bacteriophages (bacterial viruses) showing that DNA and not protein is the genetic material of the virus.

Further support for the genetic role of DNA came from two more sets of data: (1) Almost all somatic cells of any given species contain equal amounts of DNA, and the gametes contain just half that amount. (2) The proportions of nitrogenous bases are the same in the DNA of all cells of a given species, but they vary in different species.

In 1953, Watson and Crick proposed a structure for DNA. The DNA molecule, according to their model, is a double-stranded helix, shaped like a twisted ladder. The two sides of the ladder are composed of repeating subunits consisting of a phosphate group and the five-carbon sugar deoxyribose. The "rungs" are made up of paired nitrogenous bases, one purine base pairing with one pyrimidine base. There are four bases in DNA—adenine (A) and guanine (G), both of which are purines, and thymine (T) and cytosine (C), which are pyrimidines. A can pair only with T, and G only with C. The four bases are the four "letters" used to spell out the genetic message. The paired bases are joined by hydrogen bonds.

When the DNA molecule replicates, the two strands come apart, breaking at the hydrogen bonds. Each strand acts as a template for the formation of a new, complementary strand from nucleotides available in the cell. The addition of nucleotides to the new strands is catalyzed by DNA polymerases. A variety of other enzymes also play key roles in the replication process.

Replication begins at a particular nucleotide sequence on the chromosome, the origin of replication. It proceeds bidirectionally, by way of two replication forks that move in opposite directions. In the course of DNA replication, DNA polymerase proofreads, backtracking when necessary to remove nucleotides that are not correctly paired with the template strand.

The nucleotides incorporated into the growing DNA strands are supplied in the form of triphosphates. The energy required to power replication is provided by the removal and splitting of the "extra" P~P groups.

On the basis of the structure of the DNA double helix, as revealed by Watson and Crick, the role of DNA as the carrier and transmitter of the genetic information was universally accepted. With the discovery of the mechanism by which the living cell replicates its DNA, the question of how the hereditary information is faithfully transmitted from parent cell to daughter cell, generation after generation, was answered.

Questions

1. What are the steps by which Griffith demonstrated the existence of the transforming factor? Can you think of any implications of Griffith's discovery for modern medicine?

2. What characteristics of bacteriophages make them a useful experimental tool?

3. When the structure of DNA was being worked out, it became apparent that one purine base must be paired with a pyrimidine base, and that the other purine base must be paired with the other pyrimidine base. The evidence for this requirement came from two types of data. What were the data, and how did they indicate this structural requirement?

4. Further consideration of the structures of the four nitrogenous bases indicated that adenine could pair only with thymine, and cytosine only with guanine. What structural features of the bases imposed this requirement on the structure of the DNA molecule?

5. One of the chief arguments for the erroneous hypothesis that proteins constitute the genetic material was that proteins are heterogeneous. Explain why the genetic material must have this property. What feature of the Watson-Crick model of DNA structure is important in this respect?

6. Shown here is the sequence of bases in the 5' to 3' di-

rection in one strand of a hypothetical DNA molecule. Identify the sequence of bases in the complementary strand.

5′ —A—A—G—T—T—T—G—G—T—T—A—C—T—T—G— 3′

3′ — — — — — — — — — — — — — — — 5′

7. Distinguish among the following terms: origin of replication, replication bubble, and replication fork.

8. Among the functions of DNA polymerase is the proofreading of the newly synthesized strands as they are formed. DNA polymerase can identify any mismatched bases, snip them out, and replace them with the correct bases. What is the biological significance of this activity?

9. Eukaryotic cells are grown for a number of generations in a medium containing thymine labeled with tritium, a radioactive isotope of hydrogen (^{3}H). The cells are then removed from the radioactive medium, placed in an ordinary, nonradioactive medium, and allowed to divide. Studies of the distribution of the radioactive isotope are made after each generation to determine the presence or absence of radioactive material in the chromatids.

Before the cells are placed in the nonradioactive medium, all the chromatids contain ^{3}H. After one generation in the nonradioactive medium, all of the chromatids still contain radioactive ^{3}H. (**a**) Assuming each chromatid contains a single DNA molecule, explain the results. (**b**) Is this consistent with the Watson-Crick hypothesis of DNA replication? (**c**) What would be the distribution of the ^{3}H after two generations in the nonradioactive medium? Why? *(Hint:* You may find it helpful to look at Figure 14–10 as you think about these questions.)

10. Suppose you are talking to someone who has never heard of DNA. How would you support an argument that DNA is the genetic material? List at least five of the strong points in such an argument.

Suggestions for Further Reading

Crick, Francis: *What Mad Pursuit: A Personal View of Scientific Discovery,* Basic Books, Inc., New York, 1988.*

> *This book is one of a series of autobiographies by major figures in twentieth-century science, commissioned by the Sloan Foundation and intended for the general reader. Crick's account of the discovery of the structure of the DNA molecule provides an interesting counterpoint to Watson's* The Double Helix *(see below). As a physicist who became a biologist, Crick has many valuable insights into the significant differences between biology and the physical sciences.*

Felsenfeld, Gary: "DNA," *Scientific American,* October 1985, pages 58–67.

Judson, Horace F.: *The Eighth Day of Creation: Makers of the Revolution in Biology,* Simon and Schuster, New York, 1979.*

> *A comprehensive study of the human and scientific aspects of molecular biology from the 1930s through the mid-1970s. As events unfold, the story is told from each participant's point of view. We are treated to an enlightening and personal glimpse of the development of scientific thought.*

Kornberg, Arthur: *For the Love of Enzymes: The Odyssey of a Biochemist,* Harvard University Press, Cambridge, Mass., 1991.*

> *Kornberg, who began his scientific career with studies of jaundice in World War II army recruits, went on to discover the DNA polymerases and to make major contributions to our understanding of DNA replica-tion. He is a wonderful writer, and this book is enriched by the clarity of his explanations, by his wise reflections on the scientific enterprise, and by his sense of humor.*

Luria, S. E.: *A Slot Machine, a Broken Test Tube,* Harper & Row, Publishers, Inc., New York, 1985.*

> *Luria was one of the founders of the "phage group" in the 1940s, whose work with the bacteriophages of* E. coli *led to the beginnings of molecular genetics. A highly reflective book, this autobiography recounts not only his scientific experiences but also many aspects of his personal life, including his flight from Italy through France to the United States at the beginning of World War II.*

Olby, Robert: *The Path to the Double Helix,* University of Washington Press, Seattle, 1975.

> *An account, written by a professional historian of science, of twentieth-century genetics. Olby is interested not only in the scientific concepts and experiments but also in the various personalities involved and their effects on one another and on the course of scientific discovery.*

Radman, Miroslav, and Robert Wagner: "The High Fidelity of DNA Duplication," *Scientific American,* August 1988, pages 40–46.

Watson, James D.: *The Double Helix,* Atheneum Publishers, New York, 1968.*

> *"Making out" in molecular biology. A brash and lively book about how to become a Nobel laureate.*

* Available in paperback.

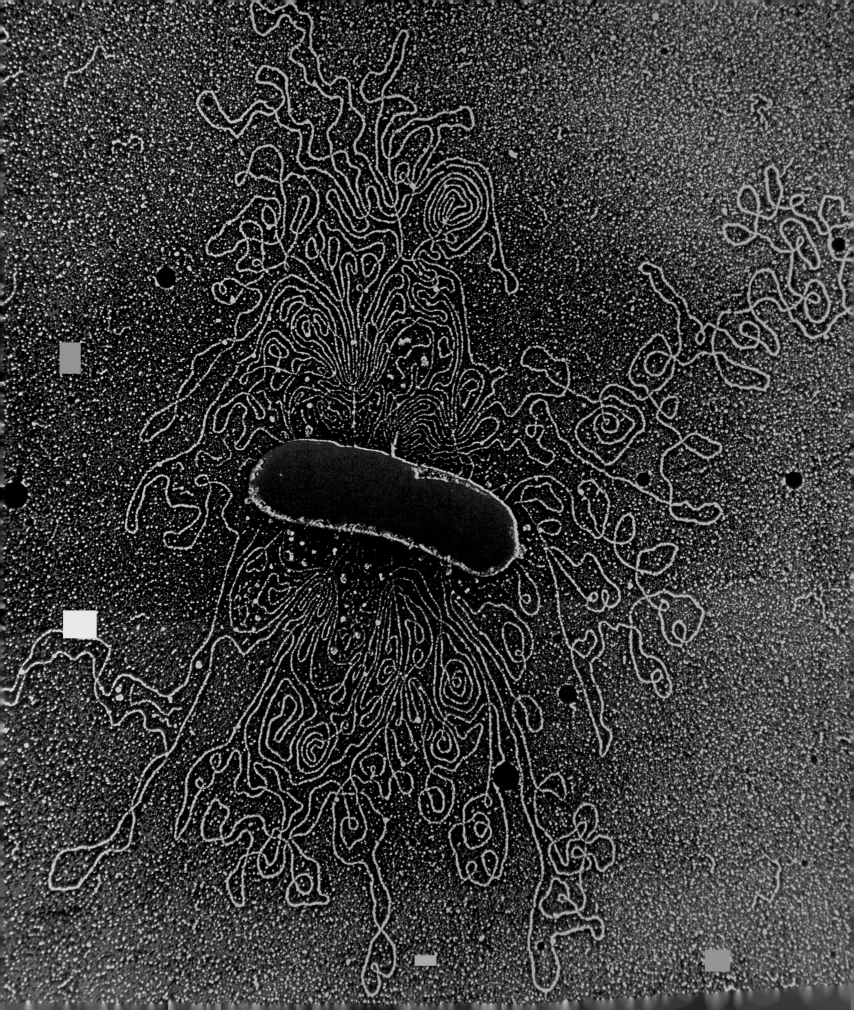

The Genetic Code and Its Translation

Escherichia coli, or *E. coli,* for short, is one of the most common of microscopic organisms. Its preferred residence is your intestinal tract. A prokaryote, it is much smaller than the eukaryotic cells of your intestinal lining, with which it cohabits. In fact, it is only about the size of a mitochondrion in one of those cells. At any given moment, however, a single *E. coli* cell contains some 5,000 different macromolecules. One of these is a single long molecule of DNA—the bacterial chromosome. In this electron micrograph, an *E. coli* cell has been exploded by osmotic shock to release its DNA, which can be seen surrounding it.

The revelation by Watson and Crick of the double helical structure of DNA led immediately to an understanding, in principle, of the way in which the molecule duplicates itself. ("It has not escaped our notice," they stated slyly in their original report, "that the special pairing we have postulated immediately suggests a possible copying mechanism for the genetic material.") The next big question, however, remained wide open: How does the genetic material, now identified as DNA, direct the day-to-day activities of living cells?

In the search for the answer to this question, *E. coli* has turned out to be a major player, comparable in status to the garden pea, *Drosophila,* and the bacteriophages. Rising from its humble origins, *E. coli,* it is now generally recognized, is the best understood of all living things on this planet.

The Watson-Crick model both established the structure of DNA and indicated the mechanism for its replication. The DNA molecule carries the instructions for the structure and function of the cell and also transmits these instructions to new cells and organisms. Yet two fundamental questions were left unanswered: How are instructions encoded in the DNA molecule? And, how are they carried out?

The studies that led to the answers to these questions had their beginnings in a concept first proposed in 1908. In a series of lectures in that year, an English physician, Sir Archibald Garrod, set forth a new concept of human diseases, which he called "inborn errors of metabolism." Garrod postulated that certain diseases are caused by the body's inability to perform particular chemical processes and are hereditary in nature. With a leap of the imagination that spanned almost half a century, he hypothesized that such diseases are the result of enzyme deficiencies. Implicit in this hypothesis was the idea that genes act by influencing the production of enzymes.

Genes and Proteins

One Gene–One Enzyme

By the 1940s, biologists had come to realize that all of the biochemical activities of the living cell, including the multitude of biosynthetic reactions that produce all of its constituent molecules, depend upon different specific enzymes. Even the synthesis of enzymes depends on enzymes. Further, it was becoming clear that the specificity of different enzymes is a result of their primary structure, the linear sequence of amino acids in the protein molecule.

Meanwhile, George Beadle, a geneticist, was working with the eye-color mutants of *Drosophila* that had been discovered in Morgan's laboratory (page 228). As a result of his studies, Beadle formulated the hypothesis that each of the various eye colors observed in the mutants is the result of a change in a single enzyme in a biosynthetic pathway. To test this notion—that genes control enzymes—on a broader scale, he teamed up in 1941 with Edward L. Tatum, a biochemist. The organism they chose for their studies was the red bread mold *Neurospora*.

Utilizing x-rays to increase the mutation rate, Beadle and Tatum analyzed the resultant mutants by genetic mapping studies analogous to those performed with *Drosophila*. A number of the mutations proved to affect the activity of single enzymes, and Beadle and Tatum were able to show that a mutation in one gene could be correlated with the loss of function of one enzyme. On the basis of these experiments, they formulated the hypothesis that a single gene specifies a single enzyme.

"One gene–one enzyme" turned out to be an oversimplification, for there are many proteins that are not enzymes. Some, for instance, are hormones, such as insulin, others are structural proteins, such as collagen, and still others are membrane proteins, such as the sodium-potassium pump. These proteins, too, are specified by genes. This expansion of the original concept did not modify it in principle. "One gene–one enzyme" was simply amended to "one gene–one protein." Subsequently, with the realization that many proteins consist of more than one polypeptide chain, it was modified once more, to the less memorable but more precise "one gene–one polypeptide chain."

The Structure of Hemoglobin

What causes the change or loss of function in an enzyme or other protein specified by a gene that has undergone a mutation? Linus Pauling was one of the first to see the implications of the work of Beadle and Tatum with regard to this question. Perhaps, Pauling reasoned, human disorders involving hemoglobin, such as sickle-cell anemia, could be traced to a variation from the normal protein structure of the hemoglobin molecule.

To test this hypothesis, Pauling took samples of hemoglobin from people with sickle-cell anemia (a homozygous recessive condition), from others heterozygous for the allele (Figure 15–1), and from still others homozygous for the normal allele. To try to detect differences in these proteins, Pauling used a process known as **electrophoresis,** in which molecules in solution are exposed to a weak electric field. Very small differences, even in very large molecules, may be reflected in the electric charges of the molecules. These charge differences cause the molecules to move at different rates in the electric field.

Figure 15–2 shows the results of Pauling's experiment. A person who has sickle-cell anemia makes a different sort of hemoglobin than a person who does not have the disorder. A person who is heterozygous (carrying one copy of the allele for sickling and one copy of the allele for normal hemoglobin) makes both kinds of hemoglobin molecules. However, enough normal molecules are produced to prevent anemia. (Notice that the terms "dominant" and "recessive" are becoming more complicated.)

A few years later, it was shown that the actual difference between the normal and the sickle-cell hemoglobin molecules is two amino acids in 600, as we noted previously (page 65).

The Virus Coat

Additional evidence that DNA specifies the structure of proteins came from the bacteriophage studies described in Chapter 14. You will recall that the introduction of viral DNA into a bacterial cell results in the production not only of more viral DNA, but also of the proteins of the virus coat. Clearly, the viral DNA carries the information for the synthesis of the coat proteins.

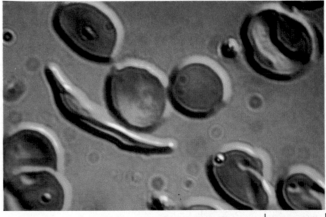

20 µm

15–1   *Although individuals heterozygous for the sickle-cell allele have no symptoms of anemia, they can be identified by the fact that they have children with sickle-cell anemia. They can also be identified by means of a simple laboratory test in which a sample of blood is treated to remove oxygen from the hemoglobin molecules. As this scanning electron micrograph of deoxygenated blood from a heterozygous individual reveals, most of the red blood cells are normal, but an occasional cell is sickled under the artificial conditions of the test.*

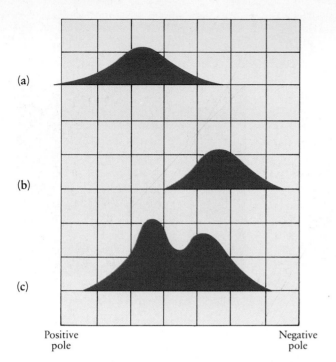

(a)

(b)

(c)

Positive pole

Negative pole

15–2 *The results, as reported by Linus Pauling in 1949, of the electrophoresis of* (a) *normal hemoglobin,* (b) *the hemoglobin of a person with sickle-cell anemia, and* (c) *the hemoglobin of a person who is heterozygous for the sickle-cell allele. The height of the curves indicates the amount of hemoglobin at each point. Because of slight differences in electric charge, normal and sickle-cell hemoglobins move at different rates in an electric field. The normal hemoglobin is more negatively charged. Hence it moves closer to the positive pole than does the sickle-cell hemoglobin. The hemoglobin from the heterozygote separates into the two different positions, indicating that it consists of both normal and sickle-cell hemoglobins.*

15–3 *Chemically, RNA is very similar to DNA, but there are two differences in its nucleotides.* (a) *One difference is in the sugar component. Instead of deoxyribose, RNA contains ribose, which has an additional oxygen atom.* (b) *The other difference is that instead of thymine, RNA contains the closely related pyrimidine uracil (U). Uracil, like thymine, pairs only with adenine.*

A third, and very important, difference between the two nucleic acids is that RNA is usually single-stranded and does not form a regular helical structure.

From DNA to Protein: The Role of RNA

As a result of all of these studies, there was general agreement that the DNA molecule contains a coded message with instructions for biological structure and function. Moreover, these instructions are carried out by proteins, which also contain a highly specific biological "language." As we saw in Chapter 3, the linear sequence of amino acids in a polypeptide chain determines the three-dimensional structure of the completed protein molecule, and it is the three-dimensional structure that, in turn, determines function. The question thus became one of translation. How did the order of bases in DNA specify the sequence of amino acids in a protein molecule?

The search for the answer to this question led to **ribonucleic acid (RNA)**, a close chemical relative of DNA. As you can see in Figure 15–3, there are three principal differences between RNA and DNA:

1. In the nucleotides of RNA, the sugar component is ribose, rather than deoxyribose.

2. The nitrogenous base thymine, found in DNA, does not occur in RNA. Instead, RNA contains a closely related pyrimidine, **uracil (U)**. Uracil, like thymine, pairs only with adenine.

3. Most RNA is single-stranded.

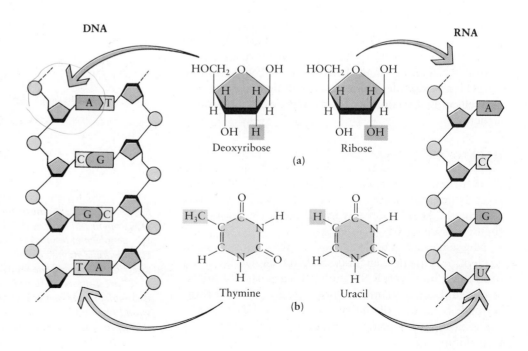

DNA

RNA

Deoxyribose

Ribose

(a)

Thymine

Uracil

(b)

15–4   (a) *Diagram of a very small portion of tobacco mosaic virus (TMV). This virus has a central core not of DNA but rather of RNA. Its outer coat is composed of 2,150 identical protein molecules, each consisting of 158 amino acids. If the RNA is separated from its protein coat and rubbed into scratches on a tobacco leaf, new TMV particles are formed, complete with new protein coats. Studies with TMV were among the first to suggest that RNA could direct the assembly of proteins. (b) When viewed under the electron microscope, TMV particles appear as rod-shaped structures.*

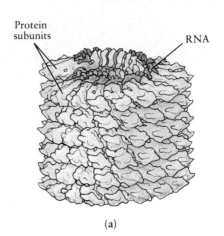

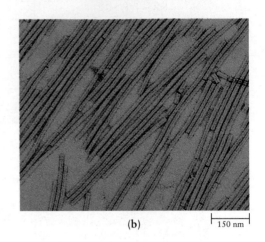

(a)

(b)

150 nm

There were several clues that RNA might play a role in the translation of genetic information from DNA into a sequence of amino acids. For instance, cells that are making large amounts of protein are rich in RNA. In eukaryotic cells, the RNA is found mostly in the cytosol, and it is here that most protein synthesis takes place. Also, both prokaryotic and eukaryotic cells making large amounts of protein have numerous ribosomes, and ribosomes are rich in RNA.

Additional evidence came from viruses. When a bacterial cell is infected by a DNA-containing bacteriophage, RNA is synthesized from the viral DNA before viral protein synthesis begins. Also, some viruses consist of only RNA and protein. Tobacco mosaic virus (Figure 15–4) is an example. These clues all indicated that RNA as well as DNA contained information about protein structure.

RNA as Messenger

As it turned out, not one but three kinds of RNA play roles as intermediaries in the steps that lead from DNA to protein. At this point, we shall consider just one of them: **messenger RNA (mRNA).**

Messenger RNA molecules are copies (transcripts) of nucleotide sequences encoded in the DNA. Unlike DNA molecules, however, RNA molecules are usually single-stranded. Each new mRNA molecule is copied, or transcribed, from one of the two strands of DNA by the same base-pairing principle that governs DNA replication (Figure 15–5).

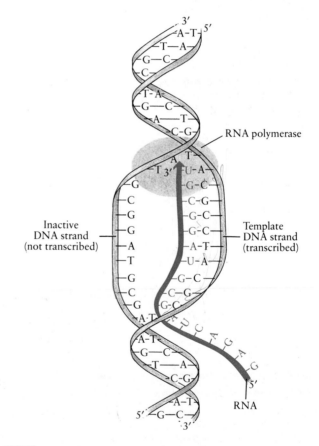

15–5   *A schematic representation of RNA transcription. At the point of attachment of the enzyme RNA polymerase, the DNA opens up, and, as the RNA polymerase moves along the DNA molecule, the two strands of the molecule separate. Nucleotide building blocks are assembled into RNA in a 5' to 3' direction as the enzyme reads the template DNA strand in a 3' to 5' direction. Note that the base sequence of the RNA strand is complementary—not identical—to that of the template strand from which it is transcribed. It is, however, identical to that of the inactive (untranscribed) DNA strand, except for the replacement of thymine (T) by uracil (U).*

The Elusive Messenger

The cytosol of cells that are synthesizing proteins is full of RNA. This observation was the major clue that RNA plays a role in directing the assembly of proteins. Although the existence of RNA molecules that carry the genetic information from DNA to protein was hypothesized, confirmation of the hypothesis required the detection and isolation of the messenger molecules. The problem was complicated, however, by the fact that most of any cell's RNA is bound into ribosomes and is therefore unlikely to be the carrier of genetic information. This paradox puzzled molecular biologists for almost a decade.

Escherichia coli and its bacteriophages once again provided the tools of discovery. As you know, the genetic material of these phages is DNA, and their coats are made of protein. When they infect a bacterial cell, new coat proteins are synthesized. If the messenger hypothesis were true, new RNA should be formed between the time of cell infection and the synthesis of new virus coats. To test the hypothesis, *E. coli* cells were infected with phage and then exposed briefly to uracil labeled with radioactive carbon. Short-lived RNA molecules with radioactive labels were detected in the cells. They were associated with ribosomes but they were not part of the ribosomes. Were they the long-sought messengers?

To answer this question it was necessary to show that the radioactive RNA was complementary to the phage DNA. The method used was simple but ingenious. If dissolved DNA molecules are heated gently, the hydrogen bonds break apart and the two strands of the double helix separate. Subsequently, when the solution is slowly cooled, complementary strands pair up again and the hydrogen bonds re-form. If the newly formed, radioactive RNA were complementary to the phage DNA, the investigators reasoned, it too should form a duplex (a double-stranded molecule) with that DNA.

As a control, the radioactive RNA molecules were first mixed in a beaker with a solution of *E. coli* DNA and heated. When this mixture was cooled, no duplexes containing radioactivity could be detected. The fact that no hybrid RNA-DNA molecules had formed

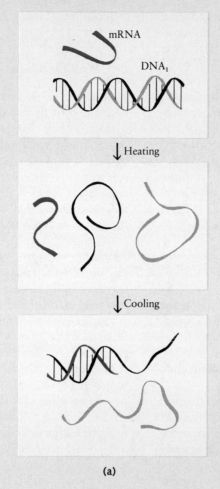

(a)

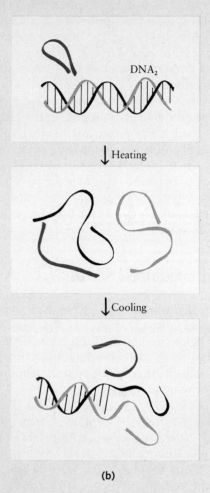

(b)

RNA-DNA hybrids can be used to show the complementarity in nucleotide sequence between an RNA molecule and the DNA molecule from which it was transcribed. **(a)** *The formation of a hybrid molecule between a radioactive RNA molecule (red) and its template DNA strand (black).* **(b)** *If the same RNA molecule is mixed with unrelated DNA, no hybrid molecules containing radioactivity are formed.*

indicated that the RNA was not complementary to the *E. coli* DNA.

Then the radioactive RNA molecules were mixed with a solution of phage DNA. When this mixture was heated and then cooled, the results were clear-cut. Duplexes containing radioactivity had formed, indicating that the RNA had bound to its complementary strand of phage DNA. The messenger had been disclosed.

Hybridization of DNA with DNA and of RNA with DNA has since become an enormously powerful tool in both molecular genetics and evolutionary biology. As we shall see in later chapters, it is used in a great variety of studies, ranging from the detection of specific genes responsible for particular human disorders to the resolution of evolutionary enigmas.

Like a strand of DNA, each RNA molecule has a 5′ end and a 3′ end. The nucleotides, which are present in the cell as triphosphates, are added, one at a time, to the 3′ end of the growing RNA chain. This process, known as **transcription,** is catalyzed by the enzyme **RNA polymerase.** The enzyme moves along the template DNA strand in a 3′ to 5′ direction, synthesizing a new complementary strand of nucleotides in a 5′ to 3′ direction.

DNA is the master copy of the genetic information, kept permanently "on file" in the chromosomes. Messenger RNA, by contrast, is the working copy of the genetic information. Incorporating the instructions encoded in the DNA molecule, mRNA dictates the sequence of amino acids in proteins. When its job is done, it is degraded into its constituent nucleotides, which are then available for use in the synthesis of other mRNA molecules.

The Genetic Code

The identification of mRNA as the working copy of the genetic instructions still left the big question unresolved. Proteins contain 20 different amino acids, but DNA and RNA each contain only four different nucleotides. Somehow, these nucleotides constituted a **genetic code** for the amino acids.

As it turned out, the idea of a code was useful not only as a dramatic metaphor but also as a working analogy. Scientists, seeking to understand how the sequence of nucleotides stored in the double helix could specify the quite dissimilar structures of protein molecules, approached the problem with methods used by cryptographers in deciphering codes.

As we saw in Chapter 3, the primary structure of each particular kind of protein molecule consists of a specific linear arrangement of the 20 different amino acids. Similarly, there are four different nucleotides, arranged in a specific linear sequence in a DNA molecule. If each nucleotide "coded" for one amino acid, only four amino acids could be specified by the four bases. If two nucleotides specified one amino acid, there could be a maximum number, using all possible arrangements of the nucleotides, of 4×4, or 16—still not quite enough to code for all 20 amino acids. Therefore, following the code analogy, at least three nucleotides in sequence must specify each amino acid. This would provide for $4 \times 4 \times 4$, or 64, possible combinations, or **codons**—clearly, more than enough.

The three-nucleotide, or triplet, codon was widely and immediately adopted as a working hypothesis. Its existence, however, was not actually demonstrated un-til the code was finally broken, a decade after Watson and Crick first presented their DNA structure. The scientists who performed the initial, crucial experiments toward breaking the code were Marshall Nirenberg and his colleague Heinrich Matthaei, both of the National Institutes of Health.

Breaking the Code

Messenger RNA, then newly discovered, gave Nirenberg the tool he needed. He broke apart *E. coli* cells, extracted their contents, and added radioactively labeled amino acids and crude samples of RNA from a variety of sources to the extracts. All of the RNA samples stimulated protein synthesis. The amounts of radioactively labeled protein produced were small but measurable. In other words, the material extracted from the *E. coli* cells would start producing protein molecules even when the RNA "orders" it received were from a "complete stranger." Even the RNA from tobacco mosaic virus, which naturally multiplies only in cells of the leaves of tobacco plants, could be read as an mRNA by the machinery of the bacterial cell.

Nirenberg and Matthaei then tried an artificial RNA. Perhaps if the *E. coli* extracts could read a foreign message and translate it into protein, they could read a totally synthetic message, one dictated by the scientists themselves. Severo Ochoa of New York University had developed a process for linking nucleotides into a long strand of RNA. With this process, carried out in a test tube, he had produced an RNA molecule that contained only one nitrogenous base, uracil, repeated over and over again. It was called "poly-U."

Nirenberg and Matthaei prepared 20 different test tubes, each of which contained extracts from *E. coli* cells. These extracts included ribosomes, ATP, the necessary enzymes, and all of the amino acids. In each test tube, one of the amino acids, and only one, carried a radioactive label. Synthetic poly-U was added to each test tube. In 19 of the test tubes, no radioactive polypeptides were produced. In the twentieth tube, however, to which radioactive phenylalanine had been added, the investigators were able to detect newly formed, radioactive polypeptide chains.

When these polypeptides were analyzed, they were found to consist only of phenylalanines, one after another. Nirenberg and Matthaei had dictated the message "uracil . . . uracil . . . uracil . . . uracil . . . uracil . . . uracil . . .," and a clear answer had come back, "phenylalanine . . . phenylalanine. . . ." The experiment not only defined the first code word (UUU = phe) but also made available a method for defining the others.

Second letter

	U	C	A	G	
U	UUU ⎫ phe UUC ⎭ UUA ⎫ leu UUG ⎭	UCU ⎫ UCC ⎪ ser UCA ⎪ UCG ⎭	UAU ⎫ tyr UAC ⎭ UAA stop UAG stop	UGU ⎫ cys UGC ⎭ UGA stop UGG trp	U C A G
C	CUU ⎫ CUC ⎪ leu CUA ⎪ CUG ⎭	CCU ⎫ CCC ⎪ pro CCA ⎪ CCG ⎭	CAU ⎫ his CAC ⎭ CAA ⎫ gln CAG ⎭	CGU ⎫ CGC ⎪ arg CGA ⎪ CGG ⎭	U C A G
A	AUU ⎫ AUC ⎪ ile AUA ⎭ AUG met	ACU ⎫ ACC ⎪ thr ACA ⎪ ACG ⎭	AAU ⎫ asn AAC ⎭ AAA ⎫ lys AAG ⎭	AGU ⎫ ser AGC ⎭ AGA ⎫ arg AGG ⎭	U C A G
G	GUU ⎫ GUC ⎪ val GUA ⎪ GUG ⎭	GCU ⎫ GCC ⎪ ala GCA ⎪ GCG ⎭	GAU ⎫ asp GAC ⎭ GAA ⎫ glu GAG ⎭	GGU ⎫ GGC ⎪ gly GGA ⎪ GGG ⎭	U C A G

First letter (5′ end) — Third letter (3′ end)

15–6 *The genetic code, consisting of 64 triplet combinations (codons) and the corresponding amino acids, indicated here by their abbreviations (see page 60). The codons shown here are the ones that would appear in the mRNA molecule. Of the 64 codons, 61 specify particular amino acids. The other three codons are stop signals, which cause the synthesis of a polypeptide chain to terminate.*

Since 61 triplets code for only 20 amino acids, there are "synonyms," as many as six different codons for leucine, for example. Most of the synonyms, as you can see, differ only in the third nucleotide. Each codon, however, specifies only one amino acid.

Similar experiments were subsequently performed in a number of laboratories, using artificial RNAs in which either two or three different nucleotides were repeated over and over again in a known sequence. As a result of these experiments, the mRNA codons for all of the amino acids were soon worked out. Of the 64 possible triplet combinations, 61 specify particular amino acids and 3 are stop signals. With 61 combinations coding for only 20 amino acids, you can see that there must be more than one codon for many of the amino acids. As shown in Figure 15–6, codons specifying the same amino acid often differ only in the third nucleotide.

The Universality of the Genetic Code

Three decades have passed since the genetic code was first broken, and the DNA, mRNA, and proteins of many more organisms have been examined. The evidence is now overwhelming: for virtually all organisms, from *Escherichia coli* to *Homo sapiens,* the genetic code is universal. UUA, for example, codes for the amino

acid leucine not only in prokaryotes but also in protists, fungi, plants, and animals. The genetic code evolved early, remained constant, and determines the underlying unity of all living things.

A few exceptions to the code shown in Figure 15–6 have been found, however. Most of them involve mitochondria, which contain their own DNA, transcribe their own RNA molecules, and carry out some protein synthesis. In several instances, the mitochondrial code differs from that carried in the chromosomes of both prokaryotes and eukaryotes.

Protein Synthesis

With a knowledge of the genetic code, we can now turn our attention to the question of how the information encoded in the DNA and transcribed into mRNA is subsequently translated into a specific sequence of amino acids in a polypeptide chain. The basic principles of protein synthesis are the same in both prokaryotic and eukaryotic cells, but there are some differences in detail, which will be described in the next chapter. Here we shall focus on the process as it takes place in prokaryotes, particularly *E. coli.*

As we have seen, instructions for protein synthesis are encoded in sequences of nucleotides in the DNA of a cell and are transcribed into mRNA molecules following the same base-pairing rules that govern DNA replication. The single-stranded mRNA molecules produced in this process range from 500 to 10,000 nucleotides in length. Specific nucleotide sequences of the DNA, called **promoters,** are the binding sites for RNA polymerase and thus are the start signals for RNA synthesis. Other nucleotide sequences, called **terminators,** are the stop signals for RNA synthesis, marking the locations at which transcription ends.

The synthesis of proteins requires, in addition to mRNA molecules, two other types of RNA: ribosomal RNA and transfer RNA. These molecules, which are transcribed from their own specific genes in the DNA of the cell, differ both structurally and functionally from mRNA.

Ribosomal RNA and Transfer RNA

Ribosomes are, as you know, the sites of protein synthesis. They are about one-third protein and two-thirds RNA. The form of RNA they contain is called **ribosomal RNA (rRNA).** As shown in Figure 15–7, each ribosome is composed of two subunits, each with its characteristic rRNAs and proteins. The smaller subunit has a binding site for messenger RNA; in *E. coli* and

15–7   (a) *One view of the three-dimensional structure of the* E. coli *ribosome, as revealed by electron micrographs.* (b) *A schematic diagram of another view of the* E. coli *ribosome. In both prokaryotes and eukaryotes, ribosomes consist of two subunits, one large and one small. Each subunit is composed of specific rRNA and protein molecules. In protein synthesis, the ribosome moves along an mRNA molecule that is threaded between the two subunits.*

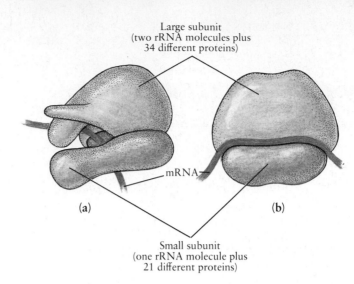

Large subunit
(two rRNA molecules plus
34 different proteins)

mRNA

(a)   (b)

Small subunit
(one rRNA molecule plus
21 different proteins)

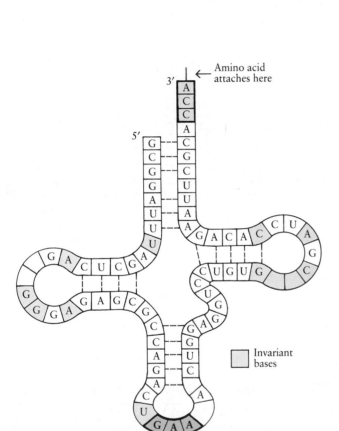

Amino acid
attaches here

3′

5′

Invariant
bases

Anticodon

15–8   *The structure of a tRNA molecule. Such molecules consist of about 80 nucleotides linked together in a single chain. The chain always terminates in a CCA sequence at its 3′ end. An amino acid links to its specific tRNA molecule at this end. Some nucleotides are the same in all tRNAs; these are shown in light green. The other nucleotides vary according to the particular tRNA. The unlabeled boxes represent unusual modified nucleotides characteristic of tRNA molecules.*

Some of the nucleotides are hydrogen-bonded to one another, as indicated by the dashed lines. In some regions, the unpaired nucleotides form loops. The loop on the right in this diagram is thought to play a role in binding the tRNA molecule to the surface of the ribosome. Three of the unpaired nucleotides in the loop at the bottom of the diagram (bright green) form the anticodon. They serve to "plug in" the tRNA molecule to an mRNA codon.

other prokaryotes, the leading (5′) end of the mRNA molecule attaches to this binding site even as the rest of the molecule is still being transcribed. The larger subunit has two binding sites for transfer RNAs.

Transfer RNA (tRNA) molecules are, in effect, a bilingual dictionary by which the language of nucleic acids is translated into the language of proteins. Cells contain more than 20 different kinds of tRNA molecules, at least one for each of the different kinds of amino acids. These molecules, which are comparatively small, all have a characteristic cloverleaf shape (Figure 15–8). At the 3′ end of the molecule is the attachment site for its specific amino acid. A second attachment site, at the opposite end of the three-dimensional structure, is exposed by a loop within the molecule. This site consists of three nucleotides that form an **anticodon,** which is complementary to a specific codon in mRNA.

A third region of the tRNA molecule functions as the recognition site for an enzyme known as an **aminoacyl-tRNA synthetase.** There are at least 20 different aminoacyl-tRNA synthetases in the cell, one or more for each amino acid. Each of these enzymes has a binding site for a particular amino acid and for its matching tRNA molecule. In reactions powered by the energy released in ATP hydrolysis, aminoacyl-tRNA synthetases catalyze the attachment of specific amino acids to specific tRNA molecules—in effect, packaging the amino acids for delivery to the correct location in a growing polypeptide chain.

Translation

The synthesis of proteins is known as **translation,** since it is the transfer of information from one language (nucleic acids) to another (proteins). Translation takes place in three stages: initiation, elongation, and termination (Figure 15–9).

15–9 *The three stages of protein synthesis. (a) Initiation. The smaller ribosomal subunit attaches to the 5' end of the mRNA molecule. The first tRNA molecule, bearing the modified amino acid fMet, plugs into the AUG initiator codon on the mRNA molecule. The larger ribosomal subunit then locks into place, with the tRNA occupying the P (peptide) site. The A (aminoacyl) site is vacant. The initiation stage is now complete.*

(b) Elongation. A second tRNA with its attached amino acid moves into the A site, and its anticodon plugs into the mRNA. A peptide bond is formed between the two amino acids brought together at the ribosome. At the same time, the bond between the first amino acid and its tRNA is broken. The ribosome moves along the mRNA chain in a 5' to 3' direction, and the second tRNA, with the dipeptide attached, is moved to the P site from the A site as the first tRNA is released from the ribosome. A third tRNA moves into the A site, and another peptide bond is formed. The growing peptide chain is always attached to the tRNA that is moving from the A site to the P site, and the incoming tRNA bearing the next amino acid always occupies the A site. This step is repeated over and over until the polypeptide is complete.

(c) Termination. When the ribosome reaches a termination codon (in this example, UGA), the polypeptide is cleaved from the last tRNA and the tRNA is released from the P site. The A site is occupied by a release factor that triggers the dissociation of the two subunits of the ribosome.

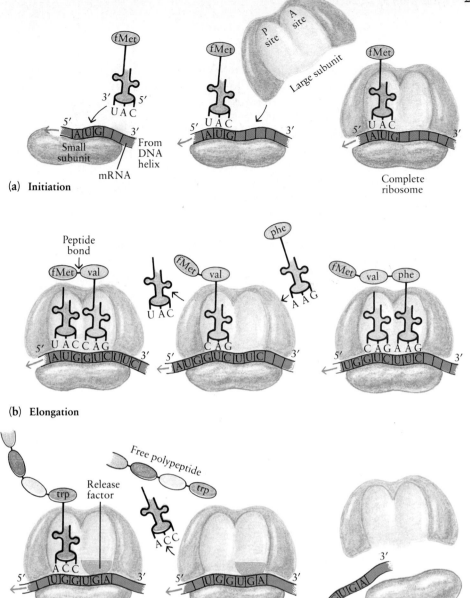

(a) Initiation

(b) Elongation

(c) Termination

The first stage, **initiation,** begins when the smaller ribosomal subunit attaches to the strand of messenger RNA near its 5' end, exposing its first, or initiator, codon. Next, the first, or initiator, tRNA comes into place to pair with the initiator codon of mRNA. This initiator codon is usually (5')–AUG–(3'), which is complementary to the tRNA anticodon (3')–UAC–(5'). In prokaryotes, the tRNA with this anticodon carries a modified form of the amino acid methionine, known as fMet. Thus, fMet is the first amino acid of the newly synthesized polypeptide chain, but it is usually later removed from the completed polypeptide.

The combination of the small ribosomal subunit, mRNA, and the initiator tRNA is known as the **initiation complex.** Once it has formed, the larger subunit of the ribosome attaches to the smaller subunit, and the initiator tRNA with its attached fMet is locked into the P (peptide) site of the larger subunit—one of the two sites for binding tRNA molecules. The energy for this step, which completes the initiation stage, is provided by the hydrolysis of guanosine triphosphate (GTP).

At the beginning of the **elongation** stage, the second codon of the mRNA is positioned opposite the A (aminoacyl) site of the large subunit. A tRNA with an anticodon complementary to the second mRNA codon plugs into the mRNA molecule and, with its attached amino acid, occupies the A site of the ribosome. When both sites are occupied, a peptide bond (see page 60) is forged between the two amino acids, attaching the first (fMet) to the second. The first tRNA is released. The ribosome then moves one codon down the mRNA molecule. Consequently, the second tRNA, to which is now attached fMet and the second amino acid, is transferred from the A position to the P position. A third tRNA–amino acid moves into the now vacant A position opposite the third codon on the mRNA, and the step is repeated. Over and over, the P position accepts the tRNA bearing the growing polypeptide chain, and the A position accepts the tRNA bearing the next amino acid that will be added to the chain.

Toward the end of the coding sequence of the mRNA molecule are one or more of the three codons that serve as **termination** signals. No tRNAs exist with anticodons that "match" these codons, and so no tRNAs will enter the A site in response to them. When a termination codon is reached, translation stops, the polypeptide chain is freed, and the two ribosomal subunits separate.

As the ribosome moves along the mRNA molecule during the elongation process, the initiator portion of the mRNA is freed, and another ribosome can form an initiation complex with it. A group of ribosomes reading the same mRNA molecule is known as a **polysome** (Figure 15–10). Polysomes make possible the rapid synthesis of multiple copies of a polypeptide from the instructions carried by a single mRNA molecule.

The working out of the details of the precise and elegant process of translation was an awe-inspiring achievement. Even more awe-inspiring is the knowledge that at this very moment a similar process is taking place in virtually every cell of our own bodies.

Biological Implications

Mutations Revisited

With the process of protein synthesis in mind, we can now consider some of the broader implications of the genetic code and its translation. For example, let us take another look at sickle-cell anemia. The mutant allele that, in the homozygous recessive condition, is responsible for this disorder apparently originated in Africa. For reasons that we shall discuss in Chapter 19, it has been maintained by natural selection at a very high frequency in the populations of certain regions of Africa. About 9 percent of African-Americans are heterozygous for the sickle-cell allele, and about 0.2 percent are homozygous for the allele and therefore have the symptoms of sickle-cell anemia.

These symptoms are triggered by low oxygen concentrations in the blood, which cause the sickle-cell hemoglobin molecules to become insoluble. The molecules stick together, forming bundles of stiff fibers that distort the shape of the red blood cells and make them more fragile. Premature degradation of these fragile cells causes the anemia. Also, the loss of flexibility of the red blood cells makes it difficult for them to make their way through small blood vessels. Blocking of the blood vessels in the joints and in vital organs by these abnormal red cells is both painful and life-threatening.

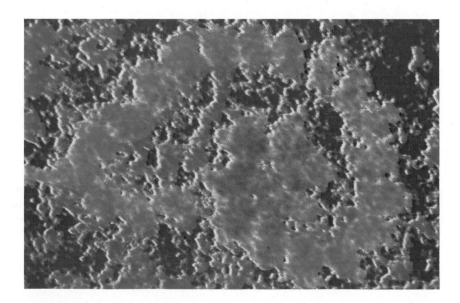

15–10 *This scanning electron micrograph shows a series of ribosomes (highlighted in green) reading a single mRNA molecule in a human brain cell. Such groups of ribosomes are called polyribosomes, or polysomes.*

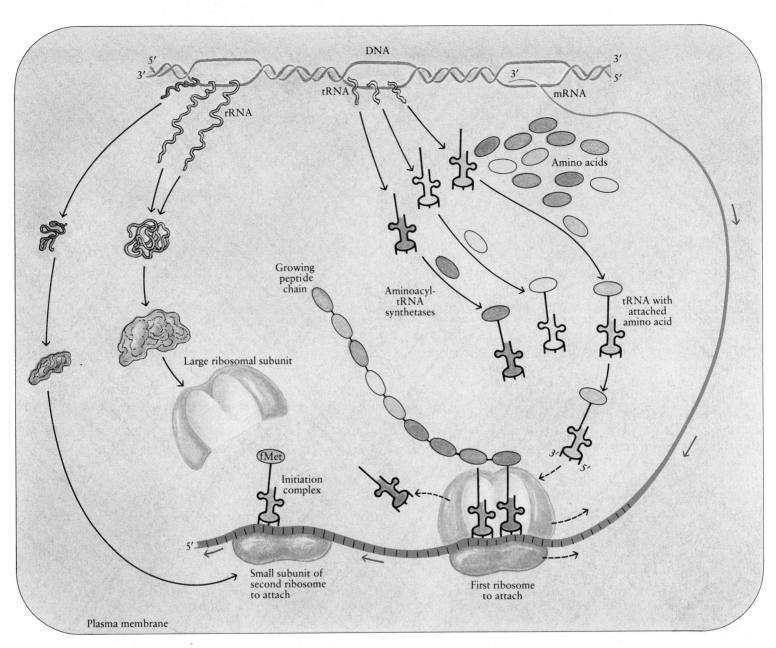

15–11   *A summary of protein synthesis in a bacterial cell. Three different rRNA molecules are transcribed from the DNA of the chromosome. These combine with specific proteins to form the ribosomal subunits. At least 32 different kinds of tRNA molecules are also transcribed from the DNA. These molecules are so structured that each can be attached at one end (by an aminoacyl-tRNA synthetase) to a specific amino acid. Each contains an anticodon that is complementary to an mRNA codon for that particular amino acid.*

The process of protein synthesis begins as an mRNA strand is being transcribed from its DNA template. When the leading (5') end of the mRNA strand attaches to a small ribosomal subunit, an initiator tRNA carrying fMet plugs into the initiator

codon of the mRNA. Addition of a large ribosomal subunit completes the initiation complex. The second amino acid is brought in by a tRNA molecule with an anticodon complementary to the next codon in the mRNA strand. The tRNA plugs in momentarily to that codon, a peptide bond is formed between the first and second amino acids, and the first tRNA molecule is released. As the ribosome moves along the mRNA strand, this process is repeated over and over, as the amino acids are brought into line one by one, following the exact order originally dictated by the DNA from which the mRNA was transcribed.

It is estimated that E. coli *can synthesize as many as 3,000 proteins, each different and each assembled in this same way.*

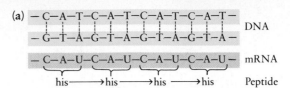

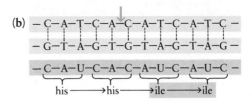

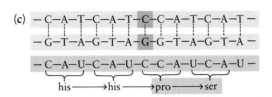

15–12   *Deletion or addition of nucleotides within a gene leads to changes in the protein produced. The original DNA molecule, the mRNA transcribed from it, and the resulting peptide are shown in* (**a**).

In (**b**) *we see the effect of the deletion of a nucleotide pair (T–A) from the location indicated by the arrow. The reading frame for the gene is altered, and a different sequence of amino acids occurs in the peptide.*

A similar change results from the addition of a nucleotide pair (bright green), as shown in (**c**).

The hemoglobin molecule, as you will recall, contains four polypeptide chains—two alpha chains and two beta chains. Normal beta chains contain glutamic acid at a particular position; sickle-cell beta chains contain valine at that position (see Figure 3–23, page 65). As Figure 15–6 reveals, the difference between the mRNA codons for glutamic acid and valine is a single nucleotide. In mRNA, GAA and GAG specify glutamic acid (glu), and GUA and GUG are among the codons that specify valine (val). So the difference between the two is the replacement of one adenine by one uracil. In the template strand of DNA from which the mRNA is transcribed, the difference is the replacement of one thymine by one adenine in a sequence of nucleotides that, since it dictates a polypeptide that contains more than 150 amino acids, must contain more than 450 nucleotides. In other words, the tremendous functional difference—literally a matter of life and death—can be traced to a single "misprint" in over 450 nucleotides.

De Vries, some 90 years ago, defined mutation in terms of characteristics appearing in the phenotype. In the light of more recent knowledge, the current definition is somewhat different: A mutation is a change in the sequence or number of nucleotides in the nucleic acid of a cell. Mutations that occur in gametes or the cells that give rise to gametes are transmitted to future generations. Mutations that occur in somatic cells are transmitted to the daughter cells produced by mitosis and cytokinesis.

Many mutations involve only a single nucleotide substitution and are called **point mutations.** As in the case of sickle-cell anemia, such a substitution can lead to changes in the protein produced by a gene. A mutation that has this effect is known as a **missense** mutation. More than 100 missense mutations involving the genes coding for the alpha and beta chains of the hemoglobin molecule are now known. About 20 of these result in serious disorders. A point mutation can also produce what is known as a **nonsense** mutation. In such cases, the result of the nucleotide substitution is a "stop" codon. This causes protein synthesis to end before the complete polypeptide has been translated.

Changes in the amino acid sequence of a protein can result not only from nucleotide substitutions but also from the deletion or addition of nucleotides within a gene (Figure 15–12). When this occurs, the reading frame of the gene may shift—that is, the way in which the nucleotides are grouped into triplets changes—resulting in the production of an entirely new protein. These **frame-shifts,** as they are known, almost invariably lead to nonfunctional proteins.

Among the agents known to cause mutations are x-rays, ultraviolet rays, radioactive materials, and a variety of other chemical substances. Most mutations occur "spontaneously"—meaning simply that we do not know the chemical factors that trigger them. The rate of spontaneous mutation is generally low. For mutations detectable in the phenotype of eukaryotes, the rate varies from 1 in 1,000 to 1 in 1,000,000 gametes per generation, depending on the gene involved. Different genes—and even different alleles of the same gene—have different rates of mutation. These differences are thought to be related to both the chemical composition of the gene (or allele) in question and its position in the chromosome.

Classical Genetics Revisited

The realization that genes code for polypeptides provided a simple and clear explanation for the phenomenon of dominance observed in many traits of pea plants, fruit flies, humans, and other organisms. In heterozygotes, enough of the protein coded by the normal allele is produced to compensate for the recessive allele, which codes either for a defective, poorly functioning

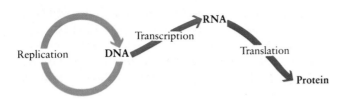

15–13   *A summary of the information flow from DNA to RNA to protein. Replication of the DNA occurs only once in each cell cycle (page 179), during the S phase prior to mitosis or meiosis. Transcription and translation, however, occur repeatedly throughout the interphase portions of the cell cycle.*

polypeptide or for none at all. Only when two recessive alleles come together in the same individual—and none of the normal polypeptide is synthesized—is the recessive phenotype expressed.

The discovery of the genetic code and the new understanding of the nature and consequences of mutations also clarified some of the apparent exceptions to Mendelian principles. Incomplete dominance, for example, results from the combined effect of the proteins translated from both the dominant and the recessive alleles of a gene.

Other exceptions, such as epistasis (page 224), are more complicated. As you know, most of the biochemical reactions in the cell occur in series—biochemical pathways—in which each reaction of the series is catalyzed by a specific enzyme (Figure 7–13, page 137). For the complete series of reactions in a biochemical

pathway to occur, at least one copy of the normal allele coding for each enzyme involved in the pathway must be present. If, for any given enzyme, two copies of a defective allele are present, all steps of the pathway beyond that point will be shut down—and the presence of normal alleles coding for other enzymes of the pathway will be masked. Moreover, the end product of the reaction series will not be formed, and harmful intermediates may accumulate. Such is the case, for example, in PKU and in Tay-Sachs disease (page 212).

Mutations of different genes coding for different enzymes in a pathway can have different effects. In the pathway for the breakdown of phenylalanine, for example, a mutation in the gene for one enzyme leads to PKU, and a mutation in the gene for another enzyme leads to a disorder known as alkaptonuria, in which the urine of affected individuals is jet black. A mutation in the gene for still another enzyme in the same pathway blocks the synthesis of the brown pigment melanin. The result is albinism—the lack of pigmentation in skin, hair, and eyes.

Although the discovery of the mechanism of protein synthesis and the working out of the genetic code shed new light on mutations and answered many questions, they raised many other questions. The genetic material, the DNA, is an enormously long sequence of nucleotides, undifferentiated in form. Yet the information it contains is compartmentalized in the units we call genes. How is this accomplished? And, how does the cell regulate its genes so they produce the right amounts of the right proteins at the right times? In the next chapter, we shall focus on these questions.

Summary

Genetic information is encoded in the sequence of nucleotides in molecules of DNA, and these, in turn, determine the sequence of amino acids in molecules of protein.

The way in which the information carried by DNA is translated into protein has been worked out in considerable detail. The information is transcribed from the DNA into a long, single strand of RNA (ribonucleic acid). This type of RNA molecule is known as messenger RNA, or mRNA. Synthesis of mRNA, catalyzed by the enzyme RNA polymerase, uses one strand of the DNA as a template, following the principles of base

pairing first suggested by Watson and Crick. The mRNA therefore is complementary to the template DNA strand. Each group of three nucleotides in the mRNA molecule is the codon for a particular amino acid.

The genetic code has now been "broken," that is, it is known which amino acid is called for by a given mRNA codon. Of the 64 possible triplet combinations of the four-letter nucleotide code, 61 combinations specify amino acids. Each of these combinations codes for one—and only one—of the 20 amino acids that make up protein molecules. The other three triplets serve as stop signals, terminating protein synthesis. The genetic code is universal, the same in all living organisms.

Protein synthesis—translation—takes place at the ribosomes. A ribosome is formed from two subunits, one large and one small, each consisting of characteristic ribosomal RNAs (rRNAs) complexed with specific proteins. Also required for protein synthesis is another group of RNA molecules, known as transfer RNA (tRNA). These small molecules can carry an amino acid on one end, and they have a triplet of bases, the anticodon, on a central loop at the opposite end of the three-dimensional structure. The tRNA molecule is the adapter that pairs the correct amino acid with each mRNA codon during protein synthesis. There is at least one kind of tRNA molecule for each kind of amino acid found in proteins.

In bacteria, even as the mRNA strand is still being transcribed, ribosomes are attaching, one after another, near its free end. At the point where the strand of mRNA is in contact with a ribosome, tRNAs are bound temporarily to the mRNA strand. This binding takes place by complementary base pairing between the mRNA codon and the tRNA anticodon. Each tRNA molecule carries the specific amino acid called for by the mRNA codon to which the tRNA attaches. Thus, following the sequence originally dictated by the DNA, the amino acid units are brought into line one by one and, as peptide bonds form between them, are linked into a polypeptide chain.

Mutations are now defined as changes in the sequence or number of nucleotides in the nucleic acid of a cell or organism. They may take the form of substitutions of one nucleotide for another, or deletions or additions of nucleotides. Nucleotide substitutions can result in missense mutations, in which one amino acid is substituted for another, or in nonsense mutations, in which protein synthesis is terminated prematurely. Deletions or additions of nucleotides can result in frame-shift mutations and the production of entirely new—and usually nonfunctional—proteins.

The fact that genes code for proteins—and that mutations result in defective proteins—makes clear the underlying basis of the phenomenon of dominance, first documented by Mendel. It also explains many of the apparent exceptions to Mendelian principles, such as incomplete dominance and epistasis.

Questions

1. Distinguish among the following: mRNA/tRNA/rRNA; code/codon/anticodon; transcription/translation; P site/A site; initiation/elongation/termination; point mutation/missense mutation/nonsense mutation/frame-shift mutation.

2. A person heterozygous for the allele for sickle-cell anemia makes the variant hemoglobin molecules but does not suffer from anemia. In what respect is this situation different from the definitions of "dominant" and "recessive" given in Chapter 12?

3. Most of the bacterial DNA codes for messenger RNA, and most of the RNA produced by the cell is mRNA. Yet analysis of the RNA content of a cell reveals that, typically, rRNA is about 80 percent of the cellular RNA and tRNA makes up most of the rest. Only about 2 percent is normally mRNA. How do you explain these findings? What do you think the functional explanation might be?

4. Given the details of protein synthesis, what further amendment would you make to the principle "one gene—one polypeptide"? (*Hint:* See page 261, and look again at Figure 15–11.)

5. Explain the term "genetic code." In what respects is it a useful analogy?

6. In a hypothetical segment of one strand of a DNA molecule, the sequence of nucleotides is

$$(3')-AAGTTTGGTTACTTG-(5')$$

(a) What would be the sequence in an mRNA strand transcribed from this DNA segment? (b) What would be the sequence of amino acids coded by the mRNA? (c) Does it matter at what point on the template strand the transcription from DNA to mRNA begins? Explain your answer.

7. Fill in the missing letters.

T G T	_ _ _	_ _ _	
_ _ A	C _ _	_ _ _	DNA
U _ _	C A	_ _ _	mRNA codons
_ _ _	_ _ _	G C A	tRNA anticodons

Specify the amino acid for which each mRNA triplet codes.

8. What amino acid is carried by the tRNA molecule shown in Figure 15–8? How do you know?

9. Suppose you have the peptide arg–lys–pro–met, and you know that the tRNA molecules used in its synthesis had the following anticodons:

$$(3')\text{–GGU–}(5')$$
$$(3')\text{–GCU–}(5')$$
$$(3')\text{–UUU–}(5')$$
$$(3')\text{–UAC–}(5')$$

What is the DNA nucleotide sequence for the template strand of the gene that codes for this peptide? *(Hint:* Determine the mRNA codons complementary to each of the tRNA codons and their sequence in the mRNA strand from which the peptide was translated.)

10. Deletion or addition of nucleotides within a gene leads to changes in the protein produced. The original DNA molecule, the mRNA transcribed from it, and the resulting peptide are:

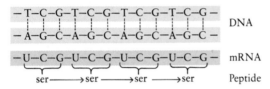

Deleting the second T–A pair, as indicated by the arrow, yields the following DNA molecule:

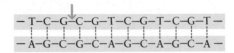

How is the resulting amino acid sequence altered? How does the addition of a C–G pair to the original molecule,

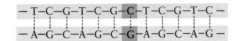

affect the amino acid sequence?

11. In prokaryotes, transcription and translation are "linked" processes. As we shall see in the next chapter, however, this is not the case in eukaryotes. On the basis of your knowledge of cell structure, propose a likely explanation.

Suggestions for Further Reading

Amato, Ivan: "Expanding the Genetic Alphabet," *Science News,* February 10, 1990, pages 88–90, 94.

Darnell, James E., Jr.: "RNA," *Scientific American,* October 1985, pages 68–87.

Dickerson, Richard E.: "The DNA Helix and How It Is Read," *Scientific American,* December 1983, pages 94–111.

Lake, James A.: "The Ribosome," *Scientific American,* August 1981, pages 84–97.

Loupe, Diane E.: "Breaking the Sickle Cycle," *Science News,* December 2, 1989, pages 360–362.

Rich, Alexander, and Sung Hou Kim: "The Three-Dimensional Structure of Transfer RNA," *Scientific American,* January 1978, pages 52–62.

Ross, Jeffrey: "The Turnover of Messenger RNA," *Scientific American,* April 1989, pages 48–55.

Scientific American: *Molecules to Living Cells,* W. H. Freeman and Company, New York, 1980.*

> More than half of the articles in this collection from Scientific American *are accounts of major breakthroughs in our understanding of the nucleic acids and their functions. Written for the general reader by the scientists who made the key discoveries, they are highly recommended.*

* Available in paperback.

This cat is a tortoiseshell, so called because of the patches of yellow and black in its fur. You can bet that a tortoiseshell is a female, because yellow and black coat colors are determined by alleles of a gene on the X chromosome. A male cat, with only one X chromosome, can be either black or yellow—but not both. However, if a female, with her two X chromosomes, is heterozygous for the gene, she can be a tortoiseshell.

If you were to examine the somatic cells of a tortoiseshell cat at interphase, you would see a dark spot of chromatin at the edge of the nucleus. This spot, known as a Barr body, is a highly condensed, inactivated X chromosome. Early in development, one X chromosome or the other is inactivated in all of the embryonic cells except those that will later give rise to the gametes. The same X chromosome is inactivated in all of the cells descended from each embryonic cell. Which is why a particular patch of fur is either yellow or black.

Barr bodies are of interest to geneticists as well as to cat fanciers. Found in the somatic cells of all female mammals, they represent one of the ways in which genes can be "turned off." Inactivation of the X chromosome occurs at random. Thus, the somatic cells of a female mammal are not identical but are of two types, depending on which X chromosome is active and which is inactive. Human females who are carriers of the allele for hemophilia, for example, show wide variability in the amount of clotting factor in their blood, ranging from 20 to 100 percent of normal levels. There also appear to be slight variations in the timing. Female carriers of the allele for color blindness are sometimes color blind in one eye but not in the other. In this cat, X-chromosome inactivation occurred relatively late, and the black and yellow patches are small and closely intermingled. When the inactivation occurs earlier in embryonic development, the patches are larger and more distinct—in which case the cats are known as calicos.

You should know that you might lose your bet about the sex of a tortoiseshell. An occasional tortoiseshell is male. Such males, as you might guess, are XXY and are sterile.

16

Chromosome Structure and the Regulation of Gene Expression

As we saw in the last chapter, the genetic code—three specific nucleotides coding for each amino acid—is universal. It is the same in bacteria, bread molds, pea plants, fruit flies, cats, humans, and all other organisms—awesome evidence that all living things are descended from a common ancestor. However, a variety of studies in molecular genetics have shown that, despite the basic similarities of prokaryotic and eukaryotic chromosomes, there are important differences in their structure. There are also important differences in the mechanisms that turn genes "on" and "off" and thus regulate gene expression. In this chapter, we shall explore some of these differences.

The Prokaryotic Chromosome

The prokaryotic chromosome is a single, continuous (circular) thread of double-stranded DNA, only 2 nanometers wide. In *Escherichia coli*, it contains some 4.7 million base pairs and is approximately 1 millimeter long when fully extended (see page 254). An *E. coli* cell is less than 2 micrometers long, so the chromosome is some 500 times longer than the cell itself. Within the cell, the chromosome is folded into an irregularly shaped body known as the nucleoid (see Figure 4–7, page 78).

Regulation of Gene Expression in Prokaryotes

Transcription in *E. coli* and other prokaryotes takes place, as we saw in the last chapter, by the synthesis of a molecule of mRNA along a template strand of DNA. The process begins when the enzyme RNA polymerase attaches to the DNA at a specific site known as the **promoter.** The RNA polymerase molecule binds tightly to the promoter and causes the DNA double helix to open,

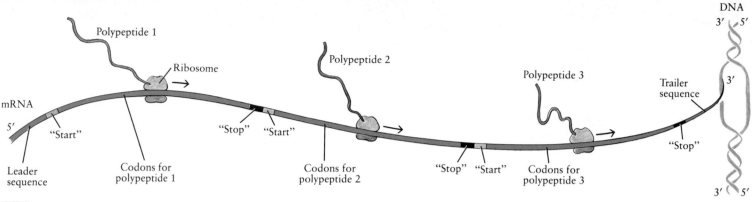

16–1 *In prokaryotes, transcription often results in an mRNA molecule that contains coding sequences for several different polypeptide chains, with the sequences separated by stop and start codons. In this diagram, the stop and start codons are adjacent, but sometimes they are separated by as many as 100 to 200 nucleotides. The 5' end of the mRNA* *molecule has a short leader sequence, and the 3' end has a trailer sequence; neither of these sequences codes for protein. Translation, which starts at the leading (5').end of the mRNA, generally begins while the rest of the molecule is still being transcribed.*

initiating transcription. The growing RNA strand remains hydrogen-bonded to the DNA template only briefly, and then it peels off in a single strand.

A segment of DNA that codes for a polypeptide is known as a **structural gene.** In the bacterial chromosome, structural genes coding for polypeptides with related functions often occur together in sequence. Such functional groups might include, for instance, two polypeptide chains that together constitute a particular enzyme or three enzymes that work in a single biochemical pathway. Groups of genes coding for such molecules are typically transcribed into a single mRNA strand (Figure 16–1). Thus, a group of polypeptides that are needed by the cell at the same time and in the same quantity can be synthesized simultaneously, a simple and efficient inventory-control system.

A cell does not make all of its possible proteins all of the time, but only when they are needed and only in the amounts needed. For example, cells of *E. coli* supplied with lactose as a nutrient require the enzyme beta-galactosidase to split the lactose, a disaccharide, into its two constituent monosaccharides, glucose and galactose. Cells growing on lactose have approximately 3,000 molecules of beta-galactosidase per cell. In the absence of lactose, there is an average of one molecule of the enzyme per cell. In short, the presence of lactose leads to the production of the enzyme molecules needed to break it down.

In other cases, the presence of a particular nutrient may inhibit the synthesis of particular proteins. Like most other bacteria, *E. coli* can synthesize each of its amino acids from ammonium ions (NH_4^+) and a carbon source. The enzymes needed for the biosynthesis of the amino acid tryptophan, for instance, are synthesized continuously in growing cells—unless tryptophan is present. In the presence of tryptophan, production of the enzymes ceases.

Mutants of *E. coli* sometimes occur that are unable to regulate enzyme production. These cells produce beta-galactosidase even in the absence of lactose, for example, or the enzymes that synthesize tryptophan even when tryptophan is present. These and similar mutants are generally at a disadvantage because they are squandering their energies and resources producing enzymes they don't need. Normal *E. coli* cells rapidly outmultiply them.

Although regulation of protein synthesis could theoretically take place at many points in the process, in prokaryotes it occurs mostly at the level of transcription. Regulation involves interactions between the chemical environment of the cell and special regulatory proteins, coded by regulatory genes. These proteins can work either as negative controls, repressing mRNA transcription, or as positive controls, enhancing transcription. The fact that mRNA is translated into protein so immediately (before transcription is even completed) and then broken down very rapidly further increases the efficiency of this strategy of regulation.

The Operon

Our current understanding of the regulation of transcription in prokaryotes rests upon a model, known as the **operon** model, proposed some years ago by the French scientists François Jacob and Jacques Monod. According to this model, an operon (Figure 16–2) comprises the promoter, one or more structural genes, and another DNA sequence known as the **operator.** The operator is a sequence of nucleotides located between the promoter and the structural gene or genes.

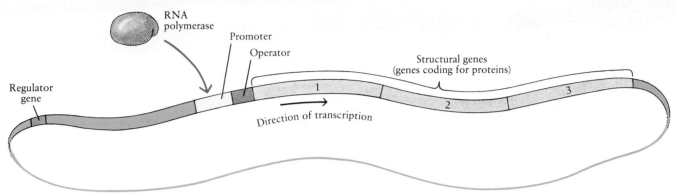

RNA polymerase

Promoter

Operator

Regulator gene

Structural genes (genes coding for proteins)

1

2

3

Direction of transcription

DNA (bacterial chromosome)

16–2 *A schematic representation of an operon. An operon consists of a promoter, an operator, and structural genes (that is, genes that code for proteins, often enzymes that work sequentially in a particular biochemical pathway). The promoter, which precedes the operator, is the binding site for RNA polymerase. The operator is the site at which a repressor can bind; it may overlap the promoter, the first structural gene, or both.*

Another gene involved in operon function is the regulator, which codes for the repressor. Although the regulator may be adjacent to the operon, in most cases it is located elsewhere on the bacterial chromosome.

Transcription of the structural genes often depends on the activity of still another gene, the **regulator,** which may be located anywhere on the bacterial chromosome. This gene codes for a protein called the **repressor,** which binds to the operator. When a repressor is bound to the operator, it obstructs the promoter. As a consequence, RNA polymerase either cannot bind to the DNA molecule or, if bound, cannot begin its movement along the molecule. The result in either case is the same: no mRNA transcription occurs. However, when the repressor is removed, transcription may begin.

The capacity of the repressor to bind to the operator and thus to block protein synthesis depends, in turn, on yet another molecule that either activates or inactivates the repressor for that particular operon (Figure 16–3).

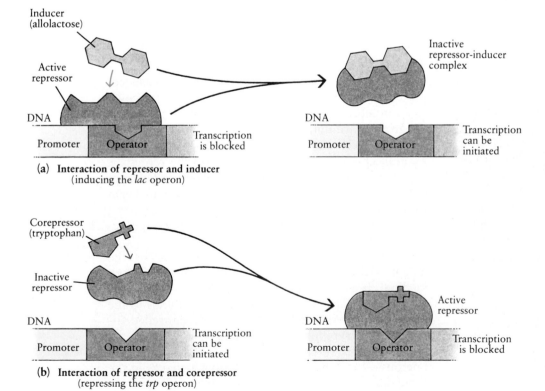

Inducer (allolactose)

Active repressor

DNA

Promoter

Operator

Transcription is blocked

Inactive repressor-inducer complex

DNA

Promoter

Operator

Transcription can be initiated

(a) Interaction of repressor and inducer
(inducing the *lac* operon)

Corepressor (tryptophan)

Inactive repressor

DNA

Promoter

Operator

Transcription can be initiated

Active repressor

DNA

Promoter

Operator

Transcription is blocked

(b) Interaction of repressor and corepressor
(repressing the *trp* operon)

16–3 *In an operon system, the transcription of mRNA—and thus the synthesis of proteins—is regulated by interactions involving either a repressor and an inducer or a repressor and a corepressor. (a) In some systems, such as the lac operon, the repressor molecule is active until it combines with the inducer. (b) In other systems, such as the trp operon, the repressor is not active until it combines with the corepressor.*

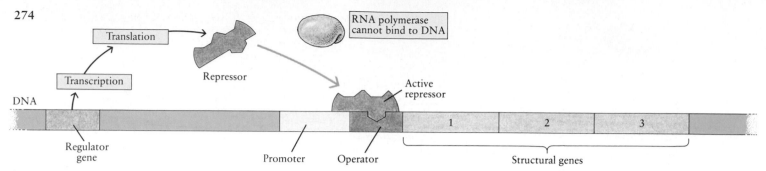

(a) *Lac* operon, operator "off"

16–4 *Regulation of the* lac *operon. (a) In the absence of lactose, the repressor binds to DNA at the operator and so prevents the RNA polymerase from initiating transcription. (b) In the presence of lactose, the repressor is inactivated. Thus RNA polymerase can bind to the promoter, and synthesis of mRNA can proceed. The genes of the operon are transcribed onto a single mRNA molecule directing the synthesis of three proteins: the enzyme beta-galactosidase, a transport protein that brings lactose from the external medium into the cell, and the enzyme transacetylase, which transfers an acetyl group from acetyl CoA (page 149) to galactose.*

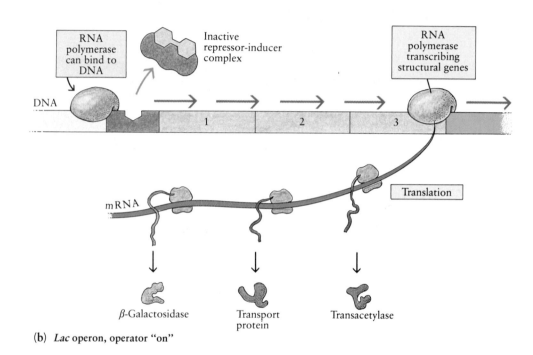

(b) *Lac* operon, operator "on"

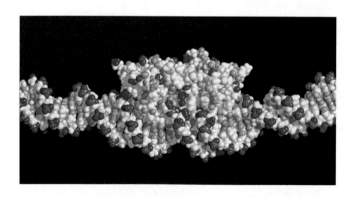

16–5 *A computer-generated model of a repressor bound to the DNA double helix. The three-dimensional shape of the repressor mirrors the shape of the nucleotide sequence that forms the operator. When the repressor is bound to the operator, it blocks a sequence of some 10 to 20 base pairs. Repressors function as molecular switches, turning off the synthesis of proteins that are not needed.*

A molecule that activates a repressor is known as a **corepressor,** and one that inactivates a repressor is known as an **inducer.** For example, when lactose is present in the growth medium, the first step in its metabolism produces a closely related sugar, allolactose. Allolactose is an inducer—it binds to and inactivates the repressor, removing it from the operator of the lactose *(lac)* operon. As a consequence, RNA polymerase can begin its movement along the DNA molecule, transcribing the structural genes of the operon into mRNA (Figure 16–4).

An example of a corepressor is provided by the amino acid tryptophan. When present in the growth medium, it activates the repressor for the tryptophan *(trp)* operon. The activated repressor then binds to the operator and blocks the synthesis of the unneeded enzymes. Both tryptophan and allolactose—as well as the molecules that interact with the repressors of other operons—exert their effects by causing a change in the three-dimensional shape of the repressor molecule.

Some 75 different operons have now been identified in *E. coli*, comprising 260 structural genes. As we shall see in Chapter 18, manipulating operons is an essential part of the scientific trick of inducing bacterial cells to produce proteins of medical importance, such as human insulin.

The Eukaryotic Chromosome

In the early days of molecular genetics (the late 1950s and into the 1960s), it was tempting to think that the eukaryotic chromosome might turn out to be simply a large-scale version of the *E. coli* chromosome. However, it soon became clear that there are many important differences between the bacterial chromosome and the chromosomes of eukaryotes—some expected and some very surprising. These differences include (1) a far greater quantity of DNA in eukaryotic chromosomes; (2) a great deal of repetition in this DNA, with much of it lacking any apparent function; (3) a close association of the DNA with proteins that play a major role in chromosome structure; and (4) considerably more complexity in the organization of the protein-coding sequences of the DNA and the regulation of their expression.

DNA is an "exquisitely thin filament," in the words of E. J. DuPraw, who calculated that a length sufficient to reach from the Earth to the sun would weigh only half a gram. The DNA of each eukaryotic chromosome is thought to be in the form of a single, linear molecule. If fully extended, the DNA forming each human chromosome would be from 3 to 4 centimeters long. Each diploid cell, with its 46 chromosomes, thus contains about 2 meters of DNA, and the entire human body contains some 25 billion kilometers of DNA double helix.

The Structure of the Chromosome

In the nucleus of the eukaryotic cell, the DNA is always found combined with proteins. This combination, as we noted in Chapter 5, is known as **chromatin** ("colored threads") because of its staining properties. Chromatin is more than half protein, and the most abundant proteins, by weight, belong to a class of small polypeptides known as **histones.** Histones are positively charged (basic) and so are attracted to the negatively charged (acidic) DNA. They are always present in chromatin and are synthesized in large amounts during the S phase of the cell cycle. The histones are primarily responsible for the folding and packaging of DNA. In a human cell, for instance, the approximately 2 meters of DNA are

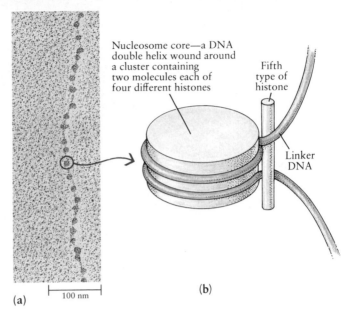

16–6 (a) *Chromatin that has been unfolded and stretched out as a result of chemical treatment, revealing the beadlike nucleosomes. The distance between nucleosomes is about 10 to 11 nanometers, and the diameter of each bead is about 7 nanometers.*

(b) *The structure of a nucleosome. The negatively charged DNA coils twice around the protein core, composed of eight positively charged histone molecules. Another histone molecule (also positively charged) binds to the outer surface of the nucleosome. The DNA of the nucleosome core contains about 140 base pairs. The strand of linker DNA between the nucleosome cores contains another 30 to 60 base pairs.*

packed into 46 cylinders (the 46 chromosomes) that, when condensed at metaphase, have a combined length of only 200 micrometers.

There are five distinct types of histones. They are present in enormous quantities—about 60 million molecules per cell of each of four of the types, and about 30 million molecules per cell of the fifth type. With the exception of the fifth type, the amino acid sequences of the histones are very similar in widely diverse groups of organisms. One histone of the garden pea, for instance, differs from the corresponding histone of the cow by only two amino acids out of a total of 102.

The fundamental packing units of chromatin are the **nucleosomes** (Figure 16–6), which resemble beads on a string. Each nucleosome consists of a core of eight histone molecules, around which the DNA filament is wrapped twice, like thread around a spool. Another histone molecule lies on the DNA, outside the nucleosome core. When a fragment of DNA is tied up in a nucleosome, it is about one-sixth the length it would be if fully extended.

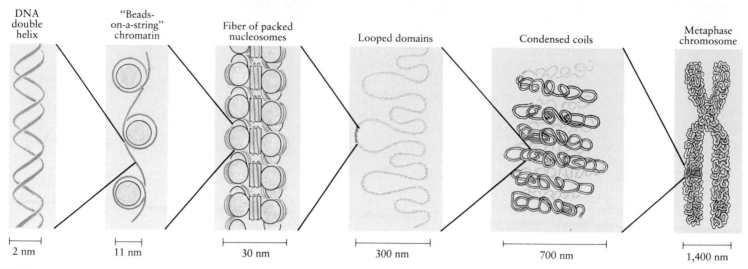

DNA double helix	"Beads-on-a-string" chromatin	Fiber of packed nucleosomes	Looped domains	Condensed coils	Metaphase chromosome
2 nm	11 nm	30 nm	300 nm	700 nm	1,400 nm

16–7   *Stages in the folding of chromatin, culminating in a fully condensed metaphase chromosome. This model is derived from electron micrographs of chromatin at different degrees of condensation. According to present evidence, each chromatid of a replicated chromosome contains a single molecule of double-stranded DNA and is, by weight, about 60 percent protein.*

As shown in Figure 16–7, additional packing of the nucleosomes produces a fiber that is about 30 nanometers in diameter. Further condensation of this fiber produces a series of loops, known as looped domains. The looped domains also coil until, ultimately, clusters of neighboring looped domains condense into the compact chromosomes that become visible during mitosis and meiosis.

Other proteins associated with the chromosome are the enzymes involved in DNA and RNA synthesis, regulatory proteins, and a large number and variety of molecules that have not yet been isolated and identified. Unlike the histones, these molecules vary from one cell type to another.

Regulation of Gene Expression in Eukaryotes

As we saw earlier, regulation of gene expression in prokaryotes typically involves turning genes on and off in response to changes in the nutrients available in the environment. In eukaryotes, especially multicellular eukaryotes, the problems of regulation are very different. A multicellular organism usually starts life as a fertilized egg, the zygote. The zygote divides repeatedly by mitosis and cytokinesis, producing many cells. At some stage these cells begin to differentiate, becoming, in humans, for example, muscle cells, nerve cells, blood cells, intestinal lining cells, and so forth. Each cell type, as it differentiates, begins to produce characteristic proteins that distinguish it structurally and functionally from other types of cells.

This is nicely illustrated by mammalian red blood cells. In the early stages of fetal life, developing red blood cells synthesize one type of fetal hemoglobin. Red blood cells produced at later stages contain a second type of fetal hemoglobin. Then, sometime after the birth of the organism, the developing red blood cells begin to produce the polypeptide chains characteristic of adult hemoglobin. The genes are expressed in a precisely timed sequence, one after the other. Moreover, the DNA segments that code for these hemoglobin molecules are expressed only in developing red blood cells.

There is evidence, however, that all of the genetic information originally present in the zygote is also present in every diploid cell of the organism (Figure 16–8). Thus, for example, the DNA segments that code for hemoglobin (both the fetal types and the adult type) are present in skin cells and heart cells and liver cells and nerve cells and, indeed, in every one of the nearly 200 different types of cells in the body. Similarly, the DNA sequence that codes for the hormone insulin is present not only in the specialized cells of the pancreas that manufacture insulin but also in all the other cells. Since each type of cell produces only its characteristic proteins—and not the proteins characteristic of other cell types—it becomes apparent that differentiation of the cells of a multicellular organism depends on the inactivation of certain groups of genes and the activation of others.

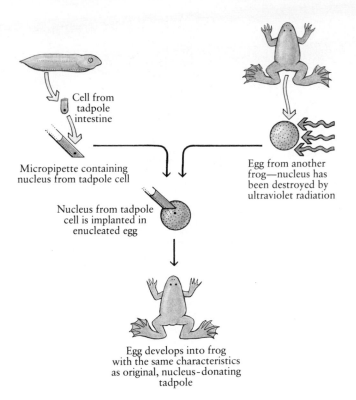

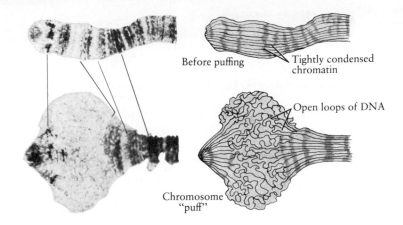

16–9 *Observations of chromosome puffs support the concept that the DNA is somehow unwound to make it available for RNA transcription. This puff was observed in a chromosome of the Brazilian gnat, which, like the fruit fly, has giant chromosomes in some of its cells. The red lines between the two views of the chromosome indicate corresponding bands of the chromosome. Puffs occur normally but can also be induced experimentally. This puff occurred in response to a hormone that causes molting.*

16–8 *A number of experiments have shown that early development does not result in permanent inactivation of genes or the loss of functional DNA. In experiments by J. B. Gurdon, illustrated here, nuclei were removed from intestinal cells of a tadpole and implanted into egg cells in which the nucleus had been destroyed. In some cases, the embryo developed normally, indicating that the tadpole intestinal cell nucleus contained all the information required for all the cells of the organism.*

In other experiments, F. C. Steward demonstrated that under certain conditions, a single differentiated cell of a carrot can be persuaded to divide repeatedly, ultimately reconstituting an entire carrot plant.

Condensation of the Chromosome and Gene Expression

Many lines of evidence indicate that the degree of condensation of the DNA of the chromosome, as shown by chromatin staining, plays a major role in the regulation of gene expression in eukaryotic cells. Staining reveals two types of chromatin: **euchromatin,** the more open chromatin, which stains weakly, and **heterochromatin,** the more condensed chromatin, which stains strongly. During interphase, heterochromatin remains tightly condensed, but euchromatin becomes less condensed. Transcription of DNA to RNA takes place only during interphase, when the euchromatin is less condensed—and is therefore accessible to molecules of RNA polymerase.

Some regions of heterochromatin are constant from cell to cell and are never expressed. An example is the highly condensed chromatin located in the centromere region of the chromosome. This region, which does not code for protein, is believed to play a structural role in the movement of the chromosomes during mitosis and meiosis. Similarly, little or no transcription takes place from Barr bodies (page 271), which are tightly condensed and irreversibly inactivated mammalian X chromosomes.

The degree of condensation of other regions of the chromatin, however, varies from one type of cell to another within the same organism. This reflects the fact that different types of cells synthesize different proteins and thus require the transcription of different segments of the DNA. Moreover, as a cell differentiates during embryonic development, the proportion of heterochromatin to euchromatin increases as the cell becomes more specialized. Segments of DNA that will not be needed by the differentiated cell are effectively "silenced."

Further evidence linking the degree of chromosome condensation to gene expression comes from studies of the giant chromosomes of insects (see page 235). At various stages of larval growth in insects, diffuse "puffs" occur in various regions of these chromosomes (Figure 16–9). The puffs are open loops of DNA. Studies with radioactive isotopes have shown that they are the sites of rapid RNA synthesis. When ecdysone, a hormone that produces molting in insects, is injected, the puffs occur in a definite sequence that can be related to the developmental stage of the animal.

277

Regulation by Specific Binding Proteins

In eukaryotes, as in prokaryotes, transcription is also regulated by proteins that bind to specific sites on the DNA molecule. Many of these proteins and their binding sites have now been identified, and it is increasingly clear that this level of transcriptional control is far more complex in multicellular eukaryotes than in prokaryotes.

A gene in a multicellular organism appears to respond to the sum of many different regulatory proteins, some tending to turn the gene on and others to turn it off. The sites at which these proteins bind may be hundreds or even thousands of base pairs away from the promoter sequence at which RNA polymerase binds and transcription begins. This, as you might expect, adds to the difficulty of identifying the regulatory molecules and also of understanding exactly how they exert their effects. Recent research suggests that changes in the activities of some of these regulators are linked with the development of cancer, a matter we shall explore in the next chapter.

The DNA of the Eukaryotic Chromosome

Early studies of the DNA of eukaryotic cells revealed two surprising facts. First, with a few exceptions, the amount of DNA per cell is the same for every diploid cell of any given species (which is not surprising), but the variations among different species are enormous. *Drosophila* has about 1.4×10^8 base pairs per haploid chromosome set, only about 70 times more than *E. coli*. Humans (with approximately 3.5×10^9 base pairs) have 25 times as much as *Drosophila*, somewhat more than a mouse, but about the same amount as a toad (3.32×10^9 base pairs). The largest amount of DNA so far has been found in a salamander with 8×10^{10} base pairs per haploid set of chromosomes.

Second, in every eukaryotic cell, there is what appears to be a great excess of DNA, or at least of DNA whose functions are unknown. It is estimated that in eukaryotic cells less than 10 percent of all the DNA codes for proteins; in humans, it may even be as little as 1 percent. By contrast, prokaryotes use their DNA very thriftily. Except for regulatory or signal sequences, virtually all of their DNA is expressed.

Continuing research is revealing the organization of eukaryotic DNA, as well as some of the functions of the long skeins of nucleotides that are never translated into protein.

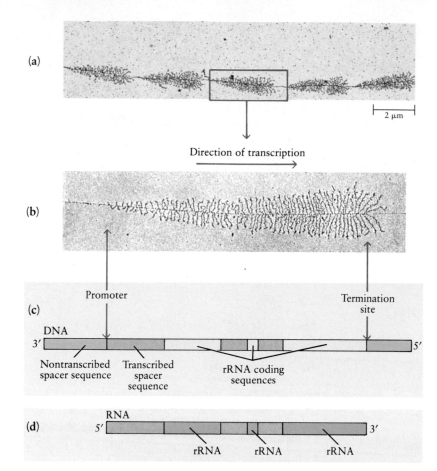

16–10   *A single gene codes for three of the four types of rRNA molecules found in the eukaryotic ribosome. This gene occurs in multiple copies, repeated in tandem (head to tail). (a) Electron micrograph of five copies of this gene separated by nontranscribed spacer sequences. (b) An enlargement of one of the genes and the RNA molecules being transcribed from it. The DNA is the thin horizontal line, and the RNA molecules are the fine fibrils perpendicular to the DNA. (c) A map of the three-rRNA gene. The portions of the DNA coding for the three rRNA molecules are shown in light green; the transcribed spacer sequences are shown as gray. (d) A map of the RNA molecule transcribed from the three-rRNA gene. When transcription is complete, enzymes cleave the molecule and remove the transcribed spacer sequences. This releases the three individual rRNA molecules.*

Classes of DNA: Repeats and Nonrepeats

Simple-Sequence DNA

The DNA of eukaryotic cells can be divided into three general classes. The first, known as **simple-sequence** DNA, consists of short repetitive sequences, arranged in tandem (head to tail). These sequences are typically 5 to 10 base pairs in length, and they are present in enormous quantities. *Drosophila virilis*, for instance, has ACAAACT repeated 12 million times. About 10 percent of the DNA of the mouse, and about 20 to 30 percent of human DNA, is made up of short, highly repetitive sequences.

Simple-sequence DNA is thought to be vital to chromosome structure. Long blocks of short repetitive sequences have been found around the centromere, and, indeed, may *be* the centromere. More recently, the tips of all human chromosomes have been found to consist of some 1,500 to 6,000 nucleotides in which a simple sequence—TTAGGG—is repeated over and over. This same simple, repeated sequence has also been found at the chromosome tips in a wide range of other mammals, birds, reptiles, and even protists. It is thought that the "caps" formed by this repeated sequence help stabilize the chromosome itself.

Intermediate-Repeat DNA

The second class of eukaryotic DNA consists of **intermediate-repeat** sequences, generally 150 to 300 nucleotides in length. Sometimes these sequences are arranged in tandem, one after another, but more often they are dispersed throughout the DNA. About 20 to 40 percent of the DNA of multicellular organisms consists of intermediate-repeat sequences.

Among the most thoroughly studied examples of this class of DNA are the genes coding for histones and for three of the four ribosomal RNAs found in eukaryotes (Figure 16–10). The histone genes are present in multiple copies (from 50 to 500) in the cells of all multicellular eukaryotes. Cells of multicellular eukaryotes, which may contain some 10 million ribosomes per cell, also have from 50 to 5,000 copies of the rRNA genes. The rRNA genes occur in tandem, head to tail; the chromosome regions in which they are located form the structure we recognize as the nucleolus (see essay).

Other intermediate-repeat sequences are similar to one another, rather than identical, and are said to form **gene families.** The best studied of these are the genes of the globin family, which code for the polypeptide chains of fetal and adult hemoglobins. Figure 16–11 summarizes the evolutionary steps that are thought to have led to this gene family. The ancestral gene is thought to have coded for a polypeptide chain resembling that of myoglobin, a relatively small molecule found in muscle cells. Myoglobin is formed by a single polypeptide chain of 153 amino acids and holds one oxygen-binding heme group. It is thought that the ancestral globin gene was accidentally duplicated several times in the course of evolutionary history. Over time, mutations led to divergence of the duplicates, eventually giving rise to the present family of genes. A corresponding divergence of function occurred in the proteins for which the globin genes code, leading to modern myoglobin and to the forerunners of the alpha and beta chains of hemoglobin.

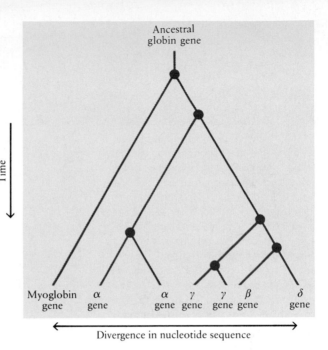

16–11 *A hypothetical evolutionary sequence that could have given rise to the genes of the globin family. The solid dots represent duplications of genes. The divergence of the genes following duplication is the result of mutations. The globin family includes the genes coding for myoglobin and for the alpha (α), beta (β), gamma (γ), and delta (δ) chains of hemoglobin. Fetal hemoglobin molecules contain two alpha chains and two gamma chains, whereas adult hemoglobin contains two alpha and two beta chains. The delta gene, which is found only in primates, codes for another variant of the beta chain.*

In addition to the protein-coding genes, the globin family contains other nucleotide sequences that are never expressed. These sequences, known as **pseudogenes,** are believed to be duplicates that have been disabled by their accumulated mutations.

Single-Copy DNA

The third class of eukaryotic DNA is **single-copy** DNA, in which each sequence is present in only one or a few copies. Depending on the species, it comprises anywhere from 50 to 70 percent of the DNA of the organism. With the exception of the histone genes and the members of gene families, all of the protein-coding genes belong to this class of DNA. However, only a small proportion of single-copy DNA—perhaps as little as 1 percent of the total DNA in the cell—appears to be translated into protein. Some, but by no means all, of the noncoding single-copy DNA is accounted for by long stretches of spacer DNA, which is never transcribed into RNA. The remainder has been found in a most unexpected location.

The Nucleolus

As you know, the most prominent feature in the nucleus of a eukaryotic cell in interphase is the nucleolus. During mitosis and meiosis, however, the nucleolus disappears, only to reappear at the end of telophase. Because of its prominence in the cell and its puzzling behavior, the nucleolus has long been an object of scrutiny by cytologists. Although some details are still missing, its structure and function are now known in broad outline.

Structurally, the nucleolus is not actually a distinct entity. It is, instead, a cluster of loops of chromatin, often from different chromosomes. For example, 10 of the 46 human chromosomes contribute chromatin loops to the nucleolus. The loops that form the basic structure of the nucleolus are the DNA segments that contain copies of the gene coding for three of the four types of rRNA molecules found in eukaryotic ribosomes. During the condensation of the chromosomes at the beginning of mitosis or meiosis, these loops are reeled back into their respective chromosomes, and the nucleolus disappears.

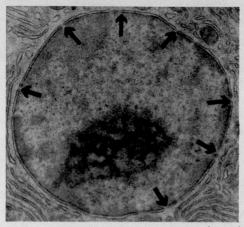

Electron micrograph of the nucleus of a type of pancreatic cell that produces and exports many of the enzymes used in digestion. The dark body in the lower center is the nucleolus, where the RNAs of the ribosomes are transcribed and the ribosomal subunits are assembled. Partially formed ribosomal subunits are around its periphery. Notice also the nuclear envelope, with its many nuclear pores (red arrows). Surrounding the nucleus are membranes of the endoplasmic reticulum, as well as a mitochondrion.

1 μm

Functionally, the nucleolus is a ribosome factory in which rRNA molecules are transcribed from the chromatin loops and ribosomal subunits are assembled. Ribosomal proteins, themselves synthesized on ribosomes in the cytosol of the cell, are transported into the nucleus and assembled into subunits. The nearly completed ribosomal subunits, containing both rRNAs and protein, are then shipped back into the cytosol. There, after a few finishing touches, the ribosomes begin to perform their essential functions in the assembly of amino acids into proteins.

Introns and Exons

One of the great surprises in the study of eukaryotic DNA was the discovery that the protein-coding sequences of genes are usually not continuous but are instead interrupted by noncoding sequences—that is, by nucleotide sequences that are not translated into protein. These noncoding interruptions within a gene are known as intervening sequences, or **introns.** The coding sequences—the sequences that *are* translated into protein—are called **exons.**

Introns were discovered in the course of mRNA-DNA hybridization experiments (page 259). Investigators found that there was not a perfect match between eukaryotic messenger RNA molecules and the genes from which they were transcribed. The nucleotide sequences of the genes were much longer than the complementary mRNA molecules found in the cytosol.

It is now known that most, but not all, structural genes of multicellular eukaryotes contain introns. The introns are transcribed onto RNA molecules, but they are snipped out before translation occurs. The number of introns per gene varies widely. For example, the gene for ovalbumin, a protein found in large quantities in vertebrate egg cells, has seven introns (Figure 16–12). In chickens, the gene for collagen, a very common structural protein, has 50 introns. Introns have also been found in genes coding for transfer RNAs and ribosomal RNAs.

The Functional Significance of Introns

The way in which introns established themselves in the DNA and their functions, if any, are not known. One suggestion is that they promote recombination. Crossing over during meiosis is more likely in genes containing introns than in genes lacking introns, just because of the distances involved.

There are also indications that, in some cases, different exons code for different structural and functional regions of the finished protein (Figure 16–13). For ex-

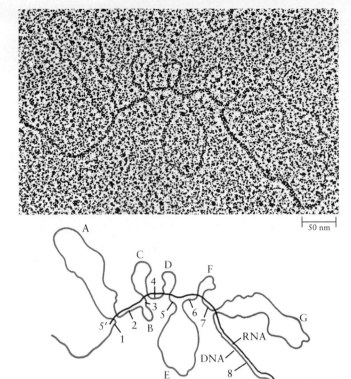

16–12 *This electron micrograph reveals the results of an experiment in which a single strand of DNA containing the gene coding for ovalbumin was hybridized with the messenger RNA for ovalbumin. The complementary sequences of the DNA and mRNA are held together by hydrogen bonds. There are eight such sequences (the exons labeled 1 through 8 in the accompanying diagram). Some segments of the DNA do not have corresponding mRNA segments and so loop out from the hybrid. These are the seven introns, labeled A through G. Only the exons are translated into protein.*

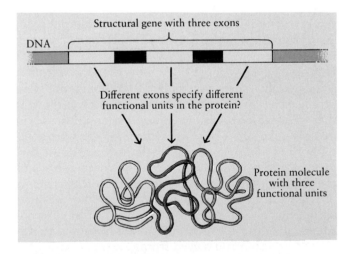

16–13 *An attractive hypothesis: Exons code for discrete functional regions in the protein for which the entire gene codes.*

ample, all of the genes of the globin family consist of three exons and two introns, one large and one small, always in the same position in the gene. The central exon of each gene codes for the portion of the polypeptide that holds the heme group, and the other two exons code for regions of the molecule that fold around this central portion (see Figure 3–21, page 64). It is hypothesized that new combinations of such regions, brought about by the reshuffling of exons, might foster the rapid evolution of new proteins.

Transcription and Processing of mRNA in Eukaryotes

Transcription in eukaryotes is the same, in principle, as in prokaryotes. It begins with the attachment of an RNA polymerase to a particular nucleotide sequence, the promoter, on the DNA molecule. The enzyme then moves along the molecule, using the 3′-to-5′ strand as a template for the synthesis of RNA molecules, as shown in Figure 15–5 (page 258). The transcribed RNA molecules (rRNAs, tRNAs, and mRNA) then play their various roles in the translation of the encoded genetic information into protein.

Despite these basic similarities, there are differences between prokaryotes and eukaryotes in transcription, translation, and the events that occur between these two processes. One important difference is that eukaryotic genes are not grouped in operons in which two or more structural genes are transcribed onto a single RNA molecule. In eukaryotes, each structural gene is transcribed separately, and its transcription is under separate controls.

Another important difference is that in eukaryotes, unlike prokaryotes, transcription and translation are separated in both time and space. After transcription is completed in the nucleus, the mRNA transcripts are extensively modified before they are transported to the cytoplasm, the site of translation.

Even before transcription is completed, while the newly forming mRNA strand is only about 20 nucleotides long, a "cap" of an unusual nucleotide is added to its leading (5′) end. This cap is necessary for the binding of the mRNA to the eukaryotic ribosome. After transcription has been completed and the molecule released from the DNA template, special enzymes add a string of adenine nucleotides to the trailing (3′) end of the molecule. This added segment, known as the poly-A tail, may contain as many as 200 nucleotides.

Before the modified mRNA molecules leave the nu-

RNA and the Origin of Life

The discovery of the role of DNA in heredity launched a debate among biologists as to whether life had its molecular beginnings in protein or in DNA. DNA was a likely choice because it is the repository of the genetic information and provides the template for its own precise replication. Proteins have neither of these qualifications. Proponents of proteins, however, have noted that virtually all of the chemical reactions of the cell depend on proteins that function as catalysts—that is, on enzymes. Even the replication of DNA depends on protein catalysts.

An important clue toward the resolution of this "chicken-and-egg" dilemma has come from an unexpected source. Using the single-celled protist *Tetrahymena,* T. C. Cech and his co-workers at the University of Colorado were studying the removal of introns and the splicing together of exons. This process must be carried out with exquisite precision, since a mistake of one nucleotide could render the entire molecule nonfunctional. In order to isolate the catalysts required for the reaction, Cech and his co-workers set up two experimental systems. One contained a *Tetrahymena* RNA molecule from which an intron was to be re-

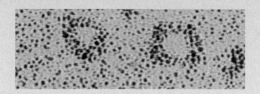

The two ringlike structures visible in this electron micrograph are introns removed from a transcribed RNA molecule of the protist Tetrahymena. *These introns have the capacity to catalyze their own removal and to splice together the exons.*

moved, plus proteins that were potential catalysts. The other system, the control, contained the RNA but no proteins. The intron was neatly removed in the first system, as expected, but, to everyone's surprise, the removal and splicing process also took place in the control.

It was subsequently shown that the intron—a segment of RNA containing some 400 nucleotides—has an enzyme-like catalytic activity that carries out its own removal and the splicing of the exons. This RNA segment folds up to form a complex surface that functions like an enzyme. Although RNA catalysts are not common, a number of them have now been found in other types of reactions and in exon-splicing in other types of cells. Such catalysts, dubbed "ribozymes," have become the subject of intense research.

The discovery that RNA can act as a catalyst makes it easier to imagine how life had its beginnings. According to cell biologist Bruce M. Alberts, "One suspects that a crucial early event was the evolution of an RNA molecule that could catalyze its own replication." These molecules then diversified into a collection of catalysts that could, for example, assemble nucleotides in RNA synthesis or accumulate lipid-like molecules to form the first primitive cell membranes. Gradually, other RNAs evolved and assembled the first proteins. Then, because the proteins were better catalysts, they gradually took over the catalytic functions. In the third step, DNA appeared on the scene, and its more stable double-stranded structure became the ultimate repository of the genetic information. Thus, researchers speculate, the catalytic intron can be regarded as a living fossil, a provocative clue to the events of almost 4 billion years ago.

cleus, the introns are removed, and the exons are spliced together to form a single, continuous molecule (Figure 16–14). The splicing mechanism is very precise, preventing the slight errors that could cause a frame-shift in the transcribed message (see page 266).

A number of instances have now been found in which identical mRNA transcripts are processed in more than one way. Such alternative splicing can result in the formation of different functional polypeptides from RNA molecules that were originally identical (Figure 16–15). In such cases, an intron may become an exon, or vice versa. Thus, the more that is learned about eukaryotic DNA and its expression, the more difficult it becomes to define "gene" or "intron" or "exon."

As we noted earlier, molecular biologists once thought that the eukaryotic chromosome might be simply a large-scale version of the prokaryotic chromosome. That turned out not to be the case. As we have seen, the structure and organization of the chromosome, the regulation of gene expression, and the processing of mRNA molecules are all much more complex in eukaryotes than in prokaryotes. For many years it also seemed reasonable to believe that the chromosomes of eukaryotes were stable. Recombinations were produced by crossing over, of course, but it was thought that—except for occasional mutations and mistakes in meiosis—the structure of each chromosome was essentially fixed and unchanging. Perhaps the greatest surprise of all has been the discovery that this too is not the case. In both prokaryotes and eukaryotes, segments of DNA can move from one place to another within a chromosome, they can move into and out of chromosomes, they can move from one chromosome to another—and sometimes even from one organism to another. In the next chapter, we shall explore some of the ways in which these movements occur and their significance for the organisms involved, including ourselves.

16–14 *A summary of the stages in the processing of an mRNA transcribed from a structural gene of a eukaryote. The genetic information encoded in the DNA (a) is transcribed into an RNA copy. This copy is then edited, with the addition of a cap at the 5′ end and a poly-A tail at the 3′ end (b). Next, the introns are snipped out, and the exons are spliced together. The mature mRNA (c) then goes to the cytoplasm, where it is translated into protein.*

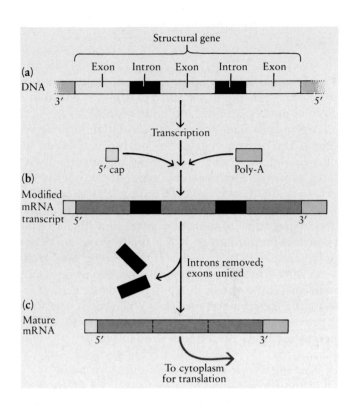

16–15 *Alternative splicing of identical mRNA transcripts results in the synthesis of different polypeptides from the information encoded by a single gene. For example, when the mRNA transcript shown here is processed (a) in the thyroid gland, segments E and F are removed in the initial processing and the poly-A tail is added to the end of exon D. The three introns are then removed, and the mature mRNA molecule is translated into the peptide hormone calcitonin. (b) In the pituitary gland, however, the poly-A tail is added to the end of exon F. Five introns are removed from this transcript, including segment D, which in the thyroid was retained as an exon. The mature mRNA, formed from five exons, is translated into a different hormone, known as calcitonin-gene-related protein (CGRP).*

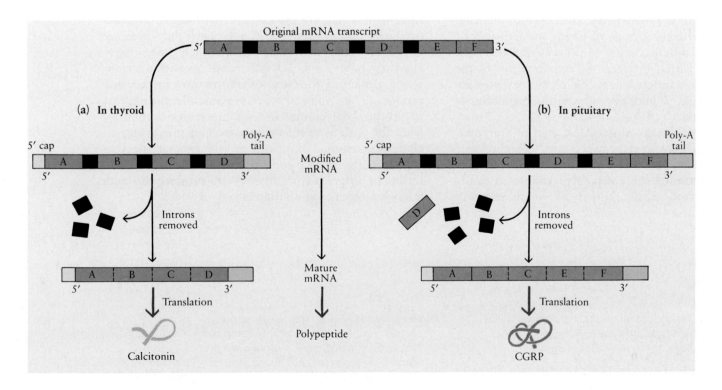

Summary

The essential genetic information of prokaryotes, of which *E. coli* is the best-studied example, is coded in a circular double-stranded molecule of DNA.

A principal means of genetic regulation in prokaryotes is the operon system. An operon is a linear sequence of genes coding for a group of functionally related proteins. The structural genes of the operon are transcribed as a single mRNA molecule. Transcription from the operon is controlled by DNA sequences that are adjacent to the structural genes and bind specific proteins. The promoter is the binding site for RNA polymerase, and the operator is the binding site for the repressor. When the repressor is attached to the DNA molecule at the operator, RNA polymerase cannot initiate the transcription of RNA. When the repressor is inactivated, RNA polymerase can attach to the DNA, permitting transcription and protein synthesis to take place.

The eukaryotic chromosome differs in many ways from the chromosome of prokaryotes. Its DNA is always associated with proteins, which constitute more than half of the chromosome. Most of these proteins are histones, which are relatively small molecules. The DNA molecule wraps around histone cores to form nucleosomes, which are the basic packaging units of eukaryotic DNA.

Regulation of gene expression is more complex in eukaryotes than in prokaryotes. During embryonic development, different groups of genes are activated or inactivated in different types of cells. Gene expression is correlated with the degree of condensation of the chromosome. A variety of specific regulatory proteins are also thought to play key roles in the regulation of gene expression.

Eukaryotes have far more DNA than prokaryotes, much of which appears to be excess and "meaningless." There are three classes of eukaryotic DNA. Multiple repeats of short nucleotide sequences, characteristically arranged in tandem, are known as simple-sequence DNA. Simple-sequence DNA is associated with the tightly condensed chromatin in the region of the centromere and with the tips of the chromosomes.

Longer repeats, usually dispersed throughout the chromosomes, are known as intermediate-repeat DNA. Intermediate-repeat DNA includes multiple copies of the genes coding for the rRNAs and histones, as well as members of gene families. The individual genes of a family differ slightly in their nucleotide sequences. As a consequence, the proteins for which they code differ slightly in structure and biological properties. Some members of gene families are not expressed; these DNA sequences are known as pseudogenes.

The third class of eukaryotic DNA, known as single-copy DNA, makes up from 50 to 70 percent of the chromosomal DNA. Current data indicate that as little as 1 percent of human DNA may be translated into protein.

In eukaryotes, the coding sequence of most structural genes is not continuous but is interrupted by sequences known as introns. Although introns are transcribed into RNA in the nucleus, they are not present in the mRNA in the cytoplasm and thus are not translated into protein. The segments that are present in the cytoplasmic mRNA and are translated into protein are known as exons.

Transcription in eukaryotes differs from that in prokaryotes in a number of respects. In eukaryotes, a multiplicity of regulatory proteins are involved. Also, structural genes are not grouped in operons as they often are in prokaryotes. The transcription of each gene is regulated separately, and each gene produces an RNA transcript containing the encoded information for a single product. RNA transcripts are processed in the nucleus to produce the mature mRNA molecules that move from the nucleus to the cytoplasm. This processing includes the removal of introns and the splicing together of exons. Alternative splicing of identical RNA transcripts in different types of cells can produce different mRNA molecules and different polypeptides.

Questions

1. Distinguish among the following: operon/promoter/operator; repressor/regulator; nucleosome/nucleolus; euchromatin/heterochromatin; intron/exon.

2. Nutrient molecules can be used by cells in two different ways. One type is broken down (usually as an energy source). Another is used as a building block for a larger molecule. Which type of molecule would you expect to function as a corepressor? As an inducer? Do the examples given in the text conform to your expectations?

3. In what ways are the chromosomes of prokaryotes and eukaryotes similar? In what ways are they different?

4. Describe the three classes of eukaryotic DNA. What are some of the functions that have been identified for each class?

5. **(a)** What might be the advantage to an organism of having multiple copies of the genes for the rRNAs and the histones? **(b)** Would you expect an organism also to have multiple copies of the genes for the tRNAs? Explain your answer.

6. Gene expression theoretically can be regulated at the level of transcription, translation, or activation of the protein. In the latter case, the polypeptide produced by protein synthesis is in an inactive form. It undergoes some structural modification (controlled by enzymes) before it can perform its function in the cell. What would be the advantages of each type of regulation, in terms of the cell? Under what circumstances might one type be more useful than another? Which is the more economical?

7. The genes for coat color in cats are carried on the X chromosome. Black (*b*) is the recessive and yellow (*B*) is the dominant. What coat colors would you expect in the offspring of a cross between a black female and a yellow male? What coat colors would you expect in the sons of a tortoiseshell female? (You may find it helpful to diagram these crosses with Punnett squares.)

Suggestions for Further Reading

Alberts, Bruce, Dennis Bray, Julian Lewis, Martin Raff, Keith Roberts, and James D. Watson: *Molecular Biology of the Cell,* 3d ed., Garland Publishing, Inc., New York, 1994.

> *This outstanding cell biology text provides an excellent discussion of chromosome structure and the regulation of gene expression.*

Cech, Thomas R.: "RNA as an Enzyme," *Scientific American,* November 1986, pages 64–75.

Chambon, Pierre: "Split Genes," *Scientific American,* May 1981, pages 60–71.

Darnell, James E., Jr.: "The Processing of RNA," *Scientific American,* October 1983, pages 90–100.

Darnell, James, Harvey Lodish, and David Baltimore: *Molecular Cell Biology,* 2d ed., W. H. Freeman and Company, New York, 1990.

> *Chapters 7–11 provide a thorough review of gene structure and function. Chapters 5 and 6 discuss the techniques used in molecular biology.*

Eigen, Manfred, William Gardiner, Peter Schuster, and Ruthild Winkler-Oswatitsch: "The Origin of Genetic Information," *Scientific American,* April 1981, pages 88–118.

Felsenfeld, Gary: "DNA," *Scientific American,* October 1985, pages 58–67.

Grunstein, Michael: "Histones as Regulators of Genes," *Scientific American,* October 1992, pages 68–74B.

Holliday, Robin: "A Different Kind of Inheritance," *Scientific American,* June 1989, pages 60–73.

Jacob, François: *The Statue Within: An Autobiography,* Basic Books, Inc., New York, 1988.

> *In this gracefully written memoir, François Jacob describes his privileged boyhood in Paris, his turbulent years of military service during World War II, and his postwar return to the laboratory, where, along with André Lwoff and Jacques Monod, he undertook his Nobel Prize–winning work on the operon. The book is rich with the insights of a brilliant and eloquent scientist-poet.*

Kornberg, Roger D., and Aaron Klug: "The Nucleosome," *Scientific American,* February 1981, pages 52–64.

McKnight, Steven Lanier: "Molecular Zippers in Gene Regulation," *Scientific American,* April 1991, pages 54–64.

Moyzis, Robert K.: "The Human Telomere," *Scientific American,* August 1991, pages 48–55.

Ptashne, Mark: "How Gene Activators Work," *Scientific American,* January 1989, pages 40–47.

Rhodes, Daniela, and Aaron Klug: "Zinc Fingers," *Scientific American,* February 1993, pages 56–65.

Steitz, Joan A.: "Snurps," *Scientific American,* June 1988, pages 56–63.

Watson, James D., Nancy H. Hopkins, Jeffrey W. Roberts, Joan A. Steitz, and Alan M. Weiner: *Molecular Biology of the Gene,* 4th ed., The Benjamin/Cummings Publishing Company, Redwood City, Calif., 1987.

> *Molecular Biology of the Gene is a classic. Now in its Fourth Edition, it continues to serve as a celebration of the extraordinary achievements of biology during the second half of the twentieth century. Well written and richly illustrated, the book covers an incredible array of topics, ranging from molecular biology of prokaryotes to that of eukaryotes and multicellular organisms.*

Genes on the Move

Barbara McClintock, shown here at Cold Spring Harbor Laboratory on Long Island, New York, spent more than half a century analyzing color differences and other variations in corn kernels (each of which is actually an embryonic corn plant). On the basis of mapping studies, analogous to those carried out with *Drosophila* in Morgan's laboratory, she deduced that the changes she observed were not the result of ordinary mutations. Instead, she postulated, the changes came about as a result of the movement of genes, or "controlling elements," that jumped from one place to another on a chromosome—or even from one chromosome to another.

McClintock's findings, first published in 1951, were largely ignored. Like Mendel before her, she was a lonely traveler with an odd tale that did not fit into the scheme of things as then understood. James Watson, who has also worked at Cold Spring Harbor (first as a nineteen-year-old graduate student, then as its director from 1968 to 1993, and now as its president), once described McClintock as "fiercely independent, beholden to no one." Despite the lack of recognition, she stubbornly pursued her research, sometimes working without pay. "It was fun," she said, "I could hardly wait to get up in the morning."

McClintock was more than vindicated by a rush of discoveries, some 30 years later, of a host of movable genetic elements such as those she had first postulated. "Jumping genes," or transposons, as they came to be known, move from place to place on the chromosome and influence the expression of other genes. When McClintock was 81 years old, she received the Nobel Prize.

Her life did not change much, however. She continued her work at Cold Spring Harbor until shortly before her death in 1992 at the age of 90. "There are really three main figures in the history of genetics," Watson noted at a tribute to her the year before. "The three M's: Mendel, Morgan, and Mc-Clintock. Gregor Mendel and Thomas Hunt Morgan showed us how regular the genome is, and Barbara McClintock showed us how irregular it is."

In the mid-1970s, when it seemed as if the mysteries of heredity had been so satisfactorily explained that all that was left was the tidying up of a few details, genetics underwent another revolution. Quite suddenly, biologists became able to manipulate genes in ways never before imagined—copying, dissecting, analyzing, modifying, and recombining portions of DNA molecules from different sources. This new technology has generated an avalanche of studies that are providing new knowledge and new possibilities—particularly in medicine and agriculture—at a stunning pace.

Like the early advances in molecular genetics, the new technology resulted from work with bacteria and viruses—and specifically from the manipulation of the natural mechanisms by which genes move both within and between chromosomes. In the next chapter, we shall look at this technology and some of its most promising and intriguing applications. Before we do so, however, we must look first at the natural phenomena on which it is based. As we shall see in this chapter, the capacity of genes to "jump" has an importance for the organisms in which it occurs—and sometimes for other organisms as well—that is quite independent of the applications made possible by its deliberate manipulation.

Although the ability of genes to move from one location to another in the eukaryotic chromosome was first detected by Barbara McClintock in her studies of corn, other scientists were not convinced until such movement was demonstrated at the molecular level in far simpler organisms. We shall begin our story where they began, with some "extra" DNA molecules found in *E. coli* and other bacteria.

Plasmids and Conjugation

Although the bacterial chromosome (page 271) contains all of the genes necessary for growth and reproduction of the cell, virtually all types of bacteria have

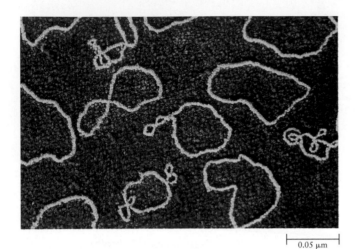

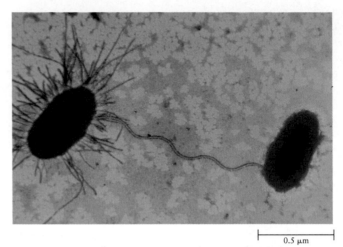

17–1   *Plasmids from an* E. coli *cell. The two plasmids that are connected, forming a figure eight, are probably about to complete replication.*

17–2   *Conjugating* E. coli *cells. The F⁺ (donor) cell at the left is connected to the F⁻ (recipient) cell at the right by a long pilus. Genes on the F plasmid are responsible for the production of these specialized pili, which are necessary for conjugation. After contact has been made, the long pilus retracts, pulling the two cells close enough to one another that a cytoplasmic bridge can form between them. Numerous shorter pili are also visible on the F⁺ cell.*

been found to carry additional DNA molecules known as **plasmids.** Plasmids, which are much smaller than the bacterial chromosome, may carry as few as two genes or as many as thirty.

Like the bacterial chromosome, plasmids are circular and self-replicating (Figure 17–1). Some plasmids replicate at the same time as the chromosome, and each daughter cell has only one copy of the plasmid. Other plasmids replicate more frequently than the chromosome, with the result that the cell may contain multiple copies. In the case of some small plasmids, as many as 50 copies have been detected in a single cell. Alternatively, if the plasmid replicates less frequently than the chromosome, some daughter cells may not receive any copies of the plasmid.

About a dozen different kinds of plasmids have been described in *E. coli* alone. Two of the most important are fertility, or F, plasmids, and drug resistance, or R, plasmids.

The F Plasmid

The first plasmid to be identified was the **F factor** of *E. coli*. This plasmid contains some 25 genes, many of which control the production of long, rod-shaped protein structures, known as **pili** (singular, pilus), that extend from the surface of cells containing the F plasmid. Such cells are known as F⁺ cells. Cells that lack the F

plasmid—and thus the pili—are known as F⁻ cells. F⁺ cells can attach themselves to F⁻ cells by the pili (Figure 17–2) and transfer a copy of the F plasmid to them through cytoplasmic bridges. Such transfer of DNA from one cell to another by cell-to-cell contact is known as **conjugation.**

Transfer of the F factor in conjugation gives the recipient cells the capacity to produce pili and to transfer the F plasmid—that is, the recipient cells become F⁺, or donor, cells. The replication of an F plasmid and its transfer from a donor cell to a recipient cell are diagrammed in Figure 17–3.

Like many other plasmids, the F factor can exist independently in the cell or it can become spliced into the bacterial chromosome. When the F factor becomes integrated into the chromosome, it then becomes possible for a portion of the chromosome itself to be transferred from one cell to another. During conjugation, the chromosome opens up within the F factor sequence and begins to replicate. As it replicates, the free end of a single strand of DNA moves into the F⁻ cell. Recombination may then occur between the chromosome of the recipient cell and those portions of the chromosome of the donor cell that have been transferred, with new material replacing old in the recipient's chromosome. Studies of this process using mutant strains of bacteria have enabled scientists to prepare detailed genetic maps of the *E. coli* chromosome.

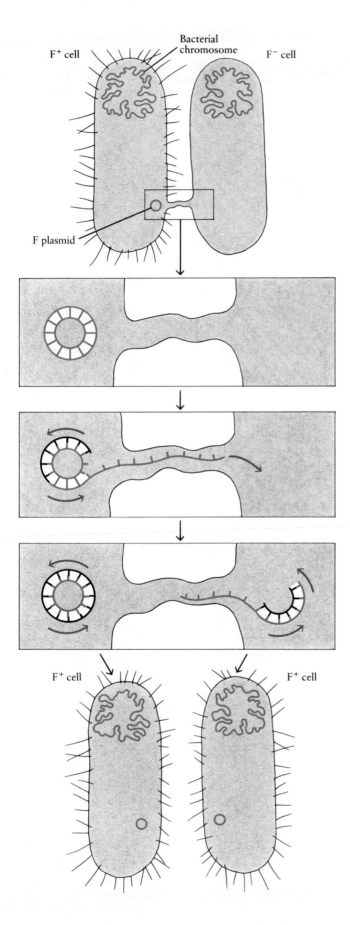

Bacterial chromosome

F⁺ cell

F⁻ cell

F plasmid

F⁺ cell

F⁺ cell

17–3 *Transfer of an F plasmid from an F⁺ cell to an F⁻ cell during conjugation. Both the bacterial chromosome and the plasmid are circular molecules of double-stranded DNA. The plasmid, however, is much smaller than the chromosome and contains far fewer base pairs.*

A single strand of DNA moves from the donor cell, through the cytoplasmic bridge, and into the recipient cell. Its complementary strand (black) is then synthesized from nucleotides available in the recipient cell. As this DNA strand is transferred, the remaining strand in the donor cell "rolls" counterclockwise, exposing its unpaired nucleotides. These serve as a template for the synthesis of a complementary DNA strand (black). As a result of this "rolling-circle" replication, the plasmid in the donor cell continues to be a circle of double-stranded DNA. The transferred plasmid converts the recipient cell to an F⁺ cell.

R Plasmids

In 1959, a group of Japanese scientists discovered that resistance to certain antibiotics and other antibacterial drugs can be readily transferred from one bacterial cell to another. Under experimental conditions, 100 percent of a population of drug-sensitive cells can become resistant within an hour after being mixed with suitable drug-resistant bacteria. It was subsequently found that the genes conveying drug resistance are often carried on plasmids that have come to be known as **R plasmids.**

Drug resistance in bacterial cells is often the result of the synthesis of enzymes that break down the drug or that set up a new enzymatic pathway, circumventing the effects of the drug. Thus, resistance may depend on the synthesis of specific enzymes in high concentrations. The fact that resistance genes are on plasmids allows many copies of those genes to be produced very rapidly within a single cell.

Resistance genes can be transferred not only from one cell to another but also from one R plasmid to another. A single plasmid may collect as many as 10 different resistance genes, making the cell it inhabits (and any cell to which it is transferred) resistant to as many as 10 different antibiotics (Figure 17–4). Typically, only a few copies of these large plasmids exist in a single

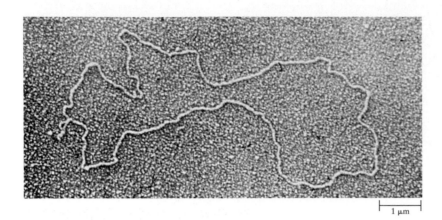

17–4   *This plasmid, known as R6, carries genes that confer resistance to six different drugs, including the antibiotics tetracycline, neomycin, and streptomycin.*

1 μm

cell. They are passed from mother to daughter cells at cell division, are transferred by conjugation, or, in another example of transformation (page 243), they may be simply passed from cell to cell through the plasma membranes.

Resistance genes can also be transferred from plasmids to the bacterial chromosome, to viruses, and to bacteria of other species, including species that cause serious human illness. Thus, for example, the usually innocuous *E. coli* can pick up R plasmids by conjugation and transfer them to *Shigella*, a bacterium capable of causing a sometimes fatal form of dysentery. Infectious drug resistance has been found among an increasing number of types of bacteria, including those responsible for typhoid fever, cholera, pneumonia, meningitis, gonorrhea, and, most disturbing of all, tuberculosis.

The incidence of tuberculosis, once thought to be essentially eradicated in the developed countries, has increased dramatically in the United States since 1985. In 1991, there were 26,000 new cases reported and, in 1992, almost 27,000. In New York City, one-third of the cases were caused by bacteria resistant to one or more of the antibiotics used to treat the disease, and drug-resistant strains of the bacteria were also identified in 35 other states. In otherwise healthy individuals, drug-resistant tuberculosis has a fatality rate of about 50 percent. In individuals with immune systems that are suppressed (often a side effect of cancer chemotherapy) or weakened (for example, by the virus that causes AIDS), the fatality rate is much higher.

For bacterial cells, the ability to transfer drug-resistance genes from one cell to another is clearly beneficial. However, as the resurgence of tuberculosis illustrates, this ability also threatens to negate the great advances that have been made in this century in the control of infectious disease.

Viruses

Some biologists consider the infectious agents known as viruses to be living organisms. Other biologists, however, take a different view. Salvador Luria, one of the pioneers in the study of bacteriophages, once described viruses as "bits of heredity looking for a chromosome." As we shall see, their quest is often successful.

The Structure of Viruses

A virus consists essentially of one or more molecules of nucleic acid enclosed in a protein coat, or **capsid.** As shown in Figure 17–5, the capsid may consist of a single protein molecule repeated over and over, or it may be formed from a number of different kinds of proteins.

The composition of the protein coat determines the specificity of the virus. A virus can infect a cell only if that type of cell has receptor sites on its surface to which the viral proteins can bind. Thus bacteriophages attack bacterial cells, tobacco mosaic virus (page 258) attacks leaf cells of the tobacco plant, and the viruses responsible for the common cold invade cells in the lining of the human respiratory tract.

Viruses contain no cytoplasm or metabolic machinery and can multiply only within a living cell. In some virus infections, the protein coat is left outside the cell while the nucleic acid enters (see page 245). In others, the intact virus enters the cell, but once inside, its protein coat is destroyed by enzymes, freeing the viral nucleic acid. Within the host cell, this nucleic acid directs the production of new viruses. This is accomplished using not only the raw materials of the cell—nucleotides and amino acids—but also the cell's enzymes, tRNA molecules, ribosomes, and ATP molecules and other energy sources.

Depending on the type of virus, its nucleic acid—that

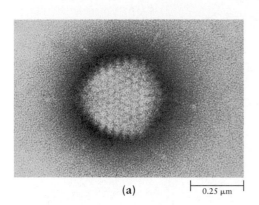

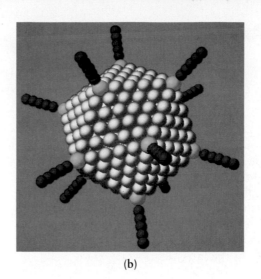

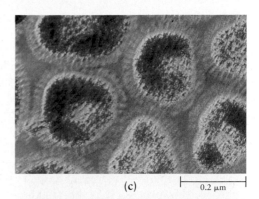

(a)   0.25 μm

(b)

(c)   0.2 μm

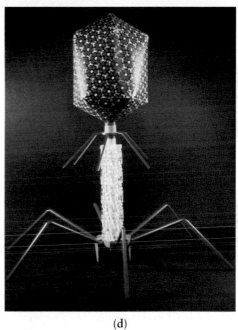

(d)

17–5   *Representative virus structures.*
*(a) An electron micrograph of the capsid of
an adenovirus, one of the many viruses that
cause colds in humans. Within the capsid is a
core of double-stranded DNA. (b) A computer-generated model of the capsid of an adenovirus. The capsid has 20 sides, each of
which is an equilateral triangle composed of
identical protein subunits. There are 252
subunits in all. Rodlike protein structures
project from the 12 subunits located at the
points where the vertices of the triangles
meet.*

  *(c) Electron micrograph of type A influ-
enza viruses. The inner layer of each capsid is
formed from helical proteins, and the outer
layer is a lipoprotein membrane through
which stubby protein spikes protrude. Within
the capsid are eight separate and dis-*

*tinct RNA molecules. For reasons that are
not understood, influenza viruses mutate frequently. The resulting changes in nucleotide
sequence alter the proteins of the outer envelope. As a consequence, previously formed
antibodies circulating in the bloodstream no
longer "recognize" the virus. New strains of
influenza are likely to arise more rapidly
than new vaccines can be produced to combat them.*

  *(d) A model of the capsid of the type of
bacteriophage used in the Hershey-Chase experiments (see Figures 14–4 and 14–5, on
pages 244 and 245). The double-stranded
DNA of the virus, located within the head
of the capsid, codes for all of the proteins
required to form its different structural
components.*

is, its chromosome—may be either DNA or RNA, single-stranded or double-stranded, circular or linear.
When DNA viruses infect a cell, two processes occur.
The viral DNA is replicated, forming more viral DNA,
and it is transcribed into mRNA that directs the synthesis of viral proteins. With most RNA viruses, the
situation is similar. The RNA replicates, forming more
viral RNA, and also serves directly as mRNA.

    The viral chromosome always codes for the coat pro-
teins and for one or more enzymes involved in replication of the viral chromosome. In at least some cases,
it codes for repressors and other regulatory molecules
as well. In most viruses, the chromosome also codes for
enzymes that, once the new virus particles are assembled, enable them to **lyse** (break apart) the host cell and
escape. Some viruses, however, insert themselves into
the plasma membrane of the host cell and are budded
off in a process similar to exocytosis (page 119).

The infection cycle of a virus is complete when the
newly synthesized viral nucleic acid molecules are packaged into the newly synthesized protein coats and the
virus particles leave the host cell.

Viruses as Vectors

The genetic makeup of bacterial cells can be altered, as
we have seen, by the introduction of DNA from other
bacterial cells by means of plasmids. Viruses can also
play a role as **vectors,** or carriers, that move pieces of
DNA from one cell to another.

    Early in the study of bacteriophages, it was noted
that a virus infection could suddenly erupt in a colony
of apparently uninfected bacterial cells. The cause of
this phenomenon, it was discovered, was the capacity
of certain viruses to set up a long-term relationship with
their host cell, remaining latent through many cell di-

visions before initiating a **lytic cycle** in which the cell is destroyed. Such viruses became known as **temperate bacteriophages.**

The DNA of temperate phages, like that of the F plasmid, may become integrated at specific sites in the host chromosome, replicating along with the chromosome. Such integrated phages are known as **prophages,** and bacteria that harbor them are known as **lysogenic bacteria.** Prophages break loose from the host chromosome spontaneously about once in every 10,000 cell divisions, initiating a lytic cycle (Figure 17–6). In the laboratory, this process may be triggered by ultraviolet light, x-rays, or other agents that damage nucleic acids.

Temperate phages resemble plasmids in that (1) they are autonomously replicating molecules of DNA, and (2) they may become integrated into the bacterial cell chromosome. Unlike plasmids, however, temperate phages can manufacture a protein coat and thus exist (although they cannot multiply) outside a host cell.

Transduction

The transfer of DNA from one host cell to another by means of viruses is known as **transduction.** During the lytic cycle of many viruses, the host DNA becomes fragmented. When the new virus particles are assembled, some of these DNA fragments, picked up at random, may be enclosed by the coat proteins. Since the amount of DNA that can be packaged within the protein coat is limited, such viruses lack some or all of their own necessary genetic information. Although they may be able to infect a new host cell, they are usually not able to complete a lytic cycle. However, the genes they carry from their previous host may become incorporated into the chromosome of the new host. This process is called **general transduction** because virtually any gene can be transferred by this mechanism.

When prophages break loose from a host chromosome to initiate a lytic cycle, they may, similarly, take a fragment of the host chromosome with them. In this situation, however, the host DNA is *not* picked up at random. It is, instead, quite specifically restricted to the portions of the host chromosome next to the insertion site of the prophage. Hence, this process is known as **specialized transduction.**

The portions of the bacterial chromosome carried by a virus are replicated along with the viral chromosome through subsequent lytic cycles. If the DNA of one of these viruses should become part of the chromosome of a new host, the genes from the previous host may be inserted into the new host's chromosome and become part of its genetic equipment. For example, the temperate bacteriophage known as lambda (Figure 17–7) can carry the bacterial genes that code for the enzymes con-

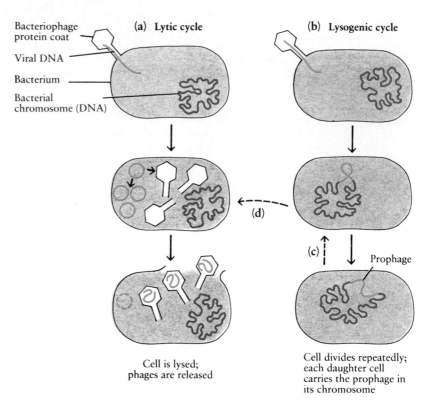

17–6   *When certain types of bacteriophages infect their host cells, one of two processes may occur.* (**a**) *The viral DNA may enter the cell and set up a lytic cycle, as described in Chapter 14. The cell is lysed as the newly formed viral particles escape from it.* (**b**) *Alternatively, the viral DNA may become part of the bacterial chromosome, replicating with it and being passed on to the daughter cells through many cell divisions. A bacterial virus integrated into the chromosome is known as a prophage, and bacteria harboring such viruses are said to be lysogenic. From time to time, however, a prophage becomes activated* (**c**) *and is released from the bacterial chromosome, setting up a new lytic cycle* (**d**). *In these diagrams, the double-stranded viral and bacterial chromosomes are depicted by single lines.*

cerned with the breakdown of galactose—in other words, the galactose operon. When a lambda phage carrying the galactose operon becomes a prophage in a mutant bacterial cell that cannot synthesize one or more of these enzymes, the infected cell gains the capacity to utilize galactose as a nutrient.

Transduction resembles conjugation in that it involves the transfer of bacterial genes from one bacterial cell to another. It differs from conjugation in that in transduction the genes are carried by viruses rather than by plasmids.

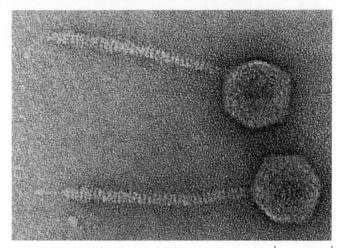

17–7   *Lambda, a temperate bacteriophage that infects* E. coli *cells. Its chromosome is a double-stranded DNA molecule that contains about 45,000 base pairs. When the chromosome is within the capsid head, it is linear (with two free ends). However, when it is released into the cytoplasm of an* E. coli *cell, it immediately forms a circle (no free ends).*

Like a plasmid, the DNA of a lambda phage may become part of a host-cell chromosome. Insertion of the lambda DNA into the host-cell chromosome is brought about by an enzyme that recognizes specific nucleotide sequences on the two DNA molecules. This enzyme, which is coded by the phage DNA, brings the two molecules together and initiates the cutting and sealing reactions.

0.05 μm

Proviruses and Retroviruses

Some viruses of eukaryotes can, like the temperate bacteriophages, become integrated into the chromosomal DNA of the host cell. When integrated, these viruses are known as **proviruses.** Such viruses are of two general types: DNA viruses (analogous to the temperate bacteriophages) and RNA retroviruses.

A number of DNA viruses are known that can, depending on the type of cell they infect, either initiate a lytic cycle or insert themselves into the chromosomal DNA of the host cell. An example is simian virus 40, or SV40 (Figure 17–8). SV40 is a virus of monkeys that was first discovered in cells, growing in tissue culture, that were being used for the development of polio vaccines. It was subsequently found to cause cancers in newborn hamsters, though not in the monkeys that are its normal hosts. The cancers are caused by specific growth-promoting proteins produced in the cells by the viral genes. In short, SV40 can introduce new, functional genes into the DNA of the host cell, as can a number of other DNA viruses.

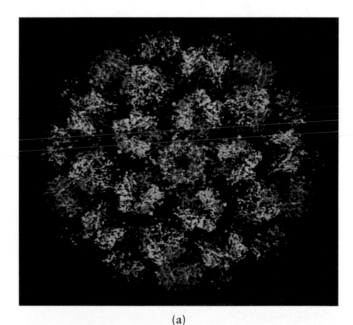

(a)

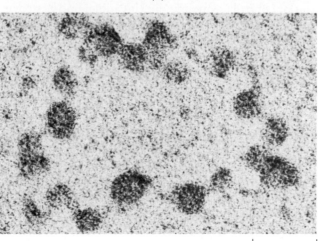

17–8   (a) *A computer-generated model of the 20-sided capsid of simian virus 40 (SV40). The capsid encloses the SV40 chromosome* (b), *a circular molecule of double-stranded DNA that, like the DNA of the eukaryotic chromosome, is complexed with histone proteins to form nucleosomes. The histones are those of the host cell. Thus, SV40 not only uses host-cell machinery for its own synthetic activities but also directly incorporates products of the host cell's synthetic activities.*

(b)   0.25 μm

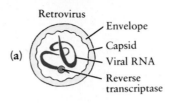

The second group of eukaryotic viruses that can become integrated into host-cell chromosomes are the **RNA retroviruses.** Human immunodeficiency virus (HIV), the deadly virus that causes AIDS, is an example of an RNA retrovirus. In Chapter 33, we shall look more closely at this virus and its effects on the human immune system.

The integration of an RNA virus into a DNA chromosome poses some special problems that are solved by a remarkable enzyme known as **reverse transcriptase.** Molecules of reverse transcriptase are carried within the capsid of an RNA retrovirus, along with its RNA. In the host cell, the viral RNA is copied by reverse transcriptase to produce, after a complex series of events, a double-stranded DNA molecule (Figure 17–9).

Once integrated into a host chromosome, the DNA derived from the viral RNA utilizes the RNA polymerases and other resources of the host cell to synthesize new viral RNA and protein molecules, which are then packaged into new virus particles. Depending on the site of insertion into the host chromosome, DNA derived from a retrovirus may cause mutations by interfering with the expression of host-cell genes, either repressing them or releasing them from repression. Characteristically, however, most retroviral insertions do not damage or destroy their host cells but become permanent additions to the host-cell DNA. If cells destined to become eggs or sperm are infected with such a retrovirus, its genetic information will be transmitted to the next generation.

Genes, Viruses, and Cancer

Cancer is a disease in which cells escape the influence of the factors, still largely unknown, that regulate normal cell growth. As a consequence, the cells multiply out of control, crowding out, invading, and destroying other tissues. Cancer is often considered a group of diseases rather than a single disease because, with few exceptions, any one of the 200 or more cell types in the human body can become malignant.

Three lines of evidence have long linked the development of cancer with mutations. First, once a cell has become cancerous, all of its daughter cells are cancerous. In other words, cancer is an inherited property of cells. Second, gross chromosomal abnormalities, such as deletions and translocations, are often visible in cancer cells. Third, most carcinogens—agents known to cause cancer, such as x-rays, ultraviolet radiation, tobacco smoke, and a variety of chemicals—are also mutagens.

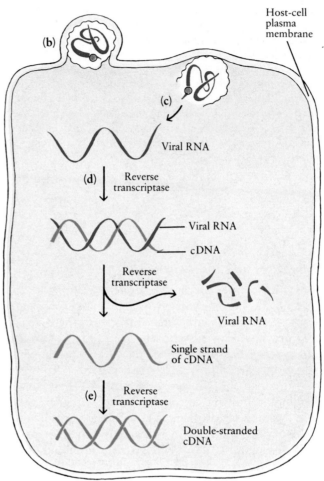

17–9   *Infection of an animal cell by a retrovirus* (**a**). *Contained within the capsid of a retrovirus are the viral RNA and one or more molecules of an enzyme, reverse transcriptase. The capsid is typically surrounded by an outer lipoprotein envelope formed from elements of the plasma membrane of its previous host. This envelope can fuse with the plasma membrane of a new host* (**b**), *allowing the virus to enter the cell by endocytosis (page 119). Once the retrovirus has gained entry to the cell, the viral RNA is* (**c**) *released from the capsid and* (**d**) *transcribed into a single strand of complementary DNA (cDNA).* (**e**) *Synthesis of the matching DNA strand follows immediately, producing a double-stranded molecule of cDNA. These two reactions, as well as the degradation of the original viral RNA molecule, are all catalyzed by reverse transcriptase.*

After the double-stranded cDNA has been inserted into a host-cell chromosome, new viral RNA molecules are transcribed from it, as are mRNA molecules that direct the synthesis of viral proteins.

Other lines of evidence, however, have linked the development of cancer with viruses. As long ago as 1911, a cancer-causing virus, the Rous sarcoma virus, was isolated from chicken tumors. Even though other cancer-causing viruses, particularly viruses affecting laboratory mice, were gradually discovered, a viral hypothesis of cancer was slow to emerge. For one thing, viruses could not be shown to be important as causes of human cancer. (Even today, after years of searching, only a few rare human cancers have been linked to viruses.) In addition, the fact that most of the known cancer-causing viruses, including the Rous sarcoma virus, are RNA viruses rather than DNA viruses seemed to make the viral hypothesis of cancer incompatible with the mutation hypothesis.

As we have seen, however, it is now known that viruses, like mutagens, can bring about changes in the cell's genetic makeup. All known cancer-causing viruses introduce information into host-cell chromosomes. These include both DNA viruses, like SV40, and RNA retroviruses. The discovery of the role of reverse transcriptase forged the crucial link between the RNA viruses and the chromosomes of eukaryotic cells.

New techniques have enabled molecular biologists to study the changes in the eukaryotic chromosome that lead to cancer. These studies are usually carried out in cells growing in tissue culture. When such cells are exposed to a carcinogenic agent, they undergo characteristic changes in their growth patterns and in their shape (Figure 17–10). Moreover, when these cells are transplanted into laboratory animals, they cause cancer.

Studies of such cells have uncovered a group of genes known as **oncogenes** (from the Greek word *onkos*, meaning "tumor"). Oncogenes closely resemble normal genes of the eukaryotic cells in which they are found.

According to the oncogene hypothesis, when something goes wrong in the expression of these normal genes, either because of mutations in the genes themselves or because of changes in gene regulation, they stimulate the uncontrolled cell division characteristic of cancer.

Other studies have revealed the presence of another group of genes that play a crucial role in determining whether cancer actually develops following oncogene activation. The function of these genes, known as **tumor suppressor genes,** is apparently to act as a brake on cell division, keeping it tightly regulated. As long as these genes are functioning normally, the development of cancer is blocked or, at the least, slowed. However, when they are disabled by mutation or actually lost from the chromosome, the way is clear for the rapid multiplication of cancer cells.

Collectively, these studies suggest that viruses can affect the development of cancer in three different ways. First, simply by their presence in the chromosome, viruses may disrupt the function of normal regulatory genes—converting normal genes into oncogenes, activating oncogenes, or disabling tumor suppressor genes. Second, viruses may encode proteins needed for viral replication that also affect the regulation of these cellular genes. Third, and most interesting of all, viruses may serve as vectors of oncogenes, enabling them to move from one cell to another or from one individual to another.

With these discoveries, the viral hypothesis and the mutation hypothesis of cancer are no longer regarded as incompatible but rather as mutually supportive. Step by step, this work is bringing us closer to the control of one of our oldest and ugliest enemies, and, at the same time, is yielding new information on the fundamental question of the regulation of cell growth.

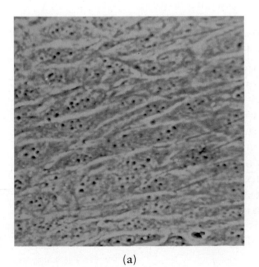
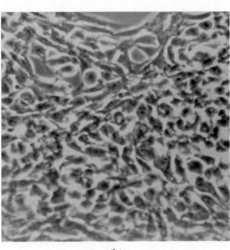

(a) (b)

17–10 (a) *Normal cells growing in tissue culture and* (b) *the same type of cells after transformation with a cancer-causing virus. The cancer cells not only show striking surface changes but also pile up on top of one another as they multiply out of control. The normal cells, by contrast, grow side by side in a single organized layer.*

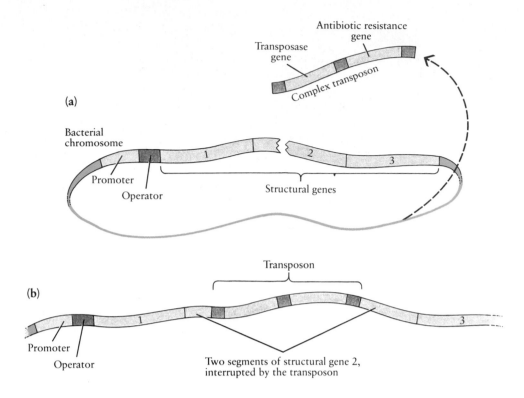

17–11   (a) *A complex transposon "jumping" from one location to another in a bacterial chromosome. The transposon shown here includes the gene that codes for the enzyme transposase and another gene that codes for an enzyme that confers resistance to an antibiotic.* (b) *This transposon has inserted itself into a structural gene, interrupting the coding sequence and inactivating the gene.*

Transposons

Both plasmids and viruses, as we have seen, can function as vectors, carrying genes from one chromosome to another. The chromosomes of both prokaryotes and eukaryotes contain yet another type of movable genetic element, one that incorporates within itself the mechanism for its movement. These movable genetic elements are known as **transposons,** and it was their activity that Barbara McClintock first detected in corn kernels more than 40 years ago.

Transposons are simply segments of DNA that move from one location to another in the chromosomes. They are characterized by a gene coding for an enzyme, **transposase,** that catalyzes their insertion into a new site. In many cases, a transposon does not disappear from its original site when it appears at a new location. Instead, the parent transposon gives rise to a copy that becomes inserted elsewhere in the chromosomes.

Two kinds of transposons—simple and complex—have been identified in prokaryotes. Simple transposons are relatively short and do not carry any genes beyond those essential for the process of transposition. They are detectable because they can cause mutations. If one of these transposons becomes inserted into a gene, it inactivates it. Simple transposons also contain promoter sequences, which may lead to the inappropriate transcription of previously inactive genes of the chromosome.

Complex transposons are much larger and carry genes that code for additional proteins (Figure 17–11). As is the case with simple transposons, complex transposons may cause mutations, but they are also detectable because of their gene products. Genes that are part of a complex transposon can move from place to place on a chromosome or from chromosome to chromosome, and are therefore known as "jumping genes." The drug-resistance genes of bacteria, for example, are often part of complex transposons, which explains why they can be transferred so readily from plasmid to plasmid and from plasmid to bacterial chromosome to plasmid again.

Complex transposons often have simple transposons flanking them—one at each end. This suggests that complex transposons are formed when two simple transposons jump at the same time, taking with them everything in between.

Eukaryotic transposons resemble their bacterial counterparts in structure, and like bacterial transposons, they can cause mutations when they become inserted into structural genes or promoter regions. However, they differ from bacterial transposons in one significant and interesting feature. In eukaryotes, many transposons are first copied into RNA—and then back into DNA—before their insertion into a new location in the chromosomal DNA. This discovery was surprising, for it was previously thought that reverse transcription was unique to retroviruses.

Although nucleotide sequences that resemble genes for reverse transcriptase have been identified in the DNA of some eukaryotes, there is, thus far, no evidence of reverse transcriptase activity in cells uninfected by retroviruses. It is, however, widely accepted that the action of reverse transcriptase in the course of retroviral infections has played an important role in the evolution of eukaryotic DNA and, in particular, in the evolution of transposons. There is also evidence that many of the repeated DNA sequences in eukaryotic chromosomes originated as transposons.

Expanding Triplet Repeats

In 1991, molecular biologists discovered yet another form of instability in eukaryotic genes. Although the affected genes were not moving from one location to another, they were expanding—often dramatically—from one generation to the next. In a few cases, the reverse process was occurring, and the genes were shrinking.

The initial discovery was made in studies of a human disorder known as **fragile X syndrome.** In this disorder, which is the most common form of inherited mental retardation, the tip of the long arm of the X chromosome hangs by a slender thread of DNA (Figure 17–12). Identification and analysis of the gene involved in fragile X syndrome revealed that it contains multiple copies of the nucleotide triplet CGG repeated in tandem, head to tail. The most astonishing discovery, however, was that the number of repeats varies from generation to generation and is correlated with the presence or absence of mental retardation and with its severity. The normal gene contains 60 or fewer CGG repeats. In healthy carriers of a fragile X chromosome, the gene may contain as many as 200 copies of the triplet. In individuals with fragile X syndrome, however, the number of copies balloons to hundreds or even thousands.

More recently, similar expanding triplet repeats have been identified in the genes responsible for Huntington's disease, for myotonic dystrophy (the most common type of muscular dystrophy in adults), and for several other rare disorders. An intensive search is now underway to identify other human genes that contain such repeats. At least 50 human genes are known to contain short triplet repeats, but it is not yet known if any of these repeats can expand.

The discovery of expanding (and occasionally contracting) triplet repeats has shed light on the incomplete

17–12   *A fragile X chromosome. The tip of the long arm is attached to the rest of the chromosome by a slender thread of DNA (arrow). Triplet repeats that increase in number from generation to generation occur in this stretch of the DNA. Like other X-linked disorders, fragile X syndrome occurs most often in males.*

penetrance and variable expressivity (page 227) observed in Huntington's disease, fragile X syndrome, and a number of other human genetic disorders. Below a certain threshold, which varies according to the disorder, the triplet repeats appear to have no harmful consequences. Above that threshold, however, symptoms of the disorder begin to appear, with their severity increasing as the number of repeats increases.

Like so many other discoveries, the discovery of expanding triplet repeats has raised many more questions than it has answered. At this time, no one knows what triggers the expansion or contraction of the repeats or how it occurs. Nor does anyone know why, in Huntington's disease, the expansion occurs most often when the affected gene is transmitted by the father, but in fragile X syndrome, it occurs most often when the gene is transmitted by the mother. The answers to these questions, when they come, may lead to new understandings of many previously inexplicable variations in human inheritance patterns. And, the answers may come very quickly. Stay tuned.

Summary

In addition to the bacterial chromosome, *E. coli* and other types of bacteria contain much smaller, also circular, double-stranded DNA molecules. These molecules are known as plasmids. Most plasmids can be transferred from cell to cell, often through the process of conjugation, in which a cytoplasmic bridge forms between two cells. Some plasmids can exist either independently in the cell or integrated into the bacterial chromosome. Plasmids often carry genes for drug resistance, with serious consequences for human health.

Viruses consist of either DNA or RNA wrapped in a protein coat. Some RNA viruses of animals, known as retroviruses, contain and code for the enzyme reverse transcriptase, which can transcribe DNA from an RNA template. Viruses can multiply only within a host cell, where the viral nucleic acid can use the cell's metabolic resources to synthesize more viral nucleic acid molecules and more viral proteins. The infection cycle of a virus is completed when newly synthesized virus particles break out of the host cell, ready to infect a new host cell.

Some DNA viruses of both prokaryotes and eukaryotes, as well as DNA transcribed from RNA retroviruses, can become integrated into a host-cell chromosome and replicate with the chromosome through many cell divisions. When integrated into a host chromosome, the DNA of a prokaryotic virus is known as a prophage. Integrated DNA from a eukaryotic virus is known as a provirus. From time to time, prophages and proviruses break loose from the chromosome and set up a new infection cycle.

Viruses can serve as vectors, transporting genes from cell to cell in a process known as transduction. General transduction occurs when host DNA, fragmented in the course of the viral infection, is incorporated into new virus particles that carry these fragments to a new host cell. Specialized transduction occurs when an integrated virus, on breaking away from the host chromosome, carries with it—as part of the viral chromosome—host genes, which are then transported to a new host cell.

According to present evidence, the development of cancer involves alterations in the function of certain normal cellular genes, converting them into oncogenes, as well as the inactivation or loss of tumor suppressor genes. The alterations in gene function may be caused by mutations or by changes in the regulation of gene expression. Viruses can cause cancer by disrupting the function of normal regulatory genes, by directing the synthesis of proteins that affect the regulation of host-cell genes, or by inserting oncogenes into a chromosome.

Transposons are another type of movable genetic element, found in both prokaryotic and eukaryotic cells. They carry a gene for the enzyme transposase, which catalyzes their transposition from one location to another in the chromosome. Transposons may cause mutations by interfering with the normal expression of host-cell genes. Simple transposons contain only genes involved in their transposition; complex transposons carry additional structural genes.

Certain human genes, including the genes responsible for fragile X syndrome and for Huntington's disease, contain triplet repeats that can increase (or sometimes decrease) in number from generation to generation. The cause or causes of these changes are not known, but it is clear that the severity of the disorders involved depends on the number of triplet repeats.

Questions

1. Distinguish among the following: transformation/conjugation/transduction; F$^+$ cell/F$^-$ cell; plasmid/virus/transposon; prophage/provirus; general transduction/specialized transduction; oncogene/tumor suppressor gene.

2. As bacterial conjugation indicates, it is possible to separate the production of new genetic combinations from reproduction. Why do you think these two processes are combined in eukaryotic cells?

3. Before the structure of any virus was known, Crick and Watson predicted that the protein coats of viruses would prove to be made up of large numbers of identical subunits. Can you explain the basis of their prediction?

4. Describe two possible outcomes of infection of a bacterial cell by a temperate bacteriophage.

5. How do RNA retroviruses differ from other RNA viruses? What are the functions of the enzyme reverse transcriptase?

6. Most bacteriophages lack introns, but introns have been found in the DNA viruses of eukaryotes. The RNA viruses of eukaryotes, however, lack introns. What do these findings suggest about the origin of viruses?

7. Some biologists consider viruses to be living organisms. By what criteria might viruses be considered alive?

Suggestions for Further Reading

Alberts, Bruce, Dennis Bray, Julian Lewis, Martin Raff, Keith Roberts, and James D. Watson: *Molecular Biology of the Cell,* 3d ed., Garland Publishing, Inc., New York, 1994.

> *This outstanding cell biology text provides excellent discussions of plasmids, viruses, and transposons.*

Amábile-Cuevas, Carlos F., and Marina E. Chicurel: "Horizontal Gene Transfer," *American Scientist,* vol. 81, pages 332–341, 1993.

Bishop, J. Michael: "Oncogenes," *Scientific American,* March 1982, pages 80–92.

Butler, P. Jonathan G., and Aaron Klug: "The Assembly of a Virus," *Scientific American,* November 1978, pages 62–69.

Campbell, A. M.: "How Viruses Insert Their DNA into the DNA of the Host Cell," *Scientific American,* December 1976, pages 102–113.

Cohen, S. N., and J. A. Shapiro: "Transposable Genetic Elements," *Scientific American,* February 1980, pages 40–49.

Doolittle, Russell F., and Peer Bork: "Evolutionarily Mobile Modules in Proteins," *Scientific American,* October 1993, pages 50–56.

Fedoroff, Nina V.: "Transposable Genetic Elements in Maize," *Scientific American,* June 1984, pages 85–98.

Gallo, Robert C.: "The AIDS Virus," *Scientific American,* January 1987, pages 46–56.

Gallo, Robert C.: "The First Human Retrovirus," *Scientific American,* December 1986, pages 88–98.

Hirsch, Martin S., and Joan C. Kaplan: "Antiviral Therapy," *Scientific American,* April 1987, pages 76–85.

Keller, Evelyn: *A Feeling for the Organism: The Life and Work of Barbara McClintock,* W. H. Freeman and Company, New York, 1993.*

> *An illuminating biography of McClintock, originally published in 1983, shortly before she was awarded the Nobel Prize.*

Levine, Arnold J.: *Viruses,* Scientific American Library, New York, 1992.

> *This beautifully illustrated book begins with the bacterial viruses that have contributed so much to our understanding of genetics. Most of the book, however, is devoted to those human viruses—herpes viruses, tumor viruses, retroviruses, and the viruses responsible for influenza and hepatitis—that have, thus far, largely defied all efforts to bring them under control.*

Morrell, Virginia: "The Puzzle of the Triple Repeats," *Science,* vol. 260, pages 1422–1423, 1993.

Rennie, John: "DNA's New Twists," *Scientific American,* March 1993, pages 122–132.

Simons, Kai, Henrik Garoff, and Ari Helenius: "How an Animal Virus Gets into and out of Its Host Cell," *Scientific American,* February 1982, pages 58–66.

Varmus, Harold: "Reverse Transcription," *Scientific American,* September 1987, pages 56–64.

Weinberg, Robert A.: "Finding the Anti-Oncogene," *Scientific American,* September 1988, pages 44–51.

*Available in paperback.

New Frontiers in Genetics

This is Cindy, playing catch in May of 1993. A few years earlier, such an activity would have been inconceivable, for Cindy was born with an inherited disorder known as SCID (severe combined immunodeficiency). In this disorder, the white blood cells—the cells that defend the body against infection—are defective. Until recently, Cindy and other victims of SCID were locked in a constant and losing battle against the multitude of infectious agents that surround us all.

The most common cause of SCID is a defect in the gene coding for an enzyme known as ADA. ADA is essential for the normal development and functioning of white blood cells. In 1988, doctors began to treat SCID children with synthetic ADA. Some of the children, including Cindy, improved. Others, however, did not.

In January of 1991, Cindy was the second child to receive a treatment that made a dramatic difference—and gave her the stamina to play baseball. White blood cells were taken from her body and treated with genetically engineered viruses that acted as vectors, inserting copies of the normal gene for ADA into the cells. The altered white blood cells were then reinfused into her body, where they synthesized the missing enzyme. As effective as this treatment was, however, it had to be repeated every few months as the altered white blood cells reached the end of their normal life span. It was an ordeal, and when asked about it, Cindy's response was, "I don't like to talk about it. It is too weird."

Shortly after this photo was taken, Cindy received a different treatment. This time, the gene for ADA was inserted into stem cells (page 180), the cells in the bone marrow that divide repeatedly, giving rise to red blood cells and to the white blood cells of the immune system. It is not yet known if the altered stem cells will reproduce in Cindy's bone marrow and make it possible for her to lead a long and normal life. If they do, it will mark the first successful cure of a human hereditary disorder—and an end to repeated trips to the hospital for this courageous pioneer.

In the last few years, advances in molecular genetics have come at an astonishing pace. One after another, new discoveries are revolutionizing our understanding of plant and animal genetics—and particularly our understanding of human genetic disorders. At the same time, new methods are being developed for the diagnosis of genetic disorders, and new hopes are being raised for actual cures. Human gene therapy, which a few years ago seemed only a distant dream, is becoming reality. And, that is only one of the many practical applications flowing from the new discoveries.

This burst of discovery and application has been made possible by the technology known as **recombinant DNA.** In recombinant DNA work, also known as **genetic engineering,** portions of DNA molecules from different sources are modified, recombined, and then inserted into other cells where the genes carried by the modified DNA are expressed.

Recombinant DNA: The Technology

In order to do recombinant DNA work, scientists must (1) have segments of DNA that are small enough to be analyzed and manipulated, (2) have large quantities of these small segments, (3) know the nucleotide sequence in the segments, and (4) be able to identify the specific segments in which they are interested. These are not, however, four sequential steps. The order in which these stages of the work are accomplished depends on the goals of a particular project. In some cases, for example, scientists know in advance the nucleotide sequences in which they are interested and where to find them. In other cases, learning the sequences and determining their location within the chromosome is the goal.

Recombinant DNA technology makes use of the natural mechanisms—plasmids, viruses, and transposons—by which genes can be moved from one location to another. It all began, however, with the discovery of a group of enzymes that protect bacterial cells against viral infection.

Obtaining Short DNA Segments

In the course of studies of the bacteriophages, it was found that viruses infecting one strain of *E. coli* were sometimes unable to infect another strain. This restriction of viral infection occurs, it was learned, because of the presence in the bacteria of enzymes that cut foreign DNA molecules into small pieces before they can be replicated or transcribed. These enzymes came to be known as **restriction enzymes.**

Restriction enzymes cut through foreign DNA molecules at specific nucleotide sequences, usually four to eight base pairs in length. These sequences are known as **recognition sequences,** since they are "recognized" by specific restriction enzymes. In the bacterial chromosome itself, the nucleotides in these sequences are chemically modified and thus protected from the enzymes.

Some restriction enzymes make straight cuts through both strands of the DNA molecule. Others, however, cleave the molecule by cutting through the two strands a few nucleotides apart. As a consequence, there is a short sequence of unpaired nucleotides at each end of the cut (Figure 18–1). These ends are said to be "sticky," because they can join again with each other when hydrogen bonds form spontaneously between the complementary bases. A type of enzyme known as a **DNA ligase,** which catalyzes the reaction that links the cut ends of each strand, is required to complete the resealing process.

From a practical standpoint, the most important thing about the "sticky" ends is that they can join with any other segments of DNA that have been cut by the same restriction enzyme and, as a result, have complementary "sticky" ends. This was the discovery that opened the way for molecular biologists to carry out their own genetic manipulations.

More than 200 different restriction enzymes have now been isolated from different types of bacteria, making it possible to cut a DNA molecule at any one of more than 90 recognition sequences. Because of the number of different restriction enzymes and recognition sequences, DNA molecules from virtually any organism can be cut into segments short enough to analyze and manipulate. These segments can be separated from one another on the basis of size by electrophoresis (Figure 18–2).

The particular segments produced by a restriction enzyme are determined by the locations at which the recognition sequence for the enzyme occurs—and these locations may be anywhere. Depending on where the recognition sequences occur, a given gene may be fragmented into a number of different segments, or, conversely, a particular segment may contain several different genes. However, short segments of DNA that represent specific genes (or known portions of specific genes) can be obtained by synthesizing them from an mRNA template, using the enzyme reverse transcriptase (page 294). If their nucleotide sequences are known, short DNA segments can also be synthesized by chemical means in the laboratory.

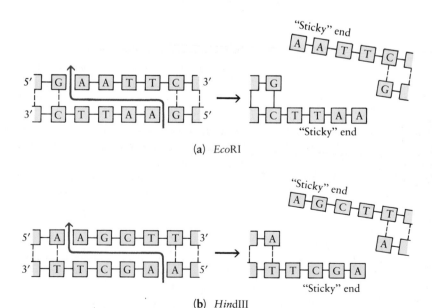

18–1 *The DNA nucleotide sequences recognized by two widely used restriction enzymes,* **(a)** *EcoRI and* **(b)** *HindIII. These two enzymes cut the DNA so that "sticky" ends result. A "sticky" end can reattach to its complementary sequence at the end of any DNA molecule that has been cut by the same restriction enzyme. Restriction enzymes are named for the bacteria from which they are obtained. EcoRI is from* E. coli, *and* HindIII *is from* Hemophilus influenzae.

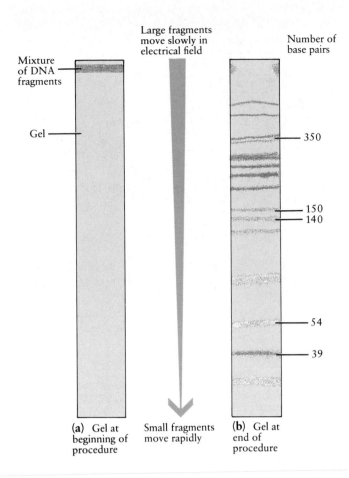

Large fragments
move slowly in
electrical field

Mixture
of DNA
fragments

Gel

Number of
base pairs

350

150
140

54

39

(a) Gel at
beginning of
procedure

Small fragments
move rapidly

(b) Gel at
end of
procedure

18–2 *Electrophoresis, the technique used by Pauling to separate normal and sickle-cell hemoglobins (page 257), can also be used to separate fragments of DNA. In electrophoresis, the electrical field separates molecules not only on the basis of their charge (the feature utilized by Pauling) but also on the basis of their size. Smaller molecules move faster than larger ones. (a) A mixture of DNA fragments containing different numbers of base pairs can be cleanly separated according to size (b). The gel can then be sliced into sections, and the separated, purified fragments washed out of the gel unharmed. This separation procedure is important in many aspects of recombinant DNA work.*

Obtaining Multiple Copies

DNA Cloning

Most recombinant DNA work requires multiple copies of the DNA segments being studied and manipulated. One way these copies can be obtained is by harnessing the replicative powers of plasmids and viruses.

Not long after the discovery of the restriction enzyme *Eco*RI, researchers isolated a small plasmid of *E. coli* that makes its host cells resistant to the antibiotic tetracycline. This plasmid has only one recognition sequence for *Eco*RI, and, as a consequence, it is cleaved at only one site by that enzyme. Thus it is possible to insert into the plasmid a small segment of foreign DNA that has also been cleaved by *Eco*RI (Figure 18–3).

Insertion of foreign DNA into the plasmid does not affect its ability to move into *E. coli* cells, its capacity to make the recipient cells tetracycline-resistant, or its ability to replicate. Typically, the plasmid replicates several times in its host cell, producing about 10 new plasmids per cell. Because the plasmids containing the foreign DNA are larger than those that do not contain it, after replication they can be readily isolated and collected. Treatment of the replicated plasmids by *Eco*RI

releases the foreign DNA, which can then be separated from the plasmid DNA by electrophoresis.

The discovery that plasmids could be used as vectors for foreign DNA opened the way to the production of large numbers of uniform, identical segments of DNA. Such multiple copies are known as **clones,** a term that has long been applied to genetically identical bacteria and other organisms produced asexually from a single parent cell or organism.

Plasmids are useful as vectors because they multiply rapidly and are easily taken up by bacteria through the plasma membrane (Figure 18–4). Modified strains of the bacteriophage lambda, which can gain ready entry to bacterial cells by way of its identifying protein coat, are also used in the laboratory to carry foreign DNA segments into bacterial cells for replication.

Polymerase Chain Reaction

In 1989, a new process was developed for producing multiple copies of DNA segments. This process, known as **polymerase chain reaction (PCR),** can take a tiny sample of DNA and, in a few hours, synthesize millions of copies of a specific segment of that DNA.

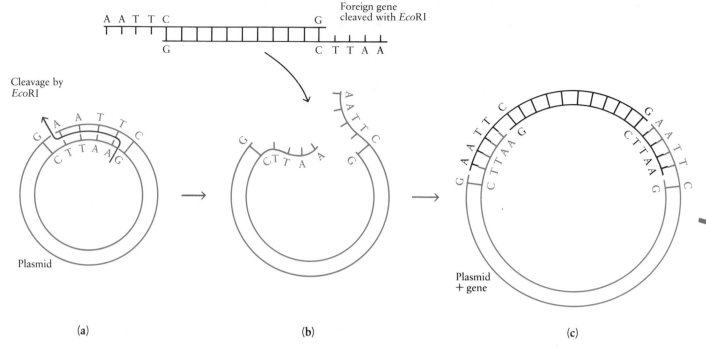

(a)   (b)   (c)

18–3   *The use of plasmids in DNA cloning. (a) The plasmid is cleaved by a restriction enzyme. In this example, the enzyme is EcoRI, which cleaves the plasmid at the sequence (5')-GAATTC-(3'), leaving "sticky" ends exposed. (b) These ends, consisting of TTAA and AATT sequences, can join with any other segment of DNA that has been cleaved by the same enzyme. Thus it is possible to splice a foreign gene into the plasmid (c). (In this illustration, the length of the GAATTC sequences is exaggerated and the lengths of the other portions of*

both the foreign gene and the plasmid are compressed.)

When plasmids incorporating a foreign gene are released into a medium in which bacteria are growing, (d) they are taken up by some of the bacterial cells. (e) As these cells multiply, the plasmids replicate. The result is an increasing number of cells, all making copies of the same plasmid. (f) The plasmids can then be separated from the other contents of the cells and treated with EcoRI to release the copies of the cloned gene.

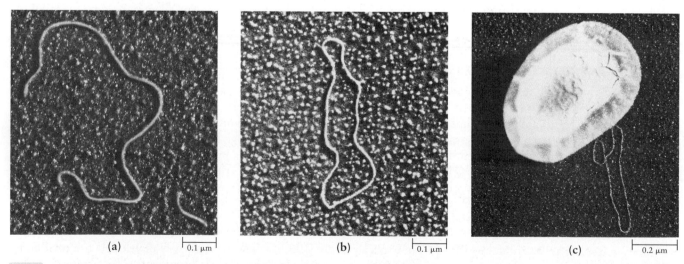

(a)   0.1 μm   (b)   0.1 μm   (c)   0.2 μm

18–4   *Insertion of a foreign gene into a bacterial cell. (a) A plasmid has been cut open with a restriction enzyme, leaving two "sticky" ends. A small segment of foreign DNA (lower right) also has "sticky" ends that can join with the "sticky" ends of the plasmid. (b) With the aid of DNA ligases, the for-*

eign DNA has been spliced into the plasmid. (c) The plasmid, now containing the foreign DNA, is about to enter a bacterial cell. (Huntington Potter and David Dressler, LIFE Magazine, 1980, Time, Inc.)

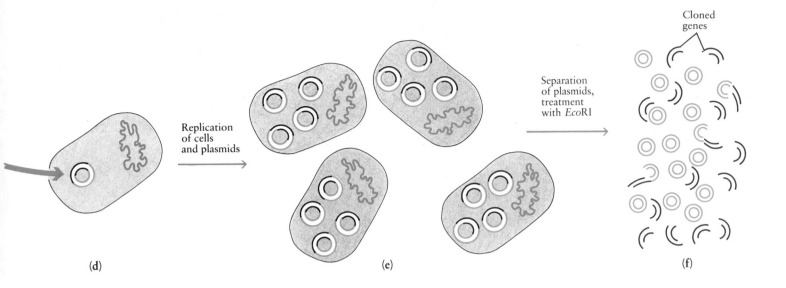

(d)

Replication of cells and plasmids

(e)

Separation of plasmids, treatment with *Eco*RI

Cloned genes

(f)

Unlike cloning, PCR requires knowledge of the nucleotide sequences at each end of the DNA segment that is to be copied. The DNA sample is added to a solution that contains DNA polymerase molecules, large quantities of the four nucleotides that occur in DNA, and "primer" molecules of DNA, about 20 nucleotides in length, that are complementary to the sequences at each end of the desired segment. When this solution is heated, the two strands of the DNA double helix separate. As the solution is cooled, the primers attach to their complementary sequences. DNA polymerase molecules "recognize" the primers and begin adding nucleotides at the locations where the primers are attached. Both strands of the DNA segment are copied simultaneously. Thus, at the end of the initial step, two copies of the segment are present, each consisting of one new strand and one old strand. The solution is then alternately heated and cooled, and with each cycle, the number of copies is doubled. If this is repeated for 20 cycles, about one million copies of the segment are produced.

Among its many applications, PCR is being used to speed up prenatal diagnosis of genetic disorders, to detect latent viral infections, and to produce large quantities of selected human DNA sequences at rates far faster than is possible with conventional DNA cloning.

Determining Nucleotide Sequences

With the development of techniques for cutting DNA molecules into smaller pieces and making multiple copies of those pieces, it is now possible, in principle, to determine the nucleotide sequence of any isolated gene. One of the most important features of restriction enzymes is that different enzymes cut DNA molecules at different sites. Cutting a DNA molecule with one restriction enzyme produces one particular set of short DNA fragments. Cutting an identical DNA molecule with a different restriction enzyme produces a different set of short DNA fragments. The fragments of each set can be separated from one another by electrophoresis and cloned into multiple copies.

These copies can then be analyzed to determine the exact nucleotide sequence of each fragment. The methods used in these analyses involve complex series of chemical or enzymatic reactions. However, the reactions have now been automated, greatly reducing the time required to determine the nucleotide sequences of short fragments.

Because the sets of fragments produced by different restriction enzymes overlap, the information obtained from sequencing the different sets can be pieced to-

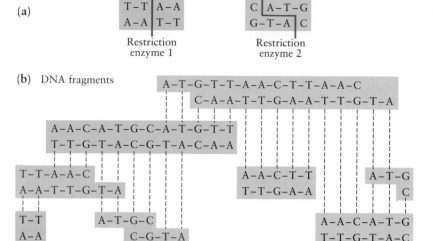

(a)

T–T | A–A
A–A | T–T
Restriction enzyme 1

C | A–T–G
G–T–A | C
Restriction enzyme 2

(b) DNA fragments

A–T–G–T–T–A–A–C–T–T–A–A–C
C–A–A–T–T–G–A–A–T–T–G–T–A

A–A–C–A–T–G–C–A–T–G–T–T
T–T–G–T–A–C–G–T–A–C–A–A

T–T–A–A–C
A–A–T–T–G–T–A

A–A–C–T–T
T–T–G–A–A

A–T–G
C

T–T
A–A

A–T–G–C
C–G–T–A

A–A–C–A–T–G
T–T–G–T–A–C

T–T–A–A–C–A–T–G–C–A–T–G–T–T–A–A–C–T–T–A–A–C–A–T–G
A–A–T–T–G–T–A–C–G–T–A–C–A–A–T–T–G–A–A–T–T–G–T–A–C

(c) Sequenced DNA molecule

18–5 *A simplified example of DNA sequencing. Identical samples of the DNA molecule to be sequenced are treated with different restriction enzymes that cut the DNA at different sites (a). One sample is treated with one enzyme, producing one set of fragments, and another sample is treated with another enzyme, producing a different set of fragments. The fragments of each set are then separated from one another, cloned, and analyzed, revealing the nucleotide sequence of each individual fragment (b). As you can see, the fragments produced by the two restriction enzymes overlap, making it possible to determine the nucleotide sequence of the molecule as a whole (c).*

gether like a puzzle to reveal the entire sequence of a DNA molecule (Figure 18–5). At the present time, the sequencing of genes from sources ranging from the smallest viruses to human cells is going forward in hundreds of laboratories around the world.

Locating Specific DNA Segments

Before a DNA segment of interest—such as a particular gene or portion of a gene—can be cloned, sequenced, or manipulated, it must first be located and isolated. The chromosomes of even the simplest eukaryotic cells contain an enormous quantity of DNA. Locating a specific segment of that DNA is like trying to find the proverbial needle in a haystack. The "magnet" that scientists use is nucleic acid hybridization, one of the earliest methods for studying DNA and RNA molecules (see essay, page 259).

The foundation of hybridization is the base-pairing properties of the nucleic acids. As you know, when dissolved DNA molecules are heated gently, the hydrogen bonds holding the two strands together are broken and the strands separate. When the solution is cooled slowly, the hydrogen bonds re-form, reconstituting the double helix.

When DNA molecules from different sources are mixed together and heated, the strands separate and undergo random collisions. If two strands with nearly complementary sequences find each other as the solution is cooled, they will form a hybrid double helix. The extent to which segments from two samples reassociate and the speed with which they do so provide an estimate of the similarity between their nucleotide sequences.

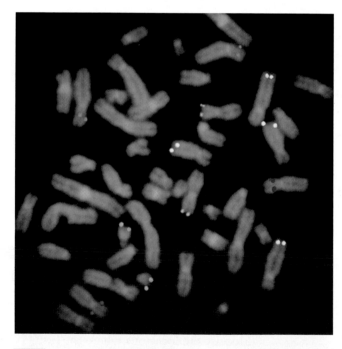

18–6 *A set of human metaphase chromosomes in which the locations of specific nucleotide sequences are revealed by nucleic acid hybridization. The probes were labeled with fluorescent dyes. Each color indicates the location of a different DNA sequence: red = the location on the X chromosome of the gene for the most common form of childhood muscular dystrophy; green = the region of chromosome 21 where the gene for Down syndrome occurs; orange = the beta-globin gene on chromosome 11; white = an oncogene on chromosome 8; violet = a gene coding for a protein found on the surface of white blood cells; yellow = an unidentified sequence on chromosome 5. Because this is a diploid set of replicated chromosomes, each probe hybridized in four locations.*

This procedure has been adapted to locate specific nucleotide sequences in either DNA or RNA. A probe can be prepared by incorporating a radioactive isotope into a short segment of single-stranded DNA or RNA that is complementary to the nucleotide sequence of interest. Alternatively, the probe can be labeled with a fluorescent dye (Figure 18–6). Probes may be mRNA molecules, DNA fragments produced by restriction enzymes, complementary DNAs synthesized from RNA by reverse transcriptase, or nucleotide sequences synthesized in the laboratory. After a probe has been prepared, it can be used to seek out segments of DNA or RNA that contain the complementary nucleotide sequence. Labeled RNA molecules are routinely used to find corresponding DNA segments, and vice versa.

Recombinant DNA: Some Applications

The "tools" of recombinant DNA technology are now being used in many different types of research studies. These studies are rapidly increasing our knowledge of chromosome structure and organization, of the regulation of gene expression, and of the ways in which things can go wrong, leading not only to inherited disorders but also to many health problems that develop later in life. An enormous number and variety of practical applications are also flowing from the new technology. In the pages that follow, we are going to look at a few representative examples.

Bacterial Synthesis of Useful Proteins

Early in the course of recombinant DNA research, biologists realized that if genes coding for proteins of medical or agricultural importance could be transferred into bacteria and expressed, the bacteria could function as "factories," producing a virtually limitless source of the proteins.

The first human protein synthesized in a bacterial cell was the hormone somatostatin (Figure 18–7). It is a small protein (only 14 amino acids long) and can be detected in very small amounts. The investigators knew the sequence of amino acids in somatostatin and thus could determine a sequence of nucleotides in DNA that would code for it. Using this information, they synthesized an artificial gene, including an initiation codon. Copies of the gene were spliced into plasmids carrying genes for drug resistance. Also spliced into the plasmids was a portion of the *lac* operon (page 274), including its regulatory sequences. The scientists then supplied the plasmids to *E. coli* cells. Only a few of the bacterial cells

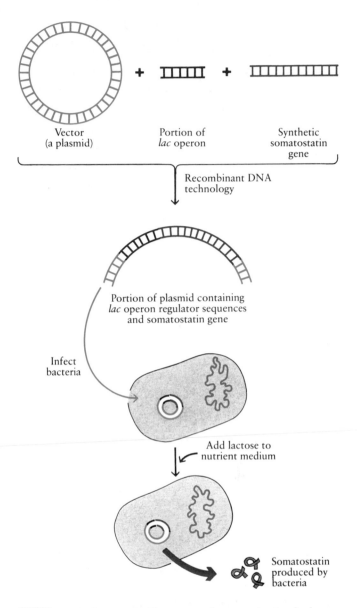

18–7   *How bacterial cells were made to synthesize the hormone somatostatin. The somatostatin gene and a portion of the* lac *operon were spliced into the plasmid. The addition of lactose to the nutrient medium turned the somatostatin gene on.*

became drug-resistant, indicating that they had taken up the plasmid. However, as these cells multiplied, the plasmids multiplied along with them. When the *lac* operon was turned on by the addition of lactose, proteins were synthesized. Chemical treatment of these proteins released pure somatostatin. When this somatostatin was tested in laboratory animals, it had the same biological activity as the natural hormone.

This experiment paved the way for the bacterial synthesis of medically useful proteins. An increasing number of medications are now produced using a similar procedure. They include human insulin (Figure 18–8), clotting factors for the treatment of hemophilia (page 231), and human growth hormone, which is used to treat some forms of dwarfism in children.

Vaccines against viral disease are another important product of this technology. As you know, all viruses consist of nucleic acid wrapped in a protein capsid. In the animal bloodstream, capsid proteins, recognized by cells of the immune system as foreign, evoke the formation of antibodies, molecules that play a crucial role in future immunity against the virus. Most vaccines are made from killed or altered forms of the virus particles. Vaccines produced from synthetic capsid proteins are safer, since without the viral nucleic acid, the vaccines cannot be contaminated by infectious particles.

The bacterial synthesis of other types of proteins is of increasing economic importance. For example, the enzyme rennin, extracted from calves' stomachs and used by the dairy industry for cheesemaking, can now be produced by recombinant DNA technology. Scientists have also succeeded in inducing bacteria to synthesize the enzyme cellulase, which is produced in nature by certain fungi. This enzyme converts cellulose, the plant polysaccharide that is indigestible by most organisms, into glucose, a food molecule of major importance in a hungry world.

Diagnosis of Genetic Disorders

Another application of recombinant DNA technology is in the diagnosis of human genetic disorders. Reliable tests have now been developed that, in conjunction with amniocentesis (page 202), make possible the prenatal diagnosis of a number of inherited disorders, including PKU, sickle-cell anemia, and the most common forms of hemophilia and muscular dystrophy. A test has also been developed that can identify those individuals at risk for Huntington's disease who are most likely to develop the disorder. All of these tests involve the use of restriction enzymes, nucleic acid probes, or both. Two examples illustrate the principles on which they are based.

Sickle-Cell Anemia

In order to develop a diagnostic test for sickle-cell anemia, radioactive copies of portions of the nucleotide sequence coding for the beta chain of hemoglobin were prepared. DNA from persons with normal hemoglobin and DNA from persons with sickle-cell anemia were

18–8   *Crystals of human insulin, produced by bacteria that have been modified by genetic engineering. Previously, human insulin was available only in very small quantities. Individuals with the type of diabetes that results from insulin deficiency were treated with insulin obtained from cows or pigs. These animal proteins were not as effective as human insulin, and they triggered allergic reactions in some individuals.*

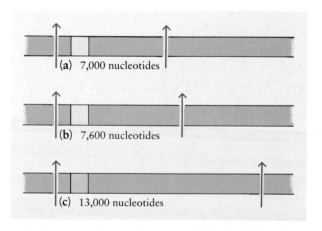

(a)   7,000 nucleotides

(b)   7,600 nucleotides

(c)   13,000 nucleotides

18–9   *A test to detect the presence of the sickle-cell allele. Treating human DNA with the restriction enzyme HpaI produces three possible restriction fragments containing the gene for the beta chain of hemoglobin. Fragments containing the normal allele (light green) are either (a) 7,000 or (b) 7,600 nucleotides long. Fragments containing the sickle-cell allele (orange) are usually (c) 13,000 nucleotides in length. A recognition sequence for the restriction enzyme HpaI—present in the DNA of persons with the normal allele—is missing in the DNA of about 87 percent of the individuals carrying the sickle-cell allele.*

Some Ethical Dilemmas

Although the rapid advances in molecular genetics are contributing greatly to our understanding of human genetic disorders, the capacity to identify individuals at risk of developing certain disorders or of transmitting them to their children is creating perplexing ethical dilemmas both for individuals and for our society as a whole. The four questions below cover situations that have either arisen already or are likely to arise by the year 2000. They were posed by Dr. Eric Lander, a human geneticist at the Whitehead Institute for Biomedical Research and the Massachusetts Institute of Technology.

1. While the right to abortion is guaranteed by the U.S. Constitution, the choice is never easy. Imagine that you learned very early in a pregnancy that the child would certainly:
   (a) die within nine months from spinal muscular atrophy, a common fatal genetic disorder;
   (b) suffer throughout life from cystic fibrosis, a painful chronic disease, and die at about age 20;
   (c) suffer from Huntington's disease at age 40 and die at about 50;
   (d) suffer from Alzheimer's disease at about 60;
   (e) be congenitally deaf;
   (f) be a dwarf, but otherwise healthy; or,
   (g) be predisposed to severe manic depression, which could be partially controlled by medication.

   Would you choose to abort? (Assume that you are young enough that you may reasonably expect to have more children if you wish.) Regardless of your own choice, would you consider it *unethical* for another couple to do so? What principles underlie your choices?

2. Suppose you could learn with certainty whether you would suffer:
   (a) from Huntington's disease at about 40;
   (b) from Alzheimer's disease at about 60.

   Would you want to know?

3. Should insurance companies have the right to:
   (a) charge higher premiums to individuals with higher risk of inherited disease?
   (b) know the results of testing for genetic predispositions?
   (c) refuse to cover a child whom prenatal tests show will suffer a severe genetic disease?

4. Suppose that 10 percent of the work force is particularly prone to cancer induced by an industrial chemical. Should an employer:
   (a) have the obligation to make the work place safe for this minority?
   (b) have the right to require pre-employment genetic screening?
   (c) have the right to refuse employment to such workers?
   (d) have the right to require such workers to pay for supplementary insurance?

   Would your answer differ if the minority involved were 1 percent of the work force or 40 percent of the work force?

---

cleaved with the restriction enzyme *Hpa*I. The fragments produced were then exposed to the radioactive probe.

In persons with normal hemoglobin, the probe consistently hybridized with a DNA fragment that was either 7,000 or 7,600 nucleotides long. By contrast, in 87 percent of persons with sickle-cell anemia, the probe hybridized with a much longer fragment (Figure 18–9). The same result is seen with cells obtained by amniocentesis, thus providing a prenatal screening test for one of the most common serious genetic disorders.

Genetic markers of this type are known as **RFLPs,** pronounced "rif-lips" and translated as **restriction-fragment-length polymorphisms.** They occur because inherited variations—that is, mutations—in the nucleotide sequences in the DNA of different individuals lead to differences in the lengths of the fragments produced by restriction enzymes. Such differences occur when the recognition sequence for a restriction enzyme has been

eliminated or otherwise altered by a mutation. In the case of the sickle-cell allele, the mutation that causes production of the abnormal hemoglobin chain is usually accompanied by another mutation, some 5,000 nucleotides away, that eliminates a cleavage site for *Hpa*I.

The principle involved here is gene linkage, the same principle that made possible the genetic mapping studies described in Chapter 13. Those studies in *Drosophila* confirmed that two genes that are close together on the same chromosome tend to stay together. The updated version is that two nucleotide sequences close together on the same DNA molecule tend to stay together. In the *Drosophila* days, a marker was a gene that produced a detectable phenotypic change, such as a different wing shape or eye color. Today, however, a marker may be "anonymous"—detectable only in the pattern produced by restriction enzyme fragments. The closer the allele and its marker are, the more accurate the diagnosis.

Who's Who: DNA Fingerprinting

Recombinant DNA techniques are now being used not only in the diagnosis of human genetic disorders but also in the identification of individuals and of their family relationships. Each individual's DNA is as distinctive as a fingerprint.

A relatively simple method for DNA fingerprinting has been devised. As we saw in Chapter 16, eukaryotic chromosomes contain many regions of simple-sequence DNA, identical short nucleotide sequences lined up in tandem and repeated thousands of times. The number of repeated units in such regions differs distinctively from individual to individual. These regions can be excised from the total DNA by the use of appropriate restriction enzymes, separated according to length by electrophoresis, and identified with radioactive probes. When the process is completed, the end result, visible on x-ray film, looks like the bar code on a supermarket package.

DNA fingerprinting is the most reliable method yet available for resolving cases of disputed parentage. It is also a precise method for identifying rapists and murderers, who do not always leave conventional fingerprints. However, semen is usually present on the person and clothing of a rape victim, and the blood of a murder victim is often on the clothing of the murderer. In some cases, samples of the murderer's blood are on the victim's body or clothing.

Often only very small samples of biological evidence are found at a crime scene—trace amounts of blood or semen, or even a single hair. Polymerase chain reaction (page 303) is used to synthesize multiple copies of the DNA present in these samples and thus make possible DNA fingerprinting.

The resulting DNA bar codes can be compared with the bar codes of blood samples from the victim and the suspects. When the tests are done correctly, the result is a reliable identification of the perpetrator, coupled with the exoneration of suspects who are innocent. To ensure the quality and admissibility in court of DNA fingerprinting evidence, uniform procedures and standards are now being developed for the laboratories that perform the tests.

The Russian royal family, photographed shortly before World War I. From left to right, Maria, Tsarina Alexandra, Tatiana, Olga, Tsar Nicholas II, Anastasia, with Alexis in front. Alexis, heir to the Romanov throne, was a victim of hemophilia (page 231).

DNA fingerprinting can also be used to identify the victims of violent death. In 1993, for example, it was used to identify the remains of Nicholas II and Alexandra, the last Tsar and Tsarina of Russia, and three of their five children. In 1918, the family, three servants, and their doctor were executed. According to the executioners' reports, the bodies of two of the children were burned, and the bodies of the others were placed in an abandoned salt mine in a forest in the Ural Mountains. Throughout the years of Communist rule, the location of the mine was a closely guarded state secret.

In 1991, however, the remains of nine bodies were recovered from the mine shaft. Forensic anthropologists in Russia tentatively identified five sets of bones as those of Nicholas, Alexandra, and their three oldest daughters, Olga, Tatiana, and Maria. The bones were then taken to the British government's Forensic Science Service Laboratory for testing. Samples of chromosomal and mitochondrial DNA from the bones were compared with samples from three living relatives—two cousins of the Tsar, who have not been publicly identified, and Prince Philip of Great Britain, whose maternal grandmother was the sister of the Tsarina. These tests confirmed that the remains are indeed those of the Tsar, the Tsarina, and three of their daughters.

In the years since the execution of the Russian royal family, rumors have circulated that the two youngest children, Anastasia and Alexis, survived. Over the years, a number of women claimed to be Anastasia. The most convincing case was made by one Anna Anderson, who was found in Berlin in a dazed state in 1920. Some believed her claim, others did not. At her death in 1984, however, she was buried in a Romanov family crypt in Europe. A lock of her hair is now being tested, and the mystery of Anna Anderson may soon be solved.

Huntington's Disease

In the early 1980s, James F. Gusella of the Massachusetts General Hospital, with a large team of co-workers, set out to develop a diagnostic test for Huntington's disease based on the RFLP technique. Huntington's presented a new problem because neither the defective gene nor its normal allele had been identified at that time. However, the workers had available a large library of cloned restriction fragments from human DNA, a family in the United States with a history of Huntington's disease, and, in Venezuela, another family of more than 3,000 individuals, also with a history of Huntington's (see page 206).

For three years, a team led by Nancy Wexler interviewed members of the Venezuelan family to obtain the information needed to construct a pedigree, performed neurological examinations of family members, and collected skin and blood samples from 570 people. As the samples were collected, they were flown to Gusella's laboratory, where the DNA was extracted, treated with restriction enzymes, and analyzed. Similar studies were conducted with the members of the family in the United States. Correlation of the restriction fragments with the family pedigrees and the results of the neurological examinations led ultimately to the identification of two RFLP patterns associated with the Huntington's gene. The presence or absence in an individual of the RFLP pattern present in a parent with Huntington's disease predicts with a high degree of certainty whether that person will or will not be similarly afflicted.

As it has turned out, however, many members of families with a history of Huntington's do not want to know their future in terms of the disease. With Huntington's, the most significant development may not be the possibility of earlier detection but rather the recent identification of the gene responsible for the disease.

Identification of Specific Human Genes

For some human genetic disorders, the underlying cause has been known for many years. In the case of sickle-cell anemia, for example, the mutation responsible for the abnormal hemoglobin molecules is known down to the last nucleotide. For the vast majority of disorders, however, the identities of the genes involved, as well as the nature and consequences of the mutations that lead to the disorders, have remained puzzles.

The search for any specific human gene is a complex and time-consuming endeavor, requiring the cooperative work of many researchers in different laboratories and the creative application of the methods of both classical and molecular genetics. One of the scientists in-

18–10   *Two members of the Venezuelan family afflicted by Huntington's disease: a man in his forties with Huntington's and his daughter, who has a 50 percent chance of developing the disorder.*

volved in the successful search for the cystic fibrosis gene, which was located in 1989, described the problem as the equivalent of looking for a broken faucet (the mutation that leads to cystic fibrosis) in a house (the gene) located somewhere in the United States (the 44 human autosomes). The problem is only slightly simpler when the gene being sought is known to be located on either the X or the Y chromosome. The enormous effort, however, is beginning to pay off.

In March of 1993, for example, some 10 years after the RFLP markers for Huntington's disease were first identified, the gene itself was located. Near the tip of chromosome 4, it consists of some 210,000 base pairs, coding for a protein that has not yet been identified. Like the gene responsible for fragile X syndrome (page 297), the Huntington's gene contains triplet repeats that vary in number. In individuals without Huntington's disease, the gene usually contains from 11 to 24 repeats of the sequence CAG. In individuals with Huntington's, the number ranges from 42 to 86 repeats. The fact that the repeats vary in number gave the gene a "now you see it, now you don't" quality that made its identification particularly difficult.

Crown Gall Disease: The Ti Plasmid

Agrobacterium tumefaciens is a common soil bacterium that infects plants, producing a lump, or tumor, of tissue known as a crown gall. Even if the bacteria are destroyed with antibiotics, the gall, once started, continues to grow. The cause of the gall, it turns out, is not the bacterium itself but rather a large plasmid it carries. A portion of this plasmid, which is known as Ti (for tumor-inducing), becomes integrated into the DNA of the host plant cell. Crown gall disease is the only naturally occurring genetic recombination yet recorded between prokaryotic and eukaryotic cells.

To try to understand how the Ti plasmid exerts its effects, recombinant DNA technology has been used to study its genes. Three of the genes direct the synthesis of plant hormones that act directly on the gall cells to promote their growth. Additional genes subvert the plant cell's metabolic machinery to produce unusual amino acids, which can be used by the gall cells but not by normal cells. Moreover, these substances appear to act as molecular aphrodisiacs, increasing bacterial conjugation and thus promoting the spread of the Ti plasmid to uninfected bacterial cells. In effect, Ti takes over and directs the activities of both of its hosts—the bacterial cells and the plant cells—to promote its own multiplication.

The Ti plasmid is of interest not only because of its remarkable powers but also because of its utility as a vector to ferry useful genes into crop plants. It is being used, for example, to transfer genes that confer resistance to major plant diseases. Other genes that are candidates for such transfer are those required for C_4 photosynthesis (page 171) and for nitrogen fixation (to be discussed in Chapter 41).

One of the technical problems in attempts at gene transfer, particularly in plants, is knowing whether a gene has actually been introduced into a new host cell and, if transferred, whether it is directing the synthesis of protein.

(a)

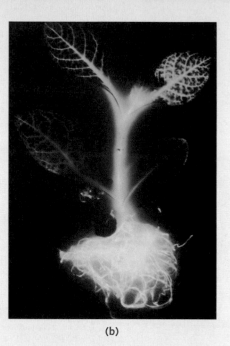

(b)

(a) *Crown galls growing on a tobacco plant. The galls, which are caused by a plasmid carried by* Agrobacterium tumefaciens, *are the large bulbous structures on the stem.*
(b) *Genes for the production of the enzyme luciferase were inserted into cells isolated from a tobacco plant, using* Agrobacterium *as a vector. After the cells multiplied and developed into a new plant, the cells that had incorporated the luciferase gene into their DNA were luminescent in the presence of luciferin, ATP, and oxygen.*

An unusual "indicator" has now been devised, using the gene for an enzyme, known as luciferase, that is found in fireflies. The substrate of luciferase is a protein called luciferin. In the presence of oxygen, luciferin plus luciferase plus ATP produces bioluminescence, as seen in the flash of the firefly.

In the first demonstration of the utility of the luciferase gene, the gene was cloned in *E. coli* and then spliced into the chromosome of a plant virus, which provided a regulatory sequence for the gene. The altered viral chromosome was then inserted into Ti plasmids, the plasmids were transferred to *Agrobacterium,* and the bacterial cells were incubated with tobacco leaf cells. The cells formed a mass of tissue from which, in a suitable growth medium, new plants were produced. The new plants were watered with a solution containing luciferin. You can see the result above: the plants shone!

Luciferase provides an extraordinary signal that gene transfer has taken place, and work is now in progress to combine its gene with other plant genes that are candidates for transfer. In the meantime, these experiments are a glowing example of the ingenuity of both molecular geneticists and the Ti plasmid.

With the gene located, work has begun to determine the exact nucleotide sequences of both the normal allele and the abnormal allele that is responsible for Huntington's disease. When these nucleotide sequences are determined, it should be possible to identify and characterize the protein coded by the gene. Ultimately, it is hoped, this work will not only reveal the underlying abnormality that causes brain cells to deteriorate some three to five decades after birth, but will also illuminate fundamental questions of the normal development and aging of the nervous system.

On the basis of previous experience, this hope seems well founded. For example, discovery of the gene responsible for the most common form of muscular dystrophy in children is shedding new light on muscle physiology. Similarly, discovery of the gene responsible for cystic fibrosis (page 212), which codes for a protein involved in the transport of chloride ions across the plasma membrane, is leading to new understandings of membrane transport processes.

Transfers of Genes between Eukaryotic Cells

As knowledge of the role of plasmids, viruses, and transposons in moving segments of DNA into and out of chromosomes increased, and as recombinant DNA techniques were refined, molecular biologists began to attempt to transfer functioning genes between eukaryotic cells. The hope was, of course, that it would someday be possible to cure human genetic disorders, substituting "good" genes for "bad" ones. As we saw at the beginning of the chapter, this hope appears on the verge of becoming reality. The shift from hope to reality has been made possible by a large body of work in other eukaryotic organisms.

The transfer of genes between eukaryotic cells is an extremely complex process. It requires, first, the preparation of a gene that will be taken up by the cells, become incorporated into a chromosome, and be expressed there—but this is only the beginning. The new gene must be established in a large number of cells of the appropriate type and be subject to the complicated regulatory controls of the normal gene.

The first stage proved easier than expected. Foreign genes will undergo recombination in eukaryotic cells growing in test tubes. In the first such experiment, SV40 virus was used as a vector to insert a rabbit gene for the beta-globin polypeptide into monkey cells. The recipient cells produced rabbit beta globin. The advantage of using viruses as vectors is that not only can viruses gain access to target cells but also they characteristically possess strong promoters (page 271), with the result that the gene is efficiently expressed. The gene coding for

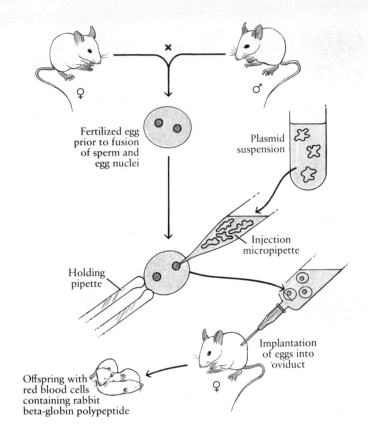

18–11 *The procedure by which the rabbit gene for the beta polypeptide chain of hemoglobin was inserted into mice. The gene was spliced into plasmids, which were then injected into the sperm nucleus within a fertilized egg before the egg and sperm nuclei fused to form a single diploid nucleus. A number of fertilized eggs were injected in this manner and then implanted in a female mouse, who subsequently gave birth. The rabbit beta-globin polypeptide was present in the red blood cells of the offspring. Nucleic acid hybridization revealed that the gene had been incorporated into their DNA.*

ADA (page 301), for example, was carried into white blood cells and, subsequently, into stem cells by a retrovirus.

Foreign genes have also been introduced into fertilized eggs and expressed in the organisms that developed from the eggs. For example, the gene for rabbit beta globin was transferred into the fertilized eggs of mice, using a plasmid as the vector (Figure 18–11). The red blood cells of the mice that developed from the eggs contained rabbit beta globin. The fact that the gene was expressed only in the red blood cells—and not in other tissues of the mouse—indicated that it had been incorporated in the "right" place and so had come under cellular control mechanisms. The gene was also passed, in a Mendelian distribution, to subsequent generations.

In the figure, the following labels appear:
- Fertilized egg prior to fusion of sperm and egg nuclei
- Plasmid suspension
- Injection micropipette
- Holding pipette
- Implantation of eggs into oviduct
- Offspring with red blood cells containing rabbit beta-globin polypeptide

18–12   *The two female mice shown here are litter mates, approximately 24 weeks old. The fertilized egg from which the mouse on the left developed was injected with a gene consisting of the promoter and regulatory sequences of a mouse gene combined with the gene for human growth hormone. Following integration of the new gene into the DNA of this mouse, it was passed on to her offspring. Mice that express the new gene grow, on average, two to three times as fast as mice lacking the gene and, as adults, they are twice the normal size. (Those unable to see the utility of a giant mouse may be interested to learn that similar procedures have been successfully performed in fish.)*

Using the same technique, the gene for human growth hormone was combined with the regulatory portion of a mouse gene and injected into fertilized mouse eggs. The resultant "transgenic" mice—mice receiving the transferred human gene—grew to twice the normal size (Figure 18–12), indicating that the human gene was incorporated into the DNA of the mouse and was producing growth hormone. Because the DNA was injected into the egg cells, it was found in all cells, including the cells destined to become gametes. Thus, it could be passed on to the next generation. This type of procedure, using plasmids or transposons as vectors, has now been carried out with a number of cloned genes. From 10 to 30 percent of the eggs survive the manipulation, and the foreign gene functions in up to 40 percent of those eggs.

Such procedures are now making it possible to introduce desirable characteristics into crop plants (see essay) and domesticated animals, such as cows and sheep. With genetic engineering, the exact identity of the gene or genes being introduced is known, providing a significant degree of control over the characteristics that will be affected. By contrast, with the conventional breeding techniques that have been used for many years, enhancing particular desirable characteristics often leads to other, less desirable characteristics. Reduced fertility, for example, is frequently a consequence of the intensive inbreeding necessary to establish a particular characteristic in a population of plants or animals.

In economically important plants and animals, the reliable transmission of the introduced genes from generation to generation is an important goal. In the case of human gene therapy, however, there are serious moral and ethical questions about the possibility of changing the course of genetic history. Because of those questions—and also because of seemingly insurmountable technical difficulties—attempts at human gene therapy are, thus far, limited to therapies that affect only somatic cells and thus only the individual receiving the therapy.

Some observers, both inside and outside the scientific community, have expressed alarm at the newfound powers being made available by the triumphs of molecular biology. It is unlikely, however, that such doubts will slow the pace of discovery and experiment. The history of our species over the past 20,000 years suggests that curiosity, the search for knowledge, and the desire to improve and to heal are, for better or worse, a large part of what it means to be human. The best we can hope for is that we will use our knowledge and newly acquired skills with wisdom and compassion.

The Human Genome Project

Work is now underway to map and sequence the entire human genome—that is, all of the DNA in human cells. This endeavor has been compared to the Manhattan Project, which brought forth the atomic bomb, and the space program, which culminated with a human footprint on the moon. Although initially the subject of some controversy and dissent, which was also true of the projects to which it is compared, the human genome project is going forward.

There is action on two fronts. One multi-institutional enterprise is focused on mapping the human chromosomes. Data for chromosome maps can be generated in a number of ways. One method involves finding genes with different and detectable alleles and documenting their rate of recombination in breeding populations. For example, scientists at the University of Utah are tracing such marker genes through three generations of 60 Mormon families. (Mormon families are large and cohesive, and their genealogical records are meticulously maintained in the Church archives.) More than 500 genes have been mapped, and other investigators have located several hundred additional markers in other families.

The techniques of recombinant DNA are also being used in chromosome mapping. In addition to analyzing RFLP patterns to locate markers, scientists are identifying the sites at which specific restriction enzymes cut the DNA of particular chromosomes. In other studies, radioactive probes are being used to locate genes for which the protein products are known.

The types of data generated by these different techniques are, unfortunately, incompatible. Thus, it has been proposed that each time a marker is identified, by whatever technique, the researchers sequence a segment of 200 to 500 base pairs at one end of the marker, and then search both ends of the segment for short, unique sequences that identify the segment. In this way, the information generated by different techniques can be integrated into one cohesive map. The goal is to create a map of the human chromosomes in which markers are located at intervals of approximately 100,000 base pairs.

The second, and more controversial, effort involves the nucleotide sequencing of the entire genome. When it was first proposed years ago, many molecular biologists were against it, mainly because of its great expense. Much of the genome is "senseless," they pointed out, most funds available for research would have to be channeled into this one project, it would take decades to complete, and, perhaps most important of all, the creative talent of almost an entire generation of young scientists in the field would be wasted on this routine enterprise. However, the prevailing winds of opinion subsequently shifted. The crucial change came about as a result of the development of polymerase chain reaction and of methods for automatic sequencing of nucleotides, which have greatly decreased the estimates of time and money required for the project.

Also, as the sequencing of the genome comes closer to realization, there is mounting excitement about the new knowledge that may be generated. What is the mysterious role of the "senseless" DNA? How are genes regulated? (The answer to this question may hold the key to understanding and treating cancers and also be important in gene therapy.) What evolutionary relationships will be revealed among the human genes themselves? And among humans and other species?

A completion date for the project has been set: the year 2004. That is little more than a hundred years after the rediscovery of Mendel's work, which marks the beginning of genetics as a science, and only 51 years after the announcement by Watson and Crick that opened up the field of molecular biology. And you will see it happen.

Summary

Recombinant DNA technology includes methods for (1) obtaining DNA segments short enough to be analyzed and manipulated, (2) obtaining large quantities of identical DNA segments, (3) determining the exact order of nucleotides in a DNA segment, and (4) locating and identifying specific DNA segments of interest.

Short DNA segments can be obtained by transcribing mRNA into DNA with the enzyme reverse transcriptase, by chemical synthesis, or by cleaving DNA molecules with restriction enzymes. These are bacterial enzymes that cleave foreign DNA molecules. The DNA segments produced by restriction enzymes can be separated from one another on the basis of their size by electrophoresis.

Different restriction enzymes cut DNA at different specific nucleotide sequences. Instead of cutting the molecule straight across, some restriction enzymes leave "sticky" ends. Any DNA cleaved by such an enzyme can be joined readily to another DNA molecule cleaved by the same enzyme. The discovery of restriction enzymes made possible the development of recombinant DNA technology.

Two techniques, DNA cloning and polymerase chain reaction, are used to produce large quantities of identical DNA segments. In cloning, the segments to be copied are introduced into bacterial cells by means of plasmids or bacteriophages, which function as vectors. Once in the bacterial cell, the vector and the foreign DNA it carries are replicated, and the multiple copies can be harvested from the cells. Polymerase chain reaction is a much more rapid process but requires a greater knowledge of the segment that is to be copied.

The availability of multiple copies makes possible, in turn, the determination of the exact order of nucleotides in a DNA segment. By combining sequencing information for sets of short segments produced by different restriction enzymes, molecular biologists can determine the complete sequence of a long DNA segment (such as an entire gene).

Segments of DNA of interest for study and manipulation can be identified by nucleic acid hybridization, using single-stranded probes labeled with radioactive isotopes or fluorescent dyes. This technique is based on the capacity of a single strand of RNA or DNA to combine, or hybridize, with another strand with a complementary nucleotide sequence.

Recombinant DNA technology has many practical applications. For example, it is possible to incorporate specific genes into suitable vectors, introduce them into bacterial cells, and induce the bacterial cells to synthesize the proteins coded by the genes. Insulin, clotting factors, and growth hormone are among the medically important proteins now produced in this way.

Recombinant DNA technology is also providing new means for early diagnosis of genetic disorders. Among the most important tools for such diagnosis are RFLPs (restriction-fragment-length polymorphisms) and radioactive probes. RFLPs are the result of mutations that eliminate or alter the recognition sequence for a restriction enzyme. When such a mutation is associated with an allele causing a genetic disorder, it can provide a diagnostic marker for that allele. Labeled probes, which bind to either the normal or the mutant allele, can be used for detection and diagnosis when the nucleotide sequence of the allele is known or can be deduced.

A variety of techniques are being used to locate and identify the genes responsible for various human disorders. Recently, the genes responsible for Huntington's disease, cystic fibrosis, and the most common form of childhood muscular dystrophy have been identified.

Increasing success is being achieved in the transfer of genes between eukaryotic cells. Such transfers have been accomplished in both plants and animals. These experiments are leading to increased understanding of the regulatory factors governing the expression of eukaryotic genes and have made possible the first attempts at human gene therapy.

A massive effort is underway to map and sequence the entire human genome. This project, which is expected to be completed by the year 2004, may answer many puzzling questions. If past experience is any guide, it is also likely to raise many new questions.

Questions

1. What are restriction enzymes? What are their uses in recombinant DNA technology?

2. Describe the role of "sticky" ends in recombinant DNA technology. How are the "sticky" ends produced? What enzyme is required to complete the recombination?

3. What is electrophoresis? Why is it of such great value in recombinant DNA studies?

4. Suppose you treated a DNA molecule with a particular restriction enzyme and obtained five fragments, which you separated and cloned into multiple copies. Using the multiple copies, you then sequenced the five fragments. What would you do next to establish their order in the original molecule?

5. Suppose you wish to locate on the chromosome the gene coding for a small protein molecule. You know the amino acid sequence of the protein, and you have the technical skill to synthesize an artificial mRNA molecule with any nucleotide sequence you choose. How would you go about locating the gene? Suppose you wish to separate the gene from the rest of the chromosome. How would you proceed?

6. Why, in the somatostatin project, did the investigators link the synthetic somatostatin gene to regulatory elements of the *lac* operon?

7. *E. coli* cells used in recombinant DNA work are "disabled." That is, they lack the capacity to synthesize key components of their cell walls, for example, or to make a nitrogenous base, such as thymine. Consequently, they are able to survive only in an enriched laboratory medium. Why is such a precaution taken by molecular biologists?

8. In the first family in which the RFLP test for the sickle-cell allele was used, both parents were known to carry the allele, and they had previously had a baby with sickle-cell anemia. The woman was pregnant again, and the parents did not want to have another child with this extremely painful disease. DNA tests from each of the parents yielded both the short and the long fragments. Tests from their child with sickle-cell anemia yielded only long fragments. Tests of the fetal cells yielded both short and long fragments. What was the status of the unborn child with respect to the sickle-cell allele?

Suggestions for Further Reading

Alberts, Bruce, Dennis Bray, Julian Lewis, Martin Raff, Keith Roberts, and James D. Watson: *Molecular Biology of the Cell,* 3d ed., Garland Publishing, Inc., New York, 1994.

This outstanding cell biology text provides an excellent discussion of recombinant DNA technology.

Anderson, W. French: "Human Gene Therapy," *Science,* vol. 256, pages 808–813, 1992.

Anderson, W. R., and E. G. Diachmakos: "Genetic Engineering in Mammalian Cells," *Scientific American,* July 1981, pages 106–121.

Ayala, Francisco J., and Bert Black: "Science and the Courts," *American Scientist,* vol. 31, pages 230–239, 1993.

Chilton, Mary-Dell: "A Vector for Introducing New Genes into Plants," *Scientific American,* June 1983, pages 50–59.

DeLisi, Charles: "The Human Genome Project," *American Scientist,* vol. 76, pages 488–493, 1988.

Erickson, Deborah: "Hacking the Genome," *Scientific American,* April 1992, pages 128–137.

Franklin-Barbajosa, Cassandra: "DNA Profiling: The New Science of Identity," *National Geographic,* May 1992, pages 112–123.

Gasser, Charles S., and Robert T. Fraley: "Transgenic Crops," *Scientific American,* June 1992, pages 62–69.

Gilbert, Walter, and Lydia Villa-Kamaroff: "Useful Proteins from Recombinant Bacteria," *Scientific American,* April 1980, pages 74–94.

Lerner, Richard A.: "Synthetic Vaccines," *Scientific American,* February 1983, pages 66–74.

Mange, Arthur P., and Elaine Johansen Mange: *Genetics: Human Aspects,* 2d ed., Sinauer Associates, Inc., Sunderland, Mass., 1990.

A thorough account of all aspects of human genetics, including applications of recombinant DNA technology to the understanding and treatment of human genetic disorders.

Morell, Virginia: "Huntington's Gene Finally Found," *Science,* vol. 260, pages 28–30, 1993.

Moses, Phyllis B., and Nam-Hai Chua: "Light Switches for Plant Genes," *Scientific American,* April 1988, pages 88–93.

Mullis, Kary B.: "The Unusual Origin of the Polymerase Chain Reaction," *Scientific American,* April 1990, pages 56–65.

Nossal, G. J. V.: *Reshaping Life,* 2d ed., Cambridge University Press, New York, 1990.

A short, excellent, easily read introduction to recombinant DNA techniques and the biotechnology industry.

Scientific American: *Recombinant DNA,* W. H. Freeman and Company, New York, 1978.[*]

The articles in this collection from Scientific American *include discussions of the critical experiments that paved the way for modern molecular genetics.*

Thompson, Larry: "At Age 2, Gene Therapy Enters a Growth Phase," *Science,* vol. 258, pages 744–746, 1992.

Torrey, John G.: "The Development of Plant Biotechnology," *American Scientist,* vol. 73, pages 354–363, 1985.

Verma, Inder M.: "Gene Therapy," *Scientific American,* November 1990, pages 68–84.

Walters, LeRoy: "The Ethics of Human Gene Therapy," *Nature,* vol. 320, pages 225–227, 1986.

Watson, James D., John Tooze, and David T. Kurtz: *Recombinant DNA,* 2d ed., W. H. Freeman and Company, New York, 1992.

A short course in genetics, organized around the central theme of recombinant DNA. Clearly written and handsomely illustrated, it is accessible to anyone who enjoys reading Scientific American.

White, Ray, and Jean-Marc Lalouel: "Chromosome Mapping with DNA Markers," *Scientific American,* February 1988, pages 40–48.

[*] Available in paperback.

Evolution

The Genetic Basis of Evolution

In Charles Darwin's time, many naturalists were abandoning the concept of special creation and moving toward some theory of evolution. An intellectual revolution was in the making—one of the greatest in human history—but there was no strong framework to support it. None of the current ideas, of which there were many, were adequate until Darwin presented his theory in *The Origin of Species*.

What Darwin (and his younger colleague, Alfred Russel Wallace) grasped was that evolution is a two-part process. First, there are differences—variations—among individuals in any population, and some of these variations can be passed from generation to generation. Second, these variations, which occur by chance, may have consequences that affect survival and reproduction. It is this second step that Darwin called "natural selection."

Darwin received primary credit for the theory because of his carefully documented, closely argued presentation, drawing on his own wide-ranging experience and his voluminous reading and correspondence. One of his many examples was camouflage in insects. As is often the case, Darwin says it best:

"Insects often resemble for the sake of protection various objects, such as green or decayed leaves, dead twigs, bits of lichen. . . . The resemblance is often wonderfully close, and is not confined to colour, but extends to form, and even to the manner in which the insects hold themselves. . . . Assuming that an insect originally happened to resemble in some degree a dead twig or a decayed leaf, and that it varied slightly in many ways, then all the variations which rendered the insect at all more like any such object, and thus favoured its escape, would be preserved, whilst other variations . . . if they rendered the insect at all less like the imitated object, they would be eliminated."

These are walkingstick insects, photographed in the Ozark Mountains of Missouri. The male is smaller than the female. Note how carefully she is lined up with the twig on which she rests. He is slightly out of line, risking his life by mating.

A vast body of evidence indicates that Earth has had a long history and that all living organisms, including ourselves, arose in the course of that history from earlier, more primitive forms. This means that all species are descended from other species; in other words, all living things share common ancestors in the distant past. The thread that links together the enormous diversity of species that live on Earth today—and that have lived in the past—is **evolution,** a process of change through time.

In this section of our text (Chapters 19–22), we shall look at the mechanisms by which evolution occurs. In the next section (Chapters 23–27), we shall turn our attention to the diverse groups of organisms that have arisen in the course of evolutionary history.

Darwin's Theory

As we saw in the Introduction,* Charles Darwin was not the first to propose that the diversity of organisms is the result of historical processes. Credit for the theory of evolution is rightly his, however, for two reasons. First, his "long argument"—as he himself described *The Origin of Species*—left little doubt that evolution had actually occurred and so marked a turning point in the science of biology. The second reason, which is closely related to the first, is that Darwin correctly perceived the general mechanism by which evolution occurs.

Although it is now more than 130 years since the publication of *The Origin of Species*, Darwin's original concept of how evolution works still provides the basic framework for our understanding of the process. This concept rested on five premises:

1. Organisms beget like organisms—in other words, there is stability in the process of reproduction.

* If you have not read pages 1–12 of the Introduction, we suggest that you do so now.

2. In most species, the number of individuals in each generation that survive and reproduce is small compared with the number initially produced.

3. In any given population, there are variations among individual organisms, and some of these variations can be inherited.

4. Which individuals will survive and reproduce and which will not are determined to a significant degree by the interaction between these variations and the environment. Some variations enable individuals to survive longer and produce more offspring than other individuals. Darwin called these variations "favorable" and argued that inherited favorable variations tend to become more and more common from one generation to the next. This is the process that Darwin called **natural selection.**

5. Given sufficient time, natural selection can lead to the accumulation of changes that differentiate groups of organisms from one another.

As originally formulated by Darwin, however, the theory of evolution had a major weakness: the absence of any valid mechanism to explain heredity. Although Mendel was at work on his experiments with pea plants at the time Darwin was writing *The Origin of Species,* his paper was not delivered until 1865 and did not enter the mainstream of biological thinking until early in the twentieth century. The subsequent development of genetics made it possible to answer three questions that Darwin was never able to resolve: (1) how inherited characteristics are transmitted from one generation to the next; (2) why inherited characteristics are not "blended out" but can disappear and then reappear in later generations (like whiteness in pea flowers); and (3) how the variations arise on which natural selection acts.

The Concept of the Gene Pool

A new branch of biology, **population genetics,** emerged from the synthesis of Mendelian principles with Darwinian evolution. A **population** can be defined as an interbreeding group of organisms of the same species, living in the same place at the same time. For instance, all of the fish of one particular species in a pond constitute a population, as do all the fruit flies in one bottle.

A population is unified and defined by its **gene pool,** which is simply the sum total of all the alleles of all the genes of all the individuals in the population. From the viewpoint of the population geneticist, each individual organism is only a temporary vessel, holding a small

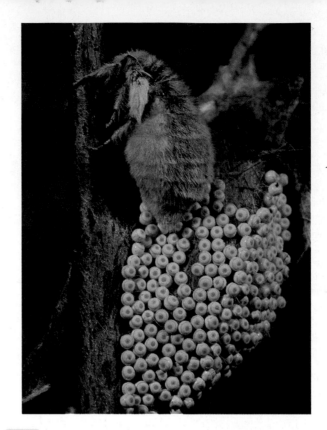

19–1   *A female rusty tussock moth, laying eggs on the empty cocoon from which she emerged after her transformation from caterpillar to adult moth. Despite the large number of eggs laid by each female who attracts a mate, there is, over a span of many generations, no increase in the number of rusty tussock moths.*

sampling of the gene pool for a moment in time. The subjects of interest to the population geneticist are gene pools, changes in their composition over time, and the forces causing these changes.

In natural populations, some alleles increase in frequency from generation to generation, and others decrease. If an individual has a favorable combination of alleles in its genotype, it is more likely to survive and reproduce. As a consequence, its alleles are likely to be present in an increased proportion in the next generation. Conversely, if the combination of alleles is not favorable, the individual is less likely to survive and reproduce. Representation of its alleles in the next generation will be reduced or perhaps eliminated. Evolution is the result of such accumulated changes in the gene pool over time.

In the context of population genetics, the **fitness** of an individual does not mean physical well-being or optimal adaptation to the environment. The only criterion—the sole measure—of an individual's fitness is the relative number of surviving offspring, that is, the extent to which the alleles in an individual's genotype are present in succeeding generations.

Survival of the Fittest

The phrase "survival of the fittest" is often used in describing the Darwinian theory. At various times in the twentieth century, the doctrine of survival of the fittest in natural populations has been used to defend gross social inequalities and ruthless competitive tactics in industry on the grounds that they are merely in accord with the "laws of nature." This philosophy is sometimes known as social Darwinism.

In fact, however, very little in the process of evolutionary change fits the concept of "nature red in tooth and claw." One plant with flowers a little brighter than those of its neighbors and so better able to catch the attention of a passing hummingbird is a more pertinent model of the struggle for survival. Fitness, as measured by population geneticists, is determined solely by the relative number of descendants of an individual in a future population.

Natural selection in operation. A female purple-throated hummingbird (Lampornis calolaema) *is making her choice. A resident of the* cloud forest of Costa Rica, this hummingbird is also known as the "mountain gem."

The Extent of Variation

Like begets like, we know now, because of the remarkable precision with which the DNA is copied and transmitted from each cell to its daughter cells. The DNA in the cells of any individual is, except for occasional mutations, a true replica of the DNA that individual received from its father and mother.

This fidelity of duplication is, of course, essential to the survival of the individual organisms of which a population is composed. However, if evolution is to occur, there must be variations among individuals. Such variations make it possible for populations to change as conditions change. They are the raw material on which evolutionary forces act.

As we have noted previously, Darwin was the first to recognize the importance of widespread, inherited variations in the process of evolution. One focus of research in population genetics has been to determine the extent of genetic variability and how it is preserved and fostered in gene pools. The extent of variation has been revealed in a number of different ways.

Artificial Selection

First-time readers of *The Origin of Species* are sometimes surprised to find themselves, in the very first chapter, caught up in a treatise on pigeon breeding. By selecting birds with particular characteristics—such as a larger beak or a tail with more feathers—for breeding, pigeon fanciers had been able, over the years, to produce a number of exotic breeds (see Figure I–9, page 10). All developed from the same wild species and still able to interbreed with one another, the new breeds differed widely in appearance, more widely in fact than many animals of different species.

Darwin coined the term **artificial selection** for this process of human choice in determining which individuals would be represented in the next generation—and which would not. He saw it as a direct analogy to the process of natural selection. As he was well aware, the selective breeding of pigeons is not an isolated example. Similar experiments in evolution have been carried out by breeders of plants and animals for centuries, producing some of our most familiar domestic varieties

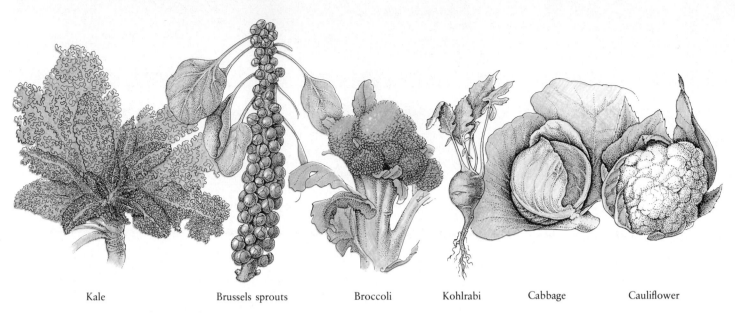

Kale   Brussels sprouts   Broccoli   Kohlrabi   Cabbage   Cauliflower

19–2   *Six vegetables produced from a single species of plant* (Brassica oleracea), *a member of the mustard family. They are the result of selection for leaves (kale), lateral buds (brussels sprouts), flowers and stem (broccoli), stem (kohlrabi), enlarged terminal buds (cabbage), and flower clusters (cauliflower). Kale most resembles the ancestral wild plant. Artificial selection, as practiced by plant and animal breeders, gave Darwin the clue to the concept of natural selection.*

(Figure 19–2). These experiments have shown that there is a large amount of variability hidden in the gene pool and that this latent variability can be expressed when selection pressures are applied.

Breeding Experiments in the Laboratory

Breeding experiments that demonstrate the extent of latent variability have also been carried out in the laboratory. In studies with *Drosophila melanogaster,* for example, an easily observable hereditary trait, the number of bristles on the ventral surface of the fourth and fifth abdominal segments (Figure 19–3), was chosen for selection.

In the starting stock, the average number of bristles was 36. From this initial population, the investigators established two separate groups of fruit flies. In one group, selection was for a decrease in the number of bristles; in the other group, it was for an increase in the number of bristles. In every generation, those individ-

19–3   *Scanning electron micrographs of* (**a**) *a fruit fly* (Drosophila melanogaster) *and* (**b**) *the ventral surface of its posterior abdominal segments. The results of selection experiments to increase and decrease the number of bristles on the ventral surface of the fourth and fifth abdominal segments are shown in Figure 19–4.*

(a)   0.5 mm

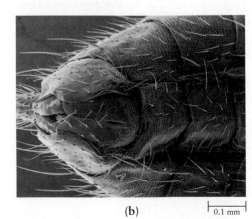

(b)   0.1 mm

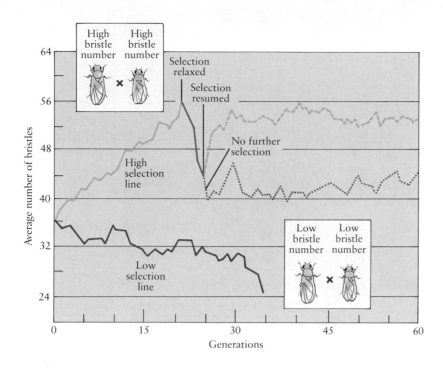

19–4 *The results of experiments with* Drosophila melanogaster, *demonstrating the extent of latent variability in a population. From a single parental stock, one group was selected for an increase in the number of bristles on the ventral surface (high selection line) and one for a decrease in the bristle number (low selection line).*

As you can see, the high selection line rapidly reached a peak at which the average number of bristles was 56, but then the stock began to become sterile. Selection was abandoned at generation 21 (dark blue line) and begun again at generation 24 (dashed line). This time, the previous high bristle number was regained, and there was no apparent loss in reproductive capacity. Note that after generation 24 the stock interbreeding without selection was also continued, as indicated by the dotted line. After 60 generations, members of the freely breeding group from the high selection line had an average of 45 bristles. The low selection line died out owing to sterility.

uals in the first group with the lowest number of bristles were selected and bred with each other. In the other group, the individuals with the highest number of bristles were selected and bred with each other.

Selection for low bristle number resulted in a drop over 30 generations from an average of 36 to an average of about 30 bristles. In the high-bristle-number line, progress was at first rapid and steady. In 21 generations, bristle number rose steadily from 36 to an average of about 56 (Figure 19–4). No new genetic material had been introduced. The potential for a wide range of bristle numbers had been already present—but hidden—within the single starting population. Subsequent experiments with *Drosophila* and other organisms have demonstrated a comparable range of natural variability in many different traits.

There is a second part to the bristle-number story. The low-bristle-number line soon died out because it became sterile. Although the flies mated, no offspring were produced. Presumably, changes in factors affecting fertility had also taken place during selection. When sterility started to become severe in the high-bristle line, the remaining flies of this line were permitted to interbreed freely without further selection. The average number of bristles fell sharply, and in five generations dropped from 56 to 40. Thereafter, as this line continued to breed without selection, the bristle number fluctuated up and down, usually between 40 and 45, which still was higher than the original 36. At generation 24, selection for high bristle number was begun again for a portion of this line. The previous high bristle number of 56 was regained, and this time there was no loss in

reproductive capacity. Apparently, the genotype had become reorganized in such a way that alleles producing a higher bristle number were present in more favorable combinations with alleles affecting fertility.

Mapping studies have shown that bristle number is controlled by a large number of genes, at least one on every chromosome and sometimes several at different sites on the same chromosome. Although we do not know how important bristle number is to the survival of the animal, we do know that selection for this trait in some way disrupted the entire genotype. Livestock breeders are well aware of this consequence of artificial selection. Loss of fertility is a major problem in virtually all circumstances in which animals have been purposely inbred for particular characteristics.

Quantifying Variability

Analysis at the molecular level provides another method for assessing variability. As you know, the amino acid sequences of proteins reflect the nucleotide sequences of the genes coding for them. In a classic study, J. L. Hubby and R. C. Lewontin ground up fruit flies from a natural population and extracted proteins from them. From these proteins, they were able to isolate 18 functionally different enzymes. Then they analyzed each of the 18 enzymes separately, to determine if its molecules existed in different structural forms in different flies or if the structure of the enzyme was uniform throughout the population. The method they used was electrophoresis—the same method used by Pauling to separate the variants of hemoglobin (see page 257).

Of the 18 enzymes studied in this way, nine were found to be composed of protein molecules that were indistinguishable by electrophoresis. In other words, the gene coding for each of these enzymes was the same throughout the entire population of fruit flies studied. However, each of the other nine enzymes was found to exist in two or more structurally different forms. Thus, without any direct analysis of the genes themselves, the investigators were able to conclude that among the fruit flies studied there were two or more alleles of the gene responsible for each of these nine enzymes. One enzyme had six slightly different structural forms (Figure 19–5). This meant that at least six alleles of the gene coding for that enzyme were present in the gene pool.

Each fruit-fly population examined was heterozygous for almost half of the genes tested. Each individual, it was estimated, was probably heterozygous for about 12 percent of its genes. Similar studies on humans, using accessible tissues such as blood or placenta, indicate that at least 25 percent of the genes in any given population are represented by two or more alleles, and individuals are heterozygous for at least 7 percent of their genes, on the average.

Now that methods for sequencing nucleotides in DNA have been developed, it is possible to make direct comparisons of the genetic material. Since all changes in nucleotides do not result in changes in amino acid sequences, and all changes in amino acid sequences are not detectable by electrophoresis, DNA sequencing is expected to reveal even more genetic variation.

Explaining the Extent of Variation

The findings of Hubby and Lewontin and others after them have, like most important scientific discoveries, raised major new questions. Many geneticists had previously thought that the individuals of a population should be close to genetic uniformity, as a result of a long history of selection for "optimal" genes. Yet, as these studies revealed, natural populations are far from uniform.

One school of geneticists, the selectionists, claim that even such small variations as those in enzyme structure are maintained by various forms of natural selection that favor some genotypes at some times and in some areas, and others at other times or in other localities. In the selectionist's view, all variation directly or indirectly affects fitness. An opposing school, the neutralists, claim that the observed variations in the protein molecules are so slight that they do not make any difference in the function of the organism and so are not acted upon by natural selection. Neutral alleles, according to this argument, accumulate as a result of random processes, including mutation. Although this difference of

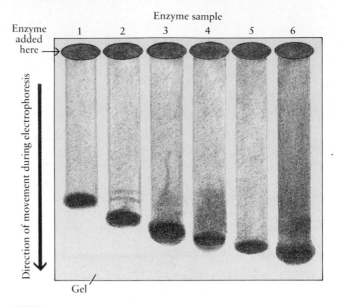

19–5   *The method used by Hubby and Lewontin to analyze* Drosophila *enzymes was electrophoresis. In this process, as you know, the sample is dissolved, placed at the edge of a sheet of gel, and exposed to a weak electric field. The rate at which the molecules move in the electric field is determined by their size and electric charge. As a result, proteins with even very slight structural differences can be separated.*

This diagram shows the electrophoretic patterns of six different forms of one enzyme. The material in each column was obtained from flies homozygous for one of the six different alleles coding for the enzyme.

opinion is not yet resolved, we do have a growing understanding of the variety of processes that can act to maintain, increase, or decrease the variability in the gene pool of a population.

A Steady State: The Hardy-Weinberg Equilibrium

In the early 1900s, biologists raised an important question about the maintenance of variability in populations. How, they asked, can both dominant and recessive alleles remain in populations? Why don't dominants simply drive out recessives? For example, given that brachydactylism (Figure 19–6) is caused by a dominant allele, why don't most or even all people have short, fat fingers? This question was answered in 1908 by G. H. Hardy, an English mathematician, and G. Weinberg, a German physician.

Working independently, Hardy and Weinberg showed that the genetic recombination that occurs at each generation in diploid organisms does not *by itself*

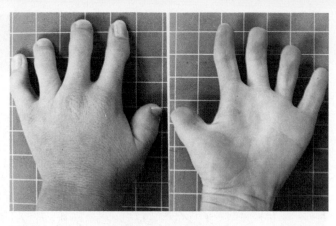

19–6 *A dominant allele is responsible for the characteristic known as brachydactylism (short fingers). In the brachydactylous hands shown here, the first bones of the fingers are of normal length, but the second and third bones are abnormally short. If brachydactylism is caused by a dominant allele, why is it such a rare characteristic?*

change the overall composition of the gene pool. To demonstrate this, they examined the behavior of alleles in an idealized population in which five conditions hold:

1. No mutations occur.

2. There is no net movement of individuals—with their genes—into the population (immigration) or out of it (emigration).

3. The population is large enough that the laws of probability apply; that is, it is highly unlikely that chance alone can alter the **frequencies,** or relative proportions, of alleles.

4. Mating is random.

5. All alleles are equally viable. In other words, there is no difference in reproductive success. The offspring of all possible matings are equally likely to survive to reproduce in the next generation.

Consider a single gene, which has only two alleles, *A* and *a*. Hardy and Weinberg demonstrated mathematically* that if the five conditions listed above are met, the frequencies of alleles *A* and *a* in the population will not change from generation to generation. Moreover, the frequencies of the three possible combinations of these alleles—the genotypes *AA, Aa,* and *aa*—will not change from generation to generation. The gene pool will be in a steady state—an equilibrium—with respect to these alleles.

This equilibrium is expressed by the following equation:

$$p^2 + 2pq + q^2 = 1$$

* This mathematical demonstration and its application in specific examples are given in Appendix A.

In this equation, the letter *p* designates the frequency of one allele (for example, *A*), and the letter *q* designates the frequency of the other allele (*a*). The sum of *p* plus *q* must always equal 1 (that is, 100 percent of the alleles of that particular gene in the gene pool). The expression p^2 designates the frequency of individuals homozygous for one allele (*AA*), q^2 the frequency of individuals homozygous for the other allele (*aa*), and $2pq$ the frequency of heterozygotes (*Aa*).

Although normally any one diploid individual has no more than two alleles of the same gene, there may, of course, be more than two alleles for a given gene in the gene pool. For instance, in the case of the gene for coat color in rabbits, there are at least four alleles. Different diploid combinations of these alleles give rise to coat colors ranging from dark gray through medium and light gray to white, as well as a variety in which most of the coat is white but the feet, ears, and muzzle are black (see Figure 13–3, page 223). The Hardy-Weinberg equilibrium applies equally well to situations such as this, in which there are multiple alleles of the same gene, although the equation representing the equilibrium is more complex.

The Significance of the Hardy-Weinberg Equilibrium

The Hardy-Weinberg equilibrium and its mathematical formulation have proved as valuable a foundation for population genetics as Mendel's principles have been for classical genetics. At first glance, this seems hard to understand, since the five conditions specified for the equilibrium are seldom likely to be met in a natural population. An analogy from physics may be useful. Newton's first law says that a body remains at rest or maintains a constant velocity when not acted upon by external force. In the real world, bodies are always acted upon by external forces, but this first law is an essential premise for examining the nature of such forces. It provides a standard against which to measure.

Similarly, the Hardy-Weinberg equation provides a standard against which we can measure the changes in allele frequencies that are always occurring in natural populations. Without the Hardy-Weinberg equation, we would not be able to detect change, determine its magnitude and direction, or uncover the forces responsible for it. If, however, we can identify and count the individuals in a population that are homozygous for a particular allele of interest, we can then calculate the frequency of that allele—and thus all the other terms of the Hardy-Weinberg equation. If we do this over a period of generations, we can chart precisely the changes that are taking place in the gene pool—and then look for the causes.

The Agents of Change

According to modern evolutionary theory, natural selection is the major force in changing allele frequencies. Because natural selection is of such great importance, we shall devote all of the next chapter to it. Here, let us look at some of the other agents that can change the frequencies of alleles in a population. There are four: mutations, gene flow, genetic drift, and nonrandom mating.

Mutations

Mutations, from the point of view of population genetics, are inheritable changes in the genotype. A mutation may involve the deletion, transposition, or duplication of a portion of a DNA molecule, or the substitution of one or more nucleotides in the molecule. Mutations can affect not only structural genes, such as the genes coding for the polypeptide chains of the hemoglobin molecule, but also regulatory genes, such as those responsible for turning on and off various processes in embryonic development.

As we noted in Chapter 15, mutations can be caused by a number of agents, including x-rays, ultraviolet rays, radioactive substances, and a variety of other chemicals. They can also be caused by the insertion of viral DNA into a chromosome or by the movement of transposons, as we saw in Chapter 17.

Most mutations occur "spontaneously"—meaning simply that we do not know the factors that trigger them. Mutations are generally said to occur at random, or by chance. This does not mean that mutations occur without cause but rather that the events triggering them are independent of their subsequent effects. Although the rate of mutation can be influenced by environmental factors, the specific mutations produced are independent of the environment—and independent of their potential for subsequent benefit or harm to the organism and its offspring.

The rate of spontaneous mutation is generally low. For mutations detectable in the phenotype, it varies from 1 in 1,000 to 1 in 1,000,000 gametes per generation, depending on the allele involved. It is estimated that each new human individual, with approximately 100,000 genes (pairs of alleles), carries two new mutations. Thus, although the incidence of mutation in any given gene or any given individual is low, the number of new mutations per population generation is very high. Mutations are regarded as the raw material for evolutionary change, since they provide the variation on which other evolutionary forces act. However, be-

19–7   *An example of the sometimes dramatic effects of mutation. The ewe in the middle is an Ancon, an unusually short-legged strain of sheep. The first Ancon on record was born in the late nineteenth century into the flock of a New England farmer. By inbreeding (the characteristic is transmitted as a recessive), it was possible to produce a strain of animals with legs too short to jump the low stone walls that traditionally enclosed New England sheep pastures. A similar strain was produced in northern Europe following an independent mutation there. At one time, it was proposed that evolution takes place in sudden, large jumps such as this—an idea sometimes referred to as the "hopeful monster" concept. One reason this idea has been abandoned is that nearly all mutations producing dramatic changes in the phenotype are harmful, as this one would be in a wild population.*

cause mutation rates are so low, mutations seldom, if ever, determine the direction of evolutionary change.

Gene Flow

Gene flow is the movement of alleles into or out of a population. It can occur as a result of the immigration or emigration of individuals of reproductive age. In the case of plants and many aquatic invertebrates, it can also occur through the movement of gametes (for example, in the form of pollen) between populations.

Gene flow can introduce new alleles into a population, or it can change existing allele frequencies. Its overall effect is to decrease the difference between populations. Natural selection, by contrast, is more likely to increase differences, producing populations more suited for different local conditions. Thus, gene flow often counteracts natural selection. As we shall see in Chapter 21, geographic barriers that prevent gene flow are very important in the formation of new species.

Genetic Drift

As we stated previously, the Hardy-Weinberg equilibrium holds true only if the population is large. This qualification is necessary because the equilibrium depends on the laws of probability. Consider, for example, an allele, say *a*, that has a frequency of 1 percent. In a population of 1 million individuals, 20,000 *a* alleles would be present in the gene pool. (Remember that each diploid individual carries two alleles for any given gene. In the gene pool of this population there are 2 million alleles for this particular gene, of which 1 percent, or 20,000, are allele *a*). If a few individuals in this population were destroyed by chance before leaving offspring, the effect on the frequency of allele *a* would be negligible.

In a population of 50 individuals, however, the situation would be quite different. In this population, it is likely that only one copy of allele *a* would be present. If the lone individual carrying this allele failed to reproduce or were destroyed by chance before leaving offspring, allele *a* would be completely lost. Similarly, if 10 of the 49 individuals without allele *a* were lost, the frequency of *a* would jump from 1 in 100 to 1 in 80.

This phenomenon, a change in the gene pool that takes place as a result of chance, is **genetic drift**. Population geneticists and other evolutionary biologists generally agree that genetic drift plays a role in determining the evolutionary course of small populations. Its relative importance, however, as compared to that of natural selection, is a matter of debate. There are at least two situations in which it has been shown to be important.

The Founder Effect

A small population that becomes separated from a larger one may or may not be genetically representative of the larger population from which it was derived (Figure 19–8). Some rare alleles may be overrepresented or, conversely, may be completely absent in the small population. As a consequence, when and if the small population increases in size, it will continue to have a different genetic composition—a different gene pool—from that of the parent group. This phenomenon, a type of genetic drift, is known as the **founder effect.**

An example of the founder effect is provided by the Old Order Amish of Lancaster, Pennsylvania (Figure 19–9). Among these people, there is an unprecedented frequency of a recessive allele that, in the homozygous state, causes a combination of dwarfism and polydactylism (extra fingers). Since the group was founded in the early 1770s, some 61 cases of this rare congenital deformity have occurred, about as many as have been

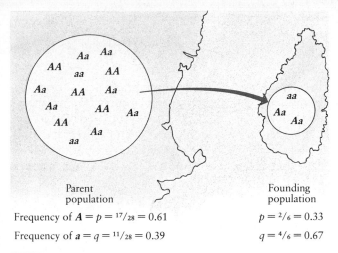

Frequency of **A** = $p = {}^{17}\!/_{28} = 0.61$ $p = {}^{2}\!/_{6} = 0.33$

Frequency of **a** = $q = {}^{11}\!/_{28} = 0.39$ $q = {}^{4}\!/_{6} = 0.67$

19–8   *When a small subset of a population founds a new colony (for example, on a previously uninhabited island), the allele frequencies within the founding group may be different from those within the parent population. Thus the gene pool of the new population will have a different composition than the gene pool of the parent population. This phenomenon is known as the founder effect.*

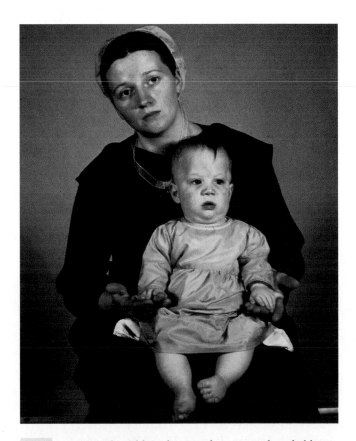

19–9   *Among the Old Order Amish, a group founded by only a few couples some 200 years ago, there is an unusually high frequency of a rare recessive allele. In its homozygous state, this allele results in extra fingers and dwarfism. This Amish child is a six-fingered dwarf.*

reported in all the rest of the world's population. Approximately 13 percent of the persons in the group, which numbers some 17,000, are estimated to carry this rare allele. The entire colony, which has kept itself virtually isolated from the rest of the world, is descended from only a few dozen individuals. By chance, one of those was a carrier of the allele.

Population Bottleneck

Population bottleneck is another type of situation that can lead to genetic drift. It occurs when a population is drastically reduced in numbers by an event that may have little or nothing to do with the usual forces of natural selection. For example, from the 1820s to the 1880s, the northern elephant seal was hunted so heavily along the coast of California and Baja California that it was rendered almost extinct, with perhaps as few as 20 individuals remaining. Since 1884, when the seal was placed under the protection of the United States and Mexican governments, the population has increased to more than 30,000, all presumably descendants of this small group. Studies of blood samples taken from 124 seal pups showed them to be homozygous for some 21 gene loci, indicating (by comparison with other mammalian groups) a drastic loss of genetic variability.

A population bottleneck is likely not only to eliminate some alleles entirely but also to cause others to become overrepresented in the gene pool. For example, the high rate of Tay-Sachs disease (page 212) among Jews of Eastern and Central European extraction has been attributed to a population bottleneck experienced by these people in the Middle Ages.

Nonrandom Mating

Disruption of the Hardy-Weinberg equilibrium can also be produced by nonrandom mating. A form of nonrandom mating particularly important in plants is self-pollination (as in the pea plants Mendel studied). In animals, nonrandom mating is often behavioral. For example, snow geese may be either white or blue (Figure 19–11). This is a type of variation known as **polymorphism,** in which two or more phenotypically distinct forms coexist in a population. White snow geese tend to mate preferentially with other white geese and blue with blue. Thus, assuming only two alleles are involved, there will be a decrease in the frequency of heterozygotes (represented as $2pq$ in the Hardy-Weinberg equation), with concomitant increases in the frequencies of the two homozygotes (p^2 and q^2).

Note that nonrandom mating can cause changes in

19–10   *Two northern elephant seal bulls fighting for supremacy of a harem, which may consist of as many as 50 females. Only a few males breed each year, and each breeding male fathers many offspring. This social system may be an additional factor contributing to the high degree of homozygosity found in the northern elephant seal population.*

19–11   *The lesser white snow goose ("the goose from beyond the north wind") and the blue goose were once believed to represent distinct species. In fact, they represent one species, with individuals of two colors. The allele for white is recessive; heterozygotes and homozygotes for blue are both the same dark blue color. However, birds mate preferentially with animals of their own color. As a result, there are more homozygotes than there would be if mating were random.*

the genotype frequencies without necessarily producing any changes in the frequencies of the alleles in question. Suppose, however, that all female snow geese, white or blue, preferred to mate only with blue snow geese. If, as a consequence, some of the blue males mated with more than one female, and some of the white males with none, changes in the frequencies of the two alleles and also of the phenotypes would occur. In fact, as we shall see in the next chapter, such nonrandom mating produced by female choice is an important agent of natural selection in some species.

Preservation and Promotion of Variability

Sexual Reproduction

By far the most important method by which eukaryotic organisms promote variation in their offspring is sexual reproduction. Sexual reproduction produces new genetic combinations in three ways: (1) by independent assortment at the time of meiosis—as diagrammed in Figure 11–16 on page 203, (2) by crossing over with genetic recombination, and (3) by the combination of two different parental genomes at fertilization. At every generation, alleles are assorted into new combinations.

In contrast, consider, for a moment, organisms that reproduce only asexually, by processes that involve mitosis and cytokinesis but not meiosis. Except when a mutation has occurred in the duplication process, the new organism will exactly resemble its only parent. In the course of time, various clones may form, each carrying one or more mutations, but potentially favorable combinations are unlikely to accumulate in one genotype.

Organisms that reproduce sexually can do so at only half the rate of asexually reproducing organisms. The only advantage to the organism of sexual reproduction, speaking strictly scientifically, appears to be the promotion of variation, the production of new combinations of alleles among the offspring. Why such variation is advantageous to the individual organism is a matter of controversy (see essay).

Mechanisms That Promote Outbreeding

Many means have evolved by which new genetic combinations are promoted in sexually reproducing populations. Among plants, a variety of mechanisms ensure that the sperm-bearing pollen from flowers on one plant is delivered to the stigmas of flowers on a different

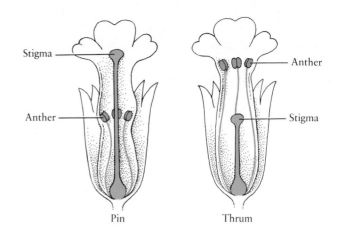

19–12 *Diagrams of two types of flowers ("pin" and "thrum") of the same species of primrose. Notice that the pollen-bearing anthers of the pin flower and the pollen-receiving stigma of the thrum flower are both situated about halfway up the length of the flower and that the pin stigma is level with the thrum anthers. An insect foraging for nectar in these flowers would collect pollen on different areas of its body, so that thrum pollen would be deposited on pin stigmas, and vice versa.*

plant. In some plants, such as the holly and the date palm, male flowers are on one tree and female flowers on another. In others, such as the avocado, the pollen of a particular plant matures at a time when its own stigma is not receptive. In some species, anatomical arrangements inhibit self-pollination (Figure 19–12).

Some plants have genes for self-sterility. Typically, such a gene has multiple alleles—s^1, s^2, s^3, and so on. A plant carrying the allele s^1 cannot fertilize a plant with an s^1 allele; one with an s^1/s^2 genotype cannot fertilize any plant with either of those alleles; and so forth. In one population of about 500 evening primrose plants, 37 different self-sterility alleles were found, and it has been estimated that there are more than 200 alleles for self-sterility in red clover. A plant with a rare self-sterility allele is more likely to be able to fertilize another plant than is a plant with a common self-sterility allele. As a consequence, the self-sterility system strongly encourages variability in a population. Selection for the rare allele makes it more common, whereas more common alleles become rarer.

Why Sex?

Sex is a very complicated and expensive way to reproduce. A sexually reproducing population wastes half of its reproductive potential on producing males and can therefore increase in numbers only half as fast as an asexually reproducing population. The same genetic penalty is paid by individual females: a sexually reproducing female will propagate her genes only half as fast as an asexually reproducing female. Moreover, sexual reproduction often involves considerable expenditure of time and effort and may expose an organism to predators while it is seeking a mate, courting, and copulating.

If sexual reproduction is so inefficient and so risky, why has it not been eliminated by natural selection? The answer appears to lie in the genetic variation promoted by sexual reproduction, but it is not easy to see how producing genetically variable offspring could compensate for producing, effectively, only half as many of them. Several hypotheses have been proposed to account for the success and persistence of sexual reproduction.

According to the "best man" hypothesis, the environment is very changeable. As a result, the offspring are likely to grow up in conditions quite different from those experienced by their parents. Offspring that are genetically identical to their parents will be poorly adapted to the changed conditions. Sexual reproduction is a way of creating a great variety of new genotypes, a few of which will happen to be well adapted to the unpredictable conditions experienced by the next generation.

The "tangled bank" hypothesis, by contrast, emphasizes that the environment is very diverse. It offers a great variety of different opportunities, but each opportunity is limited—that is, it can be exploited successfully by only a limited number of individuals of a particular genotype. Producing many uniform offspring is futile because, being very similar, they will all attempt to exploit the same opportunity. If you have 10 children, all of whom would like to stay in the same small town in which they grew up, you wouldn't want them all to be

The Red Queen from Through the Looking Glass:
"You have to run faster than that to stay in the same place."

doctors. Sexual reproduction is a way of avoiding competition between close relatives by suiting them to a variety of different ways of making a living.

Although often cited as correct, the "best man" hypothesis is, according to Graham Bell of McGill University, almost certainly wrong. If it were true, we would expect to find sexual reproduction in continually changing, disturbed, or novel conditions, with asexual reproduction prevailing in stable, long-established environments. The reverse appears to be true. For example, asexual reproduction is more common in freshwater environments, whereas sexual reproduction prevails in the more stable conditions provided by marine environments.

This ecological pattern is consistent with the "tangled bank" hypothesis but does not prove that it is correct. Other hypotheses may also be consistent with the observed pattern. For example, the "Red Queen" hypothesis—named after a character in Lewis Carroll's *Through the Looking Glass*—points out that many species (such as predators and prey, or parasites and hosts) are continually running an evolutionary race, each trying to increase its efficiency at the expense of others. Success requires a means of rapid genetic change in order to keep pace with the adaptations and

counter-adaptations of one's opponents. The continual reshuffling of alleles provided by sexual reproduction is such a means. If this hypothesis is correct, then sexual reproduction should be common wherever many species are found together, as in the sea and the tropics.

Correlation of the occurrence of sexual reproduction with ecological information appears to exclude the "best man" hypothesis. However, different or more detailed comparisons are needed to decide between the "tangled bank" and "Red Queen" hypotheses.

Other hypotheses have also been proposed to explain the success and persistence of sexual reproduction. For example, it has recently been suggested that the close pairing of homologous chromosomes at the beginning of meiosis makes possible not only the genetic recombinations that result from crossing over but also the repair of damaged DNA. Some biologists think that this "check" on the integrity of the DNA being transmitted to the next generation is of such importance that it could be a major factor in maintaining sexual reproduction.

The process of sexual reproduction, as it occurs in diverse organisms, is varied and complex. Understanding its evolution and function remains among the most challenging and perplexing problems in biology.

19–13   *Two giant African land snails* (Achatina) *in the process of mating. Like earthworms, many snails and slugs are hermaphrodites (each individual produces both sperm and egg cells). Hermaphroditism is advantageous for slow, solitary species because it doubles the chances of finding a mate. Every mature member of the same species that a hermaphrodite meets will always be of the opposite sex—and of the same sex too. As a result of the mating, each individual can produce new offspring.*

Among animals, even in those invertebrates that are **hermaphrodites,** producing both sperm and egg cells, an individual seldom fertilizes its own eggs (Figure 19–13). Among mammals, in particular, behavioral strategies promote outbreeding (mating with unrelated individuals). Often, for example, young males leave the family group as they reach reproductive age. This occurs among lions, gorillas, and baboons, to mention only a few. Among hunting dogs of the African plains, it is the young females who leave at reproductive age.

Diploidy

Another factor in the preservation of variability in eukaryotes is diploidy. In a haploid organism, genetic variations are immediately expressed in the phenotype and are therefore exposed to the selection process. In a diploid organism, however, such variations may be stored as recessives, as with the allele for white flowers in Mendel's pea plants. The extent to which a rare allele is protected is shown in Table 19–1.

As the table reveals, the lower the frequency of allele *a*, the smaller the proportion of it exposed in the *aa* homozygote becomes. The removal of the allele by natural selection slows down accordingly. This result should be of special interest to proponents of eugenics, who advocate improving the human gene pool through controlled breeding. For instance, consider a genetic disorder, such as PKU (page 212), that is expressed only in the homozygous recessive. If the frequency of allele *a* is about 0.01, the frequency of individuals with the *aa* genotype is 0.0001 (1 child for every 10,000 born). It would take 100 generations, roughly 2,500 years, of a program of sterilization of homozygous individuals with this condition to halve the allele frequency (to 0.005) and reduce the number born with this genetic disorder to 1 in 40,000, which should be enough to discourage even the most zealous eugenicist.

Heterozygote Advantage

Recessive alleles, even ones that may be harmful in the homozygous state, may not only be sheltered in the heterozygous state, but sometimes may actually be selected for. This phenomenon, in which the heterozygotes have greater reproductive success than either type of homozygote, is known as **heterozygote advantage.** It is another way that genetic variability is preserved.

One of the best-studied examples of heterozygote advantage involving a single gene locus is found in association with sickle-cell anemia. Until very recently, individuals homozygous for the sickle allele almost never lived long enough to become parents. Therefore, almost every time two sickle alleles came together in a homozygous individual, two sickle alleles were removed from the gene pool. At one time, it was thought that the sickle allele was maintained in the population by a steady in-

Frequency of Allele *a* in Gene Pool	Genotype Frequencies			Percentage of Allele *a* in Heterozygotes
	AA	*Aa*	*aa*	
0.9	0.01	0.18	0.81	10
0.1	0.81	0.18	0.01	90
0.01	0.9801	0.0198	0.0001	99

Table 19–1   Protection of Recessive Alleles by Diploidy

flux of new mutations. Yet in some African populations, as many as 45 percent of the individuals are heterozygous for the sickle allele, despite the loss of sickle alleles through homozygous individuals. To replace these alleles by mutations alone would require a rate of mutation about 1,000 times greater than any other known human mutation rate.

In the search for an alternative explanation, it was discovered that the sickle allele is maintained at high frequencies because the heterozygote has a selective advantage. In many regions of Africa, malaria is one of the leading causes of illness and death, especially among young children. Studies of malarial infections among young children showed that the severity of the illness is significantly lower in individuals heterozygous for the sickle allele than in normal homozygotes. Thus, although two alleles for sickle-cell hemoglobin are eliminated virtually every time they appear in the homozygous state, positive selection for the heterozygote maintains the allele in the gene pool. Moreover, for reasons that are not known, women who carry the sickle allele are more fertile than women who do not.

As this example makes clear, genetic variation not only is the raw material on which natural selection acts, but it can also be maintained by selection. In the next chapter, we shall consider other situations in which natural selection plays a key role in promoting and preserving genetic variation.

Summary

Population genetics is a synthesis of the Darwinian theory of evolution with the principles of Mendelian genetics. A population, for the population geneticist, is an interbreeding group of organisms, defined and united by its gene pool (the sum of all the alleles of all the genes of all the individuals in the population). Evolution is the result of accumulated changes in the composition of the gene pool.

The extent of genetic variability in a population is a major determinant of its capacity for evolutionary change. Natural populations can be shown by breeding experiments—artificial selection—to harbor a wide spectrum of genetic variations. The extent of genetic variation can be quantified by comparison of protein structures and, more recently, of DNA sequences. It has been estimated that at least 25 percent of the genes in any given human population are represented by two or more alleles.

The Hardy-Weinberg equilibrium describes the steady state in allele and genotype frequencies that would exist in an ideal population in which five conditions were met: (1) no mutations occurred, (2) no immigration or emigration occurred, (3) the population was large, (4) mating was random, and (5) there was no difference in the reproductive success of the offspring. The Hardy-Weinberg equilibrium demonstrates that the genetic recombination that results from meiosis and fertilization cannot, in itself, change the frequencies of alleles in the gene pool. The mathematical expression of the Hardy-Weinberg equilibrium provides a quantitative method for determining the extent and direction of change in allele and genotype frequencies.

The principal agent of change in the composition of the gene pool is natural selection. Other agents of change include mutation, gene flow, genetic drift, and nonrandom mating. Mutations provide the raw material for change, but mutation rates are usually so low that mutations, in themselves, do not determine the direction of evolutionary change. Gene flow, the movement of alleles into or out of the gene pool, may introduce new alleles or alter the proportions of alleles already present. It often has the effect of counteracting natural selection. Genetic drift is the phenomenon in which certain alleles increase or decrease in frequency, and sometimes even disappear, as a result of chance events. Circumstances that can lead to genetic drift, which is most likely to occur in small populations, include the founder effect and population bottleneck. Nonrandom mating causes changes in the proportions of genotypes but may or may not affect allele frequencies.

Sexual reproduction is the most important factor promoting genetic variability in populations. Mechanisms that promote outbreeding further promote variability. These mechanisms include self-sterility alleles in plants, anatomic adaptations that inhibit self-fertilization, and in animals, behavioral strategies. Variability is preserved by diploidy, which shelters rare, recessive alleles from selection. Natural selection can also act to promote and preserve variability. In cases of heterozygote advantage, for example, the heterozygote is selected over either homozygote, thus maintaining both the recessive and the dominant alleles in the population.

Questions

1. At the end of the experiment on bristle number in *Drosophila,* the flies in the high selection line had an average of 56 bristles. No fly at the beginning of the experiment had as many as 56, however. How do you explain this fact?

2. When domesticated animals, such as dogs, horses, pigs, and goats, are released into the wild, within a few generations their offspring closely resemble—in both behavior and appearance—the wild ancestors from which the animals were originally domesticated. What does this phenomenon reveal about the genetic variability latent in populations of domesticated animals? What does the speed with which the changes occur indicate about the strength of natural selection? About the long-term effectiveness of artificial selection?

3. As noted in the text, the Hardy-Weinberg equation was first formulated in response to a question: Why, if some alleles are dominant and some are recessive, don't the dominants drive out the recessives? What is the fallacy in the reasoning underlying that question? How does the Hardy-Weinberg equation answer that question?

4. What is the difference between gene flow and genetic drift? How does each affect the gene pool of a population?

5. What role might immigration play in promoting genetic variability? Compare the potential impact of immigration on the gene pool of a small island population versus that of a large mainland population.

6. How does self-fertilization, as in flowers, affect the Hardy-Weinberg equilibrium?

7. How do self-sterility alleles promote variability in a population?

8. In cases of heterozygote advantage involving a single gene with only two alleles, how does selection for the heterozygote affect the relative frequency of the two alleles? How would selection *against* the heterozygote affect this frequency?

9. Under what circumstances would two alleles be maintained in a population in exactly the same numbers?

Suggestions for Further Reading

Culotta, Elizabeth: "How Many Genes Had to Change to Produce Corn?" *Science,* vol. 252, pages 1792–1793, 1991.

Dawkins, Richard: *The Selfish Gene,* 2d ed., Oxford University Press, New York, 1989.*

> *Dawkins argues that we are "survival machines" programmed to preserve the "selfish molecules" known as genes. A wit-sharpening account and also an entertaining one.*

Koehn, Richard K., and Thomas J. Hilbish: "The Adaptive Importance of Genetic Variation," *American Scientist,* vol. 75, pages 134–141, 1987.

Lewin, Roger: "The Surprising Genetics of Bottlenecked Flies," *Science,* vol. 235, pages 1325–1327, 1987.

Li, Wen-Hsiung, and Dan Graur: *Fundamentals of Molecular Evolution,* Sinauer Associates, Inc., Sunderland, Mass., 1991.*

> *An authoritative introductory textbook on the new science that is emerging from the synthesis of evolutionary biology with the rapidly expanding discoveries of molecular genetics.*

Slatkin, Montgomery: "Gene Flow and the Geographic Structure of Natural Populations," *Science,* vol. 236, pages 787–792, 1987.

Strickberger, Monroe W.: *Evolution,* Jones and Bartlett Publishers, Boston, 1990.

> *An up-to-date textbook on evolutionary biology, written for undergraduate students who have completed a course in introductory biology.*

Travis, John: "Possible Evolutionary Role Explored for 'Jumping Genes,'" *Science,* vol. 257, pages 884–885, 1992.

Wills, Christopher: *The Wisdom of the Genes: New Pathways in Evolution,* Basic Books, Inc., New York, 1989.*

> *An introduction to modern evolutionary biology, written for a general audience and enriched with lively anecdotes and analogies. The author is particularly concerned with the evolutionary role of such features of DNA structure and organization as mobile genetic elements (transposons) and introns and exons.*

*Available in paperback.

In species after species, the females are sensible in their size, attire, and accessories, while the males are often oversized, overdressed, and burdened with awkward appendages and seemingly foolish patterns of behavior. The male peacock is perhaps the most cited example; his resplendent tail not only attracts predators but also makes it more difficult for him to escape them. The enormous antlers of the male Irish elk, as preserved in the fossil record, would have been so cumbersome that some paleontologists believe they hurried the species toward extinction. Male bowerbirds spend much of their waking life collecting ornaments for elaborate nests that are never occupied, while the females quietly prepare homes for their young, which they will raise unassisted.

Darwin concluded that "this form of selection depends, not on the struggle for existence in relation to other organic beings or to external conditions, but on a struggle between the individuals of one sex, generally the males, for the possession of the other sex."

This phenomenon of sexual selection, as Darwin called it, was one of his most controversial proposals. One reason was a lack of understanding of how such characteristics could be passed through the female to the male, generation after generation (a question that has now been answered). Another reason was that it required "a sense of beauty," as Darwin called it, on the part of the selecting females, which some of his critics found difficult to imagine.

But beauty, as we are often reminded, lies not in our eyes alone. Among proboscis monkeys *(Nasalis larvatus)*, females have, as you can see, a dainty turned-up nose. By contrast, the nose of a mature male resembles a giant cucumber, drooping down over his mouth, almost reaching his chin. In order to eat, he has to push it out of the way. Like the tail of the peacock, however, this outsized appendage is believed to be his ticket to sexual success.

These wonderfully grotesque animals are found only in the swampy mangrove forests of Borneo. They are endangered because of destruction of their habitat.

Natural Selection

According to Darwin's own account, the concept of natural selection came to him in 1838 upon reading Malthus's gloomy essay (see page 9). Darwin realized that all populations—not just the human population—are potentially doomed to exceed their resources. Only a small fraction of the individuals that might exist are born and survive to maturity. According to Darwin, those that do survive are those that are "favoured," to use his term, by reason of slight, advantageous variations. This process of survival of the "favoured" he termed natural selection, by analogy with the artificial selection practiced by breeders of domestic animals and plants.

In terms of population genetics, natural selection is now defined more rigorously as the differential rate of reproduction of different genotypes in a population. This differential reproductive success, which is a result of interactions between the individual organisms and their environment (including other organisms), can result in changes in the gene pool of a population—that is, in evolution. According to current theory, natural selection is the major force in evolution.

Natural Selection and the Maintenance of Variability

In the course of the controversies that led to the synthesis of evolutionary theory with Mendelian genetics, some biologists argued that natural selection would serve only to eliminate the "less fit." As a consequence, it would tend to reduce the genetic variation in a population and thus reduce the potential for further evolution. Modern population genetics has demonstrated this not to be true.

As we saw in the example of the sickle-cell allele (page 331), natural selection can be a critical factor in preserving and promoting variability in a population. There are many other examples of this effect. Here

335

we shall consider just one, which illustrates the fact that a variety of selective forces can be at work simultaneously.

Balanced Polymorphism: Color and Banding in Snail Shells

As you will recall (page 328), polymorphism is the co-existence within a population of two or more phenotypically distinct forms. In some cases, one phenotype gradually replaces the others, and the polymorphism is transient. In other cases, however, the phenotypes are maintained in fairly stable proportions by natural selection, and the polymorphism is said to be **balanced.**

One of the best-studied examples of a balanced polymorphism is found among land snails of the genus *Cepaea*. In one species, for instance, the shell of the snail may be yellow, brown, or any shade from pale fawn through pink and orange to red. The lip of the shell may be black or dark brown (normally) or pink or white (rarely). Up to five black or dark-brown longitudinal bands may decorate it (Figure 20–1a). Fossil evidence indicates that these different types of shells have coexisted for more than 10,000 years.

Studies among English colonies of *Cepaea* have revealed some of the selective forces at work on the snails, which occupy a variety of habitats. The snails are preyed upon by birds, among which are song thrushes. Song thrushes select snails from the colonies and take them to nearby rocks, where they break them open, eat the soft parts, and leave the shells (Figure 20–1b). By comparing the proportions of types of shells around the thrush "anvils" with the proportions in the nearby colonies, investigators have been able to correlate the shell patterns of snails seized by the thrushes with the habitats of the snails. In habitats, such as bogs, where the background is fairly uniform, unbanded snails blend in and are less likely to be preyed upon than banded ones. Conversely, in habitats, such as woodlands, where the backgrounds are mottled, unbanded snails are more likely to be the victims.

Studies of many different colonies have confirmed these correlations. In uniform environments, a higher proportion of snails are unbanded; in rough, tangled habitats, far more tend to be banded. Similarly, the greenest habitats have the highest proportion of yellow shells, but among snails living on dark backgrounds, the yellow shells are much more visible and are clearly disadvantageous, judging from the evidence conveniently assembled by the thrushes.

Many of the snail colonies studied were at distances so great from one another that the possibility of movement between populations could be ruled out. Why, then, are both shell types still present in these colonies? One would expect populations living on uniform backgrounds to be composed almost entirely of unbanded snails, and colonies on dark, mottled backgrounds to lose most of their yellow-shelled individuals. The answer to this problem is not fully worked out, but it seems that there are other factors that are correlated with the particular shell patterns and that form a part of the same group of genes that control color and banding. Experiments have shown, for instance, that unbanded snails (especially yellow ones) are more heat-resistant and cold-resistant than banded snails. In other words, several different selection pressures are at work, and they appear to maintain the genetic variations of color and banding.

20–1   (a) *Ten snails of the species* Cepaea hortensis. *The shells of snails of this species (and of other species in the genus* Cepaea) *exhibit a wide range of colors and banding patterns. Among the most common colors are yellow and pink. Shells of either color may be unbanded or may have dark bands of varying thickness and intensity. Banded snails living on dark, mottled backgrounds are less visible than unbanded snails and are therefore preyed upon less frequently.* (b) *A song thrush breaking open a yellow-shelled snail at an "anvil."*

(a)   (b)

Human Blood Types: A Puzzle

The blood types A, B, AB, and O represent the most thoroughly studied polymorphism in human populations. Apparently, the three alleles associated with these blood types (page 224) are part of our ancestral legacy, since the same blood types are also found in other primates. A great deal is known about the chemistry of the different blood types and about the allele frequencies in different populations. Yet we know very little about how this polymorphism has been maintained.

Some population biologists regard the blood types as probably neutral in selective value. Others maintain that polymorphism in human blood types is a result of selection. For example, among Caucasian males, life expectancy is greatest for those with type O and least for those with type B; exactly the opposite is true for Caucasian females with these blood types. People with type A blood run a relatively higher risk of cancer of the stomach and of pernicious anemia. People with type O blood have a higher risk of duodenal ulcers and are more likely to contract Asian flu. With

the exception of Asian flu, however, most of these conditions would not affect relative rates of reproduction—and thus act as selection forces—since they generally occur in individuals who are past reproductive age.

It has been suggested that there are correlations between the different blood types and susceptibility to such diseases as plague, leprosy, tuberculosis, syphilis, and smallpox—all diseases that, in the past, could have been powerful selection forces. For example, studies conducted in certain small towns and villages of India during a smallpox epidemic in 1965 and 1966 revealed that individuals carrying the A allele had a risk of contracting smallpox that was seven times higher than that of individuals who did not carry the A allele. Moreover, once ill with smallpox, such individuals were three times as likely to develop a severe case and twice as likely to die. These data would suggest that allele A should have been virtually eliminated in human populations long before the eradication of smallpox in the late 1970s. However, individuals with the A allele

appear to be relatively resistant to cholera, a deadly disease that still takes a very high toll in India and many other parts of the world.

The geographic distributions of the A, B, AB, and O types are irregular. For example, there is a large predominance of type O in the Western Hemisphere and an increase in the B allele as one moves from Europe toward Central Asia. The B allele is totally absent in Native Americans and Australian aborigines who have not intermarried with individuals of European descent. Although most Native Americans are type O, the Blackfoot tribe has the highest frequency of type A blood found anywhere in the world (55 percent). Even within an area as small as the British Isles, there are significant variations in allele frequency, going both from north to south and from east to west. These differences may reflect some differences in the selective forces favoring particular blood types under particular conditions, they may be the result of population migrations and genetic drift, or some combination of both. At the present time, we do not know.

What Is Selected?

As the studies with *Cepaea* illustrate, natural selection acts on the complete phenotype. In the days of fruit-fly genetics, when populations were being scanned for white eyes and short wings, phenotype became synonymous with physical appearance. In terms of evolutionary theory, however, it must be regarded as including such important characteristics as the optimum temperature at which a particular enzyme works, or the speed of response to a stimulus. In short, the phenotype includes all the physical, physiological, and behavioral attributes of an organism.

A phenotype generally is the expression of many different genes. As a corollary, any particular phenotypic characteristic may be arrived at by a number of genotypic routes. For example, selection for high and low bristle number in *Drosophila* (page 322) led, in the first series of experiments, to sterility. The second attempt to select for high bristle number led to the same number

of bristles (that is, the same phenotypic characteristic) but without a reduction in fertility. Obviously, different genotypes were responsible.

It is very rare that a single allele can determine a winning phenotype. In the peppered moth *Biston betularia* (pages 13–14), black wing color is determined by a single allele. However, the peppered coloring may be the result of several different genes, and, along with the mottled appearance, there are other characteristics important for escaping hungry predators. These include selecting the right background, lying very still while on the tree trunk, and positioning oneself correctly so as to enhance the camouflage effect, all of which involve genetic factors. Such underlying genetic complexity is typical not only of the variety of mechanisms by which animals avoid predators (Figure 20–2) but also of most characteristics of the whole organism.

As you will recall, however, the phenotype is not de-

(a)

(b)

20–2 *Coordinated anatomical and behavioral characteristics protect the peacock moth* (Automeris memusae) *from predators.* (a) *At rest, the moth positions itself in a way that enhances the camouflaging effect of its patterned wings.* (b) *Disturbed, the moth parts its wings, revealing a pair of "eyes," reported to be sufficiently menacing to scare away a bird.*

termined solely by the interactions of the multitude of alleles making up the genotype. It is also a product of the interaction of the genotype with the environment in the course of the individual's life (Figure 20–3). For example, among human identical twins (that is, twins produced from a single fertilized egg and therefore having the same genotype), noticeable differences—such as body weight—between the two are often apparent even at birth, owing to differences in the intrauterine environment.

Any particular phenotype lasts, in terms of evolutionary history, for as long as the blink of an eye. In the case of sexually reproducing organisms, the genotype is as unique and transient as the phenotype, shuffled and recombined at every generation. Only the individual genes survive. The great men and women of history have long since vanished, but each of us may carry a few of their alleles as part of our human legacy.

20–3 *Jeffrey pines usually grow tall and straight, as in* (a). *Environmental forces, however, can alter the normal growth patterns, as shown in* (b). *This tree is growing on a mountaintop in Yosemite National Park, California, where it is exposed to strong, constant winds.*

(a)

(b)

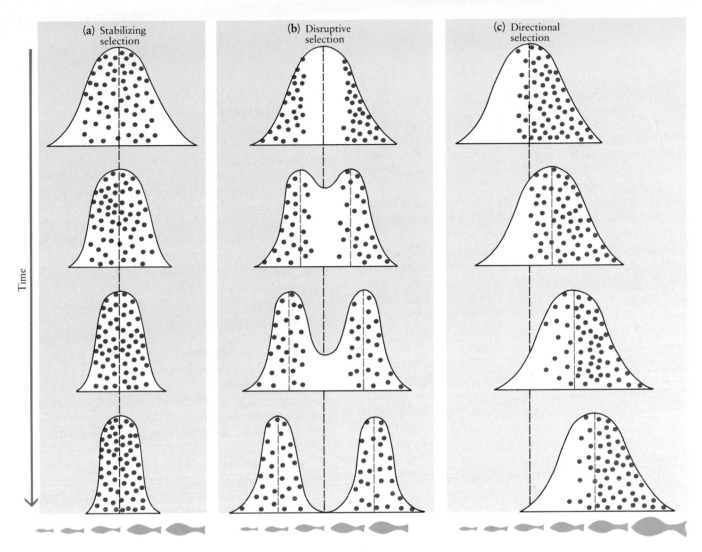

20–4 *A schematic representation of three types of natural selection acting on a trait, such as body size, that varies continuously throughout a population (see page 225 for a review of continuous variation). The horizontal axis of each small graph represents the range of values of the trait being considered, with one extreme at the left, the other extreme at the right, and intermediate states in between. The black curve in each graph summarizes the proportion of individuals in the population exhibiting a particular variant of the trait. The dots represent individuals in each generation that reproduced and left an average—or greater than average—number of offspring.*

Initially, each of the populations displays a normal bell-shaped curve. Most individuals exhibit intermediate variants of the trait, and only a few exhibit the extremes. As you can see, in (a) stabilizing selection and (b) disruptive selection, the shapes of the curves change in succeeding generations. Stabilizing selection involves the elimination of extremes, producing a more uniform population. In disruptive selection, intermediate forms are eliminated, producing two divergent populations. In (c) directional selection, one expression of the trait is gradually eliminated in favor of another. Note how the bell-shaped curve moves to the right in each succeeding generation.

Types of Selection

As we have seen, the results of natural selection in any given instance depend on the interaction of a variety of factors, both genetic and environmental. Selection processes can be classified into five broad categories on the basis of several different criteria. Depending on its effect on the distribution of characteristics within a population, natural selection can be described as stabilizing, disruptive, or directional (Figure 20–4). When selection is influenced by the relative proportions of different phenotypes within a population, it is said to be frequency-dependent. A fifth category, sexual selection, is defined by what is selected: characteristics of direct consequence in obtaining a mate and successfully reproducing.

Stabilizing Selection

Stabilizing selection, a process that is always in operation in all populations, is the elimination of individuals with extreme characteristics. Many mutant forms are probably immediately weeded out in this way, often in the zygote or embryo.

Clutch size in birds is an example of a trait affected by stabilizing selection. Clutch size (the number of eggs a bird lays) is determined genetically, although it also appears to be influenced by environmental factors. In a study of Swiss starlings (Table 20–1), the percentage of birds surviving was found to increase for each clutch size up to five. With a clutch size larger than five, a smaller percentage of birds survived—apparently because of inadequate nutrition. Female Swiss starlings whose genotypes lead to a clutch size of four or five will have more surviving young, on average, than members of the same species that lay more or fewer eggs.

Although the number of eggs laid in a clutch may appear to be a simple trait, rather like pushing the right number for copies on the Xerox machine, it involves a number of factors. They include the synthesis of proteins for the egg yolk and white, the availability of calcium for the shells, and the length of time the female will mate.

Disruptive Selection

A second type of selection, **disruptive selection,** increases the frequency of the extreme types in a population at the expense of intermediate forms. Disruptive selection in the laboratory gave rise to the high- and low-bristle-number lines of *Drosophila* described earlier.

A particularly clear instance of disruptive selection in nature has been demonstrated in studies of plants growing on soils that were previously contaminated by mining operations. The boundaries between contaminated and uncontaminated areas are often very sharp. Plants growing on uncontaminated soil are unable to survive on contaminated soil. Plants of the same species growing on contaminated soil are able to survive in the uncontaminated areas but cannot compete with those already growing there. Thus, the two extreme phenotypes have been "favoured" at the expense of intermediate forms. This has resulted in the development of very marked differences between the two groups in the 50 years or so since the mining operations were discontinued and the plants began to colonize the area. Under certain circumstances, disruptive selection of this sort may lead to the formation of new species, a subject we shall explore in the next chapter.

A more recently described example of disruptive selection is also an example of both frequency-dependent selection and sexual selection. As you may know, the coho salmon of the Pacific Northwest hatch in freshwater streams, in which they spend their first year before moving to the ocean, where they become sexually mature. Upon attaining maturity, the fish return to their home stream, breed, and die. Although females of this species attain sexual maturity and return to breed at age 3, the males may attain sexual maturity at either age 2 or age 3. The two-year-old males, known as "jacks," are about half the size of the three-year-old males, known as "hooknoses" (Figure 20–5). Whether a male will mature in two years and thus be a jack, or in three years and be a hooknose, is, in part, genetically determined.

Studies by Mart R. Gross of Simon Fraser University in British Columbia have shown that disruptive selection not only maintains these two types of males in the population but favors the smallest jacks and the largest hooknoses. When a female spawns, the males closest to the nest in which her eggs are laid are the first to deposit sperm over them. The hooknose males jockey for prox-

Table 20–1	Survival in Relation to Number of Young in Swiss Starling							
Number of young in clutch	1	2	3	4	5	6	7	8
Number of young marked	65	328	1,278	3,956	6,175	3,156	651	120
Number of marked birds recaptured after 3 months	0	6	26	82	128	53	10	1
Percentage of marked birds recaptured after 3 months	0	1.8	2.0	2.1	2.1	1.7	1.5	0.8

20–5   *Coho salmon, at the mouth of their home river, as they begin the journey upstream to the location where they will breed. The large fish at the bottom is a mature female, and the larger, more brightly colored fish above her is a hooknose male. The much smaller fish above him is a jack male. The small fish grouped together at the lower right appear to be holding back. They may be immature fish that will remain at the mouth of the river until they are mature, or they may be sexually mature jack males.*

20–6   *Scale insects on a branch of a citrus tree. With their specialized mouthparts, these insects suck fluid from the plant, debilitating it. If left unchecked, they eventually kill the plant.*

imity to the nests by fighting, and the largest males generally win. By contrast, the jacks sneak close to the nests by hiding among rocks or debris, or in shallow areas of the stream. The smaller the jack, the less likely it is to be discovered and chased away by a hooknose. Both the larger jacks and the smaller hooknoses seldom breed successfully, and the two extreme male types are thus maintained within the population.

Directional Selection

A third type of natural selection, **directional selection,** results in an increase in the proportion of individuals with an extreme phenotypic characteristic. It is therefore likely to result in the gradual replacement of one allele or group of alleles by another in the gene pool.

An example of directional selection is the development of insecticide resistance. Chemicals poisonous to insects, such as DDT, were originally hailed as major saviors of human health and property. They have fallen into disfavor not only because of their tendency to accumulate in the environment, but also because of the extraordinary increase in resistant strains of insects. At least 225 species of insects are now resistant to one or more insecticides. One species is even able to remove a chlorine atom from a DDT molecule and use the remainder as food.

A particularly striking example of insecticide resistance has been found in the scale insects (Figure 20–6) that attack citrus trees in California. In the early 1900s, a concentration of hydrocyanic gas sufficient to kill nearly 100 percent of the insects was applied to orange groves at regular intervals with great success. By 1914, orange growers near Corona, California, began to notice that the standard dose of the fumigant was no longer sufficient to destroy one type of scale insect, the red scale. A concentration of the gas that had left less than 1 survivor out of every 100 insects in the nonresistant strain left 22 survivors out of 100 in the resistant strain. By crossing resistant and nonresistant strains, it was possible to show that the difference between the two involved a single gene locus. The resistant insects were variants produced by chance in the original population. Subsequently, a new factor in the environment—a poisonous chemical—led to directional selection for these variants. Nonresistant insects died, but resistant insects survived and reproduced.

The mechanism for this insecticide resistance is not known. However, one group of experiments has shown that the resistant individual can keep its spiracles (the openings through which it draws air into its respiratory system) closed for 30 minutes under unfavorable conditions. The nonresistant insect can do so for only 60 seconds.

Another example of directional selection is provided by the peppered moth, *Biston betularia*. As we saw in the Introduction (pages 13–14), by the 1950s the black form of this moth had largely replaced the light-colored form in heavily polluted areas of the English countryside. In recent years, however, strong controls have been instituted in Great Britain on the particulate content of smoke, and the heavy soot accumulation has begun to decrease. The light-colored moths are already increasing in proportion to the black moths, but it is not yet known whether there will be a complete reversal either in pollution or in the direction of selection. There is a moral to this story. Note that the black moth is not absolutely superior to the light-colored one, or vice versa. Natural selection is, as Darwin realized, entirely a matter of time and place.

Frequency-Dependent Selection

In the examples of natural selection we have considered thus far, we have assumed that the fitness of a phenotype is independent of its relative proportion within a population. In some situations, however, a type of natural selection known as **frequency-dependent selection** acts to decrease the frequency of more common phenotypes and to increase the frequency of less common ones. Such selection is, for example, a factor in maintaining hooknose and jack males in coho salmon populations. As the frequency of either type increases, the competition between males of that type becomes more intense, allowing greater opportunities for successful reproduction by males of the other type.

Predator-prey interactions can also lead to frequency-dependent selection. If individuals of a prey species differ, for example, in color, some predators will prey disproportionately on individuals with the most common color (Figure 20–7). If, as a result, individuals with that color should become less common, selection pressure on them would be relaxed as predators switch their attention to individuals of alternative colors, which would now be more common. Thus frequency-dependent selection can be a factor in maintaining polymorphisms in prey populations.

Sexual Selection

As Darwin recognized, many of the conspicuous characteristics of animals have little to do with survival on a day-to-day basis but are instead the result of **sexual selection,** the "struggle between the members of one sex, generally the males, for the possession of the other sex."

(a)

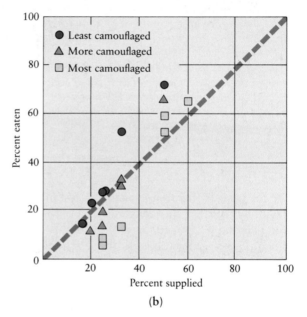

(b)

20–7 *Frequency-dependent selection has been demonstrated in the aquatic insects known as water boatmen* (a). *These insects, which are preyed upon by fish, often exist in different color forms that provide different degrees of camouflage.*
(b) *The results of experimental studies in which fish were provided with supplies of water boatmen of a species that exists in three distinct color forms. When the three forms were present in equal proportions (33 percent of each type), the least camouflaged form was the most likely to be eaten and the most camouflaged form the least likely to be eaten. However, when the proportions of the three forms were unequal, the form present in the highest proportion was the most likely to be eaten, regardless of its degree of camouflage.*

Male Ornamentation: The Role of Female Preference

Darwin hypothesized that in species in which females actively choose their mates, the preference of the females for particular characteristics plays a major role in the evolution of male ornamentation. This is, however, only one of several hypotheses that could explain the sometimes extreme characteristics observed in many male animals. For example, another hypothesis is that male ornamentation evolved as a device for signaling between competing males, with female preference having little or no influence. However, experiments performed by Malte Andersson with the long-tailed widowbirds of Kenya have provided dramatic support for Darwin's hypothesis.

Long-tailed widowbirds exhibit a striking sexual dimorphism. The females, which are smaller than the males, are mottled brown and have short tails. The larger males, by contrast, are a brilliant black, with red epaulets, and have tails that average 50 centimeters (about 20 inches) in length.

Males compete for territories, in which the females subsequently nest. Andersson observed as many as six females nesting in the territories of some males, while the territories of other males contained no females at all. As the males fly over their territories, which are in the grasslands of the African savanna, their long tails and black plumage are readily visible, presumably to both predators and other widowbirds, as well as to human observers.

Darwin suggested that selection of males by females on the basis of an ornament, such as the widowbird's tail, would produce directional selection for further elaboration of the ornament until selection in the opposite direction—as, for example, by predators—balanced the effects of the sexual selection. On the basis of this hypothesis, one would predict that, given the opportunity, females would show preference for males with still greater elaboration of the ornament, even in species in which the males are already highly ornamented.

A male long-tailed widowbird, flying over his territory.

Andersson tested this prediction by giving female widowbirds a choice between three groups of males: one group in which the tails were clipped to a length of 14 centimeters, one group in which the tails were the normal length (50 centimeters), and one group in which the tails were lengthened to 75 centimeters by gluing on feathers from the clipped tails of the first group. In the group in which the tails were the normal length, half of the birds were unaltered and half had their tails clipped and then reglued.

Before the experimental alteration of the tails, similar numbers of females were nesting in the territories of all the males. Following the alterations of tail length, the males with lengthened tails attracted four times as many females as the males with either shortened tails or tails of normal length. Moreover, the ability of the males to compete for and successfully acquire territories appeared to be unaffected by the alterations in tail length. As this study demonstrates, at least in the case of the long-tailed widowbird, female mate choice is a dominant selective agent in the evolution of ornamentation.

Andersson's experimental results were published in 1982, more than 100 years after Darwin's hypothesis about the role of female choice. The length of time between that hypothesis and these elegant experiments is indicative of the difficulty often facing evolutionary biologists as they try to devise appropriate experimental or observational tests of key hypotheses. The challenge is not only in the asking of questions but also in the framing of questions in ways that can provide clear answers.

Males produce many more gametes than do females. Thus the female has a larger investment, in terms of time, energy, and resources, in each fertilized egg. When parental care is involved, females are often the ones caring for the young. Hence, males, seeking to inseminate as many females as possible, are generally the competitors. Females, with a higher stake in each mating, seek the best possible genetic partner and so are the choosers. The competition among males may be either direct—for territories, harems, or privileges of consort—or indirect, as with nest-building and displays.

Sexual selection is thought to play a major role in establishing and maintaining **sexual dimorphism,** those differences between males and females that have to do not with the act of reproduction itself but rather with obtaining a mate (Figure 20–8). Examples of sexually selected characteristics are the nose of the male proboscis monkey (page 334), the extravagant plumage of many male birds, and the oversized antlers of male deer and elk. Among those few bird species in which the females do the courting and the males do the choosing, it is the female that is the gaudier, with the stay-at-home male the drab partner.

For those of us weary of comparisons between the resplendent peacock and the dowdy peahen, it is comforting to realize that it is the female whose size, color, and general behavior are considered optimal for the environment. The dimorphic characteristics of the male, by contrast, are useful principally for threat, display, and other bids for attention. In fact, as we noted earlier,

these characteristics may well be maladaptive with respect to factors affecting survival, such as conspicuousness to predators.

Darwin, recognizing that such males were not "fitter to survive in the struggle for existence," categorized sexual selection as a force separate from that of natural selection. However, with fitness stringently redefined in terms of relative numbers of surviving offspring, many investigators now believe that such a distinction is invalid and that sexual selection should be considered simply as one of the forms natural selection may take.

The Result of Natural Selection: Adaptation

Natural selection results in **adaptation,** a term with several meanings in biology. First, it can mean a state of being adjusted to the environment. Every living organism is adapted in this sense, just as Abraham Lincoln's legs were, as he remarked, "just long enough to reach the ground." Second, adaptation can refer to a particular characteristic, that which is adapted—such as an eye or a hand—that aids in the adjustment of an organism to its environment. Third, adaptation can mean the evolutionary process, occurring over the course of many generations, that produces organisms better suited to their environment.

(a)

(b)

20–8 *Sexual selection is caused by competition for mates and often results in differences between the two sexes. (a) In many species of birds, such as the golden pheasants shown here, a few males father most of the offspring. The males direct their energies primarily to courtship and mating, and the females invest more heavily in parental care. In such species,* *the females are colored in a way that blends with their surroundings, thus protecting them and their young, and the males have bright, conspicuous plumage. (b) In species in which mating couples form monogamous pairs, such as these albatrosses, males and females generally look alike. Both parents are involved in the care of the young.*

20–9 *The woodpecker has a number of adaptations that enable it to obtain food. These include two toes pointing backward with which the woodpecker clings to the bark of the vertical tree trunk, strong tail feathers that prop it up, a strong beak that can chisel holes in the bark, strong neck muscles that make the beak work as a hammer, air spaces in the skull that cushion the brain during hammering, and a very long tongue that can reach insects under the bark.*

20–10 *The distribution by size of male house sparrows in North America. This artist's interpretation is adapted from a map generated by a computer program that processed data on 739 birds. The higher numbers indicate larger body sizes, based on a composite of 16 different measurements of the birds' skeletons.*

Adaptation has multiple manifestations. Consider, for example, a squirrel. Note how its tail serves as a counterbalance as the animal leaps and turns; in addition, the same marvelous structure serves as a parasol, a blanket, and an aerial rudder. Pluck a burr from your clothing and consider the artful contrivances by which it clings, ensuring that you—or some other animal—will carry the seeds within to a new location, far from the parent plant. Consider the love and devotion characteristic of the domesticated dog; these adaptations, related to the procurement of food and shelter, are as stringently selected as the beak of a woodpecker (Figure 20–9). Thread a needle; your capacity to do so represents the cumulative effect of millions of years of selection pressures for digital dexterity and eye-hand coordination. (The needle itself made its appearance a mere 20,000 years ago.)

Natural selection involves interactions between individual organisms, their physical environment, and their biological environment—that is, other organisms. In many cases, the adaptations that result from natural selection can be clearly correlated with environmental factors or with the selective forces exerted by other organisms.

Adaptation to the Physical Environment: Clines and Ecotypes

Sometimes phenotypic variations within the same species follow a geographic distribution and can be correlated with gradual changes in temperature, humidity, or some other environmental condition. Such a graded variation in a trait or a complex of traits is known as a **cline**.

Many species exhibit north-south clines of various traits. House sparrows, for example, tend to have a smaller body size in the warmer parts of the range of the species, and a larger body size in the cooler parts (Figure 20–10). As size increases, the surface-to-volume ratio decreases, thereby increasing the animal's ability to conserve heat in cold weather. House sparrows were first introduced into North America between 1852 and 1860 in the form of small founder populations drawn

345

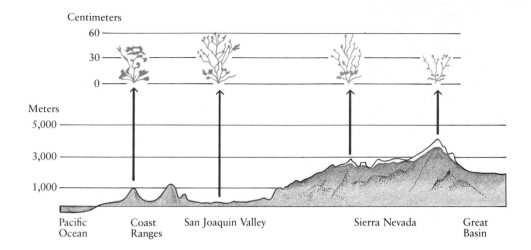

20–11   *Ecotypes of* Potentilla glandulosa, *a relative of the strawberry. Notice the correlation between the size of the plant and the altitude at which it grows. There are other phenotypic differences among the plants as well. When plants from the four geographic areas are grown under identical conditions, many of these phenotypic differences persist and are passed on to the next generation, indicating that these plants are genotypically, as well as phenotypically, different.*

principally from central England and Germany. From an evolutionary standpoint, the time that has elapsed since their introduction is very brief indeed, yet a dramatic degree of size differentiation has occurred. To cite another example, plants growing in the south often have slightly different requirements for flowering or for ending dormancy than the same kind of plants growing in the north, although they may all belong to the same species.

A species that occupies many different habitats may appear to be slightly different in each one. Each group of distinct phenotypes is known as an **ecotype.** Are the differences among ecotypes determined entirely by the environment or do these differences represent adaptation resulting from the action of natural selection on genetic variation?

Figure 20–11 illustrates a series of experiments carried out with the perennial plant *Potentilla glandulosa,* a relative of the strawberry. Experimental gardens were established at various altitudes, and wild plants collected from near each of the experimental sites were grown in all of the gardens. Because *Potentilla glandulosa* reproduces asexually, the plants collected from each area were genetically uniform. The subsequent growth of the plants in the experimental gardens revealed that many of the phenotypic differences among the ecotypes of *P. glandulosa* were due to genetic differences. It is not surprising that in these very different environments, different characteristics had resulted from selection. Over time, the genetic differences between individual plants had been translated into genetic differences between subgroups of the *P. glandulosa* population. As we shall see in the next chapter, such a process is often the first step in the formation of new species.

Adaptation to the Biological Environment: Coevolution

When populations of two or more species interact so closely that each exerts a strong selective force on the other, simultaneous adjustments occur that result in **coevolution.** One of the most important, in terms of sheer numbers of species and individuals involved, is the coevolution of flowers and their pollinators, to be described in Chapter 25. Here we shall consider other examples of coevolution, also involving plants and insects, those ancient allies and enemies.

Milkweed, Monarchs, and Mimics

Various families of plants have evolved chemical defenses that, because they are toxic, bad-tasting, or both, deter insects and other animals that try to feed on the plants. The bitter white sap of plants of the milkweed family contains a toxic substance that acts as such a deterrent. In the course of the evolutionary race to stay in the same place (see essay, page 330), some species of insects, including monarch butterflies, have evolved enzymes that enable the caterpillars to feed on the milkweeds without being poisoned (Figure 20–12). Monarch caterpillars not only utilize the plant tissues for food, but they also ingest and store up the toxic substance, which is then present in the adult forms, the butterflies. The butterflies, in turn, are distasteful and poisonous to their predators (Figure 20–13).

Tasting bad, while useful, is not an ideal defense from the point of view of the individual, since making this fact known may demand a certain amount of personal sacrifice. (Although birds usually drop the monarch butterfly after the first bite, they often inflict fatal injury in the process.) For monarchs and other animals

(a)

(b)

(c)

20–12   (a) *A monarch butterfly egg (white speck in center of photograph) clings to the bud of a milkweed.* (b) *When the egg hatches, two to four days after it is laid, a*

caterpillar emerges that feeds on the milkweed, ingesting and storing the toxic compounds produced by the plant. The monarch caterpillar and also the monarch

butterfly (c) *thus become unpalatable and poisonous. The conspicuous coloration of caterpillar and butterfly warns would-be predators.*

that have an effective protective device, it is advantageous to advertise. The more conspicuous or common such an animal is, the fewer the number of individuals that must be sacrificed before the bird or other predator learns to avoid it. Warning colors, such as those of the monarch, are common in the animal world. They are found not only among insects but also among poisonous reptiles and amphibians. The skunk is a familiar mammalian example of an animal whose distinctive markings remind us to keep our distance.

In many types of animals, especially insects, unrelated species with similar obnoxious defenses often come to resemble one another in their warning characteristics. This phenomenon is known as **Müllerian mimicry**, after F. Müller, who first described it. Bees, wasps, and hornets are probably the most familiar example of Müllerian mimicry (Figure 20–14). Even if we cannot tell which is which, we recognize them immediately as stinging insects and keep a respectful distance. Müllerian mimicry is adaptive for all the individuals involved because each benefits from a predator's experience with another. As you might expect, other insects that feed on milkweeds have the bright warning colors characteristic of the monarch.

(a)

(b)

20–13   *A blue jay that has never before tasted a monarch butterfly is fed one in a controlled experiment. Shortly after ingesting the butterfly, the blue jay begins to show signs of being uncomfortable and then vomits. Upon subsequently being offered a monarch, the bird refused it.*

Model (stings)

(a) Yellowjacket

Müllerian mimics (sting)

(b) Sand wasp

(c) Masarid wasp

(d) Anthidine bee

Batesian mimics (do not sting)

(e) Syrphid fly

(f) *Alcathae* moth

(g) Ornate checkered beetle

20–14   (a) *A yellowjacket (the model) and some of its mimics. The (b) sand wasp, (c) masarid wasp, and (d) anthidine bee are Müllerian mimics; that is, they all sting. The (e) syrphid fly, (f) Alcathae moth, and (g) ornate checkered beetle, which do not sting, are all Batesian mimics. Müllerian mimics have been compared to reputable merchants who share a common advertisement and divide its costs. Batesian mimics, by contrast, are like unscrupulous retailers who copy the advertisements of successful firms.*

Next to appear on the evolutionary scene, as events are reconstructed, were several species of butterflies that do not taste bad and are not poisonous but that have coloration similar to that of the monarch. These innocuous mimics escape predation by resembling the bad-tasting model that their would-be predators have previously learned to avoid. Such deception is known as **Batesian mimicry** because it was first described by the naturalist H. W. Bates, a friend and traveling companion of Alfred Russel Wallace (page 11).

Laboratory experiments have confirmed the selective value of Batesian mimicry. Jane Brower, working at Oxford University, made artificial models by dipping

mealworms in a solution of quinine, to give them a bitter taste, and then marking each one with a band of green cellulose paint. Other mealworms, which had first been dipped in distilled water, were painted green like the models, so as to produce mimics, and still others were painted orange to indicate another species. These colors were chosen deliberately: orange is a warning color, since it is clearly distinguishable, and green is usually found in species that are not repellent and that therefore benefit from being inconspicuous.

The painted mealworms were fed to caged starlings, which ordinarily eat mealworms voraciously. Each of the nine birds tested received models and mimics in

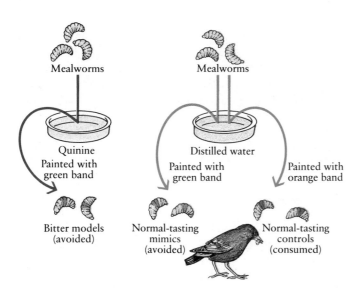

20–15   *An experimental test of Batesian mimicry. Models and mimics were produced by painting quinine-dipped worms (models) and water-dipped worms (mimics) with a green band. Other water-dipped worms were painted with an orange band, clearly distinguishing them from both the models and the mimics. After experiencing the bitter taste of a quinine-dipped model, starlings subsequently avoided not only the models but also their green-banded mimics. The starlings did, however, eagerly consume the orange-banded, water-dipped controls, demonstrating that the worms were not made unpalatable by the process of dipping or the taste of the paint.*

varying proportions. After initial tasting and violent rejection, the models were generally recognized by their appearance and avoided. In consequence, their mimics were also protected (Figure 20–15).

Batesian mimicry works only to the advantage of the mimic. The model suffers from attacks not only by inexperienced predators but also by predators who have had their first experience with the mimic rather than with the model. The mimetic pattern will be at its greatest advantage if the mimic is rare—that is, less likely to be encountered than the model. However, even when as many as 60 percent of the green-banded worms were mimics, 80 percent of the mimics escaped predation. A mimic also has a better chance if it times its emergence to appear after the model, thus reducing its chances of being encountered first. You will not be surprised to learn that in any given area, Batesian mimics do, indeed, generally emerge after the model.

The Imperfection of Adaptation

The concept of evolution has often carried with it some notion of progress. This view—regarded by most biologists as erroneous—has been reinforced by the evidence that evolution has, over time, produced organisms increasingly larger, more complex in structure, and more sophisticated in design, apparently culminating—as if it were preordained—in something as marvelous as ourselves.

Natural selection does tend to push populations toward better solutions to the particular "problems in liv-

ing" with which they are confronted. However, evolutionary change does not necessarily mean improvement by human criteria, nor does it necessarily even result in organisms that are better adapted to their immediate environment. This is because natural selection acts on the here and now. As a result, populations—particularly those engaged in intense evolutionary races—are often a generation behind. Their capacity to "keep up," and thus maintain adaptation, depends primarily on the existence of sufficient genetic variability in the population.

Natural selection has to work with what is available, not only in the chance variations in the genetic material but also in the phenotypic expression of the genotype—a constraint that has been aptly described as "the law of used parts." To take a simple example, all terrestrial vertebrates and their descendants (including ourselves) are limited structurally to a body with a long vertebral column and four limbs, one on each corner, with which our ancestors waddled up on shore. A structural engineer setting out to build a flying machine or a submarine would scrap these blueprints and start from scratch, but evolution can only build on past history. Many of our human ailments, such as unusual difficulty in childbirth and a propensity to lower back pain, can be traced directly to the conversion of this basically quadrupedal form to an upright stance. In *The Panda's Thumb*, Stephen Jay Gould points out that such absence of perfection and the presence of jerry-built contraptions (the panda's thumb, for example) provide stronger evidence that evolution has occurred than the more usually cited examples of exquisite adaptation.

Patterns of Evolution

Natural selection, as we have seen, is a complex process, operating continuously in all populations. Its outcome, in any particular instance, depends on a great many factors. Viewed from a broader perspective, however, it has the effect of producing different patterns of evolution. One of these patterns, which we have already considered, is coevolution, in which organisms of different species act as selective forces on each other. In other patterns of evolution, natural selection may produce remarkably similar phenotypes in distantly related organisms and, conversely, widely different phenotypes in closely related organisms.

Convergent Evolution

Organisms that occupy similar environments often come to resemble one another even though they may be only very distantly related. This phenomenon is known as **convergent evolution.** When organisms are subjected to similar selection pressures, they show similar adaptations. The whales, a group that includes the dolphins and porpoises, are similar to sharks and other large fish in their streamlined shape and other external features, but the fins of whales conceal the remnants of a terrestrial hand. Whales are warm-blooded, like their land-dwelling ancestors, and they have lungs rather than gills. Similarly, two families of plants invaded deserts in different parts of the world, giving rise to the cacti and the euphorbs (Figure 20–16). Both evolved large fleshy stems with water-storage tissues and protective spines, and they appear superficially similar. However, their quite different flowers reveal their widely separate evolutionary origins.

Divergent Evolution

Divergent evolution occurs when a population becomes isolated from the rest of the species and, as a result of particular selection pressures, begins to follow a different evolutionary course. For example, *Ursus arctos,* the brown bear, is distributed throughout the Northern Hemisphere, ranging from the deciduous forests up through the coniferous forests and into the tundra. As is characteristic of such widespread species, there are many local ecotypes. *Ursus arctos* was similarly widespread some 1.5 million years ago, when the Northern Hemisphere was subjected to a series of glacial advances and retreats. During one of the massive glaciations, a population of *Ursus arctos* was split off from the main group, and, according to fossil evidence, this group, under selection pressure from the harsh environment, evolved into the polar bear, *Ursus maritimus* (Figure 20–17).

(a)

(b)

20–16   *An example of convergent evolution. Members of (a) the cactus family and (b) the euphorb family have been separated for millennia of evolutionary history, with the cacti evolving in the deserts of the New World, and the euphorbs in the desert regions of Asia and Africa. Members of both families have fleshy stems adapted for water storage, protective spines, and greatly reduced leaves.*

20–17 *The polar bear is one example of divergent evolution. The white coloring of the polar bear is probably related to its need for camouflage while hunting rather than to a need to hide from predators, since it is one of the world's largest carnivores. (Its greatest enemies are commercial fishermen who do not appreciate the polar bear's fishing activities and hunters who, prizing the skins for trophies, have taken to tracking the bears by helicopter.) Unlike their southern cousins, the brown bears, polar bears (except for females with young) do not become lethargic in the winter but continue to roam the ice and icy waters in search of fish, seals, young walruses, and other prey.*

Brown bears, although they are classified as carnivores and are closely related to dogs, are mostly vegetarians, supplementing their diet only occasionally with fish and game. The polar bear, however, is almost entirely carnivorous, with seals being its staple diet. The polar bear differs physically from the other bears in a number of ways, including its white color, its streamlined head and shoulders, and the stiff bristles that cover the soles of its feet, providing insulation and traction on the slippery ice.

As the example of the polar bear illustrates, divergent evolution can lead not only to the differentiation of locally adapted ecotypes but also to the formation of new species. In the next chapter, we shall examine this process in more detail.

Summary

Natural selection is the differential reproduction of genotypes resulting from interactions between individual organisms and their environment. According to current theory, it is the major force in evolution. Natural selection can act both to produce change and to maintain variability within a population, as exemplified by the color and banding patterns in snails.

Natural selection can act only on characteristics expressed in the phenotype. The unit of selection is the entire phenotype—the whole organism. In extreme cases, a single allele may be decisive in selection. More often, however, a successful phenotype is the result of the interaction of many genes.

Three major categories of natural selection are stabilizing selection, in which extreme phenotypes are eliminated from the population; disruptive selection, in which the extreme phenotypes are selected at the expense of intermediate forms; and directional selection, in which one of the extremes is favored, pushing the population along a particular evolutionary pathway. Another type of selection is frequency-dependent selection, in which the fitness of a phenotype decreases as it becomes more common in the population and increases as it becomes less common. A fifth category, sexual selection, results from the competition for mates; it can greatly increase reproductive success without improving adaptation to other environmental factors.

The result of natural selection is the adaptation—however imperfect—of populations to their environment. Evidence of adaptation to the physical environment can be seen in gradual variations that follow a geographic distribution (clines) and in distinct groups of phenotypes (ecotypes) of the same species occupying different habitats. Adaptation to the biological environment results from the selective forces exerted by interacting species of organisms on each other (coevolution). Examples include the relationship between milkweeds and monarchs and cases of mimicry, both Müllerian and Batesian.

Evolution by natural selection does not necessarily produce a population with the best possible relationship to its environment. Because natural selection operates in the here and now, many members of a population may be at least one generation behind optimal adaptation. Moreover, the potential for evolutionary change is constrained by the extent of variability in the gene pool and by the limits on the phenotypic expression of the genotype imposed by past history.

Observed patterns of evolution resulting from natural selection include coevolution, convergent evolution, and divergent evolution. In convergent evolution, dissimilar populations, only distantly related, come to resemble one another as a result of similar selection pressures. In divergent evolution, similar, related populations become more dissimilar, a process that can lead to the formation of new species.

Questions

1. Distinguish between genotypic variability and phenotypic variability. Which is acted upon by natural selection? Which is necessary for evolution?

2. What is the difference between artificial selection and natural selection? Under what circumstances might they become indistinguishable?

3. How does stabilizing selection affect the gene pool? Disruptive selection? Directional selection?

4. What characteristics in the contemporary human population are probably being maintained by stabilizing selection? Which might be subject to directional selection?

5. Imagine a rapidly expanding population in an environment that allows every organism that is born to survive and reproduce to its maximum extent. Which organisms will make the greatest contribution to the gene pool of future generations? Is natural selection acting on such a population?

6. From an evolutionary perspective, longevity of animals past reproductive age is generally useless. Why is this so? What situations might be exceptions to this statement? Can you think of any reason why longevity might be harmful from an evolutionary standpoint?

7. In his experiments with the long-tailed widowbird, Andersson clipped and reglued the tails of half of the males in the group with tails of normal length. Why did he perform this manipulation? Why was it important?

8. If you were a poisonous butterfly, would you rather have a Batesian mimic or a Müllerian one? Why?

9. In her experiments with quinine-dipped mealworms and their Batesian mimics, Brower dipped the mimics in distilled water before painting on the distinctive green band. Similarly, the mealworms on which orange bands were painted were first dipped in distilled water. Why?

10. Many of Darwin's arguments concerning evolution focused on imperfections rather than perfections. Can you explain why?

11. Distinguish between convergent evolution and divergent evolution. Which type of selection would you expect to be the major factor in convergent evolution? In divergent evolution?

Suggestions for Further Reading

Borgia, Gerald: "Sexual Selection in Bowerbirds," *Scientific American*, June 1986, pages 92–100.

Cook, L. M., G. S. Mani, and M. E. Varley: "Postindustrial Melanism in the Peppered Moth," *Science*, vol. 231, pages 611–613, 1986.

Dawkins, Richard: *The Blind Watchmaker*, W. W. Norton & Company, New York, 1986.*

   A thorough and delightfully written explication of modern evolutionary theory, the evidence in its support, and the controversies that have surrounded it.

*Available in paperback.

Diamond, Jared: "A Pox upon Our Genes," *Natural History*, February 1990, pages 26–30.

Diamond, Jared M.: "Rapid Evolution of Urban Birds," *Nature*, vol. 324, pages 107–108, 1986.

Eldredge, Niles (ed.): *The* Natural History *Reader in Evolution*, Columbia University Press, New York, 1987.*

An anthology of articles from Natural History, *with emphasis on adaptation through natural selection, speciation, and the rise and fall of major groups of organisms.*

Gould, James L., and Carol Grant Gould: *Sexual Selection*, W. H. Freeman and Company, New York, 1989.

A clearly written, authoritative account of one of the most fascinating topics in evolutionary biology, enriched with a great variety of entertaining examples.

Gould, Stephen Jay: *Bully for Brontosaurus: Reflections in Natural History*, W. W. Norton & Company, New York, 1992.*

Gould, Stephen Jay: *Eight Little Piggies: Reflections in Natural History*, W. W. Norton & Company, New York, 1993.

Gould, Stephen Jay: *Ever Since Darwin: Reflections in Natural History*, W. W. Norton & Company, New York, 1992.*

Gould, Stephen Jay: *The Flamingo's Smile: Reflections in Natural History*, W. W. Norton & Company, New York, 1987.*

Gould, Stephen Jay: *Hen's Teeth and Horse's Toes: Further Reflections in Natural History*, W. W. Norton & Company, New York, 1984.*

Gould, Stephen Jay: *The Panda's Thumb: More Reflections in Natural History*, W. W. Norton & Company, New York, 1992.*

Collections of thoughtful, witty, and well-written essays from Natural History. *Evolution is the central theme.*

Gross, Mart R.: "Disruptive Selection for Alternative Life Histories in Salmon," *Nature*, vol. 313, pages 47–48, 1985.

Hutchinson, G. Evelyn: *The Ecological Theater and the Evolutionary Play*, Yale University Press, New Haven, Conn., 1965.

By one of the great modern experts on freshwater ecology, this is a charming and sophisticated collection of essays on the influence of environment in evolution— and also on an astonishing variety of other subjects.

May, Robert M.: "Evolution of Pesticide Resistance," *Nature*, vol. 315, pages 12–13, 1985.

Owen, Denis: *Camouflage and Mimicry*, University of Chicago Press, Chicago, 1982.*

A short, but thorough discussion of the diversity of ways in which organisms deceive other organisms. A multitude of wonderful photographs accompany the text.

Ryan, Michael J.: "Signals, Species, and Sexual Selection," *American Scientist*, vol. 78, pages 46–52, 1990.

Simpson, George Gaylord: *Penguins: Past and Present, Here and There*, Yale University Press, New Haven, Conn., 1983.*

As Simpson said, "Penguins are beautiful, interesting, inspiring, and funny." They are also excellent examples of evolution and adaptation, as this informal account demonstrates.

Small, Meredith F.: "Female Choice in Mating," *American Scientist*, vol. 80, pages 142–151, 1992.

Thornhill, Randy: "The Allure of Symmetry," *Natural History*, September 1993, pages 30–37.

Thornhill, Randy, and Darryl T. Gwynne: "The Evolution of Sexual Differences in Insects," *American Scientist*, vol. 74, pages 382–389, 1986.

Vane-Wright, R. I.: "A Case of Self-Deception," *Nature*, vol. 350, pages 460–461, 1991.

Walker, Tim: "Butterflies and Bad Taste: Rethinking a Classic Tale of Mimicry," *Science News*, June 1, 1991, pages 348–349.

*Available in paperback.

The Tasmanian devil, with its sharp teeth and powerful jaws, is both a hunter and a scavenger. Although it can kill living prey, such as lambs and chickens, it prefers carrion and can devour the entire carcass of a sheep, including the bones. Tasmanian devils are found today only on the island of Tasmania, just south of the continent of Australia. The introduction of the wild dog known as the dingo to the Australian mainland some 3,500 years ago apparently led to the extinction of the Tasmanian devil there.

Like kangaroos and koalas, Tasmanian devils are marsupials. Marsupials, unlike the more familiar placental mammals, are born at an extremely immature stage and then continue their development in a special protective pouch on the mother's body. They are native to Australia, South America, and a few neighboring islands, such as Tasmania.

The Tasmanian devil and other marsupials remind us that exotic lands and their inhabitants played an important role in raising the questions that led to the theory of evolution. Young, curious travelers, such as Darwin and Wallace, observed a profusion of different organisms and came to doubt the doctrine of special creation, which held that each species was created separately and remained fixed and immobilized in time. Madagascar, the Galapagos, Borneo—each of these faraway places has, as we have seen, its own inhabitants, many of them found nowhere else in the world.

Just as island biology raised the question of the origin of species for the early students of evolution, it has helped to answer many questions about that process for later observers. Islands are living laboratories in which organisms can become separated in time and space from their parents and their predecessors and thus set out on a new and different evolutionary adventure. Islands, as you will see, come in many different shapes, sizes, and disguises.

The Origin of Species

In the course of his voyage on H.M.S. *Beagle*, Charles Darwin was struck not only by the enormous number of species he encountered for the first time but also by their distribution. According to the doctrine of the time, accepted by Darwin when he set sail, each species had been created separately and was distinct from all other species. Moreover, each was specially created for a particular way of life and placed in the locality for which it was best suited.

It is no coincidence that Darwin first came to doubt the doctrine of special creation while traveling in the tropics, where the widest variety of species is found and where, even today, many species new to science are encountered. For example, the island continent of Australia is home to some 57 separate species of kangaroos. All of the kangaroos are marsupials. Had each kangaroo species been specially created and dropped off in Australia? And why only in Australia? Or, more plausibly, was there perhaps an ancestral marsupial that had given rise to all of these clearly related kangaroos?

Moreover, it became apparent to Darwin that species are not as distinct as had previously been thought. For example, as he traveled down the eastern coast and up the western coast of South America, he observed gradual changes in various characteristics of the plants and animals—evidence that organisms become modified with time, according to the different environments in which they live.

Darwin—like his friend and correspondent, Alfred Russel Wallace (page 11)—was also puzzled by the fact that places similar in climate and topography are often populated by very different organisms. For example, remote oceanic islands often have no terrestrial mammals at all but only peculiar species of bats. Moreover, bats are the only native placental mammals of Australia. All of the other diverse mammals indigenous to Australia are marsupials, all clearly related to one another and found only rarely elsewhere on the planet. Great Britain and continental Europe have rabbits galore, whereas similar areas in South America have only the Patagon-

(a)

(b)

(c)

21–1 *Although these three mammals are similar in appearance and have similar life styles, they are not closely related.* (**a**) *The European rabbit, a placental mammal, is classified as a lagomorph.* (**b**) *The Patagonian hare, or cavy, also a placental mammal,* *is classified as a rodent.* (**c**) *The Australian "banded hare" is actually a wallaby, a member of the kangaroo family. Like almost all of the other native mammals of Australia, it is a marsupial.* *The scientific study of the distribution of plants and animals in the various regions of the world, known as biogeography, was initially developed by Alfred Russel Wallace.*

ian hare—not a rabbit or hare at all but a rodent (Figure 21–1). A multitude of examples of this sort provided strong evidence for Darwin that living things are what and where they are because of events that occurred in the course of their previous history.

Although Darwin titled his great text *On the Origin of Species,* he was never really able to explain how species might originate. However, an enormous body of work, mostly in the twentieth century, has provided many insights into the process. A crucial starting point was the development of a clear definition of the term "species."

What Is a Species?

In Latin, *species* simply means "kind," and so species are, in the simplest sense, different kinds of organisms. A more precise definition is that a **species** is a group of natural populations whose members can interbreed with one another but cannot (or at least usually do not) interbreed with members of other such groups. The key concept in this definition is **genetic isolation:** if members of one species freely exchanged genes with members of another species, they could no longer retain those unique characteristics that identify them as different kinds of organisms.

From an evolutionary perspective, a species is a group or population of organisms, reproductively united but very probably changing as it moves through space and time. Splinter groups, reproductively isolated

from the population as a whole, can undergo sufficient change that they become new species. This process is known as **speciation.** Occurring repeatedly in the course of more than 3.5 billion years, it has given rise to the diversity of organisms that have lived in the past and that live today.

Modes of Speciation

The members of a species share a common gene pool effectively separated from the gene pools of other species. A central question, then, is how one pool of genes splits off from another to begin a separate evolutionary journey. A subsidiary question is how two species, often very similar to one another, inhabit the same place at the same time and yet remain reproductively isolated.

According to current perspectives, speciation is most commonly the result of the geographic separation of a population of organisms: this process is known as **allopatric** ("other country") speciation. Under certain circumstances, speciation may also occur without geographic isolation, in which case it is known as **sympatric** ("same country") speciation.

Allopatric Speciation

Every widespread species that has been carefully studied has been found to contain geographically representative populations that differ from each other to a greater or lesser extent. Examples are the ecotypes of

(a) (b)

21–2 *Subspecies are variants of a species that are sufficiently different to have been formally recognized with a Latin name. Two subspecies that have formed on either side of a natural geographic barrier, the Tana River in Kenya, are* (**a**) *the common giraffe,* Giraffa camelopardalis rothschildi, *and* (**b**) *the reticulated giraffe,* Giraffa camelopardalis reticulata. *The two are considered subspecies because they are phenotypically distinct (note the patterns of their coats), but they are not considered different species because they can interbreed, producing fertile hybrids. Six other subspecies of giraffes are also recognized.*

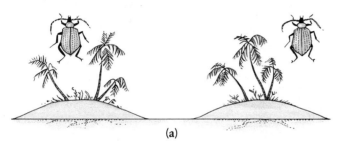

(a)

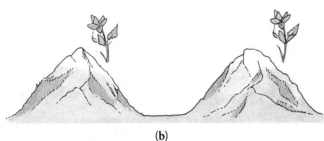

(b)

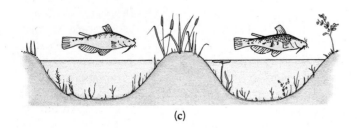

(c)

(d)

Potentilla glandulosa (page 346) and the subspecies of giraffes (Figure 21–2). A species composed of such geographic variants is particularly susceptible to speciation if geographic barriers arise, preventing gene flow.

Geographic barriers are of many different types. Islands are frequently sites for the development of new species. The breakup of Pangaea (see essay on next page) profoundly altered the course of evolution, setting the island continent of Australia adrift as a veritable Noah's ark of marsupials. Populations of many organisms can become cut off from one another by barriers less obvious than oceans (Figure 21–3). For a plant, an island may be a mist-veiled mountaintop, and for a fish, a freshwater lake. A forest grove may be an island for a small mammal, and a few meters of dry ground can isolate two populations of snails. Islands may also form by the creation of barriers between formerly contiguous geographic zones. The Isthmus of Panama, for instance, has repeatedly submerged and reemerged in the course of geologic time. With each new emergence, the Atlantic and Pacific oceans became "islands," populations of marine organisms were separated from one another, and some new species formed. Then, when the oceans joined again (with the submergence of the Isthmus), North and South America were separated and became, in turn, the "islands."

21–3 *For different populations, the genetic isolation that can lead to speciation may take different geographic forms. Genetic "islands" may be* (**a**) *true islands,* (**b**) *mountaintops,* (**c**) *ponds, lakes, or even oceans, or* (**d**) *isolated clumps of vegetation.*

The Breakup of Pangaea

In the past 30 years, continental drift has become firmly established as part of the theory of plate tectonics. According to this theory, the outermost layer of our planet is divided into a number of segments, or plates. These plates, on which the continents and sea floors rest, move in relation to one another.

We see evidence of this relative motion mainly at the boundaries between adjacent plates. Where plates collide, volcanic islands such as the Aleutians may be formed or mountain belts such as the Andes or Himalayas may be uplifted. At the boundaries where plates are separating from one another, volcanic material wells up to fill the void. It is here that ocean basins are created. Plates may also move parallel to the boundary that joins them but in opposite directions, or in the same direction but at different speeds, as with the San Andreas fault.

About 200 million years ago, all the major continents were locked together in a supercontinent, Pangaea. Several reconstructions of Pangaea have been proposed, one of which is shown here. It is generally agreed that Pangaea began to break up about 190 million years ago, about the time the dinosaurs approached their zenith and the first mammals began to appear. First, the northern group of continents (Laurasia) split apart from the southern group (Gondwana). Subsequently, Gondwana broke into three parts: Africa-South America, Australia-Antarctica, and India. India drifted northward and collided with Asia about 50 million years ago. This collision initiated the uplift of the Himalayas, which continue to rise today as India still pushes northward into Asia.

By the end of the Cretaceous period, about 65 million years ago (for a geologic timetable, see the inside of the front cover), South America and Africa had separated sufficiently to have formed half the South Atlantic. Europe, North America, and Greenland had also begun to drift apart. However, final separation between Europe and North America–Greenland did not occur until about 43 million years ago, during the Eocene epoch of the Tertiary period. Also during the Eocene, Australia finally split from Antarctica and moved northward to its present position. Later, during the Pliocene epoch, the two Americas were joined by the Isthmus of Panama, which was created by volcanic action.

From the outset, the theory of continental drift has been closely interwoven with that of evolution. One of the earliest and most impressive pieces of evidence in favor of continental drift was the discovery of fossil remains of a small, snaggle-toothed reptile, *Mesosaurus*, found in the coastal regions of Brazil and South Africa but nowhere else. Early in 1982, a team of American scientists returned from Antarctica with the first fossil of a land mammal ever found there—a marsupial. This find supports the theory that marsupials migrated by land from South America across Antarctica to Australia before the two separated, some 55 million years ago. Fossils of many extinct species of marsupials have been found in South America, but only some 40 species have survived to the present.

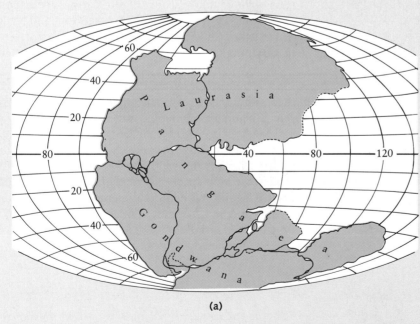

(a) Pangaea.

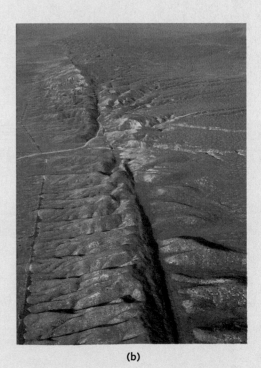

(b)

(a) Pangaea. **(b)** The San Andreas fault is a boundary between two giant moving plates. The fault, running through San Francisco and continuing southeast of Los Angeles, is responsible for California's notorious earthquakes.

If the isolated population is a small, peripheral one, it is more likely to differ from the parent population (because of the founder effect), and gene flow is less likely to occur between the two populations. Once separated, the isolated population may begin to diverge genetically under the pressure of different selective forces. If enough time elapses and the differing selective forces are sufficiently great, the isolated population may diverge so much that, even if it were reunited with the parent population, interbreeding under natural conditions would no longer occur. At this point, speciation is said to have taken place. Every isolated population does not, of course, become a species. It may rejoin the parent group or it may perish, which is probably the fate of most small, isolated populations.

Sympatric Speciation

Polyploidy

A well-documented mechanism by which new species are produced through sympatric speciation is **polyploidy,** an increase in the number of chromosomes beyond the typical diploid (2n) complement. Polyploidy may arise as a result of nondisjunction (page 198) during mitosis or meiosis, or it may be generated when the chromosomes divide properly during mitosis or meiosis but cytokinesis does not subsequently occur. Polyploid individuals can be produced deliberately in the laboratory by the use of the drug colchicine, which prevents separation of chromosomes during mitosis.

Polyploidy leading to the formation of new species sometimes occurs as a result of a doubling of the chromosome number within individual organisms of a species (Figure 21–4). This process has, for example, produced several species of whiptail lizards in the American Southwest. More frequently, however, new species are generated by a doubling of the chromosome number in hybrid organisms.

A **hybrid** organism is the offspring of parents of different species. Hybrids can occur in animals (such as the mule, which is the offspring of a donkey and a horse), but they are far more common in plants. Kentucky bluegrass, for instance, is a promiscuous hybridizer and has crossbred with many related species, producing hundreds of hybrid strains, each well adapted to the conditions of the area in which it grows. Such hybrids, spreading by means of asexual reproduction, are sometimes able to outcompete both parents. Asexually reproducing strains of hybrid plants may be considered species in that they are genetically isolated from their parents and from each other.

21–4   *Polyploidy within individual organisms can lead to the formation of new species. If the chromosomes of a diploid organism do not separate during meiosis (nondisjunction), diploid (2n) gametes may result. Union of two such gametes, produced either by the same individual or by different individuals of the same species, will produce a tetraploid (4n) individual. Although this individual may be capable of sexual reproduction, it will be reproductively isolated from the parent species.*

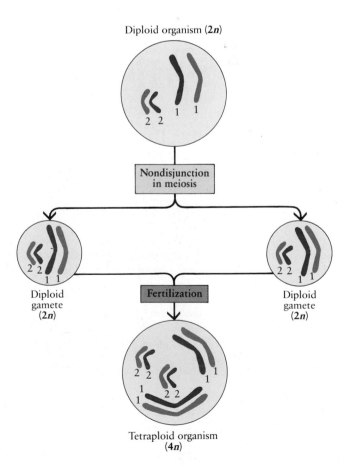

Diploid organism (**2n**)

Nondisjunction in meiosis

Diploid gamete (**2n**)

Fertilization

Diploid gamete (**2n**)

Tetraploid organism (**4n**)

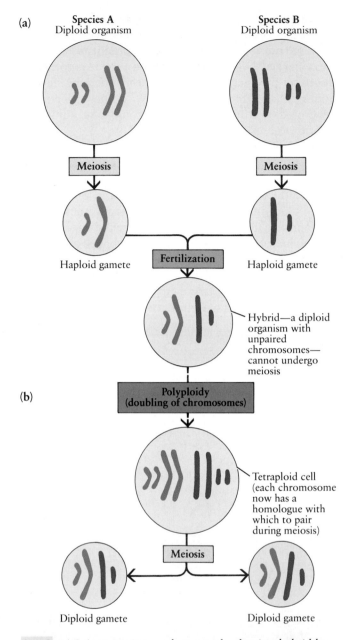

(a)

Species A
Diploid organism

Species B
Diploid organism

Meiosis

Meiosis

Haploid gamete

Fertilization

Haploid gamete

Hybrid—a diploid organism with unpaired chromosomes—cannot undergo meiosis

(b)

Polyploidy
(doubling of chromosomes)

Tetraploid cell (each chromosome now has a homologue with which to pair during meiosis)

Meiosis

Diploid gamete

Diploid gamete

21–5   (a) *An organism, such as a mule, that is a hybrid between two different species and is produced from two haploid (n) gametes, can grow normally because mitosis is normal. It cannot reproduce sexually, however, because the chromosomes cannot pair at meiosis. (b) If polyploidy subsequently occurs and the chromosome number doubles, the hybrid can then produce viable gametes. Since each chromosome will have a partner, the chromosomes can pair at meiosis. The resultant gametes will be diploid (2n).*

Hybrids in both plants and animals are often sterile because the chromosomes cannot pair at meiosis (having no homologues), a necessary step for producing viable gametes (Figure 21–5a). If, however, polyploidy occurs in such a sterile hybrid and the resulting cells divide by further mitosis and cytokinesis so that they eventually produce a new individual asexually, that individual will have twice the number of chromosomes as its parent. As a consequence, it is reproductively isolated from the parental line. However, its chromosomes—now duplicated—can pair, meiosis can occur normally, and fertility is restored (Figure 21–5b). It is a new species, capable of sexual reproduction.

Sympatric speciation through polyploidy is an important, well-established phenomenon in plants. Approximately half of the 235,000 species of flowering plants have had a polyploid origin, and many important agricultural species, including wheat, are hybrid polyploids.

Disruptive Selection

Sympatric speciation as a result of disruptive selection (page 340) is a more controversial issue. Possible cases of this type of sympatric speciation are difficult to prove because of the time factor. Either (1) the two forms are in the process of diverging, as in the plants growing on contaminated soil, in which case it is not possible to prove that complete separation will ever occur, or (2) the two forms have separated completely, and it is not possible to prove that they were not formed as a consequence of geographic separation and have since been reunited in the same locale.

One often-cited example of sympatric speciation produced by disruptive selection involves a group of species of small fruit flies of the genus *Rhagoletis*. The larvae of each species of *Rhagoletis* feed on developing fruits of only one plant family. The host fruits of the family serve not only as food but also as the rendezvous for courtship and mating, followed by the deposition of eggs (Figure 21–6). *Rhagoletis pomonella*, for example, is a species that feeds on hawthorns. In 1865, farmers in the Hudson River Valley of New York reported that some of these flies had begun attacking their apples. The infestation then spread rapidly to apple orchards in adjacent areas of Massachusetts and Connecticut. There are now two distinct varieties of *R. pomonella*, with a number of genetic differences. One variety feeds and reproduces on hawthorns, and the other, known as the apple maggot fly, on apples. The two varieties are isolated by their reproductive behavior and so may be regarded as on the path to speciation.

Many biologists, however, question whether the two varieties of *R. pomonella* are truly sympatric. They

21–6 *A female apple maggot fly, Rhagoletis pomonella. Typically, the male lies in wait on the surface of an apple. When a female arrives, attracted by the ripening fruit, he will attempt to mate with her. The female subsequently lays her eggs, one at a time, on different apples. Before laying an egg, she probes the apple, as shown here, testing for the presence of nutrients and the absence of harmful chemicals. The eggs are deposited beneath the skin, where they hatch into small, white larvae that feed on the flesh of the apples. Apples infested by the larvae soon rot and fall to the ground.*

point out that this may instead be an example of allo-patric processes operating on a greatly reduced scale. For these flies—and other small organisms that are parasitic or highly specialized for feeding and reproducing in a particular microhabitat—environmental differences that appear small to us may represent significant barriers, creating "mini-islands."

Maintaining Genetic Isolation

Once speciation has occurred, the now-separate species can live together without interbreeding, despite the fact that some are so similar phenotypically that only an expert with a microscope can tell them apart—*Dro-sophila* offers several examples. What factors operate to maintain genetic isolation of closely related species?

Isolating mechanisms may be conveniently divided into two categories: **premating mechanisms,** which prevent mating between members of different species, and **postmating mechanisms,** which prevent the production of fertile offspring from such matings as are attempted or do occur. One of the most significant things about the postmating isolating mechanisms is that, in nature, they are rarely tested. The premating mechanisms alone usually prevent any interbreeding. This is an expected consequence of sexual selection. Choosing the best mate or mates from among the members of one's own species requires a finer level of discrimination than is required to distinguish members of one's own species from those of a different species (Figure 21–7).

21–7 *A male frigate bird, a resident of the Galapagos Islands, displays his crimson pouch. Throughout the courtship period, the pouch remains bright and inflated, even when the bird is flying or sleeping. The female will not mate with a male that does not have these species-specific characteristics.*

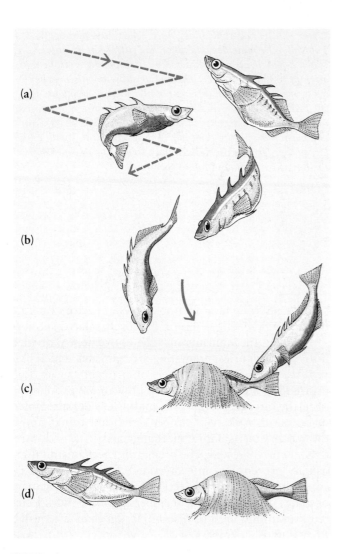

21–8 *Sticklebacks, small freshwater fish, have elaborate mating behavior. The male at breeding time, in response to increasing periods of sunlight, changes from dull brown to the radiant colors shown here (upper left). He builds a nest and (a) begins to court females, zigging toward them and zagging away from them. A female ready to lay eggs responds by displaying her swollen belly. (b) The male then leads her down to the tunnel-like nest. (c) After she enters the nest, he prods her tail and, in response, she lays her eggs and swims off. (d) He follows her through, fertilizes the eggs, and stays to tend the brood. If either partner fails in any step of this quite elaborate ritual, no young are produced.*

Premating Isolating Mechanisms

Premating isolating mechanisms in vertebrates often involve elaborate behavioral rituals, visual signals, or frequently, a combination of the two (Figure 21–8).

Visual and auditory signals are also important in many insects. Among the most familiar are the flashings of fireflies (which are actually beetles) and the chirping of crickets. In fireflies, each species has its own particular flashing pattern, which is different from that of other species both in duration of the flashes and in intervals between them. For example, one common flash pattern consists of two short pulses of light separated by about 2 seconds, with the phrase repeated every 4 to 7 seconds. The lights also vary among the species in intensity and color. The male flashes first, and the female answers, returning the species-specific signal. The predatory females of some firefly species not only produce the signal specific to their own species but also mimic the signals of other species, luring males of those species to the female's dinner table. This mimicry—and the subsequent feeding—generally occurs several days after the female has mated with a male of her own species.

Bird songs, frog calls, and the strident love notes of cicadas and crickets all serve to identify members of a species to one another. For example, studies of leopard frogs have revealed that some of the populations that can crossbreed under laboratory conditions do not do so in the wild. They are effectively isolated from one another by differences in their mating calls.

In many species of animals, chemical substances released by either the female or the male serve to bring the two sexes together. They may function as signals to attract the male, as in the case of the cecropia moth, or to trigger the release of gametes by the female, as with oysters. Because such substances are species-specific, they also act, in effect, as isolating mechanisms.

Temporal differences also play an important role in reproductive isolation. Species differences in flowering times are important isolating mechanisms in plants. Most vertebrates—we are a notable exception—have seasons for mating, often controlled by temperature or by day length. Figure 21–9 shows the mating calendar of different species of frogs and toads near Ithaca, New York.

Postmating Isolating Mechanisms

On the rare occasions when members of two different species attempt to mate or succeed in doing so, anatomical or physiological incompatibilities often maintain

Peeper

Leopard frog

Pickerel
frog and
common
toad

Tree
frog

Green
frog

Bullfrog

Wood
frog

Mating activity

March 1   April 1   May 1   June 1   July 1

21–9 *Mating timetable for various frogs
and toads that live near Ithaca, New York. In
the two cases where two different species
have mating seasons that coincide, the breeding sites differ. Peepers prefer woodland
ponds and shallow water, whereas leopard
frogs breed in swamps. Pickerel frogs mate in
upland streams and ponds, and common
toads use any ditch or puddle.*

genetic isolation. These postmating isolating mechanisms are of several different types. For instance:

1. Differences in the shape of the genitalia may prevent
   insemination, or differences in flower shape may prevent pollination.

2. The sperm may not be able to survive in the reproductive tract of the female, or the pollen tube may
   not be able to grow on the stigma.

3. The sperm cell may not be able to fuse with the egg
   cell.

4. The egg, once fertilized, may not develop.

5. The young may survive, but they may not become
   reproductively mature.

6. The mature offspring may be hardy but sterile—the
   mule, for example.

Postmating mechanisms reinforce premating ones. A
female cricket that answers to the wrong song or a frog
whose individual calendar is not synchronized with that
of the rest of the species will contribute less (or nothing)
to the gene pool. As a consequence, there is a steady
selection for premating isolating mechanisms.

An Example of Speciation:
Darwin's Finches

Admirers of Charles Darwin find it particularly appropriate that one of the best examples of speciation is provided by the finches observed by Darwin on his voyage
to the Galapagos Islands. All the Galapagos finches are
believed to have arisen from one common ancestral
group—perhaps either a single pair or even a single female bearing a fertilized egg—transported from the
South American mainland, some 950 kilometers away.

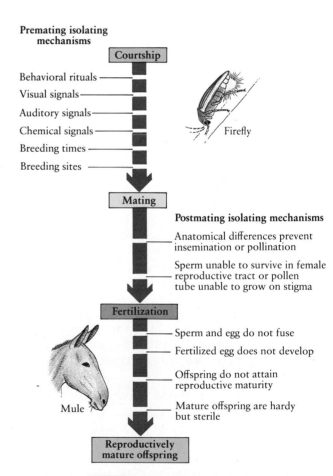

**Premating isolating
mechanisms**

Courtship

Behavioral rituals
Visual signals
Auditory signals
Chemical signals
Breeding times
Breeding sites

Firefly

Mating

Postmating isolating mechanisms

Anatomical differences prevent
insemination or pollination

Sperm unable to survive in female
reproductive tract or pollen
tube unable to grow on stigma

Fertilization

Sperm and egg do not fuse

Fertilized egg does not develop

Offspring do not attain
reproductive maturity

Mature offspring are hardy
but sterile

Mule

**Reproductively
mature offspring**

21–10 *A summary of the genetic isolating
mechanisms that usually prevent successful
interbreeding by organisms of different species. Depending on the species involved, the
complex process from courtship (or flowering, in plants) to reproductively mature offspring can be interrupted at any one of a
number of different stages.*

How she or they got there is, of course, not known, but it may have been the result of some particularly severe storm. (Periodically, for example, some North American birds and insects appear on the coasts of Ireland and England after having been blown across the North Atlantic.) It is very likely that finches were the first land birds to colonize the islands, which provide a highly diversified environment. Different groups of finches, occupying different habitats, would have been subjected to widely differing selective forces.

From the small ancestral group, 13 different species arose, plus one to the northeast on the Cocos Islands, 1,000 kilometers away. Apparently the various islands were near enough to one another that, over the years, small founding groups could emigrate. The islands were far enough apart, however, that once a group was established, there would be little or no gene flow between it and the parent group for a period of time long enough for significant differences to evolve as a result of natural selection and for genetic barriers to develop. (Finches are not very good long-distance fliers; if they were, this natural experiment in speciation would have been a failure.) Development of a new species in this way requires, it is estimated, at least 10,000 years of geographic isolation. However, since there are many islands, different species could have been evolving on different islands during the same time period.

The ancestral type was a finch, a smallish bird with a short, stout, conical bill adapted for seed-crushing. The ancestor is believed to have been a ground-feeding finch, and six of the Galapagos finches are ground finches. Four species of ground finches now live together on most of the islands. Three of them eat seeds and differ from one another mainly in the size of their beaks, which, in turn, influences the size of seeds they eat. The fourth lives largely on the prickly pear and has a much longer and more pointed beak. The other two species of ground finch are usually found only on outlying islands, where some supplement their diet with cactus.

In addition to the ground finches, there are six species of tree finches, also differing from one another mainly in beak size and shape. One has a parrotlike beak, suited to its diet of buds and fruit. Four of these tree finches have insect-eating beaks, each adapted to a different size range of insects. The sixth, and most remarkable of the insect eaters, is the woodpecker finch (Figure I–8, page 9).

By all ordinary standards of external appearance and behavior, the thirteenth species of Galapagos finch would be classified as a warbler. Its beak is thin and pointed like a warbler's, it even has the warblerlike habit of flicking its wings partly open, and, warblerlike,

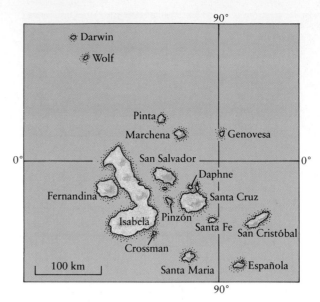

21–11   *The Galapagos archipelago, some 950 kilometers west of the coast of Ecuador, consists of 13 principal volcanic islands and many smaller islets and rocks. These islands have been called "a living laboratory of evolution." Species and subspecies of plants and animals that have been found nowhere else in the world inhabit the Galapagos. "One is astonished," wrote Charles Darwin in 1837, "at the amount of creative force . . . displayed on these small, barren, and rocky islands. . . ."*

it searches through leaves, twigs, and ground vegetation for small insects. However, its internal anatomy and other characteristics clearly place it among the finches. There is general agreement that it, too, is a descendant of the common ancestor or ancestors.

Studies directed by Peter R. Grant of Princeton University have demonstrated that reproductive isolation of the different finch species that now live together on a number of the islands depends on a combination of song and beak characteristics. It appears that young male and female finches learn to recognize their species-specific song from hearing the singing of their father and to recognize the beak of their species from observation of the beaks of both parents during feeding. Subsequently, songs, which are sung by the males only, serve as long-range cues to species identity, and beak size and shape serve as short-range cues. Females enter the territory of a courting male in response to his song but appear, at least in part, to base their decision on whether to mate on an inspection of his beak. The males are also known to actively pursue females. Sometimes a male finch will mistakenly pursue a female of another species—the finches often look very much alike from the rear—only to lose interest as soon as he sees her beak. He will rarely court a female of another species following a close inspection of her beak.

21–12   *Six of the 13 different species of Darwin's finches. Except for (a) the warbler finch (Certhidea olivacea), which resembles a warbler more than a finch, the species look very much alike. The birds are all small and dusky-brown or blackish, with stubby tails. The most obvious differences among them are in their bills, which vary from small, thin beaks to large, thick ones.*

(b) The small ground finch (Geospiza fuliginosa) and (c) the large ground finch (Geospiza magnirostris) are both seed eat- ers. G. magnirostris, *with a much larger beak than* G. fuliginosa, *is able to crack larger seeds.*

(d) The cactus finch (Geospiza scandens) lives on cactus blooms and fruit. Notice that its beak is more pointed than those of the two ground finch species.

(e) The small tree finch (Camarhynchus parvulus) and (f) the large tree finch (Camarhynchus psittacula), both insect eaters, take prey of different sizes.

The Record in the Rocks

The Earth's long history is recorded in the rocks that lie at or near its surface, layer piled upon layer, like the chapters in a book. These layers, or strata, are formed as rocks in upland areas are eroded by wind and water, and the resulting pebbles, sand, and clay are carried to the lowlands and the seas. Once deposited, they slowly become compacted and cemented into a solid form as new material is deposited above them. As continents and ocean basins change shape over the millennia, some strata sink below the surface of an ocean or a lake, others are forced upward into mountain ranges, and some are worn away, in turn, by water, wind, or ice or are deformed by heat or pressure.

Individual strata may be paper-thin or many meters thick. They can be distinguished from one another by the types of parent material from which they were laid down, the way the material was transported, and the environmental conditions under which the strata were formed, all of which leave their traces in the rock. They can be distinguished, moreover, by the types of fossils they contain. Small marine fossils, in particular, can be associated with specific periods in the Earth's history. The fossil record is seldom complete in any one place, but because of the identifying characteristics of the strata, it is possible to piece together the evidence from many different sources. It is somewhat like having many copies of the same book, all with chapters missing—but different chapters, so it is possible to reconstruct the whole.

The geologic eras—Precambrian, Paleozoic, Mesozoic, and Cenozoic—which are the major volumes of the geologic record, were identified and named in the early nineteenth century. As shown in the geologic timetable on the inside of the front cover, these eras are subdivided into periods, many of which are named, quite simply, for the areas in which the particular strata were first studied or studied most completely: the Devonian for Devonshire in southern England, the Permian for the province of Perm in Russia, the Jurassic for the Jura Mountains between France and Switzerland, and so on.

In the Badlands of South Dakota, the waters of the Missouri River and its tributaries have carved through the rock, revealing the horizontal bands of geologic strata. These formations are of particular interest because they encompass the boundary between the Cretaceous and Tertiary periods—a time of mass extinction, including that of the dinosaurs (see page 369).

Early attempts to date the various eras and periods were based on the ages of the strata in relation to each other. (A stratum occurring regularly above another is younger than the one below it.) The relative ages of the strata were then compared to the Earth's age. The first scientific estimate of the planet's age was made in the mid-1800s by the British physicist Lord Kelvin. On the basis of his calculations of the time necessary for the Earth to have cooled from its original molten state, Kelvin maintained that the planet was about 100 million years old, a calculation that posed considerable difficulties for Darwin. (Kelvin was not aware of the existence of radioactive materials under the surface that heat the planet from within.) In the last 50 years, however, new methods for determining the ages of strata have been developed using measurements of the decay of radioactive isotopes. As a result, the estimated age of the Earth has increased from 100 million years to about 4.6 billion years.

Geologic strata are now dated whenever possible by analysis of radioactive isotopes contained in crystals of igneous rock (rock formed from molten material) associated with particular strata. As we noted in Chapter 1, many naturally occurring isotopes are radioactive. All the heavier elements—atoms that have 84 or more protons in the nuclei—are unstable and, therefore, radioactive. All radioactive isotopes emit energy (as particles or rays) at a fixed rate; this process is known as radioactive "decay." The rate of decay is measured in terms of half-life: the half-life of a radioactive isotope is defined as the time in which half the atoms in a sample lose their radioactivity and become stable. Since the half-life of an isotope is constant, it is possible to calculate the fraction of decay that will take place for a given isotope in a given period of time. The radiometric clock starts to tick when the crystalline rock is formed.

Half-lives vary widely, depending on the isotope. The radioactive nitrogen isotope ^{13}N has a half-life of 10 minutes, and the most common isotope of uranium (^{238}U) has a half-life of 4.5 billion years. This uranium isotope undergoes a series of decays, eventually being transformed to an isotope of lead (^{206}Pb). Thus the proportion of ^{238}U to ^{206}Pb in a given rock sample is a good indication of how long ago that rock was formed. Five different isotopes are now commonly employed as radiometric clocks, and in thousands of instances, rocks have been dated by three or more independent clocks.

Any theory of evolution requires, as Darwin knew well, that the Earth have a long history. Thus, these radiometric clocks are doubly important to modern students of evolution. First, they demonstrate conclusively that the Earth's age is close to 5 billion years. In other words, the Earth is indeed old enough for evolution to have produced the observed diversity of organisms. Second, they provide the tools for estimating the ages of various rocks—and of the fossils within them—and so, for unraveling the details of the Earth's biological past.

The Evidence of the Fossil Record

Although we may make inferences about the course of evolution from the observation of living organisms, the proving ground of evolutionary theory is the fossil record itself. This record reveals a succession of living forms, with simpler forms generally preceding more complex forms (see essay).

Geologic studies, as well as the collecting of plant and animal specimens, were among Darwin's activities aboard the *Beagle*. The coasts of South America were of particular interest because they showed evidence of widespread upheaval, with many geologic strata exposed. In the course of his explorations, Darwin came across many fossils of extinct mammals, including those of giant armadillos. The fact that the remains of extinct armadillos were buried in the same South American plains where the only surviving species of these strange armored mammals lived provided tangible evidence of change and history.

Since Darwin's time, a steady stream of new discoveries has enormously increased our knowledge of the fossil record, which now extends back more than 3 billion years. In the case of many groups of organisms—plants and vertebrate animals, for example, as we shall see in Chapters 25 and 27—fossils have been found that exhibit a graded series of changes in anatomical characteristics, linking older forms with the modern forms and revealing pathways of divergence from common ancestors.

The fossil record provides overwhelming evidence that evolution has indeed occurred. It also reveals that the diversity of life on Earth is the result of three broad patterns of speciation, coupled with the effects of extinction.

Phyletic Change

One pattern observed in the fossil record is change within a single lineage of organisms. This is known as **phyletic change** (from the Greek *phylon,* meaning "race" or "tribe"). Under the pressures of directional selection, a species gradually accumulates changes until eventually it is so distinct from its predecessors that it may properly be considered a new species—a new kind—of organism. Darwin's concept of evolution emphasized the slow, gradual accumulation of change, and phyletic change is a large-scale version of the types of changes so ably documented by population geneticists.

Cladogenesis

A second pattern observed in the fossil record is the splitting of a lineage of organisms into two or more distinct lineages. This is known as **cladogenesis** (the forming of branches). Species formed by cladogenesis are the contemporaneous descendants of a common ancestor—like Darwin's finches.

Paleobiologists have tended to give more weight to the role of cladogenesis in evolution than to phyletic change. Ernst Mayr, an ornithologist and a leader in the formulation of modern evolutionary theory, agrees. He maintains that the formation of new species by the splitting off of small populations from the parent stock is responsible for almost all major evolutionary change. In such small populations, favorable genetic combinations, if present, can increase rapidly in number and frequency without being diluted out by gene flow to and from the larger parent population. Hence, evolution probably does not occur steadily and gradually but in spurts, accounting for the sudden increases of new species observed at many points in the fossil record. However, these two models—phyletic change and cladogenesis—have generally been viewed not as alternatives but rather as complementary to one another.

Adaptive Radiation

Another pattern observed in the fossil record is **adaptive radiation,** which is usually considered to result from the combined effects of both cladogenesis and phyletic change. Adaptive radiation is the sudden (in geologic time) diversification of a group of organisms that share a common ancestor, often itself newly evolved. It is associated with the opening up of a new biological frontier that may be as vast as the land or the air, or, as in the case of the Galapagos finches, as small as an archipelago. Adaptive radiation results in the almost simultaneous formation of many new species.

The late George Gaylord Simpson, a leading paleontologist, emphasized that adaptive radiation is the major pattern revealed by the fossil record. There are many examples. For instance, some 300 million years ago, the reptiles, liberated from an amphibian existence by the "invention" of an egg that retained its own water supply, diversified rapidly as they moved into a variety of different terrestrial environments. A similar, even more rapid burst of evolution later gave rise to the birds. As the dinosaurs became extinct, the mammals similarly burst forth upon the evolutionary scene, with many different kinds appearing simultaneously in the fossil record (Figure 21–13).

Extinction

A phenomenon particularly well documented by the fossil record is **extinction**—the complete demise not only of individual species but also of entire groups of

Placentals

Gray wolf
(*Canis lupus*)

Ocelot
(*Felis pardalis*)

Flying squirrel
(*Glaucomys volans*)

Woodchuck
(*Marmota monax*)

Giant anteater
(*Myrmecophaga
tridactyla*)

Coast mole
(*Scapanus orarius*)

House mouse
(*Mus musculus*)

Marsupials

Tasmanian wolf
(*Thylacinus
cynocephalus*)

Tiger cat
(*Dasyurus
maculatus*)

Sugar glider
(*Petaurus
breviceps*)

Wombat
(*Vombatus ursinus*)

Banded anteater
(*Myrmecobius
fasciatus*)

Marsupial mole
(*Notoryctes typhlops*)

Broadfoot marsupial mouse
(*Antechinus stuarti*)

21–13 *Early in the evolutionary history of the mammals, the lineage gave rise to two principal branches, the marsupials and the placentals, which are distinguished primarily by the stage of development at which the young are born. Although the subsequent adaptive radiation of the marsupials in Australia and of the placentals on other continents were independent events, involving different immediate ancestors, striking simi-*

larities resulted in the descendant organisms. As shown in these examples, similar adaptations are exhibited by placentals and marsupials that occupy similar habitats and have similar life styles. All of the marsupials, however, are more closely related to each other than they are to any of the placentals, and vice versa.

species. Only a small fraction of the species that have ever lived are presently in existence—certainly less than 1/10 of 1 percent, perhaps less than 1/1,000 of 1 percent.

In the 1980s, J. John Sepkoski, Jr., and David M. Raup of the University of Chicago gathered all available data on the extinction of marine organisms during the past 250 million years and analyzed the data statistically. Their analysis revealed a steady "background" extinction rate of about 180 to 300 species every one million years, interrupted every 26 million years by a period of mass extinctions. Some of these mass extinctions resulted in the elimination of enormous numbers of species. The greatest extinction of all occurred at the end of the Permian period, some 248 million years ago. It is estimated that 80 to 85 percent of all species then living became extinct, with the rate rising as high as 96 percent for marine species living in shallow waters.

Of all the organisms that once lived but live no more, the dinosaurs are perhaps the most intriguing. For some 150 million years these giant reptiles and their relatives dominated the land, air, and waters—a success story that indicates exquisite adaptation to their environments. And then, about 65 million years ago, at the end of the Cretaceous period, they vanished. Many other species of terrestrial animals died out at about the same time, including virtually all of those weighing more than 25 kilograms (about 55 pounds), as did a large proportion of the plants in temperate regions. Marine life was particularly hard hit. Paradoxically, tropical plants, small terrestrial animals, and freshwater organisms seem to have come through the general disaster relatively untouched.

In 1977, a group of scientists from the University of California at Berkeley, headed by geologist Walter Alvarez, made a most unexpected discovery. In the course of studies of sedimentary rocks in Italy, covering the transition from the Cretaceous period to the Tertiary period, they found that a layer of clay between the two sets of rocks contained unusually high levels of the metal iridium. This element, although relatively rare in the Earth's crust, is abundant in meteorites. Similar iridium deposits were subsequently found in clays at the Cretaceous-Tertiary boundary in Denmark, Spain, and New Zealand, and in deep-sea cores from both the Atlantic and Pacific Oceans.

The Berkeley scientists proposed that the cause of the iridium deposits—and the cause of the mass extinctions at the end of the Cretaceous period—was an asteroid, approximately 10 kilometers in diameter, that collided with Earth. They hypothesized that the impact and explosion of the asteroid kicked up a cloud of debris that circled Earth for a period of at least several months, producing continuous darkness, the cessation of photosynthesis, and a subsequent collapse of food supplies for heterotrophic organisms, both terrestrial and marine. Such darkness would be expected to cause significant changes in climate, putting further stress on terrestrial organisms.

Enormous interest was generated by the discovery of the iridium deposits, by the asteroid hypothesis, and by the apparent periodicity of mass extinctions. As a consequence, scientists embarked on a variety of new studies. These have included a thorough reexamination of the fossil record, studies of the boundary layers associated with the different mass extinctions, explorations of possible terrestrial sources of high levels of iridium (for example, certain types of volcanic eruptions), new analyses of geologic phenomena known to have occurred in the periods marked by high rates of extinction, as well as a search for impact craters of the right size and age. In 1990, a buried impact crater of the right size—now estimated to be 300 kilometers in diameter—was located in the Caribbean, centered on the modern town of Chicxulub in the Yucatan Peninsula of Mexico. Different dating techniques have all placed its age at almost exactly 65 million years. A much smaller crater (32 kilometers in diameter) in Iowa has also been dated at 65 million years, suggesting that more than one impact may have been involved.

Many biologists, including those most familiar with the details of the fossil record on either side of the Cretaceous-Tertiary boundary, initially had serious doubts about the asteroid hypothesis. On the whole, the fossil evidence indicates that different groups did not become extinct simultaneously at the end of the Cretaceous period, but that the extinctions occurred over a period of tens of thousands—even hundreds of thousands—of years. This suggests more gradual processes at work. Nevertheless, there is a growing consensus that although asteroid impacts may not be the whole story, they played a major role in the extinctions that occurred 65 million years ago—and perhaps in other mass extinctions as well.

Whatever the cause or causes of mass extinctions, their effects on the subsequent course of evolutionary history are clear. When entire groups of organisms have died out, apparently without regard to the success of their particular solutions to the problems facing all living systems, new opportunities have opened up for the groups of organisms that survived. As the survivors have diversified in the course of exploiting newly available living space, new sets of solutions to the common problems have appeared, based on the survivors' genetic inheritance. At each mass extinction, the course of evolution has been dramatically altered, with some branches of the evolutionary tree permanently terminated and others undergoing vast new diversifications.

Equus: A Case Study

As a case study in evolution, let us consider horses, whose history is particularly well documented in the fossil record (Figure 21–14). At the beginning of the Tertiary period, shortly after the extinction of the dinosaurs, the horse lineage was represented by *Hyracotherium,* also known as Eohippus ("dawn horse"). It was a small herbivore (25 to 50 centimeters high at the shoulders) with three toes on its hind feet, four toes on its front feet, and doglike footpads on which its weight was carried. Its eyes were halfway between the top of its head and the tip of its nose. Its teeth had small grinding surfaces and low crowns, which probably could not have stood much wear. The teeth indicate that *Hyracotherium* did not eat grass; in fact, there was probably not much grass to eat then. It probably lived in forests and browsed on succulent leaves.

Slightly higher in the fossil-bearing strata, one finds larger horses, all three-toed, still browsing. Some of these clearly had molar teeth with large crowns that continued to grow as they were worn away (as do those of modern horses).

The strata of the Oligocene yield fossils of still larger horses. In these, the middle digit of each foot had expanded and was now the weight-bearing surface. In some, the two non-weight-bearing digits had become greatly reduced. These horses are thought to have been the direct ancestors of the genus *Equus,* which includes all modern horses.

These changes can be correlated with changes in the environment. In the time of *Hyracotherium,* much of the land was marshy and the chief vegetation was succulent leaves; the teeth of *Hyracotherium* were adapted for browsing. By the Miocene, the grasslands began to spread. Horses whose teeth became adapted to grinding grasses (which are very coarse and tough) survived, whereas those that remained browsers did not. The placement of the eye higher in the head may have facilitated watching for predators while grazing. The climate became drier, and the ground became harder; reduction of the number of toes, with the development of the spring-footed gait characteristic of modern horses, was an adaptation to harder ground and to larger size. An animal twice as high as another tends to weigh about eight times as much. Little *Hyracotherium* was probably as fast as the modern horse, but a larger, heavier horse with the foot and leg structure of *Hyracotherium* would have been too slow to escape from predators. (During this same period, predators were developing adaptations that rendered them better able to catch large herbivores, including horses.)

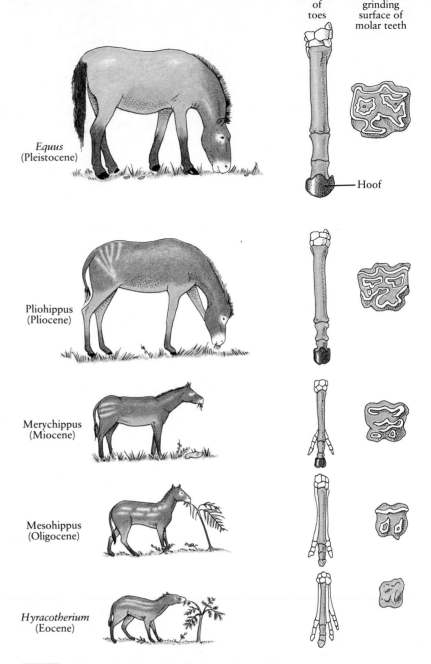

Reduction of toes

Increased grinding surface of molar teeth

Equus (Pleistocene)

Hoof

Pliohippus (Pliocene)

Merychippus (Miocene)

Mesohippus (Oligocene)

Hyracotherium (Eocene)

21–14 *The modern horse and some of its predecessors. Only one of the several branches represented in the fossil record is shown here. Over the past 60 million years, small several-toed browsers, such as* Hyracotherium, *were replaced in gradual stages by members of the genus* Equus, *characterized by, among other features, a larger size; broad molars adapted to grinding coarse grass blades; a single toe surrounded by a tough, protective keratin hoof; and a leg in which the bones of the lower leg had fused, with joints becoming more pulleylike and motion restricted to a single plane.*

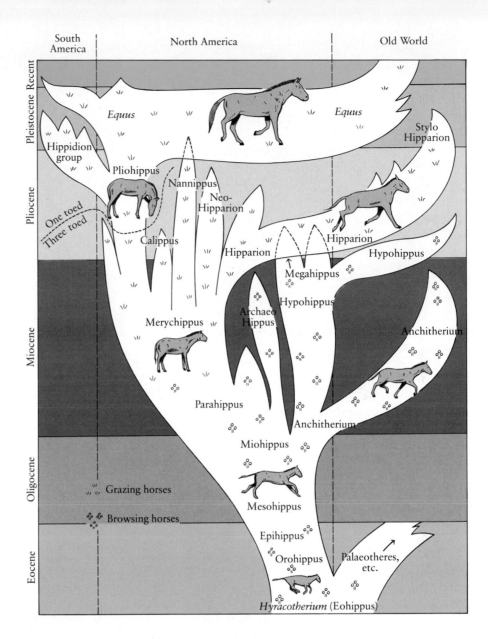

21–15 *The evolutionary history of the horse family, as summarized by George Gaylord Simpson in his classic book* Horses.

Thus the evolution of *Equus,* viewed from the long retrospective of the geologic record, represents an accumulation of adaptive changes related to pressure for increased size and for grazing. It might be considered—as it was for many years—as an example of phyletic change. As the number of fossil specimens has increased, however, it has become clear that at any given time many different species of horses coexisted, only some of which survived (Figure 21–15). Thus both phyletic change and cladogenesis are seen as contributing to the evolution of *Equus.*

Punctuated Equilibria

Although the fossil record documents many important stages in evolutionary history, there are numerous gaps. As Darwin noted in *The Origin of Species,* the geologic record is ". . . a history of the world imperfectly kept, and written in a changing dialect; of this history, we possess the last volume alone, relating only to two or three countries. Of this volume, only here and there a short chapter has been preserved; and of each page, only here and there a few lines."

Many more fossils have, of course, been discovered in the 100 years since Darwin's death. We are no longer limited to the most recent volume, and some of the volumes are now far more complete. Nevertheless, fewer examples of gradual change have been found than might have been expected. For many years, this discrepancy between the model of slow phyletic change and the poor documentation of such change in much of the fossil record was ascribed to the imperfection of the record itself.

In 1972, two young scientists, Niles Eldredge of the American Museum of Natural History and Stephen Jay

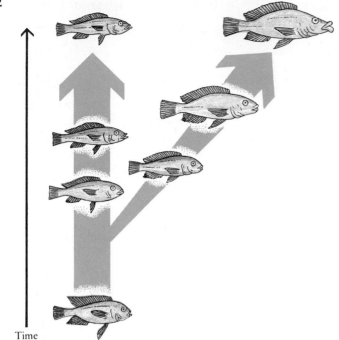

Time

(a) Gradual phyletic change

Time

(b) Punctuated equilibrium

21–16 *Two models of evolutionary change.* (a) *In the phyletic model, changes in anatomy and other characteristics within a lineage of organisms occur gradually and more or less continuously over relatively long periods of time.* (b) *In the punctuated equilibrium model, changes occur rapidly within short periods of time that are followed by much longer periods during which there are few detectable changes.*

The organisms shown here are African cichlid fishes. In Lake Victoria, which is less than 1 million years old, there are some 200 species of cichlid fishes, of many different sizes and shapes, adapted to the many different habitats and food sources available within the lake. Current evidence suggests that this diversity of species evolved from a common ancestor within the last 200,000 years.

Gould of Harvard University, ventured the proposal that perhaps the fossil record is not so imperfect after all. Both Eldredge and Gould have backgrounds in geology and invertebrate paleontology, and both were impressed with the fact that there was very little evidence of slow phyletic change in the fossil species they studied. Typically, a species would appear abruptly in fossil-bearing strata, last 5 million to 10 million years, and disappear, apparently not much different than when it first appeared. Another species, related but distinctly different, would take its place, persist with little change, and disappear equally abruptly. Suppose, Eldredge and Gould argued, that these long periods of no change, punctuated by gaps in the fossil record, are not flaws in the record but *are* the record, the evidence of what really happens.

How could new species make such sudden appearances? Eldredge and Gould found their answer in the model of allopatric speciation. If new species formed principally in small populations isolated from the parent population, if speciation occurred rapidly (by rapidly, paleobiologists mean in thousands rather than millions of years), and if the new species then outcompeted the old species, taking over their geographic range, the resulting fossil pattern would be the one observed.

As first proposed, punctuated equilibria seemed to refer principally to the tempo of evolution. Population geneticists and Darwin before them had emphasized gradual change. There was clearly room, however, for the idea that populations would change more rapidly at some times than at others, particularly in periods of environmental stress, as in the case of the peppered moth (page 13).

As this model has become more fully developed, particularly by Steven M. Stanley of Johns Hopkins University, it has become more radical and more controversial. Its proponents now argue not only that cladogenesis is the principal mode of evolutionary change (as Mayr stated some 40 years ago) but that selection occurs among species as well as among individuals. Thus, at any one time in evolutionary history it may be possible to find a number of related species coexisting, each departing in a different way from the ancestral type. For example, as the horses evolved, some species became larger, others remained small. Some became grazers, others remained browsers. The overall trends in horse evolution resulted from the differential survival of species, rather than from the phyletic change that occurred within species. Thus, in this new formulation, species take the place of individuals, and speciation and extinction substitute for birth and death. The role of speciation is analogous to the role of mutation in population genetics; it provides the variation on which selection acts. According to this proposal, there are two levels of selection: in one, selection acts on the individual, and in the other, it acts on the species.

The punctuated equilibrium model has stimulated a vigorous and continuing debate among biologists, a reexamination of evolutionary mechanisms as currently understood, and a reappraisal of the evidence. These activities have at times been misinterpreted as a sign that Darwin's theory is "in trouble." In fact, they indicate that evolutionary biology is alive and well and that scientists are doing what they are supposed to be doing—asking questions. Darwin, we think, would have been delighted.

Summary

A species is defined as a group of natural populations whose members can interbreed with one another but cannot (or at least usually do not) interbreed with members of other such groups. In order for speciation—the formation of new species—to occur, populations that formerly shared a common gene pool must be reproductively isolated from one another and subsequently subjected to different selection pressures.

Two principal modes of speciation are recognized, allopatric ("other country") and sympatric ("same country"). Allopatric speciation occurs in geographically isolated populations. Sympatric speciation, which does not require geographic isolation, occurs principally in plants through polyploidy, often coupled with hybridization. It may also take place in some cases by disruptive selection.

The key event in speciation is genetic isolation. Once species have become genetically isolated, they can once again inhabit the same geographic area without interbreeding because of numerous behavioral, anatomical, and physiological mechanisms that prevent it.

The fossil record discloses four patterns of evolutionary change: phyletic change, cladogenesis, adaptive radiation, and extinction. Phyletic change is gradual change within a single lineage over time. Cladogenesis, by contrast, is evolutionary change produced by the branching off of populations from one another to form new species. Adaptive radiation is the rapid formation of many new species from a single ancestral group, characteristically to fill a new ecological zone. Extinction is the disappearance of a species from the Earth. The fossil record reveals a low, steady rate of extinction, interrupted periodically by mass extinctions involving enormous numbers of species. In modern evolutionary theory, large-scale evolution is regarded as the product of a combination of these patterns.

Paleontologists have presented evidence for an additional pattern of evolution known as punctuated equilibrium. They propose that new species are formed during bursts of rapid speciation among small isolated populations, that the new species outcompete many of the then-existing species (which become extinct), persist for long periods with little change, and then, in turn, abruptly become extinct. The punctuationalists propose that major changes in evolution take place as a result of selection acting on species as well as on individuals.

Questions

1. Distinguish among the following: allopatric speciation/sympatric speciation; hybridization/polyploidy; premating isolating mechanisms/postmating isolating mechanisms; phyletic change/cladogenesis/adaptive radiation.

2. Define genetic isolation. Why is it such an important factor in speciation?

3. Give three possible reasons why sympatric speciation by hybridization and polyploidy is more common in plants than in animals.

4. Two species of plants, each capable of reproducing both sexually and asexually, became distinct from their parental lineages as a result of polyploidy. Individual plants of these

two species then hybridized with each other. Would the resulting plants be capable of sexual reproduction? Would they constitute a new species? Explain your answers.

5. The genetic isolation of long-established, closely related species is generally maintained almost entirely by premating isolating mechanisms. Explain, in terms of natural selection, why this is to be expected.

6. Describe the separate steps involved in the formation of the distinct species of Galapagos finches. How does this example of speciation illustrate both the founder principle and adaptive radiation?

7. Darwin emphasized that when mammals (cats, goats, sheep, rodents, and so on) were introduced onto remote islands they often thrived there, yet there were no indigenous mammals on the islands except bats. Can you explain why he considered this important?

8. Could the fossil record disprove that evolution has occurred? If so, how?

9. Recent evidence supporting the periodic occurrence of mass extinctions suggests that Cuvier and Agassiz (page 6) were correct in their assertion that a series of catastrophes led to the demise of many previously existing forms of life. Neither of these eminent scientists, however, could provide a testable explanation for the proliferation of new forms that followed each of the proposed catastrophes. How would modern evolutionary theory explain the proliferation of new forms after mass extinctions?

Suggestions for Further Reading

Alvarez, Luis W., Walter Alvarez, Frank Asaro, and Helen V. Michel: "Extraterrestrial Cause for the Cretaceous-Tertiary Extinction," *Science*, vol. 208, pages 1095–1108, 1980.

Archibald, J. David, and William A. Clemens: "Late Cretaceous Extinctions," *American Scientist*, vol. 70, pages 377–385, 1982.

Carlquist, Sherwin: *Island Life: A Natural History of the Islands of the World*, Natural History Press, Garden City, N.Y., 1965.

> *An exploration of the nature of island life and the intricate and unexpected evolutionary patterns found in island plants and animals.*

Carson, Hampton L., et al.: "Hawaii: Showcase of Evolution," *Natural History*, December 1982, pages 16–72.

> *A series of 10 articles on the evolution of the plants and animals of the Hawaiian Islands, which are, like the Galapagos Islands, a living laboratory of evolution.*

Darwin, Charles: *The Origin of Species by Means of Natural Selection, or The Preservation of Favored Races in the Struggle for Life*, New American Library, Inc., New York, 1986.*

> *Darwin's "long argument." Every student of biology should, at the very least, browse through this book to catch its special flavor and to begin to understand its extraordinary force.*

Darwin, Charles: *The Voyage of the Beagle*, New American Library, Inc., New York, 1988.*

> *Darwin's own chronicle of the expedition on which he made the discoveries and observations that eventually led him to his theory of evolution. The sensitive, eager young Darwin that emerges from these pages is very unlike the image many of us have formed of him from his later portraits.*

Eldredge, Niles: *The Miner's Canary: Unravelling the Mysteries of Extinction*, Simon and Schuster, New York, 1991.

> *An account, for the general reader, of mass extinctions and their possible causes. Eldredge maintains that environmental degradation has played the major role in triggering past mass extinctions. An important aspect of this book is his analysis of what the currently accelerating rate of extinction reveals about the state of the global environment.*

FitzGerald, Gerard J.: "The Reproductive Behavior of the Stickleback," *Scientific American*, April 1993, pages 80–85.

Gould, Stephen Jay: "Evolution and the Triumph of Homology, or Why History Matters," *American Scientist*, vol. 74, pages 60–69, 1986.

Gould, Stephen Jay: "Opus 200," *Natural History*, August 1991, pages 12–18.

Grant, Peter R.: *Ecology and Evolution of Darwin's Finches*, Princeton University Press, Princeton, N.J., 1986.*

> *A major work on the Galapagos finches, based on field studies carried out by Grant and his students over a period of some 15 years. In the course of these studies, an enormous quantity of data was gathered, which is analyzed from the perspective of modern ecological and evolutionary theory. Clearly written and beautifully illustrated, this book is already considered a classic.*

Grant, Peter R.: "Speciation and the Adaptive Radiation of Darwin's Finches," *American Scientist*, vol. 69, pages 653–663, 1981.

Kerr, Richard A.: "Huge Impact Tied to Mass Extinction," *Science*, vol. 257, pages 878–880, 1992.

---

* Available in paperback.

Lack, David: *Darwin's Finches,* Cambridge University Press, New York, 1983.*

This short, readable book, first published in 1947, gives a marvelous account of the Galapagos, their finches and other inhabitants, and of the general process of evolution. Although in many respects superseded by Grant's 1986 book, it remains an excellent introduction to the Galapagos finches.

Lewin, Roger: "A Lopsided Look at Evolution," *Science,* vol. 241, pages 291–293, 1988.

Marshall, Larry G.: "Land Mammals and the Great American Interchange," *American Scientist,* vol. 76, pages 380–388, 1988.

Mayr, Ernst: *The Growth of Biological Thought: Diversity, Evolution, and Inheritance,* Harvard University Press, Cambridge, Mass., 1982.*

This is the first of two volumes on the history of biology and its major ideas, written by one of the leading figures in the study of evolution. The introductory chapters provide an outstanding analysis of the philosophy and methodology of the biological sciences. This book, like Darwin's The Origin of Species, should at least be sampled by every serious student of biology.

Mayr, Ernst: *Populations, Species, and Evolution: An Abridgement of Animal Species and Evolution,* Harvard University Press, Cambridge, Mass., 1970.*

A masterly, authoritative, and illuminating statement of contemporary thinking about species—how they arise and their role as units of evolution.

Monastersky, Richard: "Closing in on the Killer," *Science News,* January 25, 1992, pages 56–58.

Monastersky, Richard: "Counting the Dead," *Science News,* February 1, 1992, pages 72–75.

Nance, R. Damian, Thomas R. Worsley, and Judith B. Moody: "The Supercontinent Cycle," *Scientific American,* July 1988, pages 72–79.

Prokopy, Ron, and Guy Bush: "Evolution in an Orchard," *Natural History,* September 1993, pages 4–10.

Raup, David M.: *Extinction: Bad Genes or Bad Luck?* W. W. Norton & Company, New York, 1992.*

Raup, David M.: *The Nemesis Affair: A Story of the Death of Dinosaurs and the Ways of Science,* W. W. Norton & Company, New York, 1986.*

Popular accounts of mass extinctions, the various hypotheses that have been proposed to explain them, and the debates within the scientific community over those hypotheses. Professor Raup, who has made major contributions to our knowledge of previous mass extinctions, is a lively and entertaining writer.

Russell, Dale A.: "The Mass Extinctions of the Late Mesozoic," *Scientific American,* January 1982, pages 58–65.

Sagan, Carl, and Paul R. Ehrlich: *The Cold and the Dark: The World After Nuclear War,* W. W. Norton & Company, New York, 1985.*

A discussion, for the general reader, of the hypothesis that even a limited nuclear exchange would throw such quantities of smoke and debris into the atmosphere that a "nuclear winter"—analogous to the "winter" of the asteroid hypothesis—would result, with devastating biological consequences.

Simpson, George Gaylord: *Horses,* Oxford University Press, New York, 1951.

The story of the horse family in the modern world and through 60 million years of history. Simpson was not only an outstanding vertebrate paleontologist but also a graceful and delightful writer.

Simpson, George Gaylord: *Splendid Isolation: The Curious History of South American Mammals,* Yale University Press, New Haven, Conn., 1983.*

Charles Darwin found in the animals of South America some of the major clues that led to the formulation of The Origin of Species. Today, their history can be seen as an almost ideal natural experiment in evolution, told in Simpson's engaging style.

Stanley, Steven M.: *Extinction,* W. H. Freeman and Company, New York, 1987.

A beautifully illustrated volume surveying the paleontological and geological evidence of the great extinctions of the past.

Storch, Gerhard: "The Mammals of Island Europe," *Scientific American,* February 1992, pages 64–69.

Turco, Richard P., et al.: "The Climatic Effects of Nuclear War," *Scientific American,* August 1984, pages 33–43.

Vickers-Rich, Patricia, and Thomas Hewitt Rich: "Australia's Polar Dinosaurs," *Scientific American,* July 1993, pages 50–55.

Ward, Peter Douglas: *On Methuselah's Trail: Living Fossils and the Great Extinctions,* W. H. Freeman and Company, New York, 1993.*

A fascinating account of the ancient organisms, such as Nautilus, the horseshoe crabs, and the coelacanth, that survived the great extinctions of the past. As the author makes clear, these organisms have much to teach us.

White, Mary E.: *The Flowering of Gondwana,* Princeton University Press, Princeton, N.J., 1990.

A marvelous study, enriched by some 400 photographs, of the plants of the supercontinent Gondwana and their descendant plants—both fossil and living—now found on the continent of Australia.

* Available in paperback.

In many species of birds, the first behavior to appear in nestlings is a thrusting of the head upward toward the parent, with the mouth gaping open. A complementary behavior of the parent is the regurgitation of food into the open mouth.

The big bird in this nest is a baby European cuckoo. The little bird is an adult dunnock. Programmed by her evolutionary history, the dunnock finds the gaping beak of the nestling irresistible and spends all of her daylight hours feeding the greedy chick. The chick, however, is an imposter. His mother, following her own program, kept the dunnock's nest under surveillance until it had two or three eggs in it. She then waited until the nest was momentarily unguarded, swooped in, removed one of the eggs already there, and laid her own in its place— all in 10 seconds. As she was leaving, she thriftily swallowed the stolen egg.

Although cuckoos are much bigger than dunnocks, their eggs are not much larger than dunnock eggs and their incubation time is shorter. Consequently, the baby cuckoo hatched first. Once out of his shell, still blind and naked, he hoisted the dunnock eggs one by one into a special hollow on his back, shuffled to the edge of the nest, and heaved them over the rim. The cuckoo thus ensured himself a monopoly on all of the provisions that should have gone to the rightful progeny of his foster mother, the dunnock. In the meantime, his biological mother laid eight more eggs, each containing a murderous offspring, in eight other nests.

Brood parasitism, as this phenomenon is called, is found in some 80 species of birds. The most familiar to you may be the brown-headed cowbird, whose deceitful strategies are not quite as well developed as the cuckoo's, perhaps because they have a shorter evolutionary history. Note that although the behavior of the cuckoos seems wily and deliberate—and even subject to moral judgment—the cuckoos have no more choice about what they are doing than the seemingly industrious and virtuous dunnock has about feeding the chick. Indeed, they have no more choice about this behavior than they have about the colors of their feathers or the patterns of their flight.

The Evolution of Behavior

Virtually all organisms, from the simplest bacterial cells to the most complex animals, *act*. They seek out suitable environments in which homeostasis can be maintained. They obtain nutrients and other essential molecules, such as water and oxygen. Often they successfully avoid becoming nutrients for other organisms. And, they produce offspring, a process that can involve complex patterns of courtship, mating, and parenting. These activities, along with many others, constitute **behavior**.

The behavioral characteristics of an organism—its sensitivity to particular stimuli and its patterns of response to those stimuli—are as much the products of natural selection as the size and shape of a bird's beak or the process that regulates our blood cholesterol levels (see page 58). An organism's behavior is vitally important for its survival and the successful production of offspring.

The factors governing the evolution of behavioral characteristics are the same as those that apply to any other trait. First, there are variations among individual organisms in behavioral characteristics. Second, some of these variations are genetically determined; that is, they are influenced by the presence of particular alleles or combinations of alleles. Third, individuals with certain variations tend to have greater reproductive success than individuals with certain other variations. Fourth, as a result of this greater reproductive success, certain alleles tend to increase in the gene pool of a population, generation after generation.

The study of behavior and its evolution is currently one of the most active and exciting areas of biological research. The scope of the field is enormous, ranging from laboratory studies of bacteria to long-term studies of animals in the wild. We shall, of necessity, limit our discussion to a few of the most fascinating areas of this vast subject.

The Genetic Basis of Behavior

Behavior—whether that of a *Salmonella* cell navigating toward a food source (see page 113) or yours in reading and reacting to this text—has its roots in the genetic program carried in the DNA molecules of the individual. The steps between a sequence of nucleotides in a DNA molecule and the behavior of an organism, even a very simple one, are many and complex. The process involves, at the least, the synthesis of specific molecules, their organization into particular structures capable of receiving and responding to stimuli, the development of pathways for the transmission of information within the organism, and the modification of those structures and pathways as a result of interactions of the organism with its environment.

For some behaviors, the role of genetic components can be clearly demonstrated. A classic example is the behavior of honey bees when a pupa (Figure 22–1), undergoing metamorphosis within its cell in the hive, becomes diseased and dies. In some honey-bee strains, known as hygienic, worker bees uncap the cell and remove the corpse. In other strains, known as unhygienic, this behavior does not occur.

22–1   *Two honey-bee larvae (left) and two honey-bee pupae (right) in different stages of their development. Honey bees undergo their entire development from egg to adult in individual wax cells within the hive. During the early stages of development, the cells are open, enabling worker bees to feed the larvae continuously. When a larva has reached the size at which pupation begins, the workers add a wax cap to the cell, sealing in the pupa for its metamorphosis into an adult bee. When metamorphosis is complete, the adult bee emerges from the cell by gnawing through the wax cap.*

In a series of breeding experiments performed in the early 1960s, Walter Rothenbuhler first crossed unhygienic bees with hygienic bees. All of the offspring were unhygienic, indicating that this is a dominant characteristic. Rothenbuhler then made a testcross of these hybrid offspring with the original hygienic strain. Twenty-nine colonies of bees resulted. In eight colonies, the bees were unhygienic; they left the cells capped and did not remove dead pupae. Six colonies consisted of hygienic bees that uncapped the cells and removed dead pupae. In nine of the colonies, however, worker bees uncapped cells but left the dead pupae untouched, and in the remaining six colonies, they did not uncap cells but would remove the dead pupae if the cells were uncapped for them.

These results approximate a 1:1:1:1 ratio, which is the expected ratio in a testcross involving two independently assorted genes. They indicate that one gene, designated *U*, controls the uncapping behavior, whereas another gene, designated *R*, controls the removal behavior (Figure 22–2).

As this example illustrates, a single allele may determine a particular behavioral characteristic of an organism. For most behaviors, however, the underlying genetics are considerably more complicated. For some behaviors, genes with multiple effects (that is, pleiotropic genes) play an important role. Most often, however, behavioral characteristics are the result of polygenic inheritance—that is, they depend on the integrated action of the alleles of a large number of genes.

Proximate and Ultimate Causation

Imagine a pond on a spring day, with numerous tadpoles schooling near the water's edge. If you walk quietly along the bank of the pond and then make a sudden, loud noise, you will notice that the tadpoles immediately arch their bodies and shoot away into deeper water. Similarly, if you tap on the side of an aquarium in a pet shop, the fish also arch their backs, darting to the other side of the aquarium. Why do the tadpoles and the fish exhibit this behavior? This question, and similar questions about a vast diversity of behaviors exhibited by organisms, can be approached from two different points of view. One approach seeks an answer in terms of the **proximate cause**—that is, the immediate sequence of physiological events that lead to the observed behavior. Another approach seeks to explain the **ultimate cause** of the behavior—that is, its adaptive value and its evolutionary origins.

In the example of the startle behavior exhibited by many fish and larval amphibians, such as tadpoles, the

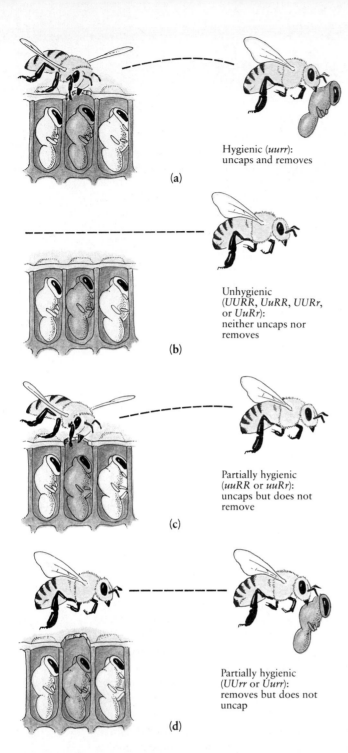

Hygienic (*uurr*):
uncaps and removes

(a)

Unhygienic
(*UURR, UuRR, UURr,*
or *UuRr*):
neither uncaps nor
removes

(b)

Partially hygienic
(*uuRR* or *uuRr*):
uncaps but does not
remove

(c)

Partially hygienic
(*UUrr* or *Uurr*):
removes but does not
uncap

(d)

22–2 *Hygienic behavior in honey bees is
determined by two independently assorted
genes,* U *and* R. **(a)** *Bees that are homozygous recessive for both genes* (uurr) *uncap
the cells containing dead pupae and remove
the corpses from the hive.* **(b)** *Bees that have
at least one copy of the dominant allele for
each gene (for example,* UuRr) *do neither.*
(c) *Bees that are* uuRR *or* uuRr *uncap the
cells but do not remove the corpses, while*
(d) *bees that are* UUrr *or* Uurr *remove dead
pupae from cells that have already been
uncapped.*

proximate cause is stimulation of two giant neurons
(nerve cells), known as Mauthner cells. Stimulation of
the Mauthner cells by signals transmitted from the
acoustic nerve leads to stimulation of other neurons on
each side of the body. This, in turn, produces powerful
contractions of the muscles of the body wall and tail
that propel the animal forward.

The ultimate cause of the behavior, however, is to be
found in its adaptive value in preserving the animal
from predation. Individuals with a highly efficient response are more likely to avoid predation—and thus
survive to reproduce—than those with a less efficient
response.

Fixed Action Patterns

The comparative study of patterns of behavior and the
construction of hypotheses concerning their evolutionary origins is known as **ethology,** from the Greek word
ethos, meaning "character" or "custom." The pioneers
in this field have been the European zoologists Konrad
Lorenz, Niko Tinbergen, and Karl von Frisch.

Some behavior patterns—for example, the gaping of
a baby bird's mouth (page 377) or the tongue flick of a
toad attempting to capture a fly—develop with a minimum of sensory experience. Although the behavior
may be subsequently refined, the pattern appears, essentially complete, the first time the organism encounters the relevant stimulus. Such an innate behavioral
pattern, which tends to be highly stereotyped, rigid, and
repetitive, is known as a **fixed action pattern.** The fixed
action patterns of the members of a given species of the
appropriate age, sex, and physiological condition are as
specific and constant as their anatomical characteristics.

Fixed action patterns are initiated by external stimuli, known as **sign stimuli.** When these stimuli are communication signals exchanged between members of a
species, they are known as **releasers.** Tinbergen hypothesized that certain circuits within the brain, which he
termed **innate releasing mechanisms,** respond to the
sign stimuli. A fixed action pattern remains blocked until the organism encounters the appropriate sign stimulus, which then stimulates the innate releasing mechanism, setting in action the sequence of movements that
constitute the behavior.

An example of a fixed action pattern studied by Lorenz and Tinbergen is the egg-rolling behavior of greylag geese. The nests of these geese are shallow depressions in the ground. If an egg rolls out of her nest, a
goose will retrieve it using a stereotyped sequence of
movements (Figure 22–3). First, she stretches her neck
toward the egg and then begins rolling it back under

22–3 *The fixed action pattern by which a greylag goose retrieves an egg that has rolled from her nest. Because an egg is not a perfect sphere, a series of lateral movements of the goose's head are required to keep the egg on a straight course back to the nest.*

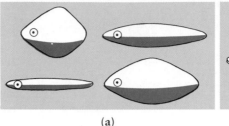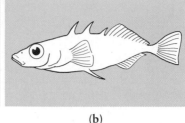

(a)   (b)

22–4 *Colors are the sign stimuli for many fixed action patterns in different animal species. In experiments by Niko Tinbergen with sticklebacks, (a) crude models, painted red on the lower surface,* *elicited much stronger reactions in both male and female sticklebacks (aggressiveness in the males and attraction in the females) than did (b) an exact replica of a male that was not colored.*

her chin. Through a series of lateral movements of her head, she keeps the egg rolling back on a straight line toward the nest. Occasionally, however, the egg slips away to the side—but the goose continues the retrieval movements all the way back to the nest, even though the egg is no longer there to retrieve. In this case, as in many others, once the fixed action pattern has been initiated, it cannot be altered but must be carried through to completion.

Through the use of physical models, ethologists have been able to identify the releasers for a number of fixed action patterns. For example, the red bellies of male sticklebacks in breeding condition are releasers not only for the stereotyped mating behavior of stickleback females (see page 362) but also for aggressive interactions between the males. Tinbergen constructed a series of models (Figure 22–4), by which he was able to show that a male stickleback reacts much more aggressively to a very crude model of a male with a red belly than to an exact replica of a stickleback without the red coloration.

Learning

All patterns of behavior, even those that seem relatively complete on their first appearance, depend not only on environmental cues but also on the normal physiological development of the animal. An enormous amount of animal behavior also requires learning—a process in which the responses of the organism are modified as a result of experience. The capacity for learning appears to be loosely correlated with the length of the life span and with the size and complexity of the brain. In small organisms, such as honey bees, with a short life span (and thus little available time for learning), most behavior appears to take the form of fixed action patterns. By contrast, in large organisms with a complex brain and a long life span, such as *Homo sapiens*, a large proportion of the behavioral repertoire is critically dependent on the prior experience of the individual.

One of the simplest forms of learning is **habituation,** in which an organism comes to ignore a persistent stimulus and go on about its other business, wasting neither time nor energy on a response. An example is provided by ground squirrels (Figure 22–5), which come to ignore the alarm calls of other ground squirrels *if* they are given frequently when there is, in fact, no danger.

Associative Learning

A type of learning familiar to us all is **association,** in which one stimulus comes to be linked, through experience, with another one. If you keep pets in your home, you will be able to cite many examples of associative learning, such as goldfish coming to the corner of the aquarium to be fed as you walk toward the tank or your dog becoming excited at the sight of a leash.

The first scientific studies of associative learning were performed in the 1920s by the Russian physiologist

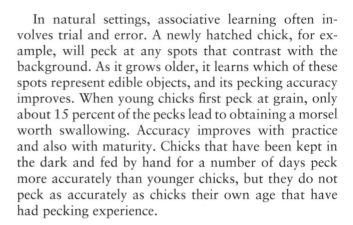

In natural settings, associative learning often involves trial and error. A newly hatched chick, for example, will peck at any spots that contrast with the background. As it grows older, it learns which of these spots represent edible objects, and its pecking accuracy improves. When young chicks first peck at grain, only about 15 percent of the pecks lead to obtaining a morsel worth swallowing. Accuracy improves with practice and also with maturity. Chicks that have been kept in the dark and fed by hand for a number of days peck more accurately than younger chicks, but they do not peck as accurately as chicks their own age that have had pecking experience.

Imprinting

Closely related to associative learning is the development of discrimination. Of vital importance for the ultimate reproductive success of many animals is the discrimination of members of one's own species from members of all other species, a discrimination that may be based on a variety of cues. In many species, particularly birds, this learning occurs very rapidly during a specific **critical period** in the early life of the individual and depends on exposure to particular characteristics of the parent or parents. This type of learning is known as **imprinting.**

The most familiar example of imprinting is the following response of birds that are physiologically mature enough to leave the nest soon after hatching (Figure 22–6). This keeps the young birds close behind and well within the protective range of the parent until the end of their juvenile period, when the following response is lost. Imprinting is also involved in song learning in birds, a process of considerable complexity.

22–5 *A female Belding's ground squirrel who has detected danger—most likely a potential predator—and is sounding the alarm. Other members of the species who hear the alarm usually retreat to the safety of their burrows. If, however, a particular squirrel repeatedly raises false alarms, the others stop responding to the calls of that squirrel.*

Ivan Pavlov. In his classic experiments, Pavlov restrained a hungry dog in a harness and offered it small portions of food at regular intervals. When he signaled the delivery of food by preceding it with an external stimulus, such as the sound of a bell or a signal light, the dog began to respond to the external stimulus by salivating. After every few trials, the external stimulus was presented without the food, and the amount of saliva produced by the dog was measured. Pavlov found that the number of drops of saliva triggered by the external stimulus alone was directly proportional to the number of previous trials in which it had been followed by food.

22–6 *Many species of precocial birds (birds that are able to walk and feed as soon as they hatch) follow the first moving object they see after they have freed themselves from the egg. The object is usually mother goose (for goslings, that is), but it can be a matchbox on a string or a member of another species, such as zoologist Konrad Lorenz, shown here. In many species, this phenomenon, an example of the type of learning known as imprinting, also influences mate selection in the adult bird. Lorenz reported many examples of ducks and geese becoming sexually fixated on objects or on members of other species, including himself.*

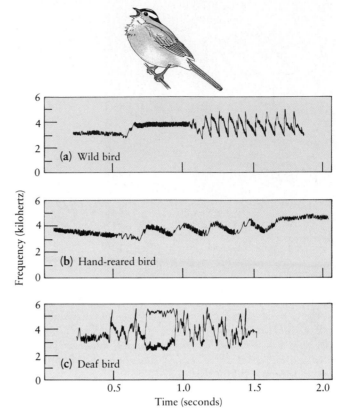

The Song of the White-Crowned Sparrow

About 150 days after hatching, a juvenile male white-crowned sparrow begins to utter a tentative, twittering call that has only a vague resemblance to the full song of the mature male. During the next 50 days or so, the juvenile's song gradually becomes more complex and sophisticated until, by the time the bird is 200 days old, he is singing a full song.

Laboratory experiments undertaken by ethologist Peter Marler and others have dissected the sequence of events in song learning in this species. Figure 22–7 shows sound spectrograms of (a) a wild bird, and (b) an adult, hand-reared bird that had never heard a song of its own species. Clearly, some form of learning is required for production of the full song. Exposing birds to tape-recorded songs at various stages of their development revealed that exposure to the full song of the white-crowned male in a critical period of 10 to 50 days after hatching is required for development of that song five to six months later. Even if a bird is kept in complete sound isolation after two months of age, he will produce a normal song if he was exposed to it during the critical period. If, however, he does not hear the song until about the time he begins to sing himself, he can never produce a full song. Exposure to the song of other sparrow species during the critical period has no effect.

If a bird is deafened after he has heard songs of his own species during the critical period—but before he himself has begun to sing—the full song is not normal (Figure 22–7c). It is, however, more complex than that of the bird that did not hear the song during the critical period. Apparently, then, for normal song, three requirements must be met: (1) the bird must have the genetic capacity to recognize and reproduce the song, (2) he must hear the song during the critical period for imprinting, and (3) he must be able to hear himself sing the song. As he sings, he apparently compares his own song to the song stored in his memory during the critical period, rehearsing until he gets it right. If a bird is deafened after he has mastered the full song, he continues to sing normally.

Imitative Learning

As the song of the white-crowned sparrow indicates, **imitation** is often a component of learning. A well-documented example of "monkey see, monkey do" is provided by the macaques (rhesus monkeys) of the island of Koshima, Japan. These animals lived and fed in the inland forest until about 40 years ago, when a group of Japanese researchers began throwing sweet potatoes on

22–7 *Sound spectrograms for adult male white-crowned sparrows:* (a) *wild;* (b) *reared by hand and isolated from songs of other white-crowned sparrows;* (c) *deafened prior to five months of age (before he himself began to sing), but after having heard normal white-crowned sparrow songs during the critical period. The unit in which the frequency (pitch) of the sound is expressed is the kilohertz; 1 kilohertz equals 1,000 cycles per second.*

the beach for them. The macaques quickly got used to venturing onto the beach, brushing sand off the potatoes, and eating them.

One year after the feeding started, a two-year-old female the scientists had named Imo was observed carrying a sweet potato to the water, dipping it in with one hand, and brushing off the sand with the other. Soon, other macaques began to wash their potatoes, too. Only macaques that had close associations with a potato washer took up the practice themselves. Thus it spread among close companions, siblings, and their mothers, but adult males, which were rarely part of these intimate groups, did not acquire the habit. However, when the young females that learned potato washing matured and had offspring of their own, those offspring learned potato washing from their mothers. Today all the macaques of Koshima dip their potatoes in the salt water to rinse them off, and many of them, having acquired a taste for salt, dip them between bites.

22–8   *A macaque of the island of Koshima, washing a sweet potato before eating it.*

This was only the beginning. Later, the scientists began scattering wheat kernels on the beach. Like the other macaques, Imo, then four years old, had been picking the grains one by one out of the sand. One day she began carrying handfuls of sand and wheat to the shore and throwing them into the water. The sand sank, the wheat kernels floated to the top, and Imo collected the wheat and ate it. The researchers were particularly intrigued by this new behavior since it involved throwing away food once collected, much less a part of the macaques' normal behavioral repertoire than holding on to food and cleaning it off. Washing wheat spread through the group in much the same way that washing potatoes had. Now the macaques, which had never even been seen on the beaches before the feeding program began, have taken up swimming. The youngsters splash in the water on hot days. Some of them dive and bring up seaweed, and at least one has left Koshima and swum to a neighboring island, perhaps as a cultural emissary.

Social Behavior: An Introduction

Of all behaviors, perhaps the most intriguing are the interactions that occur among animals, such as the macaques, living in structured societies. A **society** is a group of individuals of the same species, living together in an organized fashion, with divisions of resources, divisions of labor, and mutual dependence. Stimuli—that is, communications—exchanged among members of the group hold it together and maintain the social structure.

When biologists concerned with evolutionary theory began to analyze social behavior, some disturbing questions began to emerge. For instance, how can you explain, with a mechanism driven by differential reproductive success, the evolution of sterile castes—such as worker bees—in insects? How is it that in many vertebrate societies only a few of the males breed, with their access to the females seldom successfully challenged by other males? Why, among animals in groups, do some utter warning cries or exhibit other forms of behavior that attract attention to the individual issuing the warning, thus threatening the warner's life? These behaviors all appear to be examples of **altruism,** which is, by definition, behavior that benefits others and is performed at some risk or cost to the doer. How can such acts of apparent altruism be explained in terms of natural selection acting on the individual organism?

Not all social behavior, of course, is altruistic. Table 22–1 shows a classification of social behaviors proposed by biologist W. D. Hamilton. As you can see, each of these behaviors has a different potential effect on the fitness—that is, the reproductive success—of both the individual performing the behavior (the donor) and the recipient of the behavior. Selfish, cooperative, and altruistic behaviors have all been well documented in numerous species of animals living in a natural setting. Spiteful behavior has thus far remained unobserved except in *Homo sapiens*.

Table 22–1	A Classification of Social Behaviors	
Type of Behavior	**Effect of the Behavior on**	
	The Donor	**The Recipient**
Selfish	Increases fitness	Decreases fitness
Cooperative	Increases fitness	Increases fitness
Altruistic	Decreases fitness	Increases fitness
Spiteful	Decreases fitness	Decreases fitness

Adapted from W. D. Hamilton, "The Evolution of Social Behavior," *Journal of Theoretical Biology,* vol. 7, pages 1–52, 1964.

The evolution of selfish behavior through the action of natural selection poses no problems for evolutionary theory and, indeed, is what would be expected. Although the maintenance of cooperation by natural selection is relatively easy to understand, the mechanism by which it could have evolved initially is almost as perplexing as the mechanism underlying the evolution of altruism.

In our consideration of social behavior, we shall look first at several different types of animal societies, paying particular attention to their organization and the behavior of their individual members. Then we shall return to the question of how the observed behaviors could have evolved.

Insect Societies

Insect societies are among the most ancient of all societies and, along with modern human societies, are among the most complex. Social insects include termites, ants, wasps, and bees.

Stages of Socialization

As with other animals, the social insects evolved from forms that were originally solitary. Among bees, for example, true sociality appears to have evolved on at least eight separate occasions, and among wasps four times.

Most living species of bees and wasps are solitary, and others show varying degrees of sociality. Thus it is possible to construct a scenario of the various stages of social evolution by the analysis of present-day species. Among the solitary species, the female builds a small nest, lays her eggs in it, stocks it with a food supply, and leaves it forever (Figure 22–9). She usually dies soon after, so there is no overlap between generations.

Among subsocial or presocial species, the mother returns to feed the larvae for some period of time, and the emerging young may subsequently lay their eggs in the same nest or comb. However, the colony is not permanent (usually being destroyed over the winter), there is no division of labor, and all females are fertile.

Eusocial, or "truly social," insects are characterized by cooperation in caring for the young and a division of labor, with sterile individuals working on behalf of reproductive ones. All ants and termites and some species of wasps and bees—for example, honey bees—are eusocial.

Honey Bees

A honey-bee society usually has a population of 30,000 to 40,000 workers and one adult queen. Each worker, always a diploid female, begins life as a fertilized egg deposited by the queen in a separate wax cell. (Drones, or male bees, develop from unfertilized eggs and are therefore haploid.) The fertilized egg hatches to produce a white, grublike larva (see Figure 22–1, page 378) that is fed almost continuously by the nurse workers; each larval bee eats about 1,300 "meals" a day. After the larva has grown until it fills the cell, a matter of about six days, the nurses cover the cell with a wax lid, sealing it in. Its metamorphosis requires about 12 days, after which the adult bee emerges.

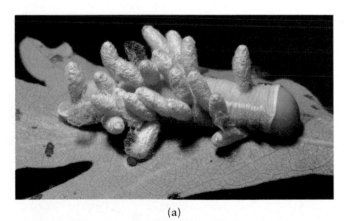

(a)

(b)

22–9 *Wasps, most of which are not truly social insects, do not tend their young, but often provide for them.* (a) *Wasps of the genus* Apanteles, *all of which are solitary and parasitic, inject their eggs under the skins of caterpillars. The resultant larvae eat the internal tissues of the caterpillar, chew their way to the surface, and spin the cocoons shown here.* (b) *Female wasps of the genus* Ammophila *hunt caterpillars, which they then deposit in previously prepared burrows in the soil. The egg is laid on the caterpillar, which, paralyzed but not dead, provides up to 40 days' food for the larva.*

(a)

(b)

(c)

22–10 *Honey bees. (a) Workers tending honey and pollen storage cells. The honey is made from nectar processed by special enzymes in the workers' bodies. (b) "Undertaker" honey bees, removing the corpse of a dead pupa from its cell. (c) A foraging worker. The first segment of each of the three pairs of legs has a patch of bristles on its inner surface. Those of the first and second pairs are pollen brushes, which gather the pollen that sticks to the bee's hairy body. On the third pair of legs, the bristles form a pollen comb that collects pollen from the brushes and the abdomen. From the comb, the pollen is forced up into the pollen basket, a concave surface fringed with hairs on the upper segment of the third pair of legs. Transfer of pollen to the pollen basket occurs in midflight. The sting is at the tip of the abdomen.*

The newly emerged adult worker rests for a day or two and then begins successive phases of employment. She is first a nurse, bringing honey and pollen from storage cells to the queen, drones, and larvae. This occupation usually lasts about a week, but it may be extended or shortened, depending on the conditions of the colony. Then she begins to produce wax, which is exuded from the abdomen, passed forward by the hind legs to the front legs, chewed thoroughly, and then used to enlarge the comb. During this stage of employment as a houseworking bee, she may also remove sick or dead comrades from the hive, clean emptied cells for reuse, or serve as a guard at the hive entrance. During this period, she begins to make brief trips outside, seemingly to become familiar with the immediate neighborhood. It is only in the third and final phase of her existence that the worker bee forages for nectar and pollen. At about six weeks of age, she dies.

22–11 *Honey-bee workers in attendance on their queen (center). The workers feed the queen and also lick queen substance, which prevents their own sexual maturation, from the surface of her body.*

The Queen

Queens are raised in special cells that are larger than the ordinary cells and shaped somewhat like a peanut shell. Although all fertilized eggs have the genetic potential to become queens, new queens develop only at certain times and under specific circumstances. According to the best recent evidence, queens become queens because of a generally more nutritious diet in the larval stage, especially rich in protein, as compared to the mostly carbohydrate (honey) diet fed the worker larvae.

The queen influences her subjects by means of chemical substances that are synthesized in and secreted from specialized glands. One of these substances, known as "queen substance," inhibits sexual maturation in the worker bees and prevents them from becoming queens or from producing rival queens.

If a hive loses its queen, workers will notice her absence very quickly and will become quite agitated. Very shortly, they begin enlarging worker cells to form emergency queen cells, and the larvae in the enlarged cells are then fed the special diet. Any diploid larva so treated will become a queen.

The Annual Cycle

One of the important differences between subsocial and eusocial bees is that colonies of eusocial bees survive the winter. Within the wintering hive, bees maintain their temperature by clustering together in a dense ball; the lower the temperature, the denser the cluster.

In the spring, when the nectar supplies are at their peak, so many young may be raised that the group separates into two colonies. The new colony is always founded by the old queen, who leaves the hive, taking about half of the workers with her. The group stays together in a swarm for a few days, gathered around the queen, after which the swarm will settle in some suitable hollow tree or other shelter found by its scouts.

As the old queen is preparing to leave the hive, the new queens are getting ready to emerge. These two events are synchronized by sound signals transmitted through the comb. As these signals are exchanged, the workers remain motionless. During this period, sexual maturation begins in some of the workers, a few of which lay eggs. The unfertilized eggs develop into males, or drones. After the old queen leaves the hive, a new young queen emerges, and any other developing queens are destroyed. The young queen then goes on her nuptial flight, exuding a chemical (apparently queen substance) that entices the drones of neighboring colonies. She mates only on this one occasion (although she may mate with more than one male) and then returns to the hive to settle down to a life devoted to egg production.

During her nuptial flight, the queen receives enough sperm to last her entire life, which may be some five to seven years. The sperm are stored in a special organ in her reproductive tract and are released to fertilize each egg as it is being laid. The queen usually lays unfertilized eggs only in the spring, at the time males are required to inseminate the new queens.

The drones' only contribution to the life of the hive is their participation in the nuptial flight. Since they are unable to feed themselves, they become an increasing liability to the social group. As nectar supplies decrease in the fall, they are stung to death by their sisters or are driven out.

Vertebrate Societies

With rare exceptions (Figure 22–12), vertebrate societies do not have the rigid caste systems characteristic of the truly social insects. Many vertebrate societies are nonetheless highly structured, with social roles and access to resources determined by specific interactions that vary according to the species and the age and sex of the individuals.

Dominance Hierarchies

In many species of birds and mammals, dominance hierarchies, maintained by species-specific patterns of behavior, determine priority of access to resources and strongly influence relative reproductive success. One type of dominance hierarchy among vertebrates that has been studied in some detail is the pecking order in chickens.

A pecking order is established whenever a flock of hens is kept together over any period of time. In any one flock, one hen usually dominates all the others; she can peck any other hen without being pecked in return. A second hen can peck all hens but the first one; a third, all hens but the first two; and so on through the flock, down to the unfortunate pullet that is pecked by all and can peck none in return.

Hens that rank high in the pecking order have privileges such as first chance at the food trough, the roost, and the nest boxes. As a consequence, they can usually be recognized on sight by their sleek appearance and confident demeanor. Low-ranking hens tend to look dowdy and unpreened and to hover timidly on the fringes of the group.

During the period when a pecking order is being established, frequent and sometimes bloody battles may ensue, but once rank is fixed in the group, a mere raising or lowering of the head is sufficient to acknowledge the

22–12   *The only vertebrates known to have a social system similar to that of the eusocial insects are the naked mole rats, which live in underground tunnels in Kenya, Ethiopia, and Somalia. In each colony, only a single dominant female (the queen rat) and one or a few males breed. The other members of the colony, both male and female, are workers who forage for food (mostly root vegetables) and dig and maintain the tunnels in which the colony lives. When they are resting, naked mole rats—which typically weigh only 20 to 30 grams—huddle together to keep warm. As many workers as possible huddle under the queen rat, who, because of her large size, gives off large quantities of body heat.*

Naked mole rats can live more than 10 years. The queen rat usually breeds four times each year, producing an average of 10 pups per litter. It is thought that reproduction of the other females in the colony is suppressed by chemicals in the queen's urine.

dominance or submission of one hen in relation to another. The overt fighting behavior has become **ritualized,** and life subsequently proceeds in harmony. If, however, new members are added to a flock, the entire pecking order must be reestablished. The resulting disorganization leads to more fighting, less eating, and less tending to the essential business of growth and egg laying.

Pecking orders have the effect of reducing the breeding population. Cocks and hens low in the pecking order copulate much less frequently than socially dominant chickens. Thus the final outcome is probably the same as it would be if the social structure did not exist: the stronger and otherwise dominant animals eat better and leave the most offspring. However, because of the social hierarchy, this comes about with a minimum expenditure of lives and energy.

Territories and Territoriality

Many vertebrates stay close to their birthplaces, occupying a home range that is likely to be the same as that occupied by their parents. Even migratory birds that travel great distances are likely to return year after year to the same areas. Often these home ranges are defended, either by individuals or by groups, against other individuals or groups of the same species or closely related species that use the same resources. Areas so defended are known as **territories,** and the behavior of defending an area against rivals is known as **territoriality.**

22–13   *Wolf packs have dominance hierarchies of both males and females. Here a subordinate wolf greets a more dominant member of the pack by licking and smelling its mouth. This same muzzle-nuzzle gesture is used by pups begging for food. Usually only the dominant male and dominant female breed, and the rest of the pack cooperate in caring for the young. Caring includes guarding the den and providing food, which is swallowed at the kill and then regurgitated for the pups. (Some domestic dogs regularly vomit at the sight of a puppy, which should be recognized not as a sign of disapproval but as a social reflex.)*

(a)

(b)

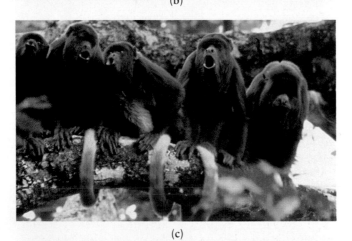

(c)

(d)

22–14 *Territories come in many shapes and sizes. (a) The male Uganda kob displays on his stamping ground, which is about 15 meters in diameter and is surrounded by similar stamping grounds on which other males display. A female signifies her choice by entering one of the stamping grounds and grazing there. Only a small proportion of males possess stamping grounds, and those that do are the only ones that breed.*

(b) A fiddler crab's territory is a burrow, from which he signals with his large claw, beckoning females and warning off other males.

(c) Howler monkeys shift their territories as they move through the jungle canopy but maintain spacing between groups by chorusing.

(d) Territoriality is common among reef fishes. For many species, a territory is a crevice in the coral, but for others, such as the clownfish, the territory is a sea anemone. The fish are covered by thick slime that partially protects them from the poison of the tentacles, but their acceptance by the anemone is chiefly a consequence of behavioral adaptations of the fish, which even mate and raise their brood among the tentacles. These two clownfish are guarding their fertilized eggs, which are on the rock just below the fish that is facing the camera. The eggs are the shiny, spherical objects that resemble small black marbles.

Territoriality in Birds

Territoriality was first recognized by an English amateur naturalist and bird watcher, Eliot Howard. Howard observed that, in the spring, female birds were attracted to areas occupied by singing males of the same species, whereas other adult males of the species avoided those areas. In general, a breeding territory is established by a male. Courtship of the female, nest building, mating, raising of the young, and often feeding are carried out within this territory. Frequently the female also participates in territory defense.

By virtue of territoriality, a mating pair has a better chance of obtaining food and nesting material in the area and a safe place to carry on all the activities associated with reproduction and care of the young. Some pairs carry out all their domestic activities within the territory. Others perform mating and nesting activities in the territories, which are defended vigorously, but gather food on a nearby communal feeding ground, where the birds congregate amicably together. A third type of territory functions only for courtship and mating, as in the bower of the bowerbird or the courtship arena of the prairie chicken. In these territories, the males prance, strut, and posture—but very rarely fight—while the females look on and eventually indicate their choice of a mate by entering his territory. Males that have not been able to secure a territory for themselves are not able to reproduce. In fact, there is

22–15 *Conflict between two male blue-striped grunts at their territorial borderline. Each grasps the other by the lips, and the loser is the one who lets go first. Although defeated, the loser swims back into his own territory unharmed. This sort of "jaw wrestling" occurs in many species of fishes.*

evidence from studies of some territorial species, such as the Australian magpie, that adult males that do not secure territories do not mature sexually.

Territorial Defense

Even though territorial boundaries may be invisible, they are clearly defined and recognized by the territory owner. With birds, for example, it is not the mere proximity of another bird of the same species that elicits aggression, but its presence within a particular area. The territory owner patrols his territory by flying from tree to tree. He will ignore a nearby rival outside his territory, but he will fly off to attack a more distant one that has crossed the border. Animals of other species are generally ignored unless they are prey or predators or are in competition for some limited resource.

Once an animal has taken possession of a territory, he or she is usually undefeatable on it. Among territory owners, prancing, posturing, scent marking, and singing and other types of calls usually suffice to repel intruders, which are at a great disadvantage. For example, a male cichlid, a tropical freshwater fish (see page 372), will dart toward a rival male within his territory, but, as he chases the rival back into its own territory, he begins to swim more slowly, the tail fin seemingly working harder and harder, just as if he were making his way against a current that increases in strength the farther he pushes into the other male's home ground. The fish know just where the boundaries are and, after chasing each other back and forth across them, will usually end up with each one trembling and victorious on his own side of the line.

Similarly, the expulsion from communal territories is typically accomplished by ritual rather than by force. For example, among the red grouse of Scotland, the males crow and threaten only very early in the morning, and then only when the weather is good. This ceremony may become so threatening that weaker members of the group leave the moor. Those that leave often starve or are killed by predators. Once the early-morning contest is over, the remaining birds flock together and feed side by side for the rest of the day.

Kin Selection

In 1962, V. C. Wynne-Edwards proposed that individuals that failed to reproduce were doing so for the benefit of the society to which they belonged. In this way, a society could maintain its population at a level always slightly below its resources, he argued, and so the entire group would benefit. Such behavior was perpetuated by the increased survival of groups whose members behaved with such altruistic self-restraint. Although Wynne-Edwards's hypothesis of group selection has now been rejected by almost all biologists, this proposal served to galvanize a whole series of extremely productive studies that have revolutionized the way modern biologists view social behavior.

Group selection was rejected on fairly simple grounds: if there were genes with alleles for breeding and alleles for restraint-of-breeding, the alleles for breeding would soon overwhelm the alleles for restraint-of-breeding in the gene pool. Rejection of group selection led to another proposal, put forward by W. D. Hamilton and based largely on studies of social insects. As Darwin himself realized, the evolution of sterile castes of insects posed a special problem for evolutionary theory. If evolution is based on the number of surviving offspring, how can natural selection—acting on the individual organism—result in the development and persistence of sterility in a large proportion of a population? Darwin concluded that, in some cases, natural selection might act not only on individuals but also on families. Conceptually the idea is a simple one: members of families share inheritable characteristics. Thus there are variations among families as well as among individuals, and families that have favorable variations are likely to leave more survivors than other families.

Hamilton elaborated Darwin's suggestion, modifying it to the concept of the gene pool. In population genetics, as we noted earlier (page 320), the measure of fitness is the extent to which the alleles in an individual's genotype are present in the gene pool of succeeding generations. Viewed from this perspective, a mother takes risks and makes sacrifices for her offspring in order to increase the representation of her alleles in the gene pool. Or, to put it another way, she is programmed genetically to take certain risks or to make certain sacrifices for her offspring. To the extent that taking these risks or making these sacrifices increases her contribution to the gene pool, the alleles dictating this altruistic behavior will increase in subsequent generations.

However, it is not just mothers and offspring who share alleles. For example, consider siblings—that is, sons and daughters of the same mother. Siblings with the same father share, on an average, half of their alleles with one another. Therefore, any allele that favorably influences altruistic behavior among siblings could similarly increase its representation in the gene pool.

Hamilton's hypothesis, based on this principle, is called **kin selection.** According to this hypothesis, within a species different groups of related individuals (that is, kin) reproduce at different rates. The critical factor in kin selection is the effect of the individual on the reproductive success of its relatives.

The concept of kin selection served to establish a new evolutionary perspective, that of **inclusive fitness.** The single criterion of Darwinian fitness is the relative number of an individual's *offspring* that survive to reproduce. The criterion of inclusive fitness is the relative number of an individual's *alleles* that are passed from generation to generation, either as a result of his or her own reproductive success or as a result of the reproductive success of related individuals.

Tests of the Hypothesis

Kin selection is a useful hypothesis because it is testable. An example of such a test is the work of Patricia Moehlman, who studied 17 litters of silver-backed jackal pups near Ndutu Lodge in Tanzania. The family unit of the silver-backed jackal consists of a monogamous pair, young pups, and, often, one to three older siblings from a previous litter that serve as helpers (Figure 22–16).

In litters with no helpers, Moehlman found that an average of only one pup was raised. With a single helper, three pups survived, and in one family with three helpers, six pups survived. Thus, by being helpers, young jackals (with the same degree of relatedness to their siblings as they would have to their own offspring) produced more copies of their own alleles than if they had reproduced themselves with no helpers.

Among Florida scrub jays, the situation is somewhat different. Here again, breeding success is positively correlated with the number of helpers, which are usually close relatives of the breeding pair. However, the success is somewhat less than would be achieved by the young jay's setting up its own nest and rearing its own young. The reason that the young jays stay on as helpers is hypothesized to be the shortage of available territories for young males. This hypothesis is supported by the fact that as soon as a territory becomes available, the male helper leaves. One of the most common ways to get a new territory is to inherit part of the parents' territory. Thus, one is led to predict that females will help less than males and that older, more dominant males will help more than the younger males, both predictions that have been shown to be true.

22–16   (a) *Silver-backed jackal parents groom a helper. Being groomed is not only pleasurable but also removes insects and other parasites from the skin. Helpers are also provided food by parents.* (b) *This young silver-backed jackal, by helping raise his younger brother, probably increases his inclusive fitness more than if he tried to raise his own family.*

(a)

(b)

Kin Recognition in Tadpoles

For kin selection to be a factor in the evolution and maintenance of altruistic behavior, animals must be able to selectively direct their altruistic acts toward related individuals—thus increasing their own inclusive fitness. In many species, dispersal from the birth site is limited, and the probability is high that any individual of the same species in the vicinity—to whom assistance might be rendered—is a relative. In other species, however, dispersal is wider, and a group may consist of both related and unrelated individuals. How, then, can an animal distinguish relatives from nonrelatives, particularly if there has been no previous opportunity to get to know the relatives?

A series of studies conducted by Andrew Blaustein and Richard O'Hara of Oregon State University have revealed the mechanism involved in one organism, the Cascades frog *(Rana cascadàe)*. Members of this species are highly gregarious, particularly in the tadpole stage, during which they maintain cohesive schools. The schooling behavior is thought to aid in the defense against predators, principally the insect larvae that feed on these tadpoles. When a Cascades frog tadpole is captured and injured by an insect larva, it releases a chemical into the water that effectively warns the other tadpoles of the danger. Release of the chemical is followed by the flight of other members of the school.

In laboratory experiments, Blaustein and O'Hara allowed fertilized eggs to develop to the free-swimming tadpole stage under four different sets of conditions: (1) Eggs from the same clutch—all laid by the same female and fertilized by the same male—developed together in the same tank; these individuals were full siblings. (2) Eggs from one clutch developed at one end of a tank, with eggs from another, unrelated clutch developing at the other end of the tank; the two groups were separated by a mesh screen that allowed water to circulate freely throughout the tank. (3) A single egg developed within a mesh cylinder in a tank, while 12 eggs from an unrelated clutch developed in the surrounding water. (4) Individual eggs developed in small tanks, in total isolation from any other individuals, related or unrelated.

When the tadpoles hatched, they were given "preference tests," using tanks that

Rana cascadae *tadpoles schooling in a pond in the Cascade Range in Oregon. The preference of individuals of this species to associate with relatives rather than with unrelated individuals persists after their metamorphosis into frogs.*

were sectioned into three parts by mesh screens. Individual tadpoles were placed in the central section of the tank, with related individuals in the section at one end of the tank and unrelated individuals in the section at the other end. Regardless of the regimen under which the tadpole had developed, it spent most of its time closest to the end of the tank that contained its relatives. When tadpoles were mixed in tanks that were not sectioned and were allowed to school freely, time after time, they chose to school with relatives rather than with nonrelatives—even when they had been raised in a tank surrounded by nonrelatives.

In field experiments, when groups of related and unrelated individuals were released into ponds and lakes, they quickly sorted out into schools composed primarily of kin. Additional experiments, in which eggs were fertilized in the laboratory, with different males fertilizing the eggs of the same female and the same male fertilizing the eggs of different females, revealed that the tadpoles could distinguish between full siblings, half siblings, and

nonsiblings, almost always preferring the company of the most closely related individuals available.

A second set of preference tests, in which the tadpoles were separated from the related and unrelated groups by glass barriers that allowed the tadpoles to see one another but prevented any water circulation between compartments, revealed that the cue that identifies its relatives to a Cascades frog tadpole is a chemical substance. In the absence of water bearing the identifying substance, the tadpoles displayed no preference for either group.

Experiments by other investigators have demonstrated varying degrees of kin recognition in a vast diversity of animals, including social insects, birds, ground squirrels, mice, and monkeys, as well as tadpoles of other frog and toad species. In most species studied, the capacity to recognize kin appears early in development. It seems to function not only in determining the distribution of favors—such as warning signals—but also in enabling animals to avoid both fighting with kin and inbreeding.

The Selfish Gene

Many years ago the English satirist Samuel Butler remarked that a chicken was just an egg's way of making another egg. Updating this concept, some biologists see an organism as just a gene's way of making more genes.

The argument is simple: The individual organism is transient. The genotype is fragmented at every generation. All that can survive from generation to generation is the gene. The way it survives is in the form of replicas. The more replicas, the better its chance of survival. The organism is the gene's survival machine, and it so programs the machine that it will turn out gene copies at a maximum rate, regardless of personal cost. Thus, for example, there is a mite in which the female has a brood of one son and 20 daughters. The son mates with his sisters while still within the mother and dies before he is born—not much of a life for the male mite but a great low-risk, high-return strategy for his genes and a wise maternal investment.

This concept of the selfish gene is admitted, even by its articulate sponsor, Richard Dawkins, to be an oversimplification. It has served, however, to bring into sharper focus questions on the evolution of behavior and of survival strategies.

Conflicts of Interest

In directing their attention to the survival of genes rather than of individuals, biologists realized that there were opportunities for serious conflicts of interest. These conflicts could occur not only between those individuals that were in obvious direct competition for some limited resource—such as territory, a food supply, or a mate—but also among individuals previously envisaged as working together for some common good—such as members of the nuclear family.

For example, a mammalian female invests in her offspring by providing nourishment in the form of milk. Nursing in mammals, however, generally prevents new pregnancies. It is to the offspring's benefit to monopolize the milk supply for as long as possible. The mother, however, can maximize her fitness by weaning the offspring at an appropriate time and beginning a new investment.

Another example of intergenerational conflict occurs when a new male takes over a female with young. The new male's evolutionary goal, as programmed by his genes, is to produce his own young as rapidly as possible. The existing offspring (fathered by another male) may interfere with this objective, either by competing with their half-siblings or by preventing a new pregnancy in the female. Sarah Blaffer Hrdy was the first to

22–17   *Mother langur holding an infant that has been fatally wounded by an infanticidal male at the time of take-over of her troop.*

report the practice of infanticide by males at the time of take-over. Her observations, which were made on Hanuman langurs in India (Figure 22–17), were greeted with considerable skepticism when first reported in 1971, but similar acts have now been observed in other primates and also in birds, rodents, and lions. Infanticide can clearly represent a successful evolutionary strategy.

Male vs. Female

Among the most closely fought conflicts of interest are those that occur between the partners of a reproductive pair. A female, by definition, is the individual that produces the larger gametes. Female gametes are relatively expensive, metabolically speaking, and if the female must also carry the embryo in her body and care for the young after their birth, her investment in each reproductive effort may be, in terms of her total reproductive potential, very large indeed. Male gametes, by contrast, are usually cheap; a single insemination costs males of most species almost nothing.

In most animal species, all that the male contributes to the next generation is his genes. In such situations, it is to the interest of the female's selfish genes to find themselves in the best possible company, thus promoting their survival in the next generation. Hence, intense competition may arise among the males to prove to the females that they are the best endowed. Such proof may take the form of elaborate courtship displays, in which strength, vigor, stamina, and other desirable traits are displayed, or it may take the form of dominance over other males. Males also attract females by offers of food or other resources in the form of a territory or a nesting

22–18 *A great bowerbird, decorating his bower with hibiscus flowers and other red objects (note the bottle cap in his beak). The male spends much of his time building this elaborate nuptial structure, which is the focus of his territory, and decorating it with brightly colored objects. Male birds call from their bowers and display to females if they approach. The female selects one male, enters his bower, copulates briefly, and departs. Then she builds her own utilitarian nest some distance from the bower of her mate, lays her eggs, and cares for the young entirely on her own.*

site. Because the male that commands the most resources, or the best resources, is likely also to be the male that is superior genetically, the offering or the territory may be less important as an actual resource than as a symbol of the male's desirability (Figure 22–18).

In some few species, males make an equal contribution to the care of the young. Selection for such a reproductive strategy would depend on at least two factors: first, that the additional parental investment would result in a significant increase in surviving young and, second, that the male has some way of being sure that the offspring he cares for are his own. Females in such situations may give up the opportunity of mating with the highest-ranking male in return for a male that will provide care for the young. Selection for such a reproductive strategy would include a means for the female to be sure that the male she has selected will remain with her after her eggs have been fertilized.

A long courtship, such as is seen among some species of birds, serves the purposes of both male and female. It gives each of them a chance to assess the qualities of the other. For her, it symbolizes commitment and keeps him away from other females. It may also bring material benefits; for example, many female birds refuse to copulate until the nest is built. For him, the long courtship provides assurance that she has not already been inseminated. She loses nothing by cheating, whereas he loses his entire investment if she does.

The Advantage of Waiting

Dominance hierarchies and territories can also be viewed from the perspective of the selfish gene. An animal that is not breeding because it is low in the dominance hierarchy or lacks a territory is not sacrificing itself for the good of the group or of the species. It is more likely waiting for its chance. If the individual is very low in a hierarchy, such an opportunity may never come, but still its chances of perpetuating its genotype are better than if it challenges a superior and is killed or badly injured.

Studies of bird territories confirm the hypothesis that individuals that lack territories are waiting for their turn. An empty territory is almost always immediately filled, as in the example of Florida scrub jays. Another example is provided by lions. As young male lions approach maturity, they are driven from the pride by the established adult males. The young males tend to stay together as nomads (Figure 22–19) until they find a pride they can take over, often by defeating the resident males. Although their period of dominance may be limited, their reproductive success is greater than if they had been killed or maimed resisting expulsion from the pride in which they were born. As we noted in Chapter 19, this social system also promotes outbreeding.

22–19 *Young male lions, living as nomads. Such lions are typically found in small groups in which all members are about the same age and are probably brothers or, at least, from the same pride. In all likelihood, they will eventually take over a pride of their own. Like the males who preceded them, they will be transients, holding the females—and the territory occupied by the females—for only a few months, or, at most, a few years.*

The females of a pride (usually sisters or half-sisters) hunt together cooperatively, become sexually receptive at about the same time, give birth at about the same time, and nurse each other's cubs. New males taking over a pride often kill the young cubs. Following the death of the cubs, the females become sexually receptive again and so are available to perpetuate the genes of the newcomers, at least during the brief period of their reign.

Reciprocal Altruism

Seemingly altruistic acts may pay off, as we have seen, because they increase the probability of survival of genes shared by the performer of the act and the beneficiary. In 1971, Robert Trivers proposed another model for seemingly self-sacrificing behavior, which he termed **reciprocal altruism**. According to this model, an altruistic act is performed with the expectation that the favor will be returned. It might involve warning of danger, removing a tick from an inaccessible portion of the anatomy, or sharing food, to cite just a few possible examples. It also might involve acts of cooperation between organisms of different species, as well as between members of the same species. Reciprocal altruism may well have become established in early human societies, which is why this model is of special interest.

An example of reciprocal altruism is provided by vampire bats (Figure 22–20), which often have a difficult time locating suitable warm-blooded victims. Many individuals return each morning to the communal nesting sites with an empty stomach after an unsuc-

cessful night of searching. A study by Gerald Wilkinson of wild vampire bats in Costa Rica showed that the bats regularly regurgitate blood to feed each other and that individuals in dire need are fed preferentially. A starved bat that receives blood reciprocates on a later day when its hunt has been successful. Not only related animals but also unrelated animals reciprocate with one another.

For reciprocal altruism to be a successful strategy, it must be resistant to exploitation by individuals who accept favors but never return them (that is, cheaters). Two conditions are necessary. The first is that individuals meet more than once and are able to recognize each other. The second is that each individual cooperates on the first encounter and, on subsequent encounters, does whatever the other individual did on the preceding encounter. Under these particular conditions, cooperation pays off and cheating is punished.

A significant number of cooperators must be present in a population for the reciprocal behavior to become established. Thus, an important question is how reciprocal altruism gets started initially. One possible explanation is that cooperation originates in a group of related individuals—that is, as kin selection—and spreads from the original cluster to include unrelated individuals.

The Biology of Human Behavior

It is tempting—indeed almost irresistible—to draw parallels between human behavior and the behavior observed in other species. The extent to which concepts concerning the evolution of behavior can be extrapolated to the human species is a matter of controversy. Some biologists maintain that the human species is basically no different from any other species, that our genes are as selfish as any, and that if we seek to modify human behavior for the common good, we need to understand its roots. Other biologists, however, maintain that while early human ancestors may have been governed by their genes, modern humans are so much a product of their culture and of their individual experience that such analyses are no longer valid. Moreover, they may be dangerous. The concept that biology determines human behavior lies at the root of all notions of racial superiority. Thus it has provided the rationale for slavery, exploitation, and genocide. More commonly, the notion that human behavior is, to some extent, biologically determined allows us to forgive ourselves for, and even to justify, violence, aggression, docility, and greed.

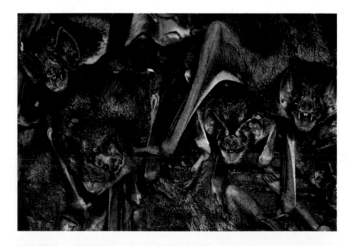

22–20   *Vampire bats, such as the females shown here, congregate during the day in communal nesting areas in caves or hollow trees. At night they emerge and begin their search for animals, preferably horses or cows, from which they can obtain a blood meal. Because of their high metabolic rate, each night these bats need to consume at least 50 percent of their body weight in blood (about 15 grams, which is a little more than half an ounce). If a bat's hunt is unsuccessful on two consecutive nights, it is in danger of death by starvation. In a striking example of reciprocal altruism, successful hunters share their food with the unsuccessful, a favor that is returned when their fortunes are reversed.*

Summary

The behavioral characteristics of organisms are the products of natural selection. Like anatomical and physiological adaptations, they affect the fitness of organisms. Although all behavior has its roots in the underlying genetic program of the individual, in only a few instances has it been possible to correlate a particular behavior with the presence or absence of specific alleles. Behaviors are typically the result of the interactions of a large number of genes, the influence of which may be further modified by interactions of the individual with its environment.

A full explanation as to why an organism performs a particular behavior involves both proximate and ultimate causation. The proximate cause of a behavior is typically a series of physiological responses to a particular stimulus. The ultimate cause, however, is the adaptive value of the behavior for the survival and successful reproduction of the individual.

Ethology is the study of the behavior of animals in their natural environment, with particular emphasis on species-specific patterns of behavior and their evolutionary origins. Patterns of behavior that appear essentially complete the first time the organism encounters the relevant stimulus are known as fixed action patterns. These patterns are highly stereotyped and repetitive and, for members of a given species of a particular age, sex, and physiological condition, are as predictable and constant as the anatomical characteristics of the species. Communication signals among members of a species that act as stimuli for fixed action patterns are known as releasers, and the brain circuits that respond to them are known as innate releasing mechanisms.

Learning is the modification of behavior as a result of experience. Various categories of learning are recognized, of which habituation is one of the simplest. In associative learning, which often takes the form of trial-and-error learning, one stimulus comes to be associated with another. Imprinting is a form of learning of particular importance in the discrimination of members of one's own species from members of other species; it occurs quickly within a critical period in early life. Another form of learning, important in many species of birds and mammals, is imitation.

Of particular interest from an evolutionary standpoint is the behavior of animals in societies—groups of individuals of the same species, living together in an organized fashion, with divisions of resources and labor and with mutual dependence. Such behavior may be selfish, cooperative, altruistic, or spiteful—with differing consequences for the fitness of the donor and the recipient.

Insect societies are among the largest and most complex of animal societies. In the honey-bee colony (and among other eusocial insects as well), the queen is the only female reproductive form. She and her brood are tended by sterile workers. The behavior and physiology of the members of the hive are controlled by the exchange of chemical signals.

Vertebrate societies are often organized in terms of social dominance. Dominance hierarchies may take the form of pecking orders, in which higher-ranking animals have priority of access to food and other resources and reproduce more. Territoriality is a system of social dominance in which only animals with territories reproduce and animals with better territories reproduce more.

A central question in the study of social behavior is the mechanism by which natural selection can result in behavior that limits the reproductive potential of individuals in the society. An explanation is provided by the hypothesis of kin selection, which includes the concept of inclusive fitness. Previous concepts of evolutionary fitness focused on the relative number of surviving offspring in future generations. Inclusive fitness focuses on the relative number of an individual's alleles that are passed from generation to generation, as a result of the reproductive success either of the individual or of its relatives.

From the viewpoint of inclusive fitness, social behavior is regulated by the "selfish gene," which programs the individual not necessarily for the individual's own well-being, or even survival, but only for the perpetuation of its alleles. Competition for representation in the future gene pool can occur not only between individuals obviously competing for the same resources but also between parents and offspring and between members of a reproductive pair. The selfish-gene concept can also be used to explain the establishment of dominance hierarchies and territorial behavior.

Altruism is behavior that carries a cost to the individual that performs it and benefits some other individual or individuals. Some acts of altruism are thought to be based on inclusive fitness, and others may be based on reciprocal altruism, the performance of an unselfish act with the expectation that the favor will be returned. Reciprocal altruism is of particular interest in terms of the evolution of cooperative behavior in the ancestors of *Homo sapiens*.

Questions

1. Distinguish among the following: proximate cause/ultimate cause; fixed action pattern/innate releasing mechanism; sign stimulus/releaser; imprinting/imitation; subsocial/eusocial; queen/worker/drone; social dominance/territoriality; group selection/kin selection; Darwinian fitness/inclusive fitness; altruism/reciprocal altruism.

2. Define behavior. What distinguishes social behavior from other forms of behavior?

3. Territoriality and social dominance achieve the same results. What are they? How do the results differ in a population in which these forms of behavior are not present?

4. Among organisms in which both sexes are diploid, the degree of relatedness between parents and children is always 0.5 (that is, each child inherits one-half, or 50 percent, of its alleles from its mother and one-half, or 50 percent, from its father). The degree of relatedness between siblings has a theoretically possible range from 0 to 1 and averages 0.5. Explain.

5. The late J. B. S. Haldane, a mathematician who made major contributions to evolutionary biology, once remarked that he would lay down his life for two brothers or eight cousins. On what basis did he make such an offer? Actually, Haldane was being too generous. Explain why.

6. Animals that live within a dominance hierarchy may never breed, yet they may be as fit as those that do breed. How can this be explained?

7. In order to protect their young against infanticidal males, females might refuse to mate with murderers. Would this be a successful strategy?

8. A hypothesis known as the "superterritory hypothesis" suggests that an animal should defend a territory larger than it really needs. Could this be spiteful behavior? Explain.

9. In what ways are human societies different from insect societies? How are they similar?

Suggestions for Further Reading

Alcock, John: *Animal Behavior: An Evolutionary Approach,* 5th ed., Sinauer Associates, Inc., Sunderland, Mass., 1993.

An outstanding text on animal behavior. Highly recommended.

Blaustein, Andrew R., and Richard K. O'Hara: "Kin Recognition in Tadpoles," *Scientific American,* January 1986, pages 108–116.

Dawkins, Richard: *The Selfish Gene,* 2d ed., Oxford University Press, New York, 1989.*

Dawkins argues that we are "survival machines" programmed to preserve the "selfish molecules" known as genes. A wit-sharpening account and also an entertaining one.

Franks, Nigel R.: "Army Ants: A Collective Intelligence," *American Scientist,* vol. 77, pages 138–145, 1989.

Goodall, Jane: *In the Shadow of Man,* 2d ed., Houghton Mifflin Company, Boston, 1988.*

A personal account, first published in 1971, of 11 years spent observing the complex social organization of a single chimpanzee community in Tanzania.

Goodall, Jane: *Through a Window: My Thirty Years with the Chimpanzees of Gombe,* Houghton Mifflin Company, Boston, 1991.*

The sequel to In the Shadow of Man, *this fascinating and insightful book carries the story of the chimpanzee community through the subsequent 20 years. Highly recommended.*

Hölldobler, Bert, and Edward O. Wilson: *The Ants,* Harvard University Press, Cambridge, Mass., 1990.

An outstandingly detailed and richly illustrated compendium of information about ants, the dominant organisms in many terrestrial environments. This delightful resource covers all aspects of these fascinating insects, from their evolution to their anatomy and physiology, as well as their social organization and ecology.

Honeycutt, Rodney L.: "Naked Mole-Rats," *American Scientist,* vol. 80, pages 43–53. 1992.

Hrdy, Sarah Blaffer: *The Langurs of Abu: Female and Male Strategies of Reproduction,* Harvard University Press, Cambridge, Mass., 1981.*

This book, which has been described as "spell-binding," is notable not only for its original observations but also for its contributions to the theoretical structure of sexual combat.

* Available in paperback.

Hrdy, Sarah Blaffer: *The Woman that Never Evolved,* Harvard University Press, Cambridge, Mass., 1981.*

Both a sociobiologist and a feminist, Hrdy examines what it means to be female. Males are almost universally dominant over females in primate species, and Homo sapiens is no exception. Yet in studying our closest living relatives, the primates, she discovers that the female is competitive, independent, sexually assertive, and has just as much at stake in the evolutionary game as her male counterpart.

Krebs, J. R., and N. B. Davies: *An Introduction to Behavioural Ecology,* 3d ed., Blackwell Scientific Publications, Boston, 1993.*

The best introduction to the general subject of social behavior. Well written with many good illustrations and an excellent balance between examples and theory.

Mech, L. David: *The Wolf: The Ecology and Behavior of an Endangered Species,* University of Minnesota Press, Minneapolis, 1983.*

The definitive book on the social history of wolves, originally published in 1970. Mech, a wildlife biologist with the U.S. Fish and Wildlife Service, devoted more than 20 years to the study of wolves in their natural habitats.

Mock, Douglas (ed.): *Behavior and Evolution of Birds,* W. H. Freeman and Company, New York, 1991.*

A collection of articles from Scientific American *on various aspects of the behavior and evolution of these familiar and fascinating animals.*

Mock, Douglas W., Hugh Drummond, and Christopher H. Stinson: "Avian Siblicide," *American Scientist,* vol. 78, pages 438–449, 1990.

Moehlman, Patricia D.: "Social Organization in Jackals," *American Scientist,* vol. 75, pages 366–375, 1987.

Moss, Cynthia: *Portraits in the Wild: Animal Behavior in East Africa,* 2d ed., The University of Chicago Press, Chicago, 1982.*

A description of some of the studies carried out on various wild species. Full of fascinating bits of information and firsthand observations.

Sapolsky, Robert M.: "Stress in the Wild," *Scientific American,* January 1990, pages 116–123.

Seeley, Thomas D.: "The Honey Bee Colony as a Superorganism," *American Scientist,* vol. 77, pages 546–553, 1989.

Seeley, Thomas D.: "How Honeybees Find a Home," *Scientific American,* October 1982, pages 158–168.

Sherman, Paul W., Jennifer U. M. Jarvis, and Stanton H. Braude: "Naked Mole Rats," *Scientific American,* September 1992, pages 78–86.

Tinbergen, Niko: *Curious Naturalists,* Doubleday & Company, Inc., Garden City, N.Y., 1984.*

Some charming descriptions of the activities and discoveries of scientists studying the behavior of animals in their natural environment, originally published in 1968.

Topoff, Howard (ed.): *The* Natural History *Reader in Animal Behavior,* Columbia University Press, New York, 1987.*

A collection of Natural History *articles covering four broad topics: sensory processes and orientation, evolution of behavior, social organization, and behavioral development.*

Topoff, Howard: "Slave-making Ants," *American Scientist,* vol. 78, pages 520–528, 1990.

Trivers, Robert: *Social Evolution,* The Benjamin/Cummings Publishing Company, Redwood City, Calif., 1985.

An excellent introductory text on the evolution of social behavior, this book assumes a minimal background on the part of the reader. Well illustrated, with many examples.

Tumlinson, James H., W. Joe Lewis, and Louise E. M. Vet: "How Parasitic Wasps Find Their Hosts," *Scientific American,* March 1993, pages 100–106.

Wilkinson, Gerald S.: "Food Sharing in Vampire Bats," *Scientific American,* February 1990, pages 76–82.

Wilson, E. O.: *The Insect Societies,* Harvard University Press, Cambridge, Mass., 1971.*

A comprehensive and fascinating account of the social insects.

Wilson, E. O.: *Sociobiology: The New Synthesis,* Harvard University Press, Cambridge, Mass., 1975.

In this extremely interesting, beautifully written, and beautifully illustrated book, Wilson undertakes to set forth the biological principles that govern social behavior in all kinds of animals. The last chapter, which concerns human sociobiology, became a focus of controversy about the inheritance of behavioral traits in Homo sapiens. Also available in an abridged, less technical version.

Winston, Mark L., and Keith N. Slessor: "The Essence of Royalty: Honey Bee Queen Pheromone," *American Scientist,* vol. 80, pages 374–385, 1992.

Woolfenden, G. E., and J. W. Fitzpatrick: "The Inheritance of Territory in Group Breeding Birds," *BioScience,* vol. 28, pages 104–108, 1978.

* Available in paperback.

The Diversity of Life

We share our planet with a multitude of other species, or kinds, of organisms. Each of these species is, like ourselves, the end product of almost 4 billion years of evolutionary history. Adapted in the course of that history to different environments and different ways of life, organisms have diversified to fill every nook and cranny of the biosphere.

Most of the Earth's species remain unknown. About 2 million have been named and classified, but far more remain unrecorded. Some estimates range as high as 50 million. Many of these undiscovered species are hidden in the upper, sunlit foliage (the canopy) of the tropical forests. Even as the forests are disappearing, increasing numbers of biologists are hoisting themselves into the treetops to seek out the many species of organisms that live nowhere else. Here, suspended high in a forest in Peru, Terry Erwin of the Smithsonian Institution is spraying an environmentally safe insecticide into the canopy. The resulting rain of spiders, insects, and other invertebrates will be caught in the sheets below. Explorers like Erwin are fighting a battle—perhaps a losing battle—against time. Tropical forests are currently being destroyed at a rate of 55,000 square miles a year.

In the past, these forests have been the source of many valuable natural products, such as rubber, quinine, vanilla, cocoa, and coffee, to name a few. Environmentalists struggling to save the forests point out that many more such treasures may be hidden in the canopy. "From a beetle without a name atop an orchid in a distant threatened forest may come a cure for cancer," notes E. O. Wilson of Harvard University, who himself found 43 species of ants in a single tree in Amazonia. Many of us believe that the tropical forests should be saved for their own sake, because they are beautiful, because we would like them to be there for our children and for their children, and because they are full of wonderful secrets, waiting to be explored. However, to quote Wilson once more, "It is as though the stars began to vanish at the moment astronomers focused their telescopes."

The Classification of Organisms

In the previous four chapters, we have considered the mechanisms by which evolutionary change occurs. Now we shall turn our attention to the products of evolution—the multitude of different kinds of organisms, past and present, that have occupied our planet since the appearance of the first living cells. In this chapter and the next four, we are going to sample the diversity of organisms that live on Earth today and, at the same time, trace the major developments in the course of evolutionary history that have given rise to this diversity.

The Need for Classification

Most people have a limited awareness of the natural world and are concerned chiefly with the organisms that influence their own lives. For example, gauchos, the cowboys of Argentina, who are famous for their horsemanship, have some 200 names for different colors of horses but generally divide plants into only four groups, based on their uses: *pasto*, or fodder; *paja*, bedding; *cardo*, wood; and *yuyos*, everything else.

Most of us are like the gauchos. Once beyond the range of common plants and animals, and perhaps a few uncommon ones that are of special interest to us, we usually run out of names and categories. Biologists, however, face the task of systematically identifying, studying, and exchanging information about the vast diversity of organisms with which we share this planet—at least 5 million different species, and perhaps many more. In order to do this, they must have a system for naming all these organisms and for grouping them together in orderly and logical ways. The problems of developing such a system are immensely complicated and begin with the basic unit of biological classification, the species.

Species Revisited

As you know, **species** in Latin simply means "kind," and so species, in the simplest sense, are different kinds of organisms. A more precise definition, as we noted in Chapter 21, is that a species is a group of natural populations whose members can interbreed with one another but cannot (or at least usually do not) interbreed with members of other such groups. The essential feature of this definition is reproductive isolation. We know that two distinct species exist when the two can occupy the same space without interbreeding.

This definition works well for animal species and is generally accepted by zoologists. Many plants, however, can reproduce asexually and also can form fertile hybrids with other species. Bacteria, with their variety of forms of genetic exchange, do not fit this definition neatly, nor do the many unicellular eukaryotes that reproduce by cell division, forming clones of identical cells. Thus, although botanists and microbiologists use the term "species," they are more likely to consider it a category of convenience, existing in the human mind rather than in the natural world.

For most practical purposes, a species is a category into which is placed an individual organism that conforms to certain fairly rigid criteria concerning its structure and other characteristics. From an evolutionary perspective, however, a species is a population of organisms, reproductively united but very probably changing as it moves through space and time. As we saw in Chapter 21, splinter groups, reproductively isolated from the population as a whole, can undergo sufficient change that they become new species.

A group of closely related species, presumably derived from a common ancestor, constitute a **genus** (plural, genera). According to the binomial system of nomenclature devised by the Swedish naturalist Linnaeus in the eighteenth century and still in use today, the scientific name of an organism consists of two parts—the name of the genus plus a **specific epithet** (an adjective or modifier) that indicates the particular species within that genus. The wolf, for example, is classified in the genus *Canis* (the Latin word for "dog"), and its scientific name is *Canis lupus*. The use of this unique combination of words indicates that we are referring to the wolf rather than such near relations as the coyote (*Canis latrans*) and the domestic dog *(Canis familiaris).*

Hierarchical Classification

One of the tasks confronting biologists as they attempt to perceive order in the diversity of organisms is to con-

23–1   *The eighteenth-century Swedish professor, physician, and naturalist Carolus Linnaeus (1707–1778), who devised the binomial system for naming species of organisms and established the major categories that are used in the hierarchical system of biological classification. When he was 25, Linnaeus spent five months exploring Lapland for the Swedish Academy of Sciences; he is shown here wearing his Lapland collector's outfit.*

struct classification schemes that are not only convenient and useful but also reflect accurately the historical relationships among species. Such classifications are, in effect, hypotheses about evolutionary history. Like other hypotheses, they can be tested (through detailed study of the fossil record and of the structure and other characteristics of living organisms) and revised as necessary.

The classification of organisms is a hierarchical system—that is, it consists of groups within groups, with each group ranked at a particular level (Figure 23–3). In such a system, a particular group is called a **taxon** (plural, taxa), and the level at which it is ranked is called a **category.** For example, genus and species are categories, and *Homo* and *Homo sapiens* are taxa within those categories.

(a)

(b)

(c)

23–2   *These orchids are all members of the genus* Cypripedium: **(a)** *the yellow lady's slipper,* Cypripedium calceolus,
(b) *the showy lady's slipper,* Cypripedium reginae, *and* **(c)** *the
pink lady's slipper,* Cypripedium acaule. *Although there is an*
*overall similarity among the plants of all three species, there
are significant differences in the leaves, in the color and size
of the flowers, and in other characteristics.*

23–3   *The hierarchical nature of biological classifications, consisting of groups
within groups, can be represented visually in
a number of different ways. This "Shaker
box" diagram shows the classification of two
species,* Homo sapiens *and* Tyto alba.

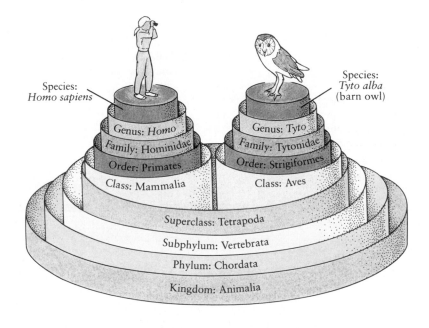

Species:
Homo sapiens

Species:
Tyto alba
(barn owl)

Genus: Homo
Family: Hominidae
Order: Primates
Class: Mammalia

Genus: Tyto
Family: Tytonidae
Order: Strigiformes
Class: Aves

Superclass: Tetrapoda

Subphylum: Vertebrata

Phylum: Chordata

Kingdom: Animalia

The Naming of Species

The often tongue-twisting binomials that constitute the scientific names of organisms are a necessary tool for clear and unambiguous communication among biologists. Many species lack common names, and, even when common names do exist for a kind of organism, more than one name may be given to the same species, such as groundhog and woodchuck, gnu and wildebeest, pill bug and sow bug and wood louse. Or the same name may be given, in different parts of the world, to different organisms. For example, a robin in North America is distinctly different from the English bird of the same name. A yam in the southern United States is a totally different vegetable from a yam several hundred kilometers away in the West Indies. When different languages are involved, the problems of communication would be virtually insurmountable without a system of nomenclature universally recognized and agreed upon by biologists.

The name of the genus is always written first, as in *Drosophila melanogaster*, and it may be used alone when one is referring to members of the entire group of species making up that genus, such as *Drosophila* or *Paramecium*. A specific epithet is meaningless when written alone, however, because many different species in different genera may have the same specific epithet. For example, *Drosophila melanogaster* is the fruit fly that has played such an important role in genetics; *Thamnophis melanogaster*, however, is a semiaquatic garter snake. Thus, by itself, the specific epithet *melanogaster* ("black stomach") would not identify either organism. For this reason, the specific epithet is always preceded by the name of the genus, or, in a context where no ambiguity is possible, the name of the genus may be abbreviated to its initial letter. Thus *Drosophila melanogaster* may be designated *D. melanogaster*.

(a)

*Although they are both called robins, **(a)** the North American robin, Turdus migratorius ("thrush, of the migratory habit"), is a distinctly*

(b)

*different bird from **(b)** the English robin, Erithacus rubecula.*

Whoever describes a species first has the privilege of naming it. It may not be named after oneself, but often it is named after a friend or colleague. *Escherichia*, for example, is named after Theodor Escherich, a German physician (*coli* simply means intestinal); and *Rhea darwinii*, an ostrichlike bird found in Patagonia, is named after Charles Darwin.

Names may be descriptive. The first discovered early fossil of a horse—which does not superficially resemble a modern horse at all—was named *Hyracotherium*, the "hyrax-like beast." (A hyrax looks like a big guinea pig and is thought, by some, to be a distant relative of the elephant.) When O. C. Marsh of Yale University began, in the 1870s, to study fossil horses, he recognized that the little dog-sized *Hyracotherium* was an early equine and gave it the charming name Eohippus ("dawn horse"). However, *Hyracotherium* remains its official designation because it was published first.

Some names are heartfelt. Thus, members of various mosquito genera have been given the specific epithets *punctor, tormentor, vexans, horrida, perfidiosus, abominator*, and *excrucians*. Others are frivolous. An English entomologist coined a whole series of generic names based on the pseudo-Greek ending *chisme*, pronounced "kiss me." Thus, there are squash bugs, stink bugs, and seed bugs known variously as *Polychisme, Peggichisme, Dolichisme*, and the promiscuous *Ochisme*. A species of wasp in the genus *Lalapa* was given the specific epithet *lusa*, and there is a (presumably treacherous) beetle named *Ytu brutus*. So, although species names may appear formidable and be unpronounceable, they are not necessarily as pompous (or even as informative) as they may seem.

Table 23–1   Biological Classifications

Corn (Zea mays)

Category	Taxon	Characteristics
Kingdom	Plantae	Multicellular organisms primarily adapted for life on land; usually have rigid cell walls and chlorophylls a and b contained in chloroplasts
Division	Anthophyta	Vascular plants (plants with conducting tissues) with seeds and flowers; ovules enclosed in ovary; seeds enclosed in fruit; the flowering plants
Class	Monocotyledones	Embryo with one seed leaf (cotyledon); flower parts usually in threes; many scattered vascular bundles in stem
Order	Commelinales	Monocots with fibrous leaves; reduction and fusion in flower parts
Family	Poaceae	Hollow-stemmed monocots with reduced greenish flowers; simple, dry, one-seeded fruits; the grasses
Genus	Zea	Robust grasses with separate male and female flower clusters; fleshy fruits
Species	Zea mays	Corn

Human Being (Homo sapiens)

Category	Taxon	Characteristics
Kingdom	Animalia	Multicellular organisms requiring complex organic substances for food; food usually ingested
Phylum	Chordata	Animals with notochord, dorsal hollow nerve cord, gill pouches in pharynx at some stage of life cycle
Subphylum	Vertebrata	Chordates with spinal cord enclosed in a vertebral column; body basically segmented; skull enclosing brain
Superclass	Tetrapoda	Four-limbed land vertebrates
Class	Mammalia	Tetrapods with milk glands that provide nourishment for the young; skin with hair; body cavity divided by a muscular diaphragm; red blood cells without nuclei; three ear bones; high body temperature
Order	Primates	Tree-dwelling mammals or their descendants, usually with fingers and flat nails; sense of smell reduced
Family	Hominidae	Primates with flat face; eyes forward; color vision; upright, bipedal locomotion
Genus	Homo	Hominids with large brain, speech, long childhood
Species	Homo sapiens	Prominent chin; high forehead; sparse body hair

In the hierarchical system of biological classification, related species are grouped into genera, related genera are grouped into **families**, related families into **orders**, related orders into **classes**, related classes into **phyla** or **divisions**, and related phyla or divisions into **kingdoms.** (The categories of division and phylum are equivalent. The term "division" is generally used in the classification of prokaryotes, algae, fungi, and plants, whereas "phylum" is used in the classification of protozoa and animals.) These categories may be further subdivided or aggregated into a number of less frequently employed categories such as subphylum and superfamily.

This system of classification makes it possible to generalize. For example, Table 23–1 shows the classification of two different organisms. Notice how the classification of an animal as a mammal, for example, or a plant as Anthophyta, provides an index to a vast amount of information. Notice also that, in progressing downward from kingdom to species, there is an increase in detail, proceeding from the general to the particular. In short, hierarchical classification is a highly useful means of organizing information.

Information Used in Classifying Organisms

The grouping of organisms into taxa from the categorical levels of genus through phylum or division is based on similarities in structure and other characteristics. From Aristotle on, however, biologists have recognized that superficial similarities are not useful criteria for classification. To take a simple example, birds and insects should not be grouped together simply because both have wings and are capable of flight. On the basis of overall structure, a wingless insect (such as an ant) is still an insect, and a flightless bird (such as the kiwi) is still a bird.

A key question in evolutionary classification is the origin of a similarity or difference. Does a similarity reflect inheritance from a common ancestor, or does it reflect adaptation to similar environments by organisms that do not share a common ancestor (convergent evolution)? A related question arises with differences between organisms: Does a difference reflect separate evolutionary histories, or does it reflect instead the

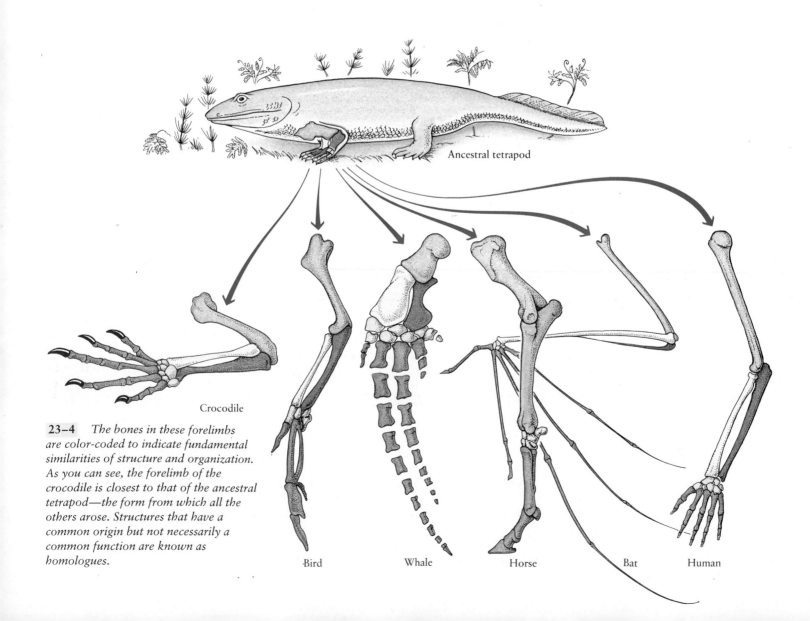

Ancestral tetrapod

Crocodile

23–4 *The bones in these forelimbs are color-coded to indicate fundamental similarities of structure and organization. As you can see, the forelimb of the crocodile is closest to that of the ancestral tetrapod—the form from which all the others arose. Structures that have a common origin but not necessarily a common function are known as homologues.*

Bird   Whale   Horse   Bat   Human

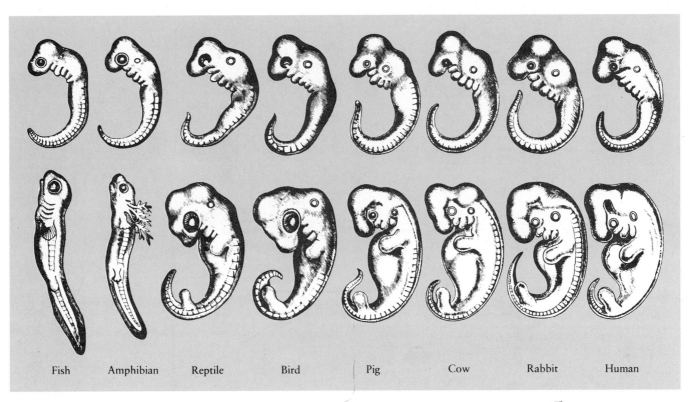

| Fish | Amphibian | Reptile | Bird | Pig | Cow | Rabbit | Human |

23–5 *The embryos of eight different vertebrates at an early stage of development (top row) and at a slightly later stage (bottom row). These drawings are based on the work of the nineteenth-century embryologist Karl Ernst von Baer. As you can see, the early embryos are almost indistinguishable. All have prominent gill pouches and fishlike tails. Even in the later stage, the human embryo still has a tail that is longer than its limbs. Such homologies in embryonic development are clear indicators of close evolutionary relationships.*

adaptations of closely related organisms to very different environments (divergent evolution)?

A classic example is the vertebrate forelimb. The wing of a bird, the flipper of a whale, the foreleg of a horse, and the human arm have quite different functions and appearances. Detailed study of the underlying bones reveals, however, the same basic structure (Figure 23–4). Such structures, which have a common origin but not necessarily a common function, are said to be **homologous.** These are the features upon which evolutionary classification systems are ideally constructed.

By contrast, other structures, which may have a similar function and superficial appearance, have an entirely different evolutionary background. Such structures are said to be **analogous.** Thus the wings of a bird and the wings of an insect would be said to be analogous, not homologous. Similarly, the spine of a cactus (a modified leaf) and the thorn of a rose (a modified branch) are analogous, not homologous.

Decisions as to homology and analogy are seldom so simple. In general, the features most likely to be homologous—and thus useful in determining evolutionary relationships—are those that are complex and detailed, consisting of a number of separate parts. This is true whether the similar feature is anatomical, as in the bones of the vertebrate forelimb, or is a biochemical pathway or a behavioral pattern. The more separate parts involved in a feature shared by several species, the less likely it is that the feature evolved independently in each.

Other types of information have also long been important in classifying organisms. For example, fossils are taken into account whenever they are available. Various stages in the life cycle and in the patterns of embryonic development are also examined (Figure 23–5). As you will see in Chapters 26 and 27, some major decisions concerning the relatedness of animal groups are based on similarities in early development.

(a)

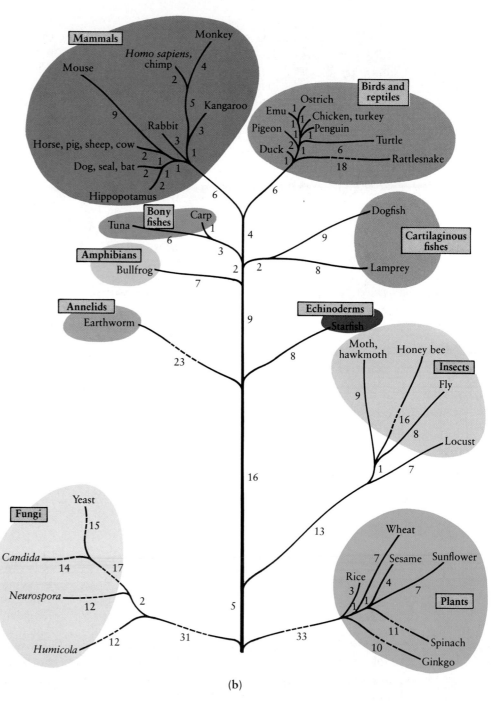

(b)

23–6 *The use of amino acid sequences of homologous proteins to determine evolutionary relationships. This method assumes that the greater the number of amino acid differences between a protein in any two organisms, the more distant their evolutionary relationship. Conversely, the smaller the number of differences, the closer their relationship.*

One of the most frequently sequenced

proteins is cytochrome c, one of the carriers of the electron transport chain (page 151). (a) In cytochrome c molecules from the more than 60 species that have been studied thus far, 27 of the amino acids are identical (dark blue-green). (b) The main branches of an evolutionary tree based on comparisons of amino acid sequences of cytochrome c molecules. The numbers indicate the number of amino acids by which each cytochrome

c differs from the cytochrome c at the nearest branch point. Dashes indicate that a line has been shortened and is therefore not to scale. Although based on comparisons of a single type of protein molecule, this tree is in fairly good agreement with evolutionary trees constructed by more conventional means, which take into account a variety of data.

Biochemical information is another, increasingly important tool in evolutionary classification. Biochemical studies can reveal, for example, similarities and differences in enzymes, reaction pathways, hormones, membrane receptors, and important structural molecules. With the development of techniques for sequencing the amino acids in proteins and the nucleotides in DNA and RNA molecules, it has become possible to compare different organisms at the most basic level of all—the gene. The results obtained by such methods (for example, Figure 23–6) conform well, but not perfectly, with classifications obtained by more traditional methods. Another recent development is the use of nucleic acid hybridization (page 259) to compare the DNA molecules of different species. Unlike sequencing studies, which are generally limited to a few proteins or genes, hybridization techniques make possible the comparison of a much larger proportion of the total genetic endowment.

Although the new molecular techniques are providing important information for use in classifying multicellular organisms (see essay), their greatest value may well be in the classification of one-celled organisms. Many of these organisms are extremely difficult to distinguish on the basis of structural features, and it is only in the last few years that biochemical data have made it possible to begin assigning them to taxa that reflect their evolutionary relationships.

A Question of Kingdoms

Until fairly recently, most biologists recognized only two kingdoms of organisms. Kingdom Animalia included those organisms that moved and ate things, and whose bodies grew to a certain size and then stopped growing. Kingdom Plantae comprised all living things that did not move or eat and that grew indefinitely. Thus the fungi, algae, and bacteria were grouped with the plants, and the protozoa—the one-celled organisms that moved and ate—were classified with the animals.

With the application of new techniques for the study of cell structure and biochemistry, a wealth of new data on the differences and similarities among organisms has accumulated. As a result, the number of groups recognized as constituting different kingdoms has increased. The new techniques revealed, for example, the fundamental differences between prokaryotic and eukaryotic cells—differences sufficiently great to warrant placing the prokaryotes in a separate kingdom, Monera.

Other studies have provided new information about the evolutionary history of the major types of organisms. As we shall see in the next chapter, there is strong evidence that different lineages of eukaryotes arose independently from different prokaryotic ancestors. Moreover, different lineages of unicellular eukaryotes appear to have given rise to the multicellular plants, fungi, and animals, and, at least in the case of photosynthetic organisms, multicellularity has arisen several times. This history makes it impossible, on the basis of current knowledge, to establish kingdoms that reflect both evolutionary history and the similarities and differences among major groups of living organisms.

Most contemporary proposals concerning kingdoms are based not on evolutionary history, but rather on the cellular organization and the mode of nutrition of the organisms. The proposal we shall follow recommends five kingdoms: Monera, Protista, Fungi, Plantae, and Animalia (Figure 23–7). In Appendix B, the major groups of organisms are classified in these five kingdoms.

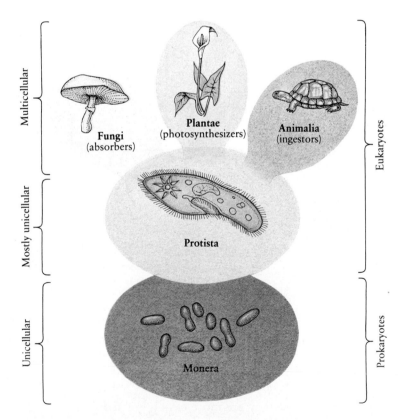

23–7   *The five kingdoms of living organisms. The organisms of kingdom Monera are all prokaryotes, whereas the organisms of the other four kingdoms are all eukaryotes. In the chapters that follow, we shall examine the diversity of organisms in each of these kingdoms.*

The Riddle of the Giant Panda

Since 1869, when the giant panda of China was discovered by the French missionary Père David, its true identity has been a riddle. It was initially classified as a member of the bear family, but biologists almost immediately began to wonder if it might instead be most closely related to another unusual mammal indigenous to China, the lesser panda. The lesser panda was clearly a member of the raccoon family, even though there were no other living members of that family in the Old World—unless, perhaps, the giant panda was also a raccoon, albeit a very odd one.

The two pandas share many unusual anatomical and behavioral characteristics, and although the giant panda does resemble the bears in certain features, it differs significantly in others. Over the years, biologists debated the question, opting in roughly equal numbers for "bear" or "raccoon." The answer seemed to be at hand in 1964 with the publication of a detailed anatomical study of the giant panda. This study demonstrated, to the satisfaction of most biologists, that the giant panda is a bear and that the features in which it resembles the lesser panda are adaptations to feeding on its exclusive energy source, bamboo.

This conclusion has now been confirmed by the application of four different techniques: DNA-DNA hybridization, comparison of the charge and size of homologous protein molecules (as determined by electrophoresis), comparison of the antibody-binding properties of homologous proteins, and detailed study of the banding patterns of the chromosomes. Each procedure provided the same answer: the giant panda is a bear, as children the world over have known all along.

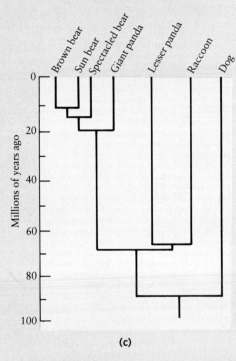

(a)　　　　　　　　　　　(b)　　　　　　　　　　　(c)

Although both native to China and sharing many unusual anatomical and behavioral adaptations, the lesser panda (a) and the giant panda (b) are not each other's closest rela- *tives. The results of DNA-DNA hybridization studies (c), as well as studies using other techniques, have demonstrated that the lesser* *panda is most closely related to the raccoons, whereas the giant panda is most closely related to the bears.*

The members of kingdom Monera, the prokaryotes, are identified on the basis of their unique cellular organization and biochemistry. Members of the kingdom Protista are eukaryotes, both autotrophs and heterotrophs, and most are unicellular. A few groups of relatively simple multicellular organisms are also included in this kingdom, because they are more closely related to unicellular forms than to superficially similar fungi, plants, or animals. All the other multicellular eukaryotes are divided into three kingdoms, based primarily on their mode of nutrition: fungi absorb organic molecules from the surrounding medium, plants manufac-

408

(a)   5 μm

(b)

(c)

(d)

(e)

23–8   *Representatives of the five king-doms.* (a) *Monera.* Cells of the bacterium Lactobacillus bulgaricus, *isolated from a yo-gurt culture. This bacterium produces lactic acid by the fermentation (page 154) of car-bohydrates. The lactic acid, in turn, acts as a preservative and is responsible for the char-acteristic flavor of cheese, yogurt, and other fermented milk products.* (b) *Protista. Two cells of* Paramecium *that are just about to* complete cell division. Before dividing, the original Paramecium *had been feasting on yeast cells, which were stained red to give them greater visibility.* (c) *Fungi. An inky cap mushroom. Fungi are characterized by a multicellular underground network and also by spores and sporangia, which, in mush-rooms, are borne in these familiar structures.* (d) *Plantae. Wildflowers in an alpine meadow in Mount Ranier National Park.*

Flowers, attractive to pollinators, are among the principal reasons for the evolutionary success of the plants of division Anthophyta. (e) *Animalia. A European grass snake* (Natrix natrix) *that has just captured a frog. A nervous system that makes possible coor-dinated and, often, very rapid movements—such as those required to catch a frog—is a chief characteristic of the animal kingdom.*

Table 23–2   Characteristics of the Five Kingdoms

	Prokaryotic	Eukaryotic			
	Monera	Protista	Fungi	Plantae	Animalia
Nuclear envelope	Absent	Present	Present	Present	Present
Mitochondria	Absent	Present	Present	Present	Present
Chloroplasts	Absent (photosynthetic membranes in some types)	Present (some forms)	Absent	Present	Absent
Cell wall	Noncellulose (polysaccharide plus amino acids)	Present in some forms, various types	Chitin and other noncellulose polysaccharides	Cellulose and other polysaccharides	Absent
Means of genetic recombination	Conjugation, transduction, transformation, or none	Fertilization and meiosis, conjugation, or none	Fertilization and meiosis, or none	Fertilization and meiosis	Fertilization and meiosis
Mode of nutrition	Autotrophic (chemosynthetic or photosynthetic) or heterotrophic	Photosynthetic or heterotrophic, or combination of these	Heterotrophic, by absorption	Photosynthetic	Heterotrophic, by ingestion
Motility	Bacterial flagella, gliding, or nonmotile	9 + 2 cilia and flagella, amoeboid, or contractile fibrils	Nonmotile	9 + 2 cilia and flagella in gametes of some forms, none in most forms	9 + 2 cilia and flagella, contractile fibrils
Multicellularity	Absent	Absent in most forms	Present	Present	Present
Nervous system	Absent	Primitive mechanisms for conducting stimuli in some forms	Absent	Absent	Present, often complex

ture them by photosynthesis, and animals ingest them in the form of other organisms. These three groups of organisms have distinct ecological roles—plants are generally producers, animals are consumers, and fungi are decomposers. Table 23–2 summarizes some of the essential similarities and differences among these five kingdoms of organisms. We shall discuss each group in turn in the chapters that follow.

Note that viruses, those ubiquitous segments of DNA or RNA enveloped in a protein capsid, are not included in this classification scheme and will not be discussed in these chapters. As we noted earlier (page 290), many biologists believe that viruses should not be regarded as living organisms at all but rather as "bits of heredity" that have set up a partially independent existence. In accord with that view, we considered viruses in Chapter 17, in the context of their genetics. Our focus here will be on the organisms of the five kingdoms.

Summary

More than 2 million different kinds of organisms are known to inhabit planet Earth, and many times that number await discovery. Scientists have sought order in this vast diversity of living things by classifying them, that is, by grouping them in meaningful ways.

The basic unit of classification is the species, most easily defined as an interbreeding group of organisms that does not breed with other groups of organisms. Species are named by a binomial system that includes the name of the genus (written first) and a descriptive word, the specific epithet, that identifies the particular species within the genus.

Biological classification is hierarchical. Genera are groups of similar species, presumably species that have recently diverged. Genera are grouped into families, families into orders, orders into classes, classes into phyla or divisions, and divisions or phyla into kingdoms. The grouping of divisions or phyla into kingdoms is based on cellular organization and mode of nutrition. The classification system followed in this text consists of five kingdoms: Monera, Protista, Fungi, Plantae, and Animalia.

In classifying organisms in the categories of genus through phylum or division, biologists seek to group the organisms in ways that reflect both their similarities and differences and their evolutionary history. A major principle of such classification is that the similarities taken into account should be homologous—that is, the result of common ancestry rather than of adaptation to similar environments (analogous).

Among the types of data used in classification are structural and biochemical characteristics, fossil evidence, stages in the life cycle, and patterns of embryonic development. In addition, new techniques in molecular biology are providing comparisons of organisms at the most basic level of all, the gene. Amino acid sequencing, nucleotide sequencing of DNA and RNA molecules, and DNA-DNA hybridization are all making valuable contributions to more accurate classification schemes and, most important, to our understanding of organisms and their evolutionary history.

Questions

1. Distinguish between the following: category/taxon; genus/species; division/phylum; homology/analogy.

2. Identify which of the following are categories and which are taxa: undergraduates; the faculty of the University of New Mexico; the Washington Redskins; major league baseball teams; the U.S. Marine Corps; Mozart's symphonies.

3. The use of the terms "division" and "phylum" for the same category in biological classification is an accident of history, resulting from the human tendency to place different subjects of study in separate compartments. What compartmentalization produced these two terms?

4. The differences in the nucleotide and amino acid se-

quences of different lineages of organisms are the result of the accumulation of random mutations. Based on your knowledge of DNA structure, the genetic code, and protein structure, what sorts of random mutations would you expect to persist in a lineage of organisms, generation after generation, unaffected by natural selection? What sorts of mutations would you expect to be harmful to the organisms and thus suppressed by natural selection?

5. It is generally thought that similarities and differences in homologous nucleotide sequences provide a more sensitive indicator of the evolutionary distance between organisms than similarities and differences in the amino acid sequences of homologous proteins. Explain why this should be so.

6. What are the major identifying characteristics of each of the five kingdoms?

7. Homologies among organisms constitute strong evidence that evolution has occurred. As we have seen in earlier chapters, biochemical analyses and electron microscopy have uncovered many homologies that were unknown in Darwin's time. Name five.

Suggestions for Further Reading

Barnes, R. S. K.: *A Synoptic Classification of Living Organisms*, Blackwell Scientific Publications, Boston, 1984.

> *A concise presentation of the diversity of life on Earth, organized according to a five-kingdom system. A comparison of the decisions made in this book, which incorporates the judgment of many British biologists, with those made by Margulis and Schwartz (see below) reveals many of the current disagreements about the classification of particular groups of organisms.*

Kessel, R. G., and C. Y. Shih: *Scanning Electron Microscopy in Biology: A Students' Atlas on Biological Organization*, Springer-Verlag, New York, 1976.

> *An atlas filled with marvelous scanning electron micrographs of specialized structures of prokaryotes, protists, fungi, plants, and animals, as well as of whole organisms.*

Lowenstein, Jerold M.: "Molecular Approaches to the Identification of Species," *American Scientist*, vol. 73, pages 541–547, 1985.

Margulis, Lynn, and Karlene V. Schwartz: *Five Kingdoms: An Illustrated Guide to the Phyla of Life on Earth*, 2d ed., W. H. Freeman and Company, New York, 1987.*

> *A concise presentation of the diversity of life on Earth, organized according to a five-kingdom system. Although many biologists disagree with some of the decisions made by the authors, this book contains a wealth of fascinating information about organisms and their life styles, coupled with outstanding micrographs and diagrams.*

May, Robert M.: "How Many Species Inhabit the Earth?" *Scientific American*, October 1992, pages 42–48.

Mayr, Ernst: "Biological Classification: Toward a Synthesis of Opposing Methodologies," *Science*, vol. 214, pages 510–516, 1981.

Mayr, Ernst: *The Growth of Biological Thought: Diversity, Evolution, and Inheritance*, Harvard University Press, Cambridge, Mass., 1982.*

> *This is the first of two volumes on the history of biology and its major ideas, written by one of the leading figures in the study of evolution. Following an outstanding introductory analysis of the philosophy and methodology of the biological sciences, the first major section of the book provides a thorough treatment of the history and current status of taxonomy.*

O'Brien, Stephen J.: "The Ancestry of the Giant Panda," *Scientific American*, November 1987, pages 102–107.

Perry, Donald: *Life Above the Jungle Floor*, Simon and Schuster, New York, 1986.*

> *Many, and perhaps most, of Earth's undiscovered organisms reside high in the trees of tropical forests. In a lively book, Perry, who devised the climbing and suspension apparatus that is making possible the exploration of this life zone, describes his adventures in the treetops.*

Schaller, George B.: *The Last Panda*, The University of Chicago Press, Chicago, 1993.

> *A vivid account of the five years that Schaller and his wife spent in Sichuan province studying the giant panda. Their research, a joint project of the Chinese government and the World Wildlife Fund, was marked by many wonderful discoveries as well as by numerous bureaucratic frustrations, as they worked to devise conservation plans to protect the panda. Unfortunately, their proposals have not been implemented, and the outlook for the panda, of which fewer than 1,000 animals survive in the wild, remains grim.*

* Available in paperback.

Sibley, Charles G., and Jon E. Ahlquist: "Reconstructing Bird Phylogeny by Comparing DNA's," *Scientific American,* February 1986, pages 82–92.

Terborgh, John: *Diversity and the Tropical Rain Forest,* W. H. Freeman and Company, New York, 1992.

This beautifully illustrated volume explores the incredible diversity of organisms found in the tropical forests, as well as the complex interactions of ecology and evolution that have given rise to that diversity. The author, who is the director of Duke University's Center for Tropical Conservation, also discusses current scientific efforts to counteract the devastating effects of the destruction of the tropical forests.

Wilson, Edward O.: *The Diversity of Life,* Harvard University Press, Cambridge, Mass., 1992.

A presentation of the diversity of life and of the evolutionary processes that have produced it, by one of the world's leading authorities on both diversity and evolution. Filled with marvelous stories of many fascinating organisms, Wilson's text is also infused with examples of how rapidly species are being driven to extinction throughout the world and the obstacles that must be overcome if the loss of diversity is to be stemmed.

Wilson, Edward O.: "Threats to Biodiversity," *Scientific American,* September 1989, pages 108–116.

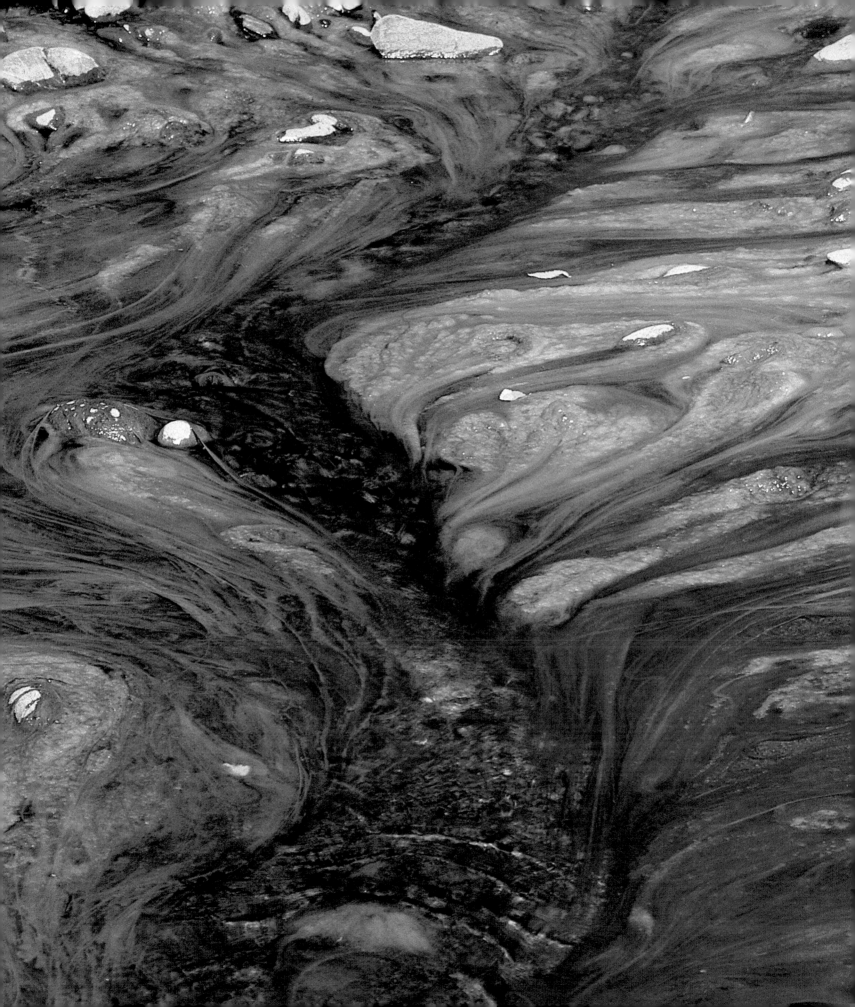

For four billion years, the Earth's surface was bare. To an observer from outer space, it would have looked much like the surface of the moon today. Yet during those four billion years, the most important events in the evolution of life took place.

The first, of course, was the origin of life itself—the appearance of the earliest self-contained, self-replicating units (cells). There are two astonishing facts concerning this event. One is that it took place very early in the Earth's history. The other is that it took place only once, or, alternatively, only one of the earliest beginnings was successful. All things living on Earth today are related to one another and to those earliest successful cells.

The next milestone in life's history was the evolution, some 3.5 billion years ago, of cells with complex pigment systems that could trap the sun's rays and convert their radiant energy to chemical energy. These photosynthetic prokaryotes and the heterotrophs that depended on them were (and are) highly successful organisms. They survived and prevailed as the only life forms on Earth for 2.5 billion years, more than half of our planet's existence. The first eukaryotes, according to the fossil record, made their appearance some 1.5 billion years ago. They have since dominated the planet in mass, though not in numbers and versatility.

Photosynthesis had consequences. The most common photosynthetic mechanism splits water and releases oxygen. This oxygen slowly changed the atmosphere of the planet and the forms of life that inhabit it. By 450 million years ago, a layer of ozone, produced by the action of ultraviolet light on the accumulating oxygen, had formed in the upper atmosphere. This layer, which absorbs large quantities of ultraviolet light, shields the surface of the Earth from those high-energy radiations. Its formation enabled cells to survive near the surface of the waters and prepared the way for the invasion of the land.

In this new environment, photosynthetic eukaryotes, such as the green algae in this flowing stream, crowded the ocean shores, pushed their way up onto the continents, and began the transition to land.

The Prokaryotes and the Protists

In this chapter, we shall consider the two kingdoms that contain principally single-celled organisms: **Monera,** which includes all of the prokaryotes, and **Protista,** which includes the single-celled eukaryotes and some relatively simple multicellular eukaryotes.

As we have seen in previous chapters, prokaryotic and eukaryotic cells are fundamentally distinct. Prokaryotes are by far the simplest of all cells. The prokaryotic cell contains a long, circular chromosome of DNA and, often, smaller molecules of DNA known as plasmids (page 288). The genetic material is not isolated from the rest of the cell by a membrane, and cell division is not accompanied by mitosis. The cytoplasm contains ribosomes but not complex membrane-bound organelles, and it is surrounded by a plasma membrane and a cell wall. This wall, which is a combination of unique polysaccharides and proteins, is different in composition from the cell walls of organisms in any of the other kingdoms.

In eukaryotic cells, by contrast, the DNA is combined with histone proteins (page 275), organized into a number of distinct chromosomes, and enclosed within a nuclear envelope. Eukaryotic cells contain an elaborate network of endoplasmic reticulum and numerous membrane-bound organelles. Many eukaryotic cells have cilia or flagella with the characteristic $9 + 2$ structure (page 103).

The greater complexity of the eukaryotic cell brings with it a number of advantages that ultimately made possible the evolution of multicellular organisms. The eukaryotic cell is capable of carrying vastly more genetic information than the prokaryotic cell—enough, for example, to specify an oak tree or a human being. Because of the compartmentalization of functions by membranes, eukaryotic cells are more efficient metabolically and can be larger. Although a larger cell requires more energy, it can obtain it more easily, either because of an increased surface area for photosynthesis or the absorption of food molecules or because of a greater ability to catch and subdue prey organisms.

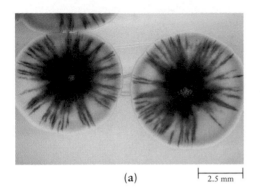

(a) |— 2.5 mm —|

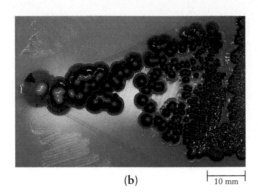

(b) |— 10 mm —|

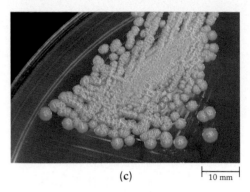

(c) |— 10 mm —|

24–1   *One method of identifying bacteria is by the appearance of their colonies. The colonies are produced when a sample of a bacterial culture is streaked across a solidi-* *fied agar surface and allowed to incubate overnight or longer. Shown here are colonies of* (a) Chromobacterium violaceum, (b) Ser- *ratia marcescens, and* (c) Pseudomonas cepa- *cia. The bright colors result from the presence of pigments synthesized by these particular bacterial cells.*

Moreover, larger cells are often better able to make necessary adjustments to potentially life-threatening environmental changes, for example, in temperature or available water.

The Prokaryotes

The prokaryotes are, in evolutionary terms, the oldest group of organisms on Earth. And, despite their relative simplicity, contemporary prokaryotes are the most abundant organisms in the world. Although there are sometimes difficulties in defining prokaryotic species, about 2,700 distinct species are currently recognized. Prokaryotes are the smallest cellular organisms; a single gram (about 1/28 of an ounce) of fertile soil can contain as many as 2.5 billion individuals.

The success of the prokaryotes is undoubtedly due to their great metabolic diversity and their rapid rate of cell division. Under optimum conditions, a population of *Escherichia coli* can double in size every 20 minutes. Prokaryotes can survive in many environments that can support no other form of life. They have been found in the icy wastes of Antarctica, the dark depths of the oceans, and even in the near-boiling waters of natural hot springs. When conditions are unfavorable, many kinds can take the form of hard, resistant spores, which may lie dormant for years until conditions become more favorable for growth.

The Classification of Prokaryotes

Until quite recently, the possibility of classifying the prokaryotes on an evolutionary basis seemed remote. Most of the characteristics used to determine evolutionary relationships among eukaryotes—for example, in-

tricate structural features and complex patterns of reproduction, development, and growth—simply do not exist in prokaryotes. Biologists studying the prokaryotes were forced to rely upon such features as the size and shape of individual cells, the general appearance of the colonies formed by some prokaryotes (Figure 24–1), mode of nutrition, the presence or absence of spores, and the characteristics of the cell wall, as revealed by its ability to retain certain chemical dyes. Many of these characteristics do not reflect evolutionary relationships, and classifications derived from them are not hierarchical. Nonetheless, such classifications are of immense practical importance in medical and veterinary diagnosis and treatment, agriculture, and industrial microbiology.

In recent years, detailed studies of cell structure and biochemistry have begun to unravel the evolutionary relationships of the prokaryotes. Crucial breakthroughs have come with the development of techniques for amino acid sequencing, nucleotide sequencing, and nucleic acid hybridization. These techniques are leading to a revolution in our understanding of the relationships among the different groups of prokaryotes.

One of the most striking discoveries thus far has been that a few types of prokaryotes, not previously classified together, differ very dramatically from all other types of cells in the details of their metabolic pathways. Some of these prokaryotes live in swamps and synthesize methane from carbon dioxide and hydrogen gas, others live in very salty environments (Figure 24–2), and still others are found in acidic environments at high temperatures.

These unusual bacteria, which would have been excellently suited to the conditions prevailing on Earth during the earliest stages of biological evolution, also

24–2   *These evaporating ponds of sea water, seen in an aerial photograph taken near San Francisco Bay, are filled with salt-loving (halophilic) bacteria. The evaporation of sea water yields table salt, as well as other salts of commercial value. As the water evaporates and the salt concentration increases, the halophilic bacteria multiply, forming massive "blooms" that turn the water red or reddish-purple.*

have transfer RNAs and ribosomal RNAs with nucleotide sequences markedly different from those in all other organisms. It appears that they may be the few surviving representatives of an evolutionary lineage that diverged very early from the lineage that gave rise to all the other prokaryotes. It has been proposed that they be placed in a new kingdom, **Archaebacteria.** In this proposal, all of the other prokaryotes, collectively known as the **Eubacteria,** would remain in kingdom Monera.

At the present time, molecular studies of the prokaryotes are proceeding rapidly in a number of laboratories, and the probable evolutionary relationships of the different groups are being worked out. An overview of the emerging evolutionary classification of the prokaryotes is shown in Figure 24–3.

Modes of Nutrition

Prokaryotes exhibit more metabolic diversity than all other organisms put together, obtaining energy in a great variety of ways. Most bacteria are heterotrophs, and some of them obtain their energy-rich organic molecules from the tissues or body fluids of other living

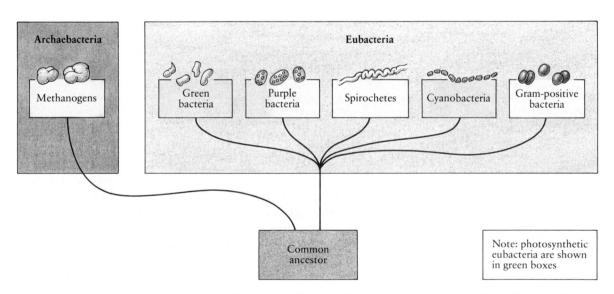

24–3   *The evolutionary relationships of the principal groups of prokaryotes. The archaebacteria include the methanogens (Figure 4–5, page 76), the salt-loving bacteria (Figure 24–2), and certain bacteria that thrive in high-temperature acidic environments. Among the eubacteria, the cyanobacteria, the green bacteria, and many of the*

purple bacteria are photosynthetic. Closely related to the purple bacteria are many familiar nonphotosynthetic forms, including E. coli *and the nitrogen-fixing bacteria of genus* Rhizobium, *which live in close association with the roots of plants. Among the spirochetes are a number of pathogens, including the organisms that cause syphilis*

and Lyme disease. The gram-positive bacteria are distinguished from the other types on the basis of the staining properties of their cell walls. They include a number of economically important species, such as the lactic acid bacteria (Figure 23–8a, page 409) that are used in the production of yogurt and other fermented foods.

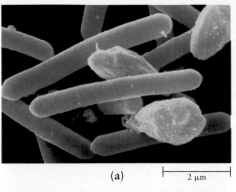

(a)   2 μm

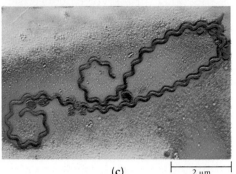

(c)   2 μm

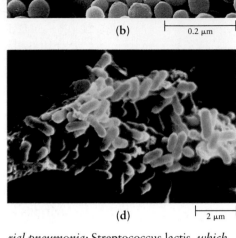

(b)   0.2 μm

(d)   2 μm

(e)   0.25 mm

24–4   *Some heterotrophic prokaryotes.*
(a) Clostridium botulinum, *the source of the toxin that causes the deadly food poisoning botulism, is a rod-shaped bacterium. The saclike structures are dormant spores, which can be killed only by boiling in the presence of oxygen. Other rod-shaped bacteria include the organisms that cause lockjaw* (Clostridium tetani), *diphtheria* (Corynebacterium diphtheriae), *and tuberculosis* (Mycobacterium tuberculosis), *as well as the familiar* E. coli.

(b) *Many bacteria, such as* Micrococcus luteus, *shown here, take the shape of spheres. Among the spherical bacteria, which may form pairs, clusters, or chains, are* Streptococcus pneumoniae, *a cause of bacte-*

rial pneumonia; Streptococcus lactis, *which is used in the commercial production of cheese; and* Nitrosococcus, *soil bacteria that oxidize ammonia to nitrates.*

(c) *A single spirochete of the genus* Leptospira. *Spirochetes of this genus infect many wild animals and can be transmitted to humans by rats, causing a disease known as leptospirosis. The spirochetes range in size up to 500 micrometers long, which is an enormous size for prokaryotes. They move by means of a rapid spinning or whirling of the cell about its long axis.*

(d) Rickettsiae *are the smallest known cells. Typhus is caused by* Rickettsia prowazekii, *shown here. It is transmitted by human body lice, and under crowded condi-*

tions, large numbers of people can be infected in a very short time. More human lives have been taken by rickettsial diseases than by any other infection except malaria. At the siege of Granada in 1489, 17,000 Spanish soldiers were killed by typhus, but only 3,000 in combat.

(e) *A fruiting body of* Chondromyces crocatus, *a myxobacterium. The myxobacteria secrete slime tracks, along which the cells glide. Each fruiting body produces enormous numbers of spores, which are contained in clustered sacs branching off from a central stalk. Although most bacteria reproduce by simple cell division, virtually all exhibit extremely rapid rates of reproduction.*

organisms. The disease-causing (pathogenic) bacteria belong to this group, as do a number of nonpathogenic forms. Some of these bacteria have little effect on their hosts, and some are actually beneficial. Cows and other ruminants can utilize cellulose only because their stomachs contain bacteria and certain protozoans that have cellulose-digesting enzymes.

By far the largest numbers of heterotrophic bacteria, however, are **saprobes** (from the Greek, *sapros,* "rotten" or "putrid"), feeding on dead organic matter. Bacteria and other microorganisms are responsible for the decay and recycling of organic material in the soil. Typically, different groups of bacteria play different roles— such as the digestion of cellulose, starches, or other polysaccharides, the hydrolysis of specific peptide bonds,

or the breakdown of amino acids. Because of their high degree of nutritional specialization, bacteria are able to live in large numbers in the same small area with little competition and, indeed, with mutual assistance, as the activities of one group make food molecules available to another group. These combined activities release the nutrients and make them available to plants and, through plants, to animals. Thus, bacteria are an essential part of ecological systems.

Other bacteria are autotrophs; that is, they are able to synthesize energy-rich organic molecules from simple inorganic substances. Some of the autotrophs are chemosynthetic. They obtain their energy from the oxidation of inorganic molecules such as certain compounds of nitrogen, sulfur, and iron. These bacteria are the only

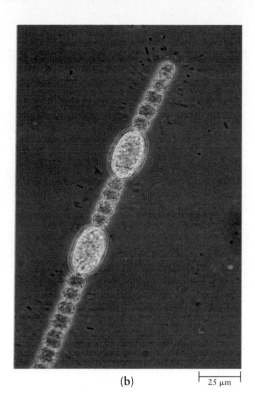

(a)   (b)

50 μm   25 μm

24–5   (a) Oscillatoria, *a filamentous cyanobacterium. The cyanobacteria were formerly regarded as algae, because of the similarity between their photosynthetic processes and those of the eukaryotic algae and the plants. Like the photosynthetic eukaryotes, they contain chlorophyll* a *and produce oxygen during photosynthesis. Biochemical and electron microscope studies, however, have revealed their prokaryotic nature.*

(b) A filament of cells of the freshwater cyanobacterium Anabaena. *When nitrate levels are low, some of the cells—such as the two large cells in this filament—differentiate into specialized cells that do not carry out photosynthesis but instead fix nitrogen into organic compounds. In Southeast Asia, rice can often be grown on the same land continuously without the addition of fertilizers because of nitrogen-fixing cyanobacteria. Most of the cyanobacteria live not in the water but rather in the tissues of a small water fern that grows in the rice paddies.*

organisms able to use inorganic compounds as energy sources.

Among the autotrophic prokaryotes are three distinct types of photosynthetic bacteria: green bacteria, photosynthetic purple bacteria, and cyanobacteria. Like plants, these bacteria capture light energy with the aid of specialized pigments. In the green and purple bacteria, however, the pigments differ from those found in algae or plants. Unlike algae and plants, these two types of photosynthetic bacteria split hydrogen sulfide (H_2S) and other compounds rather than water and do not release oxygen.

The cyanobacteria resemble plants and algae in that they contain chlorophyll *a* and split water during photosynthesis, releasing oxygen (see page 168). In the cyanobacteria, however, chlorophyll and the other photosynthetic pigments (including phycocyanin, from which the group derives its name) are not enclosed in chloroplasts, as they are in eukaryotic cells. Instead, they are part of a membrane system distributed in the peripheral portion of the cell (see Figure 4–8, page 78).

Cyanobacteria grow for the most part in fresh water. They are sometimes found as separate cells, but more often the cells form clusters, threads, or chains (Figure 24–5). Some species are able to incorporate atmospheric nitrogen into organic compounds. These cyanobacteria have the simplest nutritional requirements of any living thing, needing only nitrogen and carbon di-

oxide (which are always present in the atmosphere), a few minerals, and water. In Southeast Asia, such cyanobacteria make the productivity of rice paddies about 10 times greater than that of other types of farmland.

The Origin and Evolution of Eukaryotes

The step from the prokaryotes to the first eukaryotes was one of the major evolutionary transitions, preceded in importance only by the origin of life itself and the evolution of photosynthetic cells. The question of how it came about is a matter of current and lively discussion. One interesting hypothesis, gaining increasing acceptance, is that larger, more complex cells evolved when certain prokaryotes took up residence inside other cells.

About 2.5 billion years ago, oxygen began to accumulate in the atmosphere as a result of the photosynthetic activity of the cyanobacteria. Those prokaryotes that were able to use oxygen in ATP production gained a strong advantage, and such forms began to prosper and increase. Some of these cells evolved into modern forms of aerobic bacteria. Others, according to this hypothesis, moved into larger cells and evolved into mitochondria.

Infectious disease can be caused by bacteria, protists, and fungi, as well as by the detached bits of genetic information that we know as viruses. Most of the time, potentially pathogenic microbes live within their host organisms without apparent ill effect. Disease is generally the result of a sudden change in the microbe, in the host, or in their relationship. For instance, many people harbor small numbers of *Mycobacterium tuberculosis* without any symptoms of disease. However, factors such as malnutrition, fatigue, or other diseases may weaken host defenses so that the signs of tuberculosis appear. Similarly, the herpes simplex viruses that cause fever blisters or genital lesions may remain latent for months or years at a time, with an outbreak occurring only in response to some change in the condition of the host.

The pathogenic effects of microbes are produced in a variety of ways. Viruses, as we saw in Chapter 17, enter particular types of cells and often destroy them. Bacteria may also produce cell destruction. Frequently, however, the effects we recognize as disease are caused not by the direct action of the pathogens but by toxins, or poisons, produced by them. For instance, diphtheria is caused by *Corynebacterium diphtheriae*. The organisms are inhaled and establish infection in the upper respiratory tract, where they produce a powerful toxin that is transported through the bloodstream to body cells. This toxin, which is made only when the bacterium is harboring a particular prophage (page 292), inhibits protein synthesis.

Some diseases are the result of the body's reaction to the pathogen. In pneumonia caused by *Streptococcus pneumoniae,* the infection causes a tremendous outpouring of fluid and cells into the lungs, thus interfering with breathing. The symptoms caused by fungus infections of the skin similarly result from inflammatory responses.

A single pathogen can cause a variety of diseases. Skin infections of *Streptococcus pyogenes* cause the disease known as impetigo. Throat infections by the same bacterium are the familiar strep throat. Throat infections with strains of the bacterium that produce toxins

5 nm

Polysaccharide capsule
Cell wall
Plasma membrane
Cytoplasm

N. meningitidis cell

Meningitis is a potentially life-threatening infection of the membranes covering the brain and spinal cord. Most cases in individuals between the ages of 5 and 40 are caused by the bacterium Neisseria meningitidis. In this electron micrograph of the periphery of an N. meningitidis cell, you can distinguish the plasma membrane, the cell wall, and a fuzzy-appearing polysaccharide capsule. Different strains of N. meningitidis are characterized by polysaccharide capsules of different composition. Scientists at the Walter Reed Army Institute of Research and Rockefeller University have used purified capsule polysaccharides to develop effective vaccines against two of the three major strains of this bacterium.

(again as a result of a bacteriophage) are known as scarlet fever. Among persons with untreated strep throat or scarlet fever, about 0.5 percent develop rheumatic fever, which is characterized by inflammatory changes in the joints, heart, and other tissues, apparently as a result of reactions involving the body's own immune system. Conversely, many agents may cause the same symptoms. The "common cold" can result from infection with any one of a large number of viruses.

About 100 years ago, microbes were recognized as disease agents. This opened the way to control measures, among the most important of which was the introduction of sterile procedures in hospitals. Even more important was the institution of public health measures—for example, the eradication of fleas, lice, mosquitoes, and other agents that carry disease; disposal of sewage and other wastes; protection of public water supplies; pasteurization of milk; and quarantine. Not long ago, for instance, infant mortality during the first years of life was often as high as 50 percent in some localities owing to infant diarrhea caused by contaminated milk and water. In some regions in the developing countries, it still remains quite high.

Many infectious diseases, both bacterial and viral, can be prevented by immunization (to be discussed in Chapter 33). Bacteria, in particular, are also susceptible to antimicrobial drugs, such as sulfa and penicillin, for which the battlefields of World War II were the proving grounds. Penicillin, which is synthesized by the fungus *Penicillium,* was the first known antibiotic—by definition, a chemical that is produced by a living organism and is capable of inhibiting the growth of microorganisms. Many antibiotics are produced by bacteria, and some are formed by fungi. Many, including penicillin, can now be synthesized in the laboratory. Antibiotics and other chemotherapeutic agents are effective because they interfere with some essential process of the pathogen without affecting the cells of the host. Penicillin, for example, blocks a key reaction in the synthesis of the cell wall in many types of bacteria.

With a few exceptions, viruses have been impervious to attack by chemotherapeutic agents. Drugs that successfully disrupt viral replication generally have devastating effects on cellular processes. However, new approaches to the control of both bacterial and viral diseases are likely to emerge from the wealth of information being generated by recombinant DNA studies. Almost as a byproduct, these studies are shedding new light on disease processes, for example, by identifying bacterial and viral genes that affect virulence. If previous patterns of biological history hold, it is reasonable to expect that as molecular biologists learn more about the organisms themselves—and about the genes dictating their properties—medical scientists will apply that knowledge in practical ways.

Several lines of evidence support the idea that mitochondria are descended from specialized bacteria. Mitochondria contain their own DNA, and this DNA is present in a single, circular molecule, like the DNA of bacteria. Many of the enzymes contained in the bacterial plasma membrane are also found in mitochondrial membranes. Mitochondria contain ribosomes that resemble those of bacteria both in their small size and in details of their chemical composition, including some of the nucleotide sequences in their rRNA molecules. Further, mitochondria appear to be produced only by other mitochondria, which divide within their host cell (Figure 24–6). However, to make the situation more complex, there is not nearly enough DNA in mitochondria to code for all the mitochondrial proteins. Host-cell DNA is also required for the synthesis of mitochondrial proteins.

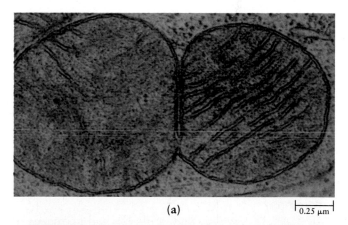

(a)   0.25 μm

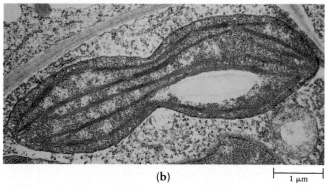

(b)   1 μm

24–6 *It is hypothesized that both mitochondria and chloroplasts originated from prokaryotes that took up residence inside other cells. Both of these organelles contain their own DNA. Moreover, new mitochondria and new chloroplasts are produced only by the replication and division of existing mitochondria and chloroplasts. (a) A dividing mitochondrion. (b) A dividing chloroplast from the leaf of a pea plant* (Pisum sativum). *The large white area is a starch storage body. The smaller white areas, scattered among the photosynthetic membranes, contain the chloroplast DNA.*

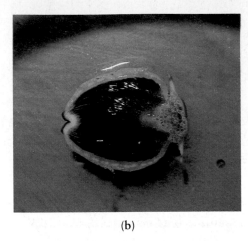

24-7 Plakobranchus, *a marine mollusk. The tissues of this animal contain chloroplasts, which it obtains by eating certain green algae.* (**a**) *Ordinarily, flaplike structures called parapodia are folded over the animal's back, hiding the chloroplast-containing tissues.* (**b**) *When the parapodia are spread apart, however, the deep green tissues become visible. The chloroplasts carry on photosynthesis so efficiently that within a 24-hour cycle of light and darkness some of the animals produce more oxygen than they consume.*

(a) |———| 1 cm   (b)

It is thought that mitochondria are most likely derived from a lineage of purple bacteria in which the capacity for photosynthesis had been lost. Little is known, however, about the original cells in which these bacteria first set up housekeeping—or, indeed, if they actually existed. Contemporary prokaryotes provide few clues about the origins of two key features of all eukaryotic cells—a nuclear envelope and chromosomes containing both DNA and histone proteins. On the basis of nucleotide sequences in rRNA and tRNA molecules, it has been suggested that very early in the history of life a common ancestral lineage gave rise not only to the two lineages (Archaebacteria and Eubacteria) shown in Figure 24–3 (page 417) but also to a third lineage. Members of this third lineage are hypothesized to have been the host cells in the evolution of the eukaryotes.

If such cells existed, they probably had no means of using oxygen for cellular respiration. Thus they were dependent entirely on glycolysis and fermentation, anaerobic processes that are, as we have seen (page 145), relatively inefficient. Cells with oxygen-utilizing respiratory assistants would have been more efficient than those lacking them and so would have reproduced at a faster rate.

In an analogous fashion, photosynthetic prokaryotes ingested by larger, nonphotosynthetic cells are believed to be the forerunners of chloroplasts. Such associations are thought to have occurred independently in a number of lineages, giving rise to the various groups of modern photosynthetic eukaryotes.

This hypothesis accounts for the presence in eukaryotic cells of complex organelles not found in the far simpler prokaryotes. It gains support from the fact that many modern organisms contain bacteria, cyanobacteria, or algae within their cells (Figure 24–7), indicating that such associations are not difficult to establish and maintain. In these associations, whether ancient or modern, the smaller cells gain nutrients and protection, and the larger cells are given a new energy source.

The Protists

An extremely diverse group of eukaryotic organisms are included in the kingdom Protista. This kingdom contains heterotrophs, photosynthetic autotrophs, and a few versatile organisms that are both heterotrophic and photosynthetic. Although most protists are unicellular, they are far from simple. The members of one group, the ciliates, are probably the most complex of all cells, with an astonishing variety of highly specialized structures (Figure 24–8). That single cells should be so complicated is actually not surprising. Among the protists, each cell is a self-sufficient organism, as capable of meeting all the requirements of living as any multicellular plant or animal.

Most of the modern protists probably bear little or no resemblance to the first eukaryotes. As their complexity reminds us, they are not beginnings. They, like the modern prokaryotes, are the end products of millions of years of evolution, and in that time, they have undergone tremendous diversification. Those that have not changed much—amoebas might be an example—have survived over the millennia because they are exquisitely adapted to the relatively unchanging environment in which they live. Biologists generally agree (1) that the modern protists represent a number of quite different evolutionary lineages, and (2) that all other eukaryotic organisms—fungi, plants, and animals—originated from primitive protists, as is shown in Figure 24–9.

The classification of the protists, like that of the prokaryotes, is currently undergoing a major upheaval, as the electron microscope and modern biochemical and molecular techniques make available a vast amount of new information about these organisms. For our purposes, the protists can be divided into three broad groups—the protozoa, the slime molds and water molds, and the algae. Each of these informal groups encompasses a number of phyla or divisions, which may or may not be closely related to one another.

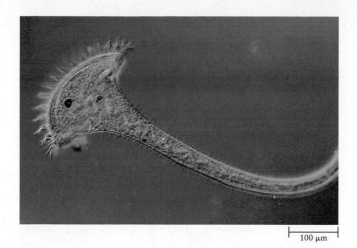

100 µm

24–8 Stentor, *a ciliated protist. Extended, it looks like a trumpet (this genus is named after "bronze-voiced" Stentor of the* Iliad, *"who could cry out in as great a voice as 50 other men"). Stentor is crowned with a wreath of membranelles, brushlike structures formed by cilia that adhere to each other in rows. The membranelles beat in rhythm, creating a powerful vortex that draws edible particles into the funnel-like groove leading to the "mouth." Elastic protein threads run the length of the cell. When contracted, Stentor is an almost perfect sphere.*

Protozoa

The protozoa are a collection of varied and interesting organisms, most of which are unicellular heterotrophs that ingest their food. They usually reproduce asexually, by simple cell division. Many also have sexual cycles, involving meiosis and the fusion of haploid gametes, which results in a diploid ($2n$) zygote. The zygote is often a thick-walled, resistant resting cell, formed during periods of drought or cold. Three phyla of protozoans—Mastigophora, Sarcodina, and Ciliophora—contain both free-living and parasitic forms and are distinguished principally on the basis of their mode of locomotion. Two additional phyla—Opalinida and Sporozoa—are entirely parasitic.

The Mastigophora, which are thought to be the most primitive of the heterotrophic protists, are characterized by their flagella. Smaller cells typically have one or two flagella, and larger cells frequently have many. Some mastigophorans are free-living, but most live in or on other organisms from which they obtain nutrients. Examples of such parasites are *Trypanosoma gambiense,* a flagellate that causes African sleeping sickness; *Trichonympha,* which lives in the digestive tract of termites (Figure 24–10a); and *Giardia lamblia,* the cause of "hiker's diarrhea" (Figure 24–10b).

Plants

Fungi

Animals

Green algae

Brown algae

Red algae

Mastigophorans

Water molds

Sarcodines

Dinoflagellates

Euglenoids

Ciliates

Slime molds

Diatoms

Sporozoans

Protists

Archaebacteria

Eubacteria

Ancestral eukaryote

Prokaryotes

Ancestral prokaryote

24–9 *A simplified scheme of the possible evolutionary relationships among the principal groups of protists. According to current hypotheses, the fungi and the different groups of modern protists evolved separately from single-celled eukaryotes. The origins of the plants have been traced to one group of photosynthetic protists, the green algae. The animals are presumed to have arisen from a single-celled eukaryotic heterotroph, although direct evidence is lacking.*

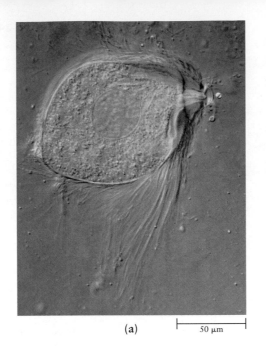

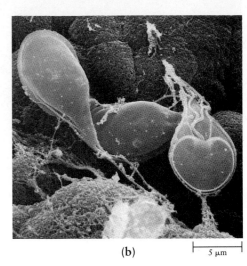

24–10   (a) Trichonympha, *one of the few organisms that can digest cellulose and thus utilize its glucose molecules for energy. This flagellate is entirely responsible for its termite hosts' well-known proficiency at digesting wood.* (b) *Three cells of* Giardia lamblia, *one of the many protists that cause diarrhea in humans. These cells are characterized by four pairs of flagella and an adhesive disk with which they adhere to the lining of the small intestine. Encysted* Giardia *are shed in the feces and can be transmitted to new hosts through contaminated water.*

(a)   50 μm

(b)   5 μm

The Sarcodina include the amoebas and several additional groups of amoeba-like organisms. Although there are similarities among these organisms, their evolutionary relationships are still poorly understood. Generally, the sarcodines move and feed by the formation of pseudopodia, which are extensions of the cytoplasm (see Figure 1–12a, page 33). Some sarcodines, the heliozoans ("sun animals"), resemble pincushions, with fine pseudopodia stiffened by microtubules radiating out from their bodies. Other sarcodines have outer shells, often brightly colored. The white cliffs of Dover and similar chalky deposits throughout the world are the result of the long accumulation of shells from one group of sarcodines, the foraminiferans (Figure 24–11).

The Ciliophora, or ciliates, are the most specialized and complicated of the protozoans. They are characterized by cilia and by the presence of two kinds of nuclei, known as macronuclei and micronuclei. They have a complex process for the exchange of genetic information, in which cells conjugate and haploid micronuclei, formed by meiosis, are exchanged. *Paramecium,* shown in Figure 24–12, is an example of a ciliate.

The Opalinida and the Sporozoa are all parasites. Opalinids, which move by means of cilia, have been found most often in the digestive tracts of frogs and toads, and, occasionally, in fishes and reptiles. Sporozoans are characterized by the absence of cilia or flagella and by complex life cycles (Figure 24–13). The best known sporozoans are the members of the genus *Plasmodium* that cause malaria in many species of birds and mammals. Malaria has generally been controlled by drugs that act on the parasite at various stages of its

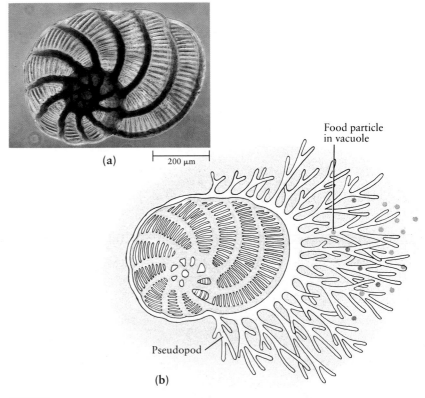

(a)   200 μm

Food particle in vacuole

Pseudopod

(b)

24–11   (a) *The shell of a foraminiferan, formed from calcium carbonate extracted from sea water. In addition to as many as 7,000 living species of foraminiferans, there are about 30,000 extinct species, known only from their fossilized shells.* (b) *In the living foraminiferan, pseudopods extend out through numerous small openings in the shell, as well as through the large opening at the "mouth" of the shell. The exterior surface of the shell is, in effect, coated with living cytoplasm. The pseudopods function in both feeding and locomotion.*

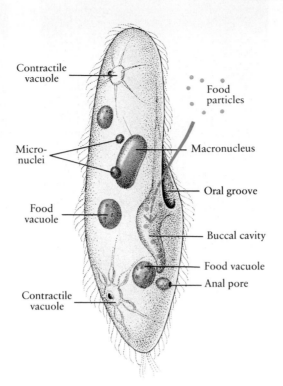

Contractile vacuole

Food particles

Micro-nuclei

Macronucleus

Oral groove

Food vacuole

Buccal cavity

Food vacuole

Anal pore

Contractile vacuole

24–12   Paramecium, *a ciliated protist. The body of this single-celled organism is completely covered by cilia, although only a relative few are shown here. Like other ciliates,* Paramecium *feeds largely on bacteria, smaller microorganisms, and other particulate matter. The beating of specialized cilia drives particles into the buccal cavity ("mouth"), where they are enclosed in food vacuoles that enter the cytoplasm. The food is digested in the vacuoles, and the undigested matter, still in vacuoles, is emptied out through the anal pore. As we saw in Figure 6–5 (page 115), the contractile vacuoles serve to eliminate excess water from the cell.* Paramecium *and other ciliates contain one or more copies of each of two types of nuclei; the much larger macronucleus is thought to contain multiple copies of the DNA present in the micronuclei.*

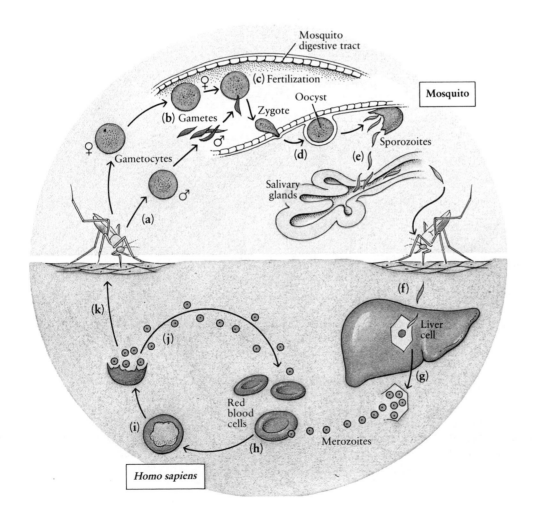

Mosquito digestive tract

(c) Fertilization

Oocyst

Mosquito

(b) Gametes

Zygote

♀

Gametocytes

Sporozoites

♂

Salivary glands

(a)

(d)   (e)

(f)

(k)

Liver cell

(j)

(g)

Red blood cells

(i)

Merozoites

(h)

Homo sapiens

24–13   *Life cycle of* Plasmodium vivax, *one of the sporozoans that cause malaria in humans. The cycle begins (a) when a female* Anopheles *mosquito "bites" a person with malaria, and, along with the blood, sucks up gametocytes of the sporozoan. In the digestive tract of the mosquito, the gametocytes differentiate into gametes (b), which unite (c), forming a zygote. The zygote invades the wall of the digestive tract and develops into a multinucleate structure called an oocyst (d). Within a few days, the oocyst divides into thousands of very small, spindle-shaped cells, the sporozoites. The sporozoites then migrate (e) to the mosquito's salivary glands. When the mosquito "bites" another victim (f), she infects the person with sporozoites. These travel through the bloodstream to liver cells, where they undergo multiple divisions (g). The products of these divisions (merozoites) enter the red blood cells (h), where again they divide repeatedly (i). They break out of the blood cells (j) at regular intervals of about 48 hours, producing the recurring episodes of fever characteristic of the disease and infecting other red blood cells. After a period of asexual reproduction, some of these merozoites become gametocytes (k). If they are ingested by a mosquito at this stage, the cycle begins anew.*

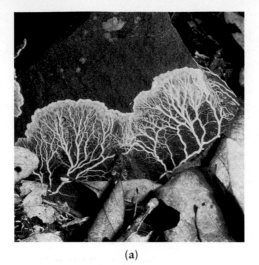

24–14   (a) *The plasmodium—a streaming mass of protoplasm—of a slime mold. Such a plasmodium, with its multiple nuclei, can pass through a piece of silk or filter paper and come out the other side apparently unchanged.* (b) *Sporangia of a plasmodial slime mold on a rotting log. The insect resting on the stalk of one sporangium is a springtail.*

(a)

(b)

life cycle, by insecticides that kill the mosquitoes that transmit it, and by draining the swamps in which the mosquitoes breed. Recently, however, both the parasite and the mosquitoes have begun to develop resistance to the chemicals used to attack them, and malaria remains a major cause of human death and disability. It is estimated that some 200 to 400 million people, principally in tropical regions of the world, are infected with the *Plasmodium* parasite. Recombinant DNA techniques are now being used to identify the genes coding for the surface proteins of the cells at various stages of the life cycle, as well as the regulatory mechanisms governing their expression. The hope is to prepare synthetic versions of key proteins that will function as effective vaccines, stimulating the immune system to attack the parasite.

Slime Molds and Water Molds

The slime molds (divisions Myxomycota and Acrasiomycota) are a group of curious organisms, usually classified with the protists because of their similarity to amoebas (see essay on page 122). However, during some or all of their life cycle, the slime molds are multicellular or, at least, **multinucleate**—that is, the cytoplasm contains many nuclei but is not partitioned by plasma membranes.

Slime molds reproduce by means of spores. A **spore,** as we noted in Chapter 11 (page 192), is a cell that is capable of developing into a new individual without fusion with another cell. The spores of slime molds, like the spores of many other organisms, are produced by meiosis in structures known as **sporangia** (Figure 24–14). Sporangia are often elevated, which facilitates the dispersal of the spores by air currents.

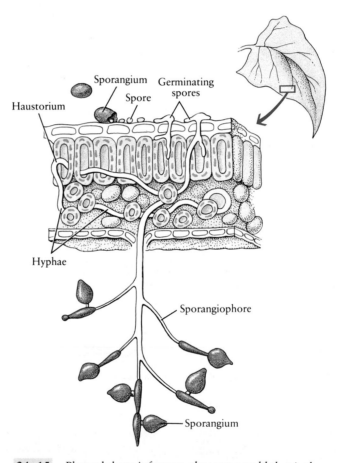

24–15   Phytophthora infestans, *the water mold that is the cause of potato blight* (phytophthora *literally means "plant destroyer"). Infection begins when an airborne sporangium lands on a leaf, releasing spores that move about in the film of water on the leaf's surface. These spores germinate, producing a network of filaments, known as hyphae. Specialized hyphae, called haustoria, penetrate the leaf epidermis and attack the photosynthetic cells in the interior of the leaf. Eventually aerial hyphae—sporangiophores—break through the epidermis, bearing sporangia from which a new generation of asexual spores will be released.*

The water molds (divisions Chytridiomycota and Oomycota) are multinucleate organisms, many of which resemble fungi in their basic structure. Until quite recently they were usually classified with the fungi, but their biochemical characteristics and the presence of flagellated reproductive cells (which do not occur in any of the fungi) indicate that they represent distinct lineages. Most water molds are saprobes, living on dead organic matter. Some, however, are parasites, including *Phytophthora infestans* (Figure 24–15), the cause of the "late blight" of potatoes, which produced the great potato famines in Ireland.

Algae

By the time of the earliest known eukaryotes—about 1.5 billion years ago—a number of distinct lines of simple photosynthetic eukaryotes had already evolved. All modern algae have chlorophyll *a* as their principal photosynthetic pigment, but six different divisions (representing different evolutionary lineages) can be distinguished on the basis of the accessory photosynthetic pigments, the nature of stored food reserves, the presence or absence of a cell wall, and the composition of the cell wall, if present (Table 24–1). Three of these divisions (Euglenophyta, Chrysophyta, and Dinoflagellata) consist almost entirely of unicellular organisms. The other three divisions (Chlorophyta, Phaeophyta, and Rhodophyta) include groups that are multicellular.

Unicellular Algae

The unicellular algae are usually found floating near the surface of the oceans and inland waters, where light is abundant. Each cell is a totally self-sufficient individual, dependent only on light from the sun and carbon dioxide and minerals from the surrounding water. Together with small invertebrate animals and the immature forms of larger animals, the unicellular algae form the **plankton** (from *planktos*, the Greek word for "wanderer"). The photosynthetic members of the plankton community—sometimes called the **phytoplankton**—carry out most of the photosynthesis that takes place in the oceans. They are the ultimate energy source for most of the oceans' other inhabitants and, as a byproduct of their photosynthetic activities, contribute enormous amounts of oxygen to the atmosphere.

Multicellular Algae

Most of the multicellular algae are adapted to living in shallow waters and along the shores. Here the waters

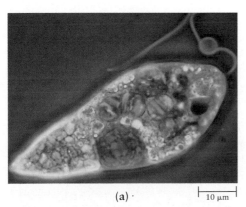

(a) · |— 10 μm

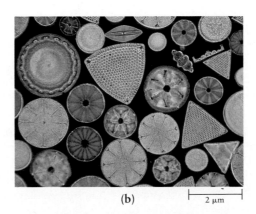

(b) |— 2 μm

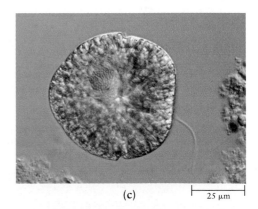

(c) |— 25 μm

24–16 *Representatives of the three divisions of unicellular algae.* (a) Euglena *(division Euglenophyta), one of the most versatile of all one-celled organisms. Containing numerous chloroplasts, it is photosynthetic, but it can also absorb organic nutrients from the surrounding medium and can live without light.* Euglena *is pulled through the water by the whiplike motion of its flagellum.*

(b) *The fossilized shells of a variety of diatoms (division Chrysophyta). Such intricately*
marked, silicon-containing shells characterize most members of this division. The piled-up shells of diatoms, which have collected over millions of years, form the fine, crumbly substance known as "diatomaceous earth." It is used as an abrasive in silver polish, as a filtering material in water treatment plants, and as insulation.

(c) Glenodinium, *a dinoflagellate (division Dinoflagellata). Dinoflagellates move through the water with a spinning motion,*
propelled by their two flagella, one of which is visible here. In dinoflagellates, unlike other eukaryotic cells, the chromosomes are always condensed. If you look closely, you can see them within the nucleus (near the center of the cell). Many dinoflagellates have thick cellulose walls and bizarre shapes, and some are bioluminescent. The infamous red tides, in which thousands of fish die, are caused by great blooms of red dinoflagellates containing a powerful nerve toxin.

Table 24–1   Characteristics of the Algae

Division		Photosynthetic Pigments	Food Reserve	Cell Wall Components	Other Characteristics
Euglenophyta (euglenoids)		Chlorophylls *a* and *b*, carotenoids	Fats and a unique polysaccharide	No cell wall; flexible protein strips inside plasma membrane	Unicellular; mostly freshwater
Chrysophyta (diatoms, golden-brown algae)		Chlorophylls *a* and *c*, carotenoids	Oils and a unique polysaccharide	Cellulose, frequently impregnated with silicon	Unicellular; marine and freshwater.
Dinoflagellata (dinoflagellates)		Chlorophylls *a* and *c*, carotenoids	Starch and oils	Cellulose	Unicellular; mostly marine
Chlorophyta (green algae)		Chlorophylls *a* and *b*, carotenoids	Starch	Polysaccharides, cellulose in some	Unicellular, colonial, or multicellular; freshwater and marine
Phaeophyta (brown algae)		Chlorophylls *a* and *c*, carotenoids	Oils and a unique polysaccharide	Cellulose and algin (a polysaccharide)	Multicellular; almost all marine, flourish in cold ocean waters
Rhodophyta (red algae)		Chlorophyll *a*, carotenoids, phycocyanin	Unusual form of starch	Cellulose, pectin compounds, impregnated with calcium carbonate in some	Multicellular; mostly marine; many species tropical

are usually rich in nutrients, washed down from the land or swept up in currents from the deeper waters. Living conditions are otherwise difficult, however. Along a rocky shore, for instance, multicellular algae—the seaweeds—are subject to great fluctuations of humidity, temperature, salinity, and light; to the pounding of the surf; and to the abrasive action of sand particles churned up by the waves.

One important adaptation found among many coastal seaweeds is the holdfast, a special anchoring structure that adheres to the rocks (Figure 24–17). The upper portions of a multicellular alga characteristically

(a)

(b)

24–17   (a) Laminaria digitata, *a brown alga (division Phaeophyta). In many brown algae, the body is differentiated into a blade, the photosynthetic structure that is exposed to sunlight; a stipe, through which the products of photosynthesis are transported downward; and a holdfast, which anchors the alga to the substrate. Among the most familiar brown algae are the kelps and the rockweeds.*

(b) Unlike the brown algae, many red algae (division Rhodophyta) are made up of branched filaments. In some red algae, the filaments are hooked, enabling them to cling to other algae. Some types of red algae, known as coralline algae, have the capacity to deposit calcium carbonate in their cell walls and play an important role in building coral reefs.

are thin and spread out. They produce enough sugars to nourish the other parts of the organism, which may be 30 meters or more below the surface of the water and overshadowed by the upper portions of the alga. Some of the large brown algae have specialized conducting tissues that transport the products of photosynthesis downward.

Life close to the shore is also crowded and competitive, and different groups of algae have become specialists at exploiting particular areas along tidal shores. Not only have these algae taken on specialized shapes, but they have also developed specialized accessory pigments. Water filters out light, removing first the longer wavelengths, the reds and oranges; at the deeper levels, only a faint blue light penetrates. The red algae capture the energy of this blue light. The brown algae absorb blue-green rays, and the green algae, whose principal pigments are the chlorophylls, make fullest use of the longer-waved red light near the surface. As a consequence of the evolution of specialized pigments, the various multicellular algae use virtually the entire spectrum of light as well as the living space along the rocky shore.

Green Algae: The Increase of Complexity

The green algae (division Chlorophyta), of which *Chlamydomonas* (page 79) is an example, are of particular interest to students of evolution. Multicellular organisms have evolved from unicellular organisms in at least

three different ways, and among the various green algae are not only unicellular and multicellular forms but also a number of intermediate forms. Moreover, the green algae are believed to include the group from which the plants arose. Evidence for this relationship is the fact that both the plants and the green algae—and no other type of algae—contain chlorophylls *a* and *b* and carotenoids as their photosynthetic pigments and store their reserve food as starch. In addition, the cell walls of certain green algae contain cellulose, as do the cell walls of all plants.

One form intermediate between the unicellular and the multicellular involves the association of individual cells in **colonies**. Colonies differ from true multicellular organisms in that, in colonies, the individual cells preserve a high degree of independent function. The cells are often connected by cytoplasmic strands, which integrate the colony sufficiently that it may be regarded as a single organism (Figure 24–18a).

Another type of intermediate form results from repeated nuclear divisions that are not accompanied by a corresponding division of the cytoplasm and the formation of cell walls (Figure 24–18b). True multicellularity is found among other green algae that consist of filaments (Figure 24–18c), sheets, or a three-dimensional body (Figure 24–18d). In these organisms, nuclear division is followed by cytokinesis and the formation of cell walls, but the daughter cells do not separate from each other.

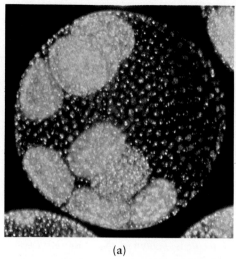

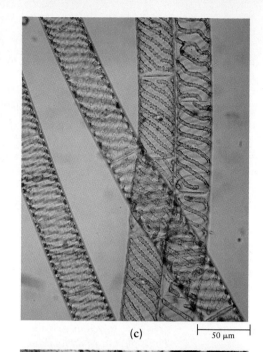

(a)

(b)

(c)

50 μm

24–18   *Among the green algae are unicellular, colonial, multinucleate, and multicellular forms. The different types of green algae represent distinct evolutionary lineages.* (a) Volvox *is a colonial green alga in which there is some specialization of function for reproduction. Hundreds or thousands of single cells, resembling the single-celled alga* Chlamydomonas, *are united by fine strands of cytoplasm into a hollow sphere. New daughter colonies can be seen forming inside the mother colonies.*

(b) Valonia, *which is an enormous single cell—about the size of a hen's egg—has* many nuclei but no partitions separating them.

(c) In Spirogyra, *a freshwater alga, the cells elongate and then divide in one plane only, remaining strung together in long, fine filaments. The chloroplasts form spirals that look like strips of green tape within each cell.*

(d) Ulva, *or sea lettuce, is a marine alga in which the cells divide both longitudinally and laterally, with a single division in the third plane. This produces a broad body that is only two cells thick.*

The green algae also exhibit a diversity of life cycles. The relatively simple life cycle of *Chlamydomonas* (page 191) is typical of many unicellular green algae. A more complex life cycle, characterized by alternation of generations, is found in some multicellular green algae (Figure 24–19). As we saw in Chapter 11 (page 190), in this type of life cycle, a diploid form that reproduces asexually (by spores) alternates with a haploid form that reproduces sexually (by gametes). The diploid, spore-producing form is known as the **sporophyte,** and the haploid, gamete-producing form is known as the **gametophyte.** In some algae, all of the gametes are identical in appearance. In other algae, one type of gamete is nonmotile and usually larger than the other type; the nonmotile gamete is the egg, and the motile gamete is the sperm. This is where sexual differentiation, with its mixed blessings and many complications, all began.

A life cycle involving alternation of generations and gametes that are differentiated into egg and sperm is found in all members of the plant kingdom. Its presence among the green algae provides further evidence of the evolutionary link between the two groups.

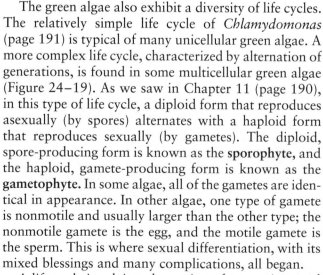

(d)

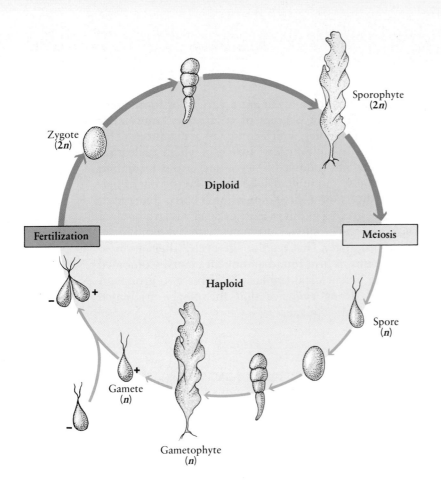

24–19 *In the sea lettuce,* Ulva, *we can see the reproductive pattern known as alternation of generations, in which one generation produces spores, the other gametes. The haploid (n) gametophyte produces haploid gametes, which are identical in appearance. These gametes fuse to form a diploid (2n) zygote. A sporophyte, a multicellular body in which all the cells are diploid, develops from the zygote. The sporophyte produces haploid spores by meiosis. The haploid spores develop into haploid gametophytes, and the cycle begins again.*

In Ulva, *the two generations look alike. In some other species of multicellular algae and in plants, the sporophyte and the gametophyte do not resemble one another and the two types of gametes are dissimilar.*

Summary

Prokaryotes are small, relatively simple cells surrounded by cell walls of unique composition. Their genetic material is in the form of a circular DNA molecule, and they lack a membrane-bound nucleus and organelles. In eukaryotic cells, by contrast, the DNA is associated with proteins and organized into chromosomes. These are contained within a nucleus bounded by a double membrane, the nuclear envelope. Eukaryotic cells also have membrane-bound organelles, including mitochondria and, in the case of photosynthetic cells, chloroplasts. They have cilia and flagella with the characteristic 9 + 2 structure.

The prokaryotes (kingdom Monera) are both the most ancient group of organisms and the most abundant. The application of molecular techniques is enabling biologists to begin determining the evolutionary relationships in this diverse kingdom. Two distinct lineages of prokaryotes, the Archaebacteria and the Eubacteria, have been identified. Prokaryotes are extremely diverse metabolically, including autotrophs (both chemosynthetic and photosynthetic) and heterotrophs. Among the heterotrophs are the pathogenic (disease-causing) bacteria and a variety of highly specialized forms that feed on dead organic matter. These bacteria

are of great importance in recycling nutrients through ecological systems. The photosynthetic green and purple bacteria split hydrogen sulfide and other compounds, instead of water, and do not release oxygen. The cyanobacteria, by contrast, have a type of photosynthesis essentially like that in plants, and some forms can also incorporate atmospheric nitrogen into organic compounds.

It is hypothesized that the first eukaryotes evolved as certain prokaryotes began living inside other prokaryotic cells and, internalized, eventually became specialized as mitochondria and chloroplasts. It is thought that the protists represent a number of distinct evolutionary lineages and, moreover, that all other eukaryotic organisms are derived from primitive protists.

The kingdom Protista comprises an enormous variety of eukaryotic organisms, mostly unicellular with some relatively simple multicellular forms. These organisms can be grouped informally into the protozoa, the slime molds and water molds, and the algae. The protozoa are unicellular heterotrophs that ingest their food; some are free-living and others are parasitic. They are classified into five phyla, principally on the basis of their mode of locomotion. The slime molds are multinucleate or multicellular heterotrophs that resemble amoebas in certain key characteristics. The water molds are multinucleate heterotrophs that superficially resemble fungi but are thought to represent distinct evolutionary lineages.

The algae, simple eukaryotic photosynthetic organisms, are classified into six divisions on the basis of their photosynthetic pigments, reserve food supply, and cell wall composition. Unicellular algae, including euglenoids, diatoms, dinoflagellates, and many green algae, form the photosynthetic component of the plankton, the food source for many aquatic organisms. Multicellular algae, including brown algae, red algae, and some green algae, are the common marine seaweeds.

Among the green algae are a variety of forms, representing different degrees of complexity: unicellular organisms; colonies, which consist of associations of individual cells; multinucleate forms, in which nuclear divisions are not followed by cytokinesis and formation of cell walls; and multicellular forms consisting of filaments, sheets, or a three-dimensional body. Some multicellular green algae have a life cycle in which a haploid form, the gametophyte, alternates with a diploid form, the sporophyte. This type of life cycle, alternation of generations, is also found among all plants. Combined with the biochemical similarities of the two groups, it provides strong evidence that the plants originated among the green algae.

Questions

1. Distinguish between the following: prokaryotes/eukaryotes; Archaebacteria/Eubacteria; pathogen/toxin; protist/protozoan; colonial organism/multicellular organism; spore/gamete; sporophyte/gametophyte.

2. Prokaryotes are considered more primitive than eukaryotes. Does this mean they are all identical to forms of life that existed before eukaryotes arose? Explain your answer.

3. A whimsical biologist has referred to the fruiting bodies of the myxobacteria (Figure 24–4e) as "multicellularity by committee." To which protists could such a description also apply?

4. Many microorganisms produce antibiotics. What do you think their function might be for the organisms that produce them?

5. Name the five phyla of protozoans, and give the distinguishing characteristics of each.

6. Consider the life cycle of *Plasmodium* (Figure 24–13). At what stages in the cycle do its numbers increase? Why might a parasite that requires several hosts find it advantageous to evolve a life cycle in which its numbers increase at several stages? Why might it be advantageous to a parasite to have a second host, such as the mosquito?

7. Describe three different pathways to multicellularity, as exemplified in organisms discussed in this chapter.

8. Explain alternation of generations, using as your example the sea lettuce, *Ulva*.

9. In some classification schemes, the green algae, the brown algae, and the red algae are placed in the plant kingdom. What similarities between the plants and these three divisions of algae might justify such a placement? What differences between the plants and these divisions justify their placement in the kingdom Protista? Which of these similarities and differences are most likely to be homologous and which analogous?

Suggestions for Further Reading

Blakemore, Richard P., and Richard B. Frankel: "Magnetic Navigation in Bacteria," *Scientific American*, December 1981, pages 58–65.

Brock, Thomas D.: "Life at High Temperatures," *Science*, vol. 230, pages 132–138, 1985.

Brock, Thomas D., and Michael T. Madigan: *Biology of Microorganisms*, 6th ed., Prentice-Hall, Inc., Englewood Cliffs, N.J., 1991.

An interesting and often entertaining presentation of microbiology, including algae and protozoa, with an emphasis on the whole cell and its ecology.

Brock, Thomas D. (ed.): *Microorganisms: From Smallpox to Lyme Disease*, W. H. Freeman and Company, 1990.*

A collection of articles from Scientific American, *covering a diversity of pathogenic microorganisms.*

* Available in paperback.

Burnet, Macfarlane, and David O. White: *Natural History of Infectious Disease,* 4th ed., Cambridge University Press, New York, 1972.*

> *A general introduction to the ecology of infectious diseases and their influence on human activities.*

Corliss, John O.: "The Kingdom Protista and Its 45 Phyla," *BioSystems,* vol. 17, pages 87–126, 1984.

Costerton, J. W., G. G. Geesey, and K. J. Cheng: "How Bacteria Stick," *Scientific American,* January 1978, pages 86–95.

Curtis, Helena: *The Marvelous Animals,* Natural History Press, Garden City, N.Y., 1968.

> *An informal introduction to one-celled eukaryotes.*

Ewald, Paul W.: "The Evolution of Virulence," *Scientific American,* April 1993, pages 86–93.

Fischetti, Vincent A.: "Streptococcal M Protein," *Scientific American,* June 1991, pages 58–65.

Fox, G. E., et al.: "The Phylogeny of Prokaryotes," *Science,* vol. 209, pages 457–463, 1980.

Godson, G. Nigel: "Molecular Approaches to Malaria Vaccines," *Scientific American,* May 1985, pages 52–59.

Goodenough, Ursula W.: "Deception by Pathogens," *American Scientist,* vol. 79, pages 344–355, 1991.

Gould, James L., and Carol Grant Gould (eds.): *Life at the Edge,* W. H. Freeman and Company, New York, 1989.*

> *A collection of articles from* Scientific American *on organisms that thrive in extraordinarily difficult environments—such as the frigid waters of the Antarctic or boiling volcanic vents in the ocean depths.*

Kabnick, Karen S., and Debra A. Peattie: "*Giardia:* A Missing Link between Prokaryotes and Eukaryotes," *American Scientist,* vol. 79, pages 34–43, 1991.

Kessin, Richard H., and Michiel M. Van Lookeren Campagne: "The Development of a Social Amoeba," *American Scientist,* vol. 80, pages 556–565, 1992.

Koch, Arthur L.: "Growth and Form of the Bacterial Cell Wall," *American Scientist,* vol. 78, pages 327–341, 1990.

Lee, J. J., S. H. Hutner, and E. C. Bovee (eds.): *An Illustrated Guide to the Protozoa,* Society of Protozoologists, Lawrence, Kansas, 1985.

> *The most recent comprehensive reference on the protozoan protists. This well-illustrated guide to an enormous diversity of fascinating organisms also includes a glossary of terms used by students of the protists.*

Margulis, Lynn: *Symbiosis in Cell Evolution: Microbial Communities in the Archean and Proterozoic Eons,* 2d ed., W. H. Freeman and Company, New York, 1993.*

> *A fascinating discourse on the origin of eukaryotic cells by serial symbiotic events.*

Penny, David: "What Was the First Living Cell?" *Nature,* vol. 331, pages 111–112, 1988.

Rietschel, Ernst Theodor, and Helmut Brade: "Bacterial Endotoxins," *Scientific American,* August 1992, pages 54–61.

Round, F. E.: *The Ecology of Algae,* Cambridge University Press, New York, 1984.*

> *A comprehensive account of the ecology of both freshwater and marine algae.*

Schopf, J. William: "The Evolution of the Earliest Cells," *Scientific American,* September 1978, pages 110–138.

Schwartz, Robert M., and Margaret O. Dayhoff: "Origins of Prokaryotes, Eukaryotes, Mitochondria, and Chloroplasts," *Science,* vol. 199, pages 395–403, 1978.

Shapiro, James A.: "Bacteria as Multicellular Organisms," *Scientific American,* June 1988, pages 82–89.

Vidal, Gonzalo: "The Oldest Eukaryotic Cells," *Scientific American,* February 1984, pages 48–57.

Woese, Carl R.: "Archaebacteria," *Scientific American,* June 1981, pages 98–122.

Yates, George T.: "How Microorganisms Move through Water," *American Scientist,* vol. 74, pages 358–365, 1986.

Zinsser, Hans: *Rats, Lice, and History,* The Atlantic Monthly Press/Little, Brown and Company, Boston, 1984.*

> *A classic popular account of the influence of infectious diseases and their animal vectors on the course of human history, first published in 1935.*

* Available in paperback.

25

The Fungi and the Plants

The transition to land was a remarkable and improbable event. Cells need water. In fact, cells *are* water—as much as 95 percent water. How is it possible to reconcile this most basic requirement of living systems with an existence on the bare, arid surface of the land?

The plant body can be regarded as a blueprint for meeting the challenge of life on the land. Unlike the life forms that preceded them, plants are complex multicellular organisms with tissues that are specialized for different roles. They have roots, which not only anchor them to the substrate (as do the holdfasts of brown algae) but also reach down below the surface, forming a complex moisture-collecting network. Plants also have sophisticated two-way plumbing systems that transport water up from the ground and transport sugars, dissolved in water, to the roots and other nonphotosynthetic parts of the plant. Leaves, the specialized photosynthetic surfaces of plants, are covered with a waxy cuticle that limits the escape of this precious commodity. Finally, plants have a unique molecule, lignin, that stiffens and supports the plant body.

Because of these features, plants—such as these magnificent Western hemlocks and Douglas firs—can reach as high as 250 feet into the air, with massive trunks that are as much as 12 feet in diameter. Among the oldest and largest plants of the North American continent, Douglas firs are the dominant trees of the forests that grow along the western side of the Cascade Mountains in Oregon, Washington, and British Columbia.

Once on the land—a transition that was, as we shall see, aided by the fungi—plants diversified rapidly to fill most of the planet's barren spaces. The great variety of plant communities that developed became the home for everything else that lives on the land. A forest, or a prairie, or a meadow or a marsh is not just plants but includes all the birds and deer and dragonflies and foxes and fungi and turtles—and more organisms than you can name or than even have names—that the plants nourish and provide with shelter and nests and hiding places.

As we saw in the last chapter, the eukaryotic cell, by virtue of its size and complexity, has a number of properties that made possible a great diversification of the unicellular eukaryotes in both structure and mode of life. These properties include the capacity to carry a great deal of genetic information and to transmit it reliably from generation to generation; the compartmentalization and specialization of different parts of the cell for different functions, leading to greater efficiency; the ability to acquire more food; and greater adaptability to life-threatening environmental changes.

There are, however, limits to the size that a single cell can attain and still function efficiently. One critical factor is the surface-to-volume ratio (page 88). The larger the cell, the greater the quantity of materials that must be moved in and out, but this movement can occur only through the cell surface. Another critical factor is the capacity of the nucleus to regulate a large amount of cytoplasm and the diverse functions of a complex cell. These problems can be minimized to some extent in a cell that is flattened or spread out and by the presence of multiple nuclei in a common cytoplasm, solutions that are exemplified most strikingly in the fungi. A more effective solution is multicellularity: the repetition of individual units—cells—each with an efficient surface-to-volume ratio and each with its own nucleus.

Multicellularity makes possible a vast increase in the potential size of organisms, as well as specialization on a broad scale. As we saw in Chapter 16, the selective activation or inactivation of particular genes during development results in cells or groups of cells specialized for particular functions. Although each nucleus carries a complete set of genetic instructions, the number of functions it must control simultaneously in a given cell is limited.

With these advantages, however, come new problems to be solved: the organism must acquire increased amounts of food to supply its numerous cells with energy; larger quantities of materials must be moved into and out of the organism and transported to and from the individual cells; a larger body requires more physical support; the activities of all of the cells must be integrated for the smooth functioning of the entire organism; and a greater length of time is required for the full development of the organism, often requiring protection and nourishment of the immature stages.

As we shall see in this chapter and the next two, the diversity of multicellular organisms represents a variety of natural experiments at solving these problems. Some of the solutions have been of limited success and are utilized by only a few types of organisms. Others have been enormously successful, as measured by the number and variety of organisms employing them.

The Fungi

The fungi are so unlike any other organisms in their biochemical characteristics and reproductive patterns that, although they were long classified with the plants,

biologists now assign them to a separate kingdom. Although some fungi, including the yeasts, are unicellular, most species are multinucleate or multicellular organisms composed of masses of filaments (Figure 25–1).

A fungal filament is called a **hypha,** and all the hyphae of a single organism are collectively called a **mycelium.** The walls of the hyphae contain chitin, a polysaccharide that is never found in plants. (It is, however, the principal component of the exoskeleton—the hard outer covering—of insects and other arthropods.) The visible structures of most fungi (Figure 25–2) represent only a small portion of the organism. These structures, such as mushrooms, are tightly packed hyphae, specialized for reproduction.

All fungi are heterotrophs that obtain food by absorbing dissolved organic molecules from the surrounding environment. Typically a fungus secretes digestive enzymes onto a food source and then absorbs the smaller molecules released by the action of the enzymes. Consequently, fungi live in soil, in water, or in some other medium containing organic substances. Growth is their only form of motility, except for spores, which may travel considerable distances through air or water.

(a)

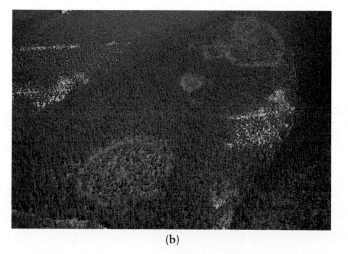

(b)

25–1   (a) *A fungus growing on a fallen tree trunk. The bulk of its body consists of masses of thin filaments through which nutrients are absorbed. In many fungi, the cells are incompletely separated by perforated cell walls, and both cytoplasm and nuclei flow through the filaments. Most fungi are nonmotile and can obtain additional food supplies only by growing. Some fungi can grow a mass of new filaments in 24 hours that, if placed end-to-end, would total more than a kilometer in length.*

(b) The growth of fungal filaments has produced what are believed to be the largest, and possibly the oldest, living organ-

isms on Earth today. This aerial view of a virgin coniferous forest in Montana reveals three large areas, identified by their dull color, in which the roots of the trees are infected by a fungus of the genus Armillaria. The largest area encompasses 8 hectares (about 20 acres). A similar area in the state of Washington is more than 600 hectares (1,500 acres). Genetic studies have demonstrated that, within each infected area, Armillaria filaments are genetically identical—that is, all part of a single individual. One growth of Armillaria in Michigan is estimated to be between 1,500 and 10,000 years old.

Predaceous Fungi

Among the most highly specialized of the fungi are the predaceous fungi that have developed a number of mechanisms for capturing small animals that they use as food. Some secrete a sticky substance on the surface of their hyphae in which passing protists, small insects, or other animals become glued. More than 50 species of fungi capture small roundworms (nematodes) that abound in the soil. In the presence of a population of roundworms (or even of water in which the worms have been growing), the hyphae of the fungi produce loops that swell rapidly, closing the opening like a noose when a nematode rubs against its inner surface. The stimulation of the cell walls is thought to increase the amount of osmotically active material in the cells, causing water to enter the cells and expand them rapidly.

The predaceous fungus shown below, *Arthrobotrys anchonia*, has trapped a nematode. The traps consist of rings, each comprising three cells, which swell rapidly to about three times their original size and garrote the nematode. Once the worm has been trapped, fungal hyphae grow into its body and digest it. When triggered, the ring cells can expand completely in less than a tenth of a second. This species was appropriately called the "nefarious noose fungus" by the late W. H. Weston of Harvard University, who made vast contributions to our present knowledge of the fungi. Another nematode-trapping fungus, *Dactylella drechsleri*, traps the worms with small adhesive knobs and was dubbed the "lethal lollipop fungus" by Weston.

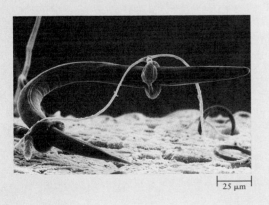

25 μm

Yeasts are of great economic importance in the production of wine, beer, and bread. Other fungi are important in the preparation of cheeses, such as Roquefort and Camembert, and antibiotics, including penicillin. Fungi cause many plant diseases, including chestnut blight and Dutch elm disease, which have had devastating effects on the chestnut and elm populations of North America. Fungi also cause a number of diseases in humans, such as ringworm (including "athlete's foot") and thrush (to which infants are particularly susceptible), and are important as destroyers of food and clothing (especially cotton and leather).

The fungi, together with the bacteria, are the principal decomposers of organic matter. As we shall see in Chapter 45, their activities are as vital to the continued function of the Earth's ecosystems as are those of the food producers.

Reproduction in the Fungi

Fungi are classified into four divisions (Figure 25–3), principally on the basis of their patterns of reproduction. Most fungi reproduce both asexually and sexually.

(a)

(b)

25–2   *Two fungi.* (a) *A common morel,* Morchella esculenta. *These (and the truffles) are among the most prized of the edible fungi.* (b) Amanita muscaria. *Members of the genus* Amanita *include the most beautiful and also the most poisonous of the mushrooms. The spots on the cap, the "skirt" near the top of the stalk, and the cup around the base are among the identifying characteristics of this genus.*

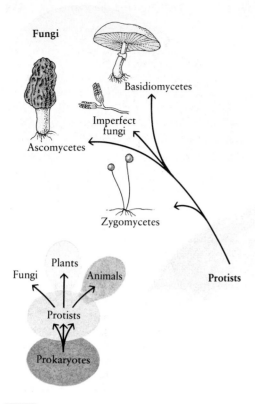

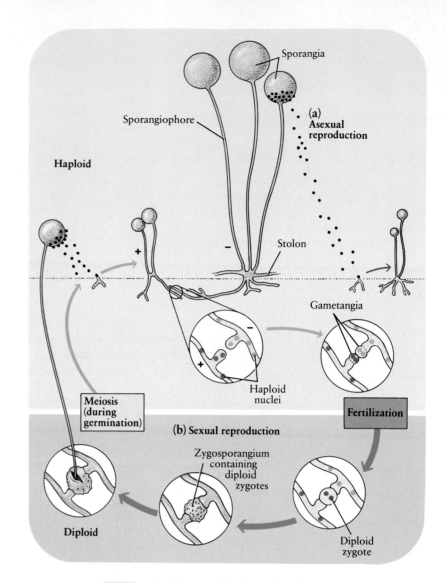

25–3   *Four divisions of fungi are distinguished on the basis of structural features and patterns of reproduction, particularly sexual reproduction. It is thought that the ascomycetes and the basidiomycetes are derived from a common ancestor, and that this ancestor and the zygomycetes evolved from an earlier common ancestor. Fungi in which sexual reproduction is unknown—either because it has been lost in the course of evolution or because it has not been observed—are classified as imperfect fungi.*

Asexual reproduction takes place either by the fragmentation of the hyphae (with each fragment becoming a new individual) or by the production of spores. In some of the fungi, asexual spores are produced in **sporangia,** which are borne on specialized hyphae called **sporangiophores.** These hyphae typically elevate the sporangia above the mycelium. The bright colors and powdery textures associated with many types of molds are the colors and textures of spores and sporangia.

Fungi are generally haploid through much of the life cycle, and sexual reproduction is often initiated by the coming together of hyphae of different mating strains (Figure 25–4). Either before or after they come in contact, portions of the hyphae may develop into specialized reproductive structures called **gametangia.** In some fungi, such as the black bread molds, the gametangia fuse and many pairs of haploid nuclei unite, forming diploid zygotes. The resulting multinucleate cell, containing a number of zygotes, forms a hard, warty wall and becomes a dormant zygosporangium. At the end of dormancy, germination occurs and the diploid nuclei undergo meiosis, producing haploid spores.

25–4   *The life cycle of a black bread mold. During most of the life cycle, the organism is haploid. The mycelium of a bread mold consists of branched hyphae that anchor the organism and absorb nutrients. Hyphae that run above the surface of the bread are known as stolons, and the specialized hyphae that elevate the sporangia are known as sporangiophores.*

(a) Asexual reproduction occurs when the sporangia mature and their fragile walls disintegrate, releasing the asexual spores, which are carried away by air currents. Under suitable conditions of warmth and moisture, the spores germinate (red arrow), giving rise to new masses of hyphae.

(b) Sexual reproduction occurs when two hyphae from different mating strains (designated here as + and −) come together, forming gametangia. After fusion of the haploid nuclei, the gametangia develop into a thick-walled, resistant structure, the zygosporangium, which contains a number of diploid zygotes. After a period of dormancy, the zygotes undergo meiosis, and the zygosporangium germinates, producing a new sporangium from which the haploid spores formed through sexual reproduction are released.

In other fungi, the life cycle is more complex. In many species, the fusion of hyphae is not followed immediately by the fusion of nuclei. The fungal mycelium thus has two or more genetically distinct kinds of haploid nuclei operating simultaneously. Ultimately, a complex reproductive structure forms, the nuclei fuse, and meiosis occurs, producing the haploid spores that give rise to a new generation.

Symbiotic Relationships of Fungi

Symbiosis ("living together") is a close and long-term association between organisms of different species. Some relationships, typically disease-causing, are parasitic. One species (the parasite) benefits from the association and the other (the host) is harmed. Although many fungi are parasites, other fungi are involved in symbiotic relationships that are of mutual benefit to both organisms. Two of these—lichens and mycorrhizae—have been and are of extraordinary importance in enabling photosynthetic organisms to become established in previously barren terrestrial environments.

The Lichens

A **lichen** is a combination of a specific fungus and a green alga or a cyanobacterium. The product of such a combination (Figure 25–5) is very different from either the photosynthetic organism or the fungus growing alone, as are the physiological conditions under which the lichen can survive.

The lichens are widespread in nature, and they are often the first colonists of bare rocky areas (Figure 25–6). Lichens do not need an organic food source, as

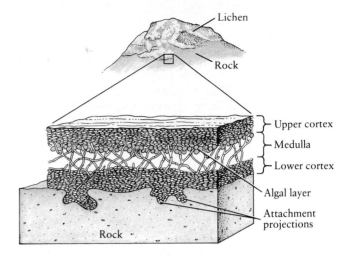

25–5   *The structure of the lichen* Lobaria verrucosa. *This lichen consists of four layers: the upper cortex, which is a protective layer of heavily gelatinized fungal hyphae; an algal layer, in which the algal cells are interspersed among loosely interwoven, thin-walled hyphae; the medulla, a thick layer of loosely packed, lightly gelatinized hyphae; and the lower cortex, which is covered with fine projections that attach the lichen to its substrate.*

25–6   (a) *Crustose ("encrusting") lichens growing on bare rock in the Wallowa Mountains of Oregon.* (b) *A foliose ("leafy") lichen growing on a rock surface in North Carolina.* (c) *British soldier lichen (*Cladonia cristatella*), a fruticose ("shrubby") lichen. Each soldier (so called because of the scarlet color) is 1 to 2 centimeters tall.*

(a)   (b)   (c)

do their component fungi, and unlike many free-living algae and cyanobacteria, they can remain alive even when dried out. They require only light, air, and a few minerals. Because lichens absorb substances from rainwater, they are particularly susceptible to airborne toxic compounds. Thus, the presence or absence of lichens is a sensitive index of air pollution.

Lichens reproduce most commonly by the breaking off of fragments containing both fungal hyphae and photosynthetic cells. New individuals can also be formed by the capture of an appropriate alga or cyanobacterium by a lichen fungus. Sometimes the captured photosynthetic cells are destroyed by the fungus, in which case the fungus also dies. If the photosynthetic cells survive, a lichen is produced.

Mycorrhizae: "Fungus-Roots"

Mycorrhizae are symbiotic associations between fungi and the roots of vascular plants and play a crucial role in the growth of the plants. If the seedlings of forest trees are grown in nutrient solutions and then transplanted to prairie or other grassland soils, they generally fail to grow. Eventually they may die from malnutrition, even if analysis shows that there are abundant nutrients in the soil. If a small amount (0.1 percent by volume) of forest soil containing fungi is added to the soil around the roots of the seedlings, however, they will grow promptly and normally. Mycorrhizae are now thought to occur in more than 90 percent of all families of plants.

In some mycorrhizal associations, the fungal hyphae penetrate the cells of the root, forming coils, swellings, or branches (Figure 25–7a). The hyphae also extend out into the surrounding soil. In other associations, the hyphae form a sheath around the root but do not actually penetrate its cells (Figure 25–7b).

The exact relationship between roots and fungi is not known. Apparently the roots secrete sugars, amino acids, and possibly some other organic substances that are used by the fungi. Although the evidence is still accumulating, it appears that the fungi convert minerals in the soil and decaying material into an available form and transport them into the roots. There is also evidence that the fungi facilitate water uptake by the roots. One of the most intriguing recent observations is that, under certain circumstances, mycorrhizae appear to function as a bridge through which phosphates, carbohydrates, and probably other substances pass from one plant to another.

A study of the fossils of early vascular plants has revealed that mycorrhizae occurred as frequently in them as they do in modern vascular plants. This has led to the interesting suggestion that the evolution of my-

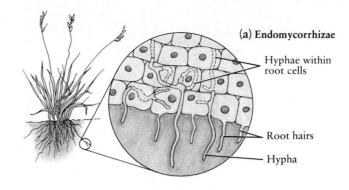

(a) Endomycorrhizae

Hyphae within root cells

Root hairs

Hypha

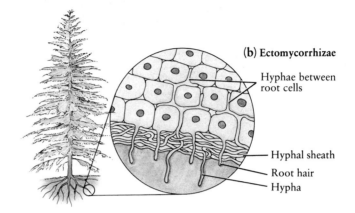

(b) Ectomycorrhizae

Hyphae between root cells

Hyphal sheath

Root hair

Hypha

25–7   (a) *Mycorrhizal associations in which the fungal hyphae penetrate the cells of the root are known as endomycorrhizae. They are common in grasses growing on nutrient-poor soils or at high elevations.* (b) *Mycorrhizal associations in which the hyphae form a sheath around the root but do not actually penetrate its cells are known as ectomycorrhizae. They are found on the roots of a wide variety of trees. In both types of mycorrhizae, hyphae penetrate the surrounding soil and function as extensions of the root system.*

corrhizal associations may have been the critical step allowing plants to make the transition to the bare and relatively sterile soils of the then-unoccupied land.

The Plants

Plants are multicellular photosynthetic organisms primarily adapted for life on land. Their characteristics are best understood in terms of the transition from water to land, an event that occurred some 500 million years ago. The land offered a wealth of advantages to photosynthetic organisms. On land, light is abundant from dawn to dusk and is not blocked by turbulent water. Carbon dioxide, needed for photosynthesis, is plentiful in the atmosphere and circulates more freely in air than in water. And, most important, the land was then unoccupied by competing forms of life.

25–8 *Although plants are primarily adapted for life on land, some, such as the hardy water lily,* Nymphaea fabiola, *have returned to an aquatic existence. Like whales and dolphins,* Nymphaea *retains the traces of its ancestors' terrestrial sojourn. These include a water-resistant cuticle, stomata (openings through which gas exchange occurs), and a highly developed internal transport system.*

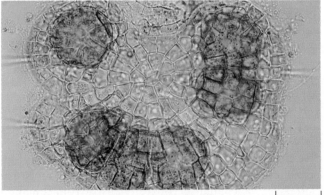

100 μm

25–9 Coleochaete, *a green alga that grows on the surface of submerged, freshwater plants in shallow water. This multicellular alga takes the form of a disk, generally one cell thick. The large, dark cells in this micrograph are diploid zygotes, which are retained on the alga and protected by a layer of smaller, haploid cells of the parent alga. The hair cells that extend outward from the disk are ensheathed at its base and are the source of the organism's name;* Coleochaete *means "sheathed hair." The hairs are thought to discourage aquatic animals from feeding on the alga.*

The Ancestral Alga

As we noted in the last chapter, the plants appear to have arisen from among the green algae. Like the plants, the green algae contain chlorophylls *a* and *b* and beta-carotene as their photosynthetic pigments, and they accumulate their food reserves in the form of starch. Beyond these similarities, however, the green algae exhibit a great diversity of characteristics, some of which are shared with the plants and some of which are not. The constellation of characteristics that could have given rise to the plants is found in only a few contemporary green algae, most strikingly in members of the genus *Coleochaete.*

Although *Coleochaete* itself does not seem to have been the alga from which the plants evolved, it is thought to be closely related to it. The cell walls of *Coleochaete,* like those of plants, contain cellulose. Additional evidence of a close relationship is found in the pattern of cytokinesis. In plants and in *Coleochaete,* the cytoplasm is divided by the formation of a cell plate at the equator of the spindle (see Figure 10–10, page 185). By contrast, in almost all other organisms, including most green algae, division of the cytoplasm takes place by constriction and pinching off of the plasma membrane.

One important characteristic shared by all plants, but absent in *Coleochaete,* is a well-defined alternation of generations. This type of life cycle is found not only in many multicellular green algae but also in the red algae and the brown algae, which biochemical evidence indicates had a totally different origin than the green algae. This suggests that alternation of generations has arisen independently on several occasions, and *Coleochaete* provides a significant clue as to how it may have occurred.

Coleochaete, like *Chlamydomonas* and a number of other green algae, is haploid for most of its life cycle. It produces clearly differentiated egg and sperm that fuse to form a zygote that subsequently undergoes meiosis, producing haploid cells from which new individuals develop. In *Coleochaete,* however, fusion of the gametes occurs not in the open waters but rather on the surface of the parent organism, and neighboring cells grow around the zygote, enclosing and protecting it (Figure 25–9). Prior to meiosis, additional rounds of DNA replication occur in the zygote, with the result that anywhere from 8 to 32 haploid cells are ultimately released. Only slight modifications of this life cycle—division of the diploid cells of the zygote by mitosis and cytokinesis *prior to* meiosis, followed by meiotic division of some, but not all, of the diploid cells—would be required to produce alternation of generations.

The Transition to Land

By the time the immediate ancestor of the plants moved from shallow waters onto the land, it had apparently evolved a well-defined alternation of generations. After the transition to land, new adaptations began to evolve. These adaptations were critical to the ultimate success of the plants on land and must have occurred early in their evolutionary history—most modern plants, even though very diverse, share them. They include a waxy coating, the cuticle, that covers the aboveground parts of the plant body and retards water evaporation, and stomata (Figure 9–10, page 169), specialized openings in leaves and green stems through which the gas exchanges necessary for photosynthesis can take place.

Another adaptation was the development of multicellular reproductive organs (gametangia and sporangia) in which the reproductive cells are protected from drying out by surrounding layers of covering cells. A related adaptation was the retention of the fertilized egg (the zygote) within the female gametangium and its development there into an embryo. Thus, during its critical early stages of development, the embryo, or young sporophyte, is protected by the tissues of the female gametophyte.

Not long after the transition to land, the plants diverged into at least two separate lineages. One gave rise to the **bryophytes,** a group that includes the modern mosses, hornworts, and liverworts, and the other to the **vascular plants,** a group that includes all of the larger land plants. A principal difference between the bryophytes and the vascular plants is that the latter, as their name implies, have a well-developed vascular system that transports water, minerals, sugars, and other nutrients throughout the plant body. The bryophytes first appear in the fossil record during the Devonian period, about 370 million years ago. (For the major events in the evolution of plants and animals, see the table on the inside of the front cover.) The oldest fossils of vascular plants are from the Silurian period, about 430 million years ago. The vascular plants, as we shall see, subsequently underwent a great diversification.

The Bryophytes

Most bryophytes are comparatively simple in their structure and are relatively small, usually less than 20 centimeters in length. Although bryophytes do not have true roots, they are generally attached to the substrate by means of **rhizoids,** which are elongate single cells or filaments of cells. Many bryophytes also have small leaf-like structures in which photosynthesis takes place. These structures do not have the specialized cells of the

25–10   *A bryophyte, a haircap moss with spore capsules. As shown in the diagram, the lower green structures are the haploid* (n) *gametophytes. The nonphotosynthetic stalks and capsules are the diploid* (2n) *sporophytes. The bryophytes are the only plants in which the gametophyte is the dominant, nutritionally independent generation.*

true leaves of the vascular plants and are only one or a few cell layers thick. Lacking water-gathering roots and specialized vascular tissues to transport water up the plant body, the bryophytes must absorb moisture through aboveground structures. As a consequence, they grow most successfully in moist, shady places and in bogs.

Bryophytes, like all plants, have a life cycle with alternation of generations. In contrast to the vascular plants, however, the bryophytes are characterized by a haploid gametophyte that is usually larger than the diploid sporophyte (Figure 25–10).

The life cycle of a moss (Figure 25–11) begins when a haploid spore germinates to form a network of horizontal filaments. Individual gametophytes grow up like branches from this network. When sufficient moisture is present, flagellated sperm are released from the male gametangium (the **antheridium**) and swim to the female gametangium (the **archegonium**). Fusion of sperm and egg takes place within the archegonium. Inside the archegonium, the zygote develops into a sporophyte, which remains attached to the gametophyte and is nutritionally dependent upon it. Typically the sporophyte consists of a foot, a stalk, and a single, large sporangium (or capsule), from which the spores are discharged.

Asexual reproduction, often by fragmentation, is also common among the bryophytes.

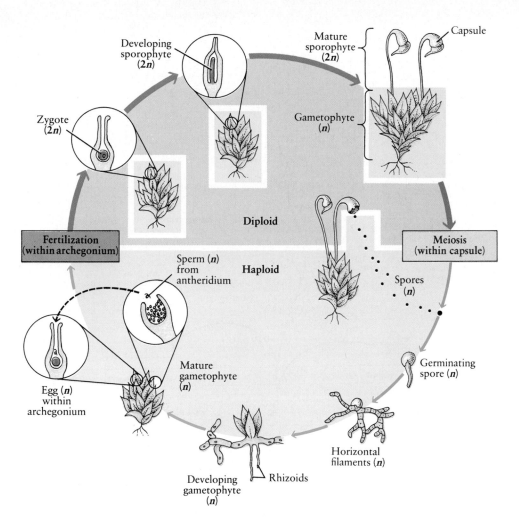

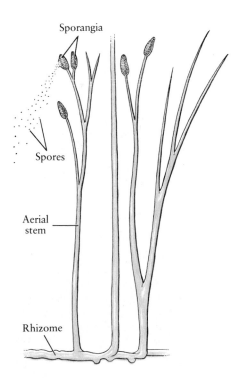

25–11 *The life cycle of a moss. The mature diploid sporophyte (upper right) consists of a capsule, which may be raised on a stalk, and a foot. Meiosis occurs within the capsule, and the resulting haploid spores are released when a small lid on the capsule bursts. The spore germinates to form a branched network of horizontal filaments, from which a leafy gametophyte develops. Sperm and egg cells are formed in specialized reproductive structures—the antheridia and the archegonia—of the gametophyte. Sperm cells, dispersed from the mature antheridium by raindrops or dew, are attracted into the archegonium, where one fuses with the egg cell to produce the diploid zygote. The zygote divides mitotically to form the sporophyte, which grows up from the gametophyte.*

25–12 *Rhynia major is one of the earliest known vascular plants, dating back some 400 million years. It lacked leaves and roots. Its aerial stems, which were photosynthetic, were attached to an underground stem, or rhizome. The aerial stems were covered with a cuticle and contained stomata. The dark structures at the tips of the stems were sporangia, which apparently released their spores by splitting lengthwise.*

The Vascular Plants

Rhynia major (Figure 25–12), now extinct, is an example of the earliest known vascular plants. It does not look much like any modern vascular plant and, indeed, it appears to have been much simpler than a bryophyte. However, it differed from the bryophytes in one important respect: within its stem was a central cylinder of vascular tissue, specialized for conducting water and dissolved substances up the plant body and products of photosynthesis down.

This evolutionary development gave rise to the great variety of plants that dominate the modern landscape, as well as to a number of groups that are now extinct. The living vascular plants are classified into nine divisions, each of which is thought to represent a single evolutionary lineage (Figure 25–13). These divisions are frequently grouped, for convenience, in ways that may or may not reflect evolutionary history. For instance, the vascular plants, as a group, are often referred to as tracheophytes. They can be grouped into

25–13   *An overview of the evolutionary relationships of the plants and of the ways in which they are usually grouped. Each of the nine divisions of living vascular plants is thought to represent a distinct evolutionary lineage. Other lineages of seedless vascular plants and of gymnosperms, now extinct, are known from the fossil record.*

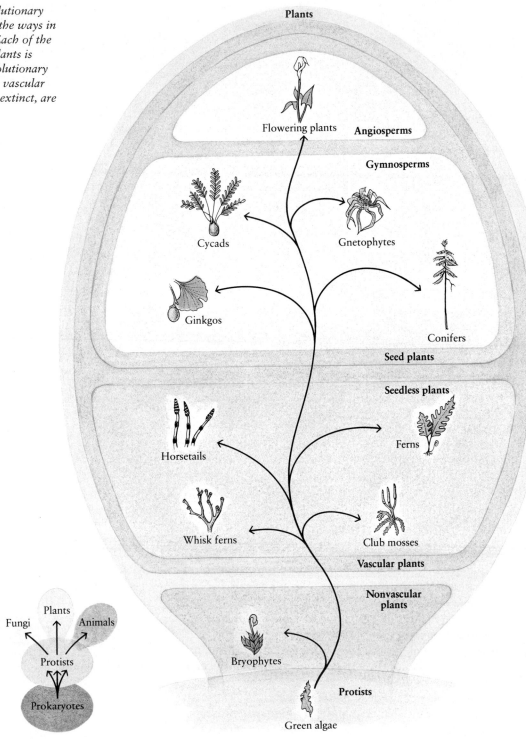

(a)   (b)   (c)   (d)

25–14 *Representatives of the four divisions of living seedless vascular plants.*
(a) *The whisk fern,* Psilotum, *one of the two living genera of division Psilotophyta. The yellow bulbous structures are the sporangia, which occur in fused groups of three. Psilotum is unique among living vascular plants in that it lacks roots and leaves. If you look closely, however, you can see small, scalelike outgrowths below the sporangia.*

(b) *The club mosses of the genus* Lycopodium *are the most familiar members of division Lycophyta. In this genus, the sporangia are borne on specialized leaves, sporophylls,* which are aggregated into a cone at the apex (top) of the branches, as shown in the broom moss, Lycopodium clavatum. *The airborne, waxy spores give rise to small subterranean gametophytes that are independent of the sporophyte. The sperm, which are biflagellate, swim to the archegonium, where fertilization occurs and the young sporophyte, or embryo, develops.*

(c) *The horsetails, division Sphenophyta, of which there is only one living genus (Equisetum), are easily recognized by their jointed, finely ribbed stems, which contain silica. At* each node, there is a circle of small, scalelike leaves. Spore-bearing structures are clustered into a cone at the apex of the stem. The gametophytes are independent, and the sperm are coiled, with numerous flagella.

(d) *A cinnamon fern. The ferns (division Pterophyta) are the most abundant of the seedless plants, with about 12,000 living species. Among ferns, the leaf, or frond, is commonly divided into leaflets, or pinnae. Spores are borne in sporangia, on the margins or undersides of leaves or on separate stalks, as shown here.*

those without seeds (Figure 25–14) and those with seeds. The seed plants also form two informal groups, the gymnosperms and the angiosperms. The **gymnosperms** (from the Greek *gymnos,* "naked" and *sperma,* "seed") are those with seeds that are not enclosed by protective tissues. The **angiosperms** (from the Greek *angio,* "vessel"—literally, a seed borne in a vessel) are those with enclosed, protected seeds.

Evolutionary Developments in the Vascular Plants

Better Conducting Systems   Beginning with a very simple vascular plant, such as *Rhynia,* it is possible to trace some major evolutionary trends, as well as several key innovations. One early innovation was the root, a structure specialized for anchorage and the absorption of water. Another was the leaf, a structure specialized for photosynthesis. One of the most striking trends in plant evolution has been the development of increasingly efficient conducting systems between these two portions of the plant body. One system, the **xylem** (pronounced "zy-lem"), transports water and ions from the roots to the leaves, a distance of more than 100 meters in some trees. The other system, the **phloem** (pronounced "flow-em"), carries dissolved sucrose and other products of photosynthesis from the leaves to the nonphotosynthetic cells of the plant.

With the development of roots, leaves, and efficient conducting systems, the plants effectively solved the most basic problems confronting multicellular photosynthetic organisms on land—acquiring adequate supplies of water and food and delivering them to all of the cells making up the organism.

Reduction of the Gametophyte   Another pronounced trend has been the reduction in size of the gametophyte. In all vascular plants, the gametophyte is smaller than the sporophyte. However, in most ferns and other seedless vascular plants, the gametophyte is separate and nutritionally independent of the sporophyte. Ferns usually have only one type of gametophyte, producing both male and female gametes (see Figure 11–5, the life cycle

25–15   *Male cones of the Monterey pine* (Pinus radiata) *shedding pollen, which is blown about by the wind. The pollen grains are immature male gametophytes, which complete their maturation when they reach the ovules, located at the base of the scales of the female cones. The female gametophytes develop within the ovules. When pollen grains land in the vicinity of the ovules, they produce pollen tubes that carry the nonmotile sperm cells into the female gametophytes, where they fertilize the egg cells.*

of a fern, on page 192). In a few species, however, the gametophytes exist as separate sexes, with some gametophytes producing male gametes and others producing female gametes.

In the seed plants—the gymnosperms and the angiosperms—the gametophyte has been reduced to microscopic size. In both gymnosperms and angiosperms, there are two types of gametophytes, one male and one female. These gametophytes are entirely heterotrophic, being dependent on the parent sporophyte for their nutrition and development. The female gametophyte, in fact, never leaves the protection of the sporophyte, and the egg cell is fertilized there by a sperm cell from the male gametophyte. In the angiosperms, the mature female gametophyte typically consists of only seven cells with a total of eight haploid nuclei. The male gametophyte is reduced to two or three cells, each of which contains a haploid nucleus. And yet the ancestral pattern of alternation of generations persists, and it is not possible to understand angiosperm reproduction, which we shall examine in detail in Chapter 39, without knowledge of its legacy from the past.

The Seed   The final innovation among the vascular plants, and perhaps the most important to their enormous success on land, was the seed. The seed is a complex structure in which the young sporophyte, or embryo, is contained within a protective outer covering, the seed coat (Figure 25–16). The seed coat, which is derived from tissues of the parent sporophyte, protects the embryo while it remains dormant, sometimes for years, until conditions are favorable for its germination. The earliest known seeds were fossilized in late Devonian deposits some 360 million years ago.

25–16   *The structure of a pine seed. The outer layers of the ovule have hardened into a seed coat, enclosing the female gametophyte and the embryo (the young sporophyte). The embryo consists of a root and a number of embryonic leaves, the cotyledons.*

When the seed germinates, the root will emerge from the seed coat and penetrate the soil. When the root absorbs water, the tightly packed cotyledons will elongate and swell with the moisture, rising above ground on the lengthening stem and forcing off the seed coat. During this period, the cotyledons absorb nutrients that are stored in the gametophyte tissue and are essential for the growth of the young sporophyte into a seedling.

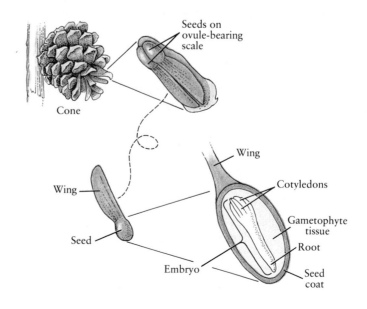

(a)

(b)

(c)

25–17 *Representatives of the three less familiar divisions of gymnosperms.* (a) *Encephalartos woodii, a cycad (division Cycadophyta) native to Africa. Most cycads are rare, and this species is one of the most severely endangered. It continues to exist only through offshoots, nurtured in many botanical gardens, that were originally obtained from a single plant.*

(b) *Leaves and fleshy seeds of* Ginkgo biloba, *the only surviving species of division Ginkgophyta, a lineage that extends back to the late Paleozoic era. Ginkgo is especially resistant to air pollution and is commonly cultivated in urban parks and along city streets. The fleshy coating of the seeds has a putrid smell, similar to that of rancid butter. However, the inner "kernel" of the seed, which has a fishy taste, is a much-prized delicacy in the Orient.*

(c) *A large seed-producing plant of* Welwitschia mirabilis, *growing in the Namib Desert of southern Africa. Welwitschia, a member of division Gnetophyta, produces only two adult leaves, which continue to grow for the life of the plant. As growth continues, the leaves break off at the tips and split lengthwise; hence, older plants appear to have numerous leaves.*

The Seed Plants

The humid Carboniferous period, which ended some 286 million years ago, was the age when most of the Earth's coal deposits were formed from lush vegetation that sank so swiftly into the warm, marshy soil that there was no chance for much of it to decompose. Seed plants were in existence by this period. According to the fossil record, some of the fernlike plants and even some of the club mosses had seedlike structures.

In the Permian period (248 to 286 million years ago), there were worldwide changes of climate, with the advent of widespread glaciers and drought. Terrestrial plants and animals were under strong selection pressures for specialized structures that would enable them to survive during periods when no water was available. Those organisms in which water-conserving, protective structures had previously evolved were at a great advantage. For example, the amphibians gave way to the reptiles with their scaly skins and shelled eggs, which may have been better suited to the harsh, dry climate. Similarly, by the close of this period, which ended the Paleozoic era, plants with seeds gained a major evolutionary advantage and came to be the dominant plants of the land.

Gymnosperms   It was during the Permian period that the gymnosperms—the "naked seed" plants—diversified. Four groups of gymnosperms have living representatives—three small divisions (Figure 25–17) and one large and familiar division, Coniferophyta. The conifers ("cone-bearers") include the pines, firs, spruces, hemlocks, cypresses, junipers, and the giant coastal redwoods of California and Oregon. In these plants, the female gametophyte develops on the scale of a cone, and the egg is fertilized there. The life cycle of a familiar conifer is shown in Figure 25–18.

448

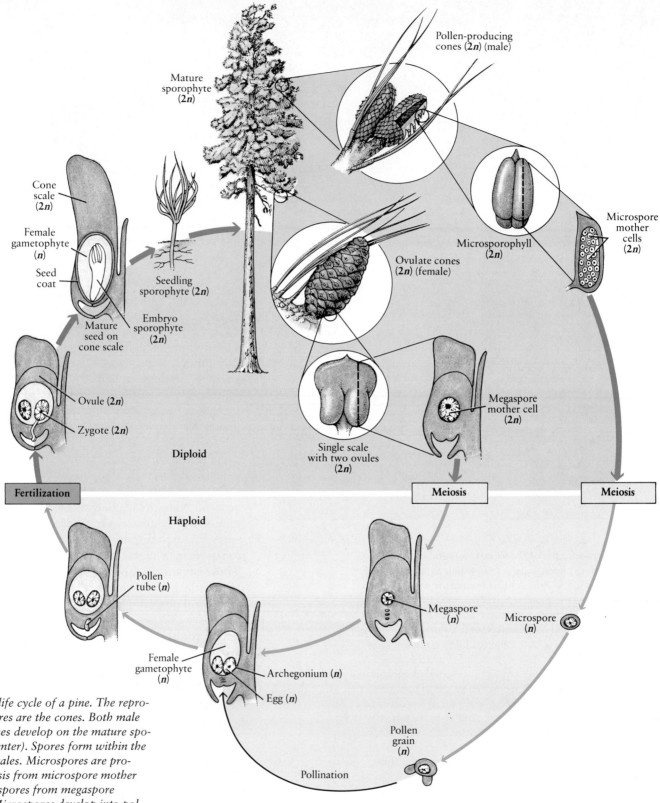

Pollen-producing
cones (**2n**) (male)

Mature
sporophyte
(**2n**)

Microspore
mother
cells
(**2n**)

Microsporophyll
(**2n**)

Cone
scale
(**2n**)

Female
gametophyte
(**n**)

Seed
coat

Seedling
sporophyte (**2n**)

Embryo
sporophyte
(**2n**)

Mature
seed on
cone scale

Ovulate cones
(**2n**) (female)

Megaspore
mother cell
(**2n**)

Ovule (**2n**)

Zygote (**2n**)

Single scale
with two ovules
(**2n**)

Diploid

Fertilization

Meiosis

Meiosis

Haploid

Pollen
tube (**n**)

Megaspore
(**n**)

Microspore
(**n**)

Female
gametophyte
(**n**)

Archegonium (**n**)

Egg (**n**)

Pollen
grain
(**n**)

Pollination

25–18 *The life cycle of a pine. The reproductive structures are the cones. Both male
and female cones develop on the mature sporophyte (top center). Spores form within the
cones on the scales. Microspores are produced by meiosis from microspore mother
cells and megaspores from megaspore
mother cells. Microspores develop into pollen grains, which are immature male gametophytes. Within the ovules, megaspores develop into female gametophytes. Each female
gametophyte contains several archegonia,
each with one egg cell. Although more than
one egg may be fertilized, usually only one*

*embryo develops fully in each female game-
tophyte.*

*The nonmotile sperm cells are carried to
the archegonium by the pollen tube, and the
egg is fertilized. After fertilization, the ovule
matures into the seed. The seed consists of*

*the embryonic sporophyte, surrounding nu-
tritive tissue of the female gametophyte, and
an outer coat derived from the protective
layers of the ovule. As the seed matures,
the cone opens, releasing the winged seeds,
which germinate, producing the seedling.*

(a)

(b)

(c)

(d)

25–19   *Flowers and their pollinators.*
(a) *A blister beetle feeding on the pollen of a California poppy. Primitive flowers are believed to have been—like the poppy—radially symmetrical, with numerous separate floral parts.*

(b) *A passion vine butterfly* (Heliconius hecale), *sipping nectar from lantana flowers in the rain forest of Costa Rica. Notice the long, sucking tongue of the butterfly.*

(c) *A male Anna's hummingbird* (Calyptae anna), *probing for nectar in a fuchsia. His head is collecting pollen, which he will carry to another flower. Flowers pollinated by birds are usually scentless, bright red or orange, and have copious nectar that makes the visit worthwhile.*

(d) *A lesser long-nosed bat* (Leptonycteris curasoae), *thrusting its face into the center of a saguaro cactus flower and lapping up nectar with its long, bristly tongue. Pollen grains clinging to its face and neck are transferred to the next flower visited by the bat. Bats of this species migrate each spring from central and southern Mexico to the deserts of the southwestern United States and then back again in the fall. They feed on the nectar and pollen of saguaro and organ-pipe cacti on their northward journey and on the flowers of the agave on the return trip. This species, which is endangered, is the only known pollinator of the flowers of these three plants.*

Angiosperms: The Flowering Plants   It is believed that the angiosperms—plants with enclosed, protected seeds—evolved from a now-extinct group of gymnosperms. They appear in the fossil record in abundance during the Cretaceous period, about 120 million years ago, as the dinosaurs were declining. Of the numerous angiosperm genera that appeared at that time, many seem to have been very similar to our modern genera.

Angiosperms have two new, interrelated structures that distinguish them from all other plants: the flower and the fruit. Both are devices by which animals are induced, rewarded, tricked, and even seduced into carrying out the plants' reproductive strategies. Plants are, generally speaking, immobile. Thus, uniting the sperm of one individual with the egg of another individual of the same species presents a problem, to which the flower—that apt symbol of spring and romance—is a solution. The fruit, which is derived from parts of the flower, is a solution to the related problem of dispersing the seeds into new environments.

The early gymnosperms from which the angiosperms evolved were probably wind-pollinated, as are modern gymnosperms (see Figure 25–15). And, as in the modern gymnosperms, the ovule probably exuded droplets of sticky sap in which pollen grains were caught and drawn to the female gametophyte. Insects, principally beetles and flies, that fed on plants must have come across the protein-rich pollen grains and the sticky, sugary droplets. As they began to depend on these newfound food supplies, they inadvertently carried pollen from plant to plant.

Insect pollination must have been more efficient than wind pollination for some plant species because, clearly, selection began to favor those plants that had insect pollinators. The more attractive the plants were to the insects, the more frequently they would be visited and the more seeds they would produce. Any chance variations that made the visits more frequent or that made pollination more efficient offered immediate advantages. More seeds would be produced, and thus more offspring would be likely to survive. Nectaries (nectar-secreting structures) evolved, which lured the pollinators. Plants developed white or brightly colored flowers that called attention to the nectar and other food supplies.

By the beginning of the Cenozoic era, some 65 million years ago, the first bees, wasps, butterflies, and moths had appeared. These are insects for which flowers are often the only source of nutrition for the adult forms. From this time on, flowers and certain insect groups have had a profound influence on one another's history, each shaping the other as they evolved together.

A flower that attracts only a few kinds of animal visitors and attracts them regularly has an advantage over flowers visited by more promiscuous pollinators: its pollen is less likely to be wasted on a plant of another species. In turn, it is an advantage for an animal to have a "private" food supply that is relatively inaccessible to competing species. Many of the distinctive features of modern flowers are adaptations that encourage regular visits (constancy) by particular pollinators. The varied shapes, colors, and odors allow sensory recognition by pollinators. The diverse, sometimes bizarre, structures such as deep nectaries and complex landing platforms that are found, for example, in orchids, snapdragons, and irises, represent ways of excluding indiscriminate pollinators and of increasing the precision of pollen deposits on the animal messengers.

Like the flower, the fruit evolved as a payment to an animal visitor for transportation services. A great variety of fruits, adapted for many different dispersal mechanisms, have evolved in the course of angiosperm history (Figure 25–20). In most cases, the chief requirement is that the seed be transported some distance from the parent plant, where it is more likely to find open ground and sunlight. A familiar example is provided by the many edible fleshy fruits that become sweet and brightly colored as they ripen, attracting the attention of birds and mammals, including ourselves. The seeds within the fruits pass through the digestive tract hours later and are often deposited some distance away. In some species the seed coat's exposure to digestive juices is a prerequisite for germination. The seeds themselves may be bitter or toxic, as in apples, discouraging animals from grinding up and digesting them.

Another factor in the dominance of the flowering plants has been the evolution of characteristic bad-tasting or toxic compounds. The mustard family, for example, is distinguished by the pungent taste and odor associated with cabbage, horseradish, and mustard. Milkweeds (page 346) contain substances that act as heart poisons in vertebrates, potential predators of this group of plants. Bitter-tasting quinine is derived from tropical trees and shrubs of the genus *Cinchona*. Nicotine, caffeine, mescaline, tetrahydrocannabinol, opium, and cocaine are all plant products. All of these substances were, at one time, considered by-products of plant metabolism and were termed "secondary plant substances." They are now recognized as products of

(a)

(b)

25–20 *Angiosperms are characterized by fruits. As we shall see in Chapter 39, a fruit is a mature ovary, enclosing the seed or seeds. It often also includes accessory parts of the flower. The fruit aids in dispersal of the seed. Some fruits are blown by the wind, some are carried from one place to another by animals, some float on water, and some are even forcibly ejected by the parent plant.*

(a) *When the fruit of the musk thistle (Carduus nutans) is ripe, it releases seeds bearing tufts of silky hair that aid in their dispersal by the wind. This thistle is native to Europe and western Asia but is now found throughout the United States.*

(b) *The tough seeds in these blackberries will pass unharmed through the digestive tract of the dormouse. If they are deposited in a suitable environment, they will germinate, giving rise to new blackberry plants.*

angiosperm evolution that provide powerful defenses against animal predators.

About 235,000 different species of angiosperms are known, all included in a single division, Anthophyta (literally, the flowering plants). They dominate the tropical and temperate regions of the world, occupying well over 90 percent of the Earth's vegetative surface. The angiosperms include not only the plants with conspicuous flowers but also the great hardwood trees, all the fruits, vegetables, nuts, herbs, and grains and grasses that are the staples of the human diet and the basis of agricultural economy all over the world. In Section 7, we shall consider the structure and physiology of the angiosperms, paying particular attention to the processes of reproduction, growth, and development, the transport of materials by the vascular system, and the integration of the activities of the diverse types of cells forming the plant body.

The Role of Plants

The only forms of life on land that do not depend on plants for their existence are a few kinds of autotrophic prokaryotes and protists. For all other terrestrial organisms, the chloroplast of the plant cell is the "needle's eye" through which the sun's energy is channeled into the biosphere. Even those animals that eat only other animals—the carnivores—could not exist if their prey, or their prey's prey, had not been nourished by plants.

Moreover, plants are the channels by which many of the simple inorganic substances vital to life enter the biosphere. Carbon in the form of carbon dioxide is taken from the atmosphere and incorporated into organic compounds during photosynthesis. Elements such as nitrogen and sulfur are taken from the soil in the form of simple inorganic compounds and incorporated into proteins, vitamins, and other essential organic compounds within green plant cells. Animals cannot make these organic compounds from inorganic materials and so are entirely dependent on plants for these compounds as well as for their energy supply.

It was only after the plants had successfully invaded the land that members of a number of different animal groups could, in turn, carry out their own invasions of this vast new environment. As we shall see in the next two chapters, both invertebrates and vertebrates underwent great diversification as they took advantage of the variety of habitats and food supplies made available by the plants.

Summary

Multicellularity made possible the evolution of larger organisms, with different portions of the body specialized for different functions, but it also brought with it new problems to be solved. These include obtaining adequate food, supplying the individual cells, supporting the body, integrating the activities of the component cells, and protecting and nourishing the immature stages. Different groups of multicellular organisms have solved these problems in different ways.

Fungi are heterotrophs that absorb organic molecules from the substrate on which they live. The body of a fungus is typically a mycelium—a mass of multinucleate or multicellular filaments called hyphae. The cell walls contain chitin.

Most fungi reproduce both asexually and sexually. Asexual spores are often formed in sporangia. The sexual cycle is initiated by the fusion of hyphae of different mating strains. In some groups of fungi, the nuclei in the fused hyphae combine immediately, and a zygote is formed. In other groups, the mycelium exists with two genetically distinct types of nuclei for a period of time before specialized reproductive structures form. Once the nuclei fuse, meiosis follows immediately, producing sexual spores.

Fungi have an important ecological role as decomposers of organic material. They are also parasitic on many types of organisms, particularly plants, in which they often cause serious disease. Fungi participate in two additional types of symbioses that are of ecological importance—lichens and mycorrhizae ("fungus-roots").

Lichens are combinations of fungi and green algae or cyanobacteria that are structurally and physiologically different from either organism as it exists separately. They are able to survive under adverse environmental conditions where neither partner could exist independently.

Mycorrhizae are associations between soil-dwelling

fungi and plant roots. Mycorrhizal associations facilitate the uptake of minerals and water by the roots of the plant and provide organic molecules for the fungus. They are thought to have played a key role in enabling plants to make the transition to land.

Plants are multicellular photosynthetic organisms adapted for life on land. Among their adaptations are a waxy cuticle; stomata, through which gases are exchanged; protective layers of cells surrounding the reproductive cells; and retention of the young sporophyte within the female gametophyte during embryonic development.

The ancestor of the plants is believed to have been a multicellular green alga, similar to the modern genus *Coleochaete*. Its life cycle was characterized by alternation of generations, and its gametes were differentiated into egg and sperm. From this common ancestor, two principal lineages diverged: the bryophytes and the vascular plants.

The bryophytes are relatively small plants that generally lack specialized vascular tissues. They include the mosses, liverworts, and hornworts. In the bryophytes—and in no other members of the plant kingdom—the haploid gametophyte is dominant and the diploid sporophyte is smaller, attached, and often nutritionally dependent.

Although the bryophytes seem to have changed little in the course of their history, the vascular plants have undergone a great diversification. Major developments in their evolution include better conducting systems, a progressive reduction in the size of the gametophyte, and the "invention" of the seed. The vascular plants can be informally grouped into the seedless vascular plants and the seed plants. The seed plants, in turn, can be grouped into the gymnosperms and the angiosperms. In the gymnosperms, the seed is "naked"—that is, it is not enclosed by outer layers. In the angiosperms—the flowering plants—the seed is enclosed in an outer protective structure, the fruit, formed from parts of the flower.

Flowers attract pollinators, and fruits enhance the dispersal of seeds. The various shapes and colors of flowers evolved under selection pressures for more efficient pollinating mechanisms. In addition to the flower and the fruit, a third factor in the success of the angiosperms has been the evolution of chemicals that discourage the predations of foraging animals. Angiosperms are the dominant plants of the modern landscape, providing a diversity of habitats and foods for terrestrial animals.

Questions

1. Distinguish between the following: hypha/mycelium; sporangia/gametangia; lichens/mycorrhizae; bryophytes/vascular plants; xylem/phloem; gymnosperm/angiosperm.

2. Multinucleate organisms, such as the fungi, show little differentiation. When differentiation does occur, as in the formation of a sporangium or gametangium, it is preceded by construction of a partition. In your opinion, why?

3. How have fungi solved the problem of supplying the individual cells with food and water? What ways do they have of obtaining new food supplies when they have exhausted a particular source?

4. How are fungi thought to have aided photosynthetic organisms in the transition to land?

5. Consider the green alga *Coleochaete*. What advantages does the alga gain from the production of "extra" haploid cells, made possible by additional replication of the DNA of the zygote prior to meiosis? What might be the advantages—to *Coleochaete*—of retaining the zygote on the body of the parent alga?

6. Bryophytes, among the plants, and amphibians, among the animals, often live in habitats intermediate between fresh water and dry land, rather than between salt water and land. Propose an argument (referring back to Chapter 6) to explain why invasions of the land were more likely by organisms previously adapted for life in fresh water.

7. Vascular plants are characterized by a molecule known as lignin, which stiffens and supports the plant body as it grows upward toward the light. What is the primary source of support for the photosynthetic portions of the multicellular seaweeds?

8. How have plants solved the problem of obtaining adequate food and water? Of supplying food and water to the individual cells? Of protecting and nourishing the immature stages?

9. Sketch a pine seed and label it. Indicate the origin of each of its components, and whether the component is haploid or diploid. How many generations are represented in the tissues of the seed?

Suggestions for Further Reading

Ahmadjian, Vernon: "The Nature of Lichens," *Natural History,* March 1982, pages 30–37.

Ahmadjian, V., and S. Paracer: *Symbiosis: An Introduction to Biological Associations,* University Press of New England, Hanover, N.H., 1986.

> *In this small but fascinating book, Ahmadjian and Paracer describe the nature of the relationship between the fungal and algal components of a lichen.*

Barth, Friedrich G.: *Insects and Flowers: The Biology of a Partnership,* Princeton University Press, Princeton, N.J., 1991.*

> *A well-written and beautifully illustrated introduction to the interrelationships of flowers and insects. Incorporating many recent discoveries, the text considers not only the diverse structures of flowers but also the sensory, navigational, and communication abilities of pollinating insects.*

Cox, Paul Alan: "Water-Pollinated Plants," *Scientific American,* October 1993, pages 68–74.

Crepet, W. L.: "Ancient Flowers for the Faithful," *Natural History,* April 1984, pages 38–45.

Dilcher, D., and P. R. Crane: "In Pursuit of the First Flower," *Natural History,* March 1984, pages 56–61.

Fleming, Theodore H.: "Plant-Visiting Bats," *American Scientist,* vol. 81, pages 460–467, 1993.

Gensel, Patricia G., and Henry N. Andrews: "The Evolution of Early Land Plants," *American Scientist,* vol. 75, pages 478–489, 1987.

Graham, Linda E.: "The Origin of the Life Cycle of Land Plants," *American Scientist,* vol. 73, pages 178–186, 1985.

Gray, Jane, and William Shear: "Early Life on Land," *American Scientist,* vol. 80, pages 444–455, 1992.

Handel, Steven N., and Andrew J. Beattie: "Seed Dispersal by Ants," *Scientific American,* August 1990, pages 76–83A.

Knoll, Andrew H.: "End of the Proterozoic Eon," *Scientific American,* October 1991, pages 64–73.

Large, E. C.: *The Advance of the Fungi,* Dover Publications, Inc., New York, 1962.*

> *A fascinating popular account of the closely interwoven histories of fungi and humans, first published in 1940.*

Litten, W.: "The Most Poisonous Mushrooms," *Scientific American,* March 1975, pages 90–101.

Matossian, Mary K.: "Ergot and the Salem Witchcraft Affair," *American Scientist,* vol. 70, pages 355–357, 1982.

Mulcahy, David L.: "Rise of the Angiosperms," *Natural History,* September 1981, pages 30–35.

Newhouse, Joseph R.: "Chestnut Blight," *Scientific American,* July 1990, pages 106–111.

Niklas, Karl J.: "Aerodynamics of Wind Pollination," *Scientific American,* July 1987, pages 90–95.

Niklas, Karl J.: "Computer-Simulated Plant Evolution," *Scientific American,* March 1986, pages 78–86.

Norstog, Knut: "Cycads and the Origin of Insect Pollination," *American Scientist,* vol. 75, pages 270–279, 1987.

Raghavan, V.: "Germination of Fern Spores," *American Scientist,* vol. 80, pages 176–185, 1992.

Raven, Peter H., Ray F. Evert, and Susan E. Eichhorn: *Biology of Plants,* 5th ed., Worth Publishers, Inc., New York, 1992.

> *This general botany text contains an excellent presentation of the evolution of plants and related organisms, as well as a wealth of information on prokaryotes, protists, and fungi.*

Rosenthal, Gerald A.: "The Chemical Defenses of Higher Plants," *Scientific American,* January 1986, pages 94–99.

Smith, A. H.: *The Mushroom Hunter's Field Guide,* The University of Michigan Press, Ann Arbor, Mich., 1980.

> *A clear, concise, well-illustrated guide to edible mushrooms, enlivened with good advice and pertinent anecdotes.*

Strobel, Gary A., and Gerald N. Lanier: "Dutch Elm Disease," *Scientific American,* August 1981, pages 56–66.

Valentine, James W.: "The Evolution of Multicellular Plants and Animals," *Scientific American,* September 1979, pages 140–158.

White, Mary E.: "Plant Life between Two Ice Ages Down Under," *American Scientist,* vol. 78, pages 253–262, 1990.

* Available in paperback.

Of the 1.5 million named species of animals, two-thirds—about a million—are insects. The insects emerged from the water along with the plants and took off, evolutionarily speaking, with the innovation of flight, about 325 million years ago. It has been calculated that for every pound of you and me and other *Homo sapiens* on this planet, there are 300 pounds of ants, bees, beetles, crickets, earwigs, fireflies, grasshoppers, fleas, lice, termites, mosquitoes, and other insects. Almost 10,000 new species are discovered each year, mostly in tropical forests, and about the same number, it is estimated, become extinct.

To your left is an assassin bug *(Arilus cristatus)* about to kill a caterpillar that would have, if left undisturbed, metamorphosed into a spicebush swallowtail butterfly *(Papilio troilus)*. The bug will use its hollow beak to stab the caterpillar and will then suck out its tissues.

Together, bug and caterpillar illustrate several of the secrets of the insects' success—a success measured by endurance (the modern cockroach is 200,000 years old), numbers of species, and multitudes of individuals. For a start, insects are small, so thousands can live in a space that might accommodate only a single vertebrate. And, they have wings and can fly, an ability that opened many habitats, or living spaces, to them. Even more important, they have many different, highly specialized ways of making a living. The outward evidence is the intricacy and versatility of their mouthparts. They can stab and suck like the assassin bug, or they can chew, sip, siphon, bite, grasp, cut, tear, or sift. As a result, different species generally do not compete with one another. Moreover, among many insects, the immature forms, such as the caterpillar, are very different from the adult forms, such as the butterfly, so parents and offspring do not compete with each other. Finally, if insects were human, they would be called wily and ingenious. Note, for instance, the eyespots on the doomed caterpillar, which are not eyes at all but only pigmented spots. Although ignored by the assassin bug, these spots might well have warded off an attack by a bird.

The Animals I: Invertebrates

Animals are many-celled heterotrophs, and their principal mode of nutrition is ingestion. They depend directly or indirectly for their nourishment on photosynthetic autotrophs—algae or plants. Typically they digest their food in an internal cavity and store food reserves as glycogen or fat. Their cells, unlike those of most other eukaryotes, do not have walls.

In animals, reproduction is usually sexual, and the gametes are the only haploid stages in the life cycle. As adults, most animals are fixed in size and shape, in contrast to plants, in which growth often continues for the lifetime of the organism. Generally, animals move by means of contractile cells (muscle cells) containing characteristic proteins. The most complex animals—the octopuses, the insects, and the vertebrates—have many kinds of specialized tissues, including elaborate sensory and neuromotor mechanisms not found in any other kingdom.

For most of us, animal means mammal. However, the mammals, or even the vertebrates as a whole, represent only a small fraction of the animal kingdom. More than 1.5 million different species of animals have been described, of which more than 95 percent are **invertebrates**—that is, animals without backbones.

The unifying characteristic among all the animals is their mode of nutrition. Unlike plants, which receive their energy from the sun, animals must actively seek out food sources or, alternatively, devise strategies for ensuring that the food comes to them. Thus motility—of the entire organism, or its parts, or both—is a requirement for animal survival. Motility in general and the hunting of prey in particular require efficient systems of integration and control of the multitude of cells that make up the organism. In short, muscle and nerve, distinguishing features of the animal kingdom, have their origin in heterotrophy.

Paradoxically, these unifying characteristics of the animal kingdom are also the key to its diversity. Given the properties of living tissue and the characteristics of our planet—particularly the force of gravity and the

26-1 *Some animals, such as the assassin bug, actively seek their food, whereas others, such as these goose barnacles (Lepas anserifera), arrange to have it delivered. Adult barnacles, which are attached to a substrate and are therefore nonmotile, depend on six pairs of bristly appendages to deliver their food supply. Louis Agassiz, America's leading nineteenth-century naturalist, characterized a barnacle as "nothing more than a little shrimp-like animal, standing on its head in a limestone house and kicking food into its mouth."*

26-2 *Fossils of many ancient animals have been found in deposits in Australia and Canada. Among the unusual organisms discovered in the Ediacaran Hills of Australia are (a) Tribrachidium, a segmented animal that lacked a clearly defined head and appendages, and (b) Spriggina floundersi, a segmented animal with a prominent "head shield." Animals from the Burgess Shale in British Columbia include (c) Hatlothrentia carinatus, which is thought to be a mollusk.*

physical properties of water and air—there are only a few basic ways that locomotion, the capture of food (Figure 26–1), self-defense, and coordination can be accomplished. In the course of evolution, however, as animals have adapted to new or changing environments, those few basic ways have been "reinvented," refined, and elaborated upon, resulting in the great diversity of structural and functional detail that we see today.

The Origin and Classification of Animals

Animals, like plants, presumably had their origins among the protists. In the case of animals, however, we have fewer clues about which protists most closely resemble the ancestral ones, and multicellularity may well have arisen more than once.

By the Cambrian period (590 to 505 million years ago), two major diversifications of animal life had already occurred, only one of which was to survive. The earliest evidences of animal life are provided by the fossilized animals (Figure 26–2) found in greatest abundance in the Ediacaran Hills of Australia and a rock formation known as the Burgess Shale in the Canadian Rockies. The Ediacaran animals, which arose some 670 million years ago and flourished for about 100 million years, were quite unlike later animals and apparently were not their ancestors. The fundamental structure of most living animals is a tube within a tube, modified and elaborated in a variety of ways that maintain a high ratio of surface to volume. Ediacaran animals, by contrast, lacked tubular internal structures and were flat, leaflike, or, occasionally, quilted like an air mattress. Although this evolutionary solution to the problem of maintaining an adequate surface-to-volume ratio was ultimately unsuccessful, it produced a diversity of exquisitely beautiful organisms. The Burgess Shale, which is dated at about 530 million years ago, includes members of at least 10 extinct phyla as well as fossil representatives of virtually all living phyla.

(a)

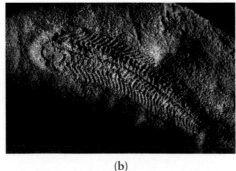

(b)

(c)

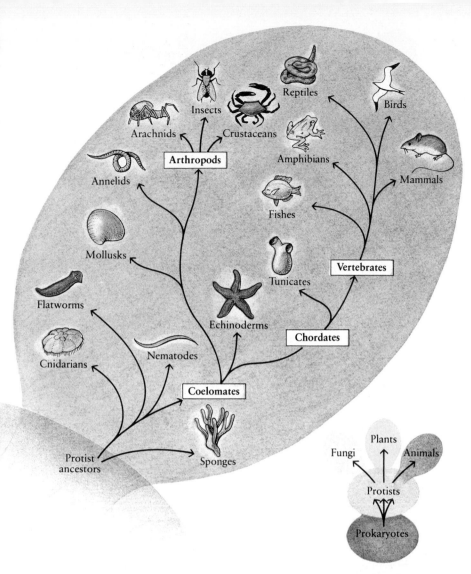

26–3 *The evolutionary relationships of the most familiar groups in the animal kingdom. These relationships are deduced from structural similarities, such as segmentation of the body in annelids and arthropods, and also from resemblances among larval forms and in developmental patterns. By the time the earliest known fossils ancestral to modern animals were deposited, the major phyla had already diverged. Thus, although later fossils illuminate evolutionary history within phyla (for example, within the arthropods or the chordates), we lack clear documentary evidence of the relationships among the different phyla.*

Table 26–1	An Outline of Animal Classification*

I. Subkingdom Parazoa ("beside the animals"): phylum **Porifera**

II. Subkingdom Mesozoa ("middle animals"): phylum Mesozoa

III. Subkingdom Eumetazoa ("true animals")

  A. Radially symmetrical animals: phyla **Cnidaria** and Ctenophora

  B. Bilaterally symmetrical animals

    1. Acoelomates (animals that lack a body cavity): phyla **Platyhelminthes,** Gnathostomulida, and **Rhynchocoela**

    2. Pseudocoelomates (animals with the type of body cavity known as a pseudocoelom): phyla **Nematoda,** Nematomorpha, Acanthocephala, Kinorhyncha, Gastrotricha, Loricifera Rotifera, and Entoprocta

    3. Coelomates (animals with the type of body cavity known as a coelom)

      a. Protostomes (animals in which the mouth appears at or near the first opening that forms in the developing embryo): phyla **Mollusca, Annelida,** Sipuncula, Echiura, Priapulida, Pogonophora, Pentastomida, Tardigrada, Onychophora, and **Arthropoda**

      b. Lophophorates (protostomes but with some deuterostome characteristics): phyla Bryozoa, Phoronida, and Brachiopoda

      c. Deuterostomes (animals in which the anus appears at or near the first opening that forms in the developing embryo): phyla **Echinodermata,** Chaetognatha, Hemichordata, and **Chordata**

* A summary of the distinctive characteristics of each phylum can be found in Appendix B.

Although the fossils of the Burgess Shale and other Cambrian formations around the world attest to the ancient origins of the modern phyla of animals, they shed little light on the chronological order of their appearance. Most of the evidence for diagrams such as Figure 26–3 comes from studies of living animals.

Modern animals are classified in about 30 phyla. The members of each phylum are thought to be descended from a common ancestral species. A number of criteria are used in classifying an animal and in attempting to determine the evolutionary relationships among the different phyla. Among the most important are the number of tissue layers into which the cells are organized, the basic plan of the body and the arrangement of its parts, the presence or absence of body cavities and the manner in which they form, and the pattern of development from fertilized egg to adult animal. Table 26–1 provides a short outline of animal classification and indicates the key features that are used in grouping the various phyla. In our examination of the diversity of animals in this chapter and the next, we shall discuss the phyla indicated in boldface type, which exemplify the most significant developments and adaptations in animal evolution.

Sponges: Phylum Porifera

Sponges may have had a different origin from other members of the animal kingdom, and they are generally regarded as an evolutionary dead end—that is, no other groups were derived from them. As adults, all sponges are **sessile** (attached to a substrate), and they are essentially water-filtering systems, made up of one or more chambers. Sponges are common on ocean floors throughout the world. Most live along the coasts in shallow water, but some are found at great depths, where currents are relatively slow.

A sponge represents a level of organization somewhere between a colony of cells and a true multicellular organism. The cells are not organized into tissues or organs, yet there is a form of recognition among the cells that holds them together and organizes them. As in colonial forms, if the individual cells of a living sponge are separated (as by squeezing the body through a fine sieve) and then mixed together, the cells will reassemble, resuming their previous organization. In contrast to colonial forms, however, the cells of the sponge are to a certain extent differentiated and specialized (Figure 26–4b). The different cell types include flagellated **collar cells,** which are feeding cells; **epithelial,** or covering, cells, some of which contain contractile fibers; and **amoebocytes,** which carry food particles from the collar cells to the epithelial and other nonfeeding cells. Amoebocytes also play several roles in reproduction, and, in addition, they produce stiffening structures. These structures, which may be either inorganic spicules or organic fibers, form the supporting skeleton for the animal. When a sponge is dried out and cleaned, only this skeleton remains.

Because all the digestive processes of sponges are carried out within single cells, even a giant sponge—and some are more than 2 meters in height—can consume nothing larger than microscopic particles. The large sponges, which need more food, have highly folded body walls that greatly increase the filtering and feeding surfaces. We have already encountered this evolutionary stratagem for increasing biological work surfaces at the cellular level—as in the inner membrane of the mitochondrion. It is a stratagem that occurs repeatedly throughout the animal kingdom.

The reproduction of sponges exhibits two features characteristic of both sessile and slow-moving animals.

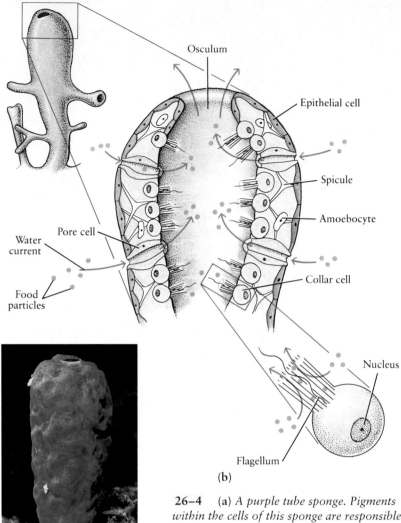

(a)

26–4   (a) *A purple tube sponge. Pigments within the cells of this sponge are responsible for its brilliant color, which may also be enhanced by the refraction, or bending, of light as it passes through the water.* (b) *The body of a simple sponge is dotted with tiny pores, from which the phylum derives its name (Porifera, or "pore-bearers"). Water, containing food particles, is drawn into the internal cavity of the sponge through these pores and is forced out through the osculum. The flow is created by the sucking effect of external water currents moving across the osculum and by the beating of the flagella protruding from the collar cells. Each collar cell has about 20 retractile filaments surrounding a single flagellum, the lashing of which directs a current of water through the filaments. Minute particles are filtered out and cling to one or more filaments and are then drawn into the cell. A sponge 10 centimeters high filters more than 20 liters of water a day.*

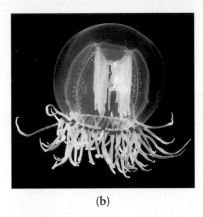

(a)   (b)

26–5 *Adult cnidarians are usually radially symmetrical and may take the form of polyps or medusas.* (a) *The tentacles of* Hydra, *a freshwater polyp, have just made contact with a small crustacean (*Daphnia*). The* Hydra *will wrap its tentacles around the* Daphnia *and then push it into its mouth. These activities are carried out by specialized contractile cells and are coordinated by the animal's simple but effective nervous system.* (b) *This jellyfish, the medusa form of* Polyorchis, *is swimming actively, its bell contracted by muscles around its margin. The specialized muscle cells contain assemblies of contractile protein fibers. A bright red eyespot is located at the base of each tentacle. The white strands suspended within the bell are the reproductive structures in which meiosis takes place and from which eggs or sperm are released into the water.*

First, asexual reproduction is quite common, usually by fragments that break off from the parent animal. Second, most kinds of sponges are **hermaphrodites**—that is, the same individual has both male and female reproductive structures and produces both sperm and egg cells. As we saw in Chapter 19, this is a great advantage for animals with little or no motility. In animals with separate sexes, a given individual can mate only with members of the opposite sex, but for a hermaphrodite any partner—or the gametes of any partner—will suffice.

Radially Symmetrical Animals: Phylum Cnidaria

The cnidarians, which include jellyfishes, *Hydra*, sea anemones, and corals, are a large group of aquatic animals in which the adult form is generally radially symmetrical (Figure 26–5). In radial symmetry, the body parts are arranged around a central axis, like spokes around the hub of a wheel. Any plane that passes through the central axis divides the body into halves that are mirror images of one another.

The cells of cnidarians, unlike those of sponges, are organized into distinct tissues. As you can see in Figure 26–6a, the basic body plan is simple: the animal is essentially a hollow container, which may be either a vase-shaped **polyp** or a bowl-shaped **medusa.** The polyp is usually sessile, and the medusa, motile. Both consist of two layers of tissue: **epidermis** and **gastrodermis** (from the Greek *epi,* "on" or "over," and *gaster,* "stomach," plus *derma,* "skin"). Between these two layers is a gelatinous filling, the **mesoglea** ("middle jelly"), most conspicuous in the medusa. The epidermis is derived from an embryonic tissue known as the **ectoderm,** and

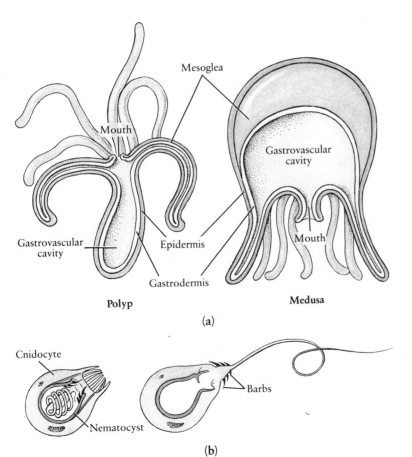

Mesoglea

Mouth

Gastrovascular cavity

Gastrovascular cavity

Epidermis

Mouth

Gastrodermis

Polyp

Medusa

(a)

Cnidocyte

Barbs

Nematocyst

(b)

26–6 (a) *Among cnidarians, there are two basic body forms: the vase-shaped polyp (left) and the bowl-shaped medusa (right). The gastrovascular cavity has a single opening. The cnidarian body has two tissue layers, epidermis and gastrodermis, with gelatinous mesoglea between them.*

(b) *Cnidocytes, specialized cells located in the tentacles and body wall, are a distinguishing feature of cnidarians. The interior of the cnidocyte is filled by a nematocyst, which consists of a capsule containing a coiled tube, as shown on the left. A trigger on the cnidocyte, responding to chemical or mechanical stimuli, causes the tube to shoot out, as shown on the right. The capsule is forced open and the tube turns inside out, exploding to the outside. The cnidocyte cannot be "reloaded." It is absorbed and a new cell grows to take its place.*

The Coral Reef

The most diverse of all marine communities—the coral reef—depends for its very existence on tiny cnidarians. Corals, which are cnidarians in which the medusa stage of the life cycle has been lost, form extensive colonies of polyps that reproduce both asexually and sexually.

Each polyp in the colony secretes its own calcium-containing skeleton, which then becomes part of the reef. The photosynthetic activity of the reef is carried out almost entirely by symbiotic dinoflagellates and green algae living within the tissues of the corals. In fact, as much as half the living substance of a coral reef may consist of green algae. Carbon, oxygen, and dissolved minerals flow over the reef as a result of the movement of waves and the ocean currents. The reef furnishes both food and shelter for other sea animals, including numerous species of reef fishes and a tremendous variety of invertebrates, such as sponges, sea urchins, marine worms, and crustaceans.

The coral polyps and algae that form the reef can grow only in warm, well-lighted surface waters, where the temperature seldom falls below 21°C. The largest coral-created land masses in the world are the 2,000-kilometer-long Great Barrier Reef, off the northeast shores of Australia, and the Marshall Islands in the Pacific. Other reefs are found throughout tropical waters and as far north as Bermuda, which is warmed by the Gulf Stream.

(a)

(b)

(a) *Living alcyonarian coral, photographed in Indonesia. Notice the individual polyps, extended from their limestone skeletons.* **(b)** *A portion of a coral reef in the Solomon Islands, located in the Pacific Ocean east of Australia. Alcyonarian coral is in the foreground. The large, deep red corals are known as tropical sea fans. Numerous surgeon fish can be seen in the background.*

the gastrodermis is derived from the embryonic **endoderm** (from the Greek *ektos*, "outer," and *endon*, "inner"). Highly specialized stinging cells, the **cnidocytes** (Figure 26–6b), characterize these animals and give the phylum its name.

In both polyps and medusas, food is captured by means of tentacles armed with cnidocytes and is then pushed into the **gastrovascular cavity.** Within this cavity, which has only one opening, enzymes are released that partially digest the food. The food particles are then taken up by the cells lining the cavity. These cells complete the digestive process and pass the products on to the other cells of the animal. Water circulating through the cavity supplies dissolved oxygen to the lining cells and carries carbon dioxide, other waste products, and the inedible remains of food particles out through the single opening.

Cnidarians have primitive nervous systems. The polyp form—as in *Hydra*—has a continuous network of nerve cells just below the outer surface. This network links the body into a functional whole and makes possible coordinated movements. The medusa form—exemplified by the jellyfish—has two rings of nerve cells that circle the margin of the bell. Medusas also have two simple types of multicellular sense organs (Figure 26–7).

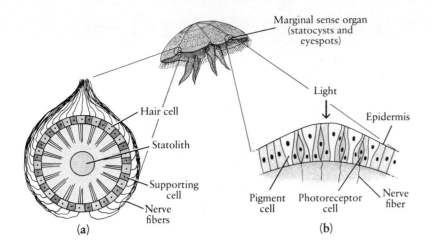

26-7 *Cnidarians have two types of simple sensory organs that are also found in more complex animals.* (**a**) *The statocyst is a specialized receptor organ that orients the animal with respect to gravity. When the bell of a jellyfish tilts, gravity pulls the statolith, a grain of hardened calcium salts, down against the hair cells. This stimulates the nerve fibers and signals the animal to right itself.* (**b**) *An eyespot, the simplest type of photoreceptor organ. In some cnidarians, marginal sense organs contain both statocysts and eyespots, as shown here. In other cnidarians, the eyespots are located at the bases of the tentacles (see Figure 26-6b).*

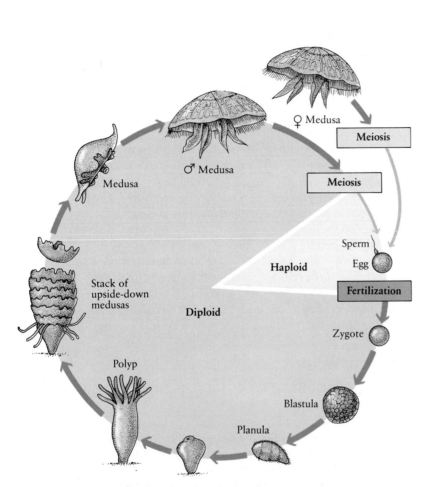

26-8 *The life cycle of the cnidarian* Aurelia. *Sperm and egg cells are released from adult medusas (jellyfish) into the surrounding water (far right). Fertilization takes place, and the resulting zygote develops first into a hollow sphere of cells, the blastula. It then elongates and becomes a ciliated larva called a planula. After dispersal, the planula settles to the bottom, attaches by one end to some object, and develops a mouth and tentacles at the other end, thus transforming into the polyp stage. The body of the polyp grows and, as it grows, begins to form medusas, stacked upside down like saucers. In a phase of rapid asexual reproduction, these bud off, one by one, and grow into full-sized medusas.*

The cnidarian life cycle is characterized by an immature larval form, known as the **planula,** which is a small, free-swimming ciliated organism. Following this larval stage, some cnidarians go through both a polyp and a medusa stage in their life cycle (Figure 26-8). In such species, the planula attaches to a surface and gives rise to a polyp, which reproduces asexually and may form extensive colonies. From the polyp, young medusas may bud off. The medusas are sexually reproducing forms that give rise to planulas again. This life cycle allows for rapid asexual reproduction (by the polyp), dispersal and genetic recombination (by the medusa), and habitat selection (by the planula larva).

Some biologists think that the wormlike animals, the next big step in the order of complexity, evolved from the cnidarians by way of ciliated planula larvae that became sexually mature without going through polyp or medusa stages. Other biologists think the evidence points to independent origins for the wormlike animals from among the ciliates. The new molecular techniques for determining evolutionary relationships (page 407) may ultimately resolve this question.

Flatworms: Phylum Platyhelminthes

The flatworms are the simplest bilaterally symmetrical animals. In such animals, the body is organized along a longitudinal axis, with the right half an approximate mirror image of the left half (Figure 26-9). Bilateral symmetry makes possible more efficient locomotion than does radial symmetry, which is typically found in animals that move slowly or are sedentary. A bilaterally symmetrical animal also has a top and a bottom or, in more precise terms, a **dorsal** and a **ventral** surface. These terms are applicable even when the organism is turned upside down, or, as with humans, stands upright, in which case dorsal means back and ventral means front.

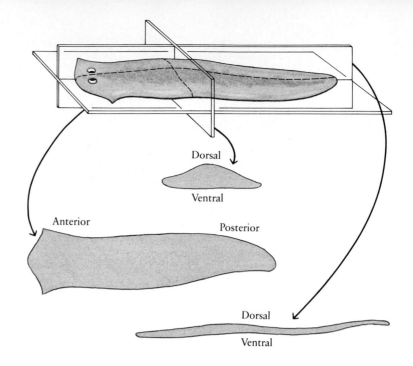

26–9 *In a bilaterally symmetrical organism, such as the planarian shown here, the right and left halves of the body are mirror images of one another. The upper and lower (or back and front) surfaces are known as dorsal and ventral. The end that goes first is termed anterior, and the one that brings up the rear, posterior.*

Like most bilateral organisms, the flatworm also has distinct "head" and "tail" ends, **anterior** and **posterior.** Having one end that goes first is characteristic of actively moving animals. In such animals, many of the sensory cells are collected into the anterior end, enabling the animal to assess an area before entering it. With the aggregation of sensory cells, there came a concomitant gathering of nerve cells into clusters at the anterior end of the animal. These clusters, known as **ganglia** (singular, ganglion), are the forerunner of the brain.

Flatworms, like all animals above the cnidarian level of organization, have three embryonic tissue layers. The third layer is the **mesoderm,** located between the ectoderm and the endoderm. These three layers can be detected very early in development and give rise to the various specialized tissues of the adult animal. Covering and lining tissues, as well as nerve tissues, are generally derived from ectoderm, digestive structures from endoderm, and muscles and most other parts of the body from mesoderm. Not only are the tissues of flatworms specialized for various functions, but also two or more types of tissue cells may combine to form organs. Thus, while cnidarians are largely limited to the tissue level of organization, flatworms can be said to exemplify the organ level of complexity.

The free-living flatworms, of which the freshwater planarian is a good example (Figure 26–10), suck bits of dead animals into their highly branched digestive cavities. The food is digested by cells lining the cavity, and indigestible residues are ejected through the mouth. Two light-sensitive spots at the anterior end resemble a pair of crossed eyes, and two projections on either side of the head are sensitive to chemical stimuli such as those from meat or other food. A fairly complex mus-

culature enables the animal, when disturbed, to move rapidly with a sort of loping motion. Planarians have an extensive excretory system and complex reproductive organs.

This phylum also includes the parasitic flukes and tapeworms. Flukes and tapeworms, like all parasites, are believed to have originated as free-living forms and to have lost certain tissues and organs (such as the digestive cavity) as a secondary effect of their parasitic existence, while developing adaptations of advantage to the parasitic way of life (Figure 26–11). Such adaptations often include a complex life cycle involving two or more hosts.

Ribbon Worms: Phylum Rhynchocoela

The ribbon worms, although a small phylum, are of special interest to biologists attempting to reconstruct the evolution of the invertebrates. They appear to be closely related to the flatworms, but they exhibit two significant new features. Ribbon worms have a one-way digestive tract beginning with a mouth and ending with an anus (Figure 26–12). This is a far more efficient arrangement than the digestive cavity of the cnidarians and flatworms, with its single opening. In the two-opening tract, food moves assembly-line fashion, always in the same direction, with the consequent possibilities (1) that eating can be continuous and (2) that various segments of the tract can become specialized for different stages of digestion. These worms also have a circulatory system, typically consisting of one dorsal and two lateral vessels that carry the colorless blood.

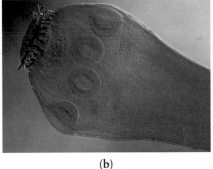

(a)

(b)

(a)

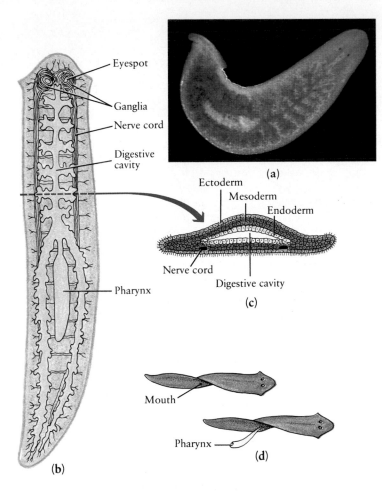

Eyespot

Ganglia

Nerve cord

Digestive cavity

Pharynx

(a)

Ectoderm

Mesoderm

Endoderm

Nerve cord

Digestive cavity

(c)

Mouth

Pharynx

(d)

(b)

26–10 (a) *A flatworm, a freshwater planarian.* (b) *The structure of a planarian. The digestive cavity is shown in yellow, and the nervous system in fine black line. Some of the nerve fibers are aggregated into two cords, one on each side of the body, and, at the anterior, nerve cells are clustered into ganglia. The light-sensitive eyespots enable a planarian to see about as well as you can with your eyes closed.* (c) *Planarians, like other flatworms, but unlike cnidarians, have three layers of body tissues, derived from the three embryonic tissue layers. The only body cavity is the digestive cavity.* (d) *A carnivore, the planarian feeds by means of its extensible pharynx.*

26–11 (a) Dipylidium caninum, *a common tapeworm of dogs and cats. This specimen was removed from a cat. Tapeworms are intestinal parasites that lack any digestive system of their own. They cling by their heads, which, as shown in* Taenia pisiformis (b), *another common tapeworm of dogs, are equipped with hooks and suckers. They absorb food molecules—digested by their hosts—through their body walls.*

26–12 (a) *A ribbon worm of the genus* Lineus, *photographed in Panama. The worm is the long, dull blue ribbon decorated with brilliant gold stripes and blue-green dots. Ribbon worms range in length from less than 2 centimeters to 30 meters and are virtually all the colors of the spectrum. All members of the phylum have thin bodies (seldom more than 0.5 centimeter thick).* (b) *The structure of a ribbon worm. These animals are characterized by a mouth-to-anus digestive tract, a simple circulatory system, and a long, muscular proboscis that can be thrust out to grasp prey and then retracted back into the body.*

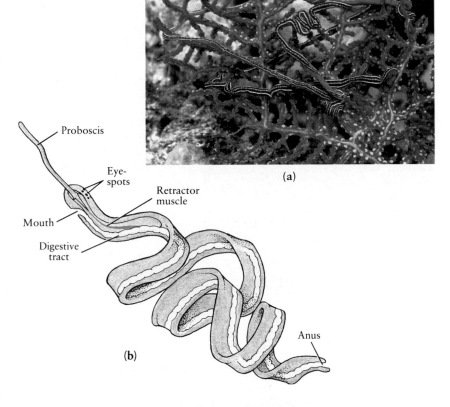

(a)

Proboscis

Eye-spots

Retractor muscle

Mouth

Digestive tract

Anus

(b)

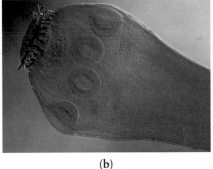

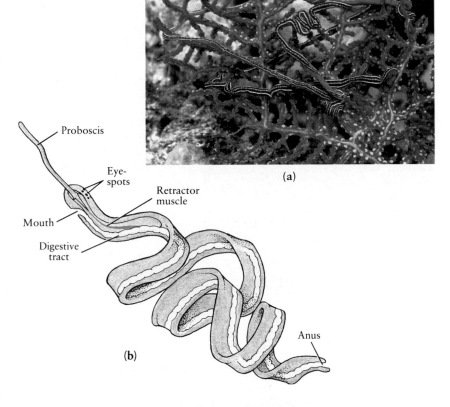

The Politics of Parasites

Throughout the world, parasites are a major cause of death and disability. Estimates of the numbers of people affected are staggering: 150 million new cases of malaria each year (caused, as we saw in Chapter 24, by protists of the genus *Plasmodium*); 200 to 300 million people infected by the blood fluke *Schistosoma*, a flatworm; 650 million people infected by *Ascaris* and 450 million people by *Ancylostoma* (hookworm), both intestinal nematodes; 250 million people infected by *Filaria* and 50 million by *Onchocerca*, both nematodes with minute larvae known as microfilariae. The diseases caused by these organisms and others are epidemic in parts of China, the Middle East, Africa, Latin America, and the Caribbean. Pockets of these diseases, as well as diseases caused by other parasites, also exist in the United States.

All of these parasites have complex life cycles involving two or more different hosts. In the case of three species of the genus *Schistosoma*, the hosts are freshwater snails and humans. The life cycle begins when small, swimming larvae are released from a snail; a single infected snail can shed some 100,000 larvae in its six-month lifetime. The larvae, if

successful, attach to human skin and penetrate it. They feed, mature, and mate in the bloodstream. The female lays her eggs, which have sharp spines on the surface, in the capillaries of the bladder wall or the intestine (depending on the species). The symptoms of schistosomiasis (also sometimes called bilharzia, or snail fever) are caused by the eggs: they lodge in the liver and spleen, blocking blood vessels, and their sharp spines tear the surrounding tissues, causing hemorrhages.

The damage inflicted by *Schistosoma* eggs on the human host can be fatal, but not before the parasite has ensured its return passage to a snail and, ultimately, to its next human host. The eggs leave the human host by passing through the blood vessel wall into the bladder or intestinal tract, where they are flushed out with the urine or feces. If the eggs are deposited in fresh water, they hatch immediately into ciliated larvae that seek out the particular species of snail in which they multiply asexually.

Ironically, western-style progress has often contributed to the spread of parasitic diseases. Snails, for example, thrive in the still waters of irrigation canals and artificial lakes. Schistoso-

miasis spread rapidly through Upper Egypt following the construction of the Aswan High Dam and the digging of permanent irrigation canals. Before the creation of the huge man-made Lake Volta in Ghana, the schoolchildren of the villages along the riverbank had an incidence of schistosomiasis of 1 percent. Now the prevalence in some lakeside villages is 100 percent.

Medical progress has not followed technology into the tropics. When the colonial period ended, the major research programs into tropical diseases ended too. Western medical research is geared overwhelmingly to the conquest of diseases affecting mainly rich, urban, older men and women. For instance, in 1987 the U.S. government alone spent more than $1.4 billion on research on cancer, a disease that afflicts some 10 million people. By contrast, only $8 million was spent worldwide on schistosomiasis and $24 million on malaria. The challenge of schistosomiasis and the other tropical parasitic diseases lies not so much in the disease process itself as in providing incentives (and therefore funding) to cure—or, better yet, prevent—the diseases of the politically powerless rural poor.

(a)

(b)

(a) *A scanning electron micrograph of male and female worms of* Schistosoma mansoni, *one of the three species that cause schistosomiasis in humans. The slender female is embraced in a groove in the much larger body of the male. The two worms remain intertwined except when the female moves away to lay her eggs— about 300 per day. Adult worms live from 3 to* 30 years, during which egg production continues unabated. **(b)** In the snail-infested waters near Aswan, Egypt, bathing children are rapidly infected by Schistosoma larvae. During the years that pass before the symptoms of disease become apparent, large quantities of eggs are shed by the children, continually infecting new generations of snails.

This phylum is called Rhynchocoela ("snout" plus "hollow") because these worms are characterized by a long, retractile, slime-covered tube (proboscis). The proboscis, sometimes armed with a barb, seizes prey and draws it to the mouth, where it is engulfed.

Roundworms: Phylum Nematoda

Roundworms, or nematodes, are unsegmented worms, often microscopic in size (Figure 26–13). They are covered by a thick, continuous cuticle, which is molted periodically as they grow. There are probably several hundred thousand species, but only a small percentage of these have been described and named. Most nematodes are free-living, and it has been estimated that a spadeful of good garden soil usually contains about a million individuals. Some are parasites, and most species of plants and animals are parasitized by at least one species of nematode. Humans are hosts to about 50 species.

Like the ribbon worms, the nematodes have a three-layered body plan and a one-way digestive tract. They also have a new feature—an additional body cavity, known as the **pseudocoelom,** that develops between the endoderm and the mesoderm (Figure 26–14). The pseu-

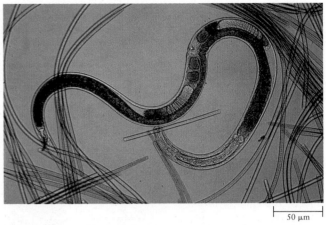

50 μm

26–13 *A free-living freshwater nematode of genus* Monochromadora, *surrounded by filamentous cyanobacteria (Oscillatoria). In addition to a one-way digestive tract, nematodes have the type of body cavity known as a pseudocoelom. The pseudocoelom assists the circulation of body fluids and functions as an internal hydrostatic skeleton. Nematodes have longitudinal muscles but no circular muscles and move with a characteristic whipping motion. This nematode is a female and contains eggs.*

26–14 *Among animals that have attained the tissue level of organization, there are four basic arrangements of the tissues. These arrangements are distinguished by the number of tissue layers and the presence or absence of a body cavity in addition to the central digestive cavity.*

(a) A body that consists of only two tissue layers separated by a gelatinous mesoglea is characteristic of the cnidarians.

(b) Flatworms and ribbon worms have three-layered bodies, with the layers closely packed on one another.

(c) Nematodes and the animals of seven other phyla have three-layered bodies in which a pseudocoelom ("false coelom") develops between the endoderm and mesoderm.

(d) Mollusks, annelids, and most other animals, including vertebrates, have bodies that are three-layered with a cavity, the coelom, that develops within the middle layer (mesoderm). Double layers of mesoderm, known as mesenteries, suspend the gut and other organs within the body wall.

(a) Two-layered, no coelom
(cnidarians)

Ectoderm
Mesoglea
Endoderm
Gastrovascular cavity

(b) Three-layered, no coelom
(flatworms, ribbon worms)

Ectoderm
Mesoderm
Endoderm
Digestive cavity

(c) Three-layered, pseudocoelom
(nematodes, etc.)

Ectoderm
Mesoderm
Pseudocoelom
Endoderm
Digestive cavity

(d) Three-layered, coelom
(mollusks, annelids, arthropods, chordates, etc.)

Ectoderm
Mesoderm
Coelom
Endoderm
Mesentery
Digestive cavity

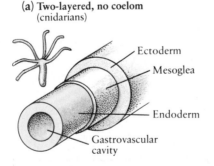
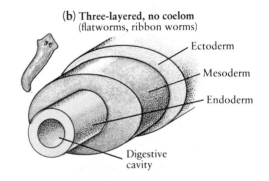
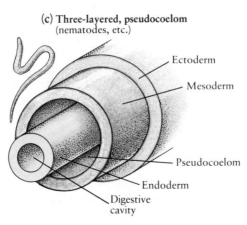
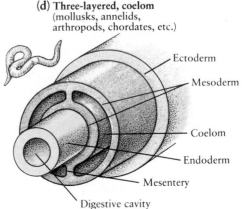

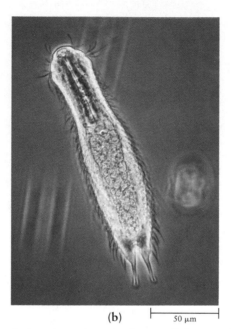

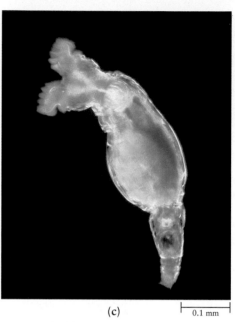

(a) ┠─────┨ 100 μm

(b) ┠───┨ 50 μm

(c) ┠──┨ 0.1 mm

26–15   *Among the residents of the sand and silt of shorelines are members of three pseudocoelomate phyla: Kinorhyncha, Gastrotricha, and Rotifera. (a) Centroderes spinosus, a kinorhynch. Unable to swim, a kinorhynch burrows by forcing fluid into its head. When the head is anchored in the mud* *by its spines, the animal can pull the rest of its body forward. (b) Chaetonotus, a common gastrotrich. Gastrotrichs can both swim and crawl, clinging to surfaces by means of adhesive tubes that project from the sides of their bodies. (c) Philodina gregaria, a freshwater rotifer that lives in the Antarctic. It* *survives the long winter in a protective cyst. When pools and lakes of meltwater form in the summer, it hatches and reproduces in great numbers. Rotifers are sometimes called "wheel animalcules" because the beating of a crown of cilia around the mouth causes them to spin through the water like tiny wheels.*

docoelom, which is essentially a sealed, fluid-filled tube, increases the effectiveness of the animal's muscular contractions. In addition to working against the water or a substrate, muscles must also have something to work against in an animal's body—otherwise the body just bends in the direction of contraction and a floppy, uncoordinated motion results. Because it resists bending, the pseudocoelom functions as a **hydrostatic skeleton** within the animal's body, causing the body to return to its original shape after the muscles have contracted. It makes possible a significant advance over the simple, rather flaccid movements of the animals we have considered previously.

Seven other phyla, quite diverse, share a body plan based upon the pseudocoelom. Representatives of three of these phyla are shown in Figure 26–15.

Mollusks: Phylum Mollusca

The mollusks constitute one of the largest phyla of animals, both in number of species and in number of individuals. They are characterized by soft bodies within a hard, calcium-containing shell. In some forms, however, the shell has been lost in the course of evolution, as in slugs and octopuses, or greatly reduced in size and internalized, as in squids.

Mollusks exhibit a tremendous diversity of form and behavior. The three major classes are (1) the bivalves, including the clams, oysters, and mussels, which have two shells joined by a hinge ligament; (2) the gastropods, such as the snails, whose shells are generally in one piece; and (3) the cephalopods, the most active and intelligent of the mollusks, including the cuttlefish, squids, and octopuses.

Structurally, the mollusks are quite distinct from all other animals. The basic body plan is shown in Figure 26–16a. There are three distinct body zones: a **head-foot,** which contains both the sensory and motor organs; a **visceral mass,** which contains the organs of digestion, excretion, and reproduction; and a **mantle,** a specialized tissue that hangs over and enfolds the visceral mass and that secretes the shell. The **mantle cavity,** a space between the mantle and the visceral mass, houses the gills. The digestive, excretory, and reproductive systems discharge into the mantle cavity.

26–16 *Mollusks are characterized by soft bodies composed of a head-foot, a visceral mass, and a mantle, which can secrete a shell. They exchange gases with the surrounding water through gills, which are protected within the mantle cavity. (In land snails, the mantle cavity has become modified for air breathing.) The hypothetical primitive mollusk (a) illustrates the basic body plan.*

The three major modern classes are the bivalves, the gastropods, and the cephalopods. (b) The bivalves, such as the clam shown here, are generally sedentary and feed by filtering water currents, created by beating cilia, through large gills. (c) In the gastropods, exemplified here by an aquatic snail, the visceral mass has become coiled and rotated through 180° so that mouth, anus, and gills all face forward and the head can be withdrawn into the mantle cavity. (d) In the cephalopods, such as the squid, the head is modified into a circle of arms, and part of the head-foot forms a tubelike siphon through which water can be forcibly expelled, providing for locomotion by jet propulsion. The blue arrows indicate the direction of water movement.

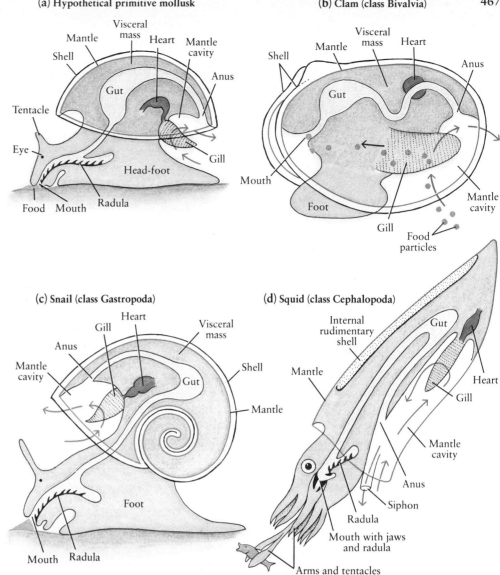

(a) Hypothetical primitive mollusk

(b) Clam (class Bivalvia)

(c) Snail (class Gastropoda)

(d) Squid (class Cephalopoda)

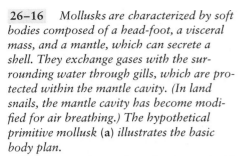

(a)

(b)

(c)

26–17 *Representatives of the three major classes of mollusks. (a) A bivalve, the calico scallop. Its bright yellow eyes, visible among its tentacles, allow the animal to detect light and dark, including passing shadows cast by other moving organisms.*

(b) A land-dwelling gastropod. The shell of this tree snail (Orthaliculus) is secreted by the mantle and grows as the soft body grows. It covers and protects the visceral mass. The head contains sensory organs, including two eyes at the tips of the longer tentacles.

(c) A cephalopod, the lesser octopus, Eledone cirrhosa. Its well-developed eyes are similar in structure to those of vertebrates, and its vision is as acute as ours. Its eight arms, which have numerous suckers on the undersurface, are extremely sensitive to touch and to chemicals in the water. When an octopus becomes aware of a crab, its favorite food, it becomes so excited that its arms weave about and its skin changes color.

A characteristic organ, found only in this phylum, is the **radula**. This tooth-bearing, movable chitinous strap is used both to scrape off algae and other food materials and to convey them backward to the digestive tract. In some species, it is also used in combat. The radula is present in all mollusks except the bivalves, which are filter-feeders.

Throughout the molluscan phylum, there is a wide range of development of the nervous system. The nervous systems of the generally sedentary bivalves and the slow-moving gastropods are relatively simple. By contrast, those of the extremely active cephalopods are much more complex. The octopus, in particular, has a highly developed nervous system and a brain and pair of eyes that rival those of the vertebrates in complexity and are strikingly similar in both structure and function.

Supply Systems

In the protists and in the smaller and simpler animals, oxygen and food molecules are supplied to cells—and waste products removed from them—largely by diffusion, aided by the movement of external fluids. For larger, thicker animals, a more effective method of providing each cell with a direct and rapid line of supply is a circulatory system that propels extracellular fluid—blood—around the body in a systematic fashion.

The molluscan circulatory system consists of a muscular pumping organ, the heart, and vessels that carry the blood to and from the heart. The heart usually has three chambers: two of them (atria) receive blood from the gills, and the third (the ventricle) pumps it to the other body tissues. Except for the cephalopods, mollusks have what is known as an open circulation; that is, the blood does not circulate entirely within vessels. The blood is collected from the gills, pumped through the heart, and released directly into spaces in the tissues, from which it returns to the gills and then to the heart. Cephalopods, whose active lives require that the cells be supplied with large quantities of oxygen and food molecules, have a closed circulatory system of continuous vessels and accessory hearts that propel the blood into the gills.

Oxygen enters the body of a mollusk through the moist surface of the mantle and the gills. A **gill** is an external structure with an increased amount of surface area, through which gases can diffuse, and a rich blood supply for the transport of the gases to and from the rest of the body. Oxygen travels inward by diffusion, down the concentration gradient. The gradient exists because the surface film, exposed to the dissolved oxygen in the passing water, contains more oxygen than does the blood within the gills, which was depleted of oxygen as it passed through the body tissues. Similarly, carbon dioxide, produced by cellular respiration, moves out to the surface film and then into the surrounding water, diffusing down its concentration gradient. In fact, all gas exchange in animals, whether water-dwelling or land-dwelling, takes place by diffusion across moist membranes.

The digestive tract of all mollusks is extensively ciliated and has many different working areas. Food is taken up by cells lining the digestive glands in the anterior portions of the tract, and the products of digestion are passed into the blood. Undigested materials are compressed into mucus-coated fecal pellets, which are discharged through the anus into the mantle cavity and are carried away from the animal in the water currents. This packaging of digestive wastes in solid form prevents fouling of the water passing over the gills.

Nitrogenous wastes produced by the metabolic activities of the cells are removed by one or two tubular structures known as **nephridia**. One opening of each nephridium is in the area surrounding the heart, and the other opening discharges into the mantle cavity. Fluid is forced, under pressure, into the opening near the heart. As the fluid passes through the tubule of the nephridium, water, sugar, salts, and other needed materials are returned to the tissues through its walls, while waste products are secreted into the tubule for excretion. Thus the excretory system is concerned not only with the problem of water balance but also with the regulation of the chemical composition of body fluids.

Segmented Worms: Phylum Annelida

This phylum includes almost 9,000 species of marine, freshwater, and terrestrial worms, including the familiar earthworms. The term annelid means "ringed" and refers to the most distinctive feature of this group: the division of the body into segments, visible as rings on the outside and separated by partitions on the inside (Figure 26–18). This segmented pattern is found in a modified form in arthropods, too, such as millipedes, crustaceans, and insects, which are thought to have evolved from the same ancestors that gave rise to the modern annelids.

Although the annelids differ significantly in their basic body plan from the mollusks, there are some important similarities between the two groups that suggest an evolutionary link. One similarity is the trochophore larva (Figure 26–19). Most marine mollusks (except the cephalopods) and marine annelids pass through this very distinctive larval form during their development.

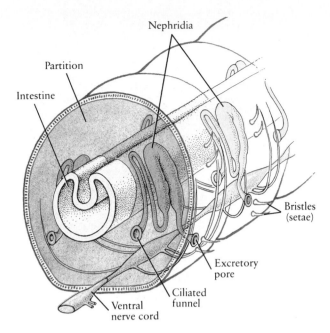

Nephridia

Partition

Intestine

Bristles
(setae)

Excretory
pore

Ciliated
funnel

Ventral
nerve cord

26–18 *The segmented structure of an earthworm, an annelid. On each segment are four pairs of bristles (setae), which are extended and retracted by special muscles. These are used by the worm to anchor one part of its body while it moves another part forward. Two excretory tubes, or nephridia, are in each segment (except the first three and the last). Each nephridium really occupies two segments, since it opens externally by a pore in one segment and internally by a ciliated funnel in the segment immediately in front of it. The intestine, nephridia, and other internal organs are suspended in the large fluid-filled coelom, which also serves as a hydrostatic skeleton.*

A second similarity, shared by the mollusks, the annelids, and all of the remaining animal phyla, is the **coelom** (pronounced "see-loam"), a fluid-filled cavity that develops within the mesoderm (see Figure 26–14d). Although the opening of a cavity within the mesoderm may seem less dramatic than other evolutionary innovations, it is extremely important. Within this cavity, organ systems—suspended by double layers of mesoderm known as **mesenteries**—can bend, twist, and fold back on themselves, increasing their functional surface areas and filling, emptying, and sliding past one another, surrounded by lubricating coelomic fluid. Consider the human lung, constantly expanding and contracting in the chest cavity, or the 6 or 7 meters of coiled human intestine; neither of these could have evolved before the coelom. The coelom, like the pseudocoelom, also constitutes a hydrostatic skeleton, stiffening the body in somewhat the same way water pressure stiffens and distends a fire hose.

In the mollusks, the coelom is greatly reduced and its functions are performed by the blood-filled spaces in the tissues and by the external shell. The annelid coelom, by contrast, occupies a significant proportion of the body and is the ultimate in a hydrostatic skeleton. Annelids have two sets of segmental muscles, one set running lengthwise and the other encircling the segments. These muscles can be contracted independently to shorten or lengthen the body. Moreover, in each segment, the coelom is divided into left and right compartments, which are separated from those ahead and

(a)

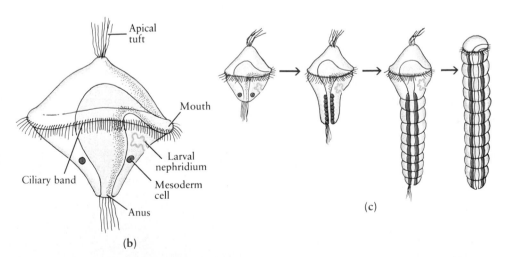

Apical
tuft

Mouth

Larval
nephridium

Mesoderm
cell

Ciliary band

Anus

(b)

(c)

26–19 *(a) The trochophore larva of a marine annelid from the Caribbean. This larva will develop into a polychaete worm. (b) Diagram of a trochophore larva. Although their adult forms are very different, certain annelids and mollusks have larvae of this type,* *suggesting an evolutionary link. (c) Development of an annelid trochophore into a segmented worm. The process begins with the elongation of the lower part of the trochophore. The elongated region then becomes constricted into segments, which soon de-* *velop bristles. The apical tuft disappears, and the upper part of the trochophore becomes the head. The worm's growth will continue throughout its lifetime by the addition of new segments just in front of the rear segment.*

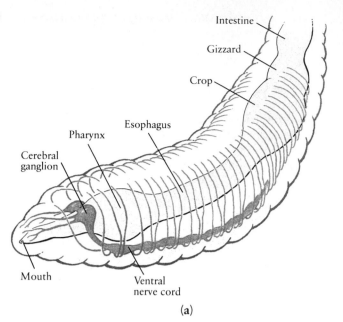

Intestine
Gizzard
Crop
Esophagus
Pharynx
Cerebral
ganglion
Mouth
Ventral
nerve cord

(a)

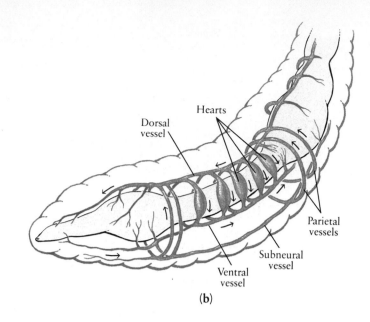

Hearts
Dorsal
vessel
Parietal
vessels
Subneural
vessel
Ventral
vessel

(b)

26–20 (a) *The digestive tract of an earthworm. The mouth leads into a muscular pharynx, which sucks in decaying vegetation and other materials. These are stored in the crop and ground up in the gizzard with the help of soil particles. The rest of the tract is a long intestine (gut) in which food is digested and absorbed.*

(b) *The circulatory system of an earthworm is made up of longitudinal vessels running the entire length of the animal, one dorsal and several ventral. Smaller vessels (the parietal vessels) in each segment collect the blood from the tissues and from the subneural vessel and feed it into the muscular dorsal vessel, through which it is pumped forward. In the anterior segments are five pairs of hearts—muscular pumping areas in the blood vessels—whose irregular contractions force the blood downward to the ventral vessel, from which it returns to the posterior segments. The arrows indicate the direction of blood flow.*

behind. This arrangement allows exquisite control over the movements of small parts of the body. In earthworms, lengthening the body places the bristles of the extended part of the body ahead of the rest of the animal. These bristles anchor the worm while contractions that shorten the posterior segments pull the rest of the body forward. Many of the marine annelids, or polychaetes, augment hydrostatic movements with swimming motions of the lateral appendages of their bodies, the parapodia.

The annelids have a tubular gut and a closed circulatory system that transports oxygen (diffused through the skin) and food molecules (from the gut) to all parts of the body (Figure 26–20). The excretory system consists of paired nephridia (see Figure 26–18) that occur in each segment of the body except the first three and the last. Annelids have a chainlike nervous system that makes possible precise control of the movements of each segment of the body. They also have a number of special sensory cells, including touch cells, taste receptors, light-sensitive cells, and cells concerned with the detection of moisture.

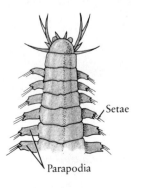

Setae
Parapodia

(a)

(b)

26–21 (a) *A deep-sea polychaete worm. As shown in the diagram (b), this annelid has a well-differentiated head with sensory appendages. The lateral parapodia ("side feet") have numerous bristles (setae). The parapodia are used in locomotion and also provide increased surface area for gas exchange with the surrounding water.*

Arthropods: Phylum Arthropoda

Arthropods, the "joint-footed" animals, are, as we noted earlier, by far the most numerous and diverse animals on Earth. More than 1 million species of insects and other arthropods have been classified to date, and estimates of the total number are as high as 50 million. As to the number of individuals, it has been calculated that of insects alone, as many as 10^{18}—a billion billion—are alive at any one time. Arthropods are abundant in virtually all habitats. It has been estimated that over every square kilometer in the temperate zone there are, at certain seasons, some 20 million individual arthropods, layered in the atmosphere like plankton.

Characteristics of the Arthropods

All arthropods are segmented, a characteristic that strongly suggests a common ancestry with the annelids. In the course of arthropod evolution, however, the body has become shorter, and it has fewer segments, which have become fixed in number and more specialized. In many arthropods, segments have fused to form distinct body regions—a head, a thorax (sometimes fused with the head to form a cephalothorax), and an abdomen. But the basic segmented pattern is often still clearly evident in the immature stages (witness the caterpillar) and can be discerned in the adult by examination of the appendages, the musculature, and the nervous system.

At some point well after the lineage leading to the arthropods diverged from that leading to the annelids, further major branchings occurred (see Figure 26–3, page 457). These branchings gave rise to the three major classes of arthropods: arachnids (spiders, ticks, mites, and scorpions), crustaceans (crabs and lobsters), and insects. There are also six smaller classes including such organisms as centipedes, millipedes, and horseshoe crabs.

26–22   *Some arthropods. (a) A centipede of the genus Scolopendra, photographed in southern Africa. Arthropods, like the annelids, are segmented. In the more primitive forms, the segmented pattern remains clearly visible in the adult animal. However, unlike annelids, adult arthropods have relatively rigid, jointed exoskeletons and appendages. In centipedes, the appendages of the first segment are modified as poison claws.*

(b) A female hunting spider, an arachnid, photographed in a rain forest in Costa Rica as she guarded her cocoon of fertilized eggs. In spiders and other arachnids, the segments of the body have become fused into two regions—cephalothorax ("head-thorax") and abdomen.

(c) This small crustacean, a female copepod, is an inhabitant of the plankton. The two large structures at the posterior (to the right in the photograph) are egg sacs in which the zygotes are developing; they will be released as free-swimming larvae. Copepods are the most numerous animals not only in the plankton but also in the world. The individuals of a single genus (Calanus) are thought to outnumber all other animals put together.

(d) The most numerous of the insects are the beetles, of which some 275,000 species have been described. This is a golden beetle (Plusiotis), photographed in the cloud forest of Costa Rica. Note the three pairs of jointed legs, the jointed antennae, and the segmented body. Beetles have two pairs of wings. When a beetle is at rest, as shown here, the hardened outer wings are held close to the body, protecting the delicate inner wings that are used in flight. The golden beetles of the cloud forest, as well as the related silver beetles, command a high price from collectors, which does not bode well for their future.

(a)

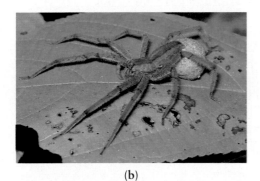

(b)

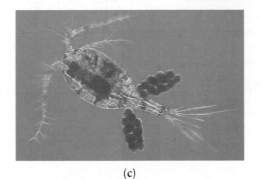

(c)

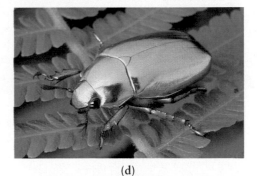

(d)

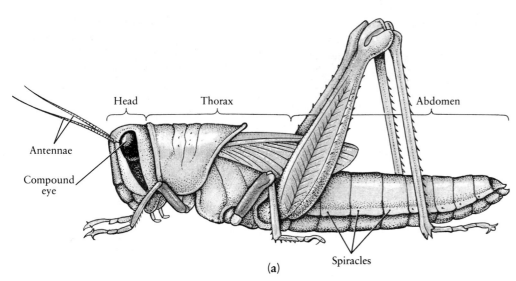

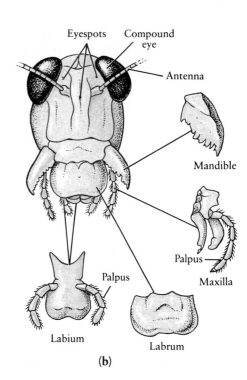

26–23 (a) *In a grasshopper, an insect, the head consists of six fused segments that have appendages specialized for tasting and biting. Each of the three segments of the thorax carries a pair of legs (three pairs in all), and two of them carry wings (in grasshoppers, as in beetles, the forewings are hardened as protective covers). The spiracles in the abdomen open into a network of chitin-lined tubules through which air circulates to various tissues of the body. This sort of tubular breathing system is found only among insects and some other land-dwelling arthropods.*
(b) The complex mouthparts of a grasshopper are modified appendages. The mandibles are crushing jaws. The labium and the labrum are the lower and upper lip. The maxillae move food into the mouth, and the palpi assist in tasting.

The members of the different classes are distinguished by the degree of fusion of their body segments and by characteristic differences in their jointed appendages. In some of the arthropods, such as the centipedes, which have a pair of legs on almost every segment of the body, the jointed appendages are uniform in size and structure. More typically, the appendages are highly differentiated and specialized for walking, swimming, biting, chewing, drinking, mating, and other related functions, depending on the species (Figure 26–23).

Arthropods have a hard outer covering, or **exoskeleton,** containing chitin. It resists predators and, in the land forms, protects the animal from drying out. The exoskeleton is many-jointed and has muscles attached to it. When the muscles contract, the exoskeleton moves at its joints. Such movements can be exquisitely precise because the force is brought to bear on very small areas, for example, on the different sections of an insect's leg.

The exoskeleton covers the animal completely and does not grow. As a consequence, arthropods must molt, a process by which the old exoskeleton is discarded and a new and larger one is formed (see Figure 3–7, page 55). At molting time, the animal secretes an enzyme that dissolves the inner layer of the exoskeleton, and a new skeleton, not yet hardened, is formed beneath the old one. The animal wriggles out from the old skeleton, which splits open. After emerging, the arthropod expands rapidly by taking in air or water, stretching the new exoskeleton before it hardens.

The exoskeleton not only covers the surface of the animal but also extends inward at both ends of the tubular digestive tract and, in insects, lines the **tracheae** (air ducts) as well. These ducts pipe air directly into various parts of the body, with the flow regulated by the opening and closing of special pores (spiracles) in the exoskeleton. Terrestrial arthropods that do not have tracheae have book lungs, which resemble the leaves of a partially opened book. Excretion in terrestrial forms is by means of tubes, known as **Malpighian tubules,** attached to and emptying into the hindgut. Tracheae, book lungs, and Malpighian tubules are found almost exclusively in this phylum.

In arthropods, as in mollusks, the coelom is markedly reduced. The circulatory system is an open one, in which a tubular heart pumps blood through vessels into

26–24 (a) *The arthropod nervous system, as exemplified in a bee. The brain consists of three fused pairs of dorsal ganglia (aggregations of nerve cells) at the anterior end of a double chain of ganglia. The ganglia are interconnected by two bundles of nerve fibers running along the ventral surface. Because of this segmental type of nervous system, many arthropod activities are controlled at the local level, and a number of species can carry on some of their normal activities after the brain has been removed.*
(b) *The head of a Mediterranean fruit fly (medfly),* Ceratitis capitata, *as revealed by the scanning electron microscope. Note the large compound eyes on either side of the head. Although insect eyes cannot change focus, they can define objects only a millimeter from the lens, a useful adaptation for an insect.*

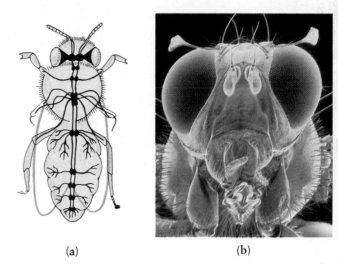

(a)   (b)

spaces in the tissues. Blood returns to the heart through valved openings.

The nervous system of arthropods is complex (Figure 26–24a), making possible the intricate and finely tuned movements involved in activities such as flight, mating in midair, and building webs and hives. Arthropods also have a number of extremely sensitive sensory organs, such as the compound eye (Figure 26–24b) found among crustaceans and insects. They also have a complex hormone-producing system, which plays a major role in molting.

The Success of the Insects

As we noted at the beginning of this chapter, of all the animals, the insects have been the most spectacularly successful. The diversity of insects and the high specificity of the diet and other requirements of each species may be, in part, an evolutionary response to the great diversity of microenvironments provided by the vascular plants. Not only are there 235,000 species of angiosperms alone, but also the structure of each individual plant is so complex that it provides a variety of resources that can be exploited by insects with different requirements and adaptations. The resources provided by the vascular plants were, in effect, an evolutionary laboratory in which great numbers of variations could be tested, leading to the diversity of adaptations we see today. This process, of course, continues.

Another factor in insect success has been the pattern of their life histories, which are fundamentally different from those of marine invertebrates. Although many ma-

rine invertebrate larvae feed, their most important biological function seems to be the invasion and selection of new habitats at some distance from the parental habitat. Such larvae may migrate great distances, resulting in the dispersal of the species over a wide area. The immature forms of most insects, by contrast, have limited mobility and usually pass through the various stages of their development very close to the location where the adult female originally laid the eggs. Insect young are voracious feeders, acquiring the resources needed for their own growth and development and, in some species, storing up reserves for an adult stage in which they will not feed. The capacity to fly has, of course, given most adult insects great mobility, and they have little difficulty finding mates or locating new habitats in which to lay their eggs—thereby ensuring dispersal of the species.

Young growing insects change not only in size but often in form, a phenomenon known as **metamorphosis.** The extent of change varies. In some species, the young, although sexually immature, look like small adults; they grow larger by a series of molts until they reach full size. In others, like the grasshopper, the newly hatched young are wingless and somewhat different in proportions from the adults, but they are otherwise similar. In almost 90 percent of the insect species, however, a complete metamorphosis occurs, and the adults are drastically different from their immature larval forms. Following the larval period, the insect undergoing complete metamorphosis enters a pupal stage, in which extensive remodeling of the organism occurs. The adult insect emerges from the pupa. Such an insect exists in four different forms in the course of its life history: the egg, the larva, the pupa, and the adult (Figure 26–25).

The use of chemicals for communication is common among organisms, and the substances employed range from the sex attractants of the alga *Chlamydomonas* to Chanel No. 5. Many insects and other animals communicate by chemicals known as **pheromones.** These chemical messengers, usually produced in special glands, are discharged into the environment, where they act on other members of the same species.

Pheromones often serve as sex attractants, drawing males to females. Among the best studied are the mating substances of moths. Female gypsy moths, by the emission of minute amounts of a pheromone commonly known as disparlure, can attract male moths that are several kilometers downwind. One female produces about one-millionth of a gram of disparlure, enough to attract more than a billion males if it were distributed with maximum efficiency. The male moth characteristically flies upwind, and the pheromone, of course, disperses downwind. When a male moth detects the pheromone of a female of the species, he will fly toward the source. Since the male can detect as little as a few hundred molecules per milliliter of the attractant, disparlure is still potent even when it has become widely diffused. If the male loses the scent, he flies about at random until he either picks it up

again or abandons the search. It is not until he is quite close to the female that he can fly "up the gradient" and use the intensity of the odor as a locating device.

Pheromones play a variety of other roles in insect communication. Ants lay down pheromones as trail markers, signposts to a food source. When a honey bee stings, she leaves not only the stinger in the victim's skin but also a chemical substance that recruits other honey bees to the attack. Similarly, worker ants of many species release pheromones as alarm substances when they are threatened by an invader; the pheromone spreads through the air to alarm and recruit other workers. If these ants, too, encounter the invader, they will release the pheromone, so the signal will either die out or build up, depending on the magnitude of the threat.

Pheromones have also been found among mammals. The most familiar are the scent-marking substances in the urine of male dogs and cats, which serve as a warning signal to other males. Male mice, it has been found, release substances in their urine that alter the reproductive cycles of the females. The smell of urine from a strange male, for example, can alter hormone balance and interrupt the pregnancy of a female, leaving her free to mate with the newcomer.

A male emperor gum moth (Antheraea eucalypti), *a native of Australia. His lavishly plumed antennae are receptors for the alluring pheromones released by females of the species.*

Sex attractant pheromones have also been found among the primates. It is suspected that sex attractants are exchanged among members of the human species, but no human substance has yet been isolated and definitely identified as a pheromone. However, the fact that the apocrine sweat glands, the sources of "body odor," begin to function at puberty strongly suggests that the chemicals they produce originally played such a role. Similarly, striking differences in the capacity of men and women to detect certain odors may be a clue to the existence of human pheromones.

Both eggs and pupae (which are nonfeeding) can endure lengthy cold or dry seasons, a critical adaptation for terrestrial organisms.

Echinoderms: Phylum Echinodermata

The echinoderms ("spiny skins") are characterized by an internal skeleton, or **endoskeleton,** that bears projecting spines. In most echinoderm species, the endoskeleton is made up of tiny, separate calcium-containing plates held together by skin tissues and muscles.

The echinoderms include the sea lilies, the sea cucumbers, and the more familiar sea urchins, sand dollars, and starfish. A unique feature of these animals is a **water vascular system** (Figure 26–26). This system, which is a modified coelomic cavity, creates a hydrostatic support for the **tube feet,** structures with which many echinoderms cling, move, and attack their prey.

Most adult echinoderms are radially symmetrical, with the body parts arranged in fives. The fossil record, however, shows that radial symmetry is a late, secondary development in the evolution of this group. Echinoderm larvae are bilaterally symmetrical. As they drift in the ocean currents, feeding all the while, different parts of their bodies grow at different rates, preparing the way for the change in body symmetry. At the time of metamorphosis, the larvae temporarily attach to a solid surface and then rapidly transform into adults with a fivefold radial symmetry.

(a)

(b)

(c)

26–25 *Developmental stages of the cecropia moth (Hyalophora cecropia): (a) eggs; (b) caterpillar, the larval form; (c) pupa, the resting form, protected within a cocoon, in which metamorphosis takes place; (d) an adult moth, the reproductive form. This moth is a female.*

(d)

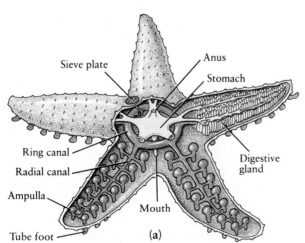

(a)

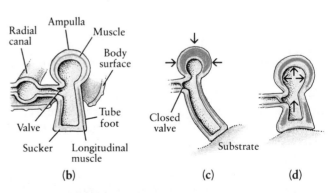

(b)   (c)   (d)

26–26 *(a) Locomotion of the starfish and other echinoderms is supported by the water vascular system, shown here in red. Five radial canals, one for each arm, connect the ring canal with many pairs of tube feet, which are hollow, thin-walled cylinders ending in suckers (b). At the other end of each tube foot is a rounded muscular sac, known as an ampulla. When an ampulla contracts (c), the water in it, prevented by a valve from flowing back into the radial canal, is forced under*

pressure into the tube foot. This stiffens the tube, making it rigid enough to walk on, and extends the foot until it attaches to the substrate by its sucker. Muscles at the base of the foot then contract, forcing the water back into the ampulla (d) and creating the suction that holds the foot to the surface. If the tube feet are planted on a hard surface, such as a rock or a clamshell, their combined force will be great enough to pull the starfish forward or to open the clam.

(a)

(b)

26–27   *The five-part body plan of the echinoderms is readily apparent in the starfish and in the sand dollar (a). Like other echinoderms, the sand dollar has a water vascular system and tube feet. The mouth is on the lower surface.*

In other echinoderms, the basic body plan is harder to see. The protective spines of the sea urchin, for example, hide the

characteristic five rows of tube feet from view. The sea cucumber (b), which exhibits a partial return to bilateral symmetry, also has five rows of tube feet on its body. Depending on the species, sea cucumbers range in length from about 5 millimeters to more than 1 meter. Some tropical species attain a length of 3 meters.

The embryonic development of the echinoderms is significantly different from that of all the animals we have considered thus far. In echinoderms, the anus appears at or near the first opening that forms in the developing embryo, while the mouth breaks through secondarily at the other end of the embryonic digestive tract. Animals with such a pattern of early development are known as **deuterostomes** ("second the mouth"), in contrast to the **protostomes** ("first the mouth"), such as mollusks, annelids, and arthropods. In the protostomes,

the mouth develops at or near the first opening in the early embryo.

The chordates, of which the vertebrates constitute the largest group, are also deuterostomes and share other similarities in embryonic development with the echinoderms. For this reason, the vertebrates, to which we shall turn our attention in the next chapter, are believed to have a closer evolutionary relationship with the echinoderms than with the other invertebrate phyla (see Figure 26–3, page 457).

Summary

Animals are multicellular heterotrophs that depend directly or indirectly on plants or algae as their source of food energy. Almost all digest their food in an internal cavity. Most are motile. Reproduction is usually sexual. More than 95 percent of the animal species are invertebrates, animals without backbones.

Modern animals are classified into about 30 phyla. Criteria for classification include the number of embryonic tissue layers, the basic body plan and the arrangement of body parts, the presence or absence of body cavities, and the pattern of development from fertilized egg to adult.

The sponges (phylum Porifera) are essentially water-filtering systems composed of several different cell types, including flagellated collar cells, epithelial cells, and amoebocytes. Most sponges, like many other sessile or slow-moving animals, are hermaphrodites.

The jellyfishes, sea anemones, and corals of phylum Cnidaria are characterized by radial symmetry; a two-layered body plan in which the two tissue layers, epidermis and gastrodermis, are separated by a jellylike mesoglea; a gastrovascular cavity, in which food is partially digested and oxygen-bearing water is circulated; and special stinging cells, the cnidocytes. Adult cnidarians may take the form of either polyps or medusas. In many species, the life cycle includes an asexually reproducing polyp, a sexually reproducing medusa, and a ciliated planula larva.

The animals in all of the remaining phyla have primary bilateral symmetry, with distinct "head" and "tail" ends and a concomitant clustering of nerve cells in the anterior region. They also have three distinct embryonic tissue layers—ectoderm, endoderm, and mesoderm. The simplest of the bilateral animals are the flatworms (phylum Platyhelminthes), which are characterized by a flattened body and a branched digestive cavity with only one opening. The ribbon worms (phylum Rhynchocoela) are the most primitive animals with a one-way (mouth-to-anus) digestive tract and a circulatory system.

The roundworms (phylum Nematoda) and the animals of seven small phyla have a body plan based on the pseudocoelom, a fluid-filled cavity that develops between the endoderm and the mesoderm. The pseudocoelom functions as a firm hydrostatic skeleton, enabling these animals to move more efficiently than the cnidarians, flatworms, and ribbon worms. Nematodes and other pseudocoelomates have a one-way digestive tract, but they lack a circulatory system.

One of the most significant innovations in the course of animal evolution was the coelom, a fluid-filled cavity that develops within the mesoderm. The coelom not only functions as a hydrostatic skeleton but also provides space within which the internal organs can be suspended.

In the mollusks (phylum Mollusca), soft-bodied animals that typically have an external shell, the coelom has been greatly reduced. The basic body plan of all mollusks is the same—a head-foot, a visceral mass, and a shell-secreting mantle—and they have efficient systems for gas exchange, digestion, circulation, and excretion.

On the basis of similarities in their trochophore larvae, the mollusks and the segmented worms of phylum Annelida are thought to have diverged from a common ancestor. Annelids are characterized by bodies that are conspicuously segmented, well-developed coeloms, tubular digestive tracts, paired excretory structures (nephridia) in each segment, and closed circulatory systems.

Phylum Arthropoda, which includes the arachnids, crustaceans, and insects, is the largest animal phylum in both number of species and number of individuals. Arthropods are segmented animals with jointed chitinous exoskeletons and a variety of highly specialized appendages. In most groups, the segments are combined, forming a head, a thorax (sometimes fused with the head as a cephalothorax), and an abdomen. The arthropods are also characterized by an open circulatory system and a nervous system consisting of a series of ganglia, a pair per segment, interconnected by a double ventral nerve cord. Among the factors contributing to the extraordinary success of the arthropods are their exoskeleton, their generally small size, and their great specialization in both diet and habitat. Additional factors in the success of the insects are the capacity for flight and complete metamorphosis, which reduces competition between adults and immature forms.

The animals of phylum Echinodermata are characterized by an internal, calcium-containing skeleton that bears projecting spines, a five-part body plan with radial symmetry as adults, and a water vascular system. On the basis of an important similarity in embryonic development—the anus appears at or near the first opening that forms in the embryo, with the mouth forming secondarily at the opposite end of the digestive tract—the echinoderms are thought to be the invertebrates to which the chordates are most closely related.

Questions

1. Distinguish among the following: radial symmetry/bilateral symmetry; polyp/medusa; endoderm/mesoderm/ectoderm; pseudocoelom/coelom; planula/trochophore; exoskeleton/endoskeleton; protostome/deuterostome.

2. In which phylum or phyla do you find each of the following and what is its function: collar cell, gastrovascular cavity, cnidocyte, statocyst, mantle, radula, siphon, nephridia, Malpighian tubules, tracheae, spiracles, compound eyes, tube feet?

3. What advantages does radial symmetry provide to sessile organisms? What advantages does bilateral symmetry provide to motile organisms?

4. A pseudocoelom or a coelom provides hydrostatic support for an animal. Why is a gastrovascular cavity less likely to fulfill this same function?

5. How do the parasitic flatworms of phylum Platyhelminthes differ from the free-living flatworms? What general fea-

tures are characteristic of adaptation to a parasitic way of life?

6. Consider the life cycle of the parasitic flatworms of the genus *Schistosoma* (see page 464). At what stages in the cycle do their numbers increase? Why might it be advantageous for a parasite that requires several hosts to have a life cycle in which its numbers increase at several stages? Why might it be advantageous to a parasite to have a second host, such as a freshwater snail?

7. Many mollusks have lost or may be in the evolutionary process of losing their shells. What are the advantages to an organism of having a shell? Of losing one?

8. Smallness may also be an advantage to an organism. What are some of the advantages of smallness?

9. Nematodes have only longitudinal muscles in their body walls but have very high internal fluid pressures. Earthworms, with both longitudinal and circular muscles, have low fluid pressures. Can you suggest a mechanism by which internal fluid at high pressure can circumvent the need for certain muscles?

10. Note that both the mollusks and the arthropods have a greatly reduced coelom. What two features—one obvious, the other not—are also shared by most members of these two groups? What is the correlation between these structural similarities?

11. The arthropods, which include some of the most active animals, have open circulatory systems, often considered inefficient. Annelids, which may share a common ancestor with arthropods, have closed circulatory systems. Presumably, half of the annelid system has been lost in the course of evolution. How could the acquisition of a relatively rigid

exoskeleton make superfluous the vessels returning blood to the heart?

12. Describe gas exchange in a clam, a terrestrial snail, an earthworm, and an insect. How does gas exchange in these animals differ from that in a cnidarian? How is it similar?

13. Describe the water vascular system of a starfish. What are its similarities to and differences from the hydrostatic skeleton of an earthworm?

14. Label the drawing below.

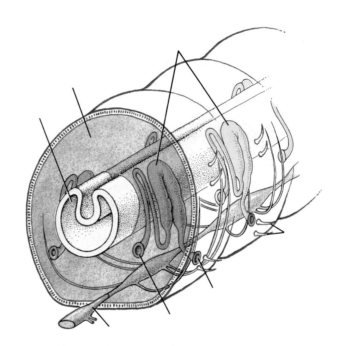

Suggestions for Further Reading

Barnes, Robert D.: *Invertebrate Zoology,* 5th ed., Saunders College Publishing, Philadelphia, 1987.

> *One of the best general introductions to protozoans and invertebrates.*

Barth, Friedrich G.: *Insects and Flowers: The Biology of a Partnership,* Princeton University Press, Princeton, N.J., 1991.*

> *A well-written and beautifully illustrated introduction to the interrelationships of flowers and insects. Incorporating many recent discoveries, the text considers not only the diverse structures of flowers but also the sensory, navigational, and communication abilities of pollinating insects.*

Birkeland, Charles: "The Faustian Traits of the Crown-of-Thorns Starfish," *American Scientist,* vol. 77, pages 154–163, 1989.

Briggs, Derek E. G.: "Extraordinary Fossils," *American Scientist,* vol. 79, pages 130–141, 1991.

Brown, Barbara E., and John C. Ogden: "Coral Bleaching," *Scientific American,* January 1993, pages 64–70.

Buchsbaum, Ralph, Mildred Buchsbaum, John Pearse, and

* Available in paperback.

Vicki Pearse: *Animals without Backbones,* 3d ed., University of Chicago Press, Chicago, 1987.*

> *A delightful introduction to the invertebrates, for the general reader, with a multitude of photographs.*

Cameron, James N.: "Molting in the Blue Crab," *Scientific American,* May 1985, pages 102–109.

Evans, Howard E.: *Life on a Little-Known Planet,* University of Chicago Press, Chicago, 1984.*

> *Professor Evans is the author of many popular articles and books on insects. This book profits from his wide knowledge, clarity, and humor.*

Field, Katharine G., et al.: "Molecular Phylogeny of the Animal Kingdom," *Science,* vol. 239, pages 748–753, 1988.

Ghiselin, Michael T.: "A Movable Feaster," *Natural History,* September 1985, pages 54–61.

Gosline, John M., and M. Edwin DeMont: "Jet-propelled Swimming in Squids," *Scientific American,* January 1985, pages 96–103.

Gould, Stephen Jay: *Wonderful Life: The Burgess Shale and the Nature of History,* W. W. Norton & Company, New York, 1989.*

> *The Burgess Shale, a rock formation in British Columbia, contains one of the world's largest fossil assemblages of soft-bodied invertebrates from the Cambrian period. In this delightfully written book, the reader is introduced not only to these fascinating organisms but also to the scientists who have assessed them—and reassessed them—since they were first discovered in 1909.*

Gray, Jane, and William Shear: "Early Life on Land," *American Scientist,* vol. 80, pages 444–455, 1992.

Hadley, Neil F.: "The Arthropod Cuticle," *Scientific American,* July 1986, pages 104–112.

Horridge, G. A.: "The Compound Eye of Insects," *Scientific American,* July 1977, pages 108–122.

Klots, Alexander B., and Elsie B. Klots: *Living Insects of the World,* Doubleday & Company, Inc., Garden City, N.Y., 1975.

> *A spectacular gallery of insect photos. The text is informal but informative, written by experts for the general public.*

Koehl, M. A. R.: "The Interaction of Moving Water and Sessile Organisms," *Scientific American,* December 1982, pages 124–134.

LaBarbera, Michael, and Steven Vogel: "The Design of Fluid Transport Systems in Organisms," *American Scientist,* vol. 70, pages 54–60, 1982.

Lenhoff, Howard M., and Sylvia G. Lenhoff: "Trembley's Polyps," *Scientific American,* April 1988, pages 108–113.

Levinton, Jeffrey S.: "The Big Bang of Animal Evolution," *Scientific American,* November 1992, pages 84–91.

Macurda, D. Bradford, Jr., and David L. Meyer: "Sea Lilies and Feather Stars," *American Scientist,* vol. 71, pages 354–365, 1983.

McMahon, Thomas A., and John Tyler Bonner: *On Size and Life,* W. H. Freeman and Company, New York, 1985.

> *Written by an engineer and a biologist, this beautifully illustrated book explores the relationship of the size of an organism to its shape, its speed, its physiological functions, its evolution, and its ecology.*

McMenamin, Mark A. S.: "The Emergence of Animals," *Scientific American,* April 1987, pages 94–102.

Morris, Simon Conway: "The Search for the Precambrian-Cambrian Boundary," *American Scientist,* vol. 75, pages 156–167, 1987.

Morse, Aileen N. C.: "How Do Planktonic Larvae Know Where to Settle?" *American Scientist,* vol. 79, pages 154–167, 1991.

Perry, Donald: *Life Above the Jungle Floor,* Simon and Schuster, New York, 1986.*

> *Many, and perhaps most, of Earth's undiscovered animals reside high in the trees of tropical forests. In a lively book, Perry, who devised the climbing and suspension apparatus that is making possible the exploration of this life zone, describes his adventures in the treetops.*

Richardson, Joyce R.: "Brachiopods," *Scientific American,* September 1986, pages 100–106.

Sebens, K. P.: "The Anemone Below," *Natural History,* November 1986, pages 48–53.

Shear, William A.: "One Small Step for an Arthropod," *Natural History,* March 1993, pages 46–51.

Smith, Kathleen K., and William M. Kier: "Trunks, Tongues, and Tentacles: Moving with Skeletons of Muscle," *American Scientist,* vol. 77, pages 28–35, 1989.

Vollrath, Fritz: "Spider Webs and Silks," *Scientific American,* March 1992, pages 70–76.

Wood, Rachel: "Reef-Building Sponges," *American Scientist,* vol. 78, pages 224–235, 1990.

Wootton, Robin J.: "The Mechanical Design of Insect Wings," *Scientific American,* November 1990, pages 114–120.

* Available in paperback.

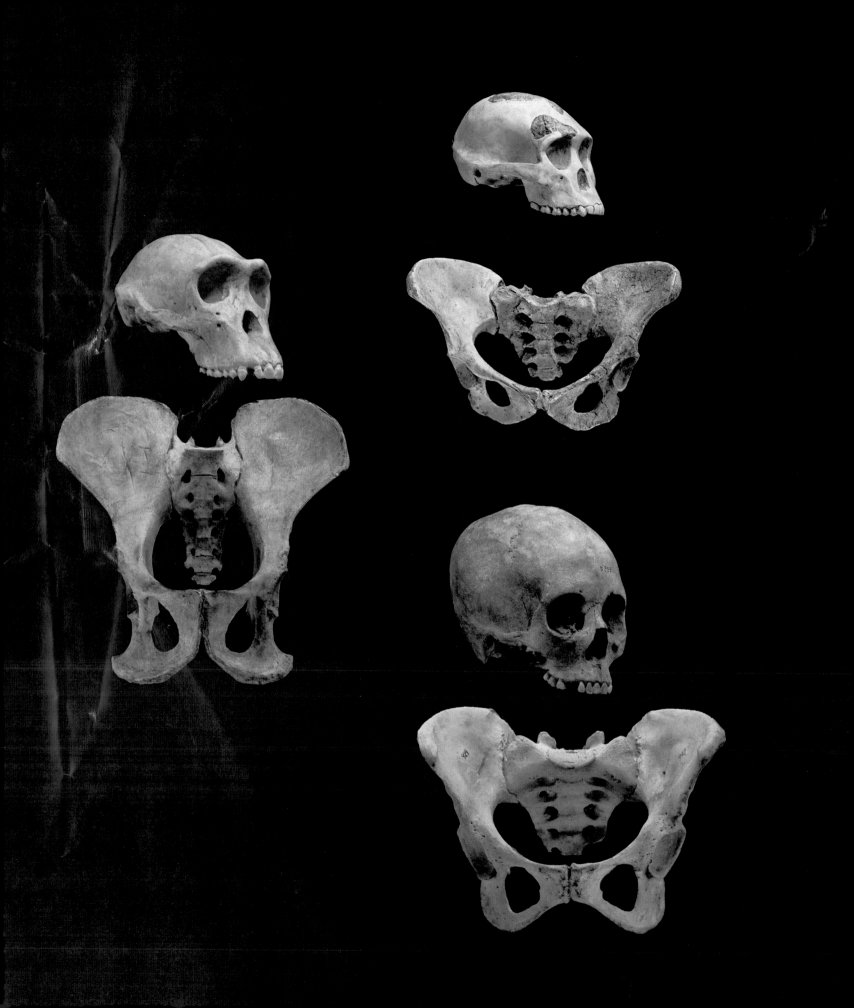

In 1959, Mary Leakey uncovered a fossil skull at Olduvai Gorge in Tanzania. Her discovery marked the beginning of a surge of research into the origins of the vertebrates that intrigue us the most—ourselves. Many previously held concepts about human evolution—who we are and how we came into being—have now been shattered, and new ones, some controversial, have taken their place.

Shown here are the skull and pelvis of a female chimpanzee (left), a fossil fondly known as Lucy (top right), and a modern woman (bottom). Lucy's skeleton, found largely intact, is about 3 million years old. As you can see, her pelvis is much more like that of a modern human than like that of a chimp. Moreover, her arms are proportionately shorter than those of a chimp and her legs are longer. On the basis of this evidence, researchers agree that she walked upright and fully erect, unlike the apes, which stoop over and use their hands as well as their feet when traveling on the ground.

Darwin conjectured that walking upright, using tools, and an increase in brain size were interdependent and evolved together. This idea prevailed for almost 100 years, bolstered perhaps by our desire to attribute our "humanness" to our superior intelligence. Note, however, that Lucy's brain is very small. Bipedalism—walking on two feet—did, of course, eventually free our hands for other things, but it was our feet and not our brain that set us on our present course.

Another long-held belief was that human evolution proceeded in a fairly direct line from ape-man to early *Homo* to us. Now, however, with a wealth of hominid fossils, a completely different picture has emerged. Human evolution was not a ladder of progress but was, like that of *Equus,* a bush with many branches, most of which led to extinction. We are lucky to be here!

Lucy was discovered in 1973 by a group led by Donald Johanson of the Cleveland Museum of Natural History. The camp was "rocking with excitement" the night of the discovery, and a tape of the Beatles' "Lucy in the Sky with Diamonds" was played at top volume, over and over, soaring across the Ethiopian desert. Which is how she got her name.

The Animals II:
Vertebrates

As we saw in the last chapter, the many phyla of invertebrate animals exhibit an enormous diversity of form and life style. The vertebrates, by contrast, share a common body plan that has undergone relatively few modifications in the course of evolution. Those few modifications, however, have had far-reaching consequences, enabling different groups of vertebrates to invade not only the land but also the skies. They led, over millions of years, to the rise of some of the largest organisms ever to inhabit the Earth—and ultimately to the emergence of our own species.

Vertebrates are characterized by a **vertebral column,** or backbone. This flexible, usually bony support forms the structural axis of the animal. Dorsal projections of the vertebrae encircle the nerve cord along the length of the spine. The brain is similarly enclosed and protected, usually by bony skull plates. Between the vertebrae are cartilaginous disks, which give the vertebral column its flexibility. Associated with the vertebrae are segmental muscles by which sections of the vertebral column can be moved separately. This segmental pattern persists in the embryonic forms of higher vertebrates but is largely lost in the course of development.

One of the great advantages of a bony endoskeleton is that it is composed of living tissue that can grow with the animal. In the vertebrate embryo, the skeleton is largely cartilaginous, with bone gradually replacing cartilage in the course of maturation. The growing portions of the bones typically remain cartilaginous until the animal reaches its full size.

There are seven classes of living vertebrates: the fishes (comprising three classes), the amphibians, the reptiles, the birds, and the mammals. These familiar animals, however, represent only one subphylum of the phylum Chordata. To understand the vertebrate body plan and its subsequent modifications, we must look first at the characteristics the vertebrates share with the other members of the phylum. We shall then trace the evolutionary history of the vertebrates, ending with the evolution of *Homo sapiens.*

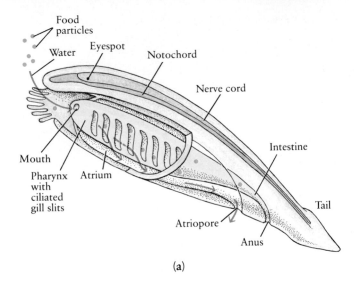

(a)

(b)

27-1 (a) Branchiostoma, *a lancelet, exemplifies the four distinctive chordate characteristics: (1) a notochord, the dorsal rod that extends the length of the body; (2) a dorsal, tubular nerve cord; (3) pharyngeal gill slits; and (4) a tail. Branchiostoma retains these four characteristics throughout its life.*

Many chordates, however, have a notochord, gill slits or pouches, and a tail only during their immature stages.

(b) Although lancelets can swim very efficiently, they spend most of their time buried in the sandy bottom of the seashore with only the anterior end protruding from the substrate.

Characteristics of the Chordates

In addition to some 44,000 living species of vertebrates, the phylum Chordata includes another 1,300 species, grouped in two subphyla: the lancelets, semitransparent animals found in shallow marine waters all over the warmer parts of the world, and the tunicates, of which the most familiar are the sea squirts.

Four features that characterize all chordates, including the vertebrates, are exemplified in *Branchiostoma,* a lancelet (Figure 27–1). The first is the **notochord,** a cartilaginous rod that extends the length of the body and serves as a firm but flexible structural support. Because of the notochord, *Branchiostoma* can swim with strong undulatory motions that move it through the water with a speed unattainable by the flatworms or aquatic annelids. In the vertebrate embryo, the vertebral column develops around the notochord and, in most species, ultimately supplants it.

The second chordate characteristic is the **dorsal, hollow nerve cord,** a fluid-filled tube that runs beneath the dorsal surface of the animal, parallel to and above the notochord. The principal nerve cords in the other phyla, by contrast, are solid and are almost always near the ventral surface.

The third characteristic is a **pharynx with gill slits.** This pharyngeal apparatus becomes highly developed in fishes, in which it serves a respiratory function, and traces of the gill pouches remain even in the human embryo. In *Branchiostoma,* the pharynx serves primarily for collecting food. The cilia on the sides of the gill slits pull in a steady current of water, which passes from the mouth through the pharyngeal slits into a chamber known as the atrium and then exits through the atriopore. Food particles are collected in the sievelike pharynx, mixed with mucus, and channeled along ciliated grooves to the intestine.

The fourth characteristic is a **tail,** posterior to the anus, consisting of blocks of muscle around an axial skeleton. Most of the body tissue of *Branchiostoma* is made up of blocks of muscles.

The chordates are thought to have arisen from a group of organisms that resembled the modern tunicates. Adult tunicates do not have all of the typical chordate features, but the larvae are clearly chordates (Figure 27–2).

Although the origins of the earliest vertebrates from their chordate ancestors are lost in the mists of time, the subsequent evolution of the vertebrates is clearly documented in the fossil record.

Fishes

The first fishes were jawless and had a strong notochord running the length of their bodies. Today the jawless fishes (class Agnatha), once a large and diverse group, are represented only by the hagfish and the lampreys. They have a notochord throughout their lives, like *Branchiostoma.* Although their ancestors had bony skeletons, modern agnaths have a cartilaginous skeleton.

482

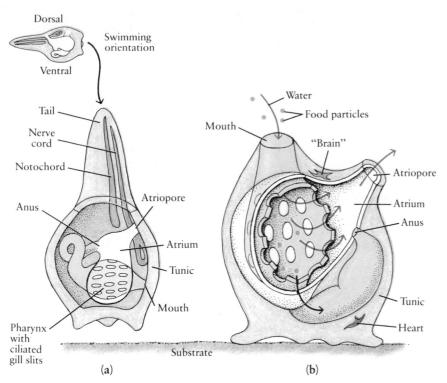

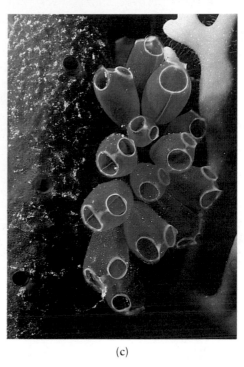

27–2   *Two stages in the life of a tunicate: (a) the larva, and (b) the adult form. In the larva, the tunic covers the mouth and atriopore, preventing the flow of water through the pharynx and its ciliated gill slits. Thus, even in species in which the larval pharynx is well developed, the larva is unable to feed. After a brief free-swimming existence, it set- tles to the bottom and attaches at the anterior end, as shown in (a). Metamorphosis then begins. The larval tail, with the notochord and dorsal nerve cord, disappears, and the animal's body is rotated 180°. The mouth is carried backward to open at the end opposite that of attachment, and all the other internal organs are also rotated back. It has been hypothesized that the ancestral vertebrates arose from tunicate larvae that became sexually mature, and thus capable of reproducing, without undergoing metamorphosis.*

(c) Living tunicates, or sea squirts.

27–3   *A sea lamprey (class Agnatha), which normally attaches to other fish and feeds on their blood, has attached itself to a rock. Respiration occurs through the prominent gill pouches, seen here as seven indentations.*

The sharks (including the dogfish) and skates, the second major class of fishes (the Chondrichthyes), also have a cartilaginous skeleton. Their ancestors, like those of the jawless fishes, were bony animals. In these fishes, the skin is covered with small, pointed teeth (denticles), which resemble vertebrate teeth structurally and give the skin the texture of coarse sandpaper.

The third major class of fishes includes those with bony skeletons, the Osteichthyes. This group of more than 21,000 species includes the trout, bass, salmon, perch, and many others—almost all of the familiar freshwater and saltwater fishes.

One of the major events in the evolution of fishes— and of the vertebrate groups descended from them— was the transformation of the anterior gill arches of filter-feeding fish into jaws. The development of powerful jaws, often armed with formidable teeth, greatly increased the range of other organisms on which the fish could feed. With more efficient feeding on larger and more concentrated sources of energy came the possibility of significant increases in size.

27–4   *A modern lungfish,* Protopterus dolloi. *When the dry seasons come, members of this African genus wriggle downward into the mud, which eventually hardens around them. Mucus glands under the skin secrete a watertight film around the body, preventing evaporation. Only the mouth is left exposed. During this period, the fish takes a breath only about once every two hours.*

27–5   *Many toads are clearly fishlike in their larval (tadpole) stages. As adults, they require water to reproduce and their moist skins are an important accessory respiratory organ. Toads have a warty skin, compact bodies, and shorter legs than frogs (with toads' legs seldom appearing on the menus of stylish restaurants). Like all adult amphibians, toads are carnivores. They catch insects with a flick of their long tongues, which are attached at the front of their mouths and which have a sticky, flypaper-like surface. This common toad (Bufo bufo) is about to capture a beetle larva (a grub).*

The Transition to Land

According to present evidence, the first fishes lived in fresh water. The cartilaginous fishes moved to the sea early in their evolution, while the bony fishes went through most of their evolution in fresh water and spread to the seas at a much later period. Fresh water is often shallow and, unlike ocean water, can become stagnant (depleted of oxygen). The primitive bony fishes seem to have had simple lungs or lunglike structures that served as accessories to the gills, enabling the fishes to survive in stagnant water.

Lunged fishes apparently evolved independently several times, and they were the most common fishes in the later Devonian period, a time of recurring drought (see the geologic timetable, inside the front cover). In most of these fishes, the lung subsequently evolved into a swim bladder, which serves as a flotation chamber. A fish alters its buoyancy—and thus maintains its position at a given depth—by adding gases to or removing them from the swim bladder via the bloodstream.

Some of the primitive lunged fishes evolved into the modern lungfishes (Figure 27–4). These fishes, which surface and gulp air into their lungs, can live in water that does not have sufficient oxygen to support other fish life. In other lunged fishes, skeletal supports evolved that served to prop up the thorax. These fishes could gulp air even when their bodies were not supported by water. It is thought that they could waddle, dragging their bellies along the muddy bottom of a drying stream bed, to seek deeper water or perhaps even make their way from one water source to another one nearby. Thus the transition to land may have begun as an attempt to remain in the water.

Amphibians

The amphibians are descended from air-breathing lunged fishes. One of the earliest amphibians, *Icthyostega,* is shown as the ancestral tetrapod in Figure 23–4 on page 404. It lived about 350 million years ago.

The 2,500 species of modern amphibians include frogs and toads (which typically lack tails as adults) and salamanders (which have tails throughout their lives). They can readily be distinguished from the reptiles by their thin, usually scaleless skins, which serve as respiratory organs. Adult frogs also have lungs, into which they gulp air, but some salamanders respire entirely through their skins and the mucous membranes of their throats. Because water evaporates rapidly through their skins, amphibians can die of desiccation in a dry environment.

Most frogs in cold climates have two life stages, one in water and the other on land (hence their name, from *amphi* and *bios,* meaning "both lives"). The eggs are laid in water and are fertilized externally. They hatch into gilled larvae (tadpoles). The tadpoles later develop into adults that lose their gills and develop lungs. The adults may live out of the water, at least in the summer.

There are, however, many variations on this theme. Some of the American salamanders fertilize their eggs internally. The males deposit sperm packets, either in water or on moist land, and these packets are picked up by the females. Many modern amphibians skip the free-living larval stage. The eggs, which may be laid on land, in a hollow log or cupped leaf, or may even be carried by the parent, hatch into miniature versions of the adult.

Reptiles

As we saw in Chapter 25, the vascular plants were freed from the water by the evolution of the seed. Analogously, the vertebrates became truly terrestrial with the evolution in the reptiles of the **amniote egg** (Figure 27–6), an egg that retains its own water supply and so can survive on land. The reptilian egg, which is much like the familiar hen's egg in basic design, contains a large yolk, the primary food supply for the developing embryo, and abundant albumen (egg white), which supplies additional nutrients and water. A membrane, the **amnion,** surrounds the developing embryo with a liquid-filled space that substitutes for the ancestral pond. The embryo passes through a gill-like stage, either in a shelled egg or in the maternal reproductive tract. In mammals also, although their eggs typically lack shells and develop internally, the embryos are enclosed in water within the amnion and pass through a stage with gill pouches before birth.

Reptiles are characteristically four-legged, although the legs are absent in snakes and some lizards. They have a thick, dry skin, usually covered with protective scales, that makes possible their terrestrial existence. Modern reptiles, of which there are about 6,000 species, include lizards, snakes, turtles, and crocodiles.

Evolution of the Reptiles

By late in the Carboniferous period, the first reptiles had begun to evolve from amphibian ancestors. During the succeeding Permian period, there was an explosive increase in the number of reptilian species. During this same period, conifers began to replace the ferns and other "amphibious" plants, suggesting that a drier climate may have been a primary selective force in both of these events.

Throughout the Permian and much of the Triassic, the dominant land vertebrates were mammal-like reptiles, an abundant and diverse group, members of which were later to give rise to the mammals. Around the beginning of the Triassic period, several of the more specialized reptile groups arose, including turtles, lizards, and the ancestors of the Archosauria, or ruling reptiles, perhaps the most spectacular of all the land's inhabitants so far.

The origin and rise of the Archosauria were associated with changes in reptilian locomotion. Vertebrates first came to the land with all four limbs sprawled far out to the side; turtles have retained this sprawling gait. Among the early Archosauria, there was a progressive tendency toward **bipedalism** (walking on two feet), with a concomitant freeing of the front legs for other purposes—flight, for example. There were three major groups of archosaurs: the pterosaurs, or flying reptiles; the crocodilians, which, reverting to (or perhaps retaining) four-legged posture, became our modern crocodiles and alligators; and the dinosaurs, a varied and splendid group of reptiles.

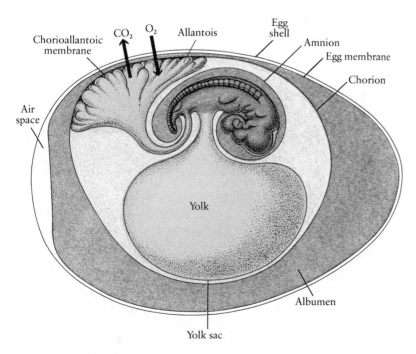

27–6 *Amniote egg. The membranes, which are produced as outgrowths from the embryo as it develops, surround and protect the embryo and the yolk (its food supply). The egg shell and egg membrane, which are waterproof but permeable to gases, are added as the early embryo passes down the maternal reproductive tract.*

(a)

(b)

27–7 (a) *A painted turtle,* Chrysemys picta. *The dorsal carapace (shell) of turtles and tortoises is partly fused to the vertebral column and the ribs, with a mosaic of horny plates on the surface. Unlike other reptiles, most tortoises do not molt but add new epidermal scales on the undersurface. This re-* sults in the addition of a growth ring each year.

(b) *Alligators and crocodiles, the largest modern reptiles, lay their eggs on land, and their skins are reinforced with horny scales. Crocodiles, which are essentially tropical an-* imals, have more slender snouts than alligators. Alligators have jaws that are broader and more rounded anteriorly. They are also reported to be less aggressive. The animal shown here is the American alligator,* Alligator mississipiensis.

The dinosaurs were long assumed to have been **ectothermic**—that is, to have maintained their body temperatures within broad limits by taking in heat from the environment or by giving it off to the environment. Some biologists contend, however, that at least some groups of dinosaurs were **endothermic**—that is, their body temperatures were maintained by heat generated internally, as are the body temperatures of birds and mammals. Only a few modern reptiles, such as leatherback turtles, show any degree of endothermy.

Throughout the long Mesozoic era, the dinosaurs dominated the life of the land, rulers of the Earth for 150 million years. Then, about 65 million years ago, they vanished, leaving only a single line of descendants, the birds. The cause of their extinction has been the subject of speculation since the first dinosaur fossils were discovered in the nineteenth century, but it is only recently that new data have emerged, making possible the formulation of testable hypotheses (see page 369).

Birds

Birds are essentially reptiles specialized for flight (Figure 27–8). Their bodies contain air sacs, and their bones are hollow. The frigate bird (page 361), a large seagoing bird with a wingspread of more than 2 meters, has a skeleton that weighs only 110 grams (about 4 ounces). The most massive bone in the bird skeleton is the sternum, or breastbone, which bears the keel, to which the huge muscles that operate the wings are attached. Flying birds have jettisoned all extra weight. For example, the female's reproductive system has been trimmed down to a single ovary, and even this becomes large enough to be functional only in the mating season.

27–8 *A cast of one of the six known specimens of* Archaeopteryx, *the most extensively studied fossil bird, which dates from the late Jurassic period, about 150 million years ago. Its skeleton is very similar to that of a small bipedal dinosaur, and it had many reptilian characteristics. The teeth and the long, jointed tail are not found in modern birds. The clearly evident feathers may have been related as much to endothermy as to flight.*

486

Birds have feathers, which is their outstanding, unique physical characteristic. They are endothermic, generating heat by internal metabolic processes and maintaining a high and constant body temperature. In modern birds, feathers make flight possible and also serve as insulation. (Only animals that are endothermic require insulation; insulation would be a disadvantage for animals that warm their bodies by exposure to the environment.) Birds also have scales on their legs, a reminder of their reptilian ancestry. Many birds hatch at a very immature stage, and virtually all birds require a long period of parental care.

Evolution of Flight

How did flight evolve? Biologists agree that evolution occurs by a series of changes, each of which, to be conserved by natural selection, must be of survival value. Being able to fly—but not very well—is a dubious advantage. One widely held hypothesis for the origin of flight is that the ancestors of the birds were tree-dwelling reptiles and that flight evolved from gliding, as a way to extend or brake jumps from branch to branch.

An alternative hypothesis is based on fossil evidence that *Archaeopteryx* was a close relative of small, bipedal, carnivorous dinosaurs known as theropods, one of the groups of dinosaurs now thought to have been endothermic. According to this hypothesis, the early stages of flight began with a feathered, endothermic, carnivorous dinosaur running after its prey, flapping its long feathered arms, and leaping. The fact that the feathered forelimbs of *Archaeopteryx* ended in claws, as did the elongated arms of the theropods, lends support to this image. Natural selection may have favored the evolution of the long "wing" feathers because they increased the speed of the running predator or because they served as cagelike traps—natural nets—for capturing prey. Some modern predatory birds use their wings in this way. Thus, in this hypothesis, flight is seen as the culmination of a long, successful predatory leap.

Whether flight evolved from the "trees down" or from the "ground up," it made available to the birds a new and vast life zone, filled with a previously inaccessible food supply—airborne insects. An enormous diversification followed, producing not only the 9,000 species of living birds but also some 14,000 species that have become extinct.

Mammals

The mammals, like the birds, are descended from the reptiles. Characteristics distinguishing these animals from other vertebrates are that mammals (1) have hair, (2) provide milk for their young from specialized glands (mammary glands), and (3) like birds, but unlike most other vertebrates, maintain a high body temperature by generating heat metabolically.

Nearly all mammalian species bear live young, as do some fish and reptiles, which retain the eggs in their bodies until they hatch. However, some very primitive mammals, the **monotremes,** such as the duckbilled platypus, lay eggs with shells but nurse their young after hatching. By contrast, the **marsupials,** which include the opossums and the kangaroos, bear live young. They differ from the largest group of mammals, however, in that the infants are born at a tiny and extremely immature stage and are often kept in a special protective pouch in which they suckle and continue their development (Figure 27–9). Most of the familiar mammals are **pla-**

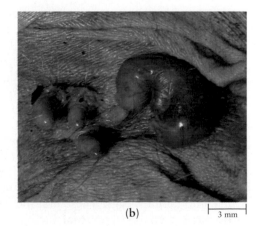

27–9   *Marsupial infants are born at a very immature stage and continue their development in a special protective pouch of the mother.* (a) *A newborn Tammar wallaby (Macropus eugenii), still attached to the umbilical cord. Within 15 seconds of its birth, this tiny infant began using its sharp claws to tear through the fetal membranes in which it was enclosed. Freed from the membranes and oriented upward, it is about to begin the long climb from just outside its mother's womb to the warmth, safety, and food supply of her pouch. As the newborn moves through her fur, it appears to be swimming.* (b) *The journey completed, the newborn has attached itself to a nipple and begun to feed.*

(a)

(b)

3 mm

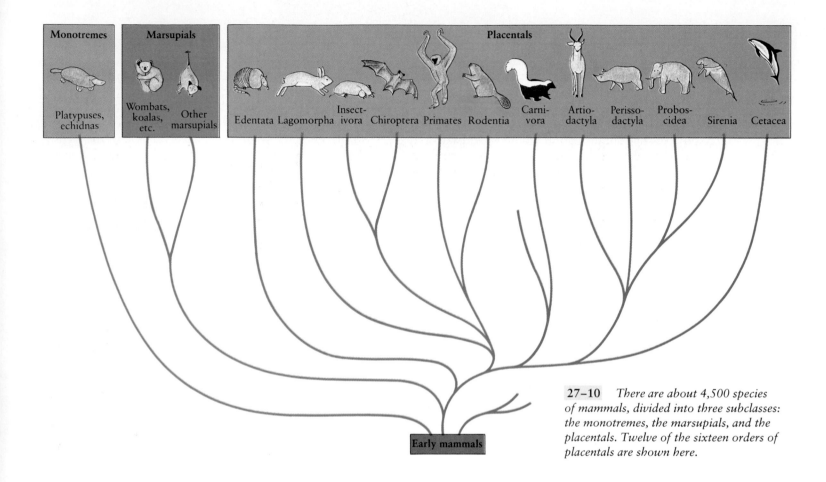

27–10   *There are about 4,500 species of mammals, divided into three subclasses: the monotremes, the marsupials, and the placentals. Twelve of the sixteen orders of placentals are shown here.*

centals, so called because they utilize their efficient nutritive connection, the placenta, between the uterus and the embryo for a relatively long period of time. As a result, the young develop to a much more advanced stage before birth and are afforded protection during their most vulnerable period.

Evolution of the Mammals

In the early Mesozoic era, some 200 million years ago—give or take a few million years—the first mammals appeared, arising from a primitive reptilian stock. Our information about these animals is very slight. The Jurassic and Cretaceous periods have left us with only a few fragments of skulls and some occasional teeth and jaws. From these scraps of evidence, we know that the first mammals were about the size of a mouse. They had sharp teeth, indicating that they were basically carnivorous. Since they were too small to attack most other vertebrates, they are assumed to have lived on insects and worms, supplementing their diet with tender buds, fruits, and perhaps eggs. These first mouse-sized mammals were probably **nocturnal** (active at night), thus avoiding the carnivorous dinosaurs that were active during daylight hours, and they were almost certainly warm-blooded. If such an animal were alive today, it would resemble an insectivore, such as a ground shrew.

For about 130 million years, these small mammals led furtive existences in a land dominated by reptiles. In the course of this time, they diverged into the three principal lineages: monotremes, marsupials, and placentals. Then suddenly, as geologic time is measured, the giant reptiles, the dinosaurs, disappeared. With their demise, a great diversity of previously occupied habitats became available. This was followed almost immediately by an explosive adaptive radiation of the mammals, giving rise to a variety of marsupials and about two dozen different lines of placentals (Figure 27–10).

Among the placentals are carnivores, ranging in size from the saber-toothed cats down to small, weasel-like creatures; herbivores, which include not only the many wild grazing animals but also most of our domesticated farm animals; the omnipresent rodents; and such groups as the whales and dolphins, the bats, the modern insectivores, and the primates. We are placental mammals and members of the primate order. The story of our evolution begins with the earliest primates.

(a)

(b)

(c)

(d)

(e)

27–11 *An assortment of mammals. (a) Lagomorphs, such as the snowshoe hare shown here in its summer coat, have two pairs of upper incisors. By contrast, rodents, such as the beaver (b), have only one pair. In both lagomorphs and rodents, the teeth grow continuously.*

(c) Carnivores, such as this young male lion chasing a herd of zebras and springbok, are adapted to hunt and kill for food. The zebras are perissodactyls (odd-toed ungulates), and the springbok are artiodactyls (even-toed ungulates). (d) Hippopotamuses, which are also artiodactyls, graze by night and spend most of the daylight hours resting in the water.

(e) Elephants, the Proboscidea, are the largest land mammals living today; some reach a weight of 7.5 metric tons. As a consequence of both the loss of habitat and the killing of elephants for the ivory in their tusks, these magnificent animals are seriously endangered. From 1979 to 1989, the number of elephants on the entire continent of Africa dropped from about 1.3 million to 609,000. An international ban on the commercial trade of ivory and all other elephant-derived products, enacted in 1989, has reduced the number of animals lost to poachers, but the long-term survival of elephant populations remains far from certain.

Trends in Primate Evolution

Primate evolution is thought to have begun when a group of small, shrewlike mammals took to the trees. Most trends in primate evolution seem to be related to various adaptations to arboreal life.

The Primate Hand and Arm

The first four-legged mammals all had five separate digits on each foot. Each digit, with the exception of the first toe, had three separate segments, making it flexible and capable of independent movement. In the course of evolution, selection pressures for greater efficiency in running, digging, and seizing prey led to specialized hooves and paws in most mammals. In other mammals, selective forces led to the modification of limbs as flippers for swimming. The primates, however, retained and elaborated on the basic five-digit pattern, and, with few exceptions, modern primates have hands in which the thumb is divergent. The divergent thumb, which can be brought into opposition to the forefinger, greatly increases gripping power and dexterity. There is an evolutionary trend among the primates toward finer manipulative ability that reaches its culmination in humans (Figure 27–12).

In the basic quadrupedal structure of the early mammals and reptiles, the forelimb has two long bones (the radius and the ulna), a pattern that provides for flexibility. Among mammals, it is the primates, in particular, that can twist the radius, the bone on the thumb side, over the ulna so that the hand can be rotated through a full semicircle without moving the elbow or the upper arm. Similarly, only a few mammals have the ability to move the upper arm freely in the shoulder socket. A dog or horse, for instance, usually moves its legs in only one plane, forward and backward. Humans, apes, South American monkeys, and some lemurs are among the few mammals that can rotate the arm widely in the socket, an advantageous characteristic for a tree-dwelling animal.

Most primates also have nails rather than claws. Nails leave the tactile surface of the digit free and so greatly increase the sensitivity of the digits for exploration and manipulation.

Visual Acuity

Another result of the move to the trees is the high premium placed on visual acuity, with a decreasing emphasis on the sense of smell, the most important of the senses among many of the other mammalian orders. (Flying produced similar evolutionary pressures among the birds, which were also evolving rapidly during this same period.) This shift from dependence on smell to dependence on sight has anatomical consequences. Among the primates, one can trace a steady evolutionary trend from laterally directed eyes to frontally directed eyes and stereoscopic vision. Almost all primates have color vision, and most primate eyes have areas of closely packed photoreceptor cells that produce sharp visual images.

Care of the Young

Another trend in primate evolution is toward increased care of the young. Because mammals, by definition, nurse their young, they tend to have longer, stronger mother-child relationships than other vertebrates (with the exception, in some cases, of birds). In the larger primates, the young mature slowly and have long periods of dependency and learning.

27–12 *Some primate hands. The hand of the tarsier has enlarged adhesive skin pads for grasping branches. In the orangutan, the fingers are lengthened and the thumb reduced, which provide for efficient brachiating (swinging arm over arm through the trees). The gorilla's hand, which is used in walking as well as handling, has shortened fingers. The human thumb is larger proportionately than that of any other primate, and opposition of thumb and fingers, on which handling ability depends, is greatest in humans.*

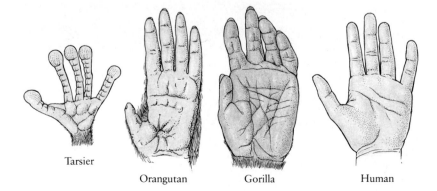

Tarsier

Orangutan   Gorilla   Human

(a)

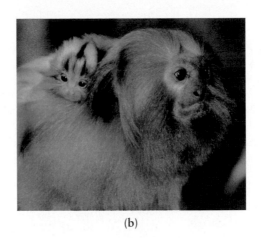

(b)

(c)

27–13   *Life in the treetops made maternal care a major factor in infant survival. Also the necessity for carrying the young for long periods resulted in strong selection pressures for reduced numbers of offspring.* **(a)** *A ring-* *tailed lemur mother and her offspring, sharing a snack. Lemurs are, as you know, native to Madagascar.* **(b)** *New World monkeys, such as the golden lion tamarin monkey, spend most of their time in the trees. They* *usually have single births.* **(c)** *A mother gorilla with her infant. Field studies suggest that bonds between mother and offspring and perhaps also among siblings last well into adulthood, perhaps for a lifetime.*

(a)

(b)

27–14   *Two prosimians.* **(a)** *A lesser bush baby. Bush babies move through the trees in a series of rapid bounds, with their bodies held upright. They sometimes leap 3 or 4 meters through the air from one tree to another. Their fingers work together against the thumb, not independently like human fingers.* **(b)** *A native of Indonesia, the little tarsier (about the size of a kitten) is specialized for leaping between vertical supports. Living entirely in trees, it has hands and feet with enlarged skin pads for grasping branches. Tarsiers have stereoscopic vision and, as you may have guessed from the owl-like eyes, are primarily nocturnal.*

Lemurs, the prosimians of Madagascar with which we began this text, are shown in Figure I–2 (page 3).

Uprightness

Another adaptation to arboreal life is the ability to adopt an upright posture. Even quadrupedal primates, such as monkeys, can sit upright. One consequence of this posture is a change in the orientation of the head, allowing the animal to look straight ahead while in a vertical position. It is this characteristic, above all others, that makes our fellow primates look so "human" to us. Vertical posture was an important precondition for the eventual evolution of the upright stance characteristic of modern humans.

Major Lines of Primate Evolution

Primates are generally divided into two major groups: the **prosimians** (lorises, bush babies, tarsiers, and lemurs) and the **anthropoids** (monkeys, apes, and humans).

Prosimians

During the Eocene epoch (about 55 to 38 million years ago), a great abundance and variety of prosimians inhabited the tropical and subtropical forests that spread much farther north and south of the Equator than they do today. Modern prosimians (Figure 27–14) are mostly small to medium-sized arboreal animals, and many are nocturnal. Insects typically form at least part of the diet of the smaller prosimians, while the larger ones eat varying combinations of leaves, fruits, and flowers.

Anthropoids

Monkeys

Monkeys, along with apes and humans, make up the higher primates, the anthropoids. Modern monkeys are generally larger than modern prosimians, and their skulls are more rounded. They are considered to be more intelligent, although this is an elusive quality to measure. They have full stereoscopic vision and also the ability to distinguish colors. Virtually all monkeys are **diurnal** (active during the day).

The monkeys probably arose from prosimian stock during the Eocene epoch. Related fossil forms are found in the New World as early as the Oligocene epoch. There are two principal groups (Figure 27–15): the New World monkeys, also known as **platyrrhines** (meaning "flat-nosed"), and the Old World monkeys, the **catarrhines** ("downward-nosed"). The separation of these groups took place with the breakup of Gondwana (see page 358), with the platyrrhines evolving in South America and the catarrhines in Africa, quite possibly during the Oligocene, some 38 to 25 million years ago.

Apes

A branching of the catarrhines gave rise to the **hominoids,** a group now represented by the apes and ourselves. Fossil apes are known in large numbers from deposits in Kenya and Uganda, ranging in age from 22 million to 14 million years ago. There were many species of these Miocene apes.

The modern apes comprise four genera: *Hylobates* (gibbons), *Pongo* (orangutans), *Pan* (chimpanzees), and *Gorilla* (gorillas). Apes, with the exception of the gibbons, are larger than monkeys, and their brain is larger in proportion to their size. They are all capable of suspending their bodies from branches when in the trees, although among modern apes, only the gibbons move primarily by brachiation—swinging from one arm and then the other with their bodies upright (Figure 27–16).

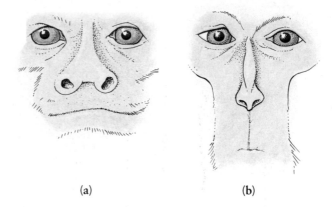

(a)   **(b)**

27–15   *Early in their evolution, the anthropoids split into two main lines, the platyrrhine, or flat-nosed (a), and the catarrhine, or downward-nosed (b). New World monkeys are platyrrhines; Old World monkeys and the hominoids are catarrhines. There are many other characteristic anatomical differences between the two groups.*

Upright suspension is thought to have played a role in the transition from the body structures associated with the horizontal position characteristic of the Old World monkeys and some lower primates to the body structure that led ultimately to our erect posture. Apes have relatively long arms and short legs, resting the weight of the front part of their bodies on their knuckles. As a result, even when they are on all fours, their bodies are partially erect.

Modern apes range widely in size. Gibbons, which are the smallest, weigh about 6 kilograms (about the size of a large house cat), with both sexes the same size.

27–16   *Brachiation, as exhibited by a gibbon, the smallest of the apes. Although gibbons can stand and walk upright, this is their usual and most efficient means of locomotion.*

(a)

(b)

(c)

(d)

(e)

(f)

27–17 *Chimpanzees are probably our closest cousins. They have a relatively fine precision grip and can* (a) *use a simple tool, such as a long twig or piece of grass, to dig out termites, which are then* (b) *nibbled off the tool. They also use leaves as blotters and branches as weapons. Chimps are* (c) *playful,* (d) *gregarious, and* (e) *noisy.* (f) *Because of their intelligence and their physiological resemblance to humans, chimpanzees are prize subjects for medical research, a practice that many find ethically troubling because of these very characteristics.*

Chimpanzees (Figure 27–17) are significantly larger; males weigh about 50 kilograms (110 pounds) and females from 35 to 40 kilograms (70 to 90 pounds). Male orangutans can weigh up to 100 kilograms (220 pounds), with the females weighing about half that. A male gorilla is only about as tall as an average-sized man, but weighs two to three times as much, from 140 to 180 kilograms (300 to 400 pounds). Gorillas have heavily ridged brows, a strong, massive jaw and, on top of the skulls of adult males, a bony crest to which the powerful jaw muscles are attached. Despite their great strength, observers in the field report that they are shy, peaceful, and rather boring: they mostly sit around eating wild celery.

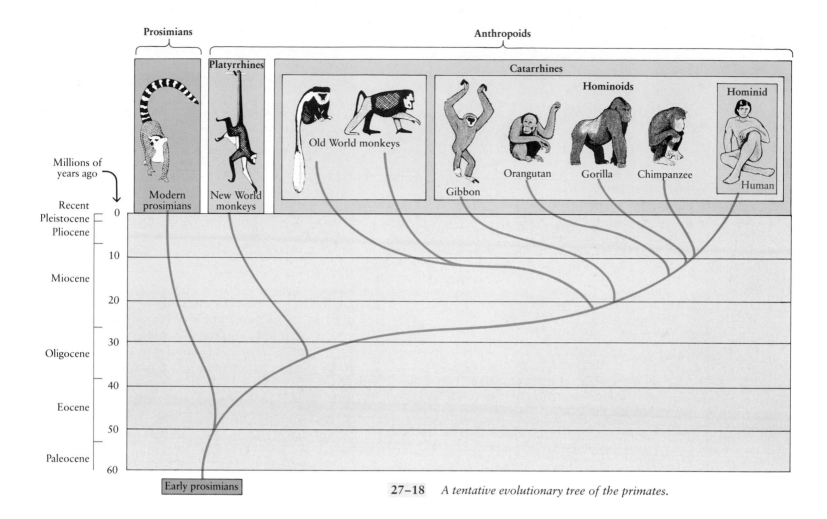

27–18   *A tentative evolutionary tree of the primates.*

The Emergence of the Hominids

The First Hominid: The Taung Child

In 1925, young Raymond Dart described a fossil skull that had been discovered the year before in a stone quarry in Taung, South Africa. He called it *Australopithecus*, the "southern ape." Dart, who was a skilled anatomist, recognized that the skull (Figure 27–19) was that of a child. He also noted that, despite the name he gave it, the child had features that set it apart from other early hominoids. His evidence for its status as a **hominid**—a member of the human family—included the rounded appearance of the skull, the size and shape of the brain case, and the shape of the teeth. More important, the point of attachment of the vertebral column to the skull indicated that the young animal walked upright.

Dart recognized the importance of his discovery, but it was largely ignored by his scientific colleagues. The brain was too small, they argued. Moreover, it was generally agreed at the time that mankind began in Asia or in Europe, where much later hominids had been discovered, rather than in Africa, so lacking in cultural advantages. It was not until 35 years later, following Mary Leakey's discovery at Olduvai Gorge (page 481), that the Taung child was accepted as a human ancestor and Africa was recognized as the cradle of humanity.

The Australopithecines

In the years since the discovery of the first *Australopithecus* specimen, fossil remains of a number of related forms have been found. The leading figures in the search for early hominid fossils have been Louis and Mary Leakey, who stubbornly explored Olduvai and the surrounding area for 20 years prior to their 1959 discovery; their son Richard, who lives and works in Kenya; and Donald Johanson and his colleagues, who have worked principally in the Afar region of Ethiopia.

The australopithecines, known only from their fossils, included at least four species. All were upright, bipedal ground walkers and had brains about one-third the size of those of modern humans. Two species were

(b)

(a)

27–19   (a) *Raymond Dart, with the skull of the Taung child, which was found in 1924 and to which he gave the name* Australopithecus africanus. *It was more than a quarter of a century later that the fossil was widely recognized as that of a hominid. In 1985, at a celebration of his ninety-second birthday and the sixtieth anniversary of his published report on his findings, Professor Dart said, "You know, I was never bitter about how I was treated. . . . I knew people wouldn't believe me. I wasn't in a hurry." Dart died in 1988, at the age of 95.* (b) *A close-up view of the skull of the Taung child.*

27–20   *The skull of the "nutcracker man," discovered by Mary Leakey, and now assigned to the species* Australopithecus boisei. *Note the bony crest atop his skull that supported his strong jaw muscles. He also had huge, grinding molar teeth.*

lightly built: *Australopithecus africanus*, represented by the Taung child, and *A. afarensis*, the species to which Lucy (page 480) belongs. Two other species—*A. boisei*, found in East Africa, and *A. robustus*, from southern Africa—had much more massive skulls and teeth. Mary Leakey's discovery at Olduvai, popularly known as the "nutcracker man" (Figure 27–20), was the first specimen of *A. boisei* to be found. The australopithecines ranged widely throughout Africa, from South Africa to northern Ethiopia. The oldest specimen so far appears to be a five-million-year-old jawbone of *A. afarensis* found in Kenya.

The australopithecines were smaller than we are. Lucy was only about 110 centimeters (3.5 feet) tall and weighed about 23 kilograms (50 pounds). Males weighed about twice as much, a fact that may give us a glimpse into australopithecine social structure. Such sexual dimorphism is found, in modern species, in animals in which the males are dominant, polygamous, and in competition for females. In species that form monogamous pairs, the two sexes are usually about the same size. (Human males today are about 20 percent larger than human females.)

The evolutionary relationships among the different species of australopithecines are still a matter of dispute. Current evidence indicates that they all became extinct about a million years ago.

Homo habilis

Somewhere in the taxonomic tangle of *Australopithecus* is the first member of the genus *Homo*, but just exactly how the australopithecines and *Homo* are related is unresolved. In 1962, Louis Leakey announced the discovery, also at Olduvai, of a hominid that had lived about 1.75 million years ago, about the same time as the "nutcracker man," *A. boisei*. Because of its larger brain size, Leakey assigned it to the genus *Homo,* and because of its possible association with early stone tools dating from the same period, he called his find *H. habilis,* "handy man." Members of this species were, according to Leakey, contemporaries of the australopithecines.

This idea was not accepted easily. However, enough specimens have now been found to secure *Homo habilis* a place on the family tree. One of the most important specimens (Figure 27–21) was discovered by Richard Leakey's group in northern Kenya. Dated at 1.9 million years ago, this individual was large and lightly built, with a brain capacity about 50 percent larger than that of the robust australopithecines. The bony crest along the skull, so prominent in the robust australopithecines, is absent.

The Footprints at Laetoli

The eruption of a volcano in the Rift Valley spewed a layer of ash over the plains of the southern Serengeti in what is now Tanzania. Soon after, there was a brief light shower and then, while the ash layer was still damp, some 20 different kinds of animals scurried, ran, and slithered over it, leaving their prints on the soft and slippery surface. These included hares, baboons, a rhinoceros, two types of giraffe, hyenas, many birds, a three-toed horse, a saber-toothed cat—and at least two hominids. At one point, one of the hominids apparently stopped, paused, turned to the left and then, perhaps reassured, continued on. Under the heat of the equatorial sun, the ash dried, setting like concrete, and soon the footprints were covered by more ash and windblown silt.

In the same spot, by a happy coincidence, some 3.6 million years later, another group of young hominids were amusing themselves by hurling dried elephant dung at one another. One of these (a scientist from Harvard who was visiting Mary Leakey's excavations at Laetoli, Tanzania) dodged, slipped, fell, and found himself on eye level with some strange indentations. And so the Laetoli footprints, one of the most significant discoveries in the study of human origins, were first observed.

Since that time, literally thousands of individual prints have been discovered in the same general area—probably more fossil animal tracks than have ever been found elsewhere in the world, an extraordinary panorama of the number and variety of ancient African animals. And, side by side with these are the arch, big toe, and heel marks that are clear proof that 3.6 million years ago there were hominids who walked fully upright with a bipedal human gait.

Hominid footprints in the volcanic ash of Laetoli. These fossil tracks extend in parallel for about 25 meters. The trail on the left was made by the smallest of the hominids, perhaps holding the hand of the one to the right. This one, the largest, may have been followed by another, smaller hominid who walked in his or her footprints, partially obliterating them. The prints on the right are those of a three-toed horse.

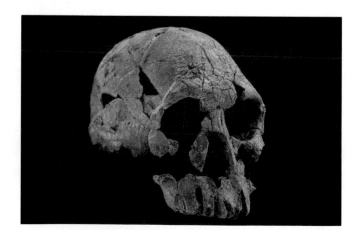

27–21 *This remarkably complete skull was reconstructed from some 300 fossil fragments found by Richard Leakey's team in northern Kenya near the Ethiopian border. It is dated at 1.9 million years. Although the skull was contemporaneous with the australopithecines, because of its much larger brain capacity and more rounded shape, it is usually assigned to the species* Homo habilis, *the most ancient human species.*

As Louis Leakey first noted, stone tools occur in abundance in the same strata as fossils of *H. habilis*. These tools include small sharp slicers, flaked off from a larger stone, and simple pounders and choppers fashioned from the stone's core (Figure 27–22). These were used, it is speculated, for preparation of vegetable foods, hunting of small game, and perhaps the butchering of larger animals. Based on the analysis of animal remains and the lack of weapons for killing larger game, it is believed that such animals were scavenged rather than hunted.

Specimens attributed to *H. habilis* are found in strata dated at 2 million to 1.5 million years ago. The origin of this species is not yet clear. Some believe that *A. afarensis* was directly ancestral to *H. habilis*. Others contend that *H. habilis* coexisted with other species at Afar and Laetoli. A third possibility is that *A. africanus* was the progenitor. And, some argue that *H. habilis* should be classified as an australopithecine.

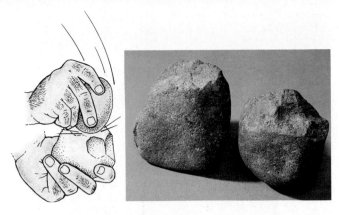

27-22   *Pebble tools have been found in fossil strata at Olduvai and other sites in East Africa. From pebbles of lava and quartz, flakes were struck off in two directions at one end, making a somewhat pointed implement, or in a row on one side, making a chopper. These tools measure up to 10 centimeters in length and may have been used to prepare plant food and to butcher game. The flakes struck from the pebbles may also have been used as tools. The oldest pebble tools have been dated at about 2.5 million years ago.*

27-23   *(a) Possible evolutionary relationships among the known hominid species, based on the available evidence. (b) As more hominid fossils have been discovered, it has become apparent that during several periods of prehistory, two or more different species coexisted. The proponents of punctuated equilibrium (page 371) point out that selection among these species fits the evidence better than gradual phyletic change from one species to another.*

New Concepts in Hominid Evolution

Although the difficulties concerning the evolutionary relationships of the early hominid species have not been resolved and, indeed, may have grown more complex as new fossils have been discovered, three new and widely accepted ideas have emerged.

First, it is now clear that hominid evolution, like that of *Equus* (page 371) and other lineages known from the fossil record, was not a ladder of progress but a bush with many branches, and most of these branches led to extinction (Figure 27–23). The fossils found thus far have decisively refuted the single-species hypothesis, which held that only one species of hominid existed at any one time and that there was a straight progression from the first ape to walk upright directly to modern humans.

Second, it was bipedalism—the capacity to walk on two feet rather than four—that set us on the path to humanness, not our intelligence.

Third, the selective pressures for bipedalism did not involve the "freeing of the hands" for tool use. Bipedalism and the splitting in the hominoid line occurred more than a million years before the appearance of simple stone tools in the fossil record.

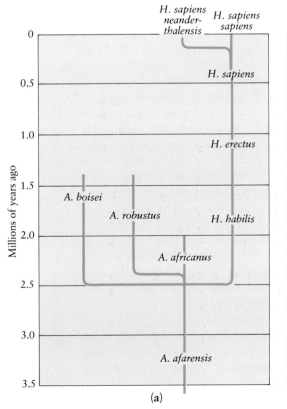

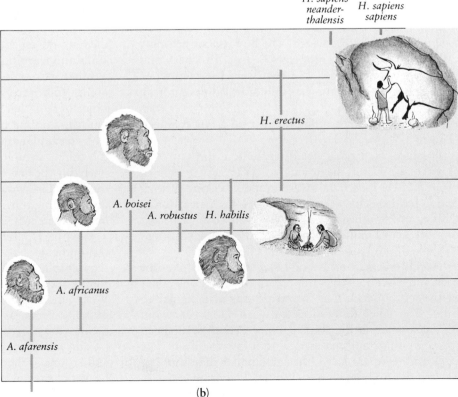

(a)

(b)

A major question now—perhaps *the* major question—is what selective pressures gave rise to bipedalism. Owen Lovejoy of Kent State University has proposed that the early appearance of bipedal locomotion in humans was due to the selective advantage it gave to males who could then procure food at a distance and carry it back to females and young, who were thus able to remain in the comparative safety of the home camp. Others, however, believe that bipedalism was triggered solely by the change in climate that set in at the end of the Miocene, shrinking the African forests and causing the spread of open grasslands. In this new habitat, bipedalism would have been a more efficient means of locomotion.

The Emergence of *Homo sapiens*

By at least 1.6 million years ago, hominids had evolved that are placed, without dispute, in the genus *Homo*. This genus includes two species, the now-extinct *Homo erectus* and our own species, *Homo sapiens*.

Homo erectus

We do not know the cause of his death, only that the body of the 12-year-old boy somehow ended up in a swamp. In 1984, it was found on the shores of a lake in northern Kenya by Kamoya Kimeu, friend and colleague of Richard Leakey. It is both the oldest—1.6 million years—and the most complete specimen of *Homo erectus* yet discovered. The boy was surprisingly tall for his age, 165 centimeters (5 feet, 5 inches), and might well have reached 183 centimeters (6 feet) when fully grown. His skeleton was only subtly different from that of modern *Homo sapiens*. His skull, however, was much heavier, with beetling brows and a low forehead (Figure 27–24).

Fossils of *H. erectus* were first found in Java in 1896 (Java man). Later, beginning in 1929, the fossils of more than 40 individuals were found in Peking (Peking man), followed by other discoveries in Africa, India, China, and Southeast Asia. The fossils cover a period of more than a million years, from 1.6 million years ago to some 400,000 to 300,000 years ago.

Homo erectus had a body skeleton much like our own and was about the same size that we are. The bones of the legs indicate that the stride was also similar to ours. The chief differences between *H. erectus* and *H. sapiens* are in the skull. *Homo erectus* skull specimens are thick and massive, with a low forehead. The jaws and teeth are large (though smaller than those of *H. habilis*), and the chin is sloping. The brain capacity

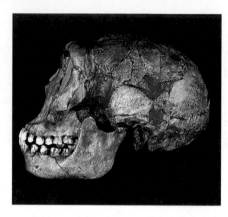

27–24   *The skull of the 12-year-old* Homo erectus *boy whose remarkably complete skeleton was discovered in 1984. Much larger than earlier hominids,* H. erectus *had a significantly larger brain. The skull walls are thick and heavy, the brow ridges are prominent, and the jaw is protruding and chinless. Strong, heavy neck muscles were attached to the skull.*

overlapped that of modern humans. Males were 10 to 20 percent larger than females.

These hominids had a new and highly distinctive tool, the hand ax (Figure 27–25). Tens of thousands of hand axes have been found throughout Africa, Asia, and Europe. They all closely resemble one another, indicating the emergence of a cultural tradition in which skills and learning were passed from one generation to another. At some point, *H. erectus* also acquired the ability to control fire. This would have extended the range of their diet, not only making meat easier to chew but also, and perhaps more important, making it possible to eat plant parts that, uncooked, would have been too tough, too bitter, or too toxic. *Homo erectus* was probably the first of the hominids to inhabit the mouths of caves. Fire would have made such habitation safer, discouraging other cave dwellers, such as bears and saber-toothed cats, and also providing a social center.

Clearly, *H. erectus* was very different from the australopithecines and *H. habilis*, yet only 200,000 years separate the 12-year-old *H. erectus* boy from northern Kenya and a number of *H. habilis* fossils from Olduvai. Was there an unprecedented spurt of evolution in those 200,000 years, as compared to the uneventfulness of the previous 1 million years? Or did *H. habilis* exist side by side with some other hominid ancestral to *H. erectus* and so to us? The answers to these questions may lie buried in the volcanic ash and dust of East Africa.

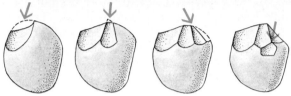

27–25 *The hand ax is a stone that has been worked on all its surfaces to provide what appears to be a gripping surface and various combinations of cutting edges, sometimes with a more or less sharp point. Hand axes came into use about 1.5 million years ago and are associated with* H. erectus.

Homo sapiens

Homo sapiens is the name we give to the species of hominid that has a body skeleton much like ours and a brain capacity similar to or approaching our own (Figure 27–26). Three varieties, or subspecies, are commonly recognized: "archaic" *Homo sapiens,* which bears many resemblances to *H. erectus* and which some believe should be regarded as a separate species; *Homo sapiens neanderthalensis,* another twig on the family bush; and *Homo sapiens sapiens,* the wisest of the wise, including, of course, ourselves.

Fossils considered early, or archaic, *H. sapiens* are dated at about 200,000 to 100,000 years ago. These individuals had prominent brow ridges but larger brains and smaller teeth than the earlier *H. erectus.* Fossil materials of this group are scarce and fragmentary, and dates are uncertain.

Homo sapiens neanderthalensis

The period from about 150,000 years ago to about 35,000 years ago abounds with specimens of what we have come to call the Neanderthals. They have been found largely in Europe but also in the Near East and central Asia. The Neanderthals stood as erect as we do but were more heavily built and more muscular. They had a brain capacity as large as ours, a long, low, massive skull, a protruding face, a low forehead, and heavy brow ridges. They are now usually classified as a variety of *H. sapiens.*

27–26 *Brain size more than tripled during the 3 million years of hominid evolution represented by these three skulls. The average brain volume of* Australopithecus afarensis *(left) was 400 cubic centimeters, that of* Homo erectus *(center) was 850 cubic centimeters, and that of modern* Homo sapiens *(right) is 1,360 cubic centimeters. Some of the increase can be correlated with the increase in body size that was also taking place during this period, but most of it is believed to represent the results of strong selection pressures for intelligence.*

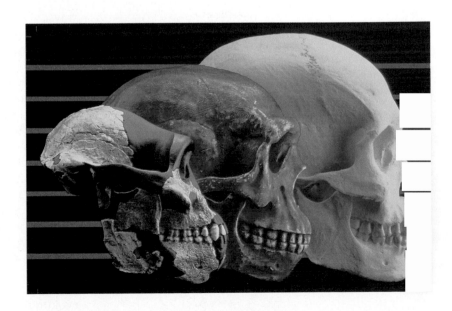

(a)

(b)

27–27   (a) *This large cave near the village of Shanidar in northern Iraq has been continuously inhabited for more than 100,000 years. Nine Neanderthal skeletons have been found in the cave, including one who was, according to analysis of fossil pollen, buried on a bed of woody branches and June flowers gathered from the hillside.*

(b) *The grave of a young Neanderthal man, excavated in 1983 on Mount Carmel in Israel. After his death, estimated to have occurred some 50,000 years ago, the flesh was allowed to decay and the brain case was removed before the remainder of the skeleton was covered with dirt.*

Neanderthals used hand-held stone tools that were much more sophisticated than those of *H. erectus*. Some of the stone tools appear to have been used for scraping hides, suggesting that Neanderthals wore clothing made of animal skins, which would certainly have been in keeping with the climate in which they lived.

Neanderthals buried their dead (Figure 27–27), sometimes with food and weapons and, in at least one instance, with spring flowers. Formal burials such as these suggest a belief in life after death.

During the relatively short span of about 100,000 years, the Neanderthals spread all across Europe, the Middle East, and western and central Asia. They were contemporaries of forms of archaic *H. sapiens* found as far away as China and South Africa. Then they all disappeared abruptly, some 30,000 years ago.

Homo sapiens sapiens

All hominid fossils of the last 30,000 years are of anatomically modern humans, *H. sapiens sapiens*. Early Europeans of this type are commonly called Cro-Magnon, after the site in southwestern France where they were first discovered. However, much earlier modern populations are known from the Middle East. For example, fossils dated at 92,000 years ago have been found at Qafzeh, in Israel.

The Cro-Magnons, when they first appeared in Europe, came bearing a new, quite different, and far better tool kit (Figure 27–28). Their stone tools were essentially flakes—which had been in use for more than 2.5 million years—but they were struck from a carefully prepared core with the aid of a punch (a tool made to make another tool). These flakes, usually referred to as blades, were longer, flatter, and narrower, and, most important, they could be shaped in a large variety of ways. They included, from the beginning, various scraping and piercing tools, flat-backed knives, awls, chisels, and a number of different engraving tools. Using these tools to work other materials, especially bone and ivory, Cro-Magnons made a variety of projectile points, barbed points for spears and harpoons, fishing hooks, and needles. They also produced some of the most exuberant and creative works of art in all of human history (see essay on page 502).

The Origin of Modern Humans

It is now generally accepted that the early hominids—all the australopithecines and *H. habilis*—evolved in Africa over a period of at least 3 million years. A major question, however, concerns the events of the last few hundred thousand years, the period during which the evolution of modern humans took place.

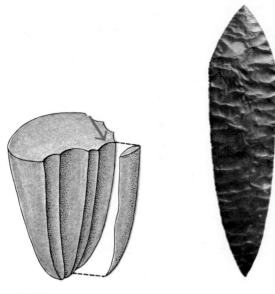

27–28 *Cro-Magnon culture was characterized by a great increase in the types of specialized tools made from long and relatively thin flakes with parallel sides, called blades. This beautifully worked "laurel leaf" blade, fashioned from flint, was used as the point of a spear. Such blades are often more than 30 centimeters long and only 0.5 centimeter thick.*

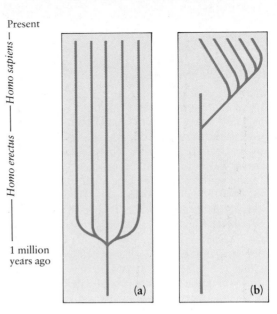

27–29 *Two models of the evolution of modern humans. (a) According to the candelabra model, populations of modern humans split off from earlier forms perhaps as much as 1 million years ago and evolved in parallel with one another. (b) The Noah's Ark model, by contrast, proposes a single, more recent source for modern humans with the extinction (for reasons unknown) of the earlier human forms.*

As we noted earlier, *H. erectus* fossils have been found not only in Africa but also abundantly, with a somewhat later appearance, in Asia and Europe. Until recently, on the basis of these fossil findings, many experts have subscribed to the so-called **candelabra model** of modern evolution (Figure 27–29a). According to this model, there were multiple early migrations from Africa, beginning perhaps as long as a million years ago, and these migrations established different populations of *H. erectus* that evolved separately into the different modern human races. Since the populations were not isolated from one another, gene flow occurred, preventing speciation.

Opposed to the candelabra hypothesis is the **Noah's Ark model** (Figure 27–29b). According to this model, a small group of already modern humans, starting in one place, colonized the entire world, like the survivors on Noah's Ark. Some fossil findings support this point of view. For instance, the oldest unequivocally modern human fossils, dated at more than 100,000 years ago, have been found in southern Africa. The next oldest are the group from Qafzeh in Israel, on the corridor to Europe and Asia, which were found in the 1930s and have recently been dated at 92,000 years. The modern humans of Europe show up about 40,000 years ago. (Either they took a long time making their way northward, or earlier evidence has yet to be discovered.) Also, no fossils have been found that appear to be unequivocally

intermediate between *H. erectus* or the Neanderthals and modern humans. Finally, the limb proportions of the first modern humans in Europe seem to be typical of equatorial people, not people adapted to cold climate, as the Neanderthals were.

A working hypothesis, based on the Noah's Ark model and supported by highly controversial studies of mitochondrial DNA, is that a population of modern humans was in existence in Africa about 200,000 years ago, began migrating through the Middle East into Asia and Europe about 100,000 years ago, and replaced the other resident human populations—*H. erectus* and *H. sapiens neanderthalensis*—upon their arrival.

The implications of this hypothesis are variously interpreted. According to Stephen Jay Gould, "It makes us realize that all human beings, despite differences in external appearance, are really members of a single entity that's had a very recent origin in one place. There is a kind of biological brotherhood that's much more profound than we ever realized." (Some of us might say sisterhood.) On the other hand, there is the unsettling question of the disappearance of *H. erectus* populations and of the Neanderthals. One possible explanation is that the migrating modern humans brought with them diseases to which *H. erectus* and the Neanderthals were not immune. Another possibility is that the appropriate biblical symbol of human evolution may not be Noah, but rather Cain.

The Art of the Caves

The cave paintings of western Spain and southern France, many surprisingly untouched by time, are part of a rich artistic tradition that endured for at least 20,000 years. Most depict animals, nearly all of them game animals, and there are also some schematic human figures and many abstract signs. The surviving paintings are deep within the caves, so they must have been viewed (as they must have been painted) by the light of crude lamps or torches.

The meaning of these drawings and paintings has long been a matter of debate. In some drawings, the animals are marked with darts or wounds (although very few appear to be seriously injured or dying). Such markings have led to the suggestion that the figures are examples of sympathetic magic, in which there is the notion that one can exert control over another creature by taking symbolic action against its image. The fact that some of the animals appear to be pregnant suggests that they may symbolize fertility. Many appear also to be in motion, an illusion greatly enhanced by the patterns of light and shadow in the dark recesses of the cave. Perhaps these animals, so vital to the hunters' welfare, were migratory in these areas, and they may have seemed to vanish at certain times in the year, mysteriously returning, heavy with young, in the springtime. This return of the animals might have been an event to be solicited or celebrated in much the same spirit as the rites of spring or Easter are celebrated by more recent peoples.

Whatever their meaning, these images touch us, like the footprints at Laetoli and the burial at Shanidar, with a sense, across time, of sharing something of what it means to be human. Paleolithic cave art in Europe came to an end perhaps 10,000 years ago, with the end of the last Ice Age. Not only were the tools and pigments laid aside, but the sacred places—for such they seem to have been— were no longer visited.

Since its beginnings, the study of human evolution has compelled us not only to look at the dry bones of our ancestors but also to examine what it means to be human. Depending on our interpretations of the past and our views of the present, we can find in our biological history the justifications for our present shortcomings or the hopes for our salvation. Alternatively, we may conclude that, based on present knowledge, human behavior in the past appears to have been as complex and contradictory as we know it to be in the present and so has little bearing on our future. Under these circumstances, the best we can do is to rely on our recently enlarged brains and make our own choices among the bewildering alternatives.

Summary

Vertebrates are distinguished by a vertebral column, a flexible and usually bony support that encloses the nerve cord. With other members of phylum Chordata, they share four identifying characteristics: a notochord, a dorsal, hollow nerve cord, a pharynx with gill slits, and a tail.

Vertebrates include the fishes (three living classes), the amphibians, the reptiles, the birds, and the mammals. Some primitive bony fishes, forerunners of the amphibians, were aided in the transition to land by the development of simple lungs and by strong skeletal supports that could prop up the body of the fish out of water. Most modern amphibians remain incompletely adapted to life on land and must spend part of the life cycle in water.

Vertebrates became truly terrestrial with the development, in the reptiles, of the amniote egg. The diversification of the reptiles that followed their conquest of the land ultimately gave rise not only to a great variety of reptiles (most of which have been extinct for about 65 million years) but also to their descendants, the birds and the mammals.

The first mammals arose from primitive reptilian stock about 200 million years ago and coexisted with the dinosaurs for 130 million years. During this period, they diversified into three principal lineages—monotremes, marsupials, and placentals. The extinction of the dinosaurs was followed by a rapid adaptive radiation of the marsupials and placentals.

The primates are an order of placental mammals that became adapted to arboreal life. The two principal groups of living primates are the prosimians and the anthropoids. Prosimians were widespread and abundant during the Eocene epoch, some 55 to 38 million years ago. Modern prosimians include lorises, bush babies, lemurs, and tarsiers. The anthropoids include the New World monkeys, the Old World monkeys, and the hominoids (apes and humans).

From more than 3.6 million years ago to at least 1.5 million years ago, groups of hominids lived that, although they were small and their skulls were apelike, walked erect. Some used simple pebble tools. At least five species are now recognized: *Australopithecus afarensis, A. africanus, A. robustus, A. boisei,* and *Homo habilis.*

Homo erectus lived from at least 1.6 million to 300,000 years ago. Individuals were tall, with body skeletons closely resembling those of modern humans, but their skulls were much heavier. The hand ax is associated with *H. erectus.* Some groups at least occasionally occupied caves and, at later stages, certainly had fire, two developments that may be related.

The species *Homo sapiens* comprises "archaic" *H. sapiens, H. sapiens neanderthalensis,* and *H. sapiens sapiens.* Archaic *H. sapiens* fossils date from 200,000 to 100,000 years ago and indicate that these individuals had larger brains and smaller teeth than the earlier *H. erectus.* Neanderthal fossils date from about 150,000 to 35,000 years ago. The majority of specimens have been found in Europe. Neanderthals had fire, inhabited caves, used stone tools of a characteristic type, hunted large animals, and probably wore some sort of clothing. They buried their dead, sometimes with food and weapons. Neanderthals disappeared some 30,000 years ago.

The Cro-Magnons, anatomically modern humans, replaced the Neanderthals. It is hypothesized that modern humans evolved in Africa and migrated from there only about 100,000 years ago, replacing previous populations of the genus *Homo* as they went.

Questions

1. Distinguish among the following: vertebrate/invertebrate; ectotherm/endotherm; monotreme/marsupial/placental; primate/prosimian; monkeys/apes; hominids/hominoids/anthropoids.

2. Describe the identifying characteristics of the phylum Chordata. What is the functional significance of each?

3. Consider the graceful swimming of a squid, a lancelet, or a fish. What is the role of the semirigid beam running the length of the animal, whether the "pen" of a squid, the notochord of a *Branchiostoma*, or the vertebral column of a fish?

4. What anatomical features make possible the flight of birds?

5. Among fish and reptiles, some species are oviparous (that is, they lay eggs from which the young hatch) and some are viviparous (giving birth to live young). Name some of the advantages of each alternative. Why are all birds oviparous?

6. Name five evolutionary trends among primates, and discuss the probable selective value of each.

7. Suppose you were to able to attend a family reunion of the following: *Homo erectus, Homo sapiens sapiens,* a Cro-Magnon, Peking man, *Homo sapiens neanderthalensis,* Lucy, *Australopithecus boisei,* and *Australopithecus africanus.* How would you distinguish one from the other?

Suggestions for Further Reading

Alexander, R. McNeill: "How Dinosaurs Ran," *Scientific American,* April 1991, pages 130–136.

Bahn, Paul G., and Jean Vertut: *Images of the Ice Age,* Facts on File, Inc., New York, 1988.

   An illustrated history of art's first 25,000 years.

Bar-Yosef, Ofer, and Bernard Vandermeersch: "Modern Humans in the Levant," *Scientific American,* April 1993, pages 94–100.

Blumenschine, Robert J., and John A. Cavallo: "Scavenging and Human Evolution," *Scientific American,* October 1992, pages 90–96.

Buffetaut, Eric: "The Evolution of the Crocodilians," *Scientific American,* October 1979, pages 130–144.

Bürgin, Toni, Olivier Rieppel, P. Martin Sander, and Karl Tschanz: "The Fossils of Monte San Giorgio," *Scientific American,* June 1989, pages 74–81.

Cavalli-Sforza, Luigi Luca: "Genes, Peoples and Languages," *Scientific American,* November 1991, pages 104–110.

de Beaune, Sophie A., and Randall White: "Ice Age Lamps," *Scientific American,* March 1993, pages 108–113.

Duellman, William E.: "Reproductive Strategies of Frogs," *Scientific American,* July 1992, pages 80–87.

Feduccia, Alan: *The Age of Birds,* Harvard University Press, Cambridge, Mass., 1980.*

   A history of the birds, tracing their evolution from reptilian ancestors through the diversifications that gave rise to the major groups of modern birds. Well-illustrated with many photographs and drawings.

Fenton, M. Brock: *Just Bats,* University of Toronto Press, Toronto, 1983.*

   A short, well-illustrated account of the fascinating lives of the bats, the only mammals capable of flight.

Gould, Stephen Jay: "Bushes All the Way Down," *Natural History,* June 1987, pages 12–19.

Gould, Stephen Jay: "A Novel Notion of Neanderthal," *Natural History,* June 1988, pages 16–21.

Griffiths, Mervyn: "The Platypus," *Scientific American,* May 1988, pages 84–91.

Hanken, James: "Development and Evolution in Amphibians," *American Scientist,* vol. 77, pages 336–343, 1989.

Harris, Marvin: *Our Kind,* Harper & Row, Publishers, Inc., New York, 1989.*

   A witty, provocative, and brilliant discussion of human evolutionary history. Recommended for anyone interested in what it means to be human and in how we got this way.

Hay, Richard L., and Mary D. Leakey: "The Fossil Footprints of Laetoli," *Scientific American,* February 1982, pages 50–57.

Heyler, Daniel, and Cecile M. Poplin: "The Fossils of Montceau-les-Mines," *Scientific American,* September 1988, pages 104–110.

* Available in paperback.

Johanson, Donald, and Maitland Edey: *Lucy: The Beginnings of Humankind*, Simon and Schuster, New York, 1981.*

> *A brash, exciting, firsthand account of the discovery of Lucy and of the conflicts and controversies among those seeking to uncover human origins.*

Johanson, Donald, and James Shreeve: *Lucy's Child: The Discovery of a Human Ancestor*, William Morrow & Company, New York, 1989.*

> *Further adventures in the search for our ancestors.*

Kirsch, John A. W.: "The Six-Percent Solution: Second Thoughts on the Adaptedness of the Marsupialia," *American Scientist*, vol. 65, pages 76–288, 1977.

Leakey, Richard E., and Roger Lewin: *Origins Reconsidered: In Search of What Makes Us Human*, Doubleday & Company, Inc., Garden City, N.Y., 1992.

> *Richard Leakey, son of Mary and Louis Leakey, has been an observer of and participant in most of the major hominid discoveries. In this thoughtful and authoritative book, he reviews our current knowledge of hominid evolution, with some speculations on human nature and the future of humankind.*

Lewin, Roger: *Bones of Contention*, Simon and Schuster, New York, 1988.*

> *A lively and highly recommended account of the personal and scientific disputes pervading the study of human evolution.*

Lewin, Roger: *In the Age of Mankind: A Smithsonian Book of Human Evolution*, Smithsonian Books, Washington, D.C., 1988.*

> *The best and most handsome introduction to this fascinating field of modern research.*

Lewontin, Richard: *Human Diversity*, W. H. Freeman and Company, New York, 1984.

> *A leading population geneticist, Lewontin analyzes the biological bases of human variation and also explores their often controversial social ramifications—such as IQ, skin color, and sex-linked differences. This is a wise and beautiful book that, transcending politics and biology, becomes a celebration of human potential.*

Lovejoy, C. Owen: "Evolution of Human Walking," *Scientific American*, November 1988, pages 118–125.

Milner, Richard: *The Encyclopedia of Evolution: Humanity's Search for Its Origins*, W. W. Norton & Company, New York, 1993.*

> *More than 600 authoritative and entertaining brief articles about many aspects of evolution. Fun to read and imaginatively illustrated.*

Milton, Katharine: "Diet and Primate Evolution," *Scientific American*, August 1993, pages 86–93.

Norell, Mark, Luis Chiappe, and James Clark: "New Limb on the Avian Family Tree," *Natural History*, September 1993, pages 38–43.

Novacek, Michael J.: "Mammalian Phylogeny: Shaking the Tree," *Nature*, vol. 356, pages 121–125, 1992.

Ostrom, John H.: "Bird Flight: How Did It Begin?" *American Scientist*, vol. 67, pages 46–56, 1979.

Pilbeam, David: "The Descent of Hominoids and Hominids," *Scientific American*, March 1984, pages 84–96.

Ross, Philip E.: "Eloquent Remains," *Scientific American*, May 1992, pages 114–125.

Simons, Elwyn L.: "Human Origins," *Science*, vol. 245, pages 1343–1350, 1989.

Stringer, Christopher B.: "The Emergence of Modern Humans," *Scientific American*, December 1990, pages 98–104.

Tattersall, Ian: "Evolution Comes to Life," *Scientific American*, August 1992, pages 80–87.

Thorne, Alan G., and Milford H. Wolpoff: "The Multiregional Evolution of Humans," *Scientific American*, April 1992, pages 76–83.

Tuttle, Russell H.: "Apes of the World," *American Scientist*, vol. 78, pages 115–125, 1990.

Walker, Alan, and Mark Teaford: "The Hunt for *Proconsul*," *Scientific American*, January 1989, pages 76–82.

Webb, Paul W.: "Form and Function in Fish Swimming," *Scientific American*, July 1984, pages 72–82.

Wellnhofer, Peter: "*Archaeopteryx*," *Scientific American*, May 1990, pages 70–77.

White, Randall: "Visual Thinking in the Ice Age," *Scientific American*, July 1989, pages 92–99.

Wilford, John Noble: *The Riddle of the Dinosaur*, Alfred A. Knopf, Inc., New York, 1985.*

> *A well-written history of dinosaur discovery and research, with balanced discussions of the many controversies that surround these fascinating animals.*

Wilson, Allan C., and Rebecca L. Cann: "The Recent African Genesis of Humans," *Scientific American*, April 1992, pages 68–73.

* Available in paperback.

Biology of Animals

The Human Animal: An Introduction

Back in the Garden of Eden, humankind, according to the Judeo-Christian tradition, was given "dominion over the fish of the sea, and over the fowl of the air, and over every living thing that moveth upon the Earth." This concept that we are special and that other living creatures are here to serve us has profoundly influenced our attitudes and our actions toward the natural world.

One of the reasons that the theory of evolution was so disquieting to Darwin's contemporaries—and indeed to many of us today—was that it makes us seem less apart from other species. Even Wallace, Darwin's partner in evolutionary theory, maintained that there must be a "spiritual" element in human evolution that makes us different in kind from other animals. Yet, as we saw in the last chapter, as the hominid story unfolds, we see how remarkably similar our own evolutionary history is to that of other species.

In the chapters that follow, we will see even more evidence that we are a part of nature. Our physical requirements are remarkably similar to those of other animals: oxygen, water, an energy source, a few kinds of atoms and molecules, a limited temperature range, a partner for reproduction. In fact, as animals, we are not particularly outstanding in most functions: eagles have better vision and they can fly; dogs have a better sense of smell; any antelope can run faster; the other primates are all better at climbing trees. Moreover, none of these animals, except a few overbred domestic dogs, have anything like our lower back problems or difficulties in childbirth.

Yet even those of us who proclaim our unity with nature know that we are indeed different from the other animals and that, for better or worse, we do have dominion. This exquisite watercolor of Adam and Eve before the Fall, from a sixteenth-century Muslim manuscript, is one kind of evidence of our difference. Evidence of our dominion is to be seen, alas, all around us. Elsewhere in the Bible, Isaiah prophesies a new Heaven and Earth so peaceable that "the wolf shall dwell with the lamb" and even the lion will be a vegetarian. Perhaps a higher goal than we can attain, but not a bad guideline.

The animal kingdom encompasses a vast and, as yet, incompletely catalogued array of organisms, ranging from relatively simple aquatic forms, often microscopic in size, to extremely complex and highly organized multicellular creatures. As we have seen in the last two chapters, the structure and function of these enormously diverse organisms can be largely understood in terms of a relatively few fundamental principles.

In this section of the book, we shall consider the principles of animal structure and function in more detail, using *Homo sapiens* as our representative organism. Because we are part of the continuum of nature, the study of the human animal (now one of the best understood) can give us a deeper understanding of animal life in general. This one species, however, will not be our only focus of attention. We shall also rely on examples from other organisms to gain further insight into important physiological principles and adaptations.

Characteristics of *Homo sapiens*

The human being is a vertebrate and as such has a bony, articulated (jointed) internal skeleton that supports the body and grows as the body grows. The dorsal, hollow nerve cord (the spinal cord) is surrounded and protected by bony segments, the vertebrae, and the brain is similarly enclosed in a protective casing, the skull.

As in other vertebrates, and most invertebrates as well, the human body contains an internal cavity, or coelom (page 469). In humans and other mammals, the coelom is divided into compartments, of which the two largest are the **thoracic cavity** and the **abdominal cavity** (Figure 28–1, on the next page). These are separated by a thin, dome-shaped muscle, the **diaphragm**. The thoracic cavity contains the heart, lungs, and esophagus (the upper portion of the digestive tract). The abdominal cavity contains a large number of organs, including the stomach, liver, and intestines.

One of the most important characteristics of humans

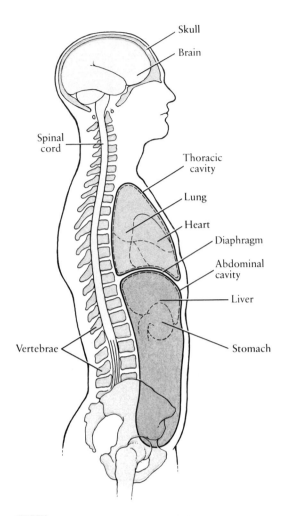

28–1 *Humans, like other vertebrates, are characterized by a dorsal central nervous system (spinal cord and brain) enclosed in vertebrae and the skull. As in other mammals, a muscular diaphragm divides the coelom into the thoracic cavity and the abdominal cavity. The dashed lines indicate the position of four of the vital organs contained within these cavities.*

and other mammals is that they are warm-blooded. More precisely, they are endotherms (page 486), generating heat internally to maintain a high and relatively constant body temperature. As a consequence, mammals (and birds, which are also endotherms) are able to achieve and sustain levels of physical activity and mental alertness generally far greater than those of ectotherms, which can regulate their body temperature only by absorbing heat from or releasing it to the environment.

Mammals have other important characteristics. They have hair or fur rather than scales or feathers. In keeping with their high levels of activity and mental alertness, they have complex systems for receiving, processing, and reacting to information from the environment. All mammals (except the monotremes, such as the duckbilled platypus) give birth to live young, as distinct from laying eggs. Mammals nurse their offspring, which entails a relatively long period of parental care and makes possible a long learning period.

The Hierarchical Organization of the Human Body

As we saw in Chapter 1 (page 30), one of the identifying properties of living organisms is that they maintain a precise structural organization. This organization is a hierarchy, in which the basic structural unit is the living cell. The human body, like that of all other complex animals, is made up of a variety of different, specialized cells. These cells are organized into **tissues,** which are groups of cells that work together to carry out a unified function. Different kinds of tissues, united structurally and coordinated in their activities, form **organs,** such as the liver and the heart. Organs are arranged into **organ systems,** such as the digestive system and the circulatory system. The circulatory system, for example, consists of the heart, the blood vessels, and the blood that flows within the vessels (Figure 28–2). Working in conjunction with one another, the organ systems make up the whole organism.

Underlying this interactive hierarchy is one of the most profound principles of biology. The structure and regulatory processes of complex organisms are such that the parts serve the whole. That is, within the hierarchy of organization, functions at lower levels are limited by requirements at higher levels. Natural selection favors this arrangement since even slight deviations in function at lower levels can result in severe impairment or even death of the organism. For example, cancer is, as you know, a group of diseases in which certain cells escape the controls that regulate normal cell division (page 180) and thus reproduce uncontrollably, often invading and destroying healthy tissues and organs. The disease process results from the failure of a lower level—cells—to operate within the constraints set by the requirements of the whole organism.

Cells and Tissues

Experts can distinguish about 200 different cell types in the human body. These different types of cells are customarily classified into only four tissue types: (1) epithelial, (2) connective, (3) muscle, and (4) nerve.

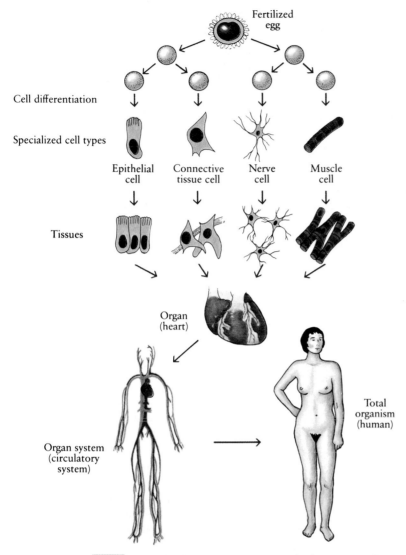

28–2 *Levels of organization within the human circulatory system.*

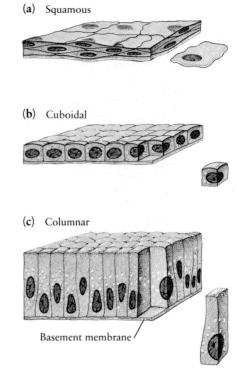

28–3 *The three types of epithelial cells that cover the inner and outer surfaces of the body. (a) Squamous cells, which are typically flat, occur in single layers (simple squamous epithelium) or piled up on one another in several layers (stratified squamous epithelium), as shown here. Simple squamous epithelium forms structures, such as the capillaries (the smallest blood vessels) and the air sacs of the lungs, across which exchanges of substances occur. Stratified squamous epithelium serves a protective function and makes up the outer layers of the skin and the lining of the mouth and other mucous membranes. (b) Cuboidal and (c) columnar cells, which line many internal passageways, are often involved in active transport and other energy-requiring processes. They also perform much of the chemical work of the body.*

Epithelial Tissues

Epithelial tissues consist of continuous sheets of cells that provide a protective covering over the whole body. They also provide a protective wrapping for individual internal organs and form the interior lining membranes of organs, cavities, and passageways. As a moment's reflection will reveal, everything that goes into and out of the body and its various organs must pass through epithelial cells, which thus play an important regulatory role in the movement of molecules and ions.

Epithelial tissues are classified according to the shape of the individual cells as **squamous, cuboidal,** or **columnar** (Figure 28–3). They may consist of only a single layer of cells (**simple epithelium**), as found in the inner lining of the circulatory system, or several layers (**stratified epithelium**), as found in the outer layer (epidermis) of the skin (Figure 28–4). One surface of the epithelial sheet is usually attached to an underlying layer, called the **basement membrane.** This layer is composed of polysaccharides and fibrous proteins produced by the epithelial cells themselves.

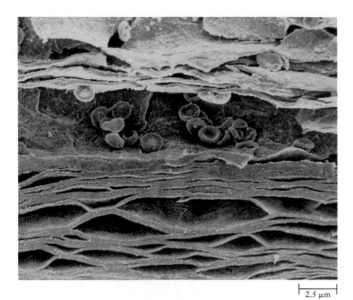

28–4 *The outer surface, or epidermis, of the skin is composed of stratified squamous epithelium. In the lower portion of this scanning electron micrograph, the layers of epithelium are clearly visible. On the skin surface, some of the older cells that have died are being sloughed off. This is a continuous process in epithelial tissue that is subjected to wear and tear. Replacement cells are produced by cell division in the underlying layers of the epidermis. A number of red blood cells can be seen just below the upper layers of epithelial cells.*

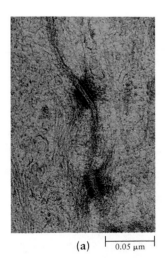
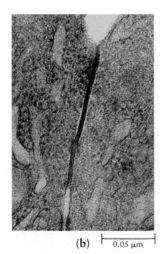

(a)   0.05 μm   (b)   0.05 μm

28–5 *Junctions between epithelial cells maintain the structural integrity of the tissue. (a) Desmosomes cement adjacent cells together. Here two desmosomes connect the plasma membranes of two adjacent epithelial cells. (b) A tight junction (the dark line) seals two intestinal epithelial cells together and prevents fluid from leaking between them.*

The epithelium of the body cavities and passageways frequently contains modified epithelial cells that secrete mucus, which lubricates the surfaces. Other epithelial cells, specialized for the synthesis and secretion of specific substances for export, are often clustered together to form **glands.** Among the substances produced by glands are perspiration, saliva, milk, hormones, and digestive enzymes. Glands are composed of cuboidal or columnar epithelial cells.

Desmosomes and Tight Junctions

In epithelial tissues that function as coverings and linings, the physical integrity of the tissue as a whole—no tears, no leaks—is often of paramount importance. Two general types of cell-cell junctions play an essential role in maintaining this integrity: **desmosomes** and **tight junctions.**

Desmosomes (Figure 28–5a) have often been compared to spot welds between cells. They consist of plaques of dense fibrous material between cells, with clusters of filaments from the cytoplasm of the neighboring cells looping in and out of them. Desmosomes attach cells to one another and give tissues mechanical strength. They are found in especially large numbers in tissues subjected to mechanical stress, such as the skin.

Tight junctions (Figure 28–5b), which appear to involve the fusion of adjacent plasma membranes, form a continuous seal around each cell in a layer of tissue, preventing leakage between cells. For example, intestinal epithelial cells are surrounded by tight junctions that keep the intestinal contents from seeping between the cells. Similar arrangements are found in epithelial cells that form the linings of other internal organs, such as the bladder and the kidney.

Connective Tissues

Connective tissues bind together, support, and protect the other three kinds of tissues. Unlike epithelial tissue cells, the cells of connective tissues are widely separated from one another by large amounts of extracellular material.

The extracellular material, synthesized by the cells of the tissue, consists of a more or less fluid and amorphous (formless) **ground substance** and, in many connective tissues, **fibers.** There are several different types of fibers, which vary according to the particular tissue: (1) connecting and supporting fibers, such as collagen fibrils, which are a major component of skin, tendons, ligaments, cartilage, and bone; (2) elastic fibers, which are found, for example, in the walls of large blood vessels; and (3) highly branched reticular fibers, which form networks inside solid organs, such as the liver.

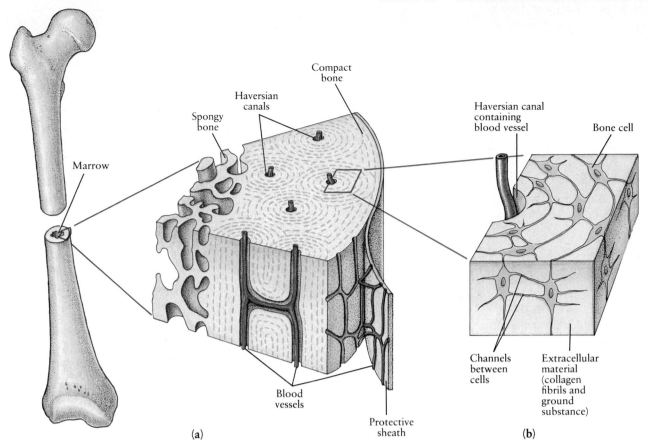

28-6 (a) *Bones are living organs, made up not only of connective tissue but of other tissue types as well. Bone includes nerve tissue and the epithelial lining of the blood vessels that are located in canals, known as Haversian canals, that run the length of the bone. Each bone is surrounded by a fibrous protective sheath containing larger blood* *vessels that supply oxygen and nutrients to the bone tissues.* (b) *A closer look reveals that the Haversian canals are encircled by living bone cells. Tiny channels containing cytoplasmic processes connect the bone cells to each other and to the blood vessels and nerves running through the Haversian canals.* *Young bone cells produce the extracellular material, which consists of collagen fibrils and ground substance. The ground substance contains calcium compounds, and it gradually hardens as those compounds crystallize.*

While epithelial tissues are classified according to cell shape and arrangement, connective tissues are grouped by the characteristics of their extracellular material.

The principal connective tissues, by volume, in the human body are bone, blood, and lymph. In blood and lymph, the ground substance is a watery fluid, called plasma, that contains numerous ions and molecules. A variety of specialized cells (to be discussed in Chapters 31 and 33) circulate through the body in this fluid. The ground substance of bone, by contrast, is impregnated with hard crystals of calcium compounds. Like other connective tissues, however, bone is living matter, consisting of cells, fibers, and ground substance (Figure 28–6). Bone tissue, despite its strength, is amazingly light. The human skeleton makes up only about 18 percent of our weight.

Muscle Tissue

Muscle cells are specialized for contraction. Every function of muscle—from running, jumping, smiling, and breathing to propelling the blood through the body and ejecting the fetus from the uterus—is carried out by the contraction of muscle cells in concert.

There are three types of muscle tissue, as shown in Figure 28–7: **skeletal muscle** and **cardiac muscle,** both of which have a striated (striped) appearance, and **smooth muscle** (no stripes). Skeletal muscles move the skeleton and are called voluntary muscles since we can move most of them at will. Cardiac muscle makes up the wall of the heart. Smooth muscle surrounds the walls of internal organs, such as the digestive organs, uterus, bladder, and blood vessels. Cardiac muscle and

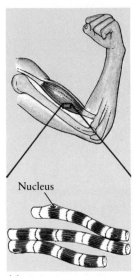

Nucleus

(a) Skeletal muscle (biceps)

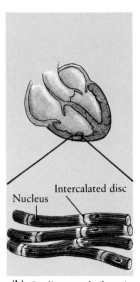

Nucleus   Intercalated disc

(b) Cardiac muscle (heart)

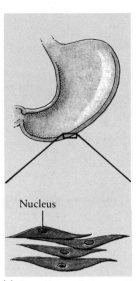

Nucleus

(c) Smooth muscle (stomach)

28–7 *Muscle cells are characterized by very thin fibrils of contractile proteins that run lengthwise through the cells and make up the bulk of the cytoplasm. These fibrils are arranged in a regular pattern in skeletal muscle and cardiac muscle, but they are irregular in smooth muscle, forming no apparent pattern.*

(a) Skeletal muscles, such as the biceps, are made up of long cells, each containing many nuclei. The tissue has a striated appearance.

(b) Cardiac muscle is made up of much shorter cells, each of which contains one or, at most, two nuclei. As in skeletal muscle, the tissue has a striated appearance. Intercalated discs join cardiac muscle cells to one another, adding strength to the tissue. They also play a role in the rapid communication between the cells that enables them to contract simultaneously, producing the heartbeat.

(c) Smooth muscle is made up of long, spindle-shaped cells. Unlike the cells of skeletal muscle, but like most cardiac muscle cells, each smooth muscle cell contains only a single nucleus.

smooth muscle are not, except in rare cases, under conscious control and are thus categorized as involuntary.

Contraction of muscle cells depends on the interaction of two proteins, actin and myosin. In skeletal and cardiac muscle, these proteins are arranged in regular, repeating assemblies, resulting in the characteristic striations. Although smooth muscle cells also contain actin and myosin, the molecules are not arranged in regular assemblies and do not form a striated pattern.

Skeletal Muscle

About 40 percent of a man's body weight is skeletal muscle; women characteristically have less, around 20 percent. A skeletal muscle is typically attached to two or more bones, either directly or, more often, indirectly, by means of the tough strands of connective tissue known as **tendons.** Some tendons, such as those that connect the finger bones with their muscles in the forearm, are very long. When the muscle contracts, the bones move around a joint, which is held together by **ligaments** and generally contains a lubricating fluid.

Most of the skeletal muscles of the body work in antagonistic groups, one flexing, or bending, the joint, and the other extending, or straightening, it (Figure 28–8). Also, two antagonistic groups may contract together to stabilize a joint. Such muscle action makes it possible for us (and others) to stand upright.

A skeletal muscle, such as the biceps, consists of bundles of muscle fibers—often hundreds of thousands of fibers—held together by connective tissue. Each fiber is a single cell with many nuclei. These fibers are often very large cells—50 to 100 micrometers in diameter and many centimeters long. Their internal structure and the mechanism by which they contract will be described in Chapter 36.

Nerve Tissue

The fourth major tissue type is **nerve tissue.** The essential functional units of nerve tissue are cells known as **neurons,** which transmit nerve impulses. Nerve tissue also contains cells of another type, known as **glial cells,** that not only support and insulate the neurons but may also supply them with nutrients.

As shown in Figure 28–9, neurons come in a diversity of sizes and shapes. Typically, a neuron consists of a **cell body,** which contains the nucleus and much of the metabolic machinery of the cell; **dendrites,** usually numerous, short, threadlike cytoplasmic extensions that, with the cell body, receive stimuli from other cells; and an **axon,** a long extension that is capable of rapidly conducting a nerve impulse over great distances. Axons are also known as **nerve fibers.**

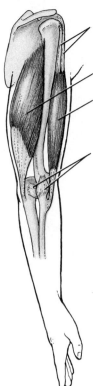

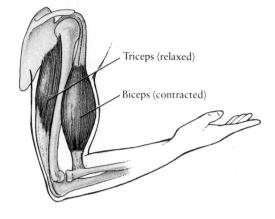

Tendons (connective tissue joining muscle to bone)

Triceps muscle (contracted) } Antagonistic pair

Biceps muscle (relaxed)

Ligaments (connective tissue joining bone to bone)

Triceps (relaxed)

Biceps (contracted)

28–8 *Muscles attached to bone move the vertebrate skeleton. Skeletal muscles often work in antagonistic pairs, with one relaxing as the other contracts. Muscles cannot lengthen spontaneously; they lengthen only when the joint moves in the opposite direction, due to contraction of the antagonistic muscles. For example, when you move your hand down, the triceps contracts and the biceps relaxes. When you move your hand toward your shoulder, the biceps contracts, while the triceps relaxes.*

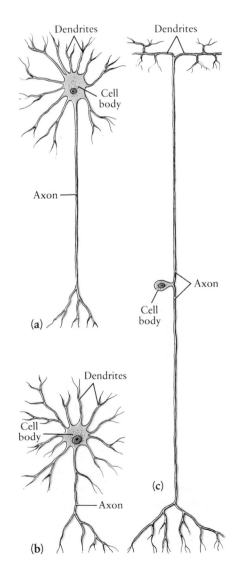

28–9 *Three of the many different forms of neurons. (a) Motor neurons and relay neurons are characterized by a cell body with numerous dendrites and a long axon that travels without interruption to its terminal, where it branches. (b) Interneurons, which are found within localized regions of the central nervous system, typically have a complex system of dendrites and, as shown here, a short axon with branches—or no axon at all. (c) In sensory neurons, which transmit impulses from sensory receptors at the ends of the dendrite branches, the cell body is off to one side of the long axon. All of these neurons form connections with other neurons.*

Dendrites   Dendrites

Cell body

Axon

Axon

Cell body

(a)

Dendrites

Cell body

(c)

Axon

(b)

Neurons are specialized to receive signals—from the external environment, the internal environment, or other neurons—to integrate the signals received, and to transmit the integrated information to other neurons, muscles, or glands. Functionally, there are four classes of neurons: **sensory neurons,** which receive sensory information and relay it to the central nervous system (brain and spinal cord); **interneurons,** which transmit signals within localized regions of the central nervous system; **relay neurons,** which relay signals between different regions of the central nervous system; and **motor neurons,** which transmit signals from the central nervous system to effectors, such as muscles or glands. These four types of neurons are linked in a variety of circuits, ranging from simple **reflex arcs** (Figure 28–10) to the extremely complex interconnections that characterize the human brain.

Neurons may reach astonishing lengths. For example, the axon of a single motor neuron may extend from the spinal cord down the whole length of the leg to the toe. Or the axon of a sensory neuron, the cell body of which is located just outside the spinal cord, may extend from the toe to the cell body and then up the entire length of the spinal cord to the lower part of the brain, where it terminates. In a giraffe, such a cell might be close to 5 meters long, and in an adult human, close to 2 meters.

Nerves are bundles of many axons from many neurons—usually hundreds and sometimes thousands. Each axon is capable of transmitting a separate message, like the wires in a telephone cable.

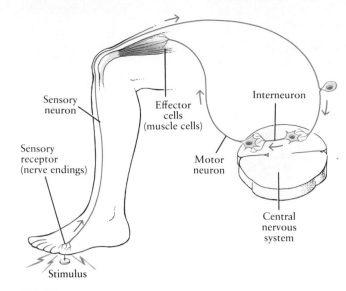

28–10   *A simple reflex arc. Sensory receptors stimulate a sensory neuron, which relays a signal to an interneuron located entirely within the central nervous system (in this example, within the spinal cord). From the interneuron, the signal is transmitted to a motor neuron, which, in turn, stimulates an effector. When stimulated, the muscle cells shown here will contract, causing the foot to move out and up, away from the tack. These basic components of the reflex arc are found in all vertebrates, from the simplest to the most complex. Reflex arcs play an essential role in the regulation of many internal processes, as well as making possible almost instantaneous responses to numerous environmental stimuli.*

Not shown in this diagram are relay neurons, which simultaneously transmit to other regions of the central nervous system (typically in the brain) the information initially received from the sensory neuron.

Organs and Organ Systems

As we have just seen, the human body comprises a variety of cells, organized into four types of tissues. At the next level of organization, different tissues unite to form organs. The stomach, for example, is an organ made up of layers—glandular epithelium (in the stomach lining), connective tissue, nerves, and smooth muscle—that collectively include tissues from each of the four basic types.

Not all organs, however, are internal and soft-structured like the stomach. The largest organ of the human body is its outer protective covering, the skin, about which we shall have more to say shortly. And, as we saw in Figure 28–6, bones are—despite superficial appearances—living organs containing several different tissue types.

Organs that work together in an integrated fashion to perform particular functions make up organ systems, the next level of organization. The digestive system, for example, is composed of the stomach, intestine, liver, pancreas, and a number of other organs, each of which carries out specific activities that contribute to the overall process of digestion. Similarly, the skeletal system consists of the 206 individual organs—bones—that together form the human skeleton (Figure 28–11). These bones are integrated through their articulations at the joints to meet the sometimes conflicting requirements of the body for support and flexibility. Bone and joint injuries resulting from the physical demands of athletic activities are reminders of both the critical functions and the vulnerability of this organ system.

Figure 28–12, on pages 516–517, provides an overview of the organ systems of the human body. As we examine the structure and physiology of the human animal in the subsequent chapters of this section, we shall do so principally in terms of these organ systems and the functions they perform.

28–11   *The skeleton of a human adult contains 206 bones. Twenty-nine are in the skull, including 14 face bones and six small bones of the ears. There are 27 bones in each hand and 26 in each foot.*

The skeletal system is not only the body's structural support but also its principal reservoir of calcium. As we shall see in subsequent chapters, calcium ions play a critical role in a number of essential physiological processes. The concentration of calcium ions in the bloodstream is maintained within relatively narrow limits by the release of calcium from mineralized bone and the deposition of calcium in newly formed bone. Thus, although the skeleton appears inert and unchanging, the bones of which it is composed undergo a continuous process of remodeling. The turnover can be rapid; the lower portion of the femur, for example, is completely replaced every five to six months.

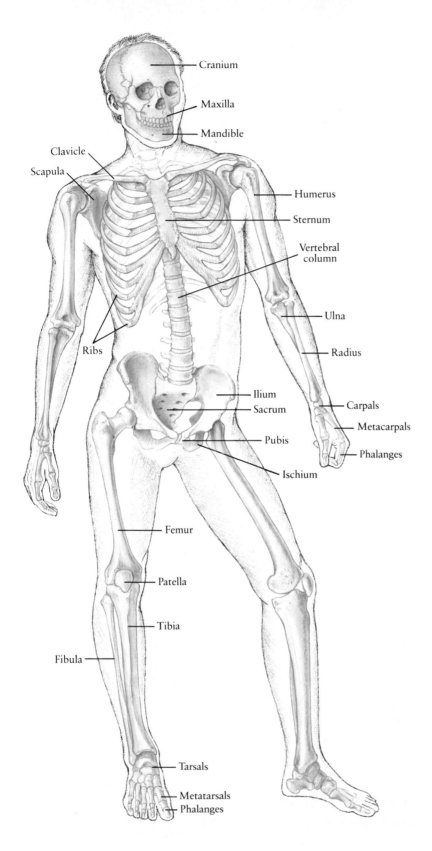

Functions of the Organism

The complex structures and physiological processes of the animal body "make sense" when seen as adaptations that solve particular problems presented by the relationship between the organism and its environment. Before we further employ this metaphor of problem and solution, however, we should stop for a moment to see what we really mean by biological problem solving.

An organism confronts its "problems" with a set of genetic instructions. If they work—that is, if they result in structures, processes, and behaviors that enable the organism to function in its environment—the organism lives and passes on these instructions to its offspring. If its instructions are better than those carried by its neighbors—enabling the organism to function more effectively—its offspring will probably be more numerous than those of its neighbors. It is as this process is repeated, generation after generation, that "problems" get "solved."

Homeostasis

Perhaps the most critical problem facing all organisms is the maintenance of a relatively constant internal environment. The chemical reactions of a living cell and the synthesis and maintenance of its constituent structures require a tightly controlled chemical environment, a fairly narrow range of temperature, and protection from foreign invaders, such as bacteria and viruses, that may feed on the substance of the cell, poison its enzymes with toxins, or subvert its genetic machinery for their own replication.

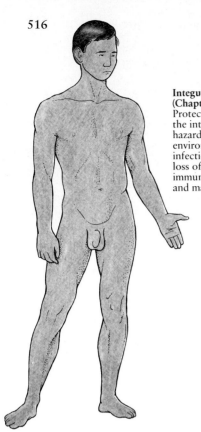

**Integumentary system
(Chapter 28):**
Protection and separation of
the internal environment from
hazards of the external
environment, such as
infectious microorganisms and
loss of moisture; contains
immunologically active cells
and many sensory receptors

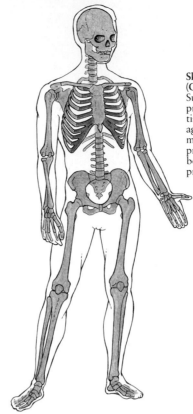

**Skeletal system
(Chapter 28):**
Support for the body and
protection of internal
tissues; provides a framework
against which muscles can act,
making movement possible;
produces blood cells (in the
bone marrow); the body's
principal reservoir of calcium

**Excretory system
(Chapter 32):**
Maintenance of water balance
and chemical composition of
blood; excretion of metabolic
wastes, particularly
nitrogenous compounds

28–12   *An overview of the principal
organ systems of the adult human and their
functions.*

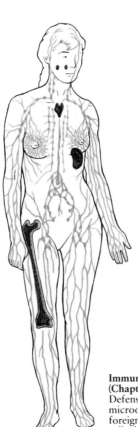

**Immune system
(Chapter 33):**
Defense against infectious
microorganisms and other
foreign invaders; removal of
cells damaged by disease or
injury

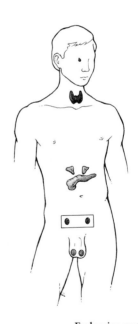

**Endocrine system
(Chapter 34):**
Regulation of physiological
processes through the
synthesis and release of
chemical messengers (hormones)
that act on specific target
tissues and organs

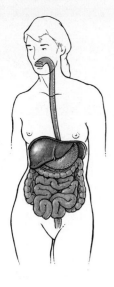

**Digestive system
(Chapter 29):**
Processing of food, absorption
of nutrients, and elimination
of undigested residues

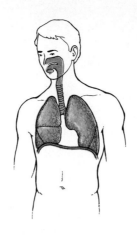

**Respiratory system
(Chapter 30):**
Uptake of oxygen and release
of carbon dioxide

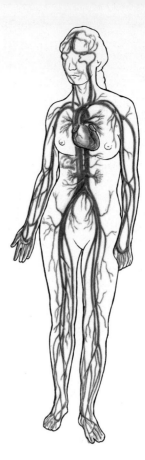

**Circulatory system
(Chapter 31):**
Transport of oxygen and
nutrients to the cells and
transport of carbon dioxide
and other metabolic wastes
away from the cells

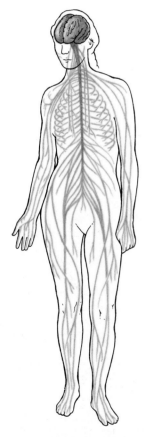

**Nervous system
(Chapters 35 and 36):**
Reception, transmission, and
integration of information
from the internal and external
environments; control and
coordination of the body's
responses to this information

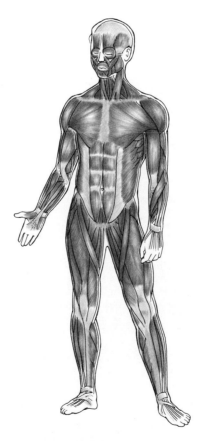

**Muscular system
(Chapter 36):**
Physical movements of the body
and of its internal components
through the alternating
contraction and relaxation of
muscle tissues

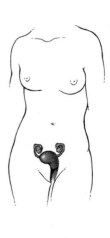
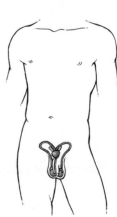

**Reproductive system
(Chapter 37):**
Production of gametes (sperm
cells in males and eggs cells
in females); mechanisms for
the transport of sperm cells
to egg cells; in females,
provision of protection and
nourishment for the developing
embryo and fetus

Skin: The Original Space Suit

Without our thin envelope of skin, we would find the Earth's environment as hostile to our bodies as the environment of outer space would be to the body of an astronaut sporting only jeans and a tee shirt. The skin is the body's first line of defense against a diversity of environmental hazards, including desiccation (drying out), abrasive materials, harmful chemicals, infectious microorganisms, and ultraviolet radiation from the sun. It is the protective "capsule" within which the body's regulatory systems can make the adjustments necessary to maintain homeostasis in the face of changing conditions.

Human skin, which is the largest organ in the body, consists of two layers, an outer epidermis and an inner dermis. As we saw in Figure 28–4, the epidermis is a stratified epithelial tissue, in which layers of cells are piled on top of one another. The structure of the dermis is considerably more complex. As shown in the illustration below, it includes connective tissue, smooth muscle, blood vessels, nerve fibers, and glands. Working together, the cells of the epidermis and the dermis provide effective physical, chemical, and biological barriers separating the body from the external environment.

Variations in the color of human skin—from almost white to a rich ebony—are the result of differences in its content of a pigment known as melanin. Melanin is produced by specialized cells, the melanocytes, that are located at the base of the epidermis. Although the number of melanocytes is approximately the same in all individuals, the type of melanin and the number, size, and distribution of the melanin granules produced by the melanocytes vary, leading to the observed variations in skin color. The type of melanin—that is, its exact chemical composition—is genetically determined. The amount of melanin produced, however, depends both on the genetic inheritance of the individual and on his or her recent history of exposure to sunlight.

Ultraviolet radiation in sunlight stimulates the melanocytes to synthesize more melanin, producing tanning. Almost all humans show a tanning response. The only exceptions are albinos, who lack one or more of the enzymes required for the synthesis of melanin. Reduced

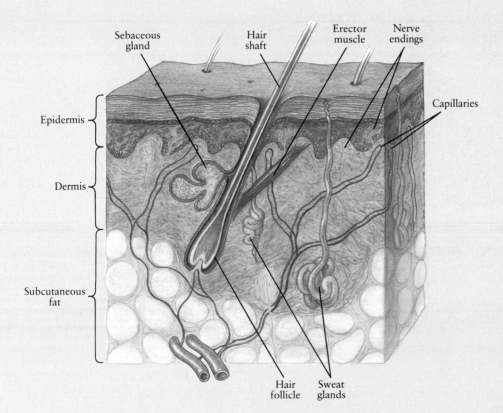

A section of human skin, showing the epidermis and the dermis, which together make up the skin, and the underlying subcutaneous fat.

The epidermis is made up mostly of epithelial cells and consists of two layers: an inner layer of living cells, and an outer layer of dead cells filled with keratin (page 63). The epidermis is a turnover system: cells produced in the inner layer migrate toward the surface and die. At the base of the epidermis are the melanocytes that produce the melanin responsible for skin color.

The dermis, consisting mostly of connective tissue, contains sensory nerve endings, small blood vessels known as capillaries, hair follicles, erector muscles that raise the hairs when contracted, and sweat and sebaceous glands, which are composed of modified epithelial cells. The sebaceous glands produce a fatty substance that lubricates the skin surface. Fatty tissue, which makes up the insulating layer below the dermis, is also a form of connective tissue, as is the blood found in the capillaries.

Sebaceous gland

Hair shaft

Erector muscle

Nerve endings

Capillaries

Epidermis

Dermis

Subcutaneous fat

Hair follicle

Sweat glands

exposure to sunlight results in a gradual fading of the skin color, as melanin production is reduced and the older layers of skin are sloughed off.

Ultraviolet radiation damages organic molecules, including DNA. Thus, overexposure can lead not only to sunburn but also to skin cancer. However, this same high-energy radiation plays a crucial role in the conversion of cholesterol molecules in the skin into vitamin D, a substance essential in the body's regulation of calcium ion (Ca^{2+}). Although large amounts of melanin in the skin protect the body from the hazards of ultraviolet light, they also reduce the production of vitamin D.

According to one hypothesis, the ancestors of modern *Homo sapiens* originated in the sun-drenched tropics and were all dark-skinned. The intensity of the sunlight throughout the year allowed sufficient amounts of vitamin D to be synthesized by the skin, despite the large amounts of melanin that protected the cells from damage. In those populations that moved northward, however, lower levels of sunlight, coupled with the need to cover more of the body during colder weather, inhibited the production of sufficient quantities of vitamin D in those with darker skin. In northern climates, natural selection generally favored the decreased production of melanin and, hence, lighter skin.

Understanding our melanin-mediated relationship to sunlight provides not only a key to our past but also a warning about our future. As a consequence of human activities, the amount of ultraviolet light striking the Earth's surface is increasing. This is a consequence of a decrease in the thickness of the ozone layer of the atmosphere (to be discussed in Chapter 45), which absorbs most of the ultraviolet rays in sunlight before they reach the Earth's surface. Although we can defend ourselves against this environmental change by wearing protective clothing and applying chemical sunscreens, other organisms, which must depend on the slow evolution of new adaptive mechanisms, cannot.

As we saw in Chapter 1, **homeostasis,** the maintenance of a relatively constant internal environment, is one of the identifying characteristics of living systems. In the absence of homeostasis, an organism cannot survive. Homeostasis is, in many ways, more difficult for one-celled organisms and small multicellular organisms than it is for large animals, such as *Homo sapiens*. Small organisms are extremely vulnerable to changes in the temperature or chemical composition of the medium in which they live, and it is likely that the strongest evolutionary pressures toward larger size and multicellularity were related to homeostasis. Larger organisms, with their lower surface-to-volume ratios (page 88) and, often, protective outer coverings, have a greater resistance to external change and foreign invaders. The world in which your individual cells function and flourish is decidedly different from the world around you— but not so different from the warm primordial soup in which we all began.

Maintaining a relatively constant environment throughout the body of a large animal is a complex process. It involves not only continuous monitoring and regulation of many different factors but also ready defenses against an enormous diversity of microorganisms. As we shall see, virtually all of the organ systems participate in homeostasis.

Energy and Metabolism

Organisms are strikingly different from the surrounding environment in both composition and structural organization. One aspect of homeostasis is the maintenance of the high level of organization characteristic of living systems in the face of the universal trend toward disorder (page 130). As we noted in our discussion of the second law of thermodynamics in Chapter 7, this trend can be countered only by the expenditure of energy.

Animals are, as we saw in Chapter 26, multicellular heterotrophs that ingest their food, from which energy is ultimately released by the oxygen-requiring reactions of the mitochondria. In large animals, a critical problem is transforming the ingested food into molecules that can be used by individual cells and then providing those molecules, along with oxygen, to each of the multitude of cells that make up the organism's body. A related problem is the removal of metabolic wastes from the body. The complexities of the human digestive, respiratory, circulatory, and excretory systems, to be discussed in Chapters 29 through 32, represent particular evolutionary solutions to these problems.

Defense against Disease

Another critical aspect of homeostasis is the defense of the organism against both infectious microorganisms and diseased cells of the body itself. In humans and other mammals, this defense is mounted by the immune system, to be discussed in Chapter 33. As we shall see, the cells of this vital system are produced in the bone marrow (part of the skeletal system) and move through the body in the bloodstream (part of the circulatory system). They are found in high concentrations at a number of sites in the body, including the outer protective wrapping of skin.

Integration and Control

In a multicellular organism, homeostasis requires coordination of the activities of the numerous cells that constitute the organism. Otherwise, tissues and organs could not respond to the overall physiological needs of the organism, which change with fluctuations in the environment. Moreover, animals are typically quite active, moving about as they search for and capture food, while simultaneously trying to avoid being captured by other animals. A life of active movement requires receiving and processing information from the external environment, followed by coordinated and appropriate contractions of the skeletal muscles.

Animals have two major control systems, the endocrine system (the hormone-secreting glands and their products) and the nervous system. Speaking very generally, the endocrine system is responsible for changes that take place over a relatively long time period—minutes to months—whereas the nervous system is involved with more rapid responses—milliseconds to minutes. However, the more we learn about these systems, the more they are seen to be closely interrelated. For instance, the production of sex hormones, as well as other hormones, was once believed to be under the control of the "master" pituitary gland. It is now known that the pituitary, though perhaps a master in some aspects of the endocrine world, is actually an executive secretary of the **hypothalamus,** a major brain center. And, within little more than the last decade, it has been discovered that the embryonic development of certain major brain centers is profoundly influenced by hormones regulated by the pituitary.

Even anatomically, the endocrine and nervous systems are not distinct. One of the body's most important glands, the adrenal medulla, a major source of adren-

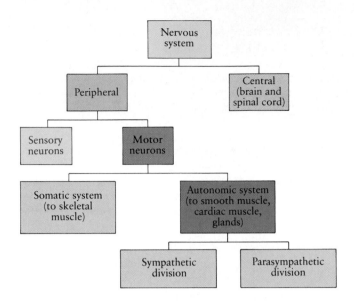

28–13   *The nervous system of vertebrates, including* Homo sapiens, *consists of a central nervous system (the brain and spinal cord) and a peripheral nervous system, a vast network of nerves connecting the central nervous system with all other parts of the body. Sensory neurons carry information to and motor neurons carry information from the central nervous system. The motor neurons are organized into the somatic and autonomic systems, and the autonomic system contains two divisions, the sympathetic and the parasympathetic.*

aline (also known as epinephrine), is not, strictly speaking, a gland. That is, it is not modified epithelial tissue but is rather a large ganglion—a collection of nerve cell bodies—whose nerve endings secrete the hormone.

Within the nervous system, there are a number of structural and functional subdivisions (Figure 28–13). One subdivision, the **somatic** system, innervates skeletal muscle. Another, the **autonomic** ("involuntary") system, innervates smooth muscle, cardiac muscle, and glands. The autonomic system is further subdivided into the **sympathetic** and the **parasympathetic** divisions, which interact with one another in a finely tuned system of checks and balances. The sympathetic division is most active in times of stress or danger. Among the major effectors of the sympathetic division is the adrenal medulla, and the overall effects of general sympathetic stimulation are those we associate with a "rush of adrenaline." The parasympathetic division plays its major role in supporting everyday activities such as digestion and excretion.

The functions of the endocrine and nervous systems will be described in greater detail in Chapters 34 through 36, but we introduce them now because every organ system to be discussed is under the influence of both the endocrine system and the sympathetic and parasympathetic divisions of the autonomic nervous system.

Feedback Control

The body's integration and control systems characteristically act through feedback loops, both negative and positive. The simplest example from everyday life of a negative feedback system is the thermostat that regulates your furnace. When the temperature in your house drops below the preset thermostat level, the thermostat turns the furnace on. When the temperature rises above the preset level, the thermostat turns the heat off.

In a living organism, systems are seldom completely on or off, and homeostatic control is much more finely tuned. The principle, however, is the same: a deviation from a "preset" condition stimulates a response that reduces the deviation. For example, endocrine cells in the pancreas, an organ of the digestive system, produce two hormones—glucagon and insulin—that are vital in maintaining a relatively constant level of glucose in the blood (Figure 28–14). Glucose is the only energy source used by brain cells, and it is a major energy source for other cells as well. When the glucose concentration of the blood falls, glucagon is produced, which causes the release of glucose from the liver. When glucose levels in the blood rise, insulin is released, causing the uptake of glucose by body cells. Thus, glucose concentration is controlled by two separate feedback systems.

In the chapters that follow, we shall see numerous examples of feedback loops, most of which involve both the nervous and the endocrine systems.

The Continuity of Life

From an evolutionary perspective, the ultimate challenge an organism faces is to multiply. The biological imperative to reproduce, following the dictates of the genes, is enormous. Reproduction may be carried out in a variety of ways, but in mammals it is always sexual, involving the formation of gametes and their union to form a zygote, or fertilized egg. The last two chapters of this section will describe the formation of human gametes, their fusion to form the fertilized egg, and the almost miraculous, though oft-repeated, phenomena by which this single cell develops into a human being.

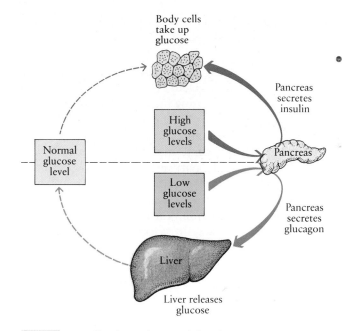

28–14   *Feedback regulation of the glucose concentration in the blood. High levels of glucose in the blood cause the pancreas to release insulin, which results in glucose uptake by cells of the body. Low levels of glucose cause the secretion of glucagon by the pancreas, which results in the release of glucose to the bloodstream from the liver. As a result of these two feedback systems, a nearly constant level of glucose is maintained in the blood.*

28–15   *Genus,* Homo; *species,* sapiens. *Embryo at 7 weeks.*

Summary

As vertebrate animals, humans are characterized by a bony, jointed internal skeleton that includes a skull and a vertebral column enclosing the central nervous system (the brain and spinal cord). The human body, like that of other mammals, contains a coelom that is divided by a muscle, the diaphragm, into two major compartments, the abdominal cavity and the thoracic cavity.

The organization of the human body is hierarchical. Cells are organized into tissues, groups of cells that carry out a unified function. Various types of tissues are grouped in different ways to form organs, and organs are, in turn, grouped to form organ systems.

The four principal tissue types in the human body are epithelial tissue, connective tissue, muscle, and nerve. Epithelium serves as a covering or lining for the body and its cavities. Glands are composed of specialized epithelial cells. Their secretions include mucus, perspiration, milk, saliva, hormones, and digestive enzymes.

Connective tissues are characterized by their capacity to secrete extracellular substances, such as collagen and other fibers. They serve to support, strengthen, and protect the other tissues of the body.

Muscle cells are specialized for contraction, which is accomplished by assemblies of two proteins, actin and myosin. In striated muscle, which includes skeletal muscle and cardiac muscle, these assemblies form a striped pattern. In smooth muscle, no such pattern is apparent.

The essential functional units of nerve tissue, the neurons, are specialized for the reception, processing, and transmission of information. Neurons typically consist of a cell body, dendrites, and an axon. They are surrounded and supported by glial cells.

A number of functions are essential to the ongoing life of a multicellular animal. The most fundamental is homeostasis, the maintenance of a relatively constant internal environment. Virtually all of the organ systems and physiological processes of the animal play a role in homeostasis. Food must be obtained and processed to yield energy-rich molecules that can be used by individual cells. These molecules, as well as oxygen, must be delivered to the cells, and metabolic wastes must be removed from them. Defenses must be mounted against infectious microorganisms and other foreign invaders. The contractions of skeletal muscles in locomotion and the physiological activities of the many tissues and organs of the body must be coordinated in response to changes in both the internal and external environments. Integration and control involve both the endocrine and nervous systems and characteristically function through feedback loops. The ultimate challenge to the animal, dictated by its genes, is to reproduce.

Questions

1. Distinguish among the following terms: animal/vertebrate/mammal; coelom/diaphragm; striated muscle/smooth muscle; skeletal muscle/cardiac muscle; neurons/glial cells; dendrite/axon; sensory neuron/interneuron/relay neuron/motor neuron; nerve fiber/nerve; somatic/autonomic; sympathetic/parasympathetic.

2. What is the functional significance of each of the four tissue types? Give an example of each type.

3. What are the three types of epithelial tissue? What is the basis for classification of epithelial cells?

4. In addition to protection, what are the other major functions of epithelial tissue?

5. What is the major structural difference between connective tissue and epithelial tissue? What functions of connective tissue account for this structural difference?

6. Identify the tissue shown in the micrograph below. What structures are visible in the micrograph?

7. What is meant by antagonistic paired muscles?

8. A young child who has not yet learned the meaning of "hot" touches a dish that has just come out of the oven. Describe the series of events that lead to the almost instantaneous removal of the hand from the dish, followed by crying.

9. Identify the major organ systems of the human body. What are the principal functions of each system?

10. How can you tell that homeostasis is occurring in each of the following situations? (a) The air temperature is 70°F, but your body temperature is 98.6°F. (b) You drink several cans of soda with pH of 3; the pH of your blood remains at about 7.5. (c) You swim in a lake that contains pathogenic bacteria, but you do not become ill.

11. Modern medical research has given us not only vastly improved health but also numerous insights into human physiological processes that cannot, ethically or morally, be experimentally manipulated. The success of this research has depended heavily on the extrapolation from other organisms to *Homo sapiens*. On what foundation does this extrapolation rest?

Suggestions for Further Reading

Caplan, Arnold I.: "Cartilage," *Scientific American*, October 1984, pages 84–94.

Eckert, Roger, and David Randall: *Animal Physiology: Mechanisms and Adaptations*, 3d ed., W. H. Freeman and Company, New York, 1988.

>   The emphasis is on basic principles. Well-written and handsomely illustrated.

Green, Howard: "Cultured Cells for the Treatment of Disease," *Scientific American*, November 1991, pages 96–103.

Kessel, Richard G., and Randy H. Kardon: *Tissues and Organs: A Text-Atlas of Scanning Electron Microscopy*, W. H. Freeman and Company, New York, 1979.*

>   A collection of more than 700 outstanding scanning electron micrographs of vertebrate tissues and organs, beautifully reproduced. The accompanying text summarizes our knowledge of each organ system and its component parts.

Luciano, Dorothy S., Arthur J. Vander, and James H. Sherman: *Human Anatomy and Physiology*, 2d ed., McGraw-Hill Publishing Company, New York, 1983.

>   An introductory anatomy and physiology textbook. The text is clearly written, and the detailed illustrations—particularly of the skeletal, muscular, nervous, and circulatory systems—are outstanding.

Marieb, Elaine N.: *Human Anatomy and Physiology*, 2d ed., The Benjamin/Cummings Publishing Company, Inc., Redwood City, Calif., 1992.

>   This well-illustrated text by an experienced educator is accessible in its organization and presentation. Up-to-date, with numerous clinical applications.

Poole, Robert M. (ed.): *The Incredible Machine*, The National Geographic Society, Washington, D.C., 1986.

>   A spectacular photographic exploration of the development and function of the human body, accompanied by clear, engaging text.

Rusting, Ricki L: "Why Do We Age?" *Scientific American*, December 1992, pages 130–141.

Schmidt-Nielsen, Knut: *Animal Physiology: Adaptation and Environment*, 4th ed., Cambridge University Press, New York, 1990.

>   Schmidt-Nielsen is concerned with underlying principles of animal physiology—the problems animals have to solve in order to survive. The emphasis is on comparative physiology, and the lucid exposition is illuminated by many interesting examples.

Vander, Arthur J., Dorothy S. Luciano, and James H. Sherman: *Human Physiology: The Mechanisms of Body Function*, 5th ed., McGraw-Hill Publishing Company, New York, 1990.

>   Most highly recommended. The text is a model of clarity, and the diagrams are splendid.

* Available in paperback.

Digestion

Odd arrangements and cumbersome solutions are, as we noted earlier, strong evidence of evolution. A cumbersome but highly successful solution to a nutritional problem is found among ruminants, a group that includes cattle, sheep, goats, deer, antelope, and giraffes. The problem is cellulose, by far the most common energy-containing molecule on this planet. Most animals cannot use it because they lack the enzymes—cellulases—to break it down.

The solution is the rumen, a vast fermentation vat, which contains an enormous number of symbiotic bacteria and protozoa. In the rumen, the microorganisms break down cellulose and other structural polysaccharides into useable materials. In return, they receive a warm, wet home, free food, and protection—except from their host. The host derives amino acids and other necessities from digesting the recipients of its hospitality.

Rumination, physiologically speaking, is a major activity. It involves huge quantities of liquids and gases. A cow, for example, secretes 60 liters of saliva a day and produces 2 liters of carbon dioxide and methane per minute, most of which escapes in sweet, chlorophyll-scented belches. A cow's rumen and its contents weigh as much as 100 kilograms. Apparently, it is all worth it, for ruminants are *the* major herbivores. They appear prominently in the fossil record some 200,000 years ago and have dominated the grasslands ever since.

"Ruminate" in everyday English means to meditate, or to chew something over. This meaning has come about because animals with rumens are also cud chewers. Cud is compacted balls of tough vegetation that are regurgitated and chewed to prepare them for processing in the rumen. This also is a useful adaptation to an herbivorous way of life, enabling wild ruminants to graze rapidly while exposed in the open and then to retire to chew their cud in comparative safety from their predators.

Rumination is often accompanied by a look of dreamy satisfaction sometimes associated with humans, who, although less specialized than ruminants in digestive activities, reportedly excel at thinking things over.

The healthy animal body is a complex, precisely organized biological machine that requires substantial energy and a supply of building materials to maintain its structure, to carry out the functions necessary for homeostasis, and to fuel its many activities. Reserves of energy and building materials are acquired through the ingestion of food. A fundamental problem for most animals, however, is that their foods—whether a leaf for a caterpillar, a fly for a spider, or a hamburger on a bun for a human—consist of complex plant and animal tissues. These tissues cannot supply energy or useful structural materials until they are broken down into their component molecules.

Digestion is the breakdown of foods into molecules that can be delivered to and used by the individual cells of the animal body. These molecules serve a variety of functions. For example, they may be energy sources or building blocks; they may provide essential chemical elements, such as calcium or iron; or they may be molecules—certain amino acids, fatty acids, and vitamins—that animal cells need but cannot synthesize for themselves.

In Chapter 26, we traced the evolution of digestive systems from the feeding cells of sponges (page 458), through the gastrovascular cavity of jellyfish (page 459) and the digestive cavity of flatworms (page 463), to the major evolutionary breakthrough—the one-way digestive tract—exemplified by the ribbon worms (page 463). As animals became larger and more complex, the tract became more convoluted, providing increased working surfaces, and different portions became specialized for different stages of the digestive process.

Digestive Tract in Vertebrates

In vertebrates, the **digestive tract**—known less elegantly but just as accurately as the gut—consists of a long convoluted tube, extending from mouth to anus (Figure 29–1). The **digestive system** also includes the salivary glands, the pancreas, the liver, and the gallbladder, accessory organs that provide the enzymes and other substances essential for digestion.

The inner surface of the gut is continuous with the outer surface of the body, and so, technically, the cavity of the gut is outside the body. Because its contents are sequestered, they can be subjected to the action of enzymes and bacteria and to conditions of pH that, although optimal for the breakdown of food, would rapidly destroy the living cells and tissues of the body proper. Nutrient molecules actually enter the body only when they pass through the epithelial lining of the digestive tract. Thus, the process of digestion involves two components: the breakdown of food molecules and their absorption into the body.

The digestive tract begins with the mouth and includes the pharynx, esophagus, stomach, small intestine, large intestine, and anus. Each of these areas is specialized for a particular phase in the overall process of digestion, but the fundamental structure of each is similar. As shown in Figure 29–2, the tube, from beginning to end, has four layers: (1) the innermost layer, the **mucosa,** which is made up of epithelial tissue, an underlying basement membrane, and connective tissue, with a thin outer coating of smooth muscle in some places; (2) the **submucosa,** which is made up of connective tissue and contains nerve fibers and blood and lymph vessels; (3) the **muscle layer;** and (4) the **serosa,** an outer coating of connective tissue.

The epithelium of the digestive tract contains many mucus-secreting cells and, in some parts of the gut, glands that secrete digestive enzymes. Along most of the digestive tract, the muscle layer consists of two layers of smooth muscle, an inner layer, in which the orientation is circular, and an outer layer, in which the cells are longitudinally arranged. Coordinated contractions of these muscles produce ringlike constrictions that mix the food, as well as the wavelike motions, known as **peristalsis,** that move food along the digestive tract (Figure 29–3).

At several points the circular layer of muscle thickens into heavy bands, called **sphincters.** These sphincters, by relaxing or contracting, act as valves to control the passage of food from one area of the digestive tract to another.

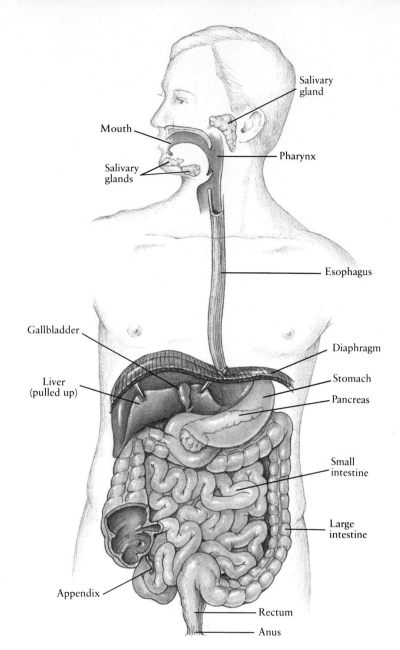

29–1   *The human digestive tract. Food passes from the mouth through the pharynx and esophagus to the stomach and small intestine, where most digestion takes place. Undigested materials pass through the large intestine, are stored briefly in the rectum, and are eliminated through the anus. Accessory organs of the digestive system are the salivary glands, pancreas, liver, and gallbladder.*

The Mouth: Initial Processing

The mechanical breakdown of food begins in the mouth. Many vertebrates, including most mammals, have teeth, which are structures adapted for the tearing and grinding of food. Modern birds, which are toothless, have gizzards containing particles of sand and gravel that serve the same function.

The tongue, also a vertebrate development, serves largely to move and manipulate food in mammals. However, there are other adaptations. The jawless hagfish and lampreys (page 483), for example, have tongues equipped with horny "teeth" that are used for

29–2   *The layers of the digestive tract include (1) the mucosa, (2) the submucosa, which contains nerves and blood and lymph vessels, (3) the muscle layer, and (4) the serosa (also known as the visceral peritoneum), which covers the outer surfaces of the organs in the abdominal cavity. Mesenteries are folds of the peritoneum that suspend the intestines from the posterior abdominal wall. Glands outside the digestive tract, principally the pancreas and liver, discharge digestive enzymes and bile into the tract through various ducts.*

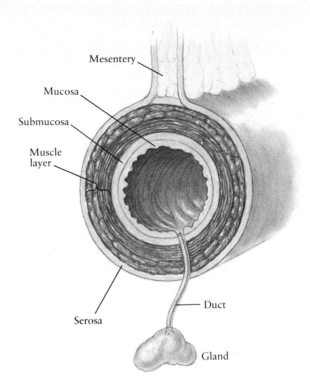

Mesentery

Mucosa

Submucosa

Muscle layer

Serosa

Duct

Gland

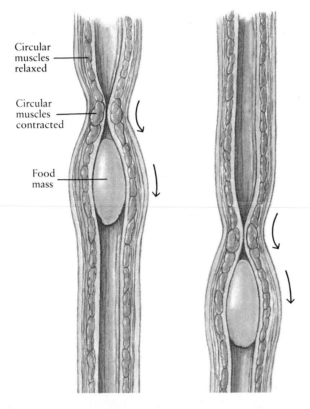

Circular muscles relaxed

Circular muscles contracted

Food mass

29–3   *Food is moved through the vertebrate digestive tract by several different patterns of contraction of the smooth muscle in its walls. One of the most important patterns, known as peristalsis, consists of progressive waves of contraction of the circular muscles.*

scraping flesh from the bodies of other fish. The sticky tongues of frogs and toads, which are attached at the front of the mouth, flip out to catch insects. Mammalian tongues carry taste buds (Figure 29–4), which contain sensory receptors for certain chemicals. In humans, the tongue has acquired a secondary function of forming sounds for communication.

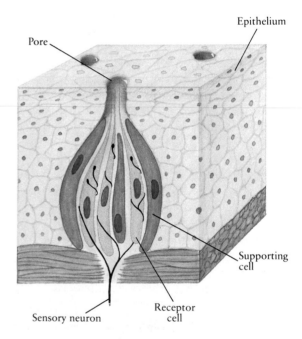

Epithelium

Pore

Supporting cell

Receptor cell

Sensory neuron

29–4   *A human taste bud. Taste buds are found primarily on the surface of the tongue but also occur on the roof of the mouth, the inner surface of the cheeks, and the pharynx. At the opening of the pore, the hairlike ends of the receptor cells are bathed in saliva containing dissolved molecules from the food. When the receptor cells are stimulated, the information is transmitted to the brain by way of the sensory neuron. The receptor cells in taste buds perform a critical function in enabling animals to distinguish what is good to eat from what is not.*

29–5 *Separation of the digestive and respiratory systems in mammals makes it possible to keep food out of the lungs. The pharynx, the common passageway of the two systems, is at the back of the mouth and connects with the trachea (windpipe) and the esophagus. (a) Prior to swallowing, the passageway from the nose to the trachea is open, allowing air to enter the lungs. The entrance to the esophagus is blocked. (b) Swallowing. As the food mass descends, the upper portion of the trachea moves upward and the epiglottis tips downward, blocking the entrance to the trachea. (c) The food mass passes into the esophagus, and the entrance to the trachea is once again opened.*

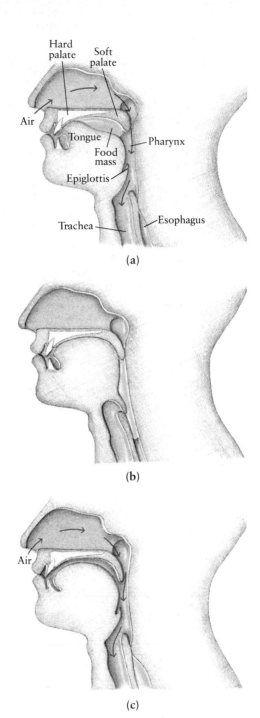

(a)

(b)

(c)

As food is chewed, it is moistened by saliva, a watery, slightly alkaline secretion produced by the **salivary glands.** The saliva, which also contains mucus, lubricates the food so that it can be swallowed easily. In humans and other mammals that chew their food, saliva also contains a digestive enzyme, **amylase,** that begins the breakdown of starches. Like all digestive enzymes, amylase works by hydrolysis; that is, the breaking of each bond involves the addition of a molecule of water (see page 53). Carnivores, such as dogs, which tear and gulp their food, have no digestive enzymes in their saliva.

The secretion of saliva is controlled by the autonomic nervous system. It can be initiated by the presence of food in the mouth, which triggers reflexes originating in taste buds and in the walls of the mouth, and also by the mere smell or anticipation of food. (Think hard, for a moment, about eating a lemon.) Fear inhibits salivation. At times of great danger or stress, the mouth may become so dry that speech is difficult. On the average, we produce 1 to 1.5 liters of saliva every 24 hours.

The Pharynx and Esophagus: Swallowing

From the mouth, food is propelled backward toward the **esophagus,** a muscular tube about 25 centimeters long in adults. Swallowing is the passing of food to the esophagus and through the esophagus on to the stomach (Figure 29–5). It begins as a voluntary action but, once under way in humans, it continues involuntarily. The upper part of the human esophagus is striated muscle, but the lower part is smooth muscle. Both liquids and solids are propelled along the esophagus by peristalsis. This process is so efficient that we can swallow water while standing on our heads.

The Heimlich Maneuver

Accidental choking on food claims the lives of almost 3,000 persons a year in this country alone, more than accidents involving firearms or airplanes. It occurs when a food mass enters the trachea rather than the esophagus (Figure 29–5). If the food becomes lodged, the person cannot speak or breathe and, if the airway is blocked for four or five minutes, will die. (The fact that the person cannot speak helps onlookers to distinguish such blockage from a heart attack. Although the symptoms are similar, heart attack victims can talk.)

The food can nearly always be dislodged by the Heimlich maneuver, a procedure so simple that it has been carried out successfully by eight-year-olds. There are three steps: (1) Stand behind the person and wrap your arms around his or her waist. (2) Make a fist with one hand, grasp it with your other hand, and then place the fist against the person's abdomen, slightly above the navel and below the rib cage. (3) Press your fist into the person's abdomen with a quick upward thrust. The sudden elevation of the diaphragm compresses the lungs and forces air up the trachea, pushing the food out. Repeat several times if necessary. If the person is sitting, the procedure can be carried out in the same way.

If the victim is lying on his or her back, face the person, and kneel astride the hips. Put the heel of one hand on the abdomen above the navel and below the rib cage, put the other hand on top of the first hand, and press with a quick upward thrust. You can even perform the Heimlich maneuver on yourself: place your hands at your waist, make a fist, and press quickly upward.

A person on whom the Heimlich maneuver has been performed should be seen by a physician immediately after the emergency treatment. This is because it is possible to break a rib or cause other internal injuries, especially if the movements are performed incorrectly. But, considering the alternative, it is well worth the risk.

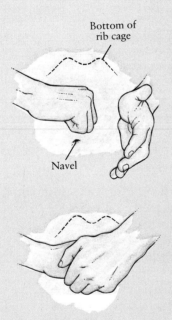

Bottom of rib cage

Navel

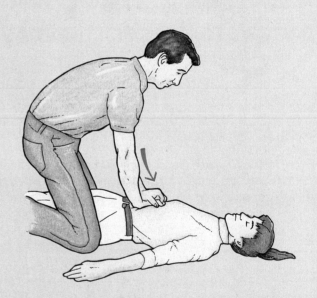

The Stomach: Storage and Liquefaction

The **stomach** is a collapsible, muscular bag, which, unless it is fully distended, lies in folds (Figure 29–6a). Stomachs vary widely in capacity. The stomach of a hyena, which may go as long as five days between kills, can hold an amount equivalent to one-third of the animal's own body weight. By contrast, a mammal that eats frequent, small meals consisting of highly nutritious food items—such as seeds or insects—typically has only a small stomach. The human stomach, distended, holds 2 to 4 liters of food.

The mucosal layer of the stomach is very thick and contains numerous pits (Figure 29–6b). The epithelial cells of the stomach mucosa secrete mucus, hydrochloric acid (HCl), and pepsinogen. These secretions, with the water in which they are dissolved, constitute **gastric juice.** As a consequence of the HCl secretion, the pH of gastric juice is normally between 1.5 and 2.5, far more acidic than any other body fluid. The burning sensation you feel if you vomit is caused by the acidity of gastric juice acting on the unprotected membranes of the esophagus. The HCl, which kills most bacteria and other living cells in the ingested food, loosens the tough, fibrous components of plant and animal tissues and erodes the cementing substances between cells. HCl also initiates the conversion of pepsinogen to its active form, **pepsin.** Pepsin, an enzyme that breaks proteins into peptides, is active only at the low pH of the normal stomach.

The stomach is influenced by both the nervous and endocrine systems. Anticipation of food and the presence of food in the mouth stimulate churning movements of the stomach and the production of gastric juice. When protein-containing food reaches the stomach, it triggers the release into the bloodstream of a hormone, **gastrin,** from specific cells of the stomach mucosa. This hormone acts on the epithelial cells of the mucosa to increase their secretion of gastric juice and on the muscle cells of the stomach wall to increase their contractions.

In the stomach, food is converted into a semiliquid mass. It is gradually moved by peristalsis through the **pyloric sphincter,** separating the stomach and small intestine. The stomach is usually empty four hours after a meal.

The Small Intestine: Digestion and Absorption

In the **small intestine,** the breakdown of food begun in the mouth and stomach is completed. The resulting nutrient molecules are then absorbed from the digestive tract into the circulatory system of the body, through which they are delivered to the individual cells.

Anatomically, the small intestine is characterized by adaptations that enormously increase its surface area. These include circular folds in the submucosa; numerous microscopic fingerlike projections of the mucosa, known as **villi;** and tiny cytoplasmic projections, known as **microvilli,** from the surface of the individual epithelial cells (Figure 29–7). If the small intestine of an adult human were to be fully extended, it would be some 6 meters long. The total surface area of the human small intestine is approximately 300 square meters, about the size of a doubles tennis court. The **duodenum,** the upper

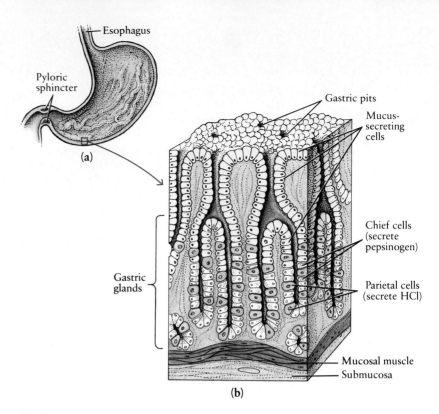

29–6   (a) *In the stomach, churning movements, combined with the action of gastric juice, convert food into a semiliquid mass that is ultimately moved by peristalsis through the pyloric sphincter into the small intestine.* (b) *A section of the stomach mucosa. The parietal cells secrete hydrochloric acid, and the chief cells produce pepsinogen, the precursor of the digestive enzyme pepsin. Mucus, secreted by other cells of the epithelium, coats the surface of the stomach and lines the gastric pits, protecting the stomach surface from digestion.*

25 centimeters of the small intestine, is the most active in the digestive process. The remaining length of the small intestine is primarily involved in the absorption of water and nutrients.

A number of substances are involved in the digestive processes of the small intestine. Mucus is secreted by cells of the intestinal mucosa. Digestive enzymes are produced by epithelial cells of both the intestinal mucosa and the pancreas. In addition to enzymes, the small intestine receives an alkaline fluid from the pancreas, which neutralizes the stomach acid, and **bile,** which is produced in the liver and stored in the gallbladder. Bile contains a mixture of salts that, like laundry detergents, emulsify fats, breaking them apart into droplets. In this form more surface area is exposed for attack by enzymes.

Fluids from the pancreas, the liver, and the gallbladder travel toward the small intestine through a series of ducts that merge to form one large duct. This duct empties into the duodenum about 10 centimeters below the pyloric sphincter (Figure 29–8).

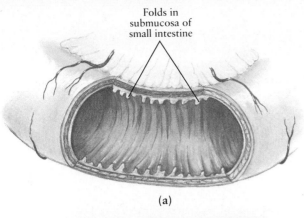

Folds in submucosa of small intestine

(a)

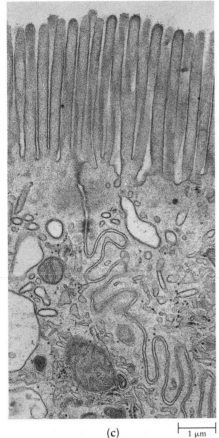

(c)

1 μm

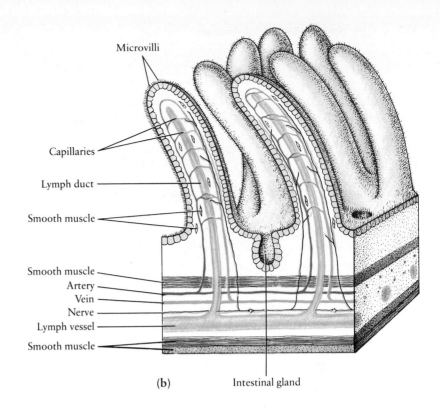

Microvilli

Capillaries

Lymph duct

Smooth muscle

Smooth muscle
Artery
Vein
Nerve
Lymph vessel
Smooth muscle

(b)

Intestinal gland

29–7 (a) *The interior of the small intestine is gathered into folds, increasing its surface area. Villi and their microvilli further increase the surface area available for absorption of water and nutrients.* (b) *A longitudinal section of villi of the small intestine. Nutrient molecules are absorbed through the walls of the villi and, with the exception of fat molecules, enter the bloodstream by way of the capillaries. After repackaging, fats are taken up by the lymph ducts. The villi can move independently of one another, and their motion increases after a meal.*

(c) *Microvilli on the surface of two adjacent epithelial cells of a villus. The plasma membrane of the microvilli contains digestive enzymes as well as transport proteins involved in the movement of nutrient molecules across the membrane. The meandering double line marks the boundary between the two cells. Note the tight junction just below the microvilli, sealing the boundary, and, a little lower, a desmosome.*

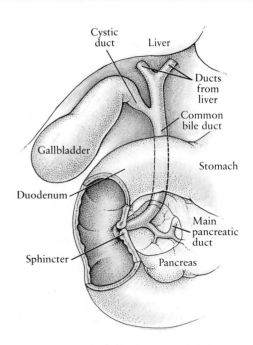

Cystic duct
Liver
Ducts from liver
Common bile duct
Gallbladder
Stomach
Duodenum
Main pancreatic duct
Sphincter
Pancreas

29–8 *Ducts from the liver, the gallbladder, and the pancreas merge just before emptying into the duodenum through a sphincter in its wall.*

531

Table 29–1	Major Gastrointestinal Hormones		
Hormone	**Source**	**Stimulus for Production**	**Action**
Gastrin	Stomach	Protein-containing food in stomach, also parasympathetic nerves	Stimulates secretion of gastric juices and muscular contractions of stomach and intestine
Secretin	Duodenum	HCl in duodenum	Stimulates secretion of alkaline pancreatic fluids and bile
Cholecystokinin	Duodenum	Fats and amino acids in duodenum	Stimulates release of enzymes of pancreas and of bile from gallbladder

In the small intestine, amylases continue the breakdown of starch begun in the mouth, producing disaccharides. **Lipases** hydrolyze fats into glycerol and fatty acids. Three types of enzymes break down proteins. The members of one group break apart the long polypeptide chains. Each enzyme in this group acts only on the bonds linking particular amino acids, so that several enzymes are required to break a large polypeptide into shorter fragments. A second type of enzyme acts only on the ends of a chain, splitting off dipeptides (pairs of amino acids). A third group of enzymes then comes into action, breaking the remaining dipeptides into single amino acids.

The digestive activities of the small intestine are coordinated and regulated by hormones (Table 29–1). In the presence of acidic gastric juice, the duodenum releases **secretin,** a hormone that stimulates the pancreas and liver to secrete alkaline fluids. Fats and amino acids in the food stimulate the production of another hormone, **cholecystokinin,** which triggers the release of enzymes from the pancreas and of bile from the gallbladder. At least eight other substances are suspected of being gastrointestinal hormones.

In addition to hormonal influences, the intestinal tract is also regulated by the autonomic nervous system. Stimulation by parasympathetic fibers increases intestinal contractions, whereas inhibition by sympathetic fibers decreases contractions. Thus, a complex interplay of stimuli and checks and balances serves to activate digestive enzymes, to adjust the chemical environment, and to regulate movements of the intestines.

Absorption of Nutrients

Water and the nutrient molecules released by the digestive processes are absorbed through the epithelial cells of the intestinal mucosa. Specialized enzymes embedded in the epithelial plasma membranes cleave disaccharides to monosaccharides, which are then rapidly absorbed

2.5 μm

29–9   *Following a high-fat meal, large droplets of fat, wrapped in a protein coating, enter the bloodstream. In this scanning electron micrograph, the fat droplets have been colored yellow. They are surrounded by numerous red blood cells.*

by active transport and facilitated diffusion (page 117). Amino acids and dipeptides are absorbed by active transport. These molecules all enter the bloodstream by way of the capillaries of the villi.

Small fatty acids also enter the blood vessels of the intestine directly, but large fatty acids, glycerol, and cholesterol travel an indirect route. These molecules enter the epithelial cells by passive diffusion. Within the cells, fatty acids and glycerol are resynthesized into fats (three molecules of fatty acid combined with one molecule of glycerol, as shown on page 56), which are then packaged into protein-coated droplets. Similarly, cholesterol is packaged into low-density lipoprotein (LDL) complexes (see page 58). Both the protein-coated droplets and the LDLs are secreted into the lymph vessels and ultimately enter the bloodstream (Figure 29–9).

The Large Intestine: Further Absorption and Elimination

The absorption of water, sodium, and other minerals, a process that occurs primarily in the small intestine, continues in the **large intestine.** In the course of digestion, large amounts of water—approximately 7 liters per day—enter the stomach and small intestine as secretions of the glands emptying into and lining the digestive tract, in the ingested food and drink, and by osmosis from body fluids. When the absorption of this water and the minerals it contains is disrupted, as in diarrhea, severe dehydration can result. Mortality from infant diarrhea, still the chief cause of infant death in many countries, is principally a consequence of water loss.

The large intestine harbors a considerable population of symbiotic bacteria, including the familiar *Escherichia coli.* The bacteria break down food substances that escaped digestion and absorption in the small intestine. Living on these food substances, largely materials we lack the enzymes to digest, the bacteria synthesize amino acids and vitamins, some of which are absorbed into the bloodstream. These bacteria are our chief source of vitamin K.

As shown in Figure 29–1, the appendix is a blind pouch off the large intestine. It is an evolutionary remnant from herbivorous ancestors and plays no known role in human digestion. However, as many individuals have learned from personal experience, the appendix may become irritated, inflamed, and then infected. If it ruptures, as a consequence of inflammation and swelling, it spills its bacterial contents into the abdominal cavity. Serious, and even fatal, infection can result.

The bulk of the fecal matter consists of water, bacteria (mostly dead cells), and cellulose fibers, along with other indigestible substances. It is lubricated by mucus, which is secreted by some of the epithelial cells lining the large intestine, stored briefly in the rectum, and then eliminated through the anus as feces. Bile pigments, released in the breakdown of hemoglobin, are responsible for the characteristic color of feces.

Regulation of Blood Glucose

The major function of digestion is, as we have seen, to provide organic molecules that can serve as energy sources and raw materials for each cell of the body. Although vertebrates rarely eat 24 hours a day, their blood glucose—the major cellular energy supply and the fundamental building-block molecule—remains extraordinarily constant.

The liver plays a central role in this critical homeostatic process. Glucose and other monosaccharides are absorbed into the blood from the intestinal tract and are passed directly to the liver by way of the hepatic portal vein. The liver converts some of these monosaccharides to glycogen and fat, storing enough glycogen to satisfy the body's needs for about four hours. The fat is stored in fat cells, which can also form fat from glucose. Similarly, the liver breaks down excess amino acids (which are not stored) and converts them to glucose. The nitrogen from the amino acids is excreted in the form of urea, and the glucose is stored as glycogen.

Whether the liver takes up or releases glucose and the amount it takes up or releases are determined primarily by the concentration of glucose in the blood. As we shall see in Chapter 34, the concentration of glucose is, in turn, regulated by a number of hormones and is influenced by the autonomic nervous system.

Some Nutritional Requirements

Because of the liver's activities in converting various types of food molecules into glucose and because most tissues can use fatty acids as an alternative fuel, the energy requirements of the body can be met by carbohydrates, proteins, or fats—the three principal types of food molecules. Energy requirements are ordinarily met by a combination of the three. Carbohydrates and proteins supply about the same number of calories per gram, and fats supply about twice as many as either of them.

In addition to calories, the cells of the body need the 20 different kinds of amino acids required for assembling proteins. When any one of the amino acids necessary for the synthesis of a particular protein is unavailable, the protein cannot be made. Vertebrates are not able to synthesize all 20 amino acids. Humans can synthesize 12, either from a simple carbon skeleton or from another amino acid. The other eight, which must be obtained in the diet, are known as essential amino acids (see essay, page 61). Plants are the original source of the essential amino acids, but it is difficult (although by no means impossible) to obtain sufficient quantities of them by eating a completely vegetarian diet. This is because most plant proteins are relatively deficient in lysine and tryptophan.

Mammals also require but cannot synthesize certain polyunsaturated fatty acids needed for the synthesis of fats and also for the synthesis of a group of hormonelike compounds known as prostaglandins (to be discussed in Chapter 34). These fatty acids can be obtained by eating plants or insects (or eating other animals that have eaten plants or insects).

Slim Chance: The Difficulty with Dieting

From a biological perspective, the slogan of the National Association to Aid Fat Americans is correct: "Fat is not a four-letter word." Fat is, in fact, an extremely efficient way to store energy.

When animals consume more calories in their food than can be stored in the form of glycogen, the excess is stored as fat. In the course of animal evolution, this capacity has become an important survival mechanism for species that eat large but infrequent meals, for species that hibernate, and for species that experience periods of famine alternating with periods of plenty—as did *Homo sapiens* throughout much of its evolutionary history. Such animals can store internal energy reserves when food is abundant and then live off their fat "pantries" when food is in short supply.

This marvelous adaptation has, however, become maladaptive to many members of our own species in the developed countries. The major nutritional problem among North Americans and Europeans is obesity. In the United States, 30 percent of middle-aged women and 15 percent of middle-aged men are obese—that is, they weigh more than 120 percent of their appropriate weight. Such excess energy storage is detrimental to more than one's image. Obesity is correlated with a significant increase in coronary artery disease, diabetes, and other serious disorders.

The approach to weight reduction taken by most obese people is to diet, that is, to reduce their caloric intake. The energy we obtain in our food is used for three main purposes—internal heat production, external work, and energy storage. Thus, it would appear that eating somewhat less than we need for internal heat production and external work, with nothing left over for storage, would enable us to live off, and lose, our fat. Unfortunately, it is not so simple.

Two important factors in the amount and location of the fat stored in our bodies are the

The physiological mechanisms by which a relatively constant body weight is maintained evolved during the long period of human history when food was often scarce and the only mode of transportation involved putting one limb in front of another. This process—whether in the form of walking, biking, or swimming— remains a key ingredient in the maintenance of an optimal weight.

number and size of our fat cells. These saclike cells can be plump and full of lipids (and may even divide if they become large enough), or they may be relatively empty and shrunken. Although dieting can reduce the size of the fat cells, it has no effect on their number. Moreover, when the content of the fat cells falls

below a certain level, signals transmitted to the brain result in increased sensations of hunger.

A further complication is that when food intake is decreased below the immediate energy requirements of the body, the metabolic rate also decreases. Energy usage becomes more efficient, and the essential physiological processes of the body are carried out with the expenditure of fewer calories. This lowering of the metabolic rate in response to a decreased caloric intake appears to be a homeostatic mechanism for maintaining a constant body weight, even when food is in short supply. When food intake is cut drastically, the body interprets the message as "famine" and takes counter measures. Moreover, upon a return to the pre-diet eating habits, the lower metabolic rate results in a weight gain—on a caloric intake that previously had simply maintained a constant weight. With the next round of dieting, the metabolic rate may decline even further. Thus, the end result of cycles of repeated dieting is generally a weight gain that exceeds the losses initially achieved by dieting.

Although weight loss may be difficult, it is not impossible. The key to success lies in the expenditure of energy in external work. During exercise, approximately 75 percent of the energy expended is dissipated in the form of heat. Moreover, as fat is replaced by lean, well-toned muscle mass, the metabolic rate either remains constant or increases. Light to moderate physical activity is also associated with reduced food intake. Studies of factory workers have revealed that sedentary jobs are associated with increased eating, while jobs involving light to moderate physical activity are associated with lower rates of food intake. These patterns are consistent with reports of reduced appetite by individuals who engage in regular moderate exercise.

Thus, a sensible dialogue with your body on the subject of weight is possible—if you understand the body's evolutionary frame of reference and its natural homeostatic responses.

Table 29–2 Vitamins			
Designation: Letter and Name	**Major Sources**	**Function**	**Deficiency Symptoms**
A, carotene	Egg yolk, green or yellow vegetables, fruits, liver, butter	Formation of visual pigments, maintenance of normal epithelial structure	Night blindness; dry, flaky skin
B-complex Vitamins:			
B_1, thiamine	Brain, liver, kidney, heart, pork, whole grains	Formation of coenzyme involved in Krebs cycle	Beri-beri, neuritis, heart failure
B_2, riboflavin	Milk, eggs, liver, whole grains	Part of electron carrier FAD	Photophobia, fissuring of skin
B_3, niacin (nicotinic acid)	Whole grains, liver and other meats, yeast	Part of electron carriers NAD, NADP, and of CoA	Pellagra, skin lesions, digestive disturbances
B_5, pantothenic acid	Present in most foods	Forms part of CoA	Neuromotor and cardiovascular disorders, gastrointestinal distress
B_6, pyridoxine	Whole grains, liver, kidney, fish, yeast	Coenzyme for amino acid metabolism and fatty acid metabolism	Dermatitis, nervous disorder
B_{12}, cyanocobalamin	Liver, kidney, brain, eggs, dairy products	Maturation of red blood cells, coenzyme in amino acid metabolism	Anemia, malformed red blood cells
Biotin	Egg yolk, synthesis by intestinal bacteria	Concerned with fatty acid synthesis, CO_2 fixation, and amino acid metabolism	Scaly dermatitis, muscle pains, weakness
Folic acid	Liver, leafy vegetables	Nucleic acid synthesis, formation of red blood cells	Failure of red blood cells to mature, anemia; spinal cord malformations in fetus
C, ascorbic acid	Citrus fruits, tomatoes, green leafy vegetables, potatoes	Vital to collagen and ground (extracellular) substance	Scurvy, failure to form connective tissue fibers
D_3, calciferol	Fish oils, liver, fortified milk and other dairy products, action of sunlight on lipids in the skin	Increases Ca^{2+} absorption from gut, important in bone and tooth formation	Rickets (defective bone formation) in children; softened bones in adults
E, tocopherol	Green leafy vegetables, wheat germ, vegetable oils	Maintains resistance of red cells to hemolysis, cofactor in electron transport chain	Increased red blood cell fragility
K, naphthoquinone	Synthesis by intestinal bacteria, leafy vegetables	Enables synthesis of clotting factors by liver	Failure of blood coagulation

Vitamins are an additional group of essential molecules that cannot be synthesized by animal cells. Many of them function as coenzymes (page 136), and they are typically required only in small amounts. Table 29–2 indicates some of the vitamins required in the human diet and their functions.

There is no clear evidence that ingesting any particular vitamin in excess of the amounts available in a well-balanced diet has any beneficial effect on a normally healthy individual. Some, including the fat-soluble vitamins A, D, E, and K, which can accumulate in body tissues, are toxic in large doses. One of the most concentrated sources of vitamin A known is polar bear liver, a half pound of which contains about 2,600 times the recommended daily allowance. For centuries, the Inuit (Eskimos) and Arctic explorers have known that consumption of polar bear liver causes illness and can be fatal—a knowledge shared by sled dogs and Arctic birds, which also refuse to eat it.

The body also has a dietary requirement for a num-

29–10   *Vitamin D is a steroid-like substance (page 59) that is produced in the skin by the action of ultraviolet rays from the sun on cholesterol. (The absorbed energy opens the second ring of the molecule.) Further chemical processing in the liver and kidneys produces the active form of the molecule shown here.*

As we discussed on page 519, high levels of melanin, the pigment responsible for human skin color, inhibit the production of vitamin D. As a consequence, selection for lighter skin color occurred in most early human populations that moved northward from the tropics. Such selection, however, did not occur among the Inuit, who eat a diet rich in fish oils, a major source of vitamin D.

Active vitamin D

ber of inorganic substances, or minerals. These include calcium and phosphorus for bone formation, iodine for thyroid hormone, iron for hemoglobin and cytochromes, and sodium, chloride, and other ions essential for ionic balance. Most of these are present in the or-

dinary diet or in drinking water. Like the vitamins, however, they must be given in supplementary form when the dietary intake is inadequate or when the individual is not able to assimilate them normally.

Summary

Digestion is the process by which food is broken down into molecules that can be taken up by the cells lining the intestine, transferred to the bloodstream, and so distributed to the individual cells of the body. It occurs in successive stages, regulated by an interplay of hormones and nervous stimuli.

In mammals, food is processed initially in the mouth, where the breakdown of starch begins in humans. It moves through the esophagus to the stomach, where gastric juices destroy bacteria and begin to break down proteins.

Most of the digestion occurs in the upper portion of the small intestine, the duodenum. Here, digestive activity, which is performed by enzymes, is almost completely under hormonal regulation. The breakdown of starch by amylases continues, fats are hydrolyzed by lipases, and proteins are reduced to dipeptides or single amino acids. Monosaccharides, amino acids, and dipeptides are absorbed into the blood vessels of the intestinal villi. Fats are absorbed into the lymph vessels and ultimately enter the bloodstream. Hormones secreted by duodenal cells stimulate the functions of the pancreas and the liver. The pancreas releases an alkaline fluid containing digestive enzymes, and the liver produces bile, which emulsifies fats.

Much of the water that enters the stomach and small intestine in the course of digestion is reabsorbed in the small intestine itself. Most of the remaining water is reabsorbed from the residue of the food mass as it passes through the large intestine. The large intestine contains symbiotic bacteria, which are the source of certain vitamins. Undigested residues are eliminated from the large intestine.

The chief energy source for cells in the mammalian body is glucose circulating in the blood. The organ principally responsible for maintaining a steady supply of glucose is the liver, which stores glucose (in the form of glycogen) when glucose levels in the blood are high and breaks down glycogen, releasing glucose, when the levels drop. These activities of the liver are regulated by a number of different hormones.

Requirements for good nutrition include molecules for fuel (which can be obtained from carbohydrates, fats, or proteins), essential amino acids, essential fatty acids, vitamins, and certain minerals.

Questions

1. Distinguish, in terms of function, between: gastrointestinal hormones/gastrointestinal enzymes; gastric juice/bile; amylases/lipases; digestion/absorption; villi/microvilli; vitamins/essential amino acids.

2. Diagram the human digestive system, including all tubes, chambers, valves, and accessory organs. Describe the function of each structure.

3. Trace the chemical processing of a hamburger on a bun, with lettuce and tomato, as it passes through your digestive tract.

4. Most of the protein-digesting enzymes are secreted in an inactive form and are themselves activated by special enzymes secreted into the digestive tract. Explain the adaptive value of this two-step process.

5. On the basis of your knowledge of the processing of the three major groups of food molecules by the digestive system, explain why carbohydrates are a better source of quick energy than either proteins or fats.

6. If you oxidize a pound of fat, whether butter or the fat in your own body cells, about 4,200 kilocalories are released. Suppose that you go on a weight-reducing diet, limiting yourself to 1,000 kilocalories a day. You spend most of your time sitting in a well-heated library (studying, of course) and so you expend only about 3,000 kilocalories a day. What is the maximum amount of weight you can expect to lose each week? Why is the actual weight loss likely to be less than you have calculated?

7. Humans are omnivores rather than strict herbivores or carnivores or specialists that use a single organism as a food source. Humans have an unusually large number of substances specifically required in their diets. How might these two phenomena be related?

Suggestions for Further Reading

Blomhoff, Rune, Michael H. Green, Trond Berg, and Kaare R. Norum: "Transport and Storage of Vitamin A," *Science*, vol. 350, pages 399–404, 1990.

Cerami, Anthony, Helen Vlassara, and Michael Brownlee: "Glucose and Aging," *Scientific American*, May 1987, pages 90–96.

Hirschhorn, Norbert, and William Greenough III: "Progress in Oral Rehydration Therapy," *Scientific American*, May 1991, pages 50–57.

Livingston, Edward H., and Paul H. Guth: "Peptic Ulcer Disease," *American Scientist*, vol. 80, pages 592–598, 1992.

Martin, Roy J., B. Douglas White, and Martin G. Hulsey: "The Regulation of Body Weight," *American Scientist*, vol. 79, pages 528–541, 1991.

McKenzie, Aline: "A Tangle of Fibers: Scientists Examine How Different Dietary Fibers Produce Their Health Benefits," *Science News*, November 25, 1989, pages 344–345.

Moog, Florence: "The Lining of the Small Intestine," *Scientific American*, November 1981, pages 154–176.

Scrimshaw, Nevin S.: "Iron Deficiency," *Scientific American*, October 1991, pages 46–53.

Uvnäs-Moberg, Kerstin: "The Gastrointestinal Tract in Growth and Reproduction," *Scientific American*, July 1989, pages 78–83.

Respiration

When you are sitting still, you take a breath 10 to 15 times a minute. Breathing transfers oxygen from the surrounding air across the delicate, moist tissues of your lungs into the underlying blood vessels. Here it is picked up by hemoglobin molecules, which carry oxygen to all the cells of your body.

Cells need oxygen, as you know, to generate ATP in their mitochondria. ATP is not stored but is produced as needed, so the uptake of oxygen is commonly used to calculate energy requirements. Of all your organs, your liver has the highest requirement: 81 liters of oxygen a day. The brain is a close second, requiring 76 liters a day. At rest, skeletal muscle requires about 74 liters a day.

Physical activity greatly increases the demand for oxygen. A top marathoner utilizes about 4 liters of oxygen per minute and about 500 liters in the course of a race. An ordinary runner uses much less oxygen per minute—about 2.5 liters—but he or she will use about the same total amount before the race is over, provided the ordinary runner reaches the finish line. A trained runner breathes more deeply, taking in more oxygen, and also uses the oxygen more efficiently.

Sprinters, such as Gail Devers, the winner of the women's 100-meter dash in the 1992 Olympics, have very high ATP requirements. For each stride in a sprint, some 10^{20} ATP molecules are required. Most of this ATP is generated anaerobically, since a sprint begins and ends before the muscle cells can be adequately supplied with oxygen-bearing blood. Lactic acid (page 154) accumulates in the muscles and is metabolized only after the race, as the sprinter continues to take in extra oxygen.

A principal physiological difference between a sedentary person and an ordinary runner, and between an ordinary runner and a trained runner, is the number of mitochondria per cell, which increases with training. Today, different individuals put different values on running. It seems safe to conjecture, however, that back when hominids first ventured out among the other animals of the grasslands, all of them began fitness training at a very early age and the mitochondria count was universally high.

Digestion, as we saw in the last chapter, is the process by which the foods consumed by an animal are broken down into smaller molecules and absorbed into the bloodstream, through which they are transported to the diverse cells of which the organism is composed. The release of energy from these molecules—the sole source of energy for heterotrophic cells—depends upon their oxidation. This process usually (although not always) requires oxygen, and when it does, it is called respiration.

Actually, respiration has two meanings in biology. At the cellular level, it refers to the oxygen-requiring chemical reactions, discussed in Chapter 8, that take place in the mitochondria and are the chief source of energy for eukaryotic cells. At the level of the whole organism, it designates the process of taking in oxygen from the environment and returning carbon dioxide to it. This exchange of gases with the environment, which is accomplished by the respiratory system, is the subject of this chapter.

Oxygen consumption is directly related to energy expenditure. The energy expenditure at rest is known as **basal metabolism.** Metabolic rates increase sharply with exercise. During exercise, a person consumes 15 to 20 times the amount of oxygen that he or she consumes at rest, with the oxygen consumption increasing in proportion to the energy expenditure.

Diffusion and Air Pressure

In every organism, from an amoeba to an elephant, the exchange of oxygen and carbon dioxide between cells and the surrounding environment takes place by diffusion. Diffusion, you will recall (page 111), is the net movement of particles from a region of higher concentration to a region of lower concentration as a result of their random motion. In describing gases, however, scientists speak of the pressure of a gas rather than its concentration.

Table 30–1	Composition of Dry Air
Gas	**% of Volume**
Oxygen	21
Nitrogen	77
Argon	1
Carbon dioxide	0.03
Other gases*	0.97

* Includes hydrogen, neon, krypton, helium, ozone, xenon, and, in some environments, radon.

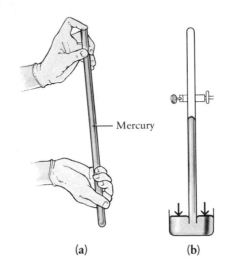

Mercury

(a)   (b)

30–1 *Atmospheric pressure is measured by means of a barometer. To avoid an inconveniently tall column, the liquid usually used is mercury, which is relatively heavy. (a) To make a simple mercury barometer, put on a pair of protective gloves and then fill a long glass tube, open at one end, with mercury. Closing the tube with your finger, (b) invert it into a dish of mercury. Remove your finger and clamp the tube to a stand. The mercury level will drop until the pressure of its weight inside the tube is equal to the atmospheric pressure (arrows) outside. At sea level, the height of the column will be about 760 millimeters (29.9 inches).*

At sea level, the air around us exerts a pressure on our skin of 1 atmosphere (about 15 pounds per square inch). This pressure is enough to support a column of water about 10 meters high or a column of mercury 760 millimeters high (Figure 30–1). The total pressure of a mixture of gases, such as air, is the sum of the pressures of the separate gases in the mixture. The pressure of each gas is proportional to its concentration. Oxygen, for instance, makes up about 21 percent by volume of dry air (Table 30–1). Thus 21 percent of the total air pressure, or 160 millimeters of mercury (mm Hg), results from the pressure of oxygen in the air. This is known as the **partial pressure** of oxygen. In diffusion, a gas moves from a region of higher partial pressure to a region of lower partial pressure.

We are so accustomed to the pressure of the air around us that we are unaware of its presence or of its effects on us. However, if you visit a place, such as Mexico City, that is at a comparatively high altitude and therefore has a lower atmospheric pressure (and thus a lower partial pressure of oxygen), you will feel lightheaded at first and will tire easily. As we shall see in an essay later in this chapter, a series of physiological adaptations are required for life at high altitude—and for successful mountain climbing.

The consequences of the opposite situation—higher gas pressures—can be seen in deep-sea divers. Early in the history of deep-sea diving, it was found that when divers come up from the bottom too quickly, they get the "bends," which are always painful and sometimes fatal. The bends develop as a result of breathing compressed air during the dive. Under high pressure, nitrogen, the principal component of air, diffuses from the air in the lungs into solution in the blood and tissues.

If the body is rapidly decompressed upon surfacing, the gas comes out of solution so quickly that nitrogen bubbles form in the blood, like the carbon dioxide bubbles that appear in a bottle of soda when you decompress it by removing the top. The nitrogen bubbles lodge in the capillaries, stopping blood flow, or they invade nerves or other tissues.

The Evolution of Respiratory Systems

Oxygen enters and moves within cells by diffusion. Within the cell, as we have noted, it takes part in the oxidation of organic compounds that serve as cellular energy sources. In this process, carbon dioxide is produced, which then diffuses out of the cell down its concentration (partial pressure) gradient. This is true of all cells, whether an amoeba, a *Paramecium,* a liver cell, or a brain cell.

As you know, substances can move efficiently by diffusion only for very short distances (less than 1 millimeter). These limits pose no problem for very small animals, in which each cell is quite close to the surface, or for animals in which much of the body mass is not metabolically active—like the "jelly" of jellyfishes. Many eggs and embryos also obtain oxygen in this simple way, particularly in the early stages of development.

Diffusion, however, cannot possibly meet the needs of large organisms in which cells in the interior may be many centimeters from the air or water serving as the oxygen source. As animals increased in size in the course of evolution, respiratory and circulatory systems evolved that transport large numbers of gas molecules by bulk flow. (Remember that whereas diffusion is the result of the random movement of individual particles, bulk flow is the overall movement of a gas or liquid in response to pressure or gravity.)

An early stage in the evolution of gas-transport systems is exemplified by the earthworm (page 470). As is the case with most other kinds of worms, an earthworm has a network of capillaries just one cell layer beneath the surface of its body. Oxygen and carbon dioxide diffuse directly through the moist body surface into and out of the blood as it travels through these capillaries. The blood picks up oxygen by diffusion as it travels near the surface of the animal and releases oxygen by diffusion as it travels past the oxygen-poor cells in the interior of the earthworm's body. Conversely, blood picks up carbon dioxide from the cells and releases it by diffusion as it travels near the surface of the animal. Thus the gases move into and out of the earthworm by diffusion but are transported within the animal by bulk flow.

This system is particularly suitable for worms because their tubular shape exposes a surface area that is relatively large compared to their body volume. Some worms can adjust their surface area in response to the oxygen supply. If you have an aquarium at home, you may be familiar with tubifex worms, which are frequently sold as fish food. When these worms are placed in water low in oxygen, such as a poorly aerated aquarium, they stretch out to as much as 10 times their normal length, increasing the surface area through which oxygen diffusion occurs and decreasing the diffusion distance.

Insects, as we have seen (page 472), have evolved a different strategy. Air is piped directly into the tissues by a network of chitin-lined tubules (Figure 30–2). In large insects, diffusion is assisted by body movements, which propel the air into and out of the spiracles by bulk flow. This system is fine for small organisms but is a major limitation on the size that can be attained by insects and other tracheal-system breathers. The planet may eventually be taken over by cockroaches or ants, but it is safe to bet they will not be the giant forms of science fiction.

The Evolution of Gills

Gills and lungs are other ways of increasing the respiratory surface. **Gills** are usually outgrowths (Figure 30–3), whereas **lungs** are ingrowths, or cavities. The respiratory surface of the gill, like that of the earthworm, is a single layer of cells, exposed to the environment on one side and to circulatory vessels on the other.

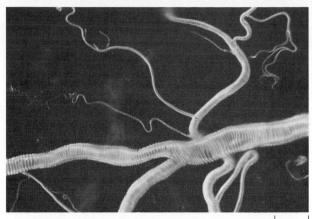

20 μm

30–2   *Insects and some other terrestrial arthropods breathe by means of tracheae. These are inward, tubular extensions of the exoskeleton, which are thickened in places, forming spiral supports that hold the tubes open. The tracheae are lined with chitin, which is molted when a new exoskeleton is formed. Gas exchange takes place at the thin, moist terminal ends of the tubules. This micrograph shows part of the tracheal system of a cockroach.*

The insect respiratory tract, like the human tract, is a fertile breeding ground for infectious organisms. For example, the tracheae of honey bees are often parasitized by mites, leaving the bees somewhat lethargic and greatly reducing their production of honey. Scientists at the U.S. Department of Agriculture have found that treatment with menthol, the soothing ingredient in most cough drops and vapor rubs, kills the mites, enabling the bees to breathe easier and get back to work.

30–3   *Gills are essentially outpocketings of the epithelium that increase the surface area exposed to water. Often gills are covered by an exoskeleton, as in crustaceans, or by a flaplike gill cover, as in fishes. In the axolotl, the amphibian shown here, the external nature of the gills is clearly evident. The source of their pink color is blood flowing through dense capillary networks just one cell layer beneath the gill surface.*

30–4   (a) *In fish, oxygen enters the blood by diffusion from water flowing through the gills.* (b) *The anatomical structure of the gills maximizes the rate of diffusion, which is proportional not only to the surface areas exposed but also to differences in concentration of the diffusing molecules. Water, carrying dissolved oxygen, flows between the gill filaments in one direction; blood flows through them in the opposite direction. Thus the blood carrying the most oxygen (that is, the oxygenated blood leaving the gill filament) meets the water carrying the most oxygen, and the blood carrying the least oxygen (the deoxygenated blood entering the gill filament) meets the water carrying the least oxygen. The result is that the oxygen concentration of the blood in any region of the filament is less than the oxygen concentration of the water flowing over that region. In fact, the concentration gradient of oxygen between the blood in the gill filament and the water flowing over it is constant along the entire length of the filament. Thus, the transfer of oxygen to the blood by diffusion (red arrows) takes place across the entire surface of the filament.*

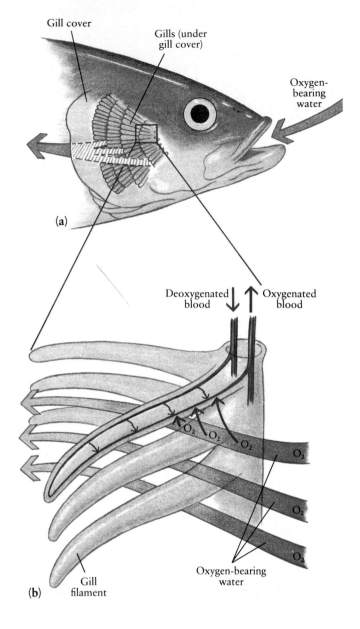

The layers of gill tissue may be spread out flat, stacked, or convoluted in various ways. The gill of a clam, for instance, is shaped like a steam-heat radiator (which is also designed to provide a high surface-to-volume ratio).

The vertebrate gill is believed to have originated primarily as a feeding device. Primitive vertebrates drew water into their mouths and expelled it through what we now call their gill slits, filtering out bits of organic matter from the water as it passed over the gills. *Branchiostoma* (page 482), which is believed to resemble closely the ancestral vertebrate, feeds in this way. Primitive vertebrates respired mostly through their skin, as does *Branchiostoma*.

In the course of time, numerous selection pressures, chiefly involved with predation, came into operation. As one consequence, there was a trend toward an increasingly thick skin, even one armored or covered with scales. As you would expect, such a skin is not useful for respiratory purposes. At the same time, related forces were operating to produce animals that were larger and swifter and so more efficient at capturing prey and escaping predators. Such animals had higher energy requirements—and consequently higher oxygen requirements. These problems were solved with the "capture" of the gill for a new purpose: respiratory exchange. The surface area and blood supply of the gill epithelium have slowly increased over the millennia. The modern gill is the result of this evolutionary process.

In most fishes, the water (in which oxygen is dissolved) is pumped in at the mouth by oscillations of the bony gill cover and flows out across the gills. In the gills of fish, the circulatory vessels are arranged so that the blood is pumped through them in a direction opposite to the movement of the oxygen-bearing water (Figure 30–4). This countercurrent arrangement results in a far more efficient transfer of oxygen to the blood than if the blood flowed in the same direction as the water.

A fish can regulate the rate of water flow, and sometimes assist it, by opening and closing its mouth. Fast swimmers, such as mackerel, obtain enough oxygen to meet their energy needs by keeping their mouths open as they swim. As a result, water moves rapidly over the gills. Such fish have become so dependent on this method of respiration that if they are kept in an aquarium or any other space where their motion is limited, they will suffocate.

The Evolution of Lungs

Lungs are internal cavities into which oxygen-containing air is taken. They have a disadvantage as compared with gills; diffusion is less efficient when there is not a continuous flow across the respiratory surface. Air, however, is a far better source of oxygen than is water. In the modern atmosphere, oxygen is 21 percent of the air, as compared to 0.5 percent by volume in water at 15°C. Not only must more water be processed to obtain a given amount of oxygen, but water is also much heavier than air. A fish spends up to 20 percent of its energy in the muscular work associated with respiration, whereas an air breather expends only 1 or 2 percent of its energy in respiration. Also, oxygen diffuses about 300,000 times more rapidly through air than through water, and so can be replenished much more quickly in air as it is used up by respiring organisms.

Lungs are not essential for air breathing. As we have seen, earthworms are air breathers. The overwhelming advantage of lungs, however, is that these interior respiratory surfaces can be kept moist without a large loss of water by evaporation. Although lungs are largely a vertebrate "invention," they are found in some invertebrates. Land-dwelling snails, for example, have independently evolved lungs that are remarkably similar to the lungs of some amphibians.

Some primitive fishes had lungs as well as gills, although their lungs were not efficient enough to serve as more than accessory respiratory structures. They were probably a special adaptation to life in fresh water, which may stagnate (become depleted of oxygen). A few species of lungfish still exist (Figure 27–4, page 484). By coming to the surface and gulping air into their lungs, they can live in water that does not have sufficient oxygen to support other fish life.

Amphibians and reptiles have relatively simple lungs, with small internal surface areas, although their lungs are far larger and more complex than those of the lungfish. The lungs of lungfish developed as simple out-pocketings of the pharynx, the posterior portion of the mouth cavity, which leads to the digestive tract. In amphibians, reptiles, and other air-breathing vertebrates, we see the evolution of the windpipe, or **trachea**, guarded by a valve mechanism, the epiglottis, and of nostrils, which make it possible for the animal to breathe with its mouth closed. Amphibians still rely largely on their skin for gas exchange, but reptiles breathe almost entirely through their lungs.

An important feature of all vertebrate lungs is that the exchange of air with the atmosphere takes place by bulk flow as a result of changes in lung volume. Such lungs are known as ventilation lungs. Frogs gulp air and force it into their lungs in a swallowing motion; then the epiglottis opens and lets the air out again. In reptiles, birds, and mammals, air is moved into and out of the lungs as a consequence of changes in the size of the thoracic cavity, brought about by muscular contractions and relaxations.

30–5 (a) *Although the lungs of birds are small, they are extraordinarily efficient. Each lung has several air sacs attached to it. With each breath, the air sacs empty and fill like balloons as they are compressed and expanded by movements of the body wall. No gas exchange takes place in the sacs. They appear rather to act as bellows, flushing fresh air through the lungs at every breath, always in the same direction. As a result, there is little residual "dead" air left in the lungs, as there is in mammals.*

(b) Scanning electron micrograph of lung tissue from a 14-day-old chicken. The tubes visible here are ventilated by air drawn in by the air sacs. Gas exchange takes place in the broad meshwork of air capillaries and blood capillaries, which make up the spongelike respiratory tissue seen here surrounding the tubes.

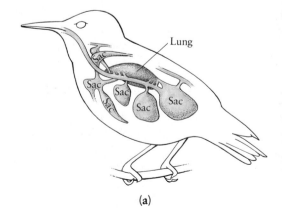

(a)

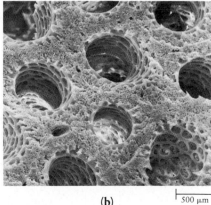

500 μm

(b)

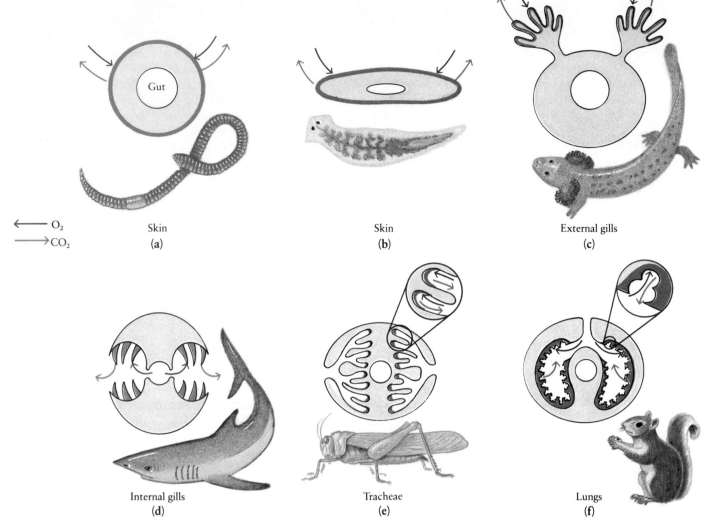

←—— O₂

——→ CO₂

Skin
(a)

Skin
(b)

External gills
(c)

Internal gills
(d)

Tracheae
(e)

Lungs
(f)

30–6   *Respiratory systems. (a) Gas exchange across the entire surface of the body is found in a wide range of small organisms from protists to earthworms. (b) Gas exchange across the surface of a flattened body is seen, for example, in flatworms. Flattening increases the surface-to-volume ratio and also decreases the distance over which diffusion has to occur within the body.*

(c) External gills increase the surface area but are unprotected and therefore easily damaged. External gills are found in polychaete worms and some amphibians. Gas exchange usually takes place across the rest of the body surface as well. (d) With internal gills, a ventilation mechanism draws water over the highly vascularized gill surfaces, as in fishes.

(e) Gas exchange at the terminal ends of fine tracheal tubes that branch through the body and penetrate all the tissues is characteristic of insects and some other arthropods. (f) Lungs are highly vascularized sacs into which air is drawn by a ventilation mechanism. Lungs are found in all air-breathing vertebrates and some invertebrates, such as terrestrial snails.

Respiration in Large Animals: Some Principles

Figure 30–6 summarizes the types of respiratory systems found among the animals. In large animals, both diffusion and bulk flow move oxygen molecules between the external environment and actively metabolizing tissues. This movement occurs in four stages:

1. Movement by bulk flow of the oxygen-containing external medium (air or water) to a thin, moist epithelium close to small blood vessels in lungs or gills.

2. Diffusion of the oxygen across this epithelium into the blood.

3. Movement by bulk flow of the circulating blood, containing dissolved oxygen, to the tissues where the oxygen will be used.

4. Diffusion of the oxygen from the blood into the extracellular fluids, from whence it diffuses into the individual cells.

Carbon dioxide, which is produced in the tissue cells, follows the reverse path as it is eliminated from the body.

The Human Respiratory System

The human respiratory system is shown in Figure 30–7a. Air enters the nasal passages, where it is warmed and cleaned, and then it passes through the pharynx and into the trachea. The trachea is strengthened by rings of cartilage that prevent it from collapsing during inspiration (breathing in).

The trachea leads into the **bronchi** (singular, bronchus), which subdivide into smaller and smaller passageways, the **bronchioles.** The bronchi and bronchioles are surrounded by thin layers of smooth muscle. Contraction and relaxation of this muscle alters resistance to air flow.

The actual exchange of gases takes place in small air sacs, the **alveoli,** which are clustered in bunches like grapes around the ends of the smallest bronchioles (Figure 30–7b). Each alveolus is about 0.1 or 0.2 millimeter in diameter, and each is surrounded by capillaries. The walls of the capillaries and of the alveoli each consist of a single layer of flattened epithelial cells, separated from one another by a thin basement membrane. Thus the barrier between the air in an alveolus and the blood in its capillaries is only about 0.5 micrometer. Gases are exchanged between the air and the blood by diffusion (Figure 30–7c). A pair of human lungs has about 300 million alveoli, providing a respiratory surface of some 70 square meters—approximately 40 times the surface area of the entire human body.

The trachea, bronchi, and bronchioles, which serve mainly to transport air by bulk flow to and from the alveoli, are lined with epithelial cells. These cells include both mucus-secreting cells and ciliated cells. The mucus coats the epithelium of the respiratory system and traps foreign particles that enter with the air. The cilia beat continuously, pushing mucus and the foreign particles embedded in it up toward the pharynx, from which it is generally swallowed. We are usually aware of the mucus production only when it increases as a result of irritation or infection.

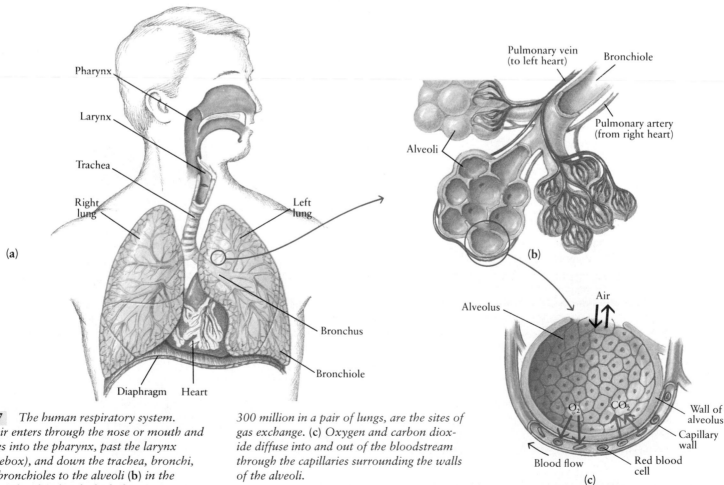

30–7   *The human respiratory system.* **(a)** *Air enters through the nose or mouth and passes into the pharynx, past the larynx (voicebox), and down the trachea, bronchi, and bronchioles to the alveoli* **(b)** *in the lungs. The alveoli, of which there are some 300 million in a pair of lungs, are the sites of gas exchange.* **(c)** *Oxygen and carbon dioxide diffuse into and out of the bloodstream through the capillaries surrounding the walls of the alveoli.*

When Smoke Gets in Your Lungs

During the period from 1950 to 1988, the death rate among men and women in the United States from cancers of the lungs and bronchi more than tripled. During that same period, a decrease occurred in the U.S. death rate from all other forms of cancer combined. However, the overall cancer death rate increased—due entirely to the increase in the number of deaths from cancers of the lungs and bronchi. These cancers are now the most common cause of death from cancer in adults. It is expected that well over 150,000 men and women will develop cancers of the lungs and bronchi in the United States this year, and more than 80 percent of them will die in less than three years. Most of these patients will be cigarette smokers.

The delicate tissues of the alveoli of the lungs are normally protected from infectious microorganisms and harmful substances in the respired air by the action of ciliated epithelial cells lining the trachea and bronchi. The cilia on the surface of these cells sweep out particles that are caught in the mucus secreted by other cells of the epithelium. Cigarette smoke paralyzes the cilia, breaching the respiratory system's first line of defense against disease and injury. Once this has occurred, both infectious microorganisms and foreign substances, including carcinogens (cancer-causing agents), can come into intimate contact with the living cells of the alveolar walls. The probability increases that the foreign invaders will actually enter the cells, initiating sequences of events that lead to serious damage and disease.

In addition to impairing the respiratory system's natural defenses, cigarette smoke exposes the tissues of the lungs to at least 43 known carcinogens. Its damaging effects increase when the lungs are simultaneously exposed to radon, a naturally occurring radioactive gas that, in some parts of the country, is released from the soil and becomes trapped in tightly insulated buildings.

The consequences of long-term exposure to cigarette smoke include not only cancer but also chronic bronchitis, characterized by a reduction in the diameter of the respiratory passageways and excessive mucus production, and emphysema. In emphysema, the fragile alveolar walls break down and are replaced by inelastic scar tissue. The result is a reduction in the surface area available for the uptake of oxygen and the release of carbon dioxide.

Although the epithelial tissues of the respiratory system are delicate, they are amazingly resilient. They are generally able to resist the assaults of cigarette smoke and other environmental hazards for many years before the cumulative damage becomes overwhelming. Most important, the damage can be halted and, in many cases, reversed, when the exposure is eliminated.

(a)

(b)   5 µm

(c)   5 µm

(a) The middle and lower lobes of a cancerous lung. The cancer is the solid grayish-white mass. Its rapid growth has replaced most of the normal lung tissue in the middle lobe. It has also probably begun to spread to other parts of the body. Notice the bronchial branch that leads into the cancer and is destroyed by it. The lung tissue remaining around the cancer is compressed, airless, and dark red. Away from the cancer, in the lower lobe, the lung is filled with air and is light pink.

Scanning electron micrographs showing (b) the normal ciliated surface of the bronchus and (c) the surface of a bronchus with cancer. The healthy, ciliated cells are shown in red, and the cancerous cells in green.

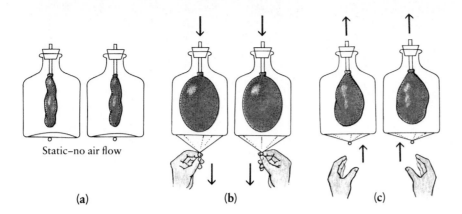

30–8   *A model illustrating the way air is taken into and expelled from the lungs, showing the action of the diaphragm.*

Static—no air flow

(a)   (b)   (c)

Mechanics of Respiration

Air flows into or out of the lungs when the air pressure within the alveoli differs from the pressure of the external air (atmospheric pressure). When alveolar pressure is greater than atmospheric pressure, air flows out of the lungs, and expiration occurs. When alveolar pressure is less than atmospheric pressure, air flows into the lungs, and inspiration occurs.

The pressure in the lungs is varied by changes in the volume of the thoracic cavity, as illustrated by the model in Figure 30–8. These changes are brought about by the contraction and relaxation of the muscular diaphragm and of the **intercostal** ("between-the-ribs") muscles. We inhale by contracting the dome-shaped diaphragm, which flattens it and lengthens the thoracic cavity, and by contracting those intercostal muscles that pull the rib cage up and out. These movements enlarge the thoracic cavity. As a result, the internal pressure falls, and air moves into the lungs. Air is forced out of the lungs as the muscles relax, reducing the volume of the thoracic cavity. Usually, only about 10 percent of the air in the lung cavity is exchanged at every breath, but as much as 80 percent can be exchanged by deliberate deep breathing.

Transport and Exchange of Gases

Oxygen is relatively insoluble in blood plasma (the liquid part of blood)—only about 0.3 milliliter of oxygen will dissolve in 100 milliliters of plasma at normal atmospheric pressure. The capacity of the blood to transport oxygen from the lungs to the body tissues is greatly increased by special oxygen-carrying protein molecules, known as **respiratory pigments.**

In vertebrates and a number of invertebrates, the respiratory pigment is **hemoglobin.** As you will recall, a hemoglobin molecule is made up of four subunits, each of which comprises a heme unit and a polypeptide chain (see Figure 3–21, page 64). The heme unit consists of a complex nitrogen-containing ring structure with one atom of iron in the center. Each heme unit can combine with one molecule of oxygen. Thus each hemoglobin molecule can carry four molecules of oxygen. Hemoglobin enables our bloodstreams to carry about 65 times as much oxygen as could be transported by an equal volume of plasma alone.

In many invertebrates, hemoglobin or other respiratory pigments circulate freely in the bloodstream. In vertebrates, hemoglobin is transported in the red blood cells. As we shall see in the next chapter, these cells are highly specialized for their transport function. A mature human red blood cell carries some 265 million molecules of hemoglobin.

Hemoglobin is a red pigment that becomes a brighter red upon oxygenation. Whether oxygen combines with hemoglobin or is released from it depends on the partial pressure of oxygen in the surrounding plasma (Figure 30–9). In the capillaries of the alveoli, where the partial pressure of oxygen is high, most of the hemoglobin is combined with oxygen. In the tissues, however, where the partial pressure is lower, oxygen is released from the hemoglobin molecules into the plasma and diffuses into the tissues. This system compensates automatically for the oxygen requirements of the tissues. Thus, because a highly active tissue uses more oxygen, the partial pressure of oxygen in the tissue is lower and more oxygen is released from the hemoglobin.

Carbon dioxide is more soluble than oxygen, and some of it is simply dissolved in the blood. Most, however, reacts with water to form carbonic acid, a weak acid that dissociates to form bicarbonate (HCO_3^-) and hydrogen (H^+) ions:

$$CO_2 + H_2O \rightleftharpoons H_2CO_3 \rightleftharpoons HCO_3^- + H^+$$

Carbon dioxide   Water   Carbonic acid   Bicarbonate ion   Hydrogen ion

As you can see by the arrows, this reaction can go in either direction. The direction it actually takes depends on the partial pressure of carbon dioxide in the blood. In the tissues, where the partial pressure of carbon dioxide is high, bicarbonate and hydrogen ions are formed. In the lungs, where the partial pressure of carbon dioxide is low, carbonic acid dissociates to form carbon dioxide and water. Once it is released, the carbon dioxide diffuses from the plasma into the alveoli and flows out of the lung with the expired air.

The reaction of carbon dioxide with water is catalyzed by the enzyme carbonic anhydrase, which is present in red blood cells along with hemoglobin. Thus the transport of carbon dioxide by the blood is also aided by the red blood cells.

Control of Respiration

The rate and depth of respiration are controlled by respiratory neurons in the brainstem. These neurons are responsible for normal breathing, which is rhythmic and automatic, like the beating of the heart. Unlike the beating of the heart, however, which few of us can control voluntarily, breathing may be brought under voluntary control within certain limits.

The respiratory neurons in the brain activate motor neurons in the spinal cord, causing the diaphragm and intercostal muscles to contract, allowing inspiration to occur. This activity of the respiratory neurons is thought to be spontaneous. Periodically, however, these neurons are inhibited, allowing expiration to occur.

In addition to their own spontaneous activity, the respiratory neurons receive signals from receptors sensitive to carbon dioxide, oxygen, and hydrogen ions, as well as from receptors sensitive to the degree of stretch of the lungs and chest. Chemoreceptor cells located in the **carotid arteries** (Figure 30–10, page 550), which supply oxygen to the brain, signal the respiratory neurons when the concentration of oxygen in the blood decreases. The concentration of dissolved carbon dioxide and of hydrogen ion is simultaneously monitored by centers in the brain and also by chemoreceptors in the carotid arteries. Thus, information is provided by a number of different, independent sensors.

This system is extremely sensitive to even the smallest change in the chemical composition of the blood, particularly to the concentration of hydrogen ion, which reflects the concentration of carbon dioxide (see equation on the preceding page). If the carbon dioxide concentration—and therefore the concentration of H^+ ions—increases only slightly, breathing immediately becomes deeper and faster, permitting more carbon di-

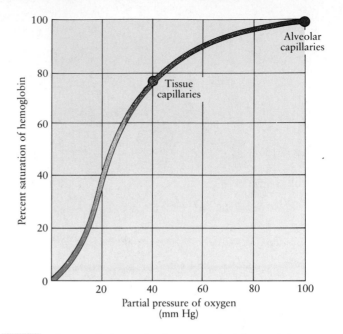

30–9 *This graph shows the degree to which hemoglobin is saturated with oxygen at different partial pressures in the surrounding medium. As the partial pressure of oxygen in the blood plasma rises, hemoglobin picks up oxygen. When it reaches 100 mm Hg—the pressure usually present in the alveoli—the hemoglobin becomes totally saturated with oxygen. As the blood travels through the tissue capillaries, the partial pressure of oxygen drops, and as it drops, oxygen is released from the hemoglobin molecules. Little oxygen is released as the partial pressure drops from 100 to 60 mm Hg. However, as the partial pressure drops below 60 mm Hg, oxygen is given up much more readily. The oxygen partial pressure in the tissue capillaries is normally about 40 mm Hg. As you can see, even after the blood has passed through these capillaries, its hemoglobin is still more than 70 percent saturated. The oxygen still carried by hemoglobin represents a reserve supply of this precious gas should the demand increase—as a result, for example, of exercise. This curve is for normal adult human hemoglobin at 38°C and at a normal pH.*

oxide to leave the blood until the concentration of H^+ ions has returned to normal. If you deliberately hyperventilate (breathe deeply and rapidly) for a few moments, you will feel faint and dizzy because of the blood's (and, therefore, the brain's) increased alkalinity.

You can, as we have noted, deliberately increase your breathing rate by contracting and relaxing your diaphragm and chest muscles, but breathing is normally involuntary. It is impossible to commit suicide by deliberately holding your breath. As soon as you lose consciousness and the concentration of carbon dioxide rises, the involuntary controls take over once more and breathing resumes.

The receptors sensitive to oxygen provide a kind of backup system to the carbon dioxide and H^+ sensors. In some cases of drug poisoning—for example, by mor-

High altitudes are harsh, dangerous environments, and their most dangerous feature is not the cold, the precipitous slopes, the threat of avalanche, or the blinding wind and snow. It is oxygen deprivation. For this reason, the ability of humans and other animals to adjust to high altitudes has long been of particular interest to physiologists.

Until recently, about 6,000 meters was thought to be the limit for human survival. In 1978, however, two European climbers reached the summit of Mt. Everest (8,848 meters) without supplemental oxygen, raising new questions about physiological adaptability.

Three years later, the American Medical Research Expedition to Everest set off with the goal of collecting the first data on human physiological function above 6,000 meters. The team included six highly experienced Himalayan climbers; a group of six "climbing scientists," all physicians with much climbing experience and an interest in high-altitude physiology; and eight physiologists who worked at the base camp at 5,400 meters and at a laboratory at 6,300 meters. Members of the expedition were able to make measurements at above 8,000 meters on human subjects (themselves), including continuous monitoring of the electrical activity of the heart and sampling of the gases in the alveoli—even at the summit itself, which two of the monitored "climbing scientists" reached.

Survival at this extreme altitude, the scientists found, depends largely on hyperventilation—extremely deep breathing. In fact, there seems to be a correlation between the capacity to hyperventilate and the capacity to be a mountain climber. This extremely deep breathing results in an astonishing decrease in the partial pressure of carbon dioxide in the lungs and bloodstream to less than one-fifth normal levels, with a corresponding increase in the pH of the blood.

Even with hyperventilation, the partial pressure of oxygen in the blood is less than one-third the partial pressure at sea level, and work capacity is greatly diminished. The European climbers, for example, reported that they moved only 2 meters per minute as they approached the summit.

There were also changes in metabolism and brain function. At 6,300 meters, there was a striking loss of body weight, with two of the expedition members each losing 15 kilograms (33 pounds). The concentration of thyroid hormone, which increases the rate of cellular respiration, increased with increasing altitude. The levels of noradrenaline, a molecule that functions both as a hormone and as a transmitter in the nervous system, were also elevated. Verbal learning and short-term memory, as measured by standard tests, declined at high altitudes but were normal one year later. In a simple brain function test, all 16 men tested showed a decrease in finger-tapping speed. In 15 of the 16, this abnormality persisted after the expedition, and in 13 of them, it was still present a year later.

In short, despite the remarkable adaptations of which the body is capable, extremely high altitudes have profound and, in some instances, permanent physiological effects. The summit of Mt. Everest is very close to the limits of human survival.

A temporary market in Nepal, south of Mt. Everest, at an altitude of 6,187 meters. The banners are Buddhist prayer flags. Inhabitants of the Himalayas and other high-altitude regions breathe more deeply than those of us who live at sea level and have more red blood cells for oxygen transport. Also, their capillaries are more dense, their cells have more mitochondria, and their muscles contain a higher concentration of myoglobin (page 279), a molecule that, like hemoglobin, takes up and releases oxygen. Individuals who have lived at high altitudes since childhood often have enlarged hearts that circulate the oxygen-carrying blood more rapidly.

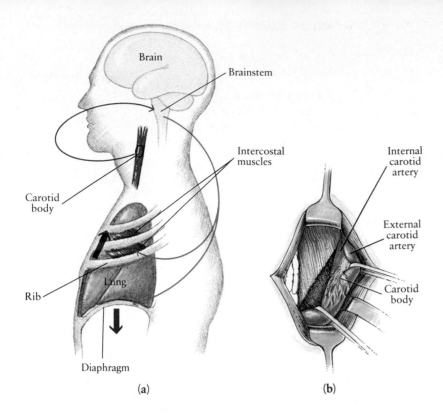

phine or barbiturates—the brainstem cells sensitive to H^+ become depressed. This causes a decrease in the breathing rate, leading ultimately to a reduction in the oxygen concentration of the blood. The oxygen sensors are then stimulated, and they maintain breathing. Massive overdoses of these drugs, however, depress the activity of the respiratory neurons themselves.

This complex system of sensors, monitoring different factors in different locations, underlines the critical importance of an uninterrupted supply of oxygen to the cells of an animal's body—particularly those of the brain.

Summary

Animals obtain energy from the oxidation of carbon-containing compounds. This process requires oxygen and releases carbon dioxide. Respiration is the means by which an animal obtains oxygen for its cells and rids itself of carbon dioxide.

Oxygen is available in both water and air. It enters cells and body tissues by diffusion, moving from regions of higher partial pressure to regions of lower partial pressure. However, efficient movement of oxygen by diffusion requires a relatively large surface area exposed to the source of oxygen and a short distance over which the oxygen has to diffuse. Selection pressures for increasingly efficient means of gas exchange led to the evolution of gills and lungs. Both gills and lungs present enormously increased surface areas for the exchange of gases. They also have a rich blood supply for transporting these gases to and from other parts of the animal's body.

Respiration in large animals involves both diffusion and bulk flow. Bulk flow brings air or water to the lungs or gills and circulates oxygen and carbon dioxide in the bloodstream. Gases are exchanged by diffusion between the blood and the air in the lungs or the water around the gills and between the blood and the tissues.

In humans, air enters the lungs through the trachea, or windpipe, and then goes into a network of increasingly smaller tubules, the bronchi and bronchioles, which terminate in small air sacs, the alveoli. Gas exchange takes place across the alveolar walls. Air moves into and out of the lungs as a result of changes in the pressure within the lungs, which, in turn, result from changes in the size of the thoracic cavity.

Respiratory pigments increase the oxygen-carrying capacity of the blood. In vertebrates, the respiratory pigment is hemoglobin, which is packed within red blood cells. Each hemoglobin molecule can bind four molecules of oxygen. Carbon dioxide is transported in the blood plasma principally in the form of bicarbonate ion.

The rate and depth of respiration are controlled by respiratory neurons in the brainstem. These neurons, which activate motor neurons in the spinal cord that cause the diaphragm and intercostal muscles to contract, respond to signals caused by very slight changes in the hydrogen ion, carbon dioxide, and oxygen concentrations of the blood.

Questions

1. Distinguish among the following: gills/lungs; trachea/pharynx; bronchi/bronchioles/alveoli.

2. What are the advantages and disadvantages of obtaining oxygen from air rather than from water? You might be able to think of several of each besides those mentioned in the text.

3. Sketch and label a diagram of the human respiratory system. When you have finished, compare your drawing to Figure 30–7a.

4. Explain the following statement: Frogs ventilate their lungs by positive pressure, whereas mammals, birds, and reptiles ventilate theirs by negative pressure.

5. One of the results of long-term smoking is the loss of bronchial cilia. What effects would you expect this to have on normal lung function?

6. Suppose a new cold remedy is guaranteed to suppress completely the secretion of mucus in the respiratory tract. Explain why you would or would not use this remedy to relieve the symptoms of a cold.

7. Carbon monoxide (CO), which is extremely poisonous, has a greater affinity for hemoglobin than does oxygen. The resulting compound, which is a brighter red than normal hemoglobin, can no longer combine with oxygen. From these facts, suggest how you might recognize and give assistance to a victim of carbon monoxide poisoning.

8. The affinity of respiratory pigments for oxygen varies among different animal species and reflects the environments to which the animals are adapted. This affinity is usually expressed in terms of the partial pressure of oxygen, in millimeters of mercury (mm Hg), required to produce 50 percent saturation of the respiratory pigment with oxygen. The partial pressures of oxygen required to produce 50 percent saturation of the hemoglobin of three freshwater fishes are as follows: catfish, 1.4 mm Hg; carp, 5 mm Hg; rainbow trout, 30 mm Hg. Which fish has the highest oxygen requirements? The lowest?

Suppose raw sewage is discharged into a river or lake in which all three species live. As the sewage decomposes and reduces the concentration of dissolved oxygen in the water, which fish species would be the first to suffocate? Which would be the last?

Suggestions for Further Reading

Bramble, Dennis M., and David R. Carrier: "Running and Breathing in Mammals," *Science*, vol. 219, pages 251–256, 1983.

Feder, Martin E., and Warren W. Burggren: "Skin Breathing in Vertebrates," *Scientific American*, November 1985, pages 126–142.

Goldberger, Ary L., David R. Rigney, and Bruce J. West: "Chaos and Fractals in Human Physiology," *Scientific American*, February 1990, pages 42–49.

Houston, Charles S.: "Mountain Sickness," *Scientific American*, October 1992, pages 58–66.

Nadel, Ethan R.: "Physiological Adaptations to Aerobic Training," *American Scientist*, vol. 73, pages 334–343, 1985.

Newsholme, Eric, and Tony Leech: *The Runner: Energy and Endurance*, Fitness Books, Roosevelt, N.J., 1984.*

> *A lively introduction to the energetics of the working human body. Equally useful as a primer for runners who want to know more about their own physiology and as an introduction to the biochemistry of carbohydrate and fat metabolism.*

West, John B.: *Everest—The Testing Place*, McGraw-Hill Publishing Company, New York, 1985.

> *An account, written for the general reader, of the adventures and findings of the American Medical Expedition to Everest. Professor West, one of the leaders of the expedition, is a lucid and delightful writer on both human physiology and mountaineering.*

*Available in paperback.

Circulation

Your mother told you not to go swimming for at least an hour after lunch. Your coach told you to warm up before running. An admirer paid you a compliment, and, if you are fair-skinned, your face turned red. But, if you are frightened, you turn pale. These phenomena are all related to the fact that the blood flow through our bodies is under micromanagement, shifting and adjusting according to internal and external events.

On a more leviathan scale, redistribution of the blood supply is a principal support system in diving mammals. Sperm whales, such as the trio shown here, are the deepest divers. They have been recorded by sonar at a depth of 1,140 meters (3,936 feet). Such deep dives are an everyday occurrence for these whales, since bottom-dwelling shrimp form the bulk of their diet. And, sperm whales can stay down for well over an hour.

Studies of diving mammals have shown that none have lungs significantly larger in proportion than ours. However, their blood volume is greater. In humans, blood is about 7 percent of body weight, whereas in diving mammals, it is 10 to 15 percent. The blood vessels are proportionately enlarged, and they appear to serve as a reservoir of oxygenated blood. Also, sperm whales store at least twice as much oxygen in their muscle cells as we do.

The major factor making possible the dives of sperm whales and other aquatic mammals is the diving reflex. During a dive, the heart rate slows and blood supply is 5 to 10 percent of normal in tissues that are tolerant of oxygen deprivation. Muscles obtain energy by anaerobic glycolysis, and most of the blood supply is shunted to the heart and brain, whose cells would begin to die in about four minutes without oxygen.

Humans also exhibit the diving reflex—if you put your face in cold water, your heart rate slows. The diving reflex seems to explain the seemingly miraculous survival of many near-drowning victims. Four minutes without oxygen is known to cause death or, at the least, irreversible brain damage. Yet of 15 persons who had been submerged for more than four minutes in the chilly waters of Lake Michigan, 11 survived without brain damage.

In all animals except those of very small size or with an extremely simple body plan, blood is the chemical "highway" interconnecting the multitude of cells that form the organism's body. It is the medium in which the nutrient molecules processed by digestion and the oxygen molecules taken in by respiration are delivered to the individual cells. Blood also carries away the waste materials, including carbon dioxide and urea, that are produced by cells in the course of their metabolic activities. As we have seen, carbon dioxide leaves the body by diffusion across the respiratory surfaces. In vertebrates, urea and other wastes carried in the blood are processed in the kidney and excreted from the body, as we shall see in the next chapter.

The blood also transports a number of other important substances, such as hormones, enzymes, and antibodies. Moreover, it has among its essential constituents the cells that defend the body against foreign invaders.

In this chapter, we shall first look briefly at the composition of the blood and then consider the structure and dynamics of the systems by which blood and another fluid, lymph, are transported in the vertebrate body.

The Blood

An individual weighing 75 kilograms (165 pounds) has about 6 liters of blood. About 55 percent of it is a straw-colored liquid called **plasma,** which is mostly water. The other 45 percent of the blood is made up of red blood cells, white blood cells, and platelets.

With the exception of the oxygen carried by hemoglobin in the red blood cells, most of the molecules needed by the individual cells of the body, as well as waste products from those cells, are dissolved in the plasma and carried along in the heavy traffic of the bloodstream. In addition, the plasma contains protein molecules, known as **plasma proteins,** that are not nu-

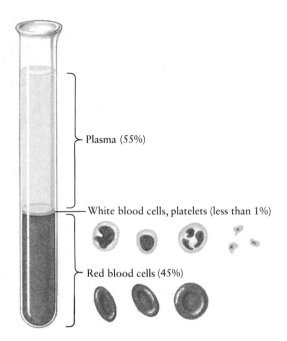

31–1   *When blood is placed in a test tube and spun in a centrifuge, its components separate from one another on the basis of their density. The two principal components, plasma and red blood cells, are layered at the top and bottom of the test tube. A thin middle layer consists of white blood cells and platelets.*

Plasma (55%)

White blood cells, platelets (less than 1%)

Red blood cells (45%)

trients or waste products, but function in the bloodstream itself. Plasma proteins are involved in blood clotting, defense of the body against foreign invaders, and maintenance of the osmotic potential of the blood, which prevents excessive loss of fluid from the bloodstream to the tissues.

Red Blood Cells

Red blood cells, or **erythrocytes** (Figure 31–2), transport oxygen to all the tissues of the body. They are among the most highly specialized of all cells. As a mammalian red blood cell matures, it extrudes its nucleus and mitochondria, and its other cellular structures dissolve. Almost the entire volume of a mature red blood cell is filled with hemoglobin.

There are about 5 million red blood cells per cubic millimeter of blood—some 25 trillion (25×10^{12}) in the adult human body. Because red blood cells, lacking a nucleus, cannot repair themselves, their life span is comparatively short, some 120 to 130 days. At this moment, in your body, red blood cells are dying at a rate of about 2 million per second. To replace them, new ones are being formed in the bone marrow at the same incredible rate.

White Blood Cells

For every 1,000 red blood cells in the human bloodstream, there are 1 or 2 white blood cells, or **leukocytes,** for a total of about 6,000 to 9,000 per cubic millimeter of blood. A number of different types of white blood

31–2   *The exchange of substances between the circulating blood and the body tissues takes place through the walls of the smallest blood vessels, the capillaries. In this micrograph, the Y-shaped structure is a branching capillary, through which red blood cells are traveling. These cells, with their oxygen-laden hemoglobin molecules, are about 7 or 8 micrometers in diameter, significantly larger than the smallest capillaries, which are only 5 micrometers in diameter. The passage of the red blood cells through the capillaries is made possible by their "donut without a hole" shape, which enables them not only to twist and turn as they move in single file but also to fold. This triumph of biological form also provides a large surface area through which oxygen can diffuse to and from the hemoglobin molecules contained within the cell.*

10 μm

cells circulate in the bloodstream. All of them are nearly colorless, contain no hemoglobin, and have a nucleus. Most are larger than red blood cells.

The chief function of the white blood cells is the defense of the body against invaders such as viruses, bacteria, and other foreign particles. Unlike red blood cells, white blood cells are not confined within the blood vessels but can migrate out into the tissues. They appear spherical in the bloodstream, but in the tissues they become flattened and amoeba-like. Like amoebas, they move by means of pseudopodia, and many are phagocytic (Figure 31–3). As we shall see in Chapter 33, certain types of white blood cells play key roles in the immune response.

White blood cells are often destroyed in the course of fighting infection. Pus is composed largely of these dead cells. New white blood cells to take the place of those that are destroyed are formed constantly in the spleen, in bone marrow, and in certain other tissues.

Platelets

Platelets, so called because they look like little plates, are colorless, oval or irregularly shaped disks less than half the size of red blood cells. Platelets are membrane-bound cytoplasmic fragments that break off from unusually large cells found in the bone marrow. They are, in effect, little bags of chemicals that play an essential role in initiating the clotting of blood and in plugging breaks in the blood vessels.

Blood Clotting

The clotting of blood is a complex phenomenon requiring platelets and at least 15 factors normally present in the bloodstream or on plasma membranes. The sequence of events begins when plasma encounters a rough surface or a protein molecule known as **tissue factor.** Tissue factor, which is thought to have many regulatory roles throughout the body, is found on the outer surface of many different cell types—but not on the cells of the inner lining of the blood vessels.

When tissue factor reacts with a specific circulating plasma protein—as can occur when a blood vessel is broken—a cascade of chemical reactions is initiated. In this cascade, the product of each step of the reaction series acts as a catalyst for the next step, and the molecules involved are, like enzymes, used over and over. Ultimately, a soluble plasma protein is converted to an insoluble protein known as **fibrin.** The fibrin molecules clump together, forming a network that enmeshes red blood cells and platelets to form a clot (Figure 31–4). The clot then contracts, pulling together the edges of the wound.

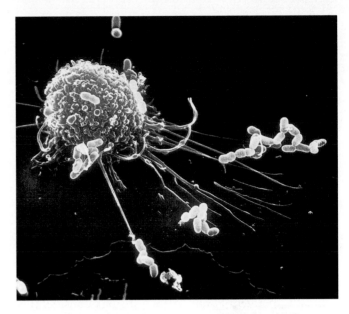

31–3 *A scanning electron micrograph of a human white blood cell entrapping bacterial cells. This type of cell defends the body against pathogens and other harmful particles by extending pseudopodia to the foreign objects, which are then engulfed within phagocytic vacuoles and destroyed with the help of enzymes from the cell's lysosomes (page 98).*

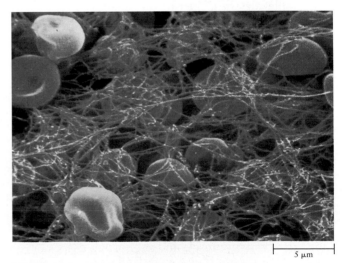

5 μm

31–4 *An early stage in the formation of a blood clot. Ultimately, the completed clot forms an impenetrable barrier, preventing both the loss of vital fluids and the entry of infectious microorganisms. In this scanning electron micrograph, the fibers formed by the insoluble protein fibrin are highlighted in pink. The red blood cells, which have become enmeshed in the fibers, are highlighted in yellow.*

The Cardiovascular System

As we saw in Chapter 26, the systems through which blood is transported in animals vary in their structure and complexity. In the earthworm, for example, blood circulates in a closed system of continuous vessels. It is propelled by the contractions of "hearts" that are little more than expanded, muscular portions of particular blood vessels. Mollusks, by contrast, have a much more complex heart, consisting of several chambers. Most mollusks, however, have an open circulatory system. Blood is pumped through vessels that open into spaces within the tissues, from which it returns to the gills and then to the heart.

The vertebrate heart, like that of the mollusks, is a muscular organ consisting of several chambers. The blood vessels, however, form a closed system, similar in principle to that of the earthworm, but considerably more elaborate. The heart and vessels together are known as the **cardiovascular system** (from *kardio*, meaning "heart," and *vasculum*, "small vessel").

The Blood Vessels

In the cardiovascular system, the heart pumps blood into the large **arteries,** from which it travels to branching, smaller arteries, then to the smallest arteries (the **arterioles**), and then into networks of very small vessels, the **capillaries.** From the capillaries, the blood passes into small veins, the **venules,** then into larger **veins,** and through them, back to the heart.

In humans, the diameter of the opening of the largest artery, the **aorta,** is about 2.5 centimeters, that of the smallest capillary is only 5 micrometers (0.0005 centimeter), and that of the largest vein, the **vena cava,** is about 3 centimeters. Arteries, veins, and capillaries differ not only in their size but also in the structure of their walls (Figure 31–5). The walls of the capillaries consist of only one layer of **endothelium,** a special type of epithelial tissue (Figure 31–6). The walls of arteries and veins, which are lined with a single layer of endothelium, also contain muscle and supporting tissues.

The Capillaries and Diffusion

The heart, arteries, and veins are, in essence, the means for getting the blood to and from the capillaries, where the actual function of the circulatory system is carried out. It is only through the thin walls of the capillaries that nutrients, oxygen, carbon dioxide, and other molecules are exchanged between the blood and the fluids surrounding the cells.

Most of the molecules that cross the capillary walls pass in or out by diffusion. Some additional molecules

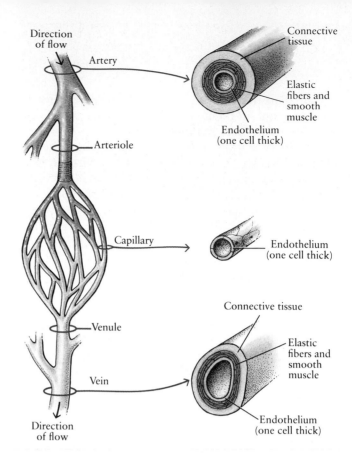

31–5   *The structure of blood vessels. Arteries have thick, tough, elastic walls that can withstand the high pressure of the blood as it leaves the heart. Arteries branch into arterioles, which have thinner walls with a reduced layer of muscle. Capillaries have walls only one cell thick. Exchange of gases, nutrients, and wastes between the blood and the cells of the body takes place through these thin capillary walls. From the capillaries, blood enters venules, which merge to form veins. Veins usually have larger lumens (passageways) than arteries and always have thinner, more readily extensible walls that minimize resistance to the flow of blood on its return to the heart.*

cross by bulk flow, because the pressure of the blood within the capillaries tends to force fluid out through the capillary walls. At the same time, because the solute concentration of the blood is higher than that of the fluids outside the capillaries (primarily because of the presence of plasma proteins), fluid tends to reenter the bloodstream by osmosis. Thus, there is normally a balance between inflow and outflow across the capillary walls. When this balance is disturbed (for example, when the endothelium is damaged by a blow) and outflow is greater than inflow, excess fluid collects in the tissues, producing the swelling known as edema.

The total length of the capillaries in a human adult is more than 80,000 kilometers (50,000 miles). No cell in the human body is farther than 130 micrometers—a distance short enough for rapid diffusion—from a capillary. Even the cells in the walls of the large veins and arteries depend on capillaries for their blood supply, as do the cells of the heart itself.

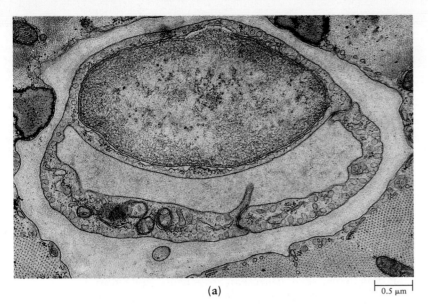

(a)

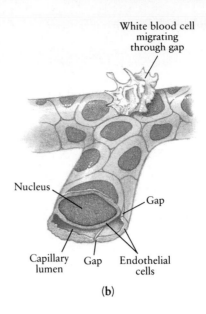

White blood cell
migrating
through gap

Nucleus

Gap

Capillary
lumen

Gap

Endothelial
cells

(b)

31–6 (a) *Electron micrograph of a cross section of a capillary, and* (b) *diagram of a capillary that has been sectioned in the same manner. Portions of two endothelial cells can be seen, fitting together to form the capillary lumen, the passageway through which blood flows. The size of the tiny gaps between the endothelial cells can be varied, and it is through these gaps that white blood cells migrate out into the tissues. The space between the capillary and the surrounding tissue contains extracellular fluids.*

The Heart

Evolution of the Heart

In the course of vertebrate evolution, the heart has undergone some structural adaptations, as shown in Figure 31–7. Fish have a single heart divided into an **atrium,** the receiving chamber for the blood, and a **ventricle,** the pumping chamber from which the blood is expelled into the vessels. The ventricle of the fish heart pumps blood directly into the capillaries of the gills, where it picks up oxygen and releases carbon dioxide. From the gills, oxygenated blood is carried to the tis-

31–7 *Vertebrate circulatory systems. Oxygen-rich blood is shown as red, and oxygen-poor blood as blue.*

(a) *In the fish, the heart has only one atrium (A) and one ventricle (V). Blood oxygenated in the gill capillaries goes straight to the capillaries of the systemic circulation without first returning to the heart.*

(b) *In amphibians, the single primitive atrium has been divided into two separate chambers. Oxygen-rich blood from the lungs enters one atrium, and oxygen-poor blood from the tissues enters the other. Little mixing of the blood occurs in the ventricle, despite its lack of a structural division. From the ventricle, oxygen-rich blood is pumped to the body tissues at the same time that oxygen-poor blood is pumped to the lungs. Some of the oxygen-poor blood is diverted from the lungs to the skin, a major respiratory organ in amphibians.*

(c) *In birds and mammals, both the atrium and the ventricle are divided into two separate chambers, so that there are, in effect, two hearts—one for pumping oxygen-poor blood through the lungs and one for pumping oxygen-rich blood through the body tissues.*

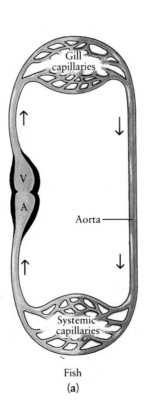

Fish
(a)

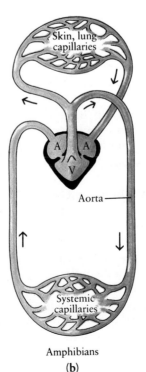

Amphibians
(b)

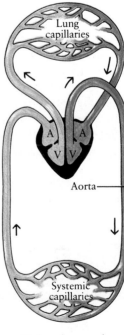

Birds and mammals
(c)

sues. By this time, however, most of the propulsive force of the heartbeat has been dissipated by the resistance of the capillaries in the gills, so that the blood flow through the rest of the tissues is relatively sluggish. In fish and other vertebrates, the flow of oxygenated blood to the body's tissues and the return trip of deoxygenated blood from the tissues is known as the **systemic circulation.**

In amphibians, there are two atria. One receives oxygenated blood from the lungs, and the other receives deoxygenated blood from the systemic circulation. Both atria empty into a single ventricle. Although the ventricle is not divided, its internal structure is ridged, ensuring that the two kinds of blood remain relatively unmixed. The oxygenated blood is pumped into the systemic circulation under high pressure at the same time that the deoxygenated blood is pumped through the lungs. Branches of the vessels leading to the lungs also carry some of the deoxygenated blood to the moist skin, a major site of gas exchange in amphibians. Blood oxygenated in the skin enters the systemic circulation before returning to the heart.

In birds and mammals, the heart is separated into two functionally distinct organs, the right heart and the left heart, each with an atrium and a ventricle. The right heart receives blood from the tissues and pumps it the short distance to the lungs, where it becomes oxygenated. From the lungs, the oxygenated blood returns to the left heart, from which it is pumped at high pressure to all of the body tissues. The flow of deoxygenated blood from the heart through the lungs and of oxygenated blood back to the heart is known as the **pulmonary circulation.**

The efficient, high-pressure circulatory system of birds and mammals, with its full separation of oxygenated and deoxygenated blood, makes possible the high metabolic rate of these animals. The high metabolic rate, in turn, makes possible their generally high levels of activity and relatively constant body temperature.

The Human Heart

Figure 31–8 shows a diagram of the human heart. Its walls are made up of a specialized type of muscle—cardiac muscle. Blood returning from the body tissues enters the right atrium through two large veins, the superior and inferior venae cavae. Blood returning from the lungs enters the left atrium through the pulmonary veins. The atria, which are thin-walled compared to the ventricles, expand as they receive the blood. Both atria then contract simultaneously, assisting the flow of blood through valves that open into the ventricles. Then the ventricles contract simultaneously. The pressure of

the blood in the ventricles closes the valves between the atria and the ventricles, preventing any backflow into the atria and ensuring that the blood flows in one direction only.

The right ventricle propels deoxygenated blood through the pulmonary arteries to the capillaries of the lungs. The left ventricle propels oxygenated blood into the aorta, from which it travels to the other body tissues. Valves between the right ventricle and the pulmonary artery and between the left ventricle and the aorta close after the ventricles contract, again preventing the backflow of blood.

If you listen to a heartbeat, you hear "lubb-dup, lubb-dup." The deeper, first sound ("lubb") is the closing of the valves between the atria and the ventricles. The second sound ("dup") is the closing of the valves between the ventricles and the arteries. If any one of the four valves is damaged, as from rheumatic fever, blood may leak back through the valve, producing the noise characterized as a "heart murmur" (a "ph-f-f-t" sound).

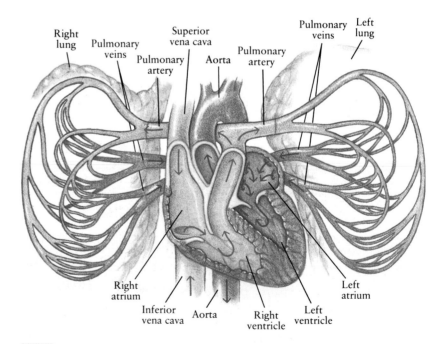

31–8   *The human heart. Blood returning from the systemic circulation through the superior and inferior venae cavae enters the right atrium and passes to the right ventricle, which propels it through the pulmonary arteries to the lungs, where it is oxygenated. Blood from the lungs enters the left atrium through the pulmonary veins, passes to the left ventricle, and then is pumped through the aorta to the body tissues.*

In a healthy adult at rest, the rhythmic beating of the heart takes place about 70 times a minute. Under strenuous exercise, the rate more than doubles. The total volume of blood pumped by the heart per minute is called the **cardiac output**. It is defined as:

$$\text{Cardiac output} \atop \text{(liters per minute)} = {\text{Heart rate} \atop \text{(beats per minute)}} \times {\text{Stroke volume} \atop \text{(liters per beat)}}$$

If, for example, the heart beats 70 times per minute and ejects 0.07 liter of blood into the aorta with each beat, the cardiac output is 4.9 liters per minute (70 beats per minute times 0.07 liter of blood per beat).

In addition to its function as a pump, the heart is also a hormone-secreting organ. In humans and other mammals, many of the cardiac muscle cells of the atria synthesize and release a molecule known as **cardiac peptide**. Receptors for cardiac peptide have been identified in the kidneys, blood vessels, adrenal glands, and brain. Although the full range of functions of this hormone remains unknown, it appears to play a major role in the regulation of blood volume and blood pressure.

Regulation of the Heartbeat

Most muscle contracts only when stimulated by a motor nerve, but the stimulation of cardiac muscle cells originates in the muscle itself. A vertebrate heart will continue to beat even after it is removed from the body if it is kept in an oxygenated nutrient solution. In vertebrate embryos, the heart begins to beat very early in development, before the appearance of any nerve supply. In fact, isolated embryonic heart cells in a test tube will beat.

The contraction of cardiac muscle is initiated by a special area of the heart, the **sinoatrial node,** which is located in the right atrium (Figure 31–9). This region of tissue functions as the pacemaker. It is composed of specialized cardiac muscle cells that can spontaneously initiate their own electrical impulse.

From the pacemaker the impulse spreads throughout the right and left atria. As it passes along the surface of the individual cardiac muscle cells, it activates their contractile machinery, causing them to contract. As we saw in Figure 28–7b (page 512), cardiac muscle cells are interconnected by structures known as **intercalated disks.** The intercalated disks make possible the rapid conduction of electrical impulses between adjacent cells and thus the rapid spread of impulses across the heart. As a consequence, many cells in both atria are activated almost simultaneously.

About 100 milliseconds after the pacemaker fires, impulses traveling through both special conducting fibers and the atrial muscle itself stimulate a second area

of nodal tissue in the right atrium, the **atrioventricular node.** From the atrioventricular node, impulses are carried by special muscle fibers, the **bundle of His** (named after its discoverer), to the walls of the right and left ventricles, which then contract almost simultaneously.

The bundle of His is the only electrical bridge between the atria and the ventricles. Although its fibers conduct impulses very rapidly, the atrioventricular node consists of slow-conducting fibers. As a consequence, a delay is imposed between the atrial and ventricular contractions, ensuring that the atrial beat is completed before the beat of the ventricles begins.

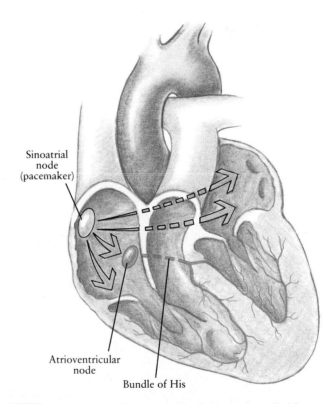

Sinoatrial node (pacemaker)

Atrioventricular node

Bundle of His

31–9 *The beat of the mammalian heart is controlled by a region of specialized muscle tissue in the right atrium, the sinoatrial node, that functions as the heart's pacemaker. Some of the nerves regulating the heart have their endings in this region. Excitation (light green lines) spreads from the pacemaker through the atrial muscle cells, causing both atria to contract almost simultaneously. When the wave of excitation reaches the atrioventricular node, its conducting fibers pass the stimulation to the bundle of His, from which excitation (dark green lines) spreads along specialized fibers of the ventricles. The result is an almost simultaneous contraction of the two ventricles. Because the fibers of the atrioventricular node conduct relatively slowly, the ventricles do not contract until after the atrial beat has been completed.*

When impulses from the conducting system travel across the heart, electric current generated on the heart's surface is transmitted to the body fluids, and from there some of it reaches the body surface. Electrodes, appropriately placed on the skin and connected to a recording instrument, can measure this current. The output, an electrocardiogram, is important in assessing the heart's ability to initiate and transmit impulses.

Although the autonomic nervous system does not initiate the vertebrate heartbeat, it does modify its rate. Fibers from the parasympathetic division travel in the vagus nerve (a large nerve that runs through the neck) to the pacemaker. Parasympathetic stimulation has a slowing effect on the pacemaker and thus decreases the rate of heartbeat. Sympathetic nerves stimulate the pacemaker, increasing the rate of heartbeat. They also stimulate cardiac muscle cells throughout the heart directly, increasing the strength of their contractions. Adrenaline, released from the adrenal glands, affects the heart in the same way as stimulation by the sympathetic nerves.

Blood Pressure

Contraction of the left ventricle of the heart propels blood into the aorta and through the systemic circulation with considerable force. **Blood pressure** is a measure of the force per unit area with which blood pushes against the walls of the blood vessels. It is generated by the pumping action of the heart and changes with the rate at which the heart contracts. The strength of these contractions, the elasticity of the arterial walls, and the rate at which blood flows from the arteries also play important roles in determining the blood pressure.

For medical purposes, blood pressure is usually measured at the artery of the upper arm with a device that uses the pressure in the artery to push up a column of mercury. Normal blood pressure in a young adult male is about 120 millimeters of mercury (120 mm Hg) when the ventricles are contracting (the **systolic** pressure) and 80 mm Hg when the ventricles relax (the **diastolic** pressure). This is stated as a blood pressure of 120/80.

The rate of blood flow is determined by both the blood pressure and the size of the vessel through which the blood is flowing. The greater the blood pressure, the greater the rate of flow. Similarly, the larger the diameter of the vessel, the larger the volume of blood that can flow through it. The diameter of the arterioles, which directly supply the capillaries, can be altered by rings of smooth muscle in the vessel walls (Figure 31–10). As the smooth muscle contracts, the opening of the

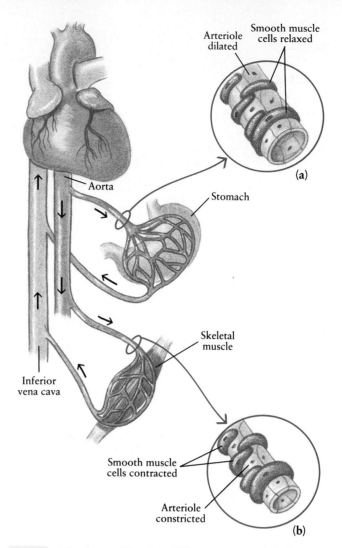

31–10   *The flow of blood to different regions of the body is regulated by the action of smooth muscle in the arteriole walls. The requirements of different tissues for oxygen and nutrients vary according to the physiological state of the organism. In the example shown here, the individual is resting after a meal. Smooth muscle ringing the arterioles that lead to capillary beds supplying the digestive tract is relaxed* (a), *allowing a high rate of blood flow that supports the digestive process. At the same time, smooth muscle ringing arterioles that lead to capillary beds supplying the skeletal muscles is contracted* (b), *reducing the blood flow to these muscles while they are at rest.*

arteriole gets smaller, a process known as **vasoconstriction,** and blood flow through the arteriole (and the capillary bed it feeds) decreases. Conversely, when the smooth muscle relaxes, the arteriole opens wider, a process known as **vasodilation,** and blood flow into the capillaries increases. These smooth muscles are influenced by autonomic nerves (chiefly sympathetic nerves), the hormones adrenaline and noradrenaline, cardiac peptide, and other chemicals that are produced locally in the tissues themselves.

Diseases of the Heart and Blood Vessels

Cardiovascular diseases cause almost as many deaths in the United States as accidents and all other diseases combined. According to recent estimates, about 70 million people in this country (about 25 percent of the total population) have some form of cardiovascular disease.

The majority of cardiovascular deaths are caused by heart attacks. A **heart attack** is the result of an insufficient supply of blood **(ischemia)** to an area of heart muscle. With their oxygen supply cut off, cardiac muscle cells in the area may die. A heart attack can be caused by a blood clot—a **thrombus**—that forms in the **coronary arteries,** the blood vessels that supply the heart itself. Or, it can be caused by a wandering clot—an **embolus**—that forms elsewhere in the body and travels to the heart and lodges in a coronary artery. A heart attack can also result from blockage of a coronary artery due to atherosclerosis. Recovery from a heart attack depends on how quickly the blockage can be cleared, how much of the heart tissue is damaged, where the damage occurs, and whether or not other blood vessels in the heart can enlarge their capacity and form new branches that supply these tissues, which then may recover to some extent.

Angina pectoris is a related condition in which the heart muscle receives an insufficient blood supply (but not so little that the muscle dies)—often as a result of a narrowing of the vessels. Its symptoms, like those of a heart attack, are pain in the center of the chest, between the shoulder blades of the back, and, often, in the left arm and shoulder.

A **stroke** is caused by interference with the blood supply to the brain. This may be the result of a thrombus, an embolus, or the bursting of a blood vessel in the brain. The effects of a stroke depend on the extent of the damage and where it occurs in the brain.

Atherosclerosis contributes to both heart attacks and strokes. In this disease, the linings of the arteries thicken due to the accumulation of abnormal smooth muscle cells, and their inner surfaces become roughened by deposits of cholesterol, fibrin, and cellular debris, as shown in the photos on page 58. These conditions pave the way for further difficulties. The arteries, becoming inelastic, no longer expand and contract, and blood moves with increasing difficulty through the narrowed ves-

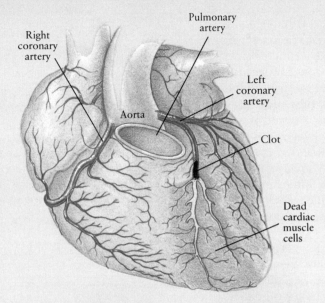

A *heart attack.* When a clot forms in one of the vessels supplying blood to the heart muscle, the cells in the area supplied by that vessel are deprived of oxygen and may die. The severity of the heart attack depends, in part, on the extent of damage to the heart muscle.

sels. Thrombi and emboli form more easily and are more likely to block the vessel.

The causes are not clear-cut. Atherosclerosis is associated with high blood pressure and with high levels of cholesterol in the blood. As we saw on page 58, a critical factor appears to be the relative proportions of the two principal forms in which cholesterol is carried in the bloodstream: high-density lipoproteins (HDLs), which appear to protect against atherosclerosis, and low-density lipoproteins (LDLs), which contribute to the disease process. Cigarette smoking, a lack of physical activity, a high dietary intake of cholesterol and saturated fats, and, in some cases, hereditary factors are associated with increased levels of LDLs and decreased levels of HDLs. Moderate, regular exercise generally increases the levels of HDLs and decreases the levels of LDLs.

Atherosclerosis is much less common in premenopausal women than it is among men of the same age, suggesting that female hormones may play a role in protecting against the disease. Paradoxically, however, men with elevated levels of female hormones appear to be at an increased risk for atherosclerosis. Whatever the underlying cause may be, the difference in susceptibility to atherosclerosis is a major reason that, in the United States, women now live, on the average, almost 10 years longer than men.

In North America, the mortality rate from ischemic heart disease reached a peak in the mid-1960s and then, quite unexpectedly, began a decline that is still continuing. The total decrease in the mortality rate has, astonishingly, exceeded 30 percent. The increase, when it occurred, was attributed to increased cigarette smoking, increased stress, and an increasingly sedentary existence. The decrease is attributed to a combination of factors, including a decrease in cigarette smoking in the population as a whole and a greater emphasis on good nutrition and exercise. Improvements in medical technology and in the treatment of heart attack victims, particularly in the first few minutes and hours after the attack, have also played a major role in the decreased mortality rate.

Hypertension—chronically increased arterial blood pressure—affects about 54 million people in the United States. It places additional strain on the arterial walls and increases the chances of emboli. Despite lack of knowledge about the causes, hypertension can be treated by drugs that act upon the autonomic nervous system to produce arteriolar dilation, and by measures that decrease blood volume, reducing the blood pressure to safer levels. In the United States, hypertension is about one-third more common among African-Americans than among individuals of European descent. Related to this disturbing statistic is the fact that the incidence of deaths among African-Americans from cardiovascular disease is significantly higher than in the population at large.

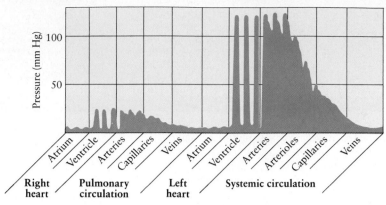

Constriction and dilation of the arterioles in different parts of the body regulate the blood flow—and thus the supply of oxygen and nutrients—according to the varying requirements of the animal. For example, blood flow through skeletal muscle increases during exercise, flow to the stomach and intestines increases during digestion, and flow through the skin increases at high temperatures and decreases at low temperatures. Of particular importance is a constant flow of blood to the brain. One of the mechanisms that prevents serious damage to human brain cells as a result of an inadequate blood supply is fainting, which causes the person to fall. As a consequence, the force of gravity does not have to be overcome for blood to flow to the brain. This response is often thwarted by well-meaning bystanders anxious to get the affected individual back on his or her feet. In fact, holding a fainting person upright can lead to severe shock and even death.

As blood flows through the circulatory system, its pressure gradually drops as a consequence of the resistance of the arterioles and capillaries. Thus, blood pressure is not the same in the various parts of the cardiovascular system (Figure 31–11). As you would expect, it is highest in the aorta and other large systemic arteries, much lower in the veins, and lowest in the right atrium.

The veins, with their relatively thin walls and large diameters, offer little resistance to flow, making possible the movement of the blood back to the heart despite its low pressure. Valves in the veins prevent backflow. The return of blood to the heart is enhanced by the contractions of skeletal muscles (Figure 31–12). For example, when you walk, your leg muscles squeeze the veins that lie between the contracting muscles. This raises the pressure within the veins and increases the flow. Also, as the thoracic cavity expands on inspiration, the elastic walls of the veins in the chest dilate, venous pressure in the region of the heart decreases, and return of blood to the heart increases.

Cardiovascular Regulating Center

Activity of the nerves controlling the smooth muscle of the blood vessels is coordinated with the activity of nerves regulating heart rate and strength of heartbeat by the **cardiovascular regulating center.** This center is located in the medulla, a small part of the brain continuous with the spinal cord. It controls the sympathetic and parasympathetic nerves to the heart as well as the nerves to the smooth muscle in the arterioles. Thus, if there is a significant increase in blood flow due to dilation of blood vessels, the heart is simultaneously stim-

31–11   *In mammals, blood goes from the right heart to the lungs and from the lungs to the left heart. From the left heart, it enters the systemic circulation, moving from arteries to arterioles to capillaries to venules to veins. Blood pressure varies in the different regions of the cardiovascular system. The fluctuations in blood pressure produced by three heartbeats are shown in each section of this diagram. Note the dramatic fall in pressure as blood traverses the arterioles of the systemic circulation.*

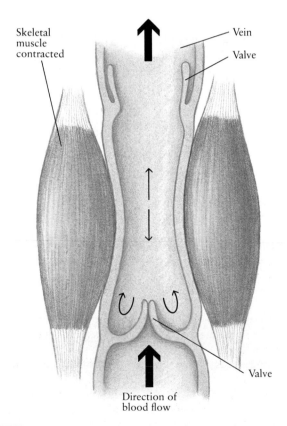

31–12   *As skeletal muscles contract, they become shorter and thicker, squeezing the veins that run between them. The resulting increased pressure in the veins assists the blood on its journey back to the heart. Valves, positioned at intervals throughout the venous network, open to allow blood flow toward the heart (upper valve) and close to prevent backflow (lower valve).*

ulated to beat faster, thus developing greater pressure to support the greater flow.

The cardiovascular regulating center integrates the reflexes that control blood pressure. It receives information about existing blood pressure from specialized stretch receptors in the carotid arteries (Figure 30–10, page 550), the venae cavae, the aorta, and the heart. The effector organs of the reflex are, as we have indicated, the heart and blood vessels.

The blood-pressure reflex is an example of negative-feedback control. When pressure falls, the activity of the heart is increased and the blood vessels are constricted, raising the pressure again. Conversely, when the pressure rises, heart activity is decreased and the blood vessels are dilated, lowering the pressure.

The Lymphatic System

Not quite all of the fluid forced out of the capillaries by the pressure of the circulating blood (page 556) reenters the capillaries by osmosis. In higher vertebrates, the fluid lost from the blood to the tissues is collected by the **lymphatic system** (Figure 31–13), which routes it back to the bloodstream. The fluid carried in this system is known as **lymph.**

The lymphatic system is like the venous system in that it consists of an interconnecting network of progressively larger vessels. An important difference, however, is that the lymph capillaries begin blindly in the tissues, rather than forming part of a continuous circuit. Fluid from the tissues seeps into the lymph capillaries, from which it travels to large ducts that empty into two veins located below the clavicles (collar bones). These veins, known as the subclavian veins, empty into the superior vena cava.

Some nonmammalian vertebrates have lymph "hearts," which help to move the fluid. In mammals, lymph is moved by contractions of the body muscles, with valves preventing backflow, as in the venous system. Also, recent studies have shown that lymph vessels contract rhythmically. These contractions may be the principal factor propelling the lymph.

Lymph nodes, which are masses of spongy tissue, are distributed throughout the lymphatic system. They have two functions: they are the sites of proliferation of lymphocytes, specialized white blood cells that are the effectors of the immune response (to be discussed in Chapter 33), and they remove cellular debris and foreign particles from the lymph before it enters the blood. The removal of chemical wastes, however, requires processing of the blood itself. This function is performed by the kidneys, as we shall see in the next chapter.

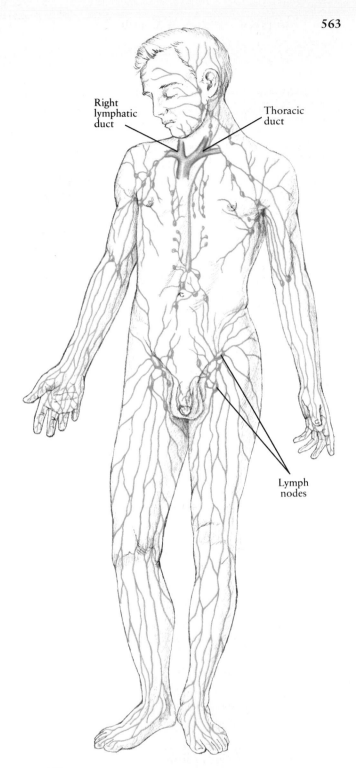

Right lymphatic duct

Thoracic duct

Lymph nodes

31–13   *The human lymphatic system consists of a network of lymph vessels and lymph nodes. Lymph reenters the bloodstream through the thoracic duct, which empties into the left subclavian vein, and through the right lymphatic duct, which empties into the right subclavian vein. These two veins empty into the superior vena cava.*

Summary

Oxygen, nutrients, and other essential molecules, as well as waste products, are carried in the blood. The blood is composed of plasma, red blood cells (erythrocytes), white blood cells (leukocytes), and platelets. Plasma, the fluid part of the blood, is chiefly water in which nutrients, waste products, ions, antibodies, hormones, enzymes, plasma proteins, and other substances are dissolved or suspended. Red blood cells contain oxygen-bearing hemoglobin, and white blood cells defend the body against foreign invaders. Platelets are involved in the clotting of blood, which occurs as the result of a reaction cascade involving at least 15 factors.

In vertebrates the blood is pumped by muscular contractions of the heart into a closed circuit of arteries, arterioles, capillaries, venules, and veins. This network ultimately services every cell in the body. The essential function of the circulatory system is performed by the capillaries, through which substances are exchanged with the fluids surrounding the individual cells of the body.

Evolutionary developments in the structure of the vertebrate heart can be correlated with changes in metabolic rates and the level of activity of the animals. Fish have a two-chambered heart, whereas amphibians have a three-chambered heart. Birds and mammals have a four-chambered heart that functions as two separate pumping organs. One side pumps deoxygenated blood to the lungs, and the other pumps oxygenated blood to the body tissues under high pressure.

Synchronization of the heartbeat is controlled by the sinoatrial node (the pacemaker), located in the right atrium, and by the atrioventricular node, which delays the stimulation of ventricular contraction until the atrial contraction is completed. The rate of heartbeat is under neural and hormonal regulation. Parasympathetic stimulation slows the heartbeat; sympathetic stimulation and adrenaline speed it up.

Blood pressure is a measure of the force per unit area with which blood pushes against the walls of the blood vessels. The magnitude of the pressure is influenced by the rate and strength of the heartbeat, the elasticity of the arterial walls, and the rate at which blood flows from the arteries into the arterioles. The rate of flow is, in turn, influenced by the degree of dilation or constriction of the arterioles. Activity of the nerves regulating the rate and strength of the heartbeat is coordinated with the activity of the nerves controlling the smooth muscle of the arterioles by the cardiovascular regulating center, located in the medulla of the brain.

Fluids that seep out of the capillaries either reenter them by osmosis or are returned to the blood by the lymphatic system. The lymph also picks up cellular debris and foreign particles, which are filtered out by the lymph nodes.

Questions

1. Distinguish between the following: blood/plasma; aorta/vena cava; atrium/ventricle; right heart/left heart; sinoatrial node/atrioventricular node; systolic/diastolic.

2. What is the function of the cardiovascular system? What are its principal components?

3. Label the diagram at the right.

4. What is the advantage of the fibers of the atrioventricular node being slow-conducting?

5. The valves of the heart are not directly controlled by nerves. Yet, in most individuals, they open and shut at precisely the right points in the cardiac cycle for efficient heart operation. How is this precise timing possible? What does determine just when the valves will open and shut?

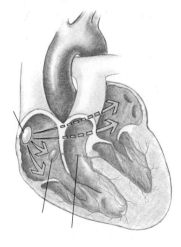

6. Trace the course of a single red blood cell from the right ventricle to the right atrium in a mammal. Trace the course of an oxygen molecule from the air to its arrival at a metabolizing cell.

7. When fair-skinned individuals are very frightened, they turn quite pale. What occurs to cause this change? Why is the underlying physiological process, which occurs in all individuals, regardless of their skin color, a useful adaptation?

8. Explain the reasons for the changes in blood pressure shown in Figure 31–11.

9. When an accident victim suffers blood loss, he or she is transfused with plasma rather than with whole blood. Why is plasma effective in meeting the immediate threat to life?

10. Individuals, particularly children, suffering from severe protein deprivation often have swollen, bloated bellies. What is the explanation for this phenomenon?

11. What are the two principal forms in which cholesterol is transported in the bloodstream? What factors affect their relative proportions? What is the effect of regular exercise?

Suggestions for Further Reading

Brown, Michael S., and Joseph L. Goldstein: "How LDL Receptors Influence Cholesterol and Atherosclerosis," *Scientific American*, November 1984, pages 58–66.

Cantin, Marc, and Jacques Genest: "The Heart as an Endocrine Gland," *Scientific American*, February 1986, pages 76–81.

Diamond, Jared: "The Saltshaker's Curse," *Natural History*, October 1991, pages 20–26.

Fackelmann, Kathy A.: "The Safer Sex? Probing a Cardiac Gender Gap," *Science News*, January 19, 1991, pages 40–41.

Golde, David W.: "The Stem Cell," *Scientific American*, December 1991, pages 86–93.

Golde, David W., and Judith C. Gasson: "Hormones that Stimulate the Growth of Blood Cells," *Scientific American*, July 1988, pages 62–70.

Harken, Alden H.: "Surgical Treatment of Cardiac Arrhythmias," *Scientific American*, July 1993, pages 68–74.

Lawn, Richard M.: "Lipoprotein (a) in Heart Disease," *Scientific American*, June 1992, pages 54–60.

Marx, Jean: "Holding the Line Against Heart Disease," *Science*, vol. 248, pages 1491–1493, 1990.

Perutz, M. F.: "Hemoglobin Structure and Respiratory Transport," *Scientific American*, December 1978, pages 92–125.

Raloff, Janet: "Beyond Oat Bran: Reaping the Benefits without Gorging on the Grain," *Science News*, May 26, 1990, pages 330–332.

Robinson, Thomas F., Stephen M. Factor, and Edmund H. Sonnenblick: "The Heart as a Suction Pump," *Scientific American*, June 1986, pages 84–91.

Stallones, Ruel A.: "The Rise and Fall of Ischemic Heart Disease," *Scientific American*, November 1980, pages 53–59.

Weiss, Rick: "First Gene-Hypertension Link Found," *Science News*, September 8, 1990, page 148.

Weiss, Rick: "Postponing Red-Cell Retirement: Can Aging Blood Cells Get a New Lease on Life?" *Science News*, December 23, 1989, pages 424–425.

Zucker, Marjorie B.: "The Functioning of Blood Platelets," *Scientific American*, June 1980, pages 86–103.

* Available in paperback.

"Warm-blooded" animals spend more than half of their energy budget maintaining their temperature at a constant level. The expenditure is higher in cold weather than in warm, and it is higher for small animals—because of the surface-to-volume ratio—than for larger ones. When food is in short supply, as in the winter months, the cost can be prohibitive.

Hibernation (from the Latin for "winter") is an obvious solution. During hibernation, body temperature drops almost to the level of the surroundings; metabolic rate, heart rate, and respiration are all greatly reduced; and the animal appears to be in a deep sleep, unresponsive to stimuli. Many animals hibernate: squirrels, hamsters, dormice, deer mice, hedgehogs, chipmunks, bats. Others become dormant but are not, strictly speaking, hibernators. Bears, for example, den up and sleep during much of the winter, but their body temperature drops only a few degrees, and they show only a moderate drop in metabolic rate. Polar bears, however, remain active all winter. They are able to do so because their food supply—seals—does not decrease in the winter. Hibernation and dormancy are adaptations not to the cold and the dark but to a seasonal food supply.

Animals prepare for hibernation or dormancy in two ways: by storing food and by storing fat. Some—chipmunks, for example—do both. Stored food is often not eaten until wake-up time in the spring. As our mothers told us, there's nothing like a good breakfast.

Emerging from a long winter's sleep requires careful timing. Early risers face the perils of late winter storms and continued food shortages. Late risers, however, run the risk of not having enough time to get themselves and their offspring ready for the following winter. Bears solve this problem by mating prior to dormancy, and their cubs are born during the winter. Thus, a female bear not only goes without eating all winter, living on her own stored fat, but also must produce milk for twins or triplets. The cubs are very small and nearly naked when they are born, but by springtime, they are fat, furry, and sturdy, more than ready to emerge with their mother. And, by then, it would be safe to say, she herself is "hungry as a bear."

Water Balance and Temperature Regulation

As we noted in Chapter 28, one of the advantages of multicellularity is the increased capacity for homeostasis—the maintenance of a tightly controlled range of internal conditions in which the cells can live and function. In animals, a great variety of activities contribute to homeostasis. Among the specific examples we have seen in previous chapters are the regulation of blood sugar levels, the uptake and distribution of oxygen to the cells, and the elimination of carbon dioxide from the body. In this chapter, we shall examine two particularly noteworthy homeostatic functions: regulation of the chemical composition of body fluids and regulation of body temperature.

Regulation of the Chemical Environment

Animals are about 70 percent water. Approximately two-thirds of this water is within the cells, and one-third is in the extracellular fluid that surrounds, bathes, and nourishes the cells. Thus the extracellular fluid serves the same purpose for the cells of an animal's body that the Precambrian seas served for the earliest organisms. As animals became multicellular in the course of evolution, they began to produce their own extracellular fluid, similar in composition to the salty fluid of the sea. As they did so, they also evolved mechanisms for regulating its composition.

Although the blood plasma constitutes only about 7 percent of total body fluids, the regulation of its composition is a key factor in regulation of the chemical environment throughout the vertebrate body. As we saw in the last chapter, blood is the supply line for chemicals taken up by the individual cells, and it carries away the wastes released by these cells. Blood can function as an efficient supply and sanitation medium only because cellular wastes are constantly removed from it. This is a very selective process of monitoring, analysis, selection, and rejection.

In many invertebrates and all vertebrates, the composition of the blood—and thus the internal chemical environment—is, to a large extent, regulated by a special **excretory system.** In terrestrial vertebrates, the most important components of this system are the **kidneys.** Although other organs, especially the liver, play important roles in regulating the chemical environment, it is possible to relate major advances in vertebrate evolution—particularly the transition to land—to increasing efficiency of kidney function.

Substances Regulated by the Kidneys

Regulation of the internal chemical environment of an animal involves solving three different—yet interwoven—problems: (1) excretion of metabolic wastes, (2) regulation of the concentrations of ions and other chemicals, and (3) maintenance of water balance.

The chief metabolic waste products that cells release into the bloodstream are carbon dioxide and nitrogenous compounds, mostly ammonia (NH_3), produced by the breakdown of amino acids. As we have seen, carbon dioxide diffuses out of the body across the respiratory surfaces, which may be skin, gills, or lungs. In simple aquatic invertebrates and most fish, ammonia also leaves the body by diffusion into the surrounding water. Ammonia, however, is highly toxic, even in low concentrations. In more complex aquatic animals—and in all terrestrial animals—rapid diffusion of ammonia from the cells into an external water supply is not possible. Thus it must be converted to nontoxic substances, which can be safely transported within the body to the excretory organs.

All birds, terrestrial reptiles, and insects convert their nitrogenous wastes into crystalline **uric acid,** with the result that very little water is required for their excretion. In birds, the uric acid is mixed with undigested wastes in the cloaca (the common exit chamber for the digestive, urinary, and reproductive tracts), and the combination is dropped as a semisolid paste, familiar to all who frequent public parks and admire outdoor statuary.

In sharks, amphibians, and mammals, the ammonia resulting from the processing of nitrogenous wastes is quickly converted in the liver to **urea,** which diffuses into the bloodstream. This relatively nontoxic compound is then carried to the kidneys. Unlike uric acid, however, urea must be dissolved in water for excretion.

Although the kidneys have an excretory function, they are more accurately regarded as regulatory organs. Chemical regulation involves not only the retention of nutrient molecules such as glucose and amino acids but also the maintenance of closely controlled concentra-

Table 32–1   Excreted Nitrogenous Compounds

Compound	Excreted by	Water Required for Excretion
Ammonia	Simple aquatic invertebrates, most fish	Copious amounts
Urea	Sharks and some other saltwater fish, amphibians, mammals	Moderate amounts
Uric acid	Insects, terrestrial reptiles, birds	Slight amount

tions of ions. Such ions as Na^+, K^+, H^+, Mg^{2+}, Ca^{2+}, and HCO_3^- play vital roles in the maintenance of protein structure, membrane permeability, and blood pH, in the propagation of nerve impulses, and in muscle contraction.

The concentration of a particular substance in the body depends not only on the absolute amount of the substance but also on the amount of water in which it is dissolved. Thus, regulation of the water content of body fluids is an important aspect of regulation of the chemical environment. The problem of water balance is such a universal one, biologically speaking, and is so important to the survival of the organism that we shall consider it in more detail before examining the structure and function of the kidney.

Water Balance: An Evolutionary Perspective

Like many small marine organisms today, the earliest organisms probably had a salt and mineral composition much like that of the seas in which they lived. They were probably also isotonic with their surroundings; that is, the organisms and their environment each had the same total effective concentration of dissolved substances. As a consequence, water did not tend to move either into or out of these organisms by osmosis.

When organisms moved to fresh water (a less con-

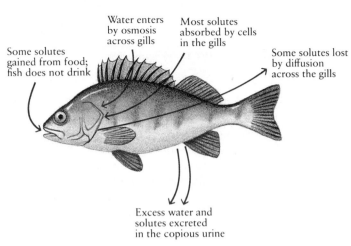

Some solutes gained from food; fish does not drink

Water enters by osmosis across gills

Most solutes absorbed by cells in the gills

Some solutes lost by diffusion across the gills

Excess water and solutes excreted in the copious urine

(a) Bony freshwater fish

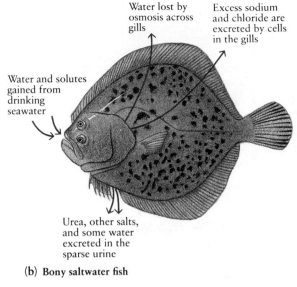

Water lost by osmosis across gills

Excess sodium and chloride are excreted by cells in the gills

Water and solutes gained from drinking seawater

Urea, other salts, and some water excreted in the sparse urine

(b) Bony saltwater fish

32–1 *Pathways by which water and solutes are gained and lost in bony fish. (a) In freshwater fish, the body fluids are more concentrated than the surrounding environment. Thus, water enters the body by osmosis across the gill epithelium. Excess water is removed from the blood by the kidneys and excreted in the urine, which is much more dilute than the body fluids. Although the kidneys reabsorb the bulk of the essential solutes, some are nonetheless lost in the urine,*

and others leave the body by diffusion across the gills. These solutes are replaced principally by the action of specialized salt-absorbing cells in the gills, and, to a lesser extent, from the diet.

(b) In bony saltwater fish, the body fluids are less concentrated than the surrounding environment. Thus, water leaves the body by osmosis across the gills. Water is also lost in the urine, where it is required to dissolve the urea removed from the blood by the kidneys.

(Although most of the nitrogenous wastes are excreted in the form of ammonia, small amounts are converted to urea by the liver.) The fish maintains its internal fluid levels by drinking sea water, which entails the ingestion of solutes. Excess sodium and chloride ions are removed from the blood and excreted by specialized cells in the gills, while other ions are removed by the kidneys and excreted in the urine.

centrated, or hypotonic, environment), water tended to move into their bodies by osmosis. Their survival depended on the development of systems for "bailing themselves out." The contractile vacuole of *Paramecium* is an example of such a bailing device (see page 115).

If the earliest vertebrates—the fish—evolved in fresh water, as is generally believed, the first function of the kidneys was probably to pump out excess water and to conserve salt and other desirable solutes, such as glucose. In freshwater fish today, the kidney works in just this way, primarily as a filter and reabsorber of solutes. The urine of these fish has a solute concentration lower than that of body fluids—that is, it is hypotonic. Some solutes are, however, inevitably lost both in the urine and by diffusion through the gills. This loss is counteracted by salt-absorbing cells in the gills, which actively transport salt back into the body (Figure 32–1a).

When fish moved to the seas, they faced a different problem: the potential loss of water to their environment, principally by osmosis across the respiratory surfaces of the gills. The hagfish, an ancient group of jawless cartilaginous fish, have solved this problem by

maintaining body fluids about as salty as the surrounding ocean waters. The body fluids of another ancient group of cartilaginous fish, the sharks, are either isotonic with or slightly hypertonic to (more concentrated than) sea water. Their isotonicity, however, is achieved in a different way. In the course of evolution, sharks developed an unusual tolerance for urea, so instead of constantly excreting it, they retain a high concentration of it in their blood.

The bony fish, which spread to the sea much later than the cartilaginous fish, have body fluids that are hypotonic to the marine environment, with a solute concentration only about one-third that of sea water. Thus they are constantly losing water to their environment by osmosis. If this water were not replaced, the solutes in their body fluids would become so concentrated that the cells would die. To compensate for their osmotic water loss, bony marine fish drink sea water. This restores their water content but leads to a new problem—how to eliminate the excess salt ingested. This problem has been solved by the evolution of special gland cells in the gills that excrete excess salt. Hence bony marine fish can drink freely from the water sur-

32–2   *Some marine animals, such as sea turtles, have special glands in their heads that can excrete sodium chloride at a concentration about twice that of sea water. Since ancient times, turtle watchers have reported that these great armored reptiles come ashore, with tears in their eyes, to lay their eggs. It is only recently that biologists have learned that this is not caused by an excess of sentiment—as is the case with Lewis Carroll's mock turtle—but is, rather, a useful solution to the problem of excess salt from ingestion of sea water. Marine birds similarly excrete a salty fluid through their nostrils.*

rounding them and still remain hypotonic to it (Figure 32–1b).

Sources of Water Gain and Loss in Terrestrial Animals

Since terrestrial animals do not always have ready access to water, they must regulate water content in other ways, balancing gains and losses. They gain water by drinking fluids, by eating water-containing foods, and as an end product of the oxidative processes that take place in the mitochondria, as we saw in Chapter 8. When 1 gram of glucose is oxidized, 0.6 gram of water is formed. When 1 gram of protein is oxidized, only about 0.3 gram of water is produced. Oxidation of 1 gram of fat, however, yields 1.1 grams of water because of the high hydrogen content of fat.

Some animals can derive all their water from food and oxidation of nutrient molecules and therefore do not require fluids. The kangaroo rat of the North American desert, for example, can live its entire existence without drinking water if it eats the right type of food. It is not surprising that the kangaroo rat selects a diet of fatty seeds, which yield a large amount of water on oxidation. If it is fed high-protein seeds, such as soybeans—the oxidation of which produces a large amount of nitrogenous waste and a relatively small amount of water—the kangaroo rat will die of dehydration unless some other source of water is available.

On average, a human takes in about 2,200 milliliters of water a day in food and drink and gains an additional 350 milliliters a day from the oxidation of nutrient molecules (Figure 32–3). Water is lost from the lungs in the form of moist exhaled air, is eliminated in the feces, is lost by evaporation from the skin, is secreted in sweat, and is removed from the blood and excreted in urine. The latter is usually the major route of water loss.

In a normal adult, the rate of water excretion in urine averages 1,500 milliliters a day. Although the actual amount of urine produced may vary from 400 to 2,500

Water gain

1. Drinking liquids (1,200 ml)
2. Eating food (1,000 ml)
3. Oxidation of nutrient molecules (350 ml)

Total: 2,550 ml

Water loss

1. Excretion of urine (1,500 ml)
2. Evaporation from skin and lungs (900 ml)
3. Elimination in feces (100 ml)
4. Secretion of sweat (50 ml)

Total: 2,550 ml

32–3   *A terrestrial animal is in water balance when the total amount of water lost in expired air, in evaporation from the skin, in sweat, and in the urine and feces equals the total amount of water gained by the intake of food and fluids and by the oxidation of food molecules.*

milliliters per day, there is a variation of less than 1 percent in the fluid content of the body. A minimum output of about 500 milliliters of water is necessary for health, since this much water is needed to remove potentially toxic waste products.

The Kidney

In vertebrates, the complex functions involved in regulating the chemical composition of body fluids are performed chiefly by the kidney. The two human kidneys (Figure 32–4) are dark red, bean-shaped organs about 10 centimeters long that lie at the back of the body, behind the stomach and liver.

The functional unit of the kidney is the **nephron.** It consists of a cluster of capillaries known as the **glomerulus** and a long narrow tube, the **renal tubule,** which originates as a bulb called **Bowman's capsule.** The renal tubule is made up of the **proximal** (near) and **distal** (far) **convoluted tubules,** which in humans and other mammals are connected by the **loop of Henle.** The nephron ends as the straight **collecting duct.** Each of the two human kidneys contains about a million nephrons with a total length of some 80 kilometers (50 miles) in an adult.

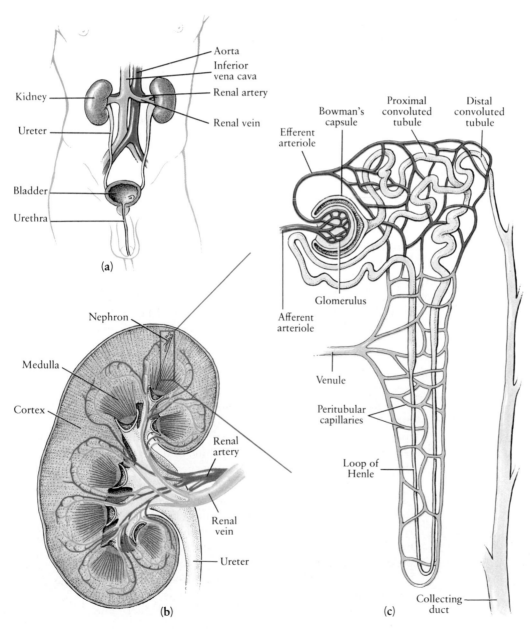

(a)

(b)

(c)

32–4   *The human excretory system.*
(a) *The kidneys regulate the chemical composition of the blood, which is carried to and from the kidneys by way of the renal arteries and veins. Waste products and water—the urine—pass from the kidneys through a pair of tubes, the ureters, to the bladder, where they are stored. Urine leaves the body by way of the urethra. In the male mammal, the urethra also serves as the passageway for semen.*

(b) *In longitudinal section, the human kidney is seen to be made up of two regions. The outer region, the cortex, contains the fluid-filtering mechanisms. The inner region, the medulla, is traversed by long loops of the renal tubules and by the collecting ducts carrying the urine. These ducts merge and empty into the funnel-shaped renal pelvis, which, in turn, empties into the ureter.*

(c) *The nephron is the functional unit of the kidney. Blood enters the nephron through the afferent arteriole leading into the glomerulus. Fluid is forced out by the pressure of the blood through the thin capillary walls of the glomerulus into Bowman's capsule. The capsule connects with the long renal tubule, which has three regions: the proximal convoluted tubule; the loop of Henle, which extends into the medulla; and the distal convoluted tubule. As the fluid travels through the tubule, almost all the water, ions, and other useful substances are reabsorbed into the bloodstream through the peritubular capillaries. Other substances are secreted from the capillaries into the tubules. Waste materials and some water pass along the entire length of the tubule into the collecting duct and are excreted from the body as urine.*

Urine is formed in the nephrons and passed from the collecting ducts into the **renal pelvis,** which is, in essence, a funnel. From this funnel, the urine is moved by peristalsis through the **ureter** to the **bladder,** which stores the urine until it is passed out of the body through the **urethra.**

Function of the Kidney

Blood enters the kidney through the renal artery, which divides into progressively smaller arteries, leading ultimately to the arterioles, each of which feeds into a glomerulus. Unlike most other capillary beds, a glomerulus lies between two arterioles—the one leading in is the **afferent arteriole,** and the one leading out is the **efferent arteriole.** The efferent arteriole then divides again into capillaries, the **peritubular** ("around-the-tubes") **capillaries,** which surround the renal tubule and then merge to form a venule that empties into a small vein, leading ultimately to the renal vein.

Constriction of the afferent and efferent arterioles keeps the blood within the glomerulus at a pressure about twice that in other capillaries. As a consequence,

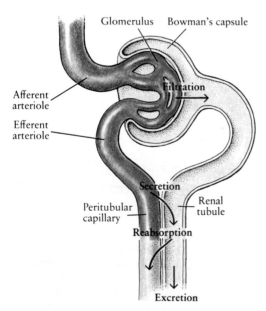

32–5 *The four basic components of kidney function: (1) filtration through the glomerular capillaries into Bowman's capsule; (2) secretion from the peritubular capillary into the renal tubule; (3) reabsorption from the renal tubule to the peritubular capillary; (4) excretion. The pressure in the glomerular capillaries—and therefore the rate of filtration—is regulated by constriction of the afferent and efferent arterioles.*

about one-fifth of the blood plasma that enters the kidney is forced through the walls of the glomerular capillaries and through the wall of Bowman's capsule into the renal tubule. This crucial first process in the formation of urine is called **filtration,** and the fluid entering the capsule is the **filtrate.** Except for the absence of large molecules, such as proteins, which cannot cross the capillary wall, the filtrate has the same chemical composition as the plasma.

The filtrate then begins its long passage through the renal tubule, the wall of which is made up of a single layer of epithelial cells specialized for active transport. In a second process in the formation of urine, **secretion,** some of the molecules remaining in the plasma after filtration are selectively removed from the peritubular capillaries and actively secreted into the filtrate. Penicillin, for example, is removed from the circulation in this way.

A third major process in urine formation, **reabsorption,** occurs simultaneously. Most of the water and solutes that initially entered the tubule during filtration are transported back into the peritubular capillaries. For example, glucose, most amino acids, and most vitamins are returned to the bloodstream.

Finally, the remaining fluid—now the urine—leaves the nephron and passes into the renal pelvis. This process is **excretion.**

Water Conservation: The Loop of Henle

The control of urinary water loss is a major mechanism by which body water is regulated. Animals with free access to fresh water typically excrete a copious urine that is less concentrated than their blood. The daily urinary output of a frog, for instance, totals 25 percent of its body weight. Most terrestrial animals cannot afford to be so profligate with water, however. In response to evolutionary pressures, birds and mammals have developed the ability to excrete urines that are more concentrated than their body fluids.

In humans and other mammals, the formation of a more concentrated (hypertonic) urine depends upon three structural features of the nephron. First, different regions of the wall of the renal tubule and the wall of the collecting duct have differing permeabilities to water, salt, and urea. Second, certain regions of the wall of the renal tubule contain membrane proteins that actively transport salt out of the tubule. Third, the pathway formed by the long renal tubule and the collecting duct is a looping one. From Bowman's capsule, the filtrate enters the proximal convoluted tubule, descends the hairpin-shaped loop of Henle, ascends the loop, passes through the distal convoluted tubule, and finally descends once more through the collecting duct. The

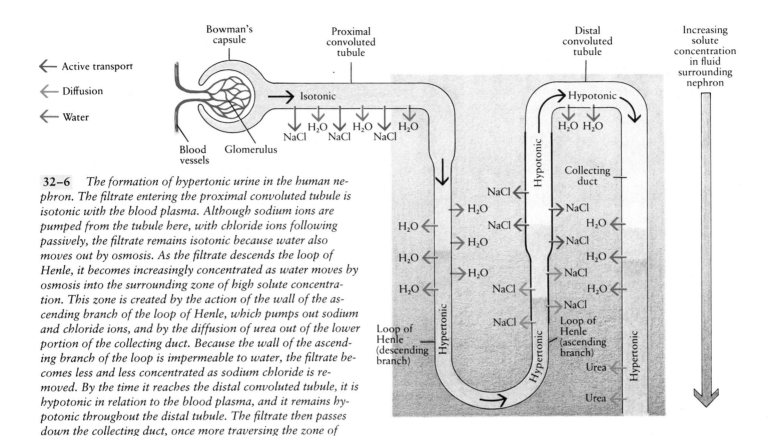

32–6 *The formation of hypertonic urine in the human nephron. The filtrate entering the proximal convoluted tubule is isotonic with the blood plasma. Although sodium ions are pumped from the tubule here, with chloride ions following passively, the filtrate remains isotonic because water also moves out by osmosis. As the filtrate descends the loop of Henle, it becomes increasingly concentrated as water moves by osmosis into the surrounding zone of high solute concentration. This zone is created by the action of the wall of the ascending branch of the loop of Henle, which pumps out sodium and chloride ions, and by the diffusion of urea out of the lower portion of the collecting duct. Because the wall of the ascending branch of the loop is impermeable to water, the filtrate becomes less and less concentrated as sodium chloride is removed. By the time it reaches the distal convoluted tubule, it is hypotonic in relation to the blood plasma, and it remains hypotonic throughout the distal tubule. The filtrate then passes down the collecting duct, once more traversing the zone of high solute concentration.*

From this point onward, the urine concentration depends on antidiuretic hormone (ADH). If ADH is absent, the wall of the collecting duct is not permeable to water, no additional water is removed, and a less concentrated urine is excreted. If ADH is present, the cells of the collecting duct are permeable to water, which moves by osmosis into the surrounding fluid, as shown in the diagram. In this case a concentrated (hypertonic) urine is passed down the duct to the renal pelvis, the ureter, the bladder, and finally out the urethra.

combined effect of these features is that the solute concentrations in the fluid surrounding different portions of the renal tubule and the collecting duct are not uniform. Instead, the solute concentration of the surrounding fluid increases steadily from the top of the loop of Henle to the bottom.

With this key fact in mind, let us follow the filtrate as it moves through the nephron (Figure 32–6). The fluid entering the proximal tubule from Bowman's capsule is isotonic with the blood plasma; that is, it has the

same solute concentration as does the plasma. In the proximal tubule, sodium ions are pumped out, with chloride ions following passively. Water also moves out of the tubule by osmosis, following after these ions. Thus, as the fluid enters the descending branch of the loop of Henle, it is still isotonic with the blood plasma. Its volume is dramatically reduced, however. Some 60 to 70 percent of the solutes and water contained within the original filtrate are removed from it during the passage through the proximal tubule. This water and the solutes it contains are taken up almost immediately by the peritubular capillaries and returned to the bloodstream.

The pathway of the loop then carries the remaining filtrate through the zone of high solute concentration. This zone is produced by the properties of both the ascending branch of the loop of Henle and the collecting duct. The wall of the ascending branch permits the movement of salt into the surrounding fluid, and the wall of the lower portion of the collecting duct is permeable to urea. (In fact, about half of the urea contained within the fluid passing through the collecting duct diffuses into the surrounding fluid, from which it is ultimately reabsorbed into the peritubular capillaries.)

The wall of the descending branch of the loop of Henle is relatively impermeable to salt but freely permeable to water. Thus, as the fluid flows through the descending branch, traversing the zone of high solute concentration, large quantities of water move by osmosis from the tubule into the surrounding fluid (and from there into the peritubular capillaries). The fluid within the tubule is now hypertonic in relation to the plasma.

As the fluid travels through the ascending branch, some salt diffuses from the lower portion of the tubule and larger quantities are removed by the action of transport proteins in the upper portion. The wall of the ascending branch, however, is impermeable to water, which must remain within the tubule. As a consequence, the fluid within the tubule becomes hypotonic (less concentrated) in relation to the plasma. In humans it remains hypotonic during its passage through the distal convoluted tubule. However, the fluid must again pass through the zone of high solute concentration as it flows down the collecting duct.

Whether additional water is removed during the passage down the collecting duct depends on the presence or absence of a hormone, **antidiuretic hormone (ADH)**. If ADH is absent, the wall of the collecting duct is impermeable to water and a hypotonic urine is excreted. If ADH is present, however, the wall is freely permeable to water. Water moves by osmosis through the wall, leaving within the duct a urine that is isotonic with the surrounding fluids but hypertonic in relation to the body fluids as a whole. In this way, mammals that need to conserve water are able to excrete a urine far more concentrated than the plasma from which it is derived.

The solutes responsible for the concentration differences on which this system depends are salt, pumped from the ascending branch of the loop of Henle, and urea, which diffuses from the lower portion of the collecting duct. Thus it is known as a **two-solute system.** A key factor in its function is the hairpin structure of the loop of Henle. The longer the loop, the greater the concentration differences that can be established. Since the primary factor limiting the concentration of the urine is the solute concentration surrounding the collecting duct, it should come as no surprise to learn that those mammals that excrete the most hypertonic urine also have the longest loops of Henle.

Control of Kidney Function: The Role of Hormones

In mammals, several different hormones act on the nephron to affect the composition of the urine. One of these is ADH, which is formed in the hypothalamus, a major regulatory center in the brain, and is stored in and released from the pituitary gland. As we have seen, ADH acts on the collecting ducts of the nephrons and increases their permeability to water, so more water moves, by diffusion, back into the blood from the nephron.

The amount of ADH released depends on the solute concentration of the blood and also on the blood pressure. Osmotic receptors that monitor the solute content of the blood are located in the hypothalamus. Pressure receptors that detect changes in blood volume are found in the walls of the heart, in the aorta, and in the carotid arteries. Stimuli received by these receptors are transmitted to the hypothalamus.

Factors that increase the concentration of solutes in the blood or decrease blood pressure, or both, increase the production of ADH and thus the conservation of water. Such factors include dehydration and hemorrhage. Pain and emotional stress also trigger ADH secretion and thus decrease urinary flow.

Table 32–2 Hormones that Affect the Nephron		
Hormone	**Source**	**Action**
ADH	Formed in hypothalamus, released from pituitary	Increases permeability of collecting ducts to water, leading to formation of hypertonic urine
Aldosterone	Adrenal glands	Increases reabsorption of sodium ions from and secretion of potassium ions into distal tubule and collecting duct
Cardiac peptide	Atria of heart	Inhibits reabsorption of sodium ions from distal tubule

Factors that decrease the concentration of solutes in the blood, such as the ingestion of large amounts of water, or that increase blood pressure—adrenaline, for instance—signal the hypothalamus to decrease ADH production, and so more water is excreted. Cold stress inhibits ADH secretion and thus increases urinary flow. Alcohol also suppresses ADH secretion and increases urinary flow, a phenomenon familiar to imbibers of beer and other alcoholic beverages.

A second hormone, **aldosterone,** which is produced by the adrenal glands, stimulates reabsorption of sodium ions from the distal tubule and collecting duct and secretion of potassium ions into them. When the adrenal glands are removed, or when they function poorly (as in Addison's disease), excessive amounts of sodium chloride and water are lost in the urine, and the tissues of the body become depleted of them. Generalized weakness results, and if a patient with Addison's disease is not given hormone replacement therapy, the fluid loss can eventually be fatal. Aldosterone production is controlled by a complex feedback circuit involving potassium ion levels in the bloodstream and processes initiated in the kidneys themselves.

Other hormones are also involved in the regulation of kidney function, particularly in response to increases in blood volume or blood pressure. The most intriguing of these substances is cardiac peptide (page 559), which inhibits the reabsorption of sodium from the distal tubule and thus increases the excretion of both sodium and water. This hormone, released from the atria of the heart, apparently exerts its effects both directly on the nephron itself and indirectly by inhibiting the release of aldosterone.

The complexity of the mechanisms regulating excretion provides additional evidence of the critical importance to the organism of maintaining a constant internal environment.

Regulation of Body Temperature

Some 200 years ago, Dr. Charles Blagden, then secretary of the Royal Society of London, went into a room that had been heated to a temperature of 126°C (260°F), taking with him a few friends, a small dog in a basket, and a steak. The entire group remained there for 45 minutes. Dr. Blagden and his friends emerged unaffected. So did the dog (the basket had kept its feet from being burned by the floor). But the steak was cooked.

The control of body temperature is an important factor in the regulation of the internal environment. Physiological processes depend on both the maintenance of

cellular structures and a multitude of biochemical reactions, virtually all controlled by enzymes. The rate at which an enzymatic reaction occurs is determined by a number of factors, one of the most important of which is temperature. In general, the reaction rate approximately doubles with every 10°C rise in temperature. However, for every protein, including every enzyme, there is a relatively narrow temperature range in which it maintains the three-dimensional shape required for its functions. Above the upper limit of this range, the protein loses its shape. Once this occurs, enzymes—and other proteins whose function depends on a specific shape—are inactivated. Low temperatures can also bring physiological processes to a halt. As a consequence of the effects of both high and low temperatures, most animals must either occupy environments in which the temperature ranges from just below freezing to between 45°C and 50°C, or through their own physiological processes create an internal environment within that range.

32–7   *Temperature regulation involves behavioral responses, as well as physiological and anatomical adaptations. In the cold, many small animals, such as these penguin chicks, huddle together for warmth. Huddling decreases the effective surface-to-volume ratio and thus the loss of heat from the body. Anatomical adaptations include an insulating layer of fat just below the surface of the skin and an external coat of heat-trapping down.*

32–8 *In laboratory studies, the internal temperature of reptiles was shown to be almost the same as the temperature of the surrounding air. It was not until observations were made of these animals in their own environment that it was found that they have behavioral means for temperature regulation that give them a surprising degree of temperature independence. By absorbing solar energy, reptiles can raise their temperature well above that of the air around them. Shown here is a horned lizard (often but not accurately known as a horned toad) that, having been overheated by the sun, has raised its body to allow cooling air currents to circulate across its underside.*

Temperature regulation, like water conservation, became a problem for those animals that, in the course of their evolution, moved from water to land. In general, large bodies of water, for the reasons discussed in Chapter 2, maintain a stable temperature. The animals that live in them, with some exceptions, maintain a body temperature that is the same as the temperature of the water. The exceptions include the aquatic mammals and certain large fish, such as tuna, that retain the heat generated by muscular activities. Because the temperatures of large bodies of water are low (seldom exceeding 5°C), the metabolic rates of most inhabitants are low, and the pace of life in the water is generally slow.

Terrestrial reptiles—snakes, lizards, and tortoises—are able to maintain remarkably stable body temperatures during their active hours by varying the amount of solar radiation they absorb (Figure 32–8). By careful selection of suitable sites, such as the slope of a hill facing the sun, and by orienting their bodies with a maximum surface exposed to sunlight, they can heat themselves rapidly. Such heating can occur as quickly as 1°C per minute, even on mornings when the air temperature is close to 0°C. By frequently changing position, these reptiles are able to keep their temperature within quite a narrow range as long as the sun is shining. When it is not, they seek the safety of their shelters, where they are protected from predators during periods when they are too sluggish to escape. As we noted in Chapter 27, such animals, which obtain their heat from the external environment, are known as ectotherms.

Homeotherms

The so-called "warm-blooded" animals, primarily mammals and birds, are endotherms (page 486). With some exceptions, endotherms are also **homeotherms,** animals that maintain a constant—and usually high—internal body temperature.

The primary source of heat in endotherms is the oxidation of glucose and other energy-yielding molecules within the body cells. Endotherms have a much higher metabolic rate than ectotherms, with a large proportion of their energy budget allocated to heat production. Also, small endotherms have a proportionally larger energy budget than large ones, because of their higher surface-to-volume ratios (page 88). Endotherms are characterized by layers of heat-retaining insulating material and by mechanisms for disposing of excess heat.

The human body maintains a remarkably constant temperature, deviating very little from 37°C. This constancy of temperature is maintained by an automatic system—a thermostat—in the hypothalamus. This thermostat receives and integrates information from widely scattered temperature receptors, compares it to the set point of the thermostat, and, on the basis of this comparison, initiates appropriate responses. Unlike a furnace thermostat, which controls a simple on-off switch, the hypothalamic thermostat has a variety of responses at its command, as summarized in Figure 32–9.

Under ordinary conditions, the skin receptors for hot and cold are probably the most important sources of information about temperature change. However, the hypothalamus itself contains receptor cells that monitor the temperature of the blood flowing through it. Information received by these hypothalamic receptors can override that from other sources, ensuring that the core body temperature remains constant.

The elevation of body temperature known as fever is due not to a malfunction of the thermostat but to a resetting. Thus, at the onset of fever, an individual typically feels cold and often has chills. Although the body temperature is rising, it is still lower than the new thermostat setting. The substance primarily responsible for the resetting of the thermostat is a protein released by white blood cells in response to pathogens. The adaptive value of fever is not known with certainty. However, current evidence indicates that moderate temperature increases stimulate the immune system, increasing the synthesis of antibodies and the production of other substances and specialized cells that work together in the body's defenses against infection.

Regulating as Body Temperature Rises

As the body temperature of a mammal rises above its thermostat setting, the blood vessels near the skin surface dilate, and the supply of blood to the skin increases. If the air is cooler than the body surface, heat can be transferred from the skin directly to the air. Animals that live in hot climates characteristically have larger exposed surface areas than animals that live in the cold (Figure 32–10).

Heat can also be lost from the surface by the evaporation of perspiration or saliva, as we saw in Chapter 2 (page 41). The heat required for evaporation comes from blood traveling in vessels just below the skin surface. In humans and most other large mammals, when the external temperature rises above body temperature,

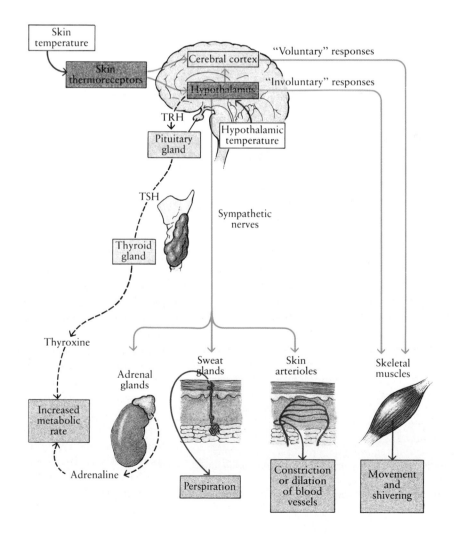

32–9 *Body temperature in mammals is regulated by a complex network of activities involving both the nervous and endocrine systems. The chief regulatory center is the hypothalamus, a brain center that controls many physiological processes. The hypothalamus receives information from thermoreceptors in the skin and in certain internal locations, including the hypothalamus itself. In this diagram, the solid arrows represent neural pathways and the dashed arrows represent hormonal pathways. In humans, the hormonal pathways play only a minor role in temperature regulation. In some animals, however, they are of major importance. TRH is thyrotropin-releasing hormone, which is produced by the hypothalamus and which stimulates the production of TSH, thyroid-stimulating hormone, by the pituitary. TSH, in turn, stimulates the production of thyroxine, the thyroid hormone. Thyroxine increases cellular metabolism, apparently by acting directly on the mitochondria. Many behavioral responses, such as seeking sun or shelter, are also involved in temperature regulation.*

(a)

(b)

(c)

32–10   *The size of the extremities in a particular type of animal can often be correlated with the climate in which it lives.* **(a)** *The fennec fox (Fennecus zerda) of the North African desert has large ears, rich with blood vessels. As blood flows through* the network of capillaries just below the skin surface, excess heat is dissipated from the body. **(b)** *The red fox (Vulpes fulva) of the eastern United States has ears of intermediate size, and* **(c)** *the Arctic fox (Alopex lagopus) has relatively small ears. As we saw in* Chapter 20 (page 345), similar correlations of characteristics—such as body size, weight, and color—with environment can sometimes be made among animals of a single species living over an extended geographic range.

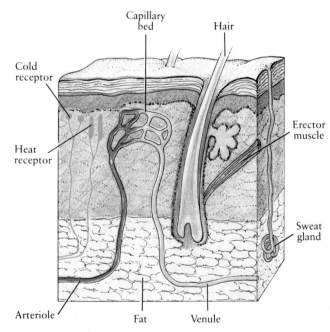

32–11   *Cross section of human skin showing structures involved in temperature regulation. In cold, the arterioles constrict, reducing the flow of blood through the capillaries, and the hairs, each of which has a small erector muscle under neural control, stand upright. Beneath the skin is a layer of fat that serves as insulation, retaining the heat in the underlying body tissues. With rising temperatures, the arterioles dilate and the sweat glands secrete a salty liquid. Evaporation of this liquid cools the skin surface, dissipating heat (approximately 540 calories for every gram of H_2O).*

perspiration begins. Horses and humans sweat from all over their body surfaces. Dogs and some other animals pant, so that air passes rapidly over their large, moist tongues, evaporating saliva. For animals that dissipate heat through evaporation, temperature regulation at high temperatures necessarily involves water loss. This, in turn, stimulates thirst and water conservation by the kidneys. Dr. Blagden and his friends were probably very thirsty.

Regulating as Body Temperature Falls

When the temperature of the circulating blood begins to fall below the thermostat setting, blood vessels near the skin surface constrict, limiting heat loss from the skin (Figure 32–11). Metabolic processes increase. Part of this increase is due to increased muscular activity, either voluntary (shifting from foot to foot) or involuntary (shivering). Part is due to direct stimulation of metabolism by the endocrine and nervous systems. Adrenaline stimulates the release and oxidation of glucose. Autonomic nerves to fat tissue increase the metabolic breakdown of fats. In some mammals exposed to prolonged cold, the thyroid gland increases its release of thyroxine, the thyroid hormone. Thyroxine appears to exert its effects directly on the mitochondria.

Most mammals have a layer of fat just below the surface of the skin that serves as insulation. Homeotherms are also typically covered with hair or feathers. As the temperature falls, these structures are pulled upright by erector muscles in the skin, trapping air that insulates the surface. All we get, as our evolutionary legacy, are goose bumps.

Capillary bed

Hair

Cold receptor

Heat receptor

Erector muscle

Sweat gland

Arteriole

Fat

Venule

Adaptations of Desert Animals

We have seen in the course of this chapter, first, that water balance is essential for the smooth functioning of the internal environment and, second, that water evaporation is a major mechanism for dissipating heat. How do animals in the hot, dry environment of the desert resolve these two conflicting demands?

Let us consider the camel, the philosophical-looking "ship of the desert." A camel has several advantages over human desert dwellers. For one thing, a camel excretes a much more concentrated urine; that is, it does not need to use as much water to dissolve its waste products. Also, a camel can lose more water proportionally than a human and still continue to function. A person who loses 10 percent of his or her body weight in water becomes delirious, deaf, and insensitive to pain. If the loss is as much as 12 percent, the individual is unable to swallow and so cannot recover without medical assistance. Laboratory rats and many other common animals can tolerate dehydration of up to 12 to 14 percent of their body weight. Camels can tolerate the loss of more than 25 percent of their body weight in water, going without drinking for one week in the summer months, three weeks in the winter.

Perhaps most important, the camel can tolerate a fluctuation in internal temperature of as much as 6°C. This tolerance means that its temperature can rise during the day (which the human thermostat would never permit) and drop during the night. The camel begins the next day at below its normal temperature—having, in effect, stored up coolness. It is estimated that the camel saves as much as 5 liters of water a day as a result of these internal temperature fluctuations.

Camels' humps were once thought to be water-storage tanks, but actually they are localized fat depos-

32–12 *The kangaroo rat, a common inhabitant of the North American desert, may spend its entire life without drinking water. The composition of its body fluids, however, remains essentially constant. A number of physiological adaptations and mechanisms contribute to this remarkable feat.*

its. Physiologists have suggested that the camel carries its fat in a dorsal hump, instead of distributed all over the body, because the hump, acting as an insulator, impedes the flow of heat from the sun into the body core.

Unlike the camel, most small desert animals are nocturnal. The avoidance of direct heat is their principal means of temperature regulation. They usually do not unload heat by sweating or panting, which, because of their relatively large surface areas, would be extravagant in terms of water loss. Small desert animals tend to be highly conservative in their water expenditures. For example, the kangaroo rat (Figure 32–12) has no sweat glands, its feces have a very low water content, and its urine is highly concentrated. Its major water loss is through respiration, and even this loss is reduced by

32–13 *Camels' bodies are insulated by fat on top, which minimizes heat gain by radiation. The underparts of their bodies and their legs, which have much less insulation, radiate heat out to the ground. Other adaptations of camels to desert life include long eyelashes, which protect their eyes from the stinging sand, and flattened nostrils, which retard water loss. These camels were photographed in the Judean desert, where they were grazing on vegetation produced by the spring rains.*

the animal's long nose in which some cooling of the expired air takes place, with condensation of water from it.

As the camel and the kangaroo rat remind us, physiological problems, even those that may seem quite distinct at first glance, are always interrelated. A successful solution to any specific physiological problem invariably involves the interactions of different systems, balancing the various demands on and needs of the animal as a whole.

Summary

Homeostasis—the maintenance of a constant internal environment—is the result of a variety of processes within the animal body. Two of the most critical homeostatic functions are regulation of the chemical composition of body fluids and regulation of body temperature.

The excretory system enables an animal to regulate its internal chemical environment. This function, which in vertebrates is carried out primarily by the kidneys, involves (1) excretion of toxic waste products, especially the nitrogenous compounds produced by the breakdown of amino acids; (2) control of the levels of ions and other solutes in body fluids; and (3) maintenance of water balance.

Animals living in saltwater, freshwater, and terrestrial environments face different problems in maintaining the composition of body fluids. Terrestrial animals generally need to conserve water. Reptiles and birds excrete nitrogenous wastes in the form of crystalline uric acid, whereas mammals excrete them as urea, which must be dissolved in water.

Maintaining water balance involves equalizing gains and losses. Although the amount of water taken in and given off may vary widely from animal to animal and also from time to time in the same animal, depending largely on environmental circumstances, the volume of water in the body remains very nearly constant.

The functional unit of the kidney is the nephron. Each nephron consists of a long tubule attached to a bulb (Bowman's capsule), which encloses a twisted cluster of capillaries, the glomerulus. Blood entering the glomerulus is under sufficient pressure to force plasma through the capillary walls into Bowman's capsule. As the filtrate makes its long passage through the nephron, cells of the renal tubule selectively reabsorb molecules from the filtrate and secrete other molecules into it. Glucose, amino acids, most ions, and a large amount of water are returned to the blood through the peritubular capillaries. Excess water and waste products are excreted from the body as urine. The loop of Henle is the portion of the mammalian nephron that makes possible the excretion of a urine that is more concentrated than the body fluids.

The function of the nephron is influenced by hormones, chiefly antidiuretic hormone (ADH), aldosterone, and cardiac peptide. ADH increases the return of water to the blood and so decreases water excretion. Aldosterone increases the reabsorption of sodium ions and the secretion of potassium ions. Cardiac peptide, by contrast, inhibits the reabsorption of sodium ions and water. All of these hormones play a role in the regulation of both blood pressure and blood volume.

Life can exist only within a very narrow temperature range, from about 0°C to about 50°C. Animals must either seek out environments with suitable temperatures or create suitable internal environments. Water-dwelling animals usually maintain a body temperature that is the same as the relatively constant temperature of the water. On land, environmental temperatures are much more variable. Ectotherms, such as terrestrial reptiles, regulate their body temperatures by adjusting the amount of heat taken in from the environment. Endotherms, principally birds and mammals, generate heat internally from the oxidation of glucose and other fuel molecules. Most endotherms are also homeotherms, maintaining a constant internal body temperature. Homeotherms in cold climates are usually insulated by fat and fur or feathers.

In mammals, temperature is regulated by a thermostat in the hypothalamus. This thermostat receives and integrates information from temperature-sensitive receptors in the skin and in the hypothalamus itself and then triggers appropriate responses. As body temperature rises, blood vessels in the skin dilate, increasing the blood flow to the surface of the body. Evaporation of water from body surfaces increases heat loss. As body temperature falls, blood vessels in the skin constrict, reducing heat loss. Energy production is increased by muscular activity and by neural and hormonal stimulation of metabolism.

Desert animals have a number of solutions to the dual problem of temperature regulation and water conservation. These include the capacity to excrete a very concentrated urine, a comparatively wide tolerance for fluctuations in temperature and total water, and a variety of behavioral adaptations.

Questions

1. Distinguish among the following: intracellular fluid/extracellular fluid; Bowman's capsule/glomerulus; ADH/aldosterone/cardiac peptide; ectotherm/endotherm/homeotherm.

2. Explain the following terms in relation to kidney function: filtration, secretion, reabsorption, and excretion.

3. Sketch a nephron, indicating the pathways of the blood and the glomerular filtrate.

4. Describe the paths of glucose and urea through the human kidney.

5. Why does a high-protein diet require an increased intake of water? (You should be able to think of two different reasons.) Why does a person lose some weight after shifting to a low-salt diet, even without reducing caloric intake? Given the fact that amino acids in excess of the body's requirements are broken down by the liver, not stored, what is the advantage of a high-protein diet? What might be a disadvantage?

6. The primary factor limiting the concentration of the urine is the solute concentration surrounding the collecting duct. Why is this so?

7. In Figure 32–6, the labels within the tubule indicate the concentration of solutes in the filtrate in relation to their concentration in the blood plasma. From the information given about the movement of ions and of water, determine, for each labeled location, the concentration of the filtrate in relation to the concentration of the surrounding fluids.

8. Viewed as a whole, the kidney selectively removes substances from the bloodstream. A closer look, however, reveals that it does so by filtering out all small molecules through the glomerulus and then, as the filtrate passes along the renal tubule, reabsorbing those that are needed. Thus, the system need not identify wastes as such; rather, it must identify useful substances. Why would such an arrangement have been beneficial to early mammals? Why is it especially beneficial to modern mammals, including ourselves?

9. Could a human being on a life raft survive by drinking sea water? By catching and eating bony saltwater fish? Explain your answers.

10. Describe what happens in the human body as its temperature rises. As it falls.

11. In terms of temperature regulation, what is the advantage of having temperature monitors that are located externally, as on the outside of a building or in the skin? Why is it also important to monitor the internal temperature? Which should be the principal source of information for your hypothalamic thermostat?

12. Compare the surface-to-volume ratio of an Eskimo igloo with that of a California ranch house. In what way is the igloo well suited to the environment in which it is found?

Suggestions for Further Reading

Crawshaw, Larry I., Brenda P. Moffitt, Daniel E. Lemons, and John A. Downey: "The Evolutionary Development of Vertebrate Thermoregulation," *American Scientist,* vol. 69, pages 543–550, 1981.

French, Alan R.: "The Patterns of Mammalian Hibernation," *American Scientist,* vol. 76, pages 568–575, 1988.

Heller, H. Craig, Larry I. Crawshaw, and Harold T. Hammel: "The Thermostat of Vertebrate Animals," *Scientific American,* August 1978, pages 102–113.

Schmidt-Nielsen, Knut: "Countercurrent Systems in Animals," *Scientific American,* May 1981, pages 118–128.

Schmidt-Nielsen, Knut: *Desert Animals,* Oxford University Press, New York, 1964.

   Although considered the definitive work on the physio-logical problems relating to heat and water, this readable book also contains numerous anecdotes—such as that about Dr. Blagden—and many fascinating personal observations.

Smith, Homer W.: *From Fish to Philosopher,* Doubleday & Company, Inc., Garden City, N.Y., 1959.*

   Smith was an eminent specialist in the physiology of the kidney. Writing for the general public, he explains the role of this remarkable organ in the story of how, in the course of evolution, organisms have increasingly freed themselves from their environments.

Stricker, Edward M., and Joseph G. Verbalis: "Hormones and Behavior: The Biology of Thirst and Sodium Appetite," *American Scientist,* vol. 76, pages 261–267, 1988.

*Available in paperback.

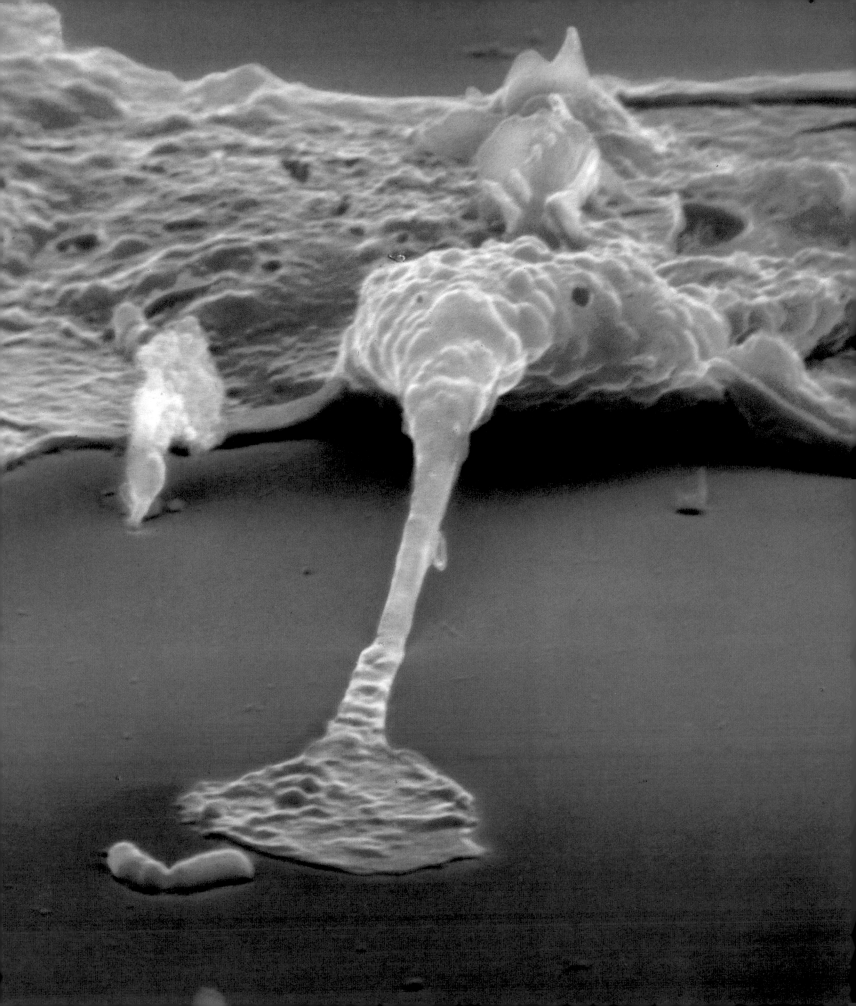

Your body is well defended. Its first line of defense is its outer wrapping of skin and inner lining of mucous membranes. Your skin, with its tough outer layer of keratin, is an impregnable barrier as long as it is intact.

Your mucous membranes lack a tough outer layer, but they have other defenses. Your tears, for example, contain lysozyme, an enzyme that destroys the cell walls of many bacteria. The cells of your respiratory tract are armed with cilia, which sweep away microorganisms, dust, soot, and debris. Your saliva, like your tears, contains lysozyme, as well as other destructive enzymes—the same enzymes that start the digestion of your food. Any intruders that accompany the food into your stomach are likely to be destroyed by the extreme acidity of your gastric juices. Survivors that find their way into the intestines encounter a host of resident bacteria, notably *E. coli,* ready to defend their territory against interlopers.

Some pathogens, however, do penetrate these defenses, through rips and punctures in the skin or mucous membranes; with the help of mosquitoes, ticks, and other blood-sucking arthropods; or by way of various disguises and other fiendish strategies. The microbes, remember, are also fighting for their lives.

Once inside, the invaders meet another army, composed of literally millions of white blood cells. Among the first members of this army to arrive at the scene when the barriers have been breached are macrophages ("big eaters"). This macrophage is extending its pseudopod toward a bacterial cell, just as its long-distant relatives, the amoebas, have been doing for about a billion years.

Your body's army of white blood cells is, like some other armies, elaborate, very complicated, hard to understand in all its details, and expensive—1 out of every 100 cells in your body belongs to this army. Mostly it wins its battles, so silently you don't even know there's a war on. Like other armies, however, it is sometimes overwhelmed and defeated. And, it sometimes makes mistakes, with lapses in military intelligence leading to friendly fire. All this and more in the pages ahead.

33

The Immune Response

Every living thing, including yourself at this moment, is surrounded by potentially harmful microorganisms. Many of these microorganisms have the capacity not only to destroy individual cells but also to disrupt the numerous interrelated processes on which the continuing life of the organism depends. Thus, the defense against such microorganisms is an essential aspect of homeostasis.

Over the long course of evolution, organisms have developed a variety of defenses that function to exclude would-be invaders or to overcome them should they gain entry. In vertebrates, such defenses have evolved into an elaborate interacting network that includes structural fortifications, such as the barrier formed by the skin; nonspecific responses that, like land mines, indiscriminately destroy any intruders that trip them; and the highly specific responses of the immune system that, like elite forces, mount an attack precisely targeted to the characteristics of each different invader.

Nonspecific Defenses

The Inflammatory Response

When microorganisms penetrate the body's first line of defense—its outer wrapping of skin and inner lining of mucous membranes—they encounter a second line of defense. This line of defense consists primarily of phagocytic white blood cells, such as the one on the facing page, carried by the circulating blood and lymph.

Suppose, for example, you cut your skin. Cells in the area immediately release **histamine** and other chemicals that increase both blood flow into the area and the permeability of nearby capillaries. Circulating white blood cells, attracted by these chemicals, move through the capillary walls, crowding into the site of the injury. These cells engulf the foreign invaders, often literally eating themselves to death. Blood clots begin to form, walling off the injured area. The local temperature often rises, creating an environment unfavorable to the multiplication of microorganisms, while increasing the motility of the white blood cells.

As a consequence of this series of events, known as the **inflammatory response** (Figure 33–1), the injured area becomes swollen, hot, red, and painful. Inflammation can also produce systemic effects (that is, effects throughout the body), as well as local effects. One systemic response is fever (page 577), caused by a protein released by certain white blood cells in the course of the inflammatory response.

Both the inflammatory response and the more specific immune response depend on the interaction of a variety of types of white blood cells. These cells, like the oxygen-carrying red blood cells, have a finite life span and must be continuously replenished. All of the different types of white blood cells, as well as the red blood cells, result from the differentiation and division of common, self-regenerating **stem cells** located in the marrow of long bones. A particularly important type of phagocytic white blood cell is the one with which we began this chapter—the **macrophage**. Macrophages are found not only at the site of an infection but also lodged in the lymph nodes, spleen, liver, lungs, and connective tissues, where they entrap any microbes or foreign particles that may have penetrated the initial defenses. These cells are also important, as we shall see, in activating **lymphocytes,** the white blood cells that play the principal roles in the immune response.

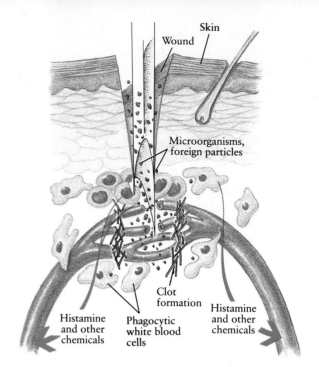

33–1   *The inflammatory response. When the body's defenses are breached, as, for example, by a shard of glass, histamine and other chemicals are released at the site of injury by both body cells and invading bacterial cells. These substances increase blood flow to the area, increase the permeability of the capillaries, and attract circulating white blood cells that migrate through the capillaries into the area. At the same time, clot formation begins, sealing off the injured area. Once the white blood cells arrive at the scene, they extend pseudopodia that engulf microorganisms and other foreign particles. Some of the white blood cells contain granules that rupture readily, releasing additional histamine that enhances the inflammatory response. Others release a protein that triggers a resetting of the body's thermostat to a higher level, producing fever.*

33–2   *The mechanism of action of interferon. (a) When a cell is infected by a virus, it releases interferon molecules, which bind to receptors on the surface of uninfected cells (b). This interaction stimulates the uninfected cells to synthesize enzymes that block the translation of viral mRNA into viral protein. If virus particles subsequently invade these cells, they are unable to reproduce. Interferon molecules also bind to receptors on the surface of white blood cells (c), stimulating both the inflammatory and immune responses.*

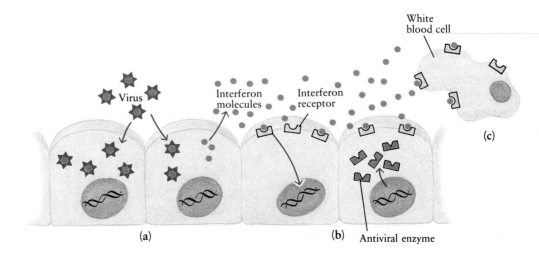

Interferons

Interferons are small proteins that differ from the other defense mechanisms of the body in two ways: (1) they are active only against viruses, and (2) they do not act directly on the invading viruses but rather stimulate the body's own cells to resist them.

When a cell is invaded by a virus, it releases interferon molecules, which then interact with receptor sites on the membranes of surrounding cells (Figure 33–2). Thus stimulated, these cells produce enzymes that block the translation of viral messenger RNA to protein. Only a very few molecules of interferon seem to be required to protect the surrounding cells from viral infection. Interferon molecules also interact with receptors on the surface of various types of white blood cells, stimulating both the inflammatory and immune responses.

Until recently, the only interferons available for research purposes were the minuscule amounts collected from mammalian cells exposed to virus in tissue culture. Now, as a result of recombinant DNA techniques, interferons are being produced in much larger quantities. These new supplies are being used to study clinical applications of interferons in the control of viral infections and to explore their effectiveness against certain forms of cancer. Among their many effects, interferons inhibit cell proliferation, which at one time led to hopes that they might be the long-sought "magic bullets" against different types of cancer. Thus far, however, only one extremely rare form of leukemia has been found to respond to an interferon.

The Immune System

As an organ system, the **immune system** (Figure 33–3) is more diffuse than the digestive system, for instance, or the excretory system. However, it resembles those organ systems in being a functional, integrated unit. It includes the bone marrow, thymus gland, lymph vessels, lymph nodes, spleen, and tonsils.

The lymph vessels, as you will recall, are the route for the return of extracellular fluid to the circulatory system. Strategically located within this system of vessels are the **lymph nodes** (Figure 33–4), which are masses of spongy tissue separated into compartments. Lymph nodes serve as filters, removing microbes, foreign particles, tissue debris, and dead cells from the circulation. They are densely populated by lymphocytes and macrophages, and it is within the lymph nodes that essential interactions among the cells involved in the immune response occur.

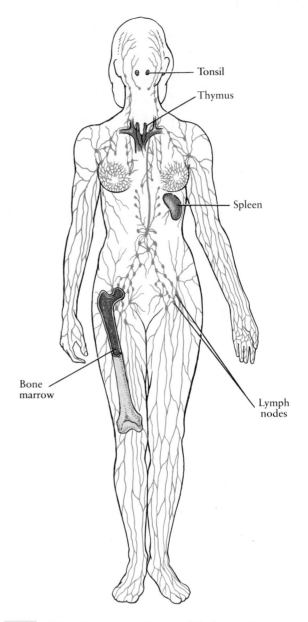

33–3   *The primary constituents of the human immune system are the bone marrow and the thymus, the sites of the initial proliferation of the B and T lymphocytes that are the principal actors in the immune response. Other major components of the system are the lymph vessels, the numerous lymph nodes, the spleen, and the tonsils.*

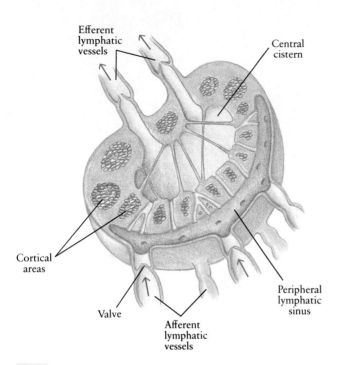

33–4   *Diagram of a lymph node. Lymph enters the node through the afferent lymphatic vessels, filters through the peripheral lymphatic sinus into the central cistern, and trickles out through the efferent lymphatic vessels. The valves within the vessels prevent backflow. The incoming lymph carries bacteria, viruses, and other potential pathogens into the node, exposing them to lymphocytes that are concentrated in the cortical areas. Those lymphocytes that are activated by this exposure then proliferate. Subsequently, they move into the lymph and circulate throughout the body in the bloodstream. Lymph nodes, which range in size from as small as a poppy seed to as large as a grape, typically increase in size during periods when the immune system is extremely active.*

Single lymph nodes are distributed throughout the body, but most are clustered in particular areas, such as the neck, armpits, and groin. The spleen and the tonsils are also rich in lymphocytes and particle-trapping cells. Other patches of lymphoid tissue, embedded in the wall of the intestine, defend the body against the billions of microorganisms that inhabit the normal intestinal tract.

The response of the immune system to foreign invaders differs from the other defenses of the body in that it is highly specific, involving recognition of the particular invader and the mounting of a precisely targeted attack against it. This **immune response** consists of two phases: a primary, short-term defense against the initial attack of an invader, and a secondary, long-term defense that provides a rapid response to any subsequent attacks by the same agent.

Two remarkable groups of lymphocytes, known as **B cells** and **T cells**, are responsible for the specificity of the immune response. In mammals, the primary sites for the differentiation and proliferation of these cells are the bone marrow (for the B cells) and the thymus gland (for the T cells).

B Cells and Antibody-Mediated Immunity

B cells are the major protagonists in one type of immune response: the production of **antibodies**, complex globular proteins that are also known as **immunoglobulins**. Antibodies make precise three-dimensional combinations with particular molecules or parts of molecules that the body recognizes as foreign, or "not-self." Any molecule or part of a molecule that can trigger the synthesis of antibodies by B cells is known as an **antigen** (short for "antibody-generating substance"). Virtually all foreign proteins and most polysaccharides can act as antigens. The surface of an invader, such as a bacterial cell, is often studded with a number of different antigens, each of which can elicit the production of a specific antibody.

The B Cell: A Life History

At any given time, about 2 trillion (2×10^{12}) B cells are on surveillance duty in the human body. Many of these cells are on patrol, circulating in the bloodstream, squeezing out between the endothelial cells that form the walls of the capillaries, and migrating through the lymphatic system. Others are clustered in the lymph nodes, spleen, and other lymphoid tissues. These cells are small, round, nondividing, and metabolically inactive.

Protruding from the plasma membrane of each B cell are antibodies with a specific three-dimensional structure. When a particular B cell meets its destiny in the form of an antigen that "fits" the antibodies on its surface, the cell enlarges and its synthesis of proteins increases. At the same time, the cell begins to divide. The proliferation of activated B cells often takes place in the lymph nodes, which may enlarge during an infection. As we shall see later, the activation and proliferation of B cells also depend on interactions with a group of T cells known as "helper" cells.

The daughter cells resulting from B-cell activation differentiate into two types (Figure 33–5), one of which is the **plasma cell**. Plasma cells, which rarely undergo further division, are, in essence, specialized antibody factories. A mature plasma cell can make 3,000 to 30,000 antibody molecules per second. These molecules are released into the bloodstream and circulate throughout the body. However, it takes about five days to produce fully mature antibody-synthesizing cells working at this maximum capacity. Thus, if the microorganism is multiplying also, it may take about this long for the immune system to catch up and for the infection to abate. Antibiotics, by suppressing the rate of multiplication of bacteria, enable antibody production to overtake the infection more rapidly.

The second type of cell produced by the antigen-stimulated B cell is the **memory cell**. Memory cells also produce antibodies, but they differ from plasma cells in their longevity: plasma cells live only a few days, whereas memory cells continue to circulate for long periods of time—up to a lifetime. Thus, the second time a particular pathogen gains entry to the body, large-scale production of antibodies to the invader begins immediately, often preventing any significant multiplication of the pathogen.

This rapid response by the memory cells is the source of immunity to many infectious diseases—such as smallpox, measles, mumps, and polio—following an infection. It is also the basis for vaccination against a number of diseases. Vaccines may be prepared using a closely related but less virulent pathogen, as in the smallpox vaccine (see essay); a killed pathogen, as in the Salk vaccine against polio; or a strain of the pathogen that has been weakened by growing in another host organism or in tissue culture, as in the widely used oral (Sabin) vaccine against polio.

The use of recombinant DNA technology to synthesize purified antigenic proteins found on the surfaces of pathogens has become a powerful new tool in the preparation of vaccines. For example, it has yielded a safe and effective vaccine against the hepatitis B virus, a major cause of liver disease throughout the world. The use of recombinant DNA technology to prepare vaccines eliminates the possibility of disease-causing infection or contaminants that might produce adverse side reactions.

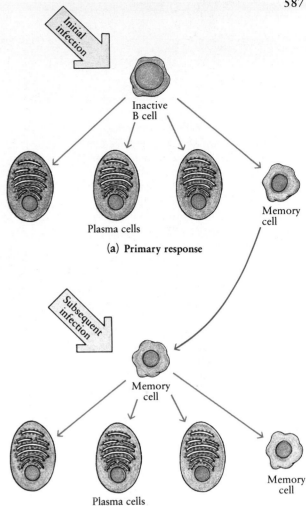

Inactive B cell

Plasma cells

Memory cell

(a) **Primary response**

Memory cell

Plasma cells

Memory cell

(b) **Secondary response**

33–5 (a) *Activation of a B cell by antigens of an infectious agent leads to the production of two types of daughter cells, plasma cells and memory cells. The plasma cells are highly specialized for the manufacture and secretion of antibodies. As a plasma cell matures, it grows larger, its nucleus becomes relatively smaller, and there is a large increase in endoplasmic reticulum and ribosomes. In its brief life span, each plasma cell synthesizes enormous quantities of antibodies that combat the infection. Although the memory cells play no role in this primary response, they persist indefinitely in the circulation.*

(b) *When these memory cells encounter the same antigen during a subsequent infection, the secondary response is triggered. The cells immediately release antibodies and also begin to divide, producing a new generation of plasma cells and new memory cells. This rapid response of the memory cells—and particularly, their immediate release of antibodies—counteracts the subsequent infection so efficiently that the individual usually has no perception of illness.*

Death Certificate for Smallpox

In the early days of medicine, the human body was the laboratory and disease itself the instructor. Smallpox was one of the great teachers. Unlike other infections, which might go unreported or misdiagnosed, smallpox left its unmistakable trace in history. It originated in the Far East and seems to have first been introduced into Europe by the returning Crusaders. Here it flourished and spread, until by the eighteenth century one in ten persons died of smallpox, and 95 percent of those who survived childhood had experienced it. About half had permanent scars and many were blinded. Young women studied their reflections in the mirror, waiting their almost inevitable turn. They learned to scratch their legs and feet at the first sign of the disease in the knowledge that the ugly lesions would then localize in these decorously concealed locations and so, perhaps, spare their faces and bosoms.

No one could miss the fact that a person who had suffered an attack was thereby protected from a future one. Smallpox scars were required of domestic servants, particularly nursemaids, as a prime certificate of employability.

It came to be recognized that some outbreaks of smallpox were more severe than others (owing, we know now, to mutations of the virus). Since one had to have the disease some time, it was reasoned, it was advantageous to choose which smallpox and when. Intentional infection of children with material preserved from a mild attack was first practiced in the Far East. The Chinese did it in the form of powdered scabs, "heavenly flowers," used as snuff. The Arabs carried matter from smallpox pustules around in nutshells and injected it under the skin on the point of a needle.

Smallpox inoculation, known as variolation, was introduced into England in 1717 by Lady Mary Wortley Montagu, wife of the British Ambassador to Turkey. In 1746, a Hospital for the Inoculation against Smallpox was established for the poor of London, where they could be confined during the course of the deliberate infection. Although the induced dis-

A child with smallpox, showing the pustules characteristic of the disease. This photograph was taken in a relief camp in Bangladesh in 1973.

ease was usually mild, it produced serious illness in some persons. Moreover, since smallpox caused by variolation was as contagious as normally contracted smallpox, it appears to have been responsible for some epidemics.

Edward Jenner was an English country doctor who, despite the derision of his colleagues, listened to the tales of country folk. They told him that smallpox never infected milkmaids or other persons who had previously had cowpox, a mild disease of farm animals that sometimes was transmitted to humans. Jenner tried variolation on several persons who had had cowpox and was unable to induce the customary infection. Then, in 1796, he performed his classic experiment on the farm boy Jamie Phipps. First he inoculated the eight-year-old with fluid taken from a pustule of a milkmaid with cowpox. Subsequently he inoculated the boy with material from a smallpox lesion. Fortunately—for himself, for Jamie, and for us all—the child did not contract smallpox.

The success of the cowpox inoculation depended on the antigenic similarity of the two naturally occurring pathogens. Jenner called the process **vaccination,** from *vacca,* the Latin word for cow. By 1800, at least 100,000 persons had been vaccinated, and smallpox began to lose its hold on the Western world.

It was not until almost 100 years later that Louis Pasteur discovered that a virus or bacterium grown in tissues other than those of its normal host would often lose its virulence while retaining its immunogenicity. In recognition of the earlier work of Jenner, Pasteur retained the term "vaccination" for inoculation with such weakened pathogens, which thus became known as **vaccines.** Pasteur's discovery provided the basis for most of our modern vaccines, such as those against polio, diphtheria, and measles.

Despite the availability of an effective vaccine for smallpox, by the 1950s there were still an estimated 2 million new cases of smallpox every year. These were largely confined to the Far East, mostly among the urban poor of India, whose crowded living conditions provided as fertile a soil as the European cities had two centuries earlier. In 1973, the World Health Organization (WHO) declared war on smallpox, which involved the mass production of vaccine and, more important, search-and-destroy missions charged with finding each new case of the disease and vaccinating every susceptible person who might be exposed to it; 150,000 persons were in the task force. On April 23, 1977, an international commission declared India free of smallpox and so it has remained.

In that same year, WHO recommended that smallpox research laboratories be closed and their supply of virus destroyed. One such laboratory was at the University of Birmingham in England. On July 25, 1978, during the last six months of the laboratory's existence, a medical photographer working on the floor below contracted smallpox. The virus, as it turned out, had made its way out of the laboratory through a duct system and the photographer was unvaccinated. The disease was diagnosed by the head of the laboratory, a prominent virologist, who was responsible for the woefully inadequate safety precautions. He committed suicide by cutting his throat. The photographer died five days later. Those appear to have been the last two deaths from smallpox on this planet.

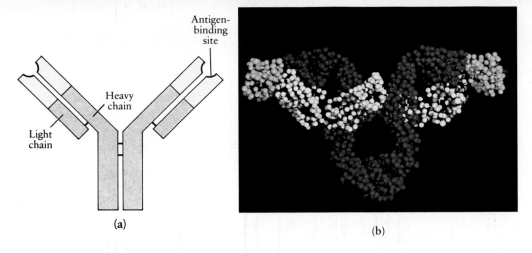

(a)

(b)

33–6 *The structure of an antibody molecule.* (a) *Each molecule consists of four polypeptide chains: two identical light (short) chains and two identical heavy (long) chains. Each chain has a variable region (yellow) and a constant region (gold). The polypeptide chains are connected to each other by disulfide bridges (page 63). The molecule has two identical antigen-binding sites, formed by the variable regions of the four chains. Although the Y-shaped structure shown here is a convenient representation of an antibody molecule, the three-dimensional shape is considerably more complex.* (b) *A model of an antibody molecule. The antigen-binding sites are shown in blue, the constant regions of the heavy chains in red, and the constant regions of the light chains in yellow.*

The Structure and Action of Antibodies

Figure 33–6 shows the structure of an antibody molecule. As you can see, it is made up of two heavy (long) polypeptide chains and two light (short) chains. The heavy and light chains both have constant and variable regions. The constant regions of the chains are characteristic of the species of organism and the general class of antibody. The variable regions form two identical, highly specific three-dimensional structures that are the binding sites for antigens with a complementary three-dimensional structure.

Antibodies commonly act against invaders in one of three ways: (1) they may coat the foreign particles and cause them to clump together in such a way that they can be taken up by phagocytic cells (Figure 33–7); (2) they may combine with them in such a way that they interfere with some vital activity—for example, by covering the protein coat of a virus at the site where the virus attaches to the plasma membrane of a host cell; or (3) they may, in combination with other blood components, known collectively as **complement**, actually lyse and destroy foreign cells. Complement is a group of at least 11 different proteins found in blood. In particular combinations, these proteins make holes in bacterial cell walls at the points where antigen and antibody combine, causing the cells to burst.

Antibody Diversity

One of the most intriguing facts about the immune response is the great variety of antigens against which a single individual can produce antibodies. It is estimated that a mouse, for instance, can form antibodies against 10 million different antigens. Moreover, antibodies can be formed not only against the natural, common invaders that an individual organism might reasonably be expected to encounter in the course of its own life and those that its ancestors might have encountered, but also against synthetic antigens that are chemically unlike any substance found in nature.

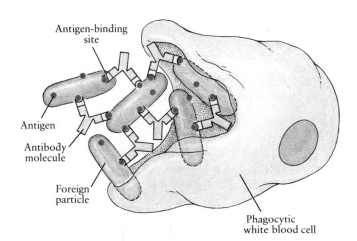

33–7 *Because each antibody molecule has two antigen-binding sites, a single antibody can bind to antigens on two different cells or particles, causing them to stick together (agglutinate). Phagocytic white blood cells then consume these larger masses of foreign particles and antibodies.*

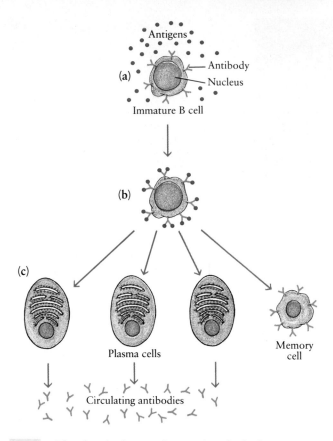

33-8 *The clonal selection theory of antibody formation.*
(a) *An immature B cell, with one specific type of antibody displayed on its surface, encounters antigen molecules with a structure complementary to the binding sites of its antibodies.*
(b) *Antigens bind to the antibodies of this B cell (and other B cells bearing the same antibody), setting in motion a series of changes within the cell. The B cell begins to divide and differentiate,* (c) *forming plasma cells and memory cells. Plasma cells secrete large quantities of circulating antibodies, all with a specificity identical to that of the antibodies on the surface of the original B cell. Memory cells bearing the same antibodies persist in the circulation indefinitely. As we saw earlier, they are responsible for the rapid secondary response to subsequent encounters with the same antigen.*

Our current understanding of antibody formation is known as the **clonal selection theory.** According to this theory, each individual has a vast variety of different B cells, each genetically equipped with the capacity to synthesize only one type of antibody, which is displayed on its surface. Any given antigen does not affect the great majority of the B cells, but only those displaying an antibody that is able to bind that specific antigen. Thus, the antigen-antibody interaction "selects" particular B cells. These B cells then proliferate, producing clones of plasma cells and memory cells, all synthesizing the antibody displayed on the original B cell (Figure 33–8).

According to the clonal selection theory, antibodies are not "tailor-made" in response to an antigen. Rather, the antibodies displayed by the B cells are like the samples in the showroom of a huge "ready-to-wear" supplier. When a customer—an antigen—comes along that fits its antibodies, the manufacturer—the B cell—gears up its factory and begins mass production of that particular size and style.

Each of us is apparently capable of making antibodies against 100 million different antigens. However, a human cell does not contain this many structural genes in its entire genome. This seeming paradox was first resolved in work with mice, and further studies have indicated that the same principles apply in other mammals. In mice, possible components of the variable regions of both the light and heavy chains of antibodies are coded by some 300 DNA sequences scattered throughout the genome. From these DNA sequences, individual segments are selected, transposed, and assembled into a specific arrangement (Figure 33–9), separated by noncoding sequences (introns). This process occurs independently in each B cell as it undergoes its initial differentiation in the bone marrow.

The completed gene is then transcribed into RNA, and the introns are excised before the RNA is translated into the protein of the final antibody molecule. Molecules of this antibody—and only this antibody—are subsequently displayed on the surface of the B cell and its descendants.

The number of possible combinations of the DNA sequences coding for components of antibody genes has been calculated at 18 billion, enough to account for the enormous diversity of antibody molecules. Incredible as it may seem, this figure is actually low. Current evidence indicates that additional combinations are generated both by slight inaccuracies in the excising and joining of the DNA sequences that form the complete variable region genes and by mutations that occur after the assembly of the genes.

Monoclonal Antibodies

In 1975, techniques for fusing cells grown in tissue culture were used to produce mouse hybrid cells that synthesized specific antibodies. One of the cell types in the hybrid was a normal B cell that had been exposed to and activated by an antigen complementary to its specific antibody. However, normal cells, including B cells, die out quickly in tissue culture. Thus, the investigators used a cancerous B cell for the other half of the hybrid, which endowed the hybrid with the capacity to survive and reproduce indefinitely. The cloned offspring of the original hybrid cell synthesized antibodies—termed

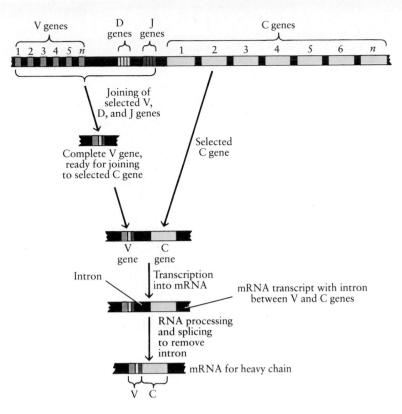

33–9 *A schematic diagram of the assembly of the gene coding for an antibody heavy chain. From the multitude of variable (V), diversity (D), and joining (J) genes present in the genome (of which only a few are shown here), specific genes are selected and transposed to form a complete variable gene. The variable gene is then joined to one of the constant (C) genes. The intron between the variable and constant sequences is removed from the RNA transcript to yield the finished mRNA molecule from which the heavy chains are translated.*

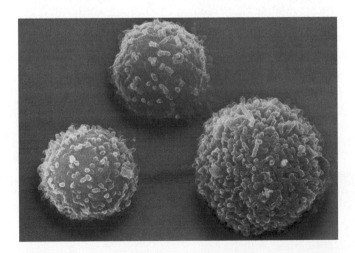

33–10 *Two T cells (blue) approaching a macrophage (yellow) that has been infected by a foreign microorganism. Note the numerous folds and microvilli that greatly increase the surface area of both the T cells and the macrophage. Cell-to-cell interactions involving B cells, T cells, macrophages, and other cells of the body depend on recognition of the pattern of carbohydrates and proteins displayed on the surface of the cells.*

monoclonal antibodies—identical to the antibodies of the original normal B cell.

Similar antibody-producing clones were subsequently developed from hybrids between cancerous human B cells and normal human B cells, each clone capable of synthesizing a different antibody. A great diversity of human monoclonal antibodies are now available in relatively large quantities. They are being used in research, medical diagnosis, and clinical trials against various diseases, including cancer. Malignant cells bear unusual antigens on their surface, unlike those found on normal cells. There is hope that by coupling chemotherapeutic agents with monoclonal antibodies to those antigens, it will be possible to target toxic chemicals directly to the cancer cells, while sparing the normal cells of the body.

T Cells and Cell-Mediated Immunity

Circulating antibodies were long believed to be the sole effectors of immunity. It is now known, however, that there is another category of highly specific immune response that is carried out by the other class of lymphocytes, the T cells. Unlike the circulating antibodies produced by B cells, which are active primarily against viruses and bacteria and the toxins they may produce, T cells interact with other eukaryotic cells, particularly the body's own cells.

Functionally, three classes of T cells are known. Two of these classes, the **helper T cells** and the **suppressor T cells,** are the principal regulators of the immune response, including the activities of the B cells. Members of the third class, the **cytotoxic T cells,** act against foreign eukaryotic cells, such as parasitic protists and fungi, and against cells of the body that are infected by viruses or other microorganisms. When a virus, for example, is multiplying within a cell, it is protected from the action of circulating antibodies. However, its presence is indicated by the appearance of new antigens on the surface of the infected cell. This makes it possible for cytotoxic T cells to find the infected cell and lyse it, exposing the viruses to antibody action.

The T Cell: A Life History

T cells, like red blood cells, B cells, and other white blood cells, are the offspring of stem cells in the bone marrow. As early as the eighth week of fetal life in humans, the future T cells creep into the embryonic thymus gland. Within the thymus gland, these cells go through a complex process of differentiation, selection, and maturation.

Differentiation involves the synthesis of several different types of membrane glycoproteins (carbohydrate-protein combinations). These molecules are ultimately displayed on the surface of the mature T cell and determine both its function and the antigens with which it can interact. One type of membrane glycoprotein exists in two forms, known as T4 and T8, and is correlated with function. Helper T cells bear the T4 molecule on their surface, whereas cytotoxic and suppressor T cells are characterized by the T8 molecule.

A second type of membrane glycoprotein is the receptor by which the T cell recognizes both the cells of the body itself and foreign antigens displayed on those cells. Although the T-cell receptor is not an antibody and is not released from the cell, there are important similarities. The receptor consists of two different polypeptide chains. During the course of differentiation of each T cell, genes coding for these two chains are assembled by the selection and splicing together of different gene segments—a process apparently identical to that occurring in differentiating B cells. The set of genes from which T-cell receptor genes are assembled is, however, distinct from the set from which antibody genes are assembled, and so are the resulting proteins.

Differentiation of T cells is followed by a process of selection. The process of gene rearrangement giving rise to the T-cell receptor is random. As a consequence, some of the receptors are incapable of recognizing "self." Others, by contrast, recognize it a little too well and thus have the potential to destroy healthy cells, wreaking havoc on the organism. Within the thymus, differentiated T cells with either of these characteristics are eliminated. T cells that survive the selection process complete their maturation within the thymus and then are released to begin their duties throughout the body.

Destiny for a T cell is like that for a B cell: the recognition and binding of antigens that fit the T-cell receptors displayed on its surface. The result is also the same: cell division and differentiation, producing clones of active cells and memory cells. The functions of the active T cells, however, are quite different from those of the plasma cells resulting from B-cell activation. Before we consider these functions, we must pause to consider the mechanism by which T cells distinguish self from "not-self."

The Major Histocompatibility Complex

Studies of T-cell functions and of tissue transplant rejections have revealed that recognition of the body's own cells depends on a group of glycoprotein molecules found on the surface of nucleated cells. The protein components of these molecules are coded by a group of genes known as the **major histocompatibility complex,** or MHC. (*Histo* is derived from the Greek word signifying "tissue.")

The major histocompatibility complex consists of at least 20 different genes, and within the human population, each of these genes has as many as 8 to 10 alleles. The total number of different combinations is astronomical, and it is predicted that no two persons—except identical twins—will ever be found to have the same major histocompatibility complex. Each is as individual as a fingerprint. Thus, these molecules provide for extremely accurate recognition of self.

During the selection process in the thymus, developing T cells are exposed to the MHC proteins displayed by the cells of the thymus. Those T cells whose receptors bind optimally to the displayed MHC molecules—tight enough, but not too tight—are the ones selected to complete their development and maturation.

Two classes of MHC molecules, known simply as Class I and Class II, have been identified. They differ in both structure and function. **Class I molecules** are found on cells throughout the body and are necessary for rec-

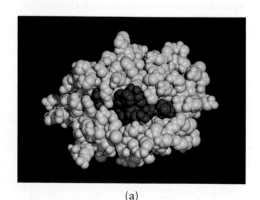

(a)

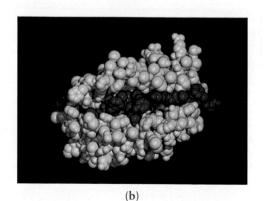

(b)

33–11 *Models of the surface of (a) a Class I MHC molecule and (b) a class II MHC molecule. Each is characterized by a deep groove in which an antigen, shown here in red, can fit. In Class I molecules, the groove is closed at both ends, and only short peptides can fit into the binding site. In Class II molecules, however, the groove is open at both ends, and longer peptides can fit into the binding site.*

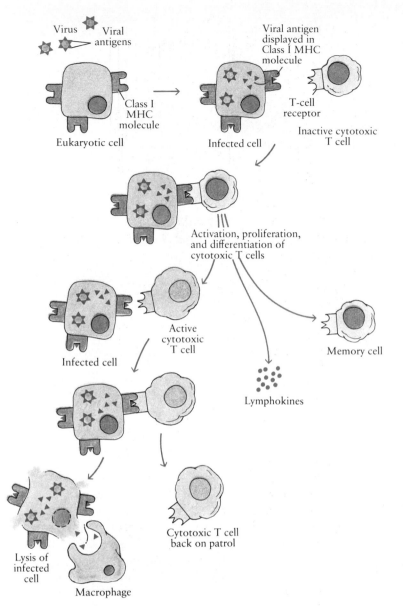

Virus   Viral antigens

Viral antigen displayed in Class I MHC molecule

Class I MHC molecule

Eukaryotic cell

Infected cell

T-cell receptor

Inactive cytotoxic T cell

Activation, proliferation, and differentiation of cytotoxic T cells

Active cytotoxic T cell

Infected cell

Lymphokines

Memory cell

Cytotoxic T cell back on patrol

Lysis of infected cell

Macrophage

33–12   *When a virus invades a eukaryotic cell, its protein coat either remains on the surface of the cell or, as shown here, breaks apart within the cytoplasm. This is an essential step in the life of the virus, releasing its nucleic acid and enabling it to begin replication. As a consequence, however, telltale markers—viral antigens—appear on the surface of the infected cell and are displayed in its Class I MHC molecules. Cytotoxic T cells whose receptors are complementary to the resulting three-dimensional display bind to the cell and are thereby activated. Activation leads to proliferation of the T cells and the differentiation of the daughter cells into clones of memory cells and clones of active cytotoxic cells that attack and destroy other infected cells. The activated T cells also release molecules, known as lymphokines, that attract macrophages and other phagocytic white blood cells to the area. These cells ingest the remains of the lysed cells, including the viruses they contained.*

The memory cells are inactive during the initial infection but are rapidly activated by any subsequent exposure to the same foreign antigen.

ognition by cytotoxic T cells. **Class II molecules** are present only on cells of the immune system and identify such cells to each other. The detailed three-dimensional structures of both Class I and Class II molecules have now been determined. Their most interesting feature is a groove in the exterior surface (Figure 33–11). The groove in Class I molecules is just the right size to hold a peptide consisting of 12 to 20 amino acids, while the groove in the Class II molecules can hold a larger peptide.

With the MHC molecules in mind, let us now turn our attention to the activities of the T cells.

The Functions of the T Cells

Functionally, the simplest of the T cells are the cytotoxic T cells. As we noted earlier, infection of a eukaryotic cell by a foreign microorganism, such as a virus, results in the appearance of new antigens on the surface of that cell. These antigens, which are short peptides derived from the antigens originally on the surface of the infectious particle, are bound to and displayed in the surface groove of the individual's Class I MHC molecules. When a cytotoxic T cell encounters a combination of Class I MHC molecule and foreign antigen to which its receptor can bind, it differentiates into active cells and memory cells (Figure 33–12). The active cells attack and lyse the infected cells. They also release powerful chemicals, known as **lymphokines,** that attract macrophages and stimulate phagocytosis. In addition, some of the cytotoxic T cells, known as "killer cells," secrete toxins that destroy target cells directly; others secrete interferon molecules (page 585). Thus, a whole battery of defenses is mobilized by the activation of cytotoxic T cells.

The activities of the cytotoxic T cells, as well as those of the B cells, are regulated by the helper and suppressor T cells. Helper T cells are characterized by receptors that recognize and bind foreign antigens displayed in conjunction with the individual's Class II MHC molecules. Such combinations are found on the surface of both macrophages that have ingested foreign microorganisms (Figure 33–13) and activated B cells. A helper T cell encountering either type of cell bearing an antigenic combination to which it can bind becomes activated and begins producing proteins known as **interleukins.** These proteins act as hormones, stimulating the differentiation and proliferation of both B cells and cytotoxic T cells following activation. The actual binding of helper T cells to antigen-displaying B cells is also an essential step in the production of plasma cells and memory cells (Figure 33–14).

33–13   *The activation of helper T cells. Phagocytosis of a virus or other foreign microorganism by a macrophage results in the display of foreign antigens in the Class II MHC molecules on the surface of the macrophage. Helper T cells with receptors that match this particular three-dimensional combination bind to the surface of the macrophage and are activated. This leads to the proliferation of the T cells and their differentiation into clones of memory cells and active helper cells. Activation of the helper T cells also triggers the release of substances known as interleukins. These molecules play a role in the stimulation of both cytotoxic T cells and B cells following activation. As shown in Figure 33–14, the active helper T cells also play an immediate and direct role in the activation of B cells.*

After its release from the helper T cell, the macrophage continues to circulate, ingesting other invading particles and displaying the foreign antigens to both inactive helper T cells and inactive B cells.

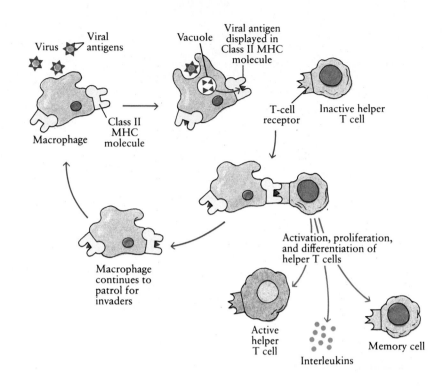

33–14   *Full activation of B cells depends not only on their exposure to antigens complementary to the antibodies displayed on their surface but also on their interaction with helper T cells. After the antibodies of an inactive B cell bind the matching antigens, some of the antigens are transferred from the antibodies to the Class II MHC molecules on the B cell surface. Helper T cells that have been previously activated by exposure to the same antigen (Figure 33–13) bind to the B cell. This binding is an essential step in the differentiation and proliferation of memory cells and plasma cells and the subsequent production of large quantities of circulating antibodies.*

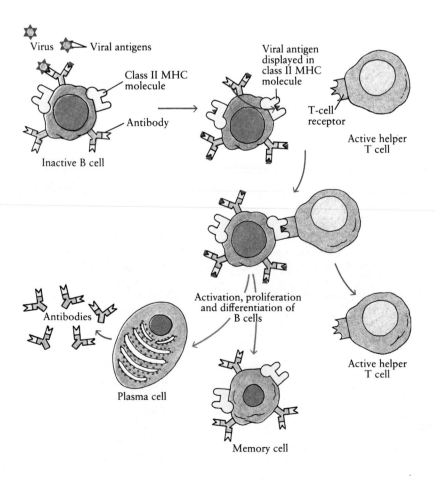

When an infection has been successfully eliminated, further activity of both B cells and T cells is suppressed. The mechanism of this suppression remains poorly understood. Although suppressor T cells are generally described as the "off" switch of the immune response, there is currently no way to distinguish such cells from other T cells bearing the T8 molecule. Thus, it is not yet known whether suppressor cells exist as a distinct class of T cells, or whether suppression is actually mediated by cytotoxic T cells and, perhaps, helper T cells.

Studies of T-cell function have provided what is perhaps only the first glimpse of a vast cellular communication network, involving far more than the immune response. As we shall see in the next two chapters, substances released by activated cells of the immune system appear to influence other cells of the body, particularly those of the endocrine and nervous systems. Conversely, the immune system seems to be influenced by the molecules responsible for communication among cells of the endocrine and nervous systems. The study of these complex interactions is one of the most intriguing areas of modern medical research, not only because of the basic biological knowledge that is being uncovered but also because of its potential bearing on many medical problems.

Cancer and the Immune Response

Cancer cells resemble an individual's normal cells in many ways. Yet, within the body, they act like foreign organisms, reproducing rapidly and invading normal tissues. Moreover, virtually all cancer cells have antigens on their cell surfaces that differ from the antigens of the normal cells of the individual and can be recognized as foreign. Does this mean that the body can mount an immune response against its own cancers?

A growing body of evidence suggests not only that cancer can induce an immune response (Figure 33–15) but also that it usually does so. In fact, according to this hypothesis, it usually does so successfully, overwhelming the cancer cells before they are ever detected. The cancers that are discovered represent occasional failures of the immune system. This conclusion suggests that bolstering the immune response may provide a means for cancer prevention or control. It also suggests that the cell-mediated immune response may have had its evolutionary origins as a defense not only against foreign invaders but also against the treacherous, malignant cells of the body itself.

33–15 *The successful destruction of a cancer cell by cytotoxic T cells. (a) Recognition of the cancer cell as abnormal depends on the presence of unusual antigens displayed on its surface in conjunction with Class I MHC molecules. (b) As the cytotoxic T cells begin lysing the membrane of the cancer cell and consuming its contents, their shape changes. (c) When their work is completed, all that remains of the cancer cell is its cytoskeleton—a mass of lifeless protein fibers.*

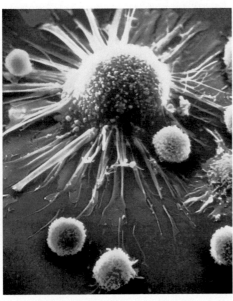
(a)

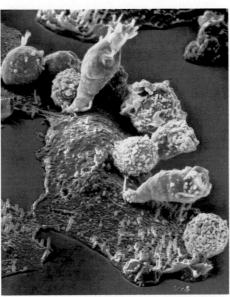
(b)

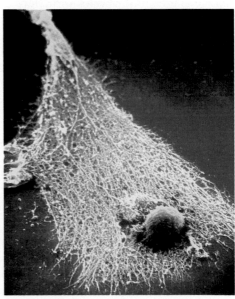
(c)

Tissue Transplants

Organ Transplants

When people suffer extensive burns, the protective barrier formed by the skin is effectively destroyed. For such persons, the most immediate and greatest threats to life are the severe infections to which they are vulnerable and the loss of body fluids from exposed areas. If, however, skin is taken from one part of the individual's body and grafted to the burned area, the new tissue adheres to the exposed area, blood vessels invade it, and the tissue grows and spreads out.

If a skin graft is taken from another individual (except the person's identical twin), the initial stages of healing take place. But then, on about the fifth to the seventh day, large numbers of the recipient's white blood cells infiltrate the transplanted tissue, and by the tenth to twelfth day it has been killed. The infiltrating cells are mainly T cells and macrophages.

The discovery and identification of the MHC molecules—a result of the search for compatible tissue grafts—is now making possible closer matches between donor and recipient in organ transplants. In the effort to further reduce rejection, transplant recipients are also generally given drugs to suppress the immune response. However, given that infection is a major complication among patients requiring skin grafts and is the leading cause of death among kidney transplant recipients, general suppression of the immune response is obviously not an ideal solution.

Cyclosporin, a drug isolated from a soil fungus, acts selectively against the T cells involved in transplant rejection. Although the mechanism of its action is still poorly understood, cyclosporin has dramatically improved the success of organ transplants. Other cells of the immune system seem to be unaffected by this drug, with the result that the patient remains protected against infection.

Blood Transfusions

The most frequently performed tissue transplants in modern medical practice are transfusions of blood. Blood transfusions are so routine that it is difficult to realize that in the past they often provoked severe—and sometimes fatal—immune responses. They became safe only when the four major human blood types (A, B, AB, and O) were identified. As we saw in Chapter 13, these blood types are determined by a gene with three alleles and are characterized by particular antigens on the surface of the red blood cells and by antibodies within the plasma (see Table 13–1, page 224).

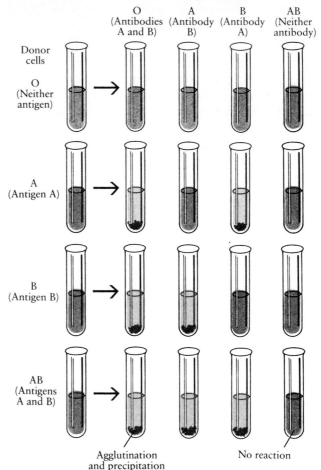

33–16   *Severe and sometimes fatal reactions can occur following transfusions of blood of a different type from the recipient's. These reactions are the result of agglutination of the donor's red blood cells caused by antibodies present in the recipient's blood. (Antibodies in the donor's blood are generally of little consequence because they are so diluted in the recipient's blood.) Blood-group reactions to transfusions can be demonstrated equally well in test tubes, as shown here. The blood serum (plasma from which the proteins involved in clotting have been removed) in which agglutination occurs has natural antibodies against the donor blood. Persons with type O blood, whose red blood cells have neither A nor B antigens, used to be called universal donors. Similarly, those with AB blood (neither A nor B antibodies in the plasma) used to be called universal recipients. Now other factors are checked as well.*

For blood transfusions to be performed safely, the blood types must be matched. If a person receives a transfusion containing red blood cells bearing a foreign antigen, the antibodies in his or her plasma will react with those cells, causing them to clump together (Figure 33–16). The resulting clumps of blood cells and antibodies can clog capillaries, blocking the vital flow of blood through the body.

The Rh Factor

Since the discovery of the major blood types at the beginning of the twentieth century, additional antigens have been identified on the surface of red blood cells. Among the most important of these antigens is the **Rh factor,** the cause of Rh disease, once a common problem in newborn babies.

During the last month before birth, the human baby usually acquires antibodies from its mother. Most of these antibodies are beneficial. An important exception, however, is found in the antibodies formed against the Rh factor. Like the other surface antigens of red blood cells, the Rh factor is genetically determined. If a woman who lacks the Rh factor (that is, an Rh-negative woman) has children fathered by a man homozygous for the Rh factor, all of the children will be Rh positive. If the man is a heterozygote, about half of the children will be Rh positive.

During the birth of an Rh-negative mother's first Rh-positive child, fetal red blood cells bearing the Rh antigen are likely to enter her bloodstream. The consequences are the same as would occur with a transfusion of Rh-positive blood: the mother's immune system produces antibodies against the foreign antigens, and these antibodies persist in her blood. In subsequent pregnancies, they may be transferred to the fetus. If that fetus is Rh positive, the antibodies will react with its red blood cells, destroying them (Figure 33–17). This reaction can be fatal either before or just after birth.

Now that its causes are recognized, Rh disease can be prevented. Within 72 hours of the delivery of her first Rh-positive child, the Rh-negative mother is injected with antibodies against the fetal Rh-positive red blood cells in her system. This destroys the fetal cells before they can trigger antibody production in the mother's body. At the birth of each Rh-positive child, this process must be repeated, in order to protect any subsequent Rh-positive fetus.

Disorders of the Immune System

Autoimmune Diseases

The immune response is a powerful bulwark against disease, but it sometimes goes awry. Usually the immune system can distinguish between self and "notself." Substances present during embryonic life, when the immune system is developing, usually do not act as antigens in later life. This recognition occasionally breaks down, however, and the immune system attacks cells of the body. Certain disorders, among them multiple sclerosis, myasthenia gravis, lupus erythematosus, and several types of anemia, have been identified as **autoimmune diseases**—that is, diseases in which an individual makes antibodies against his or her own cells. There is growing evidence that other disorders, such as juvenile-onset diabetes and some forms of rheumatoid arthritis, may have the same basis.

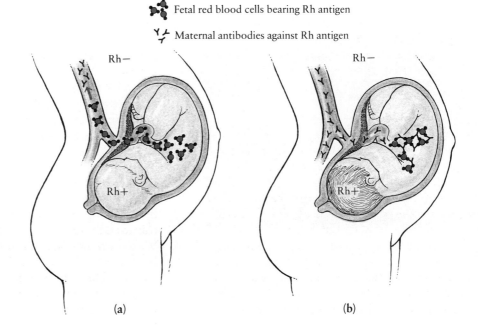

33–17 *The events leading to Rh disease.* (a) *Late in the first pregnancy—or during the birth of the child—fetal blood cells spill across the barrier that separates the maternal and fetal circulations. Antigens on the red blood cells of the Rh-positive fetus stimulate the production of antibodies by the immune system of the Rh-negative mother. These antibodies remain in the mother's bloodstream indefinitely.* (b) *Late in the second pregnancy, the antibodies pass through the barrier from the maternal blood to the blood of the fetus. If the fetus is Rh positive, the antibodies react with the antigens on its red blood cells, destroying the cells.*

Fetal red blood cells bearing Rh antigen

Maternal antibodies against Rh antigen

Rh−

Rh+

Rh−

Rh+

(a)

(b)

Allergies

Hay fever and other allergies are the result of immune responses to pollen, dust, or other substances that are weak antigens to which most people do not react. When certain individuals are exposed to particular environmental antigens, however, the production of antibodies by specific plasma cells is stimulated, as is the formation of memory cells.

Upon reexposure to the same antigen, more antibodies are formed. These antibodies circulate and attach themselves to **mast cells,** which are noncirculating white blood cells found in connective tissue. Subsequent binding of the antigen to these attached antibodies triggers the release of histamine from the cells (Figure 33–18), which, in turn, induces an inflammatory response. This reaction typically occurs on an epithelial cell surface, producing increased mucus secretion (as in hay fever), hives, or cramps and diarrhea (in the case of food allergies).

Systemic reactions may result if the mast cells release their histamine and other chemicals into the circulation. This causes dilation of the blood vessels, leading to a potentially dangerous fall in blood pressure, and constriction of the bronchioles (a syndrome known as anaphylactic shock). The evolutionary background of this decidedly maladaptive response—in which the response is far more dangerous than the antigen—remains a mystery.

Antihistamines suppress some of the symptoms of an allergic reaction. Most decongestants, on the other hand, promote the release of histamines, raising some questions about taking "allergy pills" that offer a combination of antihistamines and decongestants. Steroid hormones related to cortisone are used in more severe cases. These drugs act by suppressing the production of white blood cells and, thus, inflammation and the immune response in general.

Acquired Immune Deficiency Syndrome (AIDS)

Until fairly recently, epidemics of fatal infectious disease were a fact of life. For example, in 1918, at the close of World War I, a particularly deadly epidemic of influenza swept around the globe, leaving few families untouched. In the late 1940s and early 1950s, polio epidemics were a regular feature of summertime in the United States. However, with the widespread introduction of antibiotics following World War II and, particularly, the development of polio vaccines in the mid-1950s, we entered an era in which it seemed that the conquest of infectious disease was in sight. That this is not so—and perhaps never can be so—has been powerfully demonstrated by the appearance of a virus that attacks the human immune system, leaving its victims susceptible to numerous types of infection. The disease it causes is known as **acquired immune deficiency syndrome (AIDS).**

AIDS was first identified in 1981, when epidemiologists noticed unusual clustered occurrences of a malignancy known as Kaposi's sarcoma, which affects the endothelial linings of the blood vessels. Previously, this rare cancer had been seen only in elderly men. Its new victims, however, were young men. At about the same time—and also in young men—physicians began to observe an increasing incidence of fatal pneumonias and intestinal tract infections caused by usually innocuous protists. In the past, such infections had been observed only in cancer patients and transplant recipients whose immune systems had been suppressed in the course of their treatment.

These facts suggested that the underlying cause of the new illnesses was a massive suppression of the immune system. Because most of the early victims were homosexual men, intravenous drug users, recipients of blood transfusions, or hemophiliacs who had received clotting factors prepared from donated blood, it appeared that the agent of this suppression was a microorganism transmitted sexually or through the exchange of blood. Within a remarkably short time—three years—the virus responsible was isolated and characterized. Since then, its principal effects on the immune system have been determined, its possible origins identified, and its modes of transmission traced.

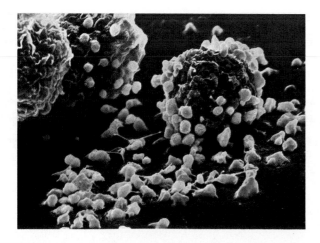

33–18   *Mast cells are "loaded" with granules filled with histamine. The cells also bear receptors to which specific antibodies are bound. When antigens complementary to those antibodies bind to them, the mast cells (shown here in red) explode, releasing their histamine-containing granules (white).*

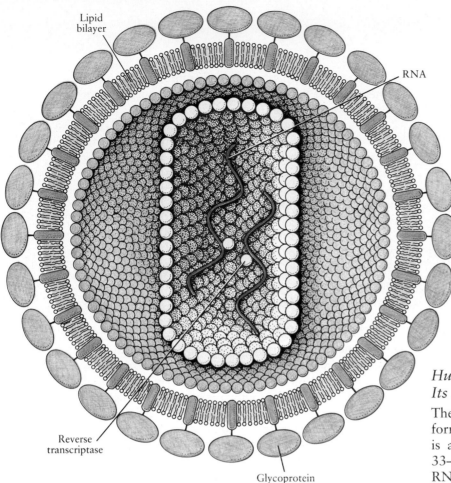

Lipid bilayer

RNA

Reverse transcriptase

Glycoprotein

33–19   *The structure of human immuno-deficiency virus (HIV), the cause of AIDS. The inner core consists of two molecules of RNA, accompanied by two or more molecules of the enzyme reverse transcriptase. Surrounding the core are envelopes formed of two distinct proteins. These are surrounded, in turn, by a lipid bilayer derived from the plasma membrane of the host cell in which the virus previously replicated. Spanning this membrane are protein molecules to which are attached glycoprotein molecules that extend out from the surface.*

Human Immunodeficiency Virus (HIV) and Its Effects

The cause of AIDS is a retrovirus (page 294), known formally as **human immunodeficiency virus (HIV)**. It is a particularly complex virus, as shown in Figure 33–19. The inner core consists of two molecules of RNA and several copies of the enzyme reverse transcriptase. Surrounding the core are two distinct protein envelopes, which are, in turn, surrounded by a lipid bilayer studded with glycoproteins. The unique—and deadly—feature of these glycoproteins is that they are a perfect three-dimensional match to the T4 molecules (page 592) that characterize helper T cells.

When HIV meets a helper T cell, the T4 molecule functions as its receptor, enabling the virus to enter the cell by receptor-mediated endocytosis (page 120). The virus is also able to infect macrophages, which are now thought to be an important reservoir for the virus within the body. The interactions between macrophages and helper T cells, described previously, may play a major role in the spread of the virus to helper T cells.

Once within a helper T cell, the viral RNA is released, and reverse transcriptase catalyzes the transcription of complementary DNA and its incorporation into a chromosome of the host cell (see Figure 17–9, page 294). In the host-cell chromosome, this complementary DNA may lie latent for some time. Sooner or later, however, replication of the virus begins—at a rate much higher than that of other known viruses. In a relatively short time, enormous numbers of new viruses burst from the infected helper T cell. These viruses invade other helper T cells (Figure 33–20), and the process is repeated.

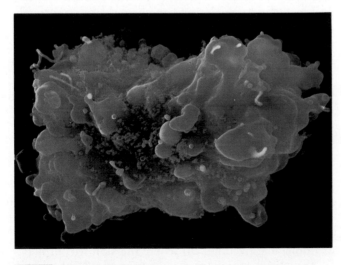

33–20   *A helper T cell being attacked by HIV viruses (blue). Glycoproteins on the surface of the HIV particles bind to the T4 glycoproteins on the surface of the helper T cell and are carried into the cell by receptor-mediated endocytosis.*

The victim is eventually left with few functional helper T cells, and, as we have seen, these cells are critical for the proliferation and activities of both B cells and cytotoxic T cells. The body cannot mount an effective immune response against cells harboring HIV, against the virus itself, against other invading microorganisms, or against any malignant cells that may be present or may develop. However, in the initial phases of the infection, before the population of helper T cells is severely depleted, B cells respond to the foreign antigens of the virus, producing circulating antibodies. Although these antibodies are apparently ineffectual in controlling the infection, they do remain in the bloodstream. Their detection is the basis of screening tests for the presence of the virus.

As the immune system becomes steadily more crippled, the victim becomes increasingly vulnerable to other diseases. Among the most common are *Pneumocystis carinii* pneumonia, parasitic gastrointestinal infections accompanied by severe diarrhea, and Kapo-

si's sarcoma and other cancers. Weight loss is extreme. Ultimately, the person dies, most often as a result of infections or heart failure.

The time span from diagnosis of full-blown AIDS to death varies from a few months to as long as five years. The time span from initial infection to the appearance of the first symptoms varies even more and may be associated with the absence or presence of other infections that trigger an immune response—an event that appears to activate latent HIV viruses. About 30 percent of infected individuals develop active disease within five years, and it appears probable that everyone infected with the virus will eventually become ill. A few cases have now been documented, however, in which some 15 years have elapsed between infection and illness.

Transmission of HIV

HIV is one of a family of primate retroviruses and appears to be most closely related to a relatively harmless virus that is endemic in the green monkeys of equatorial Africa. It has been hypothesized that this virus may have moved from the monkeys into the surrounding human population, and then undergone a series of mutations into its present deadly form. According to this hypothesis, the virus gradually spread through the population, transmitted through sexual intercourse and blood transfusions. Ultimately, it was acquired by visitors from Europe, North America, and the Caribbean and began its spread around the world.

Current evidence indicates that HIV can be transmitted through sexual intercourse, either vaginal or anal, through oral sex, and through the exchange of blood. It is present at high levels in the semen and blood of infected individuals and can enter the body through any tear in the skin or mucous membranes—including those too small to be seen. Although HIV is one of the most virulent viruses known, it is less readily transmitted than other viruses. Without a surrounding environment of blood or semen or host cells, it quickly dies.

There is no evidence that HIV can be transmitted through casual contact, hugs and light kisses, coughs or sneezes, or on toilet seats or dishes used by an infected person. Thus far, no family member has become ill as a consequence of nursing an AIDS patient, and the few medical personnel who have contracted HIV in the course of their work were either accidentally pricked with a contaminated needle or exposed to large quantities of infected blood on portions of their body (for example, ungloved hands) where the skin was broken.

The initial spread of the virus in the West among homosexual men and intravenous drug users is thought to have occurred because these are relatively self-

33–21 *An African green monkey. These monkeys are host to a virus known as simian immunodeficiency virus (SIV). Although SIV shares important similarities with human immunodeficiency virus (HIV), the cause of AIDS, it seldom produces illness in its monkey hosts.*

contained populations in which an individual might be repeatedly exposed. In equatorial Africa and in Asia, AIDS is primarily a disease of heterosexuals, and its incidence among men and women is roughly equal. Continuing studies suggest that members of the following groups in North America are presently at the greatest risk: (1) homosexual or bisexual men, (2) intravenous drug users, (3) sexual partners, whether male or female, of infected individuals, and (4) babies born to infected mothers. All available evidence indicates that once the virus has infected an individual, it remains for the rest of that person's life and can be transmitted in blood and semen—even if the person has no symptoms of illness.

The Prospects

As many as 2 million people in the United States alone are now thought to be infected by HIV. By June of 1993, 315,390 cases of AIDS had been documented in the United States, and 194,344 deaths had been attributed to the virus and its effects. Worldwide, as many as 14 million people are now thought to be infected by HIV, and it is estimated that this number will rise to between 30 million and 40 million by the turn of the century.

Although work is progressing on treatments and on the development of vaccines, any "quick fix" seems unlikely. Current treatments, which are directed at the control of infections and inhibition of reverse transcriptase (and thus of the replication of the virus), have serious side effects and are, at best, delaying actions.

The nucleotide sequences of the RNA of HIV have been determined, and the codes for the various proteins of its coat have been identified. With this information, scientists are using recombinant DNA technology in the attempt to create synthetic vaccines against the virus. However, the fact that the natural antibodies synthesized by the B cells are ineffectual suggests that development of a truly protective vaccine may be extraordinarily difficult. The task is also compounded by the high mutation rate of the genes coding for key parts of the protein coat.

Tests for antibodies to HIV have provided a means of identifying donated blood carrying the virus and now provide a high level of protection for the blood supply. As a consequence, the number of new cases of AIDS in transfusion recipients and hemophiliacs has dropped dramatically. The spread of the virus through the homosexual population has also slowed significantly, due to changes in sexual practices. Presently, the greatest increases in new cases in the United States are occurring among drug users sharing needles and among the sexual partners of infected individuals. An increasing percentage of the new cases, particularly among teenage girls and among women between the ages of 20 and 24, are attributed to heterosexual transmission.

AIDS, like smallpox in another era (page 588), is one of the great teachers on the subject of immunology. The price of its lessons—in the suffering of its victims, their families and their friends, in lives cut short in their prime, in the strains on medical and social services, and in economic terms—is extraordinarily high. Our response will say much about us as a people.

Summary

Animals have evolved a number of responses that exclude or destroy microorganisms, other foreign invaders, and cells not typically self. These responses depend on a variety of types of white blood cells, all the descendants of cells in the bone marrow.

The inflammatory response, which is nonspecific, involves the release of histamine and other chemicals and the mobilization of phagocytic white blood cells at the site of infection. Small proteins known as interferons provide another and very different type of nonspecific defense. Interferons, which are produced by virus-infected cells, stimulate nearby cells to defend themselves against viral infection. They also stimulate cells involved in the immune response.

The immune response is highly specific and involves two types of lymphocytes: B cells and T cells. B cells are the major protagonists in the formation of antibodies, large proteins whose binding sites are complementary to foreign molecules called antigens. The combination of antigen and antibody immobilizes the invader, destroying it or rendering it susceptible to phagocytosis.

Antibodies consist of two identical light chains and two identical heavy chains. Each of the four chains has a constant region and a variable region. The two antigen-binding sites on each antibody are formed by foldings of the variable regions of the molecule.

According to the clonal selection theory, differentiation of B-cell precursors, which occurs in the bone marrow, produces a variety of different B cells, each capable of synthesizing antibodies with one particular three-dimensional structure. Upon encountering an antigen that binds to the antibodies displayed on its surface, the B cell matures and divides, resulting in a clone of plasma cells all synthesizing circulating antibodies against that particular antigen. Memory cells are also produced, which persist in the bloodstream following infection and provide a rapid secondary response to any subsequent infection by the same agent.

The capacity to produce a tremendous variety of B cells, each able to synthesize one specific antibody, is accounted for by the large number of gene sequences coding for the variable regions of antibodies, by the transposition of these gene sequences in the course of B-cell differentiation, and by subsequent mutations.

T cells, which undergo differentiation and maturation in the thymus, are responsible for cell-mediated immunity. Although these cells neither display nor produce antibodies, they bear several types of glycoprotein molecules on their surface. One type, which exists in two forms, T4 and T8, identifies cells as helper T cells (T4) or as cytotoxic or suppressor T cells (T8). Cytotoxic T cells work principally against the body's own cells that are harboring viruses or other parasites. Helper T cells promote immune responses involving B cells and cytotoxic T cells, whereas suppressor T cells moderate the activities of B cells and other T cells.

The ability of the T cells to perform their functions depends upon another type of surface molecule, known as the T-cell receptor. The T-cell receptor consists of two different polypeptide chains, each coded by genes that, like those for antibodies, are rearranged in the course of differentiation. The result is an enormous diversity of T cells.

T-cell receptors recognize and bind to complementary foreign antigens presented in conjunction with genetically determined proteins found on the surface of the body's own nucleated cells. These proteins are coded by a group of genes known as the major histocompatibility complex (MHC). Class I MHC molecules, which are found on cells throughout the body, are essential in the identification of diseased cells by cytotoxic T cells. Class II MHC molecules are found on the surface of macrophages and B cells. They are essential in the presentation of foreign antigens to the helper T cells, which are, in turn, essential for the activation and proliferation of both B cells and cytotoxic T cells.

Organs transplanted between individuals other than identical twins evoke an immune response by cytotoxic T cells that can lead to rejection of the transplant. Similarly, blood transfusions can evoke an immune response by circulating antibodies to the antigens found on the surface of the donor red blood cells. Blood typing involves not only the ABO blood types but also other surface antigens of red blood cells, such as the Rh factor.

Disorders associated with the immune system include allergies, autoimmune diseases, and AIDS, a fatal infectious disease. The retrovirus responsible for AIDS (human immunodeficiency virus, or HIV) invades and destroys helper T cells, leaving the victim's immune system incapable of responding to other infections or to malignancies. HIV is present in high levels in the blood and semen of infected individuals and is transmitted by sexual contact and through the exchange of blood or blood products.

Questions

1. Distinguish among the following: macrophages/lymphocytes; B cell/T cell; antigen/antibody/T-cell receptor; antibody light chain/antibody heavy chain; T4 molecule/T8 molecule; Class I MHC molecules/Class II MHC molecules; ABO blood types/Rh factor.

2. If a live bacterium enters the body, what are the ways in which it can be destroyed? In what other ways can invading viruses be destroyed?

3. Explain the functions of the macrophages in the inflammatory and immune responses. How do these cells provide the essential link between the inflammatory response and the immune response?

4. Describe the life history of a B cell, beginning with a precursor cell in the bone marrow and ending with a clone of plasma cells and a clone of memory cells. Include its interactions with other types of white blood cells.

5. Although a human cell contains considerably fewer than 1 million structural genes, each of us is, according to immunologists, capable of making at least 100 million different kinds of antibodies. How is this possible?

6. What is the function of the intron in the assembly of an mRNA molecule coding for a polypeptide chain of an antibody molecule?

7. What prominent feature in the structure of a virus makes it more susceptible than a bacterial cell to control by vaccination? (*Hint:* see page 290)

8. Describe the life history of a cytotoxic T cell, beginning with a precursor cell in the bone marrow and ending with a clone of active cells and a clone of memory cells. Include its interactions with other types of white blood cells.

9. Predict the results of the following experiment: A mouse of strain A receives a skin graft from another strain A mouse and one from a strain B mouse. Two weeks later, the same mouse is grafted with skin from strain B and strain C. What are the fates of these grafts?

Suggestions for Further Reading

Anderson, Roy M., and Robert M. May: "Understanding the AIDS Pandemic," *Scientific American*, May 1992, pages 58–66.

Boon, Thierry: "Teaching the Immune System to Fight Cancer," *Scientific American*, March 1993, pages 82–89.

Buisseret, Paul D.: "Allergy," *Scientific American*, August 1982, pages 86–95.

Diamond, Jared: "The Mysterious Origin of AIDS," *Natural History*, September 1992, pages 24–29.

Edelson, Richard L., and Joseph M. Fink: "The Immunologic Function of Skin," *Scientific American*, June 1985, pages 46–53.

Ewald, Paul W.: "The Evolution of Virulence," *Scientific American*, April 1993, pages 86–93.

Golde, David W.: "The Stem Cell," *Scientific American*, December 1991, pages 86–93.

Johnson, Howard M., Jeffry K. Russell, and Carol H. Pontzer: "Superantigens in Human Disease," *Scientific American*, April 1992, pages 92–101.

Marx, Jean: "Key Piece Found for Immunology Puzzle?" *Science*, vol. 246, page 1561, 1989.

Marx, Jean: "Taming Rogue Immune Reactions," *Science*, vol. 249, pages 246–248, 1990.

McNeil, Paul L.: "Cell Wounding and Healing," *American Scientist*, vol. 79, pages 222–235, 1991.

Mills, John, and Henry Masur: "AIDS-Related Infections," *Scientific American*, August 1990, pages 50–57.

Nilsson, Lennart: *The Body Victorious*, Delacorte Press, New York, 1987.

> An account of the human immune system—its development, its normal functioning, and its disorders—filled with more of the stunning color photographs and micrographs for which Nilsson is noted. Moreover, if you should be in any doubt as to the physiological consequences of cigarette smoking, you owe it to yourself to examine the relevant photographs in this book.

Rennie, John: "The Body Against Itself," *Scientific American*, December 1990, pages 106–115.

Rose, Noel R.: "Autoimmune Diseases," *Scientific American*, February 1981, pages 80–103.

Schwartz, Ronald H.: "T Cell Anergy," *Scientific American*, August 1993, pages 62–71.

Scientific American: *Immunology: Recognition and Response*, W. H. Freeman and Company, New York, 1991.*

> A collection of 12 articles from Scientific American, covering the clonal selection theory, the genetics of antibody diversity, the T-cell receptor, regulation of the immune response, autoimmune disorders, and monoclonal antibodies. Highly recommended.

Scientific American: *Life, Death and the Immune System*, W. H. Freeman and Company, New York, 1994.*

> This reprint of the September 1993 issue of Scientific American, with 10 chapters by different authors, provides an up-to-date review of our understanding of the development and functioning of the immune system. Also covered are AIDS, allergies, and autoimmune diseases. Highly recommended.

Scientific American: *The Science of AIDS*, W. H. Freeman and Company, New York, 1989.*

> A reprint of the October 1988 issue of Scientific American. Its 10 chapters by different authors provide a thorough review of our knowledge of human immunodeficiency virus (HIV), the causative agent of AIDS; of the disease process and epidemiology of AIDS; and of the prospects and problems in the development of effective treatments and vaccines. Highly recommended.

Sharon, Nathan, and Halina Lis: "Carbohydrates in Cell Recognition," *Scientific American*, January 1993, pages 82–89.

Tonegawa, Susumu: "The Molecules of the Immune System," *Scientific American*, October 1985, pages 122–131.

von Boehmer, Harold, and Pawel Kisielow: "How the Immune System Learns about Self," *Scientific American*, October 1991, pages 74–81.

Welch, William J.: "How Cells Respond to Stress," *Scientific American*, May 1993, pages 56–64.

*Available in paperback.

The Endocrine System

Army ants, of which there are about 300 species, hunt in marauding swarms, driving insects and other animals before them, dismembering and eating those unable to escape. In the words of entomologist W. M. Wheeler, written almost a century ago, watching these insects suggests "the existence of a subtle, relentless and uncanny agency, directing and permeating all their activities."

Some 30 years ago, E. O. Wilson of Harvard University discovered that this "relentless and uncanny agency" is, in fact, a group of chemical messengers used by the members of a colony—which may be as large as 22 million individuals—to exchange the information that orders their complex social organization. The first ants to encounter this hapless katydid sent forth a chemical signal—a pheromone—that summoned other members of the colony. In minutes, the katydid will disappear, torn apart and carried off by the ants.

Chemical communication has powerful advantages. Chemicals can travel through air or water or be deposited in specific locations, as in the trails laid down by ant scouts and leaders. Unlike visual signals, chemical signals are effective over great distances and in the dark (in fact, all these army ants are blind). Chemicals are inexpensive, largely because of their extreme potency and durability. Also, they can be very specific: different chemicals serve as trail markers, others as alarms; one tells the army to mobilize, another rounds it up for a nightly bivouac.

Chemicals not only rule the activities of the colony but also control the activities and numbers of its individuals. Young workers—always female—engage in safe tasks, such as care and feeding of larvae. With increased age, they take on more dangerous tasks, such as food-gathering and nest defense. As Wilson points out, ". . . we send our young men to war, ants send their old ladies."

An ant colony has often been called a "superorganism." In the pages that follow, you will see the extent to which our own physiology is similarly controlled by potent chemical messengers.

As we have seen, the maintenance of homeostasis in the body of an animal is a complex process, involving the regulation of many different physiological activities. Internal and external conditions are constantly monitored, and the resulting information is communicated to centers in which it is integrated and from which appropriate responses are initiated. Throughout the living world, the communication and integration of information and the control of the subsequent responses are accomplished through chemical stimuli. Specific molecules interact with receptors that are either inside individual cells or embedded in their plasma membranes. These interactions trigger changes in the cells that lead, directly or indirectly, to the appropriate responses.

In small, simple animals and at many sites in larger, more complex animals, signalling molecules move by diffusion from the cells in which they are produced to the cells on which they act. Often, however, the target cells are a considerable distance away, and the signalling molecules are carried by bulk flow in the bloodstream—a more rapid delivery system than simple diffusion. This delivery system, although adequate for many of the processes involved in maintaining homeostasis, is nevertheless too slow for effective coordination of the numerous activities that characterize most animals. A much more rapid, direct channel of communication is provided by neurons, specialized cells that use an electrical signal—the nerve impulse—to conduct information over great distances.

For many years, the relatively slow, but long-acting regulation of the physiological processes of an animal by chemical means and the rapid conduction of information by electrical means were regarded as distinct phenomena, involving two different systems (Figure 34–1, on the next page). Chemical regulation was considered the province of the **endocrine system,** which consists of clusters of specialized secretory cells—glands—that synthesize and release chemical substances—hormones—that travel through the bloodstream. By contrast, the nervous system, the

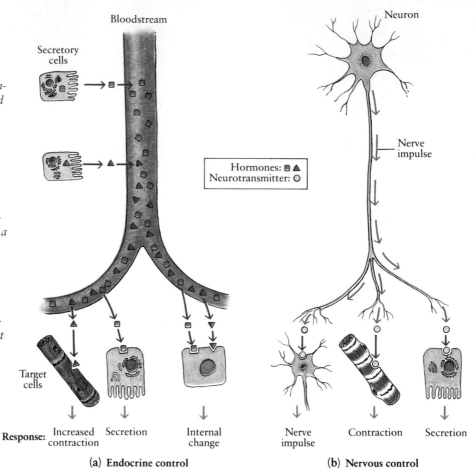

34–1 *The classical distinction between endocrine control and nervous control is based on the means by which information is carried over long distances. (a) In endocrine control, specific molecules, known as hormones, diffuse into the bloodstream, which carries them through the body to target tissues. This process may take minutes to hours, and the effects are typically long-lasting. (b) In nervous control, electrical signals—nerve impulses—are conducted along a neuron to its terminus where specific molecules, known as neurotransmitters, are released and diffuse to the target tissues. The entire process takes only a fraction of a second, and the effect is similarly short-lived. Both hormones and neurotransmitters interact with specific receptors on or in the target cells, leading to a response.*

Bloodstream

Neuron

Secretory cells

Nerve impulse

Hormones: ■ ▲
Neurotransmitter: ○

Target cells

Response: Increased contraction | Secretion | Internal change | Nerve impulse | Contraction | Secretion

(a) Endocrine control **(b) Nervous control**

interconnected network of neurons and their supporting tissues, was thought to be involved only in the conduction of nerve impulses.

It is now known, however, that there is considerable overlap between these two systems. For example, although an individual neuron conducts information electrically, it transmits that information to other cells, including other neurons, chemically, through the release of specific molecules known as **neurotransmitters.** A number of neurotransmitters are chemically identical to hormones. Moreover, the signalling molecules of some neurons, known as **neurosecretory cells,** diffuse into the bloodstream, which carries them—like hormones—to their target tissues. These examples, and others, suggest that the endocrine and nervous systems are most accurately viewed as different aspects of one overall system, the **neuroendocrine system.**

In this chapter, we shall begin our exploration of integration and control in the animal body with the hormone-secreting glands of the endocrine system. This will enable us to establish the basic principles of chemical communication among cells, which also apply to the nervous system. In Chapter 35, our focus will be the organization of the nervous system and the mechanisms by which information is transmitted within it. Chapter 36 will be devoted to sensory perception, information processing, and the subsequent motor response.

Glands and Their Products: An Overview

Hormones are, by definition, organic molecules secreted in one part of an organism that diffuse or are transported by the bloodstream to other parts of the organism, where they have specific effects on target organs or tissues. As we have seen in previous chapters, hormones are produced by a variety of different cell types: epithelial cells of the digestive tract, cardiac muscle cells, white blood cells, and even injured or infected cells. In these cell types and many others, the secretion of hormones is only one of several functions. In glandular epithelial cells and neurosecretory cells, however, the secretion of hormones is the primary function, to which all other activities are subordinated.

Glands, which are clusters of glandular epithelial cells or neurosecretory cells, are classified as exocrine or endocrine (Figure 34–2). **Exocrine glands** secrete their products into ducts that empty onto surfaces, such as the skin or the lining of the stomach. Examples are sweat glands, digestive glands, and milk glands. **Endocrine glands,** by contrast, secrete their products into the extracellular fluids, from which they diffuse into the bloodstream; thus they are sometimes referred to as "ductless" glands.

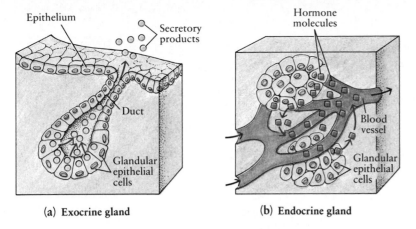

(a) **Exocrine gland**   (b) **Endocrine gland**

34–2   (a) *Exocrine glands, such as the mammary glands of the female mammal or the sweat glands of human skin, secrete their products into ducts. The products move through the ducts into body cavities, into other organs, or onto the body surface.* (b) *Endocrine glands, such as the pituitary and the thyroid, secrete their products into the extracellular fluids. From the extracellular fluids, the products—hormones—diffuse into the bloodstream and are transported through the body to their target tissues.*

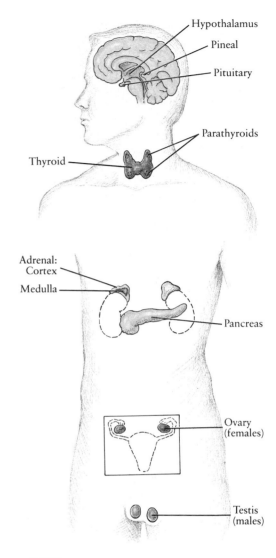

34–3   *Some of the hormone-producing (endocrine) organs. The pituitary releases hormones that, in turn, regulate the hormone secretions of the thyroid, the adrenal cortex (the outer layer of the adrenal gland), and the reproductive organs. The pituitary is itself under the regulatory control of a major brain center, the hypothalamus. The hypothalamus thus is the major link between the nervous system and the endocrine system.*

The principal endocrine glands of the vertebrate body are shown in Figure 34–3. The hormones they secrete are of three general types: steroids (see page 59), peptides or proteins, and amino acid derivatives. Hormones are characteristically active in very small amounts. It has been calculated, for example, that the concentration of adrenaline normally present in your bloodstream can be approximated by 8 milliliters (one teaspoonful) in a lake 2 meters deep and 100 meters in diameter.

As befits such potent chemicals, playing key roles in the integration and control of the body's physiological functions, hormones are themselves under tight control. One aspect of this control is the regulation of their production. With very few exceptions, hormones are under negative feedback control, as described in Chapter 28. Another, equally important aspect is that, once secreted, hormones are rapidly degraded. As a consequence, even slight increases in their concentrations—resulting from increased secretion—can trigger responses in target tissues and organs.

Table 34–1 summarizes the principal endocrine glands of vertebrates. In this chapter, we shall discuss most of these glands, the hormones they secrete, and the regulation of their secretion. We shall, however, defer discussion of the glands and hormones involved in reproduction until Chapter 37.

Table 34–1   Some of the Principal Endocrine Glands of Vertebrates and the Hormones They Produce

Gland	Hormone	Principal Action	Mechanism Controlling Secretion	Chemical Composition
Pituitary, anterior lobe	Growth hormone (somatotropin)	Stimulates growth of bone, inhibits oxidation of glucose, promotes breakdown of fatty acids	Hypothalamic hormone(s)	Protein
	Prolactin	Stimulates milk production and secretion in "prepared" gland	Hypothalamic hormone(s)	Protein
	Thyroid-stimulating hormone (TSH)	Stimulates thyroid	Thyroxine in blood; hypothalamic hormone(s)	Glycoprotein
	Adrenocorticotropic hormone (ACTH)	Stimulates adrenal cortex	Cortisol in blood; hypothalamic hormone(s)	Polypeptide (39 amino acids)
	Follicle-stimulating hormone (FSH)*	Stimulates ovarian follicle, spermatogenesis	Estrogen in blood; hypothalamic hormone(s)	Glycoprotein
	Luteinizing hormone (LH)*	Stimulates ovulation and formation of corpus luteum in female, interstitial cells in male	Progesterone or testosterone in blood; hypothalamic hormone(s)	Glycoprotein
Hypothalamus (via posterior pituitary)	Oxytocin	Stimulates uterine contractions, milk ejection	Nervous system	Peptide (9 amino acids)
	Antidiuretic hormone (ADH, vasopressin)	Controls water excretion	Osmotic concentration of blood; blood volume; nervous system	Peptide (9 amino acids)
Thyroid	Thyroxine, other thyroxinelike hormones	Stimulate and maintain metabolic activities	TSH	Iodinated amino acids
	Calcitonin	Inhibits release of calcium from bone	Concentration of Ca^{2+} ions in blood	Polypeptide (32 amino acids)
Parathyroid	Parathyroid hormone (parathormone)	Stimulates release of calcium from bone; stimulates conversion of vitamin D to active form, which promotes calcium uptake from gastrointestinal tract; inhibits calcium excretion	Concentration of Ca^{2+} ions in blood	Polypeptide (34 amino acids)
Adrenal cortex	Cortisol, other cortisol-like hormones	Affect carbohydrate, protein, and lipid metabolism	ACTH	Steroids
	Aldosterone	Affects salt and water balance	Processes initiated in the kidney; K^+ ions in blood	Steroid
Adrenal medulla	Adrenaline and noradrenaline	Increase blood sugar, dilate or constrict specific blood vessels, increase rate and strength of heartbeat	Nervous system	Amino acid derivatives
Pancreas	Insulin	Lowers blood sugar, increases storage of glycogen	Concentration of glucose and amino acids in blood; somatostatin	Polypeptide (51 amino acids)
	Glucagon	Stimulates breakdown of glycogen to glucose in the liver	Concentration of glucose and amino acids in blood; somatostatin	Polypeptide (29 amino acids)
Pineal	Melatonin	Involved in regulation of circadian rhythms	Light-dark cycles	Amino acid derivative
Ovary, follicle	Estrogens*	Develop and maintain sex characteristics in females, initiate buildup of uterine lining	FSH	Steroids
Ovary, corpus luteum	Progesterone and estrogens*	Promote continued growth of uterine lining	LH	Steroids
Testis	Testosterone*	Supports spermatogenesis, develops and maintains sex characteristics of males	LH	Steroid

* These hormones will be discussed in Chapter 37.

The Pituitary Gland

The **pituitary gland** was once considered the master gland of the body, as it is the source of hormones stimulating the reproductive organs, the adrenal cortex, and the thyroid. However, it is now known that this "master" gland is itself regulated by a key brain center, the **hypothalamus.** Hormones from the hypothalamus stimulate or, in some cases, inhibit the production of pituitary hormones. The pituitary, about the size of a kidney bean, is located at the base of the brain near the geometric center of the skull.

The anterior lobe of the pituitary is the source of at least six different hormones. One of these is **growth hor-**

mone, sometimes called **somatotropin,** which stimulates protein synthesis and promotes the growth of bone. As is the case with most of the hormones, growth hormone is best known by the effects caused by too much or too little. If the production of growth hormone is inadequate in childhood, a midget results, the so-called "pituitary dwarf." An excess of growth hormone during childhood results in a giant; most circus giants are the result of such an excess. Excessive growth hormone in the adult does not lead to giantism, since growth of the long bones has ceased. It leads instead to acromegaly, an increase in the size of the jaw and the hands and feet, adult tissues that are still sensitive to the effects of growth hormone.

Growth hormone also affects glucose metabolism, inhibiting the uptake and oxidation of glucose by some types of cells. It also stimulates the breakdown of fatty acids, thus conserving glucose. This hormone is now being produced by recombinant DNA techniques, which will extend its use in medical treatment and facilitate the study of its role in glucose metabolism.

A second hormone produced by the anterior pituitary is **prolactin,** which stimulates the secretion of milk in female mammals after they have given birth. Its production is controlled by an inhibitory hormone produced by the hypothalamus. As long as an infant continues to nurse, the nerve impulses produced by the suckling of the breast are transmitted to the hypothalamus, which decreases production of prolactin-inhibiting hormone. The pituitary then releases prolactin, which, in turn, acts upon the breast to maintain the production of milk. Once suckling ceases, the synthesis and release of prolactin decrease and so milk production stops. Thus supply is regulated by demand.

Four of the hormones secreted by the anterior pituitary are **tropic hormones**—hormones that act on other endocrine glands to regulate their secretions. One of these tropic hormones is **thyroid-stimulating hormone (TSH).** TSH stimulates cells in the thyroid gland to increase their production and release of thyroxine, the thyroid hormone. In a negative feedback loop involving both the pituitary and the hypothalamus, the increased concentration of thyroxine inhibits the further secretion of TSH by the pituitary. **Adrenocorticotropic hormone (ACTH)** has a similar regulatory relationship with the production of cortisol, a hormone produced by the adrenal cortex (the outer layer of the adrenal gland).

The other two tropic hormones secreted by the anterior pituitary are **gonadotropins**—hormones that act upon the **gonads,** or gamete-producing organs (the testes and the ovaries). These hormones, follicle-stimulating hormone (FSH) and luteinizing hormone (LH), will be discussed in Chapter 37.

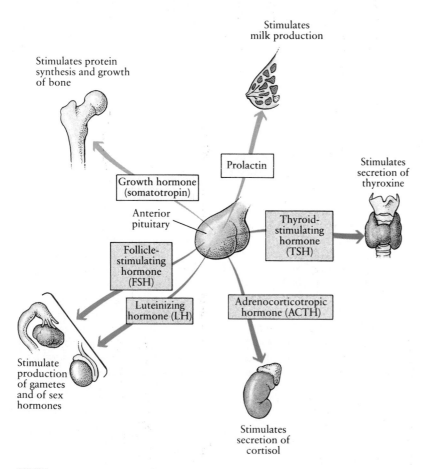

34–4   *Hormones secreted by the anterior lobe of the pituitary gland and their target tissues. Tropic hormones are indicated in blue. Each of the six different hormones produced by the anterior pituitary is synthesized by recognizably different cells. The capacity of a target tissue to respond to a particular hormone depends on the presence of receptors to which the hormone can bind.*

The Hypothalamus

The pituitary gland lies beneath the hypothalamus and is directly under its influence. The pituitary is also under the influence, by way of the hypothalamus, of other parts of the brain.

The hypothalamus is the source of at least nine hormones that act either to stimulate or inhibit the secretion of hormones by the anterior pituitary. These hormones are small peptides, one of which is only three amino acids in length. They are unusual not only for their small size but also for the way in which they reach their target gland. Produced by neurosecretory cells of the hypothalamus, they travel only a few millimeters to the pituitary, apparently never entering the general circulation. However, they make this brief passage by way of capillaries (Figure 34–5).

Figure 34–6 summarizes the negative feedback systems linking the hypothalamus and pituitary with the thyroid, adrenal cortex, and gonads. These feedback control systems, generally involving both the pituitary and the hypothalamus, provide for both homeostasis and response to changing conditions. The pituitary feedback system usually provides for constancy. However, it can be overridden by the hypothalamic system, which takes into account not only the balance between output and input but also the changes elsewhere in the body and in the external environment.

The hypothalamus is also the source of two hormones stored in and released from the posterior pituitary: **oxytocin** and **antidiuretic hormone (ADH).** Oxytocin accelerates childbirth by increasing uterine contractions during labor. These contractions also cause the uterus to regain its normal size and shape after delivery. Oxytocin is also responsible for the ejection of milk from the cells in which it is synthesized to the ducts leading to the nipple, where it is available to the nursing infant.

ADH, as we saw in Chapter 32, decreases the excretion of water by the kidneys. ADH is sometimes called vasopressin because it increases blood pressure in response to certain unusual circumstances, for example, the blood loss of a severe hemorrhage.

The Thyroid Gland

The thyroid, under the influence of thyroid-stimulating hormone from the pituitary, produces **thyroxine,** which is an amino acid combined with four atoms of iodine (Figure 34–7). Thyroxine accelerates the rate of cellular respiration. In some animals, it also plays a major role in temperature regulation (see Figure 32–9, page 577).

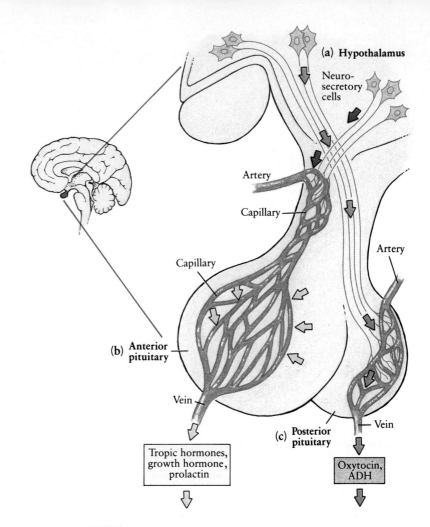

34–5   *Relationships between the hypothalamus and the pituitary.* (**a**) *The hypothalamus communicates with* (**b**) *the anterior lobe of the pituitary through a small capillary bed. Neurosecretory cells of the hypothalamus secrete releasing or inhibiting hormones (red arrows) directly onto capillaries that are linked to a second capillary bed in the anterior pituitary. Here the hypothalamic hormones affect the production of pituitary hormones (yellow arrows).*

Other hypothalamic neurosecretory cells produce oxytocin and ADH (orange arrows), which are transmitted through the nerve fibers to (**c**) *the posterior lobe of the pituitary. Following their release from the nerve endings in the posterior pituitary, these hormones diffuse into capillaries and thus enter the general circulation.*

Hyperthyroidism, the overproduction of thyroxine, results in nervousness, insomnia, and excitability; increased heart rate and blood pressure; heat intolerance and excessive sweating; and weight loss. Hypothyroidism (too little thyroxine) in infancy affects development, particularly of the brain cells. If not treated in time, it can lead to permanent mental deficiency and dwarfism. In adults, hypothyroidism is associated with dry skin, intolerance to cold, and lack of energy.

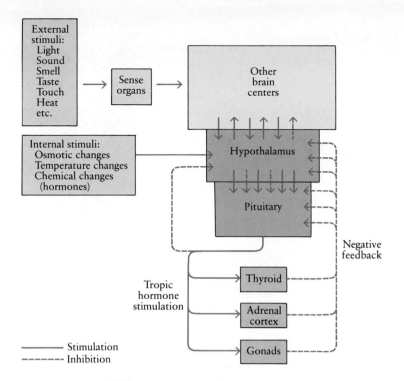

The thyroid also secretes the hormone **calcitonin** in response to rising calcium levels in the blood. Calcitonin's major action is to inhibit the release of calcium ion from bone.

The Parathyroid Glands

The pea-sized parathyroid glands, the smallest of the known endocrine glands, are located behind or within the thyroid gland (Figure 34–8). They produce **parathyroid hormone,** which is also known as **parathormone.** This hormone increases the concentration of calcium ion in blood in several different ways. Parathyroid hormone stimulates the conversion of vitamin D into its active form (see Figure 29–10, page 536). Active vitamin D, in turn, increases the absorption of calcium ion from the intestine. Parathyroid hormone also reduces excretion of calcium ion from the kidneys. In addition, it stimulates the release into the bloodstream of calcium from bone, which contains 99 percent of the body's total calcium.

34–6   *The production of many hormones is regulated by complex negative feedback systems involving the pituitary and the hypothalamus. The hypothalamus controls the pituitary's secretion of tropic hormones, and these, in turn, stimulate the secretion of hormones from the thyroid, adrenal cortex, and gonads (the testes or the ovaries). As the concentration of the hormones produced by these target glands rises in the blood, the hypothalamus decreases its production of releasing hormones, the pituitary decreases its hormone production, and production of hormones by the target glands also slows. By way of the hypothalamus, which receives information from many other parts of the brain, hormone production is also regulated in response to other changes in the external and internal environments.*

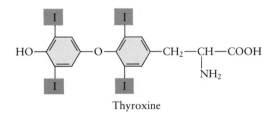

Thyroxine

34–7   *Thyroxine, the principal hormone produced by the thyroid gland. Note the four iodine atoms in its structure. Because iodine is needed for thyroxine, it is an essential component of the human diet. Insufficient iodine is often associated with goiter, an enlargement of the thyroid gland. Where iodine is present in the soil, it is available in minute quantities in drinking water and in plants. In the United States, table salt is ordinarily iodized or must be specifically labeled as being uniodized.*

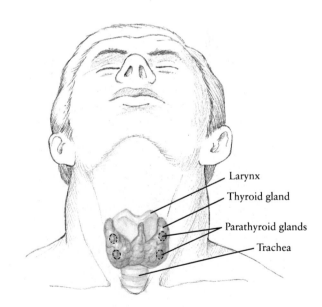

34–8   *The pea-sized parathyroid glands, the smallest of the known endocrine glands, are located behind or within the thyroid gland. They produce parathyroid hormone (parathormone), which increases concentrations of blood calcium. Calcitonin, a hormone produced by the thyroid gland, decreases blood calcium.*

Calcium ion is essential for blood coagulation, muscle contraction, nerve function, and many other processes. Parathyroid hormone and calcitonin work as a fine-tuning mechanism, regulating blood calcium, with parathyroid hormone apparently playing the principal role. The production of both hormones is regulated directly by the concentration of calcium ions in the blood (Figure 34–9).

The Adrenal Cortex

The adrenal glands are on top of the kidneys. The outer layer of the adrenal gland, the **adrenal cortex,** is the source of a number of steroid hormones. Cortisol and aldosterone are among the most active.

Cortisol and cortisol-like hormones promote the formation of glucose from protein and fat. At the same time, they decrease the utilization of glucose by most cells, with the notable exceptions of cells of the brain and the heart. Thus cortisol and related hormones favor the activities of these vital organs at the expense of other body functions. Their release increases during periods of stress, such as facing new situations, engaging in athletic competition, and taking final exams. As we shall see in the next chapter, they work in concert with the sympathetic nervous system.

In addition to their effects on glucose metabolism, cortisol and cortisol-like hormones suppress inflammatory and immune responses. Current evidence suggests that this immunosuppressive activity may be part of the normal regulatory mechanism that turns off the inflammatory and immune responses when their work is done. It may also be a factor in the increased susceptibility to illness that often accompanies stress. Because of their immunosuppressive properties, cortisol and related hormones are sometimes used in the treatment of autoimmune diseases and severe allergic reactions. However, the serious side effects of the high doses often required limit their usefulness.

Aldosterone and related hormones are involved in the regulation of ion concentrations, particularly the concentrations of sodium and potassium ions. An increase in aldosterone secretion, as we noted in Chapter 32, results in greater reabsorption of sodium ions in the distal tubule and collecting duct of the nephron and increases the secretion of potassium ions into them. A deficiency of aldosterone precipitates a critical loss of sodium ions from the body in the urine and, with it, a loss of water, leading, in turn, to a reduction in blood pressure.

In both males and females, the adrenal cortex also produces small amounts of male sex hormones. An adrenal tumor may result in increased production of these

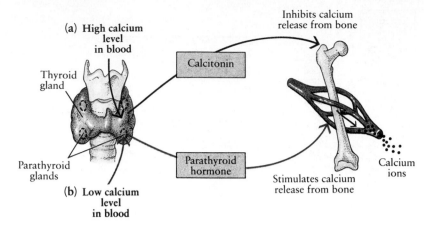

34–9   *Regulation of the level of calcium in the blood by calcitonin and parathyroid hormone.* (a) *When the level of calcium in the blood is high, calcitonin is secreted by the thyroid gland. The principal effect of this hormone is to inhibit the release of calcium ion from the reservoir provided by bone.* (b) *When the level of calcium in the blood is low, parathyroid hormone is secreted by the parathyroid glands. This hormone stimulates the release of calcium ion from bone and thus leads to an increase in the concentration of calcium ion in the blood. Parathyroid hormone also stimulates the uptake of calcium from the intestine and decreases the excretion of calcium from the kidneys.*

hormones and, consequently, in women, the production of facial hair and other masculine characteristics. The adrenal cortex is also thought to secrete small amounts of female sex hormones.

The Adrenal Medulla

The **adrenal medulla,** the central portion of the adrenal gland, is a large cluster of neurosecretory cells whose nerve endings secrete **adrenaline** and **noradrenaline,** which are taken up by the bloodstream. These hormones, known also as epinephrine and norepinephrine, increase the rate and strength of the heartbeat, raise blood pressure, stimulate respiration, and dilate the respiratory passages. They also increase the concentration of glucose in the bloodstream by promoting the activity of the enzyme that breaks down glycogen to glucose 1-phosphate. The adrenal medulla is stimulated by nerve fibers of the sympathetic division of the autonomic nervous system and so acts as an enforcer of sympathetic activity.

The Pancreas

The pancreas is both an endocrine and an exocrine gland (Figure 34–10). The **islet cells** are the source of insulin and glucagon, two hormones involved in the

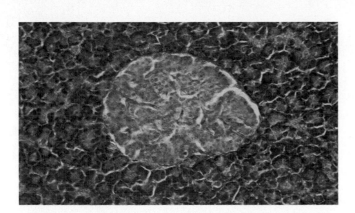

34–10 *A cross section of pancreatic tissue. The small cells clustered together in the center of the micrograph are islet cells, which are the endocrine cells of the pancreas. Different types of islet cells secrete the hormones insulin and glucagon. The meshwork of blue and white in and around the islet represents the capillaries that take up insulin and glucagon from the extracellular fluid. The surrounding exocrine cells, stained red in this tissue section, produce digestive enzymes. These enzymes are carried through the pancreatic duct to the small intestine (see Figure 29–8, page 531).*

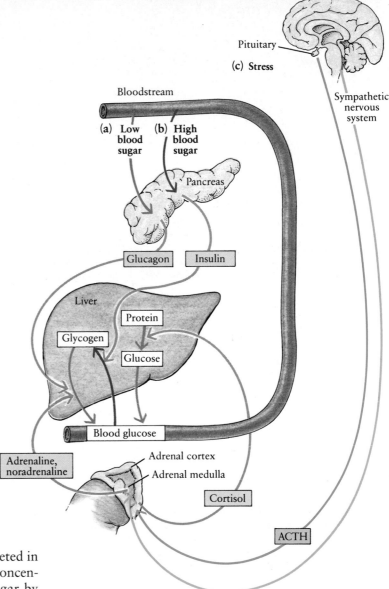

regulation of glucose metabolism. **Insulin** is secreted in response to a rise in blood sugar or amino acid concentration (as after a meal). It lowers the blood sugar by stimulating cellular uptake and utilization of glucose and by stimulating the conversion of glucose to glycogen.

Glucagon, produced by different islet cells of the pancreas, increases blood sugar. It stimulates the breakdown of glycogen to glucose in the liver and the breakdown of fats and proteins, which decreases glucose utilization.

Thus, as we have seen, at least six different hormones are involved in regulating blood sugar: growth hormone, cortisol, adrenaline and noradrenaline, insulin, and glucagon. The multiplicity of mechanisms regulating blood glucose levels, illustrated in Figure 34–11, ensures that glucose is always available for brain cells. Unlike other cells in the body, which can derive energy from the breakdown of amino acids and fats, brain cells can utilize only glucose under most circumstances. Thus they are immediately affected by low blood sugar.

34–11 *Hormonal regulation of blood glucose.* (a) *When blood sugar concentrations are low, the pancreas releases glucagon, which stimulates the breakdown of glycogen and the release of glucose from the liver.* (b) *When blood sugar concentrations are high, the pancreas releases insulin, which removes glucose from the bloodstream by increasing its uptake by cells and promoting its conversion into glycogen, the storage form.* (c) *Under conditions of stress, ACTH, produced by the pituitary, stimulates the adrenal cortex to produce cortisol and related hormones. These hormones increase the breakdown of protein and its conversion to glucose in the liver. At the same time, stimulation of the adrenal medulla by nerve fibers of the sympathetic division of the autonomic nervous system triggers the release of adrenaline and noradrenaline, which also raise blood sugar.*

Growth hormone, not shown in this diagram, also affects glucose metabolism, inhibiting the uptake and oxidation of glucose by many types of cells and stimulating the breakdown of fatty acids.

Circadian Rhythms

Virtually every living organism, from one-celled algae to plants to *Homo sapiens,* exhibits regular day-night cycles in many physiological functions. These cycles are known as **circadian rhythms,** from the Latin words *circa,* meaning "about," and *dies,* "day." Circadian rhythms are not exact. Different species and different individuals of the same species often have slightly different, but consistent, rhythms, often as much as an hour or two longer or shorter than 24 hours.

A variety of experiments have demonstrated that circadian rhythms are internal—that is, they are caused by factors within the organism itself. The mechanism by which they are controlled is known as a **biological clock.** In humans, as in other organisms, light plays a key role in setting the biological clock that governs these rhythms. The chemical nature of the clock—or whether there is just one kind of clock or many—is still not known.

Work with *Drosophila,* however, suggests that the products of one or more specific genes are involved. Mutations in a gene located on the X chromosome and known as *per* (for "period") have profound effects on the circadian rhythms of fruit flies. If the gene is deleted, the flies become insomniacs, with no discernible sleep-wake cycle. Other mutations affect the length of the daily rhythms. For example, the allele *per s* ("short periods") produces flies with daily cycles of 18 to 20 hours, and the allele *per l* ("long periods") results in daily cycles of 28 to 30 hours.

This same gene also affects the mating songs of male fruit flies, which beat their wings together in a rhythmic pattern that repeats at 60-second intervals. When the *per* gene is deleted the mating songs have no rhythms at all; *per s* results in songs at 40-second intervals, and *per l* in songs at 80-second intervals.

Investigators puzzled by how a single gene could affect both the daily 24-hour rhythm and the 60-second courtship song created mosaic flies in which cells in certain parts of the body contained the normal gene and cells in other parts of the body contained a mutant gene. They found that the location in which *per* is expressed determines its effect. In cells of the head, it affects the daily rhythm, but in cells of the thorax, it affects the song rhythms. Moreover, the greater the quantities of the *per* gene product synthesized, the faster the clock runs. The gene has now been sequenced, and it turns out to contain a series of repetitive sequences. Similar repetitive sequences have been identified in DNA from chickens, mice, and humans, but it is not yet known if these sequences are the portion of the gene controlling the rhythms.

Although the mechanism governing circadian rhythms remains mysterious, the effect of the rhythms in human physiology and behavior is not in doubt. For instance, we are more likely to be born between 3 and 4 A.M. and also to die in these same early morning hours. Body temperature fluctuates as much as 1°C (about 2°F) during the course of 24 hours, usually reaching a peak at about 4 P.M. and a low about 4 A.M. Alcohol tolerance is greatest at 5 P.M., while tolerance to pain is lowest at 6 P.M. in most persons. Respiration, heart rate, and urinary excretion of potassium, calcium, and sodium, all vary according to the time of day (heart rate by as much as 20 beats per minute). Secretion of various hormones follows circadian rhythms: serum levels of cortisol peak between 4 A.M. and 8 A.M. in people on a normal sleep-wake cycle; growth hormone levels rise about an hour after falling asleep; prolactin peaks around 3 A.M.; testosterone about 9 A.M.

Knowledge of circadian rhythms may have important practical consequences. A study by the Federal Aviation Administration showed that pilots flying from one time zone to another—from New York to Europe, for instance—exhibit "jet lag," a general decrease in mental alertness, an inability to concentrate, and an increase in decision time and physiological reaction time. Measurements of various functions show that the body may be "out of sync" for as much as a week after such a flight. This brings into question not only schedules for airline personnel but also policies of speeding diplomats to foreign capitals at times of international crisis or of air transport of troops into combat. It may be significant that the accidents at the nuclear power plants at both Three Mile Island and Chernobyl occurred in the early hours of the morning local time. In the case of Three Mile Island, the workers on duty at the time of the accident had been alternating between the day shift and the night shift every week.

Medical researchers are finding that failure to take circadian fluctuations into account can lead to mistakes in diagnosis and treatment. For example, blood pressure may vary as much as 20 percent in the course of a day, so that a person might be found to be within the normal range at one time of day and diagnosed as hypertensive at another. Numbers of white blood cells may vary as much as 50 percent in a 24-hour period, rendering the same immunosuppressive therapy ineffectual at one time of day and life-threatening at another. Studies of cancer chemotherapy in mice indicate that proper timing of drug doses can mean a doubling of survival rate. Determination of a patient's chronobiology may someday become as routine in therapy as blood typing or taking a medical history.

The Pineal Gland

The pineal gland is a small lobe lying near the center of the brain in humans. In lower vertebrates, it contains light-sensitive cells and so is sometimes called the third eye. The pineal gland secretes the hormone **melatonin.** In many species, including chickens, rats, and humans, the production of melatonin rises sharply at night and falls rapidly in the daytime. Exposure to light during the dark cycle interrupts production of the hormone.

In sparrows, injections of melatonin induce roosting and the lowering of body temperature, both of which are characteristic nighttime events in these birds. Melatonin also inhibits the development of the gonads in species as disparate as chickens and hamsters. Its decreased production in the spring, when the days are becoming longer (and the nights are becoming shorter), is believed to be associated with the seasonal enlargement of the gonads in preparation for mating. In humans, the pineal gland may be involved in sexual maturation. Tumors of the pineal have been associated with precocious puberty.

Thus there are a few tempting clues suggesting that the pineal gland may function as a biological timekeeper. However, the way in which light affects the gland and the way in which melatonin alters human physiological responses—if it does—are yet to be discovered.

Prostaglandins

Among the most potent of all substances produced by and released from cells are a group of related chemicals first detected in semen. These substances were thought at the time to be produced by the prostate gland, a structure of the male reproductive system, and so they were called **prostaglandins.** Later research revealed, however, that most of the prostaglandins in semen are synthesized in other structures, the seminal vesicles. A large number of prostaglandins have now been identified, all related structurally but with a variety of different, and sometimes directly opposite, effects.

Although prostaglandins have hormonelike properties, they differ from other hormones in several significant ways:

1. Unlike any other hormones, they are fatty acids.
2. They are produced by plasma membranes in most—if not all—organs of the body, as opposed to other hormones, which are produced by glandular epithelium or neurosecretory cells.
3. Their target tissues are generally either the same tissues in which they are produced or the tissues of another individual.
4. They produce marked effects at extremely low concentrations, much lower than those of most hormones.

Stimulation of Smooth Muscle

Among the many effects of prostaglandins, one of the most striking is their capacity to induce contractions in smooth muscle. This is believed to play several important roles in reproduction. The walls of the uterus, which are composed of smooth muscle, normally contract in continuous waves. After sexual intercourse, prostaglandins from semen are found in the female reproductive tract, where they increase the rhythmic contractions of the uterine wall and oviducts. It is believed that this action assists both the sperm on its journey to the oviduct and the oocyte as it travels from the oviduct to the uterus. The semen of some infertile males has been found to be poor in prostaglandins, and the uterus of infertile females is often unresponsive to prostaglandins.

The contractions of the uterus also increase during a menstrual period and reach their greatest strength when a woman is in labor. Prostaglandins produced in the uterine lining are believed to play a key role in triggering the onset of both menstruation and labor.

Increased understanding of prostaglandins and their capacity to stimulate smooth muscle has shed light on a long-perplexing medical problem. Between 30 and 50 percent of all women of childbearing age experience painful uterine cramps during the first day or two of each menstrual period, a condition known as dysmenorrhea. In most of these women, no abnormalities of the reproductive organs can be detected. It is now known, however, that the menstrual fluid of such women contains concentrations of prostaglandins two to three times higher than the levels found in the menstrual fluid of women without dysmenorrhea. The increased prostaglandin levels not only cause stronger, more rapid contractions of the uterine walls but also reduce the blood supply to the tissue. As a result, less oxygen is available to the actively contracting muscles, creating an oxygen debt (page 155) and its accompanying pain. It is also hypothesized that prostaglandins act directly on pain nerve endings, causing them to fire more rapidly.

Compounds that inhibit prostaglandin synthesis are highly effective in reducing or, in some cases, completely eliminating the symptoms of dysmenorrhea.

This alleviation is accompanied by a marked reduction in the levels of prostaglandins found in the menstrual fluid. The use of these compounds in preventing labor in women who are in danger of giving birth prematurely is currently under investigation.

Other Prostaglandin Effects

Prostaglandins that are closely related chemically may differ widely in their physiological effects. Although many of them stimulate contractions of smooth muscle, others inhibit smooth muscle contraction. Furthermore, one may affect the smooth muscle of bronchioles, another the smooth muscle of blood vessels. One, produced by platelets, is a potent stimulus for platelet aggregation and constriction of the blood vessels. Another, produced by the endothelial cells that line the blood vessels, is a potent inhibitor of platelet aggregation and a dilator of the vessels. The way in which prostaglandins exert these multitudinous effects is not known.

Among the prostaglandins are a group of substances known as **leukotrienes,** which are produced principally by the various white blood cells involved in the inflammatory and immune responses. The leukotrienes include the interleukins released by activated helper T cells (page 594), as well as a variety of molecules released by stimulated macrophages and mast cells. As this information might lead you to expect, increased levels of prostaglandins have been implicated in disorders of the immune system such as rheumatoid arthritis, asthma, and severe allergies.

Discovery of the prostaglandins and their involvement in inflammation has also provided a solution to another long-standing medical puzzle. Aspirin is one of the oldest and most effective medications known, but how it acts was, for many years, a mystery. Now it has been discovered that aspirin exerts its effects, at least in part, by inhibiting the synthesis of prostaglandins, thus producing its well-known soothing effects on inflammation and fever.

Mechanisms of Hormone Action

Prostaglandins, like many of the regulatory chemicals in small organisms, travel only short distances to the cells they affect. The communication is rather like that of one person talking to another in the same room. Neurotransmitters released from the nerve endings of neurons similarly travel only a short distance. As shown in Figure 34–1 (page 606), neurons make close, anatomical connections with the cells they influence, like a con-

versation on the telephone. Most hormones, by contrast, broadcast their messages. Whether these messages are received and acted upon depends upon the receptivity of the target tissue as much as on the chemical characteristics of the hormone.

The key to hormone specificity lies in receptor molecules (proteins) that have precise configurations. This allows the receptors to bind with one particular molecule but not with another that differs only slightly in structure. The study of hormones and their receptors has revealed two quite different mechanisms of action. In one, the receptors are intracellular, in the nucleus or cytoplasm. In the other, the receptors are embedded in the plasma membrane. The steroid hormones and thyroid hormone utilize the first mechanism; the peptide and protein hormones, as well as the amino acid derivatives, act through the second mechanism.

Intracellular Receptors

Steroid hormones are relatively small, lipid-soluble molecules. Thus they pass easily through plasma membranes and freely enter all cells of the body. Thyroid hormone also passes readily through plasma membranes, apparently by facilitated diffusion through a membrane protein. In their target cells—and only in their target cells—these hormones encounter specific receptor molecules with which they combine (Figure 34–12). Within the nucleus, the hormone-receptor complex binds to a specific chromosomal protein, initiating mRNA transcription from the particular DNA sequence regulated by that protein.

After appropriate processing, the mRNA moves into the cytoplasm and protein synthesis occurs. The newly synthesized proteins may be structural proteins, enzymes, or other hormones. The result is a functional change in the cell, in the substances released from it, or in the receptors displayed on its surface.

Membrane Receptors

Peptide and protein hormones, as well as amino acid derivatives such as adrenaline, cannot pass through plasma membranes. These hormones act by combining with receptors in the plasma membranes of target cells.

The existence of these receptors has important medical implications. For example, it has long been known that juvenile-onset diabetes—diabetes that typically first appears in childhood or adolescence—is caused by a deficiency of the hormone insulin, which, as we have seen, promotes the uptake of glucose by cells. It was assumed by analogy that the adult-onset diabetes that is associated with obesity had the same cause. It is now

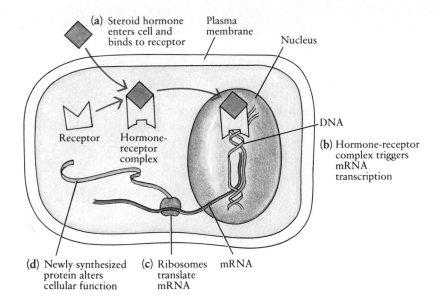

(a) Steroid hormone enters cell and binds to receptor

Plasma membrane

Nucleus

Receptor

Hormone-receptor complex

DNA

(b) Hormone-receptor complex triggers mRNA transcription

(d) Newly synthesized protein alters cellular function

(c) Ribosomes translate mRNA

mRNA

34–12 *The mechanism of action of a steroid hormone.* (a) *The lipid-soluble hormone passes through the plasma membrane into the cytoplasm. In its target cell, the hormone encounters a specific receptor to which it binds.* (b) *Within the nucleus, the hormone-receptor complex combines with a chromosomal protein, triggering the transcription of a segment of DNA into mRNA.* (c) *After processing, the mRNA is translated into protein.* (d) *The altered functions of the cell as a consequence of the newly synthesized protein constitute the cell's response to the hormone.*

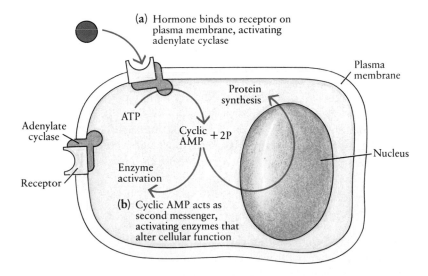

(a) Hormone binds to receptor on plasma membrane, activating adenylate cyclase

Plasma membrane

Protein synthesis

Adenylate cyclase

ATP

Cyclic AMP + 2P

Nucleus

Enzyme activation

Receptor

(b) Cyclic AMP acts as second messenger, activating enzymes that alter cellular function

34–13 *The mechanism of action of a hormone, such as adrenaline, that utilizes a second messenger.* (a) *The hormone binds to a receptor displayed on the outer surface of the plasma membrane of a target cell, activating the enzyme adenylate cyclase. This enzyme, located on the inner surface of the plasma membrane, catalyzes the conversion of ATP to cyclic AMP, with the release of two phosphate groups.* (b) *Cyclic AMP then acts as a second messenger, activating enzymes and initiating reactions that alter the functions of the cell.*

In the case of liver cells acted upon by adrenaline, the ultimate result is the release of large quantities of glucose into the bloodstream. In other types of cells, acted upon by other hormones, the action of the second messenger leads ultimately to the synthesis of new proteins. These proteins, in turn, produce changes in cellular function.

known, however, that adult-onset diabetes often results from a decrease in the number of insulin receptors in the plasma membranes of target cells rather than from a shortage of insulin. Such patients are treated most effectively by diet.

After a hormone combines with its membrane receptor, one of two events may follow, depending on the particular hormone. In some cases, the hormone-receptor complex is carried into the cytoplasm by receptor-mediated endocytosis (page 120). In other cases, the hormone never actually enters the cell. Instead, its binding to the receptor sets in motion a "second messenger" that is responsible for the sequence of events inside the cell. The second messenger for many hormones is a chemical known as **cyclic AMP**.

An example of the second-messenger mechanism is the stimulation of the release of glucose from a liver cell by adrenaline (Figure 34–13). Adrenaline molecules bind to a receptor on the outer surface of the plasma membrane. This event activates an enzyme, adenylate cyclase, that is bound on the inner surface of the membrane. Adenylate cyclase converts ATP to cyclic AMP. The cyclic AMP binds to another enzyme and activates it. This enzyme activates another enzyme, which, in turn, activates yet another. The final enzyme then breaks down glycogen at a high rate, producing glucose 1-phosphate, which is further broken down to glucose and released from the cell.

Since each enzyme molecule greatly increases the rate of the particular reaction it catalyzes and can be used over and over again, the number of molecules involved in these reactions is amplified at each step. Thus, the binding of a few molecules of adrenaline to cells in the liver initiates a reaction cascade that leads to the activation of an estimated 25 million molecules of the final enzyme. The result is the release of many grams of glucose into the blood.

Just about the same time that the role of cyclic AMP in mammalian cells was established, biologists studying that peculiar group of organisms known as the cellular slime molds (see page 122) isolated a chemical of great importance in this biological system. As you will recall, the cells of the cellular slime mold begin as individual amoebas and then come together to form a single organism. The chemical that calls them together was iden-tified as cyclic AMP. More recently, it has been found that insulin is present in fruit flies, earthworms, protists, fungi, and even *Escherichia coli*. Preliminary evidence indicates that other "mammalian" hormones, including ACTH and glucagon, are present in unicellular organisms. Their function in these organisms is not known, but they may also serve in cell-to-cell communication. These discoveries of the universality of hormones provide additional examples of the long thread of evolutionary history linking all organisms.

Summary

Hormones are signalling molecules secreted in one part of an organism that diffuse or are carried by the blood-stream to other tissues and organs, where they exert specific effects. Most hormones are secreted by glandular epithelial tissue or by neurosecretory cells. The principal endocrine glands of vertebrates include the pituitary, the hypothalamus, the thyroid, the parathyroids, the adrenal cortex and medulla, the pancreas (which is also an exocrine gland), the pineal, and the gonads (ovaries or testes).

The production of many hormones is regulated by negative feedback systems involving the anterior lobe of the pituitary gland and an area of the brain, the hypothalamus. Under the influence of hormones secreted by the hypothalamus, the pituitary produces tropic hormones that, in turn, stimulate the target glands to produce hormones. These hormones then act upon the pituitary or the hypothalamus (or both) to inhibit the production of the tropic hormones. Production of thyroid hormone and the steroid hormones of the adrenal cortex and gonads is regulated by the hypothalamus-pituitary system. The production of other hormones, such as calcitonin and parathyroid hormone, is regulated by the concentration in the bloodstream of other factors, such as ions.

Besides producing the tropic hormones, the anterior lobe of the pituitary also secretes growth hormone (somatotropin) and prolactin. In addition to producing peptide hormones that act on the anterior lobe of the pituitary, the hypothalamus produces the hormones ADH and oxytocin, which are stored in and released from the posterior lobe of the pituitary.

The islet cells of the pancreas are the source of two hormones involved in the regulation of blood glucose: insulin, which lowers blood sugar by stimulating cellular uptake of glucose, and glucagon, which raises blood sugar by stimulating the breakdown of storage forms of glucose. Blood sugar is also under the influence of adrenaline and noradrenaline, which are released from the adrenal medulla at times of stress; cortisol and related hormones, which are released from the adrenal cortex at times of stress; and growth hormone.

The pineal gland, located in the brain, is the source of melatonin and is believed to be involved with regulation of circadian and seasonal physiological changes. Its function in humans is not well understood.

Prostaglandins are a group of fatty acids that resemble other hormones in exerting effects on specific target tissues but that often act directly upon the tissues that produce them or, in some cases, on the tissues of another individual. Prostaglandins are formed in most, if not all, tissues of the body and affect such diverse functions as the contraction of smooth muscle, platelet aggregation, and the immune response.

Hormones act by at least two different mechanisms. Steroid hormones and thyroid hormone freely enter cells, where, after combining with an intracellular receptor, they exert a direct influence on the transcription of RNA. Other hormones, such as adrenaline, insulin, and glucagon, combine with receptor molecules on the plasma membranes of their target cells. The hormone-receptor combination may be carried into the cytoplasm by receptor-mediated endocytosis, or the combination may trigger the release of a "second messenger." The second messenger, in turn, sets off a series of events within the cell that is responsible for the end results of hormone activity. Cyclic AMP has been identified as the second messenger in many of these interactions.

Questions

1. Distinguish between the following: endocrine/exocrine; pituitary/hypothalamus; anterior pituitary/posterior pituitary; thyroid gland/parathyroid glands; adrenal cortex/adrenal medulla; insulin/glucagon.

2. Diagram the feedback system regulating the production and release of thyroid hormone.

3. In terms of the feedback system diagrammed in Question 2, explain why a shortage of iodine produces goiter.

4. Describe how the concentration of calcium ion in the bloodstream is regulated by the thyroid and parathyroid glands.

5. Which hormones act to increase the level of blood glucose? To decrease it? How does each hormone exert its effects?

6. How do steroid hormones and thyroid hormone exert their specific effects on target cells?

7. How do peptide or protein hormones exert their specific effects on target cells?

8. What types of functions would you expect to be controlled by the endocrine system rather than by the nervous system? Does the reality, as described in this chapter, fulfill your expectations?

Suggestions for Further Reading

Atkinson, Mark A., and Noel K. Maclaren: "What Causes Diabetes?" *Scientific American*, July 1990, pages 62–71.

Axelrod, Julius, and Terry D. Reisine: "Stress Hormones: Their Interaction and Regulation," *Science*, vol. 224, pages 452–459, 1984.

Barnes, Deborah M.: "Steroids May Influence Changes in Mood," *Science*, vol. 232, pages 1344–1345, 1986.

Berridge, Michael J.: "The Molecular Basis of Communication within the Cell," *Scientific American*, October 1985, pages 142–152.

Evans, Ronald M.: "The Steroid and Thyroid Hormone Receptor Superfamily," *Science*, vol. 240, pages 889–895, 1988.

Kolata, Gina: "Genes and Biological Clocks," *Science*, vol. 230, pages 1151–1152, 1985.

Lancaster, J. R., Jr.: "Nitric Oxide in Cells," *American Scientist*, vol. 80, pages 248–259, 1992.

Leinhard, Gustav E., Jan W. Slot, David E. James, and Mike M. Mueckler: "How Cells Absorb Glucose," *Scientific American*, January 1992, pages 86–91.

Orci, Lelio, Jean-Dominique Vassalli, and Alain Perrelet: "The Insulin Factory," *Scientific American*, September 1988, pages 85–94.

Pool, Robert: "Illuminating Jet Lag," *Science*, vol. 244, pages 1256–1257, 1989.

Snyder, Solomon H.: "The Molecular Basis of Communication between Cells," *Scientific American*, October 1985, pages 132–141.

Snyder, Solomon H., and David S. Bredt: "Biological Roles of Nitric Oxide," *Scientific American*, May 1992, pages 68–77.

Weissmann, Gerald: "Aspirin," *Scientific American*, January 1991, pages 84–90.

The Nervous System I: Structure and Function

Have you ever felt overwhelmed by fear or anger? Your heart pounds, beating faster and stronger. The blood vessels in your digestive tract and skin constrict, shunting more blood to your muscles. With these vessels constricted, more blood returns to your heart, raising your blood pressure. Your nostrils flair, and your chest heaves as you hyperventilate, bringing more oxygen into your lungs. The pupils of your eyes dilate. You may break out in a cold sweat. The muscles underlying your hair follicles contract; as a consequence, you may get "goose bumps" and feel a prickling sensation at the back of your neck. If you had a full furry coat, like Dash, the chimpanzee shown here, your hair would stand on end, making you look larger and more ferocious. The rhythmic movements of your intestines stop. The sphincter muscles of your bladder and your intestinal tract relax, which may have the disconcerting consequence of allowing involuntary urination or defecation, a not uncommon accompaniment of extreme fear. Your adrenal glands pour out a rush of adrenaline, causing the release of large quantities of glucose—an extra energy source—from your liver. Depending on how the adventure plays out, you are ready for fight or flight.

Dash was experiencing some of these feelings when this photograph was taken. He is one of about 30 chimps who were raised in captivity and have now been returned to the wild. Most, like Dash, were victims of a commercial trade in primates, a business that captures babies, usually killing their mothers in the process, and sells the babies to circuses, zoos, laboratories, or individuals who think they might make cute pets. Dash has made progress. By the age of 10, he had become top chimp in his little group of refugees, but only after many encounters such as the one that triggered the response shown here.

Fight-or-flight reactions are part of our heritage from an earlier stage of our evolutionary history. They, like other responses involving the nervous system, take place very rapidly, as we shall see in the course of this chapter.

As we saw in the last chapter, communication among the cells of a multicellular organism depends upon chemical stimuli. Molecules released from secretory cells are transported to other cells, where they interact with receptors, triggering a response. The presence in prokaryotes, protists, and fungi of signalling molecules identical to those of vertebrates and the dual function of some vertebrate molecules as both hormones and neurotransmitters provide strong evidence that the endocrine and nervous systems had a common evolutionary origin in primitive cell-cell communication systems.

The specialization that distinguishes a nervous system from other communication systems is the **neuron** (Figure 35–1, at the top of the next page), a cell that converts appropriate stimuli into electrochemical signals that are rapidly conducted through the neuron itself, often over great distances. Neurons then transmit these signals to other neurons across junctions known as **synapses,** usually by the release of specific neurotransmitter molecules, or to effector cells, such as those of muscles and glands. The rapid communication provided by neurons and their arrangement in organized networks—**nervous systems**—are the keys to the integration and control underlying the active life style of animals.

Evolution of Nervous Systems

Invertebrate nervous systems range from the very simple to the complex (Figure 35–2, at the bottom of the next page). In the cnidarian *Hydra*, the neurons form a diffuse network. They receive information from sensory receptor cells, which can be found among the epithelial cells on the inner and outer surfaces of the animal. The neurons stimulate muscular cells that cause movements in the body wall.

The nervous system of the planarian, a flatworm, is more highly organized than that of *Hydra*, providing a more efficient coordination that makes possible in-

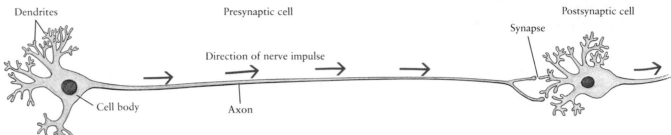

Dendrites

Presynaptic cell

Postsynaptic cell

Synapse

Direction of nerve impulse

Cell body

Axon

35–1 *The essential functional units of the nervous system are the specialized cells known as neurons. A neuron typically consists of a cell body, containing the nucleus and much of the metabolic machinery of the cell; dendrites, short threadlike cytoplasmic extensions that, with the cell body, receive stimuli from other cells; and an axon, a long extension that is capable of rapidly conducting a nerve impulse over great distances. Neurons communicate with one another across junctions known as synapses. At any given synapse, the neuron that transmits the signal across the synapse is known as the presynaptic cell, and the neuron that receives the signal is the postsynaptic cell.*

creased motility. Some of the nerve net is condensed into two cords, and there are two clusters of nerve cell bodies at the anterior end of the body. Such clusters are known as **ganglia** (singular, ganglion).

In the earthworm, the two cords have come together in a fused, double nerve cord that runs along the ventral surface of the body. Along the nerve cord are ganglia, one for each segment of the body. The nerve cord forks just below the pharynx, and the two forks meet again in the head, terminating in two large dorsal ganglia.

Arthropods, such as the crayfish, also have a double ventral nerve cord and, in addition, may have sizable clusters of nerve cell bodies in the head region. Collectively, these ganglia are large enough to be called a brain. The nervous system also contains many other ganglia interconnected by nerve fibers that run along the ventral surface. In animals with this type of nervous system, many quite complicated activities—for example, the complex movements of the finely articulated appendages—are coordinated by the nearest ganglion.

In vertebrates, the nervous system, which is dorsal rather than ventral, has been greatly elaborated. Its central processing centers—the spinal cord and the brain—are enclosed and protected by the bones of the vertebral

35–2 *In* Hydra (**a**), *a cnidarian, the nerve impulse can travel in either direction. As a consequence, it spreads out diffusely along the nerve net from the area of stimulation. In a planarian (**b**), there are two longitudinal nerve cords, with some aggregation of ganglia and sense organs at the anterior end. In annelids, such as the earthworm (**c**), the longitudinal nerve cords are fused together in a double ventral nerve cord with a ganglion in each body segment. In the crayfish (**d**), an arthropod, the nerve cord is also double and ventral, with a series of ganglia, almost as large as the brain, that control particular segments of the body.*

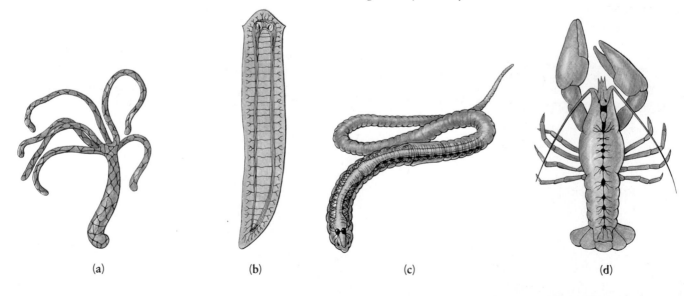

(a)   (b)   (c)   (d)

column and the skull. The trend in vertebrate evolution has been toward increased centralization of control in the brain. The precise integration that accompanies such centralization makes possible such complex behaviors as the dive of an osprey for a fish and the fashioning of a tiny lure by a fly fisherman.

Organization of the Vertebrate Nervous System

As we noted in Chapter 28, the vertebrate nervous system has a number of subdivisions (Figure 35–3) that differ in anatomy, physiology, and function. The primary and most obvious distinction is the subdivision of the system into the central nervous system, which consists of the brain and spinal cord, and the peripheral nervous system, formed by the sensory and motor pathways that carry information to and from the central nervous system. The motor pathways are further divided into the somatic system, which stimulates skeletal muscle, and the autonomic system, which relays signals to smooth muscle, cardiac muscle, and glands. The autonomic system is, in turn, subdivided into the sympathetic and parasympathetic divisions.

The functional unit of the vertebrate nervous system is, as we have noted previously, the neuron, a nerve cell characterized by a **cell body,** an **axon,** and, often, many **dendrites.** In vertebrates, as in invertebrates, nerve cell bodies are often found in clusters. Such clusters outside the central nervous system are called **ganglia;** inside the central nervous system, they are generally called **nuclei.** Axons (nerve fibers) are also grouped together, forming bundles. These bundles are known as **tracts** when they are in the central nervous system, and **nerves** when they are in the peripheral nervous system.

Neurons are surrounded and insulated by **glial cells.** The individual axons in tracts and nerves are often enveloped in insulating **myelin sheaths** formed by specialized glial cells (Figure 35–4). These sheaths are rich in lipids, giving nerves and tracts a glistening white appearance.

The Central Nervous System

The **central nervous system** is made up of the **brain** and the **spinal cord,** which provides the critical link between the brain and the rest of the body. Your spinal cord, which is a slim cylinder about as big around as your little finger, can be seen in cross section to be divided into a central area of **gray matter** and an outer area of **white matter** (Figure 35–5). The gray matter is mostly interneurons (which, as you may recall from page 514, conduct signals within localized regions of the central nervous system), the cell bodies of motor neurons, and glial cells. The white matter consists of tracts running longitudinally through the spinal cord. The fibers comprising these tracts are primarily the axons of relay neurons, which conduct signals between different regions of the central nervous system.

The spinal cord is continuous with the brainstem, which is the base of the brain. The brainstem contains tracts conducting signals to and from the spinal cord and also the cell bodies of the neurons whose axons innervate the muscles and glands of the head. In addition, within the brainstem are centers for some of the important automatic regulatory functions, such as control of respiration and blood pressure, which are also influenced by other parts of the brain.

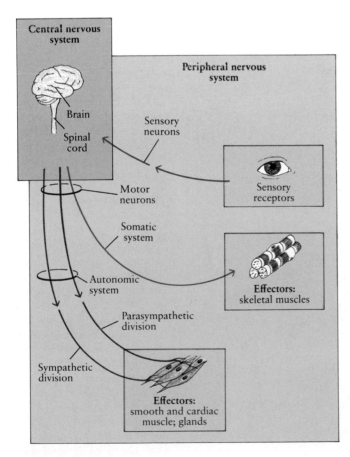

35–3 *A summary of the subdivisions of the vertebrate nervous system.*

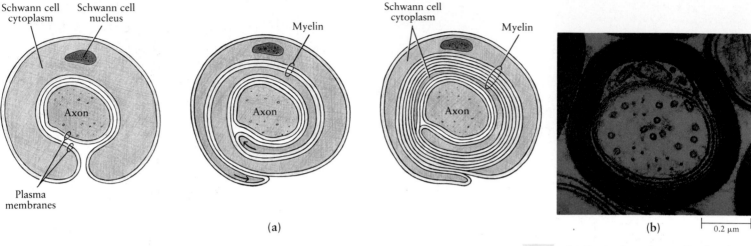

Schwann cell cytoplasm   Schwann cell nucleus   Myelin   Schwann cell cytoplasm   Myelin

Axon   Axon   Axon

Plasma membranes

(a)   **(b)**   0.2 μm

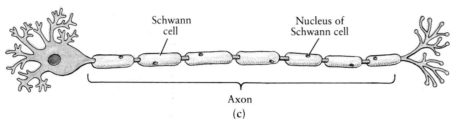

Schwann cell   Nucleus of Schwann cell

Axon

(c)

35–4   (a) *Formation of a myelin sheath by a Schwann cell, a type of glial cell found in the peripheral nervous system. As the Schwann cell grows, it wraps itself around and around the axon and gradually extrudes its cytoplasm from between the layers. The myelin sheath, which consists of layers of lipid-containing plasma membranes, insulates the nerve fiber.* (b) *Electron micrograph of a cross section of a mature myelin sheath. Its dark appearance is the result of treatment to make it opaque to electrons. Without such treatment, the myelin sheath appears white.*

(c) *A longitudinal view of a neuron, showing the myelin sheath that insulates its axon. Each segment of the sheath is produced by a single Schwann cell. Between the Schwann cells, the membrane of the axon is exposed to the surrounding medium. As we shall see later in this chapter, the alternation of insulated and uninsulated regions of the axon is the key to the rapid conduction of the nerve impulse by neurons with very long axons.*

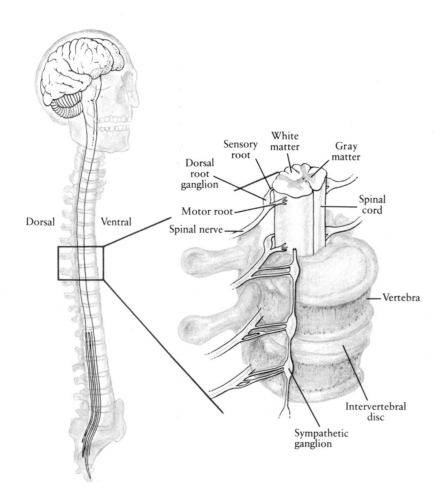

White matter
Sensory root   Gray matter
Dorsal root ganglion
Motor root   Spinal cord
Spinal nerve
Dorsal   Ventral
Vertebra
Intervertebral disc
Sympathetic ganglion

35–5   *A portion of the human spinal cord and vertebral column. Each spinal nerve divides into two fiber bundles, the sensory root and the motor root, at the vertebral column. The sensory root connects with the cord dorsally; the cell bodies of the sensory neurons are in the dorsal root ganglia. The motor root connects ventrally with the spinal cord; the cell bodies of the motor neurons are in the spinal cord itself. The sympathetic ganglia, which form a chain, are part of the autonomic nervous system.*

The butterfly-shaped gray matter within the spinal cord is composed mostly of interneurons, cell bodies of motor neurons, and glial cells. The surrounding white matter consists of ascending and descending fiber tracts.

The Peripheral Nervous System

The **peripheral nervous system** is made up of neurons whose axons extend out of the central nervous system into the tissues and organs of the body. These include both motor neurons, which carry signals out, and sensory neurons, which carry signals in. The fibers of motor and sensory neurons are bundled together into nerves. Those nerves, such as the optic nerve, that connect directly with the brain are classified as **cranial nerves.** Those that connect with the spinal cord are classified as **spinal nerves.** Pairs of spinal nerves enter and emerge from the cord through spaces between the vertebrae. The motor fibers of each pair innervate the muscles of a different area of the body, and the sensory fibers receive signals from sensory receptors in the same area. In humans there are 31 such pairs.

As you can see in Figure 35–5, the motor and sensory fibers of the spinal nerves separate from each other near the spinal cord. The cell bodies of the sensory neurons are in the dorsal root ganglia outside the spinal cord, and the sensory fibers feed into the dorsal side of the spinal cord. Here they may synapse with relay neurons, interneurons, or motor neurons, they may turn and ascend toward the brain, or they may do all of these. Fibers from the motor neurons emerge from the spinal cord on the ventral side. The cell bodies of motor neurons are located in the spinal cord, where they may receive signals from relay neurons, interneurons, and sensory neurons. Within the spinal cord, sensory neurons, interneurons, and motor neurons are often interconnected in simple reflex arcs, as shown in Figure 35–7.

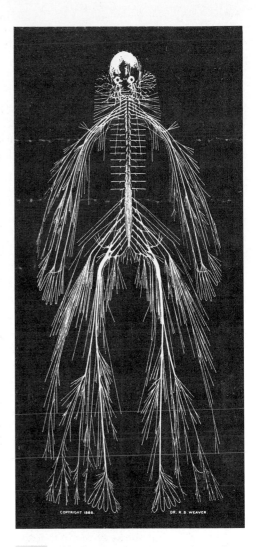

35–6   *The human nervous system, as dissected out in 1888 by Rufus B. Weaver, a Philadelphia physician. The subject, a maid employed by Dr. Weaver, had requested that her body be used to benefit science after her death. It took Dr. Weaver more than five months of daily work to remove the nerves.*

35–7   *A simple reflex arc. In this example, free nerve endings in the skin, when appropriately stimulated, transmit signals along the sensory neuron to an interneuron in the spinal cord. The interneuron transmits the signal to a motor neuron. As a result of the stimulation of the motor neuron, muscle fibers contract. Relay neurons, not shown here, are also stimulated by the sensory neuron and carry the sensory information to the brain.*

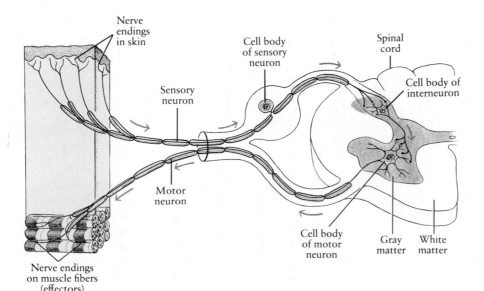

Divisions of the Peripheral Nervous System: Somatic and Autonomic

As we noted earlier, there are two subdivisions of the motor pathways of the peripheral nervous system: the somatic and the autonomic. The **somatic** ("voluntary") nervous system controls the skeletal muscles—that is, the muscles that can be moved at will. The **autonomic** ("involuntary") system consists of the motor nerves that control cardiac muscle, glands, and smooth muscle (the type of muscle found in the walls of blood vessels and in the digestive, respiratory, excretory, and reproductive systems).

You will readily recognize that the distinction here between "voluntary" and "involuntary" is not clear-cut. Skeletal muscles—part of the somatic system—often move involuntarily, as in a reflex action. On the other hand, it has been reported that some individuals, such as practitioners of yoga or those who have had biofeedback training, can control their rate of heartbeat and the contractions of some smooth muscle.

Anatomically, the motor neurons of the somatic system are distinct and separate from those of the autonomic nervous system, although axons of both types may be carried within the same nerve. The cell bodies of the motor neurons of the somatic system are located within the central nervous system, with long axons running without interruption all the way to the skeletal muscles. The pathways of the autonomic nervous system also include axons that originate in cell bodies inside the central nervous system, but these axons do not usually travel all the way to their target organs, or effectors. Instead, they synapse outside the central nervous system with motor neurons, which then innervate the effectors (Figure 35–8).

The synapses of the autonomic nervous system occur within ganglia. Thus the neurons whose axons emerge from the central nervous system and terminate in the ganglia are known as **preganglionic,** while those whose axons emerge from the ganglia and terminate in the effectors are known as **postganglionic.** This two-neuron pathway constitutes a characteristic difference between the autonomic and somatic systems.

Another major difference between these systems is that, in vertebrates, the somatic system can only stimulate or not stimulate an effector; it cannot inhibit an effector. The autonomic system, by contrast, can stimulate or inhibit the activity of an effector.

In the somatic system, sensory input often comes from neurons monitoring changes in the external environment. The autonomic nervous system receives sensory input from some of the same neurons that supply information to the somatic nervous system. Also, and most important, it receives information from sensory neurons monitoring changes in the interior of the body, such as the group of neurons signaling changes in blood pressure. These neurons are involved in reflexes similar to the one shown in Figure 35–7. In reflex arcs involving the autonomic system, however, you are usually not conscious that the reflex action has taken place.

Divisions of the Autonomic Nervous System: Sympathetic and Parasympathetic

The autonomic nervous system itself has two divisions: the **sympathetic** division and the **parasympathetic** division. These two divisions are anatomically and functionally distinct. The major anatomical differences between them are:

1. Axons of the sympathetic division originate in the thoracic (chest) and lumbar (lower back) regions of the spinal cord. Axons of the parasympathetic division emerge through the cranial (brain) region and the sacral ("tail") region of the spinal cord.

2. In the autonomic nervous system, as we have seen, there is always a relay system of two neurons connecting the central nervous system and the effector organ. These neurons synapse at a ganglion. In the sympathetic division, the ganglia are usually close to the central nervous system. In the parasympathetic division, the ganglia are close to or embedded in the target organ.

Functionally, the two divisions are generally antagonistic. As you can see in Figure 35–8, most of the internal organs are innervated by axons from both divisions. They work in close cooperation with each other and with hormones secreted by the endocrine glands for the ultimate homeostatic regulation of the body. The parasympathetic division is involved primarily in the restorative activities of the body. It is particularly active, for example, after a heavy meal. Parasympathetic stimulation slows down the heartbeat, increases the movements of the smooth muscle of the intestinal wall, and stimulates secretions of the salivary glands and the digestive glands of the stomach.

The sympathetic division, by contrast, prepares the body for action. For example, the physical characteristics of fear or anger, described on page 621, result from the increased discharge of neurons of the sympathetic division. As a consequence of the constellation of responses that follow sympathetic stimulation, the body as a whole is prepared for "fight or flight"—or, at least, for action that would have been appropriate at some earlier stage of our cultural evolution (see essay).

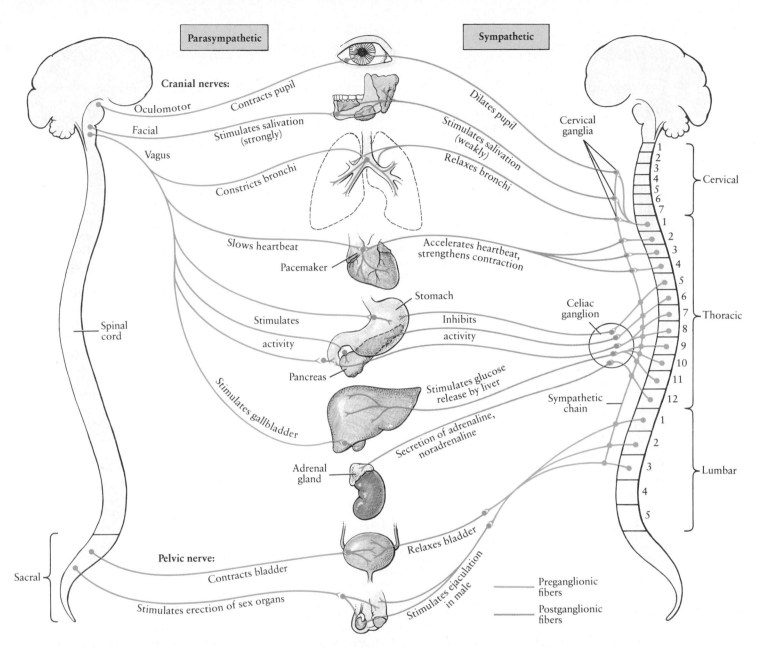

Parasympathetic   **Sympathetic**

Cranial nerves:

Oculomotor — Contracts pupil

Facial — Stimulates salivation (strongly)

Vagus

Constricts bronchi

Dilates pupil

Stimulates salivation (weakly)

Relaxes bronchi

Cervical ganglia

Slows heartbeat

Pacemaker

Accelerates heartbeat, strengthens contraction

Spinal cord

Stimulates activity

Stomach

Inhibits activity

Celiac ganglion

Pancreas

Stimulates gallbladder

Stimulates glucose release by liver

Sympathetic chain

Secretion of adrenaline, noradrenaline

Adrenal gland

Pelvic nerve:

Relaxes bladder

Contracts bladder

Sacral

Stimulates erection of sex organs

Stimulates ejaculation in male

Cervical
1
2
3
4
5
6
7

Thoracic
1
2
3
4
5
6
7
8
9
10
11
12

Lumbar
1
2
3
4
5

Preganglionic fibers

Postganglionic fibers

35–8   *The autonomic nervous system. It differs from the somatic system anatomically in that the axons emerging from the central nervous system do not travel without interruption to the effectors. Instead, they synapse outside the central nervous system with motor neurons, which then innervate the effectors. The fibers emerging from the central nervous system are known as preganglionic fibers, and those terminating in the effectors are known as postganglionic fibers.*

The autonomic nervous system consists of the sympathetic and the parasympathetic divisions. The preganglionic fibers of the parasympathetic division exit from the base of the brain and from the sacral region of the spinal cord and synapse with the postganglionic neurons at or near the target organs. The sympathetic division originates in the thoracic and lumbar regions. Preganglionic fibers of the sympathetic division synapse with postganglionic neurons in the chain

of sympathetic ganglia or in other ganglia, such as the celiac ganglion.

Most, but not all, internal organs are innervated by both divisions, which usually function in opposition to each other. In general, the sympathetic division stimulates functions involved in "fight-or-flight" reactions, and the parasympathetic division stimulates more tranquil functions, such as digestion.

Stress Points

When a firefighter breaks through a wall of flame, a camper wakes to the sound of a low growl just outside the tent, or a student delivers a speech at a graduation ceremony, the ''fight-or-flight'' response that is part of our evolutionary heritage comes into play. Each of these individuals is experiencing stress, and the physiological systems of the body, particularly the nervous and endocrine systems, mobilize for action.

Stress can result from almost any disturbance of the body's physiological systems, and it may be triggered by either internal or external events. The response to stress, which is orchestrated by the hypothalamus and mediated primarily by the sympathetic division of the autonomic nervous system, can also vary greatly. It depends not only on the nature and duration of the disturbance but also on the individual's physical and mental condition.

A full-fledged ''fight-or-flight'' response—with its pounding heart, dry mouth, and hyperventilation—often seems distinctly unhelpful as we confront the many stresses of contemporary life. However, it gives us, as it gave our ancestors, our best shot at survival in the face of severe physical danger. When the danger is past, the body's physiological systems are returned to their normal steady-state functions by stimuli from the parasympathetic division.

What happens, however, when stress is not of the acute, short-term variety but is instead chronic and prolonged? A growing body of evidence suggests that one major consequence is an increased susceptibility to illness. For example, experiments have shown that when laboratory animals are subjected to continuous stress—such as crowded living conditions or electric shocks—over which they have no control, they have a higher incidence of dis-

ease. In one particularly striking experiment, mice that were allowed merely to *see* a cat had a decreased resistance to tapeworm infections. However, when animals are given the means to control the stress, for example, by pressing a lever to stop electric shocks, they remain much healthier than those animals with no control over their environment.

One of the first intimations of the effects of prolonged stress in humans was the observation that individuals who suffer the loss of a loved one are themselves more likely to become ill or die. The causes of such effects are difficult to isolate and may involve a multiplicity of factors, such as changes in eating habits and sleep patterns. However, studies of men who had recently lost their wives to breast cancer have revealed a decrease in the responsiveness of the immune system. In these men, a decline in the activity of the lymphocytes (B cells and T cells) was detected within a month or two of the deaths of their wives and lasted as long as a year. Even the stresses of education may take their toll. Impaired activity of killer T cells and abnormally low levels of interferons have been observed in students taking exams.

Current research suggests that interactions among the nervous, endocrine, and immune systems play an important role during extended periods of stress. For example, one of the responses to stress is an increased production of the hormone cortisol by the adrenal cortex. Among its various effects, cortisol stimulates the breakdown of proteins to amino acids. The amino acids can then be used in the synthesis of enzymes needed to maintain high levels of activity, or they can be converted to glucose, the principal cellular energy supply. However, cortisol also suppresses inflammation and inhibits the immune response. As you

may recall from Chapter 33, activated helper T cells release molecules known as interleukins. These molecules function as hormones, stimulating the differentiation and proliferation of activated B cells and cytotoxic T cells. Production of one of the interleukins also appears to trigger the production of cortisol, which then inhibits further production of the interleukin. This negative feedback loop is thought to play a role in the normal regulation of the immune response. However, it is sensitive not only to cortisol produced in response to the interleukin but also to cortisol produced in response to stress. Thus, one of the consequences of stress is a suppression of interleukin production—and a general damping of the response of B cells and cytotoxic T cells to foreign invaders and diseased body cells.

Other research is uncovering intriguing new evidence of communication between the immune system and the nervous system. For example, it is now known that lymphocytes bear receptors for some neurotransmitters and for other molecules known as neuromodulators (to be discussed later in this chapter). Moreover, when activated, these white blood cells secrete certain neurotransmitters and hormones. It is not surprising, therefore, that alterations of the patterns of neural activity in the brain have been observed in the course of the immune response.

The details of the interactions among the nervous, endocrine, and immune systems are only beginning to emerge. Even at this early stage, however, the new field of psychoneuroimmunoendocrinology promises many new insights into the ways in which emotions, temperament, environment, and life events may affect our health and the ways in which the state of our health may influence our state of mind.

The Nerve Impulse

For some 200 years, it has been known that nerve conduction is associated with electrical phenomena. As you will recall, there are two types of electric charge, positive and negative. Like charges repel one another, and unlike charges attract. Thus, negatively charged particles tend to move toward a region of positive charge, and vice versa. Materials that permit the movement of charged particles, such as a copper wire or a solution containing ions, are known as **conductors.** Materials that do not permit the movement of charged particles, such as fat and rubber, are **insulators.**

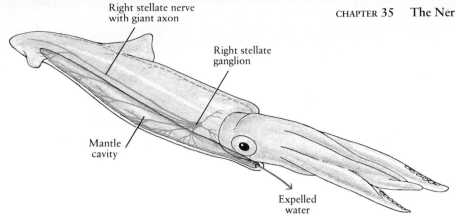

35–9   *The squid, one of the chief protagonists in research leading to an understanding of the nature of the nerve impulse. The stellate nerves contain the giant axons used in all the early studies of the nerve impulse. The giant axons innervate muscles in the wall of the mantle. Powerful contractions of those muscles result in the rapid expulsion of water from the mantle cavity, enabling the squid to escape potential predators by jet propulsion.*

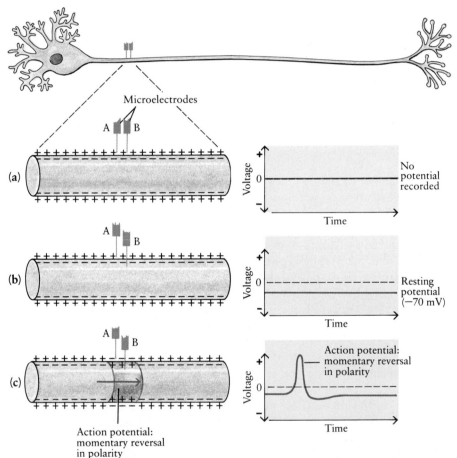

35–10   *The electric potential across the membrane of the axon is measured by microelectrodes connected to an oscilloscope.* **(a)** *When both electrodes are in the extracellular fluid outside the membrane, no potential is recorded.* **(b)** *When one electrode is inside the membrane, the oscilloscope shows that the interior is negative with respect to the exterior and that the difference between the two is about 70 millivolts. This is the resting potential.* **(c)** *If the axon is stimulated, a nerve impulse will pass along it. When the impulse reaches the region in which the microelectrodes are located, the oscilloscope shows a brief reversal of polarity—that is, the interior becomes positive in relation to the exterior. This brief reversal in polarity is the action potential.*

The difference in the amount of electric charge between a region of positive charge and a region of negative charge is called the **electric potential.** An electric potential is a form of potential energy, like the energy of a boulder at the top of a hill or of water behind a dam. This potential energy is converted to electrical energy when charged particles are allowed to move through a solution or along a wire between the two regions of differing charge. The difference in potential energy between the two regions is measured in volts or, if it is very small, in millivolts.

Major advances in understanding the nature of the nerve impulse occurred when it became possible to monitor changes in electric potential in an individual neuron. The organism that first made this possible was the squid, which has motor neurons with large, long axons (Figure 35–9). Measurements within an axon are made with microelectrodes tiny enough to penetrate a living cell without seriously injuring it. The microelectrodes are connected with a very sensitive voltmeter called an oscilloscope, which measures voltage (in millivolts) in relation to time (in milliseconds).

When both electrodes are outside the neuron, no voltage difference is recorded (Figure 35–10a). When one electrode is outside the axon and the other is inside the axon, a voltage difference of about 70 millivolts can be detected between the outside and the inside. The interior of the membrane is negatively charged with respect to the exterior (Figure 35–10b). This is the **resting potential** of the membrane. When the axon is subsequently stimulated, the oscilloscope records a very brief reversal of polarity at the point of stimulation—that is, the interior of the axon becomes positively charged in relation to the exterior (Figure 35–10c). This reversal in polarity is called the **action potential.** The action potential traveling along the membrane is the **nerve impulse.**

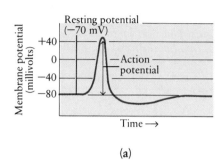

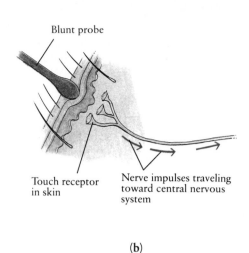

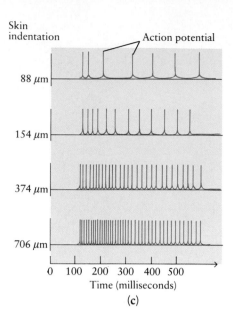

35–11   (a) *Nerve impulses can be moni-tored by electronic recording instruments. The impulses from any one neuron are all the same. That is, each impulse has the same duration and voltage change as any other.*

(b) *In a classic experiment, a blunt probe was used to stimulate a sensory neuron*

(touch receptor) in the skin of a cat. This ini-tiated nerve impulses that were conducted through the neuron to the central nervous system. (c) *The skin was touched and pressed in at varying depths, as indicated by the figures at the left. The more deeply the skin was pressed in, the more rapidly nerve*

impulses were produced. The vertical lines represent individual action potentials on a compressed time scale. As you can see, all the action potentials are the same size, but their frequency increases with the intensity of the stimulus.

The response of a neuron to stimulation is an all-or-nothing response. If the stimulus is sufficiently strong, action potentials are generated; if it is not, no action potentials are generated. Moreover, in any given neuron, the strength of the action potentials is almost always the same. Figure 35–11, for example, shows the action potentials produced by a single sensory neuron in the skin of a cat in response to pressure. As you can see, all of the action potentials are the same size. The only variation—and it is a critical variation—is the frequency with which nerve impulses are produced. The message is encoded in the number of nerve impulses generated in a given interval of time. The stronger the stimulus, the greater the number of impulses generated.

The Ionic Basis of the Action Potential

The action potential depends on the electric potential of the axon, which is, in turn, made possible by differences in the concentration of ions on the inside and the outside of the membrane. Such differences in concentration—and thus in electric potential—are characteristic of all cells, and they can be used to power a variety of cellular processes. Their use to carry information through the body of an animal is perhaps the most sophisticated application to have evolved thus far.

In axons, the critical concentration differences involve potassium ions (K^+) and sodium ions (Na^+). In the resting state, the concentration of K^+ ions in the

cytosol of an axon is about 30 times higher than in the fluid outside. Conversely, the concentration of Na^+ ions is about 10 times higher in the extracellular fluid than in the cytosol. The distribution of ions on either side of the membrane is governed by three factors: (1) the diffusion of particles down a concentration gradient (page 111), (2) the attraction of particles with opposite charges and the repulsion of particles with like charges, and (3) the properties of the membrane itself.

The lipid bilayer of the axon membrane is, like the lipid bilayer of other regions of the plasma membrane, impermeable to ions and most polar molecules. The movement of such particles through the membrane depends on the presence of integral membrane proteins that provide channels through which the particles can move, either by facilitated diffusion or active transport. The membrane of the axon is rich in proteins that provide channels for the movement of specific ions, particularly Na^+ and K^+.

An important feature of these proteins is that changes in their shape result in either the opening or the closing of the ion channels. Thus, the channels are said to be **gated.** Depending on the specific ion channel, the changes in shape that open or close its gate may be regulated by the voltage across the membrane or by the action of neurotransmitters (or drugs that mimic neurotransmitters).

Another significant feature of the axon membrane is the presence of the integral membrane protein known as the **sodium-potassium pump** (page 117). This protein

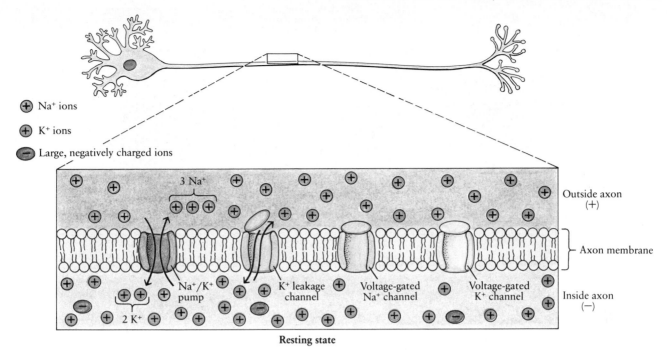

Resting state

35–12   *An axon in its resting state. Embedded in the lipid bilayer of the axon membrane are proteins that provide channels through which K⁺ and Na⁺ ions can move between the cytosol of the axon and the surrounding extracellular fluid. In this diagram, four of the principal types of channels are shown: the sodium-potassium pump, which pumps three Na⁺ ions out of the axon for every two K⁺ ions pumped in; the K⁺ leakage channel, in which the gate is open in the resting state, allowing K⁺ ions to diffuse into and out of the axon; and voltage-gated*

Na⁺ and K⁺ channels, which are closed in the resting state. Not shown in this diagram are Na⁺ leakage channels, of which there are a very small number.

In the resting state, the concentration of K⁺ ions is much higher in the cytosol of the axon than in the extracellular fluid. Thus, K⁺ ions diffuse out of the axon through the K⁺ leakage channels, moving from the region of higher concentration to the region of lower concentration. Large, negatively charged ions cannot follow the K⁺ ions out of the axon. As a consequence, the interior

of the axon is negatively charged with respect to the exterior. It is this polarization of the axon membrane—the resting potential—that makes possible the action potential.

The negative charge in the interior of the axon attracts a small number of both K⁺ and Na⁺ ions, which move into the axon through their respective leakage channels. The Na⁺ ions are promptly removed from the axon by the sodium-potassium pump—and more K⁺ ions are pumped in—maintaining the concentration differences on which the resting potential depends.

pumps Na⁺ ions out of the axon and K⁺ ions into the axon.

When the axon membrane is in its resting state (Figure 35–12), the Na⁺ gates are mostly closed. As a consequence, the membrane is almost impermeable to Na⁺ ions. The few Na⁺ ions that diffuse in through open channels—moving down their concentration gradient—are promptly removed by the sodium-potassium pump. Many K⁺ gates, however, are open, and the membrane is thus relatively permeable to K⁺ ions. Because of the concentration gradient, K⁺ ions tend to move out of the cell.

If no other forces were at work, K⁺ ions would move down the concentration gradient until their distribution were equal on either side of the membrane. However, because of the impermeability of the lipid bilayer, large, negatively charged ions cannot follow the K⁺ ions out of the cell. Thus, as K⁺ ions leave, an excess negative charge builds up inside the cell. This excess of negative

charge attracts the positive K⁺ ions, impeding their further outward movement. As a result, an equilibrium is reached at which there is no net movement of K⁺ ions across the membrane. At the point of equilibrium, there is a slight excess of negative charge inside the cell, and the membrane is said to be **polarized.** This point of equilibrium is the resting potential.

When the axon membrane is stimulated, the voltage-gated Na⁺ channels open at the site of stimulation. Na⁺ ions rush in, moving down their concentration gradient, attracted initially by the negative charge inside the axon. This influx of positively charged ions momentarily reverses the polarity of the membrane so that it becomes more positive on the inside than on the outside, producing the action potential (Figure 35–13a). This change in Na⁺ permeability lasts for only about half a millisecond. Then the Na⁺ gates close, and the stimulated region of the membrane regains its previous impermeability to Na⁺ ions.

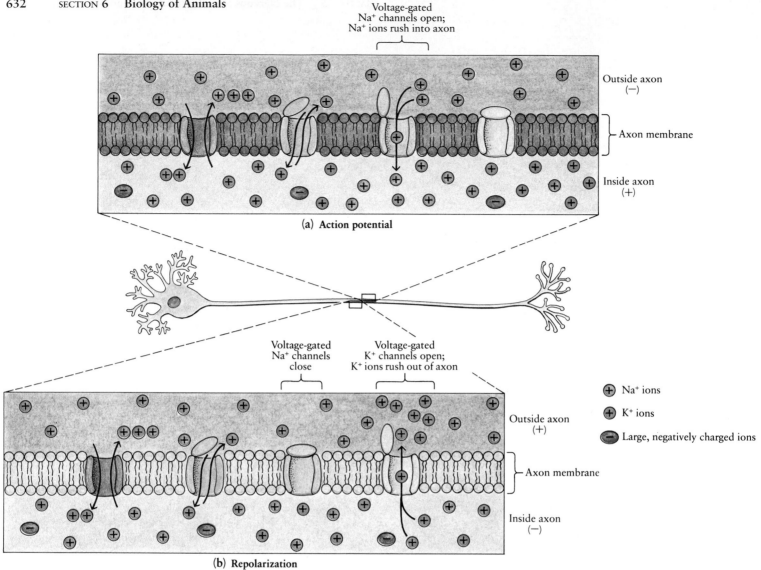

Voltage-gated
Na⁺ channels open;
Na⁺ ions rush into axon

Outside axon
(−)

Axon membrane

Inside axon
(+)

(a) Action potential

Voltage-gated
Na⁺ channels
close

Voltage-gated
K⁺ channels open;
K⁺ ions rush out of axon

Outside axon
(+)

Axon membrane

Inside axon
(−)

(b) Repolarization

⊕ Na⁺ ions

⊕ K⁺ ions

⊖ Large, negatively charged ions

35–13   (a) *The action potential. At the site of stimulation, the voltage-gated Na⁺ channels of the axon membrane open. Na⁺ ions rush into the axon, moving down their concentration gradient. The interior of the axon becomes positively charged with respect to the exterior, reversing the polarity of the membrane.*

(b) *Repolarization. Within half a millisecond, the voltage-gated Na⁺ channels close, preventing any further inward movement of Na⁺ ions. At the same time, the voltage-gated K⁺ channels open. K⁺ ions rush out of the axon, moving down their concentration gradient and restoring the resting potential. The sodium-potassium pump subsequently returns the concentrations of Na⁺ and K⁺ ions on either side of the membrane to their original levels.*

Almost immediately, the voltage-gated K⁺ channels in this same region of the membrane open, and K⁺ ions flow out of the axon (Figure 35–13b). This outward flow of positive K⁺ ions counteracts the reversal in polarity produced by the previous inward flow of positive Na⁺ ions, and the resting potential is quickly restored. Subsequently, the sodium-potassium pump shuttles Na⁺ and K⁺ ions across the axon membrane, returning the concentrations of these ions to their original levels. At the same time, the ion channels in this region of the membrane return to the shapes characteristic of the resting state.

The actual number of ions involved in the action potential is very small. Only a very few Na⁺ ions need enter to reverse the polarity of the membrane, and only the same small number of K⁺ ions need move out of the axon to restore the resting potential. As a consequence, action potentials can move along the axon in rapid fire without substantial changes occurring in the internal concentrations of Na⁺ and K⁺ ions.

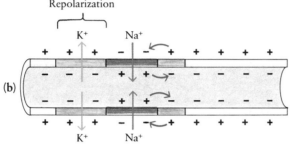

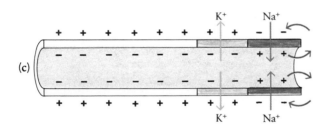

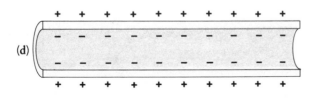

35–14 *Propagation of the nerve impulse.* (a) *In advance of the action potential, a small segment of the axon membrane becomes slightly depolarized, owing to the flow of positively charged ions along the inside of the membrane.* (b) *When a region of the membrane becomes depolarized in this way, its voltage-gated Na⁺ channels open, and Na⁺ ions move into the axon. This creates an action potential and depolarizes the next adjacent segment of the membrane. Meanwhile, in the region immediately behind the action potential, the voltage-gated K⁺ channels open and K⁺ ions move out of the axon, returning the membrane to its original polarity. During this repolarization process, the voltage-gated Na⁺ channels will not open. As a consequence,* (c) *the action potential can move in one direction only. Its passage down the axon constitutes the nerve impulse.* (d) *When the action potential has passed a given region of the membrane and repolarization has occurred, the resting potential is restored. The membrane is ready for the passage of another series of action potentials, constituting a new nerve impulse.*

membrane at the active region is comparatively positive, positively charged ions move from this region to the adjacent area inside the axon, which is still comparatively negative. As a result, the adjacent area becomes **depolarized**—that is, less negative (Figure 35–14).

This depolarization causes the voltage-gated Na⁺ channels to open and allows Na⁺ ions to rush in. The resulting increase in the internal concentration of Na⁺ ions depolarizes the next adjacent area of the membrane, causing its voltage-gated Na⁺ channels to open, and allowing the process to be repeated. As a consequence of this renewal process, repeating itself along the length of the membrane, the axon is capable of conducting a nerve impulse over a considerable distance with absolutely undiminished strength.

The nerve impulse moves in only one direction because the segment of the axon behind the action potential has a brief **refractory period** during which its voltage-gated Na⁺ channels will not open. Thus the action potential cannot go backwards.

The Role of the Myelin Sheath

As we saw in Figure 35–4 (page 624), long axons are generally enveloped in myelin sheaths formed by specialized glial cells. The myelin sheath is not just an insulator. Its most important feature is that it is interrupted at regular intervals by openings, or **nodes.** Only at the nodes is it possible for Na⁺ and K⁺ ions to move into and out of the axon. Thus, in myelinated fibers—which include all the large nerve fibers of vertebrates—

Propagation of the Impulse

An important feature of the nerve impulse is that, once initiated, this transient reversal in polarity continues to move along the axon, renewing itself continuously, just as a flame traveling along a fuse ignites the fuse ahead of it as it travels. The action potential is self-propagating because at its peak, when the inside of the

the impulse jumps from node to node (Figure 35–15), rather than moving continuously along the membrane.

This saltatory (leaping) conduction greatly increases the velocity. Some large, myelinated nerve fibers, for example, conduct impulses as rapidly as 200 meters per second, compared with velocities of only a few millimeters per second in small, unmyelinated fibers. Also, because Na^+ and K^+ ions move across only a small portion of the axon's membrane, there is an enormous saving in energy expenditure by the sodium-potassium pump.

The Synapse

Signals travel from one neuron to another across the specialized junction known as the synapse. In most synapses in the mammalian nervous system, the two neurons never touch. A space, known as the **synaptic cleft**, separates the cell transmitting information (the **presynaptic cell**) from the cell receiving information (the **postsynaptic cell**). Information is transmitted across the synaptic cleft by means of the signalling molecules known as **neurotransmitters**. Unlike the nerve impulse along the axon—an all-or-nothing proposition—signals transmitted across synapses are of varying strength and may have opposite effects. That is, some excite and some inhibit the postsynaptic cell.

Neurotransmitters are synthesized by the individual neurons, packaged into numerous small vesicles, and stored in the axon terminals. Their release is triggered by the arrival of an action potential at the axon terminal. The membrane in this region of the neuron is rich in proteins that form voltage-gated channels for the transport of calcium ions (Ca^{2+}).

Arrival of an action potential at the axon terminal alters the voltage, opening the gates and allowing Ca^{2+} ions to flow from the extracellular fluid into the axon. This influx of Ca^{2+} ions, in turn, causes the synaptic vesicles to fuse with the plasma membrane, emptying their neurotransmitter molecules into the synaptic cleft (Figure 35–16). The molecules diffuse from the presynaptic cell across the cleft and combine with receptor molecules on the membrane of the postsynaptic cell. This sets in motion a series of events that, as we shall see shortly, may or may not trigger a nerve impulse in the postsynaptic cell.

After their release, neurotransmitters are rapidly removed or destroyed, putting a halt to their effect. This is an essential feature in the control of the activities of the nervous system. The molecules may diffuse away or be broken down by specific enzymes, or they—or their breakdown products—may be taken up again by the axon terminal for recycling.

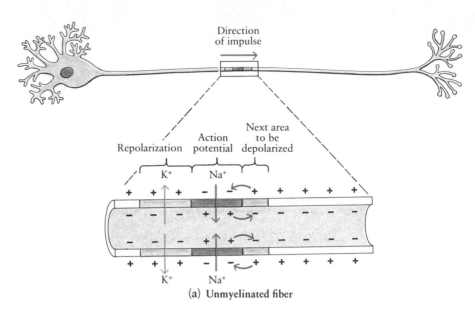

(a) **Unmyelinated fiber**

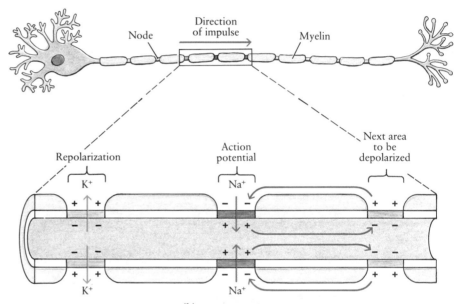

(b) **Myelinated fiber**

35–15   (a) *In an unmyelinated fiber, the entire length of the axon membrane is exposed to the extracellular fluid. All regions of the membrane are rich in ion channels and sodium-potassium pumps. Action potentials travel continuously along the axon.* (b) *In a myelinated fiber, by contrast, only the regions of the axon membrane located at the nodes are exposed to the extracellular fluid. Virtually all of the ion channels and sodium-potassium pumps are concentrated in these same regions. Thus, action potentials can be generated only at the nodes. As a consequence, the nerve impulse jumps from node to node, greatly accelerating the conduction.*

Presynaptic cell

Postsynaptic cell

Direction of nerve impulse

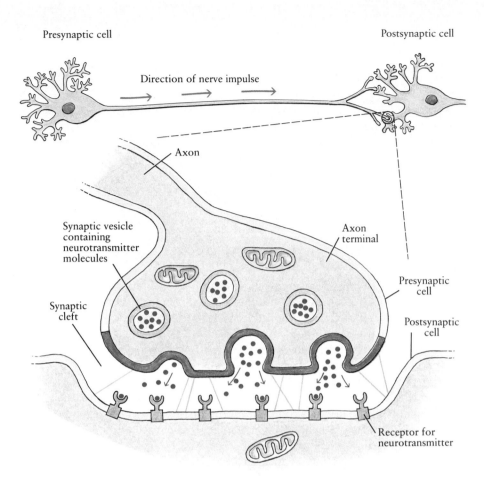

Axon

Synaptic vesicle
containing
neurotransmitter
molecules

Axon
terminal

Synaptic
cleft

Presynaptic
cell

Postsynaptic
cell

Receptor for
neurotransmitter

35–16   *A synapse. The arrival of an action
potential at the axon terminal triggers the
fusion of synaptic vesicles with the axon
membrane, releasing neurotransmitter molecules into the synaptic cleft. These molecules
rapidly diffuse across the short distance to
the postsynaptic cell, where they combine
with specific receptors in the plasma
membrane. A network of protein fibers in
the synaptic cleft anchors the presynaptic
and postsynaptic membranes and, in some
synapses, contains enzymes that rapidly degrade neurotransmitter molecules. Because
of their shape, axon terminals are known as
synaptic knobs or as "boutons" (French for
"buttons").*

The study of synapses is one of the most active areas
of contemporary neurobiological research. Although
the principal neurotransmitters have been known for
many years, researchers are discovering an enormous
number of chemicals that play a role in synaptic transmission, particularly in the central nervous system.
Moreover, the specific receptors for different neurotransmitters are being identified, their structures determined, and in some cases, their genes sequenced.
Simultaneously, the events that occur when a neurotransmitter binds to its receptor are being revealed, as
are the mechanisms by which a neuron integrates the
information that it receives from the hundreds—or even
thousands—of neurons that form synapses with it.

Neurotransmitters

A variety of chemical substances function as neurotransmitters. In the peripheral nervous system, the principal neurotransmitters are **acetylcholine** and **noradrenaline.** Acetylcholine is also found in the brain, although
at relatively few synapses. Noradrenaline is a major
neurotransmitter at some synapses in the hypothalamus
and in other specific regions of the brain, where it is

thought to play a role in arousal and attention. There
is some evidence that severe depression may be related
to an abnormally low level of noradrenaline at particular synapses. Two types of antidepressants currently
in clinical use apparently act by increasing the amount
of noradrenaline at such synapses.

Many other neurotransmitters have been found in
the central nervous system, including dopamine, serotonin, and a molecule known as GABA. **Dopamine** is a
transmitter for a relatively small group of neurons involved with muscular activity. Parkinson's disease,
which is characterized by muscular tremors and weakness, is associated with a decrease in the number of do-
pamine-producing neurons and thus with the level of
dopamine in certain areas of the brain. **Serotonin** is
found in regions of the brain associated with arousal
and attention. Increasing levels of serotonin are associated with sleep. **GABA** is a major inhibitory transmitter in the central nervous system. The loss of GABA
synapses is one of the features of Huntington's disease
(page 206).

Almost all drugs that act in the brain to alter mood
or behavior do so by enhancing or inhibiting the activity
of neurotransmitter systems. Caffeine, nicotine, and the
amphetamines, for example, stimulate brain activity by
substituting for excitatory neurotransmitters at synap-

ses. Certain types of tranquilizers block dopamine receptors, whereas LSD, a hallucinatory drug, interferes with the normal action of serotonin in the brain.

Present evidence indicates two principal mechanisms by which neurotransmitters exert their effects on postsynaptic cells. In one mechanism, the binding of a neurotransmitter to its receptor triggers a change in the shape of a membrane protein that functions as a channel for a specific ion. Depending on the receptor, binding of the neurotransmitter may open the channel, allowing ions to flow between the cytosol of the neuron and the extracellular fluid, or it may close the channel, shutting off a previously existing flow of ions. The consequence is a change in the degree of polarization across the membrane of the postsynaptic cell.

In the second mechanism, binding of the neurotransmitter to its receptor activates an enzyme in the plasma membrane and sets in motion a second messenger, generally cyclic AMP (see page 617) or a related compound, cyclic GMP (guanosine monophosphate). The events that follow activation of the second messenger are complex, but the ultimate effect is a change in the degree of polarization of the postsynaptic cell. This change, however, takes place at a slower pace than the changes triggered by the opening or closing of ion channels.

In addition to the principal neurotransmitters, other molecules also play a role in synaptic transmission. These molecules, which may be released from the same axon terminals as the principal neurotransmitters or from other cells, are known as **neuromodulators.** Although neuromodulators may move directly across the synaptic cleft, they can also diffuse over a greater distance, affecting numerous cells within a local region of the central nervous system. Like neurotransmitters,

they bind to specific membrane receptors and alter ion channels or set in motion second messengers. Their effect is often to modulate the response of the cell to a principal neurotransmitter. Over 200 different substances that function as neuromodulators have been identified thus far. They include the endorphins (see essay), interferons and interleukins, hypothalamic hormones, pituitary hormones, pancreatic hormones such as insulin, and even the digestive hormones gastrin and cholecystokinin (page 532).

The Integration of Information

The dendrites and cell body of a single neuron may receive signals—in the form of neurotransmitter or neuromodulator molecules—from hundreds, or even thousands of synapses (Figure 35–17). The binding of each molecule to its receptor has some effect on the degree of polarization of the postsynaptic cell. If the effect is to make the interior of the cell less negative (depolarization), it is said to be **excitatory.** By contrast, if the effect is to maintain the membrane at or near the resting potential, or even to make the interior more negative (hyperpolarization), it is said to be **inhibitory.**

The changes in polarity induced by neurotransmitters and neuromodulators spread from the synapses through the postsynaptic cell to a region known as the **axon hillock.** This is the region of the axon in which a nerve impulse can originate. If the collective effect is a sufficient depolarization to open the voltage-gated Na^+ channels, an action potential is initiated in the axon of the postsynaptic cell and a new message is speeded on its way to the multitude of other neurons with which the axon synapses.

35–17   *A simplified representation of the many interconnections between neurons of the central nervous system. Note how numerous synapses, representing many different presynaptic neurons, converge on an individual neuron and how its axon, in turn, diverges to synapse on a number of other neurons. As you can see, most synapses are located on the dendrites and cell body of the postsynaptic neuron. Note, however, that there are also synapses on its axon, just before some of the axon terminals. These synapses are usually inhibitory or modulatory, influencing the response of the Ca^{2+} channels to the arrival of an action potential at the terminal and thus affecting the amount of neurotransmitter released.*

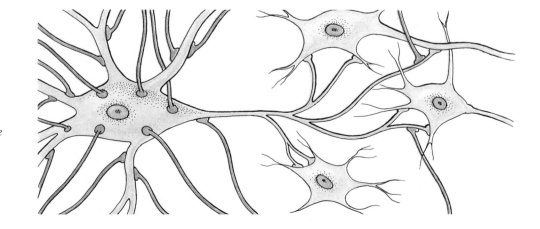

Internal Opiates

The word "opium" comes from the Greek *opion,* meaning "poppy juice." Since the time of the Greeks, poppy juice and its derivatives, such as morphine, have been used for the control of pain. They are the most potent pain-killers known, and their physiological effects are greatly enhanced by the fact that they produce euphoria. They are also highly addictive.

All substances with opiate action are related chemically and have similarities in their three-dimensional structures. Thus, it was long suspected that opiates act upon the brain by binding to specific membrane receptors. Using opium derivatives labeled with radioactive isotopes, investigators were able to show that the central nervous system does, indeed, have receptors for opiates. These receptors are located primarily in the spinal cord, brainstem, and brain regions in which drives and emotions are thought to be translated into complex actions, such as seeking food or a mate. When opiates bind to neurons bearing their receptors, they act as inhibitory neuromodulators, causing decreased production of nerve impulses by the neurons. Such receptors have been found not just in humans but in all other vertebrates tested.

Why would vertebrate brains have opiate receptors? Only one answer seemed logical: Because vertebrate brains themselves must produce opiates. This rather startling conclusion triggered a search for naturally occurring substances with opiate activity. Many such internal opiates have now been isolated, all of which act as neuromodulators. They have been given the name of **endorphins,** for endogenous morphine-like substances.

Two types of endorphins are recognized. One group, known as the enkephalins, is widespread in the central nervous system and is also abundant in the adrenal medulla. The other endorphins are produced primarily by the pituitary gland and perhaps by other tis-

Is exercise addictive? Endorphin concentrations in the blood increase substantially during exercise. This finding may explain the state of euphoria commonly known as "runner's high." Also, the depression reported by regular joggers on days when they are not able to exercise may be, in fact, a withdrawal symptom.

sues as well. Recently it has been found that the most common of these, beta-endorphin, is synthesized as part of a long peptide chain that also contains ACTH, the hormone that is released by the anterior pituitary and stimulates the adrenal cortex. Although there is a great deal of overlap in the structures of the various endorphins, the functional relationships among them are not yet known.

The endorphins are of great interest to medical researchers because of the insight they may afford into the relief of two extremely serious (and related) medical problems, opiate addiction and pain. The endorphins are believed to function as natural pain relievers. Individuals in stressful situations—soldiers in battle, athletes at crucial moments in a contest—have often reported being unaware of what later proved to be an extremely painful injury and so being able to continue to function in a life-threatening (or victory-threatening) situation. The recent discovery that macrophages are among the cell types bearing endorphin receptors suggests that endorphins may also play a role in the stimulation of inflammatory and immune responses.

Morphine, codeine, heroin, and other opiate drugs combine with the endorphin receptors, relieving stress, elevating mood, and soothing pain. However, it is hypothesized that these external opiates, acting by negative feedback, reduce the normal production of endorphins, resulting in ever-increasing dependence on the artificial source—or, in other words, in addiction.

The processing of information that occurs within the cell body of each individual neuron plays a key role in the integration and control exercised jointly by the nervous and endocrine systems. It is affected not only by the specific neurotransmitters and neuromodulators received by the cell, but also by their quantity, the precise timing of their arrival, and the locations on the neuron of the various synapses and receptors. Each neuron is a tiny computer, summing an enormous quantity of information and issuing appropriate commands carried throughout the network with which we began this chapter.

Summary

The nervous system, along with the endocrine system, integrates and controls the numerous functions that enable an animal to regulate its internal environment and to react to or deal with its external environment. The functional unit of the nervous system is the neuron. It consists of dendrites, which receive stimuli; a cell body, which contains the nucleus and metabolic machinery and also receives stimuli; and an axon, which relays stimuli to other cells.

The central nervous system consists of the brain and the spinal cord, which are encased, in vertebrates, in the skull and vertebral column. That part of the nervous system outside the central nervous system constitutes the peripheral nervous system. Cranial nerves enter and emerge from the brain in pairs. Spinal nerves enter and emerge from the vertebral column, also in pairs. Each of these pairs innervates effectors and receives signals from sensory receptors of a different and distinct area of the body.

The motor neurons of the peripheral nervous system are organized in two major divisions: (1) the somatic nervous system, which innervates the skeletal muscles, and (2) the autonomic nervous system, which controls cardiac muscle and the smooth muscles and glands involved in the digestive, circulatory, urinary, and reproductive functions. In the autonomic system, axons arising from neurons in the central nervous system synapse with motor neurons in ganglia outside the central nervous system. The postganglionic neurons stimulate or inhibit the effectors.

The autonomic system has two divisions—sympathetic and parasympathetic—which are anatomically and functionally distinct. The sympathetic division is largely responsible for reactions to the external environment, whereas the parasympathetic division controls restorative activities, such as digestion.

Information received from the internal and external environments and instructions carried to effectors, such as muscles and glands, are transmitted in the nervous system as electrochemical signals. At rest, there is a difference in electric charge between the inside and outside of the axon membrane—the resting potential. With appropriate stimulation, an action potential, a transient reversal in membrane polarity, occurs. The action potential traveling along the axon membrane is the nerve impulse. Because all action potentials are the same size, the message carried by a particular axon can be varied only by a change in the frequency or pattern of action potentials. In myelinated fibers, the nerve impulse leaps from node to node of the myelin sheath, thereby speeding conduction.

Neurons transmit signals to other neurons across junctions called synapses. In most synapses, the signal crosses the synaptic cleft in the form of a chemical, a neurotransmitter, that binds to a specific receptor on the membrane of the postsynaptic cell. Other chemicals, known as neuromodulators, can diffuse through local areas of the central nervous system as well, binding to the membrane receptors of other neurons in the area. Binding of a neurotransmitter or neuromodulator to its receptor may open or close a membrane ion channel or set a second messenger in motion. The ultimate effect is a change in the membrane voltage of the postsynaptic cell.

A single neuron may receive signals from many synapses, and, based on the summation of excitatory and inhibitory signals, an action potential will or will not be initiated in its axon. Thus individual neurons function as important relay and control centers in the integration of information by the nervous system.

Questions

1. Distinguish among the following: neuron/nerve/tract; ganglia/nuclei; central nervous system/peripheral nervous system; gray matter/white matter; resting potential/action potential/nerve impulse; presynaptic/postsynaptic; neurotransmitter/neuromodulator.

2. What are three significant differences between the somatic and autonomic nervous systems?

3. What are three significant differences between the sympathetic and parasympathetic divisions of the autonomic nervous system?

4. If a neuron is placed in a medium where ionic concentrations are the same as those of its own cytosol, what effect will this have on the resting potential?

5. Describe the way in which the nerve impulse propagates itself. Draw a diagram of this process.

6. Among the most potent of all neurotoxins (nerve poisons) are two naturally occurring substances that block the voltage-gated Na^+ channels of the axon membrane. One of these is found in the skin, ovaries, liver, and intestines of the puffer fish, or fugu, the flesh of which is regarded as a great delicacy in Japan. The other is produced by the dinoflagellates (page 427) responsible for red tides. This toxin becomes highly concentrated in shellfish that feed on the dinoflagellates. Despite the great care exercised by sushi masters in preparing fugu, and despite the warnings about contaminated shellfish that are issued whenever red tides occur, each year a number of people die as a result of ingesting these two toxins. What is the cause of death?

7. Assume that three presynaptic neurons, A, B, and C, make adjacent synapses on the same postsynaptic neuron, D. No nerve impulse is initiated in the postsynaptic neuron as a result of single nerve impulses in A, B, or C, nor is one initiated if impulses arrive simultaneously in all three, in A and B, or in A and C. Only if impulses arrive together at the synapses of B and C will D fire an impulse. Explain these results in terms of excitatory and inhibitory neurotransmitters and their effects on the membrane voltage of the postsynaptic cell.

8. If you look closely at Figure 35–17, you will notice one neuron that has only very short dendrites and no axon at all. Axon terminals of other neurons form synapses on its cell body, but it forms no axonal synapses with other cells. Such neurons, which are located primarily in the brain, are incapable of initiating an action potential and are thus known as "nonspiking." Nevertheless, they affect the firing rate of other neurons. How do you think they might do this? Why might axons be superfluous for such neurons?

9. It is early autumn, and you are walking down a desert canyon with a friend. At that time of year, temperatures are such that rattlesnakes are active at midday. You round a bend in the canyon and there, coiled up by a small water hole, is a diamondback rattler, about 1 meter in length. You see the snake immediately and note its presence to your friend. You both give it a wide berth and continue down the canyon. About 50 meters farther along, your friend almost steps on a second rattlesnake, this one about 1.5 meters in length. In the process of escaping, this snake crawls across your foot and, briefly, up your leg! Needless to say, your response to this second snake is more dramatic than your response to the first. What role has your autonomic nervous system played in the sequence of physiological events within your body? What role have your adrenal glands played? What would you expect your response to be if you were to have a third close and sudden encounter with a rattlesnake? Why?

Suggestions for Further Reading

Agnati, Luigi F., Börje Bjelke, and Kjell Fuxe: "Volume Transmission in the Brain," *American Scientist,* vol. 80, pages 362–373, 1992.

Changeux, Jean-Pierre: "Chemical Signaling in the Brain," *Scientific American,* November 1993, pages 58–62.

Cowen, Ron: "Receptor Encounters: Untangling the Threads of the Serotonin System," *Science News,* October 14, 1989, pages 248–252.

Dunant, Yves, and Maurice Israel: "The Release of Acetylcholine," *Scientific American,* April 1985, pages 58–66.

Fackelmann, Kathy A.: "Myelin on the Mend: Can Antibodies Reverse the Ravages of Multiple Sclerosis?" *Science News,* April 17, 1990, pages 218–219.

Holloway, Marguerite: "R_x for Addiction," *Scientific American,* March 1991, pages 94–103.

Kalin, Ned H.: "The Neurobiology of Fear," *Scientific American,* May 1993, pages 94–101.

Morell, Pierre, and William T. Norton: "Myelin," *Scientific American,* May 1980, pages 89–116.

Musto, David F.: "Opium, Cocaine and Marijuana in American History," *Scientific American,* July 1991, pages 40–47.

Rasmussen, Howard: "The Cycling of Calcium as an Intracellular Messenger," *Scientific American,* October 1989, pages 66–73.

Sapolsky, Robert M.: "Stress in the Wild," *Scientific American,* January 1990, pages 116–123.

Scientific American: *Progress in Neuroscience,* W. H. Freeman and Company, New York, 1985.

> *A collection of articles from* Scientific American, *originally published between 1979 and 1984, covering such topics as the role of calcium ion at the synapse, the development of the nervous system, hearing, and the genetic control of behavior.*

Tuomanen, Elaine: "Breaching the Blood-Brain Barrier," *Scientific American,* February 1993, pages 80–84.

Welch, William J.: "How Cells Respond to Stress," *Scientific American,* May 1993, pages 56–64.

Wurtman, Richard J., and Judith J. Wurtman: "Carbohydrates and Depression," *Scientific American,* January 1989, pages 68–75.

*Available in paperback.

Our senses provide us with information about the outside world. Like other primates, we receive most of this information through our sense of vision.

One way we recognize these birds as owls is by their large eyes. These eyes, with their large pupils, have some 100 times the light-gathering power of the eyes of a bird—a pigeon, for instance—that is active during the day. Owl eyes are directly in front, providing complete binocular vision. This is important for the depth perception required for hunting. Owls, however, have a visual field of only 110°. Our eyes, by contrast, are slightly to the side, so we have a visual field of about 180°. Birds that are prey rather than predator—again, pigeons are an example—have their eyes on the sides of their head, providing a visual field of as much as 340°. Owl eyes are fixed in their sockets, like the headlights of a car. In compensation, owls can rotate their heads through 270° and turn them almost upside down.

Owls' ears—long, vertical slits widely placed on either side of the head—are even more important to these expert hunters. By rapidly moving its head until a sound registers equally in both ears, an owl can pinpoint the source of the sound. These great gray owls, for instance, can locate and capture small rodents in total darkness, guided only by the noises the rodents make in scuttling across the ground or traveling through their underground tunnels. The broad facial disks are rimmed with tightly packed rows of stiff feathers that reflect these high-frequency sounds. The owl's other feathers are soft and downy, so air turbulence and sound are kept to a minimum as it swoops down on a mouse. We probably could not hear an owl in flight, but a mouse, which lives in a world dominated by smell and hearing, sometimes can. Its life depends on it.

Each of us—human, owl, mouse—receives different information from the same set of stimuli. Though we may, from time to time, share the same environment, we live, in fact, in different worlds, dictated in part by the fine-tuning of our sense organs and also—and far more important—by the universe inside our heads.

36

The Nervous System II: Sensory Perception, Information Processing, and Motor Response

The vertebrate nervous system, as we have seen, consists of a hierarchical organization of neurons, cells specialized for the conduction and transmission of electrochemical signals. Information from both the internal and external environments is received by this system and then rapidly transmitted—in the form of nerve impulses—to the appropriate control centers in the spinal cord and brain. These control centers process the information and then transmit appropriate instructions—also in the form of nerve impulses—to effectors, such as glands and muscles. The responses of the effectors—hormonal secretions and muscular contractions—subsequently alter either the internal physiology of the body or its relationship to the external environment.

In this chapter, we shall first consider how environmental stimuli, which come in a variety of forms, are able to trigger nerve impulses. Then we shall examine some of the more fascinating structures and functions of the primary vertebrate processing center, the brain. Finally, we shall consider how nerve impulses transmitted through motor neurons are translated into a concrete action, the contraction of a skeletal muscle.

Sensory Receptors and the Initiation of Nerve Impulses

The information received, processed, and transmitted by the neurons and synapses of the brain and spinal cord is carried into the central nervous system by sensory neurons. The triggering of nerve impulses in a sensory neuron depends on the conversion of the energy of a stimulus into the energy of an action potential.

Stimuli come in a variety of forms—pressure, heat or cold, vibrations, light, and chemical substances. Different types of sensory receptors are specialized to respond to different types of stimuli. In all cases, however, when a sensory receptor has been sufficiently stimulated, the membrane permeability of a sensory neuron is altered, initiating the action potentials that start the information on its way through the nervous system. The more intense the stimulus, the greater the frequency of the action potentials (see Figure 35–11, page 630).

The differences among the senses lie not in the form in which the signals are coded and conducted—the action potential—but rather in the pattern of the signals and their reception and interpretation in the central nervous system. As we shall see, information from different sensory receptors is transmitted to different regions of the brain. The particular sensation experienced—a sunset, the song of a whippoorwill, or a cooling breeze across the face—depends on the region of the brain that is stimulated.

Types of Sensory Receptors

Our sensory receptors are many and varied. Like most animals, we have mechanoreceptors (touch, hearing, body position), chemoreceptors (taste and smell), photoreceptors (vision), temperature receptors, and receptors for the sensation we recognize as pain. We apparently do not have electroreceptors or magnetoreceptors, but some animals do.

Some sensory receptors are small and relatively simple in structure. For example, consider the sensory receptors of the human skin (Figure 36–1). The simplest receptors are the free nerve endings, which are receptors for pain, temperature, and perhaps other sensations as well. Slightly more complex are the combinations of free nerve endings with a hair and its follicle. Each of these little structures is an exquisitely sensitive mechanoreceptor. When the hair is touched or bent, it causes changes in membrane permeability in the nerve endings of a sensory neuron, setting off action potentials that are carried directly to the central nervous system. Three other types of mechanoreceptors are also shown, each a combination of one or more free nerve endings with an outer layer or layers of connective tissue.

Still more complex are the taste buds of the tongue (Figure 29–4, page 527), which provide the information on which a final judgment can be made on what is and is not to be swallowed. We are able to distinguish four primary tastes: sweet, sour, salty, and bitter. While each primary taste (or, more accurately, the molecules associated with the taste) appears to stimulate a different type of taste receptor, a single taste bud may respond to more than one category of substance.

Most complex of all are the major sensory organs, the ear and the eye.

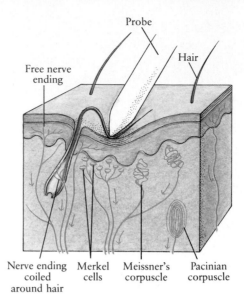

36–1   *Some sensory receptors present in human skin. The free nerve endings are largely pain and temperature receptors. Although they are the simplest type of sensory receptors, they are the most poorly understood in terms of function.*

The best understood of the skin receptors are the Pacinian corpuscles. The specialized nerve ending of a single myelinated nerve fiber is encapsulated by many concentric layers of connective tissue. Pressure on these outer layers stimulates the firing of an action potential at this nerve ending.

Merkel cells and Meissner's corpuscles also respond to touch, as do the nerve endings surrounding the hair follicles. In regions where many hairs are present, the other touch receptors are sparse or absent.

The Ear

The human ear (Figure 36–2) is an organ containing two distinct sensory receptors, each of which converts the energy of an enclosed, moving fluid into the energy of action potentials. One of these structures consists of three interlocking **semicircular canals** that provide information about the movement and orientation of the head in space. This information is critical for the maintenance of physical balance. The other structure, the **cochlea** ("snail"), is a coiled chamber in which the sensory responses involved in hearing occur. Additional structures convert vibrations of the air (sound waves) into vibrations of the fluid contained within the cochlea. In both the semicircular canals and the cochlea, the primary sensory receptors are sensitive hair cells that vibrate in response to movements of the surrounding fluid.

As shown in Figure 36–3, the cochlea consists of three fluid-filled canals, separated by membranes. Vibrations of the fluids along the outer surface of the central canal cause vibrations in the basilar membrane of the canal, which, in turn, cause vibrations in the hair cells. Vibration of a hair cell opens ion channels in its

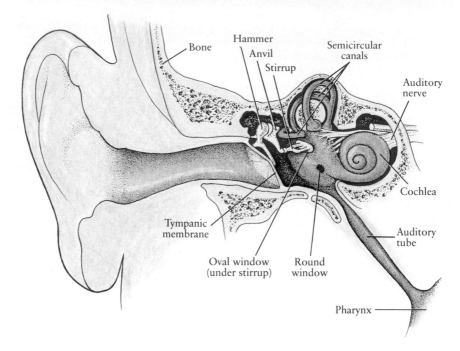

36–2   *The structure of the human ear. Sound waves entering the outer ear are funneled to the tympanic membrane (eardrum), which they cause to vibrate. These vibrations are transmitted through three small bones of the middle ear—the hammer, anvil, and stirrup—to another membrane, the oval window membrane. Vibrations of this membrane, in turn, set off vibrations in the fluids in the cochlea, the structure of the inner ear concerned with hearing.*

The three semicircular canals are additional fluid-filled chambers within the bony labyrinth of the inner ear. Each one is in a plane perpendicular to the other two. Their function is to monitor the position of the head in space and to maintain equilibrium. Movements of the head set the fluid in these canals in motion, activating sensitive hair cells and triggering action potentials in sensory neurons with which they synapse.

The middle ear, as you can see, is connected with the upper pharynx by the auditory tube (also known as the Eustachian tube). This makes it possible to equalize the air pressure in the middle ear with atmospheric pressure.

36–3   *The structures of the ear concerned with hearing. (a) Vibrations transmitted from the tympanic membrane to the stirrup cause the stirrup to push against the oval window membrane. This produces pressure waves in the fluid-filled canals of the cochlea, visible in cross section (b). The waves travel the length of the upper canal, around the far end of the cochlea, and back through the lower canal to the round window membrane. As the oval window membrane moves in, the round window membrane moves out, keeping the pressure equalized.*

(c) Within the central canal, resting on the basilar membrane, is a structure known as the organ of Corti. As the pressure waves travel through the cochlear canals, they set up vibrations in the basilar membrane and in the hair cells of the organ of Corti, pushing the cells up against the overlying tectorial membrane. This stimulation leads to the release of neurotransmitter molecules from the hair cells, initiating action potentials in the sensory neurons with which they synapse.

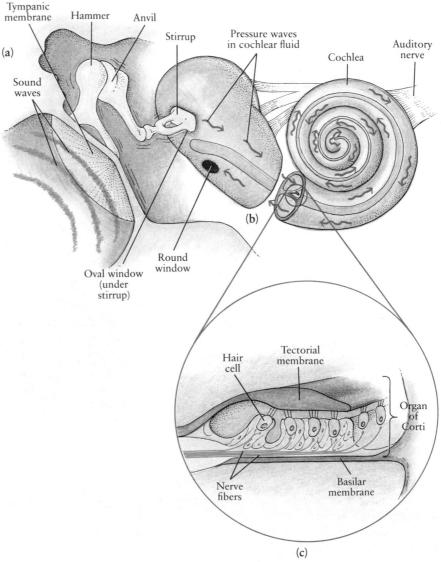

36–4   *The human eye. An image, represented by the arrow, is focused (and inverted) on the retina. The retina, which is richly supplied by a network of fine blood vessels, contains the photoreceptor cells as well as other cells that provide connections to the neurons whose axons form the optic nerve. The fovea, near the center of the retina, is the region of greatest visual acuity. The pupil is a hole in the center of the iris, which is the colored part of the eye. The cornea is the transparent outer layer that covers the iris and the pupil. Only the front of the eye is exposed. The rest of the eyeball is recessed in and protected by the bony socket of the skull.*

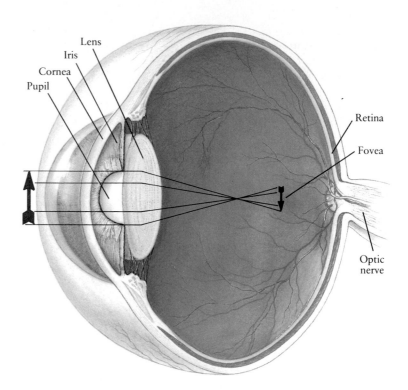

membrane, causing a depolarization that alters the release of neurotransmitter molecules from the cell. Hair cells synapse with sensory neurons, which, when adequately stimulated at these synapses, initiate action potentials. The axons of these neurons form the **auditory nerve,** which carries the information into the brain.

The basilar membrane is narrower and less elastic at one end than at the other. Hence it does not vibrate uniformly along its length. Sound at any given frequency (pitch) has its maximum effect on a specific region of the membrane. Thus, different hair cells are stimulated by different frequencies. We are capable of very fine discriminations of sound. Such discriminations are made by the brain on the basis of the signals transmitted by the different sensory neurons with which the different hair cells synapse.

The Eye

The vertebrate type of eye is often called a camera eye. In fact, it has a number of features in common with an ordinary camera equipped with several expensive accessories, such as a built-in cleaning and lubricating system, an exposure meter, and an automatic focus. Light from the object being viewed passes through the transparent **cornea** and **lens,** which focus an inverted image of the object on the light-sensitive **retina** at the back of the eyeball (Figure 36–4).

The retina of the vertebrate eye contains the **photoreceptor cells,** which capture light energy. These cells are of two types, named, because of their shapes, **rods** and **cones** (Figure 36–5). Rods are responsible for night vision; cones, for color vision.

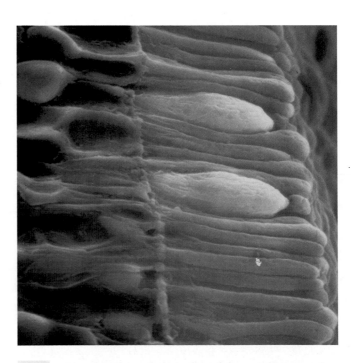

36–5   *A section through the retina of the vertebrate eye. In this scanning electron micrograph, the rods have been colored orange and the cones yellow. The other cells in the micrograph (shown in pink) are horizontal cells, bipolar cells, and amacrine cells, all of which play a role in the processing of information transmitted by the rods and the cones.*

Rods do not provide as great a degree of resolution (the amount of detail that can be distinguished) as cones do, but they are more light-sensitive than cones. Dim light does not stimulate the cones, which is why the world becomes colorless to us at night. Nocturnal animals, which include most mammals, have retinas made up almost entirely of rods and thus have virtually no color vision. Some diurnal animals, such as some reptiles and squirrels, have almost entirely cones. Higher primates, including humans, have both rods and cones.

As shown in Figure 36–6, the photoreceptor cells communicate, by way of intervening neurons, with the **ganglion cells,** whose axons form the **optic nerve.** When light is captured by the photoreceptor cells, a series of reactions occur that produce a change in their membrane polarity. This change influences their release of neurotransmitters at synapses they form with a group of cells known as **bipolar cells.** The release of neurotransmitters at these synapses causes, in turn, a change in the membrane polarity of the bipolar cells, and influences their release of neurotransmitters at synapses with the ganglion cells. The ultimate result of the stimulation by light is a change in the pattern in which action potentials fire in the axons of the ganglion cells.

There are about 125 million photoreceptors in the human retina but only about 1 million ganglion cell axons in the optic nerve, a ratio of 125 to 1. This ratio involves primarily the rods. Many rods converge on each bipolar cell, and several bipolar cells, in turn, feed into each ganglion cell. Also, other neurons in the retina—horizontal and amacrine cells—participate in these connections and interactions. Therefore, the rods do not transmit a point-to-point representation—as does a television camera, for instance—but there is processing of the information before it even leaves the retina (see essay).

The area of the retina in which the sharpest image is formed is known as the **fovea.** In the fovea, the photoreceptor cells consist entirely of closely packed cones. These cones, instead of having the 125:1 relationship of the rods and ganglion cells, make one-to-one connections with the bipolar and ganglion cells. The one-to-one connections and the close packing of the cones provide greater resolution, giving a crisper picture.

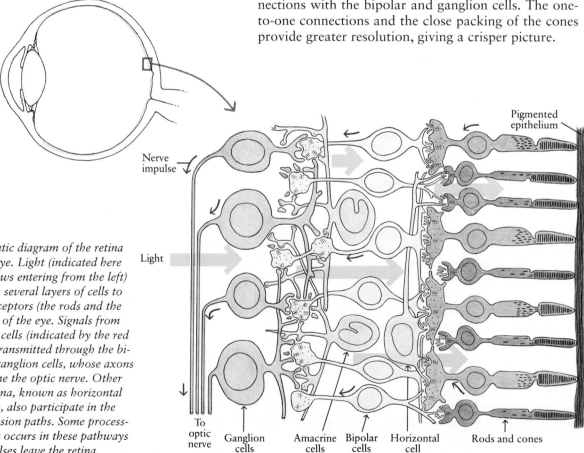

36–6   *A schematic diagram of the retina of the vertebrate eye. Light (indicated here by the yellow arrows entering from the left) must pass through several layers of cells to reach the photoreceptors (the rods and the cones) at the back of the eye. Signals from the photoreceptor cells (indicated by the red arrows) are then transmitted through the bipolar cells to the ganglion cells, whose axons converge to become the optic nerve. Other neurons in the retina, known as horizontal and amacrine cells, also participate in the elaborate transmission paths. Some processing of information occurs in these pathways before nerve impulses leave the retina.*

Nerve impulse

Light

To optic nerve

Ganglion cells

Amacrine cells

Bipolar cells

Horizontal cell

Rods and cones

Pigmented epithelium

The frog's eye differs from the human eye in that it has no central fovea and the rods and cones are distributed uniformly over the surface of the retina. It resembles the human eye, however, in that these photoreceptor cells transmit their signals to a far fewer number of ganglion cells, whose axons make up the optic nerve leading from the eye to the brain.

By inserting microelectrodes into these axons while exposing the frog's eye to various kinds of stimuli, Jerome Lettvin and coworkers at the Massachusetts Institute of Technology found that different ganglion cells responded to different stimuli—light on, light off, or a big moving shadow, for instance. Most interesting, one type of ganglion cell responded only to a small moving object. In other words, it was a bug detector. An object bigger than a bug would not stimulate these particular ganglion cells even if it was in motion, and a bug-sized object would not stimulate them if it was motionless.

The existence of the bug detector corresponds nicely with certain well-known features of the animal's behavior. A frog will strike only—and virtually always—at a small moving object and will literally starve surrounded by dead insects. Thus, the information about an extremely important aspect of the frog's world is processed right in the retina itself.

A red-eyed tree frog (Agalychnis callidryas), *leaping to capture a cricket.*

Information Processing in the Vertebrate Brain

As animals have become more complex in the course of their evolutionary history, the task of integration and control has also become increasingly complex. That it is accomplished so successfully in all vertebrates can be attributed to an accompanying evolutionary trend, the increased centralization of information processing in one dominant center, the brain.

This trend has resulted in a structure that, in humans, weighs about 1,400 grams (3 pounds), has the consistency of semisoft cheese, and is, by far, the most complex and highly organized structure on this planet. Its functions are essential not only for the integration and control of the multitude of physiological activities that occur throughout the body but also for those processes that we identify as "mind"—consciousness, perception and understanding of information from the external environment, thought, memory, and the variety of emotions that characterize human experience. Although our knowledge of the structure and function of the brain is increasing rapidly, it is such a complex system that many believe we shall never fully understand it.

The Structure of the Brain

The soft substance of the brain, upon which all of its functions depend, is well protected by the bones of the skull. Like the spinal cord, it consists of white matter—the fiber tracts, made white by their lipid-rich myelin sheaths—and gray matter. The gray matter contains not

50 μm

36–7 *In this scanning electron micrograph of tissue from a human brain, you can see the cell bodies of a number of neurons, embedded in a dense meshwork of axons. All of the information processing that occurs in the brain depends on interconnected networks of neurons, of which there are estimated to be some 100 billion in the human brain. This tissue section is from the cerebral cortex, a region of the brain that reaches its highest degree of development in* Homo sapiens. *The actions and interactions of the cells of the cerebral cortex are responsible for consciousness, intelligence, dreams, and memory.*

only the cell bodies of as many as 100 billion neurons but also numerous glial cells. In some areas of the human brain, neurons and glial cells are so densely packed that a single cubic centimeter of gray matter contains some 6 million cell bodies, with each neuron making synaptic connections to as many as 80,000 others. The result is the capacity to receive, process, and send an enormous number of different messages simultaneously.

Anatomically, the human brain consists of three principal divisions: the brainstem, the cerebellum, and the cerebrum (Figure 36–8). The **brainstem** is the "old" brain—the first region of the vertebrate brain to have evolved—and it is surprisingly similar from fish to *Homo sapiens*. Centers in the brainstem control such vital functions as heartbeat and respiration, which is why a blow to the base of the skull is so dangerous. The brainstem also contains sensory and motor neurons that serve the skin, muscles, and other structures of the head, as well as all the nerve fibers that pass between the spinal cord and the higher brain centers. Many of these fiber tracts cross over in the brainstem, so that the right side of the brain receives messages from and sends signals to the left side of the body, and vice versa.

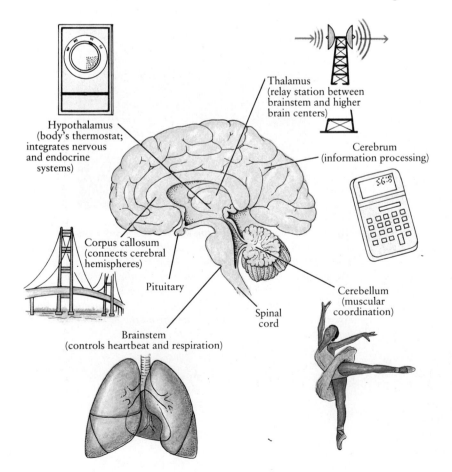

Hypothalamus (body's thermostat; integrates nervous and endocrine systems)

Thalamus (relay station between brainstem and higher brain centers)

Cerebrum (information processing)

Corpus callosum (connects cerebral hemispheres)

Pituitary

Spinal cord

Brainstem (controls heartbeat and respiration)

Cerebellum (muscular coordination)

36–8 *A diagram of the human brain, as seen in longitudinal section, showing its principal structures and some of their major functions. The cerebrum consists of two distinct cerebral hemispheres. As a consequence, the right cerebral hemisphere remains intact in the section illustrated here.*

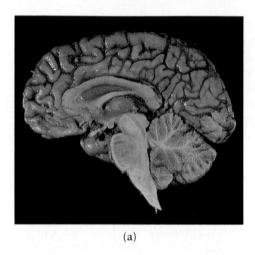

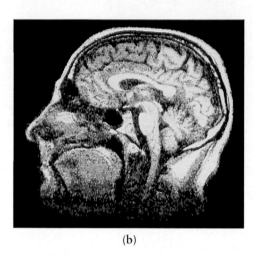

(a)   (b)

36–9   (a) *A preserved human brain, sectioned as in Figure 36–8. The light color of the corpus callosum and the brainstem reflects the fact that they are predominantly white matter, made up of myelinated axons. By contrast, the cerebral cortex is almost entirely gray matter, consisting of neuron cell bodies and dendrites, glial cells, and unmyelinated axons.*

(b) A magnetic resonance image of a human head, showing facial tissues, airways, and the intact, living brain. Magnetic resonance imaging (MRI) is one of several new diagnostic tools available to neurologists. Like CAT scans, MRI provides detailed images of soft tissues without requiring the injection of dyes or radioactive substances.

The **cerebellum** is concerned with the execution and fine-tuning of complex patterns of muscular movement. It is much larger in homeotherms than in the more slow-moving fishes and reptiles and reaches its greatest relative size in birds, in which it is associated with the exquisite coordination necessary for flight.

The **cerebrum,** which occupies 80 percent of the total brain volume in humans, forms two hemispheres (left and right) that are connected by a tightly packed mass of nerve fibers called the **corpus callosum.** The **cerebral cortex,** a thin layer of gray matter forming the outer surface of the cerebrum, is the most recent development in the evolution of the vertebrate brain. Fishes and amphibians have no cerebral cortex, and reptiles and birds have only a rudimentary indication of a cortex. More primitive mammals, such as rats, have a relatively smooth cortex. Among the primates, however, the cortex becomes increasingly extensive and complex. The human cerebral cortex is characterized by many folds and convolutions (Figure 36–10), allowing an enormous information-processing area of 2,500 square centimeters to fit within the skull.

Tucked below the cerebrum are two smaller structures, the **thalamus** and the **hypothalamus.** The thalamus consists of two egg-shaped masses of gray matter. Its neurons sort and process sensory information received from the brainstem and from the optic nerve and relay it to higher brain centers. The hypothalamus, lying just below the thalamus, contains clusters of neurons responsible for the activities associated with sex, hunger, thirst, pleasure, pain, and anger. As we have seen in earlier chapters, it contains the body's thermostat and is the source of the hormones ADH and oxytocin, which are stored in and released from the posterior pi-

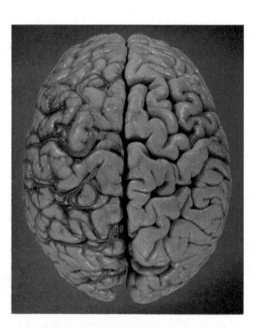

36–10   *A brain, viewed from above. The many convolutions of the cerebral cortex are clearly visible. By these, you can immediately distinguish this brain as human. The brain and spinal cord are enclosed in three layers of membranous tissue, known as the meninges. The innermost layer is visible on the left cerebral hemisphere in this photograph. Also visible are the arteries (red) and veins (blue) to and from the left cerebral hemisphere.*

tuitary. Most important, it is the major center for the integration of the nervous and endocrine systems, acting through its release of hormones that regulate the secretion of tropic hormones from the anterior pituitary.

The Cerebral Cortex

Of the approximately 100 billion nerve cells in the human brain, about 10 billion are in the cerebral cortex. This thin layer of gray matter, about 1.5 to 4 millimeters thick, is the most thoroughly studied region of the human brain. In *Homo sapiens* and other primates, each of the cerebral hemispheres is divided into lobes by two deep fissures, or grooves, in the surface. The principal fissures are the **central sulcus,** which runs down the side of each hemisphere, and the **lateral sulcus.** There are four lobes—**frontal, parietal, temporal,** and **occipital**—on each hemisphere (Figure 36–11).

Certain areas of the cortex have been mapped in terms of the functions they perform (Figure 36–12). The area just anterior to the central sulcus, in the frontal lobe, is the **motor cortex.** It is involved in the integration of the activities of the skeletal muscles. The area just posterior to the central sulcus, in the parietal lobe, is the **sensory cortex.** It is involved with the reception of tactile (touch) stimuli, as well as stimuli related to taste, temperature, and pain. Particular portions of the motor and sensory cortices are allocated to particular parts of the body, as shown in Figure 36–13.

In the temporal lobe, partially buried within the lateral sulcus, is the **auditory cortex.** This region of the cortex is the processing center for the signals relayed from the sensory neurons of the ear. Signals from sensory neurons stimulated by hair cells responding to different frequencies of sound (page 642) are relayed to different regions of the auditory cortex.

The **visual cortex** occupies the occipital lobe. By using a tiny point of light to stimulate very small regions of the retina, one after the other, investigators have shown that each region of the retina is represented by several corresponding but larger regions of the visual cortex. The fovea, which represents about 1 percent of the area of the human retina, projects to nearly 50 percent of the visual cortex. This enormous over-representation, combined with the substantial portions of the motor and sensory cortices devoted to the hands (Figure 36–13), provide evidence of the importance of eye-hand coordination in primate evolution.

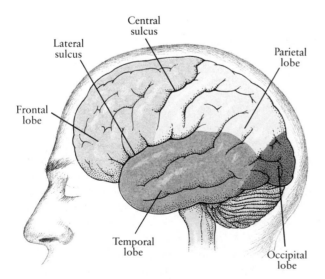

36–11   *The principal sulci (fissures) and lobes of the human cerebral cortex.*

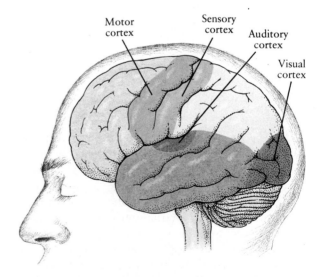

36–12   *The human cerebral cortex, showing the primary locations of the motor and sensory areas, on either side of the central sulcus, and the auditory and visual zones. The motor and sensory cortices span the brain like a headset. Functionally, the left and right motor and sensory cortices are mirror images, with the left cortex sending signals to and receiving signals from the right side of the body, and vice versa.*

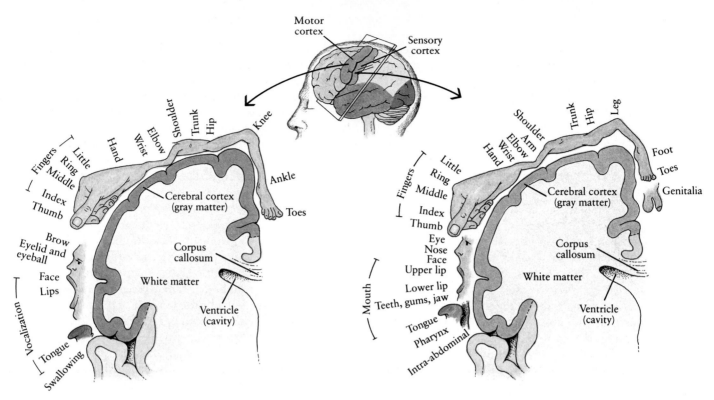

(a) Motor cortex of left cerebral hemisphere   **(b)** Sensory cortex of left cerebral hemisphere

36–13   *Functional representations of* (a) *the motor cortex and* (b) *the sensory cortex of one hemisphere of the human cerebrum. The motor and sensory cortices are located on either side of the central sulcus in each hemisphere. Stimulation of different areas of the motor cortex causes muscle contraction in different parts of the body.*   *Conversely, stimulation of different parts of the body produces electrical activity in different parts of the sensory cortex. Notice the relatively huge motor and sensory areas associated with the hand and the mouth. The motor and sensory cortices of the left cerebral hemisphere, shown here, send signals to and receive signals from the right*   *side of the body. The information in these diagrams is derived from studies of patients undergoing surgical treatment for epilepsy and from observations of individuals in whom particular areas of the cerebral cortex have been destroyed by disease or accident.*

Each of the regions of the visual cortex to which different regions of the retina project contains a variety of cells, different groups of which respond to different types of visual stimuli. As we saw earlier (page 645), considerable processing of visual information occurs in the retina before its transmission to the ganglion cells, which carry the information into the brain. Further processing occurs within the thalamus, where the ganglion cells synapse with other neurons that relay the signals to the visual cortex. The thalamic neurons apparently sort the signals representing different aspects of the visual image and then send them on to specific neurons of the visual cortex. What we see is not a direct image but rather a mental picture computed by the brain from a vast amount of coded information concerning the individual features that collectively form shapes, spatial relationships, and patterns of movement, light and dark, and color.

Left Brain/Right Brain

For more than 100 years it has been known that injury to the left side of the brain often results in impairment or loss of speech, whereas a corresponding injury to the right side of the brain usually does not. Two areas in the left cerebral hemisphere concerned with speech have been mapped, primarily through work with patients who have experienced strokes. These are known as **Broca's area** and **Wernicke's area** (Figure 36–14), each named for the nineteenth-century neurologist who first identified it. About 90 percent of all right-handed people and 65 percent of all left-handed people have these speech areas in the left cerebral cortex.

Broca's area is located just anterior to the region of the motor cortex that controls movements of the muscles of the lips, tongue, jaw, and vocal cords. Damage to this area results in slow and labored speech—if speech is possible at all—but does not affect comprehension. Wernicke's area is adjacent to and partially surrounds the auditory cortex. Localized damage in Wernicke's area results in speech that is fluent but often meaningless, and comprehension of both spoken and written words is impaired.

Because the left hemisphere in most people controls both language and the more capable hand, it has traditionally been regarded as dominant. However, perceptual and spatial disorders are now known to result from injury to the right hemisphere. For example, a person who has sustained such an injury may have difficulty with spatial orientation, may get lost easily and have difficulty finding the way in new or unfamiliar areas, and may have trouble recognizing familiar faces or voices. Musical ability also appears to reside in the right hemisphere.

The acquisition of different functions by the two cerebral hemispheres is seen as another way of increasing the functional capacity of the brain without increasing the size of the skull (which is calculated to be as large as possible in relation to the size of the birth canal). There is some evidence that this differentiation of function between the two cerebral hemispheres is part of the developmental process. It is well known, for instance, that when a young child sustains damage to the left hemisphere, the right hemisphere often takes over the functions of the damaged areas, allowing completely normal speech to develop. The ability of the right hemisphere to assume this left-hemisphere function is closely correlated with the age at which the injury occurs.

Intrinsic Processing Areas

The unmapped areas of the cerebral cortex were once known as the "association" or "silent" cortex. At the time these terms were introduced, the unmapped areas were thought to function as a sort of giant switchboard, interconnecting the motor and sensory cortices. More recent studies have shown, however, that the organization of the brain is more vertical than horizontal. For example, severing the motor cortex from the sensory cortex by deep vertical cuts appears to have little, if any, effect on an animal's behavior. Anatomical studies support the interpretation that communication between the sensory and motor areas of the cortex takes place primarily via lower brain centers, particularly the thalamus.

The regions of the cortex traditionally considered to form the association cortex—all of the areas not mapped in Figures 36–12 and 36–14—are now more accurately described as **intrinsic processing areas.** This means that although they receive and process information from neurons in other areas of the brain, they do not receive directly relayed sensory information of the type, for example, that travels from the retina to the visual cortex. Similarly, they transmit information to neurons in other areas of the brain but not directly to neurons leading out of the brain. Certain intrinsic processing areas do, however, receive information from the visual, auditory, or sensory cortices, and others transmit information to the motor cortex.

Mapping of the intrinsic processing areas is a considerably more complex undertaking than the mapping of the sensory and motor cortices, and this work is still in its earliest stages. However, neurobiologists are beginning to gain some insight into the activities of these vast areas of the cortex. For example, it is now known that the posterior region of the parietal lobe and the lower (inferior) portion of the temporal lobe receive signals transmitted by neurons of the adjacent visual cor-

36–14   *The human cerebral cortex, showing the areas associated with language. Damage to Broca's area—the more anterior area—affects speech. Damage to Wernicke's area affects comprehension of language. These two areas of the left cerebral cortex are connected by a fiber tract.*

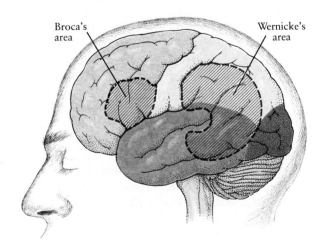

Broca's area

Wernicke's area

36–15   *The processing of visual information involves not only the visual cortex but
also adjacent intrinsic processing areas in the
parietal and temporal lobes. Information
about the components of a visual image, received in the visual cortex, is transmitted to
the temporal lobe, where complex analysis
occurs that leads to a perception of a whole
image. Simultaneously, other coded information is transmitted to the parietal lobe, where
spatial relationships of the object with other
objects in the visual field are analyzed.*

*A number of studies suggest that similar
divisions of labor may also be involved in
the processing of other types of sensory
information.*

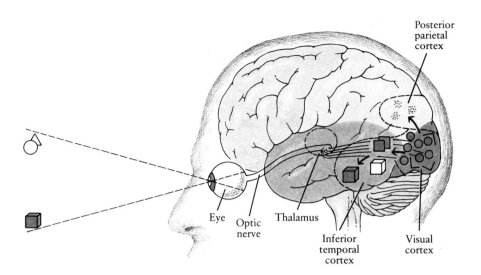

tex and are involved in the further processing of visual
information (Figure 36–15).

The proportion of the cerebral cortex devoted to intrinsic processing is much higher in primates than in
other mammals and is very large in humans. This suggests that these areas have something to do with what
is special about the human mind. Also, about half of
the total area involved is in the frontal lobes, the part
of the brain that has developed most rapidly during the
recent evolution of *Homo sapiens*. The present scientific
consensus is that the intrinsic processing areas are concerned with the integration of sensory information with
emotion and its retention in memory, with the organization of ideas, which is, of course, a key component of learning, and with long-range planning and
"intention."

Brain Circuits

In the brain, as we have seen, there is—as elsewhere in
the body—a division of labor, with different parts of
the brain performing different, specific functions. However, integration and control of the multitude of processes occurring in an animal's body depend on coordination of all of the activities occurring in the different
parts of the brain. Information is exchanged between
different regions of the brain by way of diffuse tracts of
bundled axons. Thus, a local network of neurons in one
region of the brain can have its activity modified by—
and can modify the activity of—networks of neurons
located elsewhere in the brain. Two examples of such
integrating networks are the reticular formation and the
limbic system.

The Reticular Formation

The **reticular formation** is a core of tissue, consisting of
a loose network of interneurons, running through the
brainstem (Figure 36–16). Some of the neurons in the
thalamus function as an extension of this network. The
reticular formation and its thalamic extension have numerous interconnections with the cerebral cortex.

The reticular formation is of particular interest because it is involved with arousal and with that hard-to-
define state we know as consciousness. All of the sensory systems have fibers that feed into the reticular
formation, which apparently filters incoming stimuli
and discriminates the important from the unimportant.
The existence of such a filtering system is well verified
by ordinary experience. A person may sleep through the
familiar rumble of a train or a loud radio or TV program but wake instantly at the cry of the baby or the
stealthy turn of a doorknob. Similarly, we may be unaware of the contents of a dimly overheard conversa-
tion until something important—our own name, for instance—is mentioned, and then our degree of attention
increases.

The Limbic System

The **limbic system** is a network of mostly subcortical
("below the cortex") neurons that form a loop around
the inside of the brain, linking the hypothalamus to the
cerebral cortex and to other structures as well. It is
thought to be the circuit by which drives and emotions,
such as hunger, thirst, and desire for pleasure, are translated into complex actions, such as seeking food, drinking water, or courting a mate. It is also a principal circuit in the consolidation of memory.

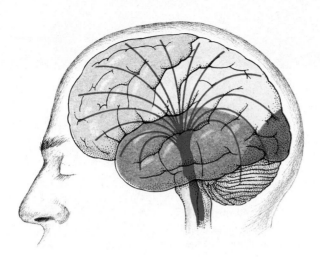

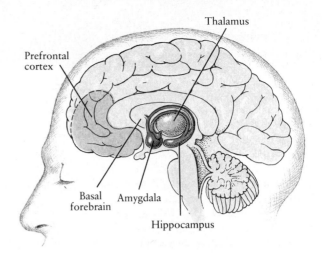

36–16 *The reticular formation, shown here in purple, is a diffuse network of neurons in the brainstem. In the reticular formation, incoming sensory stimuli of all types are monitored and analyzed, leading to modulation of the activity of other areas of the brain. The reticular formation and its thalamic extension are concerned with general alertness and the direction of attention.*

36–17 *Structures of the human brain involved in the consolidation and storage of memory, as shown in longitudinal section. Damage to any of these structures results in memory loss, the details of which vary according to the structure affected. For example, the memory loss associated with strokes typically involves damage to the prefrontal cortex, the thalamus, or the posterior portion of the hippocampus. The memory loss associated with Alzheimer's disease involves neurons in the basal forebrain. Inflammation or a temporary interruption of the oxygen supply to the brain can cause damage to the amygdala and the anterior portion of the hippocampus that also results in memory loss.*

Memory and Learning

Of all the functions of the brain, perhaps the most fascinating are the capacity to learn—that is, to modify behavior on the basis of experience—and the establishment of the memory on which learning depends.

There are two distinct types of memory, short-term and long-term. A simple example of short-term memory is looking up an unfamiliar number in a phone book. You usually remember it just long enough to dial it. If you call the number enough times, it is transferred to long-term memory. This laying down of long-term memory is analogous to the establishment of a footpath. The more frequently the path is traveled, the better established it becomes. The analogy is strengthened by the familiar experience of consciously retrieving a name, for instance, by seeking out related information that puts one "on the right track."

Studies in experimental animals and with persons who have experienced memory loss as a result of disease or injury have revealed that several regions of the brain are involved in the consolidation and storage of memory. As shown in Figure 36–17, these regions include the **hippocampus** ("seahorse") and **amygdala** ("almond"), two components of the limbic system that are located on the inner surface of the temporal lobe; the thalamus; the **basal forebrain,** one of the most ancient parts of the brain; and a portion of the frontal lobe known as the **prefrontal cortex.**

According to current hypotheses, information is transmitted along independent pathways from the various sensory cortical areas to the hippocampus and amygdala, from which independent pathways carry the information to the thalamus. Neurons of the thalamus, in turn, conduct the information to the basal forebrain and to the prefrontal cortex. Parallel circuits transmit processed information in the opposite direction, in what is thought to be a positive feedback process. The basal forebrain, which degenerates in Alzheimer's disease, is a principal source of the neurotransmitter acetylcholine in the brain. The acetylcholine that its neurons release in the feedback circuit is apparently vital for the processes that occur in other parts of the circuit, particularly the amygdala and the hippocampus.

Alzheimer's Disease

Most of us take for granted the reliability of memory, day in and day out, as we engage in the many activities—both mental and physical—that make up our lives. Alzheimer's disease, which is estimated to afflict as many as 4 million older persons in this country alone, makes clear what a precious gift memory is and how devastating its loss can be.

Alzheimer's disease was first described in 1907 by the German neurologist Alois Alzheimer, who recognized it as a distinct pathology afflicting a very small number of people in their forties and fifties. Characterized by progressive memory loss, the disease ultimately leads to severe dementia, including the inability to think, speak, or perform even the most basic tasks of personal care. Death generally follows within 3 to 10 years. By the 1970s, it had become clear that many cases of "senility" in elderly persons, previously assumed to be an inevitable consequence of age, were actually Alzheimer's disease.

The biological changes characteristic of Alzheimer's disease are revealed most strikingly in autopsy studies of brain tissue. Three abnormalities are found consistently: accumulations of tangled and twisted protein filaments within neuron cell bodies; structures known as neuritic plaques, which are clusters of degenerated axon terminals associated with a protein known as amyloid; and accumulations of this same protein adjacent to and within the walls of blood vessels. Although these abnormalities are found in various regions of the cerebral cortex, they are most apparent in the structures associated with memory—the hippocampus and the amygdala.

In addition, there is a loss of neurons whose cell bodies are located in the basal forebrain, which are a major source of acetylcholine. The axons of these neurons extend not only to the hippocampus and amygdala but also into many areas of the cerebral cortex. The death of these neurons thus reduces the supply of acetylcholine in the regions of the brain where their axon terminals are located.

Whether the abnormalities observed with Alzheimer's are the cause of the disease or, instead, its consequence is not yet known. It is also unclear whether Alzheimer's is one disease, with one underlying cause, or whether it is a family of diseases, with several different causes that lead to the same set of pathological changes. There is strong evidence that genetic factors are involved in early-onset Alzheimer's disease and less conclusive evidence that they may also be involved in some of the cases that occur much later in life.

Considerable research has been devoted to the identification of the gene coding for the amyloid protein associated with Alzheimer's disease. This protein, which consists of 42 amino acids, is synthesized as part of a much larger molecule that appears to be a membrane glycoprotein. Knowledge of the amino acid sequence has made possible the synthesis of gene probes, and these probes, along with the analysis of restriction enzyme fragments, have located the gene for the amyloid precursor protein on chromosome 21. As you will recall (page 199), Down syndrome is caused by an extra copy of chromosome 21, and perhaps significantly, many persons with Down syndrome ultimately develop Alzheimer's disease.

Now that the gene coding for the amyloid protein has been identified, studies are underway to explore its function and regulation and to determine whether the protein is a normal or abnormal gene product. The role of external factors—toxic agents, infectious agents, and immunological responses to infection—in the disease process is also being explored.

The brain, as we have seen, is an enormously complex structure, in which numerous intricate, interdependent processes occur. To even begin to understand its normal functioning is extraordinarily difficult. To determine the cause or causes of a disease such as Alzheimer's is a staggering task—but one to which investigators in many laboratories are devoting their energies.

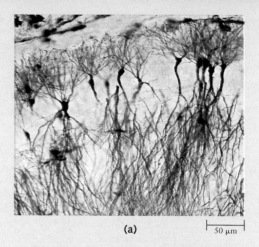

(a)

50 μm

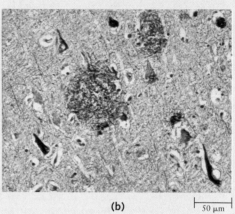

(b)

50 μm

(a) *Neurons of the hippocampus, one of the principal structures involved in the consolidation of memory.* **(b)** *A section of tissue, obtained at autopsy, from the hippocampus of an individual with Alzheimer's disease. The large round structure in the center and the oval structure at the top are neuritic plaques, aggregations of degenerating axons and amyloid protein. The dark, kite-shaped objects are the cell bodies of neurons containing abnormal tangles of protein filaments. These filaments are insoluble in water and impervious to chemical or enzymatic breakdown. They persist long after the cell containing them has died.*

Experimental studies indicate that, in addition to its role as a relay station in these circuits, the amygdala is also the region through which information from different senses is linked—so that, for example, when you think of the seashore, you can remember not only the visual image of sand and waves but also the sound of the lapping waves, the smell and taste of salt spray, and the feel of sand between your toes. And, apparently because of connections between the amygdala and the hypothalamus, the memories have emotional content—the details of which depend on your particular experiences at the seashore. Other fibers of the amygdala appear to communicate back to the primary sensory cortical areas. Current evidence indicates that the traversing of all of these pathways, including the feedback pathways, is required for the consolidation of long-term memory.

Synaptic Modification

Identification of the pathways through which information travels in the establishment of memory leaves unanswered the question of the changes at the cellular and molecular levels that form the "stuff" of memory. Possible answers to this question are beginning to emerge from studies in which invertebrate animals are used as model organisms.

A notable example of such work is that of Eric Kandel and his associates at Columbia University on the gill withdrawal reflex of the sea hare *Aplysia* (Figure 36–18). When this mollusk is touched gently on the underside, it quickly withdraws its siphon and delicate gills in a protective reaction. The stimulus, it has been found, causes 24 sensory neurons in the area to fire. These, in turn, activate interneurons and six motor neurons that control the movement of the gill and cause it to contract.

If *Aplysia* is touched repeatedly, it becomes habituated to the stimulus and ceases to withdraw. Habituation, which is regarded as a very simple form of learning (see page 380), is associated with a gradual decrease in the amount of neurotransmitter released by the repeatedly stimulated sensory neurons. This decrease is reflected, in turn, by a decline in the response of the motor neurons controlling the gill.

If, however, the stimulus is stronger—for example, a jab rather than a gentle stroking—*Aplysia* becomes sensitized to it. The motor response becomes more rapid and emphatic. At the synapse, the effect of sensitization is opposite to that of habituation; that is, there is a gradual increase in the amount of neurotransmitter released by the sensory neurons.

Numerous studies with *Aplysia*, with other mollusks, and with preparations of tissue from the hippocampus of various mammals all support the hypothesis that alterations in the strength of synaptic transmission are critical in memory and learning. These alterations are thought to depend on changes in both the presynaptic and postsynaptic cells. One important element may be the opening or blocking of ion channels that influence the release of neurotransmitter by the presynaptic cell and the degree of polarization of the postsynaptic cell in its resting state.

Various models of the possible mechanisms by which these alterations occur and are made to endure are currently being intensively investigated. Among the factors that may be involved are second messengers, including both calcium ion and cyclic AMP, changes in the pattern of protein synthesis, possibly affecting the numbers and kinds of membrane receptors, and shifts in the location of membrane receptors.

36–18   *The sea hare* Aplysia, *a shell-less mollusk that is shedding new light on the process of learning. The neurons of* Aplysia, *like those of the squid (page 629), are quite large and their axons are unmyelinated. Moreover, its nervous system has far fewer neurons than a vertebrate nervous system. Individual neurons can be identified, their pattern of organization mapped, and microelectrodes inserted into them. Thus, investigators can trace the pathways followed by nerve impulses in response to particular stimuli and monitor the modifications in transmission associated with learning.*

The Response to Information: Muscle Contraction

The activities an animal undertakes as a result of the information received and processed by its sensory receptors and brain depend upon skeletal muscles, the effectors of the somatic nervous system. Moreover, many of the adjustments in its internal environment depend upon cardiac muscle and smooth muscle, two of the effectors of the autonomic nervous system.

Muscle is the principal tissue, by weight, in the vertebrate body. However, even if there were only a single muscle cell per organism, it would be worthy of our attention and admiration. Skeletal muscle is the one tissue of the biological world whose function can actually be visualized in the structure of the very molecules of which it is composed.

The Structure of Skeletal Muscle

As we noted in Chapter 28, a skeletal muscle consists of bundles of muscle fibers—often hundreds of thousands of fibers—held together by connective tissue. Each fiber is a single, multinucleate cell 10 to 100 micrometers in diameter and, often, several centimeters long. Each muscle fiber is surrounded by a specialized plasma membrane called the **sarcolemma**. Like the membrane of the axon, the sarcolemma can propagate an action potential.

Embedded in the cytoplasm of each muscle fiber (cell) are some 1,000 to 2,000 smaller structural units, the **myofibrils** (from *myo*, the prefix for "muscle"). Tightly packed myofibrils run parallel for the length of the cell, crowding the nuclei to its periphery, where they are typically found just beneath the sarcolemma.

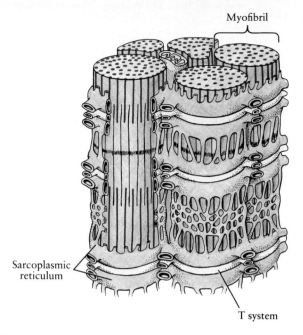

36–19   *A group of myofibrils, each of which is surrounded by the specialized endoplasmic reticulum of muscle cells, the sarcoplasmic reticulum. The sacs of the sarcoplasmic reticulum contain calcium ions, which, when released, trigger muscle contraction. Traversing the sarcoplasmic reticulum, perpendicular to the myofibrils, is a system of transverse tubules known as the T system. The membrane forming the T system is continuous with the plasma membrane (the sarcolemma). Like the sarcolemma, it can propagate an action potential.*

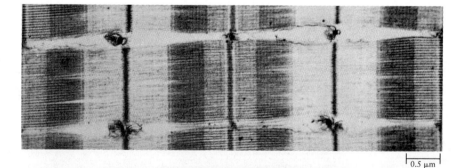

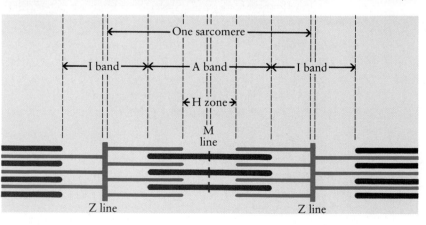

36–20   *Electron micrograph and diagram of a sarcomere, the contractile unit of muscle. Each sarcomere is composed of an array of thick and thin protein filaments arranged longitudinally along the myofibril. The Z line is a protein sheet in which thin filaments from adjoining sarcomeres are anchored. The I band is a region that contains only thin filaments. The A band is the region occupied by the thick filaments. The part of the A band where there are no thin filaments is called the H zone. The thick filaments are interconnected and held in place at the M line. Muscle contraction involves the sliding of the thin filaments between the thick ones.*

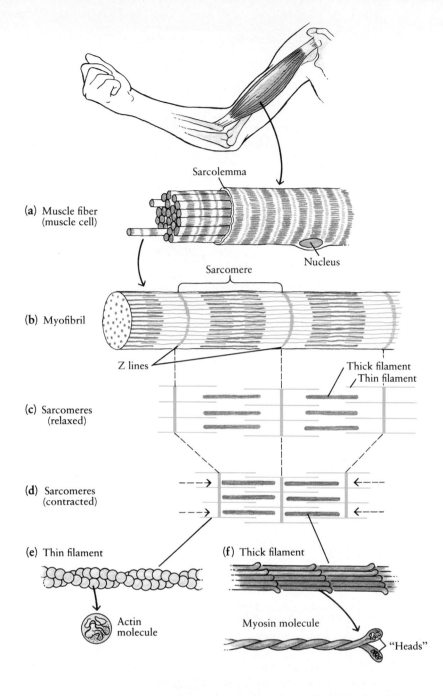

(a) Muscle fiber (muscle cell)

Sarcolemma

Nucleus

(b) Myofibril

Sarcomere

Z lines

Thick filament
Thin filament

(c) Sarcomeres (relaxed)

(d) Sarcomeres (contracted)

(e) Thin filament

Actin molecule

(f) Thick filament

Myosin molecule

"Heads"

36–21 *Skeletal muscle is composed of individual muscle cells, the muscle fibers (a). These are cylindrical cells, often many centimeters long, with numerous nuclei. Each muscle fiber is made up of many cylindrical subunits, the myofibrils (b). These fibrils, which contain contractile proteins, run from one end of the cell to the other. The myofibril is divided into segments, sarcomeres, by thin, dark partitions, the Z lines. The Z lines appear to run from myofibril to myofibril across the fiber. The sarcomeres of adjacent myofibrils are in line with each other, giving the muscle cell its striated appearance. Each sarcomere is made up of thick and thin filaments (c). When stimulated, the thick and thin filaments slide past each other, shortening the sarcomere (d).*

(e) Each thin filament consists primarily of two actin strands coiled about one another in a helical chain. Each strand is composed of globular actin molecules. (f) The thick filaments, by contrast, consist of bundles of a protein called myosin. Each individual myosin molecule is composed of two protein chains wound in a helix. The end of each chain is folded into a globular "head" structure.

Each myofibril is encased in a sleevelike membrane structure, the **sarcoplasmic reticulum,** which is a specialized endoplasmic reticulum (Figure 36–19). The sacs of the sarcoplasmic reticulum contain calcium ions (Ca^{2+}), which, as we shall see, play an essential role in muscle contraction.

Myofibrils are composed of units called **sarcomeres,** each of which is about 2 or 3 micrometers in length. The repetition of these units gives the muscle its characteristic striated pattern. Figure 36–20 shows a sarcomere as seen in a longitudinal section of muscle. As the diagram shows, each sarcomere is composed of two types of filaments running parallel to one another. The thicker filaments in the central portion of the sarcomere are composed of the protein known as **myosin.** The

thinner filaments are primarily **actin,** also a protein. When viewed in cross section, each thick filament is seen to be surrounded by six thin filaments.

The Contractile Machinery

The sarcomere is the functional unit of skeletal muscle, the mechanism by which contraction occurs. When the muscle is stimulated, the thick (myosin) filaments attach to and push the thin (actin) filaments toward the center of the sarcomere. Since the thin filaments are anchored in a protein sheet (the Z line), this causes each sarcomere to shorten, and thus the myofibril as a whole to contract (Figure 36–21).

The actin strands of the thin filaments are composed of many globular actin molecules assembled in a long chain. As shown in Figure 36–21e, each thin filament consists primarily of two such actin chains wound around one another.

The thick filaments are composed of bundles of myosin molecules. A myosin molecule consists of two long protein chains, each containing some 1,800 amino acids and each with a globular "head" at one end (Figure 36–21f). In the molecule these two chains are wound around each other, with the globular "heads" free. These heads have two crucial functions: they are the binding sites at which force is exerted on the thin filaments during contraction, and they also act as enzymes that split ATP to ADP, providing the energy for muscle contraction.

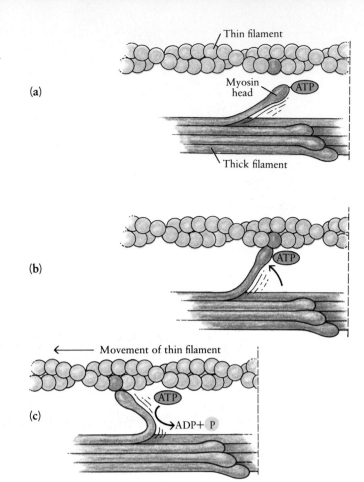

When a muscle fiber is stimulated, the heads of the myosin molecules move toward the thin (actin) filaments, to which they attach themselves, forming temporary cross bridges. The heads move with a swiveling, oarlike motion, pulling the thick filament, pushing the thin one (Figure 36–22). Rapid repeated cycles of attachment, breaking away, and reattachment move the two filaments, ratchetlike, past one another. The thin filaments on opposite sides of the sarcomere move toward one another, so the Z lines bordering the sarcomere are pulled together.

The contraction of the sarcomeres is dependent on ATP in two ways: hydrolysis of ATP by the myosin molecule provides the energy for the cycle, and combination of a new ATP molecule with the myosin molecule releases the myosin head from the binding site on the actin molecule. The stiffened muscles of a corpse—rigor mortis—are due to the absence of ATP and the subsequent locking of all the actin-myosin cross bridges.

The Regulation of Contraction

The regulation of contraction in skeletal muscle depends upon two other groups of protein molecules, **troponin** and **tropomyosin**, plus calcium ions (Ca^{2+}). As shown in Figure 36–23, tropomyosin molecules are long, thin double cables that lie along the actin molecules of the thin filament, blocking the cross-bridge binding sites on those molecules.

The troponin molecules are complexes of globular proteins that are located at regular intervals on the tropomyosin chains. When Ca^{2+} combines with the troponin molecules, they undergo changes in shape that result in shifting of the tropomyosin chains and exposure of the cross-bridge binding sites. The availability of Ca^{2+}, and thus the initiation of contraction, depend on stimulation of the muscle by a signal received from a motor neuron.

The Neuromuscular Junction

A motor neuron typically has a single long axon that branches as it reaches the muscle. At the end of each branch, the axon emerges from the myelin sheath and becomes embedded in a groove on the surface of a muscle fiber, forming the **neuromuscular junction** (Figure 36–24).

As is the case with most synapses between neurons, the signal travels across the neuromuscular junction by means of a neurotransmitter—in this case, acetylcholine. However, unlike synaptic transmission between neurons, this is a direct, one-to-one relationship involving only excitation. The acetylcholine combines with

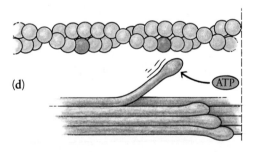

36–22   *The molecular mechanism of muscle contraction.* (a) *A globular myosin head, "charged" with ATP, protrudes from a thick filament. It serves as a hook or lever, attaching to an actin molecule of the adjacent thin filament* (b) *and pushing it toward the center of the sarcomere* (c), *shortening the sarcomere and contracting the myofibril. The energy for this process is supplied by the ATP molecule with which the myosin head was originally associated. When that ATP molecule has been "spent" and hydrolyzed to ADP and phosphate, the addition of another ATP molecule allows the myosin head to break its bond with the actin molecule* (d). *When the myosin head is "recharged" by this ATP molecule, it is once again ready to swing into position and attach to another actin molecule.*

As this process is rapidly repeated over and over along the length of the myofibrils by the heads of a multitude of myosin molecules, the myofibrils contract and the fiber as a whole shortens. When enough fibers shorten, the entire muscle shortens, producing skeletal movement.

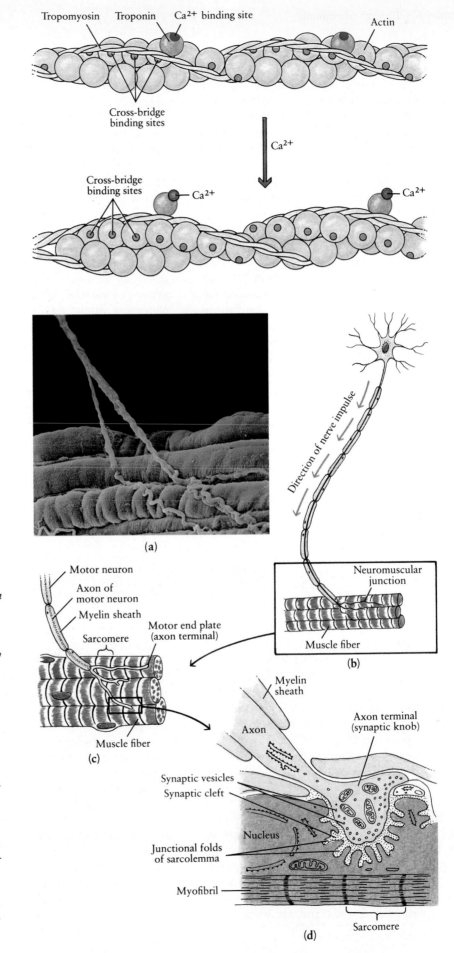

36–23 *Tropomyosin and troponin molecules, both proteins, have a regulatory role in muscle contraction. When calcium ions are not present, tropomyosin molecules, which are long, thin double cables, block the cross-bridge binding sites on the actin molecules of the thin filaments. Troponin molecules, which are globular proteins, are situated at regular intervals along the tropomyosin chain. When calcium ion, released from the sarcoplasmic reticulum, binds to troponin, the tropomyosin molecule shifts position, exposing the binding sites and permitting the cross bridges to form.*

36–24 *The neuromuscular junction. (a) A scanning electron micrograph showing the axon of a motor neuron branching to form neuromuscular junctions with three skeletal muscle fibers. (b) An action potential initiated in a motor neuron travels along its axon to the neuromuscular junctions. The axon is insulated by a myelin sheath composed of Schwann cell membranes. (c) The axon of the motor neuron divides into branches, each of which extends along the surface of a different muscle fiber (cell). A structure known as a motor end plate is found at the end of each axon branch. Extending below the motor end plates are the synaptic knobs of the axon terminals. (d) An axon terminal. The arrival of an action potential that has been conducted along the axon of the motor neuron releases acetylcholine from synaptic vesicles into the synaptic cleft. Acetylcholine combines with receptor sites on the sarcolemma, altering its membrane permeability. This initiates an action potential that sweeps along the sarcolemma of the muscle fiber, including the tubules that form the T system (Figure 36–19). This, in turn, triggers the release of Ca^{2+} ions from the sarcoplasmic reticulum and contraction of the sarcomeres.*

receptors on the sarcolemma, depolarizing the muscle plasma membrane and initiating an action potential that sweeps along the sarcolemma. This triggers the release of calcium ions (Ca^{2+}) from the sarcoplasmic reticulum. These ions continue to be released only as long as the fiber is stimulated. Once stimulation stops, the ions are pumped back into the sacs of the sarcoplasmic reticulum by active transport. Thus it is the Ca^{2+} ions that turn the contractile machinery on and off.

Many drugs act specifically on the neuromuscular junction. Curare, for instance, a plant extract once used by Indians of South America to poison their arrow tips, blocks excitation of muscle by binding to acetylcholine receptors on the sarcolemma and thus producing paralysis. Today, curare is used in very low concentrations as a muscle relaxant during surgery. The bacterial toxin of botulism—one of the most poisonous substances known—prevents nerve endings from liberating acetylcholine and so kills by paralysis of the muscles controlling breathing.

The Motor Unit

The axon of a single motor neuron and all the muscle fibers it innervates are known as a **motor unit**. Stimulation of a motor axon stimulates all of the fibers in that motor unit. The number of muscle fibers in a motor unit determines the fineness of control. In a muscle that moves the eyeball, for instance, a motor unit may contain as few as three muscle fibers, whereas in the biceps, each motor unit contains more than a thousand.

Within a given muscle, fibers of different motor units are intermingled. A slight movement may involve the contraction of only a few motor units. The strength of contraction of a muscle as a whole depends on the number of motor units that have been activated and the frequency with which they are stimulated.

Complex activities require the precisely timed contractions of different groups of fibers in antagonistic groups of muscles, often in different parts of the body. Moreover, many of an animal's most complex—and often most critical—activities are performed in relation to rapidly moving objects in its environment. Consider, for example, the swoop of an owl in pursuit of a mouse scampering for cover. Less essential from a biological point of view, but equally marvelous as an example of the exquisite integration of sensory perception, information processing, and motor response, is an event that happens numerous times each summer: a human being swings a piece of wood so that it connects with a relatively small ball 400 milliseconds after that ball was set in motion 18.4 meters away and drives it over the right-field fence.

Summary

Sensory receptors are cells or groups of cells that convert the energy of environmental stimuli into the energy of action potentials, and thus into information that can be transmitted and processed by the nervous system. Receptors may be classified according to the type of stimuli to which they respond as mechanoreceptors, chemoreceptors, photoreceptors, and temperature receptors. For example, the taste buds of the human tongue are chemoreceptors. Among the human mechanoreceptors are the touch receptors of the skin and the vastly more complex ear, which converts sound waves into nerve impulses that are carried to the brain.

The major human sense organ is the eye, which provides much of our information about the environment. Light passes through the eyeball to the retina, which contains densely packed photoreceptor cells, the rods and the cones. Rods, which are more light-sensitive than cones, are responsible for night vision. Cones provide greater resolution than rods and are responsible for color vision. Cones are most concentrated in the center of the retina in the fovea, the area of sharpest vision. The photoreceptor cells transmit signals to the bipolar cells, from which they are relayed to a network of ganglion cells, whose axons form the optic nerve. Considerable processing occurs in the retina before action potentials are triggered in the ganglion cells and transmitted to the brain.

In the course of animal evolution, the processing of information has become increasingly centralized in one dominant center, the brain. The principal structures of the human brain are the brainstem, which controls vital functions such as heartbeat and respiration and serves as the relay between the spinal cord and the rest of the brain; the cerebellum, which is associated with the coordination of fine-tuned movements; and the two hemispheres of the cerebrum, which are covered by the convoluted cerebral cortex. Two smaller structures are the thalamus, the main relay center between the brainstem and higher brain centers, and the hypothalamus, which contains clusters of neurons associated with basic drives and emotions and is the center for the integration of the nervous and endocrine systems.

The cerebral cortex is the area of the brain to which sensory information is ultimately transmitted and in which complex, conscious motor activities originate. Areas of the cortex that have been mapped include the motor cortex, sensory cortex, and parts of the cortex concerned with vision, hearing, and speech. In the motor and sensory cortices, the two cerebral hemispheres are mirror images of one another, with the right hemisphere receiving information from and controlling the left side of the body, and vice versa. However, the speech centers are found only in one hemisphere, nearly always the left, and other faculties, such as spatial orientation and musical ability, appear to be associated with the right hemisphere.

Most of the human cerebral cortex has no direct sensory or motor function. Such areas receive signals from and transmit signals to other areas of the brain. Some of them participate in the further processing of sensory information. Others function in the integration of sensory information with memory and emotion, in the organization of ideas, and in long-term planning. About half of the total area devoted to such intrinsic processing is located in the frontal lobes, the part of the brain that developed most rapidly in human evolution.

Coordination of the activities occurring in different parts of the brain depends on the exchange of information between local networks of neurons. Two major brain circuits involved in such exchanges are the reticular formation, concerned with arousal and attention, and the limbic system, concerned with translating drives and emotions into actions.

Memory and learning are thought to involve the processing of information through specific anatomical circuits. The consolidation of long-term memory appears to depend on a circuit that includes the hippocampus, amygdala, thalamus, basal forebrain, and prefrontal cortex. Studies with invertebrates indicate that, at the cellular level, memory and learning involve changes at synapses.

Skeletal muscles are the effectors of the somatic nervous system. Each muscle is made up of muscle fibers (cells), each of which is surrounded by a specialized plasma membrane, the sarcolemma. Each muscle fiber contains 1,000 to 2,000 small strands, the myofibrils, which run parallel the length of the cell. Each myofibril is surrounded by a specialized endoplasmic reticulum, the sarcoplasmic reticulum. Myofibrils are made up of units called sarcomeres, which consist of alternating thin (actin) and thick (myosin) filaments. Contraction occurs when the filaments slide past one another. The globular heads of the myosin molecules serve as binding sites linking the thick and thin filaments during contraction and as enzymes for the hydrolysis of ATP, which provides energy for muscle contraction.

Skeletal muscle fibers contract in response to stimulation by a motor neuron, whose axon terminals release acetylcholine, causing depolarization of the sarcolemma. The resulting action potential travels along the sarcolemma, triggering the release of calcium ions (Ca^{2+}) from the sarcoplasmic reticulum. The Ca^{2+} ions, in turn, permit the interaction between actin and myosin that brings about contraction of the sarcomeres.

Questions

1. Distinguish among the following: cochlea/organ of Corti; rods/cones; bipolar cell/ganglion cell; brainstem/cerebellum/cerebrum; thalamus/hypothalamus; motor cortex/sensory cortex; Broca's area/Wernicke's area; left brain/right brain; reticular formation/limbic system; muscle fiber/muscle cell/myofibrils; thick filament/thin filament; sarcomere/sarcolemma/sarcoplasmic reticulum; synapse/neuromuscular junction.

2. Why is hearing considered a form of mechanoreception?

3. Consider the process that occurs when another person speaks a word and you then hear the word. Describe the steps involved from the time the air leaves the mouth of the speaker until you are aware of the word spoken.

4. Label the drawing below.

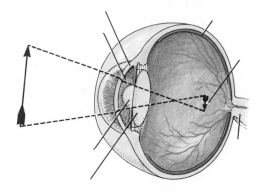

5. Why don't we see objects upside down?

6. Diagram the major structures of the human brain.

7. Sketch the human cerebral cortex and indicate on it the regions that have been mapped.

8. What functions might be affected by damage (from stroke, accident, or disease) to the cerebellum? To the reticular formation? To the dorsal portion of the cerebral cortex, anterior to the central sulcus?

9. Monkeys that have suffered damage to the amygdala are unable to remember whether an object, however familiar, is edible or inedible. Each time they encounter an object—for example, a banana—they not only look at it, but feel it, smell it, and taste it, before deciding whether to eat it or not. What is a probable explanation of this behavior?

10. Diagram and label a sarcomere that is in a relaxed state.

11. How does the structure you diagrammed in Question 10 enable muscle to perform its primary function?

12. Can you interpret the electron micrograph at the right? (You may be able to find the clue you need on page 656.)

13. In the 1960s, the advertisement for a lecture by a whimsical biologist began as follows: "Available *now*. **Linear Motor.** Rugged and dependable: design optimized by worldwide field testing over an extended period." The copy ended with "Good to eat." What was the subject of the lecture?

14. Explain the events whereby an action potential in a motor neuron causes a muscle to contract.

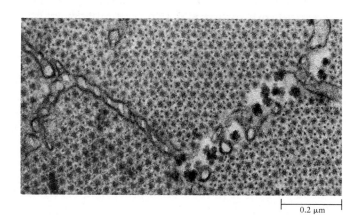

0.2 μm

Suggestions for Further Reading

Barlow, Robert B., Jr.: "What the Brain Tells the Eye," *Scientific American*, April 1990, pages 90–95.

Bloom, Floyd E., Laura Hofstadter, and Arlyne Lazerson: *Brain, Mind, and Behavior*, W. H. Freeman and Company, New York, 1988.*

   A readable companion to the television series, beautifully illustrated.

Freeman, Walter J.: "The Physiology of Perception," *Scientific American*, February 1991, pages 78–85.

Frith, Uta: "Autism," *Scientific American*, June 1993, pages 108–114.

Gazzaniga, Michael S.: "Organization of the Human Brain," *Science*, vol. 245, pages 947–952, 1989.

Glickstein, Mitchell: "The Discovery of the Visual Cortex," *Scientific American*, September 1988, pages 118–127.

Goldstein, Gary W., and A. Lorris Betz: "The Blood-Brain Barrier," *Scientific American*, September 1986, pages 74–83.

Gregory, Richard L.: *Eye and Brain: The Psychology of Seeing*, 4th ed., Princeton University Press, Princeton, N.J., 1990.*

   A vivid introduction to the science of vision.

Hinton, Geoffrey E., David C. Plaut, and Tim Shallice: "Simulating Brain Damage," *Scientific American*, October 1993, pages 76–82.

Hudspeth, A. J.: "The Hair Cells of the Inner Ear," *Scientific American*, January 1983, pages 54–64.

Jacobs, Barry L.: "How Hallucinogenic Drugs Work," *American Scientist*, vol. 75, pages 386–392, 1987.

Konishi, Masakazu: "Listening with Two Ears," *Scientific American*, April 1993, pages 66–73.

Koretz, Jane F., and George H. Handelman: "How the Human Eye Focuses," *Scientific American*, July 1988, pages 92–99.

Llinás, Rodolfo R. (ed.): *The Biology of the Brain: From Neurons to Networks*, W. H. Freeman and Company, New York, 1988.

   A collection of 10 articles from Scientific American, originally published between 1977 and 1988, that provide a state-of-the-art view of brain physiology and function.

Mishkin, Mortimer, and Tim Appenzeller: "The Anatomy of Memory," *Scientific American*, June 1987, pages 80–89.

*Available in paperback.

Poggio, Tomaso, and Christof Koch: "Synapses That Compute Motion," *Scientific American,* May 1987, pages 46–52.

Ramachandran, Vilayanur S.: "Blind Spots," *Scientific American,* May 1992, pages 86–91.

Schnapf, Julie L., and Denis A. Baylor: "How Photoreceptor Cells Respond to Light," *Scientific American,* April 1987, pages 40–47.

Scientific American: *Mind and Brain,* W. H. Freeman and Company, New York, 1993.

> *A reprint of the September 1992 issue of* Scientific American. *Its 11 articles by different authors cover such topics as the development of the brain, visual images in mind and brain, the biological basis of learning and individuality, the brain and language, working memory, sex difference in the brain, major disorders of the brain, and the aging of the brain.* *

Selkoe, Dennis J.: "Amyloid Protein and Alzheimer's Disease," *Scientific American,* November 1991, pages 68–77.

Spector, Reynold, and Conrad E. Johanson: "The Mammalian Choroid Plexus," *Scientific American,* November 1989, pages 68–74.

Stryer, Lubert: "The Molecules of Visual Excitation," *Scientific American,* July 1987, pages 42–50.

Thompson, R. F.: *Introduction to Physiological Psychology,* 2d ed., Harper & Row, Publishers, Inc., New York, 1988.

> *Intended for the undergraduate student, this text presents an up-to-date survey of the biological foundations of psychology.*

Wurtman, Richard J.: "Alzheimer's Disease," *Scientific American,* January 1985, pages 62–74.

*Available in paperback.

Reproduction

Those of us who have experienced puberty once must extend our admiration and sympathy to animals subjected to a similar torrent of hormones over and over again. Deer, such as this male reindeer, are a conspicuous example. Adults spend every summer preparing for the fall rut, the mating season. Most notable is the yearly growth of antlers. When antlers, which are made of bone, first appear, they are surrounded by a soft, nourishing tissue—the velvet. This is gradually shed and scraped away, as shown here. Despite the metabolic expense, antlers are discarded after the fall rut and regrown the following season.

Antlers grow as the result of a seasonal surge in the male hormone testosterone. If male fawns are castrated, their antlers never appear. In male deer, as in other male animals, testosterone also stimulates the sperm count, increases the sex drive, promotes aggression, and leads to prolonged episodes of deep, resonant vocalizing.

Every year the new antlers grow larger. The males with larger antlers are almost always the larger, stronger males, but it appears to be the antlers that determine mating success. If a big male loses his antlers, the females will prefer a smaller but antlered male. Similarly, a male whose antlers have been damaged in combat is at a sexual disadvantage, which may be why it is worth it, in terms of beating the competition, to regrow antlers every year. Or, it may be simply that antlers are too big and awkward to lug around all winter.

A highly successful male may corral a large harem of females and mate with all of them. Then, once the mating season is over, he retires, exhausted, for the winter. For a female deer, the situation is quite different. Her lifetime reproductive success depends almost entirely on how many of her fawns survive. Once she matures, her reproductive cycle never ends, orchestrated by a changing interplay of hormones and demanding strength, competence, and experience. She mates in the fall, gives birth in the spring, nurses her fawn all summer, and then mates again, probably with a male with a great big set of antlers.

In the preceding chapters of this section, we have been primarily concerned with the ways in which the individual vertebrate body—and particularly the individual human body—maintains itself. In this chapter and the next, we shall consider the process by which new individuals are produced, maintaining the continuity of the species, and how these individuals develop into organisms similar to their parents.

Most vertebrates—and all mammals—reproduce sexually. As you will recall, sexual reproduction involves two processes: **meiosis** and **fertilization.** In vertebrates, which are almost always diploid, meiosis produces **gametes,** the only haploid forms in the life cycle. The gametes are specialized for motility (sperm) or for the production and storage of nutrients (eggs). They are produced in the gonads of individuals of the two separate sexes, male and female.

In the following pages, we shall describe sexual reproduction in a representative mammal, *Homo sapiens.* We shall first trace the development of the male gametes, and then the development of the female gametes. Finally, we shall describe the special structures and activities that provide for fertilization and the subsequent implantation of the developing embryo.

The Male Reproductive System

The Formation of Sperm

From adolescence until old age, the human male produces an average of several hundred million sperm a day. These cells—the male gametes—are produced in the **testes** (singular, testis) of the male reproductive tract (Figure 37–1, on the next page). The testes develop in the abdominal cavity of the male embryo and, as in most male mammals, descend into an external sac, the **scrotum.**

665

37–1 *The reproductive tract of the human male, showing the penis and scrotum before (dashed lines) and during erection. Sperm cells travel from the testis, where they are formed, to the epididymis, a coiled tube overlying the testis. From there they move to the vas deferens, where most of them are stored. The vas deferens merges with a duct from the seminal vesicle and then, within the prostate gland, joins the urethra. The sperm cells are mixed with fluids, mostly from the seminal vesicles and prostate gland. The resulting mixture, the semen, is released through the urethra of the penis. The urethra is also the passageway for urine, which is stored in the bladder.*

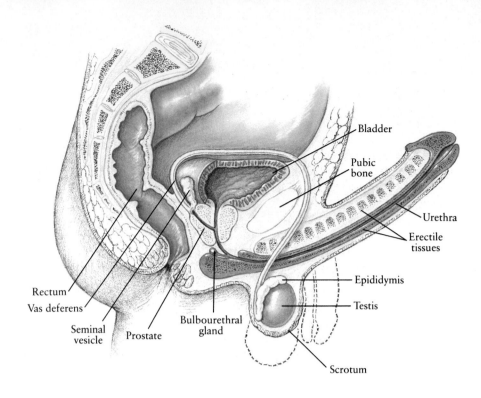

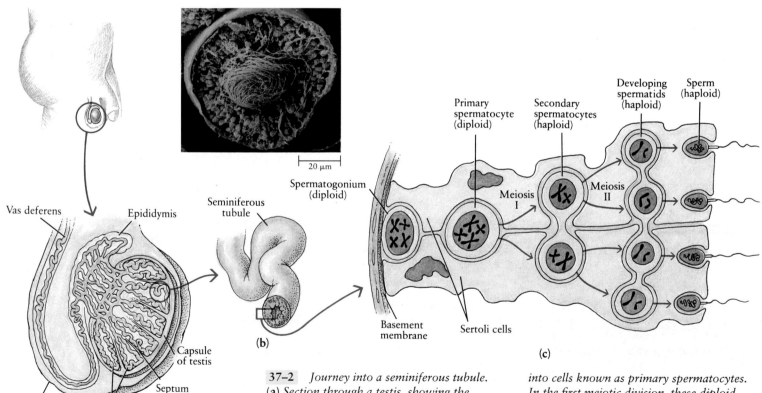

37–2 *Journey into a seminiferous tubule.* (a) *Section through a testis, showing the tightly packed coils of seminiferous tubules, where sperm cells are produced, and the epididymis, where the sperm mature and gain motility.* (b) *Scanning electron micrograph and diagram of a cross section of a seminiferous tubule.* (c) *Sperm formation, as shown in an idealized cross section of a portion of a seminiferous tubule. Spermatogonia develop into cells known as primary spermatocytes. In the first meiotic division, these diploid cells divide into two equal-sized haploid cells, the secondary spermatocytes. In the second meiotic division, four equal-sized haploid spermatids are formed. These differentiate into functional sperm cells. The Sertoli cells support and nourish the developing sperm during this process.*

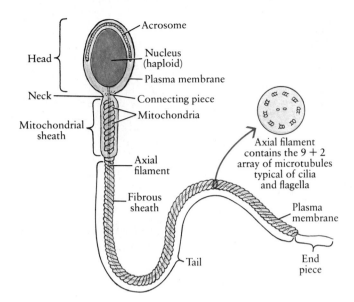

37–3   *Diagram of a human sperm cell. The mature cell consists primarily of the nucleus, carrying the "payload" of tightly condensed DNA and associated protein, the very powerful tail (flagellum), and a region between the head and the tail in which many mitochondria are concentrated. These mitochondria provide the power for sperm motility. The acrosome is a specialized lysosome (see page 98) containing enzymes that help the sperm penetrate the protective layers of an unfertilized egg cell.*

Each testis is subdivided into about 250 compartments, and each of these is packed with tightly coiled **seminiferous** ("seed-bearing") **tubules** (Figure 37–2). These are the sperm-producing regions of the testes. Each seminiferous tubule is about 80 centimeters long, and the two testes together contain a total of about 500 meters of tubules.

The sperm-producing cells of the seminiferous tubules pass through several stages of differentiation. The process begins with diploid cells known as **spermatogonia,** which line the basement membrane of each tubule and undergo repeated mitotic divisions. Some of the cells produced by these divisions remain undifferentiated and continue to divide. Others move away from the basement membrane and begin to differentiate, giving rise to cells known as primary **spermatocytes.** These diploid cells then undergo meiosis, each producing four haploid cells, the **spermatids.** (For a review of meiosis, see pages 194 to 197.) Following meiosis, the spermatids differentiate, without further division, into the highly specialized **sperm cells** (Figure 37–3).

The entire process, from one spermatogonium to four sperm cells, takes eight to nine weeks. During this time the developing cells receive nutrients from other cells of the seminiferous tubules, known as **Sertoli cells.**

The Pathway of the Sperm

The pathway of the sperm in their journey from the body can be traced in Figure 37–4. From the testis, the sperm are carried to the **epididymis,** which consists of a long, coiled tube, overlying the testis. It is surrounded by a thin, circular layer of smooth muscle fibers. The sperm are nonmotile when they enter the epididymis and gain motility only after some 12 days, on average, there.

37–4   (a) *A frontal view of the male reproductive tract, showing the pathway followed by the sperm in their journey from the body. (b) A cross section of a human penis. The penis is formed of three cylindrical masses of spongy erectile tissue that contain a large number of small spaces, each about the size of a pinhead. Erection of the penis is caused by dilation of the arteries carrying blood to the spongy tissues and compression of the veins carrying blood from these tissues.*

37–5   *In a vasectomy, the vas deferens leading from each testis is severed, and the cut ends are folded back and tied off, preventing the release of sperm from the testis. The sperm cells are reabsorbed by the body, and the ejaculate is normal except for the absence of sperm.*

This is a relatively safe procedure that does not require a general anesthetic or hospitalization. Because the nerves and blood vessels between the testes and the rest of the body are left intact, the procedure does not affect hormone levels, sexual potency, or performance. Its chief drawback is that it is generally not reversible.

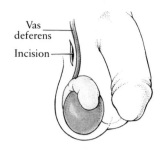
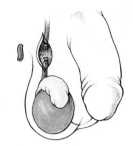
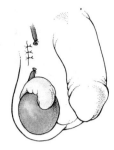

From the epididymis, the sperm pass to the **vas deferens** (plural, vasa deferentia), where most of them are stored. A vas deferens, an extension of the tightly coiled tubules of the epididymis, leads from each testis into the abdominal cavity. The vasa deferentia are covered with a heavy, three-layered coat of smooth muscle whose contractions propel the sperm along. Vasectomy, a relatively safe and almost painless means of sterilization, involves cutting and tying off the vasa deferentia, as shown in Figure 37–5.

Within the posterior wall of the abdominal cavity, the vasa deferentia loop around the bladder, where they merge with ducts of the **seminal vesicles.** The vas deferens from each testis then enters the **prostate gland** and merges with the urethra, which extends the length of the **penis.** The urethra serves both for the excretion of urine and for the rapid discharge, or **ejaculation,** of sperm, although never both at the same time.

As the sperm move along this pathway, fluids are added from the seminal vesicles and prostate gland. The seminal vesicles secrete a fructose-rich fluid that nourishes the sperm cells. This fluid also contains a high concentration of prostaglandins (page 615), which stimulate contractions in the female reproductive tract. These contractions assist the sperm in their journey to the egg. The prostate gland adds a thin, milky, alkaline fluid that helps neutralize the normally acidic fluids of the female reproductive tract. The sperm cells and the fluids in which they are suspended constitute the **semen.** About 3 to 4 milliliters of semen are released in each ejaculation.

Erection of the Penis and Orgasm in the Male

The function of the penis is to deposit sperm cells within the female reproductive tract. Erection of the penis, which can be elicited by a variety of stimuli, occurs as a consequence of an increased flow of blood that fills its spongy, erectile tissues (see Figure 37–4b). The blood flow is controlled by nerve fibers to the arteries leading into the erectile tissues. As the tissues become distended,

they compress the veins and so inhibit the flow of blood out of the tissues. With continued stimulation, the penis becomes hard and enlarged.

Erection is accompanied by the discharge into the urethra of a small amount of fluid from the **bulbourethral glands,** pea-shaped organs at the base of the penis. This fluid serves as a lubricant, aiding the movement of sperm along the urethra and the penetration of the penis into the female.

Continued stimulation of the sensory receptors of the penis sends an escalating series of nerve impulses through reflex arcs in the lower spinal cord to motor neurons innervating different muscles of the reproductive system. Contractions of smooth muscle in the walls of the vasa deferentia cause the sperm to empty into the urethra. Other muscles forcibly propel the semen through the urethra and expel it from the penis. These muscular spasms, which result in ejaculation, account for many of the sensations associated with orgasm.

The Role of Hormones

In addition to producing the sperm cells, the testes are also the major source of male hormones, known collectively as **androgens.** The principal androgen, **testosterone,** is a steroid hormone necessary for the formation of sperm cells. It is produced primarily by cells, known as **interstitial cells,** found in the tissue surrounding the seminiferous tubules of the testes. Other androgens are produced in the adrenal cortex (see page 612).

Androgens are first produced in early embryonic development, causing the male fetus to develop as a male rather than a female. After birth, androgen production continues at a very low level until the boy is about 10 years old. Then there is a surge in testosterone, resulting in the onset of sperm production (which marks the beginning of puberty) accompanied by enlargement of the penis and testes and also of the prostate and other accessory organs.

Testosterone also has effects on parts of the body not directly involved in the production and deposition of sperm. In the human male, these effects include growth of the larynx and an accompanying deepening of the voice, an increase in skeletal size, and the development of a characteristic distribution of body hair. Androgens stimulate the biosynthesis of proteins and so of muscle tissue. They also stimulate the apocrine sweat glands, whose secretions attract bacteria and so produce the body odors associated with sweat after puberty. And they may cause the oil-secreting glands of the skin to become overactive, resulting in acne. Such characteris-tics, associated with sex hormones but not directly involved in reproduction, are known as **secondary sex characteristics.**

In other animals (Figure 37–6), testosterone is responsible for such male secondary sex characteristics as the reindeer's antlers, the lion's mane, the powerful musculature and fiery disposition of the stallion, the cock's comb and spurs, and the bright plumage of many adult male birds. It is also responsible for behavior patterns such as the scent marking of dogs and for the various forms of aggression toward other males found in a great many vertebrate species.

(a)

(b)

(c)

(d)

37–6 *Secondary sex characteristics among vertebrates. The male and female animals in each of these photos are actively courting:* (a) *southern elephant seals,* (b) *sable antelope,* (c) *mandarin ducks, and* (d) *sticklebacks.*

Sexually Transmitted Diseases

From the point of view of a pathogenic microbe, always on the lookout for new hosts, it is hard to imagine a better way of life than to be the causative agent of a sexually transmitted disease. Sex is a fact of life among vertebrates, providing infectious microorganisms with a reliable route of transmission from one individual to the next. Among humans, the intimacy and privacy that surround sex often inhibit those who are infected from seeking early treatment. So it should come as no surprise that the parasitic organisms responsible for sexually transmitted diseases have enjoyed enormous success throughout much of human history.

How long have we been haunted by sexually transmitted diseases? Some authorities interpret certain Biblical passages, such as the plague of Baal-peor (*Numbers* 25:8) or the disease of the Philistines (*1 Samuel* 5 and 6), as references to syphilis. Others attribute the first clear descriptions of gonorrhea to the Greeks and Romans, whose promiscuous habits apparently left them afflicted with a frightening spectrum of diseases thought to be "venereal" (from Venus, the Roman goddess of love). Although its mode of transmission may have been poorly understood, gonorrhea was certainly commonplace by 1378, when the British surgeon John of Arderne described a "certain inward heat and excoriation of the urethra" and called it "clap"—the first known use of this nickname. An understanding of the sexual transmission of syphilis seems apparent by 1547, when the British physician Andrew

Boorde wrote that syphilis ". . . is taken when one pocky person doth synne in lechery the one with another."

Now, on the verge of the twenty-first century, the spread of sexually transmitted diseases continues unabated. In the United States, such diseases have reached epidemic proportions, especially among teenagers and young adults. While modern antibiotics can cure syphilis, gonorrhea, and other bacterial diseases, no cure exists for the 500,000 Americans who will acquire genital herpes virus infections this year. And, as we saw in Chapter 33 (pages 598–601), AIDS, with its devastating effects on the immune system, is an ultimately fatal disease. Thus, despite remarkable advances in medical science, sexually active individuals have more reason than ever to heed the advice proffered in 1564 by the Italian anatomist Gabriel Fallopius (the discoverer of the oviducts, or Fallopian tubes), who wisely urged the use of condoms to prevent the spread of venereal disease.

In the United States today, a major toll is being taken by six sexually transmitted diseases.

Chlamydia

More than 4 million Americans—most of them women—become infected with the bacterium *Chlamydia trachomatis* each year, making it the most common sexually transmitted disease. Teenage women are at the greatest risk, in part because young cervical tissues are highly susceptible to infection. The infection is

frequently without symptoms. However, if left untreated, it often spreads through the uterus to the oviducts, leading to pelvic inflammatory disease. At least 1 in 4 women with this condition subsequently experience more serious problems, such as infertility or ectopic (tubal) pregnancy.

Gonorrhea

Overall, the incidence of gonorrhea in the United States has leveled since 1976 at about 1.8 million new cases per year. But that "good news" covers a disturbing trend—rates among young people between the ages of 15 and 19 increased significantly during the 1980s. Also worrisome, the bacterium responsible, *Neisseria gonorrhoeae,* is proving increasingly resistant to antibiotics.

Human Papilloma Virus (HPV)

About 750,000 new cases of genital warts caused by human papilloma virus are reported each year—mostly in 20- to 24-year-olds. In certain populations, such as the students on some college campuses, HPV-related genital warts constitute the most prevalent sexually transmitted disease. Genital warts, however, appear in only a small fraction of the 12 million individuals estimated to be carrying and inadvertently spreading the virus—and these warts represent one of the more benign outcomes of the infection. In combination with other factors, some strains of the virus apparently play a role in the development of cancers of the cervix, vagina, and penis. For this rea-

Causative agents of the major sexually transmitted diseases in humans. **(a)** Chlamydia trachomatis, *the bacterium responsible for the most common sexually transmitted disease in women.* **(b)** Neisseria gonorrhoeae, *the bacterium responsible for gonorrhea.* **(c)** *Human papilloma virus, the cause of genital warts.* **(d)** *Herpes simplex virus 2, the cause of genital herpes infections.* **(e)** Treponema pallidum, *the bacterium responsible for syphilis. Human immunodeficiency virus (HIV), the causative agent of AIDS, is shown in Figures 33–19 and 33–20 (page 599).*

(a)

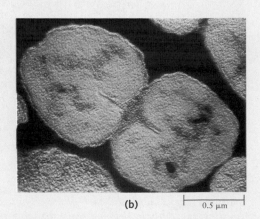

(b)　　0.5 μm

son, it is vital that HPV-infected women obtain regular Pap smears for the early detection of pre-cancerous changes in cervical cells.

Herpes Simplex Virus 2 (HSV-2)

Until the AIDS epidemic, herpes was the fastest growing sexually transmitted viral disease in the United States, with about 500,000 new cases each year—most of them in young adults aged 20 to 29. Between 15 and 20 percent of adult Americans have antibodies indicating the presence of HSV-2 infection. However, because most people remain asymptomatic, only a small proportion realize that they harbor the virus. Studies suggest that two-thirds of infected women do not know that they have the virus, and even those who do know that they are infected frequently cannot tell when they are actively shedding infectious viral particles. Besides causing periodic, painful genital sores in both men and women, the virus can be passed at birth from mother to infant. In newborns, it can cause blindness, hearing deficits, and death.

Syphilis

Credible evidence suggests that individuals as diverse as Al Capone, Florence Nightingale, and Napoleon Bonaparte all suffered from syphilis, and the disease may well have been the cause of Beethoven's deafness. But this sexually transmitted disease, with its potential to cause severe neurological damage and death, is far from glamorous. The introduction of antibiotics in the 1940s brought the number of new cases in the United States to an all-time low in 1956. However, alarming increases have occurred since 1986. With 100,000 new cases in 1989, syphilis rates have now reached their highest levels since World War II.

AIDS

This most recent and most serious sexually transmitted disease, discussed in Chapter 33, has already claimed more than 190,000 lives in the United States alone. About 2 million Americans are thought to be currently infected with human immunodeficiency virus (HIV), the retrovirus that causes AIDS. Although homosexual and bisexual men still constitute the majority of newly diagnosed cases, as infections acquired years ago—before the initiation of safer sex practices—come to fruition, the rate of transmission of the virus through heterosexual intercourse is steadily increasing. Heterosexual transmission occurring silently today will blossom into the AIDS cases of the next decade and the next century. Globally, the epidemic is increasingly one of women, with the number of HIV-infected women expected to exceed that of men by the mid-1990s. Moreover, at least half of those today infected with HIV are under the age of 25.

HIV transmission is enhanced by preexisting venereal diseases, particularly those characterized by open sores—however small—on the genitals. The risk of AIDS infection is greatly increased by the presence of genital herpes, syphilis, and chancroid, another sexually transmitted bacterial disease. Although chancroid has long been considered only a minor problem in this country, several outbreaks of this disease have occurred in U.S. cities since 1984.

With no vaccine or cure available, the only way to protect against AIDS—or to prevent any other sexually transmitted disease—remains abstinence, monogamous sex with an uninfected partner, or the use of condoms. Condom use began to increase markedly in the United States in the 1980s. Still, surveys show that individuals use condoms less frequently than might be expected, given their apparent understanding of the potentially dire consequences of unprotected sex.

Experts note that for individuals to make safety-related changes in their sexual behavior they must first feel vulnerable or susceptible to sexually transmitted diseases. This requires breaking the stereotype that such diseases affect only people with low moral character, as well as discarding the classic definition of promiscuous individuals as "people who have more sexual partners than I do." Second, individuals must perceive sexually transmitted disease as having serious consequences. The 1970s saw a growing awareness that "herpes is forever," and in the 1980s, human papilloma virus emerged as a serious risk factor for cervical cancer. Now, in the 1990s, AIDS has upped the ante even further.

It is unlikely that modern medicine will ever provide complete protection for the evolutionary niche so cleverly appropriated by sexually transmitted organisms. But, through their own behavior, individuals can do much to block the spread of these treacherous microbes.

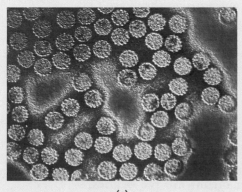

(c)

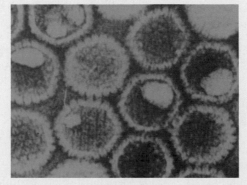

(d)

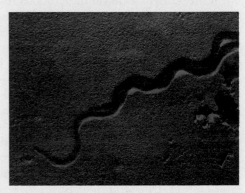

(e)

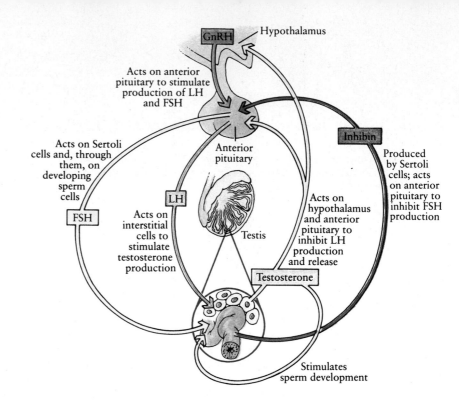

37–7 *The production of hormones affecting the testes is regulated by negative feedback. Gonadotropin-releasing hormone (GnRH), produced by the hypothalamus, stimulates the anterior pituitary to produce luteinizing hormone (LH) and follicle-stimulating hormone (FSH). LH stimulates the production and release of testosterone from the interstitial cells of the testes. Further production of GnRH by the hypothalamus is inhibited by the resulting increased concentration of testosterone. In addition, testosterone is thought to act directly on the pituitary to suppress the release of LH. As a consequence of these combined inhibitory effects, the secretion of LH by the pituitary is decreased.*

FSH acts on the Sertoli cells, which produce, in turn, the hormone inhibin. In another negative feedback loop, this hormone specifically inhibits FSH production.

The combined action of testosterone and FSH is required for the production of sperm cells.

Regulation of Hormone Production

The production of testosterone is regulated by a negative feedback system involving, among other components, a gonadotropic hormone called **luteinizing hormone (LH)**. LH is produced by the pituitary gland, under the influence of **gonadotropin-releasing hormone (GnRH)**, a hormone from the hypothalamus. LH is carried by the blood to the interstitial cells of the testes, where it stimulates the output of testosterone. As the blood level of testosterone increases, the release of LH from the pituitary is slowed (Figure 37–7).

The testes are also under the influence of a second pituitary hormone, **follicle-stimulating hormone (FSH)**. It acts on the Sertoli cells of the testes and, through them, on the developing sperm. Among the factors involved in the regulation of FSH production is a protein hormone, known as **inhibin,** that is secreted by the Sertoli cells. This hormone inhibits FSH production.

In the human male, the rate of testosterone release is fairly constant. In many animals, however, male hormone production changes seasonally, influenced by changes in temperature, daylight, or other environmental cues. Testosterone production may also be affected by social circumstances. Studies on bulls, for example, have shown that after a bull sees a cow, his blood level of LH rises as much as seventeenfold; within about half an hour, the blood level of testosterone reaches its peak. In many animal societies, including wolves and cape hunting dogs (the wild dogs of the African plains), socially inferior males may never become sexually mature, presumably because of depressed testosterone produc-

tion. Human testosterone production also may vary according to the emotional climate. In a study made among U.S. Army personnel during the Vietnam War, testosterone levels in recruits in basic training and among combat troops were markedly lower than the normal levels of testosterone found in men in behind-the-lines assignments.

Testosterone production is also dramatically affected by the synthetic compounds known as **anabolic steroids.** These drugs, which are chemical variants of testosterone, were originally developed in Germany in the 1930s in an attempt to produce the muscle-building effects of the natural hormone without its masculinizing effects. Because of their chemical similarity to testosterone, anabolic steroids function as inhibitors in the negative feedback system regulating testosterone production. In adult males, their use can reduce testosterone levels by as much as 85 percent, causing shrinkage of the testes and growth of the breasts. Long-term use of these drugs also greatly increases the risk of kidney and liver damage, liver cancer, and heart disease. In adolescents, anabolic steroids can lead to premature baldness and failure to attain full height.

The Female Reproductive System

The female reproductive system is shown in Figure 37–8. The gamete-producing organs are the **ovaries,** each a solid mass of cells about 3 centimeters long. They are suspended in the abdominal cavity by bands of

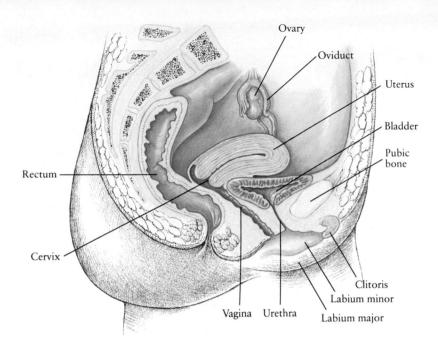

Ovary

Oviduct

Uterus

Bladder

Pubic bone

Rectum

Cervix

Clitoris

Labium minor

Vagina   Urethra   Labium major

37–8 *The female reproductive organs. Notice that the uterus lies at right angles to the vagina. This is one of the consequences of the bipedalism and upright posture of* Homo sapiens *and one of the reasons that childbirth is more difficult for the human female than for other mammals.*

connective tissue. The **oocytes,** from which the eggs develop, are in the outer layer of the ovary. Other important structures of the female reproductive system include the oviducts, the uterus, the vagina, and the vulva.

The **uterus** is a hollow, muscular, pear-shaped organ slightly smaller in size than a clenched fist (about 7.5 centimeters long and 5 centimeters wide) in the non-pregnant female. It lies almost horizontally in the abdominal cavity and is on top of the bladder. The inner lining of the uterus, the **endometrium,** has two principal layers, one of which is shed during menstruation and another, deeper layer from which the shed layer is regenerated.

The smooth muscles in the walls of the uterus move in continuous waves. This motion increases the motility both of the sperm on its journey to the oviduct and of the oocyte, from which the egg develops, as it passes from the oviduct to the uterus. These contractions increase when the endometrium is shed during a menstrual period, and they are greatest when a woman is in labor. The circular muscle at the opening of the uterus is the **cervix.** The sperm pass through this opening on their way toward the oocyte. At the time of birth, the cervix dilates to allow the fetus to emerge.

The **vagina** is a muscular tube 8 to 10 centimeters long that leads from the cervix to the outside of the body. It is the receptive organ for the penis and is also the birth canal. Its exterior opening is between the urethra (the tube leading from the bladder) and the anus.

The external genital organs of the female are collectively known as the **vulva.** The **clitoris,** most of which is embedded in the surrounding tissue, is about 2 centimeters long and is homologous with the penis of the male (in the early embryo, the structures are identical). The **labia** (singular, labium) are folds of skin. The labia majora are fleshy and, in the adult, covered with pubic hair. They enclose and protect the underlying, more delicate structures, including the thin and membranous labia minora.

The Formation of Oocytes

In human females, the primary oocytes begin to form about the third month of embryonic development. By the time an infant girl is born, her two ovaries contain some 2 million primary oocytes—all she will ever have. These diploid cells have reached prophase of the first meiotic division and will remain in prophase until the female matures sexually. Then, under the influence of hormones, the first meiotic division of a primary oocyte resumes.

Of a woman's original 2 million primary oocytes, about 300 to 400 reach maturity, usually one at a time, about every 28 days from puberty to menopause, which typically occurs at about age 50. Given this timetable, it is apparent that more than 50 years may elapse between the beginning and the end of the first meiotic division in some oocytes.

When a primary oocyte is ready to complete meiosis, its nuclear envelope fragments, and the chromosomes move to the surface of the cell. As the nucleus divides, the cytoplasm of the oocyte bulges out. One set of chromosomes moves into the bulge, which then pinches off into a small cell, the first polar body. The rest of the cellular material forms the large secondary oocyte (see Figure 11–11, page 198). The first meiotic division is completed a few hours before **ovulation** (the release of the oocyte from the ovary). The second meiotic division does not take place until after fertilization. This division produces the ovum and another small polar body. The polar bodies eventually die, their nuclei, in effect, discarded.

Maturation of the oocyte involves both the resumption of meiosis and a great increase in size. This size increase reflects the accumulation of stored food reserves and metabolic machinery, such as messenger RNA and enzymes, required for the early stages of the development of the embryo. As a consequence of the unequal cell divisions of meiosis, most of the accumulated food reserves of the oocyte are passed on to a single ovum. This cell is very large—about 100 micrometers in diameter in humans.

674

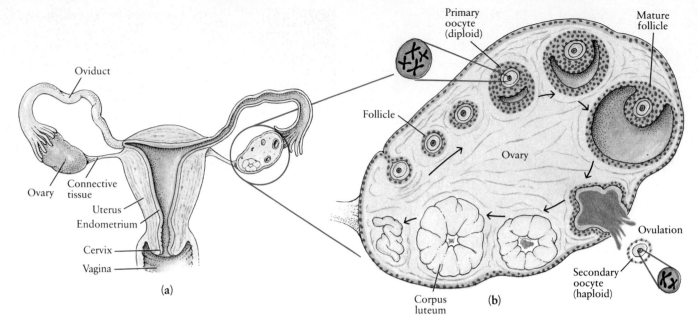

Oviduct

Ovary

Connective
tissue

Uterus
Endometrium

Cervix

Vagina

(a)

Primary
oocyte
(diploid)

Mature
follicle

Follicle

Ovary

Ovulation

Secondary
oocyte
(haploid)

Corpus
luteum

(b)

37–9   (a) *A frontal view of the internal structures of the female reproductive system.* (b) *Oocytes develop within follicles near the surface of the ovary, shown here in cross section. For clarity, the stages in the development of an oocyte and its follicle are arranged in a clockwise sequence around the periphery of the ovary. In reality, the follicle* remains in one location throughout its development.

After a secondary oocyte is discharged from a follicle (ovulation), the remaining cells of the ruptured follicle give rise to the corpus luteum, which secretes estrogens and progesterone. If the ovum is not fertilized, the corpus luteum is reabsorbed within two weeks. If the ovum is fertilized, the corpus luteum persists for about three months, continuing its production of estrogens and progesterone. These hormones, which are later produced in large quantities by the placenta, maintain the uterus during pregnancy.

Oocytes develop near the surface of the ovary. A developing oocyte and the specialized cells surrounding it are known as an **ovarian follicle** (Figure 37–9). The cells of the follicle supply nutrients to the growing oocyte and also secrete **estrogens,** the hormones that support the continued growth of the follicle and initiate the buildup of the endometrium. During the final stages of its growth, the follicle moves to the surface and produces a thin, blisterlike elevation that eventually bursts, releasing the oocyte (Figure 37–10). Usually a number of follicles begin to enlarge simultaneously, but only one becomes mature enough to release its oocyte, and the others regress.

The Pathway of the Oocyte

When the oocyte is released from the follicle at ovulation, it is swept into the adjacent **oviduct** by the movement of the funnel-shaped opening of the oviduct over the surface of the ovary and by the beating of cilia on the fingerlike projections surrounding this opening (Figure 37–11). This mechanism is so effective that women who have only one ovary and only one oviduct—and these on opposite sides of the body—have become pregnant.

The oocyte then moves slowly down the oviduct, propelled by peristaltic waves produced by the smooth muscles of the oviduct walls. The journey from the ovary to the uterus takes about three days. Although an unfertilized oocyte lives about 72 hours after it is ejected from the follicle, it is apparently capable of being fertilized for less than half of that time. Thus, fertilization, if it is to occur, must occur in an oviduct. If the egg cell is fertilized, the young embryo becomes implanted in the endometrium three to four days after it reaches the uterus, six or seven days after the egg cell was fertilized. If the oocyte is not fertilized, it dies, and the endometrial lining of the uterus is shed at menstruation.

Sterilization in women is usually carried out by severing the oviducts, thus preventing sperm from meeting the oocyte. This procedure is known as tubal ligation. Passage of the oocyte and its fertilization are also prevented when the oviducts are blocked from natural causes, such as the formation of scar tissue following infection. Until recently, it was virtually impossible for women with such blockages to bear children. In the procedure known as **in vitro fertilization,** however, the oocyte is removed surgically from the ovary just prior to ovulation, fertilized with the husband's sperm in a laboratory dish, and inserted in the uterus at the time it would have arrived had it been fertilized naturally in the oviduct. In about 15 to 20 percent of the cases in which in vitro fertilization is attempted, the fertilized egg survives the procedure and subsequent development proceeds normally.

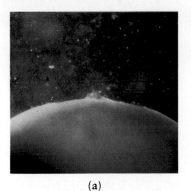 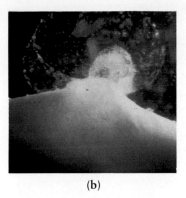 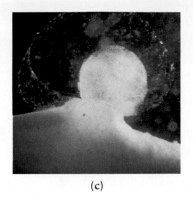 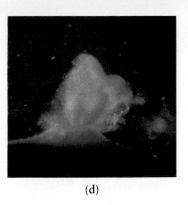

(a)   (b)   (c)   (d)

37–10 *Ovulation. (a) A secondary oocyte, visible only as a tiny speck, begins to emerge from an ovarian follicle. (b, c) With explosive force, the oocyte bursts out of the follicle, surrounded by a halo of material. (d) Completely free of the follicle and much of the material it carried out with it, the oocyte begins its journey to an oviduct.*

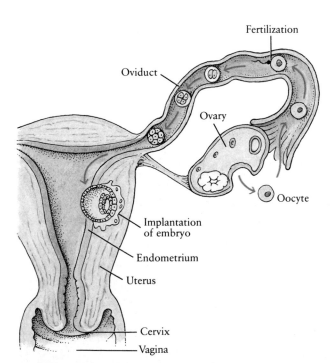

37–11 *Fertilization of the egg by a sperm cell. About once a month in the nonpregnant female of reproductive age, an oocyte is ejected from an ovary and is swept into the adjacent oviduct. Fertilization, when it occurs, normally takes place within an oviduct, after which the young embryo passes down the oviduct and becomes implanted in the lining of the uterus. Muscular movements of the oviduct, plus the beating of the cilia that line it, propel the egg cell down the oviduct toward the uterus. If the oocyte is not fertilized within about 36 hours, it dies, usually by the time it reaches the uterus. A sperm cell has a life expectancy of about 48 hours within the female reproductive tract.*

Orgasm in the Female

Under the influence of a variety of stimuli, the clitoris and its associated tissues become engorged and distended with blood, as does the penis of the male. This process, however, is somewhat slower in women than it is in men. The distension of the tissues is accompanied by the secretion into the vagina of a fluid that both lubricates its walls and neutralizes its normally acidic environment. The acidic conditions are inhospitable not only to bacteria—thereby reducing the likelihood of infections—but also to sperm cells.

Orgasm in the female, as in the male, is marked by rhythmic muscular contractions, followed by expulsion into the veins of the blood trapped in the engorged tissues. Homologous muscles produce orgasm in the two sexes, but in the female there is no ejaculation of fluid through the urethra or the vagina. At orgasm, the cervix drops down into the upper portion of the vagina, where the semen tends to form a pool. The female orgasm also may produce contractions in the oviducts that propel the sperm upward. Orgasm in the female, however, is not necessary for conception.

Hormonal Regulation in Females: The Menstrual Cycle

The production of oocytes in all vertebrate females is cyclic. It involves the interplay of hormones, changes in the follicle cells, and, in marsupial and placental mammals, changes in the endometrial lining of the uterus. In humans, this recurring pattern of varying hormone levels and tissue changes is known as the **menstrual cycle.**

The menstrual cycle is timed and controlled by the hypothalamus. The hormones that participate in the extremely complex feedback system regulating the cycle include the estrogens and progesterone (the female sex hormones), the pituitary gonadotropins FSH and LH, and gonadotropin-releasing hormone (GnRH) from the hypothalamus.

675

At the beginning of the cycle, defined as the first day of the menstrual flow, hormone levels are low (Figure 37–12). After a few days, an oocyte and its follicle begin to mature under the influence of FSH and LH. As the follicle enlarges, it secretes increased amounts of estrogens, which stimulate the regrowth of the endometrium in preparation for the implantation of a fertilized egg. The rapid rise in estrogen levels near the midpoint of the cycle triggers a sharp increase in the release of LH by the pituitary. The spurt of high LH stimulates the follicle to release the oocyte, which begins its passage to the uterus.

Under the continued stimulus of LH, the cells of the emptied follicle grow larger and fill the cavity, producing the **corpus luteum** ("yellow body"), as shown in Figure 37–9b. As the cells of the corpus luteum increase in size, they begin to synthesize significant amounts of **progesterone** as well as estrogens. As the hormone levels increase, estrogens and progesterone together inhibit the production of GnRH by the hypothalamus and thus of LH and FSH by the pituitary. If fertilization does not occur, the corpus luteum is reabsorbed, the production of ovarian hormones drops, and the blood supply to the endometrium is drastically reduced.

Deprived of hormonal support and nourishment, the endometrium can no longer sustain itself, and a portion of it is sloughed off in the menstrual fluid. Then, in response to the now low level of ovarian hormones, the levels of LH and FSH from the pituitary begin to rise again, followed by the development of a new follicle and a rise in estrogens as the next monthly cycle begins.

The menstrual cycle usually lasts about 28 days, but individual variation is common. Even in women with cycles of average length, ovulation does not always occur at the same time in the cycle (which is why the "rhythm method" is an unreliable means of birth control).

The onset of the first menstrual period marks the beginning of puberty in the human female. The average age of puberty is 12.3 years, but the normal range is very wide. The increased production of female sex hormones preceding puberty induces the development of secondary sex characteristics, such as enlargement of the hips and breasts.

Contraceptive Techniques

Using no contraceptive methods, 80 percent of women of childbearing age having regular sexual intercourse become pregnant within a year. However, a variety of contraceptive techniques is available for couples who wish to prevent or defer pregnancy. In Table 37–1, they

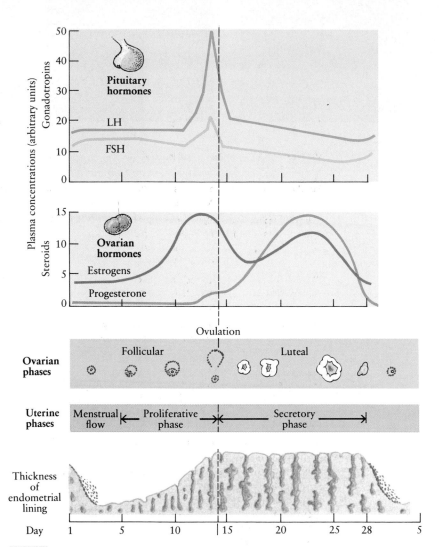

37–12   *Diagram of the events that take place during the menstrual cycle, which involves changes in hormone levels, in structures at the surface of the ovary, and in the uterine lining. The cycle begins with the first day of menstrual flow, which is caused by the shedding of the endometrium, the lining of the uterine wall. The increase of FSH and LH at the beginning of the cycle promotes the growth of the ovarian follicle and its secretion of estrogens. Under the influence of estrogens, the endometrium regrows. The sudden rise in estrogens just before midcycle triggers a sharp increase in the release of LH from the pituitary, which stimulates the release of the oocyte (ovulation). (It is not known what role, if any, is played by the simultaneous increase in FSH.) Following ovulation, LH and FSH levels drop. The follicle is converted to the corpus luteum, which secretes progesterone and also estrogens. Progesterone further stimulates the endometrium, preparing it for implantation of the fertilized egg. If fertilization does not occur, the corpus luteum degenerates, the production of progesterone and estrogens falls, the endometrium begins to slough off, FSH and LH concentrations increase once more, and the cycle begins anew.*

Table 37–1   Methods of Birth Control Currently Available

Method	Mode of Action	Effectiveness (Pregnancies per 100 Women per Year)	Action Needed at Time of Intercourse	Requires Instruction in Use	Possible Undesirable Effects
Vasectomy	Prevents release of sperm from testes	0–0.15	None	No, but does require a surgical procedure	Usually produces irreversible sterility
Tubal ligation	Prevents passage of oocyte to uterus	0–0.4	None	No, but does require a surgical procedure	Usually produces irreversible sterility
"The pill" (estrogens and progesterone)	Inhibits secretion of FSH and LH, thereby preventing follicle maturation and ovulation	0–10	None	Yes, must be taken on prescribed schedule	Early—some water retention, breast tenderness, nausea; late—increased risk of cardiovascular disease
"Minipill" (progesterone alone)	Causes changes in cervix and uterus that prevent conception	3–10	None	Yes, must be taken on prescribed schedule	Break-through bleeding, ectopic (tubal) pregnancy
"Morning-after pill" (50 × normal dose of estrogens)	Arrests pregnancy, probably by preventing implantation	?	None	Yes; prescribed only in emergencies, for example, for rape victims	Breast swelling, water retention, abdominal pain, nausea
Diaphragm with spermicidal jelly	Prevents sperm from entering uterus, jelly kills sperm	2–20	Yes, insertion before intercourse	Yes, must be inserted correctly each time; leave in at least 6 hours after	None usually, may cause irritation
Condom (worn by male)	Prevents sperm from entering vagina	3–36	Yes, male must put on after erection	Not usually	Some loss of sensation in male
Intrauterine device (wound with copper or containing progesterone that is slowly released)	Prevents fertilization or implantation	4–5	None; may be left in place for as long as 4 years	No, but must be inserted by a physician and periodically checked	Menstrual discomfort, displacement or loss of device, uterine infection or perforation, ectopic pregnancy
Cervical cap with spermicidal jelly	Prevents sperm from entering uterus, jelly kills sperm	4–?	Yes, insertion before intercourse	Yes, must be inserted correctly each time; more difficult to insert than diaphragm	None usually, may cause irritation
Contraceptive sponge (contains spermicide)	Prevents sperm from entering uterus; kills sperm	13–20	Yes, insertion before intercourse	Yes, must be inserted correctly each time; leave in at least 6 hours after	None usually, may cause irritation
Vaginal foam, jelly alone	Spermicidal, mechanical barrier to sperm	20–30	Yes, requires application before intercourse	Yes, must use within 30 minutes of intercourse; leave in at least 6 hours after	None usually, may cause irritation
Rhythm	Abstinence during probable time of ovulation	14–47	None	Yes, must know when to abstain, based on daily temperatures and/or cyclic changes in vaginal mucus	Requires abstinence during part of cycle
Douche	Washes out sperm that are still in the vagina	?–85	Yes, immediately after	Not usually	May propel sperm through cervix into uterus
Withdrawal	Removes penis from vagina before ejaculation	?–93	Yes, withdrawal	No	Frustration in some individuals

are rated in order of effectiveness, in terms of the average number of pregnancies per year among women of childbearing age using them. In most cases, two figures are given for effectiveness. The first, lower, figure is an "ideal" figure, obtained when the method is used consistently and correctly. The second figure is an average figure, reflecting actual experience.

For many years the most widely used contraceptive techniques were barrier methods, such as the diaphragm and the condom. In the 1960s and 1970s, many couples abandoned barrier methods, and "the pill," a combination of synthetic estrogens and progesterone,

came into wide use. When taken daily, it keeps the level of these hormones in the blood high enough to shut off production of the pituitary hormones FSH and LH. Without FSH the ovarian follicles do not ripen, and in the absence of LH no ovulation occurs, so pregnancy is not possible. Recently, however, the diaphragm and the condom have once more become popular. In combination with spermicidal jellies, these two methods, and particularly the latex condom, provide a barrier not only against sperm but also against many infectious microorganisms. As Table 37–1 reveals, however, their protection is far from absolute.

Summary

The male and female gametes, sperm and eggs, are formed by meiosis in the gonads (the testes and the ovaries). Sperm cells are produced in the seminiferous tubules of the testes. Cells known as spermatogonia develop into primary spermatocytes. The first meiotic division produces secondary spermatocytes, and the second meiotic division produces spermatids. The spermatids then differentiate into sperm cells.

The sperm cells enter the epididymis, a tightly coiled tube overlying the testis. The epididymis is continuous with the vas deferens, which leads through the abdominal cavity, around the bladder, and into the prostate gland. Just before entering the prostate, the two vasa deferentia merge with ducts of the seminal vesicles and then, within the prostate, with the urethra, which leads out through the penis.

The penis is composed largely of spongy erectile tissue that can become engorged with blood, enlarging and hardening. At the time of ejaculation, sperm are propelled along the vasa deferentia by contractions of smooth muscle. Secretions from the seminal vesicles, the prostate, and the bulbourethral glands are added to the sperm as they move toward the urethra. The resulting mixture, the semen, is expelled from the urethra by muscular contractions.

Production of sperm and the development of male secondary sex characteristics are under the control of hormones, including gonadotropin-releasing hormone (GnRH), the gonadotropins LH (luteinizing hormone) and FSH (follicle-stimulating hormone), and testosterone (the principal androgen). LH acts on the interstitial cells, located between the seminiferous tubules, to stimulate the production of testosterone. FSH and testosterone stimulate the production of sperm. Production of LH and FSH is regulated by a negative feedback system.

The female gamete-producing organs are the ovaries. The primary oocytes develop within nests of cells called follicles. The first meiotic division begins in the female fetus and is completed years later at ovulation. The second meiotic division is completed at fertilization. In the human female, one secondary oocyte is released every 28 days, on average, and travels down the oviduct to the uterus. If it is fertilized (which usually takes place in an oviduct), the embryo becomes implanted in the lining of the uterus (the endometrium). If it is not fertilized, it degenerates and the endometrial lining is shed at menstruation.

The production of oocytes and the preparation of the endometrium for implantation of the embryo are cyclic. The reproductive cycle, which is known in humans as the menstrual cycle, is controlled by hormones. Before ovulation, FSH and LH from the pituitary stimulate the ripening of the follicle and the secretion of estrogens. After ovulation, the corpus luteum, which forms from the emptied follicle, produces both estrogens and progesterone. Progesterone and estrogens together stimulate the growth of the endometrium.

Questions

1. Distinguish among the following: spermatocytes/spermatids/sperm cells; interstitial cells/Sertoli cells; oocyte/ovum; ovarian follicle/corpus luteum.

2. What would constitute the ejaculate of a man who has undergone a vasectomy? Would a vasectomy affect the structures associated with orgasm? Why or why not?

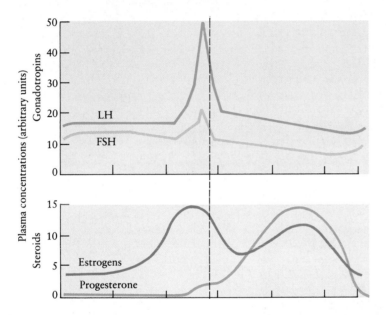

3. Describe the effects of luteinizing hormone (LH) and follicle-stimulating hormone (FSH) in the human male and female.

4. Explain the changes in the plasma concentrations of the gonadotropins and sex hormones during the menstrual cycle as shown on the graphs to the left.

5. During which days in the menstrual cycle is a woman most likely to become pregnant? (Include data on the longevity of eggs and sperm in making this calculation.) Why is the use of the calendar method of birth control much less effective than other methods?

6. What would be an appropriate method of contraception for a couple who have infrequent intercourse (once a month)? For a couple who have frequent intercourse (three times a week), with plans to have children eventually? For a couple who have frequent intercourse but do not wish to have any children? Under what circumstances, if any, would you elect to have a vasectomy or a tubal ligation?

7. A long-sought goal of contraceptive research has been the development of an effective male contraceptive, comparable to "the pill." Two candidates for such a contraceptive are the hormone inhibin, produced by the Sertoli cells of the seminiferous tubules, and testosterone. Explain the mechanism by which each of these hormones would be expected to function as a contraceptive.

Suggestions for Further Reading

Aral, Sevgi O., and King K. Holmes: "Sexually Transmitted Diseases in the AIDS Era," *Scientific American*, February 1991, pages 62–69.

Djerassi, Carl: "The Bitter Pill," *Science*, vol. 245, pages 356–361, 1989.

Djerassi, Carl: "Fertility Awareness: Jet-Age Rhythm Method?" *Science*, vol. 248, pages 1061–1062, 1990.

Frisch, Rose E.: "Fatness and Fertility," *Scientific American*, March 1988, pages 88–95.

Palca, Joseph: "The Pill of Choice?" *Science*, vol. 245, pages 1319–1323, 1989.

Riddle, John M., and J. Worth Estes: "Oral Contraceptives in Ancient and Medieval Times," *American Scientist*, vol. 80, pages 226–233, 1992.

Ulmann, André, Georges Teutsch, and Daniel Philibert: "RU 486," *Scientific American*, June 1990, pages 42–48.

Wingfield, John C., et al.: "Testosterone and Aggression in Birds," *American Scientist*, vol. 75, pages 602–608, 1987.

This western grebe is, at this very moment, serenely presenting the perfect package—her egg. Even the wrapping is an engineering marvel. The shell, mostly calcium carbonate, is riddled with thousands of tiny pores. It is almost waterproof but permeable to oxygen and carbon dioxide. The shell is strong enough to resist accidental breakage, yet fragile enough that the chick, when ready, can escape. All that is needed from the outside world is some warmth and a little oxygen.

The amniote egg, as it is known, is an evolutionary legacy from the reptiles. Before its invention, embryonic development always required the presence of water—an ocean, a stream, a puddle, a few drops of rainwater in the hollow of a leaf. For all vertebrates, development still must take place in water, but in the case of reptiles, birds, and mammals, this water is contained within a membrane, the amnion, produced by the embryo itself.

The embryo produces four different membranes that, as they grow, fuse and develop blood vessels. The most highly vascularized membrane, plastered against the inside of the shell, collects oxygen for the growing chick. Another membrane serves as a garbage bag, sequestering the chick's waste products. Yet another membrane surrounds the yolk, the embryo's primary food supply. Water is released as the yolk is digested and just the right amount escapes through the almost waterproof shell. Birds, reptiles, and mammals all have the same four extraembryonic ("outside-the-embryo") membranes, although in mammals, as you will see, some of their functions have been modified.

Another noteworthy feature of the amniote egg is its exquisite timing. Development of the embryo takes 11 days in the murderous cuckoo (page 376), 3 months in the Nile crocodile, and 22 months in the African elephant. During this interval, no matter what its duration, the events of development unfold in a precisely measured sequence, remarkably the same from species to species. And then, at the end, there emerges a most remarkable object, both ordinary and extraordinary—a whole new individual. Perhaps giving a few peeps to announce its arrival.

Development

Each of our lives began very modestly with the fusion of a tiny, motile sperm cell and a larger, but structurally simpler egg. If you were to examine the newly fertilized egg of an animal—any animal—under the electron microscope, you would find no hint of the awesome potential with which this single cell is endowed. How do the complex structures of the embryo and, subsequently, of the adult animal develop from this one, apparently simple cell? This question is one of the most fundamental in all of biology. Although it has commanded the attention of scientists for more than 100 years—and much more is known now than even a few years ago—a complete and comprehensive answer remains elusive.

During its development, the fertilized egg is transformed into a complete organism, closely resembling its parents and consisting, depending on the species, of hundreds to trillions of cells. This process involves **differentiation** (the specialization of cells, tissues, and organs), **morphogenesis** (the shaping of the adult body form), and **growth** (an increase in size). The course of development in many species has been closely observed, clarifying the sequence of developmental events. Only in recent years, however, have the tools become available to begin deciphering the underlying cellular and molecular events that give rise to and control the observed processes.

In this chapter, we shall first consider the principal stages of animal development, with examples selected from among three representative organisms—the sea urchin, the amphibian, and the chick. We shall then return to the fertilized human egg, which we left in the last chapter as it began its journey down the oviduct to the uterus, and consider the marvelous series of events that precede and lead to the birth of each new human infant.

Fertilization and Activation of the Egg

Development begins with fertilization of the egg by the sperm, a process that has been extensively studied in the sea urchin. Like the human sperm cell shown in Figure 37–3 (page 667), the sperm cell of the sea urchin consists of an acrosome, a highly condensed nucleus, a small amount of cytoplasm, and a long flagellum, all surrounded by a plasma membrane. The egg cell, which is much larger than the sperm cell (see Figure 5–8, page 93), is surrounded by an outer membrane, known as the **vitelline envelope,** attached to the plasma membrane. Embedded in the vitelline envelope and displayed on its surface are receptor molecules that participate in the binding of the sperm to the egg. Surrounding the vitelline envelope are additional layers of jelly.

When the sperm meets the egg, enzymes released from the acrosome dissolve the jelly layers at the point of contact, creating a path through which the sperm cell can move (Figure 38–1). Other molecules from the acrosome coat the outer surface of the sperm head. These molecules are the three-dimensional complement of the receptors displayed on the vitelline envelope. When they bind to those receptors, the sperm is able to penetrate the vitelline envelope. Ultimately, the plasma membranes of the sperm and the egg make contact, they fuse, and the sperm nucleus enters the cytoplasm of the egg.

Fertilization—the fusion of sperm and egg—has at least four consequences. First, changes take place on the surface of the fertilized egg to prevent the entry of additional sperm. Within a second of sperm contact, a series of reactions sweeps, wavelike, around the surface of the egg, making the plasma membrane suddenly unresponsive to the advances of other sperm and causing the vitelline envelope to lift off the surface of the egg (Figure 38–2). The protective outer membrane that forms from the vitelline envelope is known as the **fertilization membrane.**

Second, the egg is activated metabolically, as evidenced by a dramatic increase in protein synthesis and a rise in oxygen consumption.

Third, the genetic material of the male is introduced into the female gamete—the haploid sperm nucleus fuses with the haploid egg nucleus to produce a diploid zygote nucleus. The genotype of the new individual is thus established.

Fourth, the egg begins to divide by mitosis. The developmental chain of events is thus set in motion.

Cleavage and Formation of the Blastula

The initial stages of development in the sea urchin are shown in Figure 38–3. The zygote divides about once an hour for 10 hours, in the process known as **cleavage.** As the cells continue to divide, sodium ions (Na^+) are pumped from them into the extracellular spaces, followed by an osmotic flow of water. This creates a fluid-filled cavity, known as the **blastocoel,** in the center of the embryo. When the blastocoel is fully formed, the embryo—now a fluid-filled sphere of a single layer of cells—is called the **blastula.**

In eggs, such as those of the sea urchin, in which

38–1 *Fertilization in the sea urchin, as revealed by scanning electron micrographs. (a) Sperm cells on the egg surface, which is covered with microvilli. (b) Contact between the acrosome of a sperm cell and the microvilli. In the region of acrosomal contact, microvilli rise up to meet the sperm cell. Within three minutes, its head makes contact with the plasma membrane of the egg. The nucleus of the sperm cell then moves rapidly into the cytoplasm of the egg (c). One minute after the sperm cell first touches the plasma membrane of the egg, only its flagellum remains outside.*

(a) ⊢—— 5 µm

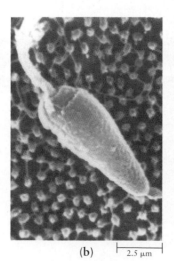

(b) ⊢—— 2.5 µm

(c) ⊢—— 2.5 µm

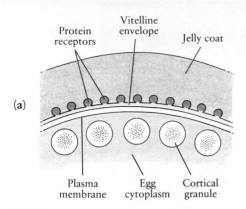

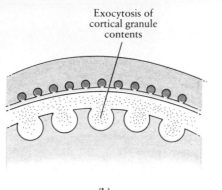

(a)

Protein receptors · Vitelline envelope · Jelly coat

Plasma membrane · Egg cytoplasm · Cortical granule

(b)

Exocytosis of cortical granule contents

(c)

Fertilization membrane

38–2 *Formation of the fertilization membrane. (a) The surface layers of an unfertilized egg consist of a jelly coat; the vitelline envelope, which bears protein receptor molecules; and the plasma membrane. Within the cytoplasm, near the plasma membrane, are membrane-bound vesicles known as cortical granules. Contact of a* sperm cell with the receptors on the vitelline envelope triggers a series of chemical reactions that cause the cortical granules to fuse with the plasma membrane (b). The granules release their contents, including enzymes and polysaccharides, into the space between the plasma membrane and the vitelline envelope. The enzymes break down the receptors on the vitelline envelope and the substances that hold it to the plasma membrane. The polysaccharides raise the osmotic potential of the space beneath the vitelline envelope, causing water to flow inward. The vitelline envelope is thus lifted off the surface of the egg (c). The vitelline envelope "hardens" to form the fertilization membrane.

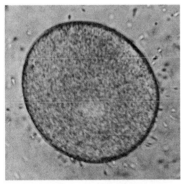

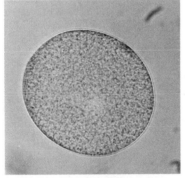

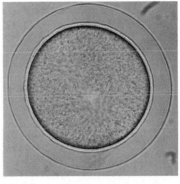

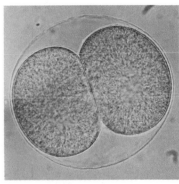

(a) *Numerous sperm cells can be seen surrounding an unfertilized egg.*

(b) *Fertilized egg. The fertilization membrane has just begun to form. The light area slightly above the center is the diploid nucleus.*

(c) *The fertilization membrane is fully formed, and the egg has begun to divide. If you look closely, you can see that there are two nuclei.*

(d) *The first division.*

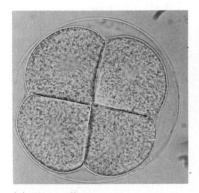

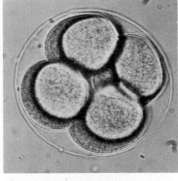

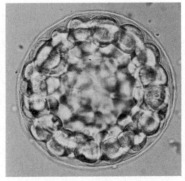

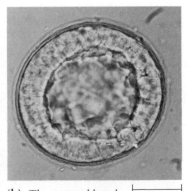

(e) *Four-cell stage.*

(f) *Eight-cell stage. The space that is beginning to form in the center will become the blastocoel.*

(g) *The embryo is now a sphere of cells, surrounding the forming blastocoel.*

(h) *The mature blastula.* 0.1 mm

38–3 *The early development of the sea urchin. Notice that as the egg divides, the* cells become progressively smaller, so that by the blastula stage they are barely distinguish- able, although the magnification has not changed.

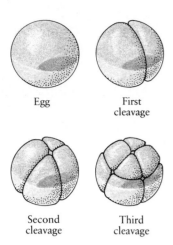

Egg

First cleavage

Second cleavage

Third cleavage

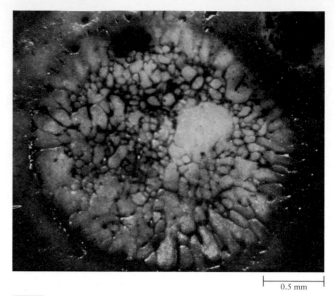

0.5 mm

38–5   *The blastodisc of a chick embryo, as viewed from above.*

38–4   *Eggs that contain a large amount of yolk concentrated in one hemisphere cleave unequally. For example, in the amphibian egg, the yolk is massed in the lower hemisphere, and the upper portion of the egg is covered by a heavily pigmented layer of cytoplasm. When the sperm penetrates the egg, the pigment cap rotates toward the point of sperm penetration, and a gray crescent appears on the side of the egg opposite the point of sperm entry. The first longitudinal cleavage usually bisects the gray crescent. The second cleavage, also longitudinal, is at right angles to the first. Thus the egg is split through the poles to produce four cells shaped like the segments of an orange. The third cleavage separates the lower, yolkier part of the embryo from the upper, less yolky part. As you can see, the four cells in the upper hemisphere are much smaller than the four in the lower hemisphere.*

stored food in the form of yolk is present only in small amounts, cleavage is uniform, producing cells of similar size. However, when larger amounts of yolk are present, as in the eggs of amphibians, the cell divisions are uneven (Figure 38–4), forming larger cells in the region of the egg in which the yolk is concentrated. Cleavage in the amphibian, as in the sea urchin, results in the formation of a blastula, but the blastocoel is small and usually off-center.

The yolk of a hen's egg, like that of the eggs of other birds, is so large and dense that cleavage does not occur in most of the egg mass. The only part of the fertilized egg that cleaves is a thin layer of cytoplasm that sits like a cap on top of the yolk and contains the nucleus. Cleavage of this thin layer produces a lozenge-shaped blastula known as a **blastodisc** (Figure 38–5). If you break open a fertilized chick egg when it is first laid, you can see the blastodisc as a white mass about 2 millimeters in diameter on top of the yolk. Microscopic examination shows that this mass is made up of close to 100,000 cells, and that the cells are in two layers. The space between the layers is the blastocoel (Figure 38–6).

38–6   *Thin sections through the center of the blastulas of (a) a sea urchin, (b) a frog, and (c) a bird. In the sea urchin and the frog, the blastula is a hollow sphere of cells. In the bird, it takes the form of a blastodisc—a flattened mass of cells—that sits on the surface of the yolk. The blastocoel is located between the upper and lower layers of cells.*

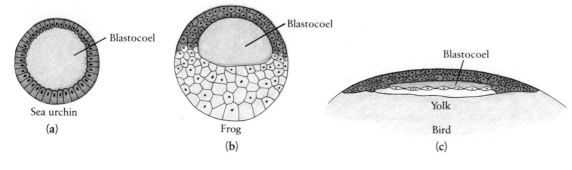

Blastocoel

Sea urchin

(a)

Blastocoel

Frog

(b)

Blastocoel

Yolk

Bird

(c)

Gastrulation and Establishment of the Body Plan

The formation of the blastula is followed by a process known as **gastrulation** (from *gaster,* the Greek word for "stomach"), which gives rise to the primitive gut.

In the sea urchin (Figure 38–7), gastrulation begins with the formation of the **blastopore,** an opening into the blastula. Cells near the blastopore break loose and, with the aid of contractile pseudopodia with sticky tips, move over the interior surface of the blastula toward the opposite pole. Next, the entire cell layer closest to the blastopore turns inward, moving through the blastocoel to the opposite pole, forming a new cavity, the **archenteron.** The archenteron will ultimately develop into the digestive tract, and the blastopore will become the anus.

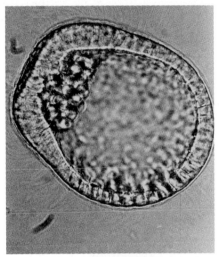

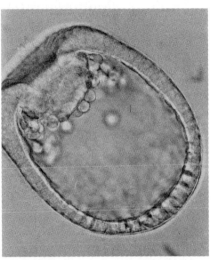

38–7 *Gastrulation in the sea urchin. Following this process, the cells of the embryo differentiate and become organized into the structures of the sea urchin larva. The entire developmental process, from fertilization to free-swimming larva, requires less than 48 hours.*

(a) *The beginning of gastrulation. The blastopore has begun to form at the upper left, and cells near the blastopore have begun to migrate into the blastocoel.*

(b) *The outer cell layer begins to fold inward at the blastopore, forming the archenteron.*

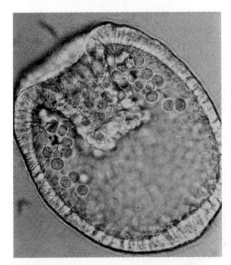

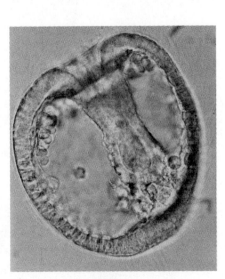

(c) *The outer layer of cells continues to move across the blastocoel, forming the archenteron.*

(d) *The mature gastrula.*

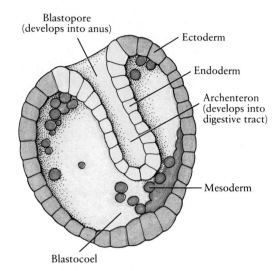

Blastopore
(develops into anus)

Ectoderm

Endoderm

Archenteron
(develops into
digestive tract)

Mesoderm

Blastocoel

38–8   *The sea urchin gastrula. Gastrula-tion produces a three-layered embryo. The archenteron becomes the digestive tract, and the blastopore becomes the anus. Ultimately, the blastocoel is almost entirely obliterated. In this and subsequent illustrations, ectoderm is blue, endoderm is yellow, and mesoderm is red.*

As a result of the movements that take place at gastrulation, three embryonic tissue layers are formed: an outer layer, the **ectoderm;** a middle layer, the **mesoderm;** and an inner layer, the **endoderm** (Figure 38–8). At the same time, the anterior-posterior axis of the embryo is established.

Gastrulation in vertebrates differs from that in the sea urchin in detail but not in principle. In the yolky frog's egg, for example, the process of inward movement is characterized by an extensive series of coordinated cellular movements—migration of single cells and sheets of cells—and the folding of cell layers (Figure 38–9). Cells at the **dorsal lip** of the blastopore change shape and sink below the surface as they move to the interior. These cells are replaced by cells that move from their original positions on the surface of the embryo to the dorsal lip. As a result, the entire sheet of cells that forms the surface of the amphibian embryo appears to stream toward and over the lip of the blastopore as if it were being hauled over a pulley.

Once inside the embryo, the migrating cells move away from the blastopore and deeper into the interior. The direction in which they move is the future anterior-posterior axis of the animal. The advancing cells form

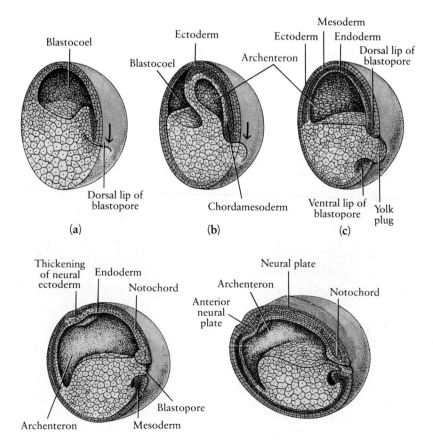

38–9   *Development of the frog gastrula.* (a) *The blastopore forms in the blastula, and cells from the outer surface begin moving across its dorsal lip to the interior.* (b) *As the cellular migrations progress, the blastocoel is obliterated and the archenteron created.* (c) *Three embryonic tissue layers are established: ectoderm, mesoderm, and endoderm.* (d) *The mesoderm directly over the roof of the archenteron, known as the chordameso-derm, differentiates to become the notochord (shown in green). The ectoderm on the dorsal surface overlying the notochord thickens and flattens, forming the neural ectoderm (shown in purple), from which the neural plate begins to form* (e).

the walls of an increasingly spacious archenteron, and the blastocoel disappears. The archenteron remains open to the outside at the blastopore, the site of the future anus. Yolk-laden cells are visible within the boundaries of the blastopore, forming the yolk plug.

In the course of gastrulation in the amphibian, as in the sea urchin, the primary embryonic tissues—endoderm, mesoderm, and ectoderm—become arranged in a three-layered pattern. The floor of the archenteron is composed of yolk-laden endodermal cells. Its roof consists of endodermal cells that have been pushed and pulled into the interior by a sheet of mesodermal cells that lies above the archenteron along the axis of the embryo. This sheet of mesoderm includes cells destined to form the notochord (page 482) and is called the **chordamesoderm.** At the sides of the archenteron, other mesodermal cells have slipped between the ectoderm and endoderm, forming the **lateral plate mesoderm.**

On the dorsal surface of the embryo, lying above the chordamesoderm, is a sheet of ectoderm that will ultimately give rise to the brain and spinal cord. This tissue is known as **neural ectoderm.** The ectoderm covering the rest of the gastrula, known as **epidermal ectoderm,** will give rise to the epidermis of the skin.

By the end of gastrulation, the first visible signs of differentiation have begun to appear. In the amphibian and other vertebrates, the chordamesoderm has formed the notochord, and the neural ectoderm has begun to thicken, forming the **neural plate** (Figure 38–10). The ridges of the neural plate curve upward and inward, forming the **neural groove.** Ultimately these ridges meet and fuse to form the **neural tube,** which pinches off from the rest of the ectoderm (Figure 38–11).

At about the same time, the two longitudinal strips of dorsal mesoderm on either side of the differentiated notochord split into segments, forming blocks of tissue called **somites.** In the lateral plate mesoderm, the coe-

38–10   *The neural plate of a frog embryo. The wide portion of the neural plate, at the anterior of the embryo, will give rise to the brain. The narrow portion will give rise to the spinal cord.*

38–11   *The formation of the neural tube from the neural plate in the frog. (a) The thickened elevations of neural ectoderm on the right and left sides of the neural plate curve inward, forming the neural groove (b). The ridges bordering the neural groove then meet and fuse, and (c) the resulting neural tube pinches off from the epidermal ectoderm.*

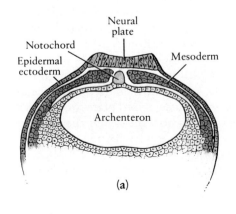

Neural
plate
Notochord
Epidermal
ectoderm
Mesoderm
Archenteron

(a)

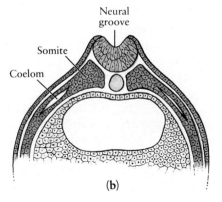

Neural
groove
Somite
Coelom

(b)

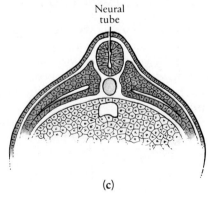

Neural
tube

(c)

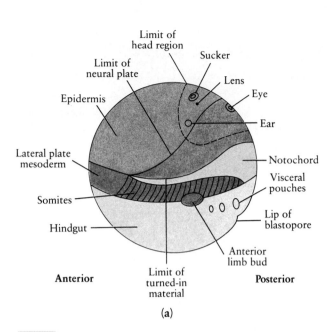

Blastula

Gastrula

Ectoderm   Endoderm

Epidermis and   Epithelium   Inner lining of   Inner lining of
associated structures   digestive tract   respiratory tract

Neural ectoderm   Glands—including liver, pancreas

Brain and   Mesoderm
nervous system

Notochord   Somites

Excretory and   Muscle   Skeleton
reproductive structures

Dermis (inner skin layer)

Lateral plate mesoderm

Outer covering   Lining
of internal organs   of coelom

Circulatory system—including heart, blood vessels, etc.

(a)   **(b)**

38–12   (a) *A fate map of the amphibian, showing the groups of cells on the surface of the blastula that, in the course of gastrulation and subsequent development, give rise to the tissues of the mature embryo (the tadpole). A comparison of the location of particular groups of cells on the blastula with their ultimate location within the tadpole makes clear the*

extent of the reorganization that occurs during development. (b) The tissue layers that form as a result of gastrulation give rise to the specialized cells and tissues of the adult animal. This pattern of differentiation is characteristic of all vertebrates, ourselves included.

lom (page 469) forms between two layers of tissue. Thus the principal features of the vertebrate have been established.

The movements of cells and tissues in gastrulation are completely regular and predictable, embryo after embryo. By applying harmless dyes to the surface of the late blastula, embryologists have developed fate maps (Figure 38–12a) that identify the groups of cells that give rise to the tissue layers of the mature gastrula. During subsequent development, these embryonic tissue layers differentiate to form the specialized tissues and organs of the adult animal (Figure 38–12b).

The Role of Tissue Interactions

Differentiation, as we saw in Chapter 16, is the result of the selective activation and inactivation of specific genes in the nucleus of a cell. These genes control the production of the specific proteins that characterize different types of cells and enable them to carry out their functions.

For some cell types, differentiation does not become irreversible until fairly late in the developmental process. Although the cells have become differentiated, they retain their full developmental potential. For many cell types, however, the developmental potential is grad-

ually limited as gastrulation proceeds and sheets of cells become properly positioned in the embryo.

The process in which the fate of a cell becomes fixed depends on a progressive series of interactions between different tissue types. This process is known as **embryonic induction.** It occurs when two different types of tissue come in contact with one another in the course of development and one tissue causes, or induces, the other to differentiate. In every vertebrate species examined thus far, the primary tissue in embryonic induction is the chordamesoderm, the tissue from which the notochord develops. The chordamesoderm induces the differentiation of the overlying ectoderm that gives rise to the neural plate and sets in motion the subsequent events of development.

Numerous experiments have shown that induction involves the exchange of chemical substances between the interacting tissue layers. Although the chemical substance or substances exchanged have not yet been identified, they are apparently of a very general nature. A number of chemicals, some of which do not occur naturally in embryos, will induce neural plate formation in embryonic ectoderm. It is not, then, that the inducing substance endows the ectoderm with the capacity to form nerve tissue; rather, it evokes a potential that is already present in the responding tissue and initiates a new pattern of gene activity.

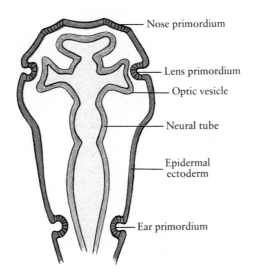

38–13   *The development of the brain and associated sense organs. At the anterior end of the neural tube, three local swellings appear, from which the complex structures of the brain will ultimately develop. The most anterior of these swellings then bulges laterally, and two saclike structures, the optic vesicles, appear. At the same time, the epidermal ectoderm folds inward to meet the optic vesicles. These inward folds are the primordia—the earliest stages—in the formation of the lens of each eye. The primordia of ears and nostrils also appear first as infoldings of the epidermal ectoderm.*

38–14   *Eyes develop as a result of interactions between the epidermal ectoderm and the optic vesicles. When the optic vesicle reaches the epidermal ectoderm, the epidermal ectoderm begins to thicken and differentiate. The optic vesicle flattens out and pushes inward, becoming the double-walled optic cup. The thickened layer of epidermal ectoderm pinches off to become the transparent lens, and the overlying epidermal ectoderm, which also becomes transparent, forms the cornea. The rim of the optic cup ultimately forms the edge of the pupil, while the interior of the optic cup becomes the retina. Note that the retina is thus a differentiated extension of the brain, with which it remains connected by the optic stalk.*

Organogenesis: The Formation of Organ Systems

The later stages of development, following cleavage and gastrulation, are known as **organogenesis,** the formation of organ systems. Organogenesis, which is essentially the same in all vertebrates, begins with the inductive interaction between ectoderm and chordamesoderm. Each of the three primary tissues formed during gastrulation then proceeds to undergo growth, differentiation, and morphogenesis. As an example of organogenesis, let us consider the interactions that give rise to the eye.

Formation of the Eye

The vertebrate brain begins as three bulges in the foremost part of the neural tube. Almost immediately, two saclike protrusions, the **optic vesicles,** appear on the sides of the anterior region of the forming brain (Figure 38–13). These spherical vesicles enlarge until they come into contact with the epidermal ectoderm. During this process, they remain connected to the neural tube by an increasingly narrow stalk, within which the optic nerve will ultimately develop.

When the optic vesicle comes into contact with the inner surface of the epidermal ectoderm, the external surface of the vesicle flattens out and pushes inward. The vesicle thus becomes a double-walled cup. The rim of the cup becomes the edge of the pupil. The opening of the cup is large at first, but the rim bends inward and converges, so that the opening of the pupil becomes smaller. At the point where the optic vesicle touches the epidermal ectoderm, the epidermal ectoderm begins to differentiate into a lens, becoming transparent (Figure 38–14).

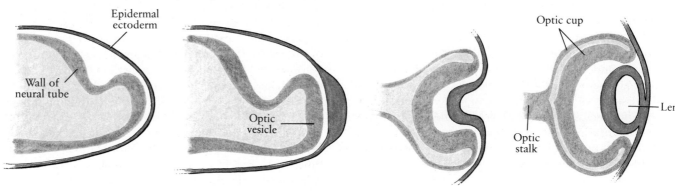

The differentiation of the lens of the eye is an example of a secondary induction. The inducing tissue (the optic vesicle) is itself the result of the primary induction of ectoderm by chordamesoderm. In fact, the formation of the complete eye involves at least six separate inductions, occurring in an orderly sequence, each one linked to the one before. Similar series of sequential inductions occur in the development of other organs. As in the eye, they ensure not only the differentiation of the specific tissues that form the organ but also that the various structures of the organ are properly positioned in relation to one another.

Morphogenesis: The Shaping of Body Form

Each part of the body has a characteristic shape and structure. For example, the spinal cord is a hollow tube, the liver is composed of lobes, the salivary glands consist of groups of secretory cells clustered around ducts, and the lungs consist of microscopic air spaces lined by extremely thin sheets of cells.

Only a few cellular processes, repeated over and over in various permutations and combinations, appear to be responsible for shaping the various structures of the body. These processes include (1) increases or decreases in the rates of cell growth and division, (2) the migration of individual cells and groups of cells from one location to another, (3) the positioning of cells by changes in their adhesion to neighboring cells, (4) the deposition of extracellular materials, (5) programmed cell death, and (6) changes in cell shape brought about by extension or contraction. For example, formation of the neural tube from the flattened neural plate is accomplished by changes in the shape of the individual cells, as shown in Figure 38–16.

These few cellular processes give rise not only to the internal structures that are similar from one vertebrate to another but also to the distinctive structures that characterize different groups of vertebrates. Among the most striking features of terrestrial vertebrates are four limbs. As we saw in Figure 23–4 (page 404), the limbs of reptiles, birds, and mammals all have the same basic structure, and yet there are obvious—and significant—differences between, for example, the wings of a bird and the arms of a human. Similar differences are apparent between the wings of a bird and its legs. The developmental process that gives rise to the shape and structure of specific body parts is known as **pattern formation.** It has been studied primarily in the developing chick wing.

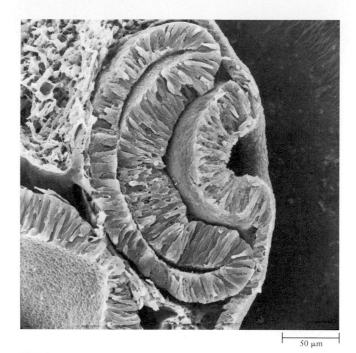

38–15   *The developing eye of a chick embryo, as revealed by the scanning electron microscope. As the cells in the interior of the optic cup differentiate into rods and cones and bipolar, horizontal, amacrine, and ganglion cells, the multiple connections that characterize the retina (see Figure 36–6, page 645) will become established.*

Development of the Chick Wing

In the chick embryo, the first visible sign of wing development is the appearance of a **wing bud,** which consists of a central mass of mesodermal cells covered by a thin layer of ectoderm. As the wing bud begins to grow outward, a ridge of specialized tissue, the **apical ectodermal ridge,** forms at its tip (Figure 38–17). Immediately beneath the apical ectodermal ridge is a region of actively dividing mesodermal cells that moves outward with the apical ridge as the wing bud elongates. As this occurs, newly divided cells are left behind and begin to differentiate into the characteristic structures of the wing from its base at the body wall (the proximal end) to its tip (the distal end).

Numerous experiments have shown that the length of time the cells spend near the apical ectodermal ridge before they are left behind is a critical factor in determining the structures into which they differentiate. The cells that are left behind first spend less time under the influence of substances secreted from the apical ectodermal ridge, and they become the proximal structures of the wing. Those that are left behind later spend more time in the region of the apical ectodermal ridge, and they become the distal structures of the wing.

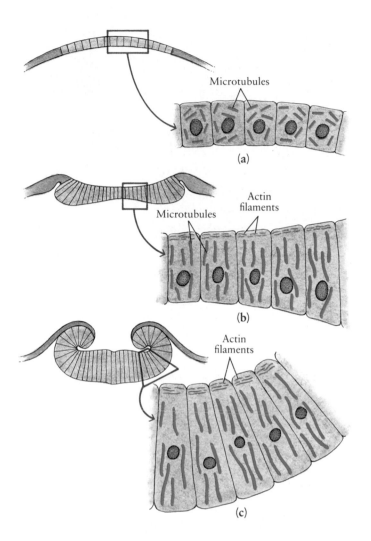

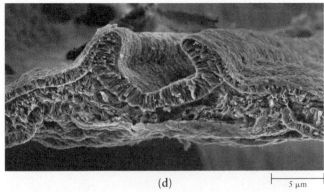

(d)

5 μm

38–16   *Formation of the vertebrate neural tube depends on changes in the shape of individual cells brought about by movements of elements of the cytoskeleton.* **(a)** *Newly differentiated cells of neural ectoderm are cuboidal. Microtubules within the cells appear to be randomly arranged.* **(b)** *As formation of the neural plate proceeds, the microtubules become arranged parallel to the dorsal-ventral axis of the embryo. As the microtubules lengthen through the addition of tubulin molecules (page 101), the cells become columnar. Just below the upper surface of each cell is a band of actin filaments. When this band contracts* **(c)**, *the cells become wedge-shaped, forcing the entire sheet of cells to curve upward. A similar process is involved in the development of other curved or tubular structures that consist of sheets of cells.*

(d) A scanning electron micrograph of the neural groove of a chick embryo. The tissue on either side of the groove is curving upward, forming the two arches that will ultimately meet to form the hollow neural tube.

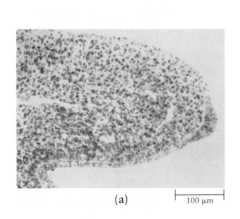

(a)   100 μm

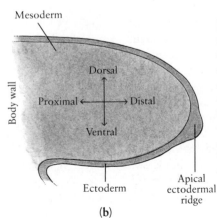

(b)

38–17   **(a)** *The wing bud of a chick embryo, viewed from the posterior. This micrograph was taken three and a half days after the wing bud began its development. Note the thickened region, the apical ectodermal ridge, at its tip.* **(b)** *The tissues of the wing bud consist of an outer layer of ectoderm, from which the skin and feathers will form, and an inner layer of mesoderm, from which all of the bones, cartilage, tendons, and muscles of the wing will develop.*

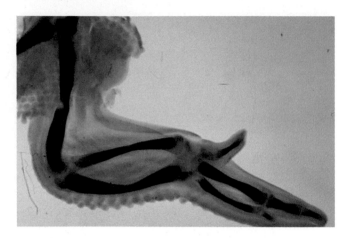

38–18   *A dorsal view of the wing of a chick embryo, 10 days after the wing bud first began its development. In the course of the evolution of the bird wing, two of the five digits of the basic vertebrate "hand" have been lost—the digits corresponding to the thumb and little finger.*

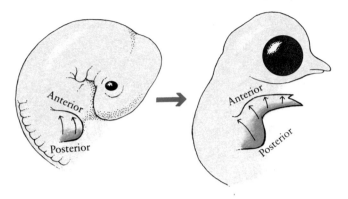

38–19   *Cells at the posterior margin of the chick wing bud release a chemical substance, now known to be retinoic acid, that diffuses from the posterior margin toward the anterior margin. This creates a concentration gradient of retinoic acid across the wing bud. The development of the correct structures on the anterior-to-posterior axis depends on their location within this concentration gradient.*

A wing is characterized not only by a specific series of structures from its proximal end to its distal end but also by a specific series of structures along the axis from anterior to posterior (Figure 38–18). Differentiation along this axis is controlled by chemical substances produced by cells at the posterior margin of the wing. For example, if a piece of tissue is transplanted from the posterior margin of one wing bud to the anterior margin of another, the wing that develops is a kind of Siamese twin in which the bones are duplicated in a mirror image of the normal wing structure.

One substance released from the posterior margin of the wing bud has been identified as **retinoic acid,** a member of the same chemical family as the carotenoids, vitamin A, and retinal, the light-sensitive pigment in the rods and cones of the vertebrate eye. As shown in Figure 38–19, retinoic acid diffuses from the posterior margin toward the anterior margin of the wing bud, creating a concentration gradient from high to low.

Receptors for retinoic acid have now been located on the surface of cells of the developing wing bud. Structurally these receptors are very similar to the receptors for steroid hormones and thyroid hormone, and it appears that their mechanism of action is also similar (see page 617). The particular sequence of cellular events that is initiated by the binding of retinoic acid to its receptors is thought to be influenced by the number of receptors activated. The number of receptors activated depends on the concentration of retinoic acid in the surrounding extracellular fluid—and that, in turn, depends on the location of the cell within the concentration gradient.

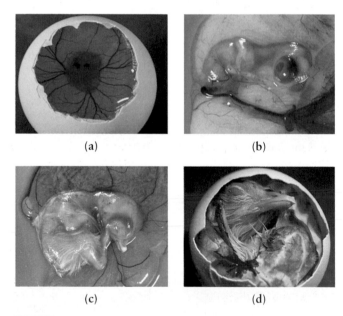

(a)   (b)

(c)   (d)

38–20   *Major stages of organogenesis and morphogenesis in the chick embryo. (a) At 5 days, large blood vessels have grown out from the embryo into the yolk sac, a membrane that encloses the yolk of the egg. Cells of the yolk sac secrete lipase, an enzyme that digests fat molecules in the yolk. The digested nutrients—the embryo's energy supply—are carried into its developing body by the blood vessels. Note the heart, above the junction of the blood vessels with the embryo. (b) By 11 days, the large, pigmented eyes are clearly visible, and the wings and legs are developing. The brain is visible through the transparent skull. (c) By 14 days, feathers are beginning to appear. (d) By 21 days, just prior to hatching, a white knob is fully formed on the end of the beak. This is the egg tooth, which is used by the chick in pecking open the shell.*

Genetic Control of Development: The Homeobox

Underlying all the processes of development are genes—sequences of information encoded by DNA molecules. The substances involved in embryonic induction and in pattern formation exert their effects by causing changes in the expression of specific genes. The first steps in identifying these genes and in determining their functions in development are being made possible by the ubiquitous fruit fly, *Drosophila*.

Drosophila, like all insects, exhibits a striking segmentation of the body, both in the larval stages and in the adult. Moreover, the adult is characterized by a series of complex appendages, including mouthparts, antennae, wings, and legs, precisely located on particular segments. Two distinct sets of genes controlling pattern formation in *Drosophila* have now been identified. One set, known as **segmentation genes,** controls the number and sequence of segments in the larva and, subsequently, in the three regions (head, thorax, and abdomen) of the adult fly. Mutations in these genes can result in larvae and flies in which certain segments are missing, or, alternatively, are duplicated.

Another set of genes, known as **homeotic genes,** controls the identity of segments. That is, these genes control the way in which a given segment develops and the appendages that it bears. Mutations of homeotic genes can result in bizarre disruptions of the normal developmental pattern. For example, mutations in the homeotic genes that control the development of antennae on the head and of legs on the thorax can result in the development of legs on the head—where antennae should have developed.

One of the most thoroughly studied groups of homeotic genes is involved in the control of

(a)

(b)

The effect of mutations in one group of homeotic genes of Drosophila. **(a)** *A normal fly with one pair of wings and one pair of halteres.* **(b)** *A mutant fly with two pairs of wings and no halteres.*

the development of both legs and wings on the three thoracic segments. In a normal fruit fly, the first thoracic segment bears a pair of legs, the second segment bears a pair of legs

and a pair of wings, and the third segment bears a pair of legs and a pair of greatly reduced wings, known as halteres. Although the halteres play no role in the propulsion required for flight, they are critical in enabling the fly to maintain its balance while airborne. Mutations in two loci of the homeotic genes produce flies in which the third thoracic segment develops as a second thoracic segment. The result is a fly with four full-size wings and no halteres.

At present, we know very little about the functions of the proteins coded by the homeotic genes. Some of these proteins undoubtedly regulate the expression of other genes that have more specific roles in development. Others may be involved in the regulation of basic cellular activities, such as cell growth and division.

Many of the homeotic genes have now been located on the *Drosophila* chromosomes and have been cloned and sequenced. In 1983, it was discovered that more than a dozen of the *Drosophila* homeotic genes contain a common DNA sequence of 180 nucleotides. This sequence is known as the **homeobox.** The nucleotide sequence of the homeobox has now been identified in the DNA of a variety of animals, ranging from worms to humans, all of which exhibit a segmental pattern at some stage of their development. The polypeptide strand dictated by the homeobox is hypothesized to function as a regulatory molecule that binds to DNA, altering the course of gene expression.

These discoveries suggest that even though the adult forms of different animals vary greatly, certain "master" genes play a critical role in the development of body organization and pattern throughout the animal kingdom.

The identification of retinoic acid as a signalling molecule in wing development and the isolation and characterization of its receptor suggest that the integration and control of development are, in principle, similar to the integration and control of other physiological processes. Developmental biologists are now seeking to identify other substances involved in the regulation of development, the mechanisms by which these substances turn genes on and off, the genes themselves, and the proteins for which they code.

Development of the Human Embryo

The patterns of embryonic development are remarkably similar throughout the animal kingdom, and particularly among vertebrates. Thus, studies of the development of sea urchins, frogs, chicks, and mammals such as mice have provided a solid foundation for our understanding of human development.

Human oocytes, as we saw in the preceding chapter, mature in the ovary and are released at approximately 28-day intervals. Fertilization (Figure 38–21) usually takes place in an oviduct. As in other species, it results in (1) changes in the outer surface of the egg that prevent the entry of other sperm cells, (2) metabolic activation of the egg, (3) introduction of the genetic material of the father, and (4) cleavage.

After fertilization, the egg continues its passage down the oviduct, where the first cell divisions take place (Figure 38–22). At about 36 hours after fertilization, the fertilized egg divides to form two cells. At 60 hours, the two cells divide to form four cells. At three days, the four cells divide to form eight. In this early stage, all of the cells are of equal size, as they are in the sea urchin. During this period, the embryo is entirely on its own, following its own genetic program and supplying its own metabolic requirements. All it needs is a suitable fluid environment containing progesterone.

By about five days after fertilization, the blastula consists of some 120 cells, and a fluid-filled blastocoel

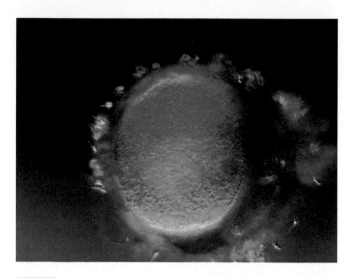

38–21   *A human egg surrounded by sperm cells. Of the 300 to 400 million sperm cells released in an ejaculation, only one can fertilize each egg cell.*

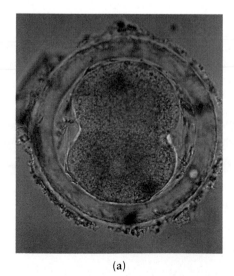

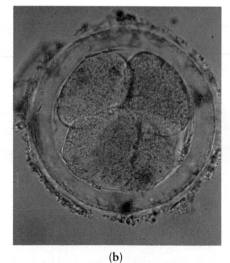

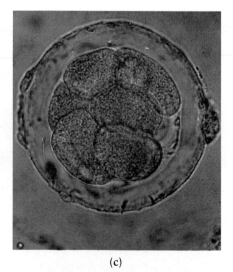

(a)   (b)   (c)

38–22   *A human embryo at (a) the two-cell stage, (b) the four-cell stage, and (c) the eight-cell stage. Although the number of cells has doubled with each cleavage, the volume of the embryo has not increased. Thus, the cells are still easily contained within the fertilization membrane. Occasionally an embryo separates into two halves during these early stages, resulting in identical twins.*

has begun to form within it. The structure of the mammalian blastula, which is known as a **blastocyst,** is rather different from that of the blastulas we have examined previously. A cross section of the blastocyst looks somewhat like a ring with the stone on the inside. The inner cell mass (the stone) is a ball of cells at one pole of the blastula that will develop into the embryo proper. The ring is called the **trophoblast** (from the Greek word *trophe,* "to nourish"). It is composed of a single layer of cells and completely encloses the developing embryo.

Implantation

About three days after the embryo reaches the uterus (about six days after fertilization), the trophoblast makes contact with the tissues of the uterus and releases a hormone known as **chorionic gonadotropin.** This hormone protects the pregnancy by stimulating the corpus luteum to continue its production of estrogens and progesterone, thus preventing menstruation (see page 675). Most pregnancy tests involve the detection of chorionic gonadotropin in blood or urine.

The trophoblast cells, multiplying rapidly by now, chemically induce changes in the endometrium and invade it. As the embryo penetrates the endometrial tissues—the process known as **implantation** (Figure 38–23)—it becomes surrounded by ruptured blood vessels and the nutrient-filled blood escaping from them. The trophoblast thickens and develops fingerlike projections that invade the uterine lining.

Extraembryonic Membranes

As we saw in Chapter 27 (page 485), one of the most significant evolutionary steps among the vertebrates was the development of the amniote egg, which contains its own water supply and so can be laid on land. As shown in Figure 38–24, the developing embryo within the amniote egg produces a series of four **extraembryonic membranes**—the **yolk sac,** the **allantois,** the **amnion,** and the **chorion.**

Although only a few mammals lay eggs (these are the monotremes, such as the duckbilled platypus), the mammalian embryo develops within a system of extraembryonic membranes that closely resemble those found in the amniote egg. Even the yolk sac is still present, enclosing a space where the yolk "used to be." Among humans and other placental mammals, the outermost membrane, the chorion, has become specialized to form part of the placenta, through which the embryo receives its nourishment.

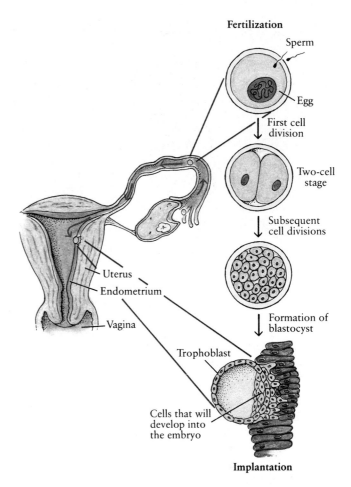

38–23 *Stages in the development of the human embryo between fertilization and implantation in the endometrial lining of the uterus. The tiny embryo, which has reached the blastocyst stage, invades the endometrium within a week after fertilization—some three to four days after it reaches the uterus. Following implantation, the placenta begins to form.*

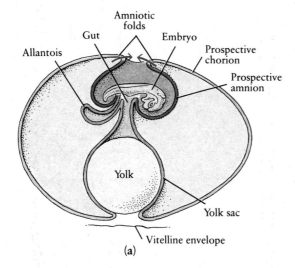

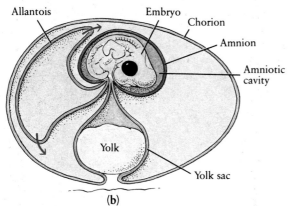

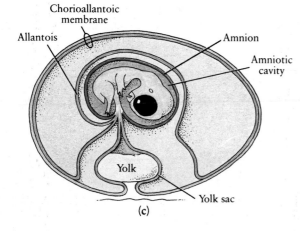

38–24 *Development of the extraembryonic membranes of the chick. In the course of gastrulation, the embryo begins to lift from the surface of the yolk. As this occurs, a series of membranes begin to form from the tissues at its base. (a) One membrane, the yolk sac, grows around and almost completely encloses the yolk. A second, the allantois, arises as an outgrowth of the rear of the gut. The third and fourth are elevated over the embryo by a folding process during which the membrane is doubled. When the folds fuse, two separate membranes are formed (b). The inner one is the amnion, and the outer one is the chorion. The chorion eventually fuses with the allantois to form the chorioallantoic membrane (c), which, in the later stages of development, encloses embryo, yolk, and all the other structures.*

A similar system of extraembryonic membranes envelopes the early human embryo. The chorion forms the embryonic component of the placenta, and the stalk of the allantois gives rise to the umbilical cord.

In **amniocentesis** (page 202), one of the two prenatal genetic testing procedures in current use, fluid samples are taken from the amniotic cavity. These samples contain cells that have been sloughed off by the embryo. The major drawback of amniocentesis is that it usually cannot be performed until the fourteenth week of pregnancy. A newer procedure, **chorionic villus biopsy,** can be performed as early as the eighth week of pregnancy. In this procedure, a small sample of tissue is removed from the chorion, which is genetically identical to the embryo itself.

The Placenta

The **placenta** is a disc-shaped mass of spongy tissue through which all exchanges between mother and embryo take place. By the end of the third week after conception, it covers 20 percent of the uterus.

The placenta is formed as a result of the interactions of a maternal tissue, the endometrium, with the extraembryonic chorion, and has a rich blood supply from both. However, the embryonic and maternal circulatory systems are not directly connected (Figure 38–25), so maternal and embryonic blood cells do not mix.

Molecules, including food and oxygen, diffuse from

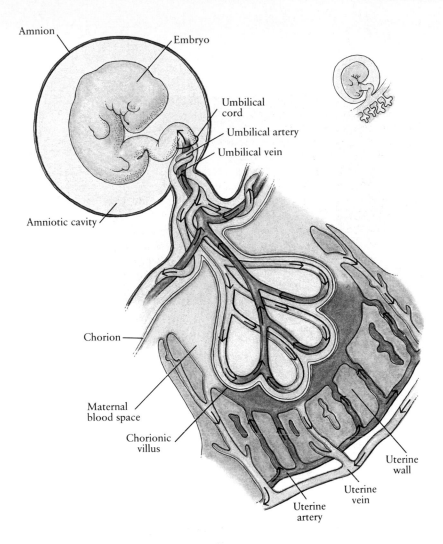

Amnion
Embryo
Umbilical cord
Umbilical artery
Umbilical vein
Amniotic cavity
Chorion
Maternal blood space
Chorionic villus
Uterine artery
Uterine vein
Uterine wall

38–25 *From the placenta, numerous fingerlike chorionic villi project into the maternal blood space in the wall of the uterus. This space is kept charged with blood from branches of the uterine artery. Across the thin barrier separating fetal from maternal blood, the exchange of materials takes place. Soluble food substances, oxygen, water, and salts diffuse into the umbilical vein from the mother's blood. Carbon dioxide and nitrogenous waste, brought to the placenta in the umbilical arteries, diffuse into the mother's blood. The placenta is thus the excretory organ of the embryo as well as its respiratory surface and its source of nourishment.*

The embryo illustrated here is 37 days old. Its actual size at this stage of development is shown in the figure at the upper right.

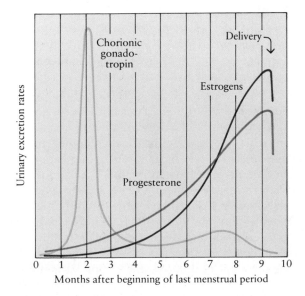

Chorionic gonadotropin

Delivery

Estrogens

Progesterone

Urinary excretion rates

0 1 2 3 4 5 6 7 8 9 10
Months after beginning of last menstrual period

38–26 *Levels of estrogens, progesterone, and chorionic gonadotropin excreted in the urine during pregnancy. Urinary excretion rates reflect the concentrations of these hormones in the blood.*

the maternal bloodstream through the placental tissue and into the blood vessels that carry them into the embryo. Similarly, carbon dioxide and other waste products from the embryo are picked up from the placenta by the maternal bloodstream and carried away for disposal through the mother's lungs and kidneys. The mother's blood volume begins to increase in response to these extra demands. Her appetite increases and the absorption of certain nutrients, such as calcium and iron, is increased.

From this point in development until birth, the embryo remains securely attached to the placenta by the **umbilical cord,** which develops from the stalk of the extraembryonic allantois. The umbilical cord permits the embryo to float freely in its sac of amniotic fluid.

As the placenta matures, it begins producing large amounts of estrogens and progesterone (Figure 38–26). By the end of the third month of pregnancy, the placenta has completely replaced the corpus luteum as a source of estrogens and progesterone. Chorionic gonadotropin is no longer produced in such large amounts, and the corpus luteum degenerates. It is hypothesized that many of the miscarriages that occur in the third month result from the degeneration of the corpus luteum before the placenta is producing adequate levels of hormones to maintain the pregnancy.

(a)

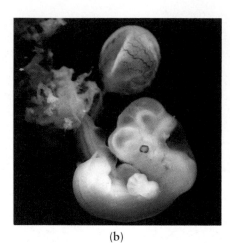

(b)

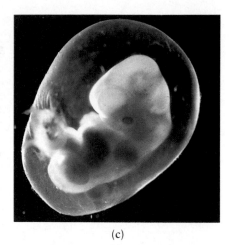

(c)

38–27   *Human embryos, early in their development. (a) In this four-week-old embryo, the neural groove is still open. The protuberances near the top are the limb buds of the future arms. The limb buds of the legs, which develop more slowly, are visible at the bottom. This embryo is 5 millimeters long. (b) By five weeks, the embryo is 10 millimeters in length. Its developing brain, an eye, hands, and a long tail can be seen.*

The yolk sac is at the top. (c) At six weeks, the embryo has grown to 15 millimeters in length. As it floats securely in the amniotic cavity, its heart beats rapidly. The brain continues to grow, and the eyes are more developed. The dark red object in the abdominal region is the liver. Skin folds have formed that will give rise to the outer ears.

The First Trimester

As implantation occurs, the extraembryonic membranes form, and the placenta becomes established, the embryo proper continues its development. In its second week of life, the embryo grows to 1.5 millimeters in length, and its major body axis begins to develop. (In this and subsequent measurements, the embryo is measured from crown to rump.) As the embryo elongates, the cell migrations of gastrulation establish its three-layered body plan.

During the third week, the embryo grows to 2.3 millimeters long, and most of its major organ systems begin to form. The neural groove is the beginning of the central nervous system, which is the first organ system to develop. By 22 days, the very rudimentary heart, still only a tube, begins to flutter and then to pulsate. From this time on, the heart will not stop its 100,000 beats per day until the death of the individual. Soon after, the eyes begin to form. Also, by this time, about 100 cells have been set aside in the yolk sac as **germ cells,** from which the egg or sperm cells of the new individual will eventually develop. These cells begin to crawl, amoeba-like, toward the site at which the gonads will develop.

By the end of the first month, the embryo is 5 millimeters in length and has increased its mass 7,000 times.

The neural groove has closed, and the embryo is now C-shaped. At this stage, it can be clearly seen that the tissues lateral to the notochord are arranged in paired somites. The embryo typically has 40 pairs of somites, from which bones, other connective tissues, and muscles will develop. The heart, even as it beats, develops from a simple set of paired, contracting tubes into a four-chambered vessel.

By 38 days, the germ cells reach their destination, the developing gonads. Although the rudimentary gonads have begun to form by this time, male and female embryos are still anatomically identical. Whether an embryo develops as a male or as a female appears to be determined by a gene located on the Y chromosome. This gene is thought to function as the master switch of sexual development, turning on other genes that lead to the differentiation of the primitive gonads into testes and of the germ cells into spermatogonia. In its absence, the gonads develop as ovaries and the germ cells become oogonia. In addition to the action of a master gene, interactions between the germ cells and the primitive gonads are necessary for development of the testes to proceed. As the testes form, they begin their secretion of androgens, and under the influence of androgens, the external genitalia and other structures become masculinized. If a female embryo is exposed to androgens, it will become similarly masculinized, although, gonadally speaking, it will be female.

During the second month, the embryo increases in mass about 500 times. By the end of this period, it weighs about 1 gram, slightly less than an aspirin tablet, and is about 3 centimeters long. Despite its small size, it is almost human-looking, and from this time on it is generally referred to as a **fetus.** Because of the early and rapid development of the brain, the head is large in relation to the rest of the body. As the growth rate of other parts of the body increases throughout the remainder of gestation (and throughout childhood as well), the apparent discrepancy in the size of the head will disappear.

Arms, legs, elbows, knees, fingers, and toes are all forming during this time (Figure 38–28). A temporary tail—another reminder of our ancestry—reaches its greatest length in the second month. Then, as its growth rate slows, it becomes surrounded by other tissues and disappears. The gallbladder and pancreas are present at this stage, and there is clear differentiation of the divisions of the digestive tract. The liver now constitutes about 10 percent of the body of the fetus and is its main blood-forming organ. By the end of the second month, the major steps in organ development are more or less complete. In fact, the rest of development is mostly concerned with growth and the maturation of physiological processes.

The first two months, when the major organ systems are forming, are the most sensitive period in human development with regard to the influence of external factors. For example, when the arms and legs are mere rudiments (fourth and fifth weeks), a number of substances can upset the normal course of events and result in limb abnormalities. Many drugs have been found to cause birth defects, and pregnant women and their doctors have become far more cautious about the use of any medication during these critical first months. Similarly, exposure to x-rays at doses that would not affect an adult or even an older fetus may produce permanent abnormalities. Alcohol consumption and smoking also affect the development of organ systems.

Infections may also affect the development of the embryo. Rubella (German measles) is a very mild disease in children and adults. Yet when contracted by the mother during the fourth through the twelfth weeks of pregnancy, it can have damaging effects on the formation of the heart, the lens of the eye, the inner ear, and the brain, depending on exactly when the infection occurs in relation to the events of development.

During the third month, the fetus begins to move its arms and kick its legs, and the mother may become aware of its movements. Reflexes, such as the startle reflex and (by the end of the third month) sucking, first appear at this time. Its face becomes expressive; the fetus can squint, frown, or look surprised. Its respiratory organs are fairly well formed by this time but are not yet functional.

By the end of the third month, the fetus is about 9 centimeters long from the top of its head to its buttocks and weighs about 15 grams (0.5 ounce). It can suck and swallow, and occasionally does swallow some of the fluid that surrounds it in the amniotic sac. The finger, palm, and toe prints are now so well developed that they can be clearly distinguished by ordinary fingerprinting methods. The kidneys and other structures of

38–28 *The development of the human hand.* (a) *In the five-week-old embryo, the future hand is a simple paddle-like structure. As development continues, the growing fingers extend outward, while a programmed process of cell death eliminates the tissues between the fingers.* (b) *By the time the embryo (now a fetus) is 11 weeks old, the hand is fully formed. The nail beds are in place, the joints of the fingers are visible, and the muscles that move the fingers and the hand have begun to function.*

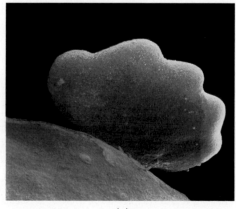

(a)

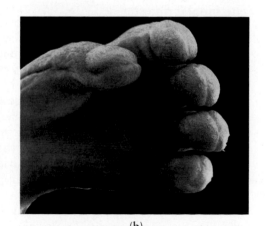

(b)

the urinary system develop rapidly, although waste products are still disposed of through the placenta. By the end of this period—the first trimester of development—all the major organ systems have been laid down.

The Second Trimester

During the fourth month, movements of the fetus become obvious to the mother. Its bony skeleton is forming and can be seen with x-rays. The body is becoming covered with a protective cheesy coating. The four-month-old fetus is about 14 centimeters long and weighs about 115 grams (4 ounces).

By the end of the fifth month, the placenta covers about 50 percent of the uterus. The fetus has grown to almost 20 centimeters and now weighs 250 grams. It has acquired hair on its head, and its body is covered with a fuzzy, soft hair called the lanugo, from the Latin word for "down." Its heart, which beats between 120 and 160 times per minute, can be heard with a stethoscope. The five-month-old fetus is already discarding some of its cells and replacing them with new ones, a process that will continue throughout its lifetime. Nevertheless, a five-month-old fetus cannot yet survive outside the uterus.

During the sixth month, the fetus has a sitting height of 30 to 36 centimeters and weighs about 680 grams. By the end of the sixth month, it could survive outside the mother's body, although probably only with respiratory assistance in an incubator. Its skin is red and wrinkled, and the cheesy body covering, which helps protect the fetus against abrasions, is now abundant. Reflexes are more vigorous. In the intestines is a pasty green mass of dead cells and bile, known as meconium, which will remain there until after birth.

The Final Trimester

During the final trimester, the fetus increases greatly in size and weight. In fact, it normally doubles in size just during the last two months. During this period, many nerve tracts are forming, and the number of brain cells is increasing at a very rapid rate. By the seventh month, brain waves can be recorded, through the abdomen of the mother, from the cerebral cortex of the fetus. Numerous studies have demonstrated that the protein intake of the mother is of critical importance during this period if the child is to have full development of its brain.

As the fetal period progresses, the physiology of the fetus becomes increasingly like the physiology of the adult, and so agents that affect the mother also threaten

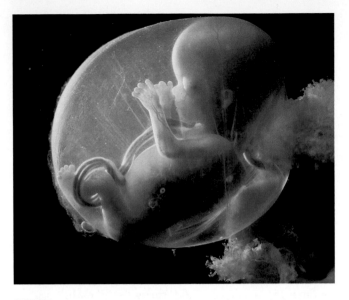

38–29   *Although the head of a 12-week-old fetus is disproportionately large, its features are clearly human. Its fingers and toes are fully developed, and outer ears and eyelids have formed. The lids are fused and will remain closed for the next three months. The umbilical cord, connecting the fetus to the placenta, contains a vein and two arteries.*

38–30   *A fetus at 18 weeks of age. When the thumb comes close to the mouth, the head turns, and the lips and tongue begin sucking motions—an important reflex for survival after birth.*

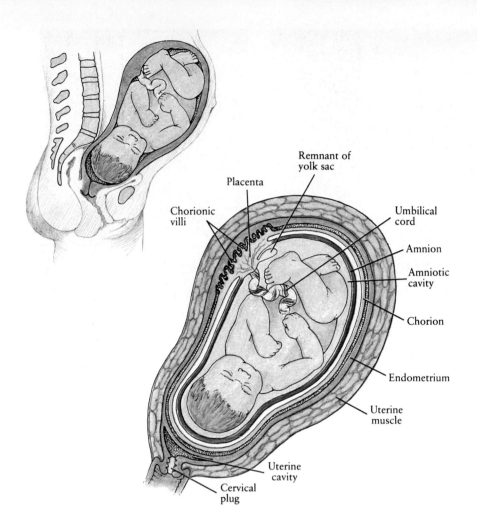

Remnant of
yolk sac

Placenta

Chorionic
villi

Umbilical
cord

Amnion

Amniotic
cavity

Chorion

Endometrium

Uterine
muscle

Uterine
cavity

Cervical
plug

38–31 *A human fetus, shortly before birth, showing the protective membranes and uterine tissues surrounding it. The cervical plug is composed largely of mucus. It develops under the influence of progesterone and serves to exclude bacteria and other infectious agents from the uterus. In 95 percent of all births, the fetus is in this head-down position.*

the maturing fetus. A depressingly familiar example is found in the infants born with cocaine, heroin, or methadone addiction.

During the last month of pregnancy, the fetus usually begins to acquire antibodies from its mother, a process that continues after birth through the mother's milk. Antibodies, which are far too large to diffuse across the placenta, are conveyed by highly selective active transport. The immunity they confer is only temporary. Within one to two months after birth, the maternal antibodies will be gradually replaced by antibodies manufactured by the baby's own immune system.

During this last month, the growth rate of the baby begins to slow down. (If it continued at the same rate as in the preceding months of development, the child would weigh 90 kilograms—about 200 pounds—by its first birthday.) The placenta begins to regress and becomes tough and fibrous.

Weight at birth is the major factor in infant mortality. Infants weighing less than 2,000 grams (4 pounds, 6 ounces) at birth are at high risk of death or severe brain damage. Infants weighing less than 2,500 grams (5 pounds, 8 ounces) are considered to be low weight, and they are 40 times as likely to die within a month of

birth as heavier infants. Two-thirds of all babies who die are low weight. The principal causes of low birth weight are poor maternal nutrition, smoking, drug use, and a lack of prenatal care. Good prenatal care and counseling—which could lead to heavier and healthier babies for the population at greatest risk, unmarried teenage mothers—cost significantly less per pregnancy than the cost per day of hospitalization of a low-weight infant in an intensive care nursery. From the 1950s until the early 1980s, the infant mortality rate in the United States declined steadily, but it has now stabilized at 8.6 deaths for each 1,000 live births, placing the United States twenty-first in infant mortality worldwide. In first place is Japan, with only 4.4 deaths for each 1,000 births.

Birth

The date of birth is calculated as about 266 days after conception or 280 days after the beginning of the last menstrual period. Babies are rarely born on the scheduled day, but some 75 percent are born within two weeks of that day.

Labor is divided into three stages: **dilation, expulsion,** and **placental** stages. Dilation, which usually lasts

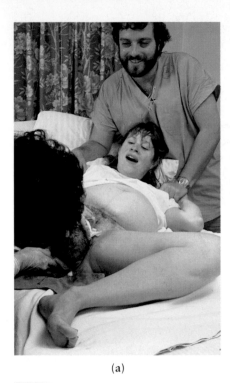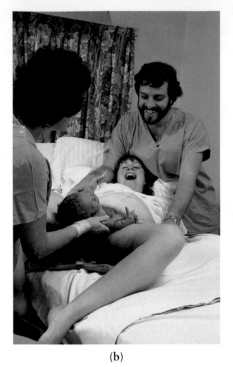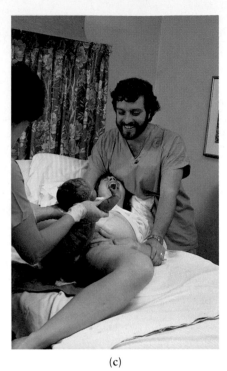

(a)   (b)   (c)

38–32   *The birth of a baby. When the umbilical cord—the baby's lifeline throughout its development—is severed, the infant will begin its separate existence.*

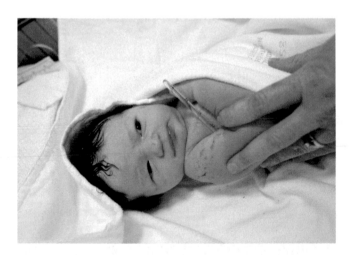

38–33   *Within five to six minutes after vaginal delivery, the human infant is alert and its pupils are dilated, even in the presence of bright lights. This alertness, which may play a role in the bonding of infant and mother in the first hour of life, is the result of a surge of adrenaline and noradrenaline similar to that in "fight-or-flight" reactions (page 621). Birth puts great stress on an infant, as it is pushed and shoved by the uterine contractions, and temporary shortages of oxygen may occur as the blood vessels of the umbilical cord are pinched closed. A series of reflexes alter the patterns of blood distribution, protecting vital organs, particularly the brain, heart, and lungs, which must begin functioning immediately after birth.*

from 2 to 16 hours (it is longer with the first baby than with subsequent births), begins with the onset of contractions of the uterus. It ends with the full dilation, or opening, of the cervix. At the beginning, uterine contractions occur at intervals of about 15 to 20 minutes and are relatively mild. By the end of the dilation stage, contractions are stronger and occur about every 1 to 2 minutes. At this point, the opening of the cervix is about 10 centimeters in diameter. Rupture of the amniotic sac, with the expulsion of fluids, usually occurs during this stage.

The second, or expulsion, stage lasts 2 to 60 minutes. It begins with the full dilation of the cervix and the appearance of the head in the cervix, called crowning. Contractions at this stage last from 50 to 90 seconds and are 1 or 2 minutes apart.

The third, or placental, stage begins immediately after the baby is born. It involves contractions of the uterus and the expelling of fluid, blood, and finally the placenta with the umbilical cord attached (also called the afterbirth). The placenta now weighs about 500 grams, about one-sixth of the weight of the infant. Minor uterine contractions continue. They help to stop the flow of blood and to return the uterus to its prepregnancy size and condition.

The baby emerges from the warm, protective enclosure in which it has been nourished and permitted to grow for nine months. The umbilical cord—until that moment, its lifeline—is severed. The baby cries as it takes its first breath, starts to breathe regularly, and so begins its independent existence.

Epilogue

When did this particular human life begin? When the sperm encountered the egg? When the embryo became a fetus, visibly human? At the first heartbeat? The first brain wave? When the infant became viable as an independent entity? In the past, these matters were discussed by philosophers and theologians who were concerned with the question of when the soul enters the body. These issues have been revived in the ethical and legal controversies concerning abortion. In the evolutionary sense, however, none of these events marks the beginning of life.

Life began more than 3 billion years ago and has been passed on since that time from organism to organism, generation after generation, to the present, and stretches on into the future, farther than the mind's eye can see. Each new organism is thus a temporary participant in the continuum of life. So is each sperm, each egg, indeed, in a sense, each living cell. Every individual, however, is a unique blend of heredity and experience, never to be duplicated and therefore irreplaceable; but from the perspective of the biological continuum, a human life lasts no longer than the blink of an eye.

Summary

In the process of development, a fertilized egg becomes a complete organism, consisting of hundreds to trillions of cells and closely resembling its parent organisms. This process involves differentiation, morphogenesis, and growth.

Development in most animals begins with fertilization—the fusion of egg and sperm. Fertilization results in (1) changes in the protective membranes of the egg that prevent the entry of other sperm cells, (2) metabolic activation of the egg, (3) introduction of the genetic material of the father into the egg, and (4) cell division.

Development takes place in three stages: cleavage, gastrulation, and organogenesis. During cleavage, the original egg cell divides, with very slight, if any, change in overall volume. When cleavage is complete the embryo consists of a cluster of cells, the blastula, with a central cavity, the blastocoel.

Gastrulation involves the movement of cells into new relative positions and results in the establishment of three embryonic tissue layers: endoderm, mesoderm, and ectoderm. At the beginning of gastrulation, an opening forms, the blastopore. Cells from the outer surface of the embryo migrate through the blastopore into the blastocoel and form a new cavity, the archenteron, which will be the primitive gut.

Each of the three primary tissue layers established during gastrulation gives rise to particular tissues and organs. The epidermis and the nervous system arise from ectoderm. The notochord, muscles, skeleton, heart and blood vessels, excretory and reproductive organs, the inner lining of the skin, and the outer lining of the internal organs arise from mesoderm. Lungs, glands, and the inner lining of the digestive tract develop from endoderm.

Organ systems differentiate and develop through a process of embryonic induction, in which one tissue induces changes in the growth or migration of an adjacent tissue with which it comes in contact. The primary inducer is the chordamesoderm, a sheet of mesoderm that lies above the roof of the archenteron and gives rise to the notochord. The chordamesoderm induces the differentiation of the overlying ectoderm that gives rise to the neural tube and sets in motion the subsequent events of development. Induction involves the exchange of chemical substances between the interacting tissue layers.

Morphogenesis, the shaping of body form, results from a few cellular processes repeated over and over. Among these processes are (1) increases or decreases in the rates of cell growth and division, (2) the migration of cells from one location to another, (3) the positioning of cells by changes in their adhesion to neighboring cells, and (4) changes in cell shape brought about by extension or contraction. Pattern formation, the developmental process that gives rise to such complex structures as the vertebrate forelimb, appears to be controlled by chemical substances exchanged between interacting cells and groups of cells. One substance known to be important in pattern formation is retinoic acid.

The earliest stages in the development of the human embryo take place in an oviduct. There is a large increase in the number of cells but little or no increase in the total size of the embryo. At the blastula stage, the embryo of humans and other mammals is known as a blastocyst. It consists of an inner cell mass and an outer layer of cells, the trophoblast. When the embryo descends into the uterus, at about the sixth day of development in humans, the trophoblast develops rapidly and invades the endometrium, in the process known as implantation.

As it develops, the human embryo produces four extraembryonic membranes—amnion, yolk sac, chorion, and allantois—similar to the extraembryonic membranes in the amniote egg of reptiles and birds. The amnion encloses the embryo in a fluid-filled cavity, and the yolk sac is the site in which germs cells are sequestered prior to their migration to the developing gonads. The chorion, which develops from the trophoblast, interacts with tissues of the endometrium to form the placenta, through which the embryo receives its food and oxygen and excretes carbon dioxide and other waste products. The allantois develops into the umbilical cord, which attaches the embryo to the placenta.

When the human embryo is about 2 weeks old, gastrulation occurs, followed by the development of the neural groove, which folds to form the neural tube. Although the embryo is still very small (about 2.5 millimeters long), most of the major organs begin to form in these early weeks, which is why damage to the embryo during this period, caused by viral infection, x-rays, or drugs, can be widespread. By the end of the second month, the embryo, now called a fetus, is almost human-looking, although it weighs only about 1 gram. By the end of the third month, all of the organ systems have been laid down. During the second trimester, development of the organ systems continues, and during the final trimester, there is a great increase in size and weight. Birth occurs, on the average, 266 days after fertilization.

Questions

1. Distinguish among the following: blastula/gastrula; blastocoel/archenteron; ectoderm/mesoderm/endoderm; chordamesoderm/lateral plate mesoderm; fertilization/implantation; extraembryonic membranes/placenta; embryo/fetus; dilation/expulsion/placental stage.

2. Describe, in general terms, the end results of each of the following events: fertilization, cleavage, gastrulation, organogenesis.

3. Describe the similarities and differences in the blastulas of a sea urchin, an amphibian, a bird, and a mammal.

4. Follow the course of a single cell from its place of origin in the fertilized egg of a frog to its position in the eye cup of an early frog embryo. Include, for each stage, the influences on this cell that might affect its history.

5. In the vertebrate eye, the lens is positioned in perfect alignment with the retina at the back of the eyeball and with the transparent cornea overlying the lens at the front of the eyeball. How do the developmental events in the formation of the eye ensure such alignment?

6. Follow the course of a single cell from its place of origin in the fertilized egg of a chick to its position in the third digit of the developing wing. Include, for each stage, the influences on this cell that might affect its history.

7. One of the most effective medications for the treatment of acne is chemically very similar to retinoic acid. Physicians strongly recommend that women who are trying to become pregnant or who have reason to believe they may be pregnant discontinue the use of this medication immediately. Why?

8. In mammals, the fetus is basically feminine and fetal androgens are necessary to produce male characteristics. In birds, however, the fetus is basically masculine and fetal estrogens are necessary to produce female characteristics. Why would the arrangement in birds be unworkable in mammals?

9. If you look closely at Figure 38–25, you will notice that the umbilical arteries carry deoxygenated blood and the umbilical vein carries oxygenated blood. Why is this the case?

10. Although the placenta allows the diffusion of substances from the bloodstream of the mother to the bloodstream of the fetus and vice versa, it does not permit blood cells of either mother or fetus to cross into the bloodstream of the other. On the basis of your knowledge of the immune system, explain why this placental barrier is essential for the well-being of both mother and fetus.

Suggestions for Further Reading

Aoki, Chiye, and Philip Siekevitz: "Plasticity in Brain Development," *Scientific American,* December 1988, pages 56–64.

Beaconsfield, Peter, George Birdwood, and Rebecca Beaconsfield: "The Placenta," *Scientific American,* August 1980, pages 95–102.

Browder, L. W., Carol Erickson, and William Jeffrey: *Developmental Biology,* 3d ed., Saunders College Publishing, Philadelphia, 1991.

>   *A general, comprehensive coverage of basic embryology and modern developmental biology.*

De Robertis, Eddy M., Guillermo Oliver, and Christopher V. E. Wright: "Homeobox Genes and the Vertebrate Body Plan," *Scientific American,* July 1990, pages 46–52.

Edelman, Gerald M.: "Cell-Adhesion Molecules: A Molecular Basis for Animal Form," *Scientific American,* April 1984, pages 118–129.

Gehring, Walter J.: "The Molecular Basis of Development," *Scientific American,* October 1985, pages 152B–162.

Gilbert, Scott F.: *Developmental Biology,* 3d ed., Sinauer Associates, Inc., Sunderland, Mass., 1991.

>   *A comprehensive, up-to-date textbook of developmental biology, with clear explanations and thorough coverage.*

Hall, Brian K.: "The Embryonic Development of Bone," *American Scientist,* vol. 76, pages 174–181, 1988.

Kalil, Ronald E.: "Synapse Formation in the Developing Brain," *Scientific American,* December 1989, pages 76–85.

Lagercrantz, Hugo, and Theodore A. Slotkin: "The 'Stress' of Being Born," *Scientific American,* April 1986, pages 100–107.

Miller, C. Arden: "Infant Mortality in the U.S.," *Scientific American,* July 1985, pages 31–37.

Nilsson, Lennart, Axel Ingleman-Sundberg, and Claes Wirsén: *A Child Is Born: The Drama of Life before Birth,* Dell Publishing, Inc., New York, 1986.[*]

>   *This is an account of the history of life before birth. The book describes in detail the development of the unborn child from the moment of fertilization and also the changes in the mother during pregnancy. There are magnificent color photographs of the developing fetus.*

Rodger, John C., and Belinda L. Drake: "The Enigma of the Fetal Graft," *American Scientist,* vol. 75, pages 51–57, 1987.

Walbot, V., and N. Holder: *Developmental Biology,* McGraw-Hill Publishing Company, New York, 1987.

>   *A modern molecular and genetic approach to developmental biology.*

Wassarman, Paul M.: "Fertilization in Mammals," *Scientific American,* December 1988, pages 78–84.

Wolpert, Lewis: "Pattern Formation in Biological Development," *Scientific American,* October 1978, pages 154–164.

[*]Available in paperback.

Biology of Plants

Animals seek food and water, find shelter from the cold, vanquish their rivals, compete for mates, and disperse their young to new territories. Plants do all these same things, but, being immobile, they do them differently—and often with what would be considered in the animal world great ingenuity.

Plants compensate for their lack of mobility by enlisting the services of other agents. This is particularly conspicuous among the angiosperms, the plants that dominate this planet and are the subject of the chapters that follow. The servants of angiosperms are the wind, all manner of insects, bats, birds, rodents, and—relative newcomers on the scene—ourselves and our clothing and our vehicles. Plants employ a variety of strategies, including, as you saw in Chapter 25, bribery, deceit (sometimes), and flashy advertising.

Fruits are conspicuous examples of bribery. Animals, such as the dwarf epauleted bat seen here picking a fig, eat the fleshy portions of fruits and spit out the seeds or, sometimes, eliminate them in their feces. In the latter case, the seed has the advantage of some accompanying fertilizer and also the removal, by the animal's digestive enzymes, of the seed's outer coating. Some seeds can't germinate without this assistance.

When the seeds are ready to go, the plants advertise. Unripe fruits are usually green, almost always inconspicuous, and often have a strong acidic taste. The changes in color that accompany ripening are the plant's announcement that the fruit is ready to be eaten.

As you can see here, plants produce fruits—and the seeds they contain—in great profusion. Yet despite this seeming overproduction and the enthusiastic cooperation of animals, in a plant's whole lifetime only one of its seeds, on the average, will grow to maturity and leave descendants. If it were otherwise, the planet would be covered with, for instance, fig trees.

The Flowering Plants: An Introduction

For most of Earth's history, the land was bare. A billion years ago, algae clung to the shores at low tide and may even have begun to cover moist surfaces farther inland. But, had anyone been there to observe it, the Earth's surface would generally have appeared almost as barren and forbidding as the bleak Martian landscape. According to the fossil record, plants began to invade the land a mere half billion years ago. Not until then did the Earth's surface truly come to life. As a film of green spread from the edges of the waters, other forms of life—heterotrophs—were able to follow. The shapes of these new forms and the ways in which they lived were determined by the plant life that preceded them. Plants supplied not only their food—their chemical energy—but also their nesting, hiding, stalking, and breeding places.

And so it is today. In almost all terrestrial communities except those created by human activities, the character of the plants still determines the character of the animals and other forms of life that inhabit a particular area. Even we members of the human species, who have seemingly freed ourselves from the life of the land and even, on occasion, from the surface of the Earth, are still dependent on the photosynthetic events that take place in the green leaves of plants.

In Chapter 25, we traced the evolution of the plants. If you have not already read pages 440–451, we recommend that you do so now. In this section, we shall focus on the group of plants that evolved most recently, the **angiosperms** (division Anthophyta). The angiosperms are, by far, the most abundant and diverse plants on Earth today, with about 235,000 living species. The group is divided into two large classes, Dicotyledones—the **dicots**—with about 170,000 species, and Monocotyledones—the **monocots**—with about 65,000 species.

Angiosperms are characterized by specialized reproductive structures—flowers—in which sexual reproduction occurs, in which the seeds are formed, and from which the fruits develop. Unlike the reproductive organs of animals, which are permanent structures that develop in the embryo, flowers are transitory, developing seasonally. After fertilization, some parts of the flower become the fruit, protecting and enclosing the seed or seeds; other parts die and are discarded.

Sexual Reproduction: The Flower

Most flowers consist of four sets of floral parts (Figure 39–1). Each floral part is thought to be, evolutionarily speaking, a modified leaf. The floral parts may be arranged spirally on a more or less elongated stalk, or similar parts—such as the petals—may be located at one level in a whorl.

The outermost parts of the flower are the **sepals,** which are usually green and leaflike. The sepals, known collectively as the **calyx,** enclose and protect the other parts of the developing flower bud. Next are the **petals,** which are collectively known as the **corolla.** Petals may also be leaflike, but they are often brightly colored. They advertise the presence of the flower among the green leaves, attracting insects or other animals that visit flowers for their nectar or for other edible substances. As these animals forage for food, they may carry pollen from flower to flower (see Figure 25–19, page 449).

Within the corolla are the **stamens.** Each stamen consists of a single elongated stalk, the **filament,** and at the end of the filament, the **anther.** The pollen grains, formed within the anther, are the immature male gametophytes. (For a review of the phenomenon of alternation of generations, see page 190.) When ripe, the pollen grains are released, often in large numbers, through slits or pores in the anther.

The centermost parts of the flower are the **carpels,** which contain the female gametophytes. A single flower may have one carpel or several carpels, which may be separate or fused together. Typically a single carpel or fused carpels consist of a **stigma,** a sticky surface to which pollen grains adhere; a stalk, the **style,** through which the pollen tubes grow; and a swollen base, the **ovary.** Within the ovary are one or more **ovules,** each of which encloses a female gametophyte, or **embryo sac,** containing a single egg cell. After the egg is fertilized, the ovule develops into a seed and the ovary into a fruit.

In some species, flowers are either male or female. Male and female flowers may be present on the same plant, as in corn, squash, oaks, and birches. Such plants are said to be **monoecious** ("one house"). Species in which the male and female flowers are on separate plants, such as the tree of heaven *(Ailanthus),* American mistletoe, and holly, are known as **dioecious** ("two houses"). As gardeners know, in order for a female

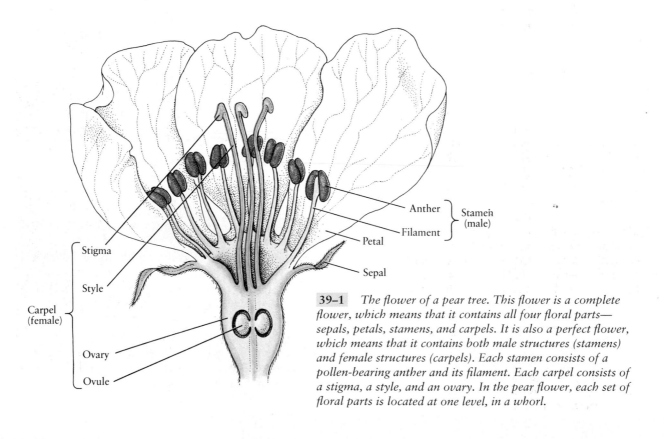

39–1   *The flower of a pear tree. This flower is a complete flower, which means that it contains all four floral parts— sepals, petals, stamens, and carpels. It is also a perfect flower, which means that it contains both male structures (stamens) and female structures (carpels). Each stamen consists of a pollen-bearing anther and its filament. Each carpel consists of a stigma, a style, and an ovary. In the pear flower, each set of floral parts is located at one level, in a whorl.*

Anther ⎫
Filament ⎬ Stamen (male)

Petal

Sepal

Stigma ⎫
Style ⎬ Carpel (female)
Ovary
Ovule

(a)

(b)

(c)

(f)

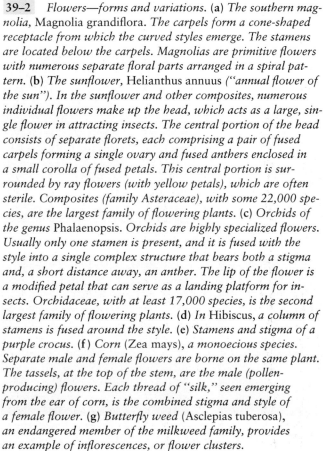

(d)

(e)

39–2 *Flowers—forms and variations. (a) The southern magnolia,* Magnolia grandiflora. *The carpels form a cone-shaped receptacle from which the curved styles emerge. The stamens are located below the carpels. Magnolias are primitive flowers with numerous separate floral parts arranged in a spiral pattern. (b) The sunflower,* Helianthus annuus *("annual flower of the sun"). In the sunflower and other composites, numerous individual flowers make up the head, which acts as a large, single flower in attracting insects. The central portion of the head consists of separate florets, each comprising a pair of fused carpels forming a single ovary and fused anthers enclosed in a small corolla of fused petals. This central portion is surrounded by ray flowers (with yellow petals), which are often sterile. Composites (family Asteraceae), with some 22,000 species, are the largest family of flowering plants. (c) Orchids of the genus* Phalaenopsis. *Orchids are highly specialized flowers. Usually only one stamen is present, and it is fused with the style into a single complex structure that bears both a stigma and, a short distance away, an anther. The lip of the flower is a modified petal that can serve as a landing platform for insects. Orchidaceae, with at least 17,000 species, is the second largest family of flowering plants. (d) In* Hibiscus, *a column of stamens is fused around the style. (e) Stamens and stigma of a purple crocus. (f) Corn* (Zea mays), *a monoecious species. Separate male and female flowers are borne on the same plant. The tassels, at the top of the stem, are the male (pollen-producing) flowers. Each thread of "silk," seen emerging from the ear of corn, is the combined stigma and style of a female flower. (g) Butterfly weed* (Asclepias tuberosa), *an endangered member of the milkweed family, provides an example of inflorescences, or flower clusters.*

(g)

holly plant to produce berries, a male holly—which never produces berries—must be planted nearby.

The Pollen Grain

By the time the **pollen grain** is released from the anther, it often consists of three haploid cells—two sperm cells contained within a larger cell, known as a **tube cell.** The tube cell, in turn, is enclosed by the thick outer wall of the pollen grain (Figure 39–3). The pollen grain contains its own nutrients and has so tough an outer coating that intact grains thousands of years old have been found in peat bogs.

As you will recall, in many multicellular algae and in the bryophytes and seedless vascular plants, there is a distinct cycle of alternation of generations. The sporophyte produces spores that produce gametophytes that produce gametes, with the gametophyte and sporophyte having separate existences. In the course of plant evolution, the gametophyte has been steadily reduced in size. In the angiosperms, all that remains of the male gametophyte is the tough, tiny pollen grain and the pollen tube that grows from it. The sperm cells are the gametes.

Fertilization

Once on the stigma, the pollen grain germinates, and, under the influence of the tube nucleus, a pollen tube grows through the style into an ovule (Figure 39–4). This may be a long distance. In corn, for instance, the pollen tube may grow to a length of 40 centimeters. The number of grains reaching the stigma is often greater than the number of ovules available for fertilization, creating intense competition among pollen tubes, the race going to the swift.

Each ovule contains a female gametophyte, which has also become reduced in size in the course of evolution. In many species, the female gametophyte consists of seven cells, with a total of eight haploid nuclei (Figure 39–5). One of the smaller cells is the egg, containing a single haploid nucleus. A large central cell contains two haploid nuclei, called the **polar nuclei** because they move to the center from each end, or pole, of the gametophyte.

The nucleus of one of the two sperm cells carried by the pollen tube unites with the egg nucleus. The fertilized egg, or zygote, develops into the embryo—the young diploid sporophyte. The nucleus of the second sperm cell unites with the two polar nuclei of the central

39–3   *Pollen grains. The walls of the pollen grain protect the male gametophyte on its journey from the anther of one flower to the stigma of another flower of the same species. These outer surfaces, which are remarkably tough and resistant, are often elaborately sculptured. As you can see, the pollen grains of different species are distinctly different:* (a) *Geranium, a dicot;* (b) *common ragweed, also a dicot (such spiny pollen grains trigger the misery of hay fever in susceptible individuals); and* (c) *cocksfoot grass, a monocot (smooth pollen grains are found in most wind-pollinated plants).*

When a pollen grain germinates, the pollen tube emerges through a pore or slit in the outer coat, clearly visible here in the cocksfoot grass pollen grain.

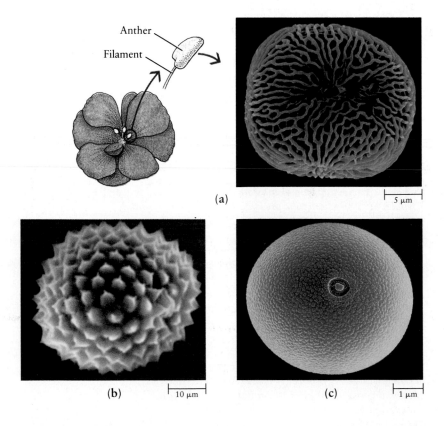

Anther

Filament

(a)   5 μm

(b)   10 μm

(c)   1 μm

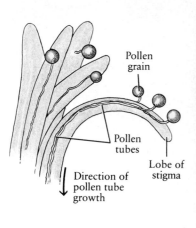

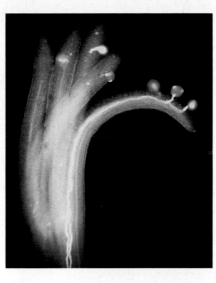

39–4   *Pollen tube growth in* Geranium maculatum, *a species in which the stigma is divided into distinct lobes. Germination of pollen grains is thought to depend on species-specific recognition, perhaps involving the interaction of chemical substances on the sticky surface of the stigma with the pollen grain. It has also been suggested that sugary substances on the stigma are used by the pollen grain to provide energy for the rapid growth of the pollen tube. In Geranium,* the pollen tube from a grain 0.1 millimeter in diameter can grow about 1 centimeter in length within 20 minutes.

cell in a process of **triple fusion.** From the resulting 3*n* (triploid) cell, a specialized tissue called the **endosperm** develops. It surrounds and nourishes the developing embryo. These extraordinary phenomena of fertilization and triple fusion—together called **double fertilization**—take place, in all the natural world, only among the flowering plants.

Figure 39–6 summarizes these events, as well as the earlier events that gave rise to the male and female gametophytes.

The Embryo

Following double fertilization, the 3*n* cell divides mitotically to produce endosperm. The zygote also divides mitotically, forming the embryo. As the embryo grows, its cells begin to **differentiate**—that is, they become different from one another. As development of the embryo proceeds, changes in its internal structure result in the formation of three distinct embryonic tissues. Gradually the embryo takes on its characteristic form, in the process known as **morphogenesis.**

In the early stages of embryonic growth, cell division takes place throughout the body of the young plant. As the embryo grows older, however, the addition of new cells becomes gradually restricted to certain parts of the plant body: the **apical meristems** (from the Greek *merizein,* "to divide"), located near the tips of the root and the shoot. During the rest of the life of the plant, primary growth—which chiefly involves the elongation of the plant body—originates from the apical meristems of roots and shoots.

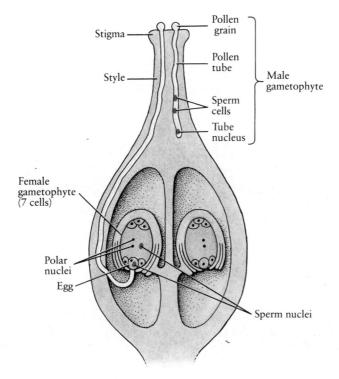

39–5   *Fertilization in angiosperms. The pollen tube of the male gametophyte, or pollen grain, grows through the style and enters an ovule, which contains a seven-celled female gametophyte (the embryo sac). One of the sperm nuclei unites with the egg, forming the zygote. The other sperm nucleus fuses with the two polar nuclei that are present in a single large cell (which in the drawing fills most of the ovule). From the resulting triploid (3n) cell, the endosperm will develop. The carpel shown here contains two ovules.*

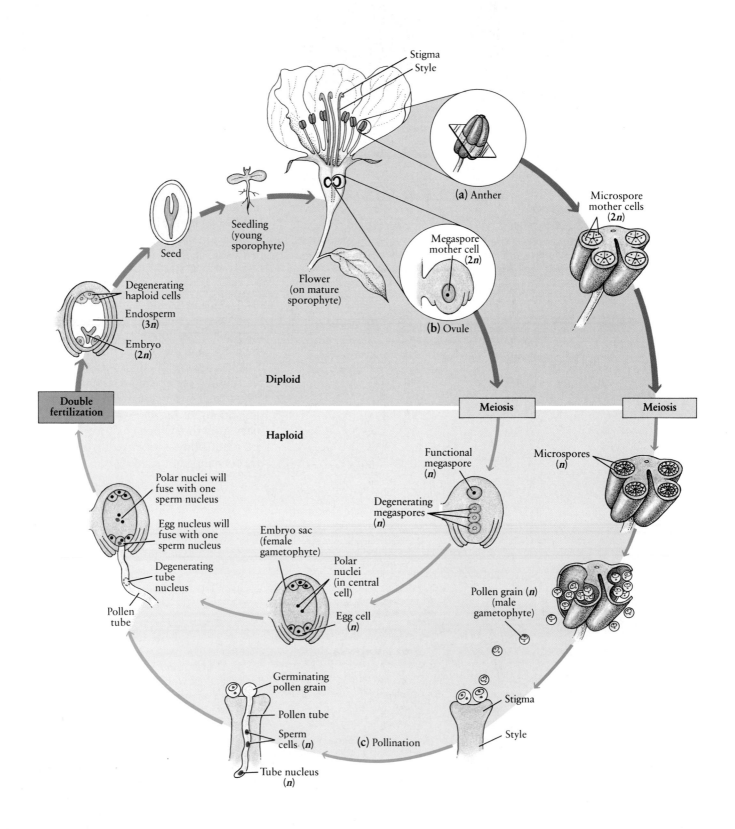

Stigma
Style

(a) Anther

Microspore
mother cells
(2n)

Seedling
(young
sporophyte)

Seed

Megaspore
mother cell
(2n)

Degenerating
haploid cells

Flower
(on mature
sporophyte)

Endosperm
(3n)

(b) Ovule

Embryo
(2n)

Diploid

Double
fertilization

Meiosis

Meiosis

Haploid

Functional
megaspore
(n)

Microspores
(n)

Polar nuclei will
fuse with one
sperm nucleus

Degenerating
megaspores
(n)

Egg nucleus will
fuse with one
sperm nucleus

Embryo sac
(female
gametophyte)

Degenerating
tube
nucleus

Polar
nuclei
(in central
cell)

Pollen
tube

Egg cell
(n)

Pollen grain (n)
(male
gametophyte)

Germinating
pollen grain

Pollen tube

Stigma

Sperm
cells (n)

Style

(c) Pollination

Tube nucleus
(n)

Grains are the small, one-seeded fruits of grasses. Because they are relatively dry, they can be stored for long periods of time. The collecting and storing of grains from wild grasses is believed to have been an important impetus to the agricultural revolution that occurred some 11,000 years ago. Today we are heavily dependent on cultivated wheat, rice, corn, rye, and other grains. In many countries, they constitute the principal component of the human diet. Wheat is about 9 to 14 percent protein, but its value as a source of protein is diminished by its deficiency in certain essential amino acids, notably lysine (see page 61).

The grain of wheat, sometimes known as the kernel, is made up of the embryo, the endosperm, and, fused together, the mature ovary wall and the remains of the seed coat. White flour is made from the endosperm, which contains 70 to 75 percent of the protein in the wheat kernel. Wheat germ, which is the embryo, is usually removed as wheat is processed because it contains oil, making the flour more likely to spoil.

Bran is the mature ovary wall (to which the remains of the seed coat are fused) and the aleurone layer (the outer part of the endosperm). It consists mostly of cellulose and is also removed when wheat is milled to make white flour. Bran somewhat decreases the caloric value of the wheat kernel, because we are unable to digest cellulose. Thus bran tends to speed the passage of food through our intestinal tracts, resulting in decreased absorption.

Until fairly recently, wheat germ and bran, which contain most of the vitamins in the wheat kernel, were sometimes used for human consumption but more often were fed to livestock. With new evidence of the importance of adequate amounts of fiber in the human diet, many more cereals and breads are made from whole grains in which the bran is retained.

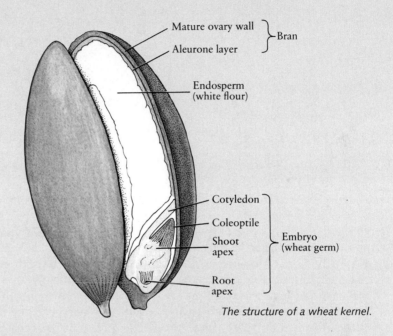

The structure of a wheat kernel.

39–6 *Life history of an angiosperm.* (a) *Within the anthers of the flower (upper right), diploid microspore mother cells divide meiotically, each giving rise to four haploid microspores. The nucleus in each microspore then divides mitotically, and the microspore develops into a two-celled pollen grain, which is an immature male gametophyte. One of the cells subsequently divides again, usually after germination, resulting in three haploid cells per pollen grain: two sperm cells and the tube cell.*

(b) *Within each ovule of the flower, a diploid megaspore mother cell divides meiotically to produce four haploid megaspores. Three of the megaspores disintegrate. The fourth, however, divides mitotically, developing into an embryo sac—the female gametophyte—consisting of seven cells with a total of eight haploid nuclei (the large central cell contains two nuclei, the polar nuclei). One of the smaller cells, containing a single haploid nucleus, is the egg.*

(c) *The pollen grain germinates on the stigma, producing a pollen tube that grows through the style into the ovary. The growing pollen tube enters an ovule through a small opening at its base. The two sperm nuclei pass through the tube into the embryo sac. Double fertilization then occurs: one sperm nucleus fertilizes the egg, and the other merges with the polar nuclei, forming a triploid (3n) cell. This triploid cell and its descendants undergo repeated mitotic divisions, producing a nutritive tissue, the endosperm. The embryo, which also undergoes repeated mitotic divisions, develops within the ovule, which becomes the seed. The ovary matures to become a fruit. The seed, released from the mother sporophyte in a dormant form, eventually germinates, forming a seedling.*

39–7 *Seeds.* (a) *In dicots such as the lima bean, the endosperm is digested as the embryo grows, and the food reserve is transferred to the two fleshy cotyledons.* (b) *In corn and other monocots, the single cotyledon absorbs food reserves from the endosperm. The coleoptile is a sheath that encloses the apical meristem of the shoot. It is the first structure to appear above ground after the seed germinates.*

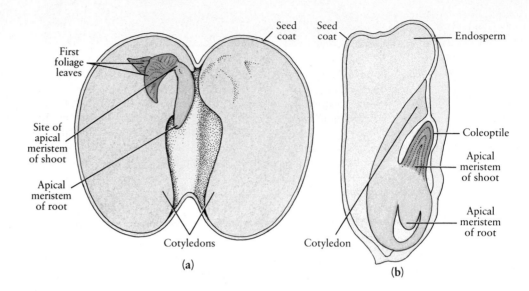

(a)

(b)

The Seed and the Fruit

The seed consists of the embryo, which develops from the fertilized egg; the stored food, which consists of or derives from the endosperm; and the seed coat, which develops from the outermost layers of the ovule (Figure 39–7). At the same time, the fruit develops from the wall of the ovary (the base of the carpel). As the ovary ripens into fruit and the seeds form, the petals, stamens, and other floral parts of the parent plant may fall away.

As you know from your own observations, fruits take many different forms. A peach, for example, develops from a flower in which the single ovary contains only one ovule. In the mature fruit, the skin, the succulent edible portion, and the stone are three distinct layers of the wall of the mature ovary. The almond-shaped structure within the stone is the seed. In a pea plant, the pod is the mature ovary wall and the peas are the seeds (the mature ovules and their contents). A raspberry is an aggregate of many small fruits from a single flower, each fruit containing a single seed and each formed from a separate carpel. In the raspberry and a number of other fruits, such as the strawberry, apple, and pear, the fleshy, edible portion is derived from the upper part of the flower stalk (Figure 39–8).

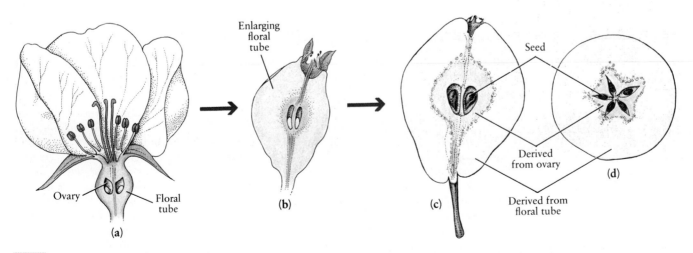

39–8 *Development and structure of the pear.* (a) *Flower of the pear. The ovary is the basal portion of the carpel.* (b) *Older flower, after the petals have fallen.* (c) *Longitudinal section and* (d) *cross section of the mature fruit. The core of the pear is the ripened ovary wall. The fleshy, edible part develops from the floral tube.*

Adaptations to Seasonal Change

The angiosperms are thought to have evolved in upland regions of the tropics during a relatively mild period in the Earth's history. As the climate grew colder, some angiosperms became extinct, while others were able to survive only near the equator. Some, presumably because of adaptations to drought (perhaps reflecting their highland origins), were able to survive in the cold, when water is locked up in ice. Chief among such adaptations is the capacity to remain dormant.

Dormancy and the Life Cycle

Depending on their characteristic patterns of active growth, dormancy, and death, modern plants are classified as annuals, biennials, and perennials. Among **annual** plants, the entire life cycle from seed to vegetative plant to flower to seed again takes place within a single growing season. Roots, stems, and leaves all die, and only the seeds, which are characteristically highly resistant to cold, desiccation, and other environmental hazards, bridge the gap between one growing season and the next. Annuals include many familiar weeds, wild flowers, garden flowers, and vegetables.

In **biennial** plants, the period from seed germination to seed formation spans two growing seasons. The first season of growth often results in a short stem, a rosette of leaves near the soil surface, and a root system. The roots are often modified for food storage; sugar beets and carrots are examples of such storage roots. During the second growing season, the stored food reserves are mobilized for flowering, fruiting, and seed formation, after which the plant dies. Species that are characteristically biennials may, in some locations, complete their cycle in a single season or may require three or more years under adverse conditions.

Perennials are plants in which the vegetative structures persist year after year. Nonwoody perennials, such as daffodils and irises, remain dormant as modified underground structures during unfavorable seasons, while woody perennials, which include vines, shrubs, and trees, survive above ground. Woody perennials in favorable climates, such as the tropical rain forest, may live year after year with little change during the annual cycle, just as did the ancestral angiosperms.

(a)

(b)

(c)

39–9   *Annual, biennial, and perennial plants.* (a) *An annual,* Linanthus dianthiflorus, *photographed in Baja California in the Sonoran desert. Many desert plants are annuals, racing through the entire life cycle from seed to flower to seed during a brief period following the seasonal rains.*

(b) *A biennial, the black-eyed Susan,* Rudbeckia hirta, *photographed on a July day in a Michigan field.*

(c) *An ancient perennial, the dawn redwood,* Metasequoia.

Twenty-five million years ago, the temperate zone of the Northern Hemisphere was covered with great forests of Metasequoia. Long thought to be extinct, living trees were discovered in western China in 1944. This tree in Philadelphia grew from seeds planted in 1947. Like the ginkgo, the bald cypress, and the larch, but unlike other gymnosperms, the dawn redwood is deciduous, shedding all its leaves in the fall.

Perennials that live in areas where part of the year is unfavorable to growth have a variety of adaptations. Some, such as cacti and most gymnosperms, undergo little apparent change, although their rates of metabolism, and therefore of growth, change with the seasons. Among dicots, many of the common vines, shrubs, and trees drop their leaves annually. Such plants are said to be **deciduous**. They are typically found in regions where there is a marked seasonal variation in available water.

Seed Dormancy

The seeds of many wild plants require a period of dormancy before they will germinate. For example, the seeds of almost all plants growing in areas with marked seasonal temperature variations require a period of cold prior to germination. This physiological requirement ensures that the seed will "wait" at least until the next favorable growth period. The seeds of a few species can remain dormant and yet viable—with the embryo in a state of suspended animation—for hundreds of years.

The seed coat often plays a major role in maintaining dormancy. In some species, it acts primarily as a mechanical barrier, preventing the entry of water and gases, without which growth is not possible. In these plants, growth is initiated when the seed coat is worn away—for example, by being abraded by sand or soil, burned away in a forest fire, or partially digested as it passes through the digestive tract of a bird or other animal.

In other species, dormancy is maintained chiefly by chemical inhibitors in the seed coat. These inhibitors undergo chemical changes in response to various environmental factors, such as light or prolonged cold or a sudden rise in temperature, which neutralize their effects, or they may be eroded or washed away by rainfall. Eventually, dormancy is broken and the stage is set for germination.

A dormant seed is usually very dry—only about 5 to 20 percent of its total weight is water. Germination cannot begin until the seed has imbibed the water required for the metabolic activities of the growing embryo. Once germination is underway, the seed coat ruptures and the young sporophyte emerges. As we shall see in the next chapter, a series of growth processes immediately begin that give rise to the plant body and continue throughout its life.

Summary

Flowers are the structures of sexual reproduction in the angiosperms. The pollen grains, which contain the male gametophytes, are produced by the anthers. The anther plus the filament is known as the stamen. The carpel typically consists of the stigma (an area on which the pollen grains germinate), a style, and, at its base, the ovary. The ovary contains one or more ovules, and within each ovule is a female gametophyte that contains an egg cell.

A new cycle of life begins when a pollen grain germinates on the stigma of a flower of the same species, sending a pollen tube through the style and into an ovule. One of the sperm nuclei of the pollen grain fertilizes the egg cell in the female gametophyte. The other sperm nucleus unites with the two polar nuclei of the female gametophyte to form a triploid ($3n$) cell. Division of this triploid cell produces a special nutritive tissue, the endosperm. The phenomena of fertilization and triple fusion, called double fertilization, occur only in the flowering plants. Following double fertilization, the ovule develops into a seed and the ovary into a fruit.

The angiosperm seed, or mature ovule, consists of the embryo, stored food, and the seed coat. The petals, stamens, and other floral parts of the parent plant may fall away as the ovary ripens into fruit and the seeds form.

Dormancy—of the seed, of vegetative parts of the plant body, or of both—enables angiosperms to bridge periods of drought or cold unsuitable for plant growth. Angiosperms are classified as annuals, biennials, or perennials, depending on whether the plant body dies at the end of one growing season (annual) or after two seasons (biennial) or whether vegetative portions of the plant body persist from year to year (perennial). In regions where there is marked variation in the availability of water, perennials are frequently deciduous, dropping their leaves at the end of each growing season.

Questions

1. Distinguish among the following: ovary/ovule/egg cell/ seed; male gametophyte/female gametophyte/embryo sac; pollination/fertilization; annual/biennial/perennial.

2. Identify the floral parts visible in the photograph below. Describe the function of each floral part. Can you name the flower?

3. In many plants, pollen production in the anthers occurs prior to or after the full development of the carpel of the same flower. What are the consequences of this shift in time frames?

4. J. B. S. Haldane, a mathematician who made major contributions to population genetics, once remarked: "A higher plant is at the mercy of its pollen grain." Explain.

5. Seeds of the jack pine (a gymnosperm) are released from the female cones only after exposure to intense heat, as in a forest fire. Following release, the tough seed coats rupture, and the seeds germinate. What might be the advantages for the jack pine of this delay in seed release and germination?

6. Sketch a dicot embryo at the time of seed release. Identify each part in terms of the future development of the plant body.

7. What are the major adaptations that enable plants to survive the periodic drought (winter) of temperate regions?

Suggestions for Further Reading

Barth, Friedrich G.: *Insects and Flowers: The Biology of a Partnership*, Princeton University Press, Princeton, N.J., 1991.*

A well-written and beautifully illustrated introduction to the interrelationships of flowers and insects. Incorporating many recent discoveries, the text considers not only the diverse structures of flowers but also the sensory, navigational, and communication abilities of pollinating insects.

Cox, Paul Alan: "Water-Pollinated Plants," *Scientific American,* October 1993, pages 68–74.

Fleming, Theodore H.: "Plant-Visiting Bats," *American Scientist,* vol. 81, pages 460–467, 1993.

Handel, Steven N., and Andrew J. Beattie: "Seed Dispersal by Ants," *Scientific American,* August 1990, pages 76–83A.

Heywood, Vernon H. (ed.): *Flowering Plants of the World,* Prentice-Hall, Inc., Englewood Cliffs, N.J., 1985.

The best available guide for students to the families of flowering plants.

Mulcahy, David L., and Gabriella B. Mulcahy: "The Effects of Pollen Competition," *American Scientist,* vol. 75, pages 44–50, 1987.

Pettitt, John, Sophie Ducker, and Bruce Knox: "Submarine Pollination," *Scientific American,* March 1981, pages 134–143.

Raven, Peter H., Ray F. Evert, and Susan E. Eichhorn: *Biology of Plants,* 5th ed., Worth Publishers, Inc., New York, 1992.

An up-to-date and handsomely illustrated general botany text, especially strong in plant evolution and ecology.

Rick, Charles M.: "The Tomato," *Scientific American,* August 1978, pages 76–87.

Robacker, David C., Bastiaan J. D. Meeuse, and Eric H. Erickson: "Floral Aroma," *BioScience,* vol. 38, pages 390–396, 1988.

Stiles, Edmund W.: "Fruit for All Seasons," *Natural History,* August 1984, pages 43–53.

Tanner, Ogden: "The Flowers That Afflict Us with 'A Sort of Madness'," *Smithsonian,* November 1985, pages 168–181.

*Available in paperback.

The Plant Body and Its Development

Plants compete for light by growing tall. This competition is particularly intense in the tropical rain forest, where water is abundant year round. Unlike their more northern counterparts, which are often shaped like pyramids, trees in the tropical forest grow straight up. Their towering trunks are bare up to the very top, where the branches, with their leaves, spread out like open umbrellas. From above, the top of the forest—the canopy—looks like a solid green carpet. Only 1 or 2 percent of the light that touches the canopy reaches the forest floor. Very few seedling trees receive sufficient light to grow tall enough to reach the canopy.

Some plants cheat. Woody vines, as long as 500 feet and as big around as your thigh, sprawl across the forest floor until they reach a tree trunk. Then they snake their way upward. Vines reach the top much more swiftly and with much less of an investment than the trees that support them.

Other cheaters, known as epiphytes, drop from the sky. These plants, unlike vines, have no connections with the ground. They collect and store water and nutrients not from their hosts but from the air, creating little patches of soil from accumulated debris. Bromeliads, among the most common of the epiphytes, have gone one step further. (You can recognize the bromeliads in the photograph because they look like the tops of pineapples; pineapples are also bromeliads, although not epiphytes.) The leaves of bromeliads merge at their bases to form watertight tanks. In some of the larger species, the tanks can hold as much as 12 gallons of rainwater. These pools of water are microcosms of bacteria, protozoa, larvae, insects, and insect-eaters. Many rain-forest mosquitoes breed exclusively in bromeliad tanks. The bromeliads absorb water from their built-in reservoirs and are also supplied with nutrients from the debris.

From time to time, weakened by age or disease or the weight of vines and epiphytes, a giant tree crashes, sunlight reaches the forest floor where a young tree is waiting, and the race to the top starts over again.

As we saw in Chapter 25, plants are multicellular photosynthetic organisms primarily adapted for life on land. The ancestor of the plants is thought to have been a multicellular green alga similar to *Coleochaete* (page 441). As in modern plants, its photosynthetic pigments were chlorophylls *a* and *b* and carotenoids, all of which were contained in chloroplasts. Each of its cells was enclosed by a plasma membrane, as well as by an external cell wall containing cellulose. Its energy source was sunlight, and it obtained oxygen, carbon dioxide, and minerals from the water in which it lived.

The photosynthetic cells of plants have the same few and relatively simple requirements: light, water, oxygen, carbon dioxide, and certain minerals. From these simple materials they, like their ancestors, make the organic substances on which all plant and animal life depends. For algae and small, simple plants living in a moist environment, each of the required materials is immediately available to every cell. Most plants, however, live in a very different environment, and their photosynthetic cells require a complex life-support system. The plant body is, in effect, that life-support system.

Figure 40–1, on the next page, diagrams the external structure of an economically important angiosperm, the potato plant. This plant, like other vascular plants, is characterized by a root system that anchors the plant in the ground and collects water and minerals from the soil; a stem or trunk that raises the photosynthetic parts of the plant toward the sun; and structures highly specialized for light capture and photosynthesis, the leaves.

The Cells and Tissues of the Plant Body

As we noted in the last chapter, the cells of the angiosperm embryo differentiate early in its development into three distinct embryonic tissues. These tissues, known as the **primary meristems,** subsequently give rise to

40–1   *The body plan of a flowering plant, the potato (Solanum tuberosum). The aboveground shoot system consists of the stem, the leaves, whose primary function is photosynthesis, and the flowers, the reproductive structures. After fertilization, the flower petals fall away and the ovaries mature to form the fruits, as we saw in the last chapter. Leaves, which may be simple or, as in the potato, compound (made up of several leaflets), appear at regions on the stem known as nodes. The portions of the stem between successive nodes are called internodes.*

In many plants, the only belowground structures are the roots, which supply water and minerals to the stem, leaves, flowers, and fruits. In the potato, however, the most conspicuous underground structures are the tubers, which are enlarged stem tips adapted for food storage.

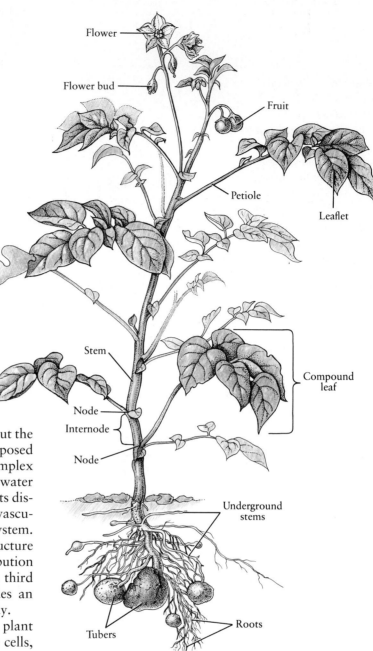

three **tissue systems** that are continuous throughout the body of the plant. The **vascular tissue** system, composed of **xylem** and **phloem** tissues (page 445), is a complex and efficient transport system. Xylem transports water and dissolved minerals, whereas phloem transports dissolved sugars and other organic molecules. The vascular tissues are embedded in the **ground tissue** system. As we shall see, the principal differences in the structure of leaves, roots, and stems lie in the relative distribution of the vascular and ground tissue systems. The third tissue system, the **dermal tissue** system, provides an outer protective covering for the entire plant body.

The most frequently encountered cells in the plant body are of a type known as **parenchyma.** These cells, which occur in all three tissue systems and predominate in the ground tissues, are typically many-sided, with thin, flexible walls. The plant cell shown in Figure 5–11 (page 95) is a photosynthetic parenchyma cell. In addition to photosynthesis, parenchyma cells perform a variety of essential functions in the plant, including respiration and storage of food and water. Each of the tissue systems also contains additional cell types, specialized for the particular functions of the tissue.

Photosynthesis is, of course, the most fundamental activity of a plant, on which all else depends. We shall therefore begin our examination of the plant body with the leaves, the primary photosynthetic organs, which are essentially the same in a seedling as in a mature plant. Following our consideration of leaves, we shall shift our attention back to the germinating seed and examine the structure and development of roots and

stems, the other two organs of the plant body. Without roots and stems, the leaves could not survive.

Leaves

Leaf Structure

The structure of a leaf (Figure 40–2) is a compromise between three conflicting evolutionary pressures: to expose a maximum photosynthetic surface to sunlight, to conserve water, and, at the same time, to provide for the exchange of gases necessary for photosynthesis.

The photosynthetic cells of leaves are parenchyma cells of two types: **palisade parenchyma,** which are columnar cells located just below the upper surface of the

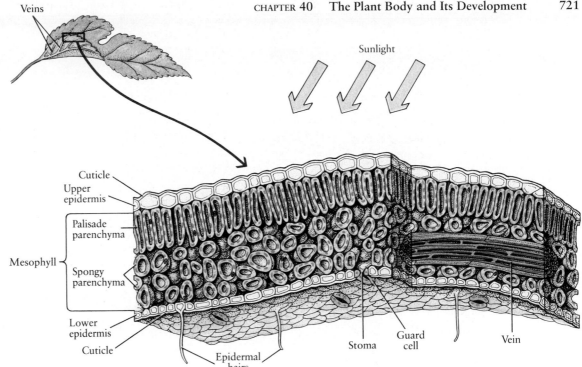

Veins

Sunlight

Cuticle

Upper epidermis

Palisade parenchyma

Mesophyll

Spongy parenchyma

Lower epidermis

Cuticle

Epidermal hairs

Stoma

Guard cell

Vein

40–2 *The structure of a leaf. Photosynthesis takes place in the palisade and spongy parenchyma cells. Note that the centers of the cells are filled with large vacuoles. These vacuoles force the cytoplasm to the periphery of the cells. The chloroplasts, indicated in bright green, move within the cytoplasm, orienting themselves to the sun. The veins carry water and solutes to and from the parenchyma cells. The interior of the leaf, the mesophyll, is enclosed by epidermal cells covered with a waxy layer, the cuticle. Openings in the epidermis, the stomata, permit the exchange of gases between the atmosphere and the cells of the mesophyll.*

leaf, and **spongy parenchyma,** which are irregularly shaped cells in the interior of the leaf, typically with large spaces between them. These spaces are filled with gases, including water vapor, oxygen, and carbon dioxide. Most of the photosynthesis occurs in the palisade cells, which are specialized for intercepting light.

Palisade and spongy parenchyma make up the ground tissue of the leaf, known as the **mesophyll,** or "middle leaf." The mesophyll is enclosed in an almost airtight wrapping formed by the cells of the **epidermis.** These cells secrete a waxy substance that forms a coating, the **cuticle,** over the outer surface of the epidermis. The epidermal cells and the cuticle are transparent, permitting light to penetrate to the photosynthetic cells.

Substances move into and out of leaves through two quite different structures: vascular bundles and stomata. Water and dissolved minerals are transported into leaves—and the sugars produced by photosynthesis are transported out—by way of the **vascular bundles.** The vascular bundles, which are known in leaves as **veins,** pass through the **petioles** (leaf stalks) and are continuous with the vascular tissues of the stem and root.

Veins form distinctive patterns in leaf blades, which are conspicuously different in monocots and dicots (Figure 40–3).

Gases—oxygen and carbon dioxide—move into and out of leaves by diffusion through **stomata** (plural of *stoma,* the Greek word for "mouth"). A stoma is a small opening, or pore. It is surrounded by two specialized cells in the epidermis, called **guard cells,** that open and close the pore (Figure 40–4). The exchange of oxygen and carbon dioxide is, as we saw in Chapter 9, necessary for photosynthesis. However, as these gases are exchanged between the atmosphere and the leaf interior, water vapor also escapes from the leaf. About 90 percent of the water loss from the plant body is through the stomata; the remaining 10 percent is through the epidermal cells and the cuticle.

Embryos	Leaves	Stems	Floral parts	Pollen grains
Dicots — Two cotyledons	Veins usually branched	Radially arranged vascular bundles	Usually occur in fours or fives	Three pores or slits
Monocots — One cotyledon	Veins usually parallel	Scattered vascular bundles	Usually occur in threes	One pore or slit

40–3 *The angiosperms are classified in two broad groups: the dicots and the monocots. The names refer to the fact that the dicot embryo has two cotyledons ("seed leaves"), and the monocot embryo has one. Other characteristic differences are visible in the plant body. The veins of dicot leaves are usually branched, producing a netted pattern. The veins of monocot leaves, by contrast, are usually parallel to one another. In dicot stems, bundles of vascular tissue are arranged around a central core of ground tissue. In monocot stems, however, the vascular bundles are usually scattered throughout the ground tissue.*

There are also characteristic differences in the reproductive structures. In dicots, the floral parts usually occur in fours or fives, but in monocots, they usually occur in threes. The pollen grains of dicots typically have three pores or slits, while those of monocots have only one.

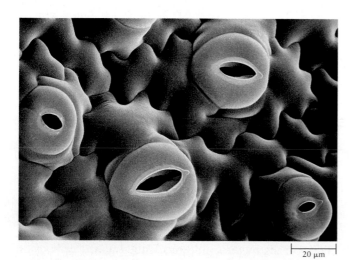

20 μm

40–4 *Open stomata on the surface of a tobacco leaf. The stomata lead into air spaces within the leaf that surround the palisade and spongy parenchyma cells. The air in these spaces, which make up 15 to 40 percent of the total volume of the leaf, is saturated with water vapor that has evaporated from the photosynthetic cells.*

Stomata are typically most abundant on the undersurface of leaves. They may be very numerous. For example, on the lower surface of tobacco leaves there are about 19,000 stomata per square centimeter. On the upper surface there are about 5,000 stomata per square centimeter.

Leaf Adaptations and Modifications

Leaves come in a variety of shapes and sizes, ranging from broad fronds to tiny scales. Some of these differences can be correlated with the environments in which the plants live. Large leaves with broad surfaces, for example, are often found in plants that grow under the canopy in a tropical rain forest, where water is plentiful but competition for light is intense.

Small, leathery leaves, in which light-capturing surface has been sacrificed for water conservation, are often associated with harsh, dry climates. This reduction in leaf surface reaches an extreme in those desert cacti in which the leaves are modified as spines—hard, dry, and nonphotosynthetic structures. (The terms "spine" and "thorn" are often used interchangeably; however,

(a)

(b)

(c)

40–5 *Modified leaves.* (a) *Spines on a giant prickly-pear cactus, photographed in the Galapagos Islands.* (b) *Tendrils of a pea plant. In the pea plant, which has compound leaves, only individual leaflets are modified as tendrils. Other leaflets of a given compound leaf are flattened, providing a broad surface for photosynthesis.* (c) *Succulent leaves of* Sedum, *adapted for water storage.*

thorns are technically modified branches.) In cacti, photosynthesis takes place in the fleshy stems, which are also water-storage organs.

In many plants, leaves are succulent—that is, they are adapted for water storage. Leaves may also be specialized for other functions, such as food storage or support. For example, a bulb, such as the onion, is a large bud consisting of a short stem with many leaves modified for storing food. The "head" of a cabbage also consists of a compressed stem bearing numerous thick, overlapping leaves. In some plants, the petioles become thick and fleshy: celery and rhubarb are two familiar examples. The tendrils of some climbing plants—the garden pea, for instance—are modified leaves or leaflets.

Characteristics of Plant Growth

As we discussed in the last chapter, the embryos of many angiosperms pass through a dormant stage prior to germination of the seed. With germination, growth resumes, the seed coat ruptures, and the young sporophyte emerges. The first foliage leaves open to the sun and begin photosynthesizing, while internally the growth processes that give rise to the plant body continue.

The **primary growth** of the plant (Figure 40–6) involves differentiation of the three tissue systems, elongation of roots and stems, and the formation of lateral roots and of aerial branches. After development of the embryo is complete, subsequent primary growth originates in the apical meristems of the root and shoot.

The existence of such meristematic areas, which add to the plant body throughout its life, is one of the principal differences between plants and many animals. Birds and mammals, for example, stop growing when they reach maturity, although the cells of certain "turnover" tissues, such as the skin or the lining of the intestine, continue to divide. Plants, however, continue to grow during their entire life span.

Growth in plants is the counterpart, to some extent, of motility in animals. Plants "move" by extending their roots and shoots, both of which involve changes in size and form. As a result of these changes, a plant modifies its relationship with the environment, for example, by curving toward the light and extending its roots toward water. The sequence of growth stages in plants thus corresponds to a whole series of motor acts in animals, especially those associated with obtaining food and water. In fact, growth in plants serves many of the functions that we group under the term "behavior" in animals.

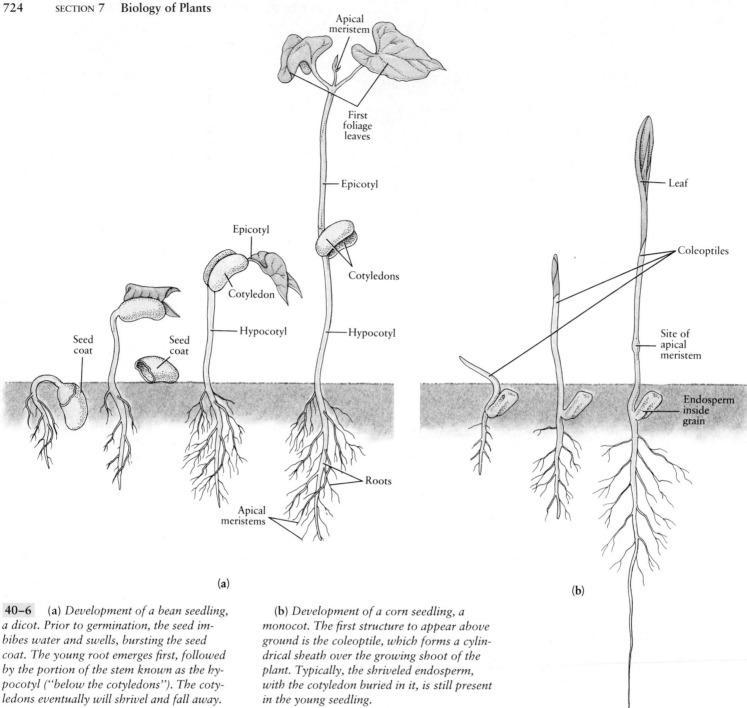

(a)

(b)

40–6   (a) *Development of a bean seedling, a dicot. Prior to germination, the seed imbibes water and swells, bursting the seed coat. The young root emerges first, followed by the portion of the stem known as the hypocotyl ("below the cotyledons"). The cotyledons eventually will shrivel and fall away.*

(b) *Development of a corn seedling, a monocot. The first structure to appear above ground is the coleoptile, which forms a cylindrical sheath over the growing shoot of the plant. Typically, the shriveled endosperm, with the cotyledon buried in it, is still present in the young seedling.*

Roots

Roots are specialized structures that anchor the plant and take up water and essential minerals from the soil. The embryonic root is the first structure to break through the seed coat, and, in an older plant, the root system may make up more than half of the plant body. The lateral spread of tree roots is usually greater than the spread of the crown of the tree. In a study made on a four-month-old rye plant, the total surface area of the root system was calculated to be 639 square meters, some 130 times the surface area of the leaves and stem.

Root Structure

The internal structure of the root is comparatively simple. In dicots and most monocots, the three tissue systems (dermal, ground, and vascular) are arranged in three concentric layers: the **epidermis**, the **cortex**, and the **vascular cylinder** (Figure 40–7).

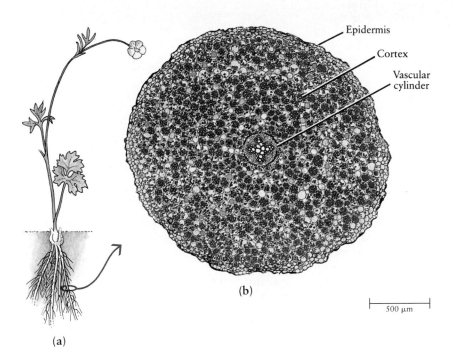

(b)

Epidermis
Cortex
Vascular cylinder

|———| 500 μm

(a)

40–7 *Root of a buttercup, a dicot, in cross section. The outer layer of cells makes up the epidermis, the inner core is the vascular cylinder, and everything in between is the cortex. Most of the parenchyma cells of the cortex contain starch grains, which are stained purple in this preparation. The vascular cylinder is shown in more detail in Figure 40–10a.*

The Epidermis

The epidermis, which covers the entire surface of the young root, absorbs water and minerals from the soil and protects the internal tissues. The cuticle is either absent or very thin compared with that found on the surface of a leaf.

The epidermal cells of the root are characterized by fine, tubular outgrowths, known as **root hairs** (Figure 40–8). Root hairs are slender extensions of the epidermal cells; in fact, the nucleus of the epidermal cell is typically found within the root hair. In the rye plant previously mentioned, the roots were estimated to have some 14 billion root hairs. Placed end to end, they would have extended more than 10,000 kilometers. Most of the water and minerals that enter the root are absorbed by these delicate outgrowths of the epidermis. In many species, however, mycorrhizal associations (page 440) substitute for root hairs.

The Cortex

As you can see in Figure 40–7, the cortex occupies by far the greatest volume of the young root. The cells of the cortex are parenchyma cells, as in the ground tissue of the leaf. However, root parenchyma usually lacks chloroplasts and is often specialized for the storage of starch and other organic substances. (In some species, the roots are highly specialized for this function; beets and carrots are examples of such roots.) There are many air spaces in the cortex, and oxygen from the soil enters these spaces through the epidermal cells and is used by the cortical cells in respiration.

40–8 *A radish seedling. Note the discarded seed coat, the cotyledons, and the primary root with its numerous root hairs. Most of the uptake of water and minerals occurs through the root hairs, which form just behind the growing tip of the root.*

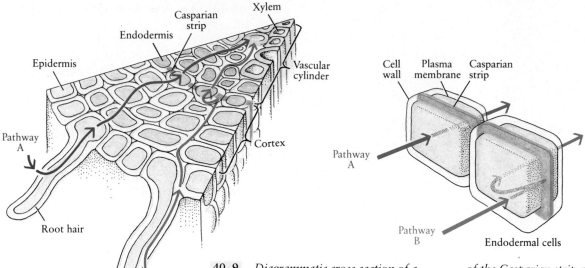

40–9 *Diagrammatic cross section of a root, showing the two pathways of uptake of water and dissolved substances. Most of the solutes and some of the water entering the root follow pathway A, indicated in red. The solutes move by active transport and diffusion and the water by osmosis through the plasma membranes and plasmodesmata (page 123) of a series of living cells. Most of the water and some of the solutes entering the root follow pathway B, indicated in blue, flowing through the cell walls and along their surfaces. Notice, however, the location* of the Casparian strip and how it blocks pathway B all around the vascular cylinder of the root. In order to pass the Casparian strip, both water and solutes must be transported through the plasma membranes of the endodermal cells, as in pathway A. After the water and solutes have crossed the endodermis, most of the solutes continue along pathway A to the conducting cells of the xylem, and most of the water returns to pathway B for the remaining distance to the conducting cells.

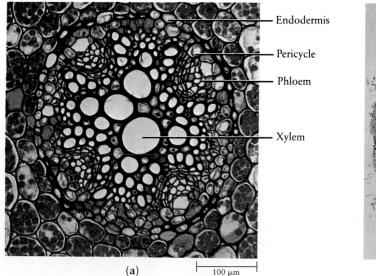

(a)

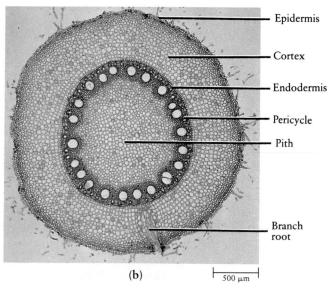

(b)

40–10 (a) *Details of the vascular cylinder of the buttercup root (a dicot) shown in Figure 40–7. This type of vascular cylinder is found in dicots and most monocots. The endodermis, which is outside the pericycle, contains the Casparian strips.*

(b) *Cross section of the root of a corn* plant *(a monocot), showing the vascular cylinder enclosing the pith. This type of root structure is found in some monocots. Part of a branch root, which emerges from the pericycle, can be seen in the lower portion of the micrograph.*

Unlike the rest of the cortex, the cells of the innermost layer, the **endodermis,** are compact and have no spaces between them. Each endodermal cell is encircled by a continuous band of wax, known as the **Casparian strip** (Figure 40–9). The Casparian strip, which is located within the cell wall and adheres tightly to the plasma membrane, is not permeable to water. Therefore, water and dissolved substances, which can move freely around the other cortical cells and through their cell walls, must pass through the plasma membranes of endodermal cells. As you will recall (page 114), water, oxygen, and carbon dioxide pass easily through plasma membranes, but many ions and other substances do not. Thus, the plasma membranes of the endodermal cells regulate the passage of such substances into the vascular tissues of the root and thereby determine what is transported to the rest of the plant body.

The Vascular Cylinder

The vascular cylinder of the root consists of the vascular tissues (xylem and phloem) surrounded by one or more layers of cells, the **pericycle.** Branch roots arise from the pericycle. In most species, the vascular tissues of the root are grouped in a solid cylinder (Figure 40–10a). In some monocots, however, the vascular tissues form a cylinder around a **pith,** a central core of ground tissue (Figure 40–10b).

Primary Growth of the Root

The first part of the embryo to break through the seed coat, in nearly all seed plants, is the primary root. Figure 40–11 diagrams the growth zone of a dicot root. At its tip is the **root cap,** which protects the apical meristem as the root is pushed through the soil. Cells of the root cap wear away and are constantly replaced by new cells from the meristem.

Certain cells in the meristem retain the capacity to produce new cells and thus to perpetuate the meristem. All the other cells in the root—which are the progeny of these relatively few meristematic cells—eventually differentiate, some becoming cells of the root cap and others forming the tissue systems of the root. The maximum rate of cell division occurs at a point well above the tip of the meristem. Then, just above the point where cell division is greatly reduced, the cells gradually elongate, growing to 10 or more times their original length, often within the span of a few hours.

As the cells elongate, they begin to differentiate. The first cells to differentiate in roots are the conducting cells of the phloem, followed by the conducting cells of the xylem. In the region of the root where the xylem first forms, the endodermis also differentiates. To the

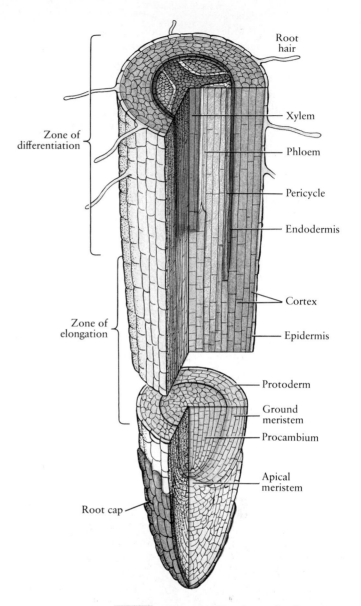

40–11 *The growth regions of a dicot root. New cells are produced by the division of cells within the apical meristem. The cells above the meristem undergo a characteristic series of changes as the distance increases between them and the root tip. First, there is a maximum rate of cell division, followed by cell elongation with little further division. This elongation accounts for most of the lengthening of the root. As the cells elongate, they differentiate into the three primary meristems that give rise to the three tissue systems of the root. The protoderm becomes the epidermis, the ground meristem becomes the cortex, and the procambium becomes the primary xylem and primary phloem. Some of the cells produced by the apical meristem differentiate to form the protective root cap.*

inside of the endodermis, the pericycle forms. At approximately the same level in the elongating root, the epidermal cells differentiate and begin to extend root hairs into crevices between the soil grains.

This sequence of growth occurs in the first root of a seedling and is repeated over and over again in all the growing root tips of a plant—even those of a tree 50 meters tall.

Patterns of Root Growth

In many dicots, the primary root develops into a **taproot,** which, in turn, gives rise to lateral, or branch, roots (Figure 40–12a). In monocots, the primary root is usually short-lived and the final root system develops from the base of the stem. Roots of this kind are called **adventitious roots.** ("Adventitious" describes any structure growing from other than its "usual" place.) These adventitious roots and their branches form a fibrous root system (Figure 40–12b).

Aerial roots are adventitious roots produced from aboveground structures. Some aerial roots, such as those of English ivy, cling to vertical surfaces and thus provide support for the climbing stem. In other plants, such as corn, aerial roots serve as prop roots (Figure 40–13a). Trees that grow in swamps, such as the red mangrove and the bald cypress, often have prop roots.

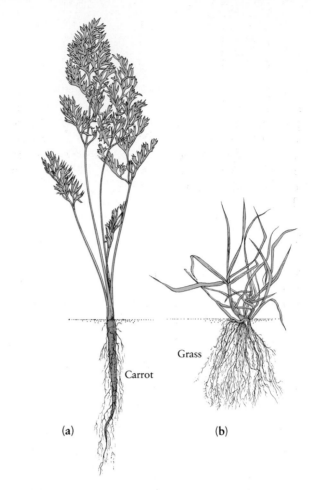

40–12   *Types of root systems.* (**a**) *Taproot system of a carrot (a dicot).* (**b**) *Fibrous root system of rye grass (a monocot).*

(a)

(b)

40–13   (a) *Prop roots of corn. These are adventitious roots, arising from the stem.* (**b**) *Air roots of a white mangrove. The* root tips grow up out of the mud in which these trees grow and take up oxygen needed by the roots for respiration.

In swampy areas the soil is usually low in oxygen. Some trees that grow in these habitats develop roots that grow out of the water and serve not only to anchor the plant but also to supply the root cells with the oxygen needed for respiration. White mangroves, for example, have air roots whose tips grow upward out of the mud and serve this aerating function (Figure 40–13b).

Stems

Stems display leaves to the light and are the pathway through which substances are transported between the roots and the leaves. They may also be adapted for storing food or water. As we saw in Figure 40–1, white potatoes are enlarged, underground stem tips, known as tubers, that are filled with starch. In some species of plants that grow in arid environments, water is stored in large parenchyma cells in the stems. The stem of a cactus, for example, may be 98 percent water by weight.

Stem Structure

The outer surface (dermal tissue) of young green stems, like that of leaves and roots, is made of epidermal cells. Like leaves, green stems are covered with a waxy cuticle, contain stomata, and are photosynthetic.

The bulk of the tissue in a young stem is ground tissue. As in leaves and roots, it is composed mostly of parenchyma cells. The turgor (page 116) of these cells provides the chief support for young green stems. The ground tissue of stems also may contain specialized supporting tissues known as **collenchyma** and **sclerenchyma** (Figure 40–14). Unlike the thin-walled parenchyma cells, collenchyma cells have primary walls (page 91) that are thickened at the corners or in some other uneven fashion. Collenchyma cells are often located just inside the epidermis, forming either a continuous cylinder or distinct vertical strips of tissue. They provide support for the growing regions of young stems and branches.

Sclerenchyma cells generally have secondary walls (page 91) impregnated with lignin, a large and complex molecule that toughens and hardens cellulose. Sclerenchyma cells are of two types: **fibers** and **sclereids.** Fibers, which are extremely elongated, somewhat elastic cells, typically occur in strands or bundles arranged in a definite pattern characteristic of the particular species. They are often associated with the vascular tissues. Plant fibers such as flax, hemp, jute, sisal, and raffia have long been used in human artifacts, including baskets, rope, and cloth. Sclereids, which are variable in form, are also common in stems. Layers of sclereids are found in seeds, nuts, and fruit stones as well, where they form the hard outer coverings.

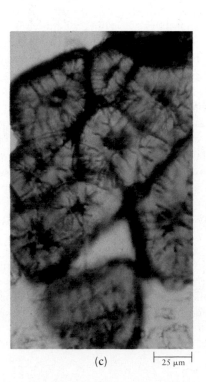

(a) — 20 μm   (b) — 20 μm   (c) — 25 μm

40–14 Some cell types found in the ground tissue of stems. (a) Collenchyma cells, viewed in longitudinal section. Their irregularly thickened cell walls are rich in pectin and contain much water. The walls are plastic (capable of expansion), and so the cell can continue to grow. (b) Longitudinal section of fibers from phloem in the stem of a linden tree. Only a portion of their length can be seen. These sclerenchyma cells have thickened, often lignified cell walls that give them strength and rigidity. Many fibers, but not all, are dead at maturity. (c) Sclereids, another type of sclerenchyma cell, have very thick lignified walls. Sclereids are often found in seeds and fruits, as well as in stems. These sclereids, called stone cells, are from a pear. They give the fruit its characteristic gritty texture.

40–15   *In angiosperms, the conducting elements of the phloem are sieve tubes, made up of individual cells, the sieve-tube members (a). These cells, which lack nuclei at maturity, are almost always found in close association with companion cells, which do have nuclei. A sieve-tube member and its companion cell arise from the same mother cell. (b) Sieve-tube members are joined to other sieve-tube members at their end walls by sieve plates.*

(c) A longitudinal view of a sieve tube in the stem of a squash plant. You can see portions of the film of protein that lines the inner surface of the cell walls of the sieve-tube members. Also present is callose, a polysaccharide deposited, often in large amounts, by the sieve-tube members in response to injury. Both callose and the protein on the inner surface of the cell walls prevent leakage from injured sieve tubes. Callose is also deposited as part of the normal aging process.

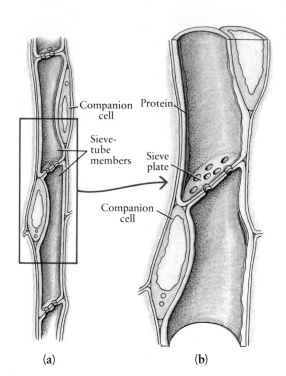
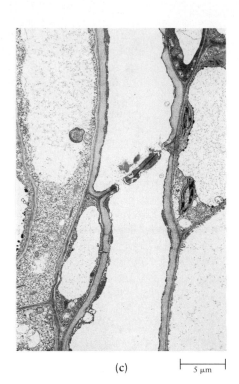

(a)   (b)   (c)   5 μm

Vascular Tissues

The vascular tissues, phloem and xylem, consist of specialized conducting cells, supporting fibers, and parenchyma cells, which store food and water. The conducting cells of the phloem transport the products of photosynthesis, chiefly in the form of sucrose, from the leaves to the nonphotosynthetic cells of the plant. In angiosperms, these cells are called **sieve-tube members** (Figure 40–15). A **sieve tube** is a vertical column of sieve-tube members joined by their end walls. These end walls, called **sieve plates,** have pores leading from one sieve-tube member to the next.

The sieve-tube members, which are alive at maturity, are filled largely with a watery fluid called sieve-tube sap. In dicots and some monocots, they also contain a protein that forms a film along the inner surface of the cell wall and is continuous from one cell to the next through the sieve plates. The function of this protein is unknown, although some botanists believe that, along with the polysaccharide callose, it seals sieve-plate pores in response to injury.

As a sieve-tube member matures, its nucleus and many of its organelles disintegrate. Sieve-tube members, however, are characteristically associated with specialized parenchyma cells called **companion cells,** which contain all of the components commonly found in living plant cells, including a nucleus. Companion cells may be responsible for the secretion of substances into and

out of the sieve-tube members and are thought also to provide nuclear functions for the sieve tubes and fulfill their energy requirements.

Specialized cells in the xylem conduct water and minerals from the roots to other parts of the plant body. It is customary to think of xylem as transporting water up and phloem as transporting sugars down, but if you think of the various shapes of plants you can see that water must also often be transported laterally, as along a tendril, or even down, as to the branches of a weeping willow. Conversely, sugars must often go upward, as into a flower or fruit.

The conducting cells of the xylem are **tracheids** and **vessel members** (Figure 40–16). Both of these cell types have thick secondary walls impregnated with lignin, and both are dead at functional maturity. Tracheids are long, thin cells that overlap one another on their tapered ends. These overlapping surfaces contain thin areas, known as **pits,** where no secondary wall has been deposited. Water passes from one tracheid to the next through the pits. Vessel members, which are much shorter and wider, also differ from tracheids in that their end walls contain perforations or are entirely absent. Thus, the vessel members form a continuous **vessel,** which is a more effective conduit than a series of tracheids. Seedless vascular plants and most gymnosperms have only tracheids; most angiosperms have both tracheids and vessel members.

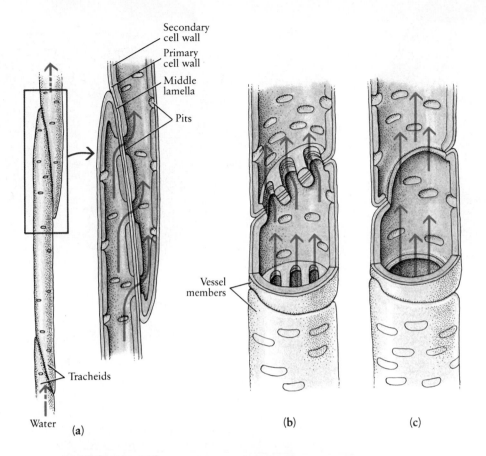

Secondary cell wall
Primary cell wall
Middle lamella
Pits
Vessel members
Tracheids
Water
(a)
(b)
(c)

40–16   *Tracheids and vessel members are the conducting cells of the xylem in angiosperms.* (a) *Tracheids are a more primitive and less efficient type of conducting cell. Water moving from one tracheid to another passes through pits. Pits are not perforations but simply areas in which there is no secondary cell wall. Thus water that is moving from one tracheid to another must pass through two primary cell walls and the middle lamella (see page 91).*

Vessel members differ from tracheids in that the primary walls and middle lamellae of vessel members are perforated at the ends where they are joined with other vessel members. (b) *There may be numerous perforations in adjoining walls of vessel members, or* (c) *the adjoining walls may dissolve completely as the cells mature, forming a single opening. Vessel members are also characteristically shorter and wider than tracheids, and their adjoining end walls are less oblique. Vessel members are connected with the members of adjacent vessels in the xylem—and also with other cells—by pits in the side walls.*

(d) *A view into a vessel in the xylem of a prop root of a corn plant.*

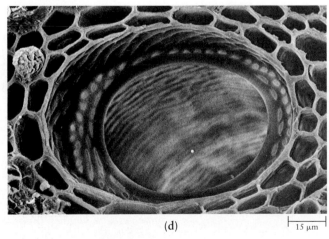

(d)

15 µm

Stem Patterns

In green stems, the xylem and the phloem are usually arranged in longitudinal parallel strands, the vascular bundles, which are embedded in the ground tissue. In young dicot stems, the vascular bundles form a ring, the vascular cylinder, around a central pith (Figure 40–17a and b). The cylinder of ground tissue outside the vascular bundles is the cortex. Within each bundle, the xylem is characteristically on the inside, adjacent to the pith, and the phloem is on the outside, adjacent to the cortex. In monocots, the vascular bundles are usually scattered throughout the ground tissue (Figure 40–17c).

The vascular tissues of the stem are continuous with those of the root, and yet, as we have seen, their arrangement in the stem is quite different from that in the root. The change from the patterns observed in the root to those found in the stem is a gradual one. The region of the plant axis between root and stem in which the change occurs is known as the transition region.

Epidermis

Cortex

Pith

Vascular
bundles

(a)

200 μm

Epidermis

Cortex

Pith

Vascular
cylinder

(b)

200 μm

Primary Growth of the Shoot System

The **shoot system** includes the stem and all of the structures that develop from it—typically, all of the aboveground parts of the plant. The pattern of growth of the developing shoot tip is similar to the pattern in the root: first, cell division takes place; next, cell elongation; and finally, differentiation. However, because of the regular occurrence of nodes (see Figure 40–1) and their appendages—leaves and buds—the growth zones are not as distinct in the shoot as they are in the root.

As in the root, the outermost layer of cells develops into the epidermis. In the shoot, these cells are covered by a relatively conspicuous cuticle. Underlying cells differentiate to form the ground tissues and the primary vascular tissues—the primary xylem and the primary phloem. The pattern of development is more complicated, however, than in the root tip since the apical meristem of the shoot is the source of tissues that give rise to new leaves, branches, and flowers.

Figure 40–18 shows the shoot tip of the familiar house plant *Coleus*. In the center is the apical meristem, which is very small, with the beginnings—**primordia**—of two leaves. Flanking the apical meristem are two previously formed leaves, which are completing their growth and development. The leaves originate by cell division in localized areas along the side of the apical meristem. As growth progresses, the vascular tissues of the stem differentiate upward into the leaf primordia, becoming part of the general vascular system that connects the plant from root to leaf tip. Leaves are formed in an orderly sequence at the shoot tip. As the internodes elongate, the young leaves become separated and thus spaced out along the stem of the plant.

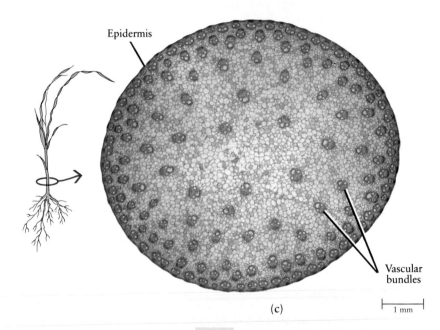

Epidermis

Vascular
bundles

(c)

1 mm

40–17   *Cross sections of two dicot stems and one monocot stem.* (a) *In alfalfa, a dicot, the vascular cylinder is made up of separate vascular bundles.* (b) *In this young stem of the linden, also a dicot, the vascular tissue forms a continuous cylinder. This stem contains mucilage ducts, which are stained red in this preparation.* (c) *In corn, a monocot, numerous vascular bundles are scattered throughout the ground tissue.*

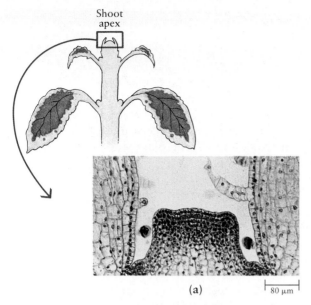

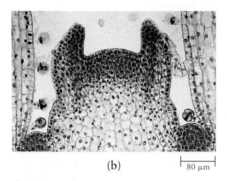

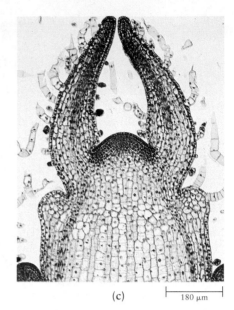

(a)   80 μm   (b)   80 μm   (c)   180 μm

40–18 *Stages in leaf development at the shoot apex of* Coleus, *as seen in longitudinal section.* Coleus *leaves develop in pairs, opposite one another, with each successive pair at right angles to the preceding pair. In these micrographs, the nuclei are stained purple; thus, the meristematic regions, densely packed with small, rapidly dividing cells, appear purple.* (a) *Two small bulges, or leaf buttresses, appear on opposite sides of the* stump-shaped apical meristem. In addition, buds are developing in the axils of two previously formed leaves. (b) *Two erect, peglike leaf primordia have developed from the leaf buttresses. Strands of vascular tissue are extending upward into the leaf primordia.* (c) *As the leaf primordia elongate, the vascular tissues continue their upward differentiation.*

As the shoot tip elongates, small masses of meristematic tissue are left just above the points at which the leaves are attached to the stem (the leaf axils). These new meristematic regions, the **axillary buds,** remain dormant until after growth of the adjacent leaf and internode is complete. In many species, the axillary buds do not develop at all unless the apical meristem of the shoot is damaged or removed. In some species, specific buds are destined to become lateral branches or specialized shoots, such as tubers or flowers. In other species, the fate of the buds is determined by environmental conditions, particularly day length, as we shall see in Chapter 42.

Modifications in the Pattern of Shoot Growth

In most plants, primary growth of the shoot proceeds from its tip, the apical meristem, and from the axillary buds. Grass plants, however, also have meristems at the bases of the leaves. The grass leaf consists of a blade (the broad part) and a sheath that fits around the plant stem (Figure 40–19). When a portion of the blade is cut off, cells at the base of the leaf are activated, producing more blade.

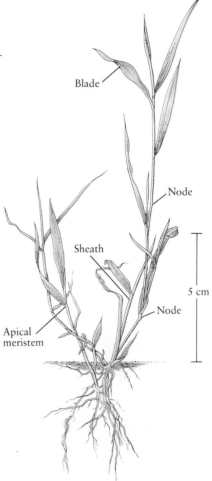

Blade

Node

Sheath

5 cm

Node

Apical
meristem

40–19 *The structure of a grass plant. The capacity of a grass leaf to grow from its base is a useful adaptation to herbivore grazing. It has also resulted, over the eons, in the piteous spectacle of thousands of hominids vibrating behind their lawnmowers from spring to fall.*

733

The shoots of some climbing plants coil themselves around the structures on which they are growing. Others produce modified branches in the form of tendrils. (As we saw previously, tendrils may also be modified leaves or leaflets.) The tendrils of grape, English ivy, and Virginia creeper are all modified branches.

Vegetative Reproduction

Among the specialized shoots that originate from the axillary buds of many species are **runners** and **rhizomes.** Runners are long, slender stems that grow along the surface of the soil. Rhizomes are also horizontal stems, growing either along or below the surface of the soil. Both runners and rhizomes develop adventitious roots and are the source of new plants, genetically identical to the parent plant.

Strawberries (Figure 40–20) are a familiar example of plants that propagate by runners, as are spider plants, commonly grown as hanging plants. Plants that reproduce by rhizomes include potatoes, many flowering garden perennials, such as lilies-of-the-valley and irises, the sod-forming grasses of lawns and pastures, and bamboos and reed grasses. In grasses, for example, rhizomes form buds that produce upright stems bearing leaves and flowers.

The production of genetically identical clones from runners and rhizomes is a very efficient way for a plant to spread quickly and invade new territory. The young plants that develop in this way have a continuous source of nourishment from the parent plant and a lower mortality than seedlings.

The capacity of plants to reproduce vegetatively has been exploited in the development of cultivated varieties of plants for food or ornamental use. Because such plants are genetically identical to the parent stock, vegetative reproduction is a way of preserving uniformity. Many plants are reproduced by stem cuttings, which involves placing young stems in soil, water, or some other planting medium and protecting them from drying out until adventitious roots appear. Rooting can often be facilitated by treating the cuttings with hormones (see page 757).

Another artificial form of plant propagation is grafting, in which a stem cutting is inserted in a slit in the stem or trunk of a rooted woody plant. The wound is sealed with tape or wax, and if the stem cutting and the rooted plant have not been damaged too much by the surgery, the grafted shoot will "take" and begin growing. Most fruit trees and rose bushes are propagated in this way.

Many economically important plants are sterile and can only be propagated vegetatively. These include pineapples, bananas, seedless grapes, navel oranges, sugar cane, and numerous ornamental plants.

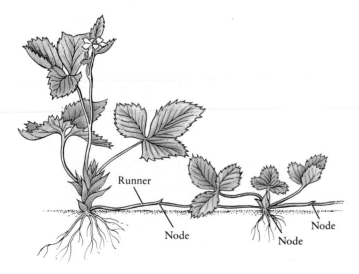

40–20   *Strawberry plants reproduce asexually by means of runners, thin horizontal stems that grow along the surface of the soil. Roots and leaves develop at every second node along the modified stems. These plants also form flowers and reproduce sexually.*

Secondary Growth

As you know from your own observations, many plants not only grow taller with age but also become thicker. The process by which woody dicots increase the thickness of their trunks, stems, branches, and roots is known as **secondary growth** (Figure 40–21). The so-called "secondary tissues" are not derived from the apical meristems. Instead, they are produced by **lateral meristems** known as the vascular cambium and the cork cambium.

The **vascular cambium** is a thin, cylindrical sheath of tissue between the xylem and the phloem. In plants with secondary growth, the cambial cells divide continually during the growing season, adding secondary xylem toward the inside of the cambium and secondary phloem toward the outside. Some cells remain as a cylinder of undifferentiated cambium, in which cell division will resume at the start of the next growing season.

As secondary growth increases the girth of stems and roots, the epidermis becomes stretched and torn. In con-

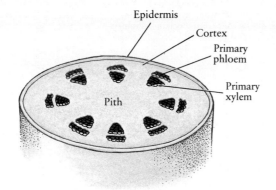

(a)

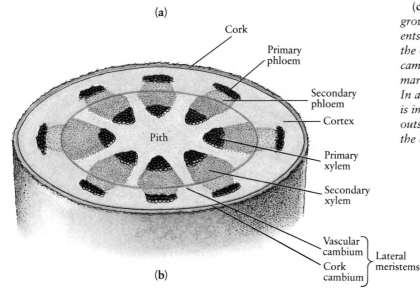

(b)

40–21   (a) *Stem of a dicot before the onset of secondary growth.*

(b) *Beginnings of secondary growth. Secondary xylem and secondary phloem are produced by the vascular cambium, a meristematic tissue formed late in primary growth. As the trunk increases in diameter, the epidermis is stretched and torn. This is accompanied by the formation of the cork cambium, from which cork is formed, replacing the epidermis.*

(c) *Cross section of a three-year-old stem, showing annual growth rings. Rays are rows of living cells that transport nutrients and water laterally (across the trunk). On the perimeter of the outermost growth layer of secondary xylem is the vascular cambium, encircled by a band of secondary phloem. The primary phloem and the cortex will eventually be sloughed off. In an older stem, the thin cylinder of active secondary phloem is immediately adjacent to the vascular cambium. The tissues outside the vascular cambium, including the phloem, constitute the bark.*

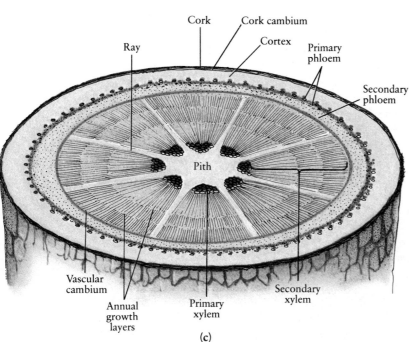

(c)

junction with this tearing process, a new type of cambium, the **cork cambium,** forms from the cortex. The cork cambium produces **cork,** which replaces the epidermis as the protective covering of woody stems and roots. Unlike the epidermis, cork is a dead tissue at maturity. The cork cambium often forms anew each year, moving further inward until, finally, there is no cortex left. The tissues outside the vascular cambium—a thin layer of phloem, the cork cambium, and the cork—constitute the **bark.**

As the plant grows older, the parenchyma cells of the xylem in the center of the stem and root die, and their neighboring vessels become clogged and cease to function. This nonconducting xylem, called **heartwood,** forms the center of the trunk and major roots of a tree, providing the support and anchorage required as the height of the tree continues to increase through primary growth. The living parenchyma cells and open vessels just inside the vascular cambium constitute the **sapwood,** through which water and minerals flow from the root tips to the leaves. Season after season the new xylem forms visible growth layers, or rings. Each growing season leaves its trace, so that the age of a tree can be estimated by counting the number of growth rings in a section near its base.

By the continuous formation of secondary xylem and (to a lesser extent) phloem, woody dicots increase their

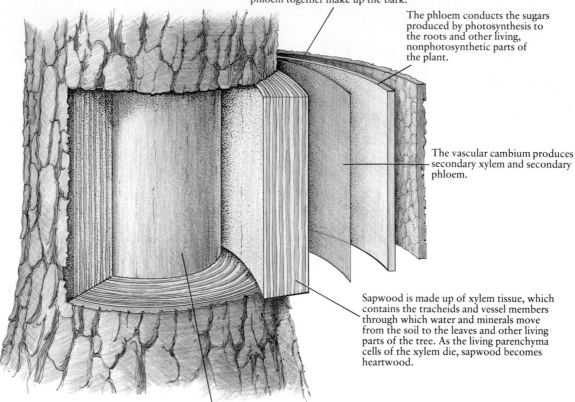

Cork, which is a dead tissue, protects the inner tissues from drying out, from mechanical injury, and from insects and other herbivores. Cork and phloem together make up the bark.

The phloem conducts the sugars produced by photosynthesis to the roots and other living, nonphotosynthetic parts of the plant.

The vascular cambium produces secondary xylem and secondary phloem.

Sapwood is made up of xylem tissue, which contains the tracheids and vessel members through which water and minerals move from the soil to the leaves and other living parts of the tree. As the living parenchyma cells of the xylem die, sapwood becomes heartwood.

Heartwood, composed entirely of dead cells, is the central supporting column of the mature tree.

40–22   *A tree trunk showing the relationships of the successive concentric layers. The heartwood is composed entirely of dead cells. The cortex and epidermis, which are outside the phloem in a green stem, are sloughed off during secondary growth.*

diameter as primary growth increases their height. Moreover, newly formed tracheids, vessel members, and sieve-tube members provide fresh conduits—undamaged by the activities of the numerous parasites and herbivores that attack plants—for the transport of water and nutrients from one part of the plant body to another. In the next chapter, we shall examine the mechanisms by which materials move through these cells.

Summary

Plants are multicellular photosynthetic organisms that are adapted to life on land. The plant body has specialized photosynthetic areas (leaves), conducting and supporting structures (stems), and organs that anchor the plant in the soil and absorb water and minerals from it (roots).

When a seed germinates, growth of the root and shoot proceeds from the apical meristems of the embryo. Certain cells within the meristem retain the ca-pacity to divide. Others elongate and then differentiate, forming, according to their position, the various specialized cells of the plant (see Table 40–1). These cells, in varying combinations, form the three tissue systems—dermal, ground, and vascular—that are continuous throughout the plant body. Parenchyma cells, which are thin-walled and many-sided, are the most common cells in plants.

The ground tissue of the leaf is composed primarily

Table 40–1 Summary of Principal Cell Types in Angiosperms

Cell Type	Location	Characteristics	Function
Meristematic (in apical meristem)	Apices of shoots and roots	Many-sided, small, thin-walled; vacuoles usually small	Origin of primary meristematic tissues and of root cap cells
Meristematic (in vascular cambium)	Lateral, between secondary phloem and secondary xylem	Elongate, often spindle-shaped	Produces secondary xylem and phloem
Epidermal	Surface of entire primary plant body	Flattened, variable in shape, overlaid by cuticle; some specialized as guard cells	Protective covering; prevents desiccation yet allows gas exchange
Parenchyma	Everywhere, usually dominant in pith, cortex, mesophyll	Many-sided, usually thin-walled; abundant air spaces between cells	Photosynthesis, respiration, storage, wound healing, among others
Collenchyma	Peripheral in cortex of stem and in leaves	Elongate, with irregularly thickened primary walls	Support for young stems and leaves
Sclereid	In pith and cortex of stems; in leaves and flesh of fruits; seed coats	Irregular; massive secondary wall; alive or dead at maturity	Produces hard texture, mechanical support
Fiber	Primary and secondary xylem and phloem; cortex	Very long, narrow cell, with secondary cell walls; often dead at maturity	Support
Tracheid	Primary or secondary xylem	Elongate, with pits in walls; dead at maturity	Conduction of water and solutes
Vessel member	Interconnected series (= vessels) in primary or secondary xylem	Elongate, with pits in walls and end walls perforated; dead at maturity	Conduction of water and solutes
Sieve-tube member	Primary or secondary phloem, usually with companion cell; cells form interconnected series (= sieve tubes)	Elongate, with specialized sieve plates; nucleus lacking at maturity	Conduction of organic solutes
Cork	Surface of stems and roots with secondary growth	Flattened cells, compactly arranged; dead at maturity, cells are often air-filled	Restricts gas exchange and water loss

of photosynthetic parenchyma cells. Palisade parenchyma cells, which are located in the upper portion of the leaf, are elongated cells with large central vacuoles. Spongy parenchyma cells, which are located below the palisade cells, are surrounded by large air spaces. The palisade and spongy parenchyma cells make up the mesophyll, or "middle leaf."

The upper and lower surfaces of the leaf consist of one or more layers of transparent epidermal cells covered with a waxy layer, the cuticle. Specialized pores, the stomata, open and close, regulating the exchange of gases and the release of water vapor. Veins, the vascular bundles of the leaf, conduct water and minerals to the

mesophyll cells of the leaf (through the xylem) and transport sugars away from them (through the phloem). The vascular tissues of leaves are continuous with those of the stem and roots.

The embryonic root is the first structure to break out of the germinating seed. Cells produced by its apical meristem form a root cap, which protects the root tip as it is pushed through the soil. Young roots have an outer layer of epidermis and, at most, a very thin cuticle. Extensions of the epidermal cells form root hairs, which greatly increase the absorptive surface of the root. Inside the epidermis is the ground tissue of the root, the cortex, composed mostly of parenchyma cells, often modified for storage. The innermost layer of the cortex is the endodermis, a single layer of specialized cells whose walls contain a waterproof zone, the Casparian strip. Just inside the endodermis is another layer of cells, the pericycle, from which branch roots arise. To the interior of the pericycle are the xylem and phloem.

Green stems, like leaves, have an outer layer of epidermal cells covered with a cuticle. The bulk of the young stem is ground tissue, which may be divided into an outer cylinder (the cortex) and an inner core (the pith). The ground tissue is composed largely of parenchyma cells but also may contain collenchyma cells and sclerenchyma cells.

The vascular tissues consist of phloem and xylem. In angiosperms, the conducting cells of the phloem are the sieve-tube members, living cells with perforated end walls that form continuous sieve tubes. Associated closely with each sieve-tube member is a companion cell. The conducting tissue of the xylem is made up of a series of tracheids or vessel members. Tracheids and vessel members characteristically have thick secondary walls and are dead at functional maturity. Phloem and xylem also contain parenchyma cells and fibers.

The height of the aboveground parts of a plant is increased through primary growth of the shoot system. Leaf primordia arise from the apical meristem of the shoot. As the nodes are separated by elongation of the internodes, small masses of meristem (buds) form in the axils of the leaves. These axillary buds may remain dormant or may give rise to branches or specialized shoots. Among the specialized shoots that may form are runners and rhizomes, through which many species reproduce asexually, producing populations of genetically identical individuals.

Secondary growth is the process by which woody plants increase their girth. Such growth arises primarily from the vascular cambium, a sheath of meristematic tissue completely surrounding the xylem and completely surrounded by phloem. The cambial cells divide during the growing season, adding new xylem cells (secondary xylem) on their inner surfaces and new phloem cells (secondary phloem) on their outer surfaces. As the trunk increases in girth, the epidermis is eventually ruptured and replaced by cork, produced by the cork cambium.

Questions

1. Distinguish among the following: dermal tissue/ground tissue/vascular tissue; epidermis/cuticle; parenchyma/collenchyma/sclerenchyma; palisade parenchyma/spongy parenchyma; xylem/phloem; sieve-tube member/tracheid/vessel member; apical meristem/lateral meristem; primary growth/secondary growth.

2. Describe the distinguishing characteristics of monocots and dicots.

3. Sketch the structure of a leaf, labeling the principal cells and tissues. Describe the function of each of the labeled parts.

4. Where might you expect to find a leaf with stomata only on the lower surface? Only on the upper surface?

5. Some leaves are specialized for functions other than photosynthesis. List three other functions for which leaves may be specialized, and give an example of each specialization.

6. We have seen that the tissue of the root cortex contains many air spaces. Also, we noted that some plants growing in swampy areas have air roots. You may have discovered that overwatering can kill house plants. What does this information indicate about roots?

7. Why are there no root hairs or lateral roots at the extreme tip of a root?

8. Sketch cross sections of (a) a root, (b) a dicot stem, and (c) a monocot stem. Label the principal tissues in each sketch, and describe the function of each of the labeled parts.

9. What two cell layers are present in roots but absent in stems? What functions do the cells of these layers perform in the roots, and why are they necessary for the survival of the plant?

10. It has often been said that the principal differences among root, stem, and leaf are quantitative in nature rather than qualitative. Explain.

11. Suppose you carve your initials 1 meter above the ground on the trunk of a mature tree that is growing vertically at an average of 15 centimeters per year. How high will your initials be at the end of 2 years? At the end of 20 years?

12. As a result of secondary growth, most of our common trees increase in girth as they increase in height. What limitations would a lack of secondary growth impose on the form and mechanical support of a tree?

13. Sketch a tree trunk with secondary growth. Compare your sketch with Figure 40–21c.

14. Porcupines often eat all of the bark off young trees, up to the height they can reach. When they do this, the tree dies. Why?

15. Consider a mature tree. Which parts of the tree are composed of living cells? Why might it be advantageous for living cells to make up such a small fraction of the mass of the tree? Similarly, why might it be advantageous for a deciduous tree to invest as little material as possible in making leaves?

16. What plant part are you eating when you eat each of the following: carrots, eggplant, lima beans, corn on the cob, green beans, celery, tomatoes, spinach, brussels sprouts, broccoli, and white potatoes?

Suggestions for Further Reading

Cutler, David F.: *Applied Plant Anatomy,* 2d ed., Longman, Inc., New York, 1983.

> *An interestingly written textbook on the fundamentals of plant anatomy, showing some of the ways in which a knowledge of plant anatomy can be applied to solve many important everyday problems.*

Hitch, Charles J.: "Dendrochronology and Serendipity," *American Scientist,* vol. 70, pages 300–305, 1982.

Langenhein, Jean H.: "Plant Resins," *American Scientist,* vol. 78, pages 16–24, 1990.

Ledbetter, M. C., and Keith Porter: *Introduction to the Fine Structure of Plant Cells,* Springer-Verlag, New York, 1970.

> *An excellent atlas of electron micrographs of plant cells, with a detailed explanation of each.*

Niklas, Karl J.: "The Cellular Mechanics of Plants," *American Scientist,* vol. 77, pages 344–349, 1989.

Shigo, Alex L.: "Compartmentalization of Decay in Trees," *Scientific American,* April 1985, pages 96–103.

Tomlinson, P. B.: "Tree Architecture," *American Scientist,* vol. 71, pages 141–149, 1983.

Troughton, John H., and F. B. Sampson: *Plants: A Scanning Electron Microscope Survey,* John Wiley & Sons, Inc., New York, 1973.*

> *A scanning electron microscope study of some anatomical features of plants and the relationship of these features to physiological processes.*

*Available in paperback.

The bodies of dead plants and animals, their excreta, and their discarded leaves and exoskeletons all return nitrogen to the soil, where a portion of this precious commodity is lost. The loss is serious. Nitrogen makes up 77 percent of the air around us but less than 1 percent of the tissues of an average plant—alfalfa, for instance. Moreover, neither plants nor animals can extract nitrogen from the atmosphere. Only a very few species of bacteria can "fix" nitrogen gas (N_2), converting it to compounds that can be taken up by the roots of plants. If these bacteria were to cease operations, there would, eventually, be no more proteins, no more nucleic acids, no more life on Earth.

Plants compete for the available nitrogen in different ways. Many species, particularly members of the legume (pea) family, enter into partnerships with nitrogen-fixing bacteria. Many more species enlist the aid of various fungi, taking advantage of their abilities to break down complex molecules into simple ones. A few species capture and devour animals.

One such carnivorous plant is *Drosera rotundifola,* the common sundew. Each plant has six to twelve modified leaves arranged in a rosette. Each leaf has as many as 260 fine filaments protruding from it (Charles Darwin counted the number of filaments on sundews he studied on the heath in Sussex). At the tip of each filament is a minute, globe-shaped gland that secretes a sticky fluid. The fluid droplets gleam in the sunlight, as the plant's name suggests, and the leaves give off an enticing odor.

Once an insect, such as the lacewing shown here, lights on one of a sundew's shimmering leaves, it cannot escape. First one leg and then another becomes trapped on the sticky glands. Neighboring filaments, each with its sticky droplet, fold slowly in toward the captive. Soon its body is immersed in a colorless mucilage. Its air passages fill up, and it suffocates. The glands then secrete a "ferment," to use Darwin's term, that digests the captive. By this means, he noted, sundews "often grow in places where hardly any other plant . . . can exist."

41

Transport Processes in Plants

As we noted in Chapter 25, the plants that have come to dominate the modern landscape are those that have evolved the best solutions (so far) to two critical problems: conveying the sperm to the egg in the absence of free water, and transporting needed materials to the various cells and tissues of the plant body. In this chapter, we shall discuss first the movement of water and minerals in the xylem; second, the transport of dissolved sugars and other organic substances in the phloem; and third, a related subject, factors affecting the availability of nitrogen and other nutrients to the plant.

You may find it helpful to review the properties of water (especially pages 37 to 42) and the ways in which it moves (pages 110 to 116) before proceeding. The transport processes of plants depend on the extraordinary properties of this very common liquid.

The Movement of Water and Minerals

Transpiration

A plant needs far more water than an animal of comparable weight. In an animal, most of its water remains in its body and recirculates. By contrast, more than 90 percent of the water entering a plant's roots is released into the air as water vapor. This loss of water vapor from the plant body is known as **transpiration.**

Transpiration is a necessary consequence of the opening of the stomata that must occur for a plant to obtain the carbon dioxide required for photosynthesis (Figure 41–1, on the next page), but the price is high. For example, a single corn plant requires 160 to 200 liters of water to grow from seed to harvest, and 1 hectare (2.47 acres) of corn requires almost 5 million liters of water a season. British ecologist John L. Harper describes the terrestrial plant as a "wick connecting the water reservoir of the soil with the atmosphere."

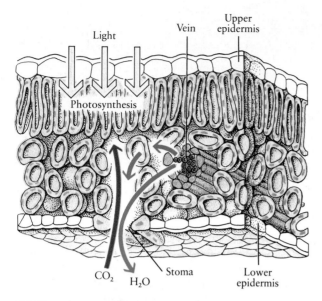

41–1   *As carbon dioxide, essential for photosynthesis, enters the leaf through the stomata, water vapor is lost. Although this water loss poses serious problems for plants, it provides both a force for the uptake of water by the roots and a mechanism for cooling the leaf. The temperature of a leaf may be as much as 10 to 15°C lower than that of the surrounding air, simply because evaporating water carries heat away (see page 41).*

41–2   *Guttation droplets on the edge of wild strawberry leaves. Guttation, the loss of liquid water, occurs when the root pressure forces water out of the xylem in the vascular bundles of the leaves. The water escapes through specialized pores located near the ends of the principal veins. Guttation, which is restricted to relatively small plants, usually occurs at night when transpiration is greatly reduced and the moist air decreases the rate of evaporation.*

The Uptake of Water

Water enters most plants almost entirely through the roots, principally through the root hairs. During periods of rapid transpiration, water may be removed from around the roots so quickly that soil water in the vicinity of the roots becomes depleted. Water then moves slowly by diffusion and capillary action through the soil toward the depleted region. Roots also obtain additional water by growing beyond the depleted region. The main roots of corn plants, for example, grow an average of 52 to 63 millimeters a day.

Root cells, like other living parts of the plant, contain a higher concentration of solutes than does soil water. As a consequence, water from the soil enters the roots by osmosis. The resulting pressure, known as **root pressure,** is sufficient to move water a short distance up the stem. **Guttation,** the loss of liquid water from a plant (Figure 41–2), is a visible consequence of root pressure. But how can water reach 20 meters high to the top of an oak tree, travel three stories up the stem of a vine, or move 125 meters up in a tall redwood?

One important clue is the observation that during times when the most rapid transpiration is taking place—which is when the flow of water up the stem must be the greatest—xylem pressures are negative (less than atmospheric pressure). This negative pressure can be demonstrated. If you peel a piece of bark from a transpiring tree and make a cut in the xylem, no liquid runs out. In fact, if you place a drop of water on the cut, the drop will be drawn in.

What is the pulling force? It is not simple suction, as the negative pressure might indicate. Suction simply removes air from a system so that the water (or other liquid) is pushed up by atmospheric pressure. But atmospheric pressure is only enough to raise water (against no resistance) approximately 10 meters at sea level, and many trees are much taller than 10 meters.

The Cohesion-Tension Theory

According to the now generally accepted theory, the explanation for the movement of water is to be found not only in the properties of plants but also in the properties of water, to which plants have become exquisitely adapted.

As we pointed out in Chapter 2, in every water molecule, two hydrogen atoms are covalently bonded to a single oxygen atom. Each hydrogen atom is also held to the oxygen atom of a neighboring water molecule by a hydrogen bond. The cohesion resulting from this secondary attraction is so great that the tensile strength in a thin column of water can be as much as 140 kilograms per square centimeter (2,000 pounds per square inch). This means that a negative pressure of more than 140 kilograms per square centimeter is required to pull the column of water apart.

In a leaf, water evaporates, molecule by molecule, from the parenchyma cells into the air spaces of the leaf. As the water potential (page 110) of a leaf cell decreases, water from the vessels and tracheids moves, molecule by molecule, into the cell. But each water molecule in a tracheid or vessel is linked to other water molecules in the tracheid or vessel. They, in turn, are linked to others, forming a long, narrow, continuous column of water reaching down all the way to a root hair. As a molecule of water moves through the stem and into the leaf, it tugs the next molecule along behind it.

Because the diameter of the vessels is relatively small and because water molecules adhere to the walls of the vessels (even as they are cohering to one another), gas bubbles, which could rupture the water column, do not usually form. The pulling action, molecule by molecule, causes the negative pressure observed in the xylem. The technical term for a negative pressure is tension, and this theory of water movement is known as the **cohesion-tension theory.**

The principle of cohesion-tension is illustrated in Figure 41–3. As indicated in the diagram, the power for this process comes not from the plant, which plays only a passive role in transpiration, but from the energy of the sun.

Factors Influencing Transpiration

Transpiration is costly, especially when water supply is limited. Several factors affect the rate of water loss. One is temperature; the rate of evaporation doubles for every 10°C increase in temperature. Humidity is also important. Water is lost much more slowly into air already laden with water vapor. Air currents also affect transpiration. Wind blows away the water vapor from leaf surfaces, which increases the steepness of the concentration gradient of water between the interior of the leaf and the air surrounding the leaf. Thus wind hastens the evaporation of water molecules from the leaf. By far the most important factor affecting transpiration, however, is the action of the stomata.

The Mechanism of Stomatal Movements

It has long been known that the osmotic movement of water is involved in the opening and closing of the sto-

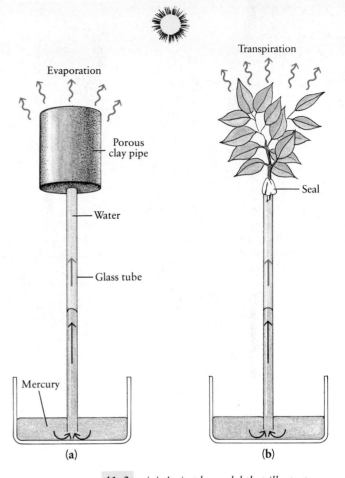

41–3   (a) *A simple model that illustrates the cohesion-tension theory. A piece of porous clay pipe, closed at both ends, is filled with water. A long, narrow glass tube, open at both ends and also filled with water, is inserted into the bottom of the clay pipe. The water-filled tube is placed with its lower end below the surface of a volume of mercury contained in a beaker. As water molecules evaporate from the pores in the pipe, they are replaced by water "pulled up" through the narrow glass tube in a continuous column. As the water evaporates, the mercury rises in the tube to replace it.* (b) *Transpiration from plant leaves results in sufficient water loss to create a similar negative pressure.*

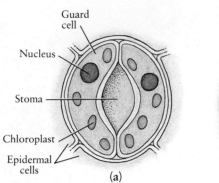

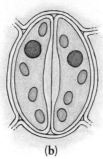

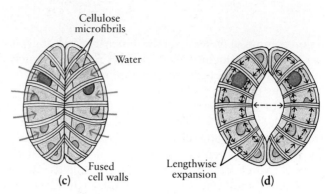

41–4   *Mechanism of stomatal movement. A stoma is bordered by two guard cells that* (a) *open the stoma when they are turgid and* (b) *close it when they lose turgor. The opening of the stoma in response to turgor is made possible by cellulose microfibrils that are arranged in hoops around the circumference of the guard cells* (c). *The arrangement is similar to that*

of the belts in radial tires. When water enters the guard cells, the microfibrils prevent radial expansion, and the only direction in which the cells can expand is lengthwise. Because the two cells are attached to each other at the ends, this lengthwise expansion forces them to bow out and the stoma to open (d).

mata. As shown in Figure 41–4, each stoma has two surrounding guard cells. Stomatal movements are caused by changes in the turgor (page 116) of these cells. When the guard cells are turgid, they bow out, opening the stoma. When they lose water, they relax, and the stoma closes.

Turgor, as you will recall, is maintained or lost due to the osmotic movement of water into or out of cells. Water moves across the plasma membrane from a solution of low solute concentration into a solution of high solute concentration. The active accumulation of solutes in the guard cells causes water to move into them by osmosis. Conversely, a decrease in the solute concentration of guard cells results in the osmotic movement of water out of the cells.

Techniques that make it possible to measure ionic concentrations within individual guard cells have revealed that the critical solute affecting the osmotic movement of water into and out of these cells is the potassium ion (K^+). With an increase in the K^+ concentration, the stomata open, and with a decrease, they close. Current evidence indicates that potassium ions are actively transported between the guard cells and the reservoir provided by the surrounding epidermal cells. Water follows by osmosis, causing the observed changes in the turgor and shape of the guard cells.

In many, but not all, species, chloride ions, Cl^-, accompany the K^+ ions across the membrane, maintaining electrical neutrality. In other, perhaps all, species, H^+ ions are transported in the opposite direction, producing a decreased hydrogen ion concentration within the guard cells of open stomata.

The energy source for the active transport of K^+ ions

between the guard cells and the surrounding epidermal cells is not yet known. On the basis of present evidence, however, one of two possibilities seems likely. Guard cells contain chloroplasts, and it may be that K^+ transport is powered by ATP produced by photophosphorylation reactions (page 168) occurring in these chloroplasts. Another possibility is that the transport of H^+ ions, in a direction opposite to that of the K^+ ions, establishes an electrochemical gradient down which the K^+ ions move. If this should be the case, the transport of K^+ ions into and out of the guard cells—and thus, the opening and closing of the stomata—would be yet another example of a vital process powered by a chemiosmotic mechanism (page 152).

Factors Influencing Stomatal Movements

A number of environmental factors affect the movement of potassium ions into and out of the guard cells and thus influence the opening and closing of the stomata. The principal one is the availability of water. When the water available to a leaf drops below a certain critical point (which varies from species to species), the stomata close, thereby limiting evaporation of the remaining water. This generally occurs *before* the leaf loses turgor and wilts.

A plant's capacity to anticipate water stress depends on the action of a hormone known as **abscisic acid.** This hormone acts by binding to specific receptors in the plasma membrane of the guard cell. The receptor-hormone complex then triggers a change in the plasma membrane, making it more permeable to potassium ions.

41–5 *Corn plants, photographed in a Wisconsin field in June of 1988, during a severe drought that affected many of the prime agricultural areas of North America. The leaves of corn plants have numerous stomata on both the upper and lower surfaces, with slightly more on the lower surface. The cuticle on the lower surface is significantly heavier than on the upper surface. When corn plants are under severe water stress, and closure of the stomata is inadequate to protect the plants against further water loss, the leaves roll up, as shown here. Leaf rolling reduces transpiration, protecting virtually all of the stomata on the upper surface, as well as many on the lower surface. All that remains exposed to the environment is a portion of the lower surface, with its heavy cuticle.*

Stomatal movement also occurs independently of the loss or gain of water by the plant. Among the other factors that affect stomatal movement are the carbon dioxide concentration, temperature, and light. In most species, an increase in carbon dioxide concentration in the intercellular spaces of a leaf causes the stomata to close. The magnitude of this response varies from species to species and with the degree of water stress that a particular plant has undergone. The site for sensing the level of carbon dioxide is inside the guard cells.

Within normal ranges, temperature changes have little effect on the stomata, but temperatures higher than 35°C (95°F) can cause stomatal closing in some species. An increase in temperature results in an increase in respiration and thus in the concentration of intercellular carbon dioxide. Many plants in hot climates close their stomata regularly at midday, apparently because of both increased water stress and the effect of temperature on the concentration of carbon dioxide in the leaves.

In most species, the stomata close regularly in the evening, when photosynthesis is no longer possible, and

open again in the morning. These movements in response to light occur even when there are no changes in the water available to the plant. Opening in the light and closing in the dark may be a result of changes in the carbon dioxide concentration. Photosynthesis uses carbon dioxide and thus reduces the internal concentration, whereas respiration causes an increase in the carbon dioxide concentration.

Light may also have a more direct effect. Blue light has long been known to stimulate stomatal opening independently of the carbon dioxide concentration. The pigment that absorbs blue light, which is thought to be located in either the plasma membrane or the vacuole membrane, promotes the uptake of K^+ ions by the guard cells. Recent experiments have demonstrated that this uptake of K^+ ions is a secondary effect. The primary effect is to stimulate the pumping of H^+ ions *out of* the guard cells. These experiments provide support for the hypothesis that an electrochemical gradient of H^+ ions powers K^+ transport, but they do not rule out the possible involvement of ATP.

Crassulacean Acid Metabolism

Although the stomata of most plants are open during the day and closed at night, the stomata of some species close in the daytime and open at night. Not only is the temperature lower at night but also the humidity is usually higher. Both factors reduce the rate of transpiration. The species that open their stomata only at night include a variety of plants adapted to hot, dry climates, for example, cacti, pineapples, and members of the stonecrop family (Crassulaceae), such as *Sedum* (Figure 40–5c, page 723).

These plants take in carbon dioxide at night, converting it to four-carbon acids. During the day, when the stomata are closed, the carbon dioxide is released from the acids and used immediately in photosynthesis. This process is known as **Crassulacean acid metabolism,** and plants that employ it are known as **CAM plants.** Crassulacean acid metabolism is analogous to the C_4 pathway in photosynthesis described on page 170, although it apparently evolved independently.

The Uptake of Minerals

In addition to water and the energy-rich sugars and other compounds produced by photosynthesis, plant cells require a number of different chemical elements. These elements are found in rocks and soil in the form of naturally occurring inorganic substances known as **minerals.** Mineral ions are taken up in water solution by the roots and travel through the xylem in the transpiration stream. As you saw in Figure 40–9 (page 726),

Table 41–1   A Summary of Mineral Elements Required by Plants

Element	Principal Form in Which Absorbed	Approximate Concentration in Whole Plants (as % of Dry Weight)	Some Functions
Macronutrients			
Nitrogen	NO_3^- (or NH_4^+)	1–4%	Component of amino acids, proteins, nucleotides, nucleic acids, chlorophyll, and coenzymes
Potassium	K^+	0.5–6%	Involved in osmosis and ionic balance and in opening and closing of stomata; activator of many enzymes
Calcium	Ca^{2+}	0.2–3.5%	Component of cell walls; enzyme cofactor; involved in plasma membrane permeability and transport of ions and hormones
Phosphorus	$H_2PO_4^-$ or HPO_4^{2-}	0.1–0.8%	Component of energy-carrying phosphate compounds (ATP and ADP), phospholipids, nucleic acids, and several essential coenzymes
Magnesium	Mg^{2+}	0.1–0.8%	Part of the chlorophyll molecule; activator of many enzymes
Sulfur	SO_4^{2-}	0.05–1%	Component of some amino acids, proteins, and coenzyme A
Micronutrients			
Iron	Fe^{2+}, Fe^{3+}	25–300 parts per million (ppm)*	Required for chloroplast development; component of cytochromes
Chlorine	Cl^-	100–10,000 ppm	Involved in osmosis and ionic balance; essential in photosynthesis in the reactions in which oxygen is produced
Copper	Cu^{2+}	4–30 ppm	Activator or component of some enzymes
Manganese	Mn^{2+}	15–800 ppm	Activator of some enzymes; required for integrity of chloroplast membrane and for oxygen release in photosynthesis
Zinc	Zn^{2+}	15–100 ppm	Activator or component of many enzymes
Boron	$B(OH)_3$ or $B(OH)_4^-$	5–75 ppm	Influences Ca^{2+} utilization, nucleic acid synthesis, and membrane integrity
Molybdenum	MoO_4^{2-}	0.1–5.0 ppm	Required for nitrogen metabolism
Elements Essential to Some Plants or to Organisms on Which Plants Depend			
Cobalt	Co^{2+}	Trace	Required by nitrogen-fixing microorganisms
Sodium	Na^+	Trace	Involved in osmosis and ionic balance; for many plants, probably not essential; required by some desert and salt-marsh species and may be required by all plants that utilize C_4 pathway of photosynthesis

* Parts per million (ppm) equal units of an element by weight per million units of oven-dried plant material; 1% equals 10,000 ppm.

(a)   (b)

41–6 *Some plants have special mineral requirements. (a) Plants of the mustard family use sulfur in the synthesis of the mustard oils that give the plants their characteristic sharp taste. These mustard plants were growing along the side of a highway near Lima, Ohio. (b) Horsetails incorporate silicon into their cell walls, making them indigestible to most herbivores but useful, at least in colonial America, for scouring pots and pans. These young horsetails were photographed in California.*

the cells of the endodermis play a major role in determining which substances enter the xylem.

The ionic composition of plant cells is far different from the ionic composition of the medium in which the plant grows, indicating that mineral ions are brought into plant cells by active transport. One way that roots move ions against a concentration gradient is by the action of transport proteins in the plasma membrane (page 117). Such proteins combine with the ions and then, utilizing the energy of ATP, carry them to the other side of the membrane and release them.

Table 41–1 lists the mineral elements required by plants, the form in which they are usually absorbed, and some of the uses plants make of them. Depending on the concentration of a mineral element in plants, it is classified as either a **macronutrient** or a **micronutrient.** The requirements of plants for a macronutrient can generally be determined by an analysis of the chemical composition of the plant cells. The requirements for micronutrients are identified by studying the capacity of the plants to grow in distilled water to which very small amounts of specific mineral ions are added. This sounds easier than it is. Sometimes, it has been found, a substance—molybdenum, for example—is needed in such small amounts that it is almost impossible to set up experimental conditions that exclude it. Thus, it is difficult to prove that its absence is lethal.

You might anticipate that organisms make use of what is most readily available. Indeed, they seem to have done this when life originated from elements in the gases of the primitive atmosphere. But Table 41–1 reveals some findings that you might not expect. Sodium, for instance, which is one of the most abundant of the elements, is apparently not required at all by most

plants. The fact that most plants evolved with no functions requiring sodium is even more striking when you consider that sodium is vital to animals. In the seas, where both plant and animal life seem to have originated, sodium is the most abundant mineral element and is far more readily available than potassium, which it closely resembles in its essential properties. Similarly, silicon and aluminum are usually present in large amounts in soils, but only a few plants require silicon and apparently none requires aluminum. Conversely, most plants need molybdenum, which is relatively rare.

The Movement of Sugars: Translocation

The photosynthetic cells of the plant, which are most abundant in the leaves, provide the food energy for all of the other cells of the plant. The process by which the products of photosynthesis are transported to other tissues is known as **translocation.**

Translocation has proved a very difficult process to study because the sieve-tube members of the phloem are quite delicate. When they are disturbed, callose and the protein that normally lines the interior of the cell walls (see Figure 40–15, page 730) plug the pores in the sieve plates, preventing the movement of substances through them.

Two investigative techniques have proved useful in getting around this difficulty. One technique is to expose photosynthesizing plants to radioactive carbon dioxide ($^{14}CO_2$). The sugars made from this carbon dioxide are radioactive, making it possible to trace their passage through the plant. Another technique involves

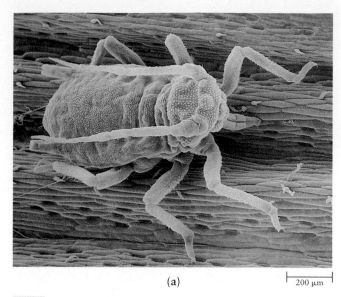

(a)   200 μm

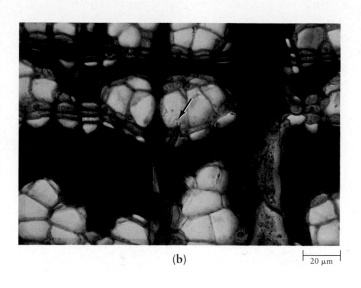

(b)   20 μm

41–7   *Assistance by aphids.* (a) *Aphids are small insects that feed on plant sap. You have probably seen them on rose bushes. The aphid drives its sharp mouthparts, or stylets, like a hypodermic needle through the epidermis of a leaf or a stem. The stylets traverse the cortical cells of the stem or the mesophyll cells of the leaf and then tap the contents of a single sieve-tube member.* (b) *In this micrograph of a basswood stem,* *the tips of the stylets (indicated by the arrow) are in a sieve-tube member in the secondary phloem. If the aphid is anesthetized, it is possible to sever its body from the stylets, leaving the stylets undisturbed in the cell. The sieve-tube sap often continues to exude through them for many hours, and pure samples can be collected for analysis without damaging the sieve tube or interfering with its function.*

the use of aphids for taking samples of sieve-tube sap (Figure 41–7).

Data gathered by these techniques indicate that sieve-tube sap contains (by weight) 10 to 25 percent solutes, more than 90 percent of which are sugars, mostly sucrose. Low concentrations of amino acids and other nitrogen-containing substances are also present. The rate of movement of the solutes along the sieve tube is remarkably fast. In one set of experiments, for example, it was estimated that the sap was moving at a rate of about 100 centimeters per hour, far faster than could be accounted for by diffusion alone. At this rate, each individual sieve-tube member was emptying and refilling every two seconds.

The Pressure-Flow Hypothesis

The movement of sugars and other organic solutes in translocation follows what is known as a **source-to-sink pattern.** The principal sources of the solutes are the photosynthesizing leaves, but storage tissues may also serve as important sources. All plant parts unable to meet their own nutritional needs may act as sinks, that is, as importers of organic solutes. Thus, storage tissues act as sinks when they are importing solutes and as sources when they are exporting solutes.

The most widely accepted explanation for the source-to-sink movement in translocation is the **pressure-flow hypothesis.** According to this hypothesis, the solutes move in solutions that move, in turn, because of differences in water potential caused by concentration gradients of sugar.

As shown in Figure 41–8, sugars from the photosynthetic cells of the leaf are moved into the sieve tubes against a concentration gradient. This active transport process is thought to involve the cotransport (page 119) of sucrose molecules and hydrogen ions by a specific protein in the sieve-tube membrane. The incoming sugar decreases the water potential in the sieve tube and causes water to move into the sieve tube from the xylem by osmosis. At a sink—for example, a storage root—sugar molecules leave the sieve tube. Water molecules follow the sugar molecules out, again by osmosis. Thus the water flows in at one end of the sieve tube and out at the other. Between these two points, the water and its solutes, including sugar, move passively by bulk flow. The speed of transport depends on the differences in concentration between source and sink. At the sink,

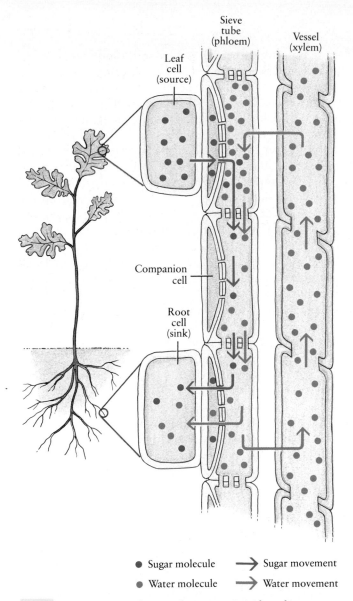

Sieve tube (phloem)

Vessel (xylem)

Leaf cell (source)

Companion cell

Root cell (sink)

● Sugar molecule ⟶ Sugar movement

● Water molecule ⟶ Water movement

41–8   *The pressure-flow mechanism as it is thought to occur in the plant body. Sugar molecules are actively loaded into the sieve tube at the source. As a consequence of the increased concentration of sugar, the water potential is decreased, and water enters the sieve tube from the xylem by osmosis. Sugar molecules are unloaded from the sieve tube at the sink, and the sugar concentration in the sieve tube falls. As a result, water moves out of the sieve tube by osmosis. Because of the active secretion of sugar molecules into the sieve tube at the source and their removal from the sieve tube at the sink, a flow of sugar solution takes place along the tube between source and sink.*

sugars may be either utilized or stored, but most of the water returns to the xylem and is recirculated in the transpiration stream.

Companion cells, which contain many mitochondria, are believed to be very active metabolically. It is hypothesized that one of their functions is to meet the energy requirements of the sieve-tube members with which they are associated. However, the actual flow of the sugar solution is a passive process, requiring no metabolic activity on the part of the phloem cells.

Factors Influencing Plant Nutrition

As we saw earlier, the mineral elements needed by plants are taken up by the roots in solution and are transported through the plant body in the transpiration stream. Although the availability of minerals depends principally on the nature of the surrounding soil, the activities of symbiotic fungi and bacteria also play a crucial role.

Soil Composition

Soil, the uppermost layer of the Earth's crust, is composed of rock fragments associated with organic material, both living and in various stages of decomposition. It typically has three layers: the A horizon, the B horizon, and the C horizon. The **A horizon,** or **topsoil,** is the zone of maximum organic accumulation (humus). The **B horizon,** or **subsoil,** consists of inorganic particles in combination with mineral ions that have leached (washed) down from the A horizon. The **C horizon** is made up of loose rock that extends down to the bedrock beneath it. As shown in Figure 41–9, the depth and composition of these three layers—and consequently, the fertility of the soil—vary considerably in different environments.

The mineral content of the soil depends in part on the parent rock from which the soil was formed. These differences in mineral content can be extremely localized, with sharp lines of demarcation. Geologists sometimes use types of vegetation or changes in color or growth patterns of plants as indicators of mineral deposits.

In most soils, however, the mineral content is more dependent on biological factors. In an undisturbed environment, most of the mineral nutrients stay within the system—the soil itself and the plants, microorganisms, and small soil animals that it supports and contains. If, however, the vegetation is repeatedly removed, as when crops are harvested or grasslands are overgrazed, or the top, humus-rich layer of the A horizon is eroded, the

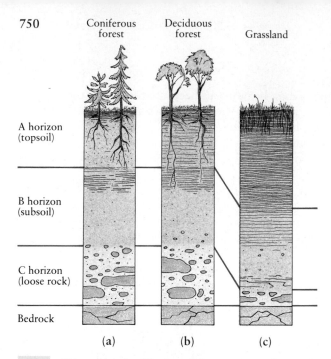

Coniferous forest   Deciduous forest   Grassland

A horizon (topsoil)

B horizon (subsoil)

C horizon (loose rock)

Bedrock

(a)   (b)   (c)

41–9 *Diagrams of soil layers of three major soil types. (a) The litter of the northern coniferous forest is acidic and slow to decay, and the soil has little accumulation of humus, is very acidic, and is leached of minerals. (b) In the cool, temperate deciduous forest, decay is somewhat more rapid, leaching less extensive, and the soil more fertile. Such soils have been widely used for agriculture, but they need to be prepared by adding lime (to reduce acidity) and fertilizer. (c) In the grasslands, almost all of the plant material above the ground dies each year as do many of the roots, and thus large amounts of organic matter are constantly returned to the soil. In addition, the finely divided roots penetrate the soil extensively. The result is highly fertile soil, often black in color, with a topsoil sometimes more than a meter in depth.*

soil rapidly becomes depleted of nutrients. It can be used for agricultural purposes only if it is heavily fertilized.

Another factor influencing the mineral content of soil is the size of the soil particles. The smaller fragments of rock are classified as sand, silt, or clay (Table 41–2). Water and mineral ions drain rapidly through soil com-

Table 41–2	Soil Classification
	Diameter of Particles (micrometers)
Coarse sand	200–2,000 (0.2–2 millimeters)
Sand	20–200
Silt	2–20
Clay	Less than 2

posed of large particles (sandy soil). Soil composed of small particles (clay) holds the water against gravity. Moreover, the small clay particles are negatively charged and therefore bind positively charged ions, such as calcium (Ca^{2+}), potassium (K^+), and magnesium (Mg^{2+}). However, a pure clay soil is poorly suited for plant growth because it is usually too tightly packed to let in enough oxygen for the respiration of plant roots, soil animals, and most soil microorganisms. Clay soils that contain enough larger particles to keep the soil from packing are known as **loams,** and these are generally the best soils for plant growth.

The pH of the soil also affects its capacity to retain nutrients. In acidic soil, positively charged ions leach out of the soil. Soil pH also affects the solubility of certain nutrient elements. Calcium, for example, is more soluble (and therefore more available to plant roots) as pH increases. Crops such as alfalfa and other legumes have high calcium requirements and grow best in alkaline soil. Rhododendrons and azaleas, on the other hand, have high requirements for iron, which is abundant only when the soil is acidic.

Soils and plant life interact. Plants constantly add to the humus, thereby changing not only the content of the soil but also its texture and its capacity to hold minerals and water. In turn, the plants are dependent upon the mineral content of the soil and its holding capacity. As these improve, plants increase in both number and size and also often change in kind, thereby producing further changes in the soil. Thus, under natural conditions, the soil—and the availability of nutrients and water to plant roots—are constantly changing.

The Role of Symbioses

Two types of symbiotic relationships, one involving fungi and the other involving bacteria, play important roles in plant nutrition.

Mycorrhizae

Mycorrhizae ("fungus-roots"), described in Chapter 25, are symbiotic associations between fungi and roots. In these associations, the fungi extract nutrients from the soil and make them available to plants, thus enabling plants to prosper on nutrient-poor soils. Recent studies indicate that the fungi may also screen out chemicals, making it possible for plants to live in soils that would otherwise be toxic.

Rhizobia and Nitrogen Fixation

All but one of the elements listed in Table 41–1 are derived principally from the weathering of rocks. The exception is nitrogen, the nutrient for which plants have

Recombinant DNA and Nitrogen Fixation

In nitrogen fixation, the key chemical reaction is the combination of nitrogen with hydrogen, producing ammonia. In nitrogen-fixing prokaryotes, this reaction is catalyzed by an enzyme complex known as **nitrogenase,** which consists of two different polypeptides, one containing iron and the other molybdenum.

Genetic mapping of a free-living nitrogen-fixing bacterium has shown that 17 genes known to be involved in nitrogen fixation are clustered on one portion of the chromosome. In experiments at the University of Sussex, biologists have succeeded in transferring this gene cluster to a plasmid (see page 288) and then introducing the plasmid into *Escherichia coli* cells. The *E. coli* cells were then able to synthesize nitrogenase and to fix nitrogen.

In other experiments at Cornell University, the gene cluster for nitrogenase has been transferred to yeast cells (which are eukaryotes) and has remained intact during repeated cell divisions. These experiments raise the hope that nitrogen-fixing genes can be transferred to the cells of plants such as corn, which could then, in effect, make their own nitrogen-containing compounds. Thus far, however, the yeast cells have not demonstrated the capacity to fix nitrogen, perhaps because of differences in the start and stop signals for protein synthesis or in the cytoplasmic makeup of prokaryotic and eukaryotic cells.

An alternative way of conferring nitrogen-fixing capability on plants would be to transfer to nonlegumes the genes involved in the leguminous plant's contribution to the symbiotic association. Because of the number of genes involved and the complexity of the factors controlling nitrogen fixation, however, such an achievement may be many years away.

Despite these difficulties, biologists in a number of laboratories are pursuing the use of recombinant DNA techniques to increase biological nitrogen fixation. The reasons are practical and simple. Each year, some 50 million tons of fixed nitrogen, in the form of chemical fertilizers, are added to the soil. These fertilizers are produced by commercial nitrogen-fixation processes in which the required energy is supplied by fossil fuels. More than one-third of the total amount of energy needed to produce a crop of corn is used in the manufacture, transportation, and application of chemical fertilizers. As the cost of the fossil fuels required for all of these processes fluctuates, so does the cost of producing the crop.

---

the greatest requirement. Although nitrogen constitutes about 77 percent of the air, most plants cannot use gaseous nitrogen (N_2). They are dependent upon nitrogen-containing ions—ammonium (NH_4^+) and nitrate (NO_3^-)—from the soil (or, in the case of carnivorous plants, such as the sundew on page 740, from their prey). Plants reduce the nitrate ions to ammonium ions, and the ammonium ions are then combined with carbon-containing compounds to form amino acids, nucleotides, chlorophyll, and other nitrogen-containing compounds.

These nitrogen-containing compounds are returned to the soil with the death of the plants (or of the animals that have eaten the plants) and are reprocessed by soil organisms, taken up by the plant roots in the form of nitrate dissolved in soil water, and reconverted to organic compounds. In the course of this cycle (to be discussed in more detail in Chapter 45), a certain amount of nitrogen is always "lost," in the sense that it becomes unavailable to plants.

The main source of nitrogen loss is the removal of plants from the soil. Soils under cultivation often show a steady decline of nitrogen content. Nitrogen may also be lost when topsoil is carried off by soil erosion or when ground cover is destroyed by fire. Nitrogen-containing ions are also leached away by water percolating down through the soil. In addition, numerous types of bacteria are present in the soil that, when oxygen is not present, break down nitrates, releasing nitrogen into the air and using the oxygen for respiration.

If the nitrogen lost from the soil were not steadily replaced, virtually all terrestrial life would finally flicker out. The "lost" nitrogen is returned to the soil by **nitrogen fixation,** the process by which atmospheric nitrogen is incorporated into organic nitrogen-containing compounds. Most nitrogen fixation is carried out by a few kinds of prokaryotes, including both free-living and symbiotic forms of cyanobacteria and heterotrophic bacteria. Of the various classes of nitrogen-fixing organisms, the symbiotic bacteria are by far the most important in terrestrial environments in terms of total amounts of nitrogen fixed.

The most common nitrogen-fixing symbiotic bacteria are members of the genus *Rhizobium* (Figure 41–10), which invade the roots of leguminous plants, such as clover, peas, beans, and alfalfa. The symbiosis between a species of *Rhizobium* and a legume is quite specific. For example, the species that invades and induces nodule formation in clover roots will not induce nodules on the roots of soybeans. The recognition

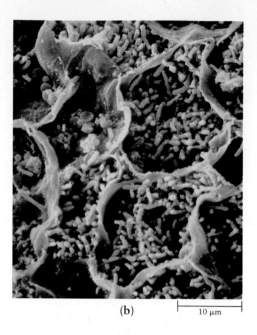

41–10   (a) *Nitrogen-fixing nodules on the roots of a yellow wax bean plant* (Phaseolus vulgaris), *a legume. These nodules are the result of a symbiotic relationship between a soil bacterium* (Rhizobium) *and root cells.* (b) *A scanning electron micrograph of the interior of cells of a nodule on the root of an alfalfa plant. Note the numerous rod-shaped bacteria. The plant supplies the bacteria with an energy source, and the bacteria supply the plant with fixed nitrogen.*

(a)   (b)   10 μm

process on which this specificity depends is thought to involve the interaction of proteins on the root surface with polysaccharides of the bacterial cell wall.

About 250 million metric tons of nitrogen are added to the soil each year, of which some 200 million tons are biological in origin. Just as all organisms, including ourselves, are ultimately dependent on photosynthesis for energy, they—and we—depend on nitrogen fixation for the nitrogen-containing molecules without which living cells cannot function.

Summary

Transpiration is the loss of water vapor by plants. As a consequence of transpiration, plants require large amounts of water. Water enters the plant from the soil through the roots and travels through the plant body in the conducting cells of the xylem. According to the cohesion-tension theory, water moves through the tracheids and vessels under negative pressure (tension). Because the molecules of water cling together (cohesion), a continuous column of water molecules is pulled from the soil solution and into the root, molecule by molecule, by the evaporation of water above.

Diffusion of gases, including water vapor, into and out of the leaf is regulated by the stomata. The stomata are opened and closed by the guard cells as a consequence of changes in turgor. Turgor is increased or decreased by the osmotic movement of water, which follows the movement of potassium ions into or out of the guard cells. The active transport of potassium ions is regulated by a variety of factors, including water stress, abscisic acid, carbon dioxide concentration, temperature, and light.

Minerals, which are naturally occurring inorganic substances, are brought into the plant from the soil in the form of ions and are carried with the transpiration stream in the xylem. They fulfill a variety of functions in plants.

The movement of organic compounds through the plant body is known as translocation. It takes place in the phloem and follows a source-to-sink pattern. According to the pressure-flow hypothesis, sugars are pumped, by active transport, into the sieve tubes in the leaves (or in tissues where they have been stored) and are removed from the sieve tubes in other parts of the plant body where they are needed for growth and energy. Water moves into and out of the sieve tubes by osmosis, following the sugar molecules. These processes create a difference in water potential along the sieve tube, which causes water and the sugars dissolved in it to move by bulk flow along the sieve tube.

Characteristics of the soil affect the availability of mineral ions to plants. These characteristics include the rock from which the soil was formed, the size of the soil particles, the amount of humus present, and the soil pH.

Two types of symbiotic relationships are important in plant nutrition: mycorrhizae (discussed in Chapter 25) and relationships involving nitrogen-fixing bacteria. The associations between nitrogen-fixing bacteria, such as *Rhizobium*, and the roots of certain plants, particularly legumes, result in the incorporation of gaseous nitrogen from the atmosphere into organic nitrogen-containing compounds.

Questions

1. Distinguish among the following: transpiration/translocation; cohesion-tension theory/pressure-flow hypothesis; source/sink; A horizon/B horizon/C horizon; rhizobia/mycorrhizae.

2. What properties of water discussed in Chapters 2 and 6 are important to the movement of water and solutes through plants?

3. Transpiration has often been described as a "necessary evil" to the plant. Why is it necessary? How is it evil?

4. Plants, like animals, are warmed above air temperature by sunlight, but, unlike animals, they cannot actively seek a shady spot. Why might the leaves in the sunlight at the top of an oak tree be smaller and more extensively lobed (more fingerlike) than leaves in the shady areas lower on the tree?

5. Gardeners advise cutting back the shoot system or removing many leaves of a plant after transplanting. How do these actions help the plant to survive?

6. In Figure 41–3, why doesn't air enter the tube through the top of the enclosed porous pipe or the leaf, even though water vapor can easily escape? How can the porous pipe (and, by analogy, the leaf) be permeable to air or water but effectively impermeable at an air-water interface?

7. Identify the cells and tissues through which a molecule of water travels from the time it enters the root until it is used in photosynthesis.

8. When K^+ ions move out of the guard cells, they move into adjacent epidermal cells. How would the influx of K^+ affect the epidermal cells? What effect would this have on the stomata?

9. Using the techniques described in this chapter for the analysis of sieve-tube sap, how would you measure the rate of movement? *(Hint:* Aphids are available.)

10. A particular sugar molecule was produced by photosynthesis in a leaf of a perennial plant late one August. Throughout the following winter, it was stored in a root of the plant. The next spring, it was oxidized in the process of respiration, providing energy for the growth of a new shoot tip. Identify the cells and tissues through which this sugar molecule traveled between its synthesis and its ultimate use.

11. As you will recall from Chapter 9, the immediate product of the Calvin cycle is glyceraldehyde phosphate, a three-carbon sugar (page 171). What chemical reactions did the sugar molecule in Question 10 probably undergo between its synthesis and its use?

12. (a) Experts in flower arranging advise recutting the stems of flowers while holding them under water. Explain why. (b) Some florists advise adding ordinary table sugar to the water in which cut flowers are placed. When this is done, some types of flowers will remain fresh for several weeks. What is the explanation?

13. (a) Nitrogen-fixation is an energy-requiring process. What is the source of the energy used by *Rhizobium*? (b) If new symbiotic associations are developed in which the nitrogen-fixing bacteria produce greater yields of nitrogen-containing compounds, what price is the host plant likely to pay? Why?

Suggestions for Further Reading

Ball, Donald M., Jeffrey F. Pedersen, and Garry D. Lacefield: "The Tall-Fescue Endophyte," *American Scientist,* vol. 81, pages 370–379, 1993.

Brill, Winston J.: "Nitrogen Fixation: Basic to Applied," *American Scientist,* vol. 67, pages 458–466, 1979.

Evert, Ray F.: "Sieve Tube Structure in Relation to Function," *BioScience,* vol. 32, pages 789–795, 1982.

Galston, Arthur W., Peter J. Davies, and Ruth L. Satter: *The Life of the Green Plant,* 3d ed., Prentice-Hall, Inc., Englewood Cliffs, N.J., 1980.*

  A comprehensive description of the functioning of the green plant, especially strong in plant physiology. Written for students without an advanced background in biology or chemistry.

Marx, Jean L.: "How Rhizobia and Legumes Get It Together," *Science,* vol. 230, pages 157–158, 1985.

Olsen, Ralph A., Ralph B. Clark, and Jesse H. Bennett: "The Enhancement of Soil Fertility by Plant Roots," *American Scientist,* vol. 69, pages 378–384, 1981.

Salisbury, Frank B., and Cleon W. Ross: *Plant Physiology,* 4th ed., Wadsworth Publishing Co., Inc., Belmont, Calif., 1991.

  A good, modern plant physiology text for more advanced students.

Walker, Dan B.: "Plants in the Hostile Atmosphere," *Natural History,* June 1978, pages 74–81.

*Available in paperback.

Responses to Stimuli and the Regulation of Plant Growth

Plants have a precise calendar of events, and flowering is a conspicuous example. All the cherry trees in Washington, D.C., flower simultaneously in the early spring, while asters and goldenrods wait till autumn. However, if one of the cherry trees flowered in the autumn, it would be immediately eliminated from the evolutionary sweepstakes.

Plants rely on environmental cues for their scheduling of events. Rainfall, soil temperature, the first fall frost are all indicators of the changing seasons. In the temperate zone, however, by far the most accurate indicator—not subject to the vagaries of wind and weather—is day length, changing on exactly the same schedule, year after year. Equally important, using day length as a signal, plants not only can respond to the changing seasons but can also anticipate the changes and prepare for them.

The loss of leaves by deciduous trees in autumn is an example. In the earliest stage of preparation, valuable ions and molecules are pumped out of the leaf and back into the stem—magnesium, for example, amino acids, and sugars. Two new layers of cells form at the base of the leaf, where the petiole is attached. The layer nearer the leaf base is made of small, weak cells with fragile cell walls, and the layer nearer the stem is tough and waxy. The waxy layer begins to clog the conduits that bring water and other supplies into the leaf. The chlorophyll molecules break down. As the green of the chlorophyll fades, the yellows and oranges of the carotenoids and the reds and purples of other pigments (the anthocyanins) become visible. Enzymes are produced that dissolve the cellulose of the walls in the weak cell layer. Leaves dangle briefly, held only by threads of vascular tissue, before they drop. The tough, waxy layer seals the breaks in the stem, forming leaf scars on the now bare branches.

Just as measuring distance requires a yardstick, measuring time requires a clock. And, as you will see in this chapter, plants have such a clock—a biological clock. You will also discover that what plants measure is not daylight at all.

The capacity to respond to stimuli—both internal and external—is one of the essential properties of living organisms. In plants, this capacity plays a critical role not only in such processes as the opening and closing of the stomata or the interaction of the roots of legumes with symbiotic nitrogen-fixing bacteria but also in the regulation of growth itself.

We saw in Chapter 40 that as a plant grows it does far more than simply increase its mass and volume. In a process that begins in the embryo and continues throughout the lifetime of the plant, cells produced in the meristematic tissues differentiate and elongate to form the specialized tissues and organs of the plant body. The rate at which this process occurs—in the plant as a whole and in different parts of the plant body—is affected by such environmental factors as temperature, water, sunlight, gravity, and physical contact with other objects, including other organisms. Growth rates, in turn, affect both the size of the whole plant and the relative sizes of its parts, thus determining its shape, or form. Moreover, many of a plant's activities—especially its cycles of active growth, reproduction, and dormancy—are finely tuned to the pattern of the changing seasons.

Application of the techniques of molecular biology to the study of plant physiology is beginning to yield new insights into the mechanisms by which plants sense and respond to environmental conditions. Nonetheless, a complete and coherent picture of the functioning plant remains elusive. It is clear, however, that regulation of the physiological activities of a plant depends on the interplay of a number of internal and external factors. Chief among the internal factors are the plant hormones.

The Role of Hormones

A **hormone** is, by definition, a substance that is produced in one tissue and transported to another, where it elicits one or more highly specific responses. Some plant hormones, however, like the prostaglandins of animals (page 615), act within the same tissues in which they are produced. In plants, as in animals, hormones are active in very small quantities. In the shoot of a pineapple plant, for example, only 6 micrograms of auxin, a common growth hormone, are found per kilogram of plant material. Hormones integrate the growth, development, and metabolic activities of the various tissues of the plant.

The term "hormone" comes from the Greek word meaning "to excite." Many hormones, however, also have inhibitory effects. So, rather than thinking of hormones as stimulators, it is perhaps more useful to consider them as chemical regulators. But this term also needs qualification. The response to a particular regulatory "message" depends not only on the content of the message (that is, its chemical structure) but also on the identity of its recipient (that is, the specific tissue) and when and how the message is received. Moreover, the response to any particular hormone is influenced by a variety of other factors in the internal environment of the plant, chief among which are likely to be other hormones.

Five principal types of plant hormones are known: auxins, cytokinins, ethylene, abscisic acid, and gibberellins.

Phototropism and the Discovery of Auxin

A **tropism** is a growth response that involves the curvature of a plant part toward or away from an external stimulus that determines the direction of movement. One of the most obvious and useful responses of plants is their **phototropism**, or curvature toward light.

Charles Darwin and his son Francis performed some of the first experiments on phototropism, which they reported in 1881. Working with grass seedlings, they noted that the curvature occurs below the tip, in a lower part of the coleoptile (the hollow sheath that surrounds the shoot tip in monocots). Then they showed that if they covered the tip of the coleoptile with a cylinder of metal foil or a hollow tube of glass blackened with India ink and exposed the plant to light coming from the side, the characteristic curvature of the seedling did not occur. If, however, light was permitted to penetrate the cylinder, curving occurred normally. Curving also occurred normally when the lightproof cylinder was placed below the tip (Figure 42–2). The Darwins con-

42–1   *This young bean seedling has just broken through the soil. Its growth into a mature plant will depend on light, water, temperature, and mineral ions from the soil and also on the interaction of many internal factors, among which are hormones.*

cluded that the exposure to lateral light caused some "influence" to be transmitted from the upper to the lower portion of the shoot tip and that this, in turn, caused the observed curvature.

In 1926, the Dutch plant physiologist Frits W. Went succeeded in separating this "influence" from the plants that produced it. Went cut off the coleoptile tips from a number of oat seedlings. He placed the tips on a slice

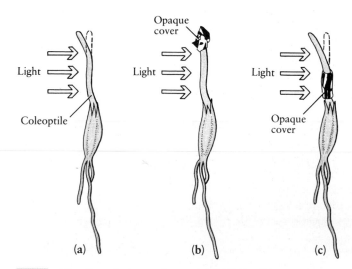

42–2   *The Darwins' experiment. (a) Light striking a growing coleoptile (such as the tip of this oat seedling) causes it to curve toward the light. (b) Placing an opaque cover over the tip of the seedling inhibits this curvature, but (c) an opaque collar placed below the tip does not. These experiments indicate that something produced in the tip of the seedling and transmitted down the coleoptile causes the curvature.*

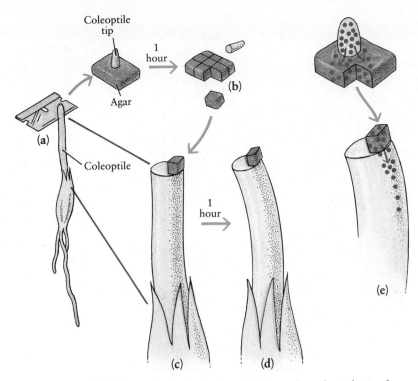

Coleoptile
tip

1
hour

(b)

Agar

(a)

Coleoptile

1
hour

(c)   (d)

(e)

42–3 *Went's experiment. (a) He cut the coleoptile tips from oat seedlings and placed them on a slice of agar. After about an hour, (b) he cut the agar in small blocks and (c) placed each block, off-center, on a decapitated seedling. (d) The seedlings curved away from the side on which the block was placed. (e) The curvature resulted from the influence of the hormone auxin, which causes plant cells to elongate. In Went's experiment, auxin molecules (red dots) produced in the coleoptile tip were first transferred to the agar, and then to one side of the seedling by means of an agar block.*

42–4 *The stalk of the African violet leaf on the left was placed in a solution containing a synthetic auxin 10 days before the picture was taken. The stalk of the leaf on the right was placed in pure water. Note the growth of adventitious roots on the stalk of the hormone-treated leaf. The commercial rooting preparations commonly used by gardeners to stimulate root growth contain synthetic auxins.*

of agar (a gelatinlike substance), with their cut surfaces touching the agar, and left them there for about an hour. He then cut the agar into small blocks and placed a block off-center on each stump of the decapitated plants, which were kept in the dark during the entire experiment. Within one hour, he observed a distinct curvature *away* from the side on which the agar block was placed (Figure 42–3).

Agar blocks that had not been previously in contact with a coleoptile tip produced either *no* curvature or only a slight curvature *toward* the side on which the block had been placed. Similarly, agar blocks that had been exposed to a section of coleoptile lower on the shoot produced no physiological effect.

Went interpreted these experiments as showing that the coleoptile tip exerted its effects by means of a chemical stimulus rather than a physical stimulus, such as an electrical impulse. This chemical stimulus came to be known as **auxin**, a term coined by Went from the Greek word *auxein*, "to increase." A few different substances with activity similar to that of auxin have now been isolated from plant tissues, and many others have been synthesized in the laboratory.

The phototropism observed by the Darwins results from the fact that, under the influence of light, auxin migrates from the light side to the dark side of the shoot tip. This movement appears to be triggered by changes in membrane permeability in the cells exposed to the light. As auxin levels increase in the cells on the dark side, they elongate more rapidly than those on the light side, causing the plant to curve toward the light—a response with high survival value for young plants.

The Mechanism of Action of Auxin

Auxin is produced principally by the apical meristems of shoots and is transported to other parts of the plant body, moving in one direction only, from shoot to root. In the shoot, auxin is essential for growth. If the auxin-producing shoot apex is removed, growth stops. Similarly, normal root growth requires very small amounts of auxin. Low concentrations of auxin also stimulate the growth of branch roots and adventitious roots (Figure 42–4). Higher concentrations of auxin, however, inhibit root growth.

The stimulatory effects of auxin on the growth of shoots and roots are the result of cell elongation. Under the influence of auxin, the plasticity of the cell wall increases and the cell expands in response to the turgor exerted by the movement of water into the cell vacuole (see page 116). Current evidence indicates that auxin activates a membrane protein that transports hydrogen ions (H^+) from the cell into the cellulose-containing cell wall. As a result, the cell wall becomes acidified and

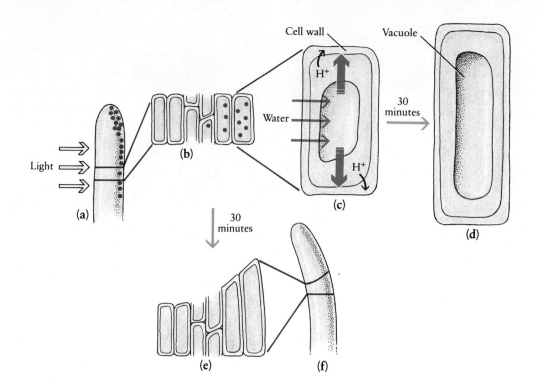

42–5 *Auxin, cell elongation, and the curvature of the shoot.* (a) *Under the influence of light, auxin moves to the dark side of the shoot,* (b) *increasing the auxin concentration in the cells on that side.* (c) *Within those cells, the increased concentration of auxin promotes the transport of H⁺ ions from the cytoplasm into the cell wall (curved black arrows). The resulting acidity activates an enzyme in the cell wall that breaks the cross-links between the cellulose molecules, increasing the plasticity of the wall. As water moves by osmosis into the vacuole of the cell, turgor (vertical blue arrows) causes the cell to elongate* (d). *The cells on the side of the shoot facing the light do not elongate* (e), *and as a result, the shoot as a whole curves toward the light* (f).

this, in turn, activates an enzyme in the cell wall that breaks the cross-links between the cellulose molecules. This allows the molecules to slide past one another as turgor acts against the cell wall (Figure 42–5). This response, known as **acid growth,** is rapid, often beginning within 3 to 5 minutes, reaching a maximum in 30 minutes, and ending after 1 to 3 hours.

Auxin also stimulates long-term growth, triggering the expression of at least 10 specific genes. Although the proteins coded by these genes have not been identified with certainty, it is hypothesized that some of them are enzymes involved in the synthesis of the new cell-wall materials necessary for continued growth.

Other Auxin Effects

In most dicot species, the growth of axillary buds is inhibited by auxin that moves down the stem from the shoot apex. This phenomenon is known as **apical dominance.** If you cut off the growing tip (apical meristem) of the stem of the house plant *Coleus,* for example, the axillary buds begin to grow vigorously, producing a plant with a bushier, more compact body and with more flowers (Figure 42–6). These buds can be repressed again by applying auxin to the cut surface of the shoot tip. Similarly, if you treat the "eyes" (actually axillary buds) of a potato with auxin, they will be inhibited from sprouting and can therefore be stored longer.

Auxin also plays a role in the seasonal initiation of activity in the vascular cambium (page 734) and the development of secondary phloem and xylem. It is also involved in the development of fruits (Figure 42–7).

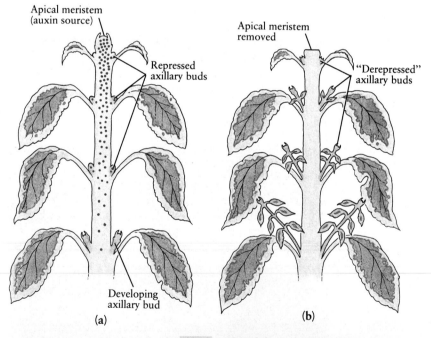

42–6 *Apical dominance in* Coleus. (a) *Auxin produced in the apical meristem of the shoot moves down the stem, repressing the growth of axillary buds. As the distance between the meristem and the axillary buds increases—and the concentration of auxin decreases—the buds are gradually freed from repression.* (b) *If the apical meristem is removed, eliminating further production of auxin, the axillary buds are "derepressed" and begin to grow vigorously.*

(a)

(b)

(c)

The Cytokinins

The isolation of auxin spurred the search for other growth-promoting factors in plants, particularly for a hormone that would stimulate cell division. Such compounds have now been found in all plants examined, especially in actively dividing tissues, such as meristems, germinating seeds, fruits, and roots. These hormones are called **cytokinins** (from "cytokinesis"). Studies of the way in which cytokinins stimulate cell division have shown that they are required for some process that takes place after DNA replication is complete but before mitosis begins.

Responses to Cytokinin and Auxin Combinations

Studies of responses to combinations of auxin and cytokinins are helping physiologists glimpse how plant hormones work together to produce the total growth pattern of the plant.

Apparently, an undifferentiated plant cell—such as a meristematic cell—has two courses open to it: either it can enlarge, divide, enlarge, and divide again, or it can elongate without cell division. The cell that divides repeatedly remains essentially undifferentiated, whereas the elongating cell tends to differentiate and become specialized. In studies of tobacco stem cells, the addition of auxin to the culture medium on which the cells were growing produced rapid cell expansion, so that giant cells were formed. A cytokinin alone had little or no effect. Auxin plus cytokinin resulted in rapid cell division, so that large numbers of relatively small cells were formed.

By slightly altering the relative concentrations of auxin and cytokinin, investigators have been able to affect the development of undifferentiated cells growing in tissue culture. At roughly equal concentrations of the

42–7 *Auxin, produced by developing embryos, promotes maturation of the ovary wall and the development of fleshy fruits. (a) Normal strawberry, (b) strawberry from which all seeds have been removed, and (c) strawberry in which a horizontal band of seeds has been removed. If a paste containing auxin were applied to (b), the strawberry would grow normally.*

42–8 *Shoots and roots growing from undifferentiated tissue (a callus) of a carrot plant following treatment with both an auxin and a cytokinin. Depending on the relative proportions of auxin and cytokinin, callus from various types of plants will continue to grow as undifferentiated tissue, will produce roots, or will produce buds and shoots.*

two hormones, the cells remain undifferentiated, forming a mass of tissue known as a **callus.** When a higher concentration of auxin is present, undifferentiated tissue gives rise to organized roots. With a higher concentration of cytokinin, buds appear. Further, careful balancing of the two hormones can produce both roots and buds and, thus, an incipient plant (Figure 42–8).

759

Other studies have shown that a third factor, the calcium ion (Ca^{2+}), can modify the action of the auxin-cytokinin combination. Auxin plus low concentrations of cytokinin favor cell enlargement, but as Ca^{2+} is added to the culture, there is a steady shift in the growth pattern from cell enlargement to cell division. High concentrations of Ca^{2+} apparently prevent the cell wall from expanding, and at such concentrations the cell switches course and divides. Thus, not only do hormones modify the effects of hormones, but these combined effects may, in turn, be modified by nonhormonal factors, such as calcium ions and, undoubtedly, many others.

Other Cytokinin Effects

Cytokinins have also been shown to reverse the inhibitory effect of auxin in apical dominance. In intact plants, cytokinins are synthesized in the roots and travel upward, primarily through the xylem, reaching the lower buds first and in highest concentration. As the dominating apex of the plant grows and moves away from these lower buds, the influence of the cytokinins overcomes that of auxin and the buds begin to grow (Figure 42–9).

Another, apparently unrelated function of cytokinins is preventing the **senescence,** or aging, of leaves. In many plants, the lower leaves turn yellow and drop off as the upper, new leaves develop. However, if a cytokinin is applied to the lower leaves, they remain green. Similarly, excised leaves remain green when maintained in a nutrient solution containing a cytokinin.

It is hypothesized that senescence in leaves, and probably in other plant parts as well, results from the progressive "turning off" of segments of DNA, with a consequent loss of messenger RNA production and protein synthesis. Recent experiments suggest that cytokinins prevent the DNA from being turned off and so promote continued enzyme synthesis and the production of such compounds as chlorophyll.

Ethylene

Ethylene is an unusual hormone in that it is a gas, a simple hydrocarbon, $H_2C{=}CH_2$. Its effects have been known for a long time. In the early 1900s, many fruit growers made a practice of improving the color and flavor of citrus fruits by "curing" them in a room with a kerosene stove. (Long before this, the Chinese used to ripen fruits in rooms where incense was being burned.) It was long believed that it was the heat that ripened the fruits. Ambitious fruit growers, who went to the expense of installing more modern heating equipment, found to their sorrow that this was not the case. As experiments showed, the incomplete combustion products of kerosene were actually responsible for ripening the fruits. The most active gas was identified as ethylene. As little as 1 part per million of ethylene in the air will speed the ripening process.

Subsequently, it was found that ethylene is produced by plants, fungi, and bacteria, as well as by kerosene stoves. The ethylene-synthesizing system is apparently located on the plasma membrane, from which the hormone is released. Ethylene can be detected just before and also during fruit ripening. It is responsible for a number of changes in color, texture, and chemical composition that take place as fruits mature. Ethylene is also involved in the senescence of floral parts that follows fertilization and precedes fruit development.

Auxin at certain concentrations causes a burst of ethylene production in some plants, and it is now believed that some of the effects attributed to auxin are related to the release of ethylene.

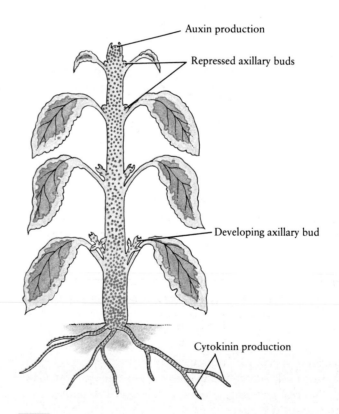

Auxin production

Repressed axillary buds

Developing axillary bud

Cytokinin production

42–9   *A schematic representation of the relative concentrations of auxin (red dots) and cytokinin (blue-green dots) in a* Coleus *plant. Auxin is produced in the shoot tip, while cytokinins are produced at the opposite end of the plant, in the root tips. Cytokinins counteract the effect of auxin in apical dominance. Thus, on the lower portions of the plant, where the cytokinin concentration is highest and the auxin concentration is lowest, the axillary buds are freed from repression and begin to develop.*

Plant Biotechnology: A New Frontier

In the millennia since the first crop plants were cultivated by *Homo sapiens,* a primary goal of agricultural scientists has been the selection and development of plant varieties with desirable characteristics. It was, as we saw in Chapter 12, in the course of plant breeding studies that the modern science of genetics had its beginnings. During this century, conventional breeding techniques have yielded a wealth of data about the genetic characteristics of plants and have led, throughout the world, to crops of much higher quality and yields.

Despite the progress that has been made, the pace of improvement has barely kept up with the increase in the human population—and with the capacity of plant pathogens and herbivorous insects to adapt to interventions designed to thwart their activities. Moreover, our knowledge of the molecular genetics of plants has lagged well behind our knowledge of the molecular genetics of bacteria, fungi, and animals. For example, it was not until 1989 that scientists learned that Mendel's wrinkled peas (page 213) result from an extra segment of DNA in the gene coding for an enzyme that converts linear starch molecules into more highly branched molecules.

Many of the most important crop plants have a genetic endowment as large and complex as our own, yet we know far less about the location and sequence of key genes, their products, and their regulation. A revolution, however, is now underway, utilizing a variety of approaches.

As long ago as the 1930s, scientists developed techniques for growing plant cells in test tubes. Tiny fragments of meristem are implanted under sterile conditions in a medium containing mineral ions and various combinations of organic compounds. Under these conditions, the meristematic cells proliferate to form clumps of undifferentiated cells. Hundreds or even thousands of subcultures of meristematic tissue can be produced in a relatively short time and in a small space. Then, by altering the hormone balance in the medium, each of these can be induced to differentiate and grow into a small but perfect plant. This technique, initially employed in the culture of orchids and other hard-to-grow plants, has now been extended to many other plants.

Within the last 20 years, it has become possible, using similar techniques, to grow isolated **protoplasts**—plant cells without cell walls—in the laboratory. The protoplasts can be grown as isolated cells, like cultures of bacteria. Alternatively, if they are grown in a suitable medium, they will regenerate cell walls, multiply, and differentiate into whole plants. The young plants produced are not only genetically uniform but also free of infectious disease.

Cultured cells are being used in a variety of ways. For instance, they are used in rapid screening tests to determine resistance to infectious disease or to detect nutritional requirements. In this way, scientists not only can work much more rapidly but can also do tests with millions of cells growing in a very small space, as compared to the far fewer number of plants that can be grown in a field or a greenhouse. Moreover, protoplasts from different species can be fused to create hybrids.

Protoplast fusion holds particular promise for combining the desirable characteristics of species that are sexually incompatible and cannot be crossbred by conventional techniques. It has been used to create both hybrids of species within the same genus and hybrids of species belonging to different genera.

Of particular interest is the fusion of the potato (*Solanum tuberosum*) and the tomato (*Lycopersicon esculentum*), both of which are members of the nightshade family (Solanaceae). Tomato plants are resistant to the water mold that causes potato blight (see page 426), a disease that remains a serious threat to potato crops. Through protoplast fusion, it will be possible to develop potato plants that incorporate the tomato plants' resistance genes. Similar work is under way with protoplast fusion involving tomato plants and strains of tobacco that are resistant to diseases to which tomato plants are particularly vulnerable.

Although protoplast fusion has become an important and useful tool, it does not allow the kind of precision that is possible with recombinant DNA procedures in which known genes are inserted into the DNA of the recipient cells. As we saw on page 312, the Ti plasmid of the soil bacterium *Agrobacterium tumefaciens* has become an important vector for ferrying genes into plants. This bacterium, however, infects only dicots, and the grains—such as rice, wheat, and corn—that feed the bulk of the

world's population are monocots. Moreover, although scientists could produce protoplasts of monocots, they were, until recently, unable to induce them to regenerate cell walls and to differentiate into whole plants. In the late 1980s, this barrier was finally overcome in both rice and corn, opening the way to genetic engineering in these vitally important crop plants. With protoplasts as the recipient cells, a diversity of plasmids and viruses can now be used as vectors to introduce desired genes. The protoplasts also provide the raw material for chromosome mapping and for detailed molecular genetic studies.

Many biologists believe that we are about to witness an extraordinary explosion of knowledge in plant biology, as the tools of biotechnology are applied to a multitude of unanswered questions. It is also hoped that rapid progress will be made in the development of new varieties of crop plants with higher nutritional values and with greater resistance to the infectious diseases and insects that destroy a large proportion of the world's food supply each year.

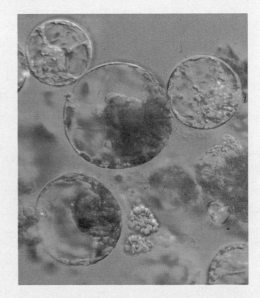

Protoplasts of wild tobacco (green) and commercial tobacco (clear). As you can see, some of the protoplasts have fused, giving rise to hybrid cells. With suitable treatment, the hybrid cells will regenerate cell walls, multiply, and differentiate into whole plants.

Ethylene and Leaf Abscission

Plants drop their leaves at regular intervals, either as a result of normal aging of the leaf or, in the case of deciduous plants, in response to environmental cues. This process of **abscission** ("cutting off") is preceded by changes in the **abscission zone,** at the base of the petiole. In woody dicots, the abscission zone consists of two cell layers: (1) a structurally weak layer in which the actual abscission occurs, and (2) a protective layer that forms a leaf scar on the stem (Figure 42–10).

Once leaf senescence begins, ethylene produced in the abscission layer is the principal regulator of leaf drop. It acts by promoting the synthesis and release of cellulase, an enzyme that breaks down plant cell walls (and which is also involved in the ripening of fruits).

Abscisic Acid

Soon after the discovery of the growth-promoting hormones, plant physiologists began to speculate that growth-inhibiting hormones would be found, since it is clearly advantageous to the plant not to grow at certain times and in certain seasons. Not long afterwards, an inhibitory hormone was isolated from dormant buds. Subsequently, the same hormone was discovered in leaves, where it was thought to promote abscission. Thus it was called **abscisic acid,** which was unfortunate since it is now known that, in most plants at least, it has little to do with abscission.

Abscisic acid may, however, induce dormancy. For example, application of abscisic acid to vegetative buds changes them to winter buds (Figure 42–11). Abscisic acid is also present in the seeds of many species, where it is a major factor in maintaining seed dormancy. Moreover, as we saw in the last chapter, it brings about the closing of stomata under conditions of impending water shortage. Thus, abscisic acid has come to be known as the **stress hormone** in recognition of its role as a protector of the plant against unfavorable environmental conditions.

The Gibberellins

The **gibberellins** were discovered slightly earlier than auxin by a Japanese scientist who was studying a disease of rice plants called "foolish seedling disease." The diseased plants grew rapidly but were spindly and tended to fall over under the weight of the developing grains. The cause of the symptoms, it was found, was a chemical produced by a fungus, *Gibberella fujikuroi,* which infected the seedlings.

This substance, which was named gibberellin, and many closely related substances were subsequently iso-

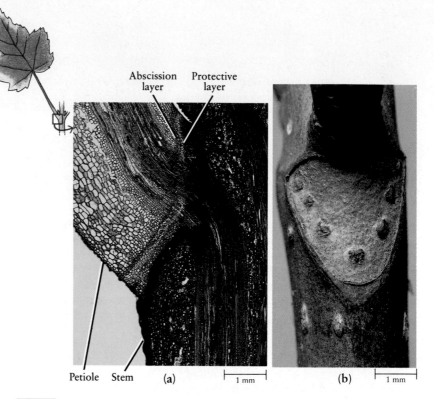

Abscission layer   Protective layer

Petiole   Stem   **(a)**   1 mm   **(b)**   1 mm

42–10   (a) *Abscission zone in a maple leaf, as seen in longitudinal section. The abscission layer, which forms across the base of the petiole, consists of structurally weak cells. Under the influence of ethylene, enzymes are produced that cause the walls of these cells to dissolve.* (b) *After the leaf drops off, the protective layer forms a covering, the leaf scar, on the stem. This leaf scar is on a branch of a horse chestnut (Aesculus hippocastanum). Immediately above the leaf scar is the base of a bud that will give rise to new growth the following spring.*

42–11   *Winter buds of an American basswood. These buds, or young shoots, form in one growing season and elongate in the next. The buds contain small amounts of water and high concentrations of proteins and lipids, which prevent the formation of ice crystals that could kill the cells. Bud scales, which are small, tough modified leaves, protect the delicate tissues of the bud from mechanical injury and desiccation. As the first shoots emerge in the spring, the bud scales are forced off. At the base of each bud, you can see the scar from the leaf of the previous growing season.*

lated not only from the fungus but also from bacteria and many species of plants. Gibberellins and auxin (to a lesser extent) control elongation in mature trees and shrubs.

In many species of plants, gibberellins characteristically produce hyperelongation of the stem. Particularly striking effects are seen in some plants that are genetic dwarfs. In some of these, application of gibberellin makes them indistinguishable from normal plants, suggesting that these dwarfs lack a gene coding for an enzyme needed for gibberellin production.

Gibberellins can also produce **bolting**, a phenomenon observed in many plants, especially biennials, that first grow as rosettes. Just before flowering, the flower stem elongates rapidly, or "bolts" (Figure 42–12). Bolting and flowering, which normally occur only after an environmental cue (such as a cold season), occur in some plants following treatment with gibberellin. Abscisic acid inhibits the effects of gibberellin in promoting bolting. This inhibition is, in turn, reversed by cytokinins.

Gibberellins can also induce cellular differentiation. In woody plants, gibberellins stimulate the vascular cambium to produce secondary phloem. Both phloem and xylem develop in the presence of gibberellins and auxins. In intact plants, interactions between the two types of hormones are thought to determine the relative rates of production of secondary phloem and secondary xylem.

The highest concentrations of gibberellins have been found in immature seeds, although they are present in varying amounts in all parts of plants. In grains (the one-seeded fruits of grasses), there is a specialized layer of cells, the **aleurone layer,** just inside the seed coat (see page 713). These cells are rich in protein. During the early stages of germination, the embryo produces gibberellin, which diffuses to the aleurone layer. In response to gibberellin, the aleurone cells produce and release enzymes that hydrolyze the starch, lipids, and proteins in the endosperm, converting them to sugars, fatty acids, and amino acids that the embryo and, later, the emerging seedling can use (Figure 42–13). In this way, the embryo itself calls forth the substances needed for its metabolism and growth at just the times they are required.

Gravitropism

As we noted earlier, phototropism—curvature toward the light—has high survival value for a young plant. Another response with high survival value is the capacity of a young plant to respond to gravity, righting itself so that the shoot grows up and the root grows down.

42–12   *In this row of cabbages, the plant in the middle has bolted and flowered naturally. Bolting and flowering can also be induced artificially by treatment with a gibberellin.*

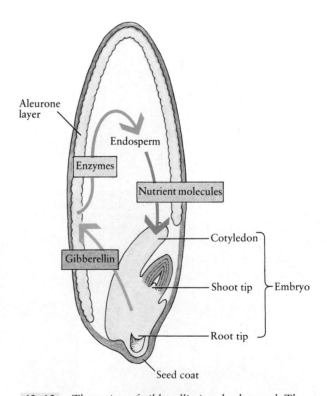

42–13   *The action of gibberellin in a barley seed. The embryo releases gibberellin, which diffuses to the aleurone layer. The gibberellin induces the aleurone cells to synthesize enzymes that digest the food reserves of the endosperm into smaller nutrient molecules. These molecules are absorbed by the cotyledon and transported to the growing regions of the embryo.*

This response is known as **gravitropism.** Like phototropism, gravitropism is thought to involve auxin.

When young monocot shoots are oriented horizontally, auxin migrates to the lower side of the coleoptile and calcium ion (Ca^{2+}) accumulates along the upper side. The increased concentration of Ca^{2+} inhibits growth along the upper side, while auxin stimulates elongation of cells on the lower side. The result is an upward curvature of the shoot. When the shoot becomes vertical, the differences in auxin and Ca^{2+} concentrations disappear, and growth continues in an upright direction. Investigators, however, have been unable to determine whether a similar mechanism is involved in the gravitropism of dicot shoots.

Roots are exquisitely sensitive to even slight increases in auxin concentration. Their downward growth is thought to result from the inhibitory effects of increased auxin concentrations along the lower side. As cells along the upper side of the root elongate more rapidly than those along the lower side, the root curves downward. The transport of auxin to the lower side of the root is dependent upon Ca^{2+}, but the mechanism by which it occurs is not known.

Another major question remains: How does a seedling "know" it is on its side? Hormones and ions are soluble, and so gravity itself should have no effect on their distribution. The answer to this question has long been thought to lie in specialized cells of the shoot and of the root cap. The inner, or core, cells of the root cap,

for example, seem to be analogous to the statocysts found in many animals (see page 461). Like statocysts, these cells contain statoliths—particles that move in response to gravity. In jellyfish, the statoliths are grains of hardened calcium salts. In the core cells of the root cap, they are **plastids** (membrane-bound organelles), filled with starch.

When a root is growing vertically, the plastids collect near the lower walls of the core cells (Figure 42–14a). If the root is placed in a horizontal position, however, the plastids slide downward and come to rest near what were previously vertically oriented walls (Figure 42–14b). Within minutes, the root begins to curve down, gradually returning the plastids to their original position. How the movement of the plastids is translated into chemical gradients of auxin and calcium ion is not yet known, but investigators think that both the detection of gravity and the translation of that information into chemical gradients probably involve alterations in both electrical properties and membrane permeability.

Photoperiodism

In many regions of the biosphere, the most important environmental changes affecting plants (and, indeed, land organisms in general) are those that result from the changing seasons. Plants are able to accommodate to these changes because of their capacity to anticipate

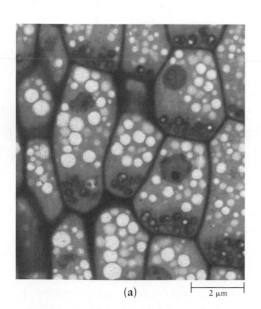
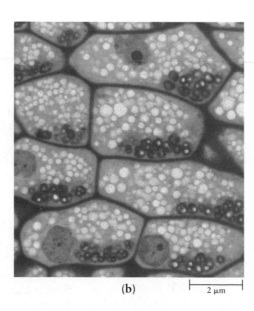

(a)   2 μm

(b)   2 μm

42–14   *Core cells of the root cap* (a) *in their normal vertical orientation and* (b) *after the root has been turned horizontally. The dark globular bodies containing white starch grains are the plastids. Note how their position within the cells changes when the orientation of the root is changed.*

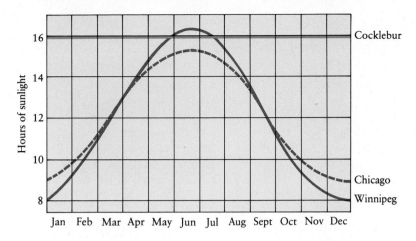

42–15 *The relative length of day and night determines when many plants flower. The red curves depict the annual change in day length in two North American cities: Chicago, at latitude 40°N, and Winnipeg, at latitude 50°N. The green line indicates the effective photoperiod of the cocklebur, a short-day plant that requires 16 hours or less of light in order to flower. In Chicago, where there is always less than 16 hours of sunlight, the cocklebur can flower as soon as it matures. In Winnipeg, however, the buds do not appear until early in August, so late that frost usually kills the plants before the seeds are mature.*

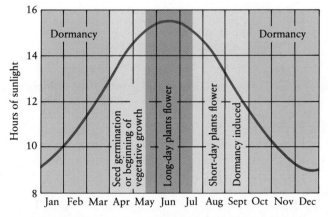

42–16 *Relationship between day length and the developmental cycle of plants in the temperate zone. The curve indicates the day length at a latitude of approximately 45°N—slightly north of Bangor, Maine.*

the yearly calendar of events: the first frost, the spring rains, long dry spells, long growing periods, and even the time that nearby plants of the same species will flower. For many plants, all of these determinations are made in the same way: by measuring the relative periods of light and darkness. This phenomenon is known as **photoperiodism.**

Photoperiodism and Flowering

The effects of photoperiodism on flowering are particularly striking. Plants are of three general types: day-neutral, short-day, and long-day. **Day-neutral** plants flower without regard to day length. **Short-day** plants flower in early spring or fall. They must have a light period *shorter* than a critical length. For instance, the cocklebur flowers when exposed to 16 hours or *less* of light (Figure 42–15). Other short-day plants are poinsettias, strawberries, primroses, ragweed, and some chrysanthemums.

Long-day plants, which flower chiefly in the summer, will flower only if the light periods are *longer* than a critical length. Spinach, potatoes, clover, henbane, and lettuce are examples of long-day plants.

The discovery of photoperiodism explained some puzzling data about the distribution of common plants. Why, for example, is there no ragweed in northern Maine? The answer, investigators found, is that ragweed starts producing flowers when the day length is 14.5 hours or less. The long summer days do not shorten to 14.5 hours in northern Maine until August (Figure 42–16), and then there is not enough time for ragweed seed to mature before the frost. For somewhat similar reasons, spinach cannot produce seeds in the tropics. Spinach needs 14 hours of light a day for a period of at least two weeks in order to flower, and days are never this long in the tropics.

Note that ragweed and spinach will both bloom if exposed to 14 hours of daylight, yet one is designated as short-day and one as long-day. The important factor is not the absolute length of the photoperiod but rather whether it is longer or shorter than a particular critical interval for that variety. Detection of the interval can be very precise. In some varieties 5 or 10 minutes' difference in exposure can determine whether a plant will flower.

The first field studies on photoperiodism in plants were performed more than 50 years ago by scientists at the U.S. Department of Agriculture. These scientists have since become known as the Beltsville group, for the small town in Maryland where they carried out their studies.

Is There a Flowering Hormone?

One of the most persistent—and still unanswered—questions in plant physiology concerns the existence of a flowering hormone. Since the 1930s, numerous experiments have suggested that such a hormone is formed in the leaves in response to an appropriate cycle of light and dark and then travels to the apical meristem of the plant and initiates flowering. For example, when certain plants, such as the cocklebur shown in the illustration, are exposed to an appropriate light cycle, **(a)** those with leaves flower and those without leaves do not. **(b)** When even one-eighth of a leaf remains on a plant, flowering occurs, and **(c)** the illumination of a single leaf—not necessarily the whole plant—suffices. These experiments indicate that chemical signals originating in the leaves cause the plant to flower.

This conclusion is supported by experiments on branched plants. **(d)** Exposure of one branch to the light induces flowering on the other, unlighted branch as well, **(e)** even when only a portion of a leaf is present on the lighted branch. **(f)** When two plants are grafted together, exposure of one of the plants to the light cycle induces flowering in both the lighted plant and the unlighted one.

Despite the evidence of these and other experiments, no specific flowering hormone has ever been isolated and identified. However, both auxin and gibberellin have been shown to induce flowering in some plants under some circumstances. These results suggest that flowering does not involve a single, unique hormone but rather a combination of hormones, probably acting in conjunction with other, as yet unknown, chemical factors.

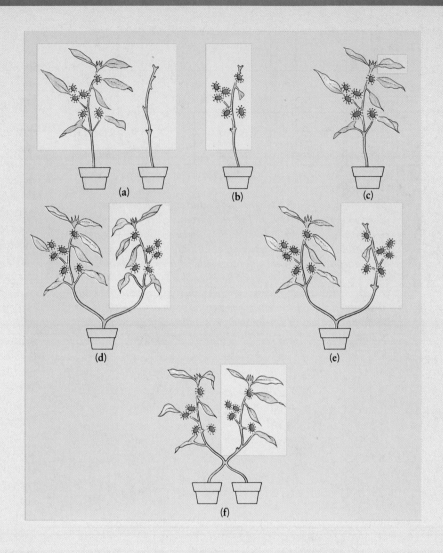

Measuring the Dark

Subsequently, other investigators, Karl C. Hamner and James Bonner, began a laboratory study of photoperiodism. They used the cocklebur (Figure 42–17) as their experimental organism. As we mentioned previously, the cocklebur is a short-day plant, requiring 16 hours or less of light per 24-hour cycle to flower. It is particularly useful for experimental purposes because a single exposure under laboratory conditions to a short-day cycle will induce flowering two weeks later, even if the plant is immediately returned to long-day conditions.

In the course of their studies, in which they tested a variety of experimental conditions, the investigators made a crucial and totally unexpected discovery. If the period of darkness is interrupted at any point by as little as a 1-minute exposure to the light of a 25-watt bulb, flowering does not occur. Interruption of the light period by darkness, on the other hand, has no effect whatsoever on flowering. Subsequent experiments with other short-day plants showed that they, too, require periods not of uninterrupted light but of uninterrupted darkness (Figure 42–18).

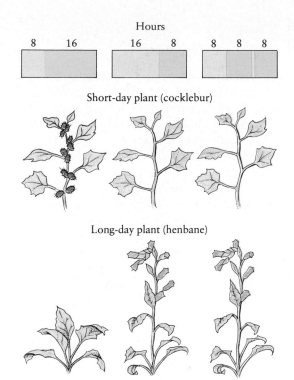

42–17   *The cocklebur, a short-day plant that has been important in experimental studies of photoperiodism. Each "burr" is an inflorescence with two flowers.*

What about long-day plants? They also measure darkness. A long-day plant that will flower if it is kept in a laboratory in which there is light for 16 hours and dark for 8 hours will also flower on 8 hours of light and 16 hours of dark if the dark is interrupted at its midpoint by even a brief exposure to light.

Photoperiodism and Phytochrome

Following up on the clues from the Hamner and Bonner experiments, the Beltsville group was able to detect and eventually to isolate the light-detecting pigment involved in photoperiodism. This pigment, which they called **phytochrome,** exists in two different forms. One form, known as P_r, absorbs red light with a wavelength of 660 nanometers; phytochrome is synthesized in the P_r form. The other form, P_{fr}, absorbs far-red light with a wavelength of 730 nanometers. P_{fr}, which is the biologically active form of the pigment (that is, it can trigger a biological response), promotes flowering in long-day plants and inhibits flowering in short-day plants.

When P_r absorbs red light, it is converted to P_{fr} (Figure 42–19). This conversion takes place in sunlight or in incandescent light. In both of these types of light, red wavelengths predominate over far-red. When P_{fr} absorbs far-red light, it is converted back to P_r. In the dark, P_{fr} is converted slowly to P_r, or it is degraded and replaced by newly synthesized P_r, or both.

42–18   *Experiments on photoperiodism showed that plants measure the length of darkness rather than that of light. Short-day plants flower only when the dark period exceeds some critical value. Thus, the cocklebur, for instance, will flower on 8 hours of light and 16 hours of darkness. If the 16-hour period of darkness is interrupted at any point—even very briefly—the plant will not flower.*

The long-day plant, on the other hand, which will not flower on 16 hours of darkness, will flower if the dark period is interrupted at its midpoint, producing two distinct 8-hour periods of darkness. Long-day plants flower only when the dark period is less than some critical value.

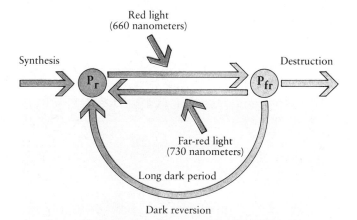

42–19   *Phytochrome is synthesized (from amino acids) in the P_r form. P_r changes to P_{fr} when exposed to red light, which is present in sunlight. P_{fr} is the active form of phytochrome that induces the biological response. P_{fr} reverts to P_r when exposed to far-red light. In darkness, P_{fr} slowly reverts to P_r or is degraded.*

Other Phytochrome Responses

The $P_r \rightleftharpoons P_{fr}$ conversion, it is now known, acts as an off-on switch for a number of plant responses to light. Many kinds of small seeds, such as lettuce, germinate only when they are in loose soil, near the surface. This ensures that the seedling will reach the light before it runs out of stored food. Red light, a sign that sunlight is present, stimulates seed germination by converting phytochrome to the active form (P_{fr}).

Phytochrome is also involved in the early development of seedlings. When a seedling develops in the dark, as it normally does underground, the stem elongates rapidly, pushing the shoot up through the soil layers. A seedling grown in the dark will be elongated and spindly with small leaves (Figure 42–20). It will also be almost colorless, because the chloroplasts do not synthesize chlorophyll until they are exposed to light. Such a seedling is said to be **etiolated.**

When the seedling tip reaches the light, normal growth begins. Phytochrome is involved in the switching from etiolated to normal growth. If a dark-grown bean seedling is exposed to only 1 minute of red (660 nanometers) light, it will respond with normal growth. If, however, the exposure to red light is followed by a 1-minute exposure to far-red light, thus negating the original exposure, etiolated growth continues.

The way in which phytochrome acts is not known. One suggestion is that it alters the permeability of the plasma membrane, permitting particular substances to enter the cell, or, perhaps, inhibiting their entry, and that these substances, which probably include hormones, regulate the cell's activities. An increase in the concentrations of all the growth-promoting hormones can be detected almost immediately after phytochrome activation.

Circadian Rhythms

How can a spinach plant distinguish a 14-hour day from a 13.5-hour day? Seeking an answer to this question leads us to another group of readily observed phenomena. Some species of plants, for instance, have flowers that open in the morning and close at dusk. Others spread their leaves in the sunlight and fold them toward the stem at night (Figure 42–21). As long ago as 1729, a French scientist noticed that these daily movements continue even when the plants are kept in constant dim light. More recent studies have shown that less evident activities, such as photosynthesis, auxin production, and the rate of cell division, also have daily rhythms.

Like the regular day-night cycles of animals (page 614), the regular day-night cycles of plants are known as **circadian rhythms.** In plants, as in animals, circadian rhythms are not exact. Different species and different individuals of the same species often have slightly different, but consistent, rhythms, often as much as an hour or two longer or shorter than 24 hours.

Biological Clocks

A variety of experiments have demonstrated that circadian rhythms are internal—that is, they are caused by factors within the organism itself. The mechanism by which they are controlled is known as a **biological clock.**

Biological clocks play a role in many aspects of plant and animal physiology, synchronizing internal and external events. For example, some flowers secrete nectar or perfume at certain specific times of the day or night. As a result, pollinators—which have their own biolog-

42–20 *Dark-grown seedlings, such as the bean plants on the left, are thin and pale with longer internodes and smaller leaves than the normal seedlings on the right. This group of characteristics, known as etiolation, has survival value because it increases the seedling's chances of reaching light before its stored energy supplies are used up.*

42–21 *Leaves of the oxalis plant, (a) during the day and (b) at dusk. Darwin believed that this folding of the leaves at night conserved heat energy. A more recent but also unconfirmed hypothesis is that the "sleep movements" prevent the leaves from absorbing moonlight on bright nights, thus protecting photoperiodic phenomena.*

(a)

(b)

42–22 *Tendrils of a bur cucumber plant* (Cucumis anguria). *Twining is caused by varying growth rates on different sides of the tendril.*

ical clocks—have become programmed to visit these flowers at these times, thereby ensuring maximum rewards for both the pollinators and the flowers. However, for most organisms, the "use" of the clock for such purposes is thought to be a secondary development. The primary function of the clock appears to be in enabling organisms to recognize the changing seasons by "comparing" external rhythms of the environment, such as changes in day length, with their own internal rhythms.

The chemical nature of the biological clock—or indeed whether there is just one kind of clock or many—is still not known. The best evidence to date indicates that the timing mechanism involves rhythmic changes in the plasma membrane, in either the protein components or the lipids, or, perhaps, in both.

Touch Responses

Many plants respond to touch. One of the most common examples is seen in tendrils, the modified stems or leaves with which many plants support themselves and climb (Figure 42–22). When the apex of a tendril touches any object, it responds to the touch by forming a tight coil. Cells touching the object shrink slightly, and those on the outer side elongate. This response can be rapid. A tendril may wrap around a support one or more times in less than an hour. Specialized epidermal cells of the tendrils are the sensors for touch, but the mechanism by which these cells induce coiling is unknown. Current evidence indicates that auxin and ethylene are probably involved. These hormones cause excised tendrils to coil even in the absence of touch.

(a)

(b)

42–23   *Sensitive plant* (Mimosa pudica).
*(a) Normal position of leaves and leaflets.
(b) Response to touch. It has been hypothesized that these reactions may prevent wilt-
ing (when they occur in response to strong
winds), startle insects, or dismay larger herbivores. Collapse of the leaflets is caused by
rapid turgor changes in cells of the petioles.
These changes are accompanied by the
release from the cells of substances known
as tannins. Tannins have an astringent taste
and are repellent to herbivores. In the
evolution of the touch response in Mimosa,
the release of tannins may have been more
important than the rapid movements.*

An even more rapid response occurs in the sensitive
plant, *Mimosa pudica.* A few seconds after a leaf is
touched, the petiole droops and the leaflets fold (Figure
42–23). This response is a result of a sudden change in
the turgor of specialized cells at the base of the leaflets
and leaves. The turgor change is preceded by an electrical signal that passes along the petiole and can be
detected by microelectrodes placed in the plant. The
electrical signal, in turn, triggers a chemical signal that
alters the permeability of the plasma membranes, leading to the observed turgor change.

Another rapid response to touch occurs in the capture of prey by the carnivorous Venus flytrap. The
leaves of the Venus flytrap have two hinged lobes, each

(a)

(b)

42–24   *The capture of prey by the hinged
leaf of a Venus flytrap. The closing of the
leaf was triggered when the fly touched one
or two of the three sensitive hairs in the middle of each leaf lobe. It was long believed
that the Venus flytrap lures its victims by exuding nectar, but current evidence indicates
that insects visit the leaves by chance. Despite its name, a Venus flytrap's usual diet in
the wild consists of crawling insects, such as
ants.*

*Unlike carnivorous animals, the Venus fly-
trap, the sundew (page 740), and other carnivorous plants do not utilize their prey for
energy but rather as a source of mineral elements, particularly nitrogen, phosphorus,
and calcium. Most of the more than 350 species of carnivorous plants are found in
swamps, bogs, and peat marshes—environments in which acids leach the soils of nutrients. Venus flytraps, for example, are found
in nature only on the coastal plain of North
and South Carolina, usually at the edges of
wet depressions and pools.*

of which is equipped with three sensitive hairs. When an insect alights on a leaf, it brushes against the hairs, setting off an electrical impulse that triggers the closing of the leaf. The toothed edges mesh like a bear trap (Figure 42–24), snapping shut in less than half a second. Once the insect is trapped, the leaf halves gradually squeeze closed, and the captive animal is pressed against digestive glands located on the inner surface of the trap.

It was long thought that the response of the Venus flytrap, like that of the sensitive plant, involved turgor changes. Recent research, however, has revealed that the response is instead another example of acid growth (page 758). The electrical impulse set off when an insect brushes against the trigger hairs activates a membrane protein that, using ATP for energy, pumps H^+ ions into the walls of epidermal cells along the outer surface of the trap's hinge. The rapid, irreversible cell expansion that results from acidification of the cell walls closes the trap. During this process, the cells on the inner surface of the hinge continue to grow at their usual, slow rate. Some 10 hours later their growth catches up with the earlier growth of the cells on the outer surface, and the trap opens. As a result of both growth processes, the leaf is slightly larger than it was before it closed.

An electrical impulse is also involved in the trapping of insects by another carnivorous plant, the sundew. As shown on page 740, the leaves of this plant are covered with tiny tentacles. A sticky droplet surrounds the tip of each tentacle. When an insect is caught on the tip of a tentacle, the surrounding tentacles bend in, carrying the prey to the center of the leaf where it is digested. Microelectrodes placed in the tentacles have revealed that the touch of an insect's foot triggers an electrical impulse that moves down the tentacle.

The electrical impulses that have been observed in plants are the same, in principle, as the nerve impulses of animals. Many botanists expect that electrical signals will be found to coordinate a variety of activities involving cells in different parts of the plant body.

Chemical Communication among Plants

As we noted in Chapter 25 (page 450), many angiosperms produce toxic or bad-tasting compounds that function as powerful defenses against herbivorous animals. In some species, the production of such compounds is initiated or increased in response to damage inflicted on the plant by herbivores, for example, chewing insects. The resulting higher concentrations of these substances deter further predation and thereby protect the plant from severe damage.

A number of studies have shown that the plants do more than protect themselves: they apparently also "warn" neighboring plants of the same species to mobilize their defenses prior to attack. This effect was first observed in Sitka willows. Three days after high levels of defensive chemicals were detected in the leaves of trees being directly attacked by tent caterpillars, high levels of the same chemicals were found in the leaves of trees that had not yet been touched by the caterpillars. Some of these trees were as far as 60 meters from the damaged trees. Presumably, the damaged leaves release an airborne substance that, upon reaching the leaves of another plant, triggers their synthesis of defensive chemicals.

Similar effects have been observed in carefully controlled laboratory studies in which mechanical damage was inflicted on the leaves of poplar and maple seedlings. Studies currently in progress are designed to isolate and identify the airborne substance or substances involved in this communication.

As we have seen in earlier chapters, chemical communication among individuals of the same species is a familiar phenomenon in animals, particularly insects. This recent research is the first hint that there may be many more interactions among individual plants than are apparent even to the well-trained eye.

Summary

Plants respond to stimuli in both their internal and external environments. Such responses enable the plant to develop normally and to remain in touch with changing external conditions.

Hormones are important factors in plant responses. A hormone is a chemical that is produced in particular tissues of an organism and carried to other tissues of the organism, where it exerts one or more specific influences. Characteristically, it is active in extremely small amounts. Table 42–1 summarizes the five major groups of hormones that have been isolated from plants: auxins, cytokinins, ethylene, abscisic acid, and gibberellins.

Auxin is produced principally in rapidly dividing tissues, such as apical meristems. It causes lengthening of the shoot, chiefly by promoting cell elongation. It is also involved in apical dominance, in which the growth of axillary buds is inhibited, restricting growth principally to the apex of the plant. At low concentrations, auxin promotes the initiation of branch roots and adventitious roots; at higher concentrations, it inhibits root growth. In developing fruits, auxin produced by the embryos stimulates growth of the ovary wall. The capacity of auxin to produce such varied effects appears to result from different responses of the various target tissues and the presence of other factors, including other hormones.

The cytokinins promote cell division. It is possible, by altering the concentrations of auxin and cytokinins, to alter patterns of growth in undifferentiated plant tissue in tissue culture.

Ethylene is a gas produced by fruits during the ripening process, which it promotes. It also plays a major role in leaf abscission. Abscisic acid, a growth-inhibiting hormone, affects stomatal closing and may be involved in the induction of dormancy in vegetative buds and the maintenance of dormancy in seeds.

Gibberellins stimulate shoot elongation, induce bolting and flowering in many plants, and are also involved in embryo and seedling growth. In the seeds of grasses, they stimulate the production of hydrolyzing enzymes that act on the stored starch, lipids, and proteins of the endosperm, converting them to sugars, fatty acids, and amino acids, which nourish the seedling.

Plants respond to a number of environmental stimuli. Two responses with high survival value for young plants are phototropism, the curvature of a plant toward the light, and gravitropism, the capacity of the shoot to grow up and the root to grow down. Phototropism is mediated by auxin and gravitropism by auxin and calcium ion (Ca^{2+}).

Photoperiodism is the response of organisms to changing periods of light and darkness in the 24-hour day. Such a response controls the onset of flowering in many plants. Some plants, known as long-day plants, will flower only when the periods of light exceed a critical length. Other plants, short-day plants, flower only when the periods of light are less than some critical period. Day-neutral plants flower regardless of photoperiod. Experiments have shown that the length of the dark period rather than that of the light period is the critical factor.

Phytochrome, a pigment commonly present in small amounts in the tissues of plants, is the receptor molecule for detecting transitions between light and darkness. The pigment exists in two forms, P_r and P_{fr}. P_r absorbs

Table 42–1	Plant Hormones and Their Effects
Hormone	**Physiological Effects and Roles**
Auxins	Stimulate cell elongation; involved in phototropism, gravitropism, apical dominance, and vascular differentiation; inhibit abscission prior to formation of abscission layer; stimulate ethylene synthesis; stimulate fruit development; induce adventitious roots on cuttings
Cytokinins	Stimulate cell division, reverse apical dominance, involved in shoot growth, delay leaf senescence
Ethylene	Stimulates fruit ripening, leaf and flower senescence, and abscission
Abscisic acid	Stimulates stomatal closure; may be necessary for dormancy in certain species
Gibberellins	Stimulate shoot elongation, stimulate bolting and flowering in biennials, regulate production of hydrolytic enzymes in grains

red light, whereas P_{fr} absorbs far-red light. P_{fr} is the biologically active form of the pigment. Among its many known effects, it promotes flowering in long-day plants, inhibits flowering in short-day plants, promotes germination in lettuce seeds, and promotes normal growth in seedlings. Its mechanism of action appears to involve changes in the permeability of the plasma membrane.

Circadian rhythms are regular cycles of growth and activity occurring approximately on a 24-hour basis. Many of these rhythms are independent of the organism's environment and are controlled by some internal regulator—a biological clock. The principal function of the biological clock is to provide the timing mechanism necessary for photoperiodic phenomena.

Some species of plants exhibit specific, rapid movements in response to touch. Examples include the winding of tendrils, the collapse of the leaves of the sensitive plant (Mimosa), the triggering of the carnivorous Venus flytrap, and the bending of the tentacles of the sundew. Plant responses to touch involve various combinations of electrical impulses, changes in turgor, and chemical changes that result in differential rates of growth.

Some species of plants respond to mechanical damage, such as may be caused by herbivorous animals, by synthesizing chemicals that deter further predation. They apparently also release airborne substances that "communicate" with other individuals of the same species, triggering in them the synthesis of defensive chemicals before any actual damage occurs.

Questions

1. Describe Went's experiments and the conclusions that can be drawn from them.

2. Describe the principal roles played by each of the following hormones: auxin, cytokinin, ethylene, abscisic acid, and gibberellin.

3. Describe the phenomenon of apical dominance. How is apical dominance reversed naturally in the intact plant? How can apical dominance be reversed by gardeners who desire bushier plants?

4. One bad apple can spoil the whole barrel. Explain.

5. Distinguish between the following: phototropism/photoperiodism; circadian rhythm/biological clock.

6. Why is a biological clock necessary for photoperiodism?

7. Plants must, in some manner, synchronize their activities with the seasons. Of the cues that they might use, day (or night) length seems to have been selected. What might be the advantage of using photoperiod rather than, say, temperature as a season detector?

8. Photoperiodic systems are not nearly as sensitive to low levels of illumination as are many visual systems. Why might extreme sensitivity be a disadvantage in a photoperiodic system? What might be the effect on plants of the widespread use of bright lights for street illumination?

9. Suppose you were given a chrysanthemum plant, in bloom, one autumn, and you decided to keep it indoors as a house plant. What precautions would you need to take the following autumn to ensure that it would bloom again?

10. What are the major survival problems facing an organism? Compare the solutions to these problems achieved by plants with those achieved by vertebrate animals.

Suggestions for Further Reading

Brady, John: *Biological Clocks, Studies in Biology, No. 104,* University Park Press, Baltimore, Maryland, 1979.*

An interesting and well-written introduction to the subject of biological clocks and their experimental study.

Chen, Ingfei: "Plants Bite Back," *Science News,* December 22, 1991, pages 408–410.

Chilton, Mary-Dell: "A Vector for Introducing New Genes into Plants," *Scientific American,* June 1983, pages 50–59.

Evans, Michael L., Randy Moore, and Karl-Heinz Hasenstein: "How Roots Respond to Gravity," *Scientific American,* December 1986, pages 112–119.

Folkerts, George W.: "The Gulf Coast Pitcher Plant Bogs," *American Scientist,* vol. 70, pages 260–267, 1982.

Gasser, Charles S., and Robert T. Fraley: "Transgenic Crops," *Scientific American,* June 1992, pages 62–69.

Goodman, Robert M., Holly Hauptli, Anne Crossway, and Vic C. Knauf: "Gene Transfer in Crop Improvement," *Science,* vol. 236, pages 48–54, 1987.

Gould, Fred: "The Evolutionary Potential of Crop Pests," *American Scientist,* vol. 79, pages 496–507, 1991.

Rosenthal, Gerald A.: "The Chemical Defenses of Higher Plants," *Scientific American,* January 1986, pages 94–99.

Schnell, Donald E.: *Carnivorous Plants of the United States and Canada,* John F. Blair, Winston-Salem, N.C., 1976.

Descriptions of 45 species of carnivorous plants with many color photographs. This book is not only interesting reading but also useful as a field guide, and it contains a chapter on growing techniques.

Strobel, Gary A.: "Biological Control of Weeds," *Scientific American,* July 1991, pages 72–78.

Torrey, John G.: "The Development of Plant Biotechnology," *American Scientist,* vol. 73, pages 354–363, 1985.

Wayne, Randy: "Excitability in Plant Cells," *American Scientist,* vol. 81, pages 140–151, 1993.

*Available in paperback.

Ecology

The California condor is North America's largest bird. This giant vulture weighs up to 25 pounds and has a wingspan of 8 to 9 feet. Although, by human standards, not very pretty up close, the condor is breathtaking to watch in its soaring flight, as it glides up to 10 miles on a single wing beat, traveling hundreds of miles in one day.

Condors do not breed until they are five or six years old. Each mating pair produces only one egg, incubates it for almost two months, and feeds and protects its solitary chick for the better part of another year. As a consequence, even under the best conditions, a mating pair produces only one offspring every two years.

The condor reminds us that just as feathers and fins, hooves and beaks, flower parts and fruit colors are shaped by an evolutionary past, so are life histories. The condor is a relic of 2 million years ago. In those days, condors ranged all over the North American continent. They had no enemies, and the giant cats, cave bears, and other carnivores of the time left ample carnage in their wake to support hundreds of pairs of condors and their chicks.

The California condor is now almost extinct. The reasons are many. They include shooting by hunters, poisoning by bait intended for coyotes, poisoning from feeding on carcasses containing lead shot, and contamination by DDT and other pesticides. The principal causes, however, are two. The first is the condor's own life history, especially its long, careful care of its young. Breeding so slowly, the condor cannot replace those young that are lost. The second is the loss of its habitat. Efforts to return condors to the wild are failing because there is no longer a "wild" for them to return to.

Ecologists are divided about the condor's future. Some support a captive breeding program, so future generations will at least be able to catch a glimpse of these homely giants. Others feel that only a free-flying condor should be preserved. In the words of writer Kenneth Brower, "For a bird that once fed on mastodons, it is too steep a fall from glory."

Population Dynamics: The Numbers of Organisms

Ecology, the study of the interactions of organisms with their physical environment and with each other, is both the oldest and the youngest of the major subdivisions of biology. It is at least as old as the inquiring human mind—clearly not a subject early hominids could safely ignore. However, as a "hard" science, ecology is young. Only recently have biologists been able to devise ways to analyze the many variables that affect organisms in their natural environment, to study them quantitatively, and to construct models, pose hypotheses, and formulate and test predictions. The great ecologists of 40 years ago, who laid the foundations of modern ecology, were sensitive observers of nature. Their successors, equally sensitive observers as they gather the data needed to test their hypotheses, are often also wizards at calculus, statistics, and computer modeling.

As a science, ecology seeks to discover how organisms affect, and are affected by, their living and nonliving environment. Ecology also seeks to define how these interactions determine the kinds and numbers of organisms found in a particular place at a particular time. Because of the pressures of the growing human population on the environment and on the other organisms with which we share it, ecology has become, from a practical standpoint, one of the most important branches of biological science. We, even more than our ancestors, cannot safely ignore it.

We are going to begin these chapters on ecology by looking at some of the factors affecting the size of populations. Later chapters will be concerned with the interactions between and among the populations of different species in communities and with the interactions between communities and their physical environment in ecosystems. We shall end with a global overview of the biosphere.

Properties of Populations

As we noted in Chapter 1, different properties emerge at different levels of organization. A molecule has different properties from its constituent atoms, a cell different properties from its constituent molecules, and a multicellular organism different properties from its constituent cells. Similarly, a population has very different properties from the individuals it comprises.

As a simple illustration, picture a patch of woods, or a field, or a lake shore to which you return from time to time, which perhaps your parents and grandparents visited and where, with any luck, your children may walk someday. If you go there in the spring, you will see, for the sake of argument, squirrels, birds, wildflowers, and beetles—the same species you saw the preceding spring and the same species you expect to see in subsequent springs. Yet, year after year, it is unlikely that you see the same individual squirrels and birds, very unlikely that you encounter the same beetles, and seldom the same individual wildflowers. The individual is transitory, but—with a little luck—the population endures.

A **population** is a group of organisms of the same species living in the same place at the same time. Among the properties of populations that are not properties of individual organisms are growth patterns, mortality patterns, age structure, density, and dispersion.

Patterns of Population Growth

As Darwin noted some 150 years ago, the reproductive potential of all species is very high. It can be calculated, for instance, that a single female housefly producing 120 eggs (with half developing into females) can, within seven generations—the expected number of generations in a single year—give rise to almost 6×10^{12} houseflies (Table 43–1). With bacteria that have a generation time of 20 minutes, a single bacterium can give rise to eight bacteria within one hour, 512 in three hours, and 262,144 in six hours. Fortunately for us, few organisms ever attain their full reproductive potential.

The **per capita rate of increase** of a population is the increase in the number of individuals in a given unit of time per individual present. Thus, the annual per capita rate of increase of the human population in the United States is currently about 0.007 (which is equivalent to 7 per thousand), and in the world as a whole it is about 0.016 (or 16 per thousand). In the absence of net immigration (the movement of other individuals of the species into the population from elsewhere) or emigration (the departure of individuals from the population), the rate of increase is equal to the birth rate minus the death rate. Thus, the rate of increase can be equal to zero, it can be a positive figure, or it can be a negative figure—as it is now for the human population of some countries.

Table 43–1   Reproductive Potential of the Housefly (*Musca domestica*)*

Generation	Numbers If All Survive and Reproduce
1	120
2	7,200
3	432,000
4	25,920,000
5	1,555,200,000
6	93,312,000,000
7	5,598,720,000,000

* In one year, about seven generations are produced. The numbers are based on the following assumptions: each female lays 120 eggs per generation, each fly survives just one generation, and half of the flies are females. Adapted from E. J. Kormondy, *Concepts of Ecology*, Prentice-Hall, Inc., Englewood Cliffs, N.J., 1969.

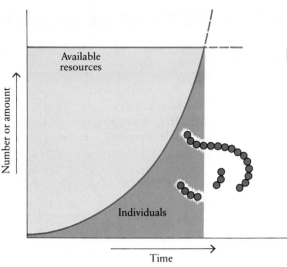

43–1   *An exponential growth curve. After an initial establishment phase, the population increases in the same fashion as a savings account with compound interest. Exponential growth is characteristic of small populations, such as laboratory cultures of bacteria, with access to abundant resources.*

The **growth rate** of a population—that is, the change in the number of individuals over time—equals the per capita rate of increase times the number of individuals *already present*. In the type of population growth known as **exponential growth,** the number of individuals increases at a constant rate. As you can see in Figure 43–1, exponential growth starts out slowly, but then it shoots up very rapidly as the number of reproducing individuals increases in each succeeding generation. The principle is the same as the compounding of interest on a savings account: the more you have, the more you get.

The exponential growth curve is most closely approximated by microorganisms cultivated in the laboratory (where resources are constantly renewed), by the initial stages of seasonal "blooms" of algae, and by the recent growth of the human population. Under most circumstances, however, a population cannot long continue to increase exponentially without reaching some environmental limits imposed by shortages of food, space, oxygen, nesting or hiding places, or accumulation of its own waste products. In nature, short-term exponential growth is characteristic of so-called fugitive or opportunistic species that invade an area, rapidly use up local resources, and then either enter a phase of dormancy or move on. Weeds and insects are typical opportunists.

Sometimes, a population that is growing very rapidly may hit an environmental limit suddenly and then "crash" to very low levels. An example can be found in the gypsy moth infestations that recurrently plague the northeastern United States (Figure 43–2). If a population grows rapidly enough, it can wipe out its food supply before the caterpillars complete their metamorphosis and reach the reproductive stages. Or, alternatively, the population may reach such a high density that it provides a suitable environment for the exponential growth of an infectious microorganism (a virus, in the case of the gypsy moth) and is wiped out by disease.

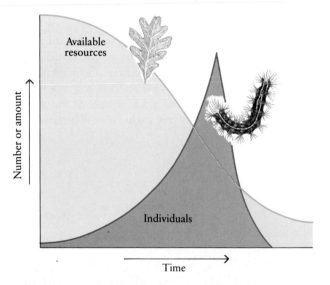

43–2   *A population "crash," as illustrated by the gypsy moth caterpillar. Infestations of these caterpillars can strip the leaves from trees so rapidly that the food supply is exhausted before the caterpillars have finished feeding and are ready to begin their metamorphosis. The result is a sudden drop in the number of individuals. Spraying the caterpillars with insecticides often reduces the size of the population enough to prevent a crash and thus maintains the population at a lower but fairly constant level.*

The Effect of the Carrying Capacity

For many populations, the number of individuals is determined not by reproductive potential but by the environment. The number of individuals of a particular population that the environment can support under a particular set of conditions is known as the **carrying capacity** of the environment. For animal species, the carrying capacity may be determined by food supply or access to sheltered sites. For plants, the determining factor may be access to sunlight or the availability of water.

The factors limiting population size may vary seasonally. When the wildebeests migrate through the Serengeti plain of East Africa, all the lions are well fed, but when the big herd animals are gone, the lion cubs often starve to death, and some of the less capable adults as well. In North America, the carrying capacity for a herd of deer is not the number counted in the springtime when new fawns appear, but the number that can survive over several winters.

The patterns of population growth observed in nature are many and complex. One of the simplest patterns, which illustrates clearly the effect of the carrying capacity, is shown in Figure 43–3. When the number of organisms is very small and resources are abundant, the curve approximates the exponential growth curve of Figure 43–1. As the number of organisms increases and resources become scarce, population growth slows. If the number of organisms exceeds the carrying capacity, the growth rate becomes negative and the population declines. Eventually, the population stabilizes and oscillates around the maximum size the environment can support. This model of population growth, represented by an S-shaped curve, is called **logistic.**

An understanding of the logistic model has practical applications. If, for instance, you want to control a population of rats, killing half of them may merely reduce the population to the point at which it increases most rapidly. A more effective approach would be to reduce the carrying capacity, which, in the case of rats, usually means tighter control of garbage disposal. Similarly, if you want to achieve maximum long-term productivity in the harvesting of a particular type of economically valuable fish, you should not harvest the population below the level of rapid growth unless you are willing to wait a long time for the population to recover.

Mortality Patterns

Another important property, affecting both the size and composition of a population, is its **mortality pattern.** Figure 43–4 shows three different patterns of mortality, represented in the form of survivorship curves. In the first, exemplified by the oyster, a very high mortality rate in early life is followed by a lower rate of mortality among older individuals that have passed some critical stage in the life cycle. In the second pattern, exemplified by *Hydra*, the rate of mortality is the same for all members of the population, regardless of age or stage in the life cycle. In the third pattern, which is characteristic of our own species, most individuals reach a fairly advanced age, after which there is a high rate of mortality.

Natural populations often show a combination of

43–3   *One of the simplest growth patterns observed in natural populations is known as logistic growth and is represented by an S-shaped curve. As with exponential growth, there is an initial establishment phase when population growth is relatively slow (1), followed by a phase of rapid acceleration (2). Then, as the population approaches the carrying capacity of the environment, the growth rate slows (3 and 4), and finally stabilizes (5), though fluctuations around the carrying capacity may continue. This graph represents the growth pattern of a herd of deer introduced into a new area with limited resources. Other growth patterns observed in natural populations are considerably more complex.*

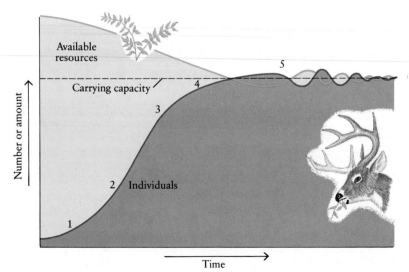

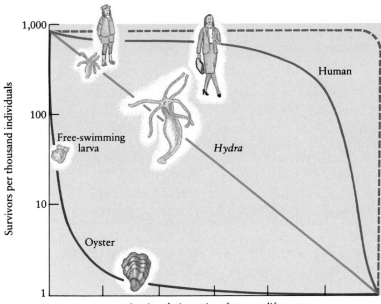

43–4   *Survivorship curves, showing three different patterns
of mortality. In the oyster, mortality is extremely high during
the free-swimming larval stage, but once the individual attaches itself to a favorable substrate, life expectancy levels off.
Among* Hydra, *the mortality rate is the same at all ages.*

*The curve at the top of the graph (dashed line) is for a hypo-
thetical population in which all individuals live out the average
life span of the species—a population, in other words, in which
all individuals die at about the same age. The fact that the
curve for humans approaches this hypothetical curve indicates
that the human population as a whole is reaching some genetically programmed uniform age of mortality.*

mortality patterns. For instance, in a study made of the
saltmarsh song sparrows of San Francisco Bay, it was
estimated that of every 100 eggs laid, 26 are lost before
hatching. Of the 74 live nestlings, only 52 leave the nest,
and of these, 42 (80 percent) die the first year. The remaining 10 breed the following season, but during the
next year, 43 percent of these die, leaving only 6 of the
original 100. Each subsequent year, mortality among
the survivors amounts to 43 percent, apparently regardless of age. Once a bird survives its first, risk-laden
year, the mortality rate for the subsequent years
remains more or less constant. Thus, the early survivorship curve for the sparrows resembles that of the
oyster, while the later curve is more *Hydra*-like.

Age Structure

The mortality pattern of a population, in turn, affects
another important property of the population, its age
structure. The **age structure** of a population is the proportion of individuals of different ages in the population.

In species in which the life span exceeds the reproductive span, knowledge of the age structure is useful
in predicting future changes in population size. For example, if a large proportion of a population is of reproductive age or younger, as in the human population of
India (Figure 43–5), population growth will continue
at a high rate even though reproductive individuals only
replace themselves. This is necessarily the case simply
because the number of reproductive individuals continues to increase as younger individuals reach reproductive age.

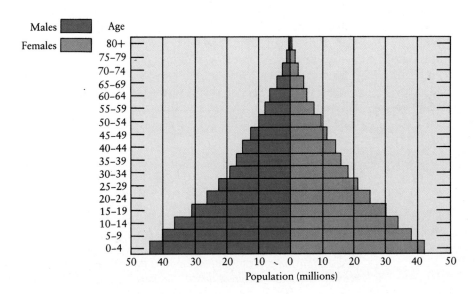

43–5   *The age structure of the human
population of India. This pyramidal shape
is characteristic of developing nations, with
half the population under 20 years of age.
In the absence of emigration, the population
can stay the same size only if death rates are
as high as birth rates. Even if members of
the present generation in India limit their
family size to only two children per couple
(which means cutting the current birth rate
in half), population growth will not level off
until about the year 2040—and at a level of
well over a billion.*

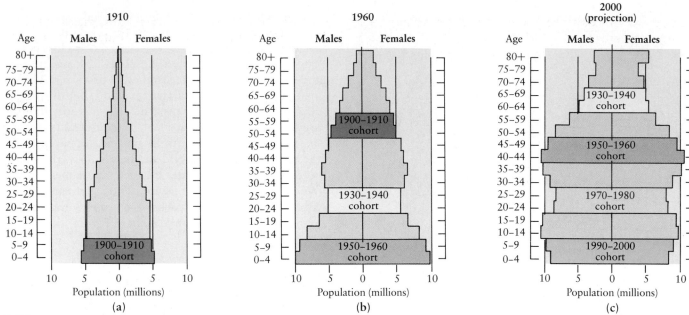

43–6 *Age structure of the United States population.* **(a)** *In 1910, the graph of age structure was shaped like a pyramid, although its base—that is, the number of persons in the youngest age groups—was not as large as that of India's. In subsequent years, as illustrated by* **(b)** *the graph of the age structure in 1960, and* **(c)** *the graph of the age structure projected for the year 2000, the proportion of the population that is more than 40 years of age has steadily increased, and the age structure of the population has become more uniform. The term*

"cohort" refers to the group of individuals born during the decade indicated.

Note the decrease in the growth of the population during the depression years of 1930 to 1940 and the bulge produced by the "baby boom" of the 1950s. Although birth rates in the United States decreased in the 1970s and have remained low, the increase in the number of individuals during the 1950s was reflected in an increase in population growth in the 1980s as the boom babies reached reproductive age.

The age structure of the United States population (Figure 43–6) is one of the reasons why the population has continued to increase despite the fact that, on the average, young couples (between the ages of 20 and 30) are having slightly fewer than two children. Another reason for the increase is continued immigration.

As population growth slows, the age structure gradually becomes more uniform. Ultimately, a population that is not increasing attains a stationary age structure—that is, the age structure of the population remains the same over long periods of time.

Density and Dispersion

Two additional, interrelated properties of populations are density and dispersion. **Population density** is the number of individual organisms per unit area or volume. For example, the number of *Paramecium* cells per cubic centimeter of pond water, of dandelions per square meter of lawn, and of field mice and oak trees per hectare (2.47 acres) are all measures of population density.

A description of the **dispersion**—the pattern of distribution of the organisms within the two- or three-dimensional space they occupy—provides additional

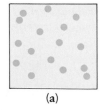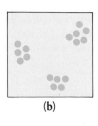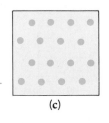

(a)   (b)   (c)

43–7 *The three basic dispersion patterns observed in natural populations are* **(a)** *random,* **(b)** *clumped, and* **(c)** *regular. The dots may represent individuals of the same species, populations of the same species, or even populations of different species. Determination of dispersion patterns requires careful observation and precise mapping, repeated in a number of different areas and at different times.*

information about the population. The three basic dispersion patterns are shown in Figure 43–7. They are (1) **random,** in which the spacing between individuals is irregular; (2) **clumped,** in which the individuals are aggregated in patches; and (3) **regular,** in which individuals are evenly spaced within the area or volume.

A number of factors, both nonliving and living, may affect the spatial distribution of a particular population. Dispersion patterns are often dependent upon the distribution of essential resources. If, for example, water is available only in patches, members of a population of a particular plant species may be found clumped around these patches. In vertebrates, dispersion patterns often reflect social behavior. Individuals of a species that is highly territorial usually exhibit a regular spacing, whereas individuals of highly gregarious species tend to be clumped.

The dispersion pattern observed for a particular population or species depends upon the scale of the observation. On a local scale—a few square meters or kilometers, for example—a species may show a random or regular distribution. However, when viewed on a larger scale, such as the entire North American continent, the same species may show a clumped distribution.

(a)

(b)

(c)

43–8   *Some examples of dispersion patterns. (a) Many species of birds exhibit a regular dispersion when perched on telephone wires, power lines, or, as shown here, a dead tree branch. These birds are carmine bee-eaters (Merops nubicoides). (b) Fish, such as this school of French grunts, photographed in a shipwreck off the coast of Mexico, and (c) quaking aspen often exhibit clumped dispersion patterns.*

Climate Change and the Movement of Populations

The probability of a major global climate change within the next century is a fairly constant topic in the news. Although the speed of the predicted change and the agent of change—the activities of *Homo sapiens*—may be news, the occurrence of climate change itself is not. Periodic warming and cooling have characterized the global climate since shortly after the first living organisms made their appearance on Earth.

During most of our planet's history, its climate appears to have been warmer than it is at the present time. However, these long periods of milder temperatures have been interrupted periodically by Ice Ages, so called because they are characterized by **glaciations,** or persistent accumulations of ice and snow. Such glaciations occur whenever the summers are not hot enough and long enough to melt ice that has accumulated during the winter. In many parts of the world, an alteration of only a few degrees in the annual average temperature is enough to begin or end a glaciation.

An early Ice Age appears to have occurred at the beginning of the Paleozoic era, some 590 million years ago. Another, marked by extensive glaciations in the Southern Hemisphere, occurred at the end of the Paleozoic, some 248 million years ago. The conifers evolved during this period and possibly the angiosperms, as older forest types disappeared. A more recent, less severe cold, dry period occurred at the end of the Mesozoic, about 65 million years ago, and may be associated with the events causing the extinction of the dinosaurs (page 369).

The most recent Ice Age began during the Pleistocene epoch, about 1.5 million years ago. The Pleistocene has been marked by four extensive glaciations that have covered large areas of North America and Europe. Between the glaciations there have been intervals, called **interglacials,** during which the climate has become warmer.

In each of the four Pleistocene glaciations,

Limit of ice sheet
6000 B.C.

4000 B.C.

6000 B.C.

8000 B.C.

0 A.D.

10,000 B.C.

Limit of ice sheet
16,000 B.C.

Hemlock

The northward migration of the eastern hemlock (Tsuga canadensis), *following the last glacial retreat. This retreat began about 18,000 years ago and ended some 8,000 years ago. The date at the right of each curve indicates the time of the first appearance of the leading edge of the hemlock population at the locations marked by the curve. This reconstruction of the northward movement of the hemlock population was compiled by Margaret B. Davis of the University of Minnesota, using records of pollen deposits that were preserved in lake sediments.*

sheets of ice, thicker than 3 kilometers in some regions, spread out from the poles, scraped their way over much of the continents—reaching as far south as southern Illinois in North America and covering Scandinavia, most of Great Britain, northern Germany, and northern Russia—and then receded again. We are living at the end of the fourth glaciation, which completed its retreat only some 8,000 years ago.

The fossil record shows that during periods

of dramatic climate change the populations of affected regions were under extraordinary ecological and evolutionary pressures. As conditions became too extreme, plant and animal populations moved, changed, or became extinct. In the interglacial periods, during which the average temperatures were at times warmer than those of today, the tropical forests and their inhabitants spread from the equatorial regions poleward through today's temperate zones. During the periods of glaciations, only animals of the polar regions could survive in these same locations.

In the colder periods, reindeer ranged as far south as southern France, while during the warmer periods, the hippopotamus reached England. Rhinoceroses, great herds of horses, large bears, and lions roamed Europe during the interglacials. In North America, as the fossil record shows, there were camels and horses, saber-toothed cats, and great ground sloths (one species as large as an elephant).

The reason for past climate changes is one of the most controversial issues in modern science. They have been variously ascribed to changes in the Earth's orbit, variations in the Earth's angle of inclination toward the sun, migration of the magnetic poles, fluctuations in solar energy, higher elevations of the continental masses, continental drift, and combinations of these and other causes. The predictions of a coming climate change are based upon a more obvious cause—the dramatic increase in the Earth's atmosphere of carbon dioxide and other by-products of human activity.

As we all wait to see the extent of this climate change, one of the greatest concerns of biologists is whether there will be time and opportunity for populations that are unable to adapt to changing conditions to move to new environments in which they can survive. If not, the resulting mass extinctions may well exceed those that marked the major climate upheavals of the past.

The Regulation of Population Size

Fifty years ago, the eminent ecologist Charles Elton noted that "no animal population remains the same for any great length of time, and . . . the numbers of most species are subject to violent fluctuations." The popular notion that "nature is in balance" and that populations generally reach an equilibrium state has also come under severe criticism from contemporary ecologists. As the two examples illustrated in Figure 43–9 show, the size of a population may vary greatly over a period of years.

(a)
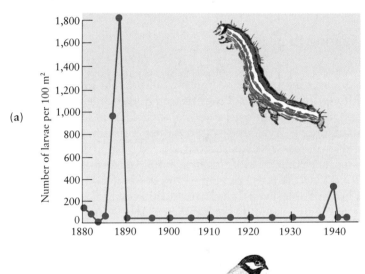

(b)
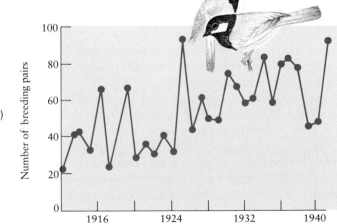

43–9 *Two examples of the sometimes extreme fluctuations that occur in population size and density.* (**a**) *The population density of dormant larvae of the moth* Dendrolimus pini, *recorded over a period of 60 years in a coniferous forest in Germany.* (**b**) *Variations over a span of 30 years in the size of the breeding population of the great tit* (Parus major), *a common European bird of the same genus as the titmice and chickadees of North America, recorded at a single location in the Netherlands.*

Although it is often difficult to understand why fluctuations in population size occur, such knowledge can be important because population fluctuations in one species can have profound effects—for good or ill—on populations of other species, including ourselves. For example, earlier in this century an enormous increase occurred in the number of field mice in Kern County, California. The population density reached thousands of mice per hectare, leading not only to the devastation of crops but also to the infestation of playgrounds and living rooms. Abrupt declines in a population can also occur with staggering speed. A field plagued with mice in June may have practically none in July. A number of factors are thought to be involved in such fluctuations.

Limiting Factors

Among the influences affecting the size and density of a population are specific limiting factors that differ for different populations. Of critical importance is the organisms' range of tolerance for such factors as light, temperature, available water, salinity, nesting space, and shortages (or excesses) of required nutrients. If any essential requirement is in short supply or any environmental feature is too extreme, growth of the population is not possible, even though all other necessities are met.

Ecologists often divide the factors that influence the growth of a population into density-dependent and density-independent factors. Factors that cause changes in either the birth rate or the mortality rate as the density of the population changes are said to be **density-dependent.** Birth rates and mortality rates, however, may also vary without regard to changes in population size or density. Factors that cause such changes are said to be **density-independent,** and they often involve weather-related events.

Numerous factors operate on populations in a density-dependent fashion. As a population increases in size, it may deplete its food supply, leading to increased competition among members of the population that ultimately leads to a higher mortality rate or a lower birth rate. Predators may be attracted to areas in which the density of prey organisms becomes high, capturing an increased proportion of the population. Similarly, infectious diseases may spread more easily when population density is high.

Environmental disturbances often act as density-independent factors. For example, when Mount St. Helens erupted in 1980, populations of many species became locally extinct. Mortality was completely independent of the size and density of the populations. In many cases, however, the identification of density-

independent factors is less clear-cut. The mortality resulting from a severe storm or cold spell may appear to be density-independent, yet animals that have secure places of shelter are more likely to survive than those that do not. And, the availability of good sheltering sites depends on the density of the population. As this example illustrates, there is often an interplay between density-dependent and density-independent factors, and populations are generally affected by both simultaneously.

Life-History Patterns

Among the most interesting and variable properties of populations are **life-history patterns,** which are groups of traits affecting reproduction and survival. Under what conditions, for example, will natural selection favor an organism that has two million almost microscopic young at one spawning, like an oyster, as opposed to an organism that has only one large infant, like an elephant?

In view of the fact that physiological constraints prevent an oyster from producing a 10-kilogram offspring, this question may sound silly. However, as a result of studies such as those on clutch size in birds (page 340), biologists have become aware that life-history patterns vary from individual to individual within a population and also from one population to another population of related organisms. In other words, these patterns involve genetically determined variations subject to natural selection.

Alternative life-history patterns have been given a variety of names: prodigal and prudent, opportunistic and equilibrium, and r-selected and K-selected. (In mathematical formulations of population growth, r represents the per capita rate of increase and K the carrying capacity.) Generally speaking, prodigal or opportunistic patterns would appear more adaptive for "weedy" species, colonizers of open fields, for instance, whereas prudent or equilibrium patterns would appear to be more adaptive for a population at the carrying capacity.

The characteristics associated with the alternative patterns are shown in Figure 43–10. Each pair of characteristics represents the extremes in a continuum of possibilities. The best choice in each continuum—the choice made by natural selection operating on a particular population—depends in large part on other properties of the population (such as its mortality pattern and its age structure) and on properties of the environment it occupies.

(a)

Prodigal or Opportunistic

Many young

Small young

Rapid maturation

Little or no parental care

Reproduction once ("Big Bang")

(b)

Prudent or Equilibrium

Few young

Large young

Slow maturation

Intensive parental care

Reproduction many times

43–10   *The characteristics associated with alternative life-history patterns.* **(a)** *Annual plants, such as the zinnia, have a prodigal, or opportunistic, life-history pattern. These plants reproduce only once, typically setting large numbers of seeds. The parent plants die soon after, and only the seeds survive.* **(b)** *Elephants, by contrast, have a prudent, or equilibrium, life-history pattern. The calves are suckled by their mothers for at least two years. The elephant social structure is a maternal hierarchy in which the young are zealously guarded by the mother and the sisters and the aunts.*

Some Examples of Life-History Patterns

Suppose you are a plant with a very short life expectancy—a flowering annual, for example. The choice that has been made for you, as a result of the action of natural selection in the past, is to put as much energy as possible, as rapidly as possible, into one reproductive effort. The result is a big burst of seeds, increasing the probability that at least a few will survive your short life span and, in turn, reproduce.

On the other hand, if you are an oak tree, once you have established a place for yourself, your life span is likely to be a very long one, more than 100 years. However, the likelihood of your offspring finding space on the forest floor is very small. You grow as tall as possible, for 20 years or more, since height aids in seed dispersal, and then put out moderate numbers of acorns per year, every year, over your entire life span. In this way, you are likely to produce more surviving offspring than if you had put the same amount of energy into one big burst of reproduction.

Bluefin tuna are like oak trees. Once a bluefin reaches a large size, its chances of continued survival are extremely good. However, the mortality among eggs and larvae is always very high, and, in years in which the plankton growth is low, there are no survivors at all. So it is better for a bluefin to concentrate its efforts first on reaching a safe size and thereafter to expend a relatively small amount of reproductive effort each season over a long period of time.

Whether breeding is early or late can greatly influence the rate of population growth. Reconsider the housefly of Table 43–1. If the generation time (that is, the interval between the time an egg is laid and the time that individual, in turn, mates and lays eggs) were shortened by only two days (from a minimum of 13 to a minimum of 11), there would be time for one more generation each season. The total number of possible offspring per season would increase by 330×10^{12}.

Early breeding does not always pay, however. Among many of the larger mammals, the rate of juvenile survival can be correlated with the experience of the mother, her size, or her social position, which is often determined by her age. In such situations, the action of natural selection favors delayed reproduction. A similar pattern is observed in the human population. Infant mortality is higher among teenage mothers in the United States than among women over 20, and higher still if the teenagers are from impoverished families, as they often are.

The Asexual Advantage

When we considered the evolution of sexual reproduction (page 329), we noted that an asexually reproducing population can increase in numbers much more rapidly than a sexually reproducing population. There are other advantages to asexual reproduction. Many plants, for example, reproduce by means of runners (page 734), and by so doing, may be able to grow to cover a very large area. All of the plants produced represent a single genotype. A young plant that develops in this way has a continued supply of resources from the parent plant and thus a much higher probability of survival.

Figure 43–11 shows the differences in survivorship in a species of buttercup between plants grown from seeds and those grown from runners. Note the very high mortality in the early stages of growth from seed—an oysterlike curve—compared to the uniform risk of death of the plantlets grown from runners.

Parthenogenesis

Another form of asexual reproduction is **parthenogenesis,** the development of an organism from an unfertilized egg. In species in which the male gamete determines the sex of the offspring, parthenogenesis always results in all female offspring. If the houseflies of Table 43–1 had reproduced parthenogenetically, each female would have had twice as many female young in every generation, and the population would have reached 358×10^{12} at the end of seven generations.

Parthenogenesis in plants lacks the advantage of the parental support system supplied by vegetative growth, which is traded off for the possibilities of larger numbers and, usually, wider dispersal of the young. Dandelions (Figure 43–12) reproduce parthenogenetically. They form conspicuous flowers and also some functionless pollen grains, which may be taken as evidence that the present asexual species of dandelions evolved from sexual ones.

Completely asexual species are also found among small invertebrates—some rotifers, for example—as well as among plants. Recently, several species of fishes and lizards have been found that apparently reproduce only parthenogenetically. Many other organisms alter-

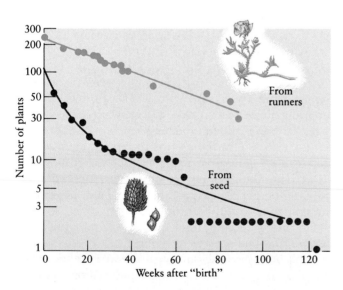

43–11   *Survivorship curves of populations of* Ranunculus repens, *a species of buttercup. Of 100 plants that started from seed (lower curve), only two (2 percent) were still alive 20 months later. Of 225 plants originating from runners, 30 (more than 15 percent) were still alive after 20 months (upper curve). Such plants receive support from the parent plant during early growth.*

43–12 *One of the most widely distributed of all the flowering plants is the dandelion. Its flowers produce enormous numbers of seeds parthenogenetically. The seeds are enclosed in fruits bearing a plumelike structure that is caught by the wind, thereby aiding dispersal. Dandelions also reproduce vegetatively. If the plant is lopped off down to the root by either hominid or herbivore, three more grow in its place.*

nate sexual and asexual phases. Freshwater *Daphnia,* for instance, multiply by parthenogenesis when the plankton on which they feed is abundant. Then, in response to some environmental cue, they start producing both males and females. Typically, the asexual phase occurs when conditions are favorable for rapid local growth, and the sexual phase when the population is facing a less certain future and less homogeneous conditions.

Some Consequences of Life-History Patterns

Opportunistic organisms, which rapidly exploit an environment and then move on, would appear to lead risky lives, both as individuals and as species. There may not always be a new environment available for exploitation. However, populations of such organisms typically possess remarkable recuperative powers, because a population can be built up quickly from only a few individuals.

By contrast, populations composed of long-lived, slow-to-mature individuals, which would appear to have a higher probability of long-term survival, are very slow to recover when their size is reduced. Both the California condor (page 774) and the whooping crane are in this category. Each does not begin to reproduce until it is about five or six years old, and each raises only a single chick in each reproductive effort. As a result of a concerted conservation effort, the whooping crane population has made a remarkable recovery from the brink of extinction. For the condor, however, as we

saw at the beginning of this chapter, the prospects are not promising.

The Human Population Explosion

The California condor and the whooping crane are only two of the thousands of species of animals and plants whose continued existence has been made precarious by the extraordinary population growth of just one species—our own.

In the period from 25,000 to 10,000 years ago, when *Homo sapiens* lived as a hunter-gatherer, the rate of population increase was low. By the end of this period, the human population probably numbered a little more than 5 million, spread over the entire world. At this point, the establishment of agricultural communities began. During the next 5,000 years, as agriculture spread around the world, population growth increased dramatically. By 3000 B.C., the human population had reached about 100 million.

Population growth continued after an agricultural way of life had become established, but at a significantly lower rate. By 1650 A.D., the population had increased to approximately 500 million. At about this time, the rapid development of science, technology, and industry began, bringing profound changes in human life and its relationship to nature. In the 150 years between 1650 and 1800, the population doubled, to 1,000 million (1 billion), and then it doubled again by 1930, to 2 billion (Figure 43–13).

43–13 *Over the last 25,000 years, the human population has grown from an estimated 3 million to more than 5 billion. A significant increase in the rate of population growth occurred as a result of the agricultural revolution, and an even more dramatic increase began with the advent of the scientific and industrial revolution in the late 1600s.*

The consequences of the rapid growth of the human population are many and varied. In our country and other parts of the developed world, they include not only the sheer numbers of people but also heavy consumption of nonrenewable fossil fuels and the resulting pollution—both as the fuels are burned and as a result of accidents, such as oil spills, during their transport. In less de-

veloped parts of the world, the consequences include malnutrition and, all too often, starvation, coupled with a continuing vulnerability to infectious diseases. The consequences for other organisms include not only the direct effects of pollution but also—and most important—the loss of habitat.

By 1991, there were 5.42 billion people on our planet, and the population is growing at a rate of 1.6 percent per year. This means a net increase in the world population of more than 160 people every minute, about 230,000 each day, and almost 84 million every year. If this rate of increase is sustained, there will be about 6.2 billion people on Earth by the year 2000.

Birth Rates and Death Rates

For *Homo sapiens,* as for other species, the rate at which the population increases is the result of both the birth rate and the death rate. The relative influence of these two factors has varied at different times in human history.

In the period between 25,000 and 10,000 years ago, the low rate of increase is believed to have resulted from a low birth rate, in turn the result of physiological factors. For example, in some primitive hunter-gatherer societies, a woman is unable to conceive a child until she is between 19 and 20. Moreover, throughout the period when she is breast-feeding a child (which may be as long as three or four years), ovulation usually does

not occur. Compared to other contemporary human societies, the age at first reproduction is later (and thus the total reproductive span of the woman is shorter) and the interval between children is longer. These factors combine to produce a much lower birth rate in the population.

It is thought that the changes in culture and nutrition that accompanied the shift to an agricultural way of life may have led to a breakdown in the mechanisms that controlled the birth rate, resulting in the dramatic population increases that occurred between 10,000 and 5,000 years ago. The reasons for the subsequent slowing of population growth are not known with certainty. One contributing factor was probably an increased vulnerability to infectious diseases as people lived in closer proximity to each other within agricultural communities. Although the population continued to grow, increased mortality—particularly among the young—had a damping effect on the growth rate.

The enormous recent increase in the world's population is primarily a result of the decline in death rates, especially among the young, as shown in Figure 43–14. Some experts believe that further reductions in death

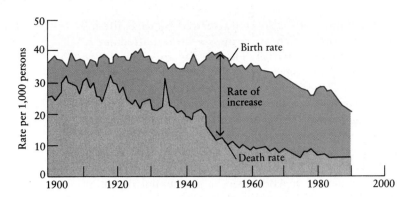

43–14   *In many tropical countries, the death rate has fallen rapidly since 1940, resulting in a rapid growth of the population. The drop in the death rate is primarily the result of increased medical services, better control of malaria-bearing mosquitoes, and the availability of new antibacterial drugs, especially the antibiotics. Note that the birth rate has also begun to fall. The area marked in color indicates the population growth in the absence of net immigration or emigration. The data shown here are for Sri Lanka.*

rates and a general increase in the standard of living will lead to a reduction in birth rates as well and thus to a reduction in the rate of population growth. Other experts, however, disagree. They point out that in many parts of the world a significant rise in the standard of living seems an almost unattainable goal and, in some cases, a rapidly receding one. Moreover, the age structure in many countries is such that even if birth rates were to drop dramatically, it would be many years before population size would begin to level off.

The current growth of the human population poses urgent and complex problems, made more difficult by the unequal distribution of growth and available resources. Births are occurring at the greatest rates in precisely those areas where the new arrivals have the least chance of an adequate diet, good housing, schools, medical care, or future employment. Moreover, because the affluent citizens of developed countries are not constantly reminded of the soaring population by experiencing firsthand the problems of hunger and crowding, they may feel it is not their problem. Yet a child born to a middle-class American will consume, in his or her lifetime, a far greater amount of the limited resources of the world—more than twice the amount of food, for instance—than a child born in a less developed country.

Bringing human population growth under control is an enormously complex undertaking, involving biological, economic, political, and social factors and requiring wisdom and commitment on the part of the diverse populations that constitute *Homo sapiens*.

Summary

Ecology is the study of the interactions of organisms with their physical environment and with each other. Ecologists seek to quantify the variables that affect organisms in nature, to construct explanatory hypotheses for the observed distribution and abundance of organisms, and to make and test predictions based on their hypotheses.

A population is a group of organisms of the same species living in the same place at the same time. Among the properties that characterize populations are patterns of growth, patterns of mortality, age structure, density, and dispersion.

In the absence of net immigration or emigration, the change in the number of individuals in a population over a given period of time—that is, the growth rate of the population—equals the per capita rate of increase (the birth rate minus the death rate) times the number of individuals present at the beginning of the period under study.

The reproductive potential of most populations is high. When the full reproductive potential of a population is achieved (a relatively infrequent occurrence in nature), exponential growth can occur. Exponential growth cannot continue long, however, without a "crash" in the population size.

The maximum number of individuals in a particular population that the local environment can support over a specified period of time is known as the carrying capacity of the environment. The logistic model, which takes the carrying capacity into account, describes one of the simpler patterns of population growth observed in nature. Growth is rapid when the population is small, gradually slows as it approaches the carrying capacity, and then oscillates as the population is maintained at or near the carrying capacity.

Populations also have characteristic mortality patterns, with varying risk of death at varying ages. A related property is the age structure of the population, that is, the proportion of individuals of different ages. Age structure is an important factor in predicting the future growth of a population.

Two additional, interrelated properties of a population are its density and its dispersion pattern. Density is the number of individual organisms per unit area or volume, whereas the dispersion pattern describes the two- or three-dimensional arrangement of the organisms in the environment.

A complex variety of environmental factors, both living and nonliving, play a role in the regulation of population size. Factors that influence birth rates or death rates independently of the population density are said to be density-independent; they often involve severe environmental disturbances. Factors that cause changes in birth rates or death rates as the density of the population changes are said to be density-dependent; these factors include numerous resources that are available in limited supply.

Populations are also characterized by their life-history patterns, which are groups of traits affecting reproduction and survival. These traits are generally genetically determined and thus subject to natural selection. The extremes of the alternative characteristics include many young versus few young, small young versus large young, rapid maturation versus slow maturation, little or no parental care versus intensive parental care, and reproducing once versus reproducing many times. In some organisms, the alternatives also include asexual reproduction versus sexual reproduction. Some combinations of these characteristics are favorable for organisms of opportunistic species undergoing exponential growth, whereas others will be selected for in more stable populations living at or near the carrying capacity.

About 25,000 years ago, the human population numbered perhaps 3 million worldwide. A surge in population growth began about 10,000 years ago with the development of agriculture. A second, more dramatic surge, still underway, began with the scientific and industrial revolution in the late 1600s. As the population now approaches 6 billion, this explosive growth is placing enormous pressure on the environment and the multitude of organisms with which we share it.

Questions

1. Distinguish between the following terms: exponential growth/logistic growth; limiting factors/carrying capacity; density-dependent factors/density-independent factors; opportunistic species/equilibrium species.

2. An old French riddle: "The pond lilies in a certain pond grow at a rate such that each day they cover twice as much of the pond as they did the day before. The pond is of a size that it will be completely covered at the end of 30 days. On what day is the pond half-covered? One-tenth covered? One-hundredth covered?"

What is the relevance of this riddle to human ecology?

3. Suppose that you have a "farm" on which you grow, harvest, and sell edible freshwater fish. The growth of the fish population is logistic. You, of course, wish to obtain maximum yields from your "farm" over a number of years. To ensure this, how large should you allow the population to become before you begin harvesting? Identify the point on the logistic growth curve (Figure 43–3) at which you should begin harvesting.

How large a population should you leave unharvested? Identify the point on the curve at which you should take no more fish from the population.

Factors in addition to the pattern of harvesting will affect the yields of fish obtained. What are some of these factors and how might you adjust them to further increase the yields?

4. What would be the shape of the survivorship curve of a population of annual plants? Of automobiles? Of salmon? Of butterflies? Of dishes in a dishwasher?

5. Note that in the three graphs in Figure 43–6 there is a marked difference in the proportion of individuals more than 80 years old, but the vertical axis has not become any longer. What do these data suggest? Does the age-structure graph of India (Figure 43–5) support your conclusion?

6. Distinguish among random, clumped, and regular dispersion patterns, and give an example of each. How does the scale from which organisms are viewed affect the dispersion pattern?

7. Explain how each of the following factors would affect the growth rate of a population: age at first reproduction; time between generations; pre-reproductive mortality; post-reproductive mortality; length of period of parental care.

8. Imagine a hypothetical species in which a particular individual lives only 48 hours and produces only two offspring. How is it possible that this individual may achieve greater fitness (see page 320) than a longer-lived individual that produces 100 offspring? Explain in evolutionary terms.

9. In an editorial entitled "Food, Overpopulation, and Irresponsibility" that appeared in a scientific journal, the authors concluded: "Because it creates a vicious cycle that compounds human suffering at a high rate, the provision of food to the malnourished nations of the world that cannot, or will not, take very substantial measures to control their own reproductive rates is inhuman, immoral and irresponsible." What is your opinion?

Suggestions for Further Reading

Begon, Michael, John L. Harper, and Colin R. Townsend: *Ecology: Individuals, Populations, and Communities*, 2d ed., Blackwell Scientific Publications, Boston, 1990.

> *An up-to-date, balanced, and thorough treatment of contemporary ecology.*

Caldwell, John C., and Pat Caldwell: "High Fertility in Sub-Saharan Africa," *Scientific American*, May 1990, pages 118–125.

Diamond, Jared: "The Worst Mistake in the History of the Human Race," *Discover*, May 1987, pages 64–66.

Ehrlich, Paul R., and J. Roughgarden: *The Science of Ecology*, The Macmillan Company, New York, 1987.

> *This textbook provides a solid introduction to the many facets of ecology. It gives a broad overview of the science and also includes a great deal of information on behavioral ecology.*

Grove, Richard H.: "Origins of Western Environmentalism," *Scientific American*, July 1992, pages 42–47.

Harper, John L.: *Population Biology of Plants*, Academic Press, Inc., New York, 1981.*

> *The world from a plant's perspective. A wonderful introduction to the ecology of populations and a must for any student of botany.*

Holloway, Marguerite, and John Horgan: "Soiled Shores," *Scientific American*, October 1991, pages 88–95.

Hutchinson, G. Evelyn: *The Ecological Theater and the Evolutionary Play*, Yale University Press, New Haven, Conn., 1965.

> *By one of the great modern experts on freshwater ecology, this is a charming and sophisticated collection of essays on the influence of environment in evolution— and also on an astonishing variety of other subjects.*

Hutchinson, G. Evelyn: *An Introduction to Population Ecology*, Yale University Press, New Haven, Conn., 1978.

> *A beautifully written and handsomely produced collection of lectures by one of the founders of the field, who was also a marvelous humanistic scholar and raconteur.*

Keyfitz, Nathan: "The Growing Human Population," *Scientific American*, September 1989, pages 118–126.

Krebs, Charles J.: *Ecology: The Experimental Analysis of Distribution and Abundance*, 4th ed., HarperCollins Publishers, New York, 1992.

> *This excellent, modern text focuses on populations and on interactions among them. Requires some knowledge of mathematics.*

Lents, James M., and William J. Kelly: "Clearing the Air in Los Angeles," *Scientific American*, October 1993, pages 32–39.

May, Robert M.: "Parasitic Infections as Regulators of Animal Populations," *American Scientist*, vol. 71, pages 36–45, 1983.

May, Robert M., and Jon Seger: "Ideas in Ecology," *American Scientist*, vol. 74, pages 256–267, 1986.

Myers, Judith H.: "Population Outbreaks in Forest Lepidoptera," *American Scientist*, vol. 81, pages 240–251, 1993.

Partridge, Linda, and Paul H. Harvey: "The Ecological Context of Life History Evolution," *Science*, vol. 241, pages 1449–1455, 1988.

Ricklefs, Robert E.: *Ecology*, 3d ed., W. H. Freeman and Company, New York, 1990.

> *This outstanding textbook provides a comprehensive introduction to the basic concepts of modern ecology. Beautifully written and rich with examples, it is highly recommended.*

Robey, Bryant, Shea O. Rutstein, and Leo Morris, "The Fertility Decline in Developing Countries," *Scientific American*, December 1993, pages 60–67.

*Available in paperback.

The Serengeti, the vast plain that spans northern Tanzania and southern Kenya, is home to the largest herds of grazing animals in the world. Zebras, wildebeest, and Thomson gazelles—affectionately known as "tommies"—move through the plains, following the annual cycle of rainfall. Ecologists have been able to demonstrate that species like these, which appear to be competitors, actually partition their resources—not only sharing but also, in some cases, facilitating each other's existence.

Zebras are the first animals to move into long-grass areas. Like horses, zebras have large incisors on both their top and bottom jaws and heavy grinding molars. They eat the long, tough grass stalks and the older leaves. By removing the coarse top stems, they make the more nutritious leaves accessible to the animals that follow.

The wildebeest—large, ungainly looking antelopes with long manes and a distinct beard—come next. A wildebeest feeds like a cow, wrapping its tongue around the long grass stems and drawing them across the saw-like ridge of its lower incisors. The great herds of wildebeest remove so much of the grass that new shoots of grass and tiny soft-stemmed dicots spring up.

The tommies come last. These exquisite little antelopes are only about 2 feet high at the shoulder and weigh about 40 pounds. With their tiny muzzles, they nibble at the new grass leaves and the small herbs that grow between them.

Grazing stimulates new growth, as do the trampling of the soil and the feces that the animals leave behind. Grazing also makes room for a variety of species, where otherwise only one or two species might dominate. Although each zebra, each wildebeest, each tommie, and every plant on the Serengeti is acting in its own self-interest, their combined activities have produced a vast, self-sustaining system that has operated to the benefit of all of its inhabitants for millennia—and perhaps will continue to do so for years to come.

44

Interactions in Communities

Populations are made up of individual organisms. Communities are made up of populations. Ecologically speaking, a **community** consists of all the populations of organisms inhabiting a common environment and interacting with one another. These interactions are major forces of natural selection. They also influence the number of individuals in each population and the number and kinds of species in the community. The interactions among different populations are enormously varied and complex, but they can generally be categorized as competitive, predatory, or symbiotic.

Competition

Competition is the interaction between individual organisms of the same species (**intraspecific competition**) or of different species (**interspecific competition**) using the same resource, often present in limited supply. As a result of competition, the overall fitness—that is, the reproductive success—of the interacting individuals may be reduced. Among the many resources for which organisms may compete are food, water, light, or living space, including nesting sites or burrows.

Competition is generally greatest among organisms that have similar requirements and life styles. Plants often compete with other plants for sunlight and water. **Herbivores,** animals that eat plants and algae, may compete with other herbivores. **Carnivores,** animals that eat animals, may compete with other carnivores. Moreover, within these categories, the more similar two species are in their requirements and life style, the more intense the competition between them is likely to be.

For many years, competition has been invoked as a major force in determining the composition and structure of communities—that is, the number and kinds of species present and their arrangement in space and time within the community. Recently, however, a number of ecologists have come to question the importance of competition as an influence on community composition and structure. The debate—at times acrimonious—that has ensued concerns not only the role of competition but also the methods to be used in testing ecological hypotheses.

We shall begin our consideration of competition—and of the controversies surrounding it—with the principle that, until very recently, dominated the study of competition and the kinds of questions that ecologists asked about it.

The Principle of Competitive Exclusion

In 1934, the Russian biologist G. F. Gause formulated what became known as the principle of **competitive exclusion.** According to this principle, if two species are in competition for the same limited resource, one or the other will be more efficient at utilizing or controlling access to this resource and will eventually eliminate the other in situations in which the two species occur together.

Gause demonstrated the validity of his principle in his own, now classic, experiments involving laboratory cultures of two species of *Paramecium: Paramecium aurelia* and *Paramecium caudatum.* When the two species were grown under identical conditions in separate containers, both grew well. *Paramecium aurelia,* however, multiplied much more rapidly than *P. caudatum,* indicating that the former used the available food supply more efficiently than the latter. When the two were grown together, *P. aurelia* rapidly outmultiplied *P. caudatum,* which soon died out (Figure 44–1).

In laboratory experiments, the fastest growing species is not always the most successful competitor, how-ever, as observed with two species of duckweed, *Lemna gibba* and *Lemna polyrrhiza.* In pure culture, *L. gibba* grows more slowly than *L. polyrrhiza,* yet *L. gibba* always replaces *L. polyrrhiza* when they are grown together. The plant bodies of *L. gibba* have air-filled sacs that serve as little pontoons, so that these plants form a mass over the other species, cutting off the light. As a consequence, the shaded *L. polyrrhiza* dies out (Figure 44–2).

It is possible to devise different culture conditions so that the outcomes of both the *Paramecium* and *Lemna* experiments are reversed. However, as long as the conditions of a particular experiment are held constant, one species always wins and the other is always eventually eliminated.

44–1   *The results of Gause's experiments with two species of* Paramecium *demonstrate the principle that if two species are in direct competition for the same limited resource— in this case, food—one eliminates the other.* (a) Paramecium aurelia *and* (b) Paramecium caudatum *were first grown separately under controlled conditions and with a constant food supply. As you can see,* P. aurelia *grew much more rapidly than* P. caudatum, *indicating that* P. aurelia *uses available food supplies more efficiently.* (c) *When the two protists were grown together, the more rapidly growing species outmultiplied and eliminated the slower-growing species.*

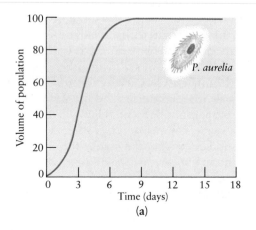

(a)

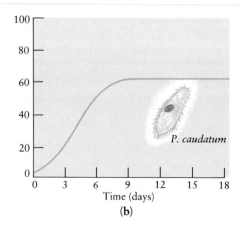

(b)

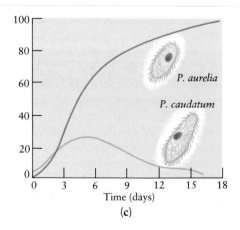

(c)

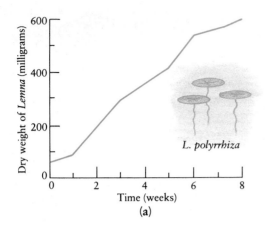
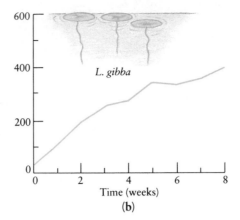
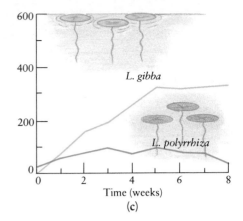

44-2   *An experiment with two species of floating duckweed, tiny angiosperms found in ponds and lakes. One species,* Lemna polyrrhiza (**a**), *grows more rapidly in pure culture than the other species,* Lemna gibba (**b**). *But* L. gibba *has tiny air-filled sacs that float it on the surface. When the two species are grown together,* L. gibba *shades* L. polyrrhiza *and is the victor in the competition for light* (**c**).

The Ecological Niche

Gause's competitive exclusion principle would lead you to expect that only dissimilar species would be found coexisting in natural communities. Yet, in fact, species with similar requirements and life styles are often found in the same community.

This observation raised the question of how similar two or more species can be and still continue to coexist in the same place at the same time, which led, in turn, to the concept of the **ecological niche.** This term is somewhat misleading because the word "niche" has the connotation of a physical space. An ecological niche, however, is not the space occupied by an organism but rather the role that it plays. The simplest analogy is that the niche is an organism's profession, as distinct from its habitat, which is its address.

A working definition of the niche is more complex, however, and includes many more factors than the way in which an organism makes its living. A niche is, in fact, the total environment and way of life of all the members of a particular species of organism in the community. Its description includes physical factors, such as the temperature limits within which the organisms can survive and their requirements for moisture. It includes biological factors, such as the nature and amount of required food sources. And, it also includes aspects of the behavior of the organisms, such as patterns of movement and daily and seasonal activity cycles. Although only a few of these factors can be studied at any one time, all are likely to influence the interactions of the members of a species with the members of other species in the community.

Resource Partitioning

The concept of the ecological niche suggested that when similar species are found coexisting together, a close examination will reveal that their niches are, in fact, different. An example is provided by the grazing animals of the Serengeti with which we began this chapter. Although the zebras, wildebeest, and gazelles appear to be sharing and competing for the same resources, they are not.

Numerous studies of different types of organisms have revealed that such dividing up, or partitioning, of resources occurs frequently among ecologically similar members of a community. The cause of the partitioning is a matter of considerable disagreement among ecologists, as we shall see shortly. First, however, let us consider two other examples.

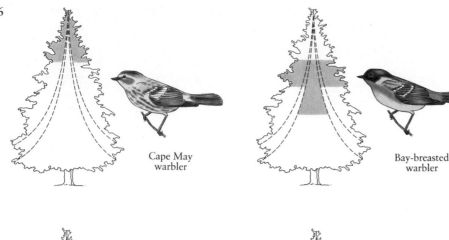

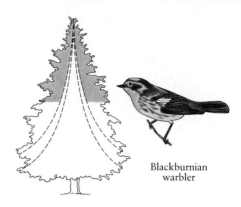

44–3 *The feeding zones in a spruce tree of five species of North American warblers. The colored areas in the tree indicate where each species spends at least half its feeding time. This partitioning of resources allows all five species to feed in the same trees.*

Cape May warbler

Bay-breasted warbler

Blackburnian warbler

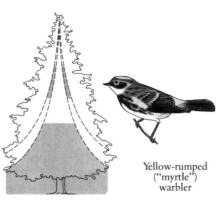

Black-throated green warbler

Yellow-rumped ("myrtle") warbler

44–4 *Mosses of the genus* Sphagnum, *which form extensive bogs in both temperate and cold regions of the world. Peat is formed from the accumulation and compression of the mosses themselves, as well as the sedges, reeds, and other grasses that grow among them. In Ireland, dried peat is widely used as an industrial fuel, as well as for home heating. It is estimated that peat bogs cover at least 1 percent of the Earth's land surface, an enormous area equivalent to half of the United States.*

Woodland Warblers

Some New England forests are inhabited by five closely related species of warblers, all about the same size and all insect eaters. The late Robert MacArthur, a brilliant and innovative young ecologist, meticulously observed and timed where the warblers fed within the trees. His data showed that the five species have different feeding zones (Figure 44–3). Because they exploit slightly different resources—that is, insects in different parts of the trees—the species can coexist.

Bog Mosses

In bogs, mosses of the genus *Sphagnum* (Figure 44–4) often appear to form a continuous cover, and several species are usually involved. How can these species, apparently very similar, continue to coexist?

When the situation is examined in detail, it is found that semiaquatic species grow along the bottoms of the wettest hollows. Other species grow in drier places on the sides of the hummocks, which they help to form. Still other species grow only in the driest conditions on the tops of the hummocks, where they are eventually succeeded by one or more species of flowering plants. Therefore, although all the species of *Sphagnum* coexist, in the sense that they are all present in the same bog, they actually occupy different microenvironments and continually replace one another as the characteristics of each microenvironment change.

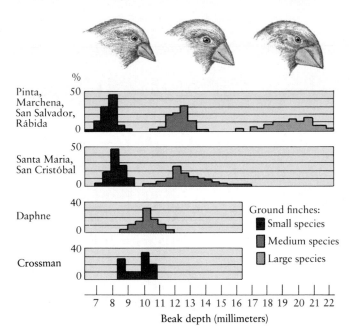

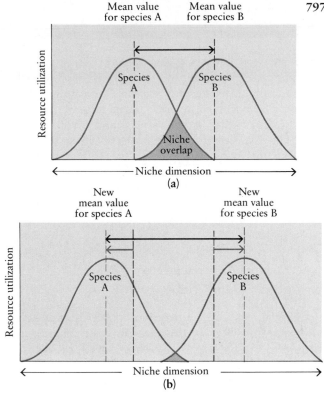

44–5 *Beak sizes in three species of ground finch found on the Galapagos Islands. Beak measurements are plotted horizontally, and the percentage of specimens of each species is shown vertically. Daphne and Crossman, which are very small islands, each have only one species of ground finch. These species have beak sizes halfway between those of the medium-sized and small finches on the larger islands.*

44–6 *Overlap in one dimension of an ecological niche. (a) The two bell-shaped curves represent utilization of a resource by two species in a community. The niche dimension might represent living space, as in the case of* Sphagnum *mosses; foraging space, as in the case of warblers; size of seeds eaten, as in the case of the Galapagos finches; and so on. Competition between the two species is potentially most intense in areas of niche overlap, leading to restriction of one or both species in living space, foraging space, or the size of seeds eaten, and so on. (b) It is hypothesized that such competition results in selection against individuals with overlapping characteristics, leading to divergence in the niches of the two species.*

The Role of Past Competition in Resource Partitioning

The resource partitioning observed among African herbivores, woodland warblers, *Sphagnum* mosses, and many other organisms was long considered to be the result of competition. In some cases, such as the warblers, the competition was thought to be occurring in the present. In other cases, such as the mosses and the herbivores, it was thought to have occurred in the evolutionary past, leading to the differing adaptations that enable the organisms to coexist. This phenomenon, in which species that live together in the same environment tend to diverge in those characteristics that overlap, is known as **character displacement.**

One of the most frequently cited examples of character displacement is provided by the beaks of Darwin's finches. As we saw in Chapter 21, the large, medium, and small ground finch species are very similar except for differences in overall body size and in the sizes and shapes of their beaks. The differences in the beaks are correlated with the sizes of the seeds the birds eat. On islands such as Pinta and Marchena (see page 364), where all three species of ground finch exist together, there are clear-cut differences in beak size (Figure 44–5). On Santa Maria and San Cristóbal Islands, the large species is not found, and the beak sizes of the medium ground finches on these islands are larger and overlap the beak sizes of the large finches found on

Pinta and Marchena. Daphne and Crossman, which are very small islands, each have only one species. Daphne has the medium-sized finch and Crossman, the small finch. These two populations have similar beak sizes, which are intermediate between those of the medium-sized and small finches on the larger islands.

These data have been interpreted in two different ways by ecologists. Some ecologists maintain that the observed differences in beak size are the result of the selection pressures exerted by interspecific competition. According to this interpretation, competition between organisms whose ecological niches overlap causes selection against individuals with overlapping characteristics, leading to the observed divergence between the species (Figure 44–6).

Other ecologists point out that it is impossible to determine if the differing beak sizes are the result of competition that occurred at times when the different species were coexisting on the same islands or if they are the result of adaptations to local conditions that occurred at times when the species were isolated from one another on different islands. Both groups of ecologists, however, agree that whatever their evolutionary cause, the differences in beak size and shape enable the different finch species to exploit different food sources and thus to coexist.

Experimental Approaches to the Study of Competition

Virtually all ecologists agree that competition does occur in nature, with the degree of intensity varying according to the particular species involved, the sizes of the interacting populations, and the abundance or scarcity of resources. Although the analysis of resource partitioning may provide clues as to the occurrence and importance of competition in a particular situation, experiments involving changes in community composition are required to demonstrate that competition is actually taking place. At their best, laboratory experiments, such as those performed by Gause, can only approximate natural conditions, which are invariably much more complex.

Barnacles in Scotland

One of the clearest experimental demonstrations of competition in a natural community was a study of barnacles performed by Joseph Connell. Barnacles are crustaceans. When they change from their free-swimming, larval form into their adult, sedentary form, they cement themselves to rocks and secrete protective calcium-containing plates. Once attached, barnacles remain fixed, so that by making careful observations over time, one can determine the dynamics of a particular population. One can identify exactly which barnacles have died and which new ones have arrived between visits to the study site.

Connell studied two barnacle species, *Chthamalus stellatus* and *Semibalanus balanoides,* that live on the coasts of Scotland. *Chthamalus* is found in the high part of the intertidal seashore. As the tides go in and out, these barnacles are exposed to wide fluctuations of temperature and salinity and to the hazards of desiccation. The other species, *Semibalanus,* occurs in the lower intertidal zone, where conditions are more constant.

Although *Chthamalus* larvae, after their period of drifting in the plankton, often attach to rocks in the lower, *Semibalanus*-occupied zone, adults are rarely

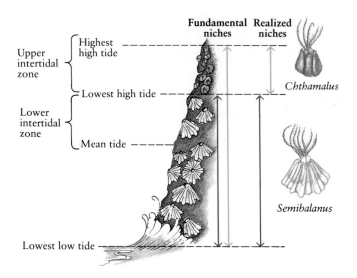

44–7   *Interspecific competition between* Semibalanus *and* Chthamalus *barnacles. The larvae of both species settle over a wide range, but the adults live in precisely restricted areas. The upper limits of the* Semibalanus *area are determined by physical factors such as exposure to air, which can lead to desiccation. Such exposure occurs at specific times during the month when the high tide covers only the lower portion of the intertidal zone, leaving the upper intertidal zone exposed. * Chthamalus *barnacles, however, are prevented from living in the* Semibalanus *area not by physical factors (they would probably thrive there since the area is less physically limiting) but by the* Semibalanus *barnacles.* Semibalanus *grows faster, and whenever it comes upon* Chthamalus *in the* Semibalanus *area, it either pries it off the rocks or grows right over it.*

found there. Connell showed that in the lower zone, *Semibalanus,* which grows faster, ousts *Chthamalus* by crowding it off the rocks and growing over it or undercutting it. When Connell experimentally removed *Semibalanus* from the lower portion of the intertidal zone, *Chthamalus* invaded the area and thrived there. In the control areas, where barnacles were not removed, each species remained in its own zone.

As this experiment demonstrates, *Chthamalus* is not restricted to the upper intertidal zone by a physiological inability to live elsewhere. It is restricted by competition with *Semibalanus.* There is no evidence, however, that competition with *Chthamalus* keeps *Semibalanus* in the lower zone. Because *Semibalanus* lacks the physiological adaptations required for life in the upper, drier zone, it cannot successfully invade it.

This study, and others like it, have generated the concepts of fundamental niche and realized niche. The **fundamental niche** describes the physiological limits of tolerance of the organism. It is the niche occupied by an organism in the absence of interactions with other organisms. The **realized niche** is that portion of the fundamental niche actually utilized. It is determined by physical factors and also by interactions with other organisms.

In its fundamental niche, *Chthamalus* can occupy both the high and low intertidal zones, but because of niche overlap with *Semibalanus, Chthamalus* actually occupies a smaller area, its realized niche (Figure 44–7). Because *Semibalanus* is restricted by physiological limits, it can occupy only the lower intertidal zone. Its fundamental niche is narrower than that of *Chthamalus* and is totally included within it. In such a situation, the species with the narrow, included fundamental niche must be a superior competitor or it will be driven to extinction. *Semibalanus* can survive in the same intertidal community as *Chthamalus* because it is the superior competitor. *Chthamalus* can survive, despite its competitive inferiority, because its fundamental niche is broader, providing it with a refuge that cannot be invaded by *Semibalanus.*

Winner Takes All

Most studies of competition have emphasized the adaptations and the partitionings of resources that make possible the coexistence of similar species within a community. This is, however, a biased view, for it is difficult to study the interactions between species after one of the protagonists has left. Just as competition within species leads to the elimination of the great majority of individual organisms, as Darwin observed, competition between species may lead to elimination of a species from the community.

One example is the disappearance from many localities of bluebirds (Figure 44–8). This is thought to have been caused, in part, by the usurpation of their nesting sites by starlings. Starlings were introduced into Central Park in New York City in 1891 and are now found throughout the United States, whereas some of us have never seen a bluebird.

Predation

Predation is the consumption of live organisms, including plants by animals, animals by animals, and even, as we have seen, animals by plants (page 741) or by fungi (page 437). Predators use a variety of techniques—known as **foraging strategies**—to obtain their food. Foraging strategies are under intense selection pressure. Those individuals that forage most efficiently are likely to leave the most offspring. From the standpoint of potential prey, those individuals that are most successful at avoiding predation are likely to leave the most offspring. Thus, predation affects the evolution of both predator and prey.

Predation also affects the number of organisms in a population and the diversity of species within a community.

Predation and Numbers

For many populations, predation is the major cause of death, yet, paradoxically, it is not at all clear that predation necessarily reduces the numbers of a prey population below the carrying capacity of its environment. However, when predation is heavier on certain age groups—juveniles versus adults, for instance—or certain life stages—such as caterpillars versus butterflies—it can alter the structure of a prey population and promote adjustments in life-history patterns.

44–8   *A male mountain bluebird at the entrance of a nesting box in Denver, Colorado. For more than 25 years, the U.S. Fish and Wildlife Service has been monitoring the breeding populations of the three bluebird species native to North America. A significant increase in the size of the populations began to occur in the mid-1980s. This increase is attributed, in large part, to a concerted effort by the North American Bluebird Society and others to encourage people to provide suitable nesting boxes for the birds in potential bluebird habitats. An important factor in the design of the nesting boxes is that the entrance holes be large enough to admit bluebirds but small enough to exclude starlings.*

44–9   (a) *Prickly-pear cactus growing on a pasture in Queensland, Australia, in November of 1926. Such rapid and environmentally destructive spread often occurs when alien organisms are introduced into a region where they have no competitors or predators.* (b) *The same pasture in October of 1929, slightly less than three years after the introduction of the cactus moth.*

(a)   (b)

Predation, especially on large herbivores, tends to cull animals in poor physical condition. Wolves, for instance, have great difficulty overtaking healthy adult moose or even healthy calves. A study of Isle Royale, an island in Lake Superior, showed that in some seasons more than 50 percent of the animals the wolves killed had lung disease, although the incidence of such individuals in the population was less than 2 percent. Thus, many of the animals killed by predators, according to this study, are animals that would have died soon anyway. (Modern human hunters, however, with their superior weapons and their desire for a "prize" specimen, are more likely to injure or destroy strong, healthy animals.)

In some situations, however, predators do limit their prey species. This has been most clearly demonstrated in cases involving the introduction of alien species into areas where they have no natural predators. When prickly-pear cactus, for instance, was brought to Australia from South America, it escaped from the garden of the gentleman who imported it and spread into fields and pastureland until more than 12 million hectares were so densely covered with prickly pears that they could support almost no other vegetation (Figure 44–9). The cactus then began to take over the rest of Australia at the rate of about 400,000 hectares a year. It was not brought under control until a natural predator was imported—a South American moth, whose caterpillars feed only on the cactus. Now only an occasional cactus and a few moths can be found. (Note, however, that the introduction of the alien moth was, in itself, risky.)

Few predator-prey relationships are this simple. Most predators have more than one prey species, al-

though one prey species may be preferred. Characteristically, when one prey species becomes less abundant, predation on other species increases so the proportions of each in the predator's diet fluctuate.

Although predators may not always limit prey populations, the availability of prey constitutes a major component of the carrying capacity for predator populations, often stringently affecting their size. This is evident in relatively simple situations, such as when a bloom of phytoplankton (mostly microscopic algae) results from an upwelling of nutrients due to ocean currents and then is followed by a corresponding increase in zooplankton (animal plankton).

A more complex example is that of the lynx and the snowshoe hare. The data (Figure 44–10) are based on pelts received yearly by the Hudson's Bay Company over a period of almost 100 years. As you can see, there are oscillations in population density that occur about every 10 years. Generally speaking, a rise in the hare population is followed by a rise in the lynx population. The hare population then plummets, and the lynx population follows.

This example, which has been studied by ecologists over the last 50 years, can be interpreted in a variety of ways. The traditional explanation is that overpredation by the lynx reduces the snowshoe hare population. The lynx population, heavily dependent on the snowshoe hare as prey, is reduced in turn. The reduction in predation then permits the snowshoe hares to increase in number, followed by an increase in the number of lynx, and so on.

A second explanation is that the hare population undergoes a regular 10-year cycle caused, perhaps, by diseases associated with crowding or the effects of its own

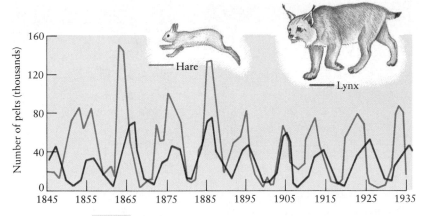

44–10 *The number of lynx and snowshoe hare pelts received yearly by the Hudson's Bay Company over a period of almost 100 years, indicating a pattern of 10-year oscillations in population density. The lynx reaches a population peak every 9 or 10 years, and these peaks are followed in each case by several years of sharp decline. The snowshoe hare follows a similar cycle, with a peak abundance generally preceding that of the lynx by a year or more.*

44–11 *Photographed off the coast of California, this starfish of the genus* Pisaster *is opening a mussel, one of its favorite foods. Studies have shown that predation by* Pisaster, *particularly on mussels and barnacles, is an important factor in maintaining species diversity in the rocky coastal communities in which it lives.*

predatory activities on the vegetation it consumes. This latter hypothesis is supported by two discoveries: (1) when overbrowsed, certain types of plants put out new shoots and leaves that contain chemicals toxic to hares, and (2) on an island where there are no lynx, the hare population undergoes similar cycles.

A third possibility is that the lynx undergo a regular cycle, independent of the hares, perhaps associated with some other factor, such as changes in the habits of their own predators, the human hunters. A decrease in the lynx population may permit growth of the hare population, or the two populations may oscillate independently. Thus, whereas the lynx and the hare used to serve as a simple model of predator-prey relationships, it is now perhaps more instructive as an example of the difficulties in dealing with ecological variables.

Predation and Species Diversity

The number and kinds of species in a community can be greatly influenced by predation. Although predation may occasionally eliminate a prey species, many experimental studies have shown that it is often an important factor in maintaining species diversity in a community.

For example, R. T. Paine studied a community on the rocky coast of Washington. In this community, the principal predator was the starfish *Pisaster* (Figure 44–11). At the beginning of the experiment, there were 15 prey species, including several species of barnacles and several kinds of mollusks, including mussels, limpets, and chitons. Paine systematically removed the starfish from one area of the community, 8 meters by 2 meters in size. By the end of the experiment, the number of prey species in the area from which the starfish were removed had declined to eight, and the community was dominated by one species of mussel. In the undisturbed community, starfish predation kept the densities of the prey populations low, reducing competition between the species and permitting all to survive. In the absence of the predator, the mussels were clear victors in the competition for living space.

Organisms, such as *Pisaster*, that are of exceptional importance in maintaining the diversity of a community are known as **keystone species.** When the keystone is removed from a stone arch, the arch falls apart. Similarly, when a keystone species is lost from a community—as a result, for example, of pollution, disease, or competition from an alien organism—the diversity of the community decreases and the structure of the community is significantly altered.

In another series of experiments, Jane Lubchenco showed that the herbivorous marine snail *Littorina lit-*

torea controls the abundance and type of algae in the higher intertidal pools on the New England coast. In such pools, the snails' preferred food (the green alga *Enteromorpha*) is competitively superior, and its removal by the snails permits the growth of other algal species (Figure 44–12). However, in areas exposed to the air at high tide as well as at low tide, *Enteromorpha* is competitively inferior. Its removal by the snails from those areas facilitates the growth of algal species that are competitively superior under exposed conditions. The result is domination by those species and a decrease in the total number of species in the exposed areas.

As this example illustrates, although predation often plays an important role in maintaining species diversity, it does not always do so. Even when the same predator and prey species are involved, the effect of predation on species diversity depends on the particular situation.

Symbiosis

As we saw in Chapter 25, **symbiosis** ("living together") is a close and long-term association between organisms of two different species. Long-continued symbiotic relationships can result in profound evolutionary changes in the organisms involved, as in the case of lichens (page 439), one of the oldest and most ecologically successful symbioses.

Although there is some disagreement among ecologists as to precisely what constitutes a symbiotic relationship, and the details of the relationship between two closely associated species are often difficult to determine, symbiotic relationships are generally considered to be of three kinds. If one species benefits and the other is harmed, the relationship is known as **parasitism.** If the relationship is beneficial to both species, it is called **mutualism.** Less common is **commensalism,** a relationship that is beneficial to one species and that neither benefits nor harms the other.

An example of commensalism is the relationship between the marine annelid *Chaetopterus* and tiny crabs of the genus *Pinnixa,* which live in the intertidal mud flats along the Atlantic coast of the United States. Each *Chaetopterus* constructs a U-shaped tube in which it lives (Figure 44–13), and the tube usually also contains two crabs, one male and one female. Both worm and crabs feed on particles of organic matter carried in water currents moved through the tube by fanlike appendages of the worm's body. The crabs obtain shelter and a steady food supply, and, as best anyone has been able to determine, their presence neither benefits nor harms the worm.

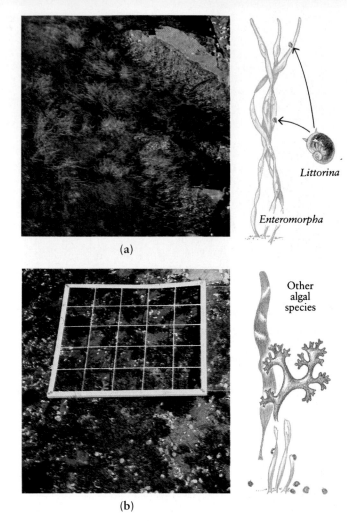

(a)

(b)

44–12   *An example of the effects of predation on species diversity.* (a) *In this high intertidal pool on the New England coast, the density of the herbivorous marine snail* Littorina littorea *is very low (between one and five individuals per square meter). The competitively superior green alga* Enteromorpha *dominates the pool, excluding other algae.* (b) *In a neighboring pool, less than a meter away, the density of* Littorina *is much higher—more than 250 individuals per square meter. The snails have grazed heavily on* Enteromorpha, *permitting the growth of other algal species that are, for* Littorina, *inedible. The aluminum grid at the top of the photograph was used to estimate the density of the snails.*

Parasitism

Parasitism may be considered a special form of predation in which the predator is considerably smaller than the prey. The plants and animals in a natural community support hundreds of parasitic species—in fact, certainly thousands and perhaps millions if one were to count nematodes and viruses.

As with more obvious forms of predation, parasitic diseases are most likely to wipe out the very young, the very old, and the disabled—either directly or, more often, indirectly, by making them more susceptible to other predators or to the effects of climate or food

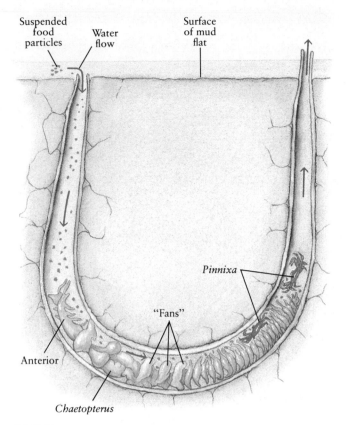

44–13 Chaetopterus, *a marine annelid, lives in a parchment tube, constructed by the worm itself, with two commensal crabs of the genus* Pinnixa. *Modified parapodia ("side feet") of the worm function as fans, pulling water through the tube. Both the worm and the crabs are nourished by food particles carried into the tube with the moving water.*

44–14 *Rabbits crowd a water hole in Australia. Imported from Europe as potentially valuable herbivores, they soon overran the countryside. The myxoma virus was introduced to control them, and now host and parasite are coexisting.*

shortages. It is predictable that a parasite-caused disease should not be too virulent or too efficient. If a parasite were to kill all the hosts for which it is adapted, it too would perish.

This principle is particularly well illustrated by a series of further misadventures on the continent of Australia. There were no rabbits in Australia until 1859, when an English gentleman imported a dozen from Europe to grace his estate. Six years later he had killed a total of 20,000 on his own property and estimated he had 10,000 remaining. In 1887, in New South Wales alone, Australians killed 20 million rabbits. By 1950, Australia was being stripped of its vegetation by the rabbit hordes (Figure 44–14).

In that year, rabbits infected with myxoma virus were released on the continent. Myxoma virus, which is carried from host to host by mosquitoes, causes only a mild disease in South American rabbits, its normal hosts, but is usually fatal to the European rabbit. At first, the effects were spectacular, and the rabbit population steadily declined, yielding a share of pasture-

land once more to the sheep herds on which much of the economy of the country depends. But then occasional rabbits began to survive the disease, and their litters also showed resistance to the myxoma virus.

A double process of selection had taken place. The original virus was so rapidly fatal that often a rabbit died before there was time for it to be bitten by a mosquito and thereby infect another rabbit. The virus strain then died with the rabbit. Strains less drastic in their effects, on the other hand, had a better chance of survival since they had a greater opportunity to spread to a new host. So selection began to work in favor of a less virulent strain of the myxoma virus. Simultaneously, rabbits that were resistant to the original virus began to proliferate. Now, as a result of coevolution, the host-parasite relationship seems to be stabilizing.

Mutualism

If the current hypothesis as to the origin of eukaryotic cells (page 419) is correct, we owe our very origins to

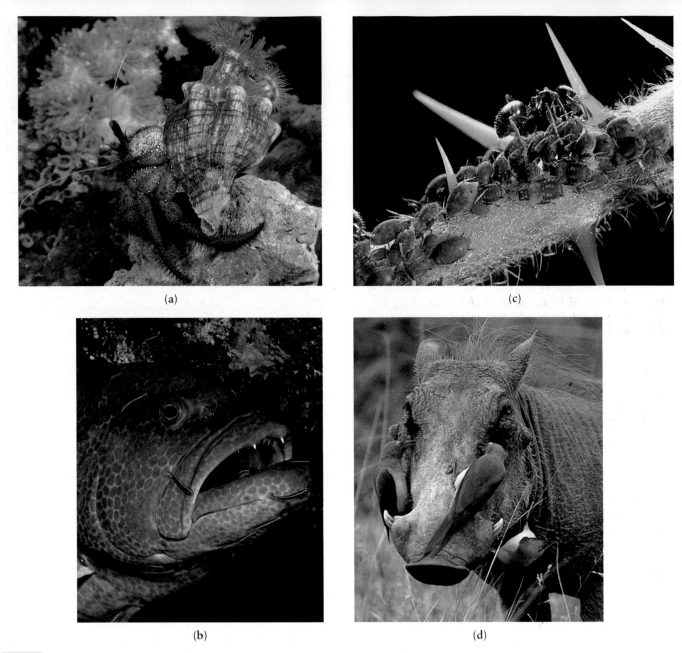

(a)

(b)

(c)

(d)

44-15   *Mutualism.* (a) *Sea anemones on the back of a whelk shell occupied by a red hairy hermit crab. The anemones protect and camouflage the crab and, in turn, gain mobility—and so a wider feeding range—from their association with the crab. Hermit crabs, which periodically move into new, larger shells, will coax their anemones to move with them.*

(b) *Cleaner fish, removing parasites from a tiger grouper. Throughout the process of having its mouth cleaned, the grouper holds its mouth perfectly still. Cleaner fish are permitted to approach larger fish with impunity because of the service they render as they feed off the algae, fungi, and other microorganisms on the fish's body. Larger fish recognize the cleaners by their brilliant colors and distinctive markings. Other species of fish, by closely resembling the cleaners, are able to get close enough to the large fish to remove big bites of flesh. What would probably happen if the number of cleaner mimics began to approach the number of true cleaners?*

(c) *Aphids suck sap from the phloem, removing certain amino acids, sugars, and other nutrients from it and excreting most of it as "honeydew," or "sugar-lerp," as it is called in Australia, where it is harvested as food by the aborigines. Some species of aphids have been domesticated by some species of ants. These aphids do not excrete their honeydew at random, but only in response to caressing movements of the ants' antennae and forelimbs. The aphids involved in this symbiotic association have lost all their own natural defenses, including even their hard outer skeletons, relying upon their hosts for protection.*

(d) *Oxpeckers live on the ticks they remove from their hosts. Like cleaner fish, they perform an essential service. An oxpecker forms an association with one particular animal, such as the warthog shown here, conducting most of its activities, including courtship and mating, on the body of its host.*

804

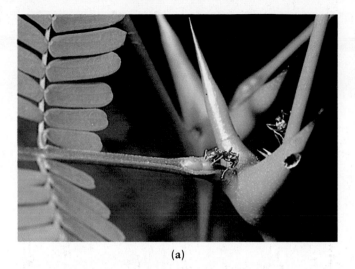

(a)

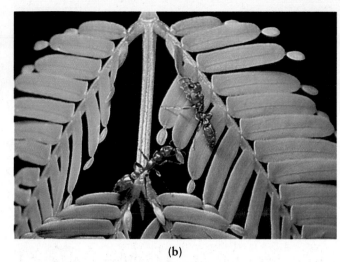

(b)

44–16 *Ants and acacias. (a) A worker ant, drinking from a nectary of a bull's-horn acacia. At the left is an entrance hole into a thorn that was cut by the queen ant. After she had hollowed out the thorn, the queen raised her brood within it. (b) Worker ants collecting Beltian bodies from the tips of acacia leaflets. Rich in proteins and oils, the Beltian bodies are an important food source for both adult and larval ants.*

mutualism. Examples of present-day mutualistic associations are so abundant that we must limit ourselves to only a few. Perhaps among the most significant are those that take place underground, between roots and nitrogen-fixing bacteria (page 751) and in mycorrhizae (page 440).

Some particularly colorful examples of mutualism are illustrated in Figure 44–15. Another example is the interaction of ants and acacias.

Ants and Acacias

Trees and shrubs of the genus *Acacia* grow throughout the tropical and subtropical regions of the world. In Africa and tropical America, where plants are preyed upon by large herbivores, acacia species are protected by thorns. (Acacias that evolved where there are no large browsers—Australia, for instance—lack thorns.)

On one of the African species of *Acacia,* ants of the genus *Crematogaster* gnaw entrance holes in the walls of the thorns and live permanently inside them. Each colony of ants inhabits the thorns on one or more trees. The ants obtain food from nectar-secreting glands on the leaves of the acacias and eat caterpillars and other small herbivores they find on the trees. Both the ants and the acacias appear to benefit from this association.

In the lowlands of Mexico and Central America, the ant-acacia relationship has been extended to even greater lengths. The bull's-horn acacia is found frequently in cutover or disturbed areas, where it grows extremely rapidly. This species of acacia has a pair of greatly swollen thorns, several centimeters in length, at the base of most leaves. The petioles bear nectaries, and at the very tip of each leaflet is a small structure, rich in oils and proteins, known as a Beltian body. Thomas

Belt, the naturalist who first described these bodies, noted that their only apparent function was to nourish the ants. Ants live in the thorns, obtain sugars from the nectaries, eat the Beltian bodies, and feed them to their larvae.

Worker ants, which swarm over the surface of the plant, are very aggressive toward other insects and, indeed, toward animals of all sizes. They become alert at the mere rustle of a passing mammal, and when their tree is brushed by an animal, they swarm out and attack at once, inflicting painful, burning bites. Moreover, and even more surprising, other plants sprouting within as much as a meter of occupied acacias are chewed and mauled, and their bark is girdled. Twigs and branches of other trees that touch an occupied acacia are similarly destroyed. Not surprisingly, acacias inhabited by these ants grow very rapidly, soon overtopping other vegetation.

Daniel Janzen, who first analyzed the ant-acacia relationship in detail, removed ants from acacias either by insecticides or by removing thorns or entire occupied branches. Acacias without their ants grew slowly and usually suffered severe damage from insect herbivores. They were soon overshadowed by competing species of plants and vines. As for the ants, according to Janzen, these particular species live only on acacias.

There is an epilogue to the ant-acacia story. Three new species—a fly, a weevil, and a spider—have been discovered that mimic the ants that inhabit the acacias. So expert is their mimicry (probably involving chemical recognition signals as well as appearance) that the patrolling ants do not recognize them as interlopers, and so they enjoy the hospitality and protection of the ant-acacia complex.

Conservation Biology and the Island Biogeography Model

As the human demand for natural resources increases, ecological communities are being fragmented at an accelerating rate. Formerly large and continuous natural communities are being reduced to isolated "islands," often surrounded by areas that are unsuitable for most of the species involved. The most dramatic examples of this destruction of natural communities are occurring in tropical forests. However, the destruction of wildlife habitat is not unique to tropical forests. Whenever a marsh is cut by a new roadway, a forest is cleared for agriculture or a housing development, or a river is interrupted by a dam, the result is a subdivision of ecological communities into smaller, increasingly isolated habitat islands.

As discussed on the facing page, the equilibrium hypothesis of island biogeography predicts higher rates of extinction on and lower rates of immigration to small, isolated islands. According to this model, as natural communities are reduced to smaller and more isolated fragments, they can be expected to support fewer and fewer species.

This prediction of reduced species numbers in small, isolated habitat fragments has been supported by research in a number of different regions. One of the most thoroughly studied sites is Barro Colorado Island, an area of tropical rain forest located in Lake Gatun in Panama. The lake was created early in the twentieth century by the completion of the Panama Canal, isolating what is now Barro Colorado Island from the previously contiguous forest. Since that time, more than 50 species of birds have disappeared from the island, although they remain abundant on the mainland, only half a kilometer away.

A similar pattern of extinction has been observed in the 86-hectare woodland of the Bogor Botanical Garden in Java, Indonesia. Fifty years ago, this woodland was isolated by the destruction of the surrounding woodlands. Since that time, it has lost 20 of the 62 species of birds originally breeding there—more than

(a)

(b)

30 percent of the original community—and another four species are close to extinction.

It has been suggested that isolation of ecological communities may be reduced—and dispersal rates increased—by maintaining natural corridors between isolated communities. This possibility is being explored in the Netherlands, where studies are examining the role of hedgerows ("paths" of trees and shrubs, ranging from 1 to 10 meters in width) in increasing the rates of colonization of forest fragments. The hedgerows appear to increase the rate of dispersal of most forest species and may play a role in maintaining animal diversity in otherwise isolated patches of forest.

One of the most ambitious ecological projects currently underway was inspired by the equilibrium hypothesis of island biogeography. Known as the Minimum Critical Size of Ecosystems Project, it is a cooperative investigation by Brazilian and North American scientists in the Amazon rain forests. The project has revealed that forest fragmentation can have substantial effects on the physical environment. Hot, dry winds blowing across surrounding cleared areas reduce the relative humidity along the edges of the forest fragments

by as much as 20 percent. Moreover, increased light penetration at the edges of the forest fragments has elevated the temperatures there as much as 4.5°C (8.1°F) above the temperature in the forest interior. These physical changes may be responsible for some of the initial biological changes in the forest fragments: more trees die, more leaves drop from the remaining trees, and both the number of bird species and their population density decrease.

Although the equilibrium hypothesis of island biogeography began as a purely theoretical exercise to explain the maintenance of species diversity on islands, it has inspired research that is providing information vital for conservation biology, for the design and management of nature reserves, and for informed land-use planning. Some of the results of this research were not predicted by the original model. However, one of the most important predictions has held up: reductions in species diversity can be expected within nature reserves as they become increasingly isolated. For most groups of organisms, the numbers of species lost will depend upon both the size of the reserves and their degree of isolation.

Community Composition and the Question of Stability

Viewed from a global perspective, ecological communities often seem to be at equilibrium, with many species persisting for many generations over large areas. However, when communities are examined on a local scale, it becomes apparent that they, like the individual populations of which they are composed, undergo many fluctuations. Two questions concerning community composition have long perplexed ecologists. First, what determines the number of species in a community? And, second, what factors underlie the changes in a community that occur with the passage of time?

The Island Biogeography Model

Because of their size and relative isolation, small islands are often excellent natural laboratories for the study of evolutionary and ecological processes. In a classic study published in 1963, Robert MacArthur and E. O. Wilson used small islands as models to explore questions of community composition and stability.

MacArthur and Wilson hypothesized that the number of species on any given island remains relatively constant through time but that the identity of the species present is continually changing. According to their proposal, known as the **equilibrium hypothesis of island biogeography,** there is a balance between the rate at which new species immigrate to an island and the rate at which species already present become locally extinct. Although the number of species is in equilibrium, the species composition is *not* in equilibrium, because when a species becomes locally extinct it is usually replaced by a different species.

The island biogeography model was tested in an ingenious way by Wilson and Daniel Simberloff. They selected a number of small mangrove islands off the Florida Keys and counted the number of species of arthropods on each. They then destroyed all the animal life (mostly insects and other small arthropods) by covering the islands with plastic tents and fumigating them. Their plant life intact, the islands were soon colonized again from the mainland, and the recolonization process was monitored.

As predicted by the model, the total number of species present on an island after recolonization tended to be the same as the total number before the island was disturbed. However, and this is an important point, the species composition was often quite different from what it had been previously. Moreover, once the equilibrium number had been reached, the species composition con-

(c)

(a) *The tropical rain forest of Barro Colorado Island, Panama, as seen from an observation tower above the forest canopy. The island is in the foreground, and the Panamanian mainland is in the background. Biologists affiliated with the Smithsonian Tropical Research Institute, assisted by Earthwatch volunteers, are conducting detailed studies of the forest and its occupants.* **(b)** *For North American visitors, one of the most familiar animals of the forest is the bay-breasted warbler, which summers in the forests of New England but winters on Barro Colorado Island— and nowhere else in the world.* **(c)** *The forest interior.*

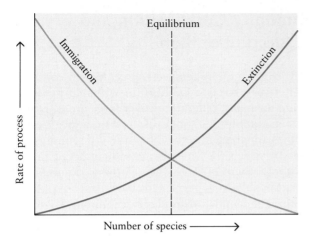

44–17   *Equilibrium model of the diversity of species on an island. The immigration rate declines as more species reach the island because species that arrived earlier have had time to become established. Thus, they are better able to compete against species that arrive later. Also, fewer immigrants will belong to new species. The extinction rate increases more rapidly at high species number because there is more competition between species. The equilibrium number of species (dashed line) is determined by the intersection of the immigration and extinction curves.*

44–18   *(a) Living sea fans, a coral that grows best in warm, clean water with a low level of nutrients. (b) The limestone remains of a sea fan of the same species that has been killed, most likely by parasites or polluted water. Reef-building corals are keystone species (page 801). When the living coral animals are destroyed, the structure of the reef community disintegrates. The diversity of other organisms that were once part of the community (page 460) must move elsewhere if they are to survive. When enormous areas of coral reef die, as is happening today in many parts of the world, there is no "elsewhere" for the other organisms.*

tinued to change, with extinction and immigration balancing one another out (Figure 44–17).

According to the island biogeography model, the two most important variables influencing species diversity are the size of the island and its distance from a source, usually the mainland, that can provide colonists. Distant islands tend to have fewer species than islands closer to the mainland. This is thought to be a result of lower rates of immigration, which appear to be a function of the distance that potential colonists must travel. Smaller islands are thought to reach equilibrium with a smaller number of species than larger islands, primarily because extinction rates are higher on smaller islands. This may be because populations tend to be smaller on such islands and thus are more susceptible to the effects of both predation and environmental disturbances.

This model applies not only to true islands but also to fragmented terrestrial areas. Thus it has important implications for conservation efforts (see essay).

The Intermediate Disturbance Hypothesis

Different types of communities vary widely in the number and diversity of species present. Among the most diverse are tropical rain forests and coral reefs. Until recently, it was thought that the species composition of these communities was relatively constant, and they were frequently cited as prime examples of an equilibrium state. Their high species diversity was thought to be a function of their stability. It now appears, however, that their diversity may be a function not of their stability but rather of the frequency and magnitude of the disturbances to which they are subjected.

Disturbances can take many forms. In the tropical forest, for example, trees are killed or severely damaged

(a)

(b)

by storms, landslides, lightning strikes, and insect out-breaks. The corals that form the basis of the coral reef community can be destroyed by predators or parasites, by the severe waves that accompany tropical storms, and by influxes of polluted water (Figure 44–18).

Soon after a disturbance, open areas—of forest or reef—are invaded by immature forms—seeds, spores, larvae, or even gametes. Initially, diversity in a newly colonized area is low. Only species that are close to the disturbed area and that are reproducing at the time are able to exploit the newly available area. If disturbances are frequent, the community will consist only of those species that can invade, mature, and reproduce before the next disturbance occurs.

According to the **intermediate disturbance hypothesis,** as the interval between disturbances increases, so does species diversity. Species that are excluded by frequent disturbances (because they are slow to mature or have limited dispersal abilities) now have an opportunity to colonize. However, if the interval between disturbances increases still further, species diversity may begin to decline (Figure 44–19). The primary factor in this decline is thought to be competition, but even if all species were competitively equal, the species most resistant to ill effects from physical extremes, predation, or disease would eventually come to dominate the community.

Among the smallest self-contained communities are those found on boulders located in the intertidal zones of rocky coasts. These communities, which are dominated by the algae growing on the rock surface, are often subjected to massive disturbances as a result of severe waves. The waves may either strip the algae away or actually overturn the boulders. As a result, bare rock becomes available for colonization. In a series of observations in intertidal zones of southern California, Wayne Sousa found that large boulders, which are infrequently overturned, and small boulders, which are frequently overturned, are typically dominated by one or a few algal species. By contrast, medium-sized boulders, which are subjected to an intermediate number of disturbances, tend to have a greater diversity of species.

Ecological Succession

Numerous observations have revealed that, if the interval between disturbances is relatively long, gradual changes occur in the composition of a community following the initial recolonization. The photosynthetic organisms that are usually the first colonists (Figure 44–20) are generally replaced in time by other types, which gradually crowd out the earliest species and

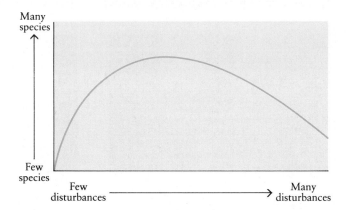

44–19   *According to the intermediate disturbance hypothesis, the diversity of species in a community is determined by the frequency of environmental disturbances. When disturbances occur either very seldom or very often, the number of species is low. By contrast, when the frequency of disturbances is somewhere in between, the number of species is high.*

44–20   *When Mt. St. Helens in the state of Washington exploded on May 18, 1980, shock waves leveled all of the trees in an area of about 21,000 hectares, and a deep layer of ash was deposited. Fireweed (Epilobium augustifolium) and grasses, as seen here four months after the eruption, were among the first plants to recolonize the area.*

(a)

(b)

44–21   (a) *Boulders in a low intertidal zone on the southern California coast. Those with bare upper surfaces have recently been overturned by the waves, while those covered with algae have remained undisturbed.* (b) *A collection of intertidal boulders covered with different successional stages of algae. The boulder in the center was overturned by wave action during the winter before this photograph was taken and has been colonized by the green alga* Ulva. *The surrounding boulders, which are larger, have not been disturbed as recently and are covered with later successional stages that are dominated by red algae.*

which may themselves be subsequently replaced. As the photosynthetic components of the system change, the accompanying animal life changes as well. This process is known as **ecological succession.**

An example of succession is provided by an abandoned field surrounded by other vegetation. Such an area is bombarded by the seeds of numerous plants and is rapidly covered by seedlings of those species whose seeds germinate the most rapidly. In an open field, the survivors among the seedlings will be those plants that can survive the sunlight and drying winds—weeds and grasses and, in many regions of North America, trees such as cedars, white pines, poplars, and birch. For a while, these plants dominate the community, but eventually they are replaced by other trees—oaks, red maples, white ash, and tulip trees. These trees are, in turn, replaced by hemlock, beech, and sugar maple, which, in the absence of major disturbances, dominate the forest indefinitely.

In many communities, the sequence of photosynthetic organisms is so regular and predictable that ecological succession was at one time viewed as analogous to the developmental processes of a single organism, with each stage "preparing the way" for the next by altering the local conditions of temperature, light, humidity, soil content, and so on. Ultimately, the com-

munity would, according to this view, reach a "mature," stable state, known as the climax community. This concept of succession, formulated in 1916 by the plant ecologist F. E. Clements, is known as the **facilitation hypothesis.** For many years, it dominated the study of ecological succession.

Recent studies, however, have suggested that there are alternative mechanisms by which succession may occur. According to the **inhibition hypothesis,** the early species prevent—rather than assist—colonization by other species. Eventually, however, the earliest colonizers are replaced by later arrivals, and those species may, in turn, prevent colonization by others, until they too are replaced or a subsequent disturbance lowers their numbers.

Another model, the **tolerance hypothesis,** suggests that the earliest species neither facilitate nor inhibit colonization by later species. The species dominant at any given time are those that can best tolerate the existing physical conditions and availability of resources.

In a continuation of his earlier studies, Wayne Sousa charted the patterns of algal succession on intertidal boulders by scraping boulders clean of algae and by adding bare rocks to the intertidal zone. The first colonists on the bare rock were the green algae *Ulva* (Figure 24–18d, page 430) and *Enteromorpha* (Figure

44–12a). These pioneer species grow rapidly and, within a short period of time, fully occupy the available space on the boulders (Figure 44–21). Later, large perennial brown and red algae become established, replacing the green algae. Ultimately, the dominant organism is *Gigartina*, a red alga.

To determine the mechanism by which this successional sequence occurred, Sousa removed algae at various stages of the process. For example, when he removed *Ulva*, he found that *Gigartina* was then able to colonize the boulders. Similarly, when he removed the middle successional species of red algae, *Gigartina* was also able to colonize. These results support the inhibition hypothesis, but they raise the question as to how the later species ever gain a foothold. The answer, it turns out, is that *Ulva* and the other early colonizers are subjected to heavy predation and have high mortality rates. As a consequence, small areas of space open up on the boulders, which are then seized by the later species.

Other experiments, however, have supported the facilitation hypothesis, suggesting that there are no simple answers that apply in all situations. Each community is a unique assemblage of organisms, the product of a unique and ever-changing history involving both physical and biological factors.

The world of living organisms is so vast and diverse—and ecologists are, relatively speaking, so few—that it may be many years before enough information is available to provide a thorough understanding of community ecology. Paradoxically, the processes that occur at the next level of ecological complexity—the ecosystem—are in many respects more clearly understood than those occurring within either populations or communities. These processes, which involve the flow of energy and the recycling of minerals, will be the subject of the next chapter.

Summary

Populations live as part of a community—an assemblage of organisms of different species inhabiting a common environment and interacting with one another. Three major types of interaction in communities are competition, predation, and symbiosis.

The more similar organisms are in their requirements and life styles, the more intense the competition between them is likely to be. As a result of competition, the overall fitness of the interacting individuals may be reduced. The importance of competition as an influence on community composition and structure is currently a matter of debate among ecologists.

The competitive exclusion principle, formulated by Gause, states that if two species are in competition for the same limited resource, one or the other will eventually be eliminated in situations in which they occur together. Similar species are able to coexist in the same community because they occupy different ecological niches. The ecological niche of a species is defined by the resource requirements and total life style of the members of that species, including their interactions with members of other species. Analyses of situations in which similar species coexist have demonstrated that resources are often subdivided, or partitioned, by the coexisting species.

Niche overlap describes the situation in which members of more than one species utilize the same limited resource. In communities in which niche overlap occurs, natural selection may result in an increase in the differences between the competing species, a phenomenon known as character displacement. Although the adaptations that enable various organisms to partition resources have often been attributed to character displacement, it is difficult, if not impossible, to distinguish between adaptations that occurred in response to competition and those that were the result of differing local conditions.

Competition has been demonstrated in numerous experimental studies. Experiments in which potential competitors were removed from a study site have given rise to the concepts of the fundamental niche and the realized niche. A fundamental niche represents the resources that would be utilized by a species in the absence of interactions with other organisms. A realized niche describes the resources actually utilized.

Predation is the consumption of live organisms. Predator-prey interactions influence population dynamics and may increase species diversity by reducing competition among prey species. The size of a predator population is often limited by the availability of prey. However, predation is not necessarily the major factor in regulating the population size of prey organisms, which may be influenced more by their own food supply.

Symbiosis is a close and long-term association between organisms of different species. It may be beneficial to one species and harmless to the other (commensalism), beneficial to one and harmful to the other (parasitism), or beneficial to both species (mutualism).

Two important questions regarding community composition and structure remain unanswered. First, what determines the number of species in a community? And, second, what factors underlie the changes in community composition with time? According to the island biogeography hypothesis, the number of species on islands reaches an equilibrium determined by the balance between immigration and extinction. Species composition may vary widely over time but the number of species remains approximately the same. For new species to gain a foothold, established species must become extinct, leading to a continual turnover in species composition.

According to the intermediate disturbance hypothesis, the greatest species diversity is found in communities, such as tropical rain forests and coral reefs, that are subjected to environmental disturbances at an intermediate frequency. Communities in which disturbances occur either very seldom or very often generally have a lower species diversity.

Following environmental disturbances, communities are recolonized by dispersal of immature forms from neighboring communities. If enough time elapses before the next major disturbance, a community typically goes through a process of ecological succession in which the earliest colonizers are replaced by other species that may, in turn, be replaced by still others. The mechanism of succession appears to vary from community to community. Communities, like the populations of which they are composed, are dynamic, continually changing as conditions change.

Questions

1. Distinguish among the following: population/community; competitive exclusion/resource partitioning; habitat/niche; niche overlap/character displacement; fundamental niche/realized niche; symbiosis/commensalism/parasitism/mutualism.

2. Compare the effects of interspecific (between species) and intraspecific (within species) competition. What is the principal reason for the differences?

3. Compare the results of MacArthur's study of the warblers with Connell's experiment with the barnacles. What step did Connell perform that MacArthur did not? Why is the step important?

4. In the American southwest, grasses and mesquite compete with each other for dominance of the landscape. Mesquite, however, was rare before cattle were introduced into the western United States. How have cattle affected the competition between the two types of plant? Suppose all cattle were removed from a large area. What change would you predict in the competition between grasses and mesquite?

5. Introducing a new species into a community can have a number of possible effects. Name some of these possible consequences both to the community and to the introduced species. What types of studies should be done before the importation of an "alien" organism? Some states and many countries have laws restricting such importations. Has your own state adopted any such laws? Are they, in your opinion, ecologically sound?

6. In the long and ruthless war between coyotes and sheep herders, studies have shown that (a) coyotes kill sheep, and (b) the percentage of sheep lost from herds in areas where coyotes have been exterminated is about the same as the percentage lost in areas where coyotes are still present. How could you explain this finding?

7. In the opinion of some ecologists, animals that eat seeds, such as the Galapagos ground finches, should be regarded as predators, while animals that eat leaves, such as deer, should be regarded as parasites. Justify this classification of herbivores as either predators or parasites.

8. A national park or a game preserve can be considered an island, from the point of view of the species living there. What lesson in the management of such a park or preserve can be learned from the experiment carried out on the mangrove islands by Simberloff and Wilson?

9. Compare the three hypotheses (facilitation, inhibition, and tolerance) that have been proposed to explain the process of ecological succession.

Suggestions for Further Reading

Arehart-Treichel, Joan: "Science Helps the Bluebird," *Science News,* June 11, 1983, pages 376–377.

Bergerud, Arthur T.: "Prey Switching in a Simple Ecosystem," *Scientific American,* December 1983, pages 130–141.

Birkeland, Charles: "The Faustian Traits of the Crown-of-Thorns Starfish," *American Scientist,* vol. 77, pages 154–163, 1989.

Brown, Barbara E., and John C. Ogden: "Coral Bleaching," *Scientific American,* January 1993, pages 64–70.

Case, Ted J., and Martin L. Cody: "Testing Theories of Island Biogeography," *American Scientist,* vol. 75, pages 402–411, 1987.

Connell, Joseph H.: "Diversity and the Coevolution of Competitors, or the Ghost of Competition Past," *Oikos,* vol. 35, pages 131–138, 1980.

Connell, Joseph H., and Ralph O. Slatyer: "Mechanisms of Succession in Natural Communities and Their Role in Community Stability and Organization," *American Naturalist,* vol. 111, pages 1119–1144, 1977.

Connor, Edward F., and Daniel Simberloff: "Competition, Scientific Method, and Null Models in Ecology," *American Scientist,* vol. 74, pages 155–162, 1986.

DeVries, Philip J.: "Singing Caterpillars, Ants and Symbiosis," *Scientific American,* October 1992, pages 76–83.

Diamond, Jared M.: "Niche Shifts and the Rediscovery of Interspecific Competition," *American Scientist,* vol. 66, pages 322–331, 1978.

Diamond, Jared M., K. David Bishop, and S. Van Balen: "Bird Survival in an Isolated Javan Woodland: Island or Mirror?" *Conservation Biology,* vol. 1, pages 132–142, 1987.

Horn, Henry S.: "Forest Succession," *Scientific American,* May 1975, pages 90–98.

Karr, James R.: "Avian Extinction on Barro Colorado Island, Panama: A Reassessment," *American Naturalist,* vol. 119, pages 220–237, 1982.

Larson, Douglas: "The Recovery of Spirit Lake," *American Scientist,* vol. 81, pages 166–177, 1993.

Lubchenco, Jane: "Plant Species Diversity in a Marine Intertidal Community: Importance of Herbivore Food Preference and Algal Competitive Abilities," *American Naturalist,* vol. 112, pages 23–29, 1978.

O'Brien, W. John, Howard I. Browman, and Barbara I. Evans: "Search Strategies of Foraging Animals," *American Scientist,* vol. 78, pages 152–160, 1990.

Owen, Jennifer: *Feeding Strategy,* University of Chicago Press, Chicago, 1982.*

This short, beautifully illustrated book explores the variety of ways in which animals obtain food. It considers not only herbivorous and carnivorous animals but also filter-feeders, parasites, and scavengers.

Peterson, Charles H.: "Intertidal Zonation of Marine Invertebrates in Sand and Mud," *American Scientist,* vol. 79, pages 236–249, 1991.

Rennie, John: "Living Together," *Scientific American,* January 1992, pages 122–133.

Rinderer, Thomas E., Benjamin P. Oldroyd, and Walter S. Sheppard: "Africanized Bees in the U.S.," *Scientific American,* December 1993, pages 84–90.

Romme, William H., and Don G. Despain: "The Yellowstone Fires," *Scientific American,* November 1989, pages 36–46.

Schaller, George: *The Serengeti Lion: A Study of Predator-Prey Relations,* University of Chicago Press, Chicago, 1976.*

One of the first, best, and most comprehensive of the modern field studies in animal ecology and behavior. Winner of the National Book Award.

Schoener, Thomas W.: "The Controversy over Interspecific Competition," *American Scientist,* vol. 70, pages 586–595, 1982.

Sousa, Wayne P.: "Experimental Investigations of Disturbance and Ecological Succession in a Rocky Intertidal Algal Community," *Ecological Monographs,* vol. 49, pages 227–254, 1979.

Storer, John H.: *The Web of Life,* New American Library, Inc., New York, 1956.*

One of the first books ever written on ecology for the general public. In its simple presentation of the interdependence of living things, it remains a classic.

Terborgh, John: "Why American Songbirds Are Vanishing," *Scientific American,* May 1992, pages 98–105.

Wilson, Edward O.: "Threats to Biodiversity," *Scientific American,* September 1989, pages 108–116.

Winkler, William G., and Konrad Bögel: "Control of Rabies in Wildlife," *Scientific American,* June 1992, pages 86–93.

Young, James A.: "Tumbleweed," *Scientific American,* March 1991, pages 82–87.

*Available in paperback.

Ecosystems

Along the North Atlantic coast, from Labrador south to Maine, a cold current of water sweeps down from the North Pole, producing an upwelling of nutrients from the ocean depths to the sunlit surface. This upwelling of nitrates, phosphates, and other compounds profoundly affects the life and times of puffins, among others. These natural fertilizers spur the growth of phytoplankton, the "meadow" of the sea. The most abundant components of the phytoplankton are diatoms (page 427), which contain droplets of oil. Oil is less dense than water, and, as a consequence, the diatoms float near the surface, where they absorb the sun's rays.

Zooplankton graze on the phytoplankton. The most numerous organisms in the zooplankton are copepods (page 471), minute crustaceans less than 2 millimeters long. Shrimp and other larger invertebrates feed on the vast microscopic herds of copepods that move through the water. Fish larvae, herring, and other small fish are next in line, and they are eaten, in turn, by larger animals. Because there are, for reasons to be discussed in this chapter, fewer large animals than small animals, these interactions can be depicted in the form of a pyramid. This puffin, its mouth stuffed with herring, is at the top of one such pyramid.

Puffins look somewhat like penguins, because, like penguins, they are upright, with their feet placed well back under them. The reason they look like penguins, of course, is that they earn their living in a similar way. On land, puffins move with ease as they scramble over the rocks, using the claws at the ends of their webbed feet to grip the surface. In the water, they are superb swimmers, powered by their short, strong wings as they dive into and breast-stroke their way through schools of fish.

During the mating season, the puffin's bill is, as you see here, ornamented with vivid red, yellow, and black plates. Puffin pairs raise their chicks in huge, multispecies colonies of sea birds. At summer's end, they head to sea and spend the long, dark winter months harvesting the ocean's riches, never touching land until the following spring.

As we saw in the last chapter, the populations within a community have numerous interactions with each other. Moreover, they interact with the **abiotic** (nonliving) environment. In all cases, these interactions have two consequences: (1) a one-way flow of energy through autotrophs (usually photosynthetic organisms) to heterotrophs, which consume either autotrophs or other heterotrophs; and (2) a cycling of materials, which move from the abiotic environment through the bodies of living organisms and back to the abiotic environment. This cycling of materials is dependent upon decomposers, organisms that break down organic materials into a form that can be used by autotrophs.

An assemblage of biotic (living) and abiotic (nonliving) components through which energy flows and materials cycle is known as an ecological system, or **ecosystem.** Taking a global view, the entire surface of the Earth can be seen as a single ecosystem. This view is useful when studying materials that are circulated on a worldwide basis, such as carbon dioxide, oxygen, and water.

A suitably stocked aquarium or terrarium is also an ecosystem, and such small-scale ecosystems can be useful models for studying certain ecological problems, such as the details of transfer of a particular mineral element. Most studies of ecosystems and their component communities have been made, however, on more or less self-contained natural units—for example, a pond, a swamp, or a meadow.

We shall begin our consideration of ecosystems with the principal physical factor affecting them—the energy received from the caldron of hydrogen and helium atoms that we know as the sun.

Solar Energy

Life on Earth is powered by the sun, which is also responsible for climate, wind, and weather. Every day, year in and year out, energy from the sun arrives at the

Earth's Threatened Ozone Shield

The living organisms of the Earth's ecosystems are shielded from the ultraviolet rays of the sun by a layer of ozone located in the stratosphere—the region of the atmosphere that begins about 10 kilometers above the Earth's surface and extends to an altitude of about 50 kilometers. Ozone (O_3) is formed when molecules of oxygen gas (O_2) are broken apart by radiant energy and then recombine.

Approximately 99 percent of the sun's ultraviolet radiation that reaches the stratosphere is converted to heat through a chemical reaction that continually recycles ozone molecules. When ultraviolet light strikes an ozone molecule, the high-energy radiation splits the molecule into highly reactive oxygen atoms. Almost immediately the atoms recombine, forming ozone once more and releasing energy in the form of heat. For hundreds of millions of years, the quantity of ozone in the atmosphere has apparently remained constant, as has the very small percentage of the ultraviolet radiation reaching the Earth's surface.

This is now changing, as a direct—although inadvertent—result of human activities. The most striking evidence of damage to the ozone layer has been the appearance of actual holes in the layer over both Antarctic and Arctic regions during their respective summers. Less dramatic, but of potentially far greater significance for living organisms, is growing evidence of a thinning of the ozone layer around the entire globe. Between 1978 and 1990, ozone levels over the temperate regions of the Northern Hemisphere dropped some 4.5 to 5 percent, and the loss since 1969 is estimated at about 10 percent.

The principal cause of the ozone loss is a group of synthetic chemicals known as chlorofluorocarbons. These chemicals are used throughout the world in refrigeration systems,

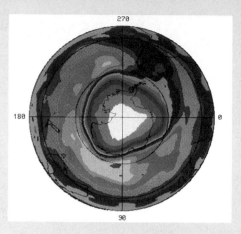

This computer-generated map of stratospheric ozone concentrations over the regions surrounding the South Pole is based on measurements made on October 6, 1993, by the Russian Meteor-3 satellite. The blue, pink, and white areas constitute the ozone hole. The ozone concentration in the center of the hole (white) is 60 percent lower than it was prior to 1975. This depletion of the ozone layer is more severe than in any year since measurements began. In this map, the outlines of the continents are shown in black and the concentric areas of color surrounding the hole indicate successively more substantial concentrations of ozone.

including home refrigerators, freezers, and air-conditioners. They are also used in fire extinguishers, particularly small household extinguishers, and they were previously used as propellants in aerosol cans.

For many years, chemists believed that chlorofluorocarbons released into the air diffused harmlessly into the upper atmosphere, where they were broken down by sunlight. In the 1970s, however, it was learned that, in the

presence of ultraviolet radiation, chlorine atoms break free of chlorofluorocarbon molecules and then react with ozone molecules. A single chlorine atom can react with—and destroy—as many as 100,000 ozone molecules. Current evidence indicates that if the addition of chlorofluorocarbons to the atmosphere were to continue unchecked, the ozone layer would eventually be entirely destroyed.

Depletion of the ozone layer leads directly and inevitably to an increase in the amount of ultraviolet radiation passing through the atmosphere. Each drop of 1 percent in the ozone layer leads to a 2 percent increase in the amount of ultraviolet radiation reaching the Earth's surface. For living systems, the consequences are many and serious. Ultraviolet radiation can trigger the dissociation of biologically important molecules, producing atoms and groups of atoms that are highly reactive.

Especially sensitive to ultraviolet radiation are the phytoplankton that form the base of aquatic food chains. Experimental studies have shown that even moderate levels of ultraviolet radiation are detrimental to these organisms and higher levels can cause mass mortality. Terrestrial photosynthetic organisms are also affected by increased levels of ultraviolet radiation. In a series of laboratory experiments in which some 100 different crop plants were exposed to slightly increased levels of ultraviolet radiation, the yields were reduced in 20 percent. Other experiments have demonstrated that the nitrogen-fixing bacteria that form symbiotic associations with the roots of legumes are killed by high levels of ultraviolet radiation.

For humans, one of the most immediate consequences of increased levels of ultraviolet radiation is an increased incidence of skin cancer. Calculations by the Environmental Protec-

upper surface of the Earth's atmosphere at an average rate of 1.94 calories per square centimeter per minute, for a total of about 1.3×10^{24} calories per year. This is known as the solar constant, and, although it is only a tiny fraction of the total energy radiated by the sun, it is a tremendous quantity of energy. However, because of the atmosphere, only a fraction of this energy reaches the surface of the Earth and becomes available to living organisms.

Of the incoming solar energy, about 30 percent is reflected back into space by clouds and atmospheric dust. Because of this reflected energy, Earth, as seen from outer space, is a shining planet, brighter than Venus. Another 20 percent of the energy is absorbed by the atmosphere.

The remaining 50 percent of the incoming radiation reaches the Earth's surface. A small amount of this is reflected from bright areas, but most is absorbed. En-tion Agency, based on 1990 satellite measurements of the ozone layer, project some 12 million additional cases of skin cancer in the United States in the next 50 years, of which more than 200,000 will be fatal. Also of concern is evidence that ultraviolet radiation can severely damage the immune system, the body's bulwark against infectious disease.

Intense efforts to find replacements for chlorofluorocarbons, particularly for refrigeration systems, are now underway. A landmark treaty, drafted in Montreal in 1987, bound its signatories to freeze chlorofluorocarbon production at 1986 levels by the end of 1989. Moreover, adherents to the treaty were required to cut production and use of these chemicals by at least half by the end of the century. It quickly became apparent, however, that the loss of stratospheric ozone is proceeding so rapidly that these measures were inadequate. Thus, in a subsequent treaty, drafted in London in 1990, the industrialized nations agreed to eliminate *all* production of chlorofluorocarbons by the turn of the century and the developing nations agreed to eliminate such production by the year 2010.

Current projections indicate that *if* the London agreement is faithfully observed, the concentration of ozone-destroying chlorine in the stratosphere should peak by the year 2005—at a level some 12 to 30 percent higher than at present. It is estimated that another 30 years would be required for the concentration of stratospheric chlorine to return to its current level, and perhaps another century for it to return to its preindustrial level. Although the London agreement cannot halt the loss of ozone that is already underway, it appears to be our best hope for slowing—and ultimately reversing—a process with potentially devastating effects for life on Earth.

Climate, Wind, and Weather

The amount of energy received by various parts of the Earth's surface is not uniform. This is the major factor determining the distribution of life on Earth. In the vicinity of the Equator, the sun's rays are almost perpendicular to the Earth's surface, and this sector receives more energy per unit area than the regions to the north and south, with the polar regions receiving the least (Figure 45–1). Moreover, because the Earth, which is tilted on its axis, rotates once every 24 hours and completes an orbit around the sun about once every 365 days, the amount of energy reaching different parts of the surface changes hour by hour and season by season (Figure 45–2).

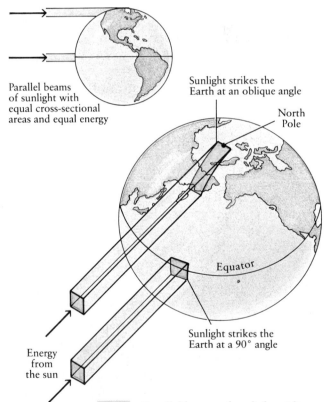

45–1 *Parallel beams of sunlight with equal cross-sectional areas contain equal amounts of energy. At the Equator, an incoming beam of sunlight strikes the Earth at a 90° angle (perpendicular) and its energy is concentrated into a small area. At higher latitudes, an identical incoming beam of sunlight strikes the curved surface of the Earth at an oblique angle. Thus, the same amount of energy is spread over a larger area. As a consequence, regions north and south of the Equator receive less solar energy per unit area than regions close to the Equator.*

ergy absorbed by the oceans warms the surface of the water, evaporating water molecules and powering the water cycle (page 45). Solar energy absorbed by the ground is reradiated from the surface in the form of infrared rays—that is, as heat. The gases of the atmosphere are transparent to visible light, but carbon dioxide and water, in particular, are not transparent to infrared rays. As a result, the heat is held in the atmosphere, warming the surface of the Earth.

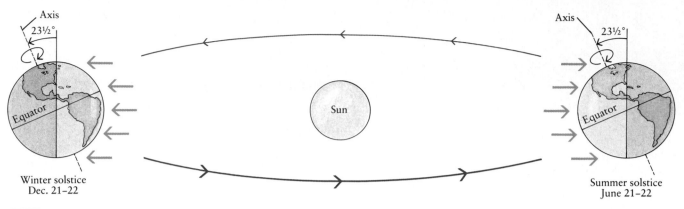

Axis
23½°

Equator

Winter solstice
Dec. 21–22

Sun

Axis
23½°

Equator

Summer solstice
June 21–22

45–2 *In the Northern and Southern Hemispheres, temperatures change in an annual cycle because the Earth is slightly
tilted on its axis of rotation in relation to its pathway around
the sun. In winter in the Northern Hemisphere, the North Pole
is tilted away from the sun, decreasing the angle at which the
sun's rays strike the surface and also decreasing the duration*
*of daylight, both of which result in lower temperatures. In
summer in the Northern Hemisphere, the North pole is tilted
toward the sun. Note that the polar region of the Northern
Hemisphere is continuously dark during the winter and continuously light during the summer.*

Temperature variations over the surface of the Earth
and the forces generated by the Earth's rotation establish major patterns of air circulation and rainfall. These
patterns depend, to a large extent, on the fact that cold
air is denser than warm air. As a consequence, warm
air rises and cold air falls. As warm air rises, it encounters lower pressure and expands, and as a gas expands,
it cools. Cooler air holds less moisture, so as warm air
rises and then cools, it tends to lose its moisture in the
form of rain or snow. As the air continues to cool, its
density increases and it falls back toward the Earth's
surface.

The air is warmest along the Equator, the region
heated most intensely by the sun. This air rises, creating
a low-pressure area (the doldrums) that draws in air
from the surface north and south of the Equator. As the
warm equatorial air rises, it cools, loses most of its
moisture, moves poleward, and then falls again at latitudes of about 30° north and south, the regions where
most of the great deserts of the world are found. This
air then either moves toward the Equator, replacing the
warm air rising there, or it moves poleward. The pole-
ward-moving air warms, picks up moisture, and rises
again at about 60° latitude (north and south). This is
the polar front, another low-pressure area. The air rising at the polar front cools, loses moisture, and descends again at the poles, producing another region of
descending air in which there is virtually no precipitation. The Earth's spinning movement twists the winds
caused by this transfer of air from Equator to poles,
creating the major wind patterns (Figure 45–3).

The worldwide patterns are modified locally by a variety of factors, such as the presence of mountains (Figure 45–4). For example, along the West Coast of the
United States, where the prevailing winds are from the
west, the western slopes of the Sierra Nevada have

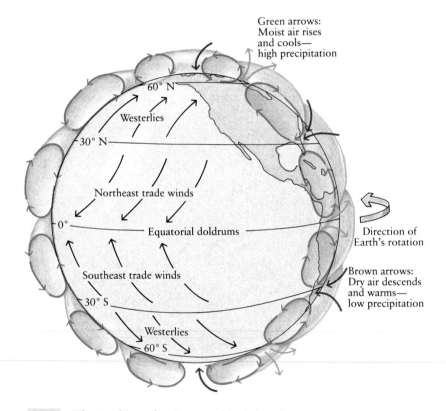

Green arrows:
Moist air rises
and cools—
high precipitation

60° N

Westerlies

30° N

Northeast trade winds

0°

Equatorial doldrums

Southeast trade winds

30° S

Westerlies

60° S

Direction of
Earth's rotation

Brown arrows:
Dry air descends
and warms—
low precipitation

45–3 *The Earth's surface is covered by belts of air currents,
which determine the major patterns of wind and rainfall. In
this diagram, the dark blue arrows indicate the direction of
movement of the air within the belts. The green arrows indicate regions of rising air, which are characterized by high precipitation, and the brown arrows indicate regions of descending air, which are characterized by low precipitation. The dry
air descending at latitudes of 30° north and south is responsible for the great deserts of the world.*

*The prevailing winds on the Earth's surface, indicated by
the black arrows, are produced by the twisting effect of the
Earth's rotation on the air currents within the individual belts.*

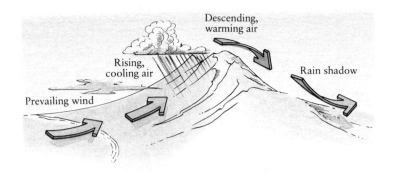

45–4   *The effect of coastal mountains on patterns of precipitation. As winds come off the water and the air is forced upward by the contour of the land, it cools and releases its moisture in the form of rain or snow. As the air descends on the far side of the mountains and becomes warmer, its capacity to hold water increases, producing a rain shadow.*

abundant rainfall, while the eastern slopes are dry and desertlike. As air from the ocean hits the western slope, it rises, is cooled, and releases its water. Then, after passing the crest of the mountain range, the air descends again. As it becomes warmer, its water-holding capacity increases, producing a "rain shadow" on the eastern slope.

The Flow of Energy

The flow of energy through ecosystems is the most important factor in their organization. Of the solar energy that reaches the Earth's surface, only a very small fraction—an estimated 0.1 percent on a worldwide basis—is diverted into living systems. Even when light falls

where vegetation is abundant, as in a forest, a cornfield, or a marsh, only 1 to 3 percent of that light (calculated on an annual basis) is used in photosynthesis. Yet this fraction, as small as it is, may result in the production—from carbon dioxide, water, and a few minerals—of several thousand grams (dry weight) of organic matter per year in a single square meter of field or forest. Worldwide, it results in the production of about 120 billion metric tons of organic matter per year.

Trophic Levels

The passage of energy from one organism to another takes place along a particular **food chain**—that is, a sequence of organisms related to one another as prey

45–5   *The energy on which life depends enters the living world in the form of light. It is converted to chemical energy by photosynthetic organisms, such as (a) wild barley, a terrestrial plant, and (b) giant kelp, a marine alga. Such photosynthetic organisms are, in turn, the energy source for all heterotrophs, including ourselves.*

(a)   (b)

and predator. The first is eaten by the second, the second by the third, and so on, in a series of feeding levels, or **trophic levels.** In most ecosystems, food chains are linked together in complex **food webs,** with many branches and interconnections (Figure 45–6).

A food web may involve more than 100 different species, with predators characteristically taking more than one type of prey, and each type of prey being exploited by several different species of predators. The relation of any species to others in its food web is an important dimension of its ecological niche.

Producers

The first trophic level of a food web is always occupied by a **primary producer.** On land, the primary producer is usually a plant; in aquatic ecosystems, it is usually an alga. These photosynthetic organisms use light energy to make carbohydrates and other compounds, which are sources of chemical energy. Producers far outweigh consumers. Some 99 percent of all the organic matter in the living world is made up of plants and algae. All heterotrophs combined account for only 1 percent of the organic matter.

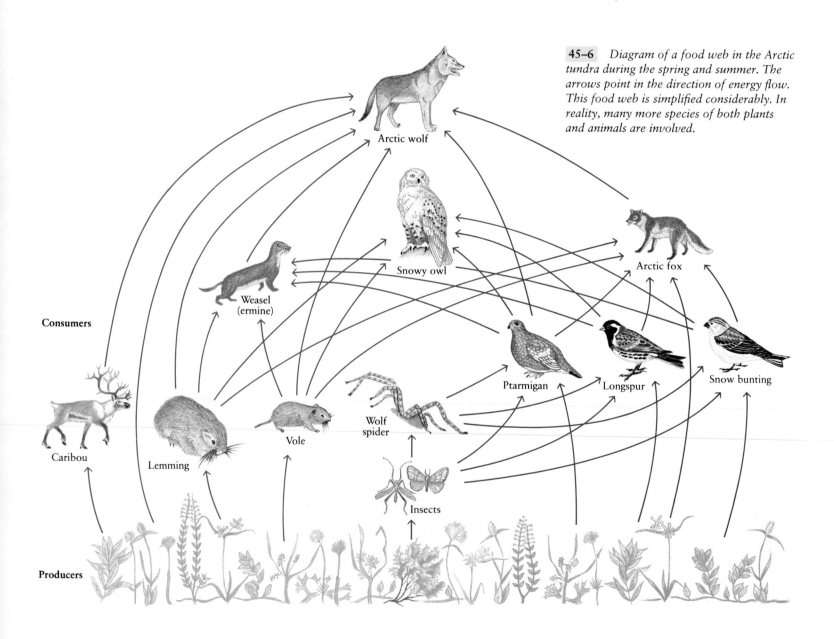

45–6   *Diagram of a food web in the Arctic tundra during the spring and summer. The arrows point in the direction of energy flow. This food web is simplified considerably. In reality, many more species of both plants and animals are involved.*

Consumers

Producers

Arctic wolf

Snowy owl

Arctic fox

Weasel (ermine)

Ptarmigan

Longspur

Snow bunting

Caribou

Lemming

Vole

Wolf spider

Insects

Life without the Sun

Throughout this text, we have emphasized that life on Earth depends on radiant energy from the sun. However, a series of discoveries beginning in 1977 have revealed the existence of entire ecosystems in which the primary producers are chemosynthetic bacteria (page 75) that are not dependent on the energy of sunlight to power their carbon-fixing reactions. Such ecosystems, which are typically located in the ocean depths, far beyond the reach of sunlight, are powered by geothermal energy—heat energy from within the Earth itself.

The first known chemosynthetic ecosystem was discovered by the *Alvin,* a research submarine from Woods Hole Oceanographic Institution, and its crew of scientific investigators. The place was the Galapagos Rift, a boundary between adjacent tectonic plates (see page 358). Sea water seeps into the fissures in the volcanic rocks erupting along this boundary, becomes heated, and rises again. As a consequence, oases of warmth are created in the near-freezing waters 2.5 kilometers below the surface. More important, the water reacts with the rocks deep within the Earth's crust. Here, under extreme heat and pressure, sulfate in the sea water is reduced to hydrogen sulfide, with heat supplying the necessary energy for the reaction. Chemosynthetic bacteria then oxidize the sulfide, obtaining the energy needed to extract carbon from carbon dioxide in the sea water and fix it in organic molecules.

More than a dozen of these deep-sea oases have now been located in the Pacific, each with chemosynthetic bacteria as the primary producers in a complex food web. The specific complement of animals varies from one chemosynthetic ecosystem to another and is thought to be the result of whatever floating larvae happened to colonize the particular area during its early stages of development.

Among the most spectacular inhabitants of the deep-sea oases are the clams, measuring some 20 centimeters in diameter, that cover the lava floor. Other permanent inhabitants include mussels, crabs, and octopuses. The bivalves feed on the bacteria, and the crustaceans and octopuses on the bivalves. One oasis, known as the Garden of Eden, is dominated by huge tube worms. These worms, which lack mouths and digestive tracts, apparently obtain energy from organic molecules synthesized by chemosynthetic bacteria living within their tissues. The bacteria have also been found living in the gills of some clams, suggesting that symbiosis plays an essential role in these ecosystems.

These discoveries triggered not only a search for other deep-sea chemosynthetic ecosystems but also a closer look at the biological interactions of chemosynthetic bacteria found in more accessible surroundings. Such ecosystems have been found deep in the Gulf of Mexico in cold waters and off the Palos Verde peninsula of California in very shallow waters.

The latter discovery was made by a young, skin-diving scientist who was puzzled by the black abalone he found grazing on mats of bacteria in sulfur-rich waters that would normally be toxic to animal life. Although the fissures through which the hot, sulfide-laden water escapes are only the size of teacups, they are surrounded by mats of chemosynthetic bacteria averaging 1 to 2 square meters in area. These mats are grazed not only by the black abalone but by other gastropods as well, including the California sea hare and the giant keyhole limpet.

Most recently, clams living in Los Angeles sewage outfalls have been found to contain symbiotic chemosynthetic bacteria, as have worms living in Bermuda. And, scientists have not yet examined the organisms of such sul-

Giant tube worms, 1.5 meters tall, growing in the depths of the Pacific Ocean at the Garden of Eden oasis. These tube worms are dependent on chemosynthetic bacteria as their energy source. Also visible are yellow-shelled clams and an eyeless, colorless crab, also participants in the food web for which the bacteria are the primary producers. Unlike the near-freezing temperatures normally found at these depths, it was 17°C (63°F) in this area.

fur-rich environments as pulpmill effluent zones, mangrove swamps, and eel grass beds. It is already clear, however, that although most life on Earth is based on radiant energy from the sun, there are highly viable alternatives—not only on this planet, but perhaps on others as well.

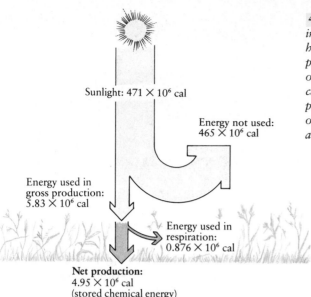

Sunlight: 471 × 10⁶ cal

Energy not used:
465 × 10⁶ cal

Energy used in
gross production:
5.83 × 10⁶ cal

Energy used in
respiration:
0.876 × 10⁶ cal

Net production:
4.95 × 10⁶ cal
(stored chemical energy)

45–7   *Calculation of the productivity of a field in Michigan in which the vegetation was mostly perennial grasses and herbs. Measurements are in terms of calories per square meter per year. In this field, the net primary production—the amount of chemical energy stored in plant material—was 4,950,000 calories per square meter per year. Thus, slightly more than 1 percent of the 471,000,000 calories per square meter per year of sunlight reaching the field was converted to chemical energy and stored in the plant tissues.*

Ecologists speak of the **productivity** of a trophic level, a community, or an ecosystem. **Gross productivity** is a measure of the rate at which energy is assimilated by the organisms in, for example, a particular trophic level. It might be considered analogous to the rate of gross income of a business.

A more useful—and often more easily measured—quantity is the **net productivity** (Figure 45–7). Net productivity is the gross productivity minus the cost of all the metabolic activities of the organisms in question. This cost might be considered analogous to the cost of doing business. Net productivity is thus comparable to the rate of net profit. It is usually expressed as the amount of energy (measured in calories) stored in chemical compounds or as the increase in biomass (measured in grams or metric tons) in a particular period of time. (**Biomass** is a convenient shorthand term meaning the total dry weight of all the organisms being measured at any one time.) Net productivity is a measure of the rate at which organisms store energy, which then becomes available to the organisms of the next trophic level.

At the first trophic level—the primary producers—the increase in the amount of plant or algal material between the beginning and the end of a specified time period represents the **net primary production** for that period. In agricultural ecosystems, for example, the standing crop at the end of the growing season represents the net primary production for that season.

As the data in Table 45–1 reveal, primary production varies enormously from one ecosystem type to another.

In terrestrial ecosystems, the key factors influencing productivity are the duration and intensity of sunlight, temperature, and precipitation. In aquatic ecosystems, the availability of essential mineral elements is more often the principal factor affecting productivity.

Consumers

Energy enters the animal world through the activities of herbivores, animals that eat plants or algae and are thus **primary consumers.** A herbivore may be a caterpillar, an elephant, a sea urchin, a snail, or a field mouse. Each type of ecosystem has its characteristic complement of herbivores.

Of the organic material consumed by herbivores, much is eliminated undigested. Most of the chemical energy from the digested food is used to maintain the metabolic processes of the animal and to power its daily activities: searching for food, eating and digesting it, mating and caring for the young, fleeing predators, and so on. Although this energy is generally described as "lost" through respiration, it is important to realize that, for the individual organism, this is the essential energy on which its life depends.

A fraction of the chemical energy consumed by the herbivore is converted to new animal biomass. The increase in animal biomass is the sum of the increase in the mass of individual animals plus the mass of new offspring. It represents energy available to the next trophic level.

Table 45-1	Average Primary Productivity of Selected Ecosystems	
Ecosystem	**Net Primary Production (grams per meter² per year)**	**Biomass (kilograms per meter²)**
Open ocean	125	0.003
Algal beds and reefs	2,500	2
Estuaries	1,500	1
Lakes and streams	250	0.02
Swamps and marshes	2,000	15
Tropical forest	2,000	12
Temperate forest	1,250	32
Taiga	800	20
Woodland and shrubland	700	6
Tropical grassland	900	4
Temperate grassland	600	1.6
Tundra and alpine	140	0.6
Desert and semidesert scrub	90	0.7
Cultivated land	650	1

Adapted from R. H. Whittaker and G. E. Likens, Table 5.2, in H. Lieth and R. H. Whittaker (eds.), *Primary Productivity of the Biosphere,* Springer-Verlag, Inc., New York, 1975. Tropical forest estimates based on Sandra Brown and Ariel E. Lugo, "Biomass of Tropical Forests: A New Estimate Based on Forest Volume," *Science,* vol. 223, pages 1290–1293, 1984.

This next level, the **secondary consumer** level, is made up of carnivores, animals that eat other animals. The carnivore that devours the herbivore may be a lion, a minnow, a starfish, a robin, or a spider. In every case, only a small part of the organic substance present in the body of the herbivore becomes incorporated into the body of the carnivore.

Some food chains have third and fourth consumer levels, but five links are usually the limit, regardless of the ecosystem. A study of 102 top predators (animals themselves free of predation) showed that there are usually only three links (four levels) in a food chain—plant to herbivore to carnivore to carnivore. For only one top predator were more than five links (six species) involved. At each higher trophic level, there is a decrease in the total amount of energy stored in animal biomass and therefore available to other consumers.

Detritivores

Detritivores are organisms that live on the refuse, or detritus, of a community—dead leaves, branches, and tree trunks, the roots of annual plants, feces, carcasses, even the discarded exoskeletons of insects. They include **scavengers,** such as vultures, jackals, crabs, and earthworms, as well as **decomposers,** such as fungi and bacteria.

Scavengers can be regarded as consumers that feed on dead prey rather than living prey. Decomposers are also consumers, but with a difference: they have evolved

(a)

(b)

45–8 *Representative consumers.* (a) *The hog deer is a herbivore of the Indian subcontinent. Its name derives from its relatively short stature. Adult hog deer generally range from 60 to*

72 centimeters in height. (b) *An Indian python, a carnivore, begins the four-hour process of swallowing an adult hog deer that it has killed by constriction.*

(a)

(b)

(c)

45–9 *Many birds, such as these vultures, are scavengers, living on dead animals. In East Africa, several species of vulture are often seen sharing the same carcass. (a) The lappet-faced vulture—the largest and most powerful, with a huge, heavy beak—is the one that can first break into a carcass. Other vultures are often seen standing by a carcass awaiting the arrival of a lappet-faced vulture. (b) Ruppell's griffon, also a vulture, specializes in reaching its long, snaky, featherless neck far inside the carcass to feed on the intestines and other internal organs. (c) The smaller Egyptian vulture, with its fine beak, can tear off scraps of flesh, such as from the skull, that the coarser beaks cannot grasp.*

specializations that enable them to exploit sources of chemical energy, such as cellulose and nitrogenous waste products, that cannot be used by animals.

In a forest community, more than 90 percent of the net primary production is eventually consumed by detritivores rather than by herbivores. Some of this energy flows through the food web by way of consumers that feed on detritivores, while the rest of it is used in the metabolic processes of the detritivores themselves. As a result, essentially all of the energy stored by plants (and, in aquatic communities, by algae) is ultimately used to support life. Energy stored in organic matter goes unused only when it is trapped in an environment, such as a highly acidic peat bog, in which most detritivores cannot live.

Efficiency of Energy Transfer

The shortness of food chains is usually attributed to the inefficiency involved in the transfer of energy from one trophic level to another. In general, only about 10 percent of the energy stored in a plant is converted to animal biomass in the herbivore that eats the plant. A similar relationship is found at each succeeding level.

If, for instance, an average of 1,500 kilocalories of light energy per square meter fall per day on a land surface covered with plants, about 15 kilocalories are converted to plant material. (Remember that only 1 to 3 percent of the sunlight falling on vegetation is actually used in photosynthesis.) Of these 15 kilocalories, about 1.5 kilocalories are incorporated into the bodies of the herbivores that eat the plants, and about 0.15 kilocalorie is incorporated into the bodies of the carnivores that prey on the herbivores. Although meat is a more concentrated source of calories and nutrients than vegetation, carnivores must usually expend more energy in foraging than herbivores do, and so the net productivity of carnivores and herbivores may be roughly equivalent.

To give a concrete example, in studies of Cayuga Lake in western New York, Lamont Cole calculated that for every 1,000 calories of light energy utilized by algae in the lake, about 150 calories are reconstituted as the biomass of small aquatic animals. Of these 150 calories, 30 calories are converted to smelt biomass. If trout eat the smelt, about 6 calories are converted to trout biomass, and if we eat the trout, we gain about 1.2 calories from each 1,000 calories of light energy originally stored in organic compounds. (Note that if we ate smelt instead of trout, we would derive about five times as much food energy from the original 1,000 calories. This, in short, is the argument for living lower on the food chain.)

As these figures suggest, the 10 percent "rule of thumb" is only a crude estimate. Actual measurements show wide variations in transfer efficiencies, from less than 1 percent to over 20 percent, depending on the species involved. The flow of energy with large losses at each successive level can be depicted as a pyramid such as that shown in Figure 45–10.

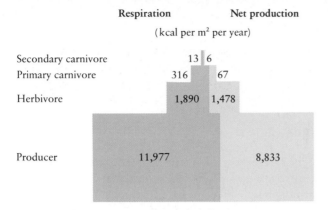

	Respiration	Net production
	(kcal per m² per year)	
Secondary carnivore	13	6
Primary carnivore	316	67
Herbivore	1,890	1,478
Producer	11,977	8,833

45–10 *Pyramid of energy flow for a river ecosystem in Florida. A relatively small proportion of the energy in the system is transferred at each trophic level. Much of the energy is used metabolically and is measured as calories lost in respiration.*

Energy Transfer and Ecosystem Structure

The energy relationships between the trophic levels determine the structure of an ecosystem in terms of both the numbers of organisms and the amount of biomass present. Figure 45–11a, for example, shows a pyramid of numbers for a grassland ecosystem. In this type of ecosystem, the primary producers (the individual grass plants) are small, and so a large number of them is required to support the primary consumers (the herbivores). In an ecosystem in which the primary producers are large (for instance, trees), one primary producer may support many herbivores, as indicated in Figure 45–11b.

Most pyramids of biomass take the form of an upright pyramid, as shown in Figure 45–12a, whether the producers are large or small. Pyramids of biomass are inverted only when the producers have very high reproduction rates. For example, in the ocean, the standing crop of phytoplankton may be smaller than the biomass of the zooplankton that feeds upon it (Figure 45–12b). Because the reproduction rate of the phytoplankton population equals or exceeds the rate of grazing by the zooplankton population, a small biomass of phytoplankton can supply food for a larger biomass of zooplankton.

Like pyramids of numbers, pyramids of biomass indicate only the quantity of organic material present at one time. They do not give the total amount of material produced or, as do pyramids of energy, the rate at which it is produced.

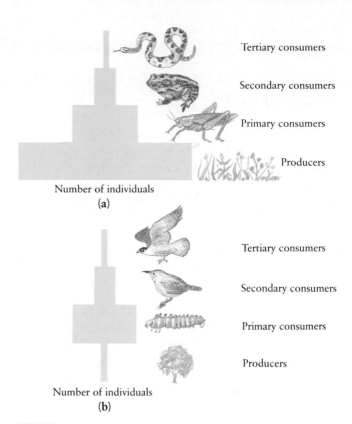

45–11 *Pyramids of numbers for (a) a grassland ecosystem, in which the number of primary producers (grass plants) is large, and (b) a temperate forest, in which a single primary producer, a tree, can support a large number of herbivores.*

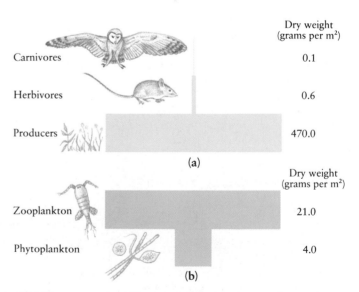

	Dry weight (grams per m²)
Carnivores	0.1
Herbivores	0.6
Producers	470.0
(a)	

	Dry weight (grams per m²)
Zooplankton	21.0
Phytoplankton	4.0
(b)	

45–12 *Pyramids of biomass for (a) the plants and animals in a field in Georgia and (b) plankton in the English Channel. Such pyramids reflect the mass present at any one time. The seemingly paradoxical relationship between the biomass of phytoplankton and zooplankton exists because the high reproduction rate of the smaller phytoplankton population is sufficient to support a larger zooplankton population.*

Biogeochemical Cycles

Energy takes a one-way course through an ecosystem, but many substances cycle through the system. Such substances include water, nitrogen, carbon, phosphorus, potassium, sulfur, magnesium, calcium, sodium, chlorine, and also a number of other minerals, such as iron and cobalt, that are required by living organisms in only very small amounts. The water cycle is shown on page 46, and the carbon cycle is on page 173.

Movements of inorganic substances are referred to as **biogeochemical cycles** because they involve geological as well as biological components of the ecosystem. The geological components are (1) the atmosphere, which is made up largely of gases, including water vapor; (2) the solid crust of the Earth; and (3) the oceans, lakes, and rivers, which cover three-fourths of the Earth's surface.

The biological components of biogeochemical cycles include the producers, consumers, and detritivores (both scavengers and decomposers). As a result of the metabolic work of the decomposers, inorganic substances are released from organic compounds and returned to the soil or water. From the soil or water, inorganic substances are taken back into the tissues of producers (plants or algae), passed along to consumers, and then released to the detritivores, from which they enter the producers again, repeating the cycle.

The cycling through the ecosystem of one important mineral, phosphorus, is shown in Figure 45–13. Other minerals in the ecosystem undergo similar cycles, differing only in details.

The Nitrogen Cycle

The chief reservoir of nitrogen is the atmosphere; in fact, nitrogen gas (N_2) makes up about 77 percent of the atmosphere. Since most living things, however, cannot use elemental atmospheric nitrogen to make amino acids and other nitrogen-containing compounds, they are dependent on nitrogen present in soil minerals. So, despite the abundance of nitrogen in the atmosphere, a shortage of nitrogen in the soil is often the major limiting factor in plant growth.

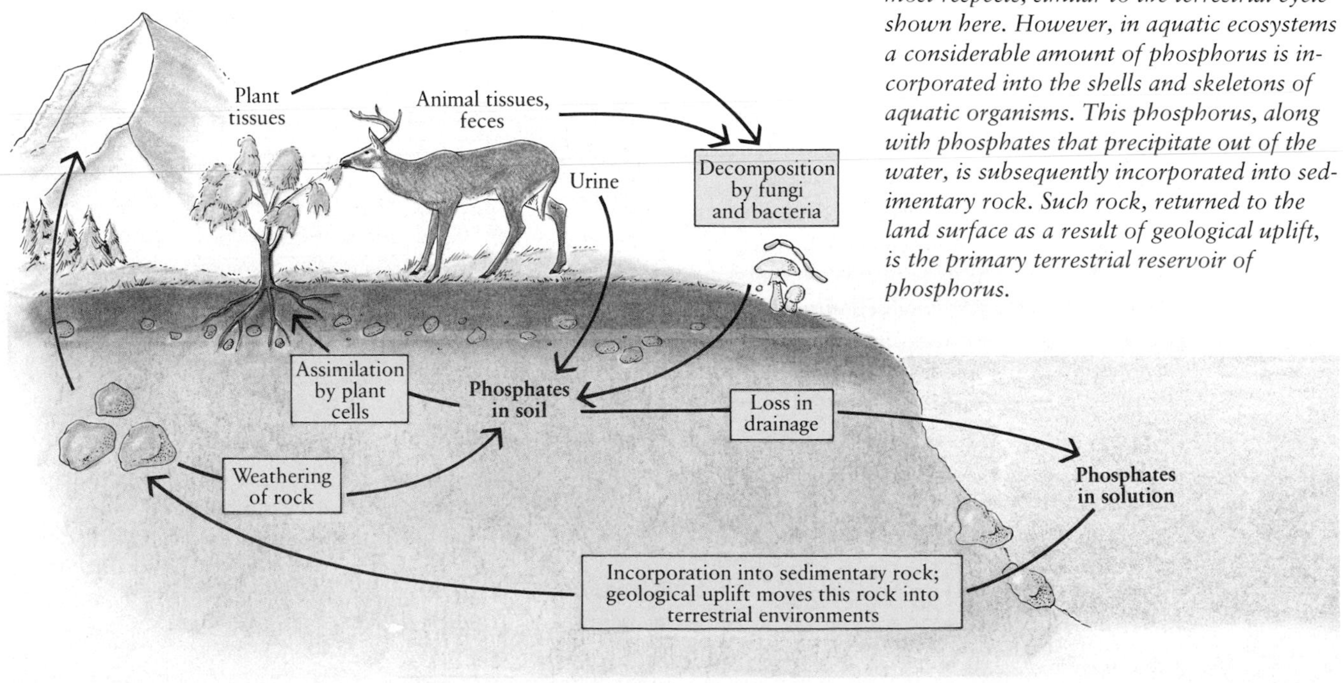

45–13   *The phosphorus cycle in a terrestrial ecosystem. Phosphorus is essential to all living organisms as a component of energy-carrier molecules, such as ATP, and also of the nucleotides of DNA and RNA. Like other minerals, it is released from dead tissues by the activities of decomposers, taken up from soil by plants, and cycled through the ecosystem.*

The phosphorus cycle in an aquatic ecosystem involves different organisms but is, in most respects, similar to the terrestrial cycle shown here. However, in aquatic ecosystems a considerable amount of phosphorus is incorporated into the shells and skeletons of aquatic organisms. This phosphorus, along with phosphates that precipitate out of the water, is subsequently incorporated into sedimentary rock. Such rock, returned to the land surface as a result of geological uplift, is the primary terrestrial reservoir of phosphorus.

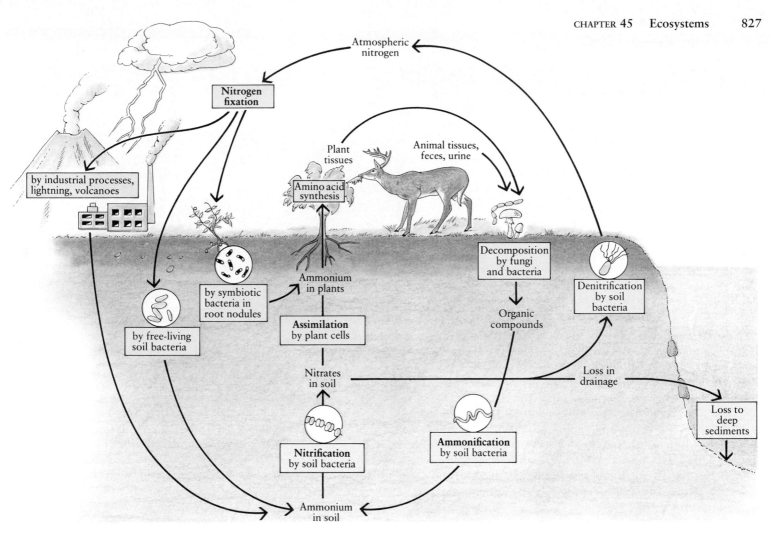

45–14 *The nitrogen cycle in a terrestrial ecosystem. The primary reservoir of nitrogen is the atmosphere, where it makes up about 77 percent of dry air. Only a few microorganisms, some symbiotic and others free-* *living, are capable of fixing nitrogen gas into inorganic compounds that can be used by plants in the synthesis of amino acids and other nitrogenous compounds.*

The nitrogen cycle in aquatic ecosystems is *similar to the terrestrial cycle. Although the specific organisms differ, the processes are essentially the same, with specific groups of bacteria carrying out the various chemical reactions on which the cycle depends.*

The process by which this limited amount of nitrogen is circulated and recirculated throughout the world of living organisms is known as the **nitrogen cycle.** The three principal stages of this cycle are (1) ammonification, (2) nitrification, and (3) assimilation.

Much of the nitrogen found in the soil is the result of the decomposition of organic materials. Initially, it is in the form of complex organic compounds, such as proteins, amino acids, nucleic acids, and nucleotides. These nitrogenous compounds are usually rapidly decomposed into simple compounds by soil-dwelling organisms, chiefly bacteria and fungi. The microorganisms use the amino acids to build their own needed proteins and release the excess nitrogen in the form of ammonia (NH_3) or ammonium ion (NH_4^+). This process is known as **ammonification.**

Several species of bacteria common in soils are able to oxidize ammonia or ammonium ion. This oxidation,

known as **nitrification,** is an energy-yielding process, and the energy released is used by these bacteria as their primary energy source. One group of bacteria oxidizes ammonia (or ammonium) to nitrite (NO_2^-):

$$2NH_3 \ + \ 3O_2 \ \longrightarrow \ 2NO_2^- \ + \ 2H^+ \ + \ 2H_2O$$

Ammonia   Oxygen   Nitrite ion   Hydrogen   Water
                                         ion

Nitrite is toxic to many plants, but it rarely accumulates. Members of another group of bacteria oxidize the nitrite to nitrate (NO_3^-), again with a release of energy:

$$2NO_2^- \ + \ O_2 \ \longrightarrow \ 2NO_3^-$$

Nitrite ion   Oxygen   Nitrate ion

Although plants can utilize ammonium directly, nitrate is the form in which most nitrogen moves from the soil into the roots. Hence, these bacteria play a crucial role in the recycling of nitrogen.

Once the nitrate is within the plant cell, it is reduced back to ammonium. In contrast to nitrification, this **assimilation** process requires energy. The ammonium ions thus formed are transferred to carbon-containing compounds to produce amino acids and other nitrogenous organic compounds needed by the plant.

Although the nitrogen cycle appears complete and self-sustaining, nitrates are steadily lost from the soil and thus removed from the cycle. As we noted in Chapter 41, a variety of processes—the harvesting of plants, soil erosion, fire, and water percolating down through the soil—can reduce the amount of nitrate available to plants.

Nitrates are also lost as a result of the activities of certain types of soil bacteria. In the absence of oxygen, these bacteria break down nitrates, releasing nitrogen gas back into the atmosphere and using the oxygen for their own respiration. This process, known as **denitrification,** takes place in poorly drained (hence, poorly aerated) soils.

The nitrogen cycle is maintained, despite the losses, primarily by the activities of the nitrogen-fixing bacteria (see pages 750–751), which incorporate gaseous nitrogen from the air into organic nitrogen-containing compounds. Just as all organisms are ultimately dependent on photosynthesis for energy, they all depend on nitrogen fixation for their nitrogen.

Recycling in a Forest Ecosystem

Studies of a deciduous forest ecosystem in the White Mountains of New Hampshire have shown that the plant life of a community plays a major role in its retention of mineral elements. Investigators at the Hubbard Brook Experimental Forest first established a procedure for determining the mineral budget—input and output, "profit" and "loss"—of areas in the forest. By analyzing the content of rain and snow, they were able to estimate input, and by constructing concrete weirs (Figure 45–15) that channeled the water flowing out of selected areas, they were able to calculate output.

From these studies, the investigators discovered that the natural forest was extremely efficient in conserving its mineral elements. For example, the annual net loss of calcium represented only about 0.3 percent of the calcium in the ecosystem. In the case of nitrogen and potassium, the ecosystem was actually accumulating mineral nutrients.

In the winter of 1965–1966, all of the trees, saplings, and shrubs in one area of the forest were cut down completely. No organic materials were removed, however, and the soil was undisturbed. During the following summer, the area was sprayed with a herbicide to

45–15   *Weir at Hubbard Brook Experimental Forest in New Hampshire. Water from each of six experimental ecosystems was channeled through weirs, built where the water leaves the watershed, and was analyzed for chemical elements. The watershed behind the weir in this photograph has been stripped of vegetation. The experiments showed that such deforestation greatly increases the loss of nutrient elements from the system.*

inhibit regrowth. During the four months from June to September, 1966, the runoff of water from the area was four times higher than in previous years. Net losses of calcium were 10 times higher than in the undisturbed forest and of potassium, 21 times higher.

The most severe disturbance was seen in the nitrogen cycle. Plant and animal tissues continued to be decomposed to ammonia or ammonium, which then were acted upon by nitrifying bacteria to produce nitrates, the form in which nitrogen is usually taken up by plants. However, no plants were present, and the nitrate was leached from the soil. As a side effect, the stream draining the area became polluted with algal blooms, and its nitrate concentration exceeded the levels established by the U.S. Public Health Service for drinking water. Trees are now beginning to grow again on the devastated site, and the runoff of nutrients has dramatically decreased.

Studies in the tropical forests of Central and South America have further underlined the importance of plant life in mineral retention. In the tropical rain forest, in particular, virtually all mineral nutrients are held within the living components of the ecosystem—and none are retained by the soil for any length of time.

Agricultural Ecosystems and a Hungry World

The earliest traces of agriculture, dating back about 11,000 years, are found in areas of the Near East that are now parts of Iran, Iraq, and Turkey. Here were the raw materials required by an agricultural economy: cereals, which are grasses with seeds capable of being stored for long periods of time without serious deterioration, and herbivorous herd animals, which can be readily domesticated. The grasses in these areas were wild wheats and barley, which still grow wild in the foothills. The animals were wild sheep and goats.

By 8,100 years ago, agricultural communities were established in eastern Europe. By 7,000 years ago, agriculture had spread to the western Mediterranean and into central Europe. During this same period, agriculture originated independently in Central and South America, and perhaps slightly later in the Far East.

With the advent of widespread agriculture, *Homo sapiens* became a major actor in ecosystem dynamics. An area under cultivation is, in effect, an artificial ecosystem, maintained at a very early stage of ecological succession by human activities. Like other ecosystems subjected to frequent, large-scale disturbances, agricultural ecosystems are characterized by a small number of species, a relatively low total biomass, a high net productivity in relation to biomass, and a limited capacity to trap and retain nutrients.

In the earliest periods of agriculture, the most important of these features was undoubtedly the high net productivity, which provided the primitive farmers with a larger and more reliable food supply than they had previously known. As we saw in Chapter 43, an immediate consequence was a significant increase in the human population, a growth that has now become exponential. The consequences of this increase are many and serious—pollution, depletion of fuel supplies, destruction of natural resources, and extinction of other species as the human population expands into previously untouched ecosystems.

Ironically, by far the most difficult problem and also the most urgent is hunger and starvation. Of the world's more than 5.4 billion people, at least 1 billion are inadequately nourished. It is estimated that some 20 million people die each year as the direct result of malnutrition, and many more deaths are thought to be caused by its indirect effects. Perhaps even more important, in terms of the world's future, are the effects, both physical and psychological, of the prolonged chronic hunger of so large a proportion of our population.

A major effort to increase world food supplies by increasing agricultural productivity, known as the Green Revolution, has been under way for the past 40 years. The principal emphasis of this effort has been the development of new crop plants, especially grains.

Enormous progress has been made. From 1950 to 1970, for example, the production of wheat in Mexico increased from 270,000 metric tons per year to 2.35 million. In 1985, wheat production in Mexico reached an all-time high of 5.2 million metric tons, but by 1987, it had dropped back to 4.4 million. In the period from 1950 to 1970, the corn harvest in Mexico increased a more modest 250 percent, with yields per hectare almost doubling. In subsequent years, the corn harvest has continued to increase, reaching a record 14 million metric tons in 1985.

Similarly, between 1950 and the present, India has increased its production of food grains about 2.8 percent each year. (Its population during this period has increased an average of 2.1 percent a year, which is significantly less.) Since 1971, China, the most populous nation in the world, has become agriculturally self-sustaining. Most of this success—in Mexico, India, China, and elsewhere—has come about as a result of improved varieties of crop plants, combined with better techniques of irrigation and fertilization.

Despite its acknowledged success, the Green Revolution has come under criticism in recent years. Modern agricultural ecosystems consist of huge stands of single species, an open invitation to insects, weeds, and disease organisms. Such opportunistic species can be kept at bay only by constant attention and the application of pesticides.

This susceptibility to invasion was tragically illustrated by the great potato famine of 1845–1847 in Ireland, which was caused by a water mold (see page 426). This famine was responsible for more than a million deaths by starvation and initiated large-scale emigration from Ireland. Within a decade, the population of Ireland dropped from 8 million to 4 million. Virtually the entire Irish potato crop was wiped out in a single week in the summer of 1846. A number of plant geneticists have warned that the new strains of wheat and rice, which have made major contributions toward feeding the growing human population, are particularly susceptible to similar disasters because of their genetic uniformity and widespread distribution.

A related criticism concerns the enormous amounts of fertilizer required by the new plant varieties in order to achieve their high yields—fertilizer that must be repeatedly applied as nutrients are removed from the ecosystem by the harvesting of the crops. Fertilizer, pesticides, farm equipment, and fuel for the equipment are all increasingly expensive. Because large landowners are able to afford the necessary investment, whereas small-scale farmers are not, the Green Revolution is seen as accelerating the consolidation of farm lands into a few large holdings by the very wealthy.

Another problem is that although food production is still outstripping population growth, increased agricultural productivity will not be able to keep pace indefinitely with the rapid growth of the world's population.

Finally, there is a more fundamental though more elusive reason for the dissatisfaction with the Green Revolution. When it first began, it appeared to many to be an almost magical solution to problems so enormous and distressing that they had seemed insoluble. It is now clear, however, that poverty and famine—and the unrest and violence they may bring—will not be solved by a "technological fix." The Green Revolution must, of course, go forward. At the same time, we must recognize that the broader solutions are social, political, and ethical, involving not only the growth of crops but their distribution, not only the limiting of populations but the raising of living standards worldwide to tolerable levels.

When inorganic nutrients are released by the action of decomposers, they are immediately taken up again by the roots of plants. This has the effect of preventing leaching of nutrients from the soil by the frequent, and often torrential, rains. When forest is cleared to make way for farming, mineral nutrients are rapidly leached from the soil. The result is that it is often impossible to grow a second crop after the first has been harvested.

Concentration of Elements

The elements needed by living organisms often are present in their tissues in higher concentrations than in the surrounding air, soil, or water. This concentration of elements comes about as a result of the selective uptake of substances by living cells, amplified by the channeling effects of food chains.

Under natural circumstances, the concentration of elements is usually valuable. Animals generally have a greater requirement for minerals than do plants because so much of the biomass of plants is cellulose. In some special cases, the effect of the accumulation may be quite dramatic. For example, the thousands of kilometers of coral reefs in the world's warmer waters are composed of calcium compounds formed from ions extracted from sea water over the millennia by one tiny polyp after another (see page 460).

Foreign substances can also get caught up in biogeochemical cycles and, as they are passed from one living organism to another, reach high concentrations as they approach the top of the food chain. The insecticide DDT is probably the best known of the toxic substances whose effects were amplified in this way (Figure 45–16).

Another was strontium-90, a radioactive element released into the atmosphere by aboveground nuclear testing in the 1950s. Strontium is closely related to calcium and can take its place in many biochemical reactions. Strontium-90 from the fallout made its way through grasses into dairy cows and into milk; from there it became concentrated in the bones and teeth of children. By 1959 the bones of children in North America and Europe contained about six times as much strontium-90 as the bones of adults. Although the level of radioactivity present has not been proved dangerous, exposure to radioactive elements is known to cause leukemia, bone cancers, and genetic abnormalities and generally to shorten the life span. And, most important, the minimum exposure that can produce such effects is not established. Since the half-life of strontium-90 is 28 years (in other words, it takes 28 years for half of the atoms present to lose their radioactivity), the exposure continues.

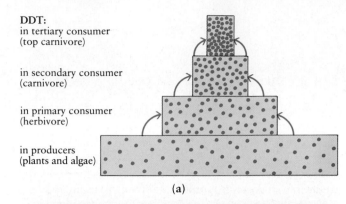

DDT:
in tertiary consumer
(top carnivore)

in secondary consumer
(carnivore)

in primary consumer
(herbivore)

in producers
(plants and algae)

(a)

(b)

45–16   (a) *Concentration of DDT residues being passed along a simple food chain. As organic matter is transferred from one level to the next in the chain, about 90 percent is usually respired or excreted. Only the remaining 10 percent forms new biomass. The losses of DDT residue, however, are small in proportion to the loss of other organic matter through respiration and excretion. Consequently, the concentration of DDT increases as the material passes up the chain, and high concentrations occur in the carnivores.*

(b) *This dead peregrine falcon embryo, its development almost completed, was found with two broken, infertile eggs in a nest in southwestern Scotland in May of 1971. When its body was analyzed, it was found to contain one of the highest concentrations of DDT residue ever recorded in a bird in Great Britain. It is not known if the developing falcon died as a direct consequence of the levels of DDT residue in its body or as a result of the collapse of its shell. DDT causes a bird's liver to break down the hormones that, in females, mobilize calcium at the time of egg production. The result is thin-shelled, fragile eggs. Birds at the top of food chains, such as the peregrine falcon, the osprey, and the bald eagle, were the principal victims. With the banning of DDT in many countries, dramatic recoveries have been observed in populations of all three species.*

The channeling of strontium-90 along the food chain could have been anticipated. Other consequences of the atomic tests were less predictable. For example, in the Arctic only slight radiation exposure was expected because the amount of fallout that reaches the ground at the poles is much less than it is in the temperate zones.

However, the Inuit peoples in the Arctic were discovered to have concentrations of radioactivity in their bodies that were much higher than those found in the inhabitants of temperate regions. The key link in the chain was the lichens. Lichens, which obtain their minerals largely from the rain, had absorbed a large amount of fallout material, little of which had time to decay and none of which was dissipated by absorption into the soil. In the winter, caribou live almost exclusively on lichens, and, at the top of the food chain, the Inuit live largely on caribou.

The Lessons of Chernobyl

In the early hours of April 26, 1986, the operators of a nuclear power plant at Chernobyl in what was then the Soviet Union began a series of tests of the operating capabilities of their newest reactor. According to official Soviet reports, without authorization the operators overrode the safety systems and deviated from standard procedures in order to conduct their tests. At some point, they lost control of the reactions occurring within the core of the reactor. Its output suddenly rose to 120 times its rated capacity, rupturing portions of the core. The cooling water in the system was instantly turned to steam, producing a powerful explosion that blew the 1,000-ton reactor cover aside and spewed hot nuclear fuel into the sky. The reactor was engulfed in fires that required more than 10 days to bring under control.

During this period, an enormous quantity of radioactive material was released into the environment—more than 1,000 times the amount released in the 1979 accident at Three Mile Island, near Harrisburg, Pennsylvania. A principal component of the material released in the Chernobyl accident was cesium-137, with a half-life of 30 years.

The consequences of the accident were most severe in the areas closest to Chernobyl, which included parts of what are now Belarus, Russia, and Ukraine (where Chernobyl is located). According to one unofficial estimate, as many as 10,000 people may have died in the five years following the accident, as a direct result of the radiation exposure. In 1986, some 135,000 people were evacuated from the most heavily contaminated areas, and in 1991 another 200,000 were still awaiting evacuation from other areas with significant contamination.

The radioactive cloud from the accident was moved in a northwesterly direction by the prevailing winds, and, when the rains subsequently came, the radioactive material fell back to the Earth. The fallout reached far beyond the Soviet Union, ultimately affecting some 100 million people in more than 20 European countries.

A substantial portion of the fallout was deposited

45–17 *In the winter of 1986–1987, as in other winters, the Sami peoples of Norway rounded up their reindeer herds in preparation for the annual slaughter—a process that previously provided them not only with meat for their own use but also with income for the next year. In central Norway, which was in the path of the radioactive cloud from the Chernobyl disaster, the levels of radioactivity in the reindeer averaged more than 10 times the level allowed in meat that is to be marketed. In some cases, the levels were more than 20 times the allowed limit. The contaminated carcasses were buried in large pits—in effect, nuclear waste disposal sites—in uninhabited regions in the far north. Although the Norwegian government has provided compensation for the financial losses, preventing economic collapse, the Sami people face an interval of some 20 to 30 years before they can hope to use and sell the milk and meat of their reindeer once more.*

more than 2,000 kilometers away in Norway, a country with no nuclear power plants. In the alpine tundra of Norway, the radioactive cesium-137 passed from rainwater to lichens and then to reindeer, in which its concentration increased to levels far exceeding those considered safe for human consumption. The highest concentrations occurred in the milk, muscles, and bones of the reindeer—the traditional subsistence for the Sami, or Lapp, peoples of central and southern Norway (Figure 45–17).

The accident at Chernobyl and its aftermath remind us of several important lessons. The first, and most obvious, is that the biological concentration of elements is a very real phenomenon with potentially severe consequences, especially for organisms at the top of the food chain, including ourselves. The second is that we dare not become complacent about the safeguards surrounding the use of potentially hazardous materials or technologies. The third, and perhaps most important, lesson is that the consequences of our misadventures do not respect international borders or local environmental regulations, no matter how well conceived or faithfully followed. We—and all other living organisms—are interconnected in one global ecosystem.

Summary

An ecosystem is a unit of biological organization made up of all the organisms in a given area and the environment in which they live. It is characterized by interactions between the living (biotic) and nonliving (abiotic) components that result in (1) a one-way flow of energy from the sun through autotrophs to heterotrophs, and (2) a cycling of mineral elements and other inorganic materials.

The ultimate source of energy for most ecosystems is the sun. Of the solar energy reaching the outer limits of the atmosphere, approximately half is reflected from or absorbed by the atmosphere. The distribution of the solar energy that reaches the Earth's surface is affected by the position and movements of the Earth in relation to the sun and by the movements of air and water over the surface. These factors can cause wide differences in temperature and rainfall from place to place and from season to season.

Within an ecosystem, there are at least three trophic (feeding) levels: primary producers, which are usually plants or algae; primary consumers, which are usually animals; and detritivores, which live on animal wastes and dead plant and animal tissues. The primary producers (the autotrophs) convert a small proportion (over 1 percent) of the sun's energy into chemical energy. The primary consumers (herbivores) eat the primary producers. A carnivore that eats a herbivore is a secondary consumer, and so on.

On average, about 10 percent of the energy transferred at each trophic level is stored in the body tissue of the consumers. Of the remaining 90 percent, part is used in the metabolism of the consumers and part is unassimilated. This unassimilated energy is ultimately utilized by detritivores.

The movements of water, carbon, nitrogen, and mineral elements through ecosystems are known as biogeochemical cycles. In such cycles, inorganic materials from the air, water, or soil are taken up by primary producers, passed on to consumers, and eventually transferred to the detritivores known as decomposers—bacteria and fungi. In the course of their metabolism, decomposers release the inorganic materials to the soil or water in a form in which they can again be taken up by primary producers.

The nitrogen cycle is of critical importance to all organisms. It involves three principal stages: ammonification, the breakdown of nitrogenous organic compounds to ammonia or ammonium ion; nitrification, the oxidation of ammonia or ammonium to nitrates, which are taken up by plants; and assimilation, the conversion of nitrates to ammonia and its incorporation into organic compounds. Nitrogen-containing organic compounds are eventually returned to the soil or water, completing the cycle. Nitrogen lost from the ecosystem is replaced by nitrogen fixation, the incorporation of elemental nitrogen into organic compounds.

In forest ecosystems, the principal reservoir of mineral elements is the vegetation. When a forest is stripped of its vegetation, the loss of mineral nutrients from the ecosystem increases dramatically.

Synthetic chemicals or radioactive elements released into the environment can be caught up in biogeochemical cycles, becoming concentrated in the tissues of organisms at higher trophic levels.

Questions

1. Distinguish among the following: community/ecosystem; biotic/abiotic; food chain/food web; gross productivity/net productivity; producer/consumer/detritivore/decomposer; ammonification/nitrification/assimilation/denitrification.

2. Describe what happens to the light energy striking a temperate forest ecosystem. What happens when it strikes a cornfield? A pond? A field on which cattle are grazing?

3. Describe what happens to a mineral nutrient in each of the ecosystems in Question 2.

4. Arrange the following ecosystems in order of decreasing net primary production: coral reef, cultivated land, desert, estuaries, open ocean, temperate forest, temperate grassland, tropical forest, and tropical grassland. What environmental factors have the greatest influence on the net primary production in terrestrial ecosystems? In marine ecosystems?

5. Consider each of the organisms below and list the effects of each on its ecosystem. Consider how the organism receives its inputs of energy and nutrients, where its outputs

(metabolic wastes, offspring, dead carcasses) go, and its effects on other organisms.

(a) Earthworm

(b) Heterotrophic soil bacterium

(c) Oak tree or grass plant

(d) Deer or grasshopper

(e) Lion or wolf

6. Explain the different kinds of information provided by a pyramid of numbers, a pyramid of biomass, and a pyramid of energy flow. For what purposes might each type be more useful than the other two?

7. Among the highest energy-transfer efficiencies known occur when reptiles consume warm-blooded prey, such as birds or small mammals. Explain, in terms of characteristics of both prey and predator, why a high energy-transfer efficiency would be expected for this particular step in a food chain.

8. Although a carnivore at the top of a food chain is free of visible predation, it is—during its lifetime—a source of energy for many different species, representing as many as four kingdoms. Explain.

9. What are the implications for mineral cycling of the human practices of fertilization of land and harvesting of crops? How are these implications different for nutrients whose major inorganic reservoir is the atmosphere rather than the soil?

Suggestions for Further Reading

Barringer, Felicity: "Chernobyl: Five Years Later the Danger Persists," *The New York Times Magazine*, April 14, 1991, pages 28–39, 74.

Carson, Rachel: *Silent Spring*, Houghton Mifflin Company, Boston, 1987.*

This is the book, originally published in 1962, that awakened the nation to the dangers of pesticides. Once seen as highly controversial, it is now regarded as a classic.

Childress, James J., Horst Felbeck, and George N. Somero: "Symbiosis in the Deep Sea," *Scientific American*, May 1987, pages 114–120.

Clark, William C.: "Managing Planet Earth," *Scientific American*, September 1989, pages 46–54.

Colinvaux, Paul: *Why Big Fierce Animals Are Rare*, Princeton University Press, Princeton, N.J., 1978.*

A series of essays for the general reader.

Crosson, Pierre R., and Norman J. Rosenberg: "Strategies for Agriculture," *Scientific American*, September 1989, pages 128–135.

Firor, John: *The Changing Atmosphere: A Global Challenge*, Yale University Press, New Haven, Conn., 1992.*

This short book, by the director of the Advanced Study Program at the National Center for Atmospheric Research, provides a clear, accessible, and nonsensational introduction to the problems of acid rain, ozone depletion, and global warming. The author considers not only the origins of these problems but also practical approaches to their solution. Highly recommended.

Gosz, J. R., R. T. Holmes, G. E. Likens, and F. H. Bormann: "The Flow of Energy in a Forest Ecosystem," *Scientific American*, March 1978, pages 92–102.

Graedel, Thomas E., and Paul J. Crutzen: "The Changing Atmosphere," *Scientific American*, September 1989, pages 58–68.

Lappé, Frances Moore: *Diet for a Small Planet*, Ballantine Books, Inc., New York, 1991.*

In this classic book, first published in 1971, Lappé established the feasibility of eating "low on the food chain," thereby greatly increasing the availability of proteins to other human populations.

Nichol, Stephen, and William de la Mare: "Ecosystem Management and the Antarctic Krill," *American Scientist*, vol. 81, pages 36–47, 1993.

Power, J. F., and R. F. Follett: "Monoculture," *Scientific American*, March 1987, pages 78–86.

Reganold, John P., Robert I. Papendick, and James F. Parr: "Sustainable Agriculture," *Scientific American*, June 1990, pages 112–120.

Rowland, F. Sherwood: "Chlorofluorocarbons and the Depletion of Stratospheric Ozone," *American Scientist*, vol. 77, pages 36–45, 1989.

Simons, Ted, Steve K. Sherrod, Michael W. Collopy, and M. Alan Jenkins: "Restoring the Bald Eagle," *American Scientist*, vol. 76, pages 252–260, 1988.

Stephens, Sharon: "Lapp Life after Chernobyl," *Natural History*, December 1987, pages 32–41.

Stolarski, Richard S.: "The Antarctic Ozone Hole," *Scientific American*, January 1988, pages 30–36.

Toon, Owen B., and Richard P. Turco: "Polar Stratospheric Clouds and Ozone Depletion," *Scientific American*, June 1991, pages 68–75.

Tunnicliffe, Verena: "Hydrothermal-Vent Communities of the Deep Sea," *American Scientist*, vol. 80, pages 336–349, 1992.

* Available in paperback.

CHAPTER

46

The Biosphere

The **biosphere,** that part of the Earth in which life exists, is only a thin film on the surface of this little planet. It extends about 8 to 10 kilometers above sea level and a few meters down into the soil, as far as roots penetrate and microorganisms are found. It includes the surface waters and the ocean depths. The biosphere is patchy, differing in both the depth and the density of life.

In this, the final chapter of our text, we shall survey the marvelously diverse communities of organisms that are found in different regions of the biosphere.

Life in the Waters

Life began in water, and although living organisms have long since conquered the land, by far the largest proportion of the biosphere consists of aquatic environments and their inhabitants. Freshwater environments can be conveniently classified as running water (rivers and streams) and standing water (lakes and ponds). Marine environments can be classified as oceans and seashores. Within these broad categories, there are a diversity of habitats, each with its own characteristic complement of organisms.

Rivers and Streams

Rivers and streams are characterized by continuously moving water. They may begin as outlets of ponds or lakes, as runoffs from melting ice or snow, or they may arise from springs (flows of groundwater emerging from below the surface of the land).

The character of life in a stream is determined to a large extent by the swiftness of the current, which generally changes as a stream moves downward from its source and, fed by tributaries, increases in volume. In swift streams, most organisms live in the riffles, or shallows, where small photosynthetic organisms—algae and mosses—cling to the rock surfaces. Many insects, both adult and immature forms, live on the underside of rocks and gravel in the riffles. For those small organ-

46–1 *A fast-moving stream in Michigan. The boulders within and beside the stream are constantly wetted by spray, making possible the luxuriant growth of mosses on their surfaces. What is a probable explanation for the total absence of mosses on some of the smaller boulders?*

46–2 *Some freshwater environments and their inhabitants. (a) A pond in the Pocono Mountains of Pennsylvania, covered with water lilies. (b) A great blue heron, with a bass that it has just caught. (c) A hungry muskrat.*

isms that can survive the swiftly moving current, there is an abundance of oxygen and nutrients swept along by the flowing waters.

As the stream travels along its course, the riffles are often interrupted by quieter pools, where organic materials may collect and be decomposed. Few plants can gain footholds on the shifting bottoms of stream pools, but some invertebrates, such as dragonflies and water striders, are typically found in or about the pools. Some organisms, notably trout, move back and forth between the riffles and the pools.

As streams broaden, they generally include peripheral habitats that have many of the characteristics of lakes and ponds.

Lakes and Ponds

Bodies of standing water vary in size from small ponds to very large lakes, covering thousands of square kilometers. They contain three distinct zones: littoral, limnetic, and profundal.

The **littoral zone,** at the edge of the lake, is the most richly inhabited. Here the most conspicuous plants are angiosperms, such as cattails and rushes, rooted to the bottom. Water lilies may grow farther out from the shore. There is often a green blanket of duckweed, a small, free-floating angiosperm. Other pond weeds grow entirely beneath the water. These plants lack a waxy outer cuticle and so can absorb minerals through their epidermis as well as through their roots. The tangle of submerged plant surfaces harbors large numbers of small organisms.

Snails, small arthropods, and mosquito larvae feed upon the plants and algae. Other insects that live among the submerged plants, such as the larvae of the dragonfly and damselfly, and the water scorpion, are carnivorous. Clams, worms, and the larvae of other insects

(a)

(b)

(c)

burrow in the mud. Frogs, salamanders, turtles, and water snakes are found almost exclusively in the littoral zone. The tadpoles of frogs and salamanders feed on algae and insect larvae, and the adults on insects. Most aquatic turtles feed on insects when young and on plants when fully grown. Water snakes are carnivores, feasting on amphibians and fishes, such as largemouth bass and bluegills, which are found in greater numbers along the lake margins. Ducks, geese, and herons feed on the plants, insects, mollusks, fishes, and amphibians abundant in this zone.

The shallow margins of some lakes and ponds are marshy. Among the inhabitants of the marshes are such animals as snails, frogs, ducks, herons, bitterns, muskrats, otters, and beavers.

In the **limnetic zone,** the zone of open water, small, floating algae—phytoplankton—are usually the dominant photosynthetic organisms. This zone, which extends down to the limits of light penetration, is the habitat, for example, of walleye, yellow perch, and, in colder waters, trout.

The deepwater **profundal zone,** extending down from the limnetic zone, has no plant life. Its principal occupants are detritivores, such as scavenging fishes, aquatic worms, insect larvae, crustaceans, fungi, and bacteria, that consume the organic debris filtering down from the overlying water. The nutrients released by bottom-dwelling decomposers are recirculated to the upper levels of the lake by the twice-yearly overturn of the water (see page 43).

The Oceans

The oceans cover almost three-fourths of the surface of the Earth. Although life extends to their deepest portions, photosynthesizing organisms are restricted to the upper, lighted zones. The sea has an average depth of nearly 4 kilometers and, except for a relatively shallow surface layer, is dark and cold. Most of its volume, therefore, is inhabited by bacteria, fungi, and animals, rather than by plants.

Sea water absorbs light readily. Even in clear water, less than 40 percent of the sunlight striking the surface reaches a depth of 1 meter, and less than 1 percent penetrates below 50 meters. Red, orange, and yellow wavelengths of light are absorbed first, so that only the shorter (higher energy) wavelengths, specifically blue and green, penetrate deeply. Thus, below depths of a few meters, only those photosynthetic organisms capable of utilizing the shorter wavelengths of light can grow.

There are two main divisions of life in the open ocean: **pelagic** (free-floating) and **benthic** (bottom-dwelling). A major component of the pelagic division is plankton. It is composed of rapidly dividing algae, intermingled with heterotrophic protists, small shrimp and other crustaceans, gelatinous invertebrates (such as jellyfish), and the eggs and larval forms of many fishes and invertebrates. These planktonic forms provide food for fishes and other relatively large pelagic animals, including some whales.

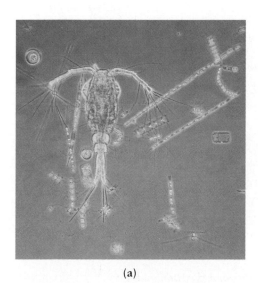

(a)

(b)

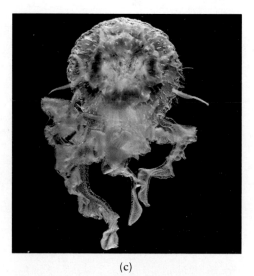

(c)

46–3 *Some ocean inhabitants.* (a) *This mixture of phytoplankton and zooplankton was photographed in the waters east of the Antarctic Peninsula. Planktonic organisms* form the base of the food webs in the pelagic division of the sea. (b) A school of marine catfish, photographed off the Philippine Islands. (c) A purple stinger jellyfish (Pelagia noctiluca), swimming off Heron Island, Australia.

The benthic division contains the sessile animals, such as sponges, sea anemones, and clams, and many motile animals, such as worms, starfish, snails, crustaceans, and fishes. A variety of fungi and bacteria also inhabit the benthic zone, subsisting on the accumulation of debris steadily drifting down from the more populated levels of the ocean.

Despite the fact that oceans cover three times more surface area of the planet than does the land, the total productivity of the open ocean—as measured by the amount of carbon converted to organic compounds by photosynthesis—is only about one-third as great. In fact, the open ocean is thought to be only slightly more productive per square meter than the desert (see Table 45–1, page 823), presumably because of the low concentration of mineral nutrients in those zones of the ocean where light penetration allows photosynthetic organisms to survive.

The major ocean currents, which are produced by a combination of winds and the Earth's rotation, profoundly affect life in the oceans and alter the climate along the ocean coasts (see essay). These patterns of water circulation—clockwise in the Northern Hemisphere and counterclockwise in the Southern Hemisphere—move currents of warm water north and south from the Equator (Figure 46–4). One such current, the Gulf Stream, warms a portion of the eastern coast of North America and the western shores of Europe, and another warms the eastern coast of South America. The same patterns of circulation bring cold waters to the western coasts of North and South America.

In regions where the winds move the surface water continuously away from the shores, as off the coasts of California, Portugal, and Peru, cold water rich in nutrients is brought to the surface from below, a process referred to as **upwelling.** Such areas contain high densities of pelagic life and traditionally have supported highly profitable fishing industries.

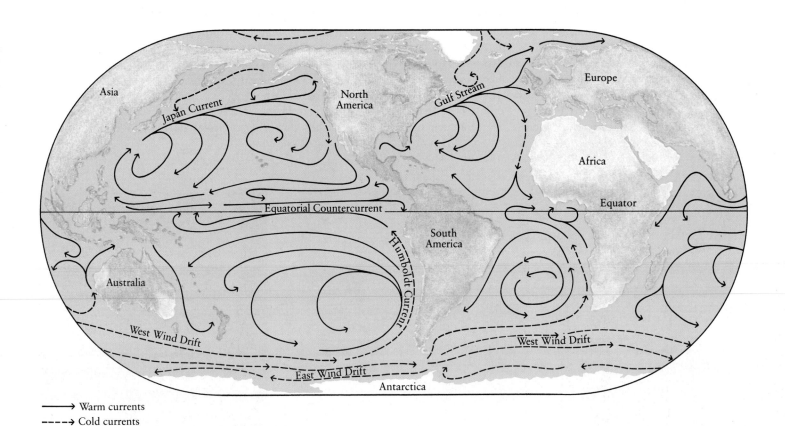

→ Warm currents

---→ Cold currents

46–4 *The major currents of the ocean have profound effects on climate. Because of the warming effects of the Gulf Stream, Europe is milder in temperature than is North America at similar latitudes. The eastern coast of South America is warmed by water from the Equator, and the Humboldt Current brings cooler weather to the western coast of South America.*

El Niño

The importance of ocean currents to weather patterns and to both marine and terrestrial organisms is dramatically illustrated by the disastrous consequences that can follow disturbance of the normal patterns of water circulation. The most frequent disturbance occurs in the Pacific Ocean and is known as El Niño (Spanish for "the Child") because its effects are usually first felt in South America around Christmastime. Major El Niño events occur approximately every 8 to 10 years, and smaller ones every 2 to 3 years. The most severe El Niño in the past century occurred between the fall of 1982 and the summer of 1983.

Normally, the patterns of weather and ocean currents in the Pacific are determined by a large high-pressure system in the eastern Pacific and a large low-pressure system over Indonesia. Air moves from the region of high pressure to the region of low pressure, creating winds that blow from east to west. These winds move warm surface water toward the western Pacific, and cold water moves in along the western coasts of North and South America as an undercurrent, as shown in Figure 46-4.

An El Niño is apparently triggered by a rise in the surface temperature of equatorial regions of the eastern Pacific, causing changes in the location and strength of the atmospheric pressure systems. These changes, in turn, cause a reversal in the wind direction and an accompanying reversal in the surface current. Over a period of two to three months, warm water surges back toward North and South America, trapping the cold undercurrent well below the surface. With higher water temperatures and a cutoff of nutrient upwelling, planktonic organisms die off—followed by the fish that feed on the plankton, followed by the seabirds that feed on the fish.

The 1982–1983 El Niño led to a collapse of the plankton-fish-seabird food chain and thus a collapse of the fishing industries in Peru, Ecuador, and the Pacific Northwest. It also apparently disrupted weather patterns in many parts of the world. Massive rainstorms battered the usually dry coasts of California, Ecuador, and Peru that winter, and record snows fell in the Sierra Nevada and Rocky Mountains, followed by record flooding in the spring. Indonesia, the Philippines, India, Australia, Mexico, and southern Africa were stricken with droughts that destroyed vegetation and millions upon millions of animals, both domestic and wild, that had neither food nor water.

The complex, interacting events occurring in an El Niño are still poorly understood, but they appear to be part of a natural, self-sustaining cycle of warming and cooling. In a 10-year international study begun in 1985, satellites, ships, and ground stations across the Pacific are continuously monitoring air pressure, water and air temperatures, and the direction and velocity of winds and water currents. This study made possible the accurate prediction, a year in advance, of a moderate El Niño in 1986–1987 and of another, somewhat stronger El Niño that occurred in 1991–1992. With each succeeding El Niño, our knowledge of the forces involved increases, as does our understanding of the pattern that occurs as an El Niño becomes established and then declines.

The power of the oceans and the atmosphere is far beyond our meager capacity to control. However, better understanding of El Niños may make possible more accurate predictions and thus action to minimize their impact on both human activities and natural ecosystems.

The Seashore

Depending on the specific area, the submerged edges of the continents extend anywhere from 10 to several hundred kilometers out into the sea. Along these shallow borders, known as the **continental shelves,** nutrients are washed out from the land, and life is much denser than in the open seas. In temperate latitudes at the edge of the sea, where the large primary producers are brown algae (such as the rockweed *Fucus* and the kelp *Laminaria),* net primary productivity is as high as anywhere else in the biosphere. Primary productivity is also extremely high in the coral reefs (page 460) of tropical regions.

Sessile animals, such as sponges and anemones, are found all over the ocean bottom, but they are most abundant in relatively shallow areas that are close to the shores. Predators, such as mollusks, echinoderms, crustaceans, and many kinds of fishes, roam over the bottoms of the continental shelves. Eel grass, turtle grass, and seaweeds provide shelter for many animals, increase the supply of oxygen, and produce large amounts of detritus that forms the base of many food webs. Snails, sea slugs, and worms crawl over the surfaces of the plants and algae, feeding as they go.

Seashores—where the sea and the land join—are of three general types along most of the shores of the temperate zones: rocky, sandy, and muddy. Organisms that live on rocky coasts, like those that live in the riffles of fast-moving streams, often have adaptations for clinging to rocks. The algae have strong holdfasts. The starfishes of the rocky coast lie spread-eagled on the rocks,

(a)

(b)

(c)

(d)

(e)

46–5   *Some examples of life at the seashore.* (a) *Railroad vine, photographed on a Georgia beach, plays a role in stabilizing the shifting sand.* (b) *Sanderlings, seen here basking in the winter sun in Santa Cruz, California, are familiar residents of both the Atlantic and Pacific coasts of North America. When not sleeping, they race along the tide line, snatching up mollusks and crustaceans exposed as the waves retreat.* (c) *Sally lightfoot crabs are common inhabitants of sandy tropical beaches. Also known as red lava crabs, they run sideways at high speeds across the sand and are even able to skip across short stretches of water.* (d) *Among the inhabitants of this rocky tidal pool on Rialto Beach, Washington, are kelp, sea anemones, and starfish.* (e) *California sea lions, sunning themselves after a swim.*

clinging with their tube feet. The abalone holds tight with its well-developed muscular foot. Mussels secrete coarse, ropelike strands that anchor them to rocky surfaces.

The organisms of the rocky coast face the additional problem of the rising and falling tides. The supratidal zone, which is wetted only by waves and spray, is a zone of dark algal and lichen growth. The intertidal zone, alternately submerged and exposed by the tides, is commonly characterized by *Fucus,* often intermixed with many other species of brown and red algae. Animal life includes barnacles, oysters, mussels, limpets, and periwinkles. The subtidal zone, always submerged, often contains forests of kelp *(Laminaria* and *Macrocystis),* sea squirts, starfish, and other invertebrates. This zonation of the organisms of rocky shores is due, in part, to gradients of light, temperature, and wave action, partly to competitive interactions, and also to predation by both herbivores, such as sea urchins, and carnivores, such as starfish.

Sandy beaches have fewer inhabitants because of the constantly shifting sands. Clams, ghost crabs, sand fleas, lugworms, and other small invertebrates live below the surface of the sand, feeding on the debris washed in and out by the tide. Beach grasses, which spread by means of underground stems, are important for stabilizing the shifting dunes behind the sandy beaches.

The mud flat, while not so rich in species as the rocky coast, supports a large number of organisms, with many animals living not only on but also beneath its surface. A mud flat can support tens of thousands of individuals per cubic meter.

Mud flats, salt marshes, and **estuaries** (areas where the fresh water of streams and rivers drains into the sea) are the receiving grounds for a constant flow of nutrients drained off from the land. As a consequence, they are extremely rich in animal life. They function as the spawning places and nurseries for many forms of marine life, including many commercially important species of fish and crustaceans. In the tropics and subtropics (including parts of Florida, Puerto Rico, and Hawaii), mangrove forests are important tideland communities, serving as spawning grounds for marine organisms and exporting minerals and nutrients to the sea.

Mud flats, salt marshes, and mangrove forests are often located in prime recreational and commercial areas. Because they cannot be directly exploited for agriculture or lumbering, each year thousands of square kilometers of these wetlands are destroyed as they are filled and paved for human habitation—and thereby rendered sterile for many other species. Their protection is of special importance because of their role in nurturing life in the oceans.

(a)

(b)

46–6 (a) *A salt marsh on the North Carolina coast. The bright green areas, which are flooded at every high tide, are primarily the marsh grass* Spartina. *The dark areas, which are* flooded only during the highest high tides, consist principally of black needlerush. (b) A snowy egret in a salt marsh in New Jersey in the autumn.

Life on the Land

The characteristic patterns of life on the land are determined principally by physical factors. Of the most immediate consequence to terrestrial organisms are temperature and precipitation (Figure 46–7). As we saw in the last chapter, these critical factors are influenced by the angle of the Earth's axis in relation to its orbit around the sun and by the Earth's rotation, which affect the duration and intensity of sunlight, the prevailing wind patterns, and the major ocean currents. Temperature and precipitation are also affected by the structure of the continents themselves.

The surfaces of the continents change slowly but constantly. They are crumpled by contractions and collisions as the continents rise, sink, and collide because of the motion of the plates on which they are carried (page 358). As a consequence, the Earth's surface is not at all uniform but varies widely from place to place in its composition and in its height above sea level. The mineral content of the Earth's surface is a major factor affecting the growth of plants and other living organisms, and, as we have seen (page 819), the mountain ranges of the continents do much to determine the patterns of rainfall.

Average temperatures decrease about 0.5°C for each degree of increase in latitude. A similar effect occurs with each rise of about 100 meters in elevation. One consequence of this relationship is that plants and animals characteristic of Arctic regions may also be found in mountain ranges near the Equator (Figure 46–8).

The Concept of the Biome

The land surface of the Earth can be seen as divided into a number of geographic areas distinguished by particular types of dominant vegetation. Each major continent has, for example, deserts, grasslands, and deciduous forests. These categories of characteristic plant life are called **biomes.**

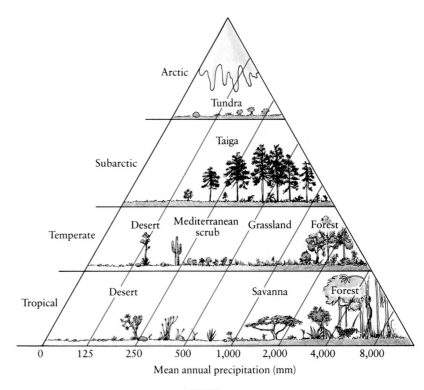

46–7   *The characteristic patterns of plant life in different regions of the biosphere are determined principally by temperature, which decreases with increasing latitude, and precipitation.*

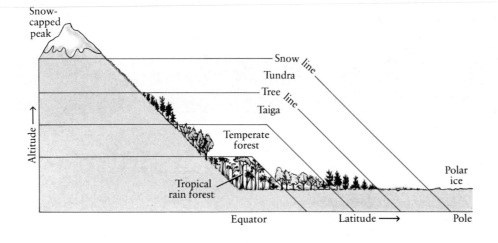

46–8   *In the Northern Hemisphere, we can experience a similar sequence of dominant plant life and its associated animal life by either traveling north for hundreds of kilometers or ascending a mountain. To experience a similar sequence in the Southern Hemisphere, we could ascend a mountain. However, by simply traveling south, we would never encounter vegetation corresponding to the taiga and the tundra of the Northern Hemisphere. Can you explain why?*

Acid Rain and Forest Decline

The average pH of normal rainfall is about 5.6 (mildly acidic), a result of the combination of carbon dioxide with water vapor to produce carbonic acid. In the 1920s, however, the pH of rain and snow in Scandinavia began to drop, and by the 1950s, the pH had begun to drop elsewhere in Europe and in the northeastern United States. It was subsequently learned that, in certain geographic areas, the average annual pH of precipitation is between 4.0 and 4.5. Occasional storms release rain with a pH as low as 2.1, which is extremely acidic.

The low pH results primarily from two acids in the rainfall: sulfuric acid and nitric acid, both of which ionize almost completely in water, releasing hydrogen ions. These acids are formed when the gaseous oxides of sulfur and nitrogen react with water vapor and other gases in the air. Sulfur and nitrogen oxides are released into the atmosphere by some natural processes, but far greater quantities are released as a result of human activities. Sulfur oxides are produced by the combustion of high-sulfur coal and oil and by the smelting of sulfur-containing ores. Nitrogen oxides are byproducts of gasoline combustion and of some generating processes for electricity.

That sulfur oxides could be damaging to vegetation was evident in the early 1900s, when a large copper smelter was opened in a mountainous area of Tennessee. Within a few years, all vegetation had been killed in the formerly luxuriant forest surrounding the smelter. The solution devised for the problem—still used today—was to build very tall smokestacks, so the wind would carry the pollutants away from the immediate area. It was assumed, incorrectly as it turned out, that the pollutants would be so widely dispersed that they would be rendered harmless.

By the 1960s, it was clear that sulfur oxides released from tall smokestacks are transported hundreds or thousands of miles by the prevailing winds and then returned to the Earth in rain and snow. Nitrogen oxides released from automobiles are also carried off by the wind. What was once a local problem has now become an international problem, in which the pollutants respect no boundaries.

The biological consequences of acid rain

(a)

(b)

(a) In the mountains of New England, red spruce trees are dying at a high rate. Only their skeletons remain, standing silent sentinel over the forest. Such scenes are also an increasingly familiar sight in the forests of central and northwestern Europe. (b) In a forest in Vermont, a student sets up an apparatus to collect rainwater that will be tested for acidity.

are numerous. In plants, it causes reduced germination of seeds, a decrease in the number of seedlings that mature, reduced growth, and lowered resistance to disease. Moreover, acid rain lowers the pH of soil, increasing the leaching of essential nutrients from the soil. Recently, it has become apparent that the forests of the eastern United States, from Maine through Georgia, are in serious decline. A dramatic slowing of growth has occurred during the past 20 years, and in some locations at high altitude, red spruce trees are dying in large numbers. Current evidence suggests that, among its other effects, acid rain reduces the cold tolerance of this species.

Acid rain also affects freshwater ecosystems, particularly lakes in mountainous regions. A 1977 study of the lakes at high elevations in the western Adirondack Mountains of New York found that 51 percent had a pH below 5.0, of which 90 percent were devoid of fish life. By contrast, a similar study carried out from 1929 to 1937 found that only 4 percent of the lakes were acidic and without fish. More recent studies indicate increasingly acidic conditions in lakes at lower levels in the Adirondacks (which were previously unaffected) and in many of the lakes in the Cas- cade Mountains in the Pacific Northwest. The effects of low pH on fish include the depletion of calcium in their bodies, leading to weakened and deformed bones; the failure of many eggs to hatch and deformed fish from those that do hatch; and the clogging of the gills by aluminum, which is released from the soil by acid.

The accumulating evidence indicates that acid rain is one of the more serious worldwide pollution problems confronting us today. The potential consequences of its effects on biological systems include lowered crop yields, decreased timber production, the need for greater amounts of fertilizer to compensate for nutrient losses, the loss of important freshwater fishing areas, and, possibly, of the eastern forests as well.

Scientists from many fields are engaged in research to gain a greater understanding of the causes and effects of acid rain and the likely consequences and costs of proposed solutions. As is the case with the other environmental problems we face, scientists can provide information on which decisions can be based, but the choices that lie ahead are essentially social and economic, to be made through political processes.

844

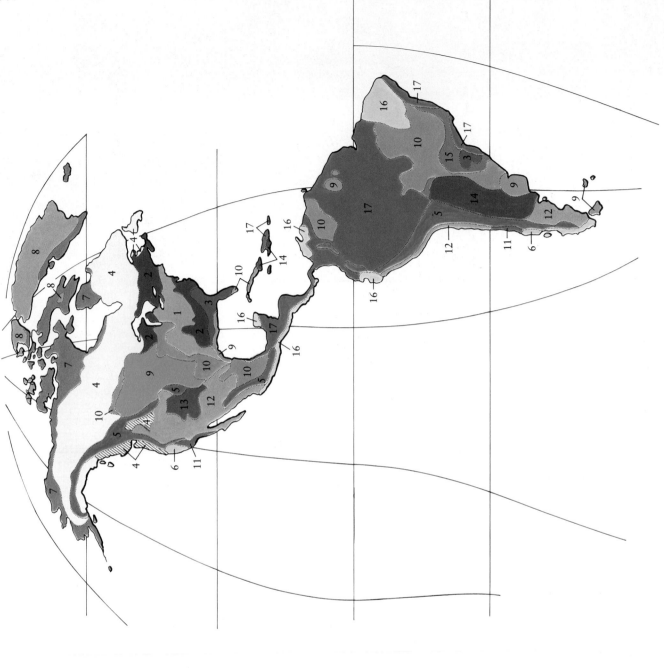

46–9 *Biomes of the world. The information in these maps and the accompanying
key was originally supplied by A. W.
Küchler of the University of Kansas, one of
the leading authorities on the distribution
of biomes. Because of the global coverage of
the maps, the scale is relatively small and the
content is generalized. Any given biome is
not always uniform, and all of the biomes include considerable variations in vegetation.
The boundaries between the biomes may be
sharp, but more often they are blurred, consisting of broad zones of transition from one
type of vegetation to another.*

1	Temperate deciduous forests
2	Temperate mixed forests
3	Subtropical mixed forests
4	Taiga
4	Northwestern coniferous forest
5	Mountain forests and alpine tundra
6	Mixed west-coast forests
7	Arctic tundra
8	Ice desert
9	Grasslands
10	Savannas
11	Mediterranean scrub
12	Deserts and semideserts
13	Juniper savanna
14	Southern woodland and scrub
15	Tropical mixed forests
16	Monsoon forests
17	Rain forests

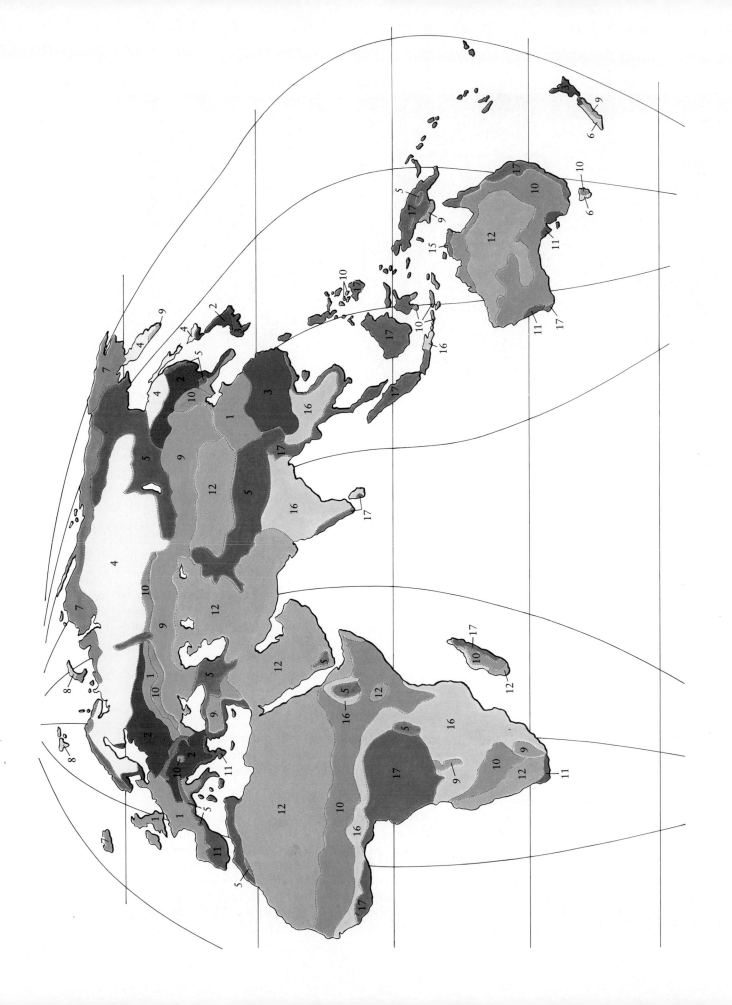

(a)

(b)

46–10   *The temperate deciduous forest of North America grades into temperate mixed forest in northern and upland regions and into subtropical mixed forest in the southeastern United States. (a) A beech and maple forest in Michigan, photographed in the spring. The forest floor is carpeted with large-flowered trillium (Trillium grandiflorum). (b) A beech and maple forest in the autumn. (c) Scarlet flycap mushrooms. The*

spots on the caps and the "skirt" near the top of each stalk identify these mushrooms as members of the deadly genus Amanita. Primary consumers in the forest include (d) chipmunks, and (e) white-tailed deer. (f) This red fox, which has just killed a cottontail rabbit, is a secondary consumer. All of these organisms are common inhabitants of the North American deciduous forest.

As Figure 46–9 on pages 844–845 shows, the communities of plants and associated animal life that make up a biome are discontinuous, but a community may closely resemble another community on the other side of the planet. For instance, many of the deserts of the world look remarkably similar. However, when you look more closely, you see that although the physical features of the environment—temperature and rainfall—are the same, the particular organisms are not. But they look and act alike. The same is true of the Mediterranean scrub of California and Spain, the grasslands of North and South America, and so on.

A biome is a class, or category, of similar communities—not a place. When we speak of the tropical forest biome, we are not speaking of one particular geographic region, but rather of all the tropical forests on the planet. As with most abstractions, important details are omitted. For example, the boundaries are not as sharp as shown on biome maps nor are all areas of the world easy to categorize. However, the biome concept emphasizes one important truth: Where the climate is the same, the organisms are also very similar, even

though they may not be related genetically and may be far apart in their evolutionary history. Organisms of geographically separate patches of the same biome provide many examples of convergent evolution (page 350).

Temperate Forests

Temperate forests once covered most of eastern North America, as well as most of Europe, part of Japan and Australia, and the tip of South America. In the United States, only scattered patches of the original forest remain.

Temperate deciduous forests occupy areas where there is a warm, mild growing season with moderate precipitation, followed by a colder period less suited to plant growth. Leaf-shedding in the deciduous forest evolved as a protection against water loss. (As you will recall, most of the water lost in transpiration escapes through the stomata of the leaves.) However, discarding leaves is expensive, in terms of both energy and nu-

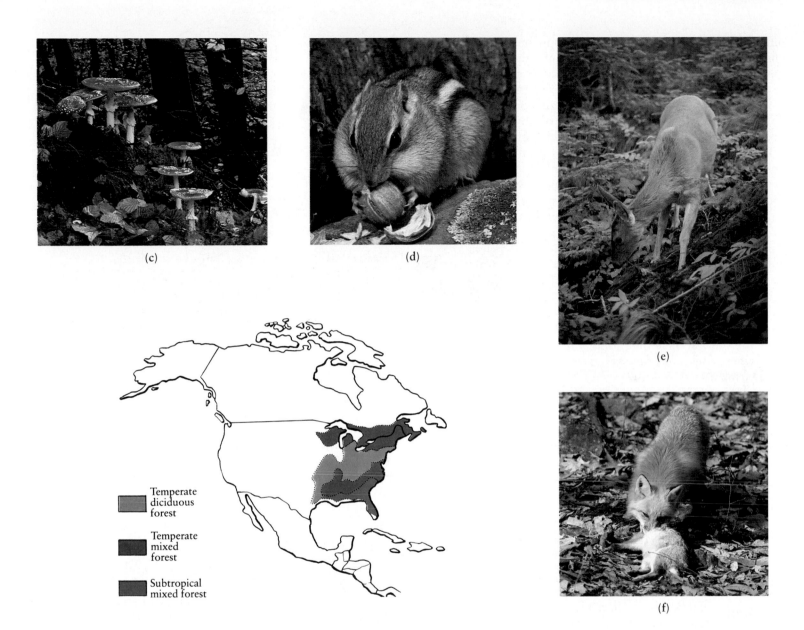

(c)

(d)

(e)

(f)

Temperate
diciduous
forest

Temperate
mixed
forest

Subtropical
mixed forest

trients. For the water-saving benefits of defoliation to exceed the costs in energy and lost nutrients, deciduous trees must have a growing season of at least four months.

In deciduous woodlands there are up to four layers of plant growth: (1) a tree layer, in which the crowns form a continuous canopy that is usually between 10 and 35 meters high; (2) a shrub layer, which grows to a height of about 5 meters; (3) a field layer, made up of ferns, grasses, and herbaceous plants, including annual flowering plants that typically bloom in the spring before the trees regain their leaves; and (4) a ground layer, which consists of mosses and liverworts and is often covered with leaf litter.

The dominant trees of temperate forests vary from region to region, depending largely on the local rainfall. In the northern and upland regions of the United States, oak, birch, beech, and maple are the most prominent trees. Maple and basswood predominate in Wisconsin and Minnesota, and maple and beech in southern Michigan, becoming mixed with hemlock and white pine as the deciduous forest gives way to mixed forest. The southern and lowland regions have forests of oak and hickory. Along the southeastern coast of the United States, the warm, seasonally wet climate supports a mixed subtropical forest of pine, oak, and magnolia.

Temperate forests support an abundance of animal life. In North America, smaller mammals, such as chipmunks, voles, squirrels, raccoons, opossums, and white-footed mice, live mainly on nuts and other fruits, mushrooms, and insects. Deer typically live on the forest borders, where they browse on shrubs and seedlings. Wolves, bobcats, gray foxes, and mountain lions, in the areas where they have not been driven out by the encroachments of civilization, feed on the other mammals.

Beneath the ground layer is the soil of the forest, often a rich, gray-brown topsoil. Such soil is composed largely of organic material—decomposing leaves and other plant parts and decaying insects and other animals—and the bacteria, protists, fungi, worms, and arthropods that live on this organic matter. The roots of plants penetrate the soil to depths measurable in meters and add organic matter to the soil when they die. Carnivorous arthropods carry fragments of their prey to considerable depths in the soil. The myriad passageways left by dead roots and fungi and by the earthworms and other small animals that inhabit the forest make the soil a sponge that holds water and nutrients. Land where deciduous forests have stood often makes good farmland.

Coniferous Forests

Most conifers are evergreens, with small, compact leaves protected against water loss by a thick cuticle. They usually cannot compete successfully with deciduous trees in temperate zones with adequate summer rainfall and rich soil. Coniferous forest biomes include the **taiga** (also known as the northern boreal forest), the northwestern coniferous forest (located in the Pacific Northwest of the United States and Canada), mountain forests, and the mixed west-coast forests found along portions of the coastlines of Chile and California.

The Taiga

The boundary between the temperate mixed forest and the taiga occurs where summers become too short and winters too long for deciduous trees to grow well. The taiga is characterized by long, severe winters and a constant cover of winter snow. It is composed chiefly of evergreen needle-leaved trees such as pine, fir, spruce, and hemlock. A thick layer of needles and dead twigs, matted together by fungal mycelia, covers the ground. Along the stream banks grow deciduous trees, such as tamarack, willow, birch, alder, and poplar. There are very few annual plants.

The principal large animals of the North American taiga are elk, moose, mule deer, black bears, and grizzlies. Among the smaller animals are porcupines, red-backed voles, snowshoe hares, wolverines, lynxes, warblers, and grouse. The small animals use the dense

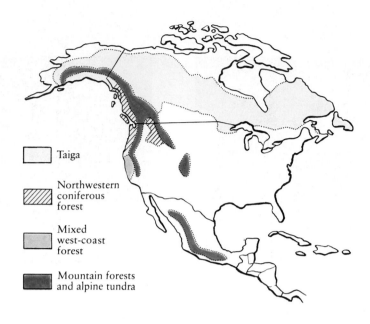

☐ Taiga

▨ Northwestern coniferous forest

▨ Mixed west-coast forest

■ Mountain forests and alpine tundra

growths of the evergreens for breeding and for shelter. Wolves feed upon the mammals, particularly the larger ones. The black bear and grizzly bear eat everything—leaves, buds, fruits, nuts, berries, fish, the supplies of campers, and occasionally the flesh of other mammals. Porcupines are bark eaters and may seriously damage trees by girdling them. Moose and mule deer are largely browsers.

The ground layer of the coniferous forest is less richly populated by invertebrates than that of the deciduous forest, and the accumulated litter is slower to decompose.

The Forests of the Pacific Northwest

The massive evergreen coniferous forests of the Pacific Northwest are adaptations to the winter-wet/summer-dry environment of that region. Because photosynthesis is limited by lack of moisture during the warm season, deciduous trees are at a disadvantage. They are usually found only along stream banks, where water is plentiful. The evergreen conifers, however, can synthesize carbohydrates all year round. Also, these trees, because of their massive size, can store water and nutrients for use during the dry season, and their thick barks and high crowns protect them from the fires characteristic of the region.

These forests, which were untouched by the Pleistocene glaciation but have been seriously disturbed by recent human activities—primarily logging—are among the most ancient in North America, in terms of both individuals and populations. Their fate, and that of the animals, such as the northern spotted owl, that live only in the old growth forest, is currently the subject of intense controversy.

(a)

(b)

(c)

(d)

(e)

(f)

46–11   *The coniferous forests of North America include the taiga, the northwestern coniferous forest, mixed west-coast forests, and mountain forests. (a) Woodland caribou, searching for food in a clearing in the taiga of Jasper National Park in Alberta, Canada. (b) The floor of Buena Vista Grove in Kings Canyon National Park in California is carpeted with the needles and cones of sweet pines. Decay is slower in the coniferous for-ests than on the warmer, wetter floors of the deciduous forest, and, because of the shade cast by the mature trees, there is no undergrowth.*

Coniferous forests support a diversity of animals, including (c) grizzly bears, (d) the northern spotted owls of the old growth forest of the Pacific Northwest, (e) moose, and (f) lynx and red squirrels.

(a)

(b)

(c)

(d)

(e)

46–12 (a) *Tundra of North America on a long Arctic day. Kettle holes, formed by chunks of glacial ice, are a common sight.* (b) *The Arctic tern is one of a number of bird species that breed in the tundra, taking advantage of the long summer days to gather food for their nestlings. The terns winter in the Antarctic, following migration routes of 13,000 to 18,000 kilometers. Within three months after hatching, the young must be ready to migrate.* (c) *A barren-ground caribou, photographed in Alaska in the autumn.* (d) *A snowy owl in its winter plumage.* (e) *An Arctic fox (foreground) that has ventured from the tundra onto the ice floes, where polar bears spend the winter.*

The Tundra

Where the climate is too cold and the winters too long even for conifers, coniferous forest grades into **tundra.** The Arctic tundra occupies one-fifth of the Earth's land surface, forming a continuous belt across northern North America, Europe, and Asia. Similar vegetation is found above the tree line in mountainous regions.

The most characteristic feature of the Arctic tundra is **permafrost,** a layer of permanently frozen subsoil. During the summer the ground thaws to a depth of a few centimeters and becomes wet and soggy; in winter it freezes again. This freeze-thaw process, which tears and crushes the roots, keeps the plants small and stunted. Drying winter winds and abrasive, driven snow further reduce the growth of tundra plants.

The virtually treeless vegetation of the tundra is dominated by herbaceous plants, such as grasses, sedges, and rushes, and woody shrubs like heather. Beneath these is a well-developed ground layer of mosses and lichens, particularly the lichen known as reindeer moss. The growing season in many areas of the tundra is less

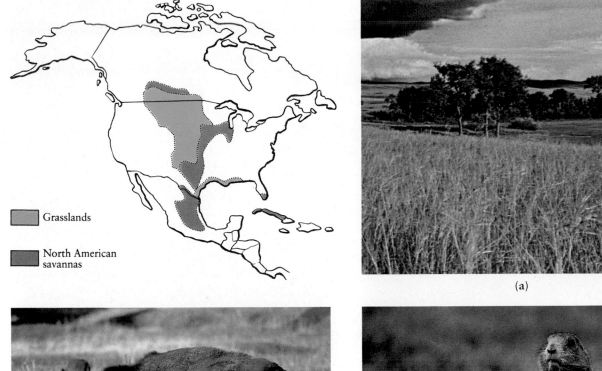

than two months, and all of the flowering plants are perennials, except for one native annual species.

The largest animals of the Arctic tundra are the herbivorous musk oxen and caribou of North America and the reindeer of the Old World. Other herbivores of the tundra include lemmings (small rodents with short tails) and ptarmigans (pigeon-sized grouse). The white fox and the snowy owl are among the principal predators, feeding largely on lemmings.

During the brief Arctic summer, insects emerge in great numbers. Migratory birds visit, taking advantage of the insect hordes and the long periods of daylight to feed their young.

Temperate Grasslands

Temperate grasslands, which are transitional areas between temperate forests and deserts, are usually found in the interior areas of continents. They are characterized by rolling to flat terrain, hot-cold seasons (rather than the wet-dry seasons of the savannas), periodic droughts, and fires. The temperate grasslands of the

46–13 *The grasslands of North America include large regions of both temperate grassland and savanna. (a) A June day on a tall-grass prairie in North Dakota. The cottonwood grove by the prairie creek is characteristic of this biome. A thunderstorm is gathering on the horizon. (b) A female bison nursing her calf in the short-grass prairie of Custer State Park, South Dakota. Among the other native animals of the grasslands are (c) the black-tailed prairie dog and (d) the red-shouldered hawk and its prey, the garter snake.*

851

46–14   *A savanna in East Africa. The giraffes are surrounded by a herd of impala. The trees are acacias.*

world include the plains and prairies of North America, the steppes of Asia, the veld of South Africa, and the pampas of Argentina.

The vegetation is largely bunch or sod-forming grasses, often mixed with legumes (clovers and wild indigo) and a variety of annuals. In North America, there is a transition from the eastern temperate deciduous woodland, through the rich, moist, tall-grass prairie (the Corn Belt), to the more desertlike, western short-grass prairie (the Great Plains). Grasslands are drier to the west, where they are in the rain shadow of the Rockies. Periodic fires serve to maintain the nature of the grasslands, destroying tree seedlings and preventing their encroachment.

The temperate grasslands of the world support small, seed-eating rodents and also large herbivores, such as the bison of early America, the gazelles and zebras of the African veld, the wild horses, wild sheep, and saiga antelope of the Asiatic steppes, and now the domestic herbivores. These large, grass-eating mammals, in turn, support carnivores, such as lions and wolves, as well as omnivorous humans.

Savannas

Savannas are grasslands with scattered clumps of trees (Figure 46–14). They occur in regions characterized by wet-dry seasons. The transition from open forest with grassy undergrowth to savanna is gradual. It is determined by the duration and severity of the dry season and, often, by fire and by grazing and browsing animals.

In the savanna, the critical competition is for water, and grasses are the victors. Grasses are well suited to a fine, sandy soil with seasonal rain because their roots form a dense network capable of extracting the maximum amount of water during the rainy period. During dry seasons, the aboveground portions of the plants die, but the deep roots are able to survive even many months of drought.

The balance between woody plants and grasses is a delicate one. If rainfall decreases, the trees die. If rainfall increases, the trees increase in number until they shade the grasses, which, in turn, die. If the grasses are overgrazed (which often happens when people begin to use the savanna for agricultural purposes and introduce livestock), enough water is left in the soil so that the woody plants can increase in number, and the grassland is eventually destroyed.

The best-known savannas are those of Africa, which are inhabited by the most abundant and diverse group of large herbivores in the world, including the gazelles, the impala, the eland, the buffalo, the giraffe, the zebra, and the wildebeest.

(a)

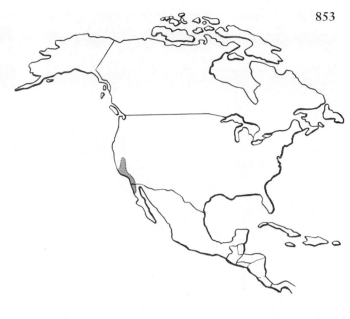

(b)

(c)

46–15 (a) *The Mediterranean scrub, or chaparral, of North America. The bushy vegetation that characterizes the chaparral grows as dense as a mat on the foothills of southern California. It is the result of long, dry summers, during which much of the plant life is semidormant, followed by a cool, rainy season. The name comes from* chaparro, *the Indian word for the scrub oak that is one of the prominent components of the chaparral.* Chaps, *the leather leggings worn by cowboys making their way through this dense, dry growth, has the same derivation.* (b) *A cacomistle, or ring-tailed cat, an inhabitant of the chaparral.*

(c) *Mediterranean scrub, or* maquis, *on the island of Crete. Note the limestone on which the plants are growing. It is characteristic of much of the Mediterranean coastline and of the islands within the Mediterranean.*

Mediterranean Scrub

Regions with mild, moist winters and long, dry summers, such as the southern coast of California and southern Spain, are dominated by small trees or, often, by spiny shrubs with broad, thick evergreen leaves.

This vegetation, known formally as **Mediterranean scrub,** has been given a variety of local names. In the United States, it is known as the **chaparral,** whereas in Mediterranean regions, it is the *maquis,* and in Chile it is the *matorral.* Mediterranean scrub is also found in southern Africa and along portions of the coast of Australia. Although the plants of these various areas are unrelated, they closely resemble one another in their growth patterns and characteristic appearance.

Mule deer live in the North American chaparral during the spring growing season, moving out to cooler regions during the summer. Many of the resident vertebrates—lizards, wren-tits, brown towhees, brush rabbits, and a diversity of rodents—are small and dull-colored, matching the dull-colored vegetation.

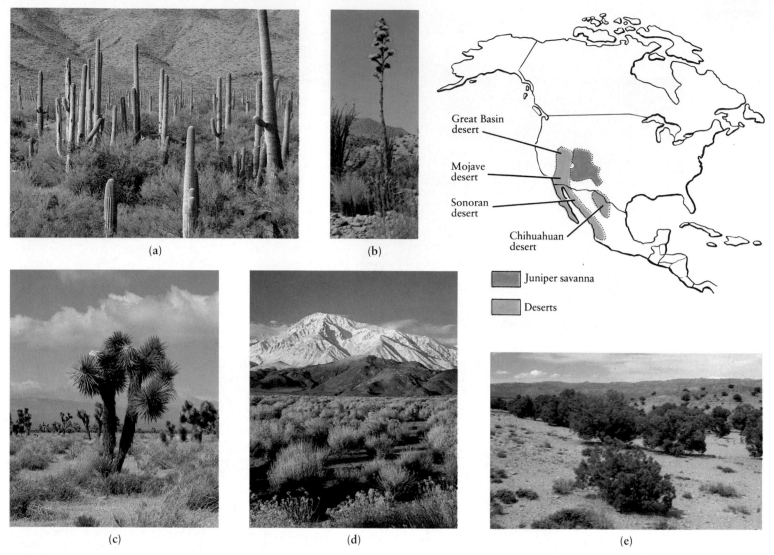

46–16 *The principal deserts of North America, adjacent to which are large areas of juniper savanna and semidesert.*

The Sonoran stretches from southern California to western Arizona and down into Mexico. A dominant plant, the giant saguaro cactus (a) is often as much as 15 meters high, with a widespreading network of shallow roots. Water is stored in a thickened stem, which expands, accordionlike, after a rainfall.

To the east is the Chihuahuan desert, one of whose principal plants is the agave (b), or century plant, a monocot.

North of the Sonoran is the Mojave, whose characteristic plant is the Joshua tree (c). This plant was named by early Mormon colonists who thought that its form resembled that of a bearded patriarch gesticulating in prayer. The Mojave contains Death Valley, the lowest point on the continent (90 meters below sea level), only 130 kilometers from Mt. Whitney, whose elevation is more than 4,000 meters.

The Mojave blends into the Great Basin, a cold desert bounded by the Sierra Nevada to the west and the Rockies to the east. It is the largest and bleakest of the American deserts. The dominant plant is sagebrush (d), shown here with the snow-covered Sierra Nevada mountains in the background.

Juniper savanna (e) is a high-altitude, cold biome named for its dominant plant species, Juniperus osteosperma.

The Desert

The great deserts of the world are located at latitudes of about 30°, both north and south, and extend poleward in the interiors of the continents. These are areas of falling, warming air and, consequently, little rainfall (see Figure 45–3, page 818). Only about 5 percent of North America is desert.

The Sahara, which stretches all the way from the Atlantic coast of Africa to the Arabian Peninsula, is the largest desert in the world (almost equal to the size of the United States) and is increasing in size, spreading along its southern boundaries. This spread is due in large part to the growth of the human population, resulting, in turn, in intensified grazing by domestic animals along the margins of the desert.

Desert regions are characterized by less than 25 centimeters of rain a year. Because there is little water vapor in the air to moderate the temperature, the nights are often extremely cold. The temperature may drop as much as 30°C at night, in contrast to the humid tropics,

(a)

(b)

(c)

(d)

46–17 *Inhabitants of the North American deserts. (a) With a few precisely targeted blows, this roadrunner will dispatch the bull snake held in its sharp beak. The roadrunner will then eat the snake headfirst. (b) A collared lizard, raising its body so cooling air currents can pass beneath it. (c) A cougar, or mountain lion. (d) A jackrabbit, which is actually a hare, resting in the shade during the hottest part of the day. Other familiar animals of the North American deserts include the horned lizard (Figure 32–8, page 576) and the kangaroo rat (Figure 32–12, page 579).*

where day and night temperatures vary by only a few degrees.

Many desert plants are annuals that race from seed to flower to seed during periods when water is available. During the brief growing seasons, the desert may be carpeted with flowers. Many of the perennials are succulents (adapted for water storage). Some are drought-deciduous (dropping their leaves in dry seasons), some have small, leathery water-conserving leaves, and some, like cacti, have leaves that are modified as hard, dry spines. C_4 and CAM photosynthesis (pages 171 and 745), both of which conserve water, are common among desert plants, as are extensive root systems that are able to trap large amounts of water during the brief periods it is available.

The animals that live in the desert are similarly adapted to this extreme climate (see page 579). Reptiles and insects, for example, have waterproof outer coverings and dry, and therefore water-conserving, excretions. Many of the mammals of the desert are small and nocturnal and obtain what little water they require from the plants they eat.

46–18   *Tropical rain forest.* (a) *The interior of a rain forest in Costa Rica. The broad-leaved plants with red flowers are* Heliconia irrasa. (b) *The diversity of the trees in the forest, which may reach several hundred species per hectare, is revealed when individual trees burst into bloom, like these trees in the coastal rain forest of Brazil.* (c) *A buttressed tree in the rain forest of Ecuador. Note the lianas on the trunk.*

(a)

(b)

(c)

Tropical Forests

In the equatorial zone, where most of the tropical forests of the world are found, the average daily temperature is the same throughout the year, and the length of daylight varies by less than one hour. Rainfall, however, is often seasonal, and variations in total rainfall from one area to another are also caused by mountains and their rain shadows.

Areas in which rainfall is limited all year are characterized by southern woodland and scrub, in which the trees have small, water-conserving leaves. In areas with distinct wet and dry seasons, tropical mixed forests and monsoon forests occur. These forests are dominated by trees that lose their leaves during the dry seasons. The tropical rain forest, the most complex of all ecosystems, is found in areas where rainfall is abundant all year round.

Tropical Rain Forest

In the tropical rain forest, total rainfall is between 200 and 400 centimeters per year, and a month with less than 10 centimeters of rain is considered relatively dry. There are more species of plants and animals in the tropical rain forest than in all the rest of the biomes of the world combined. As many as 100 species of trees can be counted on 1 hectare, but each species may be represented by only one tree. By contrast, a comparable area in a deciduous forest in the northeastern United States typically contains only a few tree species, but each species is represented by many individual trees.

The critical competition among plants of the tropical forest is for light. About 70 percent of all species of plants are trees. The upper tree story consists of solitary giants 50 to 60 meters tall. A lower story of trees characteristically forms a continuous canopy. The trees forming the canopy are remarkably similar in appearance. Their trunks are usually slender and branch only near the crowns, which are very high and relatively small as a result of crowding. Because the soil is perpetually wet, their roots do not reach deep into it, and trunks often end in thick buttresses that provide broad anchorage. Their leaves are large, leathery, and dark green; their bark is thin and smooth; and their flowers are often inconspicuous and greenish or whitish in color.

Woody vines, or **lianas,** are abundant, especially where an opening has appeared in the forest, as, for example, where a tree has fallen. Vines as long as 240 meters have been measured. There are also many **epiphytes,** which are plants that grow on other plants, often high above the forest floor. The epiphytes of the tropical rain forest germinate in the branches of trees and obtain water from the humid air of the canopy. Unlike the plants that have contact with the moist floor, epiphytes need to conserve water between rainfalls. Some epiphytes resemble desert succulents, having fleshy water-storing leaves and stems. Others have spongy roots or cup-shaped leaves that capture moisture and organic debris. Many epiphytes can take up nutrients from decaying organisms in these storage tanks. A variety of plants, including ferns, orchids, mosses, and bromeliads (page 719), have exploited the epiphyte life style.

An extraordinary variety of insects, birds, and other animals, including mammals, have moved into the forest canopy along with the vines and epiphytes to make it the most abundantly and diversely populated area of the tropical rain forest (Figure 46–19).

Little light reaches the forest floor (from 0.1 to 1 percent of the total), and the relatively few plants that are found there are adapted to growing at low light intensities. Many of these, such as the African violet, are familiar to us as house plants.

Plants also compete for nutrients. The nutrient cycles are tight, and turnover is rapid. There is almost no accumulation of leaf litter on the forest floor, such as we find in our northern forests. Everything that touches the ground disappears almost immediately—carried off, consumed, or rapidly decomposed. As a consequence, the soils of tropical rain forests are relatively infertile. Many are chiefly composed of a red clay. When they are cleared, they generally either erode rapidly or form thick, impenetrable crusts that cannot be cultivated after a season or two.

Tropical soils are generally deficient in nutrients, and what nutrients are found there are likely to be leached out by heavy rainfall. Most of the nitrogen, phosphorus, calcium, and other nutrients are found in the plants rather than the soil. Trees that most effectively store these nutrients may be the ones that win the competition for light.

Tropical Forests, Mass Extinction, and Human Responsibility

Although tropical forests currently form about half of the forested area of the Earth, they are being rapidly destroyed. If the current rate of destruction continues, almost all the tropical forests will have disappeared by the turn of the century, and with them the thousands of plants and animals found nowhere else in the biosphere.

The destruction of the forests results directly from the rapid growth of the human population in the tropics and, as we saw at the beginning of this chapter, indirectly from the nature of tropical forest soils. Because

46–19   *An enormous variety of animals live in the tropical rain forest. For example, in Colombia alone, there are at least 1,400 species of birds. (a) An* Anartia amathea *butterfly, photographed in Ecuador. (b) A Colombian cone-headed grasshopper. (c) A keel-billed toucan in Belize. (d) A green tree viper, photographed in a forest in Malaysia. (e) Squirrel monkeys in the canopy of the Amazon forest.*

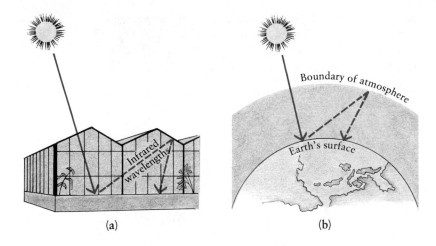

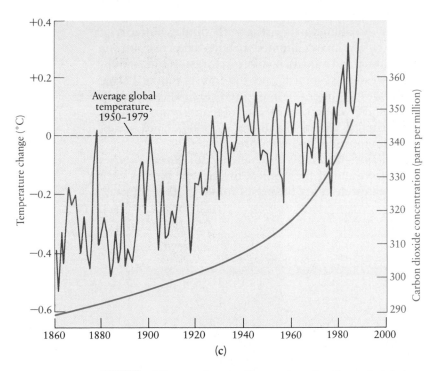

46–20   *The greenhouse effect, carbon dioxide, and global warming. (a) Light rays penetrate the glass of a greenhouse, are absorbed by the plants and soil, and are then reradiated as longer-waved infrared radiation (heat). The glass does not permit these rays to escape, and so the heat remains within the greenhouse. (b) Carbon dioxide, like glass, is transparent to light but absorbs infrared rays, preventing their escape.*

(c) Over the past 140 years, there has been a clear correlation between changes in average global temperature (red line) and the concentration of carbon dioxide in the atmosphere (blue line). As atmospheric carbon dioxide continues to increase, due to the burning of fossil fuels and forest destruction, it is expected to lead to further temperature increases. A study by the Environmental Protection Agency predicts that global temperatures could increase an average of 2°C by the year 2040 and 5°C by the year 2100.

such soils retain few nutrients, they are very poor for agricultural purposes and will support crops for only a few years after the trees have been cleared. Thus, the traditional agricultural practices in the tropics have been those of clearing and short-term cultivation. With a rapidly expanding, hungry human population, such practices have become immensely destructive as they are carried out on an ever-increasing scale. Commercial logging also contributes to the demise of the forests, as does the need of the local populations for firewood.

The loss of the organisms of the tropical forests would, in itself, represent a mass extinction on the scale of previous extinctions (see page 367). Unlike the earlier extinctions, however, it would be the direct result of the activities of a single species—our own. Moreover, many biologists are concerned that the destruction of the tropical forests will lead indirectly to extinctions in other biomes.

During the past 100 years, the concentration of carbon dioxide in the atmosphere has been rising steadily, largely as a result of increased burning of fossil fuels. Recent analyses have shown that forest destruction is also a major contributor to increased atmospheric carbon dioxide levels. As forest is cleared, the oxidation of its biomass—either by burning or by the metabolic processes of decomposers—releases large amounts of carbon dioxide. Moreover, carbon dioxide that would previously have been used by the trees and other plants of the forest in photosynthesis remains in the atmosphere.

As a result of the **greenhouse effect** (Figure 46–20), the increased carbon dioxide levels in the atmosphere are expected to result in a significant increase in global temperatures, beginning within the next 20 years. Among the expected consequences of this temperature increase are rising sea levels (as a result of the melting of polar ice), lengthened growing seasons and more precipitation in some areas, reduced precipitation and primary productivity in other areas, and the extinction of a variety of species as local environmental conditions change beyond the organisms' limits of adaptability (see page 782).

Halting, or even slowing, the destruction of the tropical forests is an enormously complex undertaking that cannot be separated from the needs of the human population of the tropics. Setting aside areas of forest as undisturbed preserves and slowing the rate of human population growth are only part of the answer. There is a pressing need for developing more productive and appropriate agricultural practices for the tropics and devising techniques for harvesting the richness of the forest—providing a long-overdue improvement in the standard of living for local populations—without simultaneously destroying the forest.

In the age of the dinosaurs, the earliest primates survived, it would appear, largely by their wits. Now, if we and other species are to survive the consequences of our own evolution, we will have to use well the contents of our own skulls. For within the human brain—that complex collection of neurons and synapses—resides the uniquely human capacity to accumulate knowledge, to plan with foresight, and so to act with enlightened self-interest and even, on occasion, with compassion for others and reverence for the diversity of life on our planet.

Summary

The biosphere is that part of the Earth that contains living organisms. It is a thin film on the surface of the planet, irregular in its thickness and its density.

The biosphere is affected by the position and movements of the Earth in relation to the sun and by the movements of air and water over the Earth's surface. These factors cause wide differences in temperature and rainfall from place to place and from season to season. There are also differences in the surfaces of the continents. These differences are reflected in the kinds of plant and animal life found in different parts of the biosphere.

Distinctive communities of organisms are found in running fresh waters (rivers and streams), standing fresh waters (lakes and ponds), at different depths of the open oceans, and along rocky, sandy, and muddy seashores.

On land, biomes are communities of organisms characterized by distinctive patterns of vegetation. They are widely distributed over areas with similar patterns of climate. The major biomes include temperate forests (deciduous and mixed), coniferous forests (taiga, mountain, west-coast mixed, and the forests of the Pacific Northwest), tundra (Arctic and alpine), temperate grasslands, savannas, Mediterranean scrub, deserts, and tropical forests (southern woodland and scrub, mixed, monsoon, and rain forest).

A major—and urgent—challenge for *Homo sapiens* is preservation of the tropical forests, the biomes with the greatest diversity of species found on this planet.

Questions

1. Mud flats are extremely rich in animal life, yet few plants are found there. What reasonable explanation can you give for the scarcity of plant life? How can the mud flats support such a profusion of animal life in the absence of plants?

2. What are the eight major biomes? Describe the principal abiotic features of each.

3. Name a plant and an animal associated with each of the eight major biomes, and describe their adaptations.

4. Although each of the biomes we have considered is sufficiently different from all the others to warrant its identification as a distinct biome, there are important similarities among some of the biomes. Consider the following groups of biomes: tropical rain forest/monsoon forest; monsoon forest/temperate deciduous forest/taiga; savanna/temperate grasslands/tundra. Describe the essential similarities and the most significant differences in the environmental factors affecting the members of each group. How do these factors affect the types of plants characterizing each biome?

5. The rate of decomposition of plant litter, animal wastes, and dead plants and animals varies from biome to biome. Describe the differences in decomposition rates in the following biomes: tropical rain forest, temperate deciduous forest, taiga. What factors in each biome are important in causing these differences? What are the consequences of these differing decomposition rates for nutrient recycling, soil quality, and the size and diversity of detritivore populations?

6. A leading ecologist has stated: "The plough is the most deadly weapon of extinction ever devised; not even thermonuclear weapons pose such a threat to the beauty and diversity of life on Earth." Explain.

Suggestions for Further Reading

Bazzaz, Fakhri A., and Eric D. Fajer: "Plant Life in a CO_2-Rich World," *Scientific American,* January 1992, pages 68–74.

Berner, Robert A., and Antonio C. Lasaga: "Modeling the Geochemical Carbon Cycle," *Scientific American,* March 1989, pages 74–81.

Bertness, Mark D.: "The Ecology of a New England Salt Marsh," *American Scientist,* vol. 80, pages 260–268, 1992.

Burney, David A.: "Recent Animal Extinctions: Recipes for Disaster," *American Scientist,* vol. 81, pages 530–541, 1993.

Colinvaux, Paul A.: "The Past and Future Amazon," *Scientific American,* May 1989, pages 102–108.

Crutzen, Paul J., and Meinrat O. Andreae: "Biomass Burning in the Tropics: Impact on Atmospheric Chemistry and Biogeochemical Cycles," *Science,* vol. 250, pages 1669–1678, 1990.

Detwiler, R. P., and C. A. S. Hall: "Tropical Forests and the Global Carbon Cycle," *Science,* vol. 239, pages 42–47, 1988.

Ellis, Gerry, and Karen Kane: *America's Rain Forest,* North Word Press, Inc., Minocqua, Wis., 1991.

> This beautiful book, filled with almost 200 outstanding color photographs, provides a valuable overview of the diversity and complexity of the forests of the Pacific Northwest. As the authors make clear, the very fabric of the ecosystem—not just individual species, such as the northern spotted owl—is endangered by logging practices that destroy the forests' capacity for renewal.

Forsyth, Adrian, and Ken Miyata: *Tropical Nature: Life and Death in the Rain Forests of Central and South America,* Charles Scribner's Sons, New York, 1984.*

> This book, for the general reader, consists of 17 essays on different aspects of the complex—and still poorly understood—ecology of tropical rain forests. It is filled with a multitude of marvelous examples of the diverse organisms of the forests and their many interactions with each other. The authors, both specialists in tropical vertebrates, spent seven years working in the forests, principally in Costa Rica and Peru, and bring vast personal experience to their essays.

Goulding, Michael: "Flooded Forests of the Amazon," *Scientific American,* March 1993, pages 114–120.

Holloway, Marguerite: "Sustaining the Amazon," *Scientific American,* July 1993, pages 90–99.

Homer-Dixon, Thomas F., Jeffrey H. Boutwell, and George W. Rathjeus: "Environmental Change and Violent Conflict," *Scientific American,* February 1993, pages 38–45.

Horn, Michael S., and Robin H. Gibson: "Intertidal Fishes," *Scientific American,* January 1988, pages 64–70.

Houghton, Richard A., and George M. Woodwell: "Global Climatic Change," *Scientific American,* April 1989, pages 36–44.

Janzen, Daniel H.: "Tropical Ecological and Biocultural Restoration," *Science,* vol. 239, pages 243–244, 1988.

Jones, Philip D., and Tom M. L. Wigley: "Global Warming Trends," *Scientific American,* August 1990, pages 84–91.

Krutch, J. W.: *The Desert Year,* Viking Press, Inc., New York, 1977.*

> A captivating, beautifully written description, first published in 1960, of the animal and plant life of the North American desert.

La Rivière, J. W. Maurits: "Threats to the World's Water," *Scientific American,* September 1989, pages 80–94.

Laws, Richard M.: "The Ecology of the Southern Ocean," *American Scientist,* vol. 73, pages 26–40, 1985.

Lopez, Barry: *Arctic Dreams: Imagination and Desire in a Northern Landscape,* Bantam Books, New York, 1987.*

> A brilliant evocation of life in the Arctic by a master nature writer. "This is a land," says Lopez, "where airplanes track icebergs the size of Cleveland and polar bears fly down out of the stars."

Mares, Michael A.: "Conservation in South America: Problems, Consequences, and Solutions," *Science,* vol. 233, pages 734–739, 1986.

Mohnen, Volker A.: "The Challenge of Acid Rain," *Scientific American,* August 1988, pages 30–38.

Moran, Joseph M., Michael D. Morgan, and James H. Wiersma: *An Introduction to Environmental Sciences,* 2d ed., W. H. Freeman and Company, New York, 1985.

> A clear, matter-of-fact presentation of the basic features of our planet and, in particular, of the biosphere. It provides a good background for the study of ecology in general and problems of environmental pollution and technological change in particular.

Newell, Reginald E., Henry G. Reichle, Jr., and Wolfgang Seiler: "Carbon Monoxide and the Burning Earth," *Scientific American,* October 1989, pages 82–88.

Perry, Donald R.: "The Canopy of the Tropical Rain Forest," *Scientific American,* November 1984, pages 138–147.

* Available in paperback.

Philander, George: "El Niño and La Niña," *American Scientist,* vol. 72, pages 451–459, 1989.

Pollack, Henry N., and David S. Chapman: "Underground Records of Changing Climate," *Scientific American,* June 1993, pages 44–50.

Post, Wilfred M., et al.: "The Global Carbon Cycle," *American Scientist,* vol. 78, pages 310–326, 1990.

Ramage, Colin S.: "El Niño," *Scientific American,* June 1986, pages 76–83.

Repetto, Robert: "Accounting for Environmental Assets," *Scientific American,* June 1992, pages 94–101.

Repetto, Robert: "Deforestation in the Tropics," *Scientific American,* April 1990, pages 36–42.

Richardson, Philip L.: "Tracking Ocean Eddies," *American Scientist,* vol. 81, pages 261–271, 1993.

Schneider, Stephen H.: "The Changing Climate," *Scientific American,* September 1989, pages 70–79.

Sebens, Kenneth P.: "The Ecology of the Rocky Subtidal Zone," *American Scientist,* vol. 73, pages 548–557, 1985.

Smith, Robert L.: *Ecology and Field Biology,* 4th ed., Harper & Row, Publishers, Inc., New York, 1990.

> *The outstanding sections of this text are those that deal with the biomes and the plants and animals within them. Although meant to accompany a course in field biology, this book is an asset and a pleasure to the amateur naturalist.*

Terborgh, John: *Diversity and the Tropical Rain Forest,* W. H. Freeman and Company, New York, 1992.

> *This beautifully illustrated volume explores the incredible diversity of organisms found in the tropical forests, as well as the complex interactions of ecology and evolution that have given rise to that diversity. The author, who is the director of Duke University's Center for Tropical Conservation, also discusses current scientific efforts to counteract the devastating effects of the destruction of the tropical forests.*

Waring, Richard H.: "Land of the Giant Conifers," *Natural History,* October 1982, pages 54–63.

White, Robert M.: "The Great Climate Debate," *Scientific American,* July 1990, pages 36–43.

Wilson, Edward O.: *The Diversity of Life,* Harvard University Press, Cambridge, Mass., 1992.

> *A presentation of the diversity of life and of the evolutionary processes that have produced it, by one of the world's leading authorities on both diversity and evolution. Filled with marvelous stories of many fascinating organisms, Wilson's text is also infused with examples of how rapidly species are being driven to extinction throughout the world and the obstacles that must be overcome if the loss of diversity is to be stemmed.*

Wilson, Edward O.: "Threats to Biodiversity," *Scientific American,* September 1989, pages 108–116.

* Available in paperback.

The Hardy-Weinberg Equation

As we noted in the text (page 325), Hardy and Weinberg demonstrated the constancy of allele and genotype frequencies in an idealized population in which five conditions hold:

1. No mutations occur.

2. There is no net movement of individuals—with their genes—into the population (immigration) or out of it (emigration).

3. The population is large enough that the laws of probability apply; that is, it is highly unlikely that chance alone can alter the frequencies, or relative proportions, of alleles.

4. Mating is random.

5. All alleles are equally viable. In other words, there is no difference in reproductive success. The offspring of all possible matings are equally likely to survive to reproduce in the next generation.

To understand how Hardy and Weinberg arrived at their equation expressing the equilibrium of allele and genotype frequencies in a population meeting these five conditions, let us look at just one gene. For simplicity, we select one that has only two alleles, which we will call A and a. We are interested in the relative proportions—that is, the frequencies—of A and a from one generation to the next. By convention, the letter p is used to designate the frequency of one allele and the letter q to designate the frequency of the other. When there are only two alleles, p and q together must equal one: $p + q = 1$.

These proportions—p and q—could be expressed in terms of fractions, as done by Mendel, but since the proportions of the two alleles will probably not be equal, as they were in Mendel's carefully controlled experiments, it is more convenient to express the numbers as decimals. For example, suppose that in a particular population 80 percent of the alleles of the gene under study are allele A. The frequency of A is 0.8, or $p = 0.8$. Given that there are only two alleles, we then know that the frequency of allele a is 0.2. In other words, $q = 1 - p$.

Let us assume that the relative frequencies of A and a are the same in both male and female (as they are for most alleles in natural populations). Now suppose that males and females mate at random with respect to alleles A and a. We can calculate the frequencies of the resulting genotypes by drawing a Punnett square. As you can see from the illustration, the genotypes in the population produced by this random mating would consist of 64 percent AA, 32 percent Aa, and 4 percent aa.

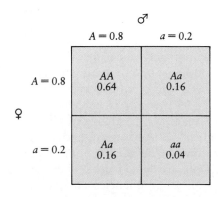

Results of random mating in a population in which the frequency (p) of allele A equals 0.8 and the frequency (q) of allele a equals 0.2. In setting up this Punnett square, we are assuming that the gene is not sex-linked and that the alleles appear in the same frequencies in males as in females.

Instead of drawing a Punnett square, we can do the same thing algebraically. Because $p + q = 1$, it follows that:

$$(p + q)(p + q) = 1 \times 1 = 1$$

Or, as you probably remember from algebra:

$$p^2 + 2pq + q^2 = 1$$

This algebraic expression of the genotype frequencies is the Hardy-Weinberg equation.

Let us apply this equation to the random mating that just occurred in our population. Taking the initial values for the frequencies of the two alleles, we obtain the following results:

$$p^2 = 0.8 \times 0.8$$
$$= 0.64 \text{ (the frequency of } AA \text{ genotypes)}$$
$$2pq = 2 \times 0.8 \times 0.2$$
$$= 0.32 \text{ (the frequency of } Aa \text{ genotypes)}$$
$$q^2 = 0.2 \times 0.2$$
$$= 0.04 \text{ (the frequency of } aa \text{ genotypes)}$$

What has happened to the frequencies of the two alleles in the gene pool as a result of this round of mating? We know from our calculations that the frequency of AA is 0.64. In addition, half of the alleles in the heterozygotes (Aa) are A, so the total frequency of allele A is 0.64 plus one-half of 0.32—that is, 0.64 plus 0.16, to give a total of 0.8. The frequency of allele A (p) has not changed. Similarly, the total frequency of allele a is 0.04 (in the homozygotes) plus 0.16 (half the alleles in the heterozygotes), or 0.2. The frequency of allele a (q) has also remained the same.

If another round of mating occurs, the proportion of AA, of Aa, and of aa genotypes in our population will again be 64 percent, 32 percent, and 4 percent, respectively. Again the frequency of allele A will be 0.8 and of allele a 0.2. And so on, and so on, generation after generation. In an ideal population in which the five conditions are met, neither the allele frequencies nor the genotype frequencies change from generation to generation.

The Hardy-Weinberg equilibrium also applies to situations in which there are more than two alleles of the same gene. However, the equation representing the equilibrium is more complex. For example, the genotype frequencies of three alleles are expressed by the algebraic expansion of $(p + q + r)^2 = 1$, with r representing the frequency in the gene pool of the third allele.

An Application of the Hardy-Weinberg Equation

As we have noted, the Hardy-Weinberg equilibrium holds only in populations in which the five stated conditions are met. What happens if one of these conditions is not met? Let us imagine another pair of alleles in which allele a has a damaging effect in the homozygous state, reducing the likelihood that a homozygous aa individual will survive to reproduce. We can estimate the number of aa genotypes in the population, perhaps by screening tests on newborn infants. Suppose, for instance, that the condition shows up in 1 in 10,000 infants. In other words, $q^2 = 1/10,000$, or 0.0001. Thus, $q = \sqrt{0.0001}$, or 0.01. If $q = 0.01$, then $p = 0.99$ and $2pq = 0.0198$, or almost 0.02. Thus about 2 percent of the population—one person in every 50—can be estimated to be a heterozygous carrier of this allele.

Now suppose that we do the same screening tests five years later and find that q is not 0.01 but 0.009, and then we repeat it a few years later and find that it has once again decreased very slightly, perhaps to 0.008. In other words, evolution is occurring. One allele is decreasing while the other is increasing. Using the Hardy-Weinberg equilibrium as our yardstick, we know not only that change has taken place and its direction but also that there must be some reason for it. We can then look for the factors causing the change.

Some Problems

1. Suppose that in a breeding experiment, 7,000 AA individuals and 3,000 aa individuals mate at random. In the first generation of offspring, what would be the frequencies of the three genotypes (AA, Aa, and aa)? What would be the frequencies of the two alleles? What would be the values in the second generation?

Solution: Initially, there are a total of 20,000 alleles in the gene pool: 14,000 A alleles and 6,000 a alleles. The frequency of allele A (that is, the value for p) is 14,000/20,000, or 0.7. The frequency of allele a (that is, the value for q) is $1 - p$, or 0.3.

In the first generation of offspring, the values of the three terms of the Hardy-Weinberg equation would be as follows:

$$p^2 = (0.7)^2$$
$$= 0.49 \text{ (frequency of } AA \text{ genotypes)}$$
$$2pq = 2(0.7)(0.3)$$
$$= 0.42 \text{ (frequency of } Aa \text{ genotypes)}$$
$$q^2 = (0.3)^2$$
$$= 0.09 \text{ (frequency of } aa \text{ genotypes)}$$

The allele frequencies would be as follows:

$$p = \sqrt{0.49} = 0.7 \text{ (frequency of allele } A)$$
$$q = \sqrt{0.09} = 0.3 \text{ (frequency of allele } a)$$

Assuming that the five conditions for the Hardy-Weinberg equilibrium continue to be met, the values will be the same in the second generation. Work through the calculations to satisfy yourself that this is the case.

2. Among African-Americans, the frequency of sickle-cell anemia (which, as you will recall, is a homozygous recessive condition) is about 0.0025. What is the frequency of heterozygotes? When one African-American marries another, what is the probability that both will be heterozygotes? If both are heterozygotes, what is the probability that their first child will have sickle-cell anemia?

B

Classification of Organisms

There are several ways to classify organisms. The one presented here follows the overall scheme described at the end of Chapter 23. Organisms are divided into five major groups, or kingdoms: Monera, Protista, Fungi, Plantae, and Animalia.

The chief taxonomic categories are kingdom, division or phylum, class, order, family, genus, and species. (The taxonomic categories of division and phylum are equivalent. The term "division" is generally used in the classification of prokaryotes, algae, fungi, and plants, whereas "phylum" is used in the classification of protozoa and animals.) The following classification includes all of the generally accepted divisions and phyla. Certain classes and orders, particularly those mentioned in this book, are also included, but the listing is far from complete. The number of species given for each group is an estimated number of living (that is, contemporary) species described and named to date.

KINGDOM MONERA

Monerans (prokaryotes) are cells that lack a nuclear envelope, chloroplasts and other plastids, mitochondria, and 9 + 2 flagella. Prokaryotes are unicellular but sometimes occur as filaments or other superficially multicellular bodies. Their predominant mode of nutrition is heterotrophic, by absorption, but some groups are autotrophic, either photosynthetic or chemosynthetic. Reproduction is primarily asexual, by cell division or budding, but genetic exchanges occur in some as a result of conjugation, transformation, transduction, and exchanges of plasmids. Motile forms move by means of bacterial flagella or by gliding.

The classification of prokaryotes is not hierarchical, and the schemes that are most widely used do not reflect evolutionary relationships. Kingdom Monera contains representatives of two distinct lineages, the Archaebacteria and the Eubacteria. Archaebacteria include the methanogens, thermoacidophiles, and extreme halophiles. Among the principal lineages of Eubacteria are the green bacteria (sulfur and nonsulfur), purple bacteria (sulfur and nonsulfur) and related forms, spirochetes, cyanobacteria, and gram-positive bacteria. About 2,700 distinct species of prokaryotes are recognized.

KINGDOM PROTISTA

Eukaryotic organisms, including unicellular or simple colonial heterotrophs (protozoa), multinucleate or multicellular heterotrophs (slime molds and water molds), and unicellular and multicellular photosynthetic autotrophs (algae). Their modes of nutrition include ingestion, absorption, and photosynthesis. Reproduction is both sexual (in some forms) and asexual. They move by 9 + 2 flagella (or cilia) or pseudopodia or are nonmotile. About 60,000 living species and another 60,000 extinct species known only from their fossils.

Phylum

Phylum Mastigophora: mastigophorans (flagellates). Unicellular heterotrophs with flagella, most of which are symbiotic forms such as *Trichonympha* and *Trypanosoma*, the cause of sleeping sickness. Some free-living species. Reproduction usually asexual by cell division. About 1,500 species.

Phylum Sarcodina: sarcodines. Unicellular heterotrophs with pseudopodia, such as amoebas. No stiffening pellicle; some produce shells. Reproduction may be asexual or sexual. About 11,500 living species and some 33,000 fossil species.

Phylum Ciliophora: ciliates. Unicellular heterotrophs with cilia, including *Paramecium* and *Stentor*. Reproduction is asexual, but genetic exchange through the phenomenon of conjugation is also common. About 8,000 species.

Phylum Opalinida: opalinids. Parasitic protists found mainly in the digestive tracts of frogs and toads. Covered uniformly with cilia or flagella. Reproduction asexual by cell division or sexual, with flagellated gametes. About 400 species.

Phylum Sporozoa: sporozoans. Parasitic protists; usually without locomotor organelles during a major part of their complex life cycle. Includes *Plasmodium,* several species of which cause malaria. About 5,000 species.

Division

Division Myxomycota: plasmodial slime molds. Heterotrophic amoeboid organisms that form a multinucleate plasmodium that creeps along as a mass and eventually differentiates into sporangia, each of which is multinucleate and eventually gives rise to many spores. Predominant mode of nutrition is by ingestion. More than 550 species.

Division Acrasiomycota: cellular slime molds. Heterotrophic organisms in which there are separate amoebas that eventually swarm together to form a mass but retain their identity within this mass, which eventually differentiates into a compound sporangium. Principal mode of nutrition is by ingestion. Seven genera and about 65 species.

Division Chytridiomycota: chytrids (water molds). Multinucleate aquatic heterotrophs with a vegetative body, or thallus, usually differentiated into rhizoids and a sporangium. Nutrition by absorption. Cell walls contain chitin. Asexual reproduction, with sexual reproduction in some forms. Both spores and gametes are flagellated. About 900 species.

Division Oomycota: oomycetes (water molds). Multinucleate filamentous heterotrophs, primarily aquatic. Nutrition by absorption. Their cell walls are composed of glucose polymers, including cellulose. Both asexual and sexual reproduction, with flagellated asexual spores and nonmotile gametes. The multinucleate filaments (hyphae) are diploid, and the haploid gametes are produced by meiosis. About 800 species.

Division

Division Euglenophyta: euglenoids. Unicellular photosynthetic (or sometimes secondarily heterotrophic) organisms with chlorophylls *a* and *b*. They store food as paramylon, an unusual carbohydrate. Euglenoids usually have a single apical flagellum and a contractile vacuole. Sexual reproduction is unknown. Euglenoids occur mostly in fresh water. There are some 1,000 species.

Division Chrysophyta: diatoms, golden-brown algae, and yellow-green algae. Unicellular photosynthetic organisms with chlorophylls *a* and *c* and the accessory pigment fucoxanthin. Food stored as the carbohydrate laminarin or as large oil droplets. Cell walls consisting mainly of cellulose, sometimes heavily impregnated with siliceous materials. There may be as many as 13,000 living species.

Class

Class Bacillariophyceae: diatoms. Chrysophyta with double siliceous shells, the two halves of which fit together like a pillbox. They are sometimes motile by the secretion of mucilage fibrils along a specialized groove, the raphe. There are nearly 10,000 living species and at least 15,000 extinct species.

Class Chrysophyceae: golden-brown algae. A diverse group of organisms, including flagellated, amoeboid, and nonmotile forms, some naked and others with a cell wall that may be ornamented with siliceous scales. About 3,000 species.

Division

Division Dinoflagellata: "spinning" flagellates. Unicellular photosynthetic organisms with chlorophylls *a* and *c.* Food is stored as starch. Cell walls contain cellulose. Most of the organisms in this division have two lateral flagella, one of which beats in a groove that encircles the organism. They probably have no form of sexual reproduction, and their mitosis is unique. Nearly 2,000 living species.

Division Chlorophyta: green algae. Unicellular, colonial, multinucleate, or multicellular, characterized by chlorophylls *a* and *b* and various carotenoids. The carbohydrate food reserve is starch. The cell walls consist of polysaccharides, including cellulose in some forms. Motile cells have two lateral or apical flagella. True multicellular genera do not exhibit complex patterns of differentiation. Multicellularity has arisen at least three times, and quite possibly more often. There are about 9,000 known species and possibly many more.

Class

Class Chlorophyceae: Unicellular, colonial, or multicellular green algae, found predominantly in fresh water. Cell division involves a system of microtubules parallel to the plane of cell division, and the nuclear envelope persists during mitosis. Sexual reproduction involves the formation of a dormant zygote that subsequently undergoes meiosis, producing the haploid cells in which the organisms spend most of the life cycle.

Class Charophyceae: Unicellular or multicellular green algae, found predominantly in fresh water. Cell division involves a system of microtubules perpendicular to the plane of cell division; the nuclear envelope breaks down during the course of mitosis. Sexual reproduction involves the formation of a dormant zygote that subsequently undergoes meiosis, producing the haploid cells in which the organisms spend most of the life cycle. Certain members of this class resemble plants more closely than do any other organisms.

Class Ulvophyceae: Multinucleate or multicellular green algae, found predominantly in salt water. Cell division involves a system of microtubules perpendicular to the plane of cell division, but the nuclear envelope persists during mitosis. Sexual reproduction often involves alternation of generations, with meiosis in the diploid sporophyte producing haploid spores that germinate to produce the haploid gametophyte.

Division

Division Phaeophyta: brown algae. Multicellular marine organisms characterized by the presence of chlorophylls *a* and *c* and the pigment fucoxanthin. Their food reserve is the carbohydrate laminarin. Motile cells are biflagellate, with one forward flagellum and one trailing flagellum. A considerable amount of tissue differentiation is found in some of the kelps, with specialized conducting cells in some genera that transport the products of photosynthesis to dimly lighted regions of the alga. There is, however, no differentiation into leaves, roots, and stem, as in plants. About 1,500 species.

Division Rhodophyta: red algae. Primarily marine organisms characterized by the presence of chlorophyll *a* and red pigments known as phycobilins. Their carbohydrate reserve is a special type of starch (floridean). No motile cells are present at any stage in the complex life cycle. The algal body is built up of closely packed filaments in a gelatinous matrix and is not differentiated into leaves, roots, and stem. It lacks specialized conducting cells. There are some 4,000 species.

KINGDOM FUNGI

Eukaryotic filamentous or, rarely, unicellular organisms. The filamentous forms consist basically of a continuous mycelium; this mycelium becomes septate (partitioned off) in certain groups and at certain stages of the life cycle. Chitin is present in the cell walls of all fungi. Fungi are saprobic or parasitic heterotrophs, with nutrition by absorption. Reproductive cycles often include both sexual and asexual phases. Most fungi are haploid, with the zygote the only diploid stage in the life cycle. No flagella or cilia at any stage of the life cycle. Some 100,000 species of fungi have been named.

Division

Division Zygomycota: terrestrial fungi, such as black bread mold, with the hyphae septate only during the formation of reproductive bodies. The division includes about 600 described species, of which about 30 occur as components of the mycorrhizae that are found in about 80 percent of all vascular plants.

Division Ascomycota: terrestrial and aquatic fungi, including *Neurospora,* powdery mildews, morels, and truffles. The hyphae are septate but the septa perforated; complete septa cut off the reproductive bodies, such as spores or gametangia. Sexual reproduction involves the formation of a characteristic cell, the ascus, in which meiosis takes place and within which spores are formed. The hyphae in many ascomycetes are packed together into complex "fruiting bodies." Yeasts are unicellular ascomycetes that reproduce asexually by budding. About 30,000 species, in addition to 25,000 species that occur in lichens.

Division Basidiomycota: terrestrial fungi, including the mushrooms and toadstools, with the hyphae septate but the septa perforated; complete septa cut off reproductive bodies. Sexual reproduction involves formation of basidia, in which meiosis takes place and on which the spores are borne. There are some 25,000 species.

Division Deuteromycota: Fungi Imperfecti. Mainly fungi in which the sexual cycle has not been observed. The deuteromycetes are classified by their asexual spore-bearing structures. There are some 25,000 species, including *Penicillium* (the original source of penicillin), fungi that cause athlete's foot and other skin diseases, and many of the molds that give cheeses, such as Roquefort and Camembert, their special flavor.

KINGDOM PLANTAE

Multicellular photosynthetic eukaryotes, primarily adapted to life on land. The photosynthetic pigment is chlorophyll *a,* with chlorophyll *b* and a number of carotenoids serving as accessory pigments. The cell walls contain cellulose. There is considerable differentiation of tissues and organs. Reproduction is primarily sexual with alternating gametophytic and sporophytic phases; the gametophytic phase has been progressively reduced in the course of evolution. The egg- and sperm-producing structures are multicellular and are surrounded by a sterile (nonreproductive) jacket layer; the zygote develops into an embryo, or young sporophyte, while encased in the archegonium (seedless plants) or embryo sac (seed plants). The living members of the plant kingdom include the bryophytes and nine divisions of vascular plants—plants with complex differentiation of the sporophyte into leaves, roots, and stems and with well-developed strands of conducting tissue for the transport of water and organic materials.

Division

Division Bryophyta: liverworts, hornworts, and mosses. Multicellular plants with photosynthetic pigments and food reserves similar to those of the green algae. They have multicellular gametangia with a sterile jacket one-cell-layer thick. The sperm are biflagellate and motile. Gametophytes and sporophytes both exhibit complex multicellular patterns of development, but conducting tissues are usually completely absent and are not well differentiated when present; true roots, leaves, and stems are absent. Most of the photosynthesis is carried out by the gametophyte, upon which the sporophyte is nutritionally dependent, at least initially. There are some 16,000 species.

Class

Class Hepaticae: liverworts. The gametophytes are either thallose (not differentiated into roots, leaves, and stem) or "leafy," and the sporophytes are relatively simple in construction. About 6,000 species.

Class Anthocerotae: hornworts. The gametophytes are thallose. The sporophyte grows from a basal meristem for as long as conditions are favorable. Stomata are present on the sporophyte. About 100 species.

Class Musci: mosses. The gametophytes are "leafy." The sporophytes have complex patterns of spore discharge. Stomata are present on the sporophyte. About 9,500 species.

Division

Division Psilotophyta: whisk ferns. Homosporous vascular plants with or without microphylls. (Homosporous plants produce only one kind of spore, giving rise to gametophytes with both male and female structures; microphylls are primitive leaves.) The sporophytes are extremely simple, and there is no differentiation between root and shoot. The sperm are motile. Two genera, with several species.

Division Lycophyta: club mosses. Homosporous and heterosporous vascular plants with microphylls; extremely diverse in appearance. (Heterosporous plants produce two kinds of spores, one kind giving rise to male gametophytes, the other to female gametophytes.) All lycophytes have motile sperm. There are five genera, with about 1,000 species.

Division Sphenophyta: horsetails. Homosporous vascular plants with jointed stems marked by conspicuous nodes and elevated siliceous ribs. Sporangia are borne in a cone at the apex of the stem. Leaves are scalelike. Sperm are motile. Although now thought to have evolved from a megaphyll (a complex leaf), the leaves of the horsetails are structurally indistinguishable from microphylls. There is one genus, *Equisetum,* with 15 living species.

Division Pterophyta: ferns. They are mostly homosporous, although some are heterosporous. All possess megaphylls. The gametophyte is more or less free-living and usually photosynthetic. Multicellular gametangia and free-swimming sperm are present. About 12,000 species.

Division Coniferophyta: conifers. Seed plants with active cambial growth and simple, needle-like leaves; the ovules are not enclosed and the sperm are not flagellated. There are some 50 genera, with about 550 species; the most familiar group of gymnosperms.

Division Cycadophyta: cycads. Seed plants with sluggish cambial growth and pinnately compound, palmlike or fernlike leaves. The ovules are not enclosed. The sperm are flagellated and motile but are carried to the ovule in a pollen tube. Cycads are gymnosperms. There are 10 genera, with about 100 species.

Division Ginkgophyta: ginkgo. Seed plants with active cambial growth and fan-shaped leaves with open dichotomous venation. The ovules are not enclosed and are fleshy at maturity. Sperm are carried to the ovule in a pollen tube but are flagellated and motile. They are gymnosperms. There is one species only.

Division Gnetophyta: gnetophytes. Seed plants with many angiospermlike features. They are the only gymnosperms in which vessels are present in the xylem. Motile sperm are absent. There are three very distinctive genera, with about 70 species.

Division Anthophyta: flowering plants (angiosperms). Seed plants in which the ovules are enclosed in a carpel (in all but a very few genera), and the seeds at maturity are borne within fruits. They are extremely diverse vegetatively but are characterized by the flower, which is basically insect-pollinated. Other modes of pollination, such as wind pollination, have been derived in a number of different lines. The gametophytes are much reduced, with the female gametophyte often consisting of only seven cells at maturity. Double fertilization involving the two nonmotile sperm nuclei of the mature male gametophyte gives rise to the zygote and to the primary endosperm nucleus; the former becomes the embryo and the latter a special nutritive tissue, the endosperm. About 235,000 species.

Class

Class Monocotyledones: monocots. Flower parts are usually in threes, leaf venation is usually parallel, vascular bundles in the young stem are scattered, true secondary growth is not present, and there is one cotyledon. About 65,000 species.

Class Dicotyledones: dicots. Flower parts are usually in fours or fives, leaf venation is usually netlike, the vascular bundles in the young stem are in a ring, there is true secondary growth with vascular cambium commonly present, and there are two cotyledons. About 170,000 species.

KINGDOM ANIMALIA

Eukaryotic multicellular organisms. Their principal mode of nutrition is by ingestion. Many animals are motile, and they generally lack the rigid cell walls characteristic of plants. Considerable cellular migration and reorganization of tissues often occur during the course of embryonic development. Their reproduction is primarily sexual, with male and female diploid organisms producing haploid gametes that fuse to form the zygote. More than 1.5 million living species have been described, and the actual number may be more than 50 million.

Phylum

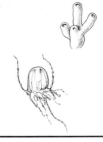

Phylum Porifera: sponges. Simple multicellular animals, largely marine, with stiff skeletons, and bodies perforated by many pores that admit water containing food particles. All have choanocytes, "collar cells." About 5,000 species.

Phylum Mesozoa: extremely simple wormlike animals, all parasites of marine invertebrates. The body consists of 20 to 30 cells, organized in two layers. About 50 species.

Phylum Cnidaria: polyps and jellyfishes. Radially symmetrical animals with a gastrovascular cavity and two-layered bodies of a jellylike consistency. Reproduction is asexual or sexual. They are the only organisms with cnidocytes, special stinging cells. All are aquatic and most are marine. About 9,000 species.

Class

Class Hydrozoa: Hydra, Obelia, and other *Hydra*-like animals. They are often colonial and frequently have a regular alternation of asexual and sexual forms. The polyp form is dominant.

Class Scyphozoa ("cup animals"): marine jellyfishes, including *Aurelia.* The medusa form is dominant. They have true muscle cells.

Class Anthozoa ("flower animals"): sea anemones, colonial corals, and related forms. They have no medusa stage.

Phylum

Phylum Ctenophora: comb jellies and sea walnuts. They are free-swimming, often almost spherical animals, with a gastrovascular cavity. They are translucent, gelatinous, delicately colored, and often bioluminescent. They possess eight bands of cilia, for locomotion. About 90 species.

Phylum Platyhelminthes: flatworms. Bilaterally symmetrical with three embryonic tissue layers. The digestive cavity has only one opening. They have no coelom or pseudocoelom and no circulatory system. They have complex hermaphroditic reproductive systems and excrete by means of special flame cells. About 13,000 species.

Class

Class Turbellaria: planarians and other nonparasitic flatworms. They are ciliated, carnivorous, and have ocelli ("eyespots").

Class Trematoda: flukes. They are parasitic flatworms with digestive cavities.

Class Cestoda: tapeworms. They are parasitic flatworms with no digestive cavities; they absorb nourishment through body surfaces.

Phylum

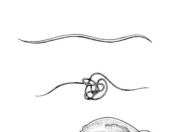

Phylum Gnathostomulida: tiny acoelomate marine worms, characterized by a unique pair of hard jaws. The digestive cavity has only one opening. About 80 species.

Phylum Rhynchocoela: proboscis, nemertine, or ribbon worms. These acoelomate worms are nonparasitic, usually marine, and have a tubelike gut with mouth and anus, a retractile proboscis armed with a hook for capturing prey, and simple circulatory and reproductive systems. About 650 species.

Phylum Nematoda: roundworms. The phylum includes minute free-living forms, such as vinegar eels, and plant and animal parasites, such as hookworms. They have a pseudocoelom and are characterized by elongated, cylindrical, bilaterally symmetrical bodies. About 12,000 species have been described and named, and there may be as many as 400,000 to 500,000 species.

Phylum Nematomorpha: horsehair worms. They are extremely slender, brown or black pseudocoelomate worms up to 1 meter long. Adults are free-living, but the larvae are parasitic in arthropods. About 230 species.

Phylum Acanthocephala: spiny-headed worms. They are parasitic pseudocoelomate worms with no digestive tract and a head armed with many recurved spines. About 500 species.

Phylum Kinorhyncha: tiny pseudocoelomate worms that burrow in muddy ocean shores. They are short-bodied, covered with spines, and have a spiny retractile proboscis. About 100 species.

Phylum Gastrotricha: microscopic, wormlike pseudocoelomate animals that move by longitudinal bands of cilia. About 400 species.

Phylum Loricifera: microscopic pseudocoelomate animals of the ocean bottom, characterized by a plate-covered body, numerous spines projecting from the head, and a retractile tube for a mouth. The first of the three known species was discovered in 1982; there may be many more.

Phylum Rotifera: microscopic, wormlike or spherical pseudocoelomate animals, "wheel animalcules." They have a complete digestive tract, flame cells, and a circle of cilia on the head, the beating of which suggests a wheel; males are minute and either degenerate or unknown in many species. About 1,500 to 2,000 species.

Phylum Entoprocta: microscopic, stalked, sessile animals that superficially resemble hydrozoans but are much more complex, having three embryonic tissue layers, a pseudocoelom, and a complete digestive tract. They were long misclassified with the coelomate bryozoans (ectoprocts), which they also resemble. About 75 species.

Phylum Mollusca: unsegmented coelomate animals, with a head, a mantle, and a muscular foot, variously modified. They are mostly aquatic; soft-bodied, often with one or more hard shells, and a heart with three chambers. All mollusks, except bivalves, have a radula (rasplike organ used for scraping or marine drilling). At least 47,000 living species, and perhaps many more; an additional 35,000 species are known only from their fossils.

Class

Class Aplacophora: solenogasters. Wormlike marine animals, with no clearly defined shell, mantle, or foot. The presence of a radula identifies them as mollusks. About 250 species.

Class Polyplacophora: chitons. The mollusks with the closest resemblance to the hypothetical primitive form, they have an elongated body covered with a mantle in which are embedded eight dorsal shell plates. About 600 species.

Class Monoplacophora: Neopilina. Mostly deep-sea mollusks with a large, single dorsal shell and multiple pairs of gills, nephridia, and retractor muscles. Two genera with eight species.

Class Scaphopoda: tooth or tusk shells. They are marine mollusks with a conical tubular shell. About 350 species.

Class Bivalvia: two-shelled mollusks, including clams, oysters, mussels, scallops. They usually have a hatchet-shaped foot and no distinct head. Generally sessile. At least 7,500 species.

Class Gastropoda: asymmetrical mollusks, including snails, whelks, slugs. They usually have a spiral shell and a head with one or two pairs of tentacles. At least 37,500 living species and about 15,000 fossil species.

Class Cephalopoda: octopuses, squids, *Nautilus.* They are characterized by a "head-foot" with eight or ten arms or many tentacles, mouth with two horny jaws, and well-developed eyes and nervous system. The shell is external *(Nautilus),* internal (squid), or absent (octopus). All except *Nautilus* have ink glands. About 600 species.

Phylum

Phylum Annelida: segmented worms. They usually have a well-developed coelom, a one-way digestive tract, head, circulatory system, nephridia, and well-defined nervous system. About 9,000 species.

Class

Class Oligochaeta: soil, freshwater, and marine annelids, including *Lumbricus* and other earthworms. They have scanty bristles and usually a poorly differentiated head. About 3,000 species.

Class Polychaeta: mainly marine worms, such as *Nereis.* They have a distinct head with tentacles, antennae, and specialized mouthparts. Parapodia are often brightly colored. About 6,000 species.

Class Hirudinea: leeches. They have a posterior sucker and usually an anterior sucker surrounding the mouth. They are freshwater, marine, and terrestrial; either free-living or parasitic. About 300 species.

Phylum

Phylum Sipuncula: peanut worms. Unsegmented marine worms with a stout body, a long, retractile proboscis, and a trochophore larva that resembles those of the polychaete annelids. About 300 species.

Phylum Echiura: spoon worms. Marine worms with a nonretractile proboscis that contracts to form a structure resembling a spoon. Their embryonic development and trochophore larvae resemble those of the polychaete annelids. About 100 species.

Phylum Priapulida: predatory burrowing marine worms, characterized by a retractile, spine-bearing proboscis. They resemble the pseudocoelomate kinorhynchs, but may have a true coelom. Only 15 species.

Phylum Pogonophora: beard worms. Unsegmented except at the posterior end, these slender marine worms live in long tubes in deep-sea sediments. Although they have no mouth or digestive tract, the anterior region of the body bears a crown of tentacles. About 100 species.

Phylum Pentastomida: wormlike parasites of vertebrate respiratory systems, sometimes with two pairs of short appendages at the anterior end of the body. They lack circulatory, respiratory, and excretory systems, but the nervous system resembles those of annelids and arthropods. About 70 species.

Phylum Tardigrada: water bears. Tiny segmented animals, with a thin cuticle and four pairs of stubby legs. They are common in fresh water and in the moisture film on mosses; when water is unavailable, they enter a state of suspended animation. About 350 species.

Phylum Onychophora: *Peripatus.* Caterpillarlike animals with many short, unjointed pairs of legs. Their relatively soft bodies, segmentally arranged nephridia, muscular body walls, and ciliated reproductive tracts resemble those of the annelids; their antennae and eyes resemble those of both the polychaete annelids and the arthropods; their jaws, protective cuticle, brain, and circulatory and respiratory systems resemble those of arthropods. Most give birth to live young, and in some species the embryo is nourished through a placenta. About 70 species.

Phylum Arthropoda: The largest phylum in the animal kingdom, arthropods are segmented animals with paired, jointed appendages, a hard jointed exoskeleton, a complete digestive tract, reduced coelom, no nephridia, a dorsal brain, and a ventral nerve cord with paired ganglia in each segment. More than 1 million species have been classified to date.

Class

Class Merostomata: horseshoe crabs. Aquatic arthropods with chelicerae (pincers or fangs), pedipalps, compound eyes, four pairs of walking legs, and book gills. Only four species.

Class Pycnogonida: sea spiders. Aquatic chelicerates with slender bodies and four or, rarely, five pairs of legs, which are often very long. About 500 species.

Class Arachnida: spiders, mites, ticks, scorpions. Most members are terrestrial, air-breathing; four pairs of walking legs; chelicerae may be pincers or fangs; pedipalps are usually sensory. About 57,000 species.

Class Crustacea: lobsters, crayfish, crabs, shrimps. Crustaceans are mostly aquatic, with compound eyes, two pairs of antennae, one pair of mandibles, and typically two pairs of maxillae. The thoracic segments have appendages, and the abdominal segments are with or without appendages. About 25,000 species.

Class Chilopoda: centipedes. They have a head and 15 to 177 trunk segments, each with one pair of jointed appendages. About 3,000 species.

Class Diplopoda: millipedes. They have a head and a trunk with 20 to 200 body rings, each with two pairs of appendages. About 7,500 species.

Class Pauropoda: tiny, soft-bodied arthropods that resemble millipedes but have only 11 or 12 segments and 9 or 10 pairs of legs. They have branched antennae. About 300 species.

Class Symphyla: garden centipedes and their relatives. Soft-bodied arthropods with one pair of antennae, 12 pairs of jointed legs with claws, and a pair of unjointed posterior appendages. About 130 species.

Class Insecta: insects, including bees, ants, beetles, butterflies, fleas, lice, flies, etc. Most insects are terrestrial and breathe by means of tracheae. The body has three distinct parts (head, thorax, and abdomen); the head bears compound eyes and one pair of antennae; the thorax bears three pairs of legs and usually two pairs of wings. About 1 million species.

Phylum

Phylum Bryozoa (Ectoprocta): "moss" animals. These microscopic aquatic organisms are characterized by the lophophore, a crown of hollow, ciliated tentacles, with which they feed and by a U-shaped digestive tract. They usually form fixed and branching colonies. Coelomates, they superficially resemble the pseudocoelomate entoprocts. Some species retain larvae in a special brood pouch. About 4,000 species.

Phylum Phoronida: sedentary, elongated worms that secrete and live in a leathery tube. They have a U-shaped digestive tract and a ring of ciliated tentacles (the lophophore) with which they feed. Marine. Only 18 species.

Phylum Brachiopoda: lamp shells. Marine animals with two hard shells (one dorsal and one ventral), they superficially resemble clams. Fixed by a stalk or one shell in adult life, they feed by means of a lophophore. About 250 living species; 30,000 extinct species.

Phylum Echinodermata: starfishes and sea urchins. Echinoderms are radially symmetrical in adult stage, with well-developed coelomic cavities, an endoskeleton of calcareous ossicles and spines, and a unique water vascular system. They have tube feet. All are marine. About 6,000 living species and 20,000 species known only from their fossils.

Class

Class Crinoidea: sea lilies and feather stars. Sessile animals, they often have a jointed stalk for attachment, and they have 10 arms bearing many slender lateral branches. Most species are fossils; only 20 living species.

Class Stelleroidea: starfishes and brittle stars. They have 5 to 50 arms, an oral surface directed downward, and rows of tube feet on each arm. Brittle stars have greatly elongated, highly flexible slender arms and rapid horizontal locomotion.

Class Echinoidea: sea urchins and sand dollars. Skeletal plates form rigid external covering that bears many movable spines.

Class Concentricycloidea: sea daisies. The microscopic members of this newly established class have five plates on the dorsal surface, two rings of tube feet on the ventral surface, and no digestive tract.

Class Holothuroidea: sea cucumbers. They have a sausage-shaped or wormlike elongated body.

Phylum

Phylum Chaetognatha: arrow worms. Free-swimming, planktonic marine worms, they have a coelom, a complete digestive tract, and a mouth with strong sickle-shaped hooks on each side. About 60 species.

Phylum Hemichordata: acorn worms. The body is divided into three regions—proboscis, collar, and trunk. The coelomic cavities provide a hydrostatic skeleton similar to the water vascular system of echinoderms, and their larvae resemble starfish larvae. They have both ventral and dorsal nerve cords, and the anterior portion of the dorsal cord is hollow in some species. They also have a pharynx with gill slits. About 80 species.

Phylum Chordata: animals having at some stage a notochord, pharyngeal gill slits (or pouches), a hollow nerve cord on the dorsal side, and a tail. More than 45,000 species.

Subphylum

Subphylum Cephalochordata: lancelets. This small subphylum contains only *Branchiostoma* and related forms. They are somewhat fishlike marine animals with a permanent notochord the whole length of the body, a nerve cord, a pharynx with gill slits, and no cartilage or bone. About 28 species.

Subphylum Urochordata: tunicates. Adults are saclike, usually sessile, often forming branching colonies. They feed by ciliary currents, have gill slits, a reduced nervous system, and no notochord. Larvae are active, with well-developed nervous system and notochord. They are marine. About 1,300 species.

Subphylum Vertebrata: vertebrates. The major subphylum of Chordata. In the vertebrates the notochord is an embryonic structure; it is typically replaced in the course of development by cartilage or bone, forming the segmented vertebral column, or backbone. A cranium, or skull, surrounds a well-developed brain. Vertebrates usually have a tail. About 44,000 species.

Class

Class Agnatha: lampreys and hagfish. These are eel-like aquatic vertebrates without limbs, with a jawless sucking mouth, and no bones, scales, or fins. About 60 species.

Class Chondrichthyes: sharks, rays, skates, and other cartilaginous fishes. They have complicated copulatory organs, scales, and no swim bladders. They are almost exclusively marine. About 625 species.

Class Osteichthyes: bony fishes, including nearly all modern freshwater fishes, such as sturgeon, trout, perch, anglerfish, lungfish, and some almost extinct groups. They usually have a swim bladder or (rarely) a lung. More than 21,000 species.

Class Amphibia: salamanders, frogs, and toads. They usually respire by gills in the larval stage and by lungs in the adult stage. They have incomplete double circulation and a usually naked skin. The limbs are legs. They were the first vertebrates to inhabit the land and were ancestors of the reptiles. Their eggs are unprotected by a shell and embryonic membranes. About 2,500 species.

Class Reptilia: turtles, lizards, snakes, crocodiles; includes many extinct species such as the dinosaurs. Reptiles breathe by lungs and have incomplete double circulation. Their skin is usually covered with scales. The four limbs are legs (absent in snakes and some lizards). Almost all are ectotherms. Most live and reproduce on land, although some are aquatic. The embryo is enclosed in an egg shell and has protective membranes. About 6,000 species.

Class Aves: birds. Birds are homeothermic animals with complete double circulation and a skin covered with feathers. The forelimbs are wings. The embryo is enclosed in an egg shell with protective membranes. About 9,000 living species; an estimated 14,000 species are extinct.

Class Mammalia: mammals. Mammals are homeothermic animals with complete double circulation. Their skin is usually covered with hair. The young are nourished with milk secreted by the mother. They have four limbs, usually legs (forelimbs sometimes arms, wings, or fins), a diaphragm used in respiration, a lower jaw made up of a single pair of bones, three bones in each middle ear connecting eardrum and inner ear, and almost always seven vertebrae in the neck. About 4,500 species.

Subclass

Subclass Prototheria: monotremes. These are the oviparous (egg-laying) mammals with imperfect temperature regulation. There are only three living species: the duckbilled platypus and spiny anteaters of Australia and New Guinea.

Subclass Metatheria: marsupials, including kangaroos, opossums, and others. Marsupials are viviparous mammals, usually with a yolk-sac placenta; the young are born in a very immature state and are carried in an external pouch of the mother for some time after birth. They are found chiefly in Australia and South America. About 260 species.

Subclass Eutheria: mammals with a well-developed chorioallantoic placenta. This subclass comprises the great majority of living mammals. There are 16 orders of Eutheria:

Order

Insectivora: shrews, moles, hedgehogs, etc.

Chiroptera: bats. Aerial mammals with the forelimbs wings.

Dermoptera: "flying" lemurs.

Edentata: toothless mammals—anteaters, sloths, armadillos, etc.

Lagomorpha: rabbits and hares.

Carnivora: carnivorous animals—cats, dogs, bears, weasels, seals, etc.

Tubulidentata: aardvarks.

Pholidota: pangolins.

Rodentia: rodents—rats, mice, squirrels, etc.

Artiodactyla: even-toed ungulates (hoofed mammals)—cattle, deer, camels, hippopotamuses, etc.

Perissodactyla: odd-toed ungulates—horses, zebras, rhinoceroses, etc.

Proboscidea: elephants.

Hyracoidea: hyraxes.

Sirenia: manatees, dugongs, and sea cows. Large aquatic mammals with the forelimbs finlike, the hind limbs absent.

Cetacea: whales, dolphins, and porpoises. Aquatic mammals with the forelimbs fins, the hind limbs absent.

Primates: prosimians, monkeys, apes, and humans.

Glossary

This list does not include units of measure, which can be found inside the back cover, or names of taxonomic groups, which can be found in Appendix B, or terms that are used only once in the text and defined there.

abdomen: In vertebrates, the portion of the trunk containing visceral organs other than heart and lungs; in arthropods, the posterior portion of the body, made up of similar segments and containing the reproductive organs and part of the digestive tract.

abiotic [Gk. *a,* not, without + *bios,* life]: Nonliving; specifically, the nonliving components of an ecosystem, such as temperature, humidity, the mineral content of the soil, etc.

abscisic acid [L. *ab,* away, off + *scissio,* dividing]: A plant hormone with a variety of inhibitory effects; brings about dormancy in buds, maintains dormancy in many types of seeds, and effects stomatal closing; also known as the "stress hormone."

abscission [L. *ab,* away, off + *scissio,* dividing]: In plants, the dropping of leaves, flowers, fruits, or stems at the end of a growing season, as the result of formation of a two-layered zone of specialized cells (the abscission zone) and the action of a hormone (ethylene).

absorption [L. *absorbere,* to swallow down]: The movement of water and dissolved substances into a cell, tissue, or organism.

acetylcholine (asset-ill-**coal**-een): One of the principal chemicals (neurotransmitters) responsible for the transmission of nerve impulses across synapses.

acid [L. *acidus,* sour]: A substance that causes an increase in the relative number of hydrogen ions (H^+) in a solution and a decrease in the relative number of hydroxide ions (OH^-); having a pH of less than 7; the opposite of a base.

ACTH: Abbreviation of adrenocorticotropic hormone.

actin [Gk. *aktis,* a ray]: A protein, composed of globular subunits, that forms filaments that are among the principal components of the cytoskeleton. Also one of the two major proteins of muscle (the other is myosin); the principal constituent of the thin filaments.

action potential: A transient change in electric potential across a membrane; in nerve cells, results in conduction of a nerve impulse; in muscle cells, results in contraction.

activation energy: The energy that must be possessed by atoms or molecules in order to react.

active site: The region of an enzyme surface that binds the substrate during the reaction catalyzed by the enzyme.

active transport: The energy-requiring transport of a solute across a plasma membrane (or a membrane of an organelle) from a region of lower concentration to a region of higher concentration (that is, against a concentration gradient).

adaptation [L. *adaptare,* to fit]: (1) The evolution of features that make a group of organisms better suited to live and reproduce in their environment. (2) A peculiarity of structure, physiology, or behavior that aids the organism in its environment.

adaptive radiation: The evolution from a primitive and unspecialized ancestor of a number of divergent forms, each specialized to fill a distinct ecological niche; associated with the opening up of a new biological frontier.

adenosine diphosphate (ADP): A nucleotide consisting of adenine, ribose, and two phosphate groups; formed by the removal of one phosphate from an ATP molecule.

adenosine monophosphate (AMP): A nucleotide consisting of adenine, ribose, and one phosphate group; can be formed by the removal of two phosphates from an ATP molecule; in its cyclic form, functions as a "second messenger" for a number of vertebrate hormones and neurotransmitters.

adenosine triphosphate (ATP): The nucleotide that provides the energy currency for cell metabolism; composed of adenine, ribose, and three phosphate groups. On hydrolysis, ATP loses one phosphate group to become adenosine diphosphate (ADP), releasing energy in the process. ATP is formed from ADP and phosphate in an enzymatic reaction that traps the energy released by the breakdown of glucose and other molecules or the energy captured in photosynthesis.

ADH: Abbreviation of antidiuretic hormone.

adhesion [L. *adhaerere,* to stick to]: The holding together of molecules of different substances.

ADP: Abbreviation of adenosine diphosphate.

adrenal gland [L. *ad,* near + *renes,* kidney]: A vertebrate endocrine gland. The cortex (outer surface) is the source of cortisol, aldosterone, and other steroid hormones; the medulla (inner core) secretes adrenaline and noradrenaline.

adrenaline: A hormone, produced by the medulla of the adrenal gland, that increases the concentration of glucose in the blood, raises blood pressure and heartbeat rate, and increases muscular power and resistance to fatigue; also a neurotransmitter across synaptic junctions. Also called epinephrine.

adrenocorticotropic hormone (ACTH): A hormone, produced by the anterior lobe of the pituitary gland, that stimulates the production of cortisol by the adrenal cortex.

adventitious [L. *adventicius,* not properly belonging to]: Referring to a structure arising from an unusual place, such as roots growing from stems or leaves.

aerobic [Gk. *aēr,* air + *bios,* life]: Any biological process that can occur in the presence of molecular oxygen (O_2).

afferent [L. *ad,* near + *ferre,* to carry]: Bringing inward to a central part, applied to nerves and blood vessels.

agar: A gelatinous material prepared from certain red algae that is used to solidify nutrient media for growing microorganisms.

aldosterone [Gk. *aldainō,* to nourish + *stereō,* solid]: A hormone produced by the adrenal cortex that affects the concentration of ions in the blood; it stimulates the reabsorption of sodium and the excretion of potassium by the kidney.

aleurone layer [Gk. *aleuron*, flour]: The outermost cell layer of the endosperm of the grains (seeds) of wheat and other grasses; when acted upon by gibberellin, the aleurone layer releases enzymes that digest the stored food of the endosperm into small nutrient molecules that can be taken up by the embryo.

alga, *pl.* **algae** (al-gah, al-jee): A unicellular or simple multicellular eukaryotic photosynthetic organism lacking multicellular sex organs.

alkaline: Pertaining to substances that increase the relative number of hydroxide ions (OH^-) in a solution; having a pH greater than 7; basic; opposite of acidic.

allantois [Gk. *allant*, sausage]: One of the four extraembryonic membranes that form during the development of reptiles, birds, and mammals.

allele frequency: The proportion of a particular allele in a population.

alleles (al-eels) [Gk. *allelon*, of one another]: Two or more different forms of a gene. Alleles occupy the same position (locus) on homologous chromosomes and are separated from each other at meiosis.

allergic reaction: An inflammatory response triggered by a weak antigen (an allergen) to which most individuals do not react; involves the release of large amounts of histamine from mast cells.

allopatric speciation [Gk. *allos*, other + *patra*, fatherland, country]: Speciation that occurs as the result of the geographic separation of a population of organisms.

alternation of generations: A sexual life cycle in which a haploid *(n)* phase alternates with a diploid *(2n)* phase. The gametophyte *(n)* produces gametes *(n)* by mitosis. The fusion of gametes yields zygotes *(2n)*. Each zygote develops into a sporophyte *(2n)* that forms haploid spores *(n)* by meiosis. Each haploid spore forms a new gametophyte, completing the cycle.

altruism: Self-sacrifice for the benefit of others; any form of behavior that increases the fitness of the recipient while reducing the fitness of the altruistic individual.

alveolus, *pl.* **alveoli** [L. dim. of *alveus*, cavity, hollow]: One of the many small air sacs within the lungs in which the bronchioles terminate. The thin walls of the alveoli contain numerous capillaries and are the site of gas exchange between the air in the alveoli and the blood in the capillaries.

amino acids (am-ee-no) [Gk. *Ammon*, referring to the Egyptian sun god, near whose temple ammonium salts were first prepared from camel dung]: Organic molecules containing nitrogen in the form of $-NH_2$ and a carboxyl group, $-COOH$, bonded to the same carbon atom; the "building blocks" of protein molecules.

aminoacyl–tRNA synthetase: In protein synthesis, one of a class of enzymes that catalyze the attachment of an amino acid to its specific tRNA molecule.

ammonification: The process by which decomposers break down proteins and amino acids, releasing the excess nitrogen in the form of ammonia (NH_3) or ammonium ion (NH_4^+).

amnion (am-neon) [Gk. dim. of *amnos*, lamb]: One of the four extraembryonic membranes that form during the development of reptiles, birds, and mammals; it encloses a fluid-filled space, the amniotic cavity, that surrounds the developing embryo.

amniote egg: An egg that is isolated and protected from the environment during the period of its development by a series of extraembryonic membranes and, often, a more or less impervious shell; the amniote eggs of birds and many reptiles are completely self-sufficient, requiring only oxygen from the outside.

amoeboid [Gk. *amoibē*, change]: Moving or feeding by means of pseudopodia (temporary cytoplasmic protrusions from the cell body).

AMP: Abbreviation of adenosine monophosphate.

anabolic steroids: Synthetic chemical variants of the male sex hormone testosterone; they produce increased muscle mass but also suppress testosterone production, leading to shrinkage of the testes, growth of the breasts, and premature baldness; long-term use increases the risk of kidney and liver damage and of liver cancer.

anabolism [Gk. *ana*, up + *-bolism* (as in metabolism)]: Within a cell or organism, the sum of all biosynthetic reactions (that is, chemical reactions in which larger molecules are formed from smaller ones).

anaerobic [Gk. *an*, without + *aēr*, air + *bios*, life]: Applied to a process that can occur without oxygen, such as fermentation; also applied to organisms that can live without free oxygen.

analogous [Gk. *analogos*, proportionate]: Applied to structures similar in function but different in evolutionary origin, such as the wing of a bird and the wing of an insect.

anaphase (anna-phase) [Gk. *ana*, up + *phasis*, form]: In mitosis and meiosis II, the stage in which the chromatids of each chromosome separate and move to opposite poles; in meiosis I, the stage in which homologous chromosomes separate and move to opposite poles.

androgens [Gk. *andros*, man + *genos*, origin, descent]: Male sex hormones; any chemical with actions similar to those of testosterone.

angiosperms (an-jee-o-sperms) [Gk. *angeion*, vessel + *sperma*, seed]: The flowering plants. Literally, a seed borne in a vessel; thus, any plant whose seeds are borne within a matured ovary (fruit).

annual plant [L. *annus*, year]: A plant that completes its life cycle (from seed germination to seed production) and dies within a single growing season.

antennae: Long, paired sensory appendages on the head of many arthropods.

anterior [L. *ante*, before, toward, in front of]: The front end of an organism.

anther [Gk. *anthos*, flower]: In flowering plants, the pollen-bearing portion of a stamen.

antheridium, *pl.* **antheridia**: In bryophytes and some vascular plants, the multicellular sperm-producing organ.

anthropoid [Gk. *anthropos*, man, human]: A higher primate; includes monkeys, apes, and humans.

antibiotic [Gk. *anti*, against + *bios*, life]: An organic compound, inhibitory or toxic to other species, that is formed and secreted by an organism.

antibody [Gk. *anti*, against]: A globular protein, synthesized by a B lymphocyte, that is complementary to a foreign substance (antigen) with which it combines specifically.

anticodon: In a tRNA molecule, the three-nucleotide sequence that base pairs with the mRNA codon for the amino acid carried by that particular tRNA; the anticodon is complementary to the mRNA codon.

antidiuretic hormone (ADH) [Gk. *anti*, against + *diurgos*, thoroughly wet + *hormaein*, to excite]: A peptide hormone synthesized in the hypothalamus that inhibits urine excretion by inducing the reabsorption of water from the nephrons of the kidneys; also called vasopressin.

antigen [Gk. *anti*, against + *genos*, origin, descent]: A foreign substance, usually a protein or polysaccharide, that, when bound to a complementary antibody displayed on the surface of a B lymphocyte or to a complementary T-cell receptor, stimulates an immune response.

aorta (a-ore-ta) [Gk. *aeirein*, to lift, heave]: The major artery in blood-circulating systems; the aorta sends blood to the other body tissues.

apical dominance [L. *apex*, top]: In plants, the hormone-mediated influence of a terminal bud in suppressing the growth of axillary buds.

apical meristem [L. *apex*, top + Gk. *meristos*, divided]: In vascular plants, the growing point at the tip of the root or stem.

arboreal [L. *arbor*, tree]: Tree-dwelling.

archegonium, *pl.* **archegonia** [Gk. *archegonos*, first of a race]: In bryophytes and some vascular plants, the multicellular egg-producing organ.

archenteron [Gk. *arch*, first, or main + *enteron*, gut]: The main cavity within the early embryo (gastrula) of many animals; lined with endoderm, it opens to the outside by means of the blastopore and ultimately becomes the digestive tract.

artery: A vessel carrying blood from the heart to the tissues; arteries are usually thick-walled, elastic, and muscular. A small artery is known as an arteriole.

artificial selection: The breeding of selected organisms for the purpose of producing descendants with desired characteristics.

asexual reproduction: Any reproductive process, such as budding or the division of a cell or body into two or more approximately equal parts, that does not involve the union of gametes.

assimilation: The energy-requiring process by which plant cells convert nitrate ions (NO_3^-) taken up by the roots of plants into ammonium ions (NH_4^+), which can then be used in the synthesis of amino acids and other nitrogenous compounds.

atmospheric pressure [Gk. *atmos,* vapor + *sphaira,* globe]: The weight of the Earth's atmosphere over a unit area of the Earth's surface.

atom [Gk. *atomos,* indivisible]: The smallest particle into which a chemical element can be divided and still retain the properties characteristic of the element; consists of a central core, the nucleus, containing protons and neutrons, and electrons that move around the nucleus.

atomic number: The number of protons in the nucleus of an atom; equal to the number of electrons in the neutral atom.

atomic weight: The number of protons plus neutrons in the nucleus of an atom.

ATP: Abbreviation of adenosine triphosphate, the principal energy-carrying compound of the cell.

ATP synthase: The enzyme complex in the inner membrane of the mitochondrion and the thylakoid membrane of the chloroplast through which protons flow down the gradient established in the first stage of chemiosmotic coupling; the site of formation of ATP from ADP and phosphate during oxidative phosphorylation and photophosphorylation.

atrioventricular node [L. *atrium,* yard, court, hall + *ventriculus,* the stomach + *nodus,* knot]: A group of slow-conducting fibers in the atrium of the vertebrate heart that are stimulated by impulses originating in the sinoatrial node (the pacemaker) and that conduct impulses to the bundle of His, a group of fibers that stimulate contraction of the ventricles.

atrium, *pl.* **atria** (a-tree-um) [L., yard, court, hall]: A thin-walled chamber of the heart that receives blood and passes it on to a thick, muscular ventricle.

autoimmune disease: A disorder, such as multiple sclerosis, in which an individual's immune system makes antibodies against his or her own cells.

autonomic nervous system [Gk. *autos,* self + *nomos,* usage, law]: In the peripheral nervous system of vertebrates, the neurons and ganglia that are not ordinarily under voluntary control; innervates the heart, glands, visceral organs, and smooth muscle; subdivided into the sympathetic and parasympathetic divisions.

autosome [Gk. *autos,* self + *soma,* body]: Any chromosome other than the sex chromosomes. Humans have 22 pairs of autosomes and one pair of sex chromosomes.

autotroph [Gk. *autos,* self + *trophos,* feeder]: An organism that is able to synthesize all needed organic molecules from simple inorganic substances (e.g., H_2O, CO_2, NH_3) and some energy source (e.g., sunlight); in contrast to heterotroph. Plants, algae, and some groups of prokaryotes are autotrophs.

auxin [Gk. *auxein,* to increase + *in,* of, or belonging to]: One of a group of plant hormones with a variety of growth-regulating effects, including promotion of cell elongation.

axillary bud [Gk. *axilla,* armpit]: In plants, a bud located just above the point at which a leaf is (or was) attached to the stem; axillary buds give rise to lateral branches or specialized shoots, such as flowers.

axis: An imaginary line passing through a body or organ around which parts are symmetrically aligned.

axon [Gk. *axon,* axle]: A long process of a neuron, or nerve cell, that is capable of rapidly conducting nerve impulses over great distances; a nerve fiber.

B cell: A type of white blood cell capable of becoming an antibody-secreting plasma cell; a B lymphocyte.

bacteriophage [L. *bacterium* + Gk. *phagein,* to eat]: A virus that parasitizes a bacterial cell.

bark: In plants, all tissues outside the vascular cambium in a woody stem.

basal body [Gk. *basis,* foundation]: A cytoplasmic organelle of animals and some protists, from which cilia or flagella arise; identical in structure to the centriole, which is involved in mitosis and meiosis in animals and some protists.

basal metabolism [Gk. *basis,* foundation + *metabole,* change]: The energy expenditure of an animal at rest.

base: A substance that causes an increase in the relative number of hydroxide ions (OH^-) in a solution and a decrease in the relative number of hydrogen ions (H^+); having a pH of more than 7; the opposite of an acid. *See* Alkaline.

base-pairing principle: In the formation of nucleic acids, the requirement that adenine must always pair with thymine (or uracil) and guanine with cytosine.

behavior: All of the acts an organism performs, as in, for example, seeking a suitable habitat, obtaining food, avoiding predators, and seeking a mate and reproducing.

biennial [L. *biennium,* a space of two years; *bi,* twice + *annus,* year]: Occurring once in two years; a plant that requires two years to complete its reproductive cycle; vegetative growth occurs in the first year, sexual reproduction and death in the second.

bilateral symmetry [L. *bi,* twice, two + *lateris,* side; Gk. *summetros,* symmetry]: A body form in which the right and left halves of an organism are approximate mirror images of each other.

bile: A yellow secretion of the vertebrate liver, temporarily stored in the gallbladder and composed of organic salts that emulsify fats in the small intestine.

binomial system [L. *bi,* twice, two + Gk. *nomos,* usage, law]: A system of naming organisms in which the name consists of two parts, with the first designating genus and the second, species; originated by Linnaeus.

biochemical pathway [Gk. *bios,* life + *chēmeia,* alchemy]: An ordered series of chemical reactions in a living cell, in which each step is catalyzed by a specific enzyme; different biochemical pathways serve different functions in the life of the cell.

biogeochemical cycle [Gk. *bios,* life + *geō,* earth + *chēmeia,* alchemy; *kyklos,* circle, wheel]: The cyclic path of an inorganic substance, such as carbon or nitrogen, through an ecosystem. Its geological components are the atmosphere, the crust of the Earth, and the oceans, lakes, and rivers; its biological components are producers, consumers, and detritivores, including decomposers.

biological clock [Gk. *bios,* life + *logos,* discourse]: Proposed internal factor(s) in organisms that governs functions that occur rhythmically in the absence of external stimuli.

biomass [Gk. *bios,* life]: Total dry weight of all organisms (or some group of organisms) living in a particular habitat or place.

biome: One of the major types of distinctive plant formations; for example, the temperate grassland biome, the tropical rain forest biome, etc.

biosphere [Gk. *bios,* life + *sphaira,* globe]: The zones of air, land, and water at the surface of the Earth occupied by living things.

biosynthesis [Gk. *bios,* life + *synthesis,* a putting together]: Formation by living organisms of organic compounds from elements or simple compounds.

biotic [Gk. *bios*, life]: Of or pertaining to life; specifically, the living components of an ecosystem.

bipedal [L. *bi*, twice, two + *pes*, foot]: Walking upright on two feet.

blade: (1) The broad, expanded part of a leaf. (2) The broad, expanded photosynthetic part of the thallus of a multicellular alga or a simple plant.

blastocoel [Gk. *blastos*, sprout + *koilos*, a hollow]: The fluid-filled cavity in the interior of a blastula.

blastocyst [Gk. *blastos*, sprout + *kystis*, sac]: The blastula stage of a developing mammal; consists of an inner cell mass that will give rise to the embryo proper and a double layer of cells, the trophoblast, that is the precursor of the chorion.

blastodisc [Gk. *blastos*, sprout + *discos*, a round plate]: Disklike area on the surface of a large, yolky egg that undergoes cleavage and gives rise to the embryo.

blastopore [Gk. *blastos*, sprout + *poros*, a way, means, path]: In the gastrula stage of an embryo, the opening that connects the archenteron with the outside; represents the future mouth in some animals (protostomes), the future anus in others (deuterostomes).

blastula [Gk. *blastos*, sprout]: An animal embryo after cleavage and before gastrulation; usually consists of a fluid-filled sphere, the walls of which are composed of a single layer of cells.

bond strength: The strength with which a chemical bond holds two atoms together; conventionally measured in terms of the amount of energy, in kilocalories per mole, required to break the bond.

botany [Gk. *botanikos*, of herbs]: The study of plants.

Bowman's capsule: In the vertebrate kidney, the bulbous unit of the nephron, which surrounds the glomerulus. In filtration, the initial process in urine formation, blood plasma is forced from the glomerular capillaries into Bowman's capsule.

brainstem: The most posterior portion of the vertebrate brain; includes medulla, pons, and midbrain.

bronchus, *pl.* **bronchi** (bronk-us, bronk-eye) [Gk. *bronchos*, windpipe]: One of a pair of respiratory tubes branching into either lung at the lower end of the trachea; it subdivides into progressively finer passageways, the bronchioles, culminating in the alveoli.

bud: (1) In plants, an embryonic shoot, including rudimentary leaves, often protected by special bud scales. (2) In animals, an asexually produced outgrowth that develops into a new individual.

bulb: A modified bud with thickened leaves adapted for underground food storage.

bulk flow: The overall movement of a fluid induced by gravity, pressure, or an interplay of both.

bundle of His: In the vertebrate heart, a group of muscle fibers that carry impulses from the atrioventricular node to the walls of the ventricles; the only electrical bridge between the atria and the ventricles.

C₃ pathway: *See* Calvin cycle.

C₄ pathway: The set of reactions by which some plants initially fix carbon in the four-carbon compound oxaloacetic acid; the carbon dioxide is later released in the interior of the leaf and enters the Calvin cycle.

callus [L. *callos*, hard skin]: In plants, undifferentiated tissue; a term used in tissue culture, grafting, and wound healing.

calorie [L. *calor*, heat]: The amount of energy in the form of heat required to raise the temperature of 1 gram of water 1°C; in making metabolic measurements the kilocalorie (Calorie) is generally used. A Calorie is the amount of heat required to raise the temperature of 1 kilogram of water 1°C.

Calvin cycle: The set of reactions in which carbon dioxide is reduced to carbohydrate during the second stage of photosynthesis; the light-independent reactions.

calyx [Gk. *kalyx*, a husk, cup]: Collectively, the sepals of a flower.

CAM photosynthesis: *See* Crassulacean acid metabolism.

capillaries [L. *capillaris*, relating to hair]: Smallest thin-walled blood vessels through which exchanges between blood and the tissues occur; connect arteries with veins.

capillary action: The movement of water or any liquid along a surface; results from the combined effect of cohesion and adhesion.

capsid: The protein coat surrounding the nucleic acid core of a virus.

capsule (kap-sul) [L. *capsula*, a little chest]: (1) A slimy layer around the cells of certain bacteria. (2) The sporangium of a bryophyte.

carbohydrate [L. *carbo*, charcoal + *hydro*, water]: An organic compound consisting of a chain or ring of carbon atoms to which hydrogen and oxygen are attached in a ratio of approximately 2:1; carbohydrates include sugars, starch, glycogen, cellulose, etc.

carbon cycle: Worldwide circulation and reutilization of carbon atoms, chiefly due to metabolic processes of living organisms. Inorganic carbon, in the form of carbon dioxide, is incorporated into organic compounds by photosynthetic organisms; when the organic compounds are broken down in respiration, carbon dioxide is released. Large quantities of carbon are "stored" in the seas and the atmosphere, as well as in fossil fuel deposits.

carbon fixation: The light-independent reactions of the second stage of photosynthesis; energy stored in ATP and NADPH by the light-dependent reactions of the first stage is used to reduce carbon from carbon dioxide to simple sugars.

cardiovascular system [Gk. *kardio*, heart + L. *vasculum*, a small vessel]: In animals, the heart and blood vessels.

carnivore [L. *caro*, *carnis*, flesh + *voro*, to devour]: Predator that obtains its nutrients and energy by eating meat.

carotenoids [L. *carota*, carrot]: A class of pigments that includes the carotenes (yellows, oranges, and reds) and the xanthophylls (yellow); accessory pigments in photosynthesis.

carpel [Gk. *karpos*, fruit]: A leaflike floral structure enclosing the ovule or ovules of angiosperms, typically divided into ovary, style, and stigma; a flower may have one or more carpels, either single or fused.

carrying capacity: In ecology, the average number of individuals of a particular population that the environment can support under a particular set of conditions.

cartilage [L. *cartilago*, gristle]: A connective tissue in skeletons of vertebrates; forms much of the skeleton of adult lower vertebrates and immature higher vertebrates.

Casparian strip (after Robert Caspary, German botanist): In the roots of plants, a thickened, waxy strip that extends around and seals the walls of endodermal cells, restricting the diffusion of solutes across the endodermis into the vascular tissues of the root.

catabolism [Gk. *katabole*, throwing down]: Within a cell or organism, the sum of all chemical reactions in which large molecules are broken down into smaller parts.

catalyst [Gk. *katalysis*, dissolution]: A substance that lowers the activation energy of a chemical reaction by forming a temporary association with the reacting molecules; as a result, the rate of the reaction is accelerated. Enzymes are catalysts.

category [Gk. *katēgoria*, category]: In a hierarchical classification system, the level at which a particular group is ranked.

cell [L. *cella*, a chamber]: The structural unit of organisms, surrounded by a plasma membrane and composed of cytoplasm and, in eukaryotes, one or more nuclei. In most plants, fungi, and bacteria, there is a cell wall outside the plasma membrane.

cell cycle: A regular, timed sequence of the events of cell growth and division through which dividing cells pass.

cell membrane: The outer membrane of the cell; the plasma membrane.

cell plate: In the dividing cells of most plants (and in some algae), a flattened structure that forms at the equator of the mitotic spindle in early telophase; gives rise to the middle lamella.

cell theory: All living things are composed of cells; cells arise only from other cells. No exception has been found to these two principles since they were first proposed well over a century ago.

cellulose [L. *cellula,* a little cell]: The chief constituent of the cell wall in all plants and some protists; an insoluble complex carbohydrate formed of microfibrils of glucose molecules.

cell wall: A plastic or rigid structure, produced by the cell and located outside the plasma membrane in most plants, algae, fungi, and prokaryotes; in plant cells, it consists mostly of cellulose.

central nervous system: In vertebrates, the brain and spinal cord; in invertebrates it usually consists of one or more cords of nervous tissue plus their associated ganglia.

centriole (sen-tree-ole) [Gk. *kentron,* center]: A cytoplasmic organelle identical in structure to a basal body; flagellated cells and all animal cells, including those without flagella, have centrioles at the spindle poles during division.

centromere (sen-tro-mere) [Gk. *kentron,* center + *meros,* a part]: Region of constriction of chromosome that holds sister chromatids together.

cerebellum [L. dim. of *cerebrum,* brain]: A subdivision of the vertebrate brain that lies above the brainstem and behind and below the cerebrum; functions in coordinating muscular activities and maintaining equilibrium.

cerebral cortex [L. *cerebrum,* brain]: A thin layer of neurons and glial cells forming the upper surface of the cerebrum, well developed only in mammals; the seat of conscious sensations and voluntary muscular activity.

cerebrum [L., brain]: The portion of the vertebrate brain occupying the upper part of the skull, consisting of two cerebral hemispheres united by the corpus callosum; coordinates most activities.

character displacement: A phenomenon in which species that live together in the same environment tend to diverge in those characteristics that overlap; exemplified by Darwin's finches.

chemical reaction: An interaction among atoms, ions, or molecules that results in the formation of new combinations of atoms, ions, or molecules; the making or breaking of chemical bonds.

chemiosmotic coupling: The mechanism by which ADP is phosphorylated to ATP in mitochondria and chloroplasts. The energy released as electrons pass down an electron transport chain is used to establish a proton gradient across an inner membrane of the organelle; when protons subsequently flow down this electrochemical gradient, the potential energy released is captured in the terminal phosphate bonds of ATP.

chemoreceptor: A sensory cell or organ that responds to the presence of a specific chemical stimulus; includes smell and taste receptors.

chemosynthetic: Applied to autotrophic bacteria that use the energy released by specific inorganic reactions to power their life processes, including the synthesis of organic molecules.

chitin (kye-tin) [Gk. *chitōn,* a tunic, undergarment]: A tough, resistant, nitrogen-containing polysaccharide present in the exoskeleton of arthropods, the epidermal cuticle or other surface structures of many other invertebrates, and in the cell walls of fungi.

chlorophyll [Gk. *chloros,* green + *phyllon,* leaf]: A class of green pigments that are the receptors of light energy in photosynthesis.

chloroplast [Gk. *chloros,* green + *plastos,* formed]: A membrane-bound, chlorophyll-containing organelle in eukaryotes (algae and plants) that is the site of photosynthesis.

chorion (core-ee-on) [Gk., skin, leather]: The outermost extraembryonic membrane of developing reptiles, birds, and mammals; in placental mammals it contributes to the structure of the placenta.

chromatid (crow-ma-tid) [Gk. *chrōma,* color]: Either of the two strands of a replicated chromosome, which are joined at the centromere.

chromatin [Gk. *chrōma,* color]: The deeply staining complex of DNA and histone proteins of which eukaryotic chromosomes are composed.

chromosome [Gk. *chrōma,* color + *soma,* body]: The structure that carries the genes. Eukaryotic chromosomes are visualized as threads or rods of chromatin, which appear in a contracted form during mitosis and meiosis, and are otherwise enclosed in a nucleus. Prokaryotic chromosomes consist of a closed circle of DNA with which a variety of proteins are associated. Viral chromosomes are linear or circular molecules of DNA or RNA.

chromosome map: A diagram of the linear order of the genes on a chromosome.

cilium, *pl.* **cilia** (silly-um) [L., eyelash]: A short, thin structure embedded in the surface of some eukaryotic cells, usually in large numbers and arranged in rows; has a highly characteristic internal structure of two inner microtubules surrounded by nine pairs of outer microtubules; involved in locomotion and the movement of substances across the cell surface.

circadian rhythms [L. *circa,* about + *dies,* day]: Regular rhythms of growth or activity that occur on an approximately 24-hour cycle.

cladogenesis [Gk. *clados,* branch + *genesis,* origin]: The splitting of an evolutionary lineage into two or more separate lineages; one of the principal patterns of evolutionary change.

class: A taxonomic grouping of related, similar orders; category above order and below phylum.

cleavage: The successive cell divisions of a fertilized egg of an animal to form a multicellular blastula.

cline [Gk. *klinein,* to lean]: A graded series of changes in some characteristic within a species, correlated with some gradual change in temperature, humidity, or other environmental factor over the geographic range of the species.

clone [Gk. *klon,* twig]: A line of cells, all of which have arisen from the same single cell by repeated cell divisions; a population of individuals derived by asexual reproduction from a single ancestor.

cnidocyte (ni-do-site) [Gk. *knide,* nettle + *kytos,* vessel]: A stinging cell containing a nematocyst; characteristic of cnidarians.

cochlea [Gk. *kochlias,* snail]: Part of the inner ear of mammals; concerned with hearing.

codominance: In genetics, the phenomenon in which the effects of both alleles at a particular locus are apparent in the phenotype of the heterozygote.

codon (code-on): Basic unit ("letter") of the genetic code; three adjacent nucleotides in a molecule of DNA or mRNA that form the code for a specific amino acid or for polypeptide chain termination.

coelom (see-loam) [Gk. *koilos,* a hollow]: A body cavity formed between layers of mesoderm and in which the digestive tract and other internal organs are suspended.

coenzyme [L. *co,* together + Gk. *en,* in + *zyme,* leaven]: A nonprotein organic molecule that plays an accessory role in enzyme-catalyzed processes, often by acting as a donor or acceptor of a substance involved in the reaction. NAD^+, FAD, and coenzyme A are common coenzymes.

coevolution [L. *co,* together + *e-,* out + *volvere,* to roll]: The simultaneous evolution of adaptations in two or more populations that interact so closely that each is a strong selective force on the other.

cofactor: A nonprotein component that plays an accessory role in enzyme-catalyzed processes; some cofactors are ions, and others are coenzymes.

cohesion [L. *cohaerere,* to stick together]: The attraction or holding together of molecules of the same substance.

cohesion-tension theory: A theory accounting for the upward movement of water in plants. According to this theory, transpiration of a water molecule results in a negative (below 1 atmosphere) pressure in the leaf cells, inducing the entrance from the vascular tissue of another water molecule, which, because of the cohesive property of water, pulls with it a chain of water molecules extending up from the cells of the root tip.

coleoptile (coal-ee-**op**-tile) [Gk. *koleon*, sheath + *ptilon*, feather]: The sheath enclosing the apical meristem and leaf primordia of a germinating monocot.

collagen [Gk. *kolla*, glue]: A fibrous protein in bones, tendons, and other connective tissues.

collenchyma [Gk. *kolla*, glue]: In plants, a type of supporting cell with an irregularly thickened primary cell wall; alive at maturity.

colony: A group of organisms of the same species living together in close association.

commensalism [L. *com*, together + *mensa*, table]: See Symbiosis.

community: All of the populations of organisms inhabiting a common environment and interacting with one another.

companion cell: In angiosperms, a specialized parenchyma cell associated with a sieve-tube member and arising from the same mother cell as the sieve-tube member.

competition: Interaction between members of the same population or of two or more populations using the same resource, often present in limited supply.

competitive exclusion: The hypothesis that two species with identical ecological requirements cannot coexist stably in the same locality and the species that is more efficient in utilizing the available resources will exclude the other; also known as Gause's principle, after the Russian biologist G. F. Gause.

complementary DNA (cDNA): DNA molecules synthesized by reverse transcriptase from an RNA template.

compound [L. *componere*, to put together]: A chemical substance composed of two or more kinds of atoms in definite ratios.

compound eye: In arthropods, a complex eye composed of many separate elements, each with light-sensitive cells and a lens that can form an image.

condensation: *See* Dehydration synthesis.

cone: (1) In plants, the reproductive structure of a conifer. (2) In vertebrates, a type of photoreceptor cell in the retina, concerned with the perception of color and with the most acute discrimination of detail.

conjugation [L. *conjugatio*, a joining, connection]: The sexual process in some unicellular organisms by which genetic material is transferred from one cell to another by cell-to-cell contact.

connective tissues: Supporting or packing tissues that lie between groups of nerves, glands, and muscle cells, and beneath epithelial cells, in which the cells are irregularly distributed through a relatively large amount of extracellular material; include bone, cartilage, blood, and lymph.

consumer, in ecological systems: A heterotroph that derives its energy from living or freshly killed organisms or parts thereof. Primary consumers are herbivores; higher-level consumers are carnivores.

continental drift: The gradual movement of the Earth's continents that has occurred over hundreds of millions of years.

continuous variation: A gradation of small differences in a particular trait, such as height, within a population; occurs in traits that are controlled by a number of genes.

convergent evolution [L. *convergere*, to turn together; *evolutio*, to unfold]: The independent development of similarities between unrelated groups, such as porpoises and sharks, resulting from adaptation to similar environments.

cork [L. *cortex*, bark]: A secondary tissue that is a major constituent of bark in woody and some herbaceous plants; made up of flattened cells, dead at maturity; restricts gas and water exchange and protects the vascular tissues from injury.

cork cambium [L. *cortex*, bark + *cambium*, exchange]: The lateral meristem that produces cork.

corolla (ko-**role**-a) [L. dim. of *corona*, wreath, crown]: Petals, collectively; usually the conspicuously colored flower parts.

corpus callosum [L., callous body]: In the vertebrate brain, a tightly packed mass of myelinated nerve fibers connecting the two cerebral hemispheres.

corpus luteum [L., yellowish body]: An ovarian structure that secretes estrogens and progesterone, which maintain the uterus during pregnancy. It develops from the remaining cells of the ruptured follicle following ovulation.

cortex [L., bark]: (1) The outer, as opposed to the inner, part of an organ, as in the adrenal gland. (2) In a stem or root, the primary tissue bounded externally by the epidermis and internally by the central cylinder of vascular tissue.

cortisol: A steroid hormone, produced by the adrenal cortex, that promotes the formation of glucose from protein and fat; also suppresses the inflammatory and immune responses.

cotyledon (cottle-ee-don) [Gk. *kotyledon*, a cup-shaped hollow]: A leaf-like structure of the embryo of a seed plant; contains stored food used during germination.

countercurrent exchange: An anatomical device for manipulating gradients so as to maximize uptake (or minimize loss) of O_2, heat, etc.

coupled reactions: In cells, the linking of endergonic (energy-requiring) reactions to exergonic (energy-releasing) reactions that provide enough energy to drive the endergonic reactions forward.

covalent bond [L. *con*, together + *valere*, to be strong]: A chemical bond formed as a result of the sharing of one or more pairs of electrons.

Crassulacean acid metabolism: A process by which some species of plants in hot, dry climates take in carbon dioxide during the night, fixing it in organic acids; the carbon dioxide is released during the day and used immediately in the Calvin cycle.

cristae: The "shelves" formed by the intricate folding of the inner membrane of the mitochondrion.

cross-fertilization: Fusion of gametes formed by different individuals; as opposed to self-fertilization.

crossing over: During meiosis, the exchange of genetic material between paired chromatids of homologous chromosomes.

cuticle (**ku**-tik-l) [L. *cuticula*, dim. of *cutis*, the skin]: (1) In plants, a layer of waxy substance (cutin) on the outer surface of epidermal cell walls. (2) In animals, the noncellular, outermost layer of many invertebrates.

cyclic AMP: A form of adenosine monophosphate (AMP) in which the atoms of the phosphate group form a ring; functions in chemical communication in slime molds and as a "second messenger" for a number of vertebrate hormones and neurotransmitters.

cytochromes [Gk. *kytos*, vessel + *chrōma*, color]: Heme-containing proteins that participate in electron transport chains; involved in cellular respiration and photosynthesis.

cytokinesis [Gk. *kytos*, vessel + *kinesis*, motion]: Division of the cytoplasm of a cell following nuclear division.

cytokinin [Gk. *kytos*, vessel + *kinesis*, motion]: One of a group of chemically related plant hormones that promote cell division, among other effects.

cytoplasm (**sight**-o-plazm) [Gk. *kytos*, vessel + *plasma*, anything molded]: The living matter within a cell, excluding the genetic material.

cytoskeleton: A network of filamentous protein structures within the cytoplasm that maintains the shape of the cell, anchors its organelles, and is involved in cell motility; includes microtubules, actin filaments, and intermediate filaments.

cytosol: A concentrated solution of ions, small molecules (such as amino acids, sugars, and ATP), and proteins that constitutes the fluid portion of the cytoplasm.

day-neutral plant: A plant that flowers without regard to the length of the light period to which it is exposed.

deciduous [L. *decidere*, to fall off]: Refers to plants that shed their leaves at a certain season.

decomposers: Specialized detritivores, usually bacteria or fungi, that consume such substances as cellulose and nitrogenous waste products. Their metabolic processes release inorganic nutrients, which are then available for reuse by plants and other organisms.

dehydration synthesis [L. *de,* from + *hydro,* water; Gk. *synthesis,* a putting together]: A type of chemical reaction in which two molecules join to form one larger molecule, simultaneously splitting out a molecule of water. The biosynthetic reactions in which monomers (e.g., monosaccharides, amino acids) are joined to form polymers (e.g., polysaccharides, polypeptides) are dehydration synthesis reactions.

dendrite [Gk. *dendron,* tree]: A process of a neuron, typically branched, that receives stimuli from other cells.

denitrification: The process by which certain bacteria living in poorly aerated soils break down nitrates, using the oxygen for their own respiration and releasing nitrogen back into the atmosphere.

density-dependent factors: Factors affecting the birth rate or death rate of a population, the effects of which vary with the density (number of individuals per unit area or volume) of the population; include resources for which members of the same or different populations compete, predation, and disease.

density-independent factors: Factors affecting the birth rate or death rate of a population, the effects of which are independent of the density of the population; often involve weather-related events.

deoxyribonucleic acid (DNA) (dee-ox-y-rye-bo-new-clay-ick): The carrier of genetic information in cells, composed of two complementary chains of nucleotides wound in a double helix; capable of self-replication as well as coding for RNA synthesis.

dermis [Gk. *derma,* skin]: The inner layer of the skin, beneath the epidermis.

desmosome [Gk. *desmos,* bond + *soma,* body]: A type of cell-cell junction that provides mechanical strength in animal tissues; consists of a plaque of dense fibrous material between adjacent cells, with clusters of filaments looping in and out from the cytoplasm of the two cells.

detritivores [L. *detritus,* worn down, worn away + *voro,* to devour]: Organisms that live on dead and discarded organic matter; include large scavengers, smaller animals such as earthworms and some insects, as well as decomposers (fungi and bacteria).

deuterostome [Gk. *deuteros,* second + *stoma,* mouth]: An animal in which the anus forms at or near the blastopore in the developing embryo and the mouth forms secondarily elsewhere; echinoderms and chordates are deuterostomes.

development: The progressive production of the phenotypic characteristics of a multicellular organism, beginning with the fertilization of an egg.

diaphragm [Gk. *diaphrassein,* to barricade]: In mammals, a sheetlike tissue (tendon and muscle) forming the partition between the abdominal and thoracic cavities; functions in breathing.

dicotyledon (dye-cottle-ee-don) [Gk. *di,* double, two + *kotyledon,* a cup-shaped hollow]: A member of the class of flowering plants having two seed leaves, or cotyledons, among other distinguishing features; often abbreviated as dicot.

differentiation: The developmental process by which a relatively unspecialized cell or tissue undergoes a progressive (usually irreversible) change to a more specialized cell or tissue.

diffusion [L. *diffundere,* to pour out]: The net movement of suspended or dissolved particles down a concentration gradient as a result of the random spontaneous movements of individual particles; the process tends to distribute the particles uniformly throughout a medium.

digestion [L. *digestio,* separating out, dividing]: The breakdown of complex, usually insoluble foods into molecules that can be absorbed into the body and used by the cells.

dioecious (dye-ee-shus) [Gk. *di,* two + *oikos,* house]: In angiosperms, having the male (staminate) and female (carpellate) flowers on different individuals of the same species.

diploid [Gk. *di,* double, two + *ploion,* vessel]: The condition in which each autosome is represented twice (2n); in contrast to haploid (n).

disaccharide [Gk. *di,* two + *sakcharon,* sugar]: A carbohydrate molecule composed of two monosaccharide subunits; examples are sucrose, maltose, and lactose.

diurnal [L. *diurnus,* of the day]: Applied to organisms that are active during the daylight hours.

division: A taxonomic grouping of related, similar classes; a high-level category below kingdom and above class. Division is generally used in the classification of prokaryotes, algae, fungi, and plants, whereas an equivalent category, phylum, is used in the classification of protozoa and animals.

DNA: Abbreviation of deoxyribonucleic acid.

dominant allele: An allele whose phenotypic effect is the same in both the heterozygous and homozygous conditions.

dormancy [L. *dormire,* to sleep]: A period during which growth ceases and metabolic activity is greatly reduced; dormancy is broken when certain requirements, for example, of temperature, moisture, or day length, are met.

dorsal [L. *dorsum,* the back]: Pertaining to or situated near the back; opposite of ventral.

double fertilization: A phenomenon unique to the angiosperms, in which the egg and one sperm nucleus fuse (resulting in a 2n fertilized egg, the zygote) and simultaneously the second sperm nucleus fuses with the two polar nuclei (resulting in a 3n endosperm nucleus).

duodenum (duo-dee-num) [L. *duodeni,* twelve each—from its length, about 12 fingers' breadth]: The upper portion of the small intestine in vertebrates, where food is digested into molecules that can be absorbed by intestinal cells.

ecological niche: A description of the roles and associations of a particular species in the community of which it is a part; the way in which an organism interacts with all of the biotic (living) and abiotic (nonliving) factors in its environment.

ecological pyramid: A graphic representation of the quantitative relationships of numbers of organisms, biomass, or energy flow between the trophic levels of an ecosystem. Because large amounts of energy and biomass are dissipated at every trophic level, these diagrams nearly always take the form of pyramids.

ecological succession: The gradual process by which the species composition of a community changes.

ecology [Gk. *oikos,* home + *logos,* a discourse]: The study of the interactions of organisms with their physical environment and with each other and of the results of such interactions.

ecosystem [Gk. *oikos,* home + *systema,* that which is put together]: The organisms in a community plus the associated abiotic (nonliving) factors with which they interact.

ecotype [Gk. *oikos,* home + L. *typus,* image]: A locally adapted variant of a species, differing genetically from other ecotypes of the same species.

ectoderm [Gk. *ecto,* outside + *derma,* skin]: One of the three embryonic tissue layers of animals; it gives rise to the outer covering of the body, the sensory receptors, and the nervous system.

ectotherm [Gk. *ecto,* outside + *therme,* heat]: An organism, such as a reptile, that maintains its body temperature by taking in heat from the environment or giving it off to the environment.

effector [L. *ex,* out of + *facere,* to make]: Cell, tissue, or organ (such as muscle or gland) capable of producing a response to a stimulus.

efferent [L. *ex,* out of + *ferre,* to bear]: Carrying away from a center, applied to nerves and blood vessels.

egg: A female gamete, which usually contains abundant cytoplasm and yolk; nonmotile and often larger than a male gamete.

electric potential: The difference in the amount of electric charge between a region of positive charge and a region of negative charge. The establishment of electric potentials across the plasma membrane and across organelle membranes makes possible a number of phenomena, including the chemiosmotic synthesis of ATP, the conduction of nerve impulses, and muscle contraction.

electron: A subatomic particle with a negative electric charge equal in magnitude to the positive charge of the proton but with a much smaller mass; the electrons of an atom move around its positively charged nucleus at almost the speed of light.

electron acceptor: Substance that accepts or receives electrons in an oxidation-reduction reaction, becoming reduced in the process.

electron carrier: A specialized molecule, such as a cytochrome, that can lose and gain electrons reversibly, alternately becoming oxidized and reduced.

electron donor: Substance that donates or gives up electrons in an oxidation-reduction reaction, becoming oxidized in the process.

electron transport: The movement of electrons down a series of electron-carrier molecules that hold electrons at slightly different energy levels; as electrons move down the chain, the energy released is used to form ATP from ADP and phosphate. Electron transport plays an essential role in the final stage of cellular respiration and in the light-dependent reactions of photosynthesis.

element: A substance composed only of atoms of the same atomic number and that cannot be decomposed by ordinary chemical means.

embryo [Gk. *en*, in + *bryein*, to swell]: The early developmental stage of an organism produced from a fertilized egg; a young organism before it emerges from the seed, egg, or body of its mother. In humans, refers to the first two months of intrauterine life. *See* Fetus.

embryo sac: The female gametophyte of a flowering plant, contained within an ovule; typically consists of seven cells with a total of eight haploid nuclei.

endergonic [Gk. *endon*, within + *ergon*, work]: Energy-requiring, as in a chemical reaction; applied to an "uphill" process.

endocrine gland [Gk. *endon*, within + *krinein*, to separate]: Ductless gland whose secretions (hormones) are released into the extracellular spaces, from which they diffuse into the circulatory system; in vertebrates, includes pituitary, sex glands, adrenal, thyroid, and others.

endocytosis [Gk. *endon*, within + *kytos*, vessel]: A cellular process in which material to be taken into the cell induces the plasma membrane to form a vacuole enclosing the material; the vacuole is released into the cytoplasm. Includes phagocytosis (endocytosis of solid particles), pinocytosis (endocytosis of liquids), and receptor-mediated endocytosis.

endoderm [Gk. *endon*, within + *derma*, skin]: One of the three embryonic tissue layers of animals; it gives rise to the epithelium that lines certain internal structures, such as most of the digestive tract and its outgrowths, most of the respiratory tract, and the urinary bladder, liver, pancreas, and some endocrine glands.

endodermis [Gk. *endon*, within + *derma*, skin]: In plants, a layer of specialized cells, one cell thick, that lies between the cortex and the vascular tissues in young roots. The Casparian strip of the endodermis prevents diffusion of solutes across the root.

endometrium [Gk. *endon*, within + *metrios*, of the womb]: The glandular lining of the uterus in mammals; thickens in response to secretion of estrogens and progesterone; one of its two principal layers is sloughed off in menstruation.

endoplasmic reticulum [Gk. *endon*, within + *plasma*, from cytoplasm; L. *reticulum*, network]: An extensive system of membranes present in most eukaryotic cells, dividing the cytoplasm into compartments and channels; often coated with ribosomes.

endorphin: One of a group of small peptides with morphine-like properties; produced by the vertebrate brain.

endosperm [Gk. *endon*, within + *sperma*, seed]: In plants, a triploid ($3n$) tissue containing stored food; develops from the union of a sperm nucleus and the two nuclei (the polar nuclei) of the central cell of the female gametophyte; found only in angiosperms.

endothelium [Gk. *endon*, + *thele*, nipple]: A type of epithelial tissue that forms the walls of the capillaries and the inner lining of arteries and veins.

endotherm [Gk. *endon*, within + *therme*, heat]: An organism, such as a bird or a mammal, that maintains its body temperature internally through metabolic processes. *See also* Homeotherm.

entropy [Gk. *en*, in + *trope*, turning]: A measure of the randomness or disorder of a system.

enzyme [Gk. *en*, in + *zyme*, leaven]: A globular protein molecule that accelerates a specific chemical reaction.

epidermis [Gk. *epi*, on or over + *derma*, skin]: In plants and animals, the outermost layers of cells.

epinephrine: *See* Adrenaline.

epistasis [Gk., a stopping]: Interaction between two nonallelic genes in which one of them interferes with or modifies the phenotypic expression of the other.

epithelial tissue [Gk. *epi*, on or over + *thele*, nipple]: In animals, a type of tissue that covers a body or structure or lines a cavity; epithelial cells form one or more regular layers with little intercellular material.

equilibrium [L. *aequus*, equal + *libra*, balance]: The state of a system in which no further net change is occurring; result of counterbalancing forward and backward processes.

equilibrium species: Species characterized by low reproduction rates, long development times, large body size, and long adult life with repeated reproductions.

erythrocyte (eh-**rith**-ro-site) [Gk. *erythros*, red + *kytos*, vessel]: Red blood cell, the carrier of hemoglobin.

estrogens [Gk. *oistros*, frenzy + *genos*, origin, descent]: Female sex hormones, which are the predominant secretions of the ovarian follicle during the preovulatory phase of the menstrual cycle; also produced by the corpus luteum and the placenta.

ethology [Gk. *ethos*, habit, custom + *logos*, discourse]: The comparative study of patterns of animal behavior, with emphasis on their adaptive significance and evolutionary origin.

ethylene: A simple hydrocarbon ($H_2C{=}CH_2$) that functions as a plant hormone; plays a role in fruit ripening and leaf abscission.

etiolation [Fr. *etioler*, to blanch]: In plants, a condition characterized by stem elongation, poor leaf development, and lack of chlorophyll; occurs in plants growing in the dark or with greatly reduced light.

eukaryote (you-**car**-ry-oat) [Gk. *eu*, good + *karyon*, nut, kernel]: A cell having a membrane-bound nucleus, membrane-bound organelles, and chromosomes in which DNA is combined with histone proteins; an organism composed of such cells.

eusocial [Gk. *eu*, good + L. *socius*, companion]: Applied to animal societies, such as those of certain insects, in which sterile individuals work on behalf of reproductive individuals.

evolution [L. *e-*, out + *volvere*, to roll]: Changes in the gene pool from one generation to the next as a consequence of processes such as mutation, natural selection, nonrandom mating, and genetic drift.

exergonic [Gk. *ex*, out of + *ergon*, work]: Energy-yielding, as in a chemical reaction; applied to a "downhill" process.

exocrine glands [Gk. *ex*, out of + *krinein*, to separate]: Glands, such as sweat glands and digestive glands, that secrete their products into ducts that empty onto surfaces, such as the skin, or into cavities, such as the interior of the stomach.

exocytosis [Gk. *ex*, out of + *kytos*, vessel]: A cellular process in which particulate matter or dissolved substances are enclosed in a vacuole and transported to the cell surface; there, the membrane of the vacuole fuses with the plasma membrane, expelling the vacuole's contents to the outside.

exon: A segment of DNA that is transcribed into RNA and expressed; dictates the amino acid sequence of part of a polypeptide.

exoskeleton: The outer supporting covering of the body; common in arthropods.

exponential growth: In populations, the increasingly accelerated rate of growth due to the increasing number of individuals being added to the reproductive base. Exponential growth is very seldom approached or sustained in natural populations.

expressivity: In genetics, the degree to which a particular genotype is expressed in the phenotype of individuals with that genotype.

extinct [L. *exstinctus,* to be extinguished]: No longer existing.

extraembryonic membranes: In reptiles, birds, and mammals, membranes formed from embryonic tissues that lie outside the embryo proper, protecting it and aiding metabolism; include amnion, chorion, allantois, and yolk sac.

F$_1$ (first filial generation): The offspring resulting from the crossing of plants or animals of a parental generation.

F$_2$ (second filial generation): The offspring resulting from crossing members of the F$_1$ generation among themselves.

facilitated diffusion: The transport of substances across the plasma membrane or across an organelle membrane from a region of higher concentration to a region of lower concentration by protein molecules embedded in the membrane; driven by the concentration gradient.

FAD: Abbreviation of flavin adenine dinucleotide, a coenzyme that functions as an electron acceptor in the Krebs cycle.

Fallopian tube: *See* Oviduct.

family: A taxonomic grouping of related, similar genera; the category below order and above genus.

fatty acid: A molecule consisting of a —COOH group and a long hydrocarbon chain; fatty acids are components of fats, oils, phospholipids, glycolipids, and waxes.

feedback systems: Control mechanisms whereby an increase or decrease in the level of a particular factor inhibits or stimulates the production, utilization, or release of that factor; important in the regulation of enzyme and hormone levels, ion concentrations, temperature, and many other factors.

fermentation: The breakdown of organic compounds in the absence of oxygen; yields less energy than aerobic processes.

fertilization: The fusion of two haploid gamete nuclei to form a diploid zygote nucleus.

fetus [L., pregnant]: An unborn or unhatched vertebrate that has passed through the earliest developmental stages; a developing human from about the second month of gestation until birth.

fibril [L. *fibra,* fiber]: Any minute, threadlike structure within a cell.

fibrous protein: Insoluble structural protein in which the polypeptide chain is coiled along one dimension. Fibrous proteins constitute the main structural elements of many animal tissues.

filament [L. *filare,* to spin]: (1) A chain of cells. (2) In flowers, the stalk of a stamen.

filtration: The first stage of kidney function; blood plasma is forced, under pressure, out of the glomerular capillaries into Bowman's capsule, through which it enters the renal tubule.

fitness: The genetic contribution of an individual to succeeding generations relative to the contributions of other individuals in the population.

fixed action pattern: A behavior that appears substantially complete the first time the organism encounters the relevant stimulus; tends to be highly stereotyped, rigid, and repetitive.

flagellum, *pl.* **flagella** (fla-**jell**-um) [L. *flagellum,* whip]: A long, threadlike organelle found in eukaryotes and used in locomotion and feeding; has an internal structure of nine pairs of microtubules encircling two central microtubules.

flower: The reproductive structure of angiosperms; a complete flower includes sepals, petals, stamens (male structures), and carpels (female structures).

food chain: A sequence of organisms related to one another as prey and predator.

food web: A set of interactions among organisms, including producers, consumers (herbivores and carnivores), and detritivores, through which energy and materials move within a community or ecosystem.

fossil [L. *fossilis,* dug up]: The remains of an organism, or direct evidence of its presence (such as tracks). May be an unaltered hard part (tooth or bone), a mold in a rock, petrification (wood or bone), unaltered or partially altered soft parts (a frozen mammoth).

founder effect: Type of genetic drift that occurs as the result of the founding of a population by a small number of individuals.

fovea [L., pit]: A small area in the center of the retina in which cones are concentrated; the area of sharpest vision.

frequency-dependent selection: A type of natural selection that decreases the frequency of more common phenotypes in a population and increases the frequency of less common phenotypes.

fruit [L. *fructus,* fruit]: In angiosperms, a matured, ripened ovary or group of ovaries and associated structures; contains the seeds.

function [L. *fungor,* to busy oneself]: Characteristic role or action of a structure or process in the normal metabolism or behavior of an organism.

gametangium, *pl.* **gametangia** [Gk. *gamein,* to marry + L. *tangere,* to touch]: A unicellular or multicellular structure in which gametes are produced.

gamete (**gam**-meet) [Gk., wife]: A haploid reproductive cell whose nucleus fuses with that of another gamete of an opposite mating type or sex (fertilization); the resulting cell (zygote) may develop into a new diploid individual or, in some protists and fungi, may undergo meiosis to form haploid somatic cells.

gametophyte: In organisms that have alternation of haploid and diploid generations (all plants and some green algae), the haploid (*n*) gamete-producing generation.

ganglion, *pl.* **ganglia** (**gang**-lee-on) [Gk. *ganglion,* a swelling]: Aggregation of nerve cell bodies; in vertebrates, refers to an aggregation of nerve cell bodies located outside the central nervous system.

gap junction: A junction between adjacent animal cells that allows the passage of materials between the cells.

gap phases: In the cell cycle, the phases that precede (G$_1$) and follow (G$_2$) the synthesis (S) phase in which DNA is replicated; in the G$_1$ phase, the cell doubles in size, and its enzymes, ribosomes, and other cytoplasmic molecules and structures increase in number; in the G$_2$ phase, the replicated chromosomes begin to condense and the structures required for mitosis or meiosis are assembled.

gastric [Gk. *gaster,* stomach]: Pertaining to the stomach.

gastrovascular cavity [Gk. *gaster,* stomach + L. *vasculum,* a small vessel]: A digestive cavity with only one opening, characteristic of the phyla Cnidaria (jellyfishes, *Hydra,* corals, etc.) and Ctenophora (comb jellies, sea walnuts); water circulating through the cavity supplies dissolved oxygen and carries away carbon dioxide and other waste products.

gastrula [Gk. *gaster,* stomach]: An animal embryo in the process of gastrulation; the stage of development during which the blastula, with its single layer of cells, turns into a three-layered embryo, made up of ectoderm, mesoderm, and endoderm, often enclosing an archenteron.

gated channel: In the plasma membrane of the axon, a protein that functions as a channel for the movement of ions into or out of the axon and that opens or closes as a result of changes in its shape.

Gause's principle: *See* Competitive exclusion.

gene [Gk. *genos,* birth, race; L. *genus,* birth, race, origin]: A unit of heredity in the chromosome; a sequence of nucleotides in a DNA molecule that performs a specific function, such as coding for an RNA molecule or a polypeptide.

gene flow: The movement of alleles into or out of a population.

gene pool: All the alleles of all the genes of all the individuals in a population.

genetic code: The system of nucleotide triplets in DNA and RNA that carries genetic information; referred to as a code because it determines the amino acid sequence in the enzymes and other protein molecules synthesized by the organism.

genetic drift: Evolution (change in allele frequencies) owing to chance processes.

genetic isolation: The absence of genetic exchange between populations or species as a result of geographic separation or of premating or postmating mechanisms (behavioral, anatomical, or physiological) that prevent reproduction.

genome: The complete set of chromosomes, with their associated genes.

genotype (jean-o-type): The genetic constitution of an individual cell or organism with reference to a single trait or a set of traits; the sum total of all the genes present in an individual.

genus, *pl.* **genera** (jean-us) [L. *genus*, race, origin]: A taxonomic grouping of closely related species.

geologic eras: See the inside of the front cover.

germ cells [L. *germinare*, to bud]: Gametes or the cells that give rise to gametes

germination [L. *germinare*, to bud]: In plants, the resumption of growth or the development from seed or spore.

gibberellins (jibb-e-rell-ins) [Fr. *gibberella*, genus of fungi]: A group of chemically related plant growth hormones, whose most characteristic effect is stem elongation in dwarf plants and bolting.

gill: The respiratory organ of aquatic animals, usually a thin-walled projection from some part of the external body surface or, in vertebrates, from some part of the digestive tract.

gland [L. *glans, glandis*, acorn]: A structure composed of modified epithelial cells specialized to produce one or more secretions that are discharged to the outside of the gland.

glial cell [Gk. *glia*, glue + L. *cella*, a chamber]: In nerve tissue, a type of cell that supports and insulates the neurons; Schwann cells, which form the myelin sheaths around axons in the peripheral nervous system of vertebrates, are glial cells.

globular protein [L. dim. of *globus*, a ball]: A polypeptide chain folded into a roughly spherical shape.

glomerulus (glom-**mare**-u-lus) [L. *glomus*, ball]: In the vertebrate kidney, a cluster of capillaries enclosed by Bowman's capsule; blood plasma minus large molecules filters through the walls of the glomerular capillaries into the renal tubule.

glucagon [Gk. *glykys*, sweet + *agō*, to lead toward]: A hormone, produced by the vertebrate pancreas, that acts to raise the concentration of glucose in the blood.

glucose [Gk. *glykys*, sweet]: A six-carbon sugar ($C_6H_{12}O_6$); the most common monosaccharide in animals.

glycerol: A three-carbon molecule with three hydroxyl (—OH) groups attached; a glycerol molecule can combine with three fatty acid molecules to form a fat or an oil.

glycogen [Gk. *glykys*, sweet + *genos*, race or descent]: A complex carbohydrate (polysaccharide); one of the main stored food substances of most animals and fungi; it is converted into glucose by hydrolysis.

glycolipids [Gk. *glykys*, sweet + *lipos*, fat]: Organic molecules similar in structure to fats, but in which a short carbohydrate chain rather than a fatty acid is attached to the third carbon of the glycerol molecule; as a result, the molecule has a hydrophilic "head" and a hydrophobic "tail." Glycolipids are important constituents of the plasma membrane and of organelle membranes.

glycolysis (gly-**coll**-y-sis) [Gk. *glykys*, sweet + *lysis*, loosening]: The process by which a glucose molecule is split anaerobically into two molecules of pyruvic acid with the liberation of a small amount of useful energy; catalyzed by enzymes in the cytosol.

glycoprotein [Gk. *glykys*, sweet + *proteios*, primary]: A large organic molecule consisting of a short carbohydrate chain covalently bonded to a protein; glycoproteins are important components of the plasma membrane and of organelle membranes; the carbohydrate chains protrude to the outside of the cell or organelle.

Golgi complex (goal-jee): An organelle present in many eukaryotic cells; consists of flat, membrane-bound sacs, tubules, and vesicles. It functions as a processing, packaging, and distribution center for substances that the cell manufactures.

gonad [Gk. *gone*, seed]: Gamete-producing organ of multicellular animals; ovary or testis.

gonadotropin [Gk. *gone*, seed + *trope*, a turning]: A vertebrate hormone, produced by the anterior lobe of the pituitary gland, that stimulates the gonads; follicle-stimulating hormone and luteinizing hormone.

granum, *pl.* **grana** [L., grain or seed]: In chloroplasts, stacked membrane-bound disks (thylakoids) that contain chlorophylls and carotenoids and are the sites of the light-dependent reactions of photosynthesis.

gravitropism [L. *gravis*, heavy + Gk. *trope*, turning]: The direction of growth or movement in which the force of gravity is the determining factor; also called geotropism.

gross productivity: A measure of the rate at which energy is assimilated by the organisms in a trophic level, a community, or an ecosystem.

ground meristem [Gk. *merizein*, to divide]: In plants, a primary meristematic tissue; gives rise to the ground tissues of the plant body.

ground tissues: In leaves and young roots and stems, all tissues other than the epidermis and the vascular tissues.

guard cells: Specialized epidermal cells surrounding a pore, or stoma, in a leaf or green stem; changes in turgor of a pair of guard cells cause opening and closing of the pore.

gymnosperm [Gk. *gymnos*, naked + *sperma*, seed]: A seed plant in which the seeds are not enclosed in an ovary; the conifers are the most familiar group.

habitat [L. *habitare*, to live in]: The place in which individuals of a particular species can usually be found.

habituation [L. *habitus*, condition]: A response to a repeated stimulus in which the stimulus comes to be ignored and a previous behavior pattern is restored; one of the simplest forms of learning.

half-life: The average time required for the disappearance or decay of one-half of any amount of a given substance.

haploid [Gk. *haploos*, single + *ploion*, vessel]: Having only one set of chromosomes (*n*), in contrast to diploid (2*n*); characteristic of eukaryotic gametes, of gametophytes in plants, and of some protists and fungi.

Hardy-Weinberg equilibrium: The steady-state relationship between relative frequencies of two or more alleles in an idealized population; both the allele frequencies and the genotype frequencies will remain constant from generation to generation in a population breeding at random in the absence of evolutionary forces.

heat of vaporization: The amount of heat required to change a given amount of a liquid into a gas; 540 calories are required to change 1 gram of liquid water into vapor.

heme [Gk. *haima*, blood]: The iron-containing group of heme proteins such as hemoglobin and the cytochromes.

hemoglobin [Gk. *haima*, blood + L. *globus*, a ball]: The iron-containing protein in vertebrate blood that carries oxygen.

hemophilia [Gk. *haima*, blood + *philios*, friendly]: A group of hereditary disorders characterized by failure of the blood to clot and consequent excessive bleeding from even minor wounds.

hepatic [Gk. *hēpatikos*, liver]: Pertaining to the liver.

herbaceous (her-**bay**-shus) [L. *herba*, grass]: In plants, nonwoody.

herbivore [L. *herba*, grass + *vorare*, to devour]: A consumer that eats plants or other photosynthetic organisms to obtain its food and energy.

heredity [L. *herres, heredis,* heir]: The transmission of characteristics from parent to offspring.

hermaphrodite [Gk. *Hermes* and *Aphrodite*]: An organism possessing both male and female reproductive organs; hermaphrodites may or may not be self-fertilizing.

heterotroph [Gk. *heteros,* other, different + *trophos,* feeder]: An organism that must feed on organic materials formed by other organisms in order to obtain energy and small building-block molecules; in contrast to autotroph. Animals, fungi, and many unicellular organisms are heterotrophs.

heterozygote [Gk. *heteros,* other + *zygōtos,* a pair]: A diploid organism that carries two different alleles at one or more genetic loci.

heterozygote advantage: The greater fitness of an organism heterozygous at a given genetic locus as compared with either homozygote.

hibernation [L. *hiberna,* winter]: A period of dormancy and inactivity, varying in length, depending on the species, and occurring in dry or cold seasons. During hibernation, metabolic processes are greatly slowed and, even in mammals, body temperature may drop to just above freezing.

histamine: A chemical, released by injured body cells and by certain types of white blood cells, that plays a role in the inflammatory response, increasing blood flow into the area and increasing the permeability of nearby capillaries; also released in large amounts in allergic reactions.

histones: A group of five relatively small, basic polypeptide molecules found bound to the DNA of eukaryotic cells.

HIV: Abbreviation of human immunodeficiency virus, the causative agent of AIDS.

homeobox: A sequence of some 180 nucleotides that has been identified in the DNA of a wide variety of animals that exhibit a segmental pattern at some stage of their development; codes for a polypeptide hypothesized to function as a regulatory molecule, altering the course of gene expression during development.

homeostasis (home-e-o-**stay**-sis) [Gk. *homos,* same or similar + *stasis,* standing]: Maintenance of a relatively stable internal physiological environment or internal equilibrium in an organism.

homeotherm [Gk. *homos,* same or similar + *therme,* heat]: An organism, such as a bird or mammal, capable of maintaining a stable body temperature independent of the environment.

hominid [L. *homo,* man]: Humans and closely related primates; includes modern and fossil forms, such as the australopithecines, but not the apes.

hominoid [L. *homo,* man]: Hominids and the apes.

homologues [Gk. *homologia,* agreement]: Chromosomes that carry corresponding genes and associate in pairs in the first stage of meiosis; each member of the pair is derived from a different parent.

homology [Gk. *homologia,* agreement]: Similarity in structure and/or position, assumed to result from a common ancestry, regardless of function, such as the wing of a bird and the foreleg of a mammal.

homozygote [Gk. *homos,* same or similar + *zygōtos,* a pair]: A diploid organism that carries identical alleles at one or more genetic loci.

hormone [Gk. *hormaein,* to excite]: An organic molecule secreted, usually in minute amounts, in one part of an organism that regulates the function of another tissue or organ.

host: (1) An organism on or in which a parasite lives. (2) A recipient of grafted tissue.

human immunodeficiency virus (HIV): The RNA retrovirus that is the causative agent of acquired immune deficiency syndrome (AIDS).

hybrid [L. *hybrida,* the offspring of a tame sow and a wild boar]: (1) Offspring of two parents that differ in one or more inheritable characteristics. (2) Offspring of two different varieties or of two different species.

hydrocarbon [L. *hydro,* water + *carbo,* charcoal]: An organic compound consisting of only carbon and hydrogen.

hydrogen bond: A weak molecular bond linking a hydrogen atom that is covalently bonded to another atom (usually oxygen, nitrogen, or fluorine) to another oxygen, nitrogen, or fluorine atom of the same or another molecule.

hydrolysis [L. *hydro,* water + Gk. *lysis,* loosening]: Splitting of one molecule into two by addition of H^+ and OH^- ions from water.

hydrophilic [L. *hydro,* water + Gk. *philios,* friendly]: Having an affinity for water; applied to polar molecules or polar regions of large molecules.

hydrophobic [L. *hydro,* water + Gk. *phobos,* fearing]: Having no affinity for water; applied to nonpolar molecules or nonpolar regions of molecules.

hypertonic [Gk. *hyper,* above + *tonos,* tension]: Of two solutions of different concentration, the solution that contains the higher concentration of solute particles; water moves across a selectively permeable membrane into a hypertonic solution.

hypha [Gk. *hyphe,* web]: A single tubular filament of a fungus or an oomycete; the hyphae together make up the mycelium, the matlike "body" of a fungus.

hypothalamus [Gk. *hypo,* under + *thalamos,* inner room]: The region of the vertebrate brain just below the thalamus; responsible for the integration of many basic behavioral patterns that involve correlation of neural and endocrine functions.

hypothesis [Gk. *hypo,* under + *tithenai,* to put]: A temporary working explanation or supposition based on accumulated facts and suggesting some general principle or relation of cause and effect; a postulated solution to a scientific problem that must be tested and if not validated, discarded.

hypotonic [Gk. *hypo,* under + *tonos,* tension]: Of two solutions of different concentration, the solution that contains the lower concentration of solute particles; water moves across a selectively permeable membrane from a hypotonic solution.

imbibition [L. *imbibere,* to drink in]: The capillary movement of water into germinating seeds and into substances such as wood and gelatin, which swell as a result.

immune response: A highly specific defensive reaction of the body to invasion by a foreign substance or organism; consists of a primary response in which the invader is recognized as foreign, or "not-self," and eliminated and a secondary response to subsequent attacks by the same invader. Mediated by two types of lymphocytes: B cells, which mature in the bone marrow and are responsible for antibody production, and T cells, which mature in the thymus and are responsible for cell-mediated immunity.

immunoglobulins: Complex, highly specific globular proteins synthesized by B cells; include both circulating antibodies and antibodies displayed on the surface of B cells prior to activation.

imprinting: A rapid and extremely narrow form of learning, common in birds and important in species recognition, that occurs during a very short critical period in the early life of an animal; depends on exposure to particular characteristics of the parent or parents.

inbreeding: The mating of individuals that are closely related genetically.

inclusive fitness: The relative number of an individual's alleles that are passed on from generation to generation, either as a result of his or her own reproductive success, or that of related individuals.

incomplete dominance: In genetics, the phenomenon in which the effects of both alleles at a particular locus are apparent in the phenotype of the heterozygote.

independent assortment: *See* Mendel's second law.

induction [L. *inducere,* to induce]: (1) In genetics, the phenomenon in which the presence of a substrate (the inducer) initiates transcription and translation of the genes coding for the enzymes required for its metabolism. (2) In embryonic development, the process in which one tissue or body part causes the differentiation of another tissue or body part.

inflammatory response: A nonspecific defensive reaction of the body to invasion by a foreign substance or organism; involves phagocytosis by white blood cells and is often accompanied by accumulation of pus and an increase in the local temperature.

innate releasing mechanism: In ethology, a circuit within an animal's brain that is hypothesized to respond to a specific stimulus, setting in motion, or "releasing," the sequence of movements that constitute a fixed action pattern.

insulin: A peptide hormone, produced by the vertebrate pancreas, that acts to lower the concentration of glucose in the blood.

interferon: A protein made by virus-infected cells that inhibits viral multiplication.

interleukin: A protein, produced by an activated helper T cell, that acts as a hormone, stimulating the differentiation and proliferation of both B cells and cytotoxic T cells following their activation.

intermediate filaments: Fibrous protein filaments that form part of the cytoskeleton; found in greatest density in cells subject to mechanical stress.

interneuron: Neuron that transmits signals from one neuron to another within a local region of the central nervous system; may receive signals from and transmit signals to many different neurons.

interphase: The portion of the cell cycle that occurs before mitosis or meiosis can take place; includes the G_1, S, and G_2 phases.

intron: A segment of DNA that is transcribed into RNA but is removed enzymatically from the RNA molecule before the mRNA enters the cytoplasm and is translated; also known as an intervening sequence.

invagination [L. *in*, in + *vagina*, sheath]: The local infolding of a layer of tissue, especially in animal embryos, so as to form a depression or pocket opening to the outside.

inversion: A chromosomal aberration in which a double break occurs and a segment is turned 180° before it is reincorporated into the chromosome.

ion (**eye**-on): Any atom or small molecule containing an unequal number of electrons and protons and therefore carrying a net positive or net negative charge.

ionic bond: A chemical bond formed as a result of the mutual attraction of ions of opposite charge.

isolating mechanisms: Mechanisms that prevent genetic exchange between individuals of different populations or species; they prevent mating or successful reproduction even when mating occurs; may be behavioral, anatomical, or physiological.

isotonic [Gk. *isos*, equal + *tonos*, tension]: Having the same concentration of solutes as another solution. If two isotonic solutions are separated by a selectively permeable membrane, there will be no net flow of water across the membrane.

isotope [Gk. *isos*, equal + *topos*, place]: Atom of an element that differs from other atoms of the same element in the number of neutrons in the atomic nucleus; isotopes thus differ in atomic weight. Some isotopes are unstable and emit radiation.

karyotype [Gk. *kara*, the head + *typos*, stamp or print]: The general appearance of the chromosomes of an organism with regard to number, size, and shape.

keratin [Gk. *karas*, horn]: One of a group of tough, fibrous proteins formed by certain epidermal tissues and especially abundant in skin, claws, hair, feathers, and hooves.

keystone species: A species that is of exceptional importance in maintaining the species diversity of a community; when a keystone species is lost, the diversity of the community decreases and its structure is significantly altered.

kidney: In vertebrates, the organ that regulates the balance of water and solutes in the blood and the excretion of nitrogenous wastes in the form of urine.

kinetic energy [Gk. *kinetikos*, putting in motion]: Energy of motion.

kinetochore [Gk. *kinetikos*, putting in motion + *choros*, chorus]: Disk-shaped protein structure within the centromere to which spindle fibers are attached during mitosis or meiosis.

kingdom: A taxonomic grouping of related, similar phyla or divisions; the highest-level category in biological classification.

kin selection: The differential reproduction of lineages of related individuals—that is, different groups of related individuals of a species reproduce at different rates; leads to an increase in the frequency of alleles shared by members of the groups with the greatest reproductive success.

Krebs cycle: Stage of cellular respiration in which acetyl groups are broken down into carbon dioxide; molecules reduced in the process can be used in ATP formation; also known as the citric acid cycle.

lamella (lah-**mell**-ah) [L. dim. of *lamina*, plate or leaf]: Layer, thin sheet.

larva [L., ghost]: An immature animal that is anatomically very different from the adult; examples are caterpillars and tadpoles.

lateral meristem [L. *latus, lateris*, side + Gk. *meristos*, divided]: In vascular plants, one of the two rings of tissue (vascular cambium and cork cambium) that produce new cells for secondary growth.

leaching: The dissolving of minerals and other elements in soil or rocks by the downward movement of water.

learning: The process that leads to modification in individual behavior as the result of experience.

leukocyte [Gk. *leukos*, white + *kytos*, vessel]: White blood cell; two principal types are phagocytic cells, such as macrophages, and lymphocytes (B cells and T cells).

leukotriene: A type of prostaglandin produced by various white blood cells involved in the inflammatory and immune responses and in allergic reactions.

lichen: Organism composed of a fungus and a green alga or a cyanobacterium that are symbiotically associated.

life cycle: The entire span of existence of any organism from time of zygote formation (or asexual reproduction) until it itself reproduces.

life-history pattern: A group of traits, such as size and number of offspring, length of maturation, age at first reproduction, and the number of times reproduction occurs, that affect reproduction, survival, and the rate of population growth.

light-dependent reactions: The reactions of the first stage of photosynthesis, in which light energy is captured by chlorophyll molecules and converted to chemical energy stored in ATP and NADPH molecules.

light-independent reactions: The carbon-fixing reactions of the second stage of photosynthesis; energy stored in ATP and NADPH by the light-dependent reactions is used to reduce carbon from carbon dioxide to simple sugars; light is not required for these reactions.

lignin: An organic molecule, found in the secondary cell walls of many vascular plants, that adds strength and rigidity to the walls.

limbic system [L. *limbus*, border]: Neuron network forming a loop around the inside of the brain and connecting the hypothalamus to the cerebral cortex; thought to be circuit by which drives and emotions are translated into complex actions and to play a role in the consolidation of memory.

linkage: The tendency for certain alleles to be inherited together because they are located on the same chromosome.

linkage group: A pair of homologous chromosomes.

lipid [Gk. *lipos*, fat]: One of a large variety of organic substances that are insoluble in polar solvents, such as water, but that dissolve readily in nonpolar organic solvents; includes fats, oils, waxes, steroids, phospholipids, glycolipids, and carotenes.

lipoprotein [Gk. *lipos*, fat + *proteios*, primary]: A large organic molecule consisting of a lipid covalently bonded to a protein.

locus, *pl.* **loci** [L., place]: In genetics, the position of a gene in a chromosome. For any given locus, there may be a number of possible alleles.

logistic growth: A pattern of population growth in which growth is rapid when the population is small, gradually slows as the population approaches the carrying capacity of its environment, and then oscillates as the population stabilizes at or near its maximum size; it is one of the simplest growth patterns observed for populations in nature.

long-day plant: A plant that must be exposed to a light period longer than some critical length for flowering to occur; long-day plants flower in spring or summer.

loop of Henle (after F. G. J. Henle, German pathologist): A hairpin-shaped portion of the renal tubule of mammals in which a hypertonic urine is formed by processes of diffusion and active transport.

lumen [L., light]: The cavity of a tubular structure, such as endoplasmic reticulum or a blood vessel.

lymph [L. *lympha*, water]: Colorless fluid derived from blood by filtration through capillary walls in the tissues; carried in special lymph ducts.

lymphatic system: The system through which lymph circulates; consists of lymph capillaries, which begin blindly in the tissues, and a network of progressively larger vessels that empty into the vena cava; also includes the lymph nodes, spleen, thymus, and tonsils.

lymph node [L. *lympha*, water + *nodus*, knot]: A mass of spongy tissues, separated into compartments; located throughout the lymphatic system, lymph nodes remove dead cells, debris, and foreign particles from the circulation; also are sites at which foreign antigens are displayed to immunologically active cells.

lymphocyte [L. *lympha*, water + Gk. *kytos*, vessel]: A type of white blood cell involved in the immune response; B cells differentiate into antibody-producing plasma cells, whereas cytotoxic T cells lyse diseased eukaryotic cells; other T cells interact with both cytotoxic T cells and with B cells.

lymphokine: A chemical, released by an activated cytotoxic T cell, that attracts macrophages and stimulates phagocytosis.

lysis [Gk. *lysis*, a loosening]: Disintegration of a cell by rupture of its plasma membrane.

lysogenic bacteria (lye-so-**jenn**-ick) [Gk. *lysis*, a loosening + *genos*, race or descent]: Bacteria carrying a bacteriophage integrated into the bacterial chromosome. The virus may subsequently set up an active cycle of infection, causing lysis of the bacterial cells.

lysosome [Gk. *lysis*, loosening + *soma*, body]: A membrane-bound organelle in which hydrolytic enzymes are sequestered.

macromolecule [Gk. *makros*, large + L. dim. of *moles*, mass]: An extremely large molecule; refers specifically to proteins, nucleic acids, polysaccharides, and complexes of these.

macronutrient [Gk. *makros*, large + L. *nutrire*, to nourish]: An inorganic nutrient required in large amounts for plant growth, such as nitrogen, potassium, calcium, phosphorus, magnesium, and sulfur.

macrophage [Gk. *makros*, large + *phagein*, to eat]: A type of phagocytic white blood cell important in both the inflammatory and immune responses.

major histocompatibility complex (MHC): In mammals, a group of at least 20 different genes, each with multiple alleles, coding for the protein components of the antigens that are displayed on nucleated cells and that serve to identify "self."

mantle: In mollusks, the outermost layer of the body wall or a soft extension of it; usually secretes a shell.

marine [L. *marini(us)*, from *mare*, the sea]: Living in salt water.

marsupial [Gk. *marsypos*, pouch, little bag]: A mammal in which the female has a ventral pouch or folds surrounding the nipples; the premature young leave the uterus and crawl into the pouch, where each one attaches itself by the mouth to a nipple until development is completed.

mast cell: A type of noncirculating white blood cell, found in connective tissue, that is the major protagonist in allergic reactions; when an allergen binds to complementary antibodies on the surface of a mast cell, large amounts of histamine are released from the cell.

matrix: The dense solution in the interior of the mitochondrion, surrounding the cristae; contains enzymes, phosphates, coenzymes, and other molecules involved in cellular respiration.

mechanoreceptor: A sensory cell or organ that receives mechanical stimuli such as those involved in touch, pressure, hearing, and balance.

medulla (med-**dull**-a) [L., the innermost part]: (1) The inner, as opposed to the outer, part of an organ, as in the adrenal gland. (2) The most posterior region of the vertebrate brain; connects with the spinal cord.

medusa: The free-swimming, bell- or umbrella-shaped stage in the life cycle of many cnidarians; a jellyfish.

megaspore [Gk. *megas*, great, large + *spora*, a sowing]: In plants, a haploid *(n)* spore that develops into a female gametophyte.

meiosis (my-o-sis) [Gk. *meioun*, to make smaller]: The two successive nuclear divisions in which a single diploid *(2n)* cell forms four haploid *(n)* nuclei, and segregation, crossing over, and reassortment of the alleles occur; gametes or spores may be produced as a result of meiosis.

memory cell: A type of white blood cell, produced by the differentiation of a B cell or a T cell following stimulation by a complementary antigen, that remains in the circulation indefinitely; enables the immune system to respond rapidly to any subsequent attack by the same invader.

Mendel's first law: The factors for a pair of alternative characters are separate, and only one may be carried in a particular gamete (genetic segregation). In modern form: Alleles segregate in meiosis.

Mendel's second law: The inheritance of a pair of factors for one trait is independent of the simultaneous inheritance of factors for other traits, such factors "assorting independently" as though there were no other factors present (later modified by the discovery of linkage). Modern form: The alleles of unlinked genes assort independently.

menstrual cycle [L. *mensis*, month]: In humans and certain other primates, the cyclic, hormone-regulated changes in the condition of the uterine lining; marked by the periodic discharge of blood and disintegrated uterine lining through the vagina.

meristem [Gk. *merizein*, to divide]: Undifferentiated plant tissue, including a mass of rapidly dividing cells, from which new tissues arise.

mesenteries [Gk. *mesos*, middle + *enteron*, gut]: Double layers of mesoderm that suspend the digestive tract and other internal organs within the coelom.

mesoderm [Gk. *mesos*, middle + *derma*, skin]: In animals, the middle layer of the three embryonic tissue layers. In vertebrates, includes the chordamesoderm, which gives rise to the notochord and skeletal muscle, and the lateral plate mesoderm, which gives rise to the circulatory system, most of the excretory and reproductive systems, the lining of the coelom, and the outer covering of the internal organs.

mesophyll [Gk. *mesos*, middle + *phyllon*, leaf]: The internal tissue of a leaf, sandwiched between two layers of epidermal cells; consists of palisade parenchyma and spongy parenchyma cells.

messenger RNA (mRNA): A class of RNA molecules, each of which is complementary to one strand of DNA and which serves to carry the genetic information from the chromosome to the ribosomes, where it is translated into protein.

metabolism [Gk. *metabole*, change]: The sum of all chemical reactions occurring within a cell or organism.

metamorphosis [Gk. *metamorphoun*, to transform]: Abrupt transition from larval to adult form, such as the transition from tadpole to adult frog.

metaphase [Gk. *meta*, middle + *phasis*, form]: The stage of mitosis or meiosis during which the chromosomes lie in the equatorial plane of the spindle.

MHC: Abbreviation of major histocompatibility complex.

microbe [Gk. *mikros*, small + *bios*, life]: A microscopic organism.

micronutrient [Gk. *mikros*, small + L. *nutrire*, to nourish]: An inorganic nutrient required in only minute amounts for plant growth, such as iron, chlorine, copper, manganese, zinc, molybdenum, and boron.

microspore [Gk. *mikros*, small + *spora*, a sowing]: In plants, a haploid (*n*) spore that develops into a male gametophyte; in seed plants, it becomes a pollen grain.

microtubule [Gk. *mikros*, small + L. dim. of *tubus*, tube]: An extremely small hollow tube composed of two types of globular protein subunits. Among their many functions, microtubules make up the internal structure of cilia and flagella.

middle lamella: In plants, distinct layer between adjacent cell walls, rich in pectins and other polysaccharides; derived from the cell plate.

mimicry [Gk. *mimos*, mime]: The superficial resemblance in form, color, or behavior of certain organisms (mimics) to other more powerful or more protected ones (models), resulting in protection, concealment, or some other advantage for the mimic.

mineral: A naturally occurring element or inorganic compound.

mitochondrion, *pl.* **mitochondria** [Gk. *mitos*, thread + *chondros*, cartilage or grain]: An organelle, bound by a double membrane, in which the reactions of the Krebs cycle, terminal electron transport, and oxidative phosphorylation take place, resulting in the formation of CO_2, H_2O, and ATP from acetyl CoA and ADP. Mitochondria are the organelles in which most of the ATP of the eukaryotic cell is produced.

mitosis [Gk. *mitos*, thread]: Nuclear division characterized by chromosome replication and formation of two identical daughter nuclei.

mole [L. *moles*, mass]: The amount of an element equivalent to its atomic weight expressed in grams, or the amount of a substance equivalent to its molecular weight expressed in grams.

molecular weight: The sum of the atomic weights of the constituent atoms in a molecule.

molecule [L. dim. of *moles*, mass]: A particle consisting of two or more atoms held together by chemical bonds; the smallest unit of a compound that displays the properties of the compound.

molting: Shedding of all or part of an organism's outer covering; in arthropods, periodic shedding of the exoskeleton to permit an increase in size.

monocotyledon [Gk. *monos*, single + *kotyledon*, a cup-shaped hollow]: A member of the class of flowering plants having one seed leaf, or cotyledon, among other distinguishing features; often abbreviated as monocot.

monoecious (mo-nee-shus) [Gk. *monos*, single + *oikos*, house]: In angiosperms, having the male and female structures (the stamens and the carpels, respectively) on the same individual but on different flowers.

monomer [Gk. *monos*, single + *meros*, part]: A simple, relatively small molecule that can be linked to others to form a polymer.

monosaccharide [Gk. *monos*, single + *sakcharon*, sugar]: A simple sugar, such as glucose, fructose, ribose.

monotreme [Gk. *monos*, single + *trēma*, hole]: A nonplacental mammal, such as the duckbilled platypus, in which the female lays shelled eggs and nurses the young.

morphogenesis [Gk. *morphe*, form + *genesis*, origin]: The development of size, form, and other structural features of organisms.

morphological [Gk. *morphe*, form + *logos*, discourse]: Pertaining to form and structure, at any level of organization.

mortality rate [L. *mors*, death]: Death rate.

motor neuron: Neuron that conducts nerve impulses from the central nervous system to an effector, which is typically a muscle or a gland.

muscle fiber: Muscle cell; a long, cylindrical, multinucleated cell containing numerous myofibrils, which is capable of contraction when stimulated.

mutagen [L. *mutare*, to change + *genus*, source or origin]: A chemical or physical agent that increases the mutation rate.

mutant [L. *mutare*, to change]: An organism carrying a gene that has undergone a mutation.

mutation [L. *mutare*, to change]: The change of a gene from one allelic form to another; an inheritable change in the DNA sequence of a chromosome.

mutualism [L. *mutuus*, lent, borrowed]: *See* Symbiosis.

mycelium [Gk. *mykes*, fungus]: The mass of hyphae forming the body of a fungus.

mycorrhizae [Gk. *mykes*, fungus + *rhiza*, root]: Symbiotic associations between particular species of fungi and the roots of vascular plants.

myelin sheath [Gk. *myelinos*, full of marrow]: A lipid-rich layer surrounding the long axons of neurons in the vertebrate nervous system; in the peripheral nervous system, made up of the plasma membranes of Schwann cells.

myofibril [Gk. *mys*, muscle + L. *fibra*, fiber]: Contractile element of a muscle fiber, made up of thick and thin filaments arranged in sarcomeres.

myoglobin [Gk. *mys*, muscle + L. *globus*, a ball]: An oxygen-binding, heme-containing globular protein found in muscles.

myosin [Gk. *mys*, muscle]: One of the principal proteins in muscle; makes up the thick filaments.

NAD: Abbreviation of nicotinamide adenine dinucleotide, a coenzyme that functions as an electron acceptor in glycolysis and the Krebs cycle.

NADP: Abbreviation of nicotinamide adenine dinucleotide phosphate, a coenzyme that functions as an electron acceptor in the light-dependent reactions of photosynthesis.

natural selection: A process of interaction between organisms and their environment that results in a differential rate of reproduction of different phenotypes in the population; can result in changes in the relative frequencies of alleles and genotypes in the population—that is, in evolution.

nectar [Gk. *nektar*, the drink of the gods]: A sugary fluid that attracts insects to plants.

negative feedback: A control mechanism whereby an increase in some substance inhibits the process leading to the increase; also known as feedback inhibition.

nematocyst [Gk. *nema, nematos*, thread + *kyst*, bladder]: A threadlike stinger, containing a poisonous or paralyzing substance, found in the cnidocyte of cnidarians.

nephridium, *pl.* **nephridia** [Gk. *nephros*, kidney]: A tubular excretory structure found in many invertebrates.

nephron [Gk. *nephros*, kidney]: The functional unit of the kidney in reptiles, birds, and mammals; a human kidney contains about 1 million nephrons.

nerve: A group or bundle of nerve fibers with accompanying connective tissue, located in the peripheral nervous system. A bundle of nerve fibers within the central nervous system is known as a tract.

nerve fiber: A filamentous process extending from the cell body of a neuron and conducting the nerve impulse; an axon.

nerve impulse: A rapid, transient, self-propagating change in electric potential across the membrane of an axon.

nervous system: All the nerve cells of an animal; the receptor-conductor-effector system; in humans, the nervous system consists of the central nervous system (brain and spinal cord) and the peripheral nervous system.

net primary production: In a community or an ecosystem, the increase in the amount of plant or algal material between the beginning and end of a specified time period, such as a growing season.

net productivity: In a trophic level, a community, or an ecosystem, the amount of energy (in calories) stored in chemical compounds or the increase in biomass (in grams or metric tons) in a particular period of time; it is the difference between gross productivity and the energy used by the organisms in respiration.

neural groove: Dorsal, longitudinal groove that forms in a vertebrate embryo; bordered by two neural folds; preceded by the neural-plate stage and followed by the neural-tube stage.

neural plate: Thickened strip of ectoderm in early vertebrate embryos that forms along the dorsal side of the body and gives rise to the central nervous system.

neural tube: Primitive, hollow, dorsal nervous system of the early vertebrate embryo; formed by fusion of neural folds around the neural groove.

neuromodulator: A chemical agent that is released by a neuron and diffuses through a local region of the central nervous system, acting on neurons within that region; generally has the effect of modulating the response to neurotransmitters.

neuromuscular junction: The junction between an axon terminal of a motor neuron and a muscle fiber innervated by that motor neuron; the axon terminal of a motor neuron is typically branched, forming neuromuscular junctions with a number of different muscle fibers.

neuron [Gk., nerve]: Nerve cell, including cell body, dendrites, and axon.

neurosecretory cell: A neuron that releases one or more hormones into the circulatory system.

neurotransmitter: A chemical agent that is released by a neuron at a synapse, diffuses across the synaptic cleft, and acts upon a postsynaptic neuron or muscle or gland cell and alters its electrical state or activity.

neutron (new-tron): An uncharged particle with a mass slightly greater than that of a proton. Found in the atomic nucleus of all elements except hydrogen, in which the nucleus consists of a single proton.

niche: *See* Ecological niche.

nitrification: The oxidation of ammonia or ammonium to nitrites and nitrates, as by nitrifying bacteria.

nitrogen cycle: Worldwide circulation and reutilization of nitrogen atoms, chiefly due to metabolic processes of living organisms; plants take up inorganic nitrogen and convert it into organic compounds (chiefly proteins), which are assimilated into the bodies of one or more animals; bacterial and fungal action on nitrogenous waste products and dead organisms return nitrogen atoms to the inorganic state.

nitrogen fixation: Incorporation of atmospheric nitrogen into inorganic nitrogen compounds available to plants, a process that can be carried out only by some soil bacteria, many free-living and symbiotic cyanobacteria, and certain symbiotic bacteria in association with legumes.

nitrogenous base: A nitrogen-containing molecule having basic properties (tendency to acquire an H^+ ion); a purine or pyrimidine.

nocturnal [L. nocturnus, of night]: Applied to organisms that are active during the hours of darkness.

node [L. nodus, knot]: (1) In plants, a joint of a stem; the place where branches and leaves are joined to the stem. (2) In the vertebrate axon, the uninsulated openings between segments of the myelin sheath; the nerve impulse jumps from node to node, greatly speeding its conduction.

nondisjunction [L. non, not + disjungere, to separate]: The failure of homologous chromosomes or of chromatids to separate during meiosis, resulting in one or more extra chromosomes in some gametes and correspondingly fewer in others.

noradrenaline: A hormone, produced by the medulla of the adrenal gland, that increases the concentration of glucose in the blood, raises blood pressure and heartbeat rate, and increases muscular power and resistance to fatigue; also one of the principal neurotransmitters; also called norepinephrine.

norepinephrine: *See* noradrenaline.

notochord [Gk. noto, back + L. chorda, cord]: A dorsal, rodlike structure that runs the length of the body and serves as the internal skeleton in the embryos of all chordates; in most adult chordates the notochord is replaced by a vertebral column that forms around (but not from) the notochord.

nuclear envelope [L. nucleus, a kernel]: The double membrane surrounding the nucleus within a eukaryotic cell.

nucleic acid (new-clay-ick): A macromolecule consisting of nucleotides; the principal types are deoxyribonucleic acid (DNA) and ribonucleic acid (RNA).

nucleoid: In prokaryotic cells, the region of the cell in which the chromosome is localized.

nucleolus (new-klee-o-lus) [L., a small kernel]: A small, dense region visible in the nucleus of nondividing eukaryotic cells; consists of rRNA molecules, ribosomal proteins, and loops of chromatin from which the rRNA molecules are transcribed.

nucleosome [L. nucleus, a kernel + Gk. soma, body]: A complex of DNA and histone proteins that forms the fundamental packaging unit of eukaryotic DNA; its structure resembles a bead on a string.

nucleotide [L. nucleus, a kernel]: A molecule composed of a phosphate group, a five-carbon sugar (either ribose or deoxyribose), and a purine or pyrimidine base; nucleotides are the building blocks of nucleic acids.

nucleus [L., a kernel]: (1) The central core of an atom, containing protons and neutrons, around which electrons move. (2) The membrane-bound structure characteristic of eukaryotic cells that contains the genetic information in the form of DNA organized into chromosomes. (3) A group of nerve cell bodies within the central nervous system.

omnivore [L. omnis, all + vorare, to devour]: An organism that "eats everything"; for example, an animal that eats both plants and meat.

oncogene [Gk. onkos, tumor + genos, birth, race]: One of a group of eukaryotic genes that closely resemble normal genes of the cells in which they are found and that are thought to play a role in the development of cancer.

oocyte (o-uh-sight) [Gk. oion, egg + kytos, vessel]: A cell that gives rise by meiosis to an ovum.

operator: A segment of DNA that interacts with a repressor protein to regulate the transcription of the structural genes of an operon.

operon [L. opus, operis, work]: In the bacterial chromosome, a segment of DNA consisting of a promoter, an operator, and a group of adjacent structural genes; the structural genes, which code for products related to a particular biochemical pathway, are transcribed onto a single mRNA molecule, and their transcription is regulated by a single repressor protein.

opportunistic species: Species characterized by high reproduction rates, rapid development, early reproduction, small body size, and uncertain adult survival.

orbital [L. orbis, circle, disk]: In the current model of atomic structure, the volume of space surrounding the atomic nucleus in which an electron will be found 90 percent of the time.

order: A taxonomic grouping of related, similar families; the category below class and above family.

organ [Gk. organon, tool]: A body part composed of several tissues grouped together in a structural and functional unit.

organelle [Gk. organon, instrument, tool]: A membrane-bound compartment in the cytoplasm of a cell.

organic [Gk. organon, instrument, tool]: Pertaining to (1) organisms or living things generally, or (2) compounds formed by living organisms, or (3) the chemistry of compounds containing carbon.

organism [Gk. organon, instrument, tool]: Any living creature, either unicellular or multicellular.

organogenesis [Gk. organon, instrument, tool + genos, origin, descent]: In vertebrates, the later stages of development, following cleavage and gastrulation, during which the organ systems form.

osmosis [Gk. osmos, impulse, thrust]: The diffusion of water across a selectively permeable membrane (a membrane that permits the free passage of water but prevents or retards the passage of a solute). In the absence of other factors that affect the water potential, the net movement of water is from the side containing a lower concentration of solute to the side containing a higher concentration.

osmotic potential [Gk. *osmos*, impulse, thrust]: The tendency of water to move across a selectively permeable membrane into a solution; it is determined by measuring the pressure required to stop the osmotic movement of water into the solution.

ovarian follicle [L. *ovum*, egg + *folliculus*, small ball]: A developing oocyte and the specialized cells surrounding it; located near the surface of the ovary; following ovulation, forms the corpus luteum.

ovary [L. *ovum*, egg]: (1) In flowering plants, the enlarged basal portion of a carpel or a fused carpel, containing the ovule or ovules; the ovary matures to become the fruit. (2) In animals, the egg-producing organ.

oviduct [L. *ovum*, egg + *ductus*, duct]: The tube serving to transport the eggs to the outside or to the uterus; also called uterine tube or Fallopian tube (in humans).

ovulation: In animals, release of an egg or eggs from the ovary.

ovule [L. dim. of *ovum*, egg]: In seed plants, a structure composed of a protective outer coat, a tissue specialized for food storage, and a female gametophyte with an egg cell; becomes a seed after fertilization.

ovum, *pl.* **ova** [L., egg]: The egg cell; female gamete.

oxidation: Gain of oxygen, loss of hydrogen, or loss of an electron by an atom, ion, or molecule. Oxidation and reduction take place simultaneously, with the electron lost by one reactant being transferred to another reactant.

oxidative phosphorylation: The process by which the energy released as electrons pass down the mitochondrial electron transport chain in the final stage of cellular respiration is used to phosphorylate (add a phosphate group to) ADP molecules, thereby yielding ATP molecules.

oxygen debt: In muscle, the cumulative deficit of oxygen that develops during strenuous exercise when the supply of oxygen is inadequate for the demand; ATP is produced anaerobically by glycolysis, and the resulting pyruvic acid is converted to lactic acid, which is subsequently metabolized when adequate oxygen is available.

pacemaker: *See* Sinoatrial node.

paleontology [Gk. *palaios*, old + *onta*, things that exist + *logos*, discourse]: The study of the life of past geologic times, principally by means of fossils.

palisade cells [L. *palus*, stake + *cella*, a chamber]: In plant leaves, the columnar, chloroplast-containing parenchyma cells of the mesophyll.

pancreas (**pang**-kree-us) [Gk. *pan*, all + *kreas*, meat, flesh]: In vertebrates, a small, complex gland located between the stomach and the duodenum, which produces digestive enzymes and the hormones insulin and glucagon.

parasite [Gk. *para*, beside, akin to + *sitos*, food]: An organism that lives on or in an organism of a different species and derives nutrients from it.

parasitism: *See* Symbiosis.

parasympathetic division [Gk. *para*, beside, akin to]: A subdivision of the autonomic nervous system of vertebrates, with centers located in the brain and in the most anterior and most posterior parts of the spinal cord; stimulates digestion; generally inhibits other functions and restores the body to normal following emergencies.

parathyroid glands [Gk. *para*, beside, akin to + *thyra*, a door]: Four pea-sized endocrine glands of vertebrates, located behind or within the thyroid gland; source of parathyroid hormone, which acts to increase the concentration of calcium in the blood.

parenchyma (pah-**renk**-ee-ma) [Gk. *para*, beside, akin to + *en*, in + *chein*, to pour]: A plant tissue composed of living, thin-walled, randomly arranged cells with large vacuoles; usually photosynthetic or storage tissue.

parthenogenesis [Gk. *parthenon*, virgin + *genesis*, birth]: The development of an organism from an unfertilized egg.

pathogen [Gk. *pathos*, suffering + *genos*, origin, descent]: An organism or a virus that causes disease.

pattern formation: The developmental process that gives rise to the shape and structure of specific body parts, such as the limbs.

penetrance: In genetics, the proportion of individuals with a particular genotype that show the phenotype ascribed to that genotype.

peptide bond [Gk. *pepto*, to soften, digest]: The type of bond formed when two amino acids are joined end to end; the acidic group (—COOH) of one amino acid is linked covalently to the basic group (—NH$_2$) of the next, and a molecule of water (H$_2$O) is removed.

perennial [L. *per*, through + *annus*, year]: A plant that persists in whole or in part from year to year and usually produces reproductive structures in more than one year.

pericycle [Gk. *peri*, around + *kyklos*, circle]: One or more layers of cells completely surrounding the vascular tissues of the root; branch roots arise from the pericycle.

peripheral nervous system [Gk. *peripherein*, to carry around]: All of the neurons and axons outside the central nervous system, including both motor neurons and sensory neurons; consists of the somatic nervous system and the autonomic nervous system.

peristalsis [Gk. *peristellein*, to wrap around]: Successive waves of muscular contraction in the walls of a tubular structure, such as the digestive tract or an oviduct; moves the contents, such as food or an egg cell, through the tube.

peritoneum [Gk. *peritonos*, stretched over]: A membrane that lines the body cavity and forms the external covering of the visceral organs.

peritubular capillaries [Gk. *peri*, around + L. *tubus*, tube]: In the vertebrate kidney, the capillaries that surround the renal tubule; water and solutes are reabsorbed into the bloodstream through the peritubular capillaries and some substances are secreted from them into the renal tubule.

permeable [L. *permeare*, to pass through]: Penetrable by molecules, ions, or atoms; usually applied to membranes that let given solutes pass through.

peroxisome: A membrane-bound organelle in which enzymes catalyzing peroxide-forming and peroxide-destroying reactions are segregated; in plant cells, the site of photorespiration.

petiole (pet-ee-ole) [Fr., from L. *petiolus*, dim. of *pes, pedis*, a foot]: The stalk of a leaf, connecting the blade of the leaf with the branch or stem.

pH: A symbol denoting the concentration of hydrogen ions in a solution; pH values range from 0 to 14; the lower the value, the more acidic a solution, that is, the more hydrogen ions it contains; pH 7 is neutral, less than 7 is acidic, more than 7 is alkaline.

phage: *See* Bacteriophage.

phagocytosis [Gk. *phagein*, to eat + *kytos*, vessel]: Cell "eating." *See* Endocytosis.

phenotype [Gk. *phainein*, to show + *typos*, stamp, print]: Observable characteristics of an organism, resulting from interactions between the genotype and the environment.

pheromone (fair-o-moan) [Gk. *phero*, to bear, carry]: Substance secreted by an animal that influences the behavior or development of other animals of the same species, such as the sex attractants of moths, the queen substance of honey bees.

phloem (flow-em) [Gk. *phloos*, bark]: Vascular tissue of higher plants; conducts sugars and other organic molecules from the leaves to other parts of the plant; in angiosperms, composed of sieve-tube members, companion cells, other parenchyma cells, and fibers.

phospholipids: Organic molecules similar in structure to fats, but in which a phosphate group rather than a fatty acid is attached to the third carbon of the glycerol molecule; as a result, the molecule has a hydrophilic "head" and a hydrophobic "tail." A double layer (bilayer) of phospholipids forms the fundamental structure of the plasma membrane and of organelle membranes.

phosphorylation: Addition of a phosphate group or groups to a molecule.

photoperiodism [Gk. *photos,* light]: The response to relative day and night length, a mechanism by which organisms measure seasonal change.

photophosphorylation [Gk. *photos,* light + *phosphoros,* bringing light]: The process by which the energy released as electrons pass down the electron transport chain between photosystems II and I during photosynthesis is used to phosphorylate ADP to ATP.

photoreceptor [Gk. *photos,* light]: A cell or organ capable of detecting light.

photorespiration [Gk. *photos,* light + L. *respirare,* to breathe]: The oxidation of carbohydrates in the presence of light and oxygen; occurs when the carbon dioxide concentration in the leaf is low in relation to the oxygen concentration.

photosynthesis [Gk. *photos,* light + *syn,* together + *tithenai,* to place]: The conversion of light energy to chemical energy; the synthesis of organic compounds from carbon dioxide and water in the presence of chlorophyll, using light energy.

photosystem [Gk. *photos,* light + *systema,* that which is put together]: A discrete unit of organization of chlorophyll and accessory pigments, embedded in the thylakoids of chloroplasts; the site of the light-dependent reactions of photosynthesis.

phototropism [Gk. *photos,* light + *trope,* turning]: Movement in which the direction of the light is the determining factor, such as the growth of a plant toward a light source; a curving response to light.

phyletic change [Gk. *phylon,* race, tribe]: The changes taking place in a single lineage of organisms over a long period of time; one of the principal patterns of evolutionary change.

phylogeny [Gk. *phylon,* race, tribe]: Evolutionary history of a taxonomic group. Phylogenies are often depicted as "evolutionary trees."

phylum, *pl.* phyla [Gk. *phylon,* race, tribe]: A taxonomic grouping of related, similar classes; a high-level category below kingdom and above class. Phylum is generally used in the classification of protozoa and animals, whereas an equivalent category, division, is used in the classification of prokaryotes, algae, fungi, and plants.

physiology [Gk. *physis,* nature + *logos,* a discourse]: The study of function in cells, organs, or entire organisms; the processes of life.

phytochrome [Gk. *phyton,* plant + *chrōma,* color]: A plant pigment that is a photoreceptor for red or far-red light and is involved with a number of developmental processes, such as flowering, dormancy, leaf formation, and seed germination.

phytoplankton [Gk. *phyton,* plant + *planktos,* wandering]: Aquatic, free-floating, microscopic, photosynthetic organisms.

pigment [L. *pigmentum,* paint]: A colored substance that absorbs light over a narrow band of wavelengths.

pinocytosis [Gk. *pinein,* to drink + *kytos,* vessel]: Cell "drinking." *See* Endocytosis.

pith: In the roots and stems of plants, the ground tissue occupying the region to the interior of the vascular cylinder; usually consists of parenchyma cells.

pituitary [L. *pituita,* phlegm]: Endocrine gland in vertebrates; the anterior lobe is the source of tropic hormones, growth hormone, and prolactin and is regulated by secretions of the hypothalamus; the posterior lobe stores and releases oxytocin and ADH produced by the hypothalamus.

placenta [Gk. *plax,* a flat object]: A tissue formed as the result of interactions between the inner lining of the mammalian uterus and the extraembryonic chorion; serves as the connection through which exchanges of nutrients and wastes occur between the blood of the mother and that of the embryo.

plankton [Gk. *planktos,* wandering]: Small (mostly microscopic) aquatic and marine organisms found in the upper levels of the water, where light is abundant; includes both photosynthetic (phytoplankton) and heterotrophic (zooplankton) forms.

planula [L. dim. of *planus,* a wanderer]: The ciliated, free-swimming type of larva formed by many cnidarians.

plasma [Gk., form or mold]: The clear, colorless fluid component of vertebrate blood, containing dissolved ions, molecules, and plasma proteins; blood minus the blood cells.

plasma cell: An antibody-producing cell resulting from the differentiation and proliferation of a B lymphocyte that has interacted with an antigen complementary to the antibodies displayed on its surface; a mature plasma cell can produce from 3,000 to 30,000 antibody molecules per second.

plasma membrane [Gk. *plasma,* form or mold + L. *membrana,* skin, parchment]: The membrane surrounding the cytoplasm of a cell; the cell membrane. Its fundamental structure consists of a bilayer of phospholipids (and a smaller number of glycolipids) in which protein molecules are embedded.

plasmid: In prokaryotes, an extrachromosomal, independently replicating, small, circular DNA molecule.

plasmodesma, *pl.* plasmodesmata [Gk. *plassein,* to mold + *desmos,* band, bond]: In plants, a minute, cytoplasmic thread that extends through pores in cell walls and connects the cytoplasm of adjacent cells.

plastid [Gk. *plastos,* formed or molded]: A membrane-bound, often pigmented, organelle in the cytoplasm of plant cells; includes chloroplasts and the starch-filled plastids in the core cells of the root cap.

platelet (plate-let) [Gk. *platus,* flat]: In mammals, a round or biconcave disk suspended in the blood and involved in the formation of blood clots.

pleiotropy (ply-o-**trop**-ee) [Gk. *pleios,* more + *trope,* a turning]: The capacity of a gene to affect a number of different phenotypic characteristics.

polar [L. *polus,* end of axis]: Having parts or areas with opposed or contrasting properties, such as positive and negative charges, head and tail.

polar body: Minute, nonfunctioning cell produced during those meiotic divisions that lead to egg cells; contains a nucleus but very little cytoplasm.

polar covalent bond: A covalent bond in which the electrons are shared unequally between the two atoms; the resulting polar molecule has regions of slightly negative and slightly positive charge.

polar nuclei: In angiosperms, the two nuclei of the central cell of the female gametophyte; they fuse with a sperm nucleus to form the triploid ($3n$) endosperm nucleus.

pollen [L., fine dust]: In seed plants, spores consisting of an immature male gametophyte and a protective outer covering.

pollination [L. *pollen,* fine dust]: The transfer of pollen from the anther to a receptive surface of a flower.

polygenic inheritance [Gk. *polus,* many + *genos,* race, descent]: The determination of a given characteristic, such as weight or height, by the interaction of many genes.

polymer [Gk. *polus,* many + *meris,* part or portion]: A large molecule composed of many similar or identical molecular subunits.

polymerase: An enzyme, such as DNA polymerase or RNA polymerase, that catalyzes the synthesis of a polymer from its subunits.

polymerase chain reaction (PCR): A recently developed chemical process in which millions of copies of a specific segment of DNA can be rapidly synthesized from a tiny sample of DNA containing that segment.

polymorphism [Gk. *polus,* many + *morphe,* form]: The presence in a single population of two or more phenotypically distinct forms of a trait.

polyp [Gk. *polus,* many + *pous,* foot]: The sessile stage in the life cycle of cnidarians.

polypeptide [Gk. *polus,* many + *pepto,* to soften, digest]: A molecule consisting of a long chain of amino acids linked together by peptide bonds.

polyploid [Gk. *polus*, many + *ploion*, vessel]: Cell with more than two complete sets of chromosomes per nucleus.

polyribosome: Two or more ribosomes together with a molecule of mRNA that they are simultaneously translating; a polysome.

polysaccharide [Gk. *polus*, many + *sakcharon*, sugar]: A carbohydrate polymer composed of monosaccharide subunits in long chains; includes starch, cellulose.

polysome: *See* Polyribosome.

population: Any group of individuals of one species that occupy a given area at the same time; in genetic terms, an interbreeding group of organisms.

population bottleneck: Type of genetic drift that occurs as the result of a population being drastically reduced in numbers by an event having little to do with the usual forces of natural selection.

population density: The number of individuals of a population per unit area or volume of living space.

posterior: Of or pertaining to the rear, or tail, end.

potential energy: Energy in a potentially usable form that is not, for the moment, being used; often called "energy of position."

predator [L. *praedari*, to prey upon; from *prehendere*, to grasp, seize]: An organism that eats other living organisms.

pressure-flow hypothesis: A hypothesis accounting for sap flow through the phloem system. According to this hypothesis, the solution containing nutrient sugars moves through the sieve tubes by bulk flow, moving into and out of the sieve tubes by active transport and diffusion.

prey [L. *prehendere*, to grasp, seize]: An organism eaten by another organism.

primary growth: In plants, growth originating in the apical meristem of the shoots and roots, as contrasted with secondary growth; results in an increase in length.

primary structure of a protein: The amino acid sequence of a protein.

primate: A member of the order of mammals that includes anthropoids and prosimians.

primitive [L. *primus*, first]: Not specialized; at an early stage of evolution or development.

primordium, *pl.* **primordia** [L. *primus*, first + *ordiri*, to begin to weave]: A cell or organ in its earliest stage of differentiation.

procambium [L. *pro*, before + *cambium*, exchange]: In plants, a primary meristematic tissue; gives rise to vascular tissues of the primary plant body and to the vascular cambium.

producer, in ecological systems: An autotrophic organism, usually a photosynthesizer, that contributes to the net primary productivity of a community.

progesterone [L. *progerere*, to carry forth or out + *steiras*, barren]: In mammals, a steroid hormone produced by the corpus luteum that prepares the uterus for implantation of the embryo; also produced by the placenta during pregnancy.

prokaryote [L. *pro*, before + Gk. *karyon*, nut, kernel]: A cell lacking a membrane-bound nucleus and membrane-bound organelles; a bacterium or a cyanobacterium.

promoter: Specific segment of DNA to which RNA polymerase attaches to initiate transcription of mRNA from an operon.

prophage: A bacterial virus (bacteriophage) integrated into a host chromosome.

prophase [Gk. *pro*, before + *phasis*, form]: An early stage in nuclear division, characterized by the condensing of the chromosomes and their movement toward the equator of the spindle. Homologous chromosomes pair up during meiotic prophase.

prosimian [L. *pro*, before + *simia*, ape]: A lower primate; includes lemurs, lorises, tarsiers, and bush babies, as well as many fossil forms.

prostaglandins [Gk. *prostas*, a porch or vestibule + L. *glans*, acorn]: A group of fatty acids that function as chemical messengers; synthesized in most, possibly all, cells of the body; thought to play key roles in fertilization and in triggering the onset of both menstruation and labor.

prostate gland [Gk. *prostas*, a porch or vestibule + L. *glans*, acorn]: A mass of muscle and glandular tissue surrounding the base of the urethra in male mammals; the vasa deferentia merge with ducts from the seminal vesicles, enter the prostate gland, and there merge with the urethra. The prostate gland secretes an alkaline fluid that has a stimulating effect on the sperm as they are released.

protein [Gk. *proteios*, primary]: A complex organic compound composed of one or more polypeptide chains, each made up of many (about 100 or more) amino acids linked together by peptide bonds.

protoderm [Gk. *protos*, first + *derma*, skin]: In plants, a primary meristematic tissue; gives rise to the epidermis.

proton: A subatomic particle with a single positive charge equal in magnitude to the charge of an electron and with a mass slightly less than that of a neutron; a component of every atomic nucleus.

protoplasm [Gk. *protos*, first + *plasma*, anything molded]: Living matter.

protostome [Gk. *protos*, first + *stoma*, mouth]: An animal in which the mouth forms at or near the blastopore in the developing embryo; mollusks, annelids, and arthropods are protostomes.

provirus: A virus of a eukaryote that has become integrated into a host chromosome.

pseudocoelom [Gk. *pseudes*, false + *koilos*, a hollow]: A body cavity consisting of a fluid-filled space between the endoderm and the mesoderm; characteristic of the nematodes.

pseudopodium [Gk. *pseudes*, false + *pous*, pod-, foot]: A temporary cytoplasmic protrusion from an amoeboid cell, which functions in locomotion or in feeding by phagocytosis.

pulmonary [L. *pulmonis*, lung]: Pertaining to the lungs.

pulmonary artery [L. *pulmonis*, lung]: In birds and mammals, an artery that carries deoxygenated blood from the right ventricle of the heart to the lungs, where it is oxygenated.

pulmonary vein [L. *pulmonis*, lung]: In birds and mammals, a vein that carries oxygenated blood from the lungs to the left atrium of the heart, from which blood is pumped into the left ventricle and from there to the body tissues.

punctuated equilibrium: A model of the mechanism of evolutionary change that proposes that long periods of no change ("stasis") are punctuated by periods of rapid speciation, with natural selection acting on species as well as on individuals.

Punnett square: The checkerboard diagram used for analysis of allele segregation.

pupa [L., girl, doll]: A developmental stage of some insects, in which the organism is nonfeeding, immotile, and sometimes encapsulated or in a cocoon; the pupal stage occurs between the larval and adult phases.

purine [Gk. *purinos*, fiery, sparkling]: A nitrogenous base, such as adenine or guanine, with a characteristic two-ring structure; one of the components of nucleic acids.

pyramid, ecological: *See* Ecological pyramid.

pyramid of energy: A diagram of the energy flow between the trophic levels of an ecosystem; plants or other autotrophs (at the base of the pyramid) represent the greatest amount of energy, herbivores next, then primary carnivores, secondary carnivores, etc.

pyrimidine: A nitrogenous base, such as cytosine, thymine, or uracil, with a characteristic single-ring structure; one of the components of nucleic acids.

quaternary structure of a protein: The overall structure of a globular protein molecule that consists of two or more polypeptide chains.

queen: In social insects (ants, termites, and some species of bees and wasps), the fertile, or fully developed, female whose function is to lay eggs.

radial symmetry [L. *radius*, a spoke of a wheel + Gk. *summetros*, symmetry]: The regular arrangement of parts around a central axis such that any plane passing through the central axis divides the organism into halves that are approximate mirror images; seen in cnidarians and adult echinoderms.

radiation [L. *radius*, a spoke of a wheel, hence, a ray]: Energy emitted in the form of waves or particles.

radioactive isotope: An isotope with an unstable nucleus that stabilizes itself by emitting radiation.

receptor: A protein or glycoprotein molecule with a specific three-dimensional structure, to which a substance (for example, a hormone, a neurotransmitter, or an antigen) with a complementary structure can bind; typically displayed on the surface of a membrane. Binding of a complementary molecule to a receptor may trigger a transport process or a change in processes occurring within the cell.

recessive allele [L. *recedere*, to recede]: An allele whose phenotypic effect is masked in the heterozygote by that of another, dominant allele.

reciprocal altruism: Performance of an altruistic act with the expectation that the favor will be returned.

recognition sequence: A specific sequence of nucleotides at which a restriction enzyme cleaves a DNA molecule.

recombinant DNA: DNA formed either naturally or in the laboratory by the joining of segments of DNA from different sources.

recombination: The formation of new gene combinations; in eukaryotes, may be accomplished by new associations of chromosomes produced during sexual reproduction or crossing over; in prokaryotes, may be accomplished through transformation, conjugation, or transduction.

reduction [L. *reducere*, to lead back]: Loss of oxygen, gain of hydrogen, or gain of an electron by an atom, ion, or molecule; oxidation and reduction take place simultaneously, with the electron lost by one reactant being transferred to another.

reflex [L. *reflectere*, to bend back]: Unit of action of the nervous system involving a sensory neuron, often one or more interneurons, and one or more motor neurons.

relay neuron: Neuron that transmits signals between different regions of the central nervous system.

releaser: In ethology, a stimulus that functions as a communication signal between members of the same species and that sets in motion, or "releases," the sequence of movements that constitute a fixed action pattern.

renal [L. *renes*, kidneys]: Pertaining to the kidney.

replication fork: In DNA synthesis, the Y-shaped structure formed at the point where the two strands of the original molecule are being separated and the complementary strands are being synthesized.

repressor [L. *reprimere*, to press back, keep back]: In genetics, a protein that binds to the operator, preventing RNA polymerase from attaching to the promoter and transcribing the structural genes of the operon; coded by a segment of DNA known as the regulator.

resolving power [L. *resolvere*, to loosen, unbind]: The ability of a lens to distinguish two lines as separate.

respiration [L. *respirare*, to breathe]: (1) In aerobic organisms, the intake of oxygen and the liberation of carbon dioxide. (2) In cells, the oxygen-requiring stage in the breakdown and release of energy from fuel molecules.

resting potential: The difference in electric potential (about 70 millivolts) across the membrane of an axon at rest.

restriction enzymes: Enzymes that cleave the DNA double helix at specific nucleotide sequences.

reticular formation [L. *reticulum*, a network]: A brain circuit involved with alertness and direction of attention to selected events; consists of a loose network of interneurons running through the brainstem, plus certain neurons in the thalamus that function as an extension of this network.

reticulum [L., network]: A fine network (e.g., endoplasmic reticulum).

retina [L. dim. of *rete*, net]: The light-sensitive layer of the vertebrate eye; contains several layers of neurons and photoreceptor cells (rods and cones); receives the image formed by the lens and transmits it to the brain via the optic nerve.

retrovirus [L., turning back]: An RNA virus that codes for an enzyme, reverse transcriptase, that transcribes the RNA into DNA and catalyzes the insertion of that DNA into a host-cell chromosome.

reverse transcriptase: An enzyme that transcribes RNA into DNA; found only in association with retroviruses.

rhizoid [Gk. *rhiza*, root]: Rootlike anchoring structure in fungi and nonvascular plants.

rhizome [Gk. *rhizoma*, mass of roots]: In vascular plants, a horizontal stem growing along or below the surface of the soil; may be enlarged for storage or may function in vegetative reproduction.

ribonucleic acid (RNA) (rye-bo-new-clay-ick): A class of nucleic acids characterized by the presence of the sugar ribose and the pyrimidine uracil; includes mRNA, tRNA, and rRNA. RNA is the genetic material of many viruses.

ribosomal RNA (rRNA): A class of RNA molecules found, along with characteristic proteins, in ribosomes; transcribed from DNA of the chromatin loops that form the nucleolus.

ribosome: A complex of specific protein and ribonucleic acid molecules; the site of translation in protein synthesis; in eukaryotic cells, often bound to the endoplasmic reticulum. Many ribosomes attached to a single strand of mRNA are called a polyribosome, or polysome.

RNA: Abbreviation of ribonucleic acid.

rod: Photoreceptor cell found in the vertebrate retina; sensitive to very dim light, responsible for "night vision."

root: The descending axis of a plant, normally below ground and serving both to anchor the plant and to take up and conduct water and dissolved minerals.

root cap: A thimblelike mass of cells that covers and protects the growing tip of a root.

root hair: An extremely fine cytoplasmic extension of an epidermal cell of a young root; root hairs greatly increase the surface area for the uptake of water and dissolved minerals.

saprobe [Gk. *sapros*, rotten, putrid + *bios*, life]: An organism that feeds on nonliving organic matter.

sarcolemma [Gk. *sarx*, the flesh + *lemma*, husk]: The specialized plasma membrane surrounding a muscle cell (muscle fiber); capable of propagating action potentials.

sarcomere [Gk. *sarx*, the flesh + *meris*, part of, portion]: Functional and structural unit of contraction in striated muscle.

sarcoplasmic reticulum [Gk. *sarx*, the flesh + *plasma*, from cytoplasm + L. *reticulum*, network]: The specialized endoplasmic reticulum that encases each myofibril of a muscle cell.

saturated fatty acid: A fatty acid in which there are no double bonds between carbon atoms; fats containing saturated fatty acids tend to be solid at room temperature.

sclerenchyma [Gk. *skleros*, hard]: In plants, a type of supporting cell with thick, often lignified, secondary walls; may be alive or dead at maturity; includes fibers and sclereids.

secondary growth: In the stems and roots of woody plants, growth originating in the lateral meristems (cork cambium and vascular cambium); as contrasted with primary growth; results in an increase in girth.

secondary sex characteristics: Characteristics of animals that distinguish between the two sexes but that do not produce or convey gametes; includes facial hair of the human male and enlarged hips and breasts of the female.

secondary structure of a protein: The simple structure (often a helix, a sheet, or a cable) resulting from the spontaneous folding of a polypeptide chain as it is formed; maintained by hydrogen bonds and other weak forces.

secretion [L. *secernere*, to sever, separate]: (1) Product of any cell, gland, or tissue that is released through the plasma membrane and that performs its function outside the cell that produced it. (2) The stage of kidney function in which, through active transport processes, molecules remaining in the blood plasma are selectively removed from the peritubular capillaries and pumped into the filtrate in the renal tubule.

seed: A complex structure formed by the maturation of the ovule of seed plants following fertilization; upon germination, a seed develops into a new sporophyte; generally consists of seed coat, embryo, and a food reserve.

segregation: *See* Mendel's first law.

selectively permeable [L. *seligere*, to gather apart + *permeare*, to go through]: Applied to membranes that permit passage of water and some solutes but block passage of most solutes; semipermeable.

self-fertilization: The union of egg and sperm produced by a single hermaphroditic organism.

self-pollination: The transfer of pollen from anther to stigma in the same flower or to another flower of the same plant, leading to self-fertilization.

semen [L., seed]: Product of the male reproductive system; includes sperm and the sperm-carrying fluids.

seminal vesicles [L. *semen*, seed + *vesicula*, a little bladder]: In male mammals, small vesicles, the ducts of which merge with the vasa deferentia as they enter the prostate gland; they produce an alkaline, fructose-containing fluid that suspends and nourishes the sperm cells.

seminiferous tubules [L. *semen*, seed + *ferre*, to bear or carry + dim. of *tubus*, tube]: The structures within the testes in which sperm cells are produced; contain spermatogonia, which give rise to the sperm cells, and Sertoli cells, which provide nourishment to the developing sperm cells.

sensory neuron: A neuron that conducts impulses from a sensory receptor to the central nervous system or a central ganglion.

sensory receptor: A cell, tissue, or organ that detects internal or external stimuli.

septum [L., fence]: A partition, or cross wall, that divides a structure, such as a fungal hypha, into compartments.

sessile [L. *sedere*, to sit]: Attached; not free to move about.

sex chromosomes: Chromosomes that are different in the two sexes and that are involved in sex determination.

sex-linked trait: An inherited trait, such as color discrimination, determined by a gene located on a sex chromosome and that therefore shows a different pattern of inheritance in males and females.

sexual reproduction: Reproduction involving meiosis and fertilization.

sexual selection: A type of natural selection that acts on characteristics of direct consequence in obtaining a mate and successfully reproducing; thought to be the chief cause of sexual dimorphism, the striking phenotypic differences between the males and females of many species.

shoot: The aboveground portions, such as the stem and leaves, of a vascular plant.

short-day plant: A plant that must be exposed to a light period shorter than some critical length for flowering to occur; short-day plants usually flower in autumn.

sieve tube: A series of sugar-conducting cells (sieve-tube members) found in the phloem of angiosperms.

sinoatrial node [L. *sinus*, fold, hollow + *atrium*, yard, court, hall + *nodus*, knot]: Area of the vertebrate heart that initiates the heartbeat; located where the superior vena cava enters the right atrium; the pacemaker.

smooth muscle: Nonstriated muscle; lines the walls of internal organs and arteries and is under involuntary control.

social dominance: A hierarchical pattern of social organization involving domination of some members of a group by other members in a relatively orderly and long-lasting pattern.

society [L. *socius*, companion]: An organization of individuals of the same species in which there are divisions of resources, divisions of labor, and mutual dependence; a society is held together by stimuli exchanged among members of the group.

sodium-potassium pump: In the plasma membrane of animal cells, a transport protein that pumps sodium ions (Na^+) out of the cell and potassium ions (K^+) into the cell in a process powered by ATP; establishes and maintains concentration gradients that are essential for osmotic balance and, in neurons, for conduction of the nerve impulse.

solution: A homogeneous mixture of the molecules of two or more substances; the substance present in the greatest amount (usually a liquid) is called the solvent, and the substances present in lesser amounts are called solutes.

somatic cells [Gk. *soma*, body]: The differentiated cells composing body tissues of multicellular plants and animals; all body cells except those giving rise to gametes.

somatic nervous system [Gk. *soma*, body]: In vertebrates, the motor and sensory neurons of the peripheral nervous system that control skeletal muscle; the "voluntary" system, as contrasted with the "involuntary," or autonomic, nervous system.

somatotropin [Gk. *soma*, body + *trope*, a turning]: A hormone, produced by the pituitary gland, that stimulates protein synthesis and promotes the growth of bone; also known as growth hormone.

specialized: (1) Of cells, having particular functions in a multicellular organism. (2) Of organisms, having special adaptations to a particular habitat or mode of life.

speciation: The process by which new species are formed.

species, *pl.* **species** [L., kind, sort]: A group of organisms that actually (or potentially) interbreed in nature and are reproductively isolated from all other such groups; a taxonomic grouping of anatomically similar individuals (the category below genus).

species-specific: Characteristic of (and limited to) a particular species.

specific: Unique; for example, the proteins in a given organism, the enzyme catalyzing a given reaction, or the antibody to a given antigen.

specific heat: The amount of heat (in calories) required to raise the temperature of 1 gram of a substance 1)C. The specific heat of water is 1 calorie per gram.

sperm [Gk. *sperma*, seed]: A mature male sex cell, or gamete, usually motile and smaller than the female gamete.

spermatid [Gk. *sperma*, seed]: Each of four haploid (*n*) cells resulting from the meiotic divisions of a spermatocyte; each spermatid becomes differentiated into a sperm cell.

spermatocytes [Gk. *sperma*, seed + *kytos*, vessel]: The diploid (2*n*) cells formed by the enlargement and differentiation of the spermatogonia; they give rise by meiotic division to the spermatids.

spermatogonia [Gk. *sperma*, seed + *gonos*, a child, the young]: The unspecialized diploid (2*n*) cells on the walls of the seminiferous tubules that, by enlargement, differentiation, and meiotic division, become spermatocytes, then spermatids, then sperm cells.

sphincter [Gk. *sphinktēr*, a band]: A circular muscle surrounding the opening of a tubular structure or the juncture of different regions of a tubular structure (e.g., the pyloric sphincter, at the juncture of the stomach and the small intestine); contraction of the sphincter closes the passageway, and relaxation opens it.

spinal cord: Part of the vertebrate central nervous system; consists of a thick, dorsal, longitudinal bundle of nerve fibers extending posteriorly from the brain.

spindle: In dividing cells, the structure formed of microtubules that extends from pole to pole; the spindle fibers appear to maneuver the chromosomes into position during metaphase and to pull the newly separated chromosomes toward the poles during anaphase.

spiracle [L. *spirare*, to breathe]: One of the external openings of the respiratory system in terrestrial arthropods.

splitting evolution: *See* Cladogenesis.

spongy parenchyma: In plant leaves, a tissue composed of loosely arranged chloroplast-containing parenchyma cells.

sporangiophore (spo-**ran**-ji-o-for) [Gk. *spora*, seed + *phore*, from *phorein*, to bear]: A specialized hypha or a branch bearing one or more sporangia.

sporangium, *pl.* **sporangia** [Gk. *spora*, seed]: A unicellular or multicellular structure in which spores are produced.

spore [Gk. *spora*, seed]: An asexual reproductive or resting cell capable of developing into a new organism without fusion with another cell; in contrast to a gamete.

sporophyte [Gk. *spora*, seed + *phytos*, growing]: In organisms that have alternation of haploid and diploid generations (all plants and some green algae), the diploid (2*n*) spore-producing generation.

stamen [L., a thread]: The male structure of a flower, which produces microspores or pollen; usually consists of a stalk, the filament, bearing a pollen-producing anther at its tip.

starch [M.E. *sterchen*, to stiffen]: A class of complex, insoluble carbohydrates, the chief food-storage substances of plants; composed of 1,000 or more glucose units and readily broken down enzymatically into these units.

statocyst [Gk. *statos*, standing + *kystis*, sac]: An organ of balance, consisting of a vesicle containing granules of sand (statoliths) or some other material that stimulates sensory cells when the organism moves.

stem: The aboveground part of the axis of vascular plants, as well as anatomically similar portions below ground (such as rhizomes).

stem cells: The common, self-regenerating cells in the marrow of long bones that give rise, by differentiation and division, to red blood cells and all of the different types of white blood cells.

stereoscopic vision [Gk. *stereōs*, solid + *optikos*, pertaining to the eye]: Ability to perceive a single, three-dimensional image from the simultaneous but separate images delivered to the brain by each eye.

steroid: One of a group of lipids having four linked carbon rings and, often, a hydrocarbon tail; cholesterol, sex hormones, and the hormones of the adrenal cortex are steroids.

stigma [Gk. *stigme*, a prick mark, puncture]: In plants, the region of a carpel serving as a receptive surface for pollen grains, which germinate on it.

stimulus [L., goad, incentive]: Any internal or external change or signal that influences the activity of an organism or of part of an organism.

stoma, *pl.* **stomata** [Gk., mouth]: A minute opening in the epidermis of leaves and stems, bordered by guard cells, through which gases pass.

strategy [Gk. *strategein*, to maneuver]: A group of related traits, evolved under the influence of natural selection, that solve particular problems encountered by living organisms; often includes anatomical, physiological, and behavioral characteristics.

striated muscle [L., from *striare*, to groove]: Skeletal voluntary muscle and cardiac muscle. The name derives from the striped appearance, which reflects the arrangement of contractile elements.

stroma [Gk., a bed, from *stronnymi*, to spread out]: A dense solution that fills the interior of the chloroplast and surrounds the thylakoids; the site of the light-independent reactions of photosynthesis.

structural gene: Any gene that codes for a protein; in distinction to regulatory genes.

style [L. *stilus*, stake, stalk]: In angiosperms, the stalk of a carpel, down which the pollen tube grows.

substrate [L. *substratus*, strewn under]: (1) The foundation to which an organism is attached. (2) A substance on which an enzyme acts.

succession: *See* Ecological succession.

sucrose: Cane sugar; a common disaccharide found in many plants; a molecule of glucose linked to a molecule of fructose.

sugar: Any monosaccharide or disaccharide.

surface tension: A tautness of the surface of a liquid, caused by the cohesion of the molecules of liquid. Water has an extremely high surface tension.

symbiosis [Gk. *syn*, together with + *bioonai*, to live]: An intimate and protracted association between two or more organisms of different species. Includes mutualism, in which the association is beneficial to both; commensalism, in which one benefits and the other is neither harmed nor benefited; and parasitism, in which one benefits and the other is harmed.

sympathetic division: A subdivision of the autonomic nervous system, with centers in the midportion of the spinal cord; slows digestion; generally excites other functions.

sympatric speciation [Gk. *syn*, together with + *patra*, fatherland, country]: Speciation that occurs without geographic isolation of a population of organisms; usually occurs as the result of hybridization accompanied by polyploidy; may occur in some cases as a result of disruptive selection.

synapse [Gk. *synapsis*, a union]: A specialized junction between two neurons where the activity in one influences the activity in the other; in mammals, transmission across the synaptic cleft is by means of chemical substances (neurotransmitters) and may be excitatory or inhibitory.

synthesis [Gk. *syntheke*, a putting together]: The formation of a more complex substance from simpler ones.

synthesis phase: In the cell cycle, the phase in which the DNA of the chromosomes is replicated and DNA-associated proteins, such as histones, are synthesized.

T cell: A type of white blood cell involved in cell-mediated immunity and in interactions with B cells; a T lymphocyte.

taxon, *pl.* **taxa** [Gk. *taxis*, arrange, put in order]: A particular group, ranked at a particular categorical level, in a hierarchical classification scheme; for example, *Drosophila* is a taxon at the categorical level of genus.

taxonomy [Gk. *taxis*, arrange, put in order + *nomos*, law]: The study of the classification of organisms; the ordering of organisms into a hierarchy that reflects their essential similarities and differences.

telophase [Gk. *telos*, end + *phasis*, form]: The last stage in mitosis and meiosis, during which the chromosomes become reorganized into two new nuclei.

temperate bacteriophage: A bacterial virus that may become incorporated into the host-cell chromosome.

template: A pattern or mold guiding the formation of a negative or complementary copy.

tentacles [L. *tentare*, to touch]: Long, flexible protrusions located about the mouth of many invertebrates; usually prehensile or tactile.

territory: An area or space occupied and defended by an individual or a group; trespassers are attacked (and usually defeated); may be the site of breeding, nesting, food gathering, or any combination thereof.

tertiary structure of a protein: A complex structure, usually globular, resulting from further folding of the secondary structure of a protein; forms spontaneously due to attractions and repulsions among amino acids with different charges on their R groups.

testcross: A mating between a phenotypically dominant individual and a homozygous recessive "tester" to determine the genetic constitution of the dominant phenotype, that is, whether it is homozygous or heterozygous for the relevant gene.

testis, *pl.* **testes** [L., witness]: The sperm-producing organ; also the source of the male sex hormone testosterone.

testosterone [Gk. *testis*, testicle + *steiras*, barren]: A steroid hormone secreted by the testes in higher vertebrates and stimulating the development and maintenance of male sex characteristics and the production of sperm; the principal androgen.

tetrad [Gk. *tetras*, four]: In genetics, a pair of homologous chromosomes that have replicated and come together in prophase I of meiosis; consists of four chromatids.

thalamus [Gk. *thalamos*, chamber]: A part of the vertebrate brain tucked below the cerebrum; the main relay center between the brainstem and the higher brain centers.

thallus [Gk. *thallos*, a young twig]: A simple plant or algal body without true roots, leaves, or stems.

theory [Gk. *theorein*, to look at]: A generalization based on many observations and experiments; a verified hypothesis.

thermodynamics [Gk. *therme*, heat + *dynamis*, power]: The study of transformations of energy. The first law of thermodynamics states that, in all processes, the total energy of a system plus its surroundings remains constant. The second law states that all natural processes tend to proceed in such a direction that the disorder or randomness of the system increases.

thorax [Gk., breastplate]: (1) In vertebrates, that portion of the trunk containing the heart and lungs. (2) In crustaceans and insects, the fused, leg-bearing segments between head and abdomen.

thylakoid [Gk. *thylakos*, a small bag]: A flattened sac, or vesicle, that forms part of the internal membrane structure of the chloroplast; the site of the light-dependent reactions of photosynthesis and of photophosphorylation; stacks of thylakoids collectively form the grana.

thyroid [Gk. *thyra*, a door]: An endocrine gland of vertebrates, located in the neck; source of an iodine-containing hormone (thyroxine) that increases the metabolic rate and affects growth.

tight junction: A junction between adjacent animal cells that prevents materials from leaking through the tissue; for example, intestinal epithelial cells are surrounded by tight junctions.

tissue [L. *texere*, to weave]: A group of similar cells organized into a structural and functional unit.

trachea, *pl.* **tracheae** (trake-ee-a) [Gk. *tracheia*, rough]: An air-conducting tube. (1) In insects and some other terrestrial arthropods, a system of chitin-lined air ducts. (2) In terrestrial vertebrates, the windpipe.

tracheid (tray-key-idd) [Gk. *tracheia*, rough]: In vascular plants, an elongated, thick-walled conducting and supporting cell of xylem, characterized by tapering ends and pitted walls without true perforations.

tract: A group or bundle of nerve fibers with accompanying connective tissue, located within the central nervous system.

transcription [L. *trans*, across + *scribere*, to write]: The enzymatic process by which the genetic information contained in one strand of DNA is used to specify a complementary sequence of bases in an RNA molecule.

transduction [L. *trans*, across + *ducere*, to lead]: The transfer of genetic material (DNA) from one cell to another by a virus.

transfer RNA (tRNA) [L. *trans*, across + *ferre*, to bear or carry]: A class of small RNAs (about 80 nucleotides each) with three functional sites: one is the attachment site for a specific amino acid, another is the nucleotide triplet (anticodon) for that amino acid, and yet another is the recognition site for the enzyme (an aminoacyl-tRNA synthetase) that catalyzes the attachment of the amino acid to the tRNA molecule. Each type of tRNA accepts a specific amino acid and transfers it to a growing polypeptide chain as specified by the nucleotide sequence of the messenger RNA being translated.

transformation [L. *trans*, across + *formare*, to shape]: A genetic change produced by the incorporation into a cell of DNA from the external medium.

translation [L. *trans*, across + *latus*, that which is carried]: The process by which the genetic information present in a strand of messenger RNA directs the sequence of amino acids during protein synthesis.

translocation [L. *trans*, across + *locare*, to put or place]: (1) In plants, the transport of the products of photosynthesis from a leaf to another part of the plant. (2) In genetics, the breaking off of a piece of chromosome with its reattachment to a nonhomologous chromosome.

transpiration [L. *trans*, across + *spirare*, to breathe]: In plants, the loss of water vapor from the stomata.

transposon [L. *transponere*, to change the position of]: A DNA sequence carrying one or more genes that is capable of moving from one location in the chromosomes to another. Simple transposons carry only the genes essential for transposition; complex transposons carry genes that code for additional proteins.

trophic level [Gk. *trophos*, feeder]: The position of a species in the food web or chain, that is, its feeding level; a step in the movement of biomass or energy through an ecosystem.

trophoblast [Gk. *trophos*, feeder + *blastos*, sprout]: In the early mammalian embryo (the blastocyst), a double layer of cells that surrounds the inner cell mass and subsequently gives rise to the chorion.

tropic [Gk. *trope*, a turning]: Pertaining to behavior or action brought about by specific stimuli, for example, phototropic ("light-oriented") motion, gonadotropic ("stimulating the gonads") hormone.

tuber [L. *tuber*, bump, swelling]: A much-enlarged, short, fleshy underground stem, such as that of the potato.

tumor suppressor gene: One of a group of eukaryotic genes that apparently act as a brake on cell division, keeping it tightly regulated; when such genes are lost or disabled by mutation, the rapid multiplication of cancer cells can occur.

turgor [L. *turgere*, to swell]: The pressure exerted on the inside of a plant cell wall by the fluid contents of the cell; the interior of the cell is hypertonic in relation to the fluids surrounding it and so gains water by osmosis.

unsaturated fatty acid: A fatty acid in which there are double bonds between some of the carbon atoms; fats containing unsaturated fatty acids tend to be liquid at room temperature.

urea [Gk. *ouron*, urine]: An organic compound formed in the vertebrate liver; principal form of disposal of nitrogenous wastes by mammals.

ureter [Gk. from *ourein*, to urinate]: The tube carrying urine from the kidney; in mammals, it empties into the bladder.

urethra [Gk. from *ourein*, to urinate]: The tube carrying urine from the bladder to the exterior of mammals.

uric acid [Gk. *ouron*, urine]: An insoluble nitrogenous waste product that is the principal excretory product in birds, reptiles, and insects.

urine [Gk. *ouron*, urine]: The liquid waste filtered from the blood by the kidney and stored in the bladder pending elimination through the urethra.

uterus [L., womb]: The muscular, expanded portion of the female reproductive tract modified for the storage of eggs or for housing and nourishing the developing embryo.

vacuole [L. *vacuus*, empty]: A membrane-bound, fluid-filled sac within the cytoplasm of a cell.

vaporization [L. *vapor*, steam]: The change from a liquid to a gas; evaporation.

vascular [L. *vasculum*, a small vessel]: Containing or concerning vessels that conduct fluid.

vascular bundle: In plants, a group of longitudinal supporting and conducting tissues (xylem and phloem).

vascular cambium [L. *vasculum*, a small vessel + *cambium*, exchange]: In plants, a cylindrical sheath of meristematic cells that divide mitotically, producing secondary phloem to one side and secondary xylem to the other, but always with a cambial cell remaining.

vas deferens, *pl.* **vasa deferentia** (vass deff-er-ens) [L. *vas*, a vessel + *deferre*, to carry down]: In mammals, the tube carrying sperm from a testis to the urethra.

vector [L., carrier]: In recombinant DNA, a small, self-replicating DNA molecule, or a portion thereof, into which a DNA segment can be spliced and introduced into a cell; generally a plasmid or a virus.

vein [L. *vena,* a blood vessel]: (1) In plants, a vascular bundle forming part of the framework of the conducting and supporting tissue of a leaf. (2) In animals, a blood vessel carrying blood from the tissues to the heart. A small vein is known as a venule.

vena cava (**vee**-na **cah**-va) [L., blood vessel + hollow]: A large vein that brings blood from the tissues to the right atrium of the four-chambered mammalian heart. The superior vena cava collects blood from the forelimbs, head, and anterior or upper trunk; the inferior vena cava collects blood from the posterior body region.

ventral [L. *venter,* belly]: Pertaining to the undersurface of an animal that holds its body in a horizontal position; to the front surface of an animal that holds its body erect.

ventricle [L. *ventriculus,* the stomach]: A muscular chamber of the heart that receives blood from an atrium and pumps blood out of the heart, either to the lungs or to the body tissues.

vertebral column [L. *vertebra,* joint]: The backbone; in nearly all vertebrates, it forms the supporting axis of the body and protects the spinal cord.

vesicle [L. *vesicula,* a little bladder]: A small, intracellular membrane-bound sac.

vessel [L. *vas,* a vessel]: A tubelike element of the xylem of angiosperms; composed of dead cells (vessel members) arranged end to end. Its function is to conduct water and dissolved minerals from the soil.

viable [L. *vita,* life]: Able to live.

villus, *pl.* **villi** [L., a tuft of hair]: In vertebrates, one of the minute, fingerlike projections lining the small intestine that serve to increase the absorptive surface area of the intestine.

virus [L., slimy, liquid, poison]: A submicroscopic, noncellular particle composed of a nucleic acid core and a protein coat (capsid); parasitic; reproduces only within a host cell.

viscera [L., internal organs]: The collective term for the internal organs of an animal.

vitamin [L. *vita,* life]: Any of a number of unrelated organic substances that cannot be synthesized by a particular organism and are essential in minute quantities for normal growth and function.

water cycle: Worldwide circulation of water molecules, powered by the sun. Water evaporates from oceans, lakes, rivers, and, in smaller amounts, soil surfaces and bodies of organisms; water returns to the Earth in the form of rain and snow. Of the water falling on land, some flows into rivers that pour water back into the oceans and some percolates down through the soil until it reaches a zone where all pores and cracks in the rock are filled with water (groundwater); the deep groundwater eventually reaches the oceans, completing the cycle.

water potential: The potential energy of water molecules; regardless of the reason (e.g., gravity, pressure, concentration of solute particles) for the water potential, water moves from a region where water potential is greater to a region where water potential is lower.

wild type: In genetics, the phenotype that is characteristic of the vast majority of individuals of a species in a natural environment.

worker: A member of the nonreproductive laboring caste of social insects.

xylem [Gk. *xylon,* wood]: A complex vascular tissue through which most of the water and dissolved minerals are conducted from the roots to other parts of the plant; consists of tracheids or vessel members, parenchyma cells, and fibers; constitutes the wood of trees and shrubs.

yolk: The stored food in egg cells that nourishes the embryo.

yolk sac: In developing reptiles and birds, the extraembryonic membrane that surrounds and encloses the yolk; performs a nutritive function. In mammals, the extraembryonic membrane in which the germ cells are set aside very early in development.

zoology [Gk. *zoe,* life + *logos,* a discourse]: The study of animals.

zooplankton [Gk. *zoe,* life + *plankton,* wanderer]: A collective term for the nonphotosynthetic organisms present in plankton.

zygote (**zi**-goat) [Gk. *zygon,* yolk, pair]: The diploid (2*n*) cell resulting from the fusion of male and female gametes (fertilization); a zygote may either develop into a diploid individual by mitotic divisions or may undergo meiosis to form haploid *(n)* individuals that divide mitotically to form a population of cells.

Illustration Credits

Page xii Frans Lanting/Minden Pictures; **Page xiv** Eric Gravé/Science Source/Photo Researchers, Inc.; **Page xvi** (left) Neil Leifer/Camera 5; (right) Dr. Gopal/MURTI–CNRI/Phototake; **Page xviii** Steven Dalton/NHPA; **Page xx** Jeff Lepore; **Page xxi** Helen Rhode; **Page xxiii** Chuck Place; **Page xxiv** Art Wolfe

Page xxvi Frans Lanting/Minden Pictures; **I–2** (a),(c)–(f) Frans Lanting/Minden Pictures; (b) Konrad Woethe/Bruce Coleman, Inc.; **I–3** George Richmond, *Portrait of Charles Darwin*, 1840, Down House, Downe, Kent, Bridgeman/Art Resource; **I–4** (a) E. R. Degginger/Bruce Coleman, Inc.; (b) George O. Poinar, Jr., University of California, Berkeley; (c) T. E. Whitely; **I–5** Frans Lanting/Minden Pictures; **I–7** (a) Mervin W. Larson/Bruce Coleman, Inc.; (b) Frans Lanting/Minden Pictures; **I–8** (a),(b) Tui De Roy **I–9** Picture Collection/New York Public Library; **Page 11** The Royal College of Surgeons; **I–10** Frans Lanting/Minden Pictures; **I–11** M. W. F. Tweedie/Bruce Coleman, Inc.; **I–14** (a) Stephen Dalton/NHPA; (b) M. K. and I. M. Morcombe/NHPA

Page 18 ©1961 California Institute of Technology and Carnegie Institution of Washington; **1–3** W. H. Hodge/Peter Arnold, Inc.; **1–6** (b) Runk and Schoenberger/Grant Heilman Photography; **1–7** Stephen Dalton/NHPA; **Pages 30–31** (a) John D. Cunningham/ Visuals Unlimited; (b) M. Walker/NHPA; (c) Mary M. Thacher/ Photo Researchers, Inc.; (d) John Cancalosi; (e) Silvestris/R. Gross/ Peter Arnold, Inc.; (f) Fred Bruemmer; (g) Stephen J. Krasemann/ DRK Photo; **1–12** (a) M. Abbey/Visuals Unlimited; (b) M. I. Walker/ Photo Researchers, Inc.; (c) David Scharf/Peter Arnold, Inc.; (d) Cabisco/Visuals Unlimited

Page 36 Michael and Patricia Fogden; **2–3** (a) G. I. Bernard/Animals Animals; (b) Runk and Schoenberger/Grant Heilman Photography; **2–5** Richard Smith/Tom Stack and Associates; **2–6** (a) Richard Kolar/Animals Animals; (b) Luciano Viti/Retna, Ltd.; (c) Anthony Bannister/Animals Animals; **2–7** (a) adapted from Roger H. Pain: "Helices of Antifreeze," *Nature*, vol. 333, page 207, 1988; (b) Steve McCutcheon

Page 48 Anthony Bannister/ABPL; **3–2** Avril Ramage/Oxford Scientific Films; **3–4** (c) Dr. Lloyd M. Beidler/Science Photo Library/ Photo Researchers, Inc.; **3–5** (b) Don W. Fawcett/Visuals Unlimited; **3–6** (b) Biophoto Associates/Photo Researchers, Inc.; **3–7** David P. Maitland/Seaphot Ltd./Planet Earth Pictures; **Page 58** (a)–(c) Lennart Nilsson/©Boehringer Ingelheim International GmbH; **3–11** Dr. Jeremy Burgess/Science Photo Library/Photo Researchers, Inc.; **3–16** sequence information from A. L. Lehninger, *Principles of Biochemistry*, Worth Publishers, Inc., New York, 1982; **3–19** (a) Stephen J. Krasemann/DRK Photo; (b) Gregory G. Dimijian/Photo Researchers, Inc.; **3–20** David Goodsell, *The Machinery of Life*, Springer-Verlag, New York, 1993; **3–21** adapted from R. E. Dickerson and I. Geis, *The Structure and Action of Proteins*, W. A. Benjamin, Inc., Menlo Park, Calif., 1969, copyright 1969 by Dickerson and Geis; **3–22** (a),(b) Stanley Flegler/Visuals Unlimited

Page 70 Courtesy of Kosuke Kamo, Sakurajima Volcanological Observatory; **4–1** Tokai University Research Center; **4–3** (a) Sidney

W. Fox; (b) David Deamer; **4–4** S. M. Awramik/Biological Photo Service; **4–5** (b) Ralph Robinson/Visuals Unlimited; **4–6** (a) Dr. Jeremy Burgess/Science Photo Library/Photo Researchers, Inc.; (b) Courtesy of the National Library of Medicine; (c) John Mais; **4–7** (b) A. Ryter; **4–8** (b) Norma J. Lang/Biological Photo Service; **4–9** (b) George Palade; **4–11** (b) Michael A. Walsh; **4–12** (b) Keith Porter; **4–14** (d) Biological Photo Service; **4–15** (a)–(c) David M. Phillips/Visuals Unlimited

Page 86 Lennart Nilsson et al., *The Body Victorious*, Delacorte Press, New York, 1987; **5–2** J. D. Robertson; **5–4** (b) Biophoto Associates/Photo Researchers, Inc.; **5–5** after P. Albersheim, *Scientific American*, April 1975; **5–6** (a) Daniel Friend; (b) Nigel Unwin; **5–7** Biophoto Associates/Science Source/Photo Researchers, Inc.; **5–8** (b) Mia Tegner and David Epel; **5–12** (b) Peter Webster; **5–13** (b) Don W. Fawcett/Visuals Unlimited; **5–14** (b) C. J. Flickinger, *Journal of Cell Biology*, vol. 49, page 221, 1975; **5–16** (b) George B. Chapman, Priscilla Devadoss/Visuals Unlimited; **5–17** (b) Bill Longcore/Photo Researchers, Inc.; **5–18** (a)–(c) Mary Osborn, Max Planck Institute for Biophysical Chemistry; **5–20** Manfred Kage/Peter Arnold, Inc.; **5–21** (a) H. Cante-Lund/NHPA; (b) Karl Aufderheide/Visuals Unlimited; **5–22** (b) Lewis Tilney

Page 108 Eric Gravé/Science Source/Photo Researchers, Inc.; **Page 113** USDA/Science Source/Photo Researchers, Inc.; **6–5** (a) Robert Brons/Biological Photo Service; (b),(c) Thomas Eisner; **6–6** after A. L. Lehninger, *Biochemistry*, 2d ed., Worth Publishers, Inc., New York, 1975; **6–11** Birgit Satir; **6–12** adapted from *Scientific American*, May 1984, page 54; **6–13** (a)–(d) M. M. Perry and A. B. Gilbert, *Journal of Cell Science*, vol. 39, pages 257–272, 1979; **Page 122** (a),(b) London Scientific Films/Oxford Scientific Films; (c)–(e) Robert Kay; **6–14** (a) Ray F. Evert; **6–15** adapted in part from James Darnell, Harvey Lodish, and David Baltimore, *Molecular Cell Biology*, W. H. Freeman and Company, New York, 1986; in part from *Scientific American*, May 1978, page 150; and in part from *Scientific American*, October 1985, page 106

Page 126 Jeff Foott/Bruce Coleman, Inc.; **7–1** Alain Eurard/Photo Researchers, Inc.; **7–5** (a)–(c) Fred Bavendam; **7–8** after A. L. Lehninger, *Biochemistry*, 2d ed, Worth Publishers, Inc., New York, 1975; **7–10** Clive Freeman, The Royal Institution/Science Photo Library/Photo Researchers, Inc.; **7–11** after A. L. Lehninger, *op. cit.*; **7–16** (a) Larry West/Bruce Coleman, Inc.; (b) Ken Lucas/Planet Earth Pictures; **Page 141** K. Amman/Bruce Coleman, Inc.

Page 142 Neil Leifer/Camera 5; **8–2** ©1989 Egyptian Expedition of the Metropolitan Museum of Art, Rogers Fund, 1915 (15.5.19e); **8–6** adapted from James Darnell et al., *Molecular Cell Biology*, W. H. Freeman and Company, New York, 1986; **8–10** adapted from A. L. Lehninger, David Nelson, and Michael Cox, *Principles of Biochemistry*, 2d ed., Worth Publishers, Inc., New York, 1993, page 554; **8–11** (a) adapted from Lehninger, Nelson, and Cox, *op. cit.*, page 557; (b) John N. Telford; **8–12** adapted from Lehninger, Nelson, and Cox, *op. cit.*, page 559; **8–14** (b) Grant Heilman Photography; **8–15** (b) Ken Regan/Camera 5; **Page 156** (a),(b) Bruce Iverson; **8–16** after A. L. Lehninger, *Biochemistry*, 2d ed., Worth Publishers, Inc., New York, 1975

Page 160 Victor Englebert; **9–1** Colin Milkins/Oxford Scientific Films; **9–4** absorption spectra determined by Govindjee; **9–6** (a) Dr. Kenneth Miller/Science Photo Library/Photo Researchers, Inc.; (b) R. Howard Berg/Visuals Unlimited; **9–9** adapted from Bruce Alberts et al., *Molecular Biology of the Cell,* 2d ed., Garland Publishing, Inc., New York, 1989, page 378; **9–10** Dr. Jeremy Burgess/Science Photo Library/Photo Researchers, Inc.

Page 176 Edward Hicks, *Noah's Ark* (detail), 1846, Philadelphia Museum of Art, bequest of Lisa Norris Elkins (50.92.7); **10–3** (a),(b) David Phillips/Visuals Unlimited; **10–4** G. Shih and R. Kessel/Visuals Unlimited; **10–5** adapted from Bruce Alberts et al., *Molecular Biology of the Cell,* Garland Publishing, Inc., New York, 1983, page 619; **10–6** (a) Andrew S. Bajer; (b),(c) adapted from Alberts et al., *op. cit.,* page 652; **10–7** (a)–(f) James Solliday/Biological Photo Service; **10–8** (a)–(d) Andrew S. Bajer; **10–9** David Phillips/Visuals Unlimited; **10–10** (d) B. A. Palevitz/E. H. Newcomb/Biological Photo Service; **10–11** M. I. Walker/Photo Researchers, Inc.

Page 188 Michael and Patricia Fogden; **11–8** (a)–(j) C. A. Hasenkampf/Biological Photo Service; **11–12** (a) Dr. Ram Verma/Phototake; **11–13** (a) Terry Mendoza/Picture Cube; **11–15** (b) Custom Medical Stock; **11–16** after J. L. Moore, *Heredity and Development,* Oxford University Press, New York, 1963; **Page 204** Ripon Microslides

Page 206 Steve Uzzell; **12–2** adapted from K. von Frisch, *Biology,* translated by Jane Oppenheimer, Harper and Row, Publishers, Inc., New York, 1964; **Page 212** John Chiasson/Gamma Liaison; **12–7** Moravian Museum, Brno

Page 220 Biophoto Associates/Photo Researchers, Inc.; **13–1** Joe McDonald/Bruce Coleman, Inc.; **13–3** (a),(c) Hans Reinhard/Bruce Coleman, Inc.; (b) Grant Heilman Photography; (d) Jane Burton/Bruce Coleman, Inc.; **13–4** (a)–(d) Ralph Somes/University of Connecticut; **13–6** Heather Angel; **13–9** Jeremy Burgess/Science Photo Library/Photo Researchers, Inc.; **Page 231** (a) Richmond Products, Boca Raton; **13–15** B. John/Cabisco/Visuals Unlimited; **13–18** P. J. Bryant/Biological Photo Service

Page 240 Manfred Kage/Peter Arnold, Inc.; **14–1** (a),(b) Bruce Iverson; **14–2** after D. D. Koob and W. E. Boggs, *The Nature of Life,* Addison-Wesley Publishing Co., Inc., Reading, Mass., 1972; **14–4** Michel Wurtz; **14–6** Lee D. Simon/Science Photo Library/Photo Researchers, Inc.; **Page 247** (a) from J. D. Watson, *The Double Helix,* Atheneum Publishers, New York, 1968; (b) Will and Deni McIntyre/Photo Researchers, Inc.; **14–11** A. B. Blumenthal, H. J. Kreigstein, and D. S. Hognes, *Cold Spring Harbor Symposium on Quantitative Biology,* vol. 38, page 205, 1973

Page 255 Dr. Gopal/MURTI-CNRI/Phototake; **15–1** Ray Nelson/Phototake; **15–2** after L. Pauling et al., "Sickle Cell Anemia, A Molecular Disease," *Science,* vol. 110, pages 543–548, 1949; **15–4** (a) after K. Namba, D. L. D. Casper, and G. J. Stubbs, *Science,* vol. 227, pages 773–776, 1985; (b) K. G. Murti/Visuals Unlimited; **15–7** adapted from J. A. Lake, *Annual Review of Biochemistry,* vol. 54, pages 507–530, 1985; **15–10** Science Photo Library/Custom Medical Stock

Page 270 M. Bruno/Jacana/Photo Researchers, Inc.; **16–5** David Goodsell, *The Machinery of Life,* Springer-Verlag, New York, 1993; **16–6** (a) Victoria Foe; **16–7** adapted from Bruce Alberts et al., *Molecular Biology of the Cell,* 2d ed., Garland Publishing, Inc., New York, 1989; **16–9** (a) George T. Rudkin; **16–10** (a),(b) Ullrich Scheer, W. W. Franke, M. F. Trendelenberg; **16–11** adapted from U. Goodenough, *Genetics,* 3d ed., Saunders College/Holt, Rinehart and Winston, Inc., New York, 1983; **Page 280** Don Fawcett/Visuals Unlimited; **16–12** (a) James German; (b) adapted from P. Chambon, "Split Genes," *Scientific American,* May 1981, pages 60–71; **Page 282** P. J. Grabowski and T. R. Cech, *Cell,* vol. 23, pages 467–476, 1981; **16–14** adapted from R. Lewin, *Science,* vol. 212, pages 28–32, 1981

Page 286 Nik Kleinberg; **17–1** K. G. Murti/Visuals Unlimited; **17–2** Dr. Dennis Kunkel/Phototake; **17–3** adapted from F. J. Ayala, and J. A. Kiger, *Modern Genetics,* Benjamin/Cummings Publishing Co. Inc., Menlo Park, Calif., 1980; **17–4** S. N. Cohen, *Nature,* 1969; **17–5** (a) M. Wurtz/Biozentrum, University of Basel/Science Photo Library/Photo Researchers, Inc.; (b) Philippe Plailly/Science Photo Library/Photo Researchers, Inc.; (c) CNRI/Phototake; (d) Biozentrum/Science Photo Library/Photo Researchers, Inc.; **17–7** Michel Wurtz/Biozentrum/University of Basel; **17–8** (a) Dr. Robert Liddington, Dana Farber Cancer Institute; (b) Jack Griffith; **17–9** adapted from A. L. Lehninger, *Principles of Biochemistry,* Worth Publishers, Inc., New York, 1982, page 863; **17–10** (a),(b) Dr. David L. Spector, Cold Spring Harbor Laboratory; **17–12** Christine J. Harrison, from Harrison et al., "The Fragile X: A Scanning Electron Microscope Study," *Journal of Medical Genetics,* vol. 20, pages 280–285, 1983

Page 300 AP/Wide World Photos (Elizabeth Fulford, 1993); **18–1** adapted from Bruce Alberts et al., *Molecular Biology of the Cell,* Garland Publishing, Inc., New York, 1983; **18–2** adapted from a photo by John T. Fiddles and Howard M. Goodman; **18–4** (a)–(c) Huntington Potter and David Dressler, LIFE Magazine, © Time Warner; **18–6** David Ward; **18–7** adapted from *City of Hope Quarterly,* vol. 7, page 2, Winter 1978; **18–8** Eli Lilly and Company; **Page 310** ITAR/TASS/Sovfoto; **18–10** Steve Uzzell III; **Page 312** (a) Eugene W. Nester; (b) Keith Wood, University of California at San Diego; **18–11** after Jon W. Gordon and Frank Ruddle; **18–12** R. L. Brinster

Page 318 Lynn Stone; **19–1** Michael Ederegger/DRK Photo; **Page 321** Michael and Patricia Fogden; **19–3** (a),(b) John Mais; **19–4** after Mather and Harrison, *Heredity,* vol. 3, page 977, 1949; **19–5** R. C. Lewontin; **19–6** Dr. Agnes Bankier, Murdoch Institute, Royal Children's Hospital, Melbourne; **19–7** Edmund Gerard; **19–9** Victor McKusick; **19–10** Kennan Ward/DRK Photo; **19–11** Jim Brandenburg/Minden Pictures; **Page 330** from Lewis Carroll, *Through the Looking Glass,* illustrated by John Tenniel; **19–13** E. S. Ross

Page 334 Mark D. DeMello/New York Zoological Society; **20–1** (a) Jane Burton/Bruce Coleman, Ltd.; (b) Kim Taylor/Bruce Coleman, Ltd.; **20–2** (a),(b) James L. Castner; **20–3** (a) M. and C. Ederegger/Peter Arnold, Inc.; (b) Kent and Donna Dannen/Photo Researchers, Inc.; **20–5** Chris Huss/The Wildlife Collection; **20–6** Steve Hopkin/Planet Earth Pictures; **20–7** (a) E. R. Degginger/Animals Animals; (b) adapted from D. Futuyma, *Evolutionary Biology,* 1979 (after Clarke 1962, based on data of Popham 1942); **Page 343** M. Andersson/VIREO; **20–8** (a) Russ Kinne/Comstock; (b) Frans Lanting/Minden Pictures; **20–9** David Kjaer/Planet Earth Pictures; **20–10** (a),(b) adapted from S. J. Gould, *Ever Since Darwin,* W. W. Norton and Company, New York, 1977, page 235; **20–11** after J. Clausen and W. M. Hiesen, Publication 615, Carnegie Institute of Washington, 1950; **20–12** (a) Lincoln Brower; (b) Frans Lanting/Minden Pictures; (c) Dwight Kuhn; **20–13** (a),(b) Lincoln Brower; **20–14** (a)–(d) E. S. Ross; (e) James L. Castner; (f) D. Wilder/Tom Stack and Associates; (g) Peter J. Bryant/Biological Photo Service; **20–16** (a) S. J. Krasemann/Peter Arnold, Inc.; (b) E. S. Ross; **20–17** E. R. Degginger/Bruce Coleman, Inc.

Page 354 Erwin and Peggy Bauer/Bruce Coleman, Inc.; **21–1** (a) C. A. Henley/Biofotos; (b) Kenneth W. Fink/Photo Researchers, Inc.; (c) Tom McHugh/Photo Researchers, Inc.; **21–2** (a) Belinda Wright/DRK Photo; (b) Stephen J. Krasemann/DRK Photo; **21–3** adapted from E. O. Wilson and W. H. Burrert, *A Primer of Population Biology,* Sinauer Associates, Sunderland, Mass., 1971; **Page 358** (b) USGS; **21–6** Ron Prokopy, University of Massachussetts–Amherst; **21–7** Sid Bahrt/Photo Researchers, Inc.; **21–9** (a) Scott Camazine/Photo Researchers, Inc.; (b) Jeff Lepore/Photo Researchers, Inc.; (c) after V. Wallace and A. Srb, *Adaptation,* Prentice-Hall, Inc., Englewood Cliffs, N. J., 1964; **21–11** after D. Lack, *Darwin's Finches,* Harper and Row, Publishers, Inc., New York, 1961; **21–12** (a)–(f) Tui De Roy; **Page 366** Michael Franis/The Wildlife Collection; **21–14**, **21–15** redrawn from G. G. Simpson, *Horses,* Oxford University Press, New York, 1951

Page 376 Steven Dalton/NHPA; **22–1** E. S. Ross; **22–5** Richard Hansen/Photo Researchers, Inc.; **22–6** Nina Leen/Time Warner, Inc.; **22–7** after Mazudazo Konish; **22–8** Heather Angel; **22–9** (a) E. S.

University Press, New York, 1983; **32–2** From Lewis Carroll, *Alice's Adventure in Wonderland*, illustrated by John Tenniel; **32–3** Gray Mortimore/Tony Stone Images; **32–7** Doug Allen/Oxford Scientific Films/Animals Animals; **32–8** John Cancalosi/Auscape International; **32–9** after A. J. Vander, J. H. Sherman and D. S. Luciano, *Human Physiology*, McGraw-Hill Publishing Company, New York, 1969; **32–10** (a) Ken Lucas/Planet Earth Pictures; (b) Robert Franz/Planet Earth Pictures; (c) Leonard Lee Rue III/Animals Animals; **32–12** John Cancalosi/DRK Photo; **32–13** Richard T. Nowitz

Page 582 Lennart Nilsson/©Boehringer Ingelheim International GmbH, *The Incredible Machine*, National Geographic Society, Washington, D.C., 1986 **Page 588** Bernard Pierre Wolff/Photo Researchers, Inc.; **33–6** (b) Leonard Lessin/Peter Arnold, Inc.; **33–10** Don Fawcett/Science Source/Photo Researchers, Inc.; **33–11** (a),(b) Paul Travers, University of London; **33–15** (a)–(c) Lennart Nilsson/©Boehringer Ingelheim International GmbH; **33–18** Lennart Nilsson/©Boehringer Ingelheim International GmbH, *The Body Victorious*, Delacorte Press, New York, 1987; **33–20** Lennart Nilsson/©Boehringer Ingelheim International GmbH, *The Incredible Machine*, National Geographic Society, Washington, D.C., 1986; **33–21** Stephen Turner/Oxford Scientific Films

Page 604 E. S. Ross; **34–1** adapted from Guttman and Hopkins; **34–10** Astrid and Hanns-Frieder Michler/Science Photo Library/Photo Researchers, Inc.; **34–13** adapted from A. J. Vander, J. H. Sherman, and D. S. Luciano, *Human Physiology*, 4th ed., McGraw-Hill Publishing Company, New York, 1985

Page 620 Janis Carter; **35–4** (b) Cedric S. Raine/Visuals Unlimited; **35–6** courtesy Dr. Peter S. Amenta, Hahneman University; **35–9** after R. Eckert and D. Randall, *Animal Physiology*, W. H. Freeman and Company, New York, 1978; **35–11** after David Hubel; **35–16** adapted from J. Darnell, H. Lodish, and D. Baltimore, *Molecular Cell Biology*, W. H. Freeman and Company, 1986; **35–17** adapted from Williams and Warwick, *Functional Neuroanatomy of Man*, W. B. Saunders, Philadelphia, 1975; **Page 637** Comstock

Page 640 Michael S. Quinton; **36–1** after W. S. Beck, *Human Design*, Harcourt Brace Jovanovich, New York, 1971; **36–5** Lennart Nilsson/©Boehringer Ingelheim International GmbH, *Behold Man*, Little, Brown and Company, Boston, 1974; **Page 646** Stephen Dalton/NHPA; **36–7** CNRI/Science Photo Library/Photo Researchers, Inc.; **36–9** (a) Martin M. Rotker/Science Source/Photo Researchers, Inc.; (b) CNRI/Science Photo Library/Photo Researchers, Inc.; **36–10** Bassett/Photo Researchers, Inc.; **36–15** adapted from *Scientific American*, June 1987, page 83; **36–17** adapted from *Scientific American*, June 1987, page 82; **Page 654** (a) Christine Gall/University of California, Irvine; (b) Daniel P. Perl; **36–18** Jim Greenfield/Planet Earth Pictures; **36–20** (a) Hugh E. Huxley; **36–24** (a) Lennart Nilsson et al., *Behold Man*, Little, Brown and Company, Boston, 1974; (c) after K. R. Porter and M. Bonneville, *Fine Structure of Cells and Tissues*, Lea and Febiger, Philadelphia, 1968; **Page 662** Hugh E. Huxley

Page 664 Helen Rhode; **37–2** (b) G. Shih and R. Kessel/Visuals Unlimited; **37–3** after M. B. V. Roberts, *Biology: A Functional Approach*, The Ronald Press Company, New York, 1971; **37–6** (a) Doug Allen/Oxford Scientific Films; (b) Jonathan Scott/Seaphot Ltd./Planet Earth Pictures; (c) Lady Philippa Scott/NHPA; (d) Oxford Scientific Films/Animals Animals; **Pages 670–671** (a) NIAID/NIH/Peter Arnold, Inc.; (b) CNRI/Science Photo Library/Photo Researchers, Inc.; (c) Institut Pasteur/CNRI/Phototake; (d) NIAID/NIH/Peter Arnold, Inc.; (e) Science VU/Visuals Unlimited; **37–10** (a)–(d) C. Edelmann/La Villette/Photo Researchers, Inc.; **37–12** after A. J. Vander, J. H. Sherman, and D. S. Luciano, *Human Physiology*, 3d ed., McGraw-Hill Publishing Company, New York, 1980

Page 680 Keith A. Szafaranski; **38–1** (a) William Byrd; (b),(c) Mia Tegner/Scripps Institution of Oceanography; **38–2** adapted from Bruce Alberts et al., *Molecular Biology of the Cell*, Garland Publishing, Inc., New York, 1983; **38–3** (a)–(h) Tryggve Gustafson; **38–4** after Huettner, 1949; **38–5** M. P. Olsen; **38–6** after E. O. Wilson et al., *Life on Earth*, 2d ed., Sinauer Associates, Sunderland, Mass., 1978; **38–7** (a)–(d) Tryggve Gustafson; **38–8** after Wilson et

al., *op. cit.*; **38–9** after Huettner, 1949; **38–10** Hans Pfletschinger/Peter Arnold, Inc.; **38–11** after Huettner, 1949; **38–12** (a) after J. J. W. Baker and G. E. Allen, *The Study of Biology*, 3d ed., Addison-Wesley Publishing Co., Inc., Reading, Mass., 1977; (b) after Maurice Sussman, *Animal Growth and Development*, 2d ed., Prentice-Hall, Inc., Englewood Cliffs, N. J., 1964; **38–13** *Ibid.*; **38–15** S. R. Hilfer; **38–16** (a)–(c) adapted from G. Karp and N. J. Berrill, *Development*, 2d ed., McGraw-Hill Publishing Company, New York, 1981; (d) Kathryn W. Tosney; **38–17** (a) Martin Raff; (b) adapted from Alberts et al., *op. cit.*; **38–18** Lewis Wolpert; **38–20** (a)–(d) Runk and Schoenberger/Grant Heilman Photography; **Page 693** (a),(b) Laboratory of Edward B. Lewis, California Institute of Technology; **38–21** Lennart Nilsson, *The Incredible Machine*, National Geographic Society, Washington, D.C., 1986; **38–22** (a)–(c) Petit Format, Nestle/Science Source/Photo Researchers, Inc.; **38–27** (a)–(c) Lennart Nilsson et al., *Behold Man*, Little, Brown and Company, Boston, 1973; **38–28** (a),(b) Lennart Nilsson/©Boehringer Ingelheim International GmbH, *The Incredible Machine*, National Geographic Society, Washington, D.C., 1986; **38–29, 38–30** Lennart Nilsson et al., *Behold Man*, Little, Brown and Company, Boston, 1974; **38–32** (a)–(c) Jeffrey Reed/Medichrome; **38–33** David Austen/Tony Stone Images

Page 706 Merlin D. Tuttle/Bat Conservation International, Inc.; **39–2** (a) Kenneth W. Fink/Bruce Coleman, Inc.; (b) E. R. Degginger/Earth Scenes; (c) Thomas C. Boyden; (d) John C. Coulter/Visuals Unlimited; (e) Doug Wechsler; (f) E. S. Ross; (g) John Tiszler/Peter Arnold, Inc.; **39–3** (a),(b) Dr. Dennis Kunkel/Phototake; (c) Dr. Jeremy Burgess/Science Photo Library/Photo Researchers, Inc.; **39–4** (b) David L. Muleahy; **39–9** (a) W. H. Hodge/Peter Arnold, Inc.; (b) Larry West; (c) Rence Purse/Photo Researchers, Inc.; **Page 717** Heather Angel

Page 718 Tom Bean/Allstock; **40–4** Dr. Jeremy Burgess/Science Photo Library/Photo Researchers, Inc.; **40–5** (a) G. J. James/Biological Photo Service; (b) Dwight R. Kuhn; (c) James L. Castner; **40–7** (b) Ray F. Evert; **40–8** G. R. Roberts; **40–9** after P. Ray, *The Living Plant*, 2d ed., Holt, Rinehart and Winston, Inc., New York, 1971; **40–10** (a),(b) Ray F. Evert; **40–11** after Ray, *op. cit.*; **40–13** (a) Jack Dermid; (b) Robert Mitchell; **40–14** (a) Biophoto Associates/Photo Researchers, Inc.; (b) Ray F. Evert; (c) Randy Moore/Visuals Unlimited; **40–15** (c) Ray F. Evert; **40–16** (d) I. N. A. Lott, McMaster University/Biological Photo Service; **40–17** (a)–(c), **40–18** (a)–(c) Ray F. Evert; **40–22** from R. Ketchum, *The Secret Life of the Forest*, American Heritage Press, 1970.

Page 740 Kim Taylor/Bruce Coleman, Inc.; **41–2** Walter H. Hodge/Peter Arnold, Inc.; **41–3** after M. Richardson, *Translocation in Plants*, St. Martin's Press, Inc., New York, 1968 (after Stoat and Hoagland, 1939); **41–4** adapted from D. K. Northington and J. R. Goodin, *The Botanical World*, Mosby-Year Book, St. Louis, 1984; **41–5** Jerry L. Davis; **41–6** (a),(b) Biological Photo Service; **41–7** (a) Science Photo; (b) M. H. Zimmerman; **41–10** (a) Runk and Schoenberger/Grant Heilman Photography; (b) C. P. Vanae/Visuals Unlimited

Page 754 Chuck Place; **42–1** Dwight Kuhn; **42–4** Runk and Schoenberger/Grant Heilman Photography; **42–7** (a)–(c) Bruce Iverson; **42–8** E. Webber/Visuals Unlimited; **Page 761** Philip Harrington; **42–10** (a) Ray F. Evert; (b) Heather Angel; **42–11** Larry West; **42–12** Carolina Biological Supply Company; **42–13** after E. O. Wilson et al., *Life: Cells, Organisms, Populations*, Sinauer Associates, Inc., Sunderland, Mass., 1977; **42–14** (a),(b) Randy Moore; **42–15, 42–16** after A. W. Naylor, "The Control of Flowering," *Scientific American*, May 1952; **42–17** John R. MacGregor/Peter Arnold, Inc.; **42–20** Breck P. Kent/Earth Scenes; **42–21** (a),(b) Jane Burton/Bruce Coleman, Inc.; **42–22** Stephen P. Parker/Photo Researchers, Inc.; **42–23** (a),(b) James L. Castner; **42–24** (a),(b) Runk and Schoenberger/Grant Heilman Photography

Page 774 Kenneth W. Fink/Photo Researchers, Inc.; **43–6** after *Nature*, vol. 298, 1982; **43–7** adapted from McNaughton and Wolfe, *General Ecology*, 2d ed., Holt, Rinehart, and Winston, Inc., New York, 1979; **43–8** (a) Peter Johnson/NHPA; (b) Doug Perrine/

Index

Page numbers in italic type indicate references to illustrations, captions, tables, and boxed essays; boldface type is used to indicate major text discussions.

Metric Measurements

	Quantity	Numerical Value	English Equivalent	Converting English to Metric
Length	kilometer (km)	1,000 (10^3) meters	1 km = 0.62 mile	1 mile = 1.609 km
	meter (m)	100 centimeters	1 m = 1.09 yards = 3.28 feet	1 yard = 0.914 m 1 foot = 0.305 m
	centimeter (cm)	0.01 (10^{-2}) meter	1 cm = 0.394 inch	1 foot = 30.5 cm 1 inch = 2.54 cm
	millimeter (mm)	0.001 (10^{-3}) meter	1 mm = 0.039 inch	1 inch = 25.4 mm
	micrometer (μm)	0.000001 (10^{-6}) meter		
	nanometer (nm)	0.000000001 (10^{-9}) meter		
	ångstrom (Å)	0.0000000001 (10^{-10}) meter		
Area	square kilometer (km²)	100 hectares	1 km² = 0.3861 square mile	1 square mile = 2.590 km²
	hectare (ha)	10,000 square meters	1 ha = 2.471 acres	1 acre = 0.4047 ha
	square meter (m²)	10,000 square centimeters	1 m² = 1.1960 square yards = 10.764 square feet	1 square yard = 0.8361 m² 1 square foot = 0.0929 m²
	square centimeter (cm²)	100 square millimeters	1 cm² = 0.155 square inch	1 square inch = 6.4516 cm²
Mass	metric ton (t)	1,000 kilograms = 1,000,000 grams	1 t = 1.103 tons	1 ton = 0.907 t
	kilogram (kg)	1,000 grams	1 kg = 2.205 pounds	1 pound = 0.4536 kg
	gram (g)	1,000 milligrams	1 g = 0.0353 ounce	1 ounce = 28.35 g
	milligram (mg)	0.001 gram		
	microgram (μg)	0.000001 gram		
Time	second (sec)	1,000 milliseconds		
	millisecond	0.001 second		
	microsecond	0.000001 second		
Volume (solids)	1 cubic meter (m³)	1,000,000 cubic centimeters	1 m³ = 1,3080 cubic yards = 35.315 cubic feet	1 cubic yard = 0.7646 m³ 1 cubic foot = 0.0283 m³
	1 cubic centimeter (cm³)	1,000 cubic millimeters	1 cm³ = 0.0610 cubic inch	1 cubic inch = 16.387 cm³
Volume (liquids)	kiloliter (kl)	1,000 liters	1 kl = 264.17 gallons	1 gal = 3.785 l
	liter (l)	1,000 milliliters	1 l = 1.06 quarts	1 qt = 0.94 l 1 pt = 0.47 l
	milliliter (ml)	0.001 liter	1 ml = 0.034 fluid ounce	1 fluid ounce = 29.57 ml
	microliter (μl)	0.000001 liter		